JACARANDA

MATHS QUEST

VICTORIAN CURRICULUM
THIRD EDITION

10+10A

JACARANDA

MATHS QUEST

VICTORIAN CURRICULUM
THIRD EDITION

10+10A

CATHERINE SMITH

JAMES SMART

GEETHA JAMES

CAITLIN MAHONY

BEVERLY LANGSFORD WILLING

CONTRIBUTING AUTHORS

Michael Sheedy | Kahni Burrows | Paul Menta

VICTORIAN
CURRICULUM
v2.0

jacaranda
A Wiley Brand

Third edition published 2023 by
John Wiley & Sons Australia, Ltd
Level 4, 600 Bourke Street, Melbourne, Vic 3000

First edition published 2014
Revised edition published 2018
Second edition published 2021

Typeset in 10.5/13 pt TimesLT Std

© John Wiley & Sons Australia, Ltd 2014, 2018, 2021, 2023

The moral rights of the authors have been asserted.

ISBN: 978-1-394-23552-0

Reproduction and communication for educational purposes

Reproduction and communication for other purposes

Trademarks

Front cover images: © virinaflora/Adobe Stock Photos;
© Marish/Shutterstock; © Irina Strelnikova/Shutterstock;
© Marina Zlochin/Adobe Stock; © Visual
Generation/Shutterstock

Illustrated by various artists, diacriTech and Wiley Composition
Services

Typeset in India by diacriTech

A catalogue record for this
book is available from the
National Library of Australia

The Publishers of this series acknowledge and pay their respects
to Aboriginal Peoples and Torres Strait Islander Peoples as the
traditional custodians of the land on which this resource was
produced.

This suite of resources may include references to (including
names, images, footage or voices of) people of Aboriginal
and/or Torres Strait Islander heritage who are deceased. These
images and references have been included to help Australian
students from all cultural backgrounds develop a better
understanding of Aboriginal and Torres Strait Islander Peoples'
history, culture and lived experience.

It is strongly recommended that teachers examine resources on
topics related to Aboriginal and/or Torres Strait Islander
Cultures and Peoples to assess their suitability for their own
specific class and school context. It is also recommended that
teachers know and follow the guidelines laid down by the
relevant educational authorities and local Elders or community
advisors regarding content about all First Nations Peoples.

All activities in this resource have been written with the safety
of both teacher and student in mind. Some, however, involve
physical activity or the use of equipment or tools. **All due care
should be taken when performing such activities**. To the
maximum extent permitted by law, the author and publisher
disclaim all responsibility and liability for any injury or loss that
may be sustained when completing activities described in this
resource.

The Publisher acknowledges ongoing discussions related to
gender-based population data. At the time of publishing, there
was insufficient data available to allow for the meaningful
analysis of trends and patterns to broaden our discussion of
demographics beyond male and female gender identification.

Printed in Singapore
M WEP283355 090524

Contents

About this resource

NEW FOR

VICTORIAN CURRICULUM v2.0

JACARANDA
MATHS QUEST 10+10A
VICTORIAN CURRICULUM
THIRD EDITION

Developed by teachers for students

Tried, tested and trusted. The third edition of the *Jacaranda Maths Quest series* continues to focus on helping teachers achieve learning success for every student — ensuring no student is left behind, and no student is held back.

Because both *what* and *how* students learn matter

Learning is personal

Whether students need a challenge or a helping hand, you'll find what you need to create engaging lessons.

Whether in class or at home, students can get unstuck and progress! Scaffolded lessons, with detailed worked examples, are all supported by teacher-led video eLessons. Automatically marked, differentiated question sets are all supported by detailed worked solutions. And brand-new Quick Quizzes support in-depth skill acquisition.

Learning is effortful

Learning happens when students push themselves. With learnON, Australia's most powerful online learning platform, students can challenge themselves, build confidence and ultimately achieve success.

Learning is rewarding

Through real-time results data, students can track and monitor their own progress and easily identify areas of strength and weakness.

And for teachers, Learning Analytics provide valuable insights to support student growth and drive informed intervention strategies.

Learn online with Australia's most

Everything you need for each of your lessons in one simple view

- Trusted, curriculum-aligned content
- Engaging, rich multimedia
- All the teacher support resources you need
- Deep insights into progress
- Immediate feedback for students
- Create custom assignments in just a few clicks.

Practical teaching advice and ideas for each lesson provided in teachON

Teaching videos for all lessons

Reading content and rich media including embedded videos and interactivities

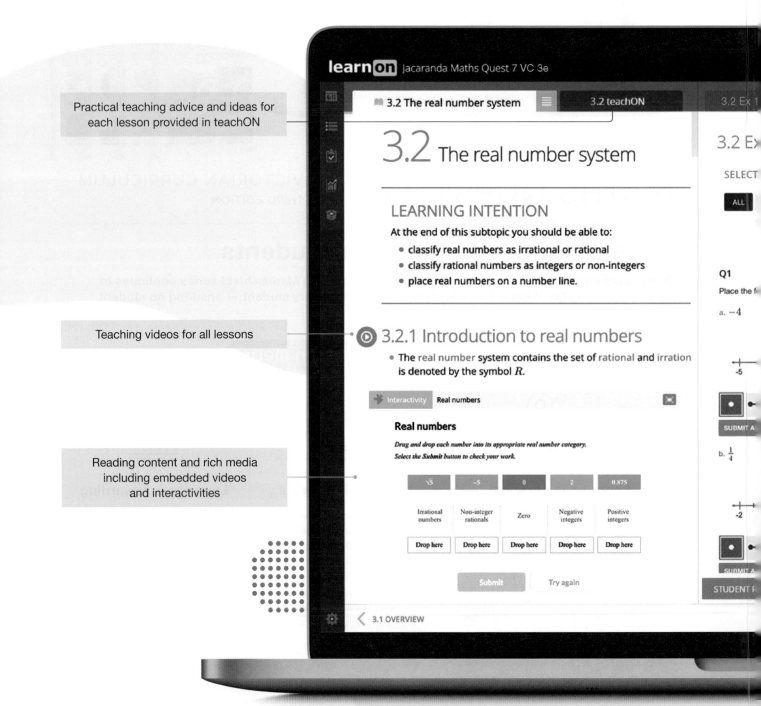

powerful learning tool, learnON

New! **Quick Quiz questions** for skill acquisition

Differentiated question sets

Teacher and student views

Textbook questions

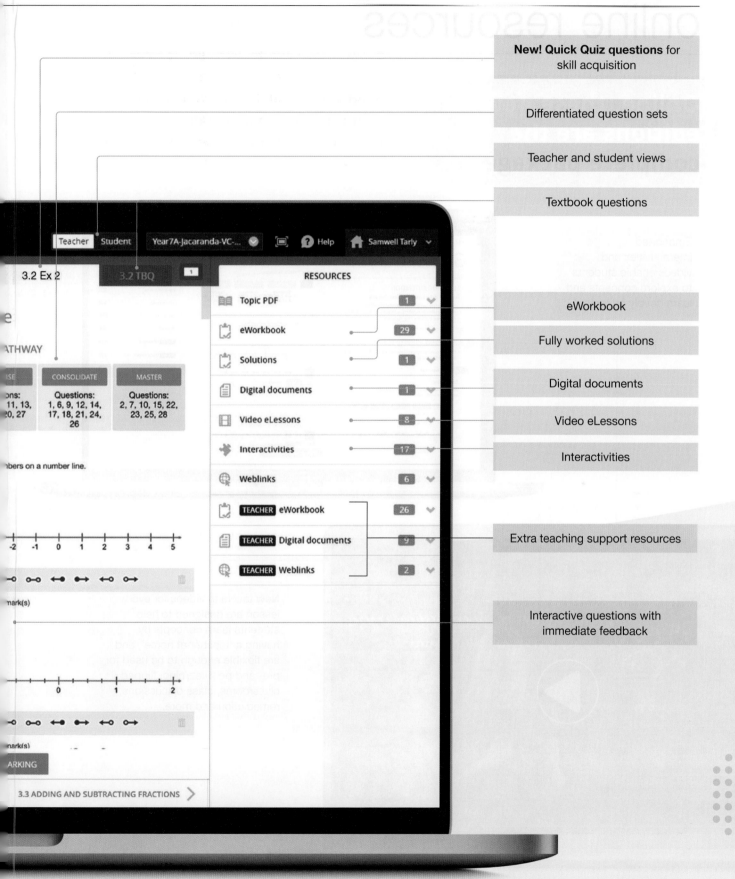

eWorkbook

Fully worked solutions

Digital documents

Video eLessons

Interactivities

Extra teaching support resources

Interactive questions with immediate feedback

Get the most from your online resources

Online, these new editions are the complete package

Trusted Jacaranda theory, plus tools to support teaching and make learning more engaging, personalised and visible.

Embedded interactivities and videos enable students to explore concepts and learn deeply by 'doing'.

New teaching videos for every lesson are designed to help students learn concepts by having a 'teacher at home', and are flexible enough to be used for pre- and post-learning, flipped classrooms, class discussions, remediation and more.

Brand new! Quick Quiz questions for skill acquisition in every lesson.

Three differentiated question sets, with immediate feedback in every lesson, enable students to challenge themselves at their own level.

Instant reports give students visibility into progress and performance.

Every question has immediate, corrective feedback to help students overcome misconceptions as they occur and get unstuck as they study independently — in class and at home.

Maths Quest 7 includes Powering up for Year 7

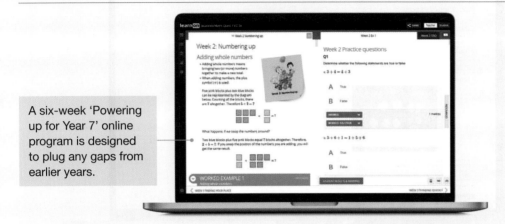

A six-week 'Powering up for Year 7' online program is designed to plug any gaps from earlier years.

NAPLAN Online Practice

Go online to complete practice NAPLAN tests. There are 6 NAPLAN-style question sets available to help students prepare for this important event. They are also useful for practising Mathematics skills in general.

Also available online is a video that provides strategies and tips to help with students' preparation.

eWorkbook

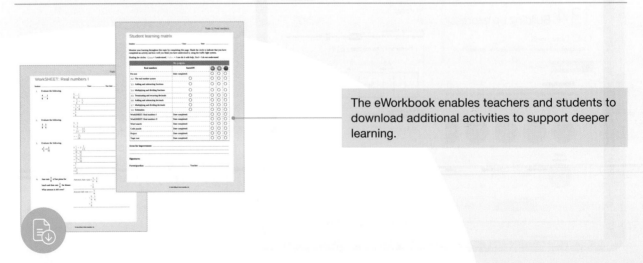

The eWorkbook enables teachers and students to download additional activities to support deeper learning.

A wealth of teacher resources

Enhanced teacher support resources for every lesson, including:
- work programs and curriculum grids
- practical teaching advice
- three levels of differentiated teaching programs
- quarantined topic tests (with solutions)

Customise and assign

An inbuilt testmaker enables you to create custom assignments and tests from the complete bank of thousands of questions for immediate, spaced and mixed practice.

Reports and results

Data analytics and instant reports provide data-driven insights into progress and performance within each lesson and across the entire course.

Show students (and their parents or carers) their own assessment data in fine detail. You can filter their results to identify areas of strength and weakness.

Acknowledgements

The authors and publisher would like to thank the following copyright holders, organisations and individuals for their assistance and for permission to reproduce copyright material in this book.

The Victorian Curriculum F–10 content elements are © VCAA, reproduced by permission. The VCAA does not endorse or make any warranties regarding this resource. The Victorian Curriculum F–10 and related content can be accessed directly at the VCAA website.

This may be a modified extract from the Victorian Curriculum F–10 and may include the work of other authors. VCAA does not endorse nor verify the accuracy of the information provided and accepts no responsibility for incomplete or inaccurate information. You can find the most up-to-date version of the Victorian Curriculum at http://victoriancurriculum.vcaa.vic.edu.au/.

Images

• © wpadington/Adobe Stock Photos: **788** • © onlyyouqj/Adobe Stock Photos: **205** • © cristianstorto/Adobe Stock Photos: **365** • © Ivan Moreno Stock/Adobe Stock Photos: **739** • © Viorel Kurnosov/Adobe Stock Photos: **898** • © ViDi Studio/Adobe Stock Photos: **788** • © Racamani/Adobe Stock Photos: **789** • © AMR Image/Getty Images: **282** • © fsachs78/Getty Images: **397, 716** • © vgajic/Getty Images: **320** • © Piter Lenk/Alamy Stock Photo: **94** • © Getty Images: **753** • © Szepy/Getty Images/iStockphoto: **314** • © Martin Ruegner/Getty Images: **536** • Shutterstock: **605** • © A.D.S.Portrait/Shutterstock: **234** • © Aarcady/Shutterstock: **687** • © Africa Studio/Shutterstock: **950, 1204** • © aiyoshi597/Shutterstock: **1136** • © Aleks Kend/Shutterstock: **1129** • © Alexander Raths/Shutterstock: **150, 871** • © alexandre zveiger/Shutterstock: **541** • © ALPA PROD/ Shutterstock: **338** • © Ana Vasileva/Shutterstock: **688** • © Balaguta Evgeniya/Shutterstock: **912** • © benjaminec/Shutterstock: **77** • © Bobex-73/Shutterstock: **765** • © Brilliantist Studio/Shutterstock: **1226** • © BSG_1974/Shutterstock: **988** • © Canoneer/Shutterstock: **687** • © Cemil Aksoy/Shutterstock: **966** • © charobnica/Shutterstock: **679** • © Chrislofotos/Shutterstock: **498** • © d13/Shutterstock: **972** • © Daniel Prudek/Shutterstock: **956** • © Dean Drobot/Shutterstock: **1164** • © del-Mar/Shutterstock: **2** • © DGLimages/ Shutterstock: **874** • © Don Mammoser/Shutterstock: **92** • © donatas1205/Shutterstock: **787** • © Dusan Petkovic/Shutterstock: **843** • © elfishes/Shutterstock: **681, 687** • © enciktepstudio/Shutterstock: **1221** • © enterphoto/Shutterstock: **756** • © Erin Alexis Randolph/Shutterstock: **808** • © ESB Professional/ Shutterstock: **106, 924, 1109** • © Evgeny Karandaev/Shutterstock: **689** • © FamVeld/Shutterstock: **771** • © FiledIMAGE/Shutterstock: **1192** • © freesoulproduction/Shutterstock: **687** • © Gena Melendrez/ Shutterstock: **957** • © Georgios Kollidas/Shutterstock: **158** • © Gordon Bell/Shutterstock: **410** • © HSNphotography/Shutterstock: **695** • © HST6/Shutterstock: **697** • © ian woolcock/Shutterstock: **1192** • © Igor Corovic/Shutterstock: **856** • © Igor Havok/Shutterstock: **948** • © Inna Ogando/Shutterstock: **682, 689** • © Irina Rogova/Shutterstock: **686** • © Iryna Imago/Shutterstock: **734** • © Kaewta Mungkung/Shutterstock: **1215** • © Kevin Wells Photography/Shutterstock: **68** • © Kotruro2/Shutterstock: **1163** • © kovop58/Shutterstock: **145** • © light poet/Shutterstock: **320** • © Lev Kropotov/Shutterstock: **698** • © Lukas Gojda/Shutterstock: **387** • © Maksym Dykha/Shutterstock: **240** • © Margarita Borodina/Shutterstock: **936** • © Maridav/Shutterstock: **387, 1044** • © Meawstory15 Production/Shutterstock: **1164** • © Mejini Neskah/Shutterstock: **936** • © mfauzisaim/Shutterstock: **864** • © Mijolanta/Shutterstock: **1044** • © miwa-in-oz/Shutterstock: **866** • © Monkey Business Images/Shutterstock: **772, 784, 862** • © MvanCaspel/Shutterstock: **696** • © Neale Cousland/Shutterstock: **512, 862** • © Nic Vilceanu/Shutterstock: **1094** • © niderlander/Shutterstock: **319** • © NigelSpiers/Shutterstock: **63** • © nuntarat eksawetanant/Shutterstock: **1226** • © oksana2010/Shutterstock: **726** • © OlegDoroshin/Shutterstock: **751** • © Oleksiy Mark/Shutterstock: **276, 544** • © Olga Danylenko/ Shutterstock: **348** • © oneSHUTTER oneMEMORY/Shutterstock: **894** • © OtmarW/Shutterstock: **697** • © P Meybruck/Shutterstock: **378** • © PabloBenii/Shutterstock: **521** • © Pavlo Baliukh/Shutterstock: **1098** • © Peterfz30/Shutterstock: **92** • © PHONT/Shutterstock: **1077** • © Photo Melon/Shutterstock: **711** • © PhotographyByMK/Shutterstock: **114** • © Pressmaster/Shutterstock: **138, 782** • © Racheal

Marish/Shutterstock, Irina Strelnikova/Shutterstock, Marina Zlochin/Adobe Stock Photos, Visual Generation/Shutterstock: **1, 219, 505, 807, 1093, 1135** • © Wiley art: **1189** • © wonlopcolors/Shutterstock: **349** • © worldswildlifewonders/Shutterstock: **53** • © Yellowj/Shutterstock: **864** • © GeoGebra., **663, 664, 705–707** • © CASIO: **511, 519, 525, 534, 1101, 1102, 1111, 1120, 1147, 1149, 1158, 1168, 1169, 1175** • © CASIO COMPUTER CO., LTD., **19, 24, 30, 32, 44, 57, 58, 71, 77, 109, 118, 123, 128, 134, 141, 225, 238, 258, 285, 286, 288, 289, 306, 343, 357, 358, 372, 389, 427, 467, 487, 573–576, 599, 611, 618, 758, 779, 815, 817, 829, 838, 847, 850, 859, 860, 868, 869, 880, 922, 923, 952, 953, 1037, 1038, 1040, 1047, 1053, 1066, 1075, 1217–1220** • © Casio screenshot: **225, 226** • © Haali/Shutterstock: **85, 147, 205, 313, 397, 450, 494, 539, 622, 720, 791, 889, 1021, 1079, 1126, 1188, 1228** • © John Wiley and Sons Australia: **470** • © Nadezda Cruzova/Shutterstock: **831** • © Shutterstock: **561** • © The TI screenshots shown are from the TI-Nspire CX/CX II handheld. For more information on this product and others, visit https://education.ti.com, **19, 24, 30, 32, 44, 57, 58, 71, 77, 109, 118, 123, 128, 134, 141, 225, 226, 238, 258, 285, 286, 288, 289, 306, 343, 357, 358, 372, 389, 416, 427, 467, 487, 511, 519, 525, 534, 573–576, 592, 599, 611, 758, 779, 815, 817, 818, 829, 838, 847, 850, 859, 860, 868, 869, 880, 922, 923, 952, 953, 954, 1037, 1038, 1040, 1047, 1053, 1066, 1075, 1101, 1102, 1111, 1120, 1147, 1149, 1158, 1168, 1169, 1175, 1217–1220** • © m.bonotto/Shutterstock: **451** • © iravgustin/Shutterstock: **556** • © John Carnemolla/Shutterstock: **561** • © Amy Johansson/Shutterstock: **561** • © Paul D Smith/Shutterstock: **561** • © XiXinXing/Shutterstock: **579** • © Orla/Shutterstock: **579** • © Michael Cola/Shutterstock: **594** • © Den Rozhnovsky/Shutterstock: **594** • © zhu difeng/Shutterstock: **595** • © Peddalanka Ramesh Babu/Shutterstock: **603** • © Elena Schweitzer/Shutterstock: **604** • © CC7/Shutterstock: **605** • © Levente Fazakas/Shutterstock: **607** • © DavideAngelini/Shutterstock: **607** • © Andrei Armiagov/Shutterstock: **608** • © kavalenkau/Shutterstock: **608** • © AlexxAn/Shutterstock: **623**

Text

• © Source: The Courier Mail, 14 Sept. 2010, p. 25.: **886** • © Source: The Great Aussie Dads survey, The Sunday Mail, 05/09/2010, David Briggs, Galaxy principal: **881** • © Source: Sponges are toxic, The Sunday Mail, 5 Sept. 2010, p. 36.: **882** • © Source: The Courier Mail, 27 Aug. 2010, pp. 40–1.: **885**

Every effort has been made to trace the ownership of copyright material. Information that will enable the publisher to rectify any error or omission in subsequent reprints will be welcome. In such cases, please contact the Permissions Section of John Wiley & Sons Australia, Ltd.

1 Exponents, surds and logarithms

LESSON SEQUENCE

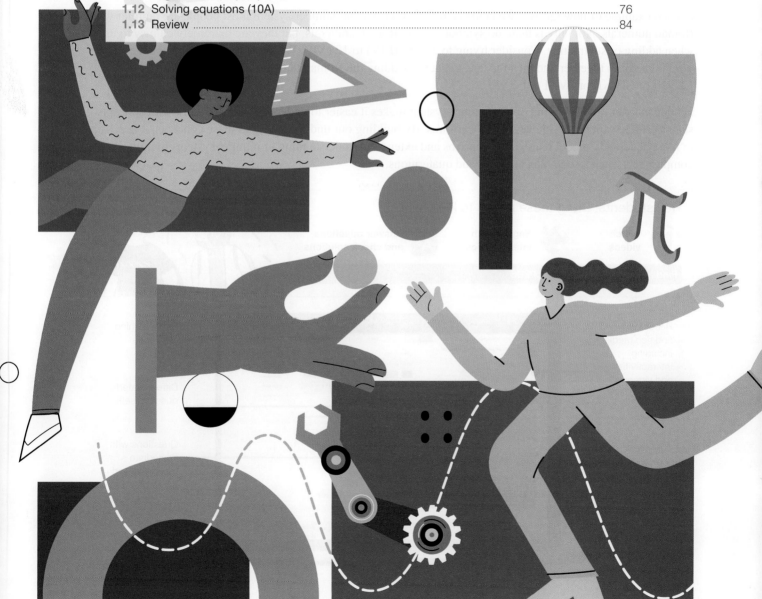

LESSON
1.1 Overview

Why learn this?

We often take for granted the amount of time and effort that has gone into developing the number system we use on a daily basis. In ancient times, numbers were used for bartering and trading goods between people. Thus, numbers were always attached to objects, for example 5 cows, 13 sheep or 20 gold coins. Consequently, it took a long time before more abstract concepts such as the number 0 were introduced and widely used. It took even longer for negative numbers or irrational numbers such as surds to be accepted as their own group of numbers. Historically, there has always been resistance to these changes and updates. In folklore, Hippasus — the man first credited with the discovery of irrational numbers — was drowned at sea for angering the gods with his discovery.

A good example of how far we have come is to look at an ancient number system most people are familiar with: Roman numerals. Not only is there no symbol for 0 in Roman numerals, but they are extremely clumsy to use when adding or subtracting. Consider trying to add 54 (LIV) to 12 (XII). We know that to determine the answer we add the ones together and then the tens to get 66. Adding the Roman numeral is more complex; do we write LXVIII or LIVXII or LVXI or LXVI?

Having a better understanding of our number system makes it easier to understand how to work with concepts such as surds, exponents (indices) and logarithms. By building our understanding of these concepts, it is possible to more accurately model real-world scenarios and extend our understanding of number systems to more complex sets, such as complex numbers and quaternions.

1. Positive numbers are also known as natural numbers. Is this statement true or false?

2. State whether $\sqrt{36}$ is a rational or irrational number.

3. Simplify the following expression: $3n^{\frac{1}{5}} \times 5n^{\frac{1}{3}}$.

4. Simplify the following expression: $\sqrt[5]{32p^{10}q^{15}}$.

5. Determine the exact value of $81^{-\frac{3}{4}}$.

6. **MC** Select which of the numbers of the set $\left\{\sqrt{0.25}, \pi, 0.\overline{261}, -5, \frac{2}{3}\right\}$ are rational.

   A. $\left\{\sqrt{0.25}, \pi, 0.\overline{261}\right\}$

   B. $\left\{0.261, -5, \frac{2}{3}\right\}$

   C. $\left\{\pi, 0.\overline{261}\right\}$

   D. $\left\{\sqrt{0.25}, 0.\overline{261}, -5, \frac{2}{3}\right\}$

7. **MC** $\dfrac{12x^8 \times 3x^7}{9x^{10} \times x^3}$ simplifies to:

   A. $\dfrac{5x^2}{3}$

   B. $4x^2$

   C. $4x^{26}$

   D. $\dfrac{5x^{26}}{3}$

8. Simplify the following expression: $3\sqrt{2} \times \sqrt{10}$.

9. Simplify the following expression: $5\sqrt{2} + 12\sqrt{2} - 3\sqrt{2}$.

10. **MC** Choose the most simplified form of the following expression: $\sqrt{8a^3} + \sqrt{18a} + \sqrt{a^5}$

    A. $5\sqrt{2a} + a\sqrt{a}$

    B. $2a\sqrt{2a^2} + 3\sqrt{2a} + a^4\sqrt{a}$

    C. $2a^2\sqrt{2a} + 2\sqrt{3a} + a^4\sqrt{a}$

    D. $2a\sqrt{2a} + 3\sqrt{2a} + a^2\sqrt{a}$

11. Simplify the following expression: $a^{\frac{2}{3}}b^{\frac{3}{2}} \div a^{\frac{3}{5}}b^{\frac{5}{7}}$

12. Solve the following equation for y: $\dfrac{1}{125} = 5^{y+2}$.

13. Solve the following equation for x: $x = \log_{\frac{1}{4}} 16$.

14. Simplify the following expression. $\log_2\left(\dfrac{1}{4}\right) + \log_2(32) - \log_2(8)$.

15. **MC** Choose the correct value for x in $3 + \log_2 3 = \log_2 x$.

    A. $x = 0$

    B. $x = 3$

    C. $x = 9$

    D. $x = 24$

LESSON
1.2 Number classification review

▶ 1.2.1 The real number system

eles-4661

- The number systems used today evolved from a basic and practical need of people to count and measure magnitudes and quantities such as livestock, people, possessions, time and so on.
- As societies grew and architecture and engineering developed, number systems became more sophisticated. Number use developed from solely whole numbers to fractions, decimals and irrational numbers.

- The real number system contains the set of rational and irrational numbers. It is denoted by the symbol R. The set of real numbers contains a number of subsets that can be classified as shown in the chart below.

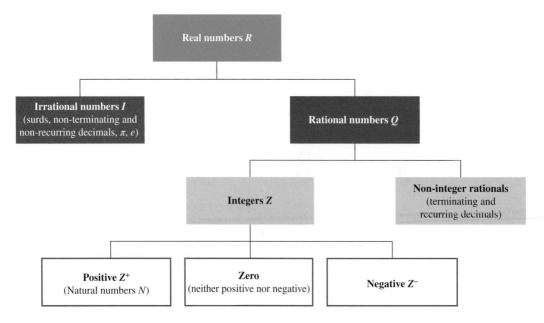

Integers (Z)

- The set of **integers** consists of whole positive and negative numbers and 0 (which is neither positive nor negative).
- The set of integers is denoted by the symbol Z and can be visualised as:

$$Z = \{..., -3, -2, -1, 0, 1, 2, 3, ...\}$$

- The set of positive integers is also known as the **natural numbers** (or counting numbers) and is denoted Z^+ or N. That is:

$$Z^+ = N = \{1, 2, 3, 4, 5, 6, ...\}$$

- The set of negative integers is denoted Z^-.

$$Z^- = \{... - 6, -5, -4, -3, -2, -1\}$$

- Integers may be represented on the number line as illustrated below.

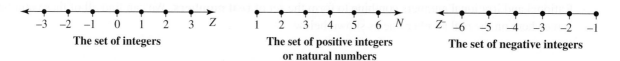

The set of integers	**The set of positive integers or natural numbers**	**The set of negative integers**

Rational numbers (Q)

- A **rational number** is a number that can be expressed as a ratio of two integers in the form $\dfrac{a}{b}$, where $b \neq 0$.
- The set of rational numbers is denoted by the symbol Q.
- Rational numbers include all whole numbers, all fractions, and all terminating and recurring decimals.
- **Terminating decimals** are decimal numbers that terminate after a specific number of digits. Examples are:

$$\frac{1}{4} = 0.25, \frac{5}{8} = 0.625, \frac{9}{5} = 1.8.$$

- **Recurring decimals** do not terminate but have a specific digit (or number of digits) repeated in a pattern. Examples are:

$$\frac{1}{3} = 0.333\,333... = 0.\dot{3} \text{ or } 0.\overline{3}$$

$$\frac{133}{666} = 0.199\,699\,699\,6... = 0.1\dot{9}9\dot{6} \text{ or } 0.1\overline{996}$$

- Recurring decimals are represented by placing a dot or line above the repeating digit(s).
- Using set notation, we can represent the set of rational numbers as:

$$Q = \left\{ \frac{a}{b} : a, b \in Z, b \neq 0 \right\}$$

- This can be read as 'Q is all numbers of the form $\dfrac{a}{b}$ given a and b are integers and b is not equal to 0'.

Irrational numbers (I)

- An **irrational number** is a number that cannot be expressed as a ratio of two integers in the form $\dfrac{a}{b}$, where $b \neq 0$.
- All irrational numbers have a decimal representation that is non-terminating and non-recurring. This means the decimals do not terminate and do not repeat in any particular pattern or order.
  For example:
  $$\sqrt{5} = 2.236\ 067\ 997\ 5 \ldots$$
  $$\pi = 3.141\ 592\ 653\ 5 \ldots$$
  $$e = 2.718\ 281\ 828\ 4 \ldots$$

- The set of irrational numbers is denoted by the symbol I. Some common irrational numbers that you may be familiar with are $\sqrt{2}, \pi, e, \sqrt{5}$.
- The symbol π (**pi**) is used for a particular number that is the circumference of a circle whose diameter is 1 unit.
- In decimal form, π has been calculated to more than 29 million decimal places with the aid of a computer.

Rational or irrational

- Rational and irrational numbers combine to form the set of **real numbers**. We can find all of these numbers somewhere on the real number line as shown below.

- To classify a number as either rational or irrational, use the following procedure:
  1. Determine whether it can be expressed as a whole number, a fraction, or a terminating or recurring decimal.
  2. If the answer is yes, the number is rational. If no, the number is irrational.

WORKED EXAMPLE 1 Classifying numbers as rational or irrational

Classify whether the following numbers are rational or irrational.

a. $\dfrac{1}{5}$ b. $\sqrt{25}$ c. $\sqrt{13}$ d. 3π

e. 0.54 f. $\sqrt[3]{64}$ g. $\sqrt[3]{32}$ h. $\sqrt[3]{\dfrac{1}{27}}$

THINK

a. $\dfrac{1}{5}$ is already a rational number.

b. 1. Evaluate $\sqrt{25}$.

   2. The answer is an integer, so classify $\sqrt{25}$.

c. 1. Evaluate $\sqrt{13}$.

   2. The answer is a non-terminating and non-recurring decimal. Classify $\sqrt{13}$.

WRITE

a. $\dfrac{1}{5}$ is rational.

b. $\sqrt{25} = 5$

   $\sqrt{25}$ is rational.

c. $\sqrt{13} = 3.605\ 551\ 275\ 46 \ldots$

   $\sqrt{13}$ is irrational.

d. **1.** Use your calculator to find the value of 3π.

    2. The answer is a non-terminating and non-recurring decimal. Classify 3π.

e. 0.54 is a terminating decimal; classify it accordingly.

f. **1.** Evaluate $\sqrt[3]{64}$.

    2. The answer is a whole number, so classify $\sqrt[3]{64}$.

g. **1.** Evaluate $\sqrt[3]{32}$.

    2. The result is a non-terminating and non-recurring decimal. Classify $\sqrt[3]{32}$.

h. **1.** Evaluate $\sqrt[3]{\dfrac{1}{27}}$.

    2. The result is a number in a rational form.

d. $3\pi = 9.424\,777\,960\,77\ldots$

    3π is irrational.

e. 0.54 is rational.

f. $\sqrt[3]{64} = 4$

    $\sqrt[3]{64}$ is rational.

g. $\sqrt[3]{32} = 3.174\,802\,103\,94\ldots$

    $\sqrt[3]{32}$ is irrational.

h. $\sqrt[3]{\dfrac{1}{27}} = \dfrac{1}{3}$.

    $\sqrt[3]{\dfrac{1}{27}}$ is rational.

Exercise 1.2 Number classification review

learn on

| 1.2 Quick quiz   on | 1.2 Exercise |

Individual pathways

■ PRACTISE	■ CONSOLIDATE	■ MASTER
1, 4, 7, 10, 13, 14, 17, 20, 23	2, 5, 8, 11, 15, 18, 21, 24	3, 6, 9, 12, 16, 19, 22, 25

Fluency

For questions **1** to **6**, classify whether the following numbers are rational (Q) or irrational (I).

1. **WE1**

  a. $\sqrt{4}$ **b.** $\dfrac{4}{5}$ **c.** $\dfrac{7}{9}$ **d.** $\sqrt{2}$

2. **a.** $\sqrt{7}$ **b.** $\sqrt{0.04}$ **c.** $2\dfrac{1}{2}$ **d.** $\sqrt{5}$

3. **a.** $\dfrac{9}{4}$ **b.** 0.15 **c.** -2.4 **d.** $\sqrt{100}$

4. **a.** $\sqrt{14.4}$ **b.** $\sqrt{1.44}$ **c.** π **d.** $\sqrt{\dfrac{25}{9}}$

5. **a.** 7.32 **b.** $-\sqrt{21}$ **c.** $\sqrt{1000}$ **d.** $7.216\,349\,157\ldots$

6. **a.** $-\sqrt{81}$ **b.** 3π **c.** $\sqrt[3]{62}$ **d.** $\sqrt{\dfrac{1}{16}}$

For questions **7** to **12**, classify the following numbers as rational (Q), irrational (I) or neither.

7. **a.** $\dfrac{1}{8}$ **b.** $\sqrt{625}$ **c.** $\dfrac{11}{4}$ **d.** $\dfrac{0}{8}$

8. **a.** $-6\dfrac{1}{7}$ **b.** $\sqrt[3]{81}$ **c.** $-\sqrt{11}$ **d.** $\sqrt{\dfrac{1.44}{4}}$

9. **a.** $\sqrt{\pi}$ **b.** $\dfrac{8}{0}$ **c.** $\sqrt[3]{21}$ **d.** $\dfrac{\pi}{7}$

10. **a.** $\sqrt[3]{(-5)^2}$ **b.** $-\dfrac{3}{11}$ **c.** $\sqrt{\dfrac{1}{100}}$ **d.** $\dfrac{64}{16}$

11. **a.** $\sqrt{\dfrac{2}{25}}$ **b.** $\dfrac{\sqrt{6}}{2}$ **c.** $\sqrt[3]{27}$ **d.** $\dfrac{1}{\sqrt{4}}$

12. **a.** $\dfrac{22\pi}{7}$ **b.** $\sqrt[3]{-1.728}$ **c.** $6\sqrt{4}$ **d.** $4\sqrt{6}$

13. **MC** Identify a rational number from the following.

    A. π **B.** $\sqrt{\dfrac{4}{9}}$ **C.** $\sqrt{\dfrac{9}{12}}$ **D.** $\sqrt[3]{3}$

14. **MC** Identify which of the following best represents an irrational number .

    A. $-\sqrt{81}$ **B.** $\dfrac{6}{5}$ **C.** $\sqrt[3]{343}$ **D.** $\sqrt{22}$

15. **MC** Select the correct statement regarding the numbers $-0.69, \sqrt{7}, \dfrac{\pi}{3}, \sqrt{49}$ from the following.

    A. $\dfrac{\pi}{3}$ is the only rational number.

    B. $\sqrt{7}$ and $\sqrt{49}$ are both irrational numbers.

    C. -0.69 and $\sqrt{49}$ are the only rational numbers.

    D. -0.69 is the only rational number.

16. **MC** Select the correct statement regarding the numbers $2\frac{1}{2}, -\frac{11}{3}, \sqrt{624}, \sqrt[3]{99}$ from the following.

    A. $-\frac{11}{3}$ and $\sqrt{624}$ are both irrational numbers.

    B. $\sqrt{624}$ is an irrational number and $\sqrt[3]{99}$ is a rational number.

    C. $\sqrt{624}$ and $\sqrt[3]{99}$ are both irrational numbers.

    D. $2\frac{1}{2}$ is a rational number and $-\frac{11}{3}$ is an irrational number.

Understanding

17. Simplify $\sqrt{\dfrac{a^2}{b^2}}$.

18. **MC** If $p < 0$, then $\sqrt{p}$ is:

    A. positive   **B.** negative

    C. rational   **D.** none of these

19. **MC** If $p < 0$, then $\sqrt{p^2}$ must be:

    A. positive   **B.** negative

    C. rational   **D.** irrational

Reasoning

20. Simplify $\left(\sqrt{p} - \sqrt{q}\right) \times \left(\sqrt{p} + \sqrt{q}\right)$. Show full working.

21. Prove that if $c^2 = a^2 + b^2$, it does not follow that $a = b + c$.

22. Assuming that x is a rational number, for what values of k will the expression $\sqrt{x^2 + kx + 16}$ always be rational? Justify your response.

Problem solving

23. Determine the value of m and n if $\dfrac{36}{11}$ is written as:

    a. $3 + \dfrac{1}{\frac{m}{n}}$ **b.** $3 + \dfrac{1}{3 + \frac{m}{n}}$

    c. $3 + \dfrac{1}{3 + \frac{1}{\frac{m}{n}}}$ **d.** $3 + \dfrac{1}{3 + \frac{1}{1 + \frac{m}{n}}}$

24. If x^{-1} means $\dfrac{1}{x}$, determine the value of $\dfrac{3^{-1} - 4^{-1}}{3^{-1} + 4^{-1}}$.

25. If $x^{-n} = \dfrac{1}{x^n}$, evaluate $\dfrac{3^{-n} - 4^{-n}}{3^{-n} + 4^{-n}}$ when $n = 3$.

LESSON
1.3 Surds (10A)

LEARNING INTENTIONS

At the end of this lesson you should be able to:
- determine whether a number under a root or radical sign is a surd
- prove that a surd is irrational by contradiction.

▶ ## 1.3.1 Identifying surds

eles-4662

- A **surd** is an irrational number that is represented by a root sign or a radical sign, for example $\sqrt{}, \sqrt[3]{}, \sqrt[4]{}$.

  Examples of surds include $\sqrt{7}, \sqrt{5}, \sqrt[3]{11}, \sqrt[4]{15}$.

- The numbers $\sqrt{9}, \sqrt{16}, \sqrt[3]{125}$, and $\sqrt[4]{81}$ are not surds as they can be simplified to rational numbers; that is, $\sqrt{9} = 3, \sqrt{16} = 4, \sqrt[3]{125} = 5, \sqrt[4]{81} = 3$.

WORKED EXAMPLE 2 Identifying surds

Determine which of the following numbers are surds.

a. $\sqrt{16}$ b. $\sqrt{13}$ c. $\sqrt{\dfrac{1}{16}}$ d. $\sqrt[3]{17}$ e. $\sqrt[4]{63}$ f. $\sqrt[3]{1728}$

THINK	WRITE
a. 1. Evaluate $\sqrt{16}$.	a. $\sqrt{16} = 4$
2. The answer is rational (since it is a whole number); state your conclusion.	$\sqrt{16}$ is not a surd.
b. 1. Evaluate $\sqrt{13}$.	b. $\sqrt{13} = 3.605\,551\,275\,46 \ldots$
2. The answer is irrational (since it is a non-recurring and non-terminating decimal); state your conclusion.	$\sqrt{13}$ is a surd.
c. 1. Evaluate $\sqrt{\dfrac{1}{16}}$.	c. $\sqrt{\dfrac{1}{16}} = \dfrac{1}{4}$
2. The answer is rational (a fraction); state your conclusion.	$\sqrt{\dfrac{1}{16}}$ is not a surd.
d. 1. Evaluate $\sqrt[3]{17}$.	d. $\sqrt[3]{17} = 2.571\,281\,590\,66 \ldots$
2. The answer is irrational (a non-terminating and non-recurring decimal); state your conclusion.	$\sqrt[3]{17}$ is a surd.

e. 1. Evaluate $\sqrt[4]{63}$.

    e. $\sqrt[4]{63} = 2.817\,313\,247\,26\ldots$

    2. The answer is irrational, so classify $\sqrt[4]{63}$ accordingly.

    $\sqrt[4]{63}$ is a surd.

f. 1. Evaluate $\sqrt[3]{1728}$.

    f. $\sqrt[3]{1728} = 12$

    2. The answer is rational; state your conclusion.

    $\sqrt[3]{1728}$ is not a surd. So **b**, **d** and **e** are surds.

▶ 1.3.2 Proof that a number is irrational

eles-4663

- In Mathematics you are required to study a variety of types of proofs. One such method is called proof by contradiction.
- This proof is so named because the logical argument of the proof is based on an assumption that leads to contradiction within the proof. Therefore, the original assumption must be false.
- An irrational number is one that cannot be expressed in the form $\dfrac{a}{b}$ (where a and b are integers). The next worked example sets out to prove that $\sqrt{2}$ is irrational.

WORKED EXAMPLE 3 Proving the irrationality of $\sqrt{2}$

Prove that $\sqrt{2}$ is irrational.

THINK

1. Assume that $\sqrt{2}$ is rational; that is, it can be written as $\dfrac{a}{b}$ in simplest form. We need to show that a and b have no common factors.

2. Square both sides of the equation.

3. Rearrange the equation to make a^2 the subject of the formula.

4. $2b^2$ is an even number and $2b^2 = a^2$.

5. Since a is even it can be written as $a = 2r$.

6. Square both sides.

7. Equate [1] and [2].

WRITE

Let $\sqrt{2} = \dfrac{a}{b}$, where a and b are integers that have no common factors and $b \neq 0$.

$2 = \dfrac{a^2}{b^2}$

$a^2 = 2b^2$ [1]

$\therefore a^2$ is an even number and a must also be even; that is, a has a factor of 2.

$\therefore a = 2r$

$a^2 = 4r^2$ [2]

But $a^2 = 2b^2$ from [1].

$\therefore 2b^2 = 4r^2$

$b^2 = \dfrac{4r^2}{2}$

$\quad = 2r^2$

$\therefore b^2$ is an even number and b must also be even; that is, b has a factor of 2.

8. Use reasoning to deduce that $\sqrt{2} = \dfrac{a}{b}$ where a and b have no common factor.

Both a and b have a common factor of 2. This contradicts the original assumption that $\sqrt{2} = \dfrac{a}{b}$, where a and b have no common factors.
Therefore, $\sqrt{2}$ is not rational.
Therefore, it must be irrational.

- *Note:* An irrational number written in surd form gives an exact value of the number; whereas the same number written in decimal form (for example, to 4 decimal places) gives an approximate value.

DISCUSSION

How can you be certain that $\sqrt{a}$ is a surd?

on Resources

eWorkbook Topic 1 Workbook (worksheets, code puzzle and project) (ewbk-2093)

Interactivity Surds on the number line (int-6029)

Exercise 1.3 Surds (10A)

learn on

1.3 Quick quiz on	**1.3 Exercise**

Individual pathways

■ PRACTISE	■ CONSOLIDATE	■ MASTER
1, 4, 7, 8, 11, 14, 17	2, 5, 9, 12, 15, 18	3, 6, 10, 13, 16, 19

Fluency

WE2 For questions **1** to **6**, determine which of the following numbers are surds.

1. a. $\sqrt{81}$ b. $\sqrt{48}$ c. $\sqrt{16}$ d. $\sqrt{1.6}$

2. a. $\sqrt{0.16}$ b. $\sqrt{11}$ c. $\sqrt{\dfrac{3}{4}}$ d. $\sqrt[3]{\dfrac{3}{27}}$

3. a. $\sqrt{1000}$ b. $\sqrt{1.44}$ c. $4\sqrt{100}$ d. $2 + \sqrt{10}$

4. a. $\sqrt[3]{32}$ b. $\sqrt{361}$ c. $\sqrt[3]{100}$ d. $\sqrt[3]{125}$

5. a. $\sqrt{6} + \sqrt{6}$ b. 2π c. $\sqrt[3]{169}$ d. $\sqrt{\dfrac{7}{8}}$

6. a. $\sqrt[4]{16}$ b. $\left(\sqrt{7}\right)^2$ c. $\sqrt[3]{33}$ d. $\sqrt{0.0001}$
    e. $\sqrt[5]{32}$ f. $\sqrt{80}$

7. **MC** The correct statement regarding the set of numbers $\left\{ \sqrt{\dfrac{6}{9}}, \sqrt{20}, \sqrt{54}, \sqrt[3]{27}, \sqrt{9} \right\}$ is:

   A. $\sqrt[3]{27}$ and $\sqrt{9}$ are the only rational numbers of the set.

   B. $\sqrt{\dfrac{6}{9}}$ is the only surd of the set.

   C. $\sqrt{\dfrac{6}{9}}$ and $\sqrt{20}$ are the only surds of the set.

   D. $\sqrt{20}$ and $\sqrt{54}$ are the only surds of the set.

8. **MC** Identify the numbers from the set $\left\{ \sqrt{\dfrac{1}{4}}, \sqrt[3]{\dfrac{1}{27}}, \sqrt{\dfrac{1}{8}}, \sqrt{21}, \sqrt[3]{8} \right\}$ that are surds.

   A. $\sqrt{21}$ only

   B. $\sqrt{\dfrac{1}{8}}$ only

   C. $\sqrt{\dfrac{1}{8}}$ and $\sqrt[3]{8}$

   D. $\sqrt{\dfrac{1}{8}}$ and $\sqrt{21}$ only

9. **MC** Select a statement regarding the set of numbers $\left\{ \pi, \sqrt{\dfrac{1}{49}}, \sqrt{12}, \sqrt{16}, \sqrt{3}, +1 \right\}$ that is *not* true.

   A. $\sqrt{12}$ is a surd.

   B. $\sqrt{12}$ and $\sqrt{16}$ are surds.

   C. π is irrational but not a surd.

   D. $\sqrt{12}$ and $\sqrt{3} + 1$ are not rational.

10. **MC** Select a statement regarding the set of numbers $\left\{ 6\sqrt{7}, \sqrt{\dfrac{144}{16}}, 7\sqrt{6}, 9\sqrt{2}, \sqrt{18}, \sqrt{25} \right\}$ that is *not* true.

    A. $\sqrt{\dfrac{144}{16}}$ when simplified is an integer.

    B. $\sqrt{\dfrac{144}{16}}$ and $\sqrt{25}$ are not surds.

    C. $7\sqrt{6}$ is smaller than $9\sqrt{2}$.

    D. $9\sqrt{2}$ is smaller than $6\sqrt{7}$.

Understanding

11. Complete the following statement by selecting appropriate words, suggested in brackets:
    $\sqrt{a}$ is definitely not a surd if a is … (any multiple of 4; a perfect square; cube).

12. Determine the smallest value of m, where m is a positive integer, so that $\sqrt[3]{16m}$ is not a surd.

13. **a.** Determine any combination of m and n, where m and n are positive integers with $m < n$, so that
    $\sqrt[4]{(m+4)(16-n)}$ is not a surd.

    b. If the condition that $m < n$ is removed, how many possible combinations are there?

Reasoning

14. Determine whether the following are rational or irrational.

   a. $\sqrt{5} + \sqrt{2}$

   b. $\sqrt{5} - \sqrt{2}$

   c. $\left(\sqrt{5} + \sqrt{2} \right)\left(\sqrt{5} - \sqrt{2} \right)$

15. **WE3** Prove that the following numbers are irrational, using proof by contradiction.

   a. $\sqrt{3}$

   b. $\sqrt{5}$

   c. $\sqrt{7}$

16. π is an irrational number and so is $\sqrt{3}$. Therefore, determine whether $\left(\pi - \sqrt{3} \right)\left(\pi + \sqrt{3} \right)$ is an irrational number.

Problem solving

17. Many composite numbers have a variety of factor pairs. For example, factor pairs of 24 are 1 and 24, 2 and 12, 3 and 8, 4 and 6.

   a. Use each pair of possible factors to simplify the following surds.

      i. $\sqrt{48}$
                       ii. $\sqrt{72}$

   b. Explain if the factor pair chosen when simplifying a surd affects the way the surd is written in simplified form.

   c. Explain if the factor pair chosen when simplifying a surd affects the value of the surd when it is written in simplified form.

18. Consider the expression $\left(\sqrt{p} + \sqrt{q} \right)\left(\sqrt{m} - \sqrt{n} \right)$. Determine under what conditions the expression will produce a rational number.

19. Solve $\sqrt{3}x - \sqrt{12} = \sqrt{3}$ and indicate whether the result is rational or irrational.

LESSON
1.4 Operations with surds (10A)

⊙ 1.4.1 Multiplying and simplifying surds

eles-4664

Multiplication of surds

- To multiply surds, multiply the expressions under the radical sign.
  For example: $\sqrt{8} \times \sqrt{3} = \sqrt{8 \times 3} = \sqrt{24}$
- If there are coefficients in front of the surds that are being multiplied, multiply the coefficients and then multiply the expressions under the radical signs.
  For example: $2\sqrt{3} \times 5\sqrt{7} = (2 \times 5)\sqrt{3 \times 7} = 10\sqrt{21}$

Multiplication of surds

In order to multiply two or more surds, use the following:

- $\sqrt{a} \times \sqrt{b} = \sqrt{a \times b}$
- $m\sqrt{a} \times n\sqrt{b} = mn\sqrt{a \times b}$

where a and b are positive real numbers.

Simplification of surds

- To simplify a surd means to make the number under the radical sign as small as possible.
- Surds can only be simplified if the number under the radical sign has a factor that is a perfect square $(4, 9, 16, 25, 36, \ldots)$.
- Simplification of a surd uses the method of multiplying surds in reverse.
- The process is summarised in the following steps:
  1. Split the number under the radical into the product of two factors, one of which is a perfect square.
  2. Write the surd as the product of two surds multiplied together. The two surds must correspond to the factors identified in step 1.
  3. Simplify the surd of the perfect square and write the surd in the form $a\sqrt{b}$.
- The example below shows the how the surd $\sqrt{45}$ can be simplified by following the steps 1 to 3.

$$\begin{aligned} \sqrt{45} &= \sqrt{9 \times 5} &\text{(Step 1)} \\ &= \sqrt{9} \times \sqrt{5} &\text{(Step 2)} \\ &= 3 \times \sqrt{5} = 3\sqrt{5} &\text{(Step 3)} \end{aligned}$$

- If possible, try to factorise the number under the radical sign so that the largest possible perfect square is used. This will ensure the surd is simplified in one step.

Simplification of surds

$$\begin{aligned} \sqrt{n} &= \sqrt{a^2 \times b} \\ &= \sqrt{a^2} \times \sqrt{b} \\ &= a \times \sqrt{b} \\ &= a\sqrt{b} \end{aligned}$$

WORKED EXAMPLE 4 Simplifying surds

Simplify the following surds. Assume that x and y are positive real numbers.

a. $\sqrt{384}$ b. $3\sqrt{405}$ c. $-\dfrac{1}{8}\sqrt{175}$ d. $5\sqrt{180x^3y^5}$

THINK

WRITE

a. 1. Express 384 as a product of two factors where one factor is the largest possible perfect square.

a. $\sqrt{384} = \sqrt{64 \times 6}$

   2. Express $\sqrt{64 \times 6}$ as the product of two surds.

$= \sqrt{64} \times \sqrt{6}$

3. Simplify the square root from the perfect square (that is, $\sqrt{64} = 8$).

$= 8\sqrt{6}$

b. 1. Express 405 as a product of two factors, one of which is the largest possible perfect square.

b. $3\sqrt{405} = 3\sqrt{81 \times 5}$

2. Express $\sqrt{81 \times 5}$ as a product of two surds.

$= 3\sqrt{81} \times \sqrt{5}$

3. Simplify $\sqrt{81}$.

$= 3 \times 9\sqrt{5}$

4. Multiply together the whole numbers outside the square root sign (3 and 9).

$= 27\sqrt{5}$

c. 1. Express 175 as a product of two factors in which one factor is the largest possible perfect square.

c. $-\dfrac{1}{8}\sqrt{175} = -\dfrac{1}{8}\sqrt{25 \times 7}$

2. Express $\sqrt{25 \times 7}$ as a product of 2 surds.

$= -\dfrac{1}{8} \times \sqrt{25} \times \sqrt{7}$

3. Simplify $\sqrt{25}$.

$= -\dfrac{1}{8} \times 5\sqrt{7}$

4. Multiply together the numbers outside the square root sign.

$= -\dfrac{5}{8}\sqrt{7}$

d. 1. Express each of 180, x^3 and y^5 as a product of two factors where one factor is the largest possible perfect square.

d. $5\sqrt{180x^3y^5} = 5\sqrt{36 \times 5 \times x^2 \times x \times y^4 \times y}$

2. Separate all perfect squares into one surd and all other factors into the other surd.

$= 5 \times \sqrt{36x^2y^4} \times \sqrt{5xy}$

3. Simplify $\sqrt{36x^2y^4}$.

$= 5 \times 6 \times x \times y^2 \times \sqrt{5xy}$

4. Multiply together the numbers and the pronumerals outside the square root sign.

$= 30xy^2\sqrt{5xy}$

WORKED EXAMPLE 5 Multiplying surds

Multiply the following surds, expressing answers in the simplest form. Assume that x and y are positive real numbers.

a. $\sqrt{11} \times \sqrt{7}$ b. $5\sqrt{3} \times 8\sqrt{5}$ c. $6\sqrt{12} \times 2\sqrt{6}$ d. $\sqrt{15x^5y^2} \times \sqrt{12x^2y}$

THINK

WRITE

a. Multiply the surds together, using $\sqrt{a} \times \sqrt{b} = \sqrt{ab}$ (that is, multiply expressions under the square root sign).
Note: This expression cannot be simplified any further.

a. $\sqrt{11} \times \sqrt{7} = \sqrt{11 \times 7}$
$\qquad\qquad\quad = \sqrt{77}$

b. Multiply the coefficients together and then multiply the surds together.

b. $5\sqrt{3} \times 8\sqrt{5} = 5 \times 8 \times \sqrt{3} \times \sqrt{5}$
$= 40 \times \sqrt{3 \times 5}$
$= 40\sqrt{15}$

c. 1. Simplify $\sqrt{12}$.

c. $6\sqrt{12} \times 2\sqrt{6} = 6\sqrt{4 \times 3} \times 2\sqrt{6}$
$= 6 \times 2\sqrt{3} \times 2\sqrt{6}$
$= 12\sqrt{3} \times 2\sqrt{6}$

2. Multiply the coefficients together and multiply the surds together.

$= 24\sqrt{18}$

3. Simplify the surd.

$= 24\sqrt{9 \times 2}$
$= 24 \times 3\sqrt{2}$
$= 72\sqrt{2}$

d. 1. Simplify each of the surds.

d. $\sqrt{15x^5y^2} \times \sqrt{12x^2y}$
$= \sqrt{15 \times x^4 \times x \times y^2} \times \sqrt{4 \times 3 \times x^2 \times y}$
$= x^2 \times y \times \sqrt{15 \times x} \times 2 \times x \times \sqrt{3 \times y}$
$= x^2y\sqrt{15x} \times 2x\sqrt{3y}$

2. Multiply the coefficients together and the surds together.

$= x^2y \times 2x\sqrt{15x \times 3y}$
$= 2x^3y\sqrt{45xy}$
$= 2x^3y\sqrt{9 \times 5xy}$

3. Simplify the surd.

$= 2x^3y \times 3\sqrt{5xy}$
$= 6x^3y\sqrt{5xy}$

- When working with surds, it is sometimes necessary to multiply surds by themselves; that is, square them. Consider the following examples:

$$\left(\sqrt{2}\right)^2 = \sqrt{2} \times \sqrt{2} = \sqrt{4} = 2$$
$$\left(\sqrt{5}\right)^2 = \sqrt{5} \times \sqrt{5} = \sqrt{25} = 5$$

- Observe that squaring a surd produces the number under the radical sign. This is not surprising, because squaring and taking the square root are *inverse operations* and, when applied together, leave the original unchanged.

Squaring surds

When a surd is squared, the result is the expression under the radical sign; that is:

$$\left(\sqrt{a}\right)^2 = a$$

where a is a positive real number.

WORKED EXAMPLE 6 Squaring surds

Simplify each of the following expressions.

a. $\left(\sqrt{6}\right)^2$

b. $\left(3\sqrt{5}\right)^2$

THINK

WRITE

a. Use $\left(\sqrt{a}\right)^2 = a$, where $a = 6$.

a. $\left(\sqrt{6}\right)^2 = 6$

b. 1. Square 3 and apply $\left(\sqrt{a}\right)^2 = a$ to square $\sqrt{5}$.

b. $\left(3\sqrt{5}\right)^2 = 3^2 \times \left(\sqrt{5}\right)^2$

$\qquad = 9 \times 5$

2. Simplify.

$\qquad = 45$

1.4.2 Addition and subtraction of surds

eles-4665

- Surds may be added or subtracted only if they are *like*.
  Examples of *like* surds include $\sqrt{7}, 3\sqrt{7}$ and $-5\sqrt{7}$.

  Examples of *unlike* surds include $\sqrt{11}, \sqrt{5}, 2\sqrt{13}$ and $-2\sqrt{3}$.
- In some cases surds will need to be simplified before you decide whether they are like or unlike, and then addition and subtraction can take place. The concept of adding and subtracting surds is similar to adding and subtracting like terms in algebra.

WORKED EXAMPLE 7 Adding and subtracting surds

Simplify each of the following expressions containing surds. Assume that a and b are positive real numbers.

a. $3\sqrt{6} + 17\sqrt{6} - 2\sqrt{6}$

b. $5\sqrt{3} + 2\sqrt{12} - 5\sqrt{2} + 3\sqrt{8}$

c. $\dfrac{1}{2}\sqrt{100a^3b^2} + ab\sqrt{36a} - 5\sqrt{4a^2b}$

THINK

WRITE

a. All 3 terms are alike because they contain the same surd ($\sqrt{6}$). Simplify.

a. $3\sqrt{6} + 17\sqrt{6} - 2\sqrt{6} = (3 + 17 - 2)\sqrt{6}$

$\qquad = 18\sqrt{6}$

b. 1. Simplify surds where possible.

b. $5\sqrt{3} + 2\sqrt{12} - 5\sqrt{2} + 3\sqrt{8}$

$\quad = 5\sqrt{3} + 2\sqrt{4 \times 3} - 5\sqrt{2} + 3\sqrt{4 \times 2}$

$\quad = 5\sqrt{3} + 2 \times 2\sqrt{3} - 5\sqrt{2} + 3 \times 2\sqrt{2}$

2. Add like terms to obtain the simplified answer.

$\quad = 5\sqrt{3} + 4\sqrt{3} - 5\sqrt{2} + 6\sqrt{2}$

$\quad = 9\sqrt{3} + \sqrt{2}$

c. 1. Simplify surds where possible.

c. $\dfrac{1}{2}\sqrt{100a^3b^2} + ab\sqrt{36a} - 5\sqrt{4a^2b}$

$= \dfrac{1}{2} \times 10\sqrt{a^2 \times a \times b^2} + ab \times 6\sqrt{a} - 5 \times 2 \times a\sqrt{b}$

$= \dfrac{1}{2} \times 10 \times a \times b\sqrt{a} + ab \times 6\sqrt{a} - 5 \times 2 \times a\sqrt{b}$

2. Add like terms to obtain the simplified answer.

$= 5ab\sqrt{a} + 6ab\sqrt{a} - 10a\sqrt{b}$

$= 11ab\sqrt{a} - 10a\sqrt{b}$

TI	THINK	DISPLAY/WRITE	CASIO	THINK	DISPLAY/WRITE
a–c.	**a–c.**	**a–c.**	**a–c.**
In a new document, on a Calculator page, complete the entry lines as: $3\sqrt{6} + 17\sqrt{6} - 2\sqrt{6}$ $5\sqrt{3} + 2\sqrt{12} - 5\sqrt{2}$ $+3\sqrt{8}$ $\dfrac{1}{2}\sqrt{100a^3b^2} + ab\sqrt{36a} - 5\sqrt{4a^2b} \mid a > 0$ and $b > 0$ Press ENTER after each entry.	$3\sqrt{6} + 17\sqrt{6} - 2\sqrt{6} = 18\sqrt{6}$ $5\sqrt{3} + 2\sqrt{12} - 5\sqrt{2} + 3\sqrt{8}$ $= 9\sqrt{3} + \sqrt{2}$ $\dfrac{1}{2}\sqrt{100a^3b^2} + ab\sqrt{36a}$ $- 5\sqrt{4a^2b} = 11a^{\frac{2}{3}}b - 10a\sqrt{b}$	On the Main screen, complete the entry lines as: $3\sqrt{6} + 17\sqrt{6} - 2\sqrt{6}$ $5\sqrt{3} + 2\sqrt{12} - 5\sqrt{2}$ $+3\sqrt{8}$ simplify $\left(\dfrac{1}{2}\sqrt{100a^3b^2} + \right.$ $a \times b \times \sqrt{36a} -$ $5\sqrt{4a^2b} \mid a > 0 \mid b > 0)$ Press EXE after each entry.	$3\sqrt{6} + 17\sqrt{6} - 2\sqrt{6} = 18\sqrt{6}$ $5\sqrt{3} + 2\sqrt{12} - 5\sqrt{2} + 3\sqrt{8}$ $= 9\sqrt{3} + \sqrt{2}$ $\dfrac{1}{2}\sqrt{100a^3b^2} + ab\sqrt{36a}$ $- 5\sqrt{4a^2b} = 11a^{\frac{2}{3}}b - 10a\sqrt{b}$

▶ 1.4.3 Dividing surds

eles-4666

- To divide surds, divide the expressions under the radical signs.
- When dividing surds it is best to simplify them (if possible) first. Once this has been done, the coefficients are divided next, and then the surds are divided.

> **Dividing surds**
>
> $$\dfrac{\sqrt{a}}{\sqrt{b}} = \sqrt{\dfrac{a}{b}}$$
>
> $$\dfrac{m\sqrt{a}}{n\sqrt{b}} = \dfrac{m}{n}\sqrt{\dfrac{a}{b}}$$
>
> **where a and b are positive real numbers.**

Divide the following surds, expressing answers in the simplest form. Assume that x and y are positive real numbers.

a. $\dfrac{\sqrt{55}}{\sqrt{5}}$
b. $\dfrac{\sqrt{48}}{\sqrt{3}}$
c. $\dfrac{9\sqrt{88}}{6\sqrt{99}}$
d. $\dfrac{\sqrt{36xy}}{\sqrt{25x^9y^{11}}}$

THINK	WRITE
a. 1. Rewrite the fraction using $\dfrac{\sqrt{a}}{\sqrt{b}} = \sqrt{\dfrac{a}{b}}$.	a. $\dfrac{\sqrt{55}}{\sqrt{5}} = \sqrt{\dfrac{55}{5}}$
2. Divide the numerator by the denominator (that is, 55 by 5). Check if the surd can be simplified any further.	$= \sqrt{11}$
b. 1. Rewrite the fraction using $\dfrac{\sqrt{a}}{\sqrt{b}} = \sqrt{\dfrac{a}{b}}$.	b. $\dfrac{\sqrt{48}}{\sqrt{3}} = \sqrt{\dfrac{48}{3}}$
2. Divide 48 by 3.	$= \sqrt{16}$
3. Evaluate $\sqrt{16}$.	$= 4$
c. 1. Rewrite the surds using $\dfrac{\sqrt{a}}{\sqrt{b}} = \sqrt{\dfrac{a}{b}}$.	c. $\dfrac{9\sqrt{88}}{6\sqrt{99}} = \dfrac{9}{6}\sqrt{\dfrac{88}{99}}$
2. Simplify the fraction under the radical by dividing both numerator and denominator by 11.	$= \dfrac{9}{6}\sqrt{\dfrac{8}{9}}$
3. Simplify surds.	$= \dfrac{9 \times 2\sqrt{2}}{6 \times 3}$
4. Multiply the whole numbers in the numerator together and those in the denominator together.	$= \dfrac{18\sqrt{2}}{18}$
5. Cancel the common factor of 18.	$= \sqrt{2}$
d. 1. Simplify each surd.	d. $\dfrac{\sqrt{36xy}}{\sqrt{25x^9y^{11}}} = \dfrac{6\sqrt{xy}}{5\sqrt{x^8 \times x \times y^{10} \times y}}$
	$= \dfrac{6\sqrt{xy}}{5x^4y^5\sqrt{xy}}$
2. Cancel any common factors — in this case $\sqrt{xy}$.	$= \dfrac{6}{5x^4y^5}$

⏵ 1.4.4 Rationalising denominators

eles-4667

- If the **denominator** of a fraction is a surd, it can be changed into a rational number through multiplication. In other words, it can be rationalised.
- As discussed earlier in this chapter, squaring a simple surd (that is, multiplying it by itself) results in a rational number. This fact can be used to rationalise denominators as follows.

Rationalising the denominator

$$\frac{\sqrt{a}}{\sqrt{b}} = \frac{\sqrt{a}}{\sqrt{b}} \times \frac{\sqrt{b}}{\sqrt{b}} = \frac{\sqrt{ab}}{b}$$

- If both numerator and denominator of a fraction are multiplied by the surd contained in the denominator, the denominator becomes a rational number. The fraction takes on a different appearance, but its numerical value is unchanged, because multiplying the numerator and denominator by the same number is equivalent to multiplying by 1.

WORKED EXAMPLE 9 Rationalising the denominator

Express the following in their simplest form with a rational denominator.

a. $\dfrac{\sqrt{6}}{\sqrt{13}}$

b. $\dfrac{2\sqrt{12}}{3\sqrt{54}}$

c. $\dfrac{\sqrt{17} - 3\sqrt{14}}{\sqrt{7}}$

THINK	WRITE
a. 1. Write the fraction.	a. $\dfrac{\sqrt{6}}{\sqrt{13}}$
2. Multiply both the numerator and denominator by the surd contained in the denominator (in this case $\sqrt{13}$). This has the same effect as multiplying the fraction by 1, because $\dfrac{\sqrt{13}}{\sqrt{13}} = 1$.	$= \dfrac{\sqrt{6}}{\sqrt{13}} \times \dfrac{\sqrt{13}}{\sqrt{13}}$ $= \dfrac{\sqrt{78}}{13}$
b. 1. Write the fraction.	b. $\dfrac{2\sqrt{12}}{3\sqrt{54}}$
2. Simplify the surds. (This avoids dealing with large numbers.)	$\dfrac{2\sqrt{12}}{3\sqrt{54}} = \dfrac{2\sqrt{4 \times 3}}{3\sqrt{9 \times 6}}$ $= \dfrac{2 \times 2\sqrt{3}}{3 \times 3\sqrt{6}}$ $= \dfrac{4\sqrt{3}}{9\sqrt{6}}$

3. Multiply both the numerator and denominator by $\sqrt{6}$. This has the same effect as multiplying the fraction by 1,

because $\dfrac{\sqrt{6}}{\sqrt{6}} = 1$.

Note: We need to multiply only by the surd part of the denominator (that is, by $\sqrt{6}$ rather than by $9\sqrt{6}$).

$$= \frac{4\sqrt{3}}{9\sqrt{6}} \times \frac{\sqrt{6}}{\sqrt{6}}$$

$$= \frac{4\sqrt{18}}{9 \times 6}$$

4. Simplify $\sqrt{18}$.

$$= \frac{4\sqrt{9 \times 2}}{9 \times 6}$$

$$= \frac{4 \times 3\sqrt{2}}{54}$$

$$= \frac{12\sqrt{2}}{54}$$

5. Divide both the numerator and denominator by 6 (cancel down).

$$= \frac{2\sqrt{2}}{9}$$

c. 1. Write the fraction.

c.

$$\frac{\sqrt{17} - 3\sqrt{14}}{\sqrt{7}}$$

2. Multiply both the numerator and denominator by $\sqrt{7}$. Use grouping symbols (brackets) to make it clear that the whole numerator must be multiplied by $\sqrt{7}$.

$$= \frac{\left(\sqrt{17} - 3\sqrt{14}\right)}{\sqrt{7}} \times \frac{\sqrt{7}}{\sqrt{7}}$$

3. Apply the Distributive Law in the numerator.
$a(b + c) = ab + ac$

$$= \frac{\sqrt{17} \times \sqrt{7} - 3\sqrt{14} \times \sqrt{7}}{\sqrt{7} \times \sqrt{7}}$$

$$= \frac{\sqrt{119} - 3\sqrt{98}}{7}$$

4. Simplify $\sqrt{98}$.

$$= \frac{\sqrt{119} - 3\sqrt{49 \times 2}}{7}$$

$$= \frac{\sqrt{119} - 3 \times 7\sqrt{2}}{7}$$

$$= \frac{\sqrt{119} - 21\sqrt{2}}{7}$$

▶ 1.4.5 Rationalising denominators using conjugate surds

eles-4668

- The product of pairs of **conjugate surds** results in a rational number.
- Examples of pairs of conjugate surds include $\sqrt{6} + 11$ and $\sqrt{6} - 11$, $\sqrt{a} + b$ and $\sqrt{a} - b$, $2\sqrt{5} - \sqrt{7}$ and $2\sqrt{5} + \sqrt{7}$.
  This fact is used to rationalise denominators containing a sum or a difference of surds.

Using conjugates to rationalise the denominator

- To rationalise a denominator that contains a sum or a difference of surds, multiply both the numerator and the denominator by the conjugate of the denominator.
  Two examples are given below.

  1. To rationalise the denominator of the fraction $\dfrac{1}{\sqrt{a}+\sqrt{b}}$, multiply it by $\dfrac{\sqrt{a}-\sqrt{b}}{\sqrt{a}-\sqrt{b}}$.

  2. To rationalise the denominator of the fraction $\dfrac{1}{\sqrt{a}-\sqrt{b}}$, multiply it by $\dfrac{\sqrt{a}+\sqrt{b}}{\sqrt{a}+\sqrt{b}}$.

- A quick way to simplify the denominator is to use the difference of two squares identity:

$$\left(\sqrt{a}-\sqrt{b}\right)\left(\sqrt{a}+\sqrt{b}\right) = \left(\sqrt{a}\right)^2 - \left(\sqrt{b}\right)^2$$
$$= a - b$$

WORKED EXAMPLE 10 Using conjugates to rationalise the denominator

Rationalise the denominator and simplify the following expressions.

a. $\dfrac{1}{4-\sqrt{3}}$

b. $\dfrac{\sqrt{6}+3\sqrt{2}}{3+\sqrt{3}}$

THINK

WRITE

a. 1. Write the fraction.

a. $\dfrac{1}{4-\sqrt{3}}$

2. Multiply the numerator and denominator by the conjugate of the denominator.
   (Note that $\dfrac{\left(4+\sqrt{3}\right)}{\left(4+\sqrt{3}\right)}=1$.)

$= \dfrac{1}{\left(4-\sqrt{3}\right)} \times \dfrac{\left(4+\sqrt{3}\right)}{\left(4+\sqrt{3}\right)}$

3. Apply the Distributive Law in the numerator and the difference of two squares identity in the denominator.

$= \dfrac{4+\sqrt{3}}{(4)^2 - \left(\sqrt{3}\right)^2}$

4. Simplify.

$= \dfrac{4+\sqrt{3}}{16-3}$

$= \dfrac{4+\sqrt{3}}{13}$

b. 1. Write the fraction.

b. $\dfrac{\sqrt{6}+3\sqrt{2}}{3+\sqrt{3}}$

2. Multiply the numerator and denominator by the conjugate of the denominator.
   (Note that $\dfrac{\left(3-\sqrt{3}\right)}{\left(3-\sqrt{3}\right)}=1$.)

$= \dfrac{\left(\sqrt{6}+3\sqrt{2}\right)}{\left(3+\sqrt{3}\right)} \times \dfrac{\left(3-\sqrt{3}\right)}{\left(3-\sqrt{3}\right)}$

3. Multiply the expressions in grouping symbols in the numerator and apply the difference of two squares identity in the denominator.

$$= \frac{\sqrt{6} \times 3 + \sqrt{6} \times \left(-\sqrt{3}\right) + 3\sqrt{2} \times 3 + 3\sqrt{2} \times - \left(\sqrt{3}\right)}{(3)^2 - \left(\sqrt{3}\right)^2}$$

4. Simplify.

$$= \frac{3\sqrt{6} - \sqrt{18} + 9\sqrt{2} - 3\sqrt{6}}{9 - 3}$$

$$= \frac{-\sqrt{18} + 9\sqrt{2}}{6}$$

$$= \frac{-\sqrt{9 \times 2} + 9\sqrt{2}}{6}$$

$$= \frac{-3\sqrt{2} + 9\sqrt{2}}{6}$$

$$= \frac{6\sqrt{2}}{6}$$

$$= \sqrt{2}$$

TI | THINK	DISPLAY/WRITE	CASIO | THINK	DISPLAY/WRITE
a, b. On a Calculator page, complete the entry lines as: $$\frac{1}{4 - \sqrt{3}}$$ $$\frac{\sqrt{6} + 3\sqrt{2}}{3 + \sqrt{3}}$$ Press ENTER after each entry.	**a, b.** $$\frac{1}{4 - \sqrt{3}} = \frac{4 + \sqrt{3}}{13}$$ $$\frac{\sqrt{6} + 3\sqrt{2}}{3 + \sqrt{3}} = \sqrt{2}$$	**a, b.** On the Main screen, complete the entry lines as: $$\text{simplify} \left(\frac{1}{4 - \sqrt{3}} \right)$$ $$\text{simplify} \left(\frac{\sqrt{6} + 3\sqrt{2}}{3 + \sqrt{3}} \right)$$ Press EXE after each entry.	**a, b.** $$\frac{1}{4 - \sqrt{3}} = \frac{4 + \sqrt{3}}{13}$$ $$\frac{\sqrt{6} + 3\sqrt{2}}{3 + \sqrt{3}} = \sqrt{2}$$

 Resources

 eWorkbook   Topic 1 Workbook (worksheets, code puzzle and project) (ewbk-2093)

Video eLessons Surds (eles-1906)
                    Rationalisation of surds (eles-1948)

 Interactivities   Simplifying surds (int-6028)
                    Addition and subtraction of surds (int-6190)
                    Multiplying surds (int-6191)
                    Dividing surds (int-6192)
                    Conjugate surds (int-6193)

1.4 Quick quiz on	1.4 Exercise

Individual pathways

■ PRACTISE	■ CONSOLIDATE	■ MASTER
1, 4, 7, 10, 12, 15, 18, 21, 24, 27, 30, 33, 36, 40	2, 5, 8, 11, 13, 16, 19, 22, 25, 28, 31, 34, 37, 38, 41	3, 6, 9, 14, 17, 20, 23, 26, 29, 32, 35, 39, 42

Fluency

WE4a For questions **1** to **3**, simplify the following surds.

1. a. $\sqrt{12}$ b. $\sqrt{24}$ c. $\sqrt{27}$ d. $\sqrt{125}$

2. a. $\sqrt{54}$ b. $\sqrt{112}$ c. $\sqrt{68}$ d. $\sqrt{180}$

3. a. $\sqrt{88}$ b. $\sqrt{162}$ c. $\sqrt{245}$ d. $\sqrt{448}$

WE4b,c For questions **4** to **6**, simplify the following surds.

4. a. $2\sqrt{8}$ b. $8\sqrt{90}$ c. $9\sqrt{80}$ d. $7\sqrt{54}$

5. a. $-6\sqrt{75}$ b. $-7\sqrt{80}$ c. $16\sqrt{48}$ d. $\frac{1}{7}\sqrt{392}$

6. a. $\frac{1}{9}\sqrt{162}$ b. $\frac{1}{4}\sqrt{192}$ c. $\frac{1}{9}\sqrt{135}$ d. $\frac{3}{10}\sqrt{175}$

WE4d For questions **7** to **9**, simplify the following surds. Assume that a, b, c, d, e, f, x and y are positive real numbers.

7. a. $\sqrt{16a^2}$ b. $\sqrt{72a^2}$ c. $\sqrt{90a^2b}$ d. $\sqrt{338a^4}$

8. a. $\sqrt{338a^3b^3}$ b. $\sqrt{68a^3b^5}$ c. $\sqrt{125x^6y^4}$ d. $5\sqrt{80x^3y^2}$

9. a. $6\sqrt{162c^7d^5}$ b. $2\sqrt{405c^7d^9}$ c. $\frac{1}{2}\sqrt{88ef}$ d. $\frac{1}{2}\sqrt{392e^{11}f^{11}}$

10. **WE5a** Simplify the following expressions containing surds. Assume that x and y are positive real numbers.
    a. $3\sqrt{5}+4\sqrt{5}$
    b. $2\sqrt{3}+5\sqrt{3}+\sqrt{3}$
    c. $8\sqrt{5}+3\sqrt{3}+7\sqrt{5}+2\sqrt{3}$
    d. $6\sqrt{11}-2\sqrt{11}$

11. Simplify the following expressions containing surds. Assume that x and y are positive real numbers.
    a. $7\sqrt{2}+9\sqrt{2}-3\sqrt{2}$
    b. $9\sqrt{6}+12\sqrt{6}-17\sqrt{6}-7\sqrt{6}$
    c. $12\sqrt{3}-8\sqrt{7}+5\sqrt{3}-10\sqrt{7}$
    d. $2\sqrt{x}+5\sqrt{y}+6\sqrt{x}-2\sqrt{y}$

WE5b For questions **12** to **14**, simplify the following expressions containing surds. Assume that a and b are positive real numbers.

12. a. $\sqrt{200}-\sqrt{300}$
    b. $\sqrt{125}-\sqrt{150}+\sqrt{600}$
    c. $\sqrt{27}-\sqrt{3}+\sqrt{75}$
    d. $2\sqrt{20}-3\sqrt{5}+\sqrt{45}$

13. a. $6\sqrt{12}+3\sqrt{27}-7\sqrt{3}+\sqrt{18}$
    b. $\sqrt{150}+\sqrt{24}-\sqrt{96}+\sqrt{108}$
    c. $3\sqrt{90}-5\sqrt{60}+3\sqrt{40}+\sqrt{100}$
    d. $5\sqrt{11}+7\sqrt{44}-9\sqrt{99}+2\sqrt{121}$

14. a. $2\sqrt{30} + 5\sqrt{120} + \sqrt{60} - 6\sqrt{135}$

b. $6\sqrt{ab} - \sqrt{12ab} + 2\sqrt{9ab} + 3\sqrt{27ab}$

c. $\dfrac{1}{2}\sqrt{98} + \dfrac{1}{3}\sqrt{48} + \dfrac{1}{3}\sqrt{12}$

d. $\dfrac{1}{8}\sqrt{32} - \dfrac{7}{6}\sqrt{18} + 3\sqrt{72}$

WE5c For questions **15** to **17**, simplify the following expressions containing surds. Assume that a and b are positive real numbers.

15. a. $7\sqrt{a} - \sqrt{8a} + 8\sqrt{9a} - \sqrt{32a}$

b. $10\sqrt{a} - 15\sqrt{27a} + 8\sqrt{12a} + 14\sqrt{9a}$

c. $\sqrt{150ab} + \sqrt{96ab} - \sqrt{54ab}$

d. $16\sqrt{4a^2} - \sqrt{24a} + 4\sqrt{8a^2} + \sqrt{96a}$

16. a. $\sqrt{8a^3} + \sqrt{72a^3} - \sqrt{98a^3}$

b. $\dfrac{1}{2}\sqrt{36a} + \dfrac{1}{4}\sqrt{128a} - \dfrac{1}{6}\sqrt{144a}$

c. $\sqrt{9a^3} + \sqrt{3a^5}$

d. $6\sqrt{a^5b} + \sqrt{a^3b} - 5\sqrt{a^5b}$

17. a. $ab\sqrt{ab} + 3ab\sqrt{a^2b} + \sqrt{9a^3b^3}$

b. $\sqrt{a^3b} + 5\sqrt{ab} - 2\sqrt{ab} + 5\sqrt{a^3b}$

c. $\sqrt{32a^3b^2} - 5ab\sqrt{8a} + \sqrt{48a^5b^6}$

d. $\sqrt{4a^2b} + 5\sqrt{a^2b} - 3\sqrt{9a^2b}$

WE6 For questions **18** to **20**, multiply the following surds, expressing answers in the simplest form. Assume that a, b, x and y are positive real numbers.

18. a. $\sqrt{2} \times \sqrt{7}$

b. $\sqrt{6} \times \sqrt{7}$

c. $\sqrt{8} \times \sqrt{6}$

d. $\sqrt{10} \times \sqrt{10}$

e. $\sqrt{21} \times \sqrt{3}$

f. $\sqrt{27} \times 3\sqrt{3}$

19. a. $5\sqrt{3} \times 2\sqrt{11}$

b. $10\sqrt{15} \times 6\sqrt{3}$

c. $4\sqrt{20} \times 3\sqrt{5}$

d. $10\sqrt{6} \times 3\sqrt{8}$

e. $\dfrac{1}{4}\sqrt{48} \times 2\sqrt{2}$

f. $\dfrac{1}{9}\sqrt{48} \times 2\sqrt{3}$

20. a. $\dfrac{1}{10}\sqrt{60} \times \dfrac{1}{5}\sqrt{40}$

b. $\sqrt{xy} \times \sqrt{x^3y^2}$

c. $\sqrt{3a^4b^2} \times \sqrt{6a^5b^3}$

d. $\sqrt{12a^7b} \times \sqrt{6a^3b^4}$

e. $\sqrt{15x^3y^2} \times \sqrt{6x^2y^3}$

f. $\dfrac{1}{2}\sqrt{15a^3b^3} \times 3\sqrt{3a^2b^6}$

WE7 For questions **21** to **23**, simplify the following expressions.

21. a. $\left(\sqrt{2}\right)^2$

b. $\left(\sqrt{5}\right)^2$

c. $\left(\sqrt{12}\right)^2$

22. a. $\left(\sqrt{15}\right)^2$

b. $\left(3\sqrt{2}\right)^2$

c. $\left(4\sqrt{5}\right)^2$

23. a. $\left(2\sqrt{7}\right)^2$

b. $\left(5\sqrt{8}\right)^2$

WE8 For questions **24** to **26**, divide the following surds, expressing answers in the simplest form. Assume that a, b, x and y are positive real numbers.

24. a. $\dfrac{\sqrt{15}}{\sqrt{3}}$

b. $\dfrac{\sqrt{8}}{\sqrt{2}}$

c. $\dfrac{\sqrt{60}}{\sqrt{10}}$

d. $\dfrac{\sqrt{128}}{\sqrt{8}}$

25. a. $\dfrac{\sqrt{18}}{4\sqrt{6}}$ b. $\dfrac{\sqrt{65}}{2\sqrt{13}}$ c. $\dfrac{\sqrt{96}}{\sqrt{8}}$ d. $\dfrac{7\sqrt{44}}{14\sqrt{11}}$

26. a. $\dfrac{9\sqrt{63}}{15\sqrt{7}}$ b. $\dfrac{\sqrt{2040}}{\sqrt{30}}$ c. $\dfrac{\sqrt{x^4 y^3}}{\sqrt{x^2 y^5}}$

d. $\dfrac{\sqrt{16xy}}{\sqrt{8x^7 y^9}}$ e. $\dfrac{\sqrt{xy}}{\sqrt{x^5 y^7}} \times \dfrac{\sqrt{12x^8 y^{12}}}{\sqrt{x^2 y^3}}$ f. $\dfrac{2\sqrt{2a^2 b^4}}{\sqrt{5a^3 b^6}} \times \dfrac{\sqrt{10a^9 b^3}}{3\sqrt{a^7 b}}$

WE9a,b For questions **27** to **29**, express the following fractions in their simplest form with a rational denominator.

27. a. $\dfrac{5}{\sqrt{2}}$ b. $\dfrac{7}{\sqrt{3}}$ c. $\dfrac{4}{\sqrt{11}}$ d. $\dfrac{8}{\sqrt{6}}$ e. $\dfrac{\sqrt{12}}{\sqrt{7}}$

28. a. $\dfrac{\sqrt{15}}{\sqrt{6}}$ b. $\dfrac{2\sqrt{3}}{\sqrt{5}}$ c. $\dfrac{3\sqrt{7}}{\sqrt{5}}$ d. $\dfrac{5\sqrt{2}}{2\sqrt{3}}$ e. $\dfrac{4\sqrt{3}}{3\sqrt{5}}$

29. a. $\dfrac{5\sqrt{14}}{7\sqrt{8}}$ b. $\dfrac{16\sqrt{3}}{6\sqrt{5}}$ c. $\dfrac{8\sqrt{3}}{7\sqrt{7}}$ d. $\dfrac{8\sqrt{60}}{\sqrt{28}}$ e. $\dfrac{2\sqrt{35}}{3\sqrt{14}}$

Understanding

WE9c For questions **30** to **32**, express the following in their simplest form with a rational denominator.

30. a. $\dfrac{\sqrt{6}+\sqrt{12}}{\sqrt{3}}$ b. $\dfrac{\sqrt{15}-\sqrt{22}}{\sqrt{6}}$ c. $\dfrac{6\sqrt{2}-\sqrt{15}}{\sqrt{10}}$ d. $\dfrac{2\sqrt{18}+3\sqrt{2}}{\sqrt{5}}$

31. a. $\dfrac{3\sqrt{5}+6\sqrt{7}}{\sqrt{8}}$ b. $\dfrac{4\sqrt{2}+3\sqrt{8}}{2\sqrt{3}}$ c. $\dfrac{3\sqrt{11}-4\sqrt{5}}{\sqrt{18}}$ d. $\dfrac{2\sqrt{7}-2\sqrt{5}}{\sqrt{12}}$

32. a. $\dfrac{7\sqrt{12}-5\sqrt{6}}{6\sqrt{3}}$ b. $\dfrac{6\sqrt{2}-\sqrt{5}}{4\sqrt{8}}$ c. $\dfrac{6\sqrt{3}-5\sqrt{5}}{7\sqrt{20}}$ d. $\dfrac{3\sqrt{5}+7\sqrt{3}}{5\sqrt{24}}$

WE10 For questions **33** to **35**, rationalise the denominator and simplify.

33. a. $\dfrac{1}{\sqrt{5}+2}$ b. $\dfrac{1}{\sqrt{8}-\sqrt{5}}$ c. $\dfrac{4}{2\sqrt{11}-\sqrt{13}}$

34. a. $\dfrac{5\sqrt{3}}{3\sqrt{5}+4\sqrt{2}}$ b. $\dfrac{\sqrt{8}-3}{\sqrt{8}+3}$ c. $\dfrac{\sqrt{12}-\sqrt{7}}{\sqrt{12}+\sqrt{7}}$

35. a. $\dfrac{\sqrt{3}-1}{\sqrt{5}+1}$ b. $\dfrac{3\sqrt{6}-\sqrt{15}}{\sqrt{6}+2\sqrt{3}}$ c. $\dfrac{\sqrt{5}-\sqrt{3}}{4\sqrt{2}-\sqrt{3}}$

Reasoning

36. Calculate the area of a triangle with base length $\dfrac{3}{\sqrt{2}+2}$ and perpendicular height $\dfrac{5}{\sqrt{8}-1}$. Express your answer with a rational denominator. Show full working.

37. Provide specific examples to show the following inequations for $a, b > 0$.

    a. $\sqrt{a+b} \neq \sqrt{a} + \sqrt{b}$ **b.** $\sqrt{a-b} \neq \sqrt{a} - \sqrt{b}$

38. Determine the average of $\dfrac{1}{2\sqrt{x}}$ and $\dfrac{1}{3-2\sqrt{x}}$, writing your answer with a rational denominator. Show full working.

39. **a.** Show that $\left(\sqrt{a} + \sqrt{b}\right)^2 = a + b + 2\sqrt{ab}$.

    b. Use this result to evaluate the following.

      i. $\sqrt{8 + 2\sqrt{15}}$ **ii.** $\sqrt{8 - 2\sqrt{15}}$ **iii.** $\sqrt{7 + 4\sqrt{3}}.$

Problem solving

40. Simplify $\dfrac{\sqrt{5} + \sqrt{3}}{\sqrt{3} + \sqrt{3} + \sqrt{5}} - \dfrac{\sqrt{5} - \sqrt{3}}{\sqrt{3} + \sqrt{3} - \sqrt{5}}.$

41. Solve the following equations for x, where x is positive.

    a. $\sqrt{9+x} - \sqrt{x} = \dfrac{5}{\sqrt{9+x}}$ **b.** $\dfrac{9\sqrt{x} - 7}{3\sqrt{x}} = \dfrac{3\sqrt{x} + 1}{\sqrt{x} + 5}$

42. Solve the following equation for x: $2 - \sqrt{2 - \sqrt{2 - \sqrt{2 - \ldots}}} = x$

LESSON
1.5 Review of exponent laws

LEARNING INTENTIONS

At the end of this lesson you should be able to:
- recall and apply the exponent laws
- simplify expressions involving multiplication and division of terms with the same base
- evaluate expressions involving powers of zero
- simplify expressions involving raising a power to another power.

1.5.1 Review of exponent laws

eles-4669

Exponent notation

- When a number or pronumeral is repeatedly multiplied by itself, it can be written in a shorter form called **exponent form**.
- Other names for an exponent are *index* or *power*.
- A number written in exponent form has two parts, the **base** and the **exponent**, and is written as shown.
- In the example shown, a is the base and x is the exponent.

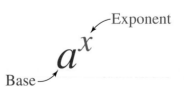

Exponent laws

• Performing operations on numbers or pronumerals written in exponent form requires application of the exponent laws. There are eight exponent or index laws; six are discussed in this lesson.

First Exponent Law

When terms with the same base are multiplied, the exponents are added.

$$a^m \times a^n = a^{m+n}$$

Second Exponent Law

When terms with the same base are divided, the exponents are subtracted.

$$a^m \div a^n = a^{m-n}$$

WORKED EXAMPLE 11 Simplifying using the first two exponent laws

Simplify each of the following expressions.

a. $m^4n^3p \times m^2n^5p^3$

b. $2a^2b^3 \times 3ab^4$

c. $\dfrac{2x^5y^4}{10x^2y^3}$

THINK	WRITE
a. 1. Write the expression.	a. $m^4n^3p \times m^2n^5p^3$
2. Multiply the terms with the same base by adding the exponents. *Note: $p = p^1$.*	$= m^{4+2}n^{3+5}p^{1+3}$ $= m^6n^8p^4$
b. 1. Write the expression.	b. $2a^2b^3 \times 3ab^4$
2. Simplify by multiplying the coefficients, then multiply the terms with the same base by adding the exponents.	$= 2 \times 3 \times a^{2+1} \times b^{3+4}$ $= 6a^3b^7$
c. 1. Write the expression.	c. $\dfrac{2x^5y^4}{10x^2y^3}$
2. Simplify by dividing both of the coefficients by the same factor, then divide terms with the same base by subtracting the exponents.	$= \dfrac{1x^{5-2}y^{4-3}}{5}$ $= \dfrac{x^3y}{5}$

TI	THINK	DISPLAY/WRITE	CASIO	THINK	DISPLAY/WRITE

a–c.

In a new document on a Calculator page, complete the entry lines as:

$m^4 \times n^3 \times p \times m^2 \times n^5 \times p^3$

$2 \times a^2 \times b^3 \times 3 \times a \times b^4$

$\dfrac{2 \times x^5 \times y^4}{10 \times x^2 \times y^3}$

Press ENTER after each entry.

Be sure to include the multiplication sign between each variable.

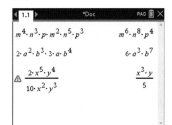

a–c.

$m^4 n^3 p \times m^2 n^5 p^3 = m^6 n^8 p^4$

$2a^2 b^3 \times 3ab^4 = 6a^3 b^7$

$\dfrac{2x^5 y^4}{10x^2 y^3} = \dfrac{x^3 y}{5}$

a–c.

On the Main screen, using the Var tab, complete the entry lines as:

$m^4 n^3 p \times m^2 n^5 p^3$

$2a^2 b^3 \times 3ab^4$

$\dfrac{2x^5 y^4}{10x^2 y^3}$

Press EXE after each entry.

a–c.

$m^4 n^3 p \times m^2 n^5 p^3 = m^6 n^8 p^4$

$2a^2 b^3 \times 3ab^4 = 6a^3 b^7$

$\dfrac{2x^5 y^4}{10x^2 y^3} = \dfrac{x^3 y}{5}$

Third Exponent Law

Any term (excluding 0) with an exponent of 0 is equal to 1.

$$a^0 = 1, \ a \neq 0$$

WORKED EXAMPLE 12 Simplifying terms with exponents of zero

Simplify each of the following expressions.

a. $\left(2b^3\right)^0$

b. $-4\left(a^2 b^5\right)^0$

THINK	WRITE
a. 1. Write the expression.	**a.** $\left(2b^3\right)^0$
2. Apply the Third Exponent Law, which states that any term (excluding 0) with an exponent of 0 is equal to 1.	$= 1$
b. 1. Write the expression.	**b.** $-4\left(a^2 b^5\right)^0$
2. The entire term inside the brackets has an exponent of 0, so the bracket is equal to 1.	$= -4 \times 1$
3. Simplify.	$= -4$

Fourth Exponent Law

When a power (a^m) is raised to a power, the exponents are multiplied.

$$(a^m)^n = a^{mn}$$

Fifth Exponent Law

When the base is a product, raise every part of the product to the exponent outside the brackets.

$$(ab)^m = a^m b^m$$

Sixth Exponent Law

When the base is a fraction, raise both the numerator and denominator to the exponent outside the brackets.

$$\left(\frac{a}{b}\right)^m = \frac{a^m}{b^m}$$

WORKED EXAMPLE 13 Simplifying terms in exponent form raised to a power

Simplify each of the following expressions.

a. $\left(2n^4\right)^3$ b. $\left(3a^2b^7\right)^3$ c. $\left(\dfrac{2x^3}{y^4}\right)^4$ d. $(-4)^3$

THINK	WRITE
a. 1. Write the expression.	a. $\left(2n^4\right)^3$
2. Apply the Fourth Exponent Law and simplify.	$= 2^{1\times3} \times n^{4\times3}$
	$= 2^3 n^{12}$
	$= 8n^{12}$
b. 1. Write the expression.	b. $\left(3a^2b^7\right)^3$
2. Apply the Fifth Exponent Law and simplify.	$= 3^{1\times3} \times a^{2\times3} \times b^{7\times3}$
	$= 3^3 a^6 b^{21}$
	$= 27a^6 b^{21}$
c. 1. Write the expression.	c. $\left(\dfrac{2x^3}{y^4}\right)^4$
2. Apply the Sixth Exponent Law and simplify.	$= \dfrac{2^{1\times4} \times x^{3\times4}}{y^{4\times4}}$
	$= \dfrac{16x^{12}}{y^{16}}$
d. 1. Write the expression.	d. $(-4)^3$
2. Write in expanded form.	$= -4 \times -4 \times -4$
3. Simplify, taking careful note of the negative sign.	$= -64$

| TI | THINK | DISPLAY/WRITE | CASIO | THINK | DISPLAY/WRITE |

TI | THINK

a–d.

On a Calculator page, use the brackets and complete the entry lines as:

$(2n^4)^3$

$(3a^2b^7)^3$

$\left(\dfrac{2x^3}{y^4}\right)^4$

$(-4)^3$

Press Enter after each entry.

DISPLAY/WRITE

a–d.

$(2n^4)^3 = 8n^{12}$

$(3a^2b^7)^3 = 27a^6b^{21}$

$\left(\dfrac{2x^3}{y^4}\right)^4 = \dfrac{16x^{12}}{y^{16}}$

$(-4)^3 = -64$

CASIO | THINK

a–d.

On the Main screen, use the brackets and complete the entry lines as:

$(2n^4)^3$

$(3a^2b^7)^3$

$\left(\dfrac{2x^3}{y^4}\right)^4$

$(-4)^3$

Press EXE after each entry.

DISPLAY/WRITE

a–d.

$(2n^4)^3 = 8n^{12}$

$(3a^2b^7)^3 = 27a^6b^{21}$

$\left(\dfrac{2x^3}{y^4}\right)^4 = \dfrac{16x^{12}}{y^{16}}$

$(-4)^3 = -64$

 Resources

 eWorkbook   Topic 1 Workbook (worksheets, code puzzle and project) (ewbk-2093)

 Video eLesson Index laws (eles-1903)

Interactivities Individual pathway interactivity: Review of index laws (int-4562)

First Index Law (int-3709)

Second Index Law (int-3711)

Third Index Law (int-3713)

Fourth Index Law — Multiplication (int-3716)

Fifth and sixth index laws (int-6063)

Exercise 1.5 Review of exponent laws

learn

| 1.5 Quick quiz on | 1.5 Exercise |

Individual pathways

■ PRACTISE	■ CONSOLIDATE	■ MASTER
1, 4, 7, 10, 13, 15, 18, 21, 22, 26	2, 5, 8, 11, 14, 16, 19, 23, 24, 27	3, 6, 9, 12, 17, 20, 25, 28

Fluency

WE11a,b For questions 1 to 3, simplify each of the following expressions.

1. a. $a^3 \times a^4$ b. $a^2 \times a^3 \times a$ c. $b \times b^5 \times b^2$ d. $ab^2 \times a^3b^5$

2. a. $m^2n^6 \times m^3n^7$ b. $a^2b^5c \times a^3b^2c^2$ c. $mnp \times m^5n^3p^4$ d. $2a \times 3ab$

3. **a.** $4a^2b^3 \times 5a^2b \times \dfrac{1}{2}b^5$

   b. $3m^3 \times 2mn^2 \times 6m^4n^5$

   c. $4x^2 \times \dfrac{1}{2}xy^3 \times 6x^3y^3$

   d. $2x^3y^2 \times 4x \times \dfrac{1}{2}x^4y^4$

WE11c For questions **4** to **6**, simplify each of the following expressions.

4. **a.** $a^4 \div a^3$
   b. $a^7 \div a^2$
   c. $b^6 \div b^3$
   d. $\dfrac{4a^7}{3a^3}$

5. **a.** $\dfrac{21b^6}{7b^2}$
   b. $\dfrac{48m^8}{12m^3}$
   c. $\dfrac{m^7n^3}{m^4n^2}$
   d. $\dfrac{2x^4y^3}{4x^4y}$

6. **a.** $7ab^5c^4 \div ab^2c^4$
   b. $\dfrac{20m^5n^3p^4}{16m^3n^3p^2}$
   c. $\dfrac{14x^3y^4z^2}{28x^2y^2z^2}$

WE12 For questions **7** to **9**, simplify each of the following expressions.

7. **a.** a^0
   b. $(2b)^0$
   c. $\left(3m^2\right)^0$

8. **a.** $3x^0$
   b. $4b^0$
   c. $-3 \times (2n)^0$

9. **a.** $4a^0 - \left(\dfrac{a}{4}\right)^0$
   b. $5y^0 - 12$
   c. $5x^0 - \left(5xy^2\right)^0$

WE13 For questions **10** to **12**, simplify each of the following expressions.

10. **a.** $\left(a^2\right)^3$
    b. $\left(2a^5\right)^4$
    c. $\left(\dfrac{m^2}{3}\right)^4$
    d. $\left(\dfrac{2n^4}{3}\right)^2$
    e. $(-7)^2$

11. **a.** $\left(a^2b\right)^3$
    b. $\left(3a^3b^2\right)^2$
    c. $\left(2m^3n^5\right)^4$
    d. $\left(\dfrac{3m^2n}{4}\right)^3$
    e. $\left(\dfrac{a^2}{b^3}\right)^2$

12. **a.** $\left(\dfrac{5m^3}{n^2}\right)^4$
    b. $\left(\dfrac{7x}{2y^5}\right)^3$
    c. $\left(\dfrac{3a}{5b^3}\right)^4$
    d. $(-3)^5$
    e. $(-2)^5$

13. **MC** **a.** $2m^{10}n^5$ is the simplified form of:

    A. $m^5n^3 \times 2m^4n^2$

    B. $\dfrac{6m^{10}n^4}{3n}$

    C. $\left(2m^5n^2\right)^2$

    D. $2n\left(m^5\right)^2 \times n^4$

    b. The value of $4 - (5a)^0$ is:

    A. -1
    B. 9
    C. 1
    D. 3

14. **MC** **a.** $4a^3b \times b^4 \times 5a^2b^3$ simplifies to:

    A. $9a^5b^8$
    B. $20a^5b^7$
    C. $20a^5b^8$
    D. $9a^5b^7$

    b. $\dfrac{15x^9 \times 3x^6}{9x^{10} \times x^4}$ simplifies to:

    A. $5x^9$
    B. $9x$
    C. $5x^{29}$
    D. $5x$

c. $\dfrac{3p^7 \times 8q^9}{12p^3 \times 4q^5}$ simplifies to:

A. $2q^4$ **B.** $\dfrac{p^4 q^4}{2}$ **C.** $\dfrac{q^4}{2}$ **D.** $\dfrac{p^4 q^4}{24}$

d. $\dfrac{7a^5 b^3}{5a^6 b^2} \div \dfrac{7b^3 a^2}{5b^5 a^4}$ simplifies to:

A. $\dfrac{49a^3 b}{25}$ **B.** $\dfrac{25a^3 b}{49}$ **C.** $a^3 b$ **D.** ab^3

Understanding

For questions **15** to **17**, evaluate the following expressions.

15. a. $2^3 \times 2^2 \times 2$ **b.** $2 \times 3^2 \times 2^2$ **c.** $\left(5^2\right)^2$

16. a. $\dfrac{3^5 \times 4^6}{3^4 \times 4^4}$ **b.** $\left(2^3 \times 5\right)^2$ **c.** $\left(\dfrac{3}{5}\right)^3$

17. a. $\dfrac{4^4 \times 5^6}{4^3 \times 5^5}$ **b.** $\left(3^3 \times 2^4\right)^0$ **c.** $4\left(5^2 \times 3^5\right)^0$

For questions **18** to **20**, simplify the following expressions.

18. a. $\left(x^y\right)^{3z}$ **b.** $a^b \times (p^q)^0$

19. a. $m^a \times n^b \times (mn)^0$ **b.** $\left(\dfrac{a^2}{b^3}\right)^x$

20. a. $\dfrac{n^3 m^2}{n^p m^q}$ **b.** $(a^{m+n})^p$

Reasoning

21. Explain why $a^3 \times a^2 = a^5$ and not a^6.

22. Is $2x$ ever the same as x^2? Explain your reasoning using examples.

23. Explain the difference between $3x^0$ and $(3x)^0$.

24. a. Complete the table for $a = 0$, 1, 2 and 3.

a	0	1	2	3
$3a^2$				
$5a$				
$3a^2 + 5a$				
$3a^2 \times 5a$				

    b. Analyse what would happen as a becomes very large.

25. Evaluate algebraically the exact value of x if $4^{x+4} = 2^{x^2}$. Justify your answer.

Problem solving

26. Binary numbers (base 2 numbers) are used in computer operations. As the name implies, binary uses only two types of numbers, 0 and 1, to express all numbers.

    A binary number such as 101 (read one, zero, one) means $(1 \times 2^2) + (0 \times 2^1) + (1 \times 2^0) = 4 + 0 + 1 = 5$ (in base 10, the base we are most familiar with).

    The number 1010 (read one, zero, one, zero) means

    $$(1 \times 2^3) + (0 \times 2^2) + (1 \times 2^1) + (0 \times 2^0) = 8 + 0 + 2 + 0 = 10.$$

    If we read the binary number from right to left, the exponent of 2 increases by one each time, beginning with a power of zero.

    Using this information, write out the numbers 1 to 10 in binary (base 2) form.

27. Solve the following equations for x.

    a. $\dfrac{7^x \times 7^{1+2x}}{(7^x)^2} = 16\,807$

    b. $2^{2x} - 5(2^x) = -4$

28. For the following working:

    a. determine the correct answer

    b. identify the error in the solution.

    $$\left(\frac{a^2 b^3 c}{a^2 b^2}\right)^3 \times \left(\frac{a^3 b^2 c^2}{a^2 b^3}\right)^2 = \left(\frac{b^3 c}{b^2}\right)^3 \times \left(\frac{ab^2 c^2}{b^3}\right)^2$$

    $$= \left(\frac{bc}{1}\right)^3 \times \left(\frac{ac^2}{b}\right)^2$$

    $$= \left(\frac{abc^3}{b}\right)^6$$

    $$= \left(\frac{ac^3}{1}\right)^6$$

    $$= a^6 c^{18}$$

LESSON
1.6 Negative exponents

▶ 1.6.1 Negative exponents and the Seventh Exponent Law

eles-4670

- Consider the expression $\dfrac{a^3}{a^5}$. This expression can be simplified in two different ways.

    1. Written in expanded form: $\dfrac{a^3}{a^5} = \dfrac{a \times a \times a}{a \times a \times a \times a \times a}$

                      $= \dfrac{1}{a \times a}$

                      $= \dfrac{1}{a^2}$

    2. Using the Second Exponent Law: $\dfrac{a^3}{a^5} = a^{3-5}$

                                $= a^{-2}$

- Equating the results of both of these simplifications, we get $a^{-2} = \dfrac{1}{a^2}$.

- In general, $\dfrac{1}{a^n} = \dfrac{a^0}{a^n} \left(1 = a^0\right)$

               $= a^{0-n}$ (using the Second Exponent Law)

               $= a^{-n}$

This statement is the Seventh Exponent Law.

Seventh Exponent Law

A term raised to a negative exponents is equivalent to 1 over the original term with a positive exponent.

$$a^{-n} = \frac{1}{a^n}$$

- The converse of this law can be used to rewrite terms with positive exponents only.

$$\frac{1}{a^{-n}} = a^n$$

- It is also worth noting that applying a negative exponent to a fraction has the effect of swapping the numerator and denominator.

$$\left(\frac{a}{b}\right)^{-n} = \frac{b^n}{a^n}$$

Note: It is proper mathematical convention for an algebraic term to be written with each variable in alphabetical order with positive exponents only.

For example, $\dfrac{b^3 a^2 c^{-4}}{y^6 x^{-5}}$ should be written as $\dfrac{a^2 b^3 x^5}{c^4 y^6}$.

WORKED EXAMPLE 14 Writing terms with positive exponents only

Express each of the following with positive exponents.

a. x^{-3} b. $2m^{-4}n^2$ c. $\dfrac{4}{a^{-3}}$

THINK	WRITE
a. 1. Write the expression.	a. x^{-3}
2. Apply the Seventh Exponent Law.	$= \dfrac{1}{x^3}$
b. 1. Write the expression.	b. $2m^{-4}n^2$
2. Apply the Seventh Exponent Law to write the expression with positive exponents.	$= \dfrac{2n^2}{m^4}$
c. 1. Write the expression and rewrite the fraction using a division sign.	c. $\dfrac{4}{a^{-3}} = 4 \div a^{-3}$
2. Apply the Seventh Exponent Law to write the expression with positive exponents.	$= 4 \div \dfrac{1}{a^3}$
3. To divide the fraction, change fraction division into multiplication.	$= 4 \times \dfrac{a^3}{1}$
	$= 4a^3$

WORKED EXAMPLE 15 Simplifying expressions with negative exponents

Simplify each of the following, expressing the answers with positive exponents.

a. $a^2 b^{-3} \times a^{-5} b$ b. $\dfrac{2x^4 y^2}{3xy^5}$ c. $\left(\dfrac{2m^3}{n^{-2}}\right)^{-2}$

THINK	WRITE
a. 1. Write the expression.	a. $a^2 b^{-3} \times a^{-5} b$
2. Apply the First Exponent Law. Multiply terms with the same base by adding the exponents.	$= a^{2+-5} b^{-3+1}$
	$= a^{-3} b^{-2}$
3. Apply the Seventh Exponent Law to write the answer with positive exponents.	$= \dfrac{1}{a^3 b^2}$
b. 1. Write the expression.	b. $\dfrac{2x^4 y^2}{3xy^5}$

2. Apply the Second Exponent Law. Divide terms with the same base by subtracting the exponents.

$$= \frac{2x^{4-1}y^{2-5}}{3}$$

$$= \frac{2x^3 y^{-3}}{3}$$

3. Apply the Seventh Exponent Law to write the answer with positive exponents.

$$= \frac{2x^3}{3y^3}$$

c. 1. Write the expression.

c. $\left(\dfrac{2m^3}{n^{-2}}\right)^{-2}$

2. Apply the Sixth Exponent Law. Multiply the exponents of both the numerator and denominator by the exponent outside the brackets.

$$= \frac{2^{-2}m^{-6}}{n^4}$$

3. Apply the Seventh Exponent Law to express all terms with positive exponents.

$$= \frac{1}{2^2 m^6 n^4}$$

4. Simplify.

$$= \frac{1}{4m^6 n^4}$$

WORKED EXAMPLE 16 Evaluating expressions containing negative exponents

Evaluate 6×3^{-3} without using a calculator.

THINK	WRITE
1. Write the multiplication.	6×3^{-3}
2. Apply the Seventh Exponent Law to write 3^{-3} with a positive exponent.	$= 6 \times \dfrac{1}{3^3}$
3. Multiply the numerator of the fraction by the whole number.	$= \dfrac{6}{3^3}$
4. Evaluate the denominator.	$= \dfrac{6}{27}$
5. Cancel by dividing both the numerator and denominator by the highest common factor (3).	$= \dfrac{2}{9}$

---

on Resources

eWorkbook   Topic 1 Workbook (worksheets, code puzzle and project) (ewbk-2093)

Video eLesson Negative indices (eles-1910)

Interactivities   Individual pathway interactivity: Negative indices (int-4563)

Negative indices (int-6064)

Exercise 1.6 Negative exponents

| 1.6 Quick quiz **on** | 1.6 Exercise |

Individual pathways

■ PRACTISE	■ CONSOLIDATE	■ MASTER
1, 4, 7, 10, 13, 15, 17, 18, 28, 31	2, 5, 8, 11, 14, 16, 19, 20, 23, 26, 29, 32	3, 6, 9, 12, 21, 22, 24, 25, 27, 30, 33

Fluency

WE14 For questions **1** to **3**, express each of the following with positive exponents.

1. a. x^{-5}
   b. y^{-4}
   c. $2a^{-9}$
   d. $\dfrac{4}{5}a^{-3}$

2. a. $3x^2 y^{-3}$
   b. $2^{-2}m^{-3}n^{-4}$
   c. $6a^3 b^{-1} c^{-5}$
   d. $\dfrac{1}{a^{-6}}$

3. a. $\dfrac{2}{3a^{-4}}$
   b. $\dfrac{6a}{3b^{-2}}$
   c. $\dfrac{7a^{-4}}{2b^{-3}}$
   d. $\dfrac{2m^3 n^{-5}}{3a^{-2}b^4}$

WE15 For questions **4** to **6**, simplify the following, expressing the answers with positive exponents.

4. a. $a^3 b^{-2} \times a^{-5} b^{-1}$
   b. $2x^{-2} y \times 3x^{-4} y^{-2}$
   c. $3m^2 n^{-5} \times m^{-2} n^{-3}$
   d. $4a^3 b^2 \div a^5 b^7$
   e. $2xy^6 \div 3x^2 y^5$

5. a. $5x^{-2} y^3 \div 6xy^2$
   b. $\dfrac{6m^4 n}{2n^3 m^6}$
   c. $\dfrac{4x^2 y^9}{x^7 y^{-3}}$
   d. $\dfrac{2m^2 n^{-4}}{6m^5 n^{-1}}$
   e. $\left(2a^3 m^4\right)^{-5}$

6. a. $4\left(p^7 q^{-4}\right)^{-2}$
   b. $3\left(a^{-2}b^{-3}\right)^4$
   c. $\left(\dfrac{2p^2}{3q^3}\right)^{-3}$
   d. $\left(\dfrac{a^{-4}}{2b^{-3}}\right)^2$
   e. $\left(\dfrac{6a^2}{3b^{-2}}\right)^{-3}$

WE16 For questions **7** to **9**, evaluate the following expressions without using a calculator.

7. a. 2^{-3}
   b. 6^{-2}
   c. 3^{-4}
   d. $3^{-2} \times 2^3$

8. a. $4^{-3} \times 2^2$
   b. 5×6^{-2}
   c. $\dfrac{6}{2^{-3}}$
   d. $\dfrac{4 \times 3^{-3}}{2^{-3}}$

9. a. $\dfrac{1}{3} \times 5^{-2} \times 3^4$
   b. $\dfrac{16^0 \times 2^4}{8^2 \times 2^{-4}}$
   c. $\dfrac{5^3 \times 25^0}{25^2 \times 5^{-4}}$
   d. $\dfrac{3^4 \times 4^2}{12^3 \times 15^0}$

10. Write each of these numbers as a power of 2.
    a. 8
    b. $\dfrac{1}{8}$
    c. 32
    d. $\dfrac{1}{64}$

11. Solve each of the following for x.
    a. $125 = 5^x$
    b. $\dfrac{1}{16} = 4^x$
    c. $\dfrac{1}{7} = 7^x$
    d. $216 = 6^x$
    e. $0.01 = 10^x$

12. Solve each of the following for x.
    a. $1 = 8^x$
    b. $64 = 4^x$
    c. $\dfrac{1}{64} = 4^x$
    d. $\dfrac{1}{64} = 2^x$
    e. $\dfrac{1}{64} = 8^x$

13. Evaluate the following expressions.

   a. $\left(\dfrac{2}{3}\right)^{-1}$

   b. $\left(\dfrac{5}{4}\right)^{-1}$

   c. $\left(3\dfrac{1}{2}\right)^{-1}$

   d. $\left(\dfrac{1}{5}\right)^{-1}$

14. Write the following expressions with positive exponents.

   a. $\left(\dfrac{a}{b}\right)^{-1}$

   b. $\left(\dfrac{a^2}{b^3}\right)^{-1}$

   c. $\left(\dfrac{a^{-2}}{b^{-3}}\right)^{-1}$

   d. $\left(\dfrac{m^3}{n^{-2}}\right)^{-1}$

15. Evaluate the following expressions using a calculator.

   a. 3^{-6}

   b. 12^{-4}

   c. 7^{-5}

16. Evaluate the following expressions using a calculator.

   a. $\left(\dfrac{1}{2}\right)^{-8}$

   b. $\left(\dfrac{3}{4}\right)^{-7}$

   c. $(0.04)^{-5}$

Understanding

17. **MC** $\dfrac{1}{a^{-4}}$ is the same as:

   A. $4a$

   B. $-4a$

   C. a^4

   D. $\dfrac{1}{a^4}$

18. **MC** $\dfrac{1}{8}$ is the same as:

   A. 2^3

   B. 2^{-3}

   C. 3^2

   D. 3^{-2}

19. **MC** Select the expression that, when simplified, gives $\dfrac{3m^4}{4n^2}$.

   A. $\dfrac{3m^{-4}n^{-2}}{4}$

   B. $3 \times 2^{-2} \times m^4 \times n^{-2}$

   C. $\dfrac{3n^{-2}}{2^{-2}m^{-4}}$

   D. $\dfrac{2^2 n^{-2}}{3^{-1}m^{-4}}$

20. **MC** When simplified, $3a^{-2}b^{-7} \div \left(\dfrac{3}{4}a^{-4}b^6\right)$ is equal to:

   A. $\dfrac{4}{a^6 b^{13}}$

   B. $\dfrac{9b}{4a^6}$

   C. $\dfrac{9a^2}{4b}$

   D. $\dfrac{4a^2}{b^{13}}$

21. **MC** When $(2x^6 y^{-4})^{-3}$ is simplified, it is equal to:

   A. $\dfrac{2x^{18}}{y^{12}}$

   B. $\dfrac{x^{18}}{8y^{12}}$

   C. $\dfrac{y^{12}}{8x^{18}}$

   D. $\dfrac{8y^{12}}{x^{18}}$

22. **MC** If $\left(\dfrac{2a^x}{b^y}\right)^3$ is equal to $\dfrac{8b^9}{a^6}$, then x and y (in that order) are:

   A. -3 and -6

   B. -6 and -3

   C. -3 and 2

   D. -2 and -3

23. Simplify the following, expressing your answer with positive exponents.

   a. $\dfrac{m^{-3}n^{-2}}{m^{-5}n^6}$

   b. $\dfrac{(m^3 n^{-2})^{-7}}{(m^{-5}n^3)^4}$

   c. $\dfrac{5(a^3 b^{-3})^2}{(ab^{-4})^{-1}} \div \dfrac{(5a^{-2}b)^{-1}}{(a^{-4}b)^3}$

24. Simplify the following, expanding any expressions in brackets.

a. $\left(r^3 + s^3\right)\left(r^3 - s^3\right)$

b. $\left(m^5 + n^5\right)^2$

c. $\dfrac{\left(x^{a+1}\right)^b \times x^{a+b}}{x^{a(b+1)} \times x^{2b}}$

d. $\left(\dfrac{p^{x+1}}{p^{x-1}}\right)^{-4} \times \dfrac{p^{8(x+1)}}{\left(p^{2x}\right)^4} \times \dfrac{p^2}{\left(p^{12x}\right)^0}$

25. Write $\left(\dfrac{2^r \times 8^r}{2^{2r} \times 16}\right)$ in the form 2^{ar+b}.

26. Write $2^{-m} \times 3^{-m} \times 6^{2m} \times 3^{2m} \times 2^{2m}$ as a power of 6.

27. Solve for x if $4^x - 4^{x-1} = 48$.

Reasoning

28. Consider the equation $y = \dfrac{6}{x}$. Clearly $x \neq 0$ as $\dfrac{6}{x}$ would be undefined.

   Explain what happens to the value of y as x gets closer to zero coming from:

   a. the positive direction
   b. the negative direction.

29. Consider the expression 2^{-n}. Explain what happens to the value of this expression as n increases.

30. Explain why each of these statements is false. Illustrate each answer by substituting a value for the pronumeral.

   a. $5x^0 = 1$
   b. $9x^5 \div (3x^5) = 3x$
   c. $a^5 \div a^7 = a^2$
   d. $2c^{-4} = \dfrac{1}{2c^4}$

Problem solving

31. Solve the following pair of simultaneous equations.

   $3^{y+1} = \dfrac{1}{9}$ and $\dfrac{5^y}{125^x} = 125$

32. Simplify $\dfrac{x^{n+2} + x^{n-2}}{x^{n-4} + x^n}$.

33. Solve for x and y if $5^{x-y} = 625$ and $3^{2x} \times 3^y = 243$.

   Hence, evaluate $\dfrac{35^x}{7^{-2y} \times 5^{-3y}}$.

LESSON
1.7 Fractional exponents (10A)

▶ 1.7.1 Fractional exponents and the Eighth Exponent Law

eles-4671

- Consider the expression $a^{\frac{1}{2}}$. Now consider what happens if we square that expression.

$$\left(a^{\frac{1}{2}}\right)^2 = a \text{ (Using the Fourth Exponent Law, } (a^m)^n = a^{m \times n})$$

- From our work on surds, we know that $\left(\sqrt{a}\right)^2 = a$.

- Equating the two facts above, $\left(a^{\frac{1}{2}}\right)^2 = \left(\sqrt{a}\right)^2$. Therefore, $a^{\frac{1}{2}} = \sqrt{a}$.

- Similarly, $b^{\frac{1}{3}} \times b^{\frac{1}{3}} \times b^{\frac{1}{3}} = \left(b^{\frac{1}{3}}\right)^3 = b$, implying that $b^{\frac{1}{3}} = \sqrt[3]{b}$.

- This pattern can be continued and generalised to produce $a^{\frac{1}{n}} = \sqrt[n]{a}$.

- Now consider: $a^{\frac{m}{n}} = a^{m \times \frac{1}{n}}$ or $a^{\frac{m}{n}} = a^{\frac{1}{n} \times m}$

$$= (a^m)^{\frac{1}{n}} \qquad = \left(a^{\frac{1}{n}}\right)^m$$

$$= \sqrt[n]{a^m} \qquad = \left(\sqrt[n]{a}\right)^m$$

Eighth Exponent Law

A term raised to a fractional exponent (index or power) $\dfrac{m}{n}$ is equivalent to the nth root of the term raised to the power m.

$$a^{\frac{m}{n}} = \sqrt[n]{a^m} = \left(\sqrt[n]{a}\right)^m$$

WORKED EXAMPLE 17 Converting fractional exponents to surd form

Write the following expressions in simplest surd form.

a. $10^{\frac{1}{2}}$

b. $5^{\frac{3}{2}}$

THINK	WRITE
a. Since an exponent of $\dfrac{1}{2}$ is equivalent to taking the square root, this term can be written as the square root of 10.	a. $10^{\frac{1}{2}} = \sqrt{10}$

b. 1. A power of $\dfrac{3}{2}$ means the square root of the number cubed.

b. $5^{\frac{3}{2}} = \sqrt{5^3}$

2. Evaluate 5^3.

$= \sqrt{125}$

3. Simplify $\sqrt{125}$.

$= 5\sqrt{5}$

WORKED EXAMPLE 18 Evaluating fractional exponents without a calculator

Evaluate the following expressions without using a calculator.

a. $9^{\frac{1}{2}}$

b. $16^{\frac{3}{2}}$

THINK	WRITE
a. 1. Rewrite the number using the Eighth Exponent Law.	**a.** $9^{\frac{1}{2}} = \sqrt{9}$
2. Evaluate.	$= 3$
b. 1. Rewrite the number using $a^{\frac{m}{n}} = \left(\sqrt[n]{a} \right)^m$.	**b.** $16^{\frac{3}{2}} = \left(\sqrt{16} \right)^3$
	$= 4^3$
2. Simplify and evaluate the result.	$= 64$

WORKED EXAMPLE 19 Evaluating exponents with a calculator

Use a calculator to determine the values of the following expressions, correct to 1 decimal place.

a. $10^{\frac{1}{4}}$

b. $200^{\frac{1}{5}}$

c. $10^{3.47}$

THINK	WRITE
a. Use a calculator to produce the answer.	**a.** $10^{\frac{1}{4}} = 1.778\,279\,41$
	≈ 1.8
b. Use a calculator to produce the answer.	**b.** $200^{\frac{1}{5}} = 2.885\,399\,812$
	≈ 2.9
c. Use a calculator to produce the answer.	**c.** $10^{3.47} = 2951.209\,227\ldots$
	≈ 2951.2

TI | THINK	DISPLAY/WRITE	CASIO | THINK	DISPLAY/WRITE

a.
In a new document on a Calculator page, complete the entry line as:

$10^{\frac{1}{4}}$

Then press ENTER.
To convert the answer to decimal, press:
- MENU
- 2: Number
- 1: Convert to Decimal

Then press ENTER.

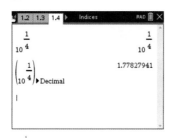

$10^{\frac{1}{4}} = 1.778\,279\,41$

≈ 1.8

b.
In a new document on a Calculator page, complete the entry line as:

$200^{\frac{1}{5}}$

Then press ENTER.
To convert the answer to decimal, press:
- MENU
- 2: Number
- 1: Convert to Decimal

Then press ENTER.

$200^{\frac{1}{5}} = 2.885\,399\,812$

≈ 2.9

a, b.
On the Main screen, complete the entry lines as:

$10^{\frac{1}{4}}$

$200^{\frac{1}{5}}$

Press EXE after each entry.

Note: Change Standard to Decimal.

$10^{\frac{1}{4}} = 1.778\,279\,41$

≈ 1.8

$200^{\frac{1}{5}} = 2.885\,399\,812$

≈ 2.9

WORKED EXAMPLE 20 Simplifying expressions with fractional exponents

Simplify the following expressions.

a. $m^{\frac{1}{5}} \times m^{\frac{2}{5}}$

b. $(a^2 b^3)^{\frac{1}{6}}$

c. $\left(\dfrac{x^{\frac{2}{3}}}{y^{\frac{3}{4}}} \right)^{\frac{1}{2}}$

THINK	WRITE
a. 1. Write the expression.	a. $m^{\frac{1}{5}} \times m^{\frac{2}{5}}$
2. Multiply numbers with the same base by adding the exponents.	$= m^{\frac{3}{5}}$
b. 1. Write the expression.	b. $(a^2 b^3)^{\frac{1}{6}}$
2. Multiply each exponent inside the grouping symbols (brackets) by the exponent on the outside.	$= a^{\frac{2}{6}} b^{\frac{3}{6}}$
3. Simplify the fractions.	$= a^{\frac{1}{3}} b^{\frac{1}{2}}$
c. 1. Write the expression.	c. $\left(\dfrac{x^{\frac{2}{3}}}{y^{\frac{3}{4}}} \right)^{\frac{1}{2}}$
2. Multiply the exponent in both the numerator and denominator by the exponent outside the grouping symbols.	$= \dfrac{x^{\frac{1}{3}}}{y^{\frac{3}{8}}}$

Resources

Exercise 1.7 Fractional exponents (10A)

learn on

1.7 Quick quiz	1.7 Exercise

Individual pathways

■ PRACTISE	■ CONSOLIDATE	■ MASTER
1, 3, 7, 10, 13, 15, 17, 20, 23, 25, 28, 31	2, 4, 5, 8, 11, 14, 18, 21, 26, 29, 32	6, 9, 12, 16, 19, 22, 24, 27, 30, 33, 34

Fluency

WE17 For questions **1** and **2**, write the following expressions in surd form.

1. a. $15^{\frac{1}{2}}$ b. $m^{\frac{1}{4}}$ c. $7^{\frac{2}{5}}$ d. $7^{\frac{5}{2}}$

2. a. $w^{\frac{3}{8}}$ b. $w^{1.25}$ c. $5^{3\frac{1}{3}}$ d. $a^{0.3}$

For questions **3** and **4**, write the following expressions in exponent form.

3. a. $\sqrt{t}$ b. $\sqrt[4]{5^7}$ c. $\sqrt[6]{6^{11}}$ d. $\sqrt[7]{x^6}$

4. a. $\sqrt[6]{x^7}$ b. $\sqrt[5]{w^{10}}$ c. $\sqrt[10]{w^5}$ d. $\sqrt[x]{11^n}$

5. **WE18** Evaluate the following expressions without using a calculator.

  a. $16^{\frac{1}{2}}$ b. $25^{\frac{1}{2}}$ c. $81^{\frac{1}{2}}$

6. Evaluate the following expressions without using a calculator.

  a. $8^{\frac{1}{3}}$ b. $64^{\frac{1}{3}}$ c. $81^{\frac{1}{4}}$

WE19 For questions **7** to **9**, use a calculator to evaluate the following expressions, correct to 1 decimal place.

7. a. $5^{\frac{1}{2}}$ b. $7^{\frac{1}{5}}$ c. $8^{\frac{1}{9}}$

8. a. $12^{\frac{3}{8}}$ b. $100^{\frac{5}{9}}$ c. $50^{\frac{2}{3}}$

9. a. $(0.6)^{\frac{4}{5}}$ b. $\left(\dfrac{3}{4}\right)^{\frac{3}{4}}$ c. $\left(\dfrac{4}{5}\right)^{\frac{2}{3}}$

For questions **10** to **19**, simplify the following expressions.

10. a. $4^{\frac{3}{5}} \times 4^{\frac{1}{5}}$ b. $2^{\frac{1}{8}} \times 2^{\frac{3}{8}}$ c. $a^{\frac{1}{2}} \times a^{\frac{1}{3}}$

11. a. $x^{\frac{3}{4}} \times x^{\frac{2}{5}}$ b. $5m^{\frac{1}{3}} \times 2m^{\frac{1}{5}}$ c. $\frac{1}{2}b^{\frac{3}{7}} \times 4b^{\frac{2}{7}}$

12. a. $-4y^2 \times y^{\frac{2}{9}}$ b. $\frac{2}{5}a^{\frac{3}{8}} \times 0.05a^{\frac{3}{4}}$ c. $5x^3 \times x^{\frac{1}{2}}$

13. a. $a^{\frac{2}{3}}b^{\frac{3}{4}} \times a^{\frac{1}{3}}b^{\frac{3}{4}}$ b. $x^{\frac{3}{5}}y^{\frac{2}{9}} \times x^{\frac{1}{5}}y^{\frac{1}{3}}$ c. $2ab^{\frac{1}{3}} \times 3a^{\frac{3}{5}}b^{\frac{4}{5}}$

14. a. $6m^{\frac{3}{7}} \times \frac{1}{3}m^{\frac{1}{4}}n^{\frac{2}{5}}$ b. $x^3 y^{\frac{1}{2}} z^{\frac{1}{3}} \times x^{\frac{1}{6}} y^{\frac{1}{3}} z^{\frac{1}{2}}$ c. $2a^{\frac{2}{5}}b^{\frac{3}{8}}c^{\frac{1}{4}} \times 4b^{\frac{3}{4}}c^{\frac{3}{4}}$

15. a. $x^3 y^2 \div \left(x^{\frac{4}{3}} y^{\frac{3}{5}} \right)$ b. $a^{\frac{5}{9}}b^{\frac{2}{3}} \div \left(a^{\frac{2}{5}}b^{\frac{2}{5}} \right)$ c. $m^{\frac{3}{8}}n^{\frac{4}{7}} \div \left(3n^{\frac{3}{8}} \right)$

16. a. $10x^{\frac{4}{5}}y \div \left(5x^{\frac{2}{3}}y^{\frac{1}{4}} \right)$ b. $\dfrac{5a^{\frac{3}{4}}b^{\frac{3}{5}}}{20a^{\frac{1}{5}}b^{\frac{1}{4}}}$ c. $\dfrac{p^{\frac{7}{8}}q^{\frac{1}{4}}}{7p^{\frac{2}{3}}q^{\frac{1}{6}}}$

17. a. $\left(2^{\frac{3}{4}} \right)^{\frac{3}{5}}$ b. $\left(5^{\frac{2}{3}} \right)^{\frac{1}{4}}$ c. $\left(7^{\frac{1}{5}} \right)^6$

18. a. $\left(a^3 \right)^{\frac{1}{10}}$ b. $\left(m^{\frac{4}{9}} \right)^{\frac{3}{8}}$ c. $\left(2b^{\frac{1}{2}} \right)^{\frac{1}{3}}$

19. a. $4\left(p^{\frac{3}{7}} \right)^{\frac{14}{15}}$ b. $\left(\dfrac{m}{x^n} \right)^{\frac{n}{p}}$ c. $\left(3m^{\frac{a}{b}} \right)^{\frac{b}{c}}$

Understanding

WE20 For questions 20 to 22, simplify the following expressions.

20. a. $\left(a^{\frac{1}{2}} b^{\frac{1}{3}} \right)^{\frac{1}{2}}$ b. $\left(a^4 b \right)^{\frac{3}{4}}$ c. $\left(x^{\frac{3}{5}} y^{\frac{7}{8}} \right)^2$

21. a. $\left(3a^{\frac{1}{3}} b^{\frac{3}{5}} c^{\frac{3}{4}} \right)^{\frac{1}{3}}$ b. $5\left(x^{\frac{1}{2}} y^{\frac{2}{3}} z^{\frac{2}{5}} \right)^{\frac{1}{2}}$ c. $\left(\dfrac{a^{\frac{3}{4}}}{b} \right)^{\frac{2}{3}}$

22. a. $\left(\dfrac{m^{\frac{4}{5}}}{n^{\frac{7}{8}}} \right)^2$ b. $\left(\dfrac{b^{\frac{3}{5}}}{c^{\frac{4}{9}}} \right)^{\frac{2}{3}}$ c. $\left(\dfrac{4x^7}{2y^{\frac{3}{4}}} \right)^{\frac{1}{2}}$

23. **MC** a. $y^{\frac{2}{5}}$ is equal to:

   A. $\left(y^{\frac{1}{2}} \right)^5$ B. $y \times \dfrac{2}{5}$ C. $\left(y^5 \right)^{\frac{1}{2}}$ D. $\left(y^{\frac{1}{5}} \right)^2$

   b. $k^{\frac{2}{3}}$ is not equal to:

   A. $\left(k^{\frac{1}{3}} \right)^2$ B. $\sqrt[3]{k^2}$ C. $\left(k^{\frac{1}{2}} \right)^3$ D. $\left(\sqrt[3]{k} \right)^2$

c. $\dfrac{1}{\sqrt[5]{g^2}}$ is equal to:

    A. $g^{\frac{2}{5}}$ **B.** $g^{-\frac{2}{5}}$ **C.** $g^{\frac{5}{2}}$ **D.** $g^{-\frac{5}{2}}$

24. **MC** **a.** If $\left(a^{\frac{3}{4}}\right)^{\frac{m}{n}}$ is equal to $a^{\frac{1}{4}}$, then m and n, in that order, could not be:

    A. 1 and 3   **B.** 2 and 6   **C.** 3 and 8   **D.** both **C** and **D**

    b. When simplified, $\left(\dfrac{a^{\frac{m}{n}}}{b^{\frac{n}{p}}}\right)^{\frac{p}{m}}$ is equal to:

    A. $\dfrac{a^{\frac{m}{p}}}{b^{\frac{n}{m}}}$ **B.** $\dfrac{a^{\frac{p}{n}}}{b^{\frac{n}{m}}}$ **C.** $\dfrac{a^{\frac{mp}{n}}}{b^{\frac{n}{m}}}$ **D.** $\dfrac{a^{p}}{b^{m}}$

25. Simplify each of the following expressions.

    a. $\sqrt{a^8}$ **b.** $\sqrt[3]{b^9}$ **c.** $\sqrt[4]{m^{16}}$

26. Simplify each of the following expressions.

    a. $\sqrt{16x^4}$ **b.** $\sqrt[3]{8y^9}$ **c.** $\sqrt[4]{16x^8y^{12}}$

27. Simplify each of the following expressions.

    a. $\sqrt[3]{27m^9n^{15}}$ **b.** $\sqrt[5]{32p^5q^{10}}$ **c.** $\sqrt[3]{216a^6b^{18}}$

Reasoning

28. The relationship between the length of a pendulum (L) in a grandfather clock and the time it takes to complete one swing (T) in seconds is given by the following rule. Note that g is the acceleration due to gravity and will be taken as 9.8 m/s^2.

$$T = 2\pi \left(\dfrac{L}{g}\right)^{\frac{1}{2}}$$

    a. Calculate the time it takes a 1-m long pendulum to complete one swing.
    b. Determine the time it takes the pendulum to complete 10 swings.
    c. Determine how many swings will be completed after 10 seconds.

29. Using the exponent laws, show that $\sqrt[5]{32a^5b^{10}} = 2ab^2$.

30. To rationalise a fraction means to remove all non-rational numbers from the denominator of the fraction. Rationalise $\dfrac{a^2}{3 + \sqrt{b^3}}$ by multiplying the numerator and denominator by $3 - \sqrt{b^3}$, and then evaluate if $b = a^2$ and $a = 2$. Show all of your working.

Problem solving

31. Simplify the following expressions.

   a. $\dfrac{x + 2x^{\frac{1}{2}}y^{\frac{1}{2}} + y - z}{\left(x^{\frac{1}{2}} + y^{\frac{1}{2}} + y^{\frac{1}{2}}\right)}$

   b. $\sqrt[5]{\dfrac{t^2}{\sqrt{t^3}}}$

32. Expand $\left(m^{\frac{3}{4}} + m^{\frac{1}{2}}n^{\frac{1}{2}} + m^{\frac{1}{4}}n + n^{\frac{3}{2}}\right)\left(m^{\frac{1}{4}} - n^{\frac{1}{2}}\right)$.

33. Simplify $\dfrac{m^{\frac{2}{5}} - 2m^{\frac{1}{5}}n^{\frac{1}{5}} + n^{\frac{2}{5}} - p^{\frac{2}{5}}}{m^{\frac{1}{5}} - n^{\frac{1}{5}} - p^{\frac{1}{5}}}$.

34. A scientist has discovered a piece of paper with a complex formula written on it. She thinks that someone has tried to disguise a simpler formula. The formula is:

$$\dfrac{\sqrt[4]{a^{13}}a^2\sqrt{b^3}}{\sqrt{a^1 b}} \times b^3 \times \left(\dfrac{\sqrt{a^3 b}}{ab^2}\right)^2 \times \left(\dfrac{b^2}{a^2\sqrt{b}}\right)^3$$

   a. Simplify the formula using exponent laws so that it can be worked with.
   b. From your simplified formula, can a take a negative value? Explain.
   c. Evaluate the smallest value for a for which the expression will give a rational answer. Consider only integers.

LESSON
1.8 Combining exponent laws

> **LEARNING INTENTION**
>
> At the end of this lesson you should be able to:
> - simplify algebraic expressions involving brackets, fractions, multiplication and division using appropriate exponent laws.

⊳ 1.8.1 Combining exponent laws

eles-4672

- When it is clear that multiple steps are required to simplify an expression, expand brackets first.
- When fractions are involved, it is usually easier to carry out all multiplications first, leaving one division as the final process.
- Make sure to simplify terms to a common base before attempting to apply the exponent laws.
  For example: $5^{2x} \times 25^3 = 5^{2x} \times \left(5^2\right)^3 = 5^{2x} \times 5^6 = 5^{2x+6}$.
- Finally, write the answer with positive exponents and variables in alphabetical order, as is convention.

WORKED EXAMPLE 21 Simplifying expressions in multiple steps

Simplify the following expressions.

a. $\dfrac{(2a)^4 b^4}{6a^3 b^2}$

b. $\dfrac{3^{n-2} \times 9^{n+1}}{81^{n-1}}$

THINK	WRITE
a. 1. Write the expression.	a. $\dfrac{(2a)^4 b^4}{6a^3 b^2}$
2. Apply the Fourth Exponent Law to remove the bracket.	$= \dfrac{16a^4 b^4}{6a^3 b^2}$
3. Apply the Second Exponent Law for each number and pronumeral to simplify.	$= \dfrac{8a^{4-3} b^{4-2}}{3}$
4. Write the answer.	$= \dfrac{8ab^2}{3}$
b. 1. Write the expression.	b. $\dfrac{3^{n-2} \times 9^{n+1}}{81^{n-1}}$
2. Rewrite each term in the expression so that it has a base of 3.	$= \dfrac{3^{n-2} \times (3^2)^{n+1}}{(3^4)^{n-1}}$
3. Apply the Fourth Exponent Law to expand the brackets.	$= \dfrac{3^{n-2} \times 3^{2n+2}}{3^{4n-4}}$
4. Apply the First and Second Exponent Laws to simplify and write your answer.	$= \dfrac{3^{3n}}{3^{4n-4}}$
	$= \dfrac{1}{3^{n-4}}$

WORKED EXAMPLE 22 Simplifying complex expressions involving multiple steps

Simplify the following expressions.

a. $(2a^3 b)^4 \times 4a^2 b^3$

b. $\dfrac{7xy^3}{(3x^3 y^2)^2}$

c. $\dfrac{2m^5 n \times 3m^7 n^4}{7m^3 n^3 \times mn^2}$

THINK	WRITE
a. 1. Write the expression.	a. $(2a^3 b)^4 \times 4a^2 b^3$
2. Apply the Fourth Exponent Law. Multiply each exponent inside the brackets by the exponent outside the brackets.	$= 2^4 a^{12} b^4 \times 4a^2 b^3$
3. Evaluate the numbers.	$= 16a^{12} b^4 \times 4a^2 b^3$
4. Multiply coefficients and multiply pronumerals. Apply the First Exponent Law to multiply terms with the same base by adding the exponents.	$= 16 \times 4 \times a^{12+2} b^{4+3}$
	$= 64a^{14} b^7$

b. 1. Write the expression.

$$\text{b.} \quad \frac{7xy^3}{(3x^3y^2)^2}$$

2. Apply the Fourth Exponent Law in the denominator. Multiply each exponent inside the brackets by the exponent outside the brackets.

$$= \frac{7xy^3}{9x^6y^4}$$

3. Apply the Second Exponent Law. Divide terms with the same base by subtracting the exponents.

$$= \frac{7x^{-5}y^{-1}}{9}$$

4. Use $a^{-m} = \dfrac{1}{a^m}$ to express the answer with positive exponents.

$$= \frac{7}{9x^5y}$$

c. 1. Write the expression.

$$\text{c.} \quad \frac{2m^5n \times 3m^7n^4}{7m^3n^3 \times mn^2}$$

2. Simplify each numerator and denominator by multiplying coefficients and terms with the same base.

$$= \frac{6m^{12}n^5}{7m^4n^5}$$

3. Apply the Second Exponent Law. Divide terms with the same base by subtracting the exponents.

$$= \frac{6m^8n^0}{7}$$

4. Simplify the numerator using $a^0 = 1$.

$$= \frac{6m^8 \times 1}{7}$$

$$= \frac{6m^8}{7}$$

WORKED EXAMPLE 23 Simplifying expressions with multiple fractions

Simplify the following expressions.

a. $\dfrac{\left(5a^2b^3\right)^2}{a^{10}} \times \dfrac{a^2b^5}{\left(a^3b\right)^7}$

b. $\dfrac{8m^3n^{-4}}{\left(6mn^2\right)^3} \div \dfrac{4m^{-2}n^{-4}}{6m^{-5}n}$

THINK

WRITE

a. 1. Write the expression.

$$\text{a.} \quad \frac{\left(5a^2b^3\right)^2}{a^{10}} \times \frac{a^2b^5}{\left(a^3b\right)^7}$$

2. Remove the brackets in the numerator of the first fraction and in the denominator of the second fraction.

$$= \frac{25a^4b^6}{a^{10}} \times \frac{a^2b^5}{a^{21}b^7}$$

3. Multiply the numerators and then multiply the denominators of the fractions. (Simplify across.)

$$= \frac{25a^6b^{11}}{a^{31}b^7}$$

4. Divide terms with the same base by subtracting the exponents. (Simplify down.)

$$= 25a^{-25}b^4$$

5. Express the answer with positive exponents.

$$= \frac{25b^4}{a^{25}}$$

b. 1. Write the expression.	**b.** $\dfrac{8m^3n^{-4}}{(6mn^2)^3} \div \dfrac{4m^{-2}n^{-4}}{6m^{-5}n}$
2. Remove the brackets.	$= \dfrac{8m^3n^{-4}}{216m^3n^6} \div \dfrac{4m^{-2}n^{-4}}{6m^{-5}n}$
3. Multiply by the reciprocal.	$= \dfrac{8m^3n^{-4}}{216m^3n^6} \times \dfrac{6m^{-5}n}{4m^{-2}n^{-4}}$
4. Multiply the numerators and then multiply the denominators. (Simplify across.)	$= \dfrac{48m^{-2}n^{-3}}{864mn^2}$
5. Cancel common factors and divide pronumerals with the same base. (Simplify down.)	$= \dfrac{m^{-3}n^{-5}}{18}$
6. Simplify and express the answer with positive exponents.	$= \dfrac{1}{18m^3n^5}$

 Resources

 eWorkbook   Topic 1 Workbook (worksheets, code puzzle and project) (ewbk-2093)

Interactivities   Individual pathway interactivity: Combining index laws (int-4565)

Combining index laws (int-6108)

Exercise 1.8 Combining exponent laws

learn on

1.8 Quick quiz on	1.8 Exercise

Individual pathways

■ PRACTISE	■ CONSOLIDATE	■ MASTER
1, 4, 7, 10, 13, 16, 17, 23, 26	2, 5, 8, 11, 14, 18, 19, 21, 24, 27	3, 6, 9, 12, 15, 20, 22, 25, 28

Fluency

 For questions **1** to **3**, simplify the following expressions.

1. a. $(3a^2b^2)^3 \times 2a^4b^3$
    b. $(4ab^5)^2 \times 3a^3b^6$
    c. $2m^3n^{-5} \times (m^2n^{-3})^{-6}$

2. a. $(2pq^3)^2 \times (5p^2q^4)^3$
    b. $(2a^7b^2)^2 \times (3a^3b^3)^2$
    c. $5(b^2c^{-2})^3 \times 3(bc^5)^{-4}$

3. a. $6x^{\frac{1}{2}}y^{\frac{1}{3}} \times \left(4x^{\frac{3}{4}}y^{\frac{4}{5}}\right)^{\frac{1}{2}}$
      b. $(16m^3n^4)^{\frac{3}{4}} \times \left(m^{\frac{1}{2}}n^{\frac{1}{4}}\right)^3$

   c. $2\left(p^{\frac{2}{3}}q^{\frac{1}{3}}\right)^{-\frac{3}{4}} \times 3\left(p^{\frac{1}{4}}q^{-\frac{3}{4}}\right)^{-\frac{1}{3}}$
      d. $\left(8p^{\frac{1}{5}}q^{\frac{2}{3}}\right)^{-\frac{1}{3}} \times \left(64p^{\frac{1}{3}}q^{\frac{3}{4}}\right)^{\frac{2}{3}}$

WE21 For questions **4** to **6**, simplify the following expressions.

4. a. $\dfrac{5a^2b^3}{(2a^3b)^3}$ b. $\dfrac{4x^5y^6}{(2xy^3)^4}$ c. $\dfrac{(3m^2n^3)^3}{(2m^5n^5)^7}$

5. a. $\left(\dfrac{4x^3y^{10}}{2x^7y^4}\right)^6$ b. $\dfrac{3a^3b^{-5}}{(2a^7b^4)^{-3}}$ c. $\left(\dfrac{3g^2h^5}{2g^4h}\right)^3$

6. a. $\dfrac{\left(5p^6q^{\frac{1}{3}}\right)^2}{25\left(p^{\frac{1}{2}}q^{\frac{1}{4}}\right)^{\frac{2}{3}}}$ b. $\left(\dfrac{3b^2c^3}{5b^{-3}c^{-4}}\right)^{-4}$ c. $\dfrac{\left(x^{\frac{1}{2}}y^{\frac{1}{4}}z^{\frac{1}{2}}\right)^2}{\left(x^{\frac{2}{3}}y^{-\frac{1}{4}}z^{\frac{1}{3}}\right)^{-\frac{3}{2}}}$

WE22c For questions **7** to **9**, simplify the following expressions.

7. a. $\dfrac{2a^2b \times 3a^3b^4}{4a^3b^5}$ b. $\dfrac{4m^6n^3 \times 12mn^5}{6m^7n^6}$ c. $\dfrac{10m^6n^5 \times 2m^2n^3}{12m^4n \times 5m^2n^3}$

8. a. $\dfrac{6x^3y^2 \times 4x^6y}{9xy^5 \times 2x^3y^6}$ b. $\dfrac{(6x^3y^2)^4}{9x^5y^2 \times 4xy^7}$ c. $\dfrac{5x^2y^3 \times 2xy^5}{10x^3y^4 \times x^4y^2}$

9. a. $\dfrac{a^3b^2 \times 2(ab^5)^3}{6(a^2b^3)^3 \times a^4b}$ b. $\dfrac{(p^6q^2)^{-3} \times 3pq}{2p^{-4}q^{-2} \times (5pq^4)^{-2}}$ c. $\dfrac{6x^{\frac{3}{2}}y^{\frac{1}{2}} \times x^{\frac{4}{5}}y^{\frac{3}{5}}}{2\left(x^{\frac{1}{2}}y\right)^{\frac{1}{5}} \times 3x^{\frac{1}{2}}y^{\frac{1}{5}}}$

WE23a For questions **10** to **12**, simplify the following expressions.

10. a. $\dfrac{a^3b^2}{5a^4b^7} \times \dfrac{2a^6b}{a^9b^3}$ b. $\dfrac{(2a^6)^2}{10a^7b^3} \times \dfrac{4ab^6}{6a^3}$ c. $\dfrac{(m^4n^3)^2}{(m^6n)^4} \times \dfrac{(m^3n^3)^3}{(2mn)^2}$

11. a. $\left(\dfrac{2m^3n^2}{3mn^5}\right)^3 \times \dfrac{6m^2n^4}{4m^3n^{10}}$ b. $\left(\dfrac{2xy^2}{3x^3y^5}\right)^4 \times \left(\dfrac{x^3y^9}{2y^{10}}\right)^2$ c. $\dfrac{4x^{-5}y^{-3}}{(x^2y^2)^{-2}} \times \dfrac{3x^5y^6}{2^{-2}x^{-7}y}$

12. (10A) a. $\dfrac{5p^6q^{-5}}{3q^{-4}} \times \left(\dfrac{5p^6q^4}{3p^5}\right)^{-2}$ b. $\dfrac{2a^{\frac{1}{2}}b^{\frac{1}{3}}}{6a^{\frac{1}{3}}b^{\frac{1}{2}}} \times \dfrac{\left(4a^{\frac{1}{4}}b\right)^{\frac{1}{2}}}{b^{\frac{1}{4}}a}$ c. $\dfrac{3x^{\frac{2}{3}}y^{\frac{1}{5}}}{9x^{\frac{1}{3}}y^{\frac{1}{4}}} \times \dfrac{4x^{\frac{1}{2}}}{x^{\frac{3}{4}}y}$

WE23b For questions **13** to **15**, simplify the following expressions.

13. a. $\dfrac{5a^2b^3}{6a^7b^5} \div \dfrac{a^9b^4}{3ab^6}$ b. $\dfrac{7a^2b^4}{3a^6b^7} \div \left(\dfrac{3ab}{2a^6b^4}\right)^3$

14. a. $\left(\dfrac{4a^9}{b^6}\right)^3 \div \left(\dfrac{3a^7}{2b^5}\right)^4$ b. $\dfrac{5x^2y^6}{(2x^4y^5)^2} \div \dfrac{(4x^6y)^3}{10xy^3}$ c. $\left(\dfrac{x^5y^{-3}}{2xy^5}\right)^{-4} \div \dfrac{4x^6y^{-10}}{(3x^{-2}y^2)^{-3}}$

15. (10A) a. $\dfrac{3m^3n^4}{2m^{-6}n^{-5}} \div \left(\dfrac{2m^4n^6}{m^{-1}n}\right)^{-2}$ b. $4m^{\frac{1}{2}}n^{\frac{3}{4}} \div \dfrac{6m^{\frac{1}{3}}n^{\frac{1}{4}}}{8m^{\frac{3}{4}}n^{\frac{1}{2}}}$ c. $\left(\dfrac{4b^3c^{\frac{1}{3}}}{6c^{\frac{1}{5}}b}\right)^{\frac{1}{2}} \div \left(2b^3c^{-\frac{1}{5}}\right)^{-\frac{3}{2}}$

Understanding

16. Evaluate the following expressions.

    a. $(5^2 \times 2)^0 \times (5^{-3} \times 2^0)^5 \div (5^6 \times 2^{-1})^{-3}$

    b. $(2^3 \times 3^3)^{-2} \div \dfrac{(2^6 \times 3^9)^0}{2^6 \times (3^{-2})^{-3}}$

17. Evaluate the following for $x = 8$. (*Hint:* Simplify first.)

    $(2x)^{-3} \times \left(\dfrac{x}{2}\right)^2 \div \dfrac{2x}{(2^3)^4}$

18. a. Simplify the following fraction: $\dfrac{a^{2y} \times 9b^y \times (5ab)^y}{(a^y)^3 \times 5(3b^y)^2}$

    b. Determine the value of y if the fraction is equal to 125.

19. (10A) **MC** Select the expression that is not the same as $(4xy)^{\frac{3}{2}}$.

    A. $8x^{\frac{3}{2}}y^{\frac{3}{2}}$
    B. $(\sqrt{4xy})^3$
    C. $\sqrt{64x^3y^3}$
    D. $4xy^{\frac{1}{2}} \times (2xy^2)^{\frac{1}{2}}$

20. **MC** The expression $\dfrac{x^2y}{(2xy^2)^3} \div \dfrac{xy}{16x^0}$ is equal to:

    A. $\dfrac{2}{x^2y^6}$
    B. $\dfrac{2x^2}{b^6}$
    C. $2x^2y^6$
    D. $\dfrac{2}{xy^6}$

21. (10A) Simplify the following expressions.

    a. $\sqrt[3]{m^2n} \div \sqrt{mn^3}$

    b. $(g^{-2}h)^3 \times \left(\dfrac{1}{n^{-3}}\right)^{\frac{1}{2}}$

    c. $\dfrac{45^{\frac{1}{3}}}{9^{\frac{3}{4}} \times 15^{\frac{3}{2}}}$

22. (10A) Simplify the following expressions.

    a. $2^{\frac{3}{2}} \times 4^{-\frac{1}{4}} \times 16^{-\frac{3}{4}}$

    b. $\left(\dfrac{a^3b^{-2}}{3^{-3}b^{-3}}\right)^{-2} \div \left(\dfrac{3^{-3}a^{-2}b}{a^4b^{-2}}\right)^2$

    c. $\left(\sqrt[5]{d^2}\right)^{\frac{3}{2}} \times \left(\sqrt[3]{d^5}\right)^{\frac{1}{5}}$

Reasoning

23. The population of the number of bacteria on a petri dish is modelled by $N = 6 \times 2^{t+1}$, where N is the number of bacteria after t days.

    a. Determine the initial number of bacteria.
    b. Determine the number of bacteria after 1 week (7 days).
    c. Calculate when the number of bacteria will first exceed 100 000.

24. In a controlled breeding program at the Melbourne Zoo, the population (P) of koalas at t years is modelled by $P = P_0 \times 10^{kt}$. Given $P_0 = 20$ and $k = 0.3$:

    a. evaluate the number of koalas after 2 years
    b. determine when the population will be equal to 1000. Show full working.

25. The decay of uranium is modelled by $D = D_0 \times 2^{-kt}$. It takes 6 years for the mass of uranium to halve. Giving your answers to the nearest whole number, determine the percentage remaining after:

a. 2 years   **b.** 5 years   **c.** 10 years.

Problem solving

26. Solve the following for x: $2^{2x+2} - 2^{2x-1} - 28 = 0$.

27. Simplify $\dfrac{7^{2x+1} - 7^{2x-1} - 48}{36 \times 7^{2x} - 252}$.

28. (10A) Simplify $\dfrac{z^4 + z^{-4} - 3}{z^2 + z^{-2} - 5^{\frac{1}{2}}}$.

LESSON
1.9 Application of exponentials: Compound interest

LEARNING INTENTION

At the end of this lesson you should be able to:
- calculate the future value of an investment earning compound interest
- calculate the amount of interest earned after a period of time on an investment with compound interest.

1.9.1 Application of exponentials: compound interest

eles-4673

- One practical application of exponentials is compound interest.
- **Compound interest** is the type of interest that is applied to savings in a bank account, term deposits and bank loans.
- Unlike simple interest, which has a fixed amount of interest added at each payment, compound interest depends on the balance (or principal) of the account. This means that the amount of interest increases with each successive payment. It is often calculated per year (per annum or p.a.)
- The following graph shows how compound interest increases over time. Each interest amount is 20% of the previous balance. As the balance grows, the interest increases and the balance growth accelerates.

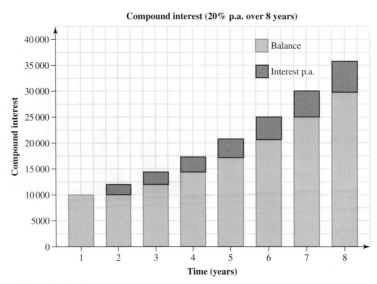

Compound interest (20% p.a. over 8 years)

- The amount to which the initial investment grows is called the **compounded value** or **future value**.
- Compound interest can be calculated by methodically calculating the amount of interest earned at each time, and adding it to the value of the investment.

WORKED EXAMPLE 24 Calculating compound interest step by step

Kyna invests $8000 at 8% p.a. for 3 years with interest paid at the end of each year. Determine the compounded value of the investment by calculating the simple interest on each year separately.

THINK	WRITE
1. Write the initial (first year) principal.	Initial principal = $8000
2. Calculate the interest for the first year.	Interest for year 1 = 8% of $8000 = $640
3. Calculate the principal for the second year by adding the first year's interest to the initial principal.	Principal for year 2 = $8000 + $640 = $8640
4. Calculate the interest for the second year.	Interest for year 2 = 8% of $8640 = $691.20
5. Calculate the principal for the third year by adding the second year's interest to the second year's principal.	Principal for year 3 = $8640 + $691.20 = $9331.20
6. Calculate the interest for the third year.	Interest for year 3 = 8% of $9331.20 = $746.50
7. Calculate the future value of the investment by adding the third year's interest to the third year's principal.	Compounded value after 3 years = $9331.20 + $746.50 = $10 077.70

- To calculate the total amount of interest received, subtract the initial value from the future value.
- In Worked example 24, the total amount of interest is $10 077.70 − $8000 = $2077.70.

1.9.2 The compound interest formula

eles-4674

- Calculating compound interest in a step by step manner is very time consuming, particularly for an investment or loan over 20 or more years.
- By investigating the process however, we are able to simplify things and create a formula to calculate compound interest over extended periods of time.
- To increase a number by a certain percentage we can multiply the number by a decimal that is greater than 1. For example, to increase a number by 8%, multiply by 1.08.
- Consider Worked example 24. Let the compounded value after n years be A_n.
  - The initial value of the investment (after 0 years) is $8000.
    Therefore, $A_0 = 8000$.
  - After 1 year, 8% interest has been added to the investment.
    Therefore, $A_1 = 8000 \times 1.08$
  - After 2 years, another 8% interest has been added to the investment.
    Therefore, $A_2 = (8000 \times 1.08) \times 1.08 = 8000 \times (1.08)^2$
  - After 3 years, another 8% interest has been added to the investment.
    Therefore, $A_3 = (8000 \times 1.08 \times 1.08) \times 1.08 = 8000 \times (1.08)^3$

- Continuing this pattern, the value of the investment after n years is given by $A_n = 8000 \times (1.08)^n$.

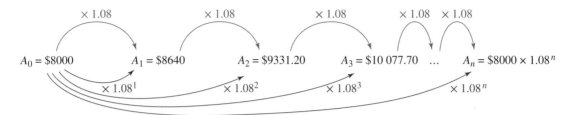

- We can simplify this so that we skip all of the values in the middle and focus on the initial value (principal) and the final (future) value.
- The compound interest formula is the repeated application of simple interest to the previous amount.

Compound interest formula

For any investment or loan, the balance after n compounding periods is given by:

$$A = P(1 + r)^n$$

where:
- A is the future value of the investment in $
- P is the principal (initial value of the investment) in $
- r is the interest rate per compounding period as a decimal. (*Note:* The pronumeral i can also be used in the formula to represent the interest rate.)
- n is the number of compounding periods.

- The interest (I) can then be calculated using the formula:

$$I = A - P$$

WORKED EXAMPLE 25 Using the compound interest formula

William has $14 000 to invest. He invests the money at 9% p.a. for 5 years with interest compounded annually.
a. Use the formula $A = P(1 + r)^n$ to calculate the amount to which this investment will grow.
b. Calculate the compound interest earned on the investment.

THINK	WRITE
a. 1. Write the compound interest formula.	a. $A = P(1 + r)^n$
2. Write down the values of P, r and n.	$P = \$14\,000$, $r = 0.09$, $n = 5$
3. Substitute the values into the formula.	$A = \$14\,000 \times 1.09^5$
4. Calculate.	$= \$21\,540.74$ The investment will grow to $21 540.74.
b. Calculate the compound interest earned.	b. $I = A - P$ $= \$21\,540.74 - \$14\,000$ $= \$7540.74$ The compound interest earned is $7540.74.

TI	THINK	DISPLAY/WRITE	CASIO	THINK	DISPLAY/WRITE

a–b.
On a Calculator page, store the value of p. To do this, complete the entry line as:
$p: = 14000$
Then press:
- MENU
- 3: Algebra
- 1: Solve

Complete the entry line as:
solve $(a = p \times (1 + r)^n, a)$
$|r = 0.09$ and $n = 5$
Then press ENTER and complete as shown.

a–b.

The investment will grow to
$21 540.74.
The compound interest earned is
$7540.74.

a–b.
On the Main screen, complete the entry line as:
solve $(A = P \times (1 + r)^n, A)|r = 0.09|n = 5|P = 14000$
Then press EXE and complete as shown.

a–b.

The investment will grow to $21 540.74
The compound interest earned is $7540.74.

1.9.3 Compounding period

eles-4675

- In Worked example 25, interest is paid annually. Interest can be paid more regularly — it may be paid six-monthly (twice a year), quarterly (4 times a year), monthly or even daily.
- The frequency of interest payment is called the **compounding period**.
- In general, the time period of a loan will be stated in years and the interest rate will be quoted as % p.a. (per annum). If the compounding period is not annual, these values must be adjusted in the compound interest formula.
- For example, an investment over 5 years at 6% p.a. compounding quarterly will have:

$$n = 5 \times 4 = 20 \text{ compounding periods}$$

$$r = \frac{6}{4}\% = 1.5\% = 0.015$$

Calculating n and r for different compounding time periods

- n is the total number of compounding time periods:

$$n = \text{number of years} \times \text{number of compounding periods per year}$$

- r is the interest rate per compounding time period:

$$r = \frac{\text{interest rate per annum}}{\text{number of compounding periods per year}}$$

Calculate the future value of an investment of $4000 at 6% p.a. for 2 years with interest compounded quarterly.

THINK	WRITE
1. Write the compound interest formula.	$A = P(1 + r)^n$
2. Write the values of P, n and r. The number of compounding periods, n, is 4 compounding periods per year for two years. The interest rate, r, is the interest rate per annum divided by the number of compounding periods per year, expressed as a decimal.	$P = \$4000$, $n = 2 \times 4 = 8$ $r = \dfrac{6}{4} \div 100 = 0.015$
3. Substitute the values into the formula.	$A = \$4000 \times 1.015^8$
4. Calculate the future value of an investment.	$= \$4505.97$ The future value of the investment is $4505.97.

TI | THINK	DISPLAY/WRITE	CASIO | THINK	DISPLAY/WRITE
Use the finance functions available on the calculator for this question. On a Calculator page, press: • MENU • 8: Finance • 2: TVM Functions • 5: Future Value Complete the entry line as: tvmFV(8, 1.5, 4000, 0) Press ENTER. Note that the number of compounding periods is 8, that is 4 times a year for 2 years, and the interest is $\dfrac{6}{4} = 1.5\%$ quarterly.	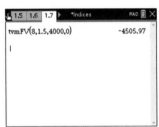 The future value is $4505.97.	On the Financial screen, press: • Compound Interest Enter the values as shown in the screenshot. The FV is left blank. Tap it and it will calculate the value. Note that the number of compounding periods is 8, that is 4 times a year for 2 years, and the interest is $\dfrac{6}{4} = 1.5\%$ quarterly.	 The future value is $4505.97.

---

  on Resources

eWorkbook   Topic 1 Workbook (worksheets, code puzzle and project) (ewbk-2093)

Interactivities   Individual pathway interactivity: Compound interest (int-4636)

Compound interest (int-6075)

Compounding periods (int-6186)

Exercise 1.9 Application of exponentials: Compound interest **learn**on

1.9 Quick quiz on	1.9 Exercise

Individual pathways

■ PRACTISE	■ CONSOLIDATE	■ MASTER
1, 4, 7, 9, 10, 14, 19, 22	2, 5, 8, 11, 15, 17, 20, 23	3, 6, 12, 13, 16, 18, 21, 24

Fluency

For questions **1** to **3**, use the formula $A = P(1 + r)^n$ to calculate the amount to which each of the following investments will grow with interest compounded annually.

1. **a.** $3000 at 4% p.a. for 2 years
    b. $9000 at 5% p.a. for 4 years

2. **a.** $16000 at 9% p.a. for 5 years
    b. $12500 at 5.5% p.a. for 3 years

3. **a.** $9750 at 7.25% p.a. for 6 years
    b. $100 000 at 3.75% p.a. for 7 years

For questions **4** and **5**, calculate the compounded value of each of the following investments.

4. **a.** $870 for 2 years at 3.50% p.a. with interest compounded six-monthly

    b. $9500 for $2\frac{1}{2}$ years at 4.6% p.a. with interest compounded quarterly

5. **a.** $148 000 $3\frac{1}{2}$ for years at 9.2% p.a. with interest compounded six-monthly

    b. $16 000 for 6 years at 8 p.a. with interest compounded monthly

6. Calculate the compounded value of each of the following investments.

    a. $130 000 for 25 years at 12.95% p.a. with interest compounded quarterly
    b. $250 000 for 8.5 years at 6.75% p.a. with interest compounded monthly

Understanding

7. **WE24** Danielle invests $6000 at 10% p.a. for 4 years with interest paid at the end of each year. Determine the compounded value of the investment by calculating the simple interest on each year separately.

8. Ben is to invest $13 000 for 3 years at 8% p.a. with interest paid annually. Determine the amount of interest earned by calculating the simple interest for each year separately.

9. **WE25** Simon has $2000 to invest. He invests the money at 6% p.a. for 6 years with interest compounded annually.

    a. Use the formula $A = P(1 + r)^n$ to calculate the amount to which this investment will grow.
    b. Calculate the compound interest earned on the investment.

10. **WE26** Calculate the future value of an investment of $14 000 at 7% p.a. for 3 years with interest compounded quarterly.

11. A passbook savings account pays interest of 0.3% p.a. Jill has $600 in such an account. Calculate the amount in Jill's account after 3 years, if interest is compounded quarterly.

12. Damien is to invest $35 000 at 7.2% p.a. for 6 years with interest compounded six-monthly. Calculate the compound interest earned on the investment.

13. Sam invests $40 000 in a one-year fixed deposit at an interest rate of 7% p.a. with interest compounding monthly.

    a. Convert the interest rate of 7% p.a. to a rate per month.

    b. Calculate the value of the investment upon maturity.

14. **MC** A sum of $7000 is invested for 3 years at the rate of 5.75% p.a., compounded quarterly. The interest paid on this investment, to the nearest dollar, is:

    A. $1208   **B.** $1308   **C.** $8208   **D.** $8308.

15. **MC** After selling their house and paying off their mortgage, Mr and Mrs Fong have $73 600. They plan to invest it at 7% p.a. with interest compounded annually. The value of their investment will first exceed $110 000 after:

    A. 5 years

    B. 6 years

    C. 8 years

    D. 10 years

16. **MC** Maureen wishes to invest $150 000 for a period of 7 years. The following investment alternatives are suggested to her. The best investment would be:

    A. simple interest at 8% p.a.

    B. compound interest at 6.7% p.a. with interest compounded annually

    C. compound interest at 6.6% p.a. with interest compounded six-monthly

    D. compound interest at 6.5% p.a. with interest compounded quarterly.

17. **MC** An amount is to be invested for 5 years and compounded semi-annually at 7% p.a. Select which of the following investments will have a future value closest to $10 000.

    A. $700   **B.** $6500   **C.** $7400   **D.** $9000

18. Jake invests $120 000 at 9% p.a. for a 1-year term. For such large investments interest is compounded daily.

    a. Calculate the daily percentage interest rate, correct to 4 decimal places. Use 1 year = 365 days.

    b. Hence, calculate the compounded value of Jake's investment on maturity.

    c. Calculate the amount of interest paid on this investment.

    d. Calculate the extra amount of interest earned compared with the case where the interest is calculated only at the end of the year.

Reasoning

19. Daniel has $15 500 to invest. An investment over a 2-year term will pay interest of 7% p.a.

    a. Calculate the compounded value of Daniel's investment if the compounding period is:

      i. 1 year   **ii.** 6 months   **iii.** 3 months   **iv.** monthly.

    b. Explain why it is advantageous to have interest compounded on a more frequent basis.

20. Jasmine invests $6000 for 4 years at 8% p.a. simple interest. David also invests $6000 for 4 years, but his interest rate is 7.6% p.a. with interest compounded quarterly.

    a. Calculate the value of Jasmine's investment on maturity.

    b. Show that the compounded value of David's investment is greater than Jasmine's investment.

    c. Explain why David's investment is worth more than Jasmine's investment despite receiving a lower rate of interest.

21. Quan has $20 000 to invest over the next 3 years. He has the choice of investing his money at 6.25% p.a. simple interest or 6% p.a. compound interest.

   a. Calculate the amount of interest that Quan will earn if he selects the simple interest option.
   b. Calculate the amount of interest that Quan will earn if the interest is compounded:
      i. annually   ii. six monthly   iii. quarterly.
   c. Clearly Quan's decision will depend on the compounding period. Explain the conditions under which Quan should accept the lower interest rate on the compound interest investment.
   d. Consider an investment of $10000 at 8% p.a. simple interest over 5 years. Use a trial-and-error method to determine an equivalent rate of compound interest over the same period.
   e. State whether this equivalent rate be the same if we change:
      i. the amount of the investment   ii. the period of the investment.

Problem solving

22. A building society advertises investment accounts at the following rates:

   i. 3.875% p.a. compounding daily   ii. 3.895% p.a. compounding monthly
   iii. 3.9% p.a. compounding quarterly.

   Peter thinks the first account is the best one because the interest is calculated more frequently. Paul thinks the last account is the best one because it has the highest interest rate. Explain whether either is correct.

23. Two banks offer the following investment packages.
   Bank west: 7.5% p. a. compounded annually fixed for 7 years.
   Bank east: 5.8% p. a. compounded annually fixed for 9 years.

   a. Determine which bank's package will yield the greatest interest.
   b. If a customer invests $20 000 with Bank west, determine how much she would have to invest with Bank east to produce the same amount as Bank west at the end of the investment period.

24. a. Consider an investment of $1 invested at 100% interest for 1 year. Calculate the value of the investment if it is compounded:
      i. quarterly   ii. monthly   iii. daily   iv. once every hour.
   b. Comment on the pattern you notice as the compounding period become more frequent. Do you notice any connection to an important mathematical constant?

LESSON
1.10 Logarithms

LEARNING INTENTIONS

At the end of this lesson you should be able to:
- convert between exponent (index) form and logarithmic form
- evaluate logarithms and use logarithms in scale measurement.

⊙ 1.10.1 Logarithms

eles-4676

- The index, power or exponent in the statement $y = a^x$ is also known as a **logarithm** (or log for short).

$$y = a^x$$

Logarithm or index or power or exponent

Base

- The statement $y = a^x$ can be written in an alternative form as $\log_a y = x$, which is read as 'the logarithm of y to the base a is equal to x'. These two statements are equivalent.

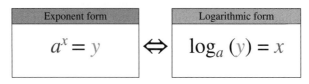

Exponent form		Logarithmic form
$a^x = y$	$\Leftrightarrow$	$\log_a (y) = x$

- For example, $3^2 = 9$ can be written as $\log_3 9 = 2$. The log form would be read as 'the logarithm of 9, to the base of 3, is 2'. In both forms, the base is 3 and the logarithm is 2.
- It helps to remember that the output of a logarithm would be the power of the equivalent expression in exponent form. Logarithms take in large numbers and output small numbers (powers).

WORKED EXAMPLE 27 Converting to logarithmic form

Write the following in logarithmic form.

a. $10^4 = 10\,000$

b. $6^x = 216$

THINK	WRITE
a. 1. Write the given statement.	a. $10^4 = 10\,000$
2. Identify the base (10) and the logarithm (4), and write the equivalent statement in logarithmic form. (Use $a^x = y \Leftrightarrow \log_a y = x$, where the base is a and the log is x.)	$\log_{10}(10\,000) = 4$
b. 1. Write the given statement.	b. $6^x = 216$
2. Identify the base (6) and the logarithm (x), and write the equivalent statement in logarithmic form.	$\log_6(216) = x$

WORKED EXAMPLE 28 Converting to exponent form

Write the following in exponent form.

a. $\log_2 (8) = 3$

b. $\log_{25}(5) = \dfrac{1}{2}$

THINK	WRITE
a. 1. Write the statement.	a. $\log_2(8) = 3$
2. Identify the base (2) and the logarithm (3), and write the equivalent statement in exponent form. Remember that the log is the same as the exponent.	$2^3 = 8$
b. 1. Write the statement.	b. $\log_{25}(5) = \dfrac{1}{2}$
2. Identify the base (25) and the logarithm $\left(\dfrac{1}{2}\right)$, and write the equivalent statement in exponent form.	$25^{\frac{1}{2}} = 5$

- In the previous examples, we found that:

$$\log_2(8) = 3 \Leftrightarrow 2^3 = 8 \quad \text{and} \quad \log_{10}(10\,000) = 4 \Leftrightarrow 10^4 = 10\,000.$$

We could also write $\log_2(8) = 3$ as $\log_2(2^3) = 3$ and $\log_{10}(10\,000) = 4$ as $\log_{10}(10^4) = 4$.

- Can this pattern be used to work out the value of $\log_3(81)$? We need to determine the power when the base of 3 is raised to that power to give 81.

WORKED EXAMPLE 29 Evaluating a logarithm

Evaluate $\log_3(81)$.

THINK	WRITE
1. Write the log expression.	$\log_3(81)$
2. Express 81 in exponent form with a base of 3.	$= \log_3(3^4)$
3. Write the value of the logarithm.	$= 4$

▶ 1.10.2 Using logarithmic scales in measurement

eles-4677

- Logarithms can be used to display data sets that cover ranges of values that vary greatly in size.
  For example, when measuring the amplitude of earthquake waves, some earthquakes will have amplitudes of around 10 000, whereas other earthquakes may have amplitudes of around 10 000 000 (1000 times greater). Rather than trying to display this data on a linear scale, we can take the logarithm of the amplitude, which gives us the magnitude of each earthquake.
- The Richter scale uses the magnitudes of earthquakes to display the difference in their power.
- The logarithm that is used in these scales is the logarithm with base 10, which means that an increase by 1 on the scale is an increase of 10 in the actual value.
- The logarithm with base 10 is often written simply as $\log(x)$, with the base omitted.

WORKED EXAMPLE 30 Real world application of logarithms

Convert the following amplitudes of earthquakes into values on the Richter scale, correct to 1 decimal place.
a. 1989 Newcastle earthquake: amplitude 398 000
b. 2010 Canterbury earthquake: amplitude 12 600 000
c. 2010 Chile earthquake: amplitude 631 000 000

THINK	WRITE
a. Use a calculator to calculate the logarithmic value of the amplitude. Round the answer to 1 decimal place. Write the answer in words.	a. $\log(398\,000) = 5.599\ldots$ $\qquad\qquad = 5.6$ The 1989 Newcastle earthquake rated 5.6 on the Richter scale.
b. Use a calculator to calculate the logarithmic value of the amplitude. Round the answer to 1 decimal place. Write the answer in words.	b. $\log(12\,600\,000) = 7.100\ldots$ $\qquad\qquad = 7.1$ The 2010 Canterbury earthquake rated 7.1 on the Richter scale.
c. Use a calculator to calculate the logarithmic value of the amplitude. Round the answer to 1 decimal place. Write the answer in words.	c. $\log(631\,000\,000) = 8.800\ldots$ $\qquad\qquad = 8.8$ The 2010 Chile earthquake rated 8.8 on the Richter scale.

Displaying logarithmic data in histograms

- If we are given a data set in which the data vary greatly in size, we can use logarithms to transform the data into more manageable values, and then group the data into intervals to provide an indication of the spread of the data.

WORKED EXAMPLE 31 Creating a histogram with a log scale

The following table displays the population of **10** different towns and cities in Victoria (using data from the 2011 census).

a. Convert the populations into logarithmic form, correct to 2 decimal places.

b. Group the data into a frequency table.

c. Draw a histogram to represent the data.

Town or city	Population
Benalla	9328
Bendigo	76 051
Castlemaine	9124
Echuca	12 613
Geelong	143 921
Kilmore	6 142
Melbourne	3 707 530
Stawell	5734
Wangaratta	17 377
Warrnambool	29 284

THINK

a. Use a calculator to calculate the logarithmic values of all of the populations. Round the answers to 2 decimal places.

WRITE

a.

Town or city	log(population)
Benalla	3.97
Bendigo	4.88
Castlemaine	3.96
Echuca	4.10
Geelong	5.16
Kilmore	3.79
Melbourne	6.57
Stawell	3.76
Wangaratta	4.24
Warrnambool	4.67

b. Group the logarithmic values into class intervals and create a frequency table.

b.

log(population)	Frequency
3 – < 4	4
4 – < 5	4
5 – < 6	1
6 – < 7	1

c. Construct a histogram of the data set.

c.

Exercise 1.10 Logarithms

learn on

1.10 Quick quiz   **1.10 Exercise**

Individual pathways

■ PRACTISE	■ CONSOLIDATE	■ MASTER
1, 4, 5, 9, 12, 14, 17, 23, 27	2, 6, 8, 10, 13, 15, 18, 20, 24, 26, 28	3, 7, 11, 16, 19, 21, 22, 25, 29

Fluency

WE27 For questions **1** to **3**, write the following in logarithmic form.

1. **a.** $4^2 = 16$　　**b.** $2^5 = 32$　　**c.** $3^4 = 81$　　**d.** $6^2 = 36$　　**e.** $1000 = 10^3$

2. **a.** $25 = 5^2$　　**b.** $4^3 = x$　　**c.** $5^x = 125$　　**d.** $7^x = 49$　　**e.** $p^4 = 16$

3. **a.** $9^{\frac{1}{2}} = 3$　　**b.** $0.1 = 10^{-1}$　　**c.** $2 = 8^{\frac{1}{3}}$　　**d.** $2^{-1} = \dfrac{1}{2}$　　**e.** $4^{\frac{3}{2}} = 8$

4. **MC** The statement $w = h^t$ is equivalent to:

   A. $w = \log_t(h)$　　**B.** $h = \log_t(w)$　　**C.** $t = \log_w(h)$　　**D.** $t = \log_h(w)$

WE28 For questions **5** to **7**, write the following in exponent form.

5. **a.** $\log_2(16) = 4$　　**b.** $\log_3(27) = 3$　　**c.** $\log_{10}(1\,000\,000) = 6$　　**d.** $\log_5(125) = 3$

6. **a.** $\log_{16}(4) = \dfrac{1}{2}$　　**b.** $\log_4(64) = x$　　**c.** $\dfrac{1}{2} = \log_{49}(7)$　　**d.** $\log_3(x) = 5$

7. **a.** $\log_{81}(9) = \dfrac{1}{2}$　　**b.** $\log_{10}(0.01) = -2$　　**c.** $\log_8(8) = 1$　　**d.** $\log_{64}(4) = \dfrac{1}{3}$

8. **MC** The statement $q = \log_r(p)$ is equivalent to:

   A. $q = r^p$　　**B.** $p = r^q$　　**C.** $r = p^q$　　**D.** $r = q^p$

WE29 For questions **9** to **11**, evaluate the following logarithms.

9. **a.** $\log_2(16)$　　**b.** $\log_4(16)$　　**c.** $\log_{11}(121)$　　**d.** $\log_{10}(100\,000)$

10. **a.** $\log_3(243)$　　**b.** $\log_2(128)$　　**c.** $\log_5(1)$　　**d.** $\log_9(3)$

11. **a.** $\log_3\left(\dfrac{1}{3}\right)$　　**b.** $\log_6(6)$　　**c.** $\log_{10}\left(\dfrac{1}{100}\right)$　　**d.** $\log_{125}(5)$

12. Write the value of each of the following logarithms.

   a. $\log_{10}(1)$ b. $\log_{10}(10)$ c. $\log_{10}(100)$

13. Write the value of each of the following logarithms.

   a. $\log_{10}(1000)$ b. $\log_{10}(10\,000)$ c. $\log_{10}(100\,000)$

Understanding

14. Use your results for questions **12** and **13** to answer the following.

   a. Between which two whole numbers would $\log_{10}(7)$ lie?
   b. Between which two whole numbers would $\log_{10}(4600)$ lie?
   c. Between which two whole numbers would $\log_{10}(85)$ lie?

15. a. Between which two whole numbers would $\log_{10}(12\,750)$ lie?
   b. Between which two whole numbers would $\log_{10}(110)$ lie?
   c. Between which two whole numbers would $\log_{10}(81\,000)$ lie?

16. **WE30** Convert the following amplitudes of earthquakes into values on the Richter scale, correct to 1 decimal place.

   a. 2016 Northern Territory earthquake: amplitude 1 260 000
   b. 2011 Christchurch earthquake: amplitude 2 000 000
   c. 1979 Tumaco earthquake: amplitude 158 000 000

17. **MC** The following graph shows the weights of animals. If a gorilla has a weight of 207 kilograms, then its weight is between that of:

   A. a monkey and a jaguar.
   B. a horse and a triceratops.
   C. a guinea pig and a monkey.
   D. a jaguar and a horse.

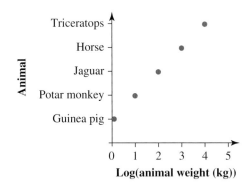

18. **MC** The following table shows a variety of top speeds.

F1 racing car	370 000 m/h
V8 supercar	300 000 m/h
Cheetah	64 000 m/h
Space shuttle	28 000 000 m/h
Usain Bolt	34 000 m/h

The correct value, to 2 decimal places, for a cheetah's top speed using a log(base 10) scale would be:

A. 5.57
B. 5.47
C. 7.45
D. 4.81

19. **MC** The following graph represents the capacity of five Victorian dams.

Capacity of Victorian dams

a. The capacity of Thomson Dam is closest to:

   A. 100 000 ML
   C. 500 000 ML

   B. 1 000 000 ML
   D. 5000 ML

b. The capacity of Sugarloaf Dam is closest to:

   A. 100 000 ML
   C. 500 000 ML

   B. 1 000 000 ML
   D. 5000 ML

c. The capacity of Silvan Dam is between the range:

   A. 1 and 10 ML
   C. 1000 and 10 000 ML

   B. 10 and 100 ML
   D. 10 000 and 100 000 ML

20. The following table shows the average weights of 10 different adult mammals.

Mammal	Weight (kg)
Black wallaroo	18
Capybara	55
Cougar	63
Fin whale	70 000
Lion	175
Ocelot	9
Pygmy rabbit	0.4
Red deer	200
Quokka	4
Water buffalo	725

Display the data in a histogram using a log (base 10) scale, using class intervals of width 1.

The following information about the flow rates of world-famous waterfalls applies to questions **21** and **22**.

Waterfall	Flow rate
Victoria Falls	$1\,088 \text{ m}^3/\text{s}$
Niagara Falls	$2\,407 \text{ m}^3/\text{s}$
Celilo Falls	$5\,415 \text{ m}^3/\text{s}$
Khane Phapheng Falls	$11\,610 \text{ m}^3/\text{s}$
Boyoma Falls	$17\,000 \text{ m}^3/\text{s}$

21. **MC** The correct value, correct to 2 decimal places, to be plotted for the flow rate of Niagara Falls using a log (base 10) scale would be:

    A. 3.38   **B.** 3.04   **C.** 4.06   **D.** 4.23

22. **MC** The correct value, correct to 2 decimal places, to be plotted for the flow rate of Victoria Falls using a log (base 10) scale would be:

    A. 3.38   **B.** 3.04   **C.** 4.06   **D.** 4.23

Reasoning

23. **a.** If $\log_{10}(g) = k$, determine the value of $\log_{10}(g^2)$. Justify your answer.
    b. If $\log_x(y) = 2$, determine the value of $\log_y(x)$. Justify your answer.
    c. By referring to the equivalent exponent statement, explain why x must be a positive number given $\log_4(x) = y$, for all values of y.

24. Calculate the following logarithms.

    a. $\log_2(64)$ **b.** $\log_3\left(\dfrac{1}{81}\right)$ **c.** $\log_{10}(0.00001)$

25. Calculate the following logarithms.

    a. $\log_3(243)$ **b.** $\log_4\left(\dfrac{1}{64}\right)$ **c.** $\log_5\left(\sqrt{125}\right)$

26. The pH scale measures acidity using a log (base 10) scale. For each decrease in pH of 1, the acidity of a substance increases by a factor of 10.
If a liquid's pH value decreases by 0.7, calculate how much the acidity of the liquid has increased.

Problem solving

27. For each of the following, determine the value of x.

    a. $\log_x\left(\dfrac{1}{243}\right) = -5$ **b.** $\log_x(343) = 3$ **c.** $\log_{64}(x) = -\dfrac{1}{2}$

28. Simplify $10^{\log_{10}(x)}$.

29. Simplify the expression $3^{2-\log_3(x)}$.

LESSON
1.11 Logarithm laws (10A)

LEARNING INTENTION

At the end of this lesson you should be able to:
- simplify expressions using logarithm laws.

▶ 1.11.1 Logarithm laws

eles-4678

- Recall the exponent (index) laws.

Exponent Law 1: $a^m \times a^n = a^{m+n}$	Exponent Law 2: $\dfrac{a^m}{a^n} = a^{m-n}$
Exponent Law 3: $a^0 = 1$	Exponent Law 4: $(a^m)^n = a^{mn}$
Exponent Law 5: $(ab)^m = a^m b^m$	Exponent Law 6: $\left(\dfrac{a}{b}\right)^m = \dfrac{a^m}{b^m}$
Exponent Law 7: $a^{-m} = \dfrac{1}{a^m}$	Exponent Law 8: $a^{\frac{m}{n}} = \sqrt[n]{a^m}$

- The exponent (index) laws can be used to produce a set of equivalent logarithm laws.

Logarithm Law 1

- If $x = a^m$ and $y = a^n$, then $\log_a x = m$ and $\log_a y = n$ (equivalent log form).

Now	$xy = a^m \times a^n$	
or	$xy = a^{m+n}$	(First Exponent Law).
So	$\log_a(xy) = m + n$	(equivalent log form)
or	$\log_a(xy) = \log_a x + \log_a y$	(substituting for m and n).

Logarithm Law 1

$$\log_a(x) + \log_a(y) = \log_a(xy)$$

- This means that the sum of two logarithms with the same base is equal to the logarithm of the product of the numbers.

WORKED EXAMPLE 32 Adding logarithms

Evaluate $\log_{10}(20) + \log_{10}(5)$.

THINK	WRITE
1. Since the same base of 10 is used in each log term, use $\log_a(x) + \log_a(y) = \log_a(xy)$ and simplify.	$\log_{10}(20) + \log_{10}(5) = \log_{10}(20 \times 5)$ $= \log_{10}(100)$
2. Evaluate. (Remember that $100 = 10^2$.)	$= 2$

Logarithm Law 2

- If $x = a^m$ and $y = a^n$, then $\log_a(x) = m$ and $\log_a(y) = n$ (equivalent log form).

Now
$$\frac{x}{y} = \frac{a^m}{a^n}$$

or
$$\frac{x}{y} = a^{m-n}$$ (Second Exponent Law).

So
$$\log_a\left(\frac{x}{y}\right) = m - n$$ (equivalent log form)

or
$$\log_a\left(\frac{x}{y}\right) = \log_a(x) - \log_a(y)$$ (substituting for m and n).

Logarithm Law 2

$$\log_a(x) - \log_a(y) = \log_a\left(\frac{x}{y}\right)$$

- This means that the difference of two logarithms with the same base is equal to the logarithm of the quotient of the numbers.

WORKED EXAMPLE 33 Subtracting logarithms

Evaluate $\log_4(20) - \log_4(5)$.

THINK	WRITE
1. Since the same base of 4 is used in each log term, use $\log_a(x) - \log_a(y) = \log_a\left(\frac{x}{y}\right)$ and simplify.	$\log_4(20) - \log_4(5) = \log_4\left(\frac{20}{5}\right)$ $= \log_4(4)$
2. Evaluate. (Remember that $4 = 4^1$.)	$= 1$

WORKED EXAMPLE 34 Simplifying multiple logarithm terms

Evaluate $\log_5(35) + \log_5(15) - \log_5(21)$.

THINK	WRITE
1. Since the first two log terms are being added, use $\log_a(x) + \log_a(y) = \log_a(xy)$ and simplify.	$\log_5(35) + \log_5(15) - \log_5(21)$ $= \log_5(35 \times 15) - \log_5(21)$ $= \log_5(525) - \log_5(21)$
2. To find the difference between the two remaining log terms, use $\log_a(x) - \log_a(y) = \log_a\left(\frac{x}{y}\right)$ and simplify.	$= \log_5\left(\frac{525}{21}\right)$ $= \log_5(25)$
3. Evaluate. (Remember that $25 = 5^2$.)	$= 2$

TI	THINK	DISPLAY/WRITE	CASIO	THINK	DISPLAY/WRITE

TI | THINK	DISPLAY/WRITE
On a Calculator page, press CTRL log (above 10^x) and complete the entry line as: $\log_5(35) + \log_5(15) - \log_5(21)$ Then press ENTER.	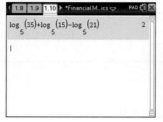 $\log_5(35) + \log_5(15) - \log_5(21) = 2$

CASIO | THINK	DISPLAY/WRITE
On the Math1 keyboard screen, tap: $\boxed{\log_\blacksquare\square}$ Complete the entry line as: $\log_5(35) + \log_5(15) - \log_5(21)$ Then press EXE.	$\log_5(35) + \log_5(15) - \log_5(21) = 2$

- Once you have gained confidence in using the first two laws, you can reduce the number of steps of working by combining the application of the laws. In Worked example 34, we could write:

$$\log_5(35) + \log_5(15) - \log_5(21) = \log_5\left(\frac{35 \times 15}{21}\right)$$
$$= \log_5(25)$$
$$= 2$$

Logarithm Law 3

- If $x = a^m$, then $\log_a(x) = m$ (equivalent log form).

Now	$x^n = (a^m)^n$	
or	$x^n = a^{mn}$	(Fourth Exponent Law)
So	$\log_a(x^n) = mn$	(equivalent log form)
or	$\log_a(x^n) = \left(\log_a(x)\right) \times n$	(substituting for m)
or	$\log_a(x^n) = n\log_a(x)$	

Logarithm Law 3

$$\log_a(x^n) = n\log_a(x)$$

- This means that the logarithm of a number raised to a power is equal to the product of the power and the logarithm of the number.

WORKED EXAMPLE 35 Simplifying a logarithm of a number raised to a power

Evaluate $2\log_6(3) + \log_6(4)$.

THINK	WRITE
1. The first log term is not in the required form to use the log law relating to sums. Use $\log_a(x^n) = n\log_a(x)$ to rewrite the first term in preparation for applying the first log law.	$2\log_6(3) + \log_6(4) = \log_6\left(3^2\right) + \log_6(4)$ $= \log_6(9) + \log_6(4)$
2. Use $\log_a(x) + \log_a(y) = \log_a(xy)$ to simplify the two log terms to one.	$= \log_6(9 \times 4)$ $= \log_6(36)$
3. Evaluate. (Remember that $36 = 6^2$.)	$= 2$

Logarithm Law 4

- As

$$a^0 = 1 \qquad \text{(Third Exponent Law)}$$
$$\log_a(1) = 0 \qquad \text{(equivalent log form)}$$

Logarithm Law 4

$$\log_a(1) = 0$$

- This means that the logarithm of 1 with any base is equal to 0.

Logarithm Law 5

- As

$$a^1 = a$$
$$\log_a(a) = 1 \qquad \text{(equivalent log form)}$$

Logarithm Law 5

$$\log_a(a) = 1$$

- This means that the logarithm of any number a with base a is equal to 1.

Logarithm Law 6

- Now

$$\log_a\left(\frac{1}{x}\right) = \log_a\left(x^{-1}\right) \qquad \text{(Seventh Exponent law)}$$

- or

$$\log_a\left(\frac{1}{x}\right) = -1 \times \log_a(x) \qquad \text{(using the fourth log law)}$$

- or

$$\log_a\left(\frac{1}{x}\right) = -\log_a(x).$$

Logarithm Law 6

$$\log_a\left(\frac{1}{x}\right) = -\log_a(x)$$

Logarithm Law 7

- Now

$$\log_a(a^x) = x\log_a(a) \qquad \text{(using the third log law)}$$

- or

$$\log_a(a^x) = x \times 1 \qquad \text{(using the fifth log law)}$$

- or

$$\log_a(a^x) = x.$$

Logarithm Law 7

$$\log_a(a^x) = x$$

Resources

Exercise 1.11 Logarithm laws (10A)

learn on

1.11 Quick quiz on	**1.11 Exercise**

Individual pathways

■ PRACTISE	■ CONSOLIDATE	■ MASTER
1, 2, 3, 6, 9, 12, 13, 15, 18, 22, 26, 29	4, 7, 10, 14, 16, 19, 21, 23, 27, 30	5, 8, 11, 17, 20, 24, 25, 28, 31

Fluency

1. Use a calculator to evaluate the following, correct to 5 decimal places.

   a. $\log_{10}(50)$ b. $\log_{10}(25)$ c. $\log_{10}(5)$ d. $\log_{10}(2)$

2. Use your answers to question **1** to show that each of the following statements is true.

   a. $\log_{10}(25) + \log_{10}(2) = \log_{10}(50)$ b. $\log_{10}(50) - \log_{10}(2) = \log_{10}(25)$

   c. $\log_{10}(25) = 2\log_{10}(5)$ d. $\log_{10}(50) - \log_{10}(25) - \log_{10}(2) = \log_{10}(1)$

WE32 For questions **3** to **5**, evaluate the following.

3. a. $\log_6(3) + \log_6(2)$ b. $\log_4(8) + \log_4(8)$

4. a. $\log_{10}(25) + \log_{10}(4)$ b. $\log_8(32) + \log_8(16)$

5. a. $\log_6(108) + \log_6(12)$ b. $\log_{14}(2) + \log_{14}(7)$

WE33 For questions **6** to **8**, evaluate the following.

6. a. $\log_2(20) - \log_2(5)$ b. $\log_3(54) - \log_3(2)$

7. a. $\log_4(24) - \log_4(6)$ b. $\log_{10}(30\,000) - \log_{10}(3)$

8. a. $\log_6(648) - \log_6(3)$ b. $\log_2(224) - \log_2(7)$

WE34 For questions **9** to **11**, evaluate the following.

9. a. $\log_3(27) + \log_3(2) - \log_3(6)$ b. $\log_4(24) - \log_4(2) - \log_4(6)$

10. a. $\log_6(78) - \log_6(13) + \log_6(1)$ b. $\log_2(120) - \log_2(3) - \log_2(5)$

11. a. $\log_7(15) + \log_7(3) - \log_7(315)$ b. $\log_9(80) - \log_9(8) - \log_9(30)$

12. Evaluate $2\log_4(8)$.

WE35 For questions **13** to **17**, evaluate the following.

13. a. $2\log_{10}(5) + \log_{10}(4)$

    b. $\log_3(648) - 3\log_3(2)$

14. a. $4\log_5(10) - \log_5(80)$

    b. $\log_2(50) + \dfrac{1}{2}\log_2(16) - 2\log_2(5)$

15. a. $\log_8(8)$

    b. $\log_5(1)$

    c. $\log_2\left(\dfrac{1}{2}\right)$

    d. $\log_4\left(4^5\right)$

16. a. $\log_6\left(6^{-2}\right)$

    b. $\log_{20}(20)$

    c. $\log_2(1)$

    d. $\log_3\left(\dfrac{1}{9}\right)$

17. a. $\log_4\left(\dfrac{1}{2}\right)$

    b. $\log_5\left(\sqrt{5}\right)$

    c. $\log_3\left(\dfrac{1}{\sqrt{3}}\right)$

    d. $\log_2\left(8\sqrt{2}\right)$

Understanding

For questions **18** to **20**, use the logarithm laws to simplify the following.

18. a. $\log_a(5) + \log_a(8)$

    b. $\log_a(12) + \log_a(3) - \log_a(2)$

    c. $4\log_x(2) + \log_x(3)$

    d. $\log_x(100) - 2\log_x(5)$

19. a. $3\log_a(x) - \log_a\left(x^2\right)$

    b. $5\log_a(a) - \log_a\left(a^4\right)$

    c. $\log_x(6) - \log_x(6x)$

    d. $\log_a\left(a^7\right) + \log_a(1)$

20. a. $\log_p\left(\sqrt{p}\right)$

    b. $\log_k\left(k\sqrt{k}\right)$

    c. $6\log_a\left(\dfrac{1}{a}\right)$

    d. $\log_a\left(\dfrac{1}{\sqrt[3]{a}}\right)$

21. **MC** *Note:* There may be more than one correct answer.

    a. The equation $y = 10^x$ is equivalent to:

    A. $x = 10^y$

    B. $x = \log_{10}(y)$

    C. $x = \log_x(10)$

    D. $x = \log_y(10)$

    b. The equation $y = 10^{4x}$ is equivalent to:

    A. $x = \log_{10}\left(\sqrt{4y}\right)$

    B. $x = \log_{10}\left(\sqrt[4]{y}\right)$

    C. $x = 10^{\frac{1}{4}y}$

    D. $x = \dfrac{1}{4}\log_{10}(y)$

    c. The equation $y = 10^{3x}$ is equivalent to:

    A. $x = \dfrac{1}{3}\log_{10}(y)$

    B. $x = \log_{10}\left(y^{\frac{1}{3}}\right)$

    C. $x = \log_{10}(y) - 3$

    D. $x = 10^{y-3}$

    d. The equation $y = ma^{nx}$ is equivalent to:

    A. $x = \dfrac{1}{n}a^{my}$

    B. $x = \log_a\left(\dfrac{m}{y}\right)^n$

    C. $x = \dfrac{1}{n}\left(\log_a(y) - \log_a(m)\right)$

    D. $x = \dfrac{1}{n}\log_a\left(\dfrac{y}{m}\right)$

For questions **22** to **24**, simplify, and evaluate where possible, each of the following without a calculator.

22. a. $\log_2(8) + \log_2(10)$ **b.** $\log_3(7) + \log_3(15)$ **c.** $\log_{10}(20) + \log_{10}(5)$ **d.** $\log_6(8) + \log_6(7)$

23. a. $\log_2(20) - \log_2(5)$ **b.** $\log_3(36) - \log_3(12)$ **c.** $\log_5(100) - \log_5(8)$ **d.** $\log_2\left(\dfrac{1}{3}\right) + \log_2(9)$

24. a. $\log_4(25) + \log_4\left(\dfrac{1}{5}\right)$ **b.** $\log_{10}(5) - \log_{10}(20)$ **c.** $\log_3\left(\dfrac{4}{5}\right) - \log_3\left(\dfrac{1}{5}\right)$

    d. $\log_2(9) + \log_2(4) - \log_2(12)$ **e.** $\log_3(8) - \log_3(2) + \log_3(5)$ **f.** $\log_4(24) - \log_4(2) - \log_4(6)$

25. **MC** **a.** The expression $\log_{10}(xy)$ is equal to:

    A. $\log_{10}(x) \times \log_{10}(y)$ **B.** $\log_{10}(x) - \log_{10}(y)$
    C. $\log_{10}(x) + \log_{10}(y)$ **D.** $y \log_{10}(x)$

    b. The expression $\log_{10}(x^y)$ is equal to:

    A. $x \log_{10}(y)$ **B.** $y \log_{10}(x)$
    C. $10 \log_x(y)$ **D.** $\log_{10}(x) + \log_{10}(y)$

    c. The expression $\dfrac{1}{3}\log_2(64) + \log_2(10)$ is equal to:

    A. $\log_2(40)$ **B.** $\log_2(80)$

    C. $\log_2\left(\dfrac{64}{10}\right)$ **D.** 1

Reasoning

26. For each of the following, write the possible strategy you intend to use.

    a. Evaluate $\left(\log_3(81)\right)\left(\log_3(27)\right)$.

    b. Evaluate $\dfrac{\log_a(81)}{\log_a(3)}$.

    c. Evaluate $5^{\log_5(7)}$.

    In each case, explain how you obtained your final answer.

27. Simplify $\log_5(10) + 2\log_5(2) - 3\log_5(10)$.

28. Simplify $\log_2\left(\dfrac{8}{125}\right) - 3\log_2\left(\dfrac{3}{5}\right) - 4\log_2\left(\dfrac{1}{2}\right)$.

Problem solving

29. Simplify $\log_a\left(a^5 + a^3\right) - \log_a\left(a^4 + a^2\right)$.

30. If $2\log_a(x) = 1 + \log_a(8x - 15a)$, determine the value of x in terms of a where a is a positive constant and x is positive.

31. Solve the following for x.

$$\log_3(x + 2) + \log_3(x - 4) = 3$$

LESSON
1.12 Solving equations (10A)

LEARNING INTENTION

At the end of this lesson you should be able to:
- simplify and solve equations involving logarithms using the logarithm laws and exponent laws.

1.12.1 Solving simple exponential equations

An exponential equation has the unknown variable as an exponent.

If exponent laws can be used to express both sides of an equation as single powers of the same base, then this allows exponents to be equated. For example, if an equation can be simplified to the form $2^{2x} = 2^6$, then for this equality to be true, $2x = 6$. Solving this equation gives the solution, $x = 3$.

WORKED EXAMPLE 36

Find the value of x in each of the following equations.
a. $3^x = 81$ **b. $4^{x-1} = 256$**

THINK	WRITE
a. 1. Write the equation.	a. $\qquad 3^x = 81$
2. Express both sides to the same base.	$\qquad 3^x = 3^4$
3. Equate the powers.	Therefore, $x = 4$,
b. 1. Write the equation.	b. $\qquad 4^{x-1} = 256$
2. Express both sides to the same base.	$\qquad 4^{x-1} = 4^4$
3. Equate the powers.	Therefore, $x - 1 = 4$,
4. Solve the linear equation for x by adding one to both sides.	$\qquad x = 5$

WORKED EXAMPLE 37

Solve $5^{3x} \times 25^{4-2x} = \dfrac{1}{125}$ for x.

THINK	WRITE
1. Use the exponent laws to express the left-hand side of the equation as a power of a single base.	$5^{3x} \times 25^{4-2x} = \dfrac{1}{125}$
	$5^{3x} \times 5^{2(4-2x)} = \dfrac{1}{125}$
	$5^{3x+8-4x} = \dfrac{1}{125}$
	$5^{8-x} = \dfrac{1}{125}$

2. Express the right-hand side as a power of the same base.

$$5^{8-x} = \frac{1}{5^3}$$
$$5^{8-x} = 5^{-3}$$

3. Equate exponents and calculate the required value of x.

Equating exponents,
$$8 - x = -3$$
$$x = 11$$

TI | THINK	WRITE	CASIO | THINK	WRITE
1. On a Calculator page, press MENU then select 3: Algebra 1: Numerical Solve Complete the entry line as nSolve $\left(5^{3x} \times 25^{4-2x} = \frac{1}{125}, x\right)$ then press ENTER.	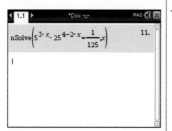	1. On the Run-Matrix screen, press OPTN then select CALC by pressing F4. Select Solve N by pressing F5 then complete the entry line as SolveN $\left(5^{3x} \times 25^{4-2x} = \frac{1}{125}, x\right)$ and press EXE.	
2. The answer appears on the screen.	$x = 11$	2. The answer appears on the screen.	$x = 11$

1.12.2 Solving equations with logarithms (10A)

eles-4679

- The equation $\log_a(y) = x$ is an example of a general **logarithmic equation**. Laws of logarithms and exponents (indices) are used to solve these equations.

WORKED EXAMPLE 38 Solving by converting to exponent form

Solve for x in the following equations.

a. $\log_2(x) = 3$ b. $\log_6(x) = -2$ c. $\log_3(x^4) = -16$ d. $\log_5(x-1) = 2$

THINK

a. 1. Write the equation.

   2. Rewrite using $a^x = y \Leftrightarrow \log_a(y) = x$.

   3. Rearrange and simplify.

b. 1. Write the equation.

   2. Rewrite using $a^x = y \Leftrightarrow \log_a(y) = x$.

   3. Rearrange and simplify.

WRITE

a. $\log_2(x) = 3$

    $2^3 = x$

    $x = 8$

b. $\log_6(x) = -2$

    $6^{-2} = x$

    $x = \dfrac{1}{6^2}$

      $= \dfrac{1}{36}$

c. **1.** Write the equation.

c. $\log_3\left(x^4\right) = -16$

    2. Rewrite using $\log_a\left(x^n\right) = n\log_a(x)$.

$4\log_3(x) = -16$

    3. Divide both sides by 4.

$\log_3(x) = -4$

    4. Rewrite using $a^x = y \Leftrightarrow \log_a(y) = x$.

$3^{-4} = x$

    5. Rearrange and simplify.

$$x = \frac{1}{3^4}$$
$$= \frac{1}{81}$$

d. **1.** Write the equation.

d. $\log_5(x-1) = 2$

    2. Rewrite using $a^x = y \Leftrightarrow \log_a(y) = x$.

$5^2 = x - 1$

    3. Solve for x.

$x - 1 = 25$
$x = 26$

WORKED EXAMPLE 39 Solving for the base of a logarithm

Solve for x in $\log_x(25) = 2$, given that $x > 0$.

THINK

1. Write the equation.

2. Rewrite using $a^x = y \Leftrightarrow \log_a(y) = x$.

3. Solve for x.
Note: $x = -5$ is rejected as a solution because $x > 0$.

WRITE

$\log_x(25) = 2$

$x^2 = 25$

$x = 5$ (because $x > 0$)

WORKED EXAMPLE 40 Evaluating logarithms

Solve for x in the following equations.

a. $\log_2(16) = x$ **b.** $\log_3\left(\dfrac{1}{3}\right) = x$ **c.** $\log_9(3) = x$

THINK

a. **1.** Write the equation.

    2. Rewrite using $a^x = y \Leftrightarrow \log_a(y) = x$.

    3. Write 16 with base 2.

    4. Equate the exponents.

WRITE

a. $\log_2(16) = x$

$2^x = 16$

$= 2^4$

$x = 4$

b. 1. Write the equation.

b. $\log_3\left(\dfrac{1}{3}\right) = x$

2. Rewrite using $a^x = y \Leftrightarrow \log_a(y) = x$.

$3^x = \dfrac{1}{3}$

$\quad\ = \dfrac{1}{3^1}$

3. Write $\dfrac{1}{3}$ with base 3.

$3^x = 3^{-1}$

4. Equate the exponents.

$x = -1$

c. 1. Write the equation.

c. $\log_9(3) = x$

2. Rewrite using $a^x = y \Leftrightarrow \log_a(y) = x$.

$9^x = 3$

3. Write 9 with base 3.

$\left(3^2\right)^x = 3$

4. Remove the grouping symbols.

$3^{2x} = 3^1$

5. Equate the exponents.

$2x = 1$

6. Solve for x.

$x = \dfrac{1}{2}$

WORKED EXAMPLE 41 Solving equations with multiple logarithm terms

Solve for x in the equation $\log_2(4) + \log_2(x) - \log_2(8) = 3$.

THINK

1. Write the equation.

2. Simplify the left-hand side.
Use $\log_a(x) + \log_a(y) = \log_a(xy)$ and
$\log_a(x) - \log_a(y) = \log_a\left(\dfrac{x}{y}\right)$.

3. Simplify.

4. Rewrite using $a^x = y \Leftrightarrow \log_a(y) = x$.

5. Solve for x.

WRITE

$\log_2(4) + \log_2(x) - \log_2(8) = 3$

$\log_2\left(\dfrac{4 \times x}{8}\right) = 3$

$\log_2\left(\dfrac{x}{2}\right) = 3$

$2^3 = \dfrac{x}{2}$

$x = 2 \times 2^3$
$\ \ = 2 \times 8$
$\ \ = 16$

| 1.12 Quick quiz   on | 1.12 Exercise |

Individual pathways

■ PRACTISE	■ CONSOLIDATE	■ MASTER
4, 7, 10, 14, 17, 20	1, 2, 3, 5, 8, 11, 14, 18, 21, 24	6, 9, 12, 15, 16, 19, 22, 25

Fluency

1. **WE36a** Solve for x in each of the following equations.

    a. $2^x = 32$
    b. $5^x = 625$
    c. $3^x = 243$

    d. $10^{-x} = \dfrac{1}{100}$
    e. $4^{-x} = 16$
    f. $6^x = \dfrac{1}{216}$

    g. $3^{-x} = \dfrac{1}{81}$
    h. $2^{-x} = 1$
    i. $8^x = 2^6$

2. **WE36b** Solve for n in each of the following equations.

    a. $2^{3n+1} = 64$
    b. $5^{2n+3} = 25$
    c. $3^{2-n} = 27$

    d. $16^{n+3} = 2^3$
    e. $49^{5-3n} = \dfrac{1}{7}$
    f. $36^{4n-3} = 216$

3. **WE37** Solve for x in each of the following.

    a. $4^{2x} = 8^{x-1}$
    b. $27^{4-x} = 9^{2x+1}$
    c. $16^{3x+1} = 128^{x-2}$

    d. $25^{2x-3} = \dfrac{1}{125}$
    e. $32^{5-x} = 4^{3x+2}$
    f. $64^{2-3x} = 16^{x+1}$

    g. $9^{3x+5} = \dfrac{1}{243}$
    h. $16^{4-3x} = \dfrac{1}{8^{x+3}}$

WE38 For questions **4** to **6**, solve for x in the following equations.

4. a. $\log_5(x) = 2$ b. $\log_3(x) = 4$ c. $\log_2(x) = -3$ d. $\log_4(x) = -2$ e. $\log_{10}\left(x^2\right) = 4$

5. a. $\log_2\left(x^3\right) = 12$ b. $\log_3(x+1) = 3$ c. $\log_5(x-2) = 3$ d. $\log_4(2x-3) = 0$ e. $\log_{10}(2x+1) = 0$

6. a. $\log_2(-x) = -5$ b. $\log_3(-x) = -2$ c. $\log_5(1-x) = 4$ d. $\log_{10}(5-2x) = 1$

WE39 For questions **7** to **9**, solve for x in the following equations, given that $x > 0$.

7. a. $\log_x(9) = 2$ b. $\log_x(16) = 4$ c. $\log_x(25) = \dfrac{2}{3}$

8. a. $\log_x(125) = \dfrac{3}{4}$ b. $\log_x\left(\dfrac{1}{8}\right) = -3$ c. $\log_x\left(\dfrac{1}{64}\right) = -2$

9. a. $\log_x\left(6^2\right) = 2$ b. $\log_x\left(4^3\right) = 3$

WE40 For questions **10** to **12**, solve for x in the following equations.

10. a. $\log_2(8) = x$ b. $\log_3(9) = x$ c. $\log_5\left(\dfrac{1}{5}\right) = x$

11. a. $\log_4\left(\dfrac{1}{16}\right) = x$ b. $\log_4(2) = x$ c. $\log_8(2) = x$

12. a. $\log_6(1) = x$ b. $\log_8(1) = x$ c. $\log_{\frac{1}{2}}(2) = x$ d. $\log_{\frac{1}{3}}(9) = x$

WE41 For questions **13** to **15**, solve for x in the following equations.

13. **a.** $\log_2(x) + \log_2(4) = \log_2(20)$ **b.** $\log_5(3) + \log_5(x) = \log_5(18)$
    c. $\log_3(x) - \log_3(2) = \log_3(5)$ **d.** $\log_{10}(x) - \log_{10}(4) = \log_{10}(2)$

14. **a.** $\log_4(8) - \log_4(x) = \log_4(2)$ **b.** $\log_3(10) - \log_3(x) = \log_3(5)$
    c. $\log_6(4) + \log_6(x) = 2$ **d.** $\log_2(x) + \log_2(5) = 1$

15. **a.** $3 - \log_{10}(x) = \log_{10}(2)$ **b.** $5 - \log_4(8) = \log_4(x)$
    c. $\log_2(x) + \log_2(6) - \log_2(3) = \log_2(10)$ **d.** $\log_2(x) + \log_2(5) - \log_2(10) = \log_2(3)$
    e. $\log_3(5) - \log_3(x) + \log_3(2) = \log_3(10)$ **f.** $\log_5(4) - \log_5(x) + \log_5(3) = \log_5(6)$

16. **MC** **a.** The solution to the equation $\log_7(343) = x$ is:

    A. $x = 2$ **B.** $x = 3$ **C.** $x = 1$ **D.** $x = 0$

    b. If $\log_8(x) = 4$, then x is equal to:

    A. 4096 **B.** 512 **C.** 64 **D.** 2

    c. Given that $\log_x(3) = \dfrac{1}{2}$, x must be equal to:

    A. 3 **B.** 6 **C.** 81 **D.** 9

    d. If $\log_a(x) = 0.7$, then $\log_a(x^2)$ is equal to:

    A. 0.49 **B.** 1.4 **C.** 0.35 **D.** 0.837

Understanding

For questions **17** to **19**, solve for x in the following equations.

17. **a.** $2^x = 128$ **b.** $3^x = 9$ **c.** $7^x = \dfrac{1}{49}$ **d.** $9^x = 1$ **e.** $5^x = 625$

18. **a.** $64^x = 8$ **b.** $6^x = \sqrt{6}$ **c.** $2^x = 2\sqrt{2}$ **d.** $3^x = \dfrac{1}{\sqrt{3}}$ **e.** $4^x = 8$

19. **a.** $9^x = 3\sqrt{3}$ **b.** $2^x = \dfrac{1}{4\sqrt{2}}$ **c.** $3^{x+1} = 27\sqrt{3}$ **d.** $2^{x-1} = \dfrac{1}{32\sqrt{2}}$ **e.** $4^{x+1} = \dfrac{1}{8\sqrt{2}}$

Reasoning

20. The apparent brightness of stars is measured on a logarithmic scale called magnitude, in which lower numbers mean brighter stars. The relationship between the ratio of the apparent brightness of two objects and the difference in their magnitudes is given by the formula:

$$m_2 - m_1 = -2.5\log_{10}\left(\frac{b_2}{b_1}\right)$$

where m is the magnitude and b is the apparent brightness.
Determine how many times brighter a magnitude 2.0 star is than a magnitude 3.0 star.

21. The decibel (dB) scale for measuring loudness, d, is given by the formula $d = 10 \log_{10}(I \times 10^{12})$, where I is the intensity of sound in watts per square metre.

a. Determine the number of decibels of sound if the intensity is 1.
b. Evaluate the number of decibels of sound produced by a jet engine at a distance of 50 metres if the intensity is 10 watts per square metre.
c. Determine the intensity of sound if the sound level of a pneumatic drill 10 metres away is 90 decibels.
d. Determine how the value of d changes if the intensity is doubled. Give your answer to the nearest decibel.
e. Evaluate how the value of d changes if the intensity is 10 times as great.
f. Determine by what factor does the intensity of sound have to be multiplied in order to add 20 decibels to the sound level.

22. The Richter scale is used to describe the energy of earthquakes.

A formula for the Richter scale is $R = \dfrac{2}{3} \log_{10}(K) - 0.9$, where R is the Richter scale value for an earthquake that releases K kilojoules (kJ) of energy.

a. Determine the Richter scale value for an earthquake that releases the following amounts of energy.
   i. 1000 kJ
   ii. 2000 kJ
   iii. 3000 kJ
   iv. 10 000 kJ
   v. 100 000 kJ
   vi. 1 000 000 kJ

b. Does doubling the energy released double the Richter scale value? Justify your answer.

c. Determine the energy released by an earthquake of:
  i. magnitude 4 on the Richter scale
  ii. magnitude 5 on the Richter scale
  iii. magnitude 6 on the Richter scale.
d. Explain the effect (on the amount of energy released) of increasing the Richter scale value by 1.
e. Explain why an earthquake measuring 8 on the Richter scale is so much more devastating than one that measures 5.

Problem solving

23. Solve the following equations for x.
  a. $3^{x+1} = 7$
  b. $3^{x+1} = 7^x$

24. Solve the following equation for x.

  $$(27 \times 3^x)^3 = 81^x \times 3^2$$

25. Solve $\left\{ x : (3^x)^2 = 30 \times 3^x - 81 \right\}$.

LESSON
1.13 Review

1.13.1 Topic summary

Number sets

- Natural numbers: $N = \{1, 2, 3, 4, 5 \ldots\}$
- Integers: $Z = \{\ldots, -2, -1, 0, 1, 2 \ldots\}$
- Rational numbers: $Q = \left\{\dfrac{1}{4}, -0.36, 2, \dfrac{9}{7}, 0.\dot{3}\dot{6}\right\}$
- Irrational numbers: $I = \{\sqrt{2}, \pi, e\}$
- Real numbers: $R = Q + I$

Exponent notation

- Exponent (index) notation is a short way of writing a repeated multiplication.
  e.g. $2 \times 2 \times 2 \times 2 \times 2 \times 2$ can be written as 2^6, which is read '2 to the power of 6'.
- The **base** is the number that is being repeatedly multiplied and the exponent or **index** is the number of times it is multiplied.
  e.g. $2^6 = 2 \times 2 \times 2 \times 2 \times 2 \times 2 = 64$

Surds

- A surd is any number that requires a $\sqrt{x}$ or $\sqrt[n]{x}$ symbol and does not simplify to a whole number.
  e.g. $\sqrt{2}$ is a surd, but $\sqrt{9} = 3$ is not.
- To simplify a surd look for the highest square factor.
  e.g. $\sqrt{48} = \sqrt{16 \times 3} = 4\sqrt{3}$
- Only like surds can be added and subtracted.
  e.g. $\sqrt{5}, 3\sqrt{5}$ and $-6\sqrt{5}$ are like surds whereas $\sqrt{7}$ and $2\sqrt{11}$ are not like surds.
- Surds are added and subtracted the same way like terms are combined in algebra. You may need to simplify first.
  e.g. $3\sqrt{2} + 7\sqrt{2} - 2\sqrt{2} = 10\sqrt{2} - 2\sqrt{2} = 8\sqrt{2}$
  e.g. $\sqrt{12} + \sqrt{75} = 2\sqrt{3} + 5\sqrt{3} = 7\sqrt{3}$

EXPONENTS, SURDS AND LOGARITHMS

Exponent laws

- 1st law: $a^m \times a^n = a^{m+n}$
- 2nd law: $a^m \div a^n = a^{m-n}$
- 3rd law: $a^0 = 1, a \neq 0$
- 4th law: $(a^m)^n = a^{m \times n} = a^{mn}$
- 5th law: $(ab)^n = a^n b^n$
- 6th law: $\left(\dfrac{a}{b}\right)^n = \dfrac{a^n}{b^n}$
- 7th law: $a^{-n} = \dfrac{1}{a^n}$
- 8th law: $a^{\frac{1}{n}} = \sqrt[n]{a}$

Multiplying and dividing surds

- $\sqrt{a} \times \sqrt{b} = \sqrt{ab}$
- $m\sqrt{a} \times n\sqrt{b} = mn\sqrt{ab}$
- $\sqrt{a} \div \sqrt{b} = \dfrac{\sqrt{a}}{\sqrt{b}} = \sqrt{\dfrac{a}{b}}$
- $m\sqrt{a} \div (n\sqrt{b}) = \dfrac{m\sqrt{a}}{n\sqrt{b}} = \dfrac{m}{n}\sqrt{\dfrac{a}{b}}$

Rationalising the denominator

- Involves rewriting a fraction with a rational denominator.
  e.g. $\dfrac{2}{\sqrt{5}} = \dfrac{2}{\sqrt{5}} \times \dfrac{\sqrt{5}}{\sqrt{5}} = \dfrac{2\sqrt{5}}{5}$
- It may be necessary to multiply by the conjugate in order to rationalise.
  e.g. $\dfrac{1}{\sqrt{6}+2} \times \dfrac{\sqrt{6}-2}{\sqrt{6}-2} = \dfrac{\sqrt{6}-2}{2}$

Logarithms

- Exponent form: $y = a^x$
- Logarithmic form: $\log_a(y) = x$
- Log laws:
  - $\log_a(x) + \log_a(y) = \log_a(xy)$
  - $\log_a(x) - \log_a(y) = \log_a\left(\dfrac{x}{y}\right)$
  - $\log_a(x)^n = n\log_a(x)$
  - $\log_a(1) = 0$
  - $\log_a(a) = 1$
  - $\log_a\left(\dfrac{1}{x}\right) = -\log_a(x)$
  - $\log_a(a^x) = x$
- Each log law is equivalent to one of the exponent laws.

Solving exponential and logarithmic equations

- Use exponent laws to express both sides of an equation as single powers of the same base, then equate the exponents.
- Simplify both sides of the equation so there is at most a single logarithm on each side.
- Switch to exponent form or log form as required.
  e.g.
  $\log_3(x) = 4 \implies x = 3^4$
  $5^x = 7 \implies \log_5(7) = x$

Compound interest

- To calculate the value of an investment earning compound interest:
  $$A = P(1 + r)^n$$
  - A = future value
  - P = principal (starting value)
  - r = interest rate as a decimal (e.g. 7.5% p.a. is equal to 0.075 p.a.)
  - n = number of periods
- To calculate the amount of interest earned: **Interest** $= I = A - P$

1.13.2 Success criteria

Tick a column to indicate that you have completed the lesson and how well you think you have understood it using the traffic light system.

(**Green:** I understand; **Yellow:** I can do it with help; **Red:** I do not understand)

Lesson	Success criteria	⬤	⬤	⬤
1.2	I can define the real, rational, irrational, integer and natural numbers.			
	I can determine whether a number is rational or irrational.			
1.3	I can determine whether the number under a root or radical sign is a surd.			
	I can prove that a surd is irrational by contradiction.			
1.4	I can multiply and simplify surds.			
	I can add and subtract like surds.			
	I can divide surds.			
	I can rationalise the denominators of fractions.			
1.5	I can recall and apply the exponent laws.			
	I can simplify expressions involving multiplication and division of terms with the same base.			
	I can evaluate expressions involving powers of zero.			
	I can simplify expressions involving raising a power to another power.			
1.6	I can evaluate expressions involving negative exponents.			
	I can simplify expressions involving negative exponents and rewrite expressions so that all exponents are positive.			
1.7	I can evaluate expressions involving fractional exponents.			
	I can simplify expressions involving fractional exponents.			
1.8	I can simplify algebraic expressions involving brackets, fractions, multiplication and division using appropriate exponent (index) laws.			
1.9	I can calculate the future value of an investment earning compound interest.			
	I can calculate the amount of interest earned after a period of time on an investment with compound interest.			
1.10	I can convert between exponent form and logarithmic form.			
	I can evaluate logarithms and use logarithms in scale measurement.			
1.11	I can simplify expressions using logarithm laws.			
1.12	I can simplify and solve equations involving exponents and logarithms using the logarithm laws and exponent laws.			

1.13.3 Project

Other number systems

Throughout history, different systems have been used to aid with counting. Ancient tribes are known to have used stones, bones and knots in rope to help them keep count. The counting system that is used around the world today is called the Hindu-Arabic system. This system had its origin in India around 300–200 BCE. The Arabs brought this method of counting to Europe in the Middle Ages.

The Hindu–Arabic method is known as the decimal or base 10 system, as it is based on counting in lots of ten. This system uses the digits 0, 1, 2, 3, 4, 5, 6, 7, 8 and 9. Notice that the largest digit is one less than the base number; that is, the largest digit in base 10 is 9. To make larger numbers, digits are grouped together. The position of the digit tells us about its value. We call this *place value*. For example, in the number 325, the 3 has a value of 'three lots of a hundred', the 2 has a value of 'two lots of ten' and the 5 has a value of 'five lots of units'. Another way to write this is:

$$3 \times 100 + 2 \times 10 + 5 \times 1 \text{ or } 3 \times 10^2 + 2 \times 10^1 + 5 \times 10^0$$

In a decimal system, every place value is based on the number 10 raised to a power. The smallest place value (units) is described by 10^0, the tens place value by 10^1, the hundreds place value by 10^2, the thousands by 10^3 and so on.

Computers do not use a decimal system. The system for computer languages is based on the number 2 and is known as the binary system. The only digits needed in the binary system are the digits 0 and 1. Can you see why?

Decimal number	0	1	2	3	4	5	6	7	8	9	10	11	12	13
Binary number	0	1	10	11	100	101	110	111	1000	1001	1010	1011	1100	1101

Consider the decimal number 7. From the table above, you can see that its binary equivalent is 111. How can you be sure this is correct?

$$111 = 1 \times 2^2 + 1 \times 2^1 + 1 \times 2^0 = 4 + 2 + 1 = 7$$

Notice that this time each place value is based on the number 2 raised to a power. You can use this technique to change any binary number into a decimal number. (The same pattern applies to other bases. For example, in base 6 the place values are based on the number 6 raised to a power.)

Binary operations

When adding in the decimal system, each time the addition is greater than 9, we need to 'carry over' into the next place value. In the example below, the units column adds to more than 9, so we need to carry over into the next place value.

$$\begin{array}{r} ^{1}17 \\ +\ 13 \\ \hline 30 \end{array}$$

The same is true when adding in binary, except we need to 'carry over' every time the addition is greater than 1.

$$\begin{array}{r} ^{1}01 \\ +\ 01 \\ \hline 10 \end{array}$$

1. Perform the following binary additions.

   a. $\begin{array}{r} 11_2 \\ +\ 01_2 \\ \hline \end{array}$

   b. $\begin{array}{r} 111_2 \\ +\ 110_2 \\ \hline \end{array}$

   c. $\begin{array}{r} 1011_2 \\ +\ 101_2 \\ \hline \end{array}$

2. Perform the following binary subtractions. Remember that if you need to borrow a number from a column on the left-hand side, you will actually be borrowing a 2 (not a 10).

   a. $\begin{array}{r} 11_2 \\ -\ 01_2 \\ \hline \end{array}$

   b. $\begin{array}{r} 111_2 \\ -\ 110_2 \\ \hline \end{array}$

   c. $\begin{array}{r} 1011_2 \\ -\ 101_2 \\ \hline \end{array}$

3. Try some multiplication. Remember to carry over lots of 2.

   a. $\begin{array}{r} 11_2 \\ \times\ 01_2 \\ \hline \end{array}$

   b. $\begin{array}{r} 111_2 \\ \times\ 110_2 \\ \hline \end{array}$

   c. $\begin{array}{r} 1011_2 \\ \times\ 101_2 \\ \hline \end{array}$

4. What if our number system had an 8 as its basis (that is, we counted in lots of 8)? The only digits available for use would be 0, 1, 2, 3, 4, 5, 6 and 7. (Remember the maximum digit is 1 less than the base value.) Give examples to show how numbers would be added, subtracted and multiplied using this base system. Remember that you would 'carry over' or 'borrow' lots of 8.

5. The hexadecimal system has 16 as its basis. Investigate this system. Explain how it would be possible to have 15, for example, in a single place position. Give examples to show how the system would add, subtract and multiply.

 Resources

eWorkbook　Topic 1 Workbook (worksheets, code puzzle and project) (ewbk-2093)

Interactivities Crossword (int-9161)

　　　　　　　　Sudoku puzzle (int-3891)

Fluency

1. **MC** Identify which of the given numbers are rational.

$$\sqrt{\frac{6}{12}}, \ \sqrt{0.81}, \ 5, \ -3.26, \ 0.5, \ \frac{\pi}{5}, \ \sqrt{\frac{3}{12}}$$

A. $\sqrt{0.81}, \ 5, \ -3.26, \ 0.5$ and $\sqrt{\frac{3}{12}}$

B. $\sqrt{\frac{6}{12}}$ and $\frac{\pi}{5}$

C. $\sqrt{\frac{6}{12}}, \ \sqrt{0.81}$ and $\sqrt{\frac{3}{12}}$

D. $5, \ -3.26$ and $\sqrt{\frac{6}{12}}$

2. For each of the following, state whether the number is rational or irrational and give the reason for your answer.

   a. $\sqrt{12}$ **b.** $\sqrt{121}$ **c.** $\frac{2}{9}$ **d.** $0.\dot{6}$ **e.** $\sqrt[3]{0.08}$

3. **MC** Identify which of the numbers of the given set are surds.

$$\left\{ 3\sqrt{2}, 5\sqrt{7}, 9\sqrt{4}, 6\sqrt{10}, 7\sqrt{12}, 12\sqrt{64} \right\}$$

A. $9\sqrt{4}, 12\sqrt{64}$

B. $3\sqrt{2}$ and $7\sqrt{12}$ only

C. $3\sqrt{2}, 5\sqrt{7}$ and $6\sqrt{10}$ only

D. $3\sqrt{2}, 5\sqrt{7}, 6\sqrt{10}$ and $7\sqrt{12}$

4. Identify which of $\sqrt{2m}, \ \sqrt{25m}, \ \sqrt{\frac{m}{16}}, \ \sqrt{\frac{20}{m}}, \ \sqrt[3]{m}, \ \sqrt[3]{8m}$ are surds:

   a. if $m = 4$ **b.** if $m = 8$.

5. Simplify each of the following.

   a. $\sqrt{50}$ **b.** $\sqrt{180}$ **c.** $2\sqrt{32}$ **d.** $5\sqrt{80}$

6. **MC** The expression $\sqrt{392x^8y^7}$ may be simplified to:

   A. $196x^4y^3\sqrt{2y}$ **B.** $2x^4y^3\sqrt{14y}$

   C. $14x^4y^3\sqrt{2y}$ **D.** $14x^4y^3\sqrt{2}$

7. Simplify the following surds. Give the answers in the simplest form.

   a. $4\sqrt{648x^7y^9}$ **b.** $-\frac{2}{5}\sqrt{\frac{25}{64}x^5y^{11}}$

8. Simplify the following, giving answers in the simplest form.

   a. $7\sqrt{12} + 8\sqrt{147} - 15\sqrt{27}$

   b. $\dfrac{1}{2}\sqrt{64a^3b^3} - \dfrac{3}{4}ab\sqrt{16ab} + \dfrac{1}{5ab}\sqrt{100a^5b^5}$

9. Simplify each of the following.

   a. $\sqrt{3} \times \sqrt{5}$ b. $2\sqrt{6} \times 3\sqrt{7}$ c. $3\sqrt{10} \times 5\sqrt{6}$ d. $\left(\sqrt{5}\right)^2$

10. Simplify the following, giving answers in the simplest form.

    a. $\dfrac{1}{5}\sqrt{675} \times \sqrt{27}$ b. $10\sqrt{24} \times 6\sqrt{12}$

11. Simplify the following.

    a. $\dfrac{\sqrt{30}}{\sqrt{10}}$ b. $\dfrac{6\sqrt{45}}{3\sqrt{5}}$ c. $\dfrac{3\sqrt{20}}{12\sqrt{6}}$ d. $\dfrac{\left(\sqrt{7}\right)^2}{14}$

12. Rationalise the denominator of each of the following.

    a. $\dfrac{2}{\sqrt{6}}$ b. $\dfrac{\sqrt{3}}{2\sqrt{6}}$ c. $\dfrac{2}{\sqrt{5}-2}$ d. $\dfrac{\sqrt{3}-1}{\sqrt{3}+1}$

13. Evaluate each of the following, correct to 1 decimal place if necessary.

    a. $64^{\frac{1}{3}}$ b. $20^{\frac{1}{2}}$ c. $10^{\frac{1}{3}}$ d. $50^{\frac{1}{4}}$

14. Evaluate each of the following, correct to 1 decimal place.

    a. $20^{\frac{2}{3}}$ b. $2^{\frac{3}{4}}$ c. $(0.7)^{\frac{3}{5}}$ d. $\left(\dfrac{2}{3}\right)^{\frac{2}{3}}$

15. Write each of the following in simplest surd form.

    a. $2^{\frac{1}{2}}$ b. $18^{\frac{1}{2}}$ c. $5^{\frac{3}{2}}$ d. $8^{\frac{4}{3}}$

16. Evaluate each of the following without using a calculator. Show all working.

    a. $\dfrac{16^{\frac{3}{4}} \times 81^{\frac{1}{4}}}{6 \times 16^{\frac{1}{2}}}$

    b. $\left(125^{\frac{2}{3}} - 27^{\frac{2}{3}}\right)^{\frac{1}{2}}$

17. Evaluate each of the following, giving your answer as a fraction.

    a. 4^{-1} b. 9^{-1} c. 4^{-2} d. 10^{-3}

18. Determine the value of each of the following, correct to 3 significant figures.

    a. 12^{-1} b. 7^{-2} c. $(1.25)^{-1}$ d. $(0.2)^{-4}$

19. Write down the value of each of the following.

    a. $\left(\dfrac{2}{3}\right)^{-1}$ b. $\left(\dfrac{7}{10}\right)^{-1}$ c. $\left(\dfrac{1}{5}\right)^{-1}$ d. $\left(3\dfrac{1}{4}\right)^{-1}$

20. **MC** a. The expression $\sqrt{250}$ may be simplified to:

   A. $25\sqrt{10}$ B. $5\sqrt{10}$ C. $10\sqrt{5}$ D. $5\sqrt{50}$

   b. When expressed in its simplest form, $2\sqrt{98} - 3\sqrt{72}$ is equal to:

   A. $-4\sqrt{2}$ B. -4 C. $-2\sqrt{4}$ D. $4\sqrt{2}$

   c. When expressed in its simplest form, $\sqrt{\dfrac{8x^3}{32}}$ is equal to:

   A. $\dfrac{x\sqrt{x}}{2}$ B. $\dfrac{\sqrt{x^3}}{4}$ C. $\dfrac{\sqrt{x^3}}{2}$ D. $\dfrac{x\sqrt{x}}{4}$

21. Determine the values of the following, giving your answer in fraction form.

   a. $\left(\dfrac{2}{5}\right)^{-1}$ b. $\left(\dfrac{2}{3}\right)^{-2}$

22. Determine the values of the following, leaving your answer in fraction form.

   a. 2^{-1} b. 3^{-2} c. 4^{-3} d. $\left(\dfrac{1}{2}\right)^{-1}$

23. **MC** $3d^{10}e^4$ is the simplified form of:

   A. $d^6e^2 \times 3d^4e^3$ B. $\dfrac{6d^{10}e^5}{2e^2}$ C. $\left(3d^5e^2\right)^2$ D. $3e\left(d^5\right)^2 \times e^3$

24. **MC** $8m^3n \times n^4 \times 2m^2n^3$ simplifies to:

   A. $10m^5n^8$ B. $16m^5n^7$ C. $16m^5n^8$ D. $10m^5n^7$

25. **MC** $8x^3 \div 4x^{-3}$ is equal to:

   A. 2 B. $2x^0$ C. $2x^6$ D. $2x^{-1}$

26. **MC** $\dfrac{12x^8 \times 2x^7}{6x^9 \times x^5}$ simplifies to:

   A. $4x^5$ B. $8x$ C. $4x$ D. $8x^5$

27. **MC** The expression $\dfrac{\left(a^2b^3\right)^5}{\left(2a^2b\right)^2}$ is equal to:

   A. $\dfrac{a^6b^{13}}{4}$ B. $2a^6b^{13}$ C. $\dfrac{a^3b^6}{2}$ D. $\dfrac{a^6b^{13}}{2}$

28. **MC** $\dfrac{\left(p^2q\right)^4}{\left(2p^5q^2\right)^3} \div \dfrac{\left(p^5q^2\right)^2}{2pq^5}$ can be simplified to:

   A. $\dfrac{1}{4p^{16}q}$ B. $\dfrac{2^2}{p^{16}q}$ C. $\dfrac{1}{4p^8}$ D. $\dfrac{1}{2p^{16}q}$

29. **MC** $16^{-\frac{3}{4}} \div 9^{\frac{3}{2}}$ can be simplified to:

A. 2 B. $\dfrac{1}{216}$ C. $\dfrac{8}{27}$ D. $3\dfrac{3}{8}$

30. **MC** $\dfrac{\left(2l^{\frac{2}{9}}m^{-1}\right)^{-3}}{8\left(\frac{1}{16}lm^{-2}\right)^2}$ can be simplified to:

A. $\dfrac{8m^7}{l^{\frac{11}{3}}}$ B. $\dfrac{2m^7}{l^{\frac{7}{3}}}$ C. $\dfrac{4m^7}{l^{\frac{8}{3}}}$ D. $\dfrac{16m^7}{l^{\frac{5}{3}}}$

31. **MC** $\sqrt[5]{32i^{\frac{10}{7}}j^{\frac{5}{11}}k^2}$ can be simplified to:

A. $\dfrac{32i^{\frac{2}{7}}j^{\frac{1}{11}}k^{\frac{2}{5}}}{5}$ B. $2i^{\frac{2}{7}}j^{\frac{1}{11}}k^{\frac{2}{5}}$ C. $\dfrac{32i^{\frac{10}{7}}j^{\frac{5}{11}}k^2}{5}$ D. $2i^{\frac{50}{7}}j^{\frac{25}{11}}k^{10}$

32. Simplify each of the following.

a. $5x^3 \times 3x^5y^4 \times \dfrac{3}{5}x^2y^6$ b. $\dfrac{26a^4b^6c^5}{12a^3b^3c^3}$ c. $\left(\dfrac{20m^5n^2}{6}\right)^3$ d. $\left(\dfrac{14p^7}{21q^3}\right)^4$

33. Evaluate each of the following.

a. $5a^0 - \left(\dfrac{2a}{3}\right)^0 + 12$ b. $-(3b)^0 - \dfrac{(4b)^0}{2}$

34. Simplify each of the following and express your answer with positive exponents.

a. $2a^{-5}b^2 \times 4a^{-6}b^{-4}$ b. $4x^{-5}y^{-3} \div \left(20x^{12}y^{-5}\right)$ c. $\left(2m^{-3}n^2\right)^{-4}$

35. Evaluate each of the following without using a calculator.

a. $\left(\dfrac{1}{2}\right)^{-3}$ b. $2 \times (3)^{-3} \times \left(\dfrac{9}{2}\right)^2$ c. $4^{-3} \times \dfrac{5}{8^{-2}} - 5$

36. Simplify each of the following.

a. $2a^{\frac{4}{5}}b^{\frac{1}{2}} \times 3a^{\frac{1}{2}}b^{\frac{3}{4}} \times 5a^{\frac{3}{4}}b^{\frac{2}{5}}$ b. $\dfrac{4^3x^{\frac{3}{4}}y^{\frac{1}{9}}}{16x^{\frac{4}{5}}y^{\frac{1}{3}}}$ c. $\left(\dfrac{4a^{\frac{1}{3}}}{b^3}\right)^{\frac{1}{2}}$

37. Evaluate each of the following without using a calculator. Show all working.

a. $\dfrac{16^{\frac{3}{4}} \times 81^{\frac{1}{4}}}{6 \times 16^{\frac{1}{2}}}$ b. $\left(125^{\frac{2}{3}} - 27^{\frac{2}{3}}\right)^{\frac{1}{2}}$

38. Simplify the following.

a. $\sqrt[3]{a^9} + \sqrt[4]{16a^8b^2} - 3\left(\sqrt[5]{a}\right)^{15}$ b. $\sqrt[5]{32x^5y^{10}} + \sqrt[3]{64x^3y^6}$

39. Simplify each of the following.

a. $\dfrac{\left(5a^{-2}b\right)^{-3}\times 4a^{6}b^{-2}}{2a^{2}b^{3}\times 5^{-2}a^{-3}b^{-6}}$

b. $\dfrac{2x^{4}y^{-5}}{3y^{6}x^{-2}}\times\left(\dfrac{4xy^{-2}}{3x^{-6}y^{3}}\right)^{-3}$

c. $\left(\dfrac{2m^{3}n^{4}}{5m^{\frac{1}{2}}n}\right)^{\frac{1}{3}}\div\left(\dfrac{4m^{\frac{1}{3}}n^{-2}}{5^{-\frac{2}{3}}}\right)^{-\frac{1}{2}}$

40. Simplify each of the following and then evaluate.

a. $\left(3\times 5^{6}\right)^{\frac{1}{2}}\times 3^{\frac{3}{2}}\times 5^{-2}+\left(3^{6}\times 5^{-\frac{1}{2}}\right)^{0}$

b. $\left(6\times 3^{-2}\right)^{-1}\div\dfrac{\left(3^{\frac{1}{2}}\times 6^{\frac{1}{3}}\right)^{6}}{-6^{2}\times\left(3^{-3}\right)^{0}}$

41. Ryan invests \$12 500 for 3 years at 8% p.a. with interest paid annually. By calculating the amount of simple interest earned each year separately, determine the amount to which the investment will grow.

42. Calculate the compound interest earned on \$45 000 at 12% p.a. over 4 years if interest is compounded:

a. annually
b. six-monthly
c. quarterly
d. monthly.

43. The quarterly incomes (in dollars) of a sample of 12 local businesses was recorded as follows:

$$\$45\,000,\ \$20\,000,\ \$360\,000,\ \$750\,000,\ \$3\,200,\ \$1\,048\,500,$$
$$\$34\,590,\ \$37\,250,\ \$65\,710,\ \$290\,060,\ \$59\,070,\ \$8450$$

Complete the frequency table below for log (base 10) of incomes using class intervals of width 1.

$\log_{10}$ (income)	Frequency

44. The following table shows the average weights of 10 different adult mammals.

Mammal	Weight (kg)
African elephant	4800
Black rhinoceros	1100
Blue whale	136 000
Giraffe	800
Gorilla	140
Humpback whale	30 000
Lynx	23
Orang-utan	64
Polar bear	475
Tasmanian devil	7

Display the data in a histogram using a log (base 10) scale.

45. Evaluate the following.
    a. $\log_{12}(18) + \log_{12}(8)$
    b. $\log_4(60) - \log_4(15)$
    c. $\log_9\left(9^8\right)$
    d. $2\log_3(6) - \log_3(4)$

46. Use the logarithm laws to simplify each of the following.
    a. $\log_a(16) + \log_a(3) - \log_a(2)$
    b. $\log_x\left(x\sqrt{x}\right)$

    c. $4\log_a(x) - \log_a\left(x^2\right)$
    d. $5\log_x\left(\dfrac{1}{x}\right)$

47. **MC** If $25^{2-x} = 125$, then x is equal to:
    A. 1
    B. $\dfrac{1}{2}$
    C. -1
    D. 2

48. Solve for x in the following equations, given that $x > 0$.
    a. $\log_2(x) = 9$
    b. $\log_5(x) = -2$
    c. $\log_x(25) = 2$
    d. $\log_x\left(2^6\right) = 6$
    e. $\log_3(729) = x$
    f. $\log_7(1) = x$

49. Solve for x in the following equations.
    a. $\log_5(4) + \log_5(x) = \log_5(24)$
    b. $\log_3(x) - \log_3(5) = \log_3(7)$

50. Solve for x in the following equations.
    a. $6^x = \dfrac{1}{36}$
    b. $7^x = \dfrac{1}{\sqrt{7}}$
    c. $2^{x+1} = 8\sqrt{2}$

51. Solve for x in the following equations, correct to 3 decimal places.
    a. $2^x = 25$
    b. $0.6^x = 7$
    c. $9^{-x} = 0.84$

Problem solving

52. Answer the following. Explain how you reached your answer.
    a. What is the hundreds digit in 3^{3^3}?
    b. What is the ones digit in 6^{704}?
    c. What is the thousands digit in 9^{1000}?

53. Solve for x: $\left(\dfrac{6}{x}\right)^{-1} + \dfrac{1}{6} = x^{-1}$

54. Simplify $\left(\left(\dfrac{\left(a^2\right)^{-1}}{b^{\frac{1}{2}}}\right)^{-1}\right)^{-1}$.

55. If $m = 2$, determine the value of:

$$\frac{6a^{3m} \times 2b^{2m} \times (3ab)^{-m}}{(4b)^m \times (9a^{4m})^{\frac{1}{2}}}$$

56. Answer the following and explain your reasoning.

a. Identify the digit in the tens of 3^{3^3}.
b. Identify the digit in the ones of 6^{309}.
c. Identify the digit in the ones of 8^{1007}.

57. For the work shown below:
a. calculate the correct answer
b. identify where the student has made mistakes.

$$\left(\frac{3a^3b^5c^3}{5a^2b}\right)^2 \div \left(\frac{2ab}{c}\right) = \frac{3a^6b^{10}c^6}{10a^4b^2} \div \frac{2ab}{c}$$

$$= \frac{3a^6b^{10}c^6}{10a^4b^2} \times \frac{c}{2ab}$$

$$= \frac{3a^6b^{10}c^7}{20a^5b^3}$$

$$= \frac{3ab^7c^7}{20}$$

58. A friend is trying to calculate the volume of water in a reservoir amid fears there may be a severe water shortage. She comes up with the following expression:

$$W = \frac{r^4u^2}{r^{\frac{3}{2}}d^2\sqrt{u}} \times \frac{ru \times d^2}{dr^3u^4}$$

where r is the amount of rain, d is how dry the area is, u is the usage of water by the townsfolk, and W is the volume of water in kL.

a. Help your friend simplify the expression by simplifying each pronumeral one at a time.
b. Explain whether the final expression contains any potential surds.
c. Express the fraction with a rational denominator.
d. List the requirements for the possible values of r, u and d to give a rational answer.
e. Calculate the volume of water in the reservoir when $r = 4$, $d = 60$ and $u = 9$. Write your answer in:
   i. kL   **ii.** L   **iii.** mL.
f. Does a high value for d mean the area is dry? Explain your reasoning.

59. The speed of a toy plane can be modelled by the equation $S = \dfrac{p^2}{2 + \sqrt{w^3}}$, where:

$w =$ wind resistance
$p =$ battery power (from 0 (empty) to 10 (full)).

a. Rationalise the denominator of the expression.
b. Using your knowledge of perfect squares, estimate the speed of a toy plane with its battery half full and a wind resistance of 2. Check your answer with a calculator.
c. How does the speed of the toy plane change with increasing wind resistance? Explain providing supportive calculations.

on   To test your understanding and knowledge of this topic, go to your learnON title at www.jacplus.com.au and complete the **post-test**.

Answers

Topic 1 Exponents, surds and logarithms

1.1 Pre-test

1. True
2. Rational
3. $15\,n^{\frac{8}{15}}$
4. $2p^2q^3$
5. $\dfrac{1}{27}$
6. D
7. B
8. $6\sqrt{5}$
9. $14\sqrt{2}$
10. D
11. $a^{\frac{1}{15}}b^{\frac{11}{14}}$
12. $y=-5$
13. $x=-2$
14. 0
15. D

1.2 Number classification review

1. a. Q b. Q c. Q d. I
2. a. I b. Q c. Q d. I
3. a. Q b. Q c. Q d. Q
4. a. I b. Q c. I d. Q
5. a. Q b. I c. I d. I
6. a. Q b. I c. I d. Q
7. a. Q b. Q c. Q d. Q
8. a. Q b. I c. I d. Q
9. a. I b. Neither
   c. I d. I
10. a. I b. Q c. Q d. Q
11. a. I b. I c. Q d. Q
12. a. I b. Q c. Q d. I
13. B
14. D
15. C
16. C
17. $\dfrac{a}{b}$
18. D
19. A
20. $p-q$
21. Sample responses can be found in the worked solutions in the online resources.
22. $8,\ -8$
23. a. $m=11, n=3$ b. $m=2, n=3$
    c. $m=3, n=2$ d. $m=1, n=2$
24. $\dfrac{1}{7}$ or 7^{-1}
25. $\dfrac{37}{91}$

1.3 Surds (10A)

1. b and d
2. b, c and d
3. a and d
4. a and c
5. a, c and d
6. c and f
7. A
8. D
9. B
10. C
11. Any perfect square.
12. $m=4$
13. a. $m=5$, $n=7$ and $m=4$, $n=14$
    b. 15
14. a. Irrational b. Irrational c. Rational
15. Sample responses can be found in the worked solutions in the online resources.
16. Irrational
17. a. i. $4\sqrt{3}$ ii. $6\sqrt{2}$
    b. Yes. If you don't choose the largest perfect square, then you will need to simplify again.
    c. No
18. $p=m$ and $q=n$
19. $x=3$, rational

1.4 Operations with surds (10A)

1. a. $2\sqrt{3}$ b. $2\sqrt{6}$
   c. $3\sqrt{3}$ d. $5\sqrt{5}$
2. a. $3\sqrt{6}$ b. $4\sqrt{7}$
   c. $2\sqrt{17}$ d. $6\sqrt{5}$
3. a. $2\sqrt{22}$ b. $9\sqrt{2}$
   c. $7\sqrt{5}$ d. $8\sqrt{7}$
4. a. $4\sqrt{2}$ b. $24\sqrt{10}$
   c. $36\sqrt{5}$ d. $21\sqrt{6}$
5. a. $-30\sqrt{3}$ b. $-28\sqrt{5}$
   c. $64\sqrt{3}$ d. $2\sqrt{2}$
6. a. $\sqrt{2}$ b. $2\sqrt{3}$
   c. $\dfrac{1}{3}\sqrt{15}$ d. $\dfrac{3}{2}\sqrt{7}$
7. a. $4a$ b. $6a\sqrt{2}$
   c. $3a\sqrt{10b}$ d. $13a^2\sqrt{2}$
8. a. $13ab\sqrt{2ab}$ b. $2ab^2\sqrt{17ab}$
   c. $5x^3y^2\sqrt{5}$ d. $20xy\sqrt{5x}$

9. a. $54c^3d^2\sqrt{2cd}$ b. $18c^3d^4\sqrt{5cd}$
   c. $\sqrt{22ef}$ d. $7e^5f^5\sqrt{2ef}$

10. a. $7\sqrt{5}$ b. $8\sqrt{3}$
   c. $15\sqrt{5}+5\sqrt{3}$ d. $4\sqrt{11}$

11. a. $13\sqrt{2}$ b. $-3\sqrt{6}$
   c. $17\sqrt{3}-18\sqrt{7}$ d. $8\sqrt{x}+3\sqrt{y}$

12. a. $10\left(\sqrt{2}-\sqrt{3}\right)$ b. $5\left(\sqrt{5}+\sqrt{6}\right)$
   c. $7\sqrt{3}$ d. $4\sqrt{5}$

13. a. $14\sqrt{3}+3\sqrt{2}$ b. $3\sqrt{6}+6\sqrt{3}$
   c. $15\sqrt{10}-10\sqrt{15}+10$ d. $-8\sqrt{11}+22$

14. a. $12\sqrt{30}-16\sqrt{15}$ b. $12\sqrt{ab}+7\sqrt{3ab}$
   c. $\dfrac{7}{2}\sqrt{2}+2\sqrt{3}$ d. $15\sqrt{2}$

15. a. $31\sqrt{a}-6\sqrt{2a}$ b. $52\sqrt{a}-29\sqrt{3a}$
   c. $6\sqrt{6ab}$ d. $32a+2\sqrt{6a}+8a\sqrt{2}$

16. a. $a\sqrt{2a}$ b. $\sqrt{a}+2\sqrt{2a}$
   c. $3a\sqrt{a}+a^2\sqrt{3a}$ d. $\left(a^2+a\right)\sqrt{ab}$

17. a. $4ab\sqrt{ab}+3a^2b\sqrt{b}$
   b. $3\sqrt{ab}\left(2a+1\right)$
   c. $-6ab\sqrt{2a}+4a^2b^3\sqrt{3a}$
   d. $-2a\sqrt{b}$

18. a. $\sqrt{14}$ b. $\sqrt{42}$ c. $4\sqrt{3}$
   d. 10 e. $3\sqrt{7}$ f. 27

19. a. $10\sqrt{33}$ b. $180\sqrt{5}$ c. 120
   d. $120\sqrt{3}$ e. $2\sqrt{6}$ f. $2\dfrac{2}{3}$

20. a. $\dfrac{2}{5}\sqrt{6}$ b. $x^2y\sqrt{y}$ c. $3a^4b^2\sqrt{2ab}$
   d. $6a^5b^2\sqrt{2b}$ e. $3x^2y^2\sqrt{10xy}$ f. $\dfrac{9}{2}a^2b^4\sqrt{5ab}$

21. a. 2 b. 5 c. 12

22. a. 15 b. 18 c. 80

23. a. 28 b. 200

24. a. $\sqrt{5}$ b. 2 c. $\sqrt{6}$ d. 4

25. a. $\dfrac{\sqrt{3}}{4}$ b. $\dfrac{\sqrt{5}}{2}$ c. $2\sqrt{3}$ d. 1

26. a. $1\dfrac{4}{5}$ b. $2\sqrt{17}$ c. $\dfrac{x}{y}$
   d. $\dfrac{\sqrt{2}}{x^3y^4}$ e. $2xy\sqrt{3y}$ f. $\dfrac{4\sqrt{a}}{3}$

27. a. $\dfrac{5\sqrt{2}}{2}$ b. $\dfrac{7\sqrt{3}}{3}$ c. $\dfrac{4\sqrt{11}}{11}$
   d. $\dfrac{4\sqrt{6}}{3}$ e. $\dfrac{2\sqrt{21}}{7}$

28. a. $\dfrac{\sqrt{10}}{2}$ b. $\dfrac{2\sqrt{15}}{5}$ c. $\dfrac{3\sqrt{35}}{5}$
   d. $\dfrac{5\sqrt{6}}{6}$ e. $\dfrac{4\sqrt{15}}{15}$

29. a. $\dfrac{5\sqrt{7}}{14}$ b. $\dfrac{8\sqrt{15}}{15}$ c. $\dfrac{8\sqrt{21}}{49}$
   d. $\dfrac{8\sqrt{105}}{7}$ e. $\dfrac{\sqrt{10}}{3}$

30. a. $\sqrt{2}+2$ b. $\dfrac{3\sqrt{10}-2\sqrt{33}}{6}$
   c. $\dfrac{12\sqrt{5}-5\sqrt{6}}{10}$ d. $\dfrac{9\sqrt{10}}{5}$

31. a. $\dfrac{3\sqrt{10}+6\sqrt{14}}{4}$ b. $\dfrac{5\sqrt{6}}{3}$
   c. $\dfrac{3\sqrt{22}-4\sqrt{10}}{6}$ d. $\dfrac{\sqrt{21}-\sqrt{15}}{3}$

32. a. $\dfrac{14-5\sqrt{2}}{6}$ b. $\dfrac{12-\sqrt{10}}{16}$ c. $\dfrac{6\sqrt{15}-25}{70}$
   d. $\dfrac{\sqrt{30}+7\sqrt{2}}{20}$

33. a. $\sqrt{5}-2$ b. $\dfrac{2\sqrt{2}+\sqrt{5}}{3}$ c. $\dfrac{8\sqrt{11}+4\sqrt{13}}{31}$

34. a. $\dfrac{15\sqrt{15}-20\sqrt{6}}{13}$ b. $12\sqrt{2}-17$
   c. $\dfrac{19-4\sqrt{21}}{5}$

35. a. $\dfrac{\sqrt{15}-\sqrt{3}-\sqrt{5}+1}{4}$
   b. $\dfrac{-6+6\sqrt{2}+\sqrt{10}-2\sqrt{5}}{2}$
   c. $\dfrac{4\sqrt{10}+\sqrt{15}-4\sqrt{6}-3}{29}$

36. $\dfrac{15\left(3\sqrt{2}-2\right)}{28}$

37. a. Answers will vary.
   A sample answer could be $\sqrt{16+9}=\sqrt{25}=5$, but $\sqrt{16}+\sqrt{9}=4+3=7$.
   b. Answers will vary.
   A sample answer could be $\sqrt{16-9}=\sqrt{7}\approx 2.646$, but $\sqrt{16}-\sqrt{9}=4-3=1$.

38. $\dfrac{9\sqrt{x}+6x}{36x-16x^2}$

39. a. Sample responses can be found in the worked solutions in the online resources.
   b. i. $\sqrt{5}+\sqrt{3}$ ii. $\sqrt{5}-\sqrt{3}$ iii. $\sqrt{3}+2$

40. $\dfrac{2}{7}$

41. a. $x = 16$ **b.** $x = 1$

42. 1

1.5 Review of exponent laws

1. a. a^7 **b.** a^6 **c.** b^8 **d.** a^4b^7

2. a. m^5n^{13} **b.** $a^5b^7c^3$ **c.** $m^6n^4p^5$ **d.** $6a^2b$

3. a. $10a^4b^9$ **b.** $36m^8n^7$ **c.** $12x^6y^6$ **d.** $4x^8y^6$

4. a. a **b.** a^5 **c.** b^3 **d.** $\dfrac{4}{3}a^4$

5. a. $3b^4$ **b.** $4m^5$ **c.** m^3n **d.** $\dfrac{1}{2}y^2$

6. a. $7b^3$ **b.** $\dfrac{5}{4}m^2p^2$ **c.** $\dfrac{1}{2}xy^2$

7. a. 1 **b.** 1 **c.** 1

8. a. 3 **b.** 4 **c.** -3

9. a. 3 **b.** -7 **c.** 4

10. a. a^6 **b.** $16a^{20}$

  c. $\dfrac{1}{81}m^8$ **d.** $\dfrac{4}{9}n^8$

  e. 49

11. a. a^6b^3 **b.** $9a^6b^4$ **c.** $16m^{12}n^{20}$

  d. $\dfrac{27}{64}m^6n^3$ **e.** $\dfrac{a^4}{b^6}$

12. a. $\dfrac{625m^{12}}{n^8}$ **b.** $\dfrac{343x^3}{8y^{15}}$ **c.** $\dfrac{81a^4}{625b^{12}}$

  d. -243 **e.** -32

13. a. D **b.** D

14. a. C **b.** D

  c. B **d.** D

15. a. 64 **b.** 72 **c.** 625

16. a. 48 **b.** 1600 **c.** $\dfrac{27}{125}$

17. a. 20 **b.** 1 **c.** 4

18. a. x^{3yz} **b.** a^b

19. a. $m^a n^b$ **b.** $\dfrac{a^{2x}}{b^{3x}}$

20. a. $n^{3-p}m^{2-q}$ **b.** a^{mp+np}

21. $a^3 = a \times a \times a$

  $a^2 = a \times a$

  $a^3 \times a^2 = a \times a \times a \times a \times a$

  $= a^5$, not a^6

  Explanations will vary.

22. They are equal when $x = 2$ or when $x = 0$. Explanations will vary.

23. $3x^0 = 3$ and $(3x)^0 = 1$. Explanations will vary.

24. a.

a	0	1	2	3
$3a^2$	0	3	12	27
$5a$	0	5	10	15
$3a^2 + 5a$	0	8	22	42
$3a^2 \times 5a$	0	15	120	405

  b. $3a^2 \times 5a$ will become much larger than $3a^2 + 5a$.

25. $x = -2$ or 4

26.
  $1 \equiv 1$
  $2 \equiv 10$
  $3 \equiv 11$
  $4 \equiv 100$
  $5 \equiv 101$
  $6 \equiv 110$
  $7 \equiv 111$
  $8 \equiv 1000$
  $9 \equiv 1001$
  $10 \equiv 1010$

27. a. $x = 4$ **b.** $x = 0, 2$

28. a. a^2bc^7

  b. The student made a mistake when multiplying the two brackets in line 3. Individual brackets should be expanded first.

1.6 Negative exponents

1. a. $\dfrac{1}{x^5}$ **b.** $\dfrac{1}{y^4}$ **c.** $\dfrac{2}{a^9}$ **d.** $\dfrac{4}{5a^3}$

2. a. $\dfrac{3x^2}{y^3}$ **b.** $\dfrac{1}{4m^3n^4}$ **c.** $\dfrac{6a^3}{bc^5}$ **d.** a^6

3. a. $\dfrac{2a^4}{3}$ **b.** $2ab^2$ **c.** $\dfrac{7b^3}{2a^4}$ **d.** $\dfrac{2m^3a^2}{3b^4n^5}$

4. a. $\dfrac{1}{a^2b^3}$ **b.** $\dfrac{6}{x^6y}$ **c.** $\dfrac{3}{n^8}$

  d. $\dfrac{4}{a^2b^5}$ **e.** $\dfrac{2y}{3x}$

5. a. $\dfrac{5y}{6x^3}$ **b.** $\dfrac{3}{m^2n^2}$ **c.** $\dfrac{4y^{12}}{x^5}$

  d. $\dfrac{1}{3m^3n^3}$ **e.** $\dfrac{1}{32a^{15}m^{20}}$

6. a. $\dfrac{4q^8}{p^{14}}$ **b.** $\dfrac{3}{a^8b^{12}}$ **c.** $\dfrac{27q^9}{8p^6}$

  d. $\dfrac{b^6}{4a^8}$ **e.** $\dfrac{1}{8a^6b^6}$

7. a. $\dfrac{1}{8}$ **b.** $\dfrac{1}{36}$ **c.** $\dfrac{1}{81}$ **d.** $\dfrac{8}{9}$

8. a. $\dfrac{1}{16}$ **b.** $\dfrac{5}{36}$ **c.** 48 **d.** $\dfrac{32}{27}$

9. a. $\dfrac{27}{25} = 1\dfrac{2}{25}$ **b.** 4 **c.** 125 **d.** $\dfrac{3}{4}$

10. a. 2^3 b. 2^{-3} c. 2^5 d. 2^{-6}

11. a. $x=3$ b. $x=-2$ c. $x=-1$ d. $x=3$ e. $x=-2$

12. a. $x=0$ b. $x=3$ c. $x=-3$ d. $x=-6$ e. $x=-2$

13. a. $\dfrac{3}{2}$ b. $\dfrac{4}{5}$ c. $\dfrac{2}{7}$ d. 5

14. a. $\dfrac{b}{a}$ b. $\dfrac{b^3}{a^2}$ c. $\dfrac{a^2}{b^3}$ d. $\dfrac{1}{m^3n^2}$

15. a. $\dfrac{1}{729}$ b. $\dfrac{1}{20\,736}$ c. $0.000\,059\,499$ or $\dfrac{1}{16\,807}$

16. a. 256 b. $\dfrac{16\,384}{2187}$ c. 9 765 625

17. C

18. B

19. B

20. D

21. C

22. D

23. a. $\dfrac{m^2}{n^8}$ b. $\dfrac{n^2}{m}$ c. $\dfrac{25}{a^7b^6}$

24. a. $r^6 - s^6$
   b. $m^{10} + 2m^5n^5 + n^{10}$
   c. 1
   d. p^2

25. 2^{2r-4}

26. 6^{3m}

27. $x=3$

28. a. As x gets closer to 0 coming from the positive direction, y gets more and more positive, approaching ∞.
   b. As x gets closer to 0 coming from the negative direction, y gets more and more negative, approaching $-\infty$.

29. $2^{-n} = \dfrac{1}{2^n}$

   A n increases, the value of 2^n increases, so the value of 2^{-n} gets closer to 0.

30. Sample responses can be found in the worked solutions of your online resources.

31. $x=-2, y=-3$

32. x^2

33. $x=3, y=-1; 7$

1.7 Fractional exponents (10A)

1. a. $\sqrt{15}$ b. $\sqrt[4]{m}$ c. $\sqrt[5]{7^2}$ d. $\sqrt{7^5}$

2. a. $\sqrt[8]{w^3}$ b. $\sqrt[4]{w^5}$ c. $\sqrt[3]{5^{10}}$ d. $\sqrt[10]{a^3}$

3. a. $t^{\frac{1}{2}}$ b. $5^{\frac{7}{4}}$ c. $6^{\frac{11}{6}}$ d. $x^{\frac{6}{7}}$

4. a. $x^{\frac{7}{6}}$ b. w^2 c. $w^{\frac{1}{2}}$ d. $11^{\frac{n}{x}}$

5. a. 4 b. 5 c. 9

6. a. 2 b. 4 c. 3

7. a. 2.2 b. 1.5 c. 1.3

8. a. 2.5 b. 12.9 c. 13.6

9. a. 0.7 b. 0.8 c. 0.9

10. a. $4^{\frac{4}{5}}$ b. $2^{\frac{1}{2}}$ c. $a^{\frac{5}{6}}$

11. a. $x^{\frac{23}{20}}$ b. $10m^{\frac{8}{15}}$ c. $2b^{\frac{5}{7}}$

12. a. $-4y^{\frac{20}{9}}$ b. $0.02a^{\frac{9}{8}}$ c. $5x^{\frac{7}{2}}$

13. a. $ab^{\frac{3}{2}}$ b. $x^{\frac{4}{5}}y^{\frac{5}{9}}$ c. $6a^{\frac{8}{5}}b^{\frac{17}{15}}$

14. a. $2m^{\frac{19}{28}}n^{\frac{2}{5}}$ b. $x^{\frac{19}{6}}y^{\frac{5}{6}}z^{\frac{5}{6}}$ c. $8a^{\frac{2}{5}}b^{\frac{9}{8}}c$

15. a. $x^{\frac{5}{3}}y^{\frac{7}{5}}$ b. $a^{\frac{7}{45}}b^{\frac{4}{15}}$ c. $\frac{1}{3}m^{\frac{3}{8}}n^{\frac{11}{56}}$

16. a. $2x^{\frac{2}{15}}y^{\frac{3}{4}}$ b. $\frac{1}{4}a^{\frac{11}{20}}b^{\frac{7}{20}}$ c. $\frac{1}{7}p^{\frac{5}{24}}q^{\frac{1}{12}}$

17. a. $2^{\frac{9}{20}}$ b. $5^{\frac{1}{6}}$ c. $7^{\frac{6}{5}}$

18. a. $a^{\frac{3}{10}}$ b. $m^{\frac{1}{6}}$ c. $2^{\frac{1}{3}}b^{\frac{1}{6}}$

19. a. $4p^{\frac{2}{5}}$ b. $x^{\frac{m}{p}}$ c. $3^{\frac{b}{c}}m^{\frac{a}{c}}$

20. a. $a^{\frac{1}{4}}b^{\frac{1}{6}}$ b. $a^3b^{\frac{3}{4}}$ c. $x^{\frac{6}{5}}y^{\frac{7}{4}}$

21. a. $3^{\frac{1}{3}}a^{\frac{1}{9}}b^{\frac{1}{5}}c^{\frac{1}{4}}$ b. $5x^{\frac{1}{4}}y^{\frac{1}{3}}z^{\frac{1}{5}}$ c. $\dfrac{a^{\frac{1}{2}}}{b^{\frac{2}{3}}}$

22. a. $\dfrac{m^{\frac{8}{5}}}{n^{\frac{7}{4}}}$ b. $\dfrac{b^{\frac{2}{5}}}{c^{\frac{8}{27}}}$ c. $\dfrac{2^{\frac{1}{2}}x^{\frac{7}{2}}}{y^{\frac{3}{8}}}$

23. a. D b. C c. B

24. a. D b. B

25. a. a^4 b. b^3 c. m^4

26. a. $4x^2$ b. $2y^3$ c. $2x^2y^3$

27. a. $3m^3n^5$ b. $2pq^2$ c. $6a^2b^6$

28. a. 2.007 s b. 20.07 s c. 4.98 swings

29. $\left(2^5a^5b^{10}\right)^{\frac{1}{5}} = 2ab^2$

30. $\dfrac{a^2\left(3-\sqrt{b^3}\right)}{9-b^3}; \dfrac{4}{11}$

31. a. $x^{\frac{1}{2}} + y^{\frac{1}{2}} - z^{\frac{1}{2}}$ b. $t^{\frac{1}{10}}$

32. $m - n^2$

33. $m^{\frac{1}{5}} - n^{\frac{1}{5}} + p^{\frac{1}{5}}$

34. a. $a^{-\frac{1}{4}} \times b^{\frac{13}{2}}$

  b. No, because you can't take the fourth root of a negative number.

  c. $a = 1$

1.8 Combining exponent laws

1. a. $54a^{10}b^9$ **b.** $48a^5b^{16}$ **c.** $\dfrac{2n^{13}}{m^9}$

2. a. $500p^8q^{18}$ **b.** $36a^{20}b^{10}$ **c.** $\dfrac{15b^2}{c^{26}}$

3. a. $12x^{\frac{7}{8}}y^{\frac{11}{15}}$ **b.** $8m^{\frac{15}{4}}n^{\frac{15}{4}}$ **c.** $\dfrac{6}{p^{\frac{7}{12}}}$ **d.** $8p^{\frac{7}{45}}q^{\frac{5}{18}}$

4. a. $\dfrac{5}{8a^7}$ **b.** $\dfrac{x}{4y^6}$ **c.** $\dfrac{27}{128m^{29}n^{26}}$

5. a. $\dfrac{64y^{36}}{x^{24}}$ **b.** $24a^{24}b^7$ **c.** $\dfrac{27h^{12}}{8g^6}$

6. a. $p^{\frac{35}{3}}q^{\frac{1}{2}}$ **b.** $\dfrac{625}{81b^{20}c^{28}}$ **c.** $x^{\frac{5}{3}}y^{\frac{1}{8}}z^{\frac{3}{2}}$

7. a. $\dfrac{3a^2}{2}$ **b.** $8n^2$ **c.** $\dfrac{m^2n^4}{3}$

8. a. $\dfrac{4x^5}{3y^8}$ **b.** $\dfrac{36x^6}{y}$ **c.** $\dfrac{y^2}{x^4}$

9. a. $\dfrac{b^7}{3a^4}$ **b.** $\dfrac{75q^5}{2p^{11}}$ **c.** $x^{\frac{17}{10}}y^{\frac{7}{10}}$

10. a. $\dfrac{2}{5a^4b^7}$ **b.** $\dfrac{4a^3b^3}{15}$ **c.** $\dfrac{n^9}{4m^9}$

11. a. $\dfrac{4m^5}{9n^{15}}$ **b.** $\dfrac{4}{81x^2y^{14}}$ **c.** $48x^{11}y^6$

12. a. $\dfrac{3p^4}{5q^9}$ **b.** $\dfrac{2b^{\frac{1}{12}}}{3a^{\frac{17}{24}}}$ **c.** $\dfrac{4x^{\frac{1}{12}}}{3y^{\frac{21}{20}}}$

13. a. $\dfrac{5}{2a^{13}}$ **b.** $\dfrac{56a^{11}b^6}{81}$

14. a. $\dfrac{1024b^2}{81a}$ **b.** $\dfrac{25}{128x^{23}y^4}$ **c.** $\dfrac{4y^{36}}{27x^{16}}$

15. a. $6m^{19}n^{19}$ **b.** $\dfrac{16m^{\frac{11}{12}}n}{3}$ **c.** $\dfrac{4b^{\frac{11}{2}}}{3^{\frac{1}{2}}c^{\frac{7}{30}}}$

16. a. $\dfrac{125}{8}$ **b.** 1

17. 1

18. a. 5^{y-1} **b.** $y = 4$

19. D

20. A

21. a. $m^{\frac{1}{6}}n^{-\frac{7}{6}}$ or $\sqrt[6]{\dfrac{m}{n^7}}$ **b.** $g^{-6}h^3n^{\frac{3}{2}}$

  c. $3^{-\frac{7}{3}} \times 5^{-\frac{7}{6}}$

22. a. 2^{-2} or $\dfrac{1}{4}$ **b.** a^6b^{-8} or $\dfrac{a^6}{b^8}$ **c.** $d^{\frac{14}{15}}$ or $\sqrt[15]{d^{14}}$

23. a. 12   **b.** 1536   **c.** 14 days

24. a. 80 koalas

  b. During the 6th year

25. a. 79%   **b.** 56%   **c.** 31%

26. $\dfrac{3}{2}$

27. $\dfrac{4}{21}$

28. $z^2 + z^{-2} + \sqrt{5}$

1.9 Application of exponentials: Compound interest

1. a. $3244.80   **b.** $10\,939.56

2. a. $24\,617.98   **b.** $14\,678.02

3. a. $14\,838.45   **b.** $129\,394.77

4. a. $932.52   **b.** $10\,650.81

5. a. $20\,2760.57

  b. $25\,816.04

6. a. $3\,145\,511.41

  b. $443\,014.84

7. $8784.60

8. $3376.26

9. a. $2837.04   **b.** $837.04

10. $17\,240.15

11. $605.42

12. $18\,503.86

13. a. 0.5833%   **b.** $42\,891.60

14. B

15. B

16. C

17. C

18. a. 0.0247%

  b. $\approx \$131\,295.85$

  c. $\approx \$11\,295.85$

  d. $\approx \$495.85$

19. a. i. $17\,745.95

   ii. $17\,786.61

   iii. $17\,807.67

   iv. $17\,821.99

  b. The interest added to the principal also earns interest.

20. a. $7920

  b. David's investment = $8108.46

  c. Because David's interest is compounded, the interest is added to the principal each quarter and earns itself interest.

21. a. $3750 interest

   b. i. $3820.32 interest

     ii. $3881.05

     iii. $3912.36

   c. Compound quarterly gives the best return.

   d. If we assume that interest is compounded annually, an equivalent return of $I = 7\%$ would be achieved.

   e. i. Yes

     ii. No

22. Neither is correct. The best option is to choose 3.895% p.a. compounding monthly.

23. a. Bank east   b. $19 976.45

24. a. i. $2.44   ii. $2.61
    iii. $2.71   iv. $2.71

   b. Compounding more frequently increases the final value, but the amount of increase becomes less and less. The final value of iv. is 2.7181 which is almost equal to iii.

1.10 Logarithms

1. a. $\log_4(16) = 2$ b. $\log_2(32) = 5$

   c. $\log_3(81) = 4$ d. $\log_6(36) = 2$

   e. $\log_{10}(1000) = 3$

2. a. $\log_5(25) = 2$ b. $\log_4(x) = 3$

   c. $\log_5(125) = x$ d. $\log_7(49) = x$

   e. $\log_p(16) = 4$

3. a. $\log_9(3) = \dfrac{1}{2}$ b. $\log_{10}(0.1) = -1$

   c. $\log_8(2) = \dfrac{1}{3}$ d. $\log_2\left(\dfrac{1}{2}\right) = -1$

   e. $\log_4(8) = \dfrac{3}{2}$

4. D

5. a. $2^4 = 16$ b. $3^3 = 27$

   c. $10^6 = 1\,000\,000$ d. $5^3 = 125$

6. a. $16^{\frac{1}{2}} = 4$ b. $4^x = 64$

   c. $49^{\frac{1}{2}} = 7$ d. $3^5 = x$

7. a. $81^{\frac{1}{2}} = 9$ b. $10^{-2} = 0.01$

   c. $8^1 = 8$ d. $64^{\frac{1}{3}} = 4$

8. B

9. a. 4   b. 2
   c. 2   d. 5

10. a. 5   b. 7   c. 0   d. $\dfrac{1}{2}$

11. a. -1 b. 1   c. -2 d. $\dfrac{1}{3}$

12. a. 0   b. 1   c. 2

13. a. 3   b. 4   c. 5

14. a. 0 and 1   b. 3 and 4   c. 1 and 2

15. a. 4 and 5   b. 2 and 3   c. 4 and 5

16. a. 6.1   b. 6.3   c. 8.2

17. D

18. D

19. a. B   b. A   c. D

20.

21. A

22. B

23. a. $\log_{10}(g) = k$ implies that $g = k$ so $g^2 = \left(10^k\right)^2$. That is, $g^2 = 10^{2k}$; therefore, $\log_{10}\left(g^2\right) = 2k$.

   b. $\log_x(y) = 2$ implies that $y = x^2$, so $x = y^{\frac{1}{2}}$ and therefore $\log_y(x) = \dfrac{1}{2}$.

   c. The equivalent exponential statement is $x = 4^y$, and we know that 4^y is greater than zero for all values of y. Therefore, x is a positive number.

24. a. 6   b. -4 c. -5

25. a. 5   b. -3 c. $\dfrac{3}{2}$

26. 5 times

27. a. 3   b. 7   c. $\dfrac{1}{8}$

28. x

29. $\dfrac{9}{x}$

1.11 Logarithm laws (10A)

1. a. 1.698 97   b. 1.397 94
   c. 0.698 97   d. 0.301 03

2. Sample responses can be found in the worked solutions in the online resources.

3. a. 1   b. 3

4. a. 2   b. 3

5. a. 4   b. 1

6. a. 2   b. 3

7. a. 1   b. 4

8. a. 3   b. 5

9. a. 2   b. $\dfrac{1}{2}$

10. a. 1   b. 3

11. a. -1 b. $-\dfrac{1}{2}$

12. 3

13. a. 2   b. 4

14. a. 3   b. 3

15. a. 1   b. 0
    c. -1 d. 5

16. a. -2 b. 1
    c. 0 d. -2

17. a. $-\dfrac{1}{2}$ b. $\dfrac{1}{2}$ c. $-\dfrac{1}{2}$ d. $\dfrac{7}{2}$

18. a. $\log_a(40)$ b. $\log_a(18)$
    c. $\log_x(48)$ d. $\log_x(4)$

19. a. $\log_a(x)$ b. 1
    c. -1 d. 7

20. a. $\dfrac{1}{2}$ b. $\dfrac{3}{2}$
    c. -6 d. $-\dfrac{1}{3}$

21. a. B   b. B, D
    c. A, B   d. C, D

22. a. $\log_2(80)$ b. $\log_3(105)$
    c. $\log_{10}(100) = 2$ d. $\log_6(56)$

23. a. $\log_2(4) = 2$ b. $\log_3(3) = 1$ c. $\log_5(12.5)$
    d. $\log_2(3)$

24. a. $\log_4(5)$ b. $\log_{10}\left(\dfrac{1}{4}\right)$
    c. $\log_3(4)$ d. $\log_2(3)$
    e. $\log_3(20)$ f. $\log_4(2) = \dfrac{1}{2}$

25. a. C   b. B   c. A

26. a. 12 (Evaluate each logarithm separately and then find the product.)
    b. 4 (First simplify the numerator by expressing 81 as a power of 3.)
    c. 7 (Let $y = 5^{\log_5(7)}$ and write an equivalent statement in logarithmic form.)

27. -2

28. $7 - 3\log_2(3)$

29. 1

30. $x = 3a, 5a$

31. 7

1.12 Solving equations (10A)

1. a. 5 b. 4 c. 5
   d. 2 e. -2 f. -3
   g. 4 h. 0 i. 2

2. a. $\dfrac{5}{3}$ b. $-\dfrac{1}{2}$ c. -1
   d. $-\dfrac{9}{4}$ e. $\dfrac{11}{6}$ f. $\dfrac{9}{8}$

3. a. -3 b. $\dfrac{10}{7}$ c. $-\dfrac{18}{5}$
   d. $\dfrac{3}{4}$ e. $\dfrac{21}{11}$ f. $\dfrac{4}{11}$
   g. $-\dfrac{5}{2}$ h. $\dfrac{25}{9}$

4. a. 25 b. 81 c. $\dfrac{1}{8}$
   d. $\dfrac{1}{16}$ e. $100, -100$

5. a. 16 b. 26 c. 127
   d. 2 e. 0

6. a. $-\dfrac{1}{32}$ b. $-\dfrac{1}{9}$ c. -624
   d. -2.5

7. a. 3 b. 2 c. 125

8. a. 625 b. 2 c. 8

9. a. 6 b. 4

10. a. 3 b. 2 c. -1

11. a. -2 b. $\dfrac{1}{2}$ c. $\dfrac{1}{3}$

12. a. 0 b. 0 c. -1
    d. -2

13. a. 5 b. 6 c. 10
    d. 8

14. a. 4 b. 2 c. 9
    d. $\dfrac{2}{5}$

15. a. 500 b. 128 c. 5
    d. 6 e. 1 f. 2

16. a. B   b. A   c. D
    d. B

17. a. 7 b. 2 c. -2
    d. 0 e. 4

18. a. $\dfrac{1}{2}$ b. $\dfrac{1}{2}$ c. $\dfrac{3}{2}$
    d. $-\dfrac{1}{2}$ e. $\dfrac{3}{2}$

19. a. $\dfrac{3}{4}$ b. $-\dfrac{5}{2}$ c. $\dfrac{5}{2}$
    d. $-\dfrac{9}{2}$ e. $-\dfrac{11}{4}$

20. Approximately 2.5 times brighter

21. a. 120 b. 130
    c. 0.001 d. 3 dB are added.
    e. 10 dB are added.   f. 100

22. a. i. 1.1 ii. 1.3 iii. 1.418
    iv. 1.77 v. 2.43 vi. 3.1
    b. No; see answers to 19a i and ii above.
    c. i. 22 387 211 KJ
       ii. 707 945 784 KJ
       iii. 22 387 211 386 KJ.
    d. The energy is increased by a factor of 31.62.
    e. It releases 31.62^3 times more energy.

23. a. $x = 0.7712$ b. $x = 1.2966$

24. $x = 7$

25. $x = 1, 3$

Project

1. a. 100_2 b. 1101_2 c. 10000_2

2. a. 10_2 b. 1_2 c. 110_2

3. a. 11_2 b. 101010_2 c. 110111_2

4. Sample responses can be found in the worked solutions in the online resources. The digits in octal math are 0, 1, 2, 3, 4, 5, 6 and 7. The value 'eight' is written as '1 eight and 0 ones', or 10_8.

5. Sample responses can be found in the worked solutions in the online resources. The numbers 10, 11, 12, 13, 14 and 15 are allocated the letters A, B, C, D, E and F respectively.

1.13 Review questions

1. A

2. a. Irrational, since it is equal to a non-recurring and non-terminating decimal.
   b. Rational, since it can be expressed as a whole number.
   c. Rational, since it is given in a rational form.
   d. Rational, since it is a recurring decimal.
   e. Irrational, since it is equal to a non-recurring and non-terminating decimal.

3. D

4. a. $\sqrt{2m}, \sqrt{\dfrac{20}{m}}, \sqrt[3]{m}, \sqrt[3]{8m}$

   b. $\sqrt{25m}, \sqrt{\dfrac{m}{16}}, \sqrt{\dfrac{20}{m}}$

5. a. $5\sqrt{2}$ b. $6\sqrt{5}$ c. $8\sqrt{2}$ d. $20\sqrt{5}$

6. C

7. a. $72x^3y^4\sqrt{2xy}$ b. $-\dfrac{1}{4}x^2y^5\sqrt{xy}$

8. a. $25\sqrt{3}$ b. $3ab\sqrt{ab}$

9. a. $\sqrt{15}$ b. $6\sqrt{42}$
   c. $30\sqrt{15}$ d. 5

10. a. 27 b. $720\sqrt{2}$

11. a. $\sqrt{3}$ b. 6
    c. $\dfrac{\sqrt{10}}{4\sqrt{3}}$ or $\dfrac{\sqrt{30}}{12}$ d. $\dfrac{1}{2}$

12. a. $\dfrac{\sqrt{6}}{3}$ b. $\dfrac{\sqrt{2}}{4}$ c. $2\sqrt{5}+4$ d. $2-\sqrt{3}$

13. a. 4 b. 4.5 c. 2.2 d. 2.7

14. a. 7.4 b. 1.7 c. 0.8 d. 0.8

15. a. $\sqrt{2}$ b. $3\sqrt{2}$
    c. $5\sqrt{5}$ d. 16

16. a. 1 b. 4

17. a. $\dfrac{1}{4}$ b. $\dfrac{1}{9}$
    c. $\dfrac{1}{16}$ d. $\dfrac{1}{1000}$

18. a. 0.0833 b. 0.0204
    c. 0.800 d. 625

19. a. $1\dfrac{1}{2}$ b. $1\dfrac{3}{7}$
    c. 5 d. $\dfrac{4}{13}$

20. a. B   b. A   c. A

21. a. $2\dfrac{1}{2}$ b. $2\dfrac{1}{4}$

22. a. $\dfrac{1}{2}$ b. $\dfrac{1}{9}$
    c. $\dfrac{1}{64}$ d. $\dfrac{2}{1}=2$

23. D

24. C

25. C

26. C

27. A

28. A

29. B

30. C

31. B

32. a. $9x^{10}y^{10}$ b. $\dfrac{13ab^3c^2}{6}$
    c. $\dfrac{1000m^{15}n^6}{27}$ d. $\dfrac{16p^{28}}{81q^{12}}$

33. a. 16 b. $-\dfrac{3}{2}$

34. a. $\dfrac{8}{a^{11}b^2}$ b. $\dfrac{y^2}{5x^{17}}$ c. $\dfrac{m^{12}}{16n^8}$

35. a. 8 b. $\dfrac{3}{2}$ c. 0

36. a. $30a^{\frac{41}{20}}b^{\frac{33}{20}}$ b. $\dfrac{4}{x^{\frac{1}{20}}y^{\frac{2}{9}}}$ c. $\dfrac{2a^{\frac{1}{6}}}{b^{\frac{3}{2}}}$

37. a. 1 b. 4

38. a. $-2a^3+2a^2b^{\frac{1}{2}}$
    b. $6xy^2$

39. a. $\dfrac{2a^{13}}{5b^2}$ b. $\dfrac{9y^4}{32x^{15}}$ c. $2^{\frac{4}{3}}m$

40. a. 46 b. $-\dfrac{1}{18}$

41. $\$15\,746.40$

42. a. $\$25\,808.37$ b. $\$26\,723.16$
    c. $\$27\,211.79$ d. $\$27\,550.17$

43.

$\log_{10}$ (income)	Frequency
$3-<4$	2
$4-<5$	6
$5-<6$	3
$6-<7$	1

44.

x-axis: Log (weight (kg))
y-axis: Frequency

45. a. 2   **b.** 1   **c.** 8   **d.** 2

46. a. $\log_a(24)$ **b.** $\dfrac{3}{2}$

   c. $\log_a(x^2)$ or $2\log_a(x)$ **d.** -5

47. B

48. a. 512   **b.** $\dfrac{1}{25}$ **c.** 5

   d. 2   **e.** 6   **f.** 0

49. a. 6   **b.** 35

50. a. -2 **b.** $-\dfrac{1}{2}$ **c.** $\dfrac{5}{2}$

51. a. 4.644   **b.** -3.809 **c.** 0.079

52. a. 9   **b.** 6   **c.** 0

53. $x = 2, -3$

54. $\dfrac{1}{a^2 b^{\frac{1}{2}}}$

55. $\dfrac{1}{36}$

56. a. 8   **b.** 6   **c.** 2

57. a. $\dfrac{9ab^7 c^7}{50}$

   b. The student has made two mistakes when squaring the left-hand bracket in line 1: $3^2 = 9, 5^2 = 25$.

58. a. $\dfrac{\sqrt{r}}{d\sqrt{u^3}}$

   b. Yes, $\sqrt{r}$, $\sqrt{u^3}$

   c. $\dfrac{\sqrt{ru^3}}{du^3}$

   d. r should be a perfect square, u should be a perfect cube and d should be a rational number.

   e. **i.** $0.001\,234\,6\,\text{kL}$

      ii. $1.2346\,\text{L}$

      iii. $1234.6\,\text{mL}$

   f. A high value for d causes the expression to be smaller, as d only appears on the denominator of the fraction. This means that when d is high there is less water in the reservoir and the area is dry.

59. a. $\dfrac{p^2\left(2 - \sqrt{w^3}\right)}{4 - w^3}$

   b. Sample responses can be found in the worked solutions in the online resources; approximately 5.

   c. Speed decreases as wind resistance increases.

2 Algebra and equations

LESSON SEQUENCE

LESSON
2.1 Overview

Why learn this?

Algebra is like the language of maths: it holds the key to understanding the rules, formulas and relationships that summarise much of our understanding of the universe. Every maths student needs this set of skills in order to process mathematical information and move on to more challenging concepts.

This explains why those who want to pursue a career in maths need algebra. Every maths teacher is faced with the question 'Why do *I* need to study algebra if I'm never going to use it?', yet no-one asks why a professional footballer would lift weights when they don't lift any weights in their sport. The obvious answer for the footballer is that they are training their muscles to be fitter and stronger for upcoming matches. Learning algebra is no different, in that you are training your mind to better handle abstract concepts.

Abstraction is the ability to consider concepts beyond what we observe. Spatial reasoning, complex reasoning, understanding verbal and non-verbal ideas, recognising patterns, analysing ideas and solving problems all involve abstract thinking to some degree. If some food were to fall on the ground, an adult would think about how long the food has been there, whether the ground is clean, and whether the food surface can be washed, whereas a young child would just pick up the food and eat it off the ground, because they lack the ability to think abstractly. Being able to think about all these considerations is just a simple example of abstract thinking. We use abstract thinking every day and develop this skill over our life. Those who have strong abstract reasoning skills tend to perform highly on intelligence tests and are more likely to be successful in later life. Algebra helps us develop our abstract reasoning skills and thus is of use to all students!

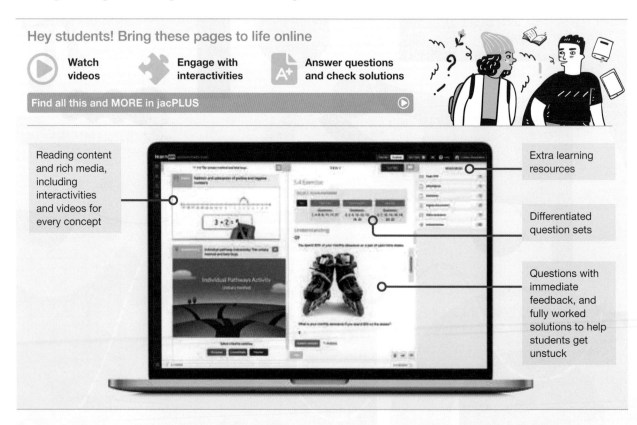

1. Evaluate $\dfrac{(-d)^2}{9c}$, if $c = \dfrac{1}{3}$ and $d = -6$.

2. If $c = \sqrt{a^2 + b^2}$, calculate c, if $a = 4$ and $b = 3$.

3. **MC** Given the integer values $x = 3$ and $y = -2$, state whether the Closure Law holds for $3y \div x$.
   A. Yes, the answer obtained is an integer value.
   B. No, the answer obtained is a negative integer.
   C. Yes, the answer obtained is a natural number.
   D. No, the answer obtained is irrational.

4. Simplify the following expression.

$$\frac{y}{5} - \frac{y}{6}$$

5. **MC** The expression $\dfrac{2}{x+1} - \dfrac{1}{(x+1)^2}$ can be simplified to:

   A. $\dfrac{1}{x+1}$ B. $\dfrac{1}{(x+1)^2}$ C. $\dfrac{x-1}{(x+1)^2}$ D. $\dfrac{2x+1}{(x+1)^2}$

6. **MC** The expression $\dfrac{5}{2x} + \dfrac{1}{3x}$ simplified is:

   A. $\dfrac{6}{5x}$ B. $\dfrac{17}{6x}$ C. $\dfrac{18}{6x}$ D. $\dfrac{5}{6x^2}$

7. Simplify the following expression.

$$\frac{x}{5} \times \frac{-10}{3y}$$

8. If the side length of a cube is x cm, then the cube's volume, V, is given by $V = x^3$. Calculate the side length, in cm, of a cube that has a volume of 1 m^3.

9. Solve the equation $\dfrac{2(4r+3)}{5} = \dfrac{3(2r+5)}{4}$.

10. Solve the equation $\dfrac{8x+3}{5} - \dfrac{3(x-1)}{2} = \dfrac{1}{2}$.

11. Solve the equation $\dfrac{\sqrt[3]{a}}{4} = -2$.

12. At a charity fundraising event, three-eighths of the profit came from sales of tickets and one-fifth came from donations. A third of the profit came from the major raffle and a pop-up stall raised $2200. Determine the amount of money raised at the event.

▶

13. **MC** If $\dfrac{x+4}{(x+1)(x-2)} = \dfrac{a}{x+1} + \dfrac{b}{x-2}$, the values of a and b respectively are:

  A. $a = x$ and $b = 4$ 　　　　　　　　 B. $a = 1$ and $b = 2$
  C. $a = -1$ and $b = 2$ 　　　　　　　 D. $a = 1$ and $b = -2$

14. **MC** Solve the literal equation $\dfrac{1}{a} + \dfrac{1}{b} = \dfrac{1}{c}$ for a.

  A. $a = \dfrac{bc}{b-c}$ 　　　 B. $a = \dfrac{1}{b-c}$ 　　　 C. $a = \dfrac{bc}{b+c}$ 　　　 D. $a = c - b$

15. **MC** Rearrange the literal equation $m = \dfrac{pa+qb}{p-q}$ to make p the subject.

  A. $p = \dfrac{qb}{-a}$ 　　　　　　　　　　 B. $p = \dfrac{q(m-b)}{m+a}$

  C. $p = \dfrac{q(m+b)}{m+a}$ 　　　　　　　 D. $p = \dfrac{q(m+b)}{m-a}$

LESSON
2.2 Substitution

LEARNING INTENTIONS

At the end of this lesson you should be able to:
  • evaluate expressions by substituting the numeric values of pronumerals
  • understand and apply the Commutative, Associative, Identity and Inverse Laws.

▶ 2.2.1 Substituting values into expressions

eles-4696

  • An expression can be evaluated by substituting the numerical value of pronumerals into an algebraic expression.
  • The substituted values are placed in brackets when evaluating an expression.

WORKED EXAMPLE 1 Substituting values into an expression

If $a = 4$, $b = 2$ and $c = -7$, evaluate the following expressions.
a. $a - b$ 　　　　　　　　　　　　　　　　 b. $a^3 + 9b - c$

THINK	WRITE
a. 1. Write the expression.	a. $a - b$
2. Substitute $a = 4$ and $b = 2$ into the expression.	$= 4 - 2$
3. Simplify and write the answer.	$= 2$
b. 1. Write the expression.	b. $a^3 + 9b - c$
2. Substitute $a = 4$, $b = 2$ and $c = -7$ into the expression.	$= (4)^3 + 9(2) - (-7)$
3. Simplify and write the answer.	$= 64 + 18 + 7$
	$= 89$

WORKED EXAMPLE 2 Substituting into Pythagoras' theorem

If $c = \sqrt{a^2 + b^2}$, calculate c if $a = 12$ and $b = -5$.

THINK	WRITE
1. Write the expression.	$c = \sqrt{a^2 + b^2}$
2. Substitute $a = 12$ and $b = -5$ into the expression.	$= \sqrt{(12)^2 + (-5)^2}$
3. Simplify.	$= \sqrt{144 + 25}$
	$= \sqrt{169}$
4. Write the answer.	$= 13$

TI | THINK	DISPLAY/WRITE	CASIO | THINK	DISPLAY/WRITE
In a new document, open a calculator page. To substitute values, use the symbol |. Press CTRL and then = to bring up the palette; use the Touchpad to select the | symbol. Then type 'and' or find it in the CATALOG. Complete the entry line as: $c = \sqrt{a^2 + b^2} \mid a = 12$ and $b = -5$ Then press ENTER.	If $a = 12$ and $b = -5$, then $c = \sqrt{a^2 + b^2} = 13$.	To type the equation, $\sqrt{}$ is on the Keyboard Math1 screen. The vertical line | is on the Keyboard Math3 screen. Complete the entry line as: $c = \sqrt{a^2 + b^2} \mid a = 12$ and $b = -5$ Then press EXE.	If $a = 12$ and $b = -5$, then $c = \sqrt{a^2 + b^2} = 13$.

2.2.2 Number laws

eles-4697

- Recall from previous studies that when dealing with numbers and pronumerals, particular rules must be obeyed. Before progressing further, let us briefly review the Commutative, Associative, Identity and Inverse Laws.
- Consider any three pronumerals, x, y and z, where x, y and z are elements of the set of real numbers.

Commutative Law

- The **Commutative Law** holds true for addition and multiplication. That is, you can add or multiply in any order, since the order in which two numbers or pronumerals are added or multiplied does not affect the result.
- The Commutative Law does not hold true for subtraction or division.

Commutative Law

$$x + y = y + x \qquad \text{For example: } 3 + 2 = 5 \text{ and } 2 + 3 = 5$$

$$x - y \neq y - x \qquad \text{For example: } 3 - 2 = 1 \text{ but } 2 - 3 = -1$$

$$x \times y = y \times x \qquad \text{For example: } 3 \times 2 = 6 \text{ and } 2 \times 3 = 6$$

$$x \div y \neq y \div x \qquad \text{For example: } 3 \div 2 = \frac{3}{2}, \text{ but } 2 \div 3 = \frac{2}{3}$$

Associative Law

- The **Associative Law** holds true for addition and multiplication, since grouping two or more numbers or pronumerals and calculating them in a different order does not affect the result.
- The Associative Law does not hold true for subtraction or division.

Associative Law

$x + (y + z) = (x + y) + z$ **For example:** $2 + (3 + 4) = 2 + 7 = 9$ and $(2 + 3) + 4 = 5 + 4 = 9$

$x - (y - z) \neq (x - y) - z$ **For example:** $2 - (3 - 4) = 2 - -1 = 3$ but $(2 - 3) - 4 = -1 - 4 = -5$

$x \times (y \times z) = (x \times y) \times z$ **For example:** $2 \times (3 \times 4) = 2 \times 12 = 24$ and $(2 \times 3) \times 4 = 6 \times 4 = 24$

$x \div (y \div z) \neq (x \div y) \div z$ **For example:** $2 \div (3 \div 4) = 2 \div \dfrac{3}{4} = 2 \times \dfrac{4}{3} = \dfrac{8}{3}$ but

$$(2 \div 3) \div 4 = \dfrac{2}{3} \div 4 = \dfrac{2}{3} \times \dfrac{1}{4} = \dfrac{2}{12} = \dfrac{1}{6}$$

Identity Law

- Under the **Identity Law**, the sum of zero and any number is the number, and the product of 1 and any number is the number. That is, x has not been changed (it has kept its identity) when zero is added to it or it is multiplied by 1.
- The Identity Law does not hold true for subtraction or division.

Identity Law

$x + 0 = 0 + x = x$ **For example:** $5 + 0 = 0 + 5 = 5$

$x - 0 \neq 0 - x$ **For example:** $5 - 0 = 5$ and $0 - 5 = -5$

$x \times 1 = 1 \times x = x$ **For example:** $7 \times 1 = 1 \times 7 = 7$

$x \div 1 \neq 1 \div x$ **For example:** $8 \div 1 = 8$ and $1 \div 8 = \dfrac{1}{8}$

Inverse Law

- The inverse of a real number x under addition is $-x$.
- The inverse of a real number x under multiplication is its reciprocal, $\dfrac{1}{x}$.
- The **Inverse Law** states that in general:
  - when the additive inverse of a number or pronumeral is added to itself, it equals 0
  - when the multiplicative inverse of a number or pronumeral is multiplied by itself, it equals 1.

Inverse Law

$x + -x = -x + x = 0$ **For example:** $5 + -5 = -5 + 5 = 0$

$x \times \dfrac{1}{x} = \dfrac{1}{x} \times x = 1$ **For example:** $7 \times \dfrac{1}{7} = \dfrac{1}{7} \times 7 = 1$

- It is worth noting that the subtraction $(5 - 2 = 3)$ is equivalent to adding an inverse $(5 + (-2) = 3)$, and the division $(10 \div 2 = 5)$ is equivalent to multiplication by an inverse $(10 \times \dfrac{1}{2} = 5)$.

Closure Law

- The **Closure Law** states that, when an operation is performed on an element (or elements) of a set, the result produced must also be an element of that set.
  For example, addition is closed on natural numbers (that is, positive integers: 1, 2, 3, …), since adding a pair of natural numbers produces a natural number.
- Subtraction is not closed on natural numbers.
  For example, 5 and 7 are natural numbers and the result of adding them is 12, a natural number. However, the result of subtracting 7 from 5 is -2, which is not a natural number.

WORKED EXAMPLE 3 Determining which operations with the integers are closed

Determine the values of the following expressions, given the integer values $x = 4$ and $y = -12$. Comment on whether the Closure Law for integers holds for each of the expressions when these values are substituted.

a. $x + y$ **b.** $x - y$ **c.** $x \times y$ **d.** $x \div y$

THINK	WRITE
a. 1. Substitute each pronumeral into the expression.	**a.** $x + y = 4 + (-12)$
2. Evaluate and write the answer.	$= -8$
3. Determine whether the Closure Law holds; that is, is the result an integer?	The Closure Law holds for these substituted values.
b. Repeat steps **1–3** of part **a.**	**b.** $x - y = 4 - (-12)$ $= 16$ The Closure Law holds for these substituted values.
c. Repeat steps **1–3** of part **a.**	**c.** $x \times y = 4 \times (-12)$ $= -48$ The Closure Law holds for these substituted values.
d. Repeat steps **1–3** of part **a.**	**d.** $x \div y = 4 \div (-12)$ $= \dfrac{4}{-12}$ $= -\dfrac{1}{3}$ The Closure Law does not hold for these substituted values since the answer obtained is a fraction, not an integer.

- It is important to note that, although a particular set of numbers may be closed under a given operation, for example multiplication, another set of numbers may not be closed under that same operation.
  For example, in part **c** of Worked example 3, integers were closed under multiplication.
- In some cases, however, the set of *irrational numbers* is not closed under multiplication; for example, $\sqrt{3} \times \sqrt{3} = \sqrt{9} = 3$.
- In this example, two irrational numbers produced a rational number under multiplication.

Resources

eWorkbook Topic 2 Workbook (worksheets, code puzzle and project) (ewbk-13282)

Video eLesson Substitution (eles-1892)

Interactivities Individual pathway interactivity: Substitution (int-4566)
Substituting positive and negative numbers (int-3765)
Commutative Law (int-6109)
Associative Law (int-6110)
Identity Law (int-6111)
Inverse Law (int-6112)

Exercise 2.2 Substitution

learn on

2.2 Quick quiz on	2.2 Exercise

Individual pathways

■ PRACTISE	■ CONSOLIDATE	■ MASTER
1, 4, 8, 10, 13, 14, 17, 22	2, 5, 9, 11, 15, 18, 19, 23	3, 6, 7, 12, 16, 20, 21, 24

Fluency

WE1 For questions **1** to **3**, if $a = 2$, $b = 3$ and $c = 5$, evaluate the following expressions.

1. a. $a + b$ b. $c - b$ c. $c - a - b$ d. $c - (a - b)$

2. a. $7a + 8b - 11c$ b. $\dfrac{a}{2} + \dfrac{b}{3} + \dfrac{c}{5}$ c. abc d. $ab(c - b)$

3. a. $a^2 + b^2 - c^2$ b. $c^2 + a$ c. $-a \times b \times -c$ d. $2.3a - 3.2b$

For questions **4** to **6**, if $d = -6$ and $k = -5$, evaluate the following expressions.

4. a. $d + k$ b. $d - k$ c. $k - d$

5. a. kd b. $-d(k + 1)$ c. d^2

6. a. k^3 b. $\dfrac{k - 1}{d}$ c. $3k - 5d$

7. If $x = \dfrac{1}{3}$ and $y = \dfrac{1}{4}$, evaluate the following expressions.

   a. $x + y$ b. $y - x$ c. xy

   d. $\dfrac{x}{y}$ e. $x^2 y^3$ f. $\dfrac{9x}{y^2}$

8. If $x = 3$, determine the values of the following expressions.

   a. x^2 b. $-x^2$ c. $(-x)^2$
   d. $2x^2$ e. $-2x^2$ f. $(-2x)^2$

9. If $x = -3$, determine the values of the following expressions.

   a. x^2 b. $-x^2$ c. $(-x)^2$
   d. $2x^2$ e. $-2x^2$ f. $(-2x)^2$

WE2 For questions **10** to **12**, calculate the unknown variables in the real-life mathematical formulas given.

10. a. If $c = \sqrt{a^2 + b^2}$, calculate c if $a = 8$ and $b = 15$.

    b. If $A = \dfrac{1}{2}bh$, determine the value of A if $b = 12$ and $h = 5$.

    c. The perimeter, P, of a rectangle is given by $P = 2L + 2W$. Calculate the perimeter, P, of a rectangle given $L = 1.6$ and $W = 2.4$.

11. a. If $T = \dfrac{C}{L}$, determine the value of T if $C = 20.4$ and $L = 5.1$.

    b. If $K = \dfrac{n+1}{n-1}$, determine the value of K if $n = 5$.

    c. Given $F = \dfrac{9C}{5} + 32$, calculate F if $C = 20$.

12. a. If $v = u + at$, evaluate v if $u = 16$, $a = 5$, $t = 6$.

    b. The area, A, of a circle is given by the formula $A = \pi r^2$. Calculate the area of a circle, correct to 1 decimal place, if $r = 6$.

    c. If $E = \dfrac{1}{2}mv^2$, calculate m if $E = 40$ and $v = 4$.

    d. Given $r = \sqrt{\dfrac{A}{\pi}}$, evaluate A to 1 decimal place if $r = 14.1$.

13. MC a. If $p = -5$ and $q = 4$, then pq is equal to:

    A. 20   **B.** 1   **C.** −1   **D.** −20

    b. If $c^2 = a^2 + b^2$, and $a = 6$ and $b = 8$, then c is equal to:

    A. 28   **B.** 100   **C.** 10   **D.** 14

    c. Given $h = 6$ and $k = 7$, kh^2 is equal to:

    A. 294   **B.** 252   **C.** 1764   **D.** 5776

Understanding

14. Knowing the length of two sides of a right-angled triangle, the third side can be calculated using Pythagoras' theorem. If the two shorter sides have lengths of 1.5 cm and 3.6 cm, calculate the length of the hypotenuse.

15. The volume of a sphere can be calculated using the formula $\dfrac{4}{3}\pi r^3$. What is the volume of a sphere with a radius of 2.5 cm?
Give your answer correct to 2 decimal places.

2.5 cm

16. A rectangular park is 200 m by 300 m. If Blake runs along the diagonal of the park, calculate how far he will run.
Give your answer to the nearest metre.

Reasoning

17. WE3 Determine the values of the following expressions, given the integer values $x = 1$, $y = -2$ and $z = -1$.
Comment on whether the Closure Law for integers holds true for each of the expressions when these values are substituted.

    a. $x + y$ **b.** $y - z$ **c.** $y \times z$

18. Determine the values of the following expressions, given the integer values $x = 1$, $y = -2$ and $z = -1$. Comment on whether the Closure Law for integers holds true for each of the expressions when these values are substituted.

    a. $x \div z$ **b.** $z - x$ **c.** $x \div y$

19. For each of the following, complete the relationship to illustrate the stated law. Justify your reasoning.

    a. $(a + 2b) + 4c =$ _____ Associative Law
    b. $(x \times 3y) \times 5c =$ _____ Associative Law
    c. $2p \div q \neq$ _____ Commutative Law
    d. $5d + q =$ _____ Commutative Law

20. Calculate the values of the following expressions, given the natural number values $x = 8$, $y = 2$ and $z = 6$. Comment on whether the Closure Law for natural numbers holds true for each of the expressions.

    a. $x + y$ **b.** $y - z$ **c.** $y \times z$
    d. $x \div z$ **e.** $z - x$ **f.** $x \div y$

21. For each of the following, complete the relationship to illustrate the stated law. Justify your reasoning.

    a. $3z + 0 =$ _____ Identity Law
    b. $2x \times$ _____ $=$ _____ Inverse Law
    c. $(4x \div 3y) \div 5z \neq$ _____ Associative Law
    d. $3d - 4y \neq$ _____ Commutative Law

Problem solving

22. $s = ut + \dfrac{1}{2}at^2$ where t is the time in seconds, s is the displacement in metres, u is the initial velocity and a is the acceleration due to gravity.

    a. Calculate s when $u = 16.5$ m/s, $t = 2.5$ seconds and $a = 9.8$ m/s^2.
    b. A body has an initial velocity of 14.7 m/s and after t seconds has a displacement of 137.2 metres. Determine the value of t if $a = 9.8$ m/s^2.

23. If $n = p\sqrt{1 + \dfrac{1}{m}}$, calculate the value of m when $n = 6$ and $p = 4$.

24. The formula for the period (T) of a pendulum in seconds is $T = 2\pi\sqrt{\dfrac{L}{g}}$, where L is the length in metres of the pendulum and $g = 9.81$ m/s^2 is the acceleration due to gravity. Determine the period of a pendulum, to 1 decimal place, in a grandfather clock with a pendulum length of 154 cm.

LESSON
2.3 Adding and subtracting algebraic fractions

LEARNING INTENTIONS

At the end of this lesson you should be able to:
- determine the lowest common denominator of two or more fractions with pronumerals in the denominators
- add and subtract fractions involving algebraic expressions.

⏵ 2.3.1 Algebraic fractions

eles-4698

- In an algebraic fraction, the denominator, the numerator or both are algebraic expressions.

  For example, $\dfrac{x}{2}$, $\dfrac{3x+1}{2x-5}$ and $\dfrac{1}{x^2+5}$ are all **algebraic fractions**.

- As with all fractions, algebraic fractions must have a common denominator if they are to be added or subtracted, so an important step is to determine the lowest common denominator (LCD).

WORKED EXAMPLE 4 Simplifying fractions with algebraic numerators

Simplify the following expressions.

a. $\dfrac{2x}{3} - \dfrac{x}{2}$

b. $\dfrac{x+1}{6} + \dfrac{x+4}{4}$

c. $\dfrac{6x-11}{2} - \dfrac{7x}{4} + \dfrac{9-5x}{3}$

THINK

WRITE

a. 1. Write the expression.

a. $\dfrac{2x}{3} - \dfrac{x}{2}$

2. Rewrite each fraction as an equivalent fraction using the LCD of 3 and 2, which is 6.

$= \dfrac{2x}{3} \times \dfrac{2}{2} - \dfrac{x}{2} \times \dfrac{3}{3}$

$= \dfrac{4x}{6} - \dfrac{3x}{6}$

3. Express as a single fraction.

$= \dfrac{4x-3x}{6}$

4. Simplify the numerator and write the answer.

$= \dfrac{x}{6}$

b. 1. Write the expression.

b. $\dfrac{x+1}{6} + \dfrac{x+4}{4}$

2. Rewrite each fraction as an equivalent fraction using the LCD of 6 and 4, which is 12.

$= \dfrac{x+1}{6} \times \dfrac{2}{2} + \dfrac{x+4}{4} \times \dfrac{3}{3}$

$= \dfrac{2(x+1)}{12} + \dfrac{3(x+4)}{12}$

3. Express as a single fraction.

$= \dfrac{2(x+1)+3(x+4)}{12}$

⏵

4. Simplify the numerator by expanding brackets and collecting like terms.

$$= \frac{2x + 2 + 3x + 12}{12}$$

5. Write the answer.

$$= \frac{5x + 14}{12}$$

c. 1. Write the expression.

c. $\dfrac{6x - 11}{2} - \dfrac{7x}{4} + \dfrac{9 - 5x}{3}$

2. Rewrite each fraction as an equivalent fraction using the LCD of 2, 4 and 3, which is 12.

$$= \frac{6x - 11}{2} \times \frac{6}{6} - \frac{7x}{4} \times \frac{3}{3} + \frac{9 - 5x}{3} \times \frac{4}{4}$$

$$= \frac{6(6x - 11)}{12} - \frac{21x}{12} + \frac{4(9 - 5x)}{12}$$

3. Express as a single fraction.

$$= \frac{6(6x - 11) - 21x + 4(9 - 5x)}{12}$$

4. Simplify the numerator by expanding brackets and collecting like terms.

$$= \frac{36x - 66 - 21x + 36 - 20x}{12}$$

5. Write the answer.

$$= \frac{-5x - 30}{12}$$

▶ 2.3.2 Pronumerals in the denominator

eles-4699

- If pronumerals appear in the denominator, the process involved in adding and subtracting the fractions is to determine a lowest common denominator as usual.
- When there is an algebraic expression in the denominator of each fraction, a common denominator can be obtained by writing the product of the denominators.
  For example, if $x + 3$ and $2x - 5$ are in the denominator of each fraction, then a common denominator of the two fractions will be $(x + 3)(2x - 5)$.

WORKED EXAMPLE 5 Simplifying fractions with algebraic denominators

Simplify $\dfrac{2}{3x} - \dfrac{1}{4x}$.

THINK

1. Write the expression.

2. Rewrite each fraction as an equivalent fraction using the LCD of $3x$ and $4x$, which is $12x$.
   Note: $12x^2$ is not the lowest LCD.

3. Express as a single fraction.

4. Simplify the numerator and write the answer.

WRITE

$\dfrac{2}{3x} - \dfrac{1}{4x}$

$= \dfrac{2}{3x} \times \dfrac{4}{4} - \dfrac{1}{4x} \times \dfrac{3}{3}$

$= \dfrac{8}{12x} - \dfrac{3}{12x}$

$= \dfrac{8 - 3}{12x}$

$= \dfrac{5}{12x}$

WORKED EXAMPLE 6 Simplifying by finding the LCD of two algebraic expressions

Simplify $\dfrac{x+1}{x+3} + \dfrac{2x-1}{x+2}$ **by writing it first as a single fraction.**

THINK	WRITE
1. Write the expression.	$\dfrac{x+1}{x+3} + \dfrac{2x-1}{x+2}$
2. Rewrite each fraction as an equivalent fraction using the LCD of $x+3$ and $x+2$, which is the product $(x+3)(x+2)$.	$= \dfrac{x+1}{x+3} \times \dfrac{x+2}{x+2} + \dfrac{2x-1}{x+2} \times \dfrac{x+3}{x+3}$ $= \dfrac{(x+1)(x+2)}{(x+3)(x+2)} + \dfrac{(2x-1)(x+3)}{(x+3)(x+2)}$
3. Express as a single fraction.	$= \dfrac{(x+1)(x+2)+(2x-1)(x+3)}{(x+3)(x+2)}$
4. Simplify the numerator by expanding brackets and collecting like terms. *Note:* The denominator is generally kept in factorised form. That is, it is not expanded.	$= \dfrac{\left(x^2+2x+x+2\right)+\left(2x^2+6x-x-3\right)}{(x+3)(x+2)}$ $= \dfrac{\left(x^2+3x+2+2x^2+5x-3\right)}{(x+3)(x+2)}$
5. Write the answer.	$= \dfrac{3x^2+8x-1}{(x+3)(x+2)}$

WORKED EXAMPLE 7 Simplification involving repeated linear factors

Simplify $\dfrac{x+2}{x-3} + \dfrac{x-1}{(x-3)^2}$ **by writing it first as a single fraction.**

THINK	WRITE
1. Write the expression.	$\dfrac{x+2}{x-3} + \dfrac{x-1}{(x-3)^2}$
2. Rewrite each fraction as an equivalent fraction using the LCD of $x-3$ and $(x-3)^2$, which is $(x-3)^2$.	$= \dfrac{x+2}{x-3} \times \dfrac{x-3}{x-3} + \dfrac{x-1}{(x-3)^2}$ $= \dfrac{(x+2)(x-3)}{(x-3)^2} + \dfrac{x-1}{(x-3)^2}$ $= \dfrac{x^2-x-6}{(x-3)^2} + \dfrac{x-1}{(x-3)^2}$
3. Express as a single fraction.	$= \dfrac{x^2-x-6+x-1}{(x-3)^2}$
4. Simplify the numerator and write the answer.	$= \dfrac{x^2-7}{(x-3)^2}$

TI | THINK	**DISPLAY/WRITE**	**CASIO | THINK**	**DISPLAY/WRITE**

TI | THINK

On a Calculator page, press CTRL and ÷ to get the fraction template, and then complete the entry line as:

$$\frac{x+2}{x-3}+\frac{x-1}{(x-3)^2}$$

Then press ENTER.

DISPLAY/WRITE

$$\frac{x+2}{x-3}+\frac{x-1}{(x-3)^2}=\frac{x^2-7}{(x-3)^2}$$

CASIO | THINK

On the Main screen, complete the entry line as:

$$\text{combine}\left(\frac{x+2}{x-3}+\frac{x-1}{(x-3)^2}\right)$$

Then press EXE.

DISPLAY/WRITE

$$\frac{x+2}{x-3}+\frac{x-1}{(x-3)^2}=\frac{x^2-7}{(x-3)^2}$$

DISCUSSION

Explain why we can't just add the numerators and the denominators of fractions, as shown in the incorrect expression below.

$$\frac{a}{b}+\frac{c}{d}=\frac{a+c}{b+d}$$

 Resources

 eWorkbook   Topic 2 Workbook (worksheets, code puzzle and project) (ewbk-13282)

Interactivities Individual pathway interactivity: Adding and subtracting algebraic fractions (int-4567)
Adding and subtracting algebraic fractions (int-6113)
Lowest common denominators with pronumerals (int-6114)

Exercise 2.3 Adding and subtracting algebraic fractions   learn

2.3 Quick quiz on	**2.3 Exercise**

Individual pathways

■ PRACTISE	■ CONSOLIDATE	■ MASTER
1, 4, 7, 10, 13, 16, 19	2, 5, 8, 11, 14, 17, 20	3, 6, 9, 12, 15, 18, 21

Fluency

For questions **1** to **3**, simplify each of the following expressions.

1. a. $\dfrac{4}{7}+\dfrac{2}{3}$ b. $\dfrac{1}{8}+\dfrac{5}{9}$ c. $\dfrac{3}{5}+\dfrac{6}{15}$

2. a. $\dfrac{4}{9}-\dfrac{3}{11}$ b. $\dfrac{3}{7}-\dfrac{2}{5}$ c. $\dfrac{1}{5}-\dfrac{x}{6}$

3. a. $\dfrac{5x}{9} - \dfrac{4}{27}$ 　　　　b. $\dfrac{3}{8} - \dfrac{2x}{5}$ 　　　　c. $\dfrac{5}{x} - \dfrac{2}{3}$

WE4 For questions 4 to 6, simplify the following expressions.

4. a. $\dfrac{2y}{3} - \dfrac{y}{4}$ 　　b. $\dfrac{y}{8} - \dfrac{y}{5}$ 　　c. $\dfrac{4x}{3} - \dfrac{x}{4}$ 　　d. $\dfrac{5x-1}{2} + \dfrac{3x}{5}$

5. a. $\dfrac{2w}{14} - \dfrac{w}{28}$ 　　b. $\dfrac{y}{20} - \dfrac{y}{4}$ 　　c. $\dfrac{12y}{5} + \dfrac{y}{7}$ 　　d. $\dfrac{3x-1}{4} + \dfrac{5x}{3} + \dfrac{2-3x}{2}$

6. a. $\dfrac{x+1}{5} + \dfrac{x+3}{2}$ 　　b. $\dfrac{x+2}{4} + \dfrac{x+6}{3}$ 　　c. $\dfrac{2x-1}{5} - \dfrac{2x+1}{6}$ 　　d. $\dfrac{x+1}{5} + \dfrac{7x}{3} - \dfrac{25-9x}{15}$

WE5 For questions 7 to 9, simplify the following expressions.

7. a. $\dfrac{2}{4x} + \dfrac{1}{8x}$ 　　b. $\dfrac{3}{4x} - \dfrac{1}{3x}$ 　　c. $\dfrac{5}{3x} + \dfrac{1}{7x}$

8. a. $\dfrac{12}{5x} + \dfrac{4}{15x}$ 　　b. $\dfrac{1}{6x} + \dfrac{1}{8x}$ 　　c. $\dfrac{9}{4x} - \dfrac{9}{5x}$

9. a. $\dfrac{2}{100x} + \dfrac{7}{20x}$ 　　b. $\dfrac{1}{10x} + \dfrac{5}{x}$ 　　c. $\dfrac{4}{3x} - \dfrac{3}{2x}$

WE6&7 For questions 10 to 12, simplify the following expressions by writing them as single fractions.

10. a. $\dfrac{2}{x+4} + \dfrac{3x}{x-2}$ 　　b. $\dfrac{2x}{x+5} + \dfrac{5}{x-1}$ 　　c. $\dfrac{5}{2x+1} + \dfrac{x}{x-2}$ 　　d. $\dfrac{2x}{x+1} - \dfrac{3}{2x-7}$

11. a. $\dfrac{4x}{x+7} + \dfrac{3x}{x-5}$ 　　b. $\dfrac{x+2}{x+1} + \dfrac{x-1}{x+4}$ 　　c. $\dfrac{x+8}{x+1} - \dfrac{2x+1}{x+2}$ 　　d. $\dfrac{x+5}{x+3} - \dfrac{x-1}{x-2}$

12. a. $\dfrac{x+1}{x+2} - \dfrac{2x-5}{3x-1}$ 　　b. $\dfrac{2}{x-1} - \dfrac{3}{1-x}$ 　　c. $\dfrac{4}{(x+1)^2} + \dfrac{3}{x+1}$ 　　d. $\dfrac{3}{x-1} - \dfrac{1}{(x-1)^2}$

Understanding

13. A classmate attempted to complete an algebraic fraction subtraction problem.

$$\dfrac{x}{x-1} - \dfrac{3}{x-2} = \dfrac{x}{x-1} \times \dfrac{(x-2)}{(x-2)} - \dfrac{3}{x-2} \times \dfrac{(x-1)}{(x-1)}$$

$$= \dfrac{x(x-2) - 3(x-1)}{(x-1)(x+2)}$$

$$= \dfrac{x^2 - 2x - 3x - 1}{(x-1)(x+2)}$$

$$= \dfrac{x^2 - 5x - 1}{(x-1)(x+2)}$$

　a. Identify the mistake she made.
　b. Determine the correct answer.

14. Simplify the following.

a. $\dfrac{y-x}{x-y}$

b. $\dfrac{3}{x-2}+\dfrac{4}{2-x}$

15. Simplify the following.

a. $\dfrac{3}{3-x}+\dfrac{3x}{(x-3)^2}$

b. $\dfrac{1}{x-2}-\dfrac{2x}{(2-x)^2}+\dfrac{x^2}{(x-2)^3}$

Reasoning

16. Simplify the following.

a. $\dfrac{1}{x+2}+\dfrac{2}{x+1}+\dfrac{1}{x+3}$

b. $\dfrac{1}{x-1}+\dfrac{4}{x+2}+\dfrac{2}{x-4}$

17. Simplify the following.

a. $\dfrac{3}{x+1}+\dfrac{2}{x+3}-\dfrac{1}{x+2}$

b. $\dfrac{2}{x-4}-\dfrac{3}{x-1}+\dfrac{5}{x+3}$

c. Explain why the process that involves determining the lowest common denominator is important in parts **a** and **b**.

18. The reverse process of adding or subtracting algebraic fractions is quite complex.

Use trial and error, or technology, to determine the value of a if $\dfrac{7x-4}{(x-8)(x+5)}=\dfrac{a}{x-8}+\dfrac{3}{x+5}$.

Problem solving

19. Simplify $\dfrac{3}{x^2+7x+12}-\dfrac{1}{x^2+x-6}+\dfrac{2}{x^2+2x-8}$.

20. Simplify $\dfrac{x^2+3x-18}{x^2-x-42}-\dfrac{x^2-3x+2}{x^2-5x+4}$.

21. Simplify $\dfrac{x^2-25}{x^2-2x-15}+\dfrac{x^2+12x+32}{x^2+4x-32}-\dfrac{2x^2}{x^2-x-12}$.

LESSON
2.4 Multiplying and dividing algebraic fractions

2.4.1 Multiplying algebraic fractions

eles-4700

- Algebraic fractions can be simplified using the index laws and by cancelling factors common to the numerators and denominators.
- A fraction can only be simplified if:
  - there is a common factor in the numerator and the denominator
  - the numerator and denominator are both written in factorised form, that is, as the *product* of two or more factors.

$$\frac{3ab}{12a} = \frac{{}^{1}\cancel{3} \times {}^{1}\cancel{a} \times b}{{}^{4}\cancel{12} \times {}^{1}\cancel{a}} \quad \leftarrow \text{product of factors} \atop \leftarrow \text{product of factors} \qquad \frac{3a + b}{12a} = \frac{3 \times a + b}{12 \times a} \quad {\leftarrow \text{not a product of factors} \atop \leftarrow \text{product of factors}}$$

$$= \frac{b}{4} \qquad\qquad\qquad\qquad\qquad \textbf{Cannot be simplified}$$

- Multiplication of algebraic fractions follows the same rules as multiplication of numerical fractions: multiply the numerators, then multiply the denominators.

WORKED EXAMPLE 8 Multiplying algebraic fractions and simplifying the result

Simplify each of the following expressions.

a. $\dfrac{5y}{3x} \times \dfrac{6z}{7y}$

b. $\dfrac{2x}{(x + 1)(2x - 3)} \times \dfrac{x + 1}{x}$

THINK	WRITE
a. 1. Write the expression.	a. $\dfrac{5y}{3x} \times \dfrac{6z}{7y}$
2. Cancel common factors in the numerators and denominators. The y can be cancelled in the denominator and the numerator. Also, the 3 in the denominator can divide into the 6 in the numerator.	$= \dfrac{5y^{1}}{{}_{1}\cancel{3}x} \times \dfrac{\cancel{6}^{2}z}{7y^{1}}$ $= \dfrac{5}{x} \times \dfrac{2z}{7}$
3. Multiply the numerators, then multiply the denominators and write the answer.	$= \dfrac{10z}{7x}$

b. 1. Write the expression.

b. $$\dfrac{2x}{(x+1)(2x-3)} \times \dfrac{x+1}{x}$$

2. Cancel common factors in the numerator and the denominator. $(x+1)$ and the x are both common in the numerator and the denominator and can therefore be cancelled.

$$= \dfrac{2x^{\cancel{1}}}{{}^{1}(\cancel{x+1})(2x-3)} \times \dfrac{\cancel{x+1}^{1}}{\cancel{x}^{1}}$$

$$= \dfrac{2}{2x-3} \times \dfrac{1}{1}$$

3. Multiply the numerators, then multiply the denominators and write the answer.

$$= \dfrac{2}{2x-3}$$

2.4.2 Dividing algebraic fractions

eles-4701

- When dividing algebraic fractions, follow the same rules as for division of numerical fractions: write the division as a multiplication and invert the second fraction.
- This process is sometimes known as multiplying by the **reciprocal**.

WORKED EXAMPLE 9 Dividing algebraic fractions

Simplify the following expressions.

a. $\dfrac{3xy}{2} \div \dfrac{4x}{9y}$

b. $\dfrac{4}{(x+1)(3x-5)} \div \dfrac{x-7}{x+1}$

THINK

WRITE

a. 1. Write the expression.

a. $\dfrac{3xy}{2} \div \dfrac{4x}{9y}$

2. Change the division sign to a multiplication sign and write the second fraction as its reciprocal.

$$= \dfrac{3xy}{2} \times \dfrac{9y}{4x}$$

3. Cancel common factors in the numerators and denominators. The pronumeral x is common to both the numerator and denominator and can therefore be cancelled.

$$= \dfrac{3y}{2} \times \dfrac{9y}{4}$$

4. Multiply the numerators, then multiply the denominators and write the answer.

$$= \dfrac{27y^2}{8}$$

b. 1. Write the expression.

b. $\dfrac{4}{(x+1)(3x-5)} \div \dfrac{x-7}{x+1}$

2. Change the division sign to a multiplication sign and write the second fraction as its reciprocal.

$$= \dfrac{4}{(x+1)(3x-5)} \times \dfrac{x+1}{x-7}$$

3. Cancel common factors in the numerators and denominators. $(x+1)$ is common to both the numerator and denominator and can therefore be cancelled.

$$= \dfrac{4}{3x-5} \times \dfrac{1}{x-7}$$

4. Multiply the numerators, then multiply the denominators and write the answer.

$$= \dfrac{4}{(3x-5)(x-7)}$$

| TI | THINK | DISPLAY/WRITE | CASIO | THINK | DISPLAY/WRITE |

TI | THINK

a.

On a Calculator page, use the fraction template twice to complete the entry line as:

$$\frac{3xy}{2} \div \frac{4x}{9y}$$

Then press ENTER.

DISPLAY/WRITE

a.

$$\frac{3xy}{2} \div \frac{4x}{9y} = \frac{27y^2}{8}$$

b.

On a Calculator page, use the fraction template twice to complete the entry line as:

$$\frac{4}{(x+1)(3x-5)} \div \frac{x-7}{x+1}$$

Then press ENTER.

b.

$$\frac{4}{(x+1)(3x-5)} \div \frac{x-7}{x+1}$$

$$= \frac{4}{(x-7)(3x-5)}$$

CASIO | THINK

a, b.

On the Main screen, use the fraction template twice to complete the entry line as:

$$\frac{3 \times xy}{2} \div \frac{4x}{9y}$$

$$\frac{4}{(x+1)(3x-5)} \div \frac{x-7}{x+1}$$

Press EXE after each entry.

DISPLAY/WRITE

a, b.

$$\frac{3xy}{2} \div \frac{4x}{9y} = \frac{27y^2}{8}$$

$$\frac{4}{(x+1)(3x-5)} \div \frac{x-7}{x+1}$$

$$= \frac{4}{(x-7)(3x-5)}$$

DISCUSSION

Explain how multiplying and dividing algebraic fractions is different to adding and subtracting them.

 Resources

Exercise 2.4 Multiplying and dividing algebraic fractions

2.4 Quick quiz on	2.4 Exercise

Individual pathways

■ PRACTISE	■ CONSOLIDATE	■ MASTER
1, 4, 7, 10, 13, 16, 19	2, 5, 8, 11, 14, 17, 20	3, 6, 9, 12, 15, 18, 21

Fluency

WE8a For questions **1** to **3**, simplify the following expressions.

1. a. $\dfrac{x}{5} \times \dfrac{20}{y}$ b. $\dfrac{x}{4} \times \dfrac{12}{y}$ c. $\dfrac{y}{4} \times \dfrac{16}{x}$ d. $\dfrac{x}{2} \times \dfrac{9}{2y}$

2. a. $\dfrac{x}{10} \times \dfrac{-25}{2y}$ b. $\dfrac{3w}{-14} \times \dfrac{-7}{x}$ c. $\dfrac{3y}{4x} \times \dfrac{8z}{7y}$ d. $\dfrac{-y}{3x} \times \dfrac{6z}{-7y}$

3. a. $\dfrac{x}{3z} \times \dfrac{-9z}{2y}$ b. $\dfrac{5y}{3x} \times \dfrac{x}{8y}$ c. $\dfrac{-20y}{7x} \times \dfrac{-21z}{5y}$ d. $\dfrac{y}{-3w} \times \dfrac{x}{2y}$

WE8b For questions **4** to **6**, simplify the following expressions.

4. a. $\dfrac{2x}{(x-1)(3x-2)} \times \dfrac{x-1}{x}$ b. $\dfrac{5x}{(x-3)(4x+7)} \times \dfrac{4x+7}{x}$

   c. $\dfrac{9x}{(5x+1)(x-6)} \times \dfrac{5x+1}{2x}$ d. $\dfrac{(x+4)}{(x+1)(x+3)} \times \dfrac{x+1}{x+4}$

5. a. $\dfrac{2x}{x+1} \times \dfrac{x-1}{(x+1)(x-1)}$ b. $\dfrac{2}{x(2x-3)} \times \dfrac{x(x+1)}{4}$

   c. $\dfrac{2x}{4(a+3)} \times \dfrac{3a}{15x}$ d. $\dfrac{15c}{12(d-3)} \times \dfrac{21d}{6c}$

6. a. $\dfrac{6x^2}{20(x-2)^2} \times \dfrac{15(x-2)}{16x^4}$ b. $\dfrac{7x^2(x-3)}{5x(x+1)} \times \dfrac{3(x-3)(x+1)}{14(x-3)^2(x-1)}$

WE9a For questions **7** to **9**, simplify the following expressions.

7. a. $\dfrac{3}{x} \div \dfrac{5}{x}$ b. $\dfrac{2}{x} \div \dfrac{9}{x}$ c. $\dfrac{4}{x} \div \dfrac{12}{x}$ d. $\dfrac{20}{y} \div \dfrac{20}{3y}$

8. a. $\dfrac{1}{5w} \div \dfrac{5}{w}$ b. $\dfrac{7}{2x} \div \dfrac{3}{5x}$ c. $\dfrac{3xy}{7} \div \dfrac{3x}{4y}$ d. $\dfrac{2xy}{5} \div \dfrac{5x}{y}$

9. a. $\dfrac{6y}{9} \div \dfrac{3x}{4xy}$ b. $\dfrac{8wx}{5} \div \dfrac{3w}{4y}$ c. $\dfrac{2xy}{5} \div \dfrac{3xy}{5}$ d. $\dfrac{10xy}{7} \div \dfrac{20x}{14y}$

WE9b For questions **10** to **12**, simplify the following expressions.

10. a. $\dfrac{9}{(x-1)(3x-7)} \div \dfrac{x+3}{x-1}$ b. $\dfrac{1}{(x+2)(2x-5)} \div \dfrac{x-9}{2x-5}$

11. a. $\dfrac{12(x-3)^2}{(x+5)(x-9)} \div \dfrac{4(x-3)}{7(x-9)}$ b. $\dfrac{13}{6(x-4)^2(x-1)} \div \dfrac{3(x+1)}{2(x-4)(x-1)}$

12. a. $\dfrac{16\,(x+5)\,(x-4)^2}{(x+3)\,(x+4)} \div \dfrac{8\,(x+3)\,(x+5)}{(x-4)\,(x+4)}$

b. $\dfrac{(x+2)\,(x+3)\,(x+4)}{(x-2)\,(x+3)^2} \div \dfrac{x^2+x-12}{x^2-4}$

Understanding

For questions **13** to **15**, determine the missing fractions.

13. a. $\dfrac{x+2}{3} \times \boxed{} = 5$

b. $\dfrac{3}{x^2} \div \boxed{} = \dfrac{1}{4}$

14. a. $\dfrac{(x+3)\,(x+2)}{(x-4)} \times \boxed{} = \dfrac{x-5}{x+2}$

b. $\dfrac{x^2\,(x-3)}{(x+4)\,(x-5)} \div \boxed{} = \dfrac{3x}{2\,(x+4)}$

15. a. $\dfrac{x^2+8x+15}{x^2-4x-21} \times \boxed{} = \dfrac{x^2-25}{x^2-11x+28}$

b. $\boxed{} \div \dfrac{x^2-2x-24}{x^2-36} = \dfrac{x^2+12x+36}{x^2}$

Reasoning

16. Explain whether $\dfrac{3}{x+2}$ is the same as $\dfrac{1}{x+2} + \dfrac{1}{x+2} + \dfrac{1}{x+2}$.

17. Does $\dfrac{12xy+16yz^2}{20xyz}$ simplify to $\dfrac{3+4z}{5}$? Explain your reasoning.

18. a. Simplify $\dfrac{(x-4)\,(x+3)}{4x-x^2} \times \dfrac{x^2-x}{(x+3)\,(x-1)}$.

b. Identify and explain the error in the following reasoning.

$$\dfrac{(x-4)\,(x+3)}{4x-x^2} \times \dfrac{x^2-x}{(x+3)\,(x-1)}$$

$$= \dfrac{(x-4)\,(x+3)}{x\,(4-x)} \times \dfrac{x\,(x-1)}{(x+3)\,(x-1)} = 1$$

Problem solving

19. Simplify $\dfrac{x^2-2x-3}{x^4-1} \times \dfrac{x^2+4x-5}{x^2-5x+6} \div \dfrac{x^2+7x+10}{x^4-3x^2-4}$.

20. Simplify $\dfrac{x+1}{x-\dfrac{x}{\dfrac{x}{a}}}$ where $a = \dfrac{x-1}{x+1}$.

21. Simplify $\left(\dfrac{\dfrac{x^2+1}{x-1}-x}{\dfrac{x^2-1}{x+1}+1}\right) \times \left(1 - \dfrac{2}{1+\dfrac{1}{x}}\right)$.

LESSON
2.5 Solving simple equations

LEARNING INTENTIONS

At the end of this lesson you should be able to:
- solve one- and two-step equations using inverse operations
- solve equations with pronumerals on both sides of the equals sign.

2.5.1 Solving equations using inverse operations

eles-4702

- **Equations** show the equivalence of two expressions.
- Equations can be solved using inverse operations.
- Determining the solution of an equation involves calculating the value or values of a variable that, when substituted into that equation, produce a true statement.
- When solving equations, the last operation performed on the pronumeral when building the equation is the first operation undone by applying inverse operations to both sides of the equation. For example, the equation $2x + 3 = 5$ is built from x by:
  first operation: multiplying by 2 to give $2x$
  second operation: adding 3 to give $2x + 3$.
- In order to solve the equation, undo the second operation of adding 3 by subtracting 3, then undo the first operation of multiplying by 2 by dividing by 2.

> **Inverse operations**
>
> **+ and − are inverse operations**
>
> **× and ÷ are inverse operations**
>
> **^{2} and $\sqrt{}$ are inverse operations**

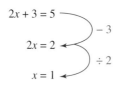

$$2x + 3 = 5$$
$$2x = 2 \quad \Big\} \; -3$$
$$x = 1 \quad \Big\} \; \div 2$$

- Equations that require one step to solve are called one-step equations.

WORKED EXAMPLE 10 Solving equations using inverse operations

Solve the following equations.

a. $a + 27 = 71$ b. $\dfrac{d}{16} = 3\dfrac{1}{4}$ c. $\sqrt{e} = 0.87$ d. $f^2 = \dfrac{4}{25}$

THINK		WRITE
a. 1. Write the equation.	**a.**	$a + 27 = 71$
2. 27 has been added to a, resulting in 71. The addition of 27 has to be reversed by subtracting 27 from both sides of the equation to obtain the solution.		$a + 27 - 27 = 71 - 27$ $a = 44$
b. 1. Write the equation.	**b.**	$\dfrac{d}{16} = 3\dfrac{1}{4}$
2. Express $3\dfrac{1}{4}$ as an improper fraction.		$\dfrac{d}{16} = \dfrac{13}{4}$

---

3. The pronumeral d has been divided by 16, resulting in $\dfrac{13}{4}$. Therefore, the division has to be reversed by multiplying both sides of the equation by 16 to obtain d.

$$\dfrac{d}{\cancel{16}} \times \cancel{16} = \dfrac{13}{\cancel{4}} \times \cancel{16}^4$$

$$d = 52$$

c. 1. Write the equation.

c. $\sqrt{e} = 0.87$

2. The square root of e has been taken, resulting in 0.87. Therefore, the square root has to be reversed by squaring both sides of the equation to obtain e.

$$\left(\sqrt{e}\right)^2 = 0.87^2$$

$$e = 0.7569$$

d. 1. Write the equation.

d. $f^2 = \dfrac{4}{25}$

2. The pronumeral f has been squared, resulting in $\dfrac{4}{25}$. Therefore, the squaring has to be reversed by taking the square root of both sides of the equation to obtain f. Note that there are two possible solutions, one positive and one negative, since two negative numbers can also be multiplied together to produce a positive result.

$$f = \pm\sqrt{\dfrac{4}{25}}$$

$$f = \pm\dfrac{2}{5}$$

▶ 2.5.2 Two-step equations

eles-4704

- Two-step equations involve the inverse of two operations in their solutions.

WORKED EXAMPLE 11 Solving two-step equations

Solve the following equations.

a. $5y - 6 = 79$

b. $\dfrac{4x}{9} = 5$

THINK	WRITE
a. 1. Write the equation.	a. $5y - 6 = 79$
2. Step 1: Add 6 to both sides of the equation.	$5y - 6 + 6 = 79 + 6$ $5y = 85$
3. Step 2: Divide both sides of the equation by 5 to obtain y.	$\dfrac{5y}{5} = \dfrac{85}{5}$
4. Write the answer.	$y = 17$
b. 1. Write the equation.	b. $\dfrac{4x}{9} = 5$
2. Step 1: Multiply both sides of the equation by 9.	$\dfrac{4x}{9} \times 9 = 5 \times 9$ $4x = 45$

3. **Step 2:** Divide both sides of the equation by 4 to obtain x.

$$\frac{4x}{4} = \frac{45}{4}$$

$$x = \frac{45}{4}$$

4. Express the answer as a mixed number.

$$x = 11\frac{1}{4}$$

TI | THINK	DISPLAY/WRITE	CASIO | THINK	DISPLAY/WRITE
a. On a Calculator page, to solve the equation, press: • MENU • 3: Algebra • 1: Solve Then complete the line as: solve $(5y - 6 = 79, y)$ The 'comma y' $(,y)$ instructs the calculator to solve for the variable y. Then press ENTER. **b.** On a Calculator page, complete the entry line as: solve $\left(\dfrac{4x}{9} = 5, x\right)$ The result is given as an improper fraction. To change to a proper fraction, press: • MENU • 2: Number • 7: Fraction Tools • 1: Proper Fraction Then complete as shown and press ENTER.	**a.** $5y - 6 = 79$ $\Rightarrow y = 17$ **b.** $\dfrac{4x}{y} = 5$ $\Rightarrow x = 11\dfrac{1}{4}$	**a, b.** On the Main screen, to solve the equation, tap: • Action • Advanced • solve Then complete the entry line as: solve $(5y - 6 = 79, y)$ Then press EXE. The 'comma y' $(,y)$ instructs the calculator to solve for the variable y. Then complete the entry line as: solve $\left(\dfrac{4x}{9} = 5\right)$ The result is given as an improper fraction. If required, to change to a proper fraction, tap: • Action • Transformation • Fraction • propFrac Then complete as shown and press EXE. If x is the only pronumeral, it is not necessary to include x at the end of the entry line.	**a, b.** $5y - 6 = 79$ $\Rightarrow y = 17$ $\dfrac{4x}{9} = 5$ $\Rightarrow x = 11\dfrac{1}{4}$

▶ 2.5.3 Equations where the pronumeral appears on both sides

eles-4705

- In solving equations where the pronumeral appears on both sides, subtract the smaller pronumeral term so that it is eliminated from both sides of the equation.

WORKED EXAMPLE 12 Solving equations with multiple pronumeral terms

Solve the following equations.

a. $5h + 13 = 2h - 2$ b. $14 - 4d = 27 - d$ c. $2(x - 3) = 5(2x + 4)$

THINK	WRITE
a. 1. Write the equation.	**a.** $\qquad 5h + 13 = 2h - 2$
2. Eliminate the pronumeral from the right-hand side by subtracting $2h$ from both sides of the equation.	$5h + 13 - 2h = 2h - 2 - 2h$ $\qquad\quad 3h + 13 = -2$

3. Subtract 13 from both sides of the equation.

$$3h + 13 - 13 = -2 - 13$$
$$3h = -15$$

4. Divide both sides of the equation by 3 and write the answer.

$$h = -5$$

b. 1. Write the equation.

b. $$14 - 4d = 27 - d$$

2. Add $4d$ to both sides of the equation.

$$14 - 4d + 4d = 27 - d + 4d$$
$$14 = 27 + 3d$$

3. Subtract 27 from both sides of the equation.

$$14 - 27 = 27 + 3d - 27$$
$$-13 = 3d$$

4. Divide both sides of the equation by 3.

$$\frac{-13}{3} = \frac{3d}{3}$$
$$-\frac{13}{3} = d$$

5. Express the answer as a mixed number.

$$-4\frac{1}{3} = d$$

6. Write the answer so that d is on the left-hand side.

$$d = -4\frac{1}{3}$$

c. 1. Write the equation.

c. $$2(x - 3) = 5(2x + 4)$$

2. Expand the brackets on both sides of the equation.

$$2x - 6 = 10x + 20$$

3. Subtract $2x$ from both sides of the equation.

$$2x - 6 - 2x = 8x + 20 - 2x$$
$$-6 = 8x + 20$$

4. Subtract 20 from both sides of the equation.

$$-6 - 20 = 8x - 20$$
$$-26 = 8x$$

5. Divide both sides of the equation by 8.

$$\frac{-26}{8} = -\frac{8x}{8}$$
$$-\frac{26}{8} = x$$

6. Simplify and write the answer with the pronumeral on the left-hand side.

$$x = -\frac{13}{4}$$

DISCUSSION

Describe in one sentence what it means to solve linear equations.

on Resources

 eWorkbook   Topic 2 Workbook (worksheets, code puzzle and project) (ewbk-13282)

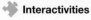 **Video eLessons** Solving linear equations (eles-1895)
Solving linear equations with pronumerals on both sides (eles-1901)

Interactivities   Individual pathway interactivity: Solving simple equations (int-4569)
Using algebra to solve problems (int-3805)
One-step equations (int-6118)
Two-step equations (int-6119)

Exercise 2.5 Solving simple equations

2.5 Quick quiz on	**2.5 Exercise**

Individual pathways

■ PRACTISE	■ CONSOLIDATE	■ MASTER
1, 4, 7, 10, 13, 16, 18, 21, 25, 26, 30, 33, 36, 40, 43, 46	2, 5, 8, 11, 14, 17, 19, 22, 23, 27, 28, 31, 34, 37, 41, 44, 47	3, 6, 9, 12, 15, 20, 24, 29, 32, 35, 38, 39, 42, 45, 48

Fluency

WE10a For questions **1** to **3**, solve the following equations.

1. **a.** $a + 61 = 85$ **b.** $k - 75 = 46$ **c.** $g + 9.3 = 12.2$

2. **a.** $r - 2.3 = 0.7$ **b.** $h + 0.84 = 1.1$ **c.** $i + 5 = 3$

3. **a.** $t - 12 = -7$ **b.** $q + \dfrac{1}{3} = \dfrac{1}{2}$ **c.** $x - 2 = -2$

WE10b For questions **4** to **6**, solve the following equations.

4. **a.** $\dfrac{f}{4} = 3$ **b.** $\dfrac{i}{10} = -6$ **c.** $6z = -42$

5. **a.** $9v = 63$ **b.** $6w = -32$ **c.** $\dfrac{k}{12} = \dfrac{5}{6}$

6. **a.** $4a = 1.7$ **b.** $\dfrac{m}{19} = \dfrac{7}{8}$ **c.** $\dfrac{y}{4} = 5\dfrac{3}{8}$

WE10c, d For questions **7** to **12**, solve the following equations.

7. **a.** $\sqrt{t} = 10$ **b.** $y^2 = 289$ **c.** $\sqrt{q} = 2.5$

8. **a.** $f^2 = 1.44$ **b.** $\sqrt{h} = \dfrac{4}{7}$ **c.** $p^2 = \dfrac{9}{64}$

9. **a.** $\sqrt{g} = \dfrac{15}{22}$ **b.** $j^2 = \dfrac{196}{961}$ **c.** $a^2 = 2\dfrac{7}{9}$

10. **a.** $\sqrt{t} - 3 = 2$ **b.** $5x^2 = 180$ **c.** $3\sqrt{m} = 12$

   d. $-2t^2 = -18$ **e.** $t^2 + 11 = 111$ **f.** $\sqrt{m} - 5 = 0$

11. **a.** $\sqrt[3]{x} = 2$ **b.** $x^3 = -27$ **c.** $\sqrt[3]{m} = \dfrac{1}{2}$

   d. $x^3 = \dfrac{27}{64}$ **e.** $\sqrt[3]{m} = 0.2$ **f.** $w^3 = 15\dfrac{5}{8}$

12. **a.** $x^3 + 1 = 0$ **b.** $3x^3 = -24$ **c.** $\sqrt[3]{m} + 5 = 6$

   d. $-2 \times \sqrt[3]{w} = 16$ **e.** $\sqrt[3]{t} - 13 = -8$ **f.** $2x^3 - 14 = 2$

WE11a For questions **13** to **20**, solve the following equations.

13. **a.** $5a + 6 = 26$ **b.** $6b + 8 = 44$ **c.** $8i - 9 = 15$

14. **a.** $7f - 18 = 45$ **b.** $8q + 17 = 26$ **c.** $10r - 21 = 33$

15. a. $6s + 46 = 75$ **b.** $5t - 28 = 21$ **c.** $8a + 88 = 28$

16. a. $\dfrac{f}{4} + 6 = 16$ **b.** $\dfrac{g}{6} + 4 = 9$ **c.** $\dfrac{r}{10} + 6 = 5$

17. a. $\dfrac{m}{9} - 12 = -10$ **b.** $\dfrac{n}{8} + 5 = 8.5$ **c.** $\dfrac{p}{12} - 1.8 = 3.4$

18. a. $6(x + 8) = 56$ **b.** $7(y - 4) = 35$ **c.** $5(m - 3) = 7$

19. a. $3(2k + 5) = 24$ **b.** $5(3n - 1) = 80$ **c.** $6(2c + 7) = 58$

20. a. $2(x - 5) + 3(x - 7) = 19$ **b.** $3(x + 5) - 5(x - 1) = 12$ **c.** $3(2x - 7) - (x + 3) = -60$

`WE11b` For questions **21** to **24**, solve the following equations.

21. a. $\dfrac{3k}{5} = 15$ **b.** $\dfrac{9m}{8} = 18$ **c.** $\dfrac{7p}{10} = -8$

22. a. $\dfrac{8u}{11} = -3$ **b.** $\dfrac{11x}{4} = 2$ **c.** $\dfrac{4v}{15} = 0.8$

23. a. $\dfrac{x - 5}{3} = 7$ **b.** $\dfrac{2m + 1}{3} = -3$ **c.** $\dfrac{3w - 1}{4} = 6$

24. a. $\dfrac{t - 5}{2} = 0$ **b.** $\dfrac{6 - x}{3} = -1$ **c.** $\dfrac{3n - 5}{4} = -6$

25. `MC` **a.** The solution to the equation $\dfrac{p}{5} + 2 = 7$ is:

   A. $p = 5$ **B.** $p = 25$ **C.** $p = 45$ **D.** $p = 10$

   b. If $5h + 8 = 53$, then h is equal to:

   A. $\dfrac{1}{5}$ **B.** 12.2 **C.** 225 **D.** 9

   c. The exact solution to the equation $14x = 75$ is:

   A. $x = 5.357\,142\,857$ **B.** $x = 5.357$ (to 3 decimal places)

   C. $x = 5\dfrac{5}{14}$ **D.** $x = 5.4$

For questions **26** to **29**, solve the following equations.

26. a. $-5h = 10$ **b.** $2 - d = 3$ **c.** $5 - p = -2$ **d.** $-7 - x = 4$

27. a. $-6t = -30$ **b.** $-\dfrac{v}{5} = 4$ **c.** $-\dfrac{r}{12} = \dfrac{1}{4}$ **d.** $-4g = 3.2$

28. a. $6 - 2x = 8$ **b.** $10 - 3v = 7$ **c.** $9 - 6l = -3$ **d.** $-3 - 2g = 1$

29. a. $-5 - 4t = -17$ **b.** $-\dfrac{3e}{5} = 14$ **c.** $-\dfrac{k}{4} - 3 = 6$ **d.** $-\dfrac{4f}{7} + 1 = 8$

`WE12a` For questions **30** to **32**, solve the following equations.

30. a. $6x + 5 = 5x + 7$ **b.** $7b + 9 = 6b + 14$ **c.** $11w + 17 = 6w + 27$

31. a. $8f - 2 = 7f + 5$ **b.** $10t - 11 = 5t + 4$ **c.** $12r - 16 = 3r + 5$

32. a. $12g - 19 = 3g - 31$　　　**b.** $7h + 5 = 2h - 6$　　　**c.** $5a - 2 = 3a - 2$

WE12b For questions **33** to **35**, solve the following equations.

33. a. $5 - 2x = 6 - x$　　　**b.** $10 - 3c = 8 - 2c$　　　**c.** $3r + 13 = 9r - 3$

34. a. $k - 5 = 2k - 6$　　　**b.** $5y + 8 = 13y + 17$　　　**c.** $17 - 3g = 3 - g$

35. a. $14 - 5w = w + 8$　　　**b.** $4m + 7 = 8 - m$　　　**c.** $14 - 5p = 9 - 2p$

WE12c For questions **36** to **38**, solve the following equations.

36. a. $3(x + 5) = 2x$　　　**b.** $8(y + 3) = 3y$　　　**c.** $6(t - 5) = 4(t + 3)$

37. a. $10(u + 1) = 3(u - 3)$　　　**b.** $12(f - 10) = 4(f - 5)$　　　**c.** $2(4r + 3) = 3(2r + 7)$

38. a. $5(2d + 9) = 3(3d + 13)$　　　**b.** $5(h - 3) = 3(2h - 1)$　　　**c.** $2(4x + 1) = 5(3 - x)$

39. MC a. The solution to $8 - 4k = -2$ is:

　A. $k = 2\dfrac{1}{2}$　　　**B.** $k = -2\dfrac{1}{2}$　　　**C.** $k = 1\dfrac{1}{2}$　　　**D.** $k = -1\dfrac{1}{2}$

　b. The solution to $-\dfrac{6n}{5} + 3 = -7$ is:

　A. $n = 3\dfrac{1}{3}$　　　**B.** $n = -3\dfrac{1}{3}$　　　**C.** $n = \dfrac{1}{3}$　　　**D.** $n = 8\dfrac{1}{3}$

　c. The solution to $p - 6 = 8 - 4p$ is:

　A. $p = \dfrac{2}{5}$　　　**B.** $p = 2\dfrac{4}{5}$　　　**C.** $p = 4\dfrac{2}{3}$　　　**D.** $p = \dfrac{2}{3}$

Understanding

40. If the side length of a cube is x cm, then its volume V is given by $V = x^3$. Calculate the side length (correct to the nearest cm) of a cube that has a volume of:

　a. $216 \, \text{cm}^3$　　　　　　　　**b.** $2 \, \text{m}^3$.

41. The surface area of a cube with side length x cm is given by $A = 6x^2$. Determine the side length (correct to the nearest cm) of a cube that has a surface area of:

　a. $37.5 \, \text{cm}^2$　　　　　　　　**b.** $1 \, \text{m}^2$.

42. A pebble is dropped down a well. In time t seconds it falls a distance of d metres, given by $d = 5t^2$.

　a. Calculate the distance the pebble falls in 1 second.
　b. Calculate the time the pebble will take to fall 40 metres. (Answer in seconds correct to 1 decimal place.)

Reasoning

43. The surface area of a sphere is given by the formula $A = 4\pi r^2$, where r is the radius of the sphere.

　a. Determine the surface area of a sphere that has a radius of 5 cm. Show your working.
　b. Evaluate the radius of a sphere that has a surface area equal to $500 \, \text{cm}^2$. (Answer correct to the nearest mm.)

44. Determine the radius of a circle of area $10 \, \text{cm}^2$. Show your working.

45. The volume of a sphere is given by the formula $V = \dfrac{4}{3}\pi r^3$, where r is the radius of the sphere. If the sphere can hold 1 litre of water, determine its radius correct to the nearest mm. Show your working.

Problem solving

46. The width of a room is three-fifths of its length. When the width is increased by 2 metres and the length is decreased by 2 metres, the resultant shape is a square. Determine the dimensions of the room.

47. Four years ago, Leon was one-third of James' age. In six years' time, the sum of their ages will be 60. Determine their current ages.

48. A target board for a dart game has been designed as three concentric circles where each coloured region is the same area. If the radius of the blue circle is r cm and the radius of the outer circle is 10 cm, determine the value of r.

LESSON
2.6 Solving multi-step equations

LEARNING INTENTIONS

At the end of this lesson you should be able to:
- expand brackets and collect like terms in order to solve multi-step equations
- solve equations involving algebraic fractions by determining the LCM of the denominators.

▶ 2.6.1 Equations with multiple brackets

eles-4706

- Equations can be simplified by expanding brackets and collecting like terms before they are solved.

WORKED EXAMPLE 13 Solving equations with brackets

Solve the following equations.
a. $6(x + 1) - 4(x - 2) = 0$ **b.** $7(5 - x) = 3(x + 1) - 10$

THINK	WRITE
a. 1. Write the equation.	**a.** $6(x + 1) - 4(x - 2) = 0$
2. Expand all the brackets. (Be careful with the −4.)	$6x + 6 - 4x + 8 = 0$
3. Collect like terms.	$2x + 14 = 0$
4. Subtract 14 from both sides of the equation.	$2x + 14 - 14 = -14$
	$2x = -14$
5. Divide both sides of the equation by 2 to obtain the value of x.	$x = -7$
b. 1. Write the equation.	**b.** $7(5 - x) = 3(x + 1) - 10$
2. Expand all the brackets.	$35 - 7x = 3x + 3 - 10$
3. Collect like terms.	$35 - 7x = 3x - 7$

▶

4. Create a single pronumeral term by adding $7x$ to both sides of the equation.

$$35 - 7x + 7x = 3x - 7 + 7x$$
$$35 = 10x - 7$$

5. Add 7 to both sides of the equation.

$$35 + 7 = 10x - 7 + 7$$
$$42 = 10x$$

6. Divide both sides of the equation by 10 to solve for x and simplify.

$$\frac{42}{10} = \frac{10x}{10}$$
$$\frac{42}{10} = x$$
$$\frac{21}{5} = x$$

7. Express the improper fraction as a mixed number fraction.

$$4\frac{1}{5} = x$$

8. Rewrite the equation so that x is on the left-hand side.

$$x = 4\frac{1}{5}$$

TI	THINK	DISPLAY/WRITE	CASIO	THINK	DISPLAY/WRITE
a, b.	**a, b.**	**a, b.**	**a, b.**
On a Calculator page, complete the entry lines as: solve $(6(x+1) - 4(x-2) = 0,\ x)$ solve $(7(5-x) = 3(x+1) - 10,\ x)$ Press ENTER after each entry. Convert **b** to a proper fraction.	 $6(x+1) - 4(x-2) = 0$ $\Rightarrow x = -7$ $7(5-x) = 3(x+1) - 10$ $\Rightarrow x = 4\frac{1}{5}$	On the Main screen, complete the entry lines as: solve $(6(x+1) - 4(x-2) = 0,\ x)$ solve $(7(5-x) = 3(x+1) - 10,\ x)$ Press EXE after each entry. Convert **b** to a proper fraction.	 $6(x+1) - 4(x-2) = 0$ $\Rightarrow x = -7$ $7(5-x) = 3(x+1) - 10$ $\Rightarrow = 4\frac{1}{5}$

⊙ 2.6.2 Equations involving algebraic fractions

eles-4707

- To solve an equation containing algebraic fractions, multiply both sides of the equation by the lowest common multiple (LCM) of the denominators. This gives an equivalent form of the equation without fractions.

WORKED EXAMPLE 14 Solving equations with algebraic fractions

Solve the equation $\dfrac{x-5}{3} = \dfrac{x+7}{4}$ **and verify the solution.**

THINK	WRITE
1. Write the equation.	$$\dfrac{x-5}{3} = \dfrac{x+7}{4}$$
2. The LCM is $3 \times 4 = 12$. Multiply both sides of the equation by 12.	$$\dfrac{^4 \cancel{12}\,(x-5)}{^1 \cancel{3}} = \dfrac{^3 \cancel{12}\,(x+7)}{^1 \cancel{4}}$$

3. Simplify the fractions.

$$4(x-5) = 3(x+7)$$

4. Expand the brackets.

$$4x - 20 = 3x + 21$$

5. Subtract $3x$ from both sides of the equation.

$$4x - 20 - 3x = 3x + 21 - 3x$$
$$x - 20 = 21$$

6. Add 20 to both sides of the equation and write the answer.

$$x - 20 + 20 = 21 + 20$$
$$x = 41$$

7. To verify, check that the answer $x = 41$ is true for both the left-hand side (LHS) and the right-hand side (RHS) of the equation by substitution.

Substitute $x = 41$ into the LHS.

$$\text{LHS} = \frac{41 - 5}{3}$$
$$= \frac{36}{3}$$
$$= 12$$

Substitute $x = 41$ into the RHS.

$$\text{RHS} = \frac{41 + 7}{4}$$
$$= \frac{48}{4}$$
$$= 12$$

8. Write the answer.

Because the LHS = RHS, the solution $x = 41$ is correct.

WORKED EXAMPLE 15 Solving equations involving algebraic fractions

Solve the following equations.

a. $\dfrac{5(x+3)}{6} = 4 + \dfrac{3(x-1)}{5}$ b. $\dfrac{4}{3(x-1)} = \dfrac{1}{x+1}$ c. $\dfrac{-2x}{3} + \dfrac{4}{9} = \dfrac{11}{5}(7x-1)$

THINK

WRITE

a. 1. Write the equation.

a.
$$\frac{5(x+3)}{6} = 4 + \frac{3(x-1)}{5}$$

2. The lowest common denominator of 5 and 6 is 30. Write each term as an equivalent fraction with a denominator of 30.

$$\frac{25(x+3)}{30} = \frac{120}{30} + \frac{18(x-1)}{30}$$

3. Multiply each term by 30. This effectively removes the denominator.

$$25(x+3) = 120 + 18(x-1)$$

4. Expand the brackets and collect like terms.

$$25x + 75 = 120 + 18x - 18$$
$$25x + 75 = 102 + 18x$$

5. Subtract $18x$ from both sides of the equation.

$$25x + 75 - 18x = 102 + 18x - 18x$$
$$7x + 75 = 102$$

▶

6. Subtract 75 from both sides of the equation.

$$7x + 75 - 75 = 102 - 75$$
$$7x = 27$$

7. Divide both sides of the equation by 7 to solve for x.

$$\frac{7x}{7} = \frac{27}{7}$$
$$x = \frac{27}{7}$$

8. Express the answer as a mixed number.

$$x = 3\frac{6}{7}$$

b. 1. Write the equation.

b. $$\frac{4}{3(x-1)} = \frac{1}{x+1}$$

2. The lowest common denominator of 3, $x + 1$ and $x - 1$ is $3(x-1)(x+1)$. Write each term as an equivalent fraction with a common denominator of $3(x-1)(x+1)$.

$$\frac{4(x+1)}{3(x-1)(x+1)} = \frac{3(x-1)}{3(x-1)(x+1)}$$

3. Multiply each term by the common denominator.

$$4(x+1) = 3(x-1)$$

4. Expand the brackets.

$$4x + 4 = 3x - 3$$

5. Subtract $3x$ from both sides of the equation.

$$4x + 4 - 3x = 3x - 3 - 3x$$
$$x + 4 = -3$$

6. Subtract 4 from both sides of the equation to solve for x.

$$x + 4 - 4 = -3 - 4$$
$$= -7$$

7. Write the answer.

$$x = -7$$

c. 1. Write the equation.

c. $$\frac{-2x}{3} + \frac{4}{9} = \frac{11}{5}(7x - 1)$$

2. The lowest common denominator of 3 and 9 is 9. Write each term as an equivalent fraction with a denominator of 9.
Express as a single fraction.

$$\frac{-2x}{3} \times \frac{3}{3} + \frac{4}{9} \times \frac{1}{1} = \frac{11}{5}(7x - 1)$$
$$\frac{-6x}{9} + \frac{4}{9} = \frac{11}{5}(7x - 1)$$
$$\frac{-6x + 4}{9} = \frac{11}{5}(7x - 1)$$

3. The lowest common denominator of 5 and 9 is 45. Write each term as an equivalent fraction with a denominator of 45.

$$\frac{-6x + 4}{9} \times \frac{5}{5} = \frac{11}{5}(7x - 1) \times \frac{9}{9}$$
$$\frac{5(-6x + 4)}{45} = \frac{99}{45}(7x - 1)$$
$$5(-6x + 4) = 99(7x - 1)$$

4. Expand the brackets and collect like terms.

$$5(-6x + 4) = 99(7x - 1)$$
$$-30x + 20 = 693x - 99$$

5. Add $30x$ to both sides of the equation.

$$-30x + 20 + 30x = 693x - 99 + 30x$$
$$20 = 723x - 99$$

6. Add 99 to both sides of the equation.	$20 + 99 = 723x - 99 + 99$
	$119 = 723x$
7. Divide both sides of the equation by 723 to solve for x.	$\dfrac{119}{723} = x$
8. Write the answer.	$x = \dfrac{119}{723}$

DISCUSSION

Do the rules for the order of operations apply to algebraic fractions? Explain.

 Resources

 eWorkbook   Topic 2 Workbook (worksheets, code puzzle and project) (ewbk-13282)

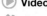 **Video eLesson** Solving linear equations with algebraic fractions (eles-1857)

Interactivities Individual pathway interactivity: Solving multi-step equations (int-4570)
Expanding brackets: Distributive Law (int-3774)

Exercise 2.6 Solving multi-step equations

learn on

| **2.6 Quick quiz** on | **2.6 Exercise** |

Individual pathways

| ■ PRACTISE | ■ CONSOLIDATE | ■ MASTER |
| 1, 4, 7, 10, 13, 17, 20 | 2, 5, 8, 11, 14, 18, 21 | 3, 6, 9, 12, 15, 16, 19, 22 |

Fluency

WE13 For questions **1** to **3**, solve the following linear equations.

1. **a.** $6(4x - 3) + 7(x + 1) = 9$ **b.** $9(3 - 2x) + 2(5x + 1) = 0$

2. **a.** $8(5 - 3x) - 4(2 + 3x) = 3$ **b.** $9(1 + x) - 8(x + 2) = 2x$

3. **a.** $6(4 + 3x) = 7(x - 1) + 1$ **b.** $10(4x + 2) = 3(8 - x) + 6$

WE14 For questions **4** to **6**, solve the following equations and verify the solutions.

4. **a.** $\dfrac{x+1}{2} = \dfrac{x+3}{3}$ **b.** $\dfrac{x-7}{5} = \dfrac{x-8}{4}$ **c.** $\dfrac{x-6}{4} = \dfrac{x-2}{2}$

5. **a.** $\dfrac{8x+3}{5} = 2x$ **b.** $\dfrac{2x-1}{5} = \dfrac{x-3}{4}$ **c.** $\dfrac{4x+1}{3} = \dfrac{x+2}{4}$

6. **a.** $\dfrac{6-x}{3} = \dfrac{2x-1}{5}$ **b.** $\dfrac{8-x}{9} = \dfrac{2x+1}{3}$ **c.** $\dfrac{2(x+1)}{5} = \dfrac{3-2x}{4}$

For questions **7** to **9**, solve the following linear equations.

7. a. $\dfrac{x}{3} + \dfrac{4x}{5} = \dfrac{1}{3}$
   b. $\dfrac{x}{4} - \dfrac{x}{5} = \dfrac{3}{4}$
   c. $\dfrac{x}{4} - \dfrac{4x}{7} = 2$
   d. $\dfrac{-3x}{5} + \dfrac{x}{8} = \dfrac{1}{4}$

8. a. $\dfrac{2x}{3} - \dfrac{x}{6} = -\dfrac{3}{4}$
   b. $\dfrac{5x}{8} - 8 = \dfrac{2x}{3}$
   c. $\dfrac{2}{7} - \dfrac{x}{8} = \dfrac{3x}{8}$
   d. $\dfrac{4}{x} - \dfrac{1}{6} = \dfrac{2}{x}$

9. a. $\dfrac{15}{x} - 4 = \dfrac{2}{x}$
   b. $\dfrac{1}{3} + \dfrac{4}{x} = \dfrac{5}{x}$
   c. $\dfrac{2x-4}{5} + 6 = \dfrac{x}{2}$
   d. $\dfrac{4x-1}{2} - \dfrac{2x+5}{3} = 0$

WE15 For questions **10** to **12**, solve the following linear equations.

10. a. $\dfrac{3(x+1)}{2} + \dfrac{5(x+1)}{3} = 4$
    b. $\dfrac{2(x+1)}{7} + \dfrac{3(2x-5)}{8} = 0$
    c. $\dfrac{2(4x+3)}{5} - \dfrac{6(x-2)}{2} = \dfrac{1}{2}$
    d. $\dfrac{8x}{5} + \dfrac{3}{2} = \dfrac{7}{3}(x+2)$

11. a. $\dfrac{5(7-x)}{2} = \dfrac{2(2x-1)}{7} + 1$
    b. $\dfrac{2(6-x)}{3} = \dfrac{9(x+5)}{6} + \dfrac{1}{3}$
    c. $\dfrac{-5(x-2)}{3} - \dfrac{6(2x-1)}{5} = \dfrac{1}{3}$
    d. $\dfrac{9x}{7} - \dfrac{1}{2} = \dfrac{4}{3}(x-5)$

12. a. $\dfrac{1}{x-1} + \dfrac{3}{x+1} = \dfrac{8}{x+1}$
    b. $\dfrac{3}{x+1} + \dfrac{5}{x-4} = \dfrac{5}{x+1}$
    c. $\dfrac{1}{x-1} - \dfrac{3}{x} = \dfrac{-1}{x-1}$
    d. $\dfrac{2x-1}{3} - \dfrac{5}{2} = \dfrac{-4}{5}(x+7)$

Understanding

13. Last week Maya broke into her money box. She spent one-quarter of the money on a birthday present for her brother and one-third of the money on an evening out with her friends, leaving her with $75.
    Determine the amount of money in her money box.

14. At work Keith spends one-fifth of his time in planning and buying merchandise. He spends seven-twelfths of his time in customer service and one-twentieth of his time training the staff. This leaves him ten hours to deal with the accounts.
    Determine the number of hours he works each week.

15. Last week's school fete was a great success, raising a good deal of money. Three-eighths of the profit came from sales of food and drink, and the market stalls recorded one-fifth of the total.
    A third of the profit came from the major raffle, and the jumping castle raised $1100.
    Determine the amount of money raised at the fete.

16. Lucy had half as much money as Mel, but since Grandma gave them each $20 she now has three-fifths as much.
    Determine the amount of money Lucy has.

Reasoning

17. Answer the following questions and justify your answers.
    a. Determine numbers smaller than 100 that have exactly 3 factors (including 1 and the number itself).
    b. Determine the two numbers smaller than 100 that have exactly 5 factors.
    c. Determine a number smaller than 100 that has exactly 7 factors.

18. To raise money for a charity, a Year 10 class has decided to organise a school lunch.

Tickets will cost $6 each. The students have negotiated a special deal for delivery of drinks and pizzas, and they have budgeted $200 for drinks and $250 for pizzas.
If they raise $1000 or more, they qualify for a special award.

    a. Write an equation to represent the minimum number of tickets they must sell to qualify for the award.
    b. Solve the equation to find the number of tickets they must sell to qualify for the award.
       Explain your answer.

19. If $\dfrac{x+7}{(x+2)(x+3)} \equiv \dfrac{a}{x+2} - \dfrac{4}{x+3}$, explain why a must be equal to 5.

    (*Note:* '$\equiv$' means identically equal to.)

Problem solving

20. Solve for x:
$$\frac{2}{9}(x-1) - \frac{5}{8}(x-2) = \frac{2}{5}(x-4) - \frac{7}{12}$$

21. If $\dfrac{2(4x+3)}{(x-3)(x+7)} \equiv \dfrac{a}{x-3} + \dfrac{b}{x+7}$, determine the values of a and b.

22. If $\dfrac{7x+20}{x^2+7x+12} = \dfrac{a}{x+3} + \dfrac{b}{x+4} + \dfrac{a+b}{x^2+7x+12}$, determine the values of a and b.

LESSON
2.7 Literal equations

LEARNING INTENTIONS

At the end of this lesson you should be able to:
- solve literal equations, which include multiple variables, by changing the subject of an equation to a particular pronumeral
- determine any restriction on a variable in an equation due to limitations imposed by the equation or the context of the question.

▶ 2.7.1 Literal equations

eles-4708

- **Literal equations** are equations that include several pronumerals or variables. Solving literal equations involves changing the subject of the equation to a particular pronumeral.
- A variable is the subject of an equation if it expressed in terms of the other variables.
  In $v = u + at$, the subject of the equation is v as it is written in terms of the variables u, a and t.
- A formula is a literal equation that records an interesting or important real-life relationship.

WORKED EXAMPLE 16 Solving literal equations

Solve the following literal equations for x.

a. $ax^2 + bd = c$　　　　　　　　　　b. $ax = cx + b$

THINK	WRITE
a. 1. Write the equation.	**a.** $\quad ax^2 + bd = c$
2. Subtract bd from both sides of the equation.	$ax^2 + bd - bd = c - bd$ $ax^2 = c - bd$
3. Divide both sides by a.	$\dfrac{ax^2}{a} = \dfrac{c - bd}{a}$ $x^2 = \dfrac{c - bd}{a}$
4. To solve for x, take the square root of both sides. This gives both a positive and a negative result for x.	$x = \pm\sqrt{\dfrac{c - bd}{a}}$
b. 1. Write the equation.	**b.** $\quad ax = cx + b$
2. Subtract cx from both sides.	$ax - cx = cx + b - cx$ $ax - cx = b$
3. Factorise by taking x as a common factor.	$x(a - c) = b$
4. To solve for x, divide both sides by $a - c$.	$\dfrac{x(a - c)}{a - c} = \dfrac{b}{a - c}$ $x = \dfrac{b}{a - c}$

TI	THINK	DISPLAY/WRITE	CASIO	THINK	DISPLAY/WRITE
a, b.	**a, b.**	**a, b.**	**a, b.**

a, b.

In a new problem on a Calculator page, complete the entry lines as:
solve $(a \times x^2 + b \times d = c, x)$
solve $(a \times x = c \times x + b, x)$
Press ENTER after each entry.

$$x = \pm \sqrt{\frac{c - bd}{a}}$$

$$x = \frac{b}{a - c}$$

a, b.

On the Main screen, complete the entry lines as:
solve $(a \times x^2 + b \times d = c, x)$
solve $(a \times x = c \times x + b, x)$
Press EXE after each entry.

$$x = \pm \sqrt{\frac{c - bd}{a}}$$

$$x = \frac{b}{a - c}$$

WORKED EXAMPLE 17 Rearranging to make a variable the subject of an equation

Make b the subject of the formula $D = \sqrt{b^2 - 4ac}$.

THINK	WRITE
1. Write the formula.	$D = \sqrt{b^2 - 4ac}$
2. Square both sides.	$D^2 = b^2 - 4ac$
3. Add $4ac$ to both sides of the equation.	$D^2 + 4ac = b^2 - 4ac + 4ac$ $D^2 + 4ac = b^2$
4. Take the square root of both sides.	$\pm \sqrt{D^2 + 4ac} = b$
5. Make b the subject of the formula by solving for b.	$b = \pm \sqrt{D^2 + 4ac}$

▶ 2.7.2 Restrictions on variables

eles-4709

- Some variables may have implicit restrictions on the values that they may be assigned in an equation or formula.

  For example:

  - if $V = \dfrac{d}{t}$, then t cannot equal zero; otherwise, the value of V would be undefined

  - if $d = \sqrt{x - 9}$, then:
    - the value of d will be restricted to positive values or 0
    - the value of $x - 9$ must be greater than or equal to zero because the square root of a negative number cannot be found.

$$x - 9 \geq 0$$
$$x \geq 9 \quad \text{(Hence, } x \text{ must be greater than or equal to 9.)}$$

- Other restrictions may arise once a formula is rearranged.

  For example, if we look at the formula $V = ls^2$, there are no restrictions on the values that the variables l and s can be assigned. (However, the sign of V must always be the same as the sign of l because s^2 is always positive.)

  If the formula is transposed to make s the subject, then:

$$V = ls^2$$
$$\frac{V}{l} = s^2$$
$$\text{or } s = \pm\sqrt{\frac{V}{l}}$$

  This shows the restrictions that $l \neq 0$ and $\frac{V}{l} \geq 0$.

- If the formula $V = ls^2$ represents the volume of the rectangular prism shown, additional restrictions become evident: the variables l and s represent lengths and must be positive numbers.

  Hence, when we make s the subject we get $s = \sqrt{\frac{V}{l}}$.

WORKED EXAMPLE 18 Identifying restrictions on variables

List any restrictions on the variables in the equations below.

a. **The literal equation:** $x = \dfrac{100\sqrt{y+4}}{z-10}$

b. **The area of a triangle:** $A = \dfrac{bh}{2}$, **where** $b =$ **base length and** $h =$ **height**

THINK	WRITE
a. We cannot substitute a negative value into a square root. This affects the possible values for y.	a. $y + 4 \geq 0$ $\quad y \geq -4$
A fraction is undefined if the denominator is equal to 0. This affects the possible values for z.	$z - 10 \neq 0$ $\quad z \neq 10$
b. In this case the restrictions do not come from the equation but from the context of the equation. Since b and h represent lengths of a shape, they must have positive values or else the shape would not exist.	b. $b > 0$ and $h > 0$ This also implies that $A > 0$.

DISCUSSION

Why is it important to consider restrictions on variables when solving literal equations?

---

 on Resources

 eWorkbook Topic 2 Workbook (worksheets, code puzzle and project) (ewbk-13282)

 Interactivities Individual pathway interactivity: Literal equations (int-4571)
Restrictions on variables (int-6120)

Exercise 2.7 Literal equations

2.7 Quick quiz on

2.7 Exercise

Individual pathways

■ PRACTISE	■ CONSOLIDATE	■ MASTER
1, 4, 7, 10, 13, 16	2, 5, 8, 11, 14, 17	3, 6, 9, 12, 15, 18

Fluency

WE16 For questions **1** to **3**, solve the following literal equations for x.

1. **a.** $\dfrac{ax}{bc} = d$ **b.** $\dfrac{x}{a} - bc = d$ **c.** $\sqrt{x} + n = m$

2. **a.** $acx^2 = w$ **b.** $\dfrac{a}{x} = \dfrac{b}{y}$ **c.** $\dfrac{x+m}{n} = w$

3. **a.** $ab(x+b) = c$ **b.** $\dfrac{a}{x} = \dfrac{b}{c} + m$ **c.** $mx = ay - bx$ **d.** $\dfrac{x}{m} + a = \dfrac{c}{d}$

WE17 For questions **4** to **6**, rearrange each of the following literal equations to make the variable in brackets the subject.

4. **a.** $V = lbh$ $[l]$ **b.** $P = 2l + 2b$ $[b]$

   c. $A = \dfrac{1}{2}bh$ $[h]$ **d.** $c = \sqrt{a^2 + b^2}$ $[a]$

5. **a.** $F = \dfrac{9C}{5} + 32$ $[C]$ **b.** $A = \pi r^2$ $[r]$ **c.** $v = u + at$ $[a]$

   d. $I = \dfrac{PRN}{100}$ $[N]$ **e.** $E = \dfrac{1}{2}mv^2$ $[m]$

6. **a.** $E = \dfrac{1}{2}mv^2$ $[v]$ **b.** $v^2 = u^2 + 2as$ $[a]$ **c.** $v^2 = u^2 + 2as$ $[u]$

   d. $\dfrac{1}{x} = \dfrac{1}{a} + \dfrac{1}{b}$ $[a]$ **e.** $x = \dfrac{nx_1 + mx_2}{m+n}$ $[x_1]$

For questions **7** to **9**, complete the following.

7. **a.** If $c = \sqrt{a^2 + b^2}$, calculate a if $c = 13$ and $b = 5$.

   b. If $A = \dfrac{1}{2}bh$, calculate the value of h if $A = 56$ and $b = 16$.

   c. If $F = \dfrac{9C}{5} + 32$, calculate the value of C if $F = 86$.

8. **a.** If $v = u + at$, calculate the value of a if $v = 83.6$, $u = 15$ and $t = 7$.
   b. If $V = ls^2$, calculate the value of s if $V = 2028$ and $l = 12$.
   c. If $v^2 = u^2 + 2as$, calculate the value of u if $v = 16$, $a = 10$ and $s = 6.75$.

9. **a.** If $A = \dfrac{1}{2}h(a + b)$, calculate the value of a if $A = 360$, $b = 15$ and $h = 18$.

   b. If $x = \dfrac{nx_1 + mx_2}{m+n}$, calculate the value of x_2 if $x = 10$, $m = 2$, $n = 1$ and $x_1 = 4$.

Understanding

10. For each of the following equations:
    i. **WE18** list any restrictions on the variables in the equation
    ii. rearrange the equation to make the variable in brackets the subject
    iii. list any new restrictions on the variables in the equation formed in part ii.

    a. $y = x^2 + 4$ $[x]$
    b. $y = \dfrac{2}{x-3}$ $[x]$
    c. $v = u + at$ $[t]$

11. For each of the following equations:
    i. list any restrictions on the variables in the equation
    ii. rearrange the equation to make the variable in brackets the subject
    iii. list any new restrictions on the variables in the equation formed in part ii.

    a. $c = \sqrt{a^2 + b^2}$ $[b]$
    b. $s = \dfrac{a}{1-r}$ $[r]$
    c. $m = \dfrac{pb + qa}{p+q}$ $[b]$

12. For each of the following equations:
    i. list any restrictions on the variables in the equation
    ii. rearrange the equation to make the variable in brackets the subject
    iii. list any new restrictions on the variables in the equation formed in part ii.

    a. $x = \dfrac{-b \pm \sqrt{b^2 - 4ac}}{2a}$ $[c]$
    b. $m = \dfrac{pb + qa}{p+q}$ $[p]$
    c. $E^2 = (pc)^2 + (mc^2)^2$ $[m]$

Reasoning

13. The area of a trapezium is given by $A = \dfrac{1}{2}(a+b)h$, where a and b are the lengths of the top and the base, and h is the height of the trapezium.

    a. State any restrictions on the variables in the formula. Justify your response.
    b. Make b the subject of the equation.
    c. Determine the length of the base of a trapezium with a height of 4 cm and top of 5 cm and a total area of 32 cm^2.
       Show your working.

14. The volume of a cylinder is given by $V = \pi r^2 h$, where r is the radius and h is the height of the cylinder.

    a. State any restrictions on the values of the variables in the formula.
       Justify your response.
    b. Make r the subject of the formula.
    c. List any new restrictions on the variables in the formula.
       Justify your response.

15. T is the period of a pendulum whose length is l and g is the acceleration due to gravity.

    The formula relating these variables is $T = 2\pi \sqrt{\dfrac{l}{g}}$.

    a. State what restrictions apply to the variables T and l. Justify your response.
    b. Make l the subject of the equation.
    c. Justify if the restrictions stated in part **a** still apply.
    d. Determine the length of a pendulum that has a period of 3 seconds, given that $g = 9.8$ m/s^2.
       Give your answer correct to 1 decimal place.

Problem solving

16. $F = 32 + \dfrac{9}{5}C$ is the formula relating degrees Celsius (C) to degrees Fahrenheit (F).

    a. Transform the equation to make C the subject.
    b. Determine the temperature when degrees Celsius is equal to degrees Fahrenheit.

17. Newton's law of universal gravitation, $F = G\dfrac{m_1 m_2}{r^2}$, tells us the gravitational force acting between two objects with masses m_1 and m_2 at a distance r metres apart. In this equation, G is the gravitational constant and has a fixed value of 6.67×10^{-11}.

    a. Transform the equation to make m_1 the subject.
    b. Evaluate the mass of the Moon, to 2 decimal places, if the value of F between Earth and the Moon is 2.0×10^{20} N and the distance between Earth and the Moon is assumed to be 3.84×10^8 m. Take the mass of Earth to be approximately 5.97×10^{24} kg.

18. Jing Jing and Pieter live on the same main road but Jing Jing lives a kilometres to the east of Pieter. Both Jing Jing and Pieter set off on their bicycles at exactly the same time and both ride in a westerly direction.
    Jing Jing rides at j kilometres per hour and Pieter rides at p kilometres per hour. It is known that $j > p$.
    Determine an equation in terms of a, j and p for the distance Jing Jing has ridden in order to catch up with Pieter.

2.8.1 Topic summary

Algebraic basics

- We can only add and subtract like terms:
  $3x + 6y - 7x + 2z = 6y + 2z - 4x$
- Multiplying algebraic terms: $10x^3y^2 \times 4x^2z = 40x^5y^2z$
- Cancelling down fractions: only cancel what is common to all terms in both the numerator and denominator.

$$\frac{3ac + 5ab}{10abc} = \frac{3ac + 5ab}{10abc} = \frac{3c + 5b}{10bc}$$

- Expanding brackets: $x(a + b) = ax + bx$

Solving equations

- Inverse operations are used to solve equations.
  - Add (+) and subtract (−) are inverses
  - Multiply (×) and divide (÷) are inverses
  - Squares (x^2) and square roots ($\sqrt{x}$) are inverses
- One-step equations can be solved using one inverse operation:
  e.g. $x + 5 = 12$
  $x + 5 - 5 = 12 - 5$
  $x = 7$

Solving complex equations

- Solving two-step and multi-step equations will involving the following.
  - Using inverse operations
  - Expanding brackets
  - Collecting like terms
  - Finding the LCM of algebraic fractions, then multiplying by the LCM to remove all denominators

Literal equations

- Literal equations are equations that involve multiple pronumerals or variables.
- The same processes (inverse operations etc.) are used to solve literal equations.
- Solving a literal equation is the same as making one variable the subject of the equation. This means it is expressed in terms of the other variables.
  e.g. P is the subject of the equation $P = \frac{nRT}{V}$.

  To make T the subject, transpose to get $T = \frac{PV}{nR}$.

Algebraic fractions: + and −

- Fractions can be added and subtracted if they have the same common denominator.
  e.g.
  $$\frac{5}{2x} + \frac{4}{3y} = \frac{15y}{6xy} + \frac{8x}{6xy} = \frac{15y + 8x}{6xy}$$
  Or
  $$\frac{3}{(x+2)} - \frac{2}{(x-2)} = \frac{3(x-2)}{x^2-4} - \frac{2(x+2)}{x^2-4}$$
  $$= \frac{3x - 6 - 2x - 4}{x^2 - 4}$$
  $$= \frac{x - 10}{x^2 - 4}$$

Substitution

- When the numeric value of a pronumeral is known, it can be substituted into an expression to evaluate the expression.
- It can be helpful to place substituted values inside brackets.
  e.g. Evaluate the expression $b^2 - 4ac$ when $a = -3$, $b = -2$ and $c = 4$:
  $$b^2 - 4ac = (-2)^2 - 4 \times (-3) \times (4)$$
  $$= 4 + 48$$
  $$= 52$$

Number laws

- **Commutative Law:** the order in which an operation is carried out does not affect the result. It holds true for:
  - Addition: $x + y = y + x$
  - Multiplication: $x \times y = y \times x$
- **Associative Law:** when calculating two or more numerals, how they are grouped does not affect the result. It holds true for:
  - Addition: $x + (y + z) = (x + y) + z$
  - Multiplication: $x \times (y \times z) = (x \times y) \times z$
- **Identity Law:** an identity is any number the when applied to another number under a specific operation does not change the result.
  - For addition, the identity is 0.
  - For multiplication, the identity is 1.
- **Inverse Law:** an inverse is any number that when applied to another number produce 0 for addition and 1 for multiplication.
  - Under addition, the inverse of x is $-x$ as $x + (-x) = 0$.
  - Under multiplication, the inverse of x is $\frac{1}{x}$ as $x \times \frac{1}{x} = 1$.

ALGEBRA & EQUATIONS

Closure

A set of numbers is closed if under a specific operation, the result produced is also an element of the set.
e.g. For integers, multiplication is closed as the product of two integers is always an integer. (i.e. $3 \times (-5) = -15$)
For integers, division is not closed as the quotient of two numbers is often not an integer.
(i.e. $3 \div (-5) = -\frac{3}{5}$)

Algebraic fractions: ×

- When multiplying fractions, multiply the numerators together and the denominators together.
- Cancel any common factors in the numerator and denominator.
  e.g.
  $$\frac{5y}{12x} \times \frac{7x^2}{15z} = \frac{(1)5y}{12x} \times \frac{7x^2}{(3)15z} = \frac{7xy}{36z}$$
  Cancel common factors from top and bottom.
  Write variables in alphabetical order.

Algebraic fractions: ÷

- When dividing two fractions, multiply the first fraction by the reciprocal of the second.
- The reciprocal of $\frac{a}{b}$ is $\frac{b}{a}$.
  e.g.
  $$\frac{10x^2}{33z} \div \frac{6x^2}{11y} = \frac{(5)10x^2}{(3)33z} \times \frac{(1)11y}{(3)6x^2} = \frac{5y}{9z}$$

2.8.2 Success criteria

Tick the column to indicate that you have completed the lesson and how well you have understood it using the traffic light system.

(**Green:** I understand; **Yellow:** I can do it with help; **Red:** I do not understand)

Lesson	Success criteria	⬤	◯	⬤
2.2	I can evaluate expressions by substituting the numeric values of pronumerals.			
	I can understand and apply the Commutative, Associative, Identity and Inverse Laws.			
2.3	I can determine the lowest common denominator of two or more fractions with pronumerals in the denominators.			
	I can add and subtract fractions involving algebraic expressions.			
2.4	I can cancel factors, including algebraic expressions, that are common to the numerators and denominators of fractions.			
	I can multiply and divide fractions involving algebraic expressions and simplify the result.			
2.5	I can solve one- and two-step equations using inverse operations.			
	I can solve equations with pronumerals on both sides of the equals sign.			
2.6	I can expand brackets and collect like terms in order to solve multi-step equations.			
	I can solve equations involving algebraic fractions by determining the LCM of the denominators.			
2.7	I can solve literal equations, which include multiple variables, by changing the subject of an equation to a particular pronumeral.			
	I can determine any restrictions on a variable in an equation due to limitations imposed by the equation or the context of the question.			

2.8.3 Project

Checking for data entry errors

When entering numbers into an electronic device, or even writing numbers down, errors frequently occur. A common type of error is a transposition error, which occurs when two digits are written in the reverse order. Take the number 2869, for example. With this type of error, it could be written as 8269, 2689 or 2896. A common rule for checking these errors is as follows.

If the difference between the correct number and the recorded number is a multiple of 9, a transposition error has occurred.

We can use algebraic expressions to check this rule. Let the digit in the thousands position be represented by a, the digit in the hundreds position by b, the digit in the tens position by c and the digit in the ones position by d. So the real number can be represented as $1000a + 100b + 10c + d$.

▶

1. If the digits in the ones position and the tens position were written in the reverse order, the number would be $1000a + 100b + 10d + c$. The difference between the correct number and the incorrect one would then be: $1000a + 100b + 10c + d - (1000a + 100b + 10d + c)$.
   a. Simplify this expression.
   b. Is the expression a multiple of 9? Explain.
2. If a transposition error had occurred in the tens and hundreds position, the incorrect number would be $1000a + 100c + 10b + d$. Perform the procedure shown in question 1 to determine whether the difference between the correct number and the incorrect one is a multiple of 9.
3. Consider, lastly, a transposition error in the thousands and hundreds positions. Is the difference between the two numbers a multiple of 9?
4. Comment on the checking rule for transposition errors.

Exercise 2.8 Review questions

learn on

Fluency

1. **MC** Given $E = \frac{1}{2}mv^2$ where $m = 0.2$ and $v = 0.5$, the value of E is:

   A. 0.000 625   B. 0.1   C. 0.005   D. 0.025

2. **MC** The expression $-6d + 3r - 4d - r$ simplifies to:
   A. $2d + 2r$ B. $-10d + 2r$ C. $-10d - 4r$ D. $2d + 4r$

3. **MC** The expression $5(2f + 3) + 6(4f - 7)$ simplifies to:
   A. $34f + 2$ B. $34f - 4$ C. $34f - 27$ D. $34f + 14$

4. **MC** The expression $7(b - 1) - (8 - b)$ simplifies to:
   A. $8b - 9$ B. $8b - 15$ C. $6b - 9$ D. $6b - 15$

5. **MC** If $14p - 23 = 6p - 7$, then p equals:
   A. -3 B. -1 C. 1 D. 2

6. Simplify the following by collecting like terms.
   a. $3c - 5 + 4c - 8$ b. $-3k + 12m - 4k - 9m$
   c. $-d + 3c - 8c - 4d$ d. $6y^2 + 2y + y^2 - 7y$

7. If $A = \frac{1}{2}bh$, determine the value of A if $b = 10$ and $h = 7$.

8. For each of the following, complete the relationship to illustrate the stated law.
   a. $(a + 3b) + 6c =$ _____ Associative Law
   b. $12a - 3b \neq$ _____ Commutative Law
   c. $7p \times$ _____ $=$ _____ Inverse Law
   d. $(x \times 5y) \times 7z =$ _____ Associative Law
   e. $12p + 0 =$ _____ Identity Law
   f. $(3p \div 5q) \div 7r =$ _____ Associative Law
   g. $9d + 11e =$ _____ Commutative Law
   h. $4a \div b \neq$ _____ Commutative Law

9. Determine the values of the following expressions given the natural number values $x = 12$, $y = 8$ and $z = 4$. Comment on whether the Closure Law holds for each of the expressions when the values are substituted.

   a. $x \times y$
   b. $z \div x$
   c. $y - x$

10. Simplify the following expressions.

    a. $\dfrac{5y}{3} - \dfrac{y}{2}$
    b. $\dfrac{x+4}{5} + \dfrac{x+2}{2}$
    c. $\dfrac{5}{3x} - \dfrac{1}{5x}$
    d. $\dfrac{x-1}{x+3} + \dfrac{2x-5}{x+2}$

11. Simplify the following expressions.

    a. $\dfrac{y}{4} \times \dfrac{32}{x}$
    b. $\dfrac{20y}{7x} \times \dfrac{35z}{16y}$
    c. $\dfrac{x+6}{(x+1)(x+3)} \times \dfrac{5(x+1)}{x+6}$

    d. $\dfrac{25}{x} \div \dfrac{30}{x}$
    e. $\dfrac{xy}{5} \div \dfrac{10x}{y}$
    f. $\dfrac{2x}{(x+8)(x-1)} \div \dfrac{9x+1}{x+8}$

12. Solve the following equations.

    a. $p - 20 = 68$
    b. $s - 0.56 = 2.45$
    c. $3b = 48$

    d. $\dfrac{r}{7} = -5$
    e. $\sqrt{x} = 12$
    f. $2(x+5) = -3$

    g. $\dfrac{y}{4} - 3 = 12$
    h. $a^2 = 36$
    i. $5 - k = -7$

13. Solve the following equations.
    a. $42 - 7b = 14$
    b. $12t - 11 = 4t + 5$
    c. $2(4p - 3) = 2(3p - 5)$

14. Solve the following equations.
    a. $5(x - 2) + 3(x + 2) = 0$
    b. $7(5 - 2x) - 3(1 - 3x) = 1$
    c. $5(x + 1) - 6(2x - 1) = 7(x + 2)$
    d. $8(3x - 2) + (4x - 5) = 7x$
    e. $7(2x - 5) - 4(x + 20) = x - 5$
    f. $3(x + 1) + 6(x + 5) = 3x + 40$

15. Solve the following equations.
    a. $\dfrac{x}{2} + \dfrac{x}{5} = \dfrac{3}{5}$
    b. $\dfrac{x}{3} - \dfrac{x}{5} = 3$
    c. $-\dfrac{1}{21} = \dfrac{x}{7} - \dfrac{x}{6}$

    d. $\dfrac{3}{x} + \dfrac{2}{5} = \dfrac{5}{x}$
    e. $\dfrac{2x-3}{2} - \dfrac{3}{5} = \dfrac{x+3}{5}$
    f. $\dfrac{2(x+2)}{3} = \dfrac{3}{7} + \dfrac{5(x+1)}{3}$

16. a. Make x the subject of $bx + cx = \dfrac{d}{2}$.
    b. Make r the subject of $V = \dfrac{4}{3}\pi r^3$.

Problem solving

17. A theatre production is in town and many parents are taking their children. An adult ticket costs $15 and a child's ticket costs $8. Every child must be accompanied by an adult, and each adult can have no more than 4 children with them. It costs the company $12 per adult and $3 per child to run the production. There is a seating limit of 300 people and all tickets are sold.
    a. Determine how much profit the company makes on each adult ticket and on each child's ticket.
    b. To maximise profit, the company should sell as many children's tickets as possible. Of the 300 available seats, determine how many should be allocated to children if there is a maximum of 4 children per adult.
    c. Using your answer to part b, determine how many adults would make up the remaining seats.
    d. Construct an equation to represent the profit that the company can make depending on the number of children and adults attending the production.
    e. Substitute your values to calculate the maximum profit the company can make.

18. You are investigating prices for having business cards printed for your new games store. A local printing company charges a flat rate of $250 for the materials used and $40 per hour for labour.
    a. If h is the number of hours of labour required to print the cards, construct an equation for the cost of the cards, C.
    b. You have budgeted $1000 for the printing job. Determine the number of hours of labour you can afford. Give your answer to the nearest minute.
    c. The printer estimates that they can print 1000 cards per hour of labour. Evaluate the number of cards that will be printed with your current budget.
    d. An alternative to printing is photocopying. The company charges 15 cents per side for the first 10 000 cards and then 10 cents per side for the remaining cards. Justify which is the cheaper option for 18 750 single-sided cards and by how much.

19. A scientist tried to use a mathematical formula to predict people's moods based on the number of hours of sleep they had the previous night. One formula that he used was what he called the 'grumpy formula', $g = 0.16(h - 8)^2$, which was valid on a 'grumpy scale' from 0 to 10 (least grumpy to most grumpy).

    a. Calculate the number of hours of sleep needed to not be grumpy.
    b. Evaluate the grumpy factor for somebody who has had:
        i. 4 hours of sleep
        ii. 6 hours of sleep
        iii. 10 hours of sleep.
    c. Determine the number of hours of sleep needed to be most grumpy.

    Another scientist already had his own grumpy formula and claims that the scientist above stole his idea and has just simplified it. The second scientist's grumpy formula was

    $$g = \frac{0.16\,(h - 8)}{8 - h} \times \frac{2\,(8 - h)}{3\,(h - 8)} \div \frac{2h}{3(h - 8)^2}$$

    d. Write the second scientist's formula in simplified form.
    e. Are the second scientist's claims justified? Explain.

 To test your understanding and knowledge of this topic, go to your learnON title at www.jacplus.com.au and complete the **post-test**.

Answers

Topic 2 Algebra and equations

2.1 Pre-test

1. 12
2. $c = \pm 5$
3. A
4. $\dfrac{y}{30}$
5. D
6. B
7. $\dfrac{-2x}{3y}$
8. $100\,\text{cm}$
9. $25\dfrac{1}{2}$
10. $x = -16$
11. $a = -512$
12. $\$24\,000$
13. C
14. A
15. D

2.2 Substitution

1. a. 5   b. 2   c. 0   d. 6
2. a. -17 b. 3   c. 30   d. 12
3. a. -12 b. 27   c. 30   d. -5
4. a. -11 b. -1 c. 1
5. a. 30   b. -24 c. 36
6. a. -125 b. 1   c. 15
7. a. $\dfrac{7}{12}$ b. $-\dfrac{1}{12}$ c. $\dfrac{1}{12}$
   d. $1\dfrac{1}{3}$ e. $\dfrac{1}{576}$ f. 48
8. a. 9   b. -9 c. 9
   d. 18   e. -18 f. 36
9. a. 9   b. -9 c. 9
   d. 18   e. -18 f. 36
10. a. 17   b. 30   c. 8
11. a. 4   b. 1.5   c. 68
12. a. 46   b. 113.1   c. 5
    d. 624.6
13. a. D   b. C   c. B
14. $3.9\,\text{cm}$
15. $65.45\,\text{cm}^3$
16. $361\,\text{m}$
17. a. -1; in this case, addition is closed on integers.
    b. -1; in this case, subtraction is closed on integers.
    c. 2; in this case, multiplication is closed on integers.

18. a. -1; in this case, division is closed on integers.
    b. -2; in this case, subtraction is closed on integers.
    c. $-\dfrac{1}{2}$; in this case, division is not closed on integers.
19. a. $(a + 2b) + 4c = a + (2b + 4c)$
    b. $(x \times 3y) \times 5c = x \times (3y \times 5c)$
    c. $2p \div q \neq q \div 2p$
    d. $5d + q = q + 5d$
20. a. 10; in this case, addition is closed on natural numbers.
    b. -4; in this case, subtraction is not closed on natural numbers.
    c. 12; in this case, multiplication is closed on natural numbers.
    d. $\dfrac{4}{3}$; in this case, division is not closed on natural numbers.
    e. -2; in this case, subtraction is not closed on natural numbers.
    f. 4; in this case, division is closed on natural numbers.
21. a. $3z + 0 = 0 + 3z = 3z$
    b. $2x \times \dfrac{1}{2x} = \dfrac{1}{2x} \times 2x = 1$
    c. $(4x \div 3y) \div 5z \neq 4x \div (3y \div 5z)$
    d. $3d - 4y \neq 4y - 3d$
22. a. $s = 71.875$ metres   b. $t = 4$ seconds
23. $m = \dfrac{4}{5}$
24. 2.5 seconds

2.3 Adding and subtracting algebraic fractions

1. a. $\dfrac{26}{21}$ or $1\dfrac{5}{21}$ b. $\dfrac{49}{72}$ c. 1
2. a. $\dfrac{17}{99}$ b. $\dfrac{1}{35}$ c. $\dfrac{6 - 5x}{30}$
3. a. $\dfrac{15x - 4}{27}$ b. $\dfrac{15 - 16x}{40}$ c. $\dfrac{15 - 2x}{3x}$
4. a. $\dfrac{5y}{12}$ b. $-\dfrac{3y}{40}$ c. $\dfrac{13x}{12}$ d. $\dfrac{31x - 5}{10}$
5. a. $\dfrac{3w}{28}$ b. $-\dfrac{y}{5}$ c. $\dfrac{89y}{35}$ d. $\dfrac{11x + 9}{12}$
6. a. $\dfrac{7x + 17}{10}$ b. $\dfrac{7x + 30}{12}$
   c. $\dfrac{2x - 11}{30}$ d. $\dfrac{47x - 22}{15}$
7. a. $\dfrac{5}{8x}$ b. $\dfrac{5}{12x}$ c. $\dfrac{38}{21x}$
8. a. $\dfrac{8}{3x}$ b. $\dfrac{7}{24x}$ c. $\dfrac{9}{20x}$
9. a. $\dfrac{37}{100x}$ b. $\dfrac{51}{10x}$ c. $-\dfrac{1}{6x}$

10. a. $\dfrac{3x^2 + 14x - 4}{(x+4)(x-2)}$ b. $\dfrac{2x^2 + 3x + 25}{(x+5)(x-1)}$

c. $\dfrac{2x^2 + 6x - 10}{(2x+1)(x-2)}$ d. $\dfrac{4x^2 - 17x - 3}{(x+1)(2x-7)}$

11. a. $\dfrac{7x^2 + x}{(x+7)(x-5)}$ b. $\dfrac{2x^2 + 6x + 7}{(x+1)(x+4)}$

c. $\dfrac{-x^2 + 7x + 15}{(x+1)(x+2)}$ d. $\dfrac{x-7}{(x+3)(x-2)}$

12. a. $\dfrac{x^2 + 3x + 9}{(x+2)(3x-1)}$ b. $\dfrac{5 - 5x}{(x-1)(1-x)} = \dfrac{5}{x-1}$

c. $\dfrac{3x + 7}{(x+1)^2}$ d. $\dfrac{3x - 4}{(x-1)^2}$

13. a. The student transcribed the denominator incorrectly and wrote $(x+2)$ instead of $(x-2)$ in line 2.
Also, the student forgot that multiplying a negative number by a negative number gives a positive number. Line 3 should have $+3$ in the numerator, not -1. They didn't multiply.

b. $\dfrac{x^2 - 5x + 3}{(x-1)(x-2)}$

14. a. -1 b. $\dfrac{-1}{(x-2)}$

15. a. $\dfrac{9}{(x-3)^2}$ b. $\dfrac{4}{(x-2)^3}$

16. a. $\dfrac{4x^2 + 17x + 17}{(x+2)(x+1)(x+3)}$ b. $\dfrac{7x^2 - 20x + 4}{(x-1)(x+2)(x-4)}$

17. a. $\dfrac{4x^2 + 17x + 19}{(x+1)(x+3)(x+2)}$ b. $\dfrac{2(2x^2 - 9x + 25)}{(x-4)(x-1)(x+3)}$

c. The lowest common denominator may not always be the product of the denominators. Each fraction must be multiplied by the correct multiple.

18. $a = 4$

19. $\dfrac{4(x-1)}{(x+3)(x+4)(x-2)}$

20. $\dfrac{2(x-1)}{(x-7)(x-4)}$

21. $\dfrac{8(x-1)}{(x-4)(x+3)}$

5. a. $\dfrac{2x}{(x+1)^2}$ b. $\dfrac{x+1}{2(2x-3)}$

c. $\dfrac{a}{10(a+3)}$ d. $\dfrac{35d}{8(d-3)}$

6. a. $\dfrac{9}{32x^2(x-2)}$ b. $\dfrac{3x}{10(x-1)}$

7. a. $\dfrac{3}{5}$ b. $\dfrac{2}{9}$ c. $\dfrac{1}{3}$ d. 3

8. a. $\dfrac{1}{25}$ b. $\dfrac{35}{6}$ or $5\dfrac{5}{6}$

c. $\dfrac{4y^2}{7}$ d. $\dfrac{2y^2}{25}$

9. a. $\dfrac{8y^2}{9}$ b. $\dfrac{32xy}{15}$ c. $\dfrac{2}{3}$ d. y^2

10. a. $\dfrac{9}{(3x-7)(x+3)}$ b. $\dfrac{1}{(x+2)(x-9)}$

11. a. $\dfrac{21(x-3)}{x+5}$ b. $\dfrac{13}{9(x-4)(x+1)}$

12. a. $\dfrac{2(x-4)^3}{(x+3)^2}$ b. $\dfrac{(x+2)^2}{(x+3)(x-3)}$

13. a. $\dfrac{15}{(x+2)}$ b. $\dfrac{12}{x^2}$

14. a. $\dfrac{(x-4)(x-5)}{(x+3)(x+2)^2}$ b. $\dfrac{2x(x-3)}{3(x-5)}$

15. a. $\dfrac{(x-5)}{(x-4)}$ b. $\dfrac{(x+4)(x+6)}{x^2}$

16. Yes, because all of the fractions have the same denominator and therefore can be added together.

17. No, x and z are not common to all terms so cannot be cancelled down.

18. a. -1
b. $4 - x$ has been considered to be the same as $x - 4$, which is incorrect.

19. 1

20. $\dfrac{(x+1)^2}{x^2 + 1}$

21. $\dfrac{-1}{x}$

2.4 Multiplying and dividing algebraic fractions

1. a. $\dfrac{4x}{y}$ b. $\dfrac{3x}{y}$ c. $\dfrac{4y}{x}$ d. $\dfrac{9x}{4y}$

2. a. $\dfrac{-5x}{4y}$ b. $\dfrac{3w}{2x}$ c. $\dfrac{6z}{7x}$ d. $\dfrac{2z}{7x}$

3. a. $\dfrac{-3x}{2y}$ b. $\dfrac{5}{24}$ c. $\dfrac{12z}{x}$ d. $\dfrac{-x}{6w}$

4. a. $\dfrac{2}{3x-2}$ b. $\dfrac{5}{x-3}$ c. $\dfrac{9}{2(x-6)}$ d. $\dfrac{1}{x+3}$

2.5 Solving simple equations

1. a. $a = 24$ b. $k = 121$ c. $g = 2.9$

2. a. $r = 3$ b. $h = 0.26$ c. $i = -2$

3. a. $t = 5$ b. $q = \dfrac{1}{6}$ c. $x = 0$

4. a. $f = 12$ b. $i = -60$ c. $z = -7$

5. a. $v = 7$ b. $w = -5\dfrac{1}{3}$ c. $k = 10$

6. a. $a = 0.425$ b. $m = 16\dfrac{5}{8}$ c. $y = 21\dfrac{1}{2}$

7. a. $t = 100$ **b.** $y = \pm 17$ **c.** $q = 6.25$

8. a. $f = \pm 1.2$ **b.** $h = \dfrac{16}{49}$ **c.** $p = \pm \dfrac{3}{8}$

9. a. $g = \dfrac{225}{484}$ **b.** $j = \pm \dfrac{14}{31}$ **c.** $a = \pm 1\dfrac{2}{3}$

10. a. $t = 25$ **b.** $x = \pm 6$ **c.** $m = 16$
   d. $t = \pm 3$ **e.** $t = \pm 10$ **f.** $m = 25$

11. a. $x = 8$ **b.** $x = -3$ **c.** $m = \dfrac{1}{8}$

   d. $x = \dfrac{3}{4}$ **e.** $m = 0.008$ **f.** $w = 2\dfrac{1}{2}$

12. a. $x = -1$ **b.** $x = -2$ **c.** $m = 1$
   d. $w = -512$ **e.** $t = 125$ **f.** $x = 2$

13. a. $a = 4$ **b.** $b = 6$ **c.** $i = 3$

14. a. $f = 9$ **b.** $q = 1\dfrac{1}{8}$ **c.** $r = 5\dfrac{2}{5}$

15. a. $s = 4\dfrac{5}{6}$ **b.** $t = 9\dfrac{4}{5}$ **c.** $a = -7\dfrac{1}{2}$

16. a. $f = 40$ **b.** $g = 30$ **c.** $r = -10$

17. a. $m = 18$ **b.** $n = 28$ **c.** $p = 62.4$

18. a. $x = 1\dfrac{1}{3}$ **b.** $y = 9$ **c.** $m = 4\dfrac{2}{5}$

19. a. $k = 1\dfrac{1}{2}$ **b.** $n = 5\dfrac{2}{3}$ **c.** $c = 1\dfrac{1}{3}$

20. a. $x = 10$ **b.** $x = 4$ **c.** $x = -7\dfrac{1}{5}$

21. a. $k = 25$ **b.** $m = 16$ **c.** $p = -11\dfrac{3}{7}$

22. a. $u = -4\dfrac{1}{8}$ **b.** $x = \dfrac{8}{11}$ **c.** $v = 3$

23. a. $x = 26$ **b.** $m = -5$ **c.** $w = \dfrac{25}{3}$

24. a. $t = 5$ **b.** $x = 9$ **c.** $n = -\dfrac{19}{3}$

25. a. B **b.** D **c.** C

26. a. $h = -2$ **b.** $d = -1$ **c.** $p = 7$
   d. $x = -11$

27. a. $t = 5$ **b.** $v = -20$ **c.** $r = -3$
   d. $g = -0.8$

28. a. $x = -1$ **b.** $v = 1$ **c.** $l = 2$
   d. $g = -2$

29. a. $t = 3$ **b.** $e = -23\dfrac{1}{3}$ **c.** $k = -36$

   d. $f = -12\dfrac{1}{4}$

30. a. $x = 2$ **b.** $b = 5$ **c.** $w = 2$

31. a. $f = 7$ **b.** $t = 3$ **c.** $r = 2\dfrac{1}{3}$

32. a. $g = -1\dfrac{1}{3}$ **b.** $h = -2\dfrac{1}{5}$ **c.** $a = 0$

33. a. $x = -1$ **b.** $c = 2$ **c.** $r = 2\dfrac{2}{3}$

34. a. $k = 1$ **b.** $y = -1\dfrac{1}{8}$ **c.** $g = 7$

35. a. $w = 1$ **b.** $m = \dfrac{1}{5}$ **c.** $p = 1\dfrac{2}{3}$

36. a. $x = -15$ **b.** $y = -4\dfrac{4}{5}$ **c.** $t = 21$

37. a. $u = -2\dfrac{5}{7}$ **b.** $f = 12\dfrac{1}{2}$ **c.** $r = 7\dfrac{1}{2}$

38. a. $d = -6$ **b.** $h = -12$ **c.** $x = 1$

39. a. A **b.** D **c.** B

40. a. 6 cm **b.** 1.26 m

41. a. 2.5 cm **b.** 41 cm

42. a. 5 m **b.** 2.8 s

43. a. 314 cm^2 **b.** 6.3 cm

44. 1.8 cm

45. 6.2 cm

46. The dimensions are 10 m by 6 m.

47. Leon is 14 and James is 34.

48. $\dfrac{10\sqrt{3}}{3}$ cm

2.6 Solving multi-step equations

1. a. $x = \dfrac{20}{31}$ **b.** $x = 3\dfrac{5}{8}$

2. a. $x = \dfrac{29}{36}$ **b.** $x = -7$

3. a. $x = -2\dfrac{8}{11}$ **b.** $x = \dfrac{10}{43}$

4. a. $x = 3$ **b.** $x = 12$ **c.** $x = -2$

5. a. $x = \dfrac{3}{2}$ **b.** $x = -\dfrac{11}{3}$ or $x = -3\dfrac{2}{3}$

   c. $x = \dfrac{2}{13}$

6. a. $x = 3$ **b.** $x = \dfrac{5}{7}$ **c.** $x = \dfrac{7}{18}$

7. a. $x = \dfrac{5}{17}$ **b.** $x = 15$

   c. $x = -6\dfrac{2}{9}$ **d.** $x = -\dfrac{10}{19}$

8. a. $x = -1\dfrac{1}{2}$ **b.** $x = -192$

   c. $x = \dfrac{4}{7}$ **d.** $x = 12$

9. a. $x = 3\dfrac{1}{4}$ **b.** $x = 3$

   c. $x = 52$ **d.** $x = 1\dfrac{5}{8}$

10. a. $x = \dfrac{5}{19}$ b. $x = 1\dfrac{31}{58}$

    c. $x = 4\dfrac{11}{14}$ d. $x = -4\dfrac{7}{22}$

11. a. $x = 5\dfrac{20}{43}$ b. $x = -1\dfrac{10}{13}$

    c. $x = 1\dfrac{2}{61}$ d. $x = 129\dfrac{1}{2}$

12. a. $x = 1.5$ b. $x = -4\dfrac{1}{3}$

    c. $x = 3$ d. $x = -1\dfrac{39}{44}$

13. $180

14. 60 hours

15. $12\,000

16. $60

17. a. $4, 9, 25, 49$ b. $16, 81$ c. 64

18. a. $6x - 450 = 1000$

    b. $241\dfrac{1}{3}$ tickets. This means they need to sell 242 tickets to qualify, as the number of tickets must be a whole number.

19. Sample responses can be found in the worked solutions in the online resources.

20. 4

21. $a = 3, b = 5$

22. $a = -8$ and $b = 15$

2.7 Literal equations

1. a. $x = \dfrac{bcd}{a}$ b. $x = a(d + bc)$ c. $x = (m - n)^2$

2. a. $x = \pm\sqrt{\dfrac{w}{ac}}$ b. $x = \dfrac{ay}{b}$ c. $x = nw - m$

3. a. $x = \dfrac{c}{ab} - b$ b. $x = \dfrac{ac}{b + mc}$

   c. $x = \dfrac{ay}{m + b}$ d. $x = \dfrac{mc - amd}{d}$

4. a. $l = \dfrac{V}{bh}$ b. $b = \dfrac{P - 2l}{2}$

   c. $h = \dfrac{2A}{b}$ d. $a = \pm\sqrt{c^2 - b^2}$

5. a. $C = \dfrac{5}{9}(F - 32)$ b. $r = \pm\sqrt{\dfrac{A}{\pi}}$

   c. $a = \dfrac{v - u}{t}$ d. $N = \dfrac{100I}{PR}$

   e. $m = \dfrac{2E}{v^2}$

6. a. $v = \pm\sqrt{\dfrac{2E}{m}}$ b. $a = \dfrac{v^2 - u^2}{2s}$

   c. $u = \pm\sqrt{v^2 - 2as}$ d. $a = \dfrac{xb}{b - x}$

e. $x_1 = \dfrac{x(m + n) - mx_2}{n}$

7. a. $a = \pm 12$ b. $h = 7$ c. $C = 30$

8. a. $a = 9.8$ b. $s = \pm 13$ c. $u = \pm 11$

9. a. $a = 25$ b. $x_2 = 13$

10. a. i. No restrictions on x
    ii. $x = \pm\sqrt{y - 4}$
    iii. $y \geq 4$
    b. i. $x \neq 3$
    ii. $x = \dfrac{2}{y} + 3$
    iii. $y \neq 0$
    c. i. No restrictions
    ii. $t = \dfrac{v - u}{a}$
    iii. $a \neq 0$

11. a. i. $c \geq 0$
    ii. $b = \pm\sqrt{c^2 - a^2}$
    iii. $|c| \geq |a|$
    b. i. $r \neq 1$
    ii. $r = \dfrac{s - a}{s}$
    iii. $s \neq 0$
    c. i. $p \neq -q$
    ii. $b = \dfrac{m(p + q) - qa}{p}$
    iii. $p \neq 0$

12. a. i. $a \neq 0, b^2 \geq 4ac$
    ii. $c = \dfrac{b^2 - (2ax + b)^2}{4a}$ or $c = -ax^2 - bx$
    iii. No new restrictions
    b. i. $p \neq -q$
    ii. $p = \dfrac{q(a - m)}{m - b}$
    iii. $m \neq b$
    c. i. $E > (pc)$
    ii. $m = \dfrac{\sqrt{E^2 - (pc)^2}}{c^2}$
    iii. $c \neq 0$

13. a. No restrictions, but all values must be positive for the trapezium to exist.
    b. $b = \dfrac{2A}{h} - a$
    c. $b = 11$ cm

14. a. No restrictions, but all values must be positive for a cylinder to exist.
    b. $r = \sqrt{\dfrac{V}{\pi h}}$
    c. $h \neq 0$, no new restrictions

15. a. T and l must be greater than zero.

b. $l = \dfrac{T^2 g}{4\pi^2}$

c. The restrictions still hold.

d. 2.2 m

16. a. $C = \dfrac{5}{9}(F - 32)$

b. $-40°$

17. a. $m_1 = \dfrac{Fr^2}{Gm_2}$ b. $7.41 \times 10^{22}\ kg$

18. The distance Jing Jing has ridden is $\dfrac{ja}{j-p}$ kilometres.

Project

1. a. $9(c - d)$

b. Yes, this is a multiple of 9 as the number that multiples the brackets is 9.

2. $90(b - c)$; 90 is a multiple of 9, so the difference between the correct and incorrect one is a multiple of 9.

3. $900(a - b)$; again, 900 is a multiple of 9.

4. If two adjacent digits are transposed, the difference between the correct number and the transposed number is a multiple of 9.

2.8 Review questions

1. D

2. B

3. C

4. B

5. D

6. a. $7c - 13$ b. $-7k + 3m$
   c. $-5c - 5d$ d. $7y^2 - 5y$

7. 35

8. a. $(a + 3b) + 6c = a + (3b + 6c)$

b. $12a - 3b \ne 3b - 12a$

c. $7p \times \dfrac{1}{7p} = \dfrac{1}{7p} \times 7p = 1$

d. $(x \times 5y) \times 7z = x \times (5y \times 7z)$

e. $12p + 0 = 0 + 12p = 12p$

f. $(3p \div 5q) \div 7r \ne 3p \div (5q \div 7r)$

g. $9d + 11e = 11e + 9d$

h. $4a \div b \ne b \div 4a$

9. a. 96; in this case, multiplication is closed on natural numbers.

b. $\dfrac{1}{3}$; in this case, division is not closed on natural numbers.

c. -4; in this case, subtraction is not closed on natural numbers.

10. a. $\dfrac{7y}{6}$ b. $\dfrac{7x + 18}{10}$

c. $\dfrac{22}{15x}$ d. $\dfrac{3x^2 + 2x - 17}{(x + 3)(x + 2)}$

11. a. $\dfrac{8y}{x}$ b. $\dfrac{25z}{4x}$ c. $\dfrac{5}{x + 3}$

d. $\dfrac{5}{6}$ e. $\dfrac{y^2}{50}$ f. $\dfrac{2x}{(x - 1)(9x + 1)}$

12. a. $p = 88$ b. $s = 3.01$ c. $b = 16$

d. $r = -35$ e. $x = 144$ f. $x = -\dfrac{13}{2}$

g. $y = 60$ h. $a = \pm 6$ i. $k = 12$

13. a. $b = 4$ b. $t = 2$ c. $p = -2$

14. a. $x = \dfrac{1}{2}$ b. $x = 6\dfrac{1}{5}$ c. $x = -\dfrac{3}{14}$

d. $x = 1$ e. $x = 12\dfrac{2}{9}$ f. $x = 1\dfrac{1}{6}$

15. a. $x = \dfrac{6}{7}$ b. $x = 22\dfrac{1}{2}$ c. $x = 2$

d. $x = 5$ e. $x = 3\dfrac{3}{8}$ f. $x = -\dfrac{16}{21}$

16. a. $x = \dfrac{d}{2(b + c)}$ b. $r = \sqrt[3]{\dfrac{3V}{4\pi}}$

17. a. $3 per adult ticket; $5 per child's ticket

b. 240

c. 60

d. $P = 3a + 5c$, where $a =$ number of adults and $c =$ number of children.

e. $1380

18. a. $C = 250 + 40h$

b. 18 hours 45 minutes

c. 18 750

d. Printing is the cheaper option by $1375.

19. a. 8 hours

b. i. 2.56

ii. 0.64

iii. 0.64

c. 0.094 hours or 15.9 hours

d. $g = \dfrac{0.16(h - 8)^2}{h}$

e. No, the formula is not the same.

3 Networks

LESSON SEQUENCE

LESSON
3.1 Overview

Why learn this?

Networks are used to show how things are connected. The study of networks and decision mathematics is a branch of graph theory. Early studies of graph theory were done in response to difficult problems. One of the most famous problems in graph theory is the Four Colour Theorem, which asked 'Is it true that any map drawn in the plane may have its regions coloured with four colours, in such a way that any two regions having a common border have different colours?' The mathematician Francis Guthrie first posed the Four Colour Theorem in 1852. In 1976, mathematicians at the University of Illinois were able to harness their knowledge of graph theory and the power of computer algorithms to test all possible combinations and prove the theorem.

Leonhard Euler

The mathematician Leonhard Euler is usually credited with being the founder of graph theory. He famously used it to solve a problem known as the 'Seven Bridges of Königsberg'. For a long time it had been pondered whether it was possible to travel around the European city of Königsberg (now called Kaliningrad) in such a way that the seven bridges would only have to be crossed once each. In the branch of mathematics known as graph theory, diagrams involving points and lines are used as a planning and analysis tool for systems and connections. Applications of graph theory include business efficiency, transportation systems, design projects, building and construction, food chains and communications networks.

Graph theory and networks are used to show how things are connected. Networks are used to organise train timetables and improve the flow of traffic. The network of airports in Australia or the kinship systems of First Nations Australians can be represented using network diagrams. Graph theory and networks have been applied to a wide range of disciplines, from social networks, where they are used to examine the structure of relationships and social identities, to biological networks, which analyse molecular networks.

Hey students! Bring these pages to life online

▶ **Watch videos**

🧩 **Engage with interactivities**

A+ **Answer questions and check solutions**

Find all this and MORE in jacPLUS ▶

Reading content and rich media, including interactivities and videos for every concept

Extra learning resources

Differentiated question sets

Questions with immediate feedback, and fully worked solutions to help students get unstuck

1. Consider the following diagram.

State how many lines directly join to point B.

2. Consider the following diagram.

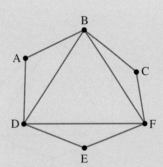

State the number of vertices.

3. Consider the following diagram.

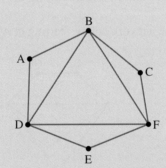

State the number of edges.

4. Consider the following diagram.

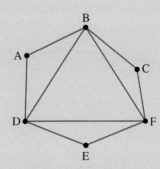

Imagine that the diagram represents walking paths in a park. Each letter represents a viewing platform. Beginning at A, trace a path that you can walk along so that you visit each viewing platform only once.

5. Consider the following diagram.

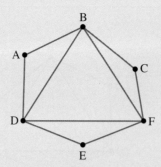

Starting from A, is it possible to walk through the park so that you travel on each path exactly once? If so, list a path you could follow.

6. Consider the following diagram.

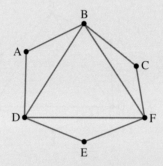

Explain whether the diagram can be traced without lifting the pen off the paper.

7. Determine an Euler circuit for the network below, starting at A.

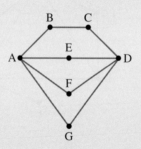

8. Redraw the following planar graph without any intersecting edges

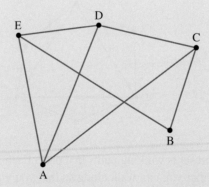

9. Consider the following diagram.

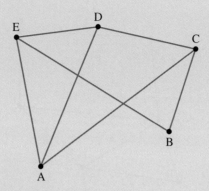

Determine the number of faces of the planar graph.

10. Consider the following diagram.

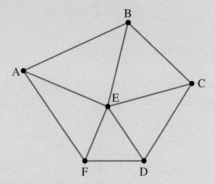

State which vertex in the network shown has the highest degree.

11. In a planar graph, the number of vertices $= 7$ and the number of edges $= 13$. Therefore, calculate the number of faces.

12. Consider the graph shown.

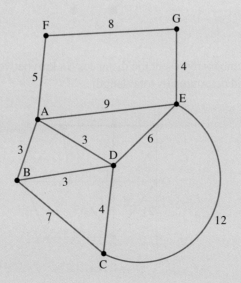

Draw the minimum spanning tree in the graph.

13. Consider the graph shown.

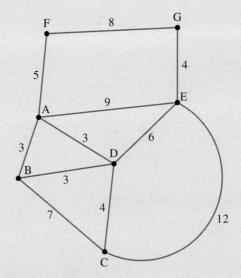

Calculate the distance of the minimum spanning tree.

14. Using the network shown, determine what two edges should be added to the network so that it has both an Euler circuit and a Hamiltonian cycle.

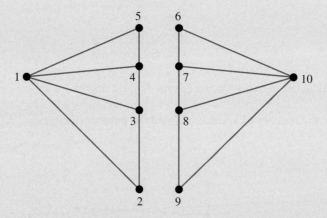

15. In the network shown, numbers represent the distances (in km) between ten towns. Calculate the minimal spanning tree and determine its total length.

LESSON
3.2 Properties of networks

LEARNING INTENTIONS

At the end of this lesson you should be able to:
- draw network diagrams to represent connections
- define edge, vertex, loop and the degree of a vertex
- identify properties and types of graphs, including connected and isomorphic graphs.

▶ 3.2.1 Networks

- A network is a collection of objects that are connected to each other in some specific way.
- The objects are called **vertices** or **nodes** and are represented by points, and the connections are called **edges** and are represented by lines. Although edges are often drawn as straight lines, they don't have to be.
- A diagram of a network can also be called a graph.

> **In a network diagram or graph, the lines are called edges and the points are called vertices (or nodes), with each edge joining a pair of vertices.**

- The network diagram shown has five edges and five vertices.

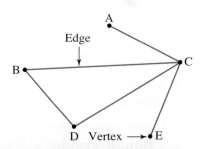

WORKED EXAMPLE 1 Identifying the number of vertices and edges

Count the number of vertices and edges in the network shown.

THINK

1. Vertices are objects that are represented by the points in the network. Count the vertices (black dots) and write the answer.

2. Edges are the lines joining the vertices (points). Count the edges (blue lines) and write the answer.

WRITE/DRAW

There are 5 vertices.

There are 6 edges.

Draw the network that represents the family tree showing Alex, his two parents Brendan and Christine, and his grandparents, David and Evelyn (paternal) and Frank and Ginny (maternal).

THINK	WRITE/DRAW
1. List the objects (people) in the network.	A for Alex, B for Brendan, C for Christine, D for David, E for Evelyn, F for Frank, G for Ginny
2. Draw them using the vertical direction to indicate the generations. Join the various people with lines representing parentage and marriage. D is married to E. F is married to G. D and E are the parents of B. F and G are the parents of C. B is married to C. B and C are the parents of A.	

3.2.2 Types and properties of graphs

eles-6153

- A graph is a series of points and lines that can be used to represent the connections that exist in various settings.

Simple graphs

A simple graph is one in which pairs of vertices are connected by at most one edge.

- The graphs below are examples of simple graphs.

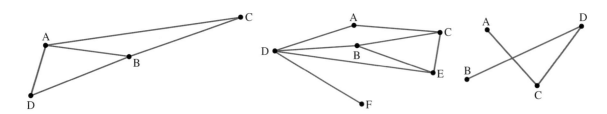

Connected graphs

If it is possible to reach every vertex of a graph by moving along the edges, it is called a connected graph; otherwise, it is a disconnected graph.

- The graph below left is an example of a connected graph, whereas the graph below right is not connected (it is disconnected).

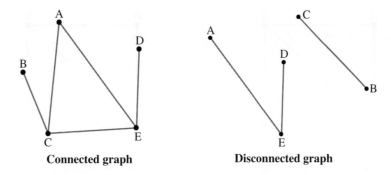

Connected graph   Disconnected graph

A bridge

A bridge is an edge that disconnects the graph when removed.

Note: **A graph can have more than one bridge.**

- In the following graph, the edge AB is a bridge, because if it is removed, the graph would become disconnected.

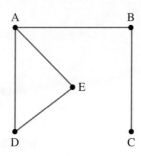

A loop

A loop is an edge that connects a vertex to itself.

Note: **A loop is only counted as one edge.**

- In the following graph, an edge drawn from a vertex D back to itself is known a loop.

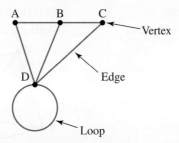

- So far, the graphs you have seen are all examples of **undirected graphs**, where we assume you can move along an edge in both directions.

- Graphs that only allow movement in a specified direction are called **directed graphs** or **digraphs**, and these edges will contain an arrow indicating the direction of travel allowed.

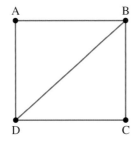

An undirected graph
(travel along an edge in any direction)

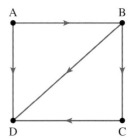

A directed graph
(arrows show the direction you must travel)

WORKED EXAMPLE 3 Drawing a graph to represent the possible ways of travelling

The diagram represents a system of paths and gates in a large park. Draw a graph to represent the possible ways of travelling to each gate in the park.

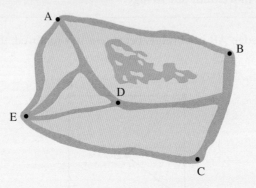

THINK	WRITE/DRAW

1. Identify, draw and label all possible vertices.	Represent each gate as a vertex.
	
	A•
	
	               •B
	
	      D•
	E•
	
	          •C

2. Draw edges to represent all the direct connections between the identified vertices.	Direct pathways exist for A–B, A–D, A–E, B–C, C–E and D–E. 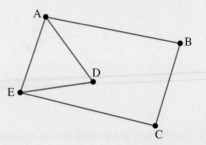

3. Identify all the other unique ways of connecting vertices.

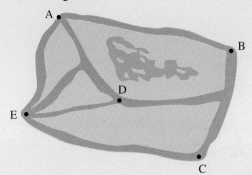

Other unique pathways exist for A–E, D–E, B–D and C–D.

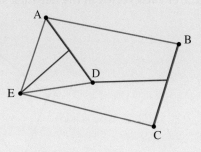

4. Draw the final graph.
That is, there are two ways to get from Gate A to Gate E and so on. The final graph is not a map but should represent the number of ways vertices connect.

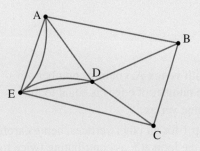

Note: The network graph could look different, but the connections should all be represented.

3.2.3 The degree of a vertex

eles-6154

- The total number of edges that are directly connected to a particular vertex is known as the **degree** of the vertex.
- The notation deg(V) is used, where V represents the vertex.
- If there is a loop at a vertex, it counts twice to the degree of that vertex.
  Remember: A loop counts as one edge but counts twice to the degree of the vertex.

The degree of a vertex

The degree of a vertex in an undirected graph is the number of edges directly connected to that vertex, *except* that a loop at a vertex contributes twice to the degree of that vertex.

- In the graph, deg(A) = 2, deg(B) = 2, deg(C) = 5, deg(D) = 2, deg(E) = 5 and deg(F) = 2.

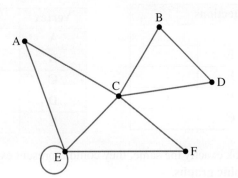

State the degree of each vertex in the network shown

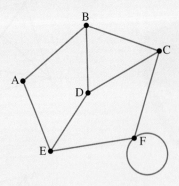

THINK	WRITE
1. Starting with vertex A, count its number of edges. The number of edges is equal to the degree of that vertex.	$\deg(A) = 2$
2. Repeat step 1 for all other vertices, being careful to include the loop at F as contributing twice to the degree.	$\deg(B) = 3$ $\deg(C) = 3$ $\deg(D) = 3$ $\deg(E) = 3$ $\deg(F) = 4$

3.2.4 Isomorphic graphs

- Consider the following graphs.

Graph 1

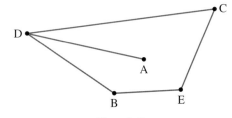

Graph 2

- For the two graphs, the connections for each vertex can be summarised as shown in the table.

Graph 1

Vertex	Connections		
A	D		
B	D	E	
C	D	E	
D	A	B	C
E	B	C	

Graph 2

Vertex	Connections		
A	D		
B	D	E	
C	D	E	
D	A	B	C
E	B	C	

- Although the graphs don't look exactly the same, they could represent exactly the same information. Such graphs are known as **isomorphic graphs**.

Isomorphic graphs

Isomorphic graphs have the same number of vertices and edges, with corresponding vertices having identical degrees and connections.

WORKED EXAMPLE 5 Identifying isomorphic graphs

Confirm whether the following two graphs are isomorphic.

Graph 1

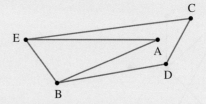

Graph 2

THINK

1. Identify the degree of the vertices for each graph.

2. Identify the number of edges for each graph.

3. Identify the vertex connections for each graph.

4. Comment on the two graphs.

WRITE

Graph	A	B	C	D	E
Graph 1	2	3	2	2	3
Graph 2	2	3	2	2	3

Graph	Edges
Graph 1	6
Graph 2	6

Vertex	Connections, Graph 1 and Graph 2		
A	B	E	
B	A	D	E
C	D	E	
D	B	C	
E	A	B	C

The two graphs are isomorphic as they have the same number of vertices and edges, with corresponding vertices having identical degrees and connections.

3.2 Quick quiz on	3.2 Exercise

Individual pathways

■ PRACTISE	■ CONSOLIDATE	■ MASTER
1, 3, 7, 9, 11, 14, 17	2, 5, 10, 12, 15, 18	4, 6, 8, 13, 16, 19

Fluency

1. **WE1** Count the number of vertices and edges in the following networks.

   a.   b.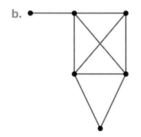

2. Count the number of vertices and edges in the following networks.

   a.   b.   c.   d.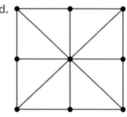

3. **WE2** Draw the network that represents the following family tree.
   Henry and Ida marry and have one child, Jane.
   Jane marries Kenneth and they have one child, Louise.
   Louise marries Mark and they have two children, Neil and Otis.

4. Four towns, Joplin, Amarillo, Flagstaff and Bairstow, are connected to each other as follows:
   Joplin to Amarillo; Joplin to Flagstaff; Amarillo to Bairstow; Amarillo to Flagstaff; Flagstaff to Bairstow.
   Draw the network represented by these connections.

5. **WE3** The diagram shows the plan of a floor of a house. Draw a graph to represent the possible ways of travelling between each room on the floor.

6. Draw a graph to represent the following tourist map.

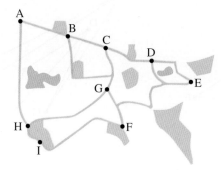

7. WE4 State the degree of each vertex in the network shown.

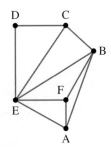

8. Identify the degree of each vertex in the following graphs.

a.

b.

c.

d.

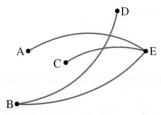

9. WE5 Confirm whether the following pairs of graphs are isomorphic.

a.

Graph 1

Graph 2

b.

Graph 1

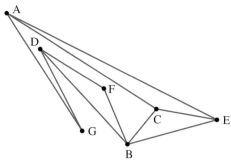

Graph 2

10. Explain why the following pairs of graphs are not isomorphic.

a.

Graph 1

Graph 2

b.

Graph 1

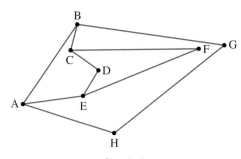

Graph 2

Understanding

11. Identify pairs of isomorphic graphs from the following.

Graph 1

Graph 2

Graph 3

Graph 4

Graph 5

Graph 6

Graph 7

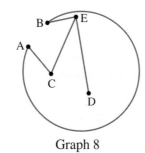

Graph 8

12. Draw a graph of:

a. a simple, connected graph with 6 vertices and 7 edges

b. a simple, connected graph with 7 vertices and 7 edges, where one vertex has degree 3 and five vertices have degree 2

c. a simple, connected graph with 9 vertices and 8 edges, where one vertex has degree 8.

13. Complete the following table for the graphs shown.

	Simple	Connected
Graph 1	Yes	Yes
Graph 2		
Graph 3		
Graph 4		
Graph 5		

Graph 1

Graph 2

Graph 3

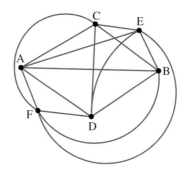

| Graph 4 | Graph 5 |

Reasoning

14. Four streets are connected to each other as follows.
Draw a network represented by these connections.
Princess Street — Bird Avenue
Princess Street — Chatlie Street
Charlie Street — Dundas Street
Princess Street — Dundas Street

15. A round robin tournament occurs when each team plays all other teams once only.
At the beginning of the school year, five schools play a round robin competition in table tennis.

   a. Draw a graph to represent the games played.
   b. State what the total number of edges in the graph indicates.

16. Consider a network of 4 vertices, where each vertex is connected to each of the other 3 vertices with a single edge (no loops, isolated vertices or parallel edges).

   a. List the vertices and edges.
   b. Construct a diagram of the network.
   c. List the degree of each vertex.

Problem solving

17. The diagram shows a map of some of the main suburbs of Beijing.

   a. Draw a graph to represent the shared boundaries between the suburbs.
   b. State which suburb has the highest degree.
   c. State the type of graph.

18. By indicating the passages with edges and the intersections and passage endings with vertices, draw a graph to represent the maze shown.

Maze

19. Jetways Airlines operates flights in South-East Asia.

The table indicates the number of direct flights per day between key cities.

From: To:	Bangkok	Manila	Singapore	Kuala Lumpur	Jakarta	Hanoi	Phnom Penh
Bangkok	0	2	5	3	1	1	1
Manila	2	0	4	1	1	0	0
Singapore	5	4	0	3	4	2	3
Kuala Lumpur	3	1	3	0	0	3	3
Jakarta	1	1	4	0	0	0	0
Hanoi	1	0	2	3	0	0	0
Phnom Penh	1	0	3	3	0	0	0

a. Draw a graph to represent the number of direct flights.
b. State whether this graph would be considered directed or undirected. Explain why.
c. State the number of ways you can travel from Hanoi to Bangkok.

LESSON
3.3 Planar graphs

▶ 3.3.1 Planar networks

eles-6155

- A **planar network** or planar graph is a network in which the edges do not cross each other.
- In some networks it is possible to redraw graphs so that they have no intersecting edges.
  For example, in the graph below it is possible to redraw one of the intersecting edges so that it still represents the same information.

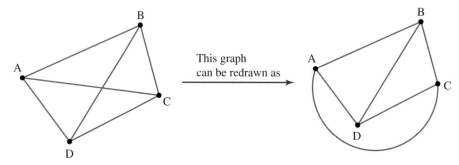

This graph can be redrawn as

WORKED EXAMPLE 6 Redrawing a graph to make it planar

Redraw the following graph so that it has no intersecting edges.

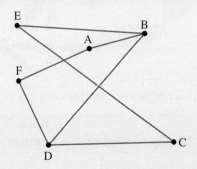

THINK	WRITE/DRAW
1. List all connections in the original graph.	Connections: AB; AF; BD; BE; CD; CE; DF
2. Draw all vertices and any section(s) of the graph that have no intersecting edges.	

3. Draw any further edges that don't create intersections. Start with edges that have the fewest intersections in the original drawing.

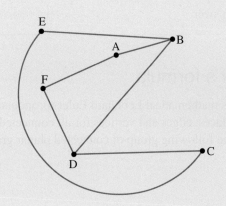

4. Identify any edges yet to be drawn and redraw the graph so that they do not intersect with the other edges.

Connections: ~~AB~~; ~~AF~~; ~~BD~~; BE; ~~CD~~; CE; ~~DF~~

5. Draw the final graph.

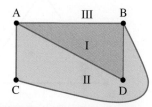

3.3.2 Vertices, edges and faces

eles-6156

- For a planar graph, the edges divide the surrounding space into regions, also known as faces.
- For the network shown, there are three faces (labelled I, II and III). The vertices (A, B, C and D) are not part of any region.
- The space outside the entire network is counted as a face. In this case it is face III.

WORKED EXAMPLE 7 Determining the number of faces in a network

Identify the number of vertices, edges and faces in the network shown.

THINK

1. Label the vertices and count them.

WRITE/DRAW

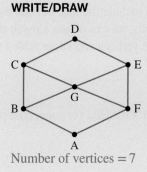

Number of vertices = 7

2. Count the edges. The easiest way to do this is to cross off every edge that has been counted with a small mark. (This will guarantee that no edge is missed and no edge is counted twice.)

Number of edges = 10

3. Colour in each face with a different colour. Count the faces. Do not forget the face outside the network.

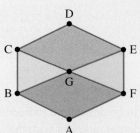

Number of faces = 5

4. Write the answer.

The network has 7 vertices, 10 edges and 5 faces.

▶ 3.3.3 Euler's formula
eles-6157

- The famous mathematical Leonhard Euler (pronounced 'oil-er') discovered the relationship between the number of faces, edges and vertices for all connected planar networks.
- Consider the following group of connected planar graphs:

Graph 1

Graph 2

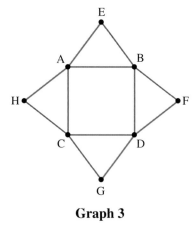

Graph 3

- The number of vertices, edges and faces for each graph is summarised in the following table.

	Vertices	Edges	Faces
Graph 1	3	3	2
Graph 2	4	5	3
Graph 3	8	12	6

Note: Do not forget to include the face outside the network.

- For each of these graphs, we can obtain a result that is well known for any connected planar graph: the difference between the vertices and edges added to the number of faces will always equal 2.

Graph 1: $3 - 3 + 2 = 2$
Graph 2: $4 - 5 + 3 = 2$
Graph 3: $8 - 12 + 6 = 2$

This is known as Euler's formula for connected planar graphs.

Euler's formula

For any connected planar graph:

$$v - e + f = 2$$

where v is the number of vertices, e is the number of edges and f is the number of faces (regions).

WORKED EXAMPLE 8 Applying Euler's formula

Determine the number of faces in a connected planar graph of 7 vertices and 10 edges.

THINK	WRITE
1. Substitute the given values into Euler's formula.	$v - e + f = 2$ $7 - 10 + f = 2$
2. Solve the equation for the unknown value.	$7 - 10 + f = 2$ $f = 2 - 7 + 10$ $f = 5$
3. Write the answer as a sentence.	There will be 5 faces in a connected planar graph with 7 vertices and 10 edges.

 Resources

 eWorkbook Topic 3 Workbook (worksheets, code puzzle and project) (ewbk-13284)

 Interactivity Individual pathway interactivity: Planar graphs (int-9164)

Exercise 3.3 Planar graphs

learn

3.3 Quick quiz **on**	3.3 Exercise

Individual pathways

■ PRACTISE	■ CONSOLIDATE	■ MASTER
1, 4, 8, 12, 15	2, 5, 6, 9, 13, 16	3, 7, 10, 11, 14, 17

Fluency

1. **WE6** Redraw the following graph so that it has no intersecting edges.

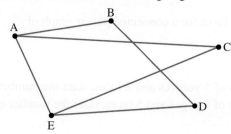

2. Convert the following graph to a planar graph.

3. Redraw the following network diagram so that it is a planar graph.

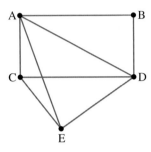

4. **WE7** Identify the number of vertices, edges and faces in the network shown.

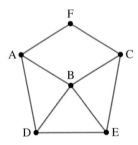

5. Identify the number of vertices, edges and faces in the networks shown.

a.

b.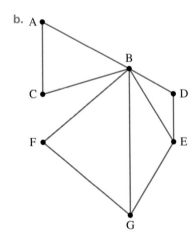

6. **WE8** Determine the number of faces for a connected planar graph of:
   a. 8 vertices and 10 edges
   b. 11 vertices and 14 edges.

7. a. For a connected planar graph of 5 vertices and 3 faces, state the number of edges.
   b. For a connected planar graph of 8 edges and 5 faces, state the number of vertices.

Understanding

8. Redraw the following graph to show that it is planar.

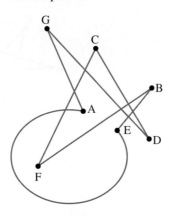

9. Identify which of the following graphs are *not* planar.

Graph 1

Graph 2

Graph 3

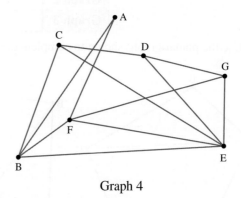

Graph 4

10. Consider the following network.

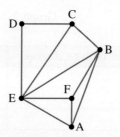

   a. Identify the number of vertices, edges and faces in the shown network.
   b. Confirm Euler's formula for the shown network.

11. For each of the following planar graphs, identify the number of faces.

a.

b.

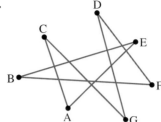

Reasoning

12. Construct a connected planar graph with:

   a. 6 vertices and 5 faces
   b. 11 edges and 9 faces.

13. Use the planar graphs shown to complete the given table.

Graph 1

Graph 2

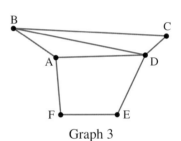

Graph 3

	Total edges	**Total degrees**
Graph 1		
Graph 2		
Graph 3		

14. a. Use the planar graphs shown to complete the given table.

Graph 1

Graph 2

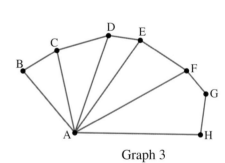

Graph 3

	Total vertices of even degree	**Total vertices of odd degree**
Graph 1		
Graph 2		
Graph 3		

   b. Comment on any pattern evident from the table.

Problem solving

15. Represent the following 3-dimensional shape as a planar graph.

Tetrahedron

16. Represent the following 3-dimensional shape as a planar graph.

Cube

17. The table displays the most common methods of communication for a group of people.

	WhatsApp	Snapchat
Adam	Ethan	Ethan
Michelle		Sophie, Ethan
Liam	Michelle	Ethan, Sophie

a. Display the information for the entire table in a graph.
b. State who would be the best person to introduce Adam.
c. Display the Snapchat information in a separate graph.

LESSON
3.4 Connected graphs

▶ 3.4.1 Traversable graphs

eles-6158

- Networks can represent many applications, such as social networks, electrical wiring, internet and communications.
- Sometimes it may be important to follow a sequence that goes through all vertices only once, for example a salesperson who wishes to visit each town once.
- Sometimes it may be important to follow a sequence that use all edges only once, such as a road repair gang repairing all the roads in a suburb.
- Movement through a simple connected graph is described in terms of starting and finishing at specified vertices by travelling along the edges.
- The definitions of the main terms used when describing movement across a network are as follows.

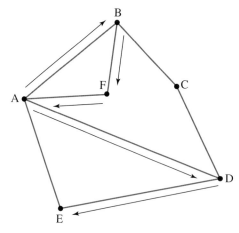

Route: ABFADE

Moving around a graph

Walk: Any route taken through a network, including routes that repeat edges and vertices

Trail: A walk with no repeated edges

Path: A walk with no repeated vertices and no repeated edges

Cycle: A path beginning and ending at the same vertex

Circuit: A trail beginning and ending at the same vertex

Walk: ABCADCB

Trail: ABCADC

Path: ABCDE

Cycle: ABCDA

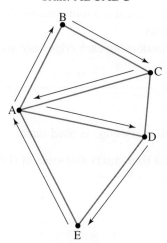

Circuit: ABCADEA

WORKED EXAMPLE 9 Identifying different routes in a graph

**In the network shown, identify two different routes:
one cycle and one circuit.**

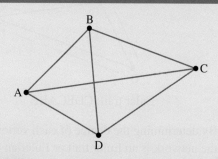

THINK

1. For a cycle, identify a route that doesn't repeat a vertex
   and doesn't repeat an edge, apart from the start/finish.

WRITE

Cycle: ABDCA

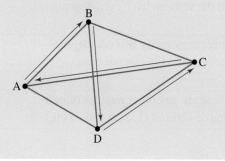

2. For a circuit, identify a route that doesn't repeat an edge and ends at the starting vertex.

Circuit: ADBCA

- In some practical situations it is most efficient if a route travels along each edge only once, for example parcel deliveries and garbage collections.
- If it is possible to move around a network using each edge only once, the route is known as a **Euler trail** or **Eulerian circuit**.

> **Euler trails and circuits**
>
> **An Euler trail is a trail in which every edge is used once.**
>
> **An Euler circuit is an Euler trail that starts and ends at the same vertex.**

Euler trail: CDECABE

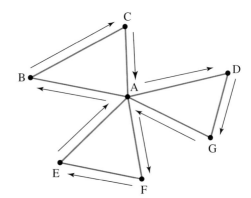

Euler circuit: ABCADGAFEA

- By determining the degree of each vertex and whether each degree is even or odd, it can be established if the network is an Euler trail or Eulerian circuit.

Degree of vertices	Euler trail	How?	Euler circuit
All the vertices have even degrees.	Yes	Start and finish at any vertex.	Yes
There are exactly *two* odd degree vertices and the rest are even.	Yes	Start and finish at the two odd degree vertices.	No
There are any other number of odd degree vertices (1, 3, 5 and so on.)	No		No

Existence conditions for Euler trails and Euler circuits

- If all of the vertices of a connected graph have an even degree, then an **Euler circuit** exists. An Euler circuit can begin at any vertex.
- If exactly 2 vertices of a connected graph have an odd degree, then an **Euler trail** exists. An Euler trail will start at one of the odd degree vertices and finish at the other.

WORKED EXAMPLE 10 Identifying an Euler trail

Determine whether there is an Euler trail through the network shown and, if so, give an example.

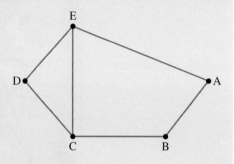

THINK

1. Determine the degree of each vertex by counting the number of edges connected to it.

2. Count the number of odd degree vertices and hence state whether there is an Euler trail through the network.

3. Since there are exactly 2 vertices with odd degrees (C and E), an Euler trail has to start and finish with these; say, begin at C and end at E. Attempt to find a trail that uses each edge. *Note:* Although each edge must be used exactly once, vertices may be used more than once.
To ensure that each edge has been used, label them as you go.

4. List the sequence of vertices along the trail.

WRITE/DRAW

Vertex A has degree = 2.
Vertex B has degree = 2.
Vertex C has degree = 3.
Vertex D has degree = 2.
Vertex E has degree = 3.

Number of odd vertices = 2
Therefore, an Euler trail exists.

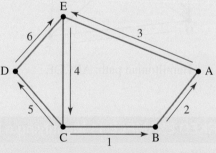

An Euler trail is C–B–A–E–C–D–E.

▶ 3.4.3 Hamiltonian paths and cycles

- In other situations, it may be more practical if all vertices can be reached without using all of the edges of the graph.
  For example, if you wanted to visit a selection of capital cities in Europe, you wouldn't want to use all of the available flight routes.
- If it is possible to travel a network using each vertex only once, the route is known as a **Hamiltonian path** or **Hamiltonian cycle**.
- There is no set method for finding these routes, except by inspection or trial and error.

TOPIC 3 Networks **187**

Hamiltonian paths and cycles

A Hamiltonian path is a path that passes through every vertex exactly once.

A Hamiltonian cycle is a Hamiltonian path that starts and finishes at the same vertex.

Hamiltonian path: ABCDE

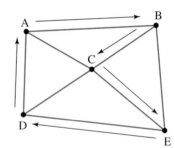

Hamiltonian cycle: ABCEDA

WORKED EXAMPLE 11 Identifying a Hamiltonian path and cycle

For the network shown, identify:
a. **a Hamiltonian path**
b. **a Hamiltonian cycle.**

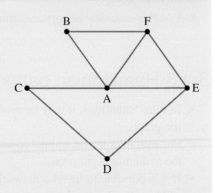

THINK	WRITE
a. Choose any starting vertex and attempt to visit all the other vertices. (Not all the edges need to be used.)	**a.** A possible Hamiltonian path is A–B–F–E–D–C.
b. Attempt to close the path obtained in part **a** by returning to the starting vertex.	**b.** A possible Hamiltonian cycle is A–B–F–E–D–C–A.

WORKED EXAMPLE 12 Describing an Eulerian trail and a Hamiltonian path

Identify an Euler trail and a Hamiltonian path in the network shown.

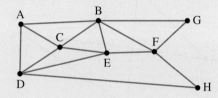

THINK	WRITE/DRAW
1. For an Euler trail to exist, there must be exactly 2 vertices with an odd degree.	$\deg(A) = 3, \deg(B) = 5, \deg(C) = 4, \deg(D) = 4, \deg(E) = 4,$ $\deg(F) = 4, \deg(G) = 2, \deg(H) = 2$ As there are only two odd-degree vertices, an Euler trail must exist.
2. Identify a route that uses each edge once. Begin the route at one of the odd-degree vertices and finish at the other.	Euler trail: ABGFHDEFBECDACB
3. Identify a route that reaches each vertex once.	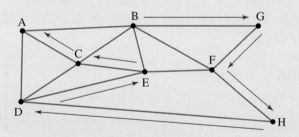 Hamiltonian path: BGFHDECA
4. Write the answer.	Euler trail: ABGFHDEFBECDACB (other answers are possible) Hamiltonian path: BGFHDECA (other answers are possible)

Exercise 3.4 Connected graphs

3.4 Quick quiz on	3.4 Exercise

Individual pathways

■ PRACTISE	■ CONSOLIDATE	■ MASTER
1, 4, 7, 9, 12, 16	2, 5, 10, 13, 14, 17	3, 6, 8, 11, 15, 18

Fluency

1. **WE9** In the network shown, identify two different routes: one cycle and one circuit.

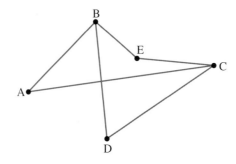

2. In the network shown, identify three different routes: one path, one cycle and one circuit.

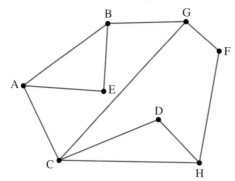

3. State which of the terms *walk*, *trail*, *path*, *cycle* and *circuit* could be used to describe the following routes on the graph shown.

   a. AGHIONMLKFGA
   b. IHGFKLMNO
   c. HIJEDCBAGH
   d. FGHIJEDCBAG

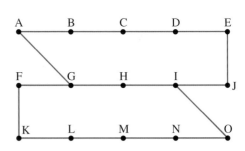

4. **WE10** Determine whether there is an Euler trail through the network shown and, if so, give an example.

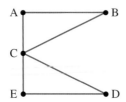

5. Starting at vertex R, determine an Euler trail for the planar graph shown.
   (*Hint:* What vertex should the trail end at?)

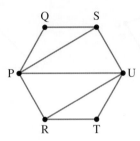

6. **WE11** For each of the following networks shown, identify:

   i. a Hamiltonian path
   ii. a Hamiltonian circuit.

   a.

   b.

   c.

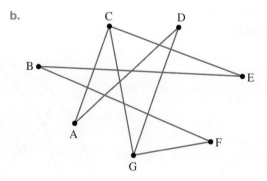

   d.

7. **WE12** Identify an Euler trail and a Hamiltonian path in each of the following graphs.

   a.

   b.

8. Identify an Euler circuit and a Hamiltonian cycle in each of the following graphs, if they exist.

a.

b.

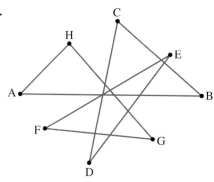

Understanding

9. Identify an Euler circuit for each of the following networks shown.

a.

b.

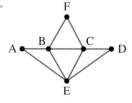

10. In the following graph, if an Euler trail commences at vertex A, identify the vertices at which it could finish.

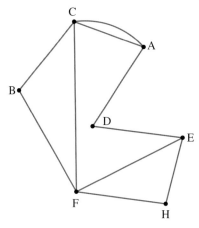

11. In the following graph, identify at which vertices a Hamiltonian path could finish if it commences by travelling from:

a. B to E

b. E to A.

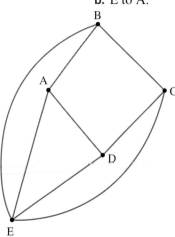

Reasoning

12. A road inspection crew must travel along each road shown on the map exactly once.

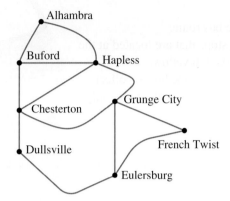

   a. State from either of which two cities the crew must begin its tour.
   b. Determine a path using each road once.
   c. State which cities are visited most often and why.

13. In a computer network, the file server (F) is connected to all the other computers as shown. Craig, the technician, wishes to test that each connecting cable is functioning properly. He wishes to route a signal, starting at F, so that it travels down each cable exactly once and then returns to the file server. Determine such a circuit for the network configuration shown.

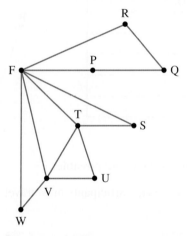

14. In the following graph, other than from G to F, identify the 2 vertices between which you must add an edge in order to create a Hamiltonian path that commences from:

   a. vertex G
   b. vertex F.

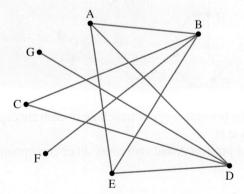

15. On the map shown, a school bus route is indicated in yellow. The bus route starts and ends at the school indicated.

   a. Draw a graph to represent the bus route.

   b. Students can catch the bus at stops that are located at the intersections of the roads marked in yellow. Determine whether it is possible for the bus to collect students by driving down each section of the route only once. Explain your answer.

Problem solving

16. a. Use the following graph to complete the table to identify all of the Hamiltonian cycles commencing at vertex A.

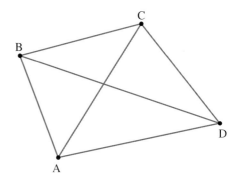

	Hamiltonian cycle
1.	ABCDA
2.	
3.	
4.	
5.	
6.	

   b. State whether any other Hamiltonian cycles are possible.

17. The map of an orienteering course is shown. Participants must travel to each of the nine checkpoints along any of the marked paths.

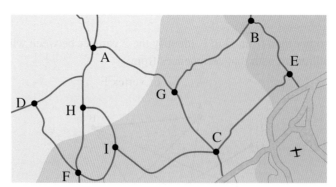

   a. Draw a graph to represent the possible ways of travelling to each checkpoint.

   b. State the degree of checkpoint H.

   c. If participants must start and finish at A and visit every other checkpoint only once, identify two possible routes they could take.

18. The graph shown outlines the possible ways a tourist bus can travel between eight locations.

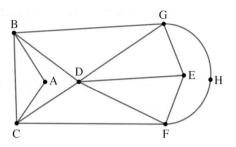

  a. If vertex A represents the second location visited, list the possible starting points.
  b. If the bus also visited each location only once, state which of the starting points listed in part **a** could not be correct.
  c. If the bus also needed to finish at vertex D, list the possible paths that could be taken.

LESSON
3.5 Trees and their applications (Extending)

LEARNING INTENTIONS

At the end of this lesson you should be able to:
  • identify the properties of trees, weighted graphs and networks
  • determine minimum spanning trees
  • apply Prim's algorithm to determine the minimum spanning tree.

▶ 3.5.1 Trees

eles-6161

  • A special kind of network is called a **tree**.
  • A tree is a simple connected graph with no circuits. It consists of the smallest number of edges necessary so that each vertex is connected to at least one other vertex.
  • This ensures that each vertex can 'communicate' to all other vertices either directly or indirectly.
    For example, the following diagrams shows trees for 2, 3, 4 and 5 vertices respectively.

2 vertices	3 vertices	4 vertices	5 vertices
A———B	A———B C	A———B C D	A———B C———E D

  • A tree cannot contain any loops, parallel (or multiple) edges or cycles.
  • The number of edges is always 1 less than the number of vertices. A tree with n vertices has $n-1$ edges.
  • The number of edges must be one fewer than the number of vertices for the network to be a tree.

For example, in the following diagram the network has one too many edges and therefore is *not* a tree.

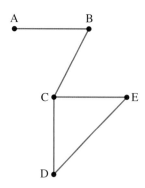

WORKED EXAMPLE 13 Identifying trees

Identify which of the following networks are trees.

a.

b.

c.

d.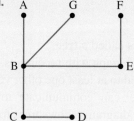

THINK	WRITE
a. 1. Count the number of vertices and edges. There should be one edge fewer than the number of vertices.	**a.** Number of vertices = 6 Number of edges = 5
2. Check whether each vertex is connected to at least one other and state your conclusion.	A is connected to B. B is connected to F. C is connected to D. D is connected to E. F is connected to B Therefore, the network shown is a tree.
b. Count the number of vertices and edges. The number of edges must be one fewer than the number of vertices for the network to be a tree.	**b.** Number of vertices = 6 Number of edges = 6 Therefore, the network shown is not a tree.
c. 1. Count the number of vertices and edges. There should be one edge fewer than the number of vertices.	**c.** Number of vertices = 7 Number of edges = 6

2.	Check whether each vertex is connected to at least one other and state your conclusion.	A is not connected to any vertex. Therefore, the network shown is not a tree.
d. 1.	Check whether the number of edges is one fewer than the number of vertices.	**d.** Number of vertices $= 7$ Number of edges $= 6$
2.	Check whether each vertex is connected to at least one other.	Each vertex is connected. Therefore, the network shown is a tree.

⏵ 3.5.2 Weighted graphs

eles-6162

- In many applications of networks, it is useful to attach a value to the edges.
- These values could represent the length of the edge in terms of time or distance, or the costs involved with moving along that section of the path. Such graphs are known as **weighted graphs**.
- Weighted graphs can help determine how to travel through a network in the shortest possible time.

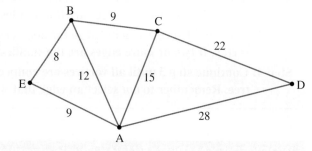

WORKED EXAMPLE 14 Identifying the shortest path between two vertices

The graph represents the distances in kilometres between eight locations.
Identify the shortest distance to travel from A to D that goes to all vertices.

THINK	**WRITE**
1. Identify the Hamiltonian paths that connect the two vertices.	Possible paths: **a.** ABGEFHCD **b.** ABCHGEFD **c.** AEGBCHFD **d.** AEFGBCHD **e.** AEFHGBCD
2. Calculate the total distances for each path to determine the shortest distance.	**a.** $3+2+2+4+2+3+4=20$ **b.** $3+4+3+6+2+4+3=25$ **c.** $3+2+2+4+3+2+3=19$ **d.** $3+4+5+2+4+3+2=23$ **e.** $3+4+2+6+2+4+4=25$
3. Write the answer as a sentence.	The shortest distance from A to D that travels to all vertices is 19 km.

▶ 3.5.3 Prim's algorithm

- **Spanning trees** are types of trees that include all the vertices of the original graph.
- A **minimum spanning tree** is a spanning tree where the total distance along the tree branches is as small as possible. That is, it has the lowest total weighting.
- **Prim's algorithm** is a set of logical steps that can be used to identify the minimum spanning tree for a weighted connected graph.

Prim's algorithm for determining a minimum spanning tree

Step 1: Select the edge that has the lowest weight. If two or more edges are the lowest, choose any of these.

Step 2: Looking at the two vertices included so far, select the smallest edge leading from either vertex. If two or more edges are the smallest, choose any of these.

Step 3: Look at all vertices included so far and select the smallest edge leading from any included vertex. If two or more edges are the smallest, choose any of these.

Step 4: Continue step 3 until all vertices are connected or until there are $n - 1$ edges in the spanning tree. Remember to *not* select an edge that would create a circuit.

WORKED EXAMPLE 15 Using Prim's algorithm to determine a minimum spanning tree

a. **Use Prim's algorithm to identify the minimum spanning tree of the graph shown.**
b. **Determine the length of the minimum spanning tree.**

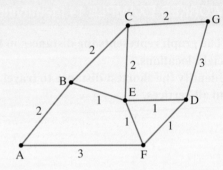

THINK

a. 1. Draw the vertices of the graph and include the edge with the lowest weight.

   Note: Since BE, EF, DE and DF all have a weight of 1, choose any of these edges.

2. Look at the two included vertices B and E. Select the smallest edge leading from B or E.

   Note: You can select edge ED or EF as they both have a weight of 1.

DRAW

a.

3. Look at the vertices included so far (B, E and D). Select the smallest edge leading from any of these vertices.

   Note: You can select EF or DF as they both have a weight of 1.

4. Look at the vertices included so far (B, E, D and F). Select the smallest edge leading from any of these vertices.

   Note: You *cannot* select edge DF as it will create a circuit. You must now select an edge with a weight of 2 (the next lowest). Choose either edge AB, BC or CE.

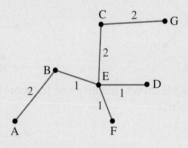

5. Repeat this process until all vertices are connected. When selecting edges, do not create any circuits.

b. 1. The length of the minimum spanning tree can be determined by adding up the weight of the edges that create the minimum spanning tree.

b. Length $= 1 + 1 + 1 + 2 + 2 + 2 = 9$

  Resources

eWorkbook Topic 3 Workbook (worksheets, code puzzle and project) (ewbk-13284)

Interactivity Individual pathway interactivity: Trees and their applications (Extending) (int-9166)

Exercise 3.5 Trees and their applications (Extending)

3.5 Quick quiz on	3.5 Exercise

Individual pathways

■ PRACTISE	■ CONSOLIDATE	■ MASTER
1, 4, 7, 10, 13	2, 6, 8, 11, 14	3, 5, 9, 12, 15

Fluency

1. **WE13** Identify which of the following networks are trees.

a.   b.   c.

d.   e.

2. In the network shown, identify one (or more) edges that, when removed, result in the remaining network becoming a tree.

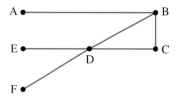

3. Identify one (or more) edges in the network shown that, when added, result in the network becoming a tree.

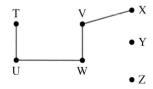

4. **WE14** Use the graph to identify the shortest distance to travel from A to D that goes to all vertices.

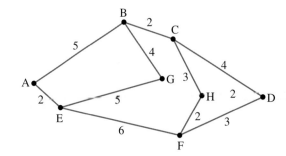

5. Use the graph to identify the shortest distance to travel from A to I that goes to all vertices.

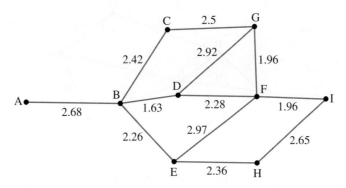

6. **WE15** Consider the graph shown.

   a. Use Prim's algorithm to identify the minimum spanning tree.

   b. State the length of the minimum spanning tree.

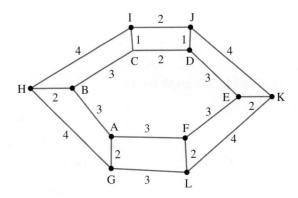

Understanding

7. A truck starts from the main distribution point at vertex A and makes deliveries at each of the other vertices before returning to A.
Determine the shortest route the truck can take.

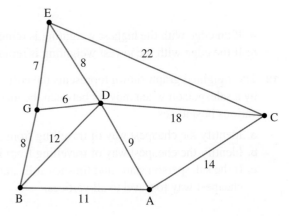

8. Identify the minimum spanning tree for the graph shown.

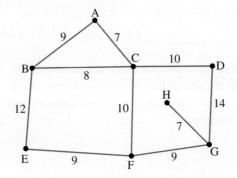

9. Identify the minimum spanning tree for the graph shown.

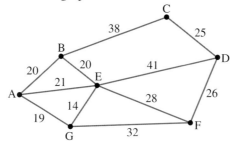

Reasoning

10. Draw diagrams to show the steps you would follow when using Prim's algorithm to identify the minimum spanning tree for the graph shown.

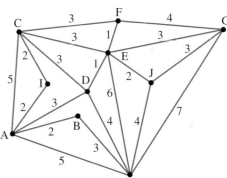

11. Consider the graph shown.

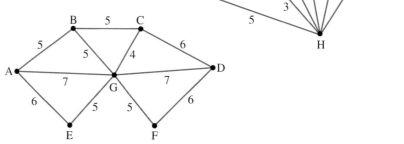

   a. If an edge with the highest weighting is removed, identify the shortest Hamiltonian path.

   b. If the edge with the lowest weighting is removed, identify the shortest Hamiltonian path.

12. The weighted graph shown represents the costs incurred by a salesperson when moving between the locations of various businesses.

   a. Identify the cheapest way of travelling from A to G.

   b. Identify the cheapest way of travelling from B to G.

   c. If the salesperson starts and finishes at E, state the cheapest way to travel to all vertices.

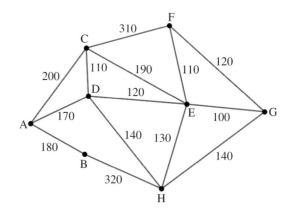

Problem solving

13. The diagram shown represents 8 cities and the roads connecting them. The distances along each road are also indicated. (The distances are in kilometres.) Determine the minimal spanning tree for this network and determine the total length of the tree.

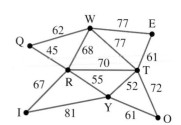

14. A group of computers are connected as shown.

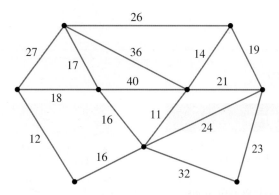

The numbers on the edges represent the cost (in cents) of sending a 1-MB mail message between the computers.

Determine the smallest possible cost to send a message to all the computers.

15. The organisers of the Tour de Vic bicycle race are using the following map to plan the event.

a. Draw a weighted graph to represent the map.

b. If they wish to start and finish in Geelong, determine the shortest route that can be taken that includes a total of nine other locations exactly once, two of which must be Ballarat and Bendigo.

LESSON
3.6 Review

3.6.1 Topic summary

Definitions and terms

- In a graph, the lines are called edges and the points are called vertices (or nodes), with each edge joining a pair of vertices.
- A simple graph is one in which pairs of vertices are connected by at most one edge.
- If it is possible to reach every vertex of a graph by moving along the edges, the graph is a *connected graph*.
- A bridge is an edge that, if removed, disconnects the graph.
- If it is only possible to move along the edges of a graph in one direction, the graph is a *directed graph* (or digraph) and the edges are represented by arrows. Otherwise, it is an *undirected graph*.
- The degree of a vertex = the number of edges directly connected to that vertex, except that a loop at a vertex contributes twice to the degree of that vertex.

Networks

Trees and weighted graphs

- Weighted graphs have values associated with each edge.
- Trees are simple connected graphs with no circuits.
- Spanning trees include all vertices of a graph.
- Minimum spanning trees have the lowest total combined weighting possible for a graph.
- Prim's algorithm:
  - **Step 1:** Select the edge that has the lowest weight. If two or more edges are the smallest, choose any of these.
  - **Step 2:** Looking at the two vertices included so far, select the smallest edge leading from either vertex. If 2 or more edges are the smallest, choose any of these.
  - **Step 3:** Look at all vertices included so far and select the smallest edge leading from any included vertex. If two or more edges are the smallest, choose any of these.
  - **Step 4:** Continue Step 3 until all vertices are selected. Remember to not select an edge that would create a circuit.
- Prim's algorithm selects an edge of lowest weight that must be connected to the edges already included in the minimum spanning tree.

Planar graphs

- If a graph can be drawn with no intersecting edges, then it is a *planar graph*.

This graph can be redrawn as

- Euler's formula identifies the relationship between the vertices (v), edges (e) and faces (f) in a planar graph:
$$v - e + f = 2$$

Connected graphs

- **Walk:** Any route taken through a network, including routes that repeat edges and vertices
- **Trail:** A walk with no repeated edges
- **Path:** A walk with no repeated vertices and no repeated edges
- **Cycle:** A path beginning and ending at the same vertex
- **Circuit:** A trail beginning and ending at the same vertex
- An Euler trail is a trail in which every edge is used once.
- An Euler circuit is a circuit in which every edge is used once.
- If all of the vertices of a connected graph are even, then an Euler circuit exists.
- If exactly two vertices of a connected graph are odd, then an Euler trail exists.
- A Hamiltonian path is a path that passes through every vertex exactly once.
- A Hamiltonian cycle is a Hamiltonian path that starts and finishes at the same vertex.

3.6.2 Success criteria

Tick the column to indicate that you have completed the lesson and how well you have understood it using the traffic light system.

(**Green:** I understand; **Yellow:** I can do it with help; **Red:** I do not understand)

Lesson	Success criteria	⬤	◯	⬤
3.2	I can draw network diagrams to represent connections.			
	I can define edge, vertex, loop and the degree of a vertex.			
	I can identify properties and types of graphs, including connected and isomorphic graphs.			
3.3	I can identify and draw planar graphs.			
	I can apply Euler's formula for planar graphs.			
3.4	I can define walks, trails, paths, cycles and circuits in the context of traversing a graph.			
3.5	I can identify the properties of trees, weighted graphs and networks.			
	I can determine minimum spanning trees.			
	I can apply Prim's algorithm to determine the minimum spanning tree.			

3.6.3 Project

Networks at school

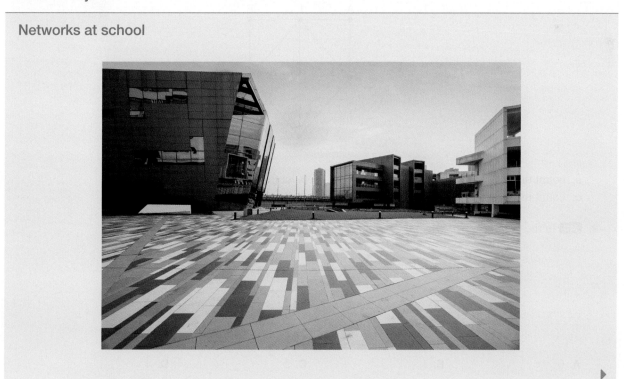

Work in a small group to analyse the pathways in your school.
1. On a map of your school, identify all the buildings that students may use at school.
2. Represent this information as a network diagram. Use edges to represent the pathways in your school.
3. As a group, determine whether a path can be found connecting each room.
4. a. If a path can be found, is it an Euler path? Add additional edges if necessary to create an Euler path.
   b. If a path cannot be found, choose a room to omit so that a path can be found. Add any paths necessary to create an Euler path.
5. If possible, identify an Euler circuit within the building you are in now (or the building you use for your Maths class).
6. Make recommendations about the pathways currently found in your school. Do you believe that any more should be created?

 Resources

 eWorkbook   Topic 3 Workbook (worksheets, code puzzle and project) (ewbk-13284)

 Interactivities Crossword (int-9167)
                    Sudoku puzzle (int-3596)

Exercise 3.6 Review questions

learn on

Fluency

1. **MC** The minimum number of edges in a connected graph with eight vertices is:
   A. 5   **B.** 6   **C.** 7   **D.** 8

2. For the network shown:

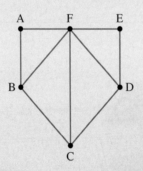

   a. count the number of vertices (V)   **b.** count the number of edges (E)
   c. count the number of faces (F)   **d.** confirm Euler's formula.

3. **MC** In the graph shown, select the number of edges that are bridges.

   A. 0   **B.** 1   **C.** 2   **D.** 4

4. **MC** A connected graph with 9 vertices has 10 faces.
The number of edges in the graph is:

A. 15   **B.** 16   **C.** 17   **D.** 18

5. a. For the network shown, determine:

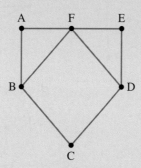

    i. an Euler path
    ii. a Hamiltonian path
    iii. a Hamiltonian circuit.
  b. Explain why is there no Euler circuit.

6. **MC** The number of faces in the planar graph shown is:

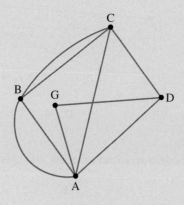

A. 6   **B.** 7   **C.** 8   **D.** 9

7. a. Identify whether the following graphs are planar or not planar.

  i.

  ii.
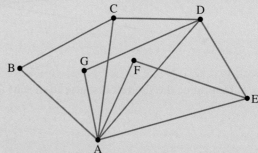

  b. Redraw the graphs that are planar without any intersecting edges.

8. Identify which of the following graphs are isomorphic.

a.

b.

c.

d.

9. **MC** Select the length of the minimum spanning tree of the graph shown.

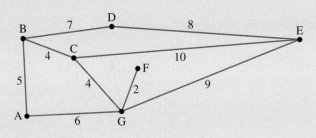

A. 33   **B.** 26   **C.** 34   **D.** 30

10. For each of the following graphs:
  i. add the minimum number of edges required to create an Euler trail
  ii. state the Euler trail created.

a.

b.

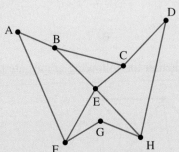

11. **a.** Determine the shortest distance from start to finish in the graph shown.

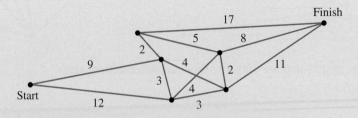

  b. Calculate the total length of the shortest Hamiltonian path from start to finish.
  c. Draw the minimum spanning tree for this graph.

Problem solving

12. The city of Kaliningrad in Russia is situated on the Pregel River. It has two large islands that are connected to each other and to the mainland by a series of bridges.

    In the early 1700s, when the city was called Königsberg, there were seven bridges. The challenge was to walk a route that crossed each bridge exactly once.

    In 1735, Swiss mathematician Leonard Euler (pronounced 'oiler') used networks to prove whether this was possible.

    a. Draw a network with four vertices (to represent the areas of land) and seven edges to represent the seven connecting bridges of Königsberg in 1735.

    b. Use your knowledge of networks to explain whether it is possible to walk a route that crosses each bridge only once.

    c. The diagrams below show the bridges in Kaliningrad in 1735 and the bridges as they are now. Use your knowledge of networks to explain whether it is possible now to walk a route that crosses each bridge only once.

Bridges as they were in 1735

Bridges as they are now

13. The graph shown represents seven towns labelled A–G in a particular district. The lines (edges) represent roads linking the towns (vertices), with distances given in km.

a. Does the graph represent a connected planar graph?
b. Determine the number of:
   i. vertices
   ii. edges
   iii. faces (regions) for the given graph.
c. Using your answers from part b, determine whether Euler's formula holds for the given graph.
d. i. Write the degree of each vertex.
   ii. Calculate the sum of the degrees of all the vertices and compare with the number of edges.
   iii. How many odd-degree vertices are there?

14. The flying distances between the capital cities of Australian mainland states and territories are listed in the following table. (Distances are in km.)

	Adelaide	Brisbane	Canberra	Darwin	Melbourne	Perth	Sydney
Adelaide		2055	1198	3051	732	2716	1415
Brisbane	2055		1246	3429	1671	4289	982
Canberra	1198	1246		4003	658	3741	309
Darwin	3051	3429	4003		3789	4049	4301
Melbourne	732	1671	658	3789		3456	873
Perth	2716	4289	3741	4049	3456		3972
Sydney	1415	982	309	4301	873	3972	

a. Draw a weighted graph to show this information.
b. If technical problems are preventing direct flights from Melbourne to Darwin and from Melbourne to Adelaide, calculate the shortest way of flying from Melbourne to Darwin.
c. If no direct flights are available from Brisbane to Perth or from Brisbane to Adelaide, calculate the shortest way of getting from Brisbane to Perth.

15. The diagram shows the streets in a suburb of a city with a section of underground tunnels shown in dark blue. Weightings indicate distances in metres.

The tunnels are used for utilities such as electricity, gas, water and drainage.

a. i. If the gas company wishes to run a pipeline that minimises its total length but reaches each vertex, calculate the total length required.
   ii. Draw a graph to show the gas lines.

b. If drainage pipes need to run from H to A, calculate the shortest path they can follow.
   Determine how long this path will be in total.

c. A single line of cable for a computerised monitoring system needs to be placed so that it starts at D and reaches every vertex once.
   Calculate the minimum length possible and determine the path it must follow.

d. If a power line has to run from D so that it reaches every vertex at least once and finishes back at the start, determine what path it must take to be a minimum.

on To test your understanding and knowledge of this topic, go to your learnON title at www.jacplus.com.au and complete the **post-test**.

Answers

Topic 3 Networks

3.1 Pre-test

1. 4
2. 6
3. 9
4. One possible path is from A is A–B–C–F–E–D
5. Yes it is possible. One possible path from A is:
   A–D–E–F–D–B–F–C–B–A
6. The diagram is able to be traced from any vertex.
7. A–B–C–D–E–A–F–D–G–A
8.

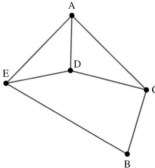

9. 4
10. E
11. 8
12.

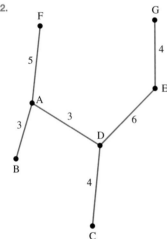

13. 25
14. Join 4 to 7 and 3 to 8
15. 184 km

3.2 Properties of networks

1. a. Vertices = 6, edges = 8
   b. Vertices = 7, edges = 9
2. a. Vertices = 5, edges = 5
   b. Vertices = 6, edges = 9
   c. Vertices = 7, edges = 11
   d. Vertices = 9, edges = 16

3.

4.

5.

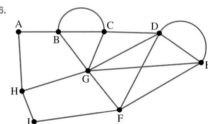

6.

7. Degree of A = 3, B = 4, C = 3, D = 2, E = 5, F = 3
8. a. deg (A) = 5; deg (B) = 3; deg (C) = 4;
      deg (D) = 1; deg (E) = 1
   b. deg (A) = 0; deg (B) = 2; deg (C) = 2;
      deg (D) = 3; deg (E) = 3
   c. deg (A) = 4; deg (B) = 2; deg (C) = 2;
      deg (D) = 2; deg (E) = 4
   d. deg (A) = 1; deg (B) = 2; deg (C) = 1;
      deg (D) = 1; deg (E) = 3
9. a. Yes   b. Yes
10. a. Different degrees and connections
    b. Different degrees and connections
11. Graphs 2 and 4; Graphs 5 and 6; Graphs 1 and 7
12. Answer will vary. Possible answers are shown.
    a.

b.

c.

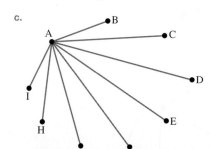

13.

Graph	Simple	Connected
Graph 1	Yes	Yes
Graph 2	Yes	Yes
Graph 3	Yes	Yes
Graph 4	No	Yes
Graph 5	Yes	Yes

14.

15. a.

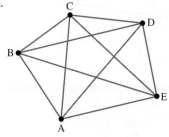

b. The total number of edges represents the total games played, as each edge represents one team playing another.

16. a. $v = \{1, 2, 3, 4\}$
$E = \{(1, 2), (1, 3), (1, 4), (2, 3), (2, 4), (3, 4)\}$

b.

c. Degree $(1) = 3$
Degree $(2) = 3$
Degree $(3) = 3$
Degree $(4) = 3$

17. a.

b. deg (Huairou) = deg (Shunyi) = 4

c. This is a simple, connected graph as there are no loops or multiple edges, and all vertices are reachable.

18.

19. a.

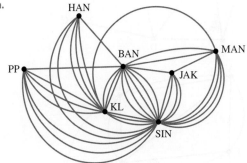

b. Directed, as it would be important to know the direction of the flight.

c. 7

3.3 Planar graphs

1.

2.

3.

4. $V = 6$, $E = 9$, $F = 5$

5. a. Vertices = 5, edges = 7, faces = 4

   b. Vertices = 7, edges = 10, faces = 5

6. a. 4   b. 5

7. a. 6   b. 5

8.

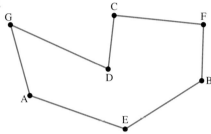

9. Graph 3 is not planar as it cannot be redrawn without intersecting edges.

10. a. Vertices = 6, edges = 10, faces = 6

    b. Please see the worked solutions.

11. a. 3   b. 2

12. a.

   b.

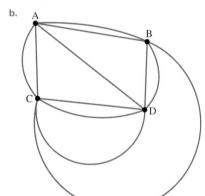

13.

Graph	Total edges	Total degrees
Graph 1	3	6
Graph 2	5	10
Graph 3	8	16

14. a.

Graph	Total vertices of even degree	Total vertices of odd degree
Graph 1	3	2
Graph 2	4	2
Graph 3	4	4

    b. No clear pattern evident

15.

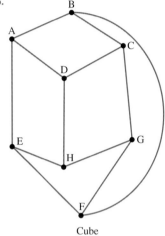

Tetrahedron

16.

Cube

17. a.

Sophie
Michelle
Liam
Adam
Ethan

   b. Ethan and Michelle, as they are in contact via more methods.

   c.

Sophie
Michelle
Liam
Adam
Ethan

3.4 Connected graphs

1. Cycle: ABECA (other exist)
   Circuit: BECDB (other exist)

2. Path: ABGFHDC (other exist)
   Cycle: DCGFHD (other exist)
   Circuit: AEBGFHDCA (other exist)

3. a. Walk
   b. Walk, trail and path
   c. Walk, trail, path, cycle and circuit
   d. Walk and trail

4. A–B–C–D–E–C–A (also a circuit)

5. R–T–U–R–P–U–S–P–Q–S

6. a. i. A–B–E–D–C
   ii. A–B–E–D–C–A
   b. i. A–B–E–D–F–C–G
   ii. A–B–E–D–F–C–G–A
   c. i. E–D–B–A–G–C–F
   ii. There is no possible circuit because both E and F have a degree of one.
   d. i. A–B–C–D–E–F
   ii. A–B–C–D–E–F–A

7. a. Euler: AFEDBECAB; Hamiltonian: BDECAF
   b. Euler: GFBECGDAC; Hamiltonian: BECADGF

8. a. Euler: AIBAHGFCJBCDEGA;
   Hamiltonian: none exist
   b. Euler: ABCDEFGHA (other exist);
   Hamiltonian: HABCDEFGH (other exist)

9. Other circuits are possible.
   a. B–A–D–B–E–C–B–F–G–B
   b. A–E–D–C–E–B–C–F–B–A

10. E

11. a. A or C   b. B or D

12. a. Buford, Eulersburg
   b. B–A–H–B–C–H–G–C–D–E–G–F–E
   c. Chesterton, Hapless, Grunge City, because they all have the highest degree

13. Other circuits are possible.
   F–P–Q–R–F–S–T–U–V–T–F–V–W–F

14. a. G to C   b. F to E

15. a.

School

   b. Yes, because the degree of each intersection or corner point is an even number.

16. a.

	Hamiltonian cycle
1.	ABCDA
2.	ABDCA
3.	ACBDA
4.	ACDBA
5.	ADBCA
6.	ADCBA

   b. Yes, commencing on vertices other than A. For example: BCDAB.

17. a.

   b. deg (H) = 4
   c. ADHFICEBGA or AHDFICEBGA

18. a. B : BCA
   C : CBA
   D : DBA or DCA
   F : FCA
   G : GBA
   Therefore, B, C, D, F or G
   b. B or C, as once A is reached, the next option is returning to the start.
   c. None possible

3.5 Trees and their applications (Extending)

1. b, d, and e

2. Remove either (B, D), (B, C), or (D, C).

3. For example: (X, Y) and (Y, Z)

4. 21

5. 20.78

6. a.

   b. The length of the minimum spanning tree is 24.

7. ABGEDCA or ACDEGBA (length 66)

8.

9.

10. Step 1   Step 2

Step 3 Step 4

 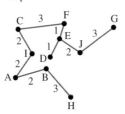

11. a. FDCGBAE (other solutions exist)

    b. FDCBAEG (other solutions exist)

12. a. ADEG

    b. BHG

    c. EGFCDABHE

13. 403

14. $1.28

15. a.

    b. Geelong → Ballarat → Ararat → Maryborough → Bridgewater → Bendigo → Castlemaine → Seymour → Sunbury → Bacchus Marsh → Geelong = 723 km

Project

The steps to analyse the pathways in your school:

1. On a map of your school, identify all the buildings that students may use at school.

- Make a list of all the buildings in your school that are accessible to students.

2. Represent this information as a network diagram. Use edges to represent the pathways in your school.

- Draw a node for each building and connect them with edges to represent the pathways between them.

3. As a group, determine whether a path can be found connecting each room.

- Try to find a path that connects all the rooms in your school.

4 a. If a path can be found, is it an Euler path? Add additional edges if necessary to create an Euler path.

- An Euler path is a path that visits every edge exactly once. If the path you found satisfies this condition, then it is an Euler path. Otherwise, you can add additional edges to create an Euler path.

b. If a path cannot be found, choose a room to omit so that a path can be found. Add any paths necessary to create an Euler path.

- If you cannot find a path that connects all the rooms, choose a room to omit and try again. You can add additional edges to create an Euler path.

5. If possible, identify an Euler circuit within the building you are in now (or the building you use for your Maths class).

- To identify an Euler circuit within a building, draw a network diagram of the rooms in the building and the pathways between them. If the building has an Euler circuit, then it is possible to walk from any room in the building and return to that room by following a sequence of pathways that visits each room exactly once.

6. Make recommendations about the pathways currently found in your school. Do you believe that any more should be created?

- Based on your analysis of the pathways in your school, you can make recommendations on whether additional pathways should be created.

  For example, if you found that there are rooms that are not connected to the rest of the buildings, you could recommend adding new pathways to connect them. Alternatively, if you found that some pathways are rarely used or lead to congestion, you could recommend creating new pathways to alleviate traffic. Your recommendations should take into account the safety, efficiency, and convenience of the pathways for students and staff.

3.6 Review questions

1. C

2. a. $V = 6$

    b. $E = 9$

    c. $F = 5$

    d. $6 = 9 - 5 + 2$

3. D

4. C

5. a. i. B–A–F–E–D–F–B–C–D

     ii. A–B–C–D–E–F (others are possible)

     iii. A–B–C–D–E–F–A (others are possible)

    b. Because there are 2 vertices of odd degree.

6. A

7. a. i. Planar   **ii.** Planar

    b. i.

ii.

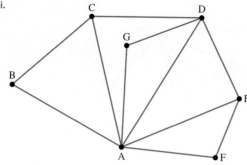

8. a and d

9. D

10. a. i.

ii. ABDBCADC

b. i.

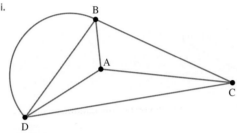

ii. BAFEHGFHDCEBC

11. a. 23

b. 34

c.

12. a.

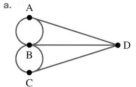

b. No. All vertices have an odd degree.

c. Yes. It is possible as there are 2 odd vertices and 2 even vertices.

13. a. Yes, because the edges do not cross each other.

b. i. 7

ii. 11

iii. 6

c. Yes, $V = E - F + 2$.

d. i. Degree A = 2, degree B = 3, degree C = 5, degree D = 3, degree E = 3, degree F = 4, degree G = 2

ii. Sum = 22

iii. 4

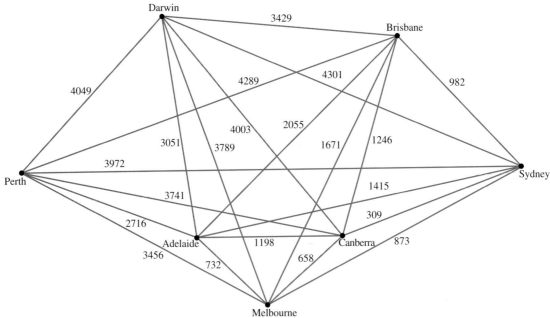

b. Via Canberra: $658 + 4003 = 4661$ km

c. Via Sydney: $982 + 3972 = 4954$ km

15. a. i. $815 + 565 + 325 + 365 + 350 + 805 + 580 = 3805$ m

   ii.

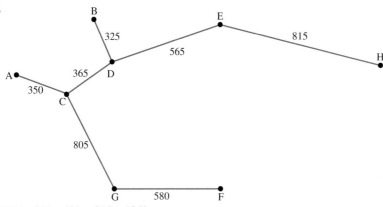

b. HEDA: $565 + 480 + 815 = 1860$ m

c. DFGCABEH: $1035 + 580 + 805 + 350 + 625 + 695 + 815 = 4905$ m

d. DEHFGCABD: $565 + 815 + 1195 + 580 + 805 + 350 + 625 + 325 = 5260$ m

4 Linear relationships

LESSON
4.1 Overview

Why learn this?

Often in life, we will be faced with a trade-off situation. This means that you are presented with multiple options and must decide on a combination of outcomes that provides you with the best result. Imagine a race with both swimming and running components, in which athletes start from a boat, swim to shore and then run along the beach to the finish line. Each athlete would have the following options:

- swim directly to shore and run a longer distance along the beach
- swim a longer distance diagonally through the ocean and reduce the distance required to run to reach the finish line
- swim directly through the ocean to the finish, covering the shortest possible distance.

Which option should an athlete take? This would depend on how far the athlete can swim or run, because reducing the swimming distance increases the running distance. To determine the best combination of swimming and running, an athlete could form equations based on speed, time and distance, and solve the equations simultaneously to find the best combination.

Just like the athletes in the scenario above, businesses face trade-offs every day, where they have to decide how much of each product they should produce in order to make the highest possible profit. As an example, a baker might make the most per-item profit from selling cakes, but if they don't produce muffins, bread and a range of other products, they will attract fewer customers and miss out on sales, reducing overall profit. Thus, a baker could use simultaneous equations to determine the best combination of baked goods to produce in order to maximise profit.

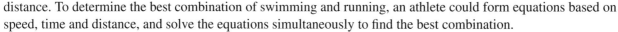

Hey students! Bring these pages to life online

▶ **Watch** videos

🧩 **Engage with** interactivities

A+ **Answer questions** and check solutions

Find all this and MORE in jacPLUS ▶

Reading content and rich media, including interactivities and videos for every concept

Extra learning resources

Differentiated question sets

Questions with immediate feedback, and fully worked solutions to help students get unstuck

1. Sammy has $35 credit from an App Store. She only buys apps that cost $2.50 each. Calculate the number of apps Sammy can buy and still have $27.50 credit.

2. Determine the equation of the line in the form $y = mx + c$.

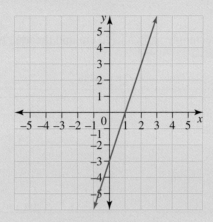

3. State the number of solutions to the pair of simultaneous equations $2x - y = 1$ and $-6x + 3y = -3$.

4. Use substitution to solve the simultaneous equations $y = 0.2x$ and $y = -0.3x + 0.5$. Give your answer as a coordinate pair.

5. **MC** Solve the inequality $2x + 3 > 5x - 6$ for x.
   A. $x < -1$ B. $x < 1$ C. $x > 1$ D. $x < 3$

6. **MC** Solve the inequality $\dfrac{5 - 2m}{3} \leq 2$ for m.

   A. $m \leq -\dfrac{1}{2}$ B. $m \geq -\dfrac{1}{2}$ C. $m \geq -\dfrac{11}{2}$ D. $m \leq -\dfrac{11}{2}$

7. Dylan received a better result for his Maths test than for his English test. If the sum of his two test results is 159 and the difference is 25, determine Dylan's maths test result.

8. **MC** Solve the pair of simultaneous equations $mx + ny = m$ and $x = y + n$ for x and y in terms of m and n.

   A. $x = \dfrac{m + n^2}{m + n}$ and $y = \dfrac{m(1 - n)}{m + n}$

   B. $x = 1$ and $y = \dfrac{m(1 - n)}{m + n}$

   C. $x = \dfrac{1}{n}$ and $y = \dfrac{m(1 - n)}{m + n}$

   D. $x = \dfrac{m + n^2}{m + n}$ and $y = \dfrac{1 - n}{n}$

9. **MC** Solve the pair of simultaneous equations $\dfrac{x}{3} - \dfrac{y}{2} = \dfrac{1}{6}$ and $\dfrac{x}{4} + \dfrac{y}{3} = \dfrac{1}{2}$.

A. $x = \dfrac{221}{6}$ and $y = \dfrac{9}{2}$

B. $x = \dfrac{13}{102}$ and $y = \dfrac{1}{34}$

C. $x = \dfrac{81}{17}$ and $y = \dfrac{17}{9}$

D. $x = \dfrac{22}{17}$ and $y = \dfrac{9}{17}$

10. **MC** If the perimeter of the rectangle shown is 22 cm and the area is 24 cm², select all possible values of x and y.

(y + 3) cm

2x cm

A. $x = 4, y = 0$

B. $x = 6.5, y = -5$

C. $x = 1\dfrac{1}{2}, y = 5$

D. $x = -\dfrac{3}{2}, y = -5$

E. $x = \dfrac{3}{2}, y = 5$

F. $x = -4, y = 0$

G. $x = 1.5, y = 5$

11. **MC** Identify the points of intersection between the line $y = x + 4$ and the hyperbola $y = \dfrac{5}{x}$.

A. $(-4, 4)$ and $(1, 5)$

B. $(-1, -5)$ and $(5, 1)$

C. $(-5, -1)$ and $(1, 5)$

D. $(5, 9)$ and $(-1, 3)$

12. **MC** Select all the point(s) of intersection between the circle $x^2 + y^2 = 8$ and the line $y = x$.

A. $(4, 4)$

B. $(2, 2)$

C. $(-4, -4)$

D. $(-2, -2)$

E. $(0, 0)$

F. $(-2, 2)$

G. $(2, -2)$

13. State at how many points the line $y = 2$ intersects with parabola $y = x^2 - 4$.

14. **MC** Identify the three inequalities that define the shaded region in the diagram.

A. $x > 5, y < 1, y < x + 2$

B. $x > 5, y < 1, y > -x + 2$

C. $x > 5, y < 1, y < -x + 2$

D. $x > 1, y < 5, y > x + 2$

15. **MC** Identify the region satisfying the systems of inequalities $2y - 3x > 1$ and $y + x < -2$.

A.

B.

C.

D.

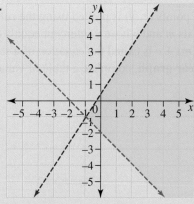

LESSON
4.2 Sketching linear graphs

LEARNING INTENTIONS

At the end of this lesson you should be able to:
- plot points on a graph using a rule and a table of values
- sketch linear graphs by determining the x- and y-intercepts
- sketch the graphs of horizontal and vertical lines
- model linear graphs from a worded context.

4.2.1 Plotting linear graphs

eles-4736

- If a series of points (x, y) is plotted using the rule $y = mx + c$, the points always lie in a straight line whose gradient equals m and whose y-intercept equals c.
- The rule $y = mx + c$ is called the equation of a straight line written in 'gradient–intercept' form.
- To plot a **linear graph**, complete a table of values to determine the points.

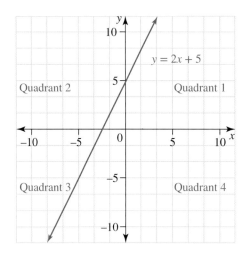

WORKED EXAMPLE 1 Plotting linear graphs

Plot the linear graph defined by the rule $y = 2x - 5$ for the x-values $-3, -2, -1, 0, 1, 2$ and 3.

THINK	WRITE/DRAW

1. Create a table of values using the given x-values.

x	-3	-2	-1	0	1	2	3
y							

2. Determine the corresponding y-values by substituting each x-value into the rule.

x	-3	-2	-1	0	1	2	3
y	-11	-9	-7	-5	-3	-1	1

3. Plot the points on a Cartesian plane and rule a straight line through them. Since the x-values have been specified, the line should only be drawn between the x-values of -3 and 3.
4. Label the graph.

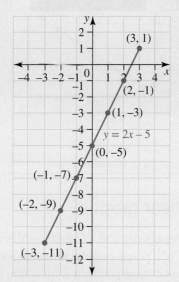

TI | THINK

1. In a new document, on a Lists & Spreadsheet page, label column A as x and label column B as y. Enter the x-values into column A.
Then in cell B1, complete the entry line as:
$= 2a1 - 5$
Then press ENTER.

DISPLAY/WRITE

2. Highlight cell B1 and press CTRL then click (the button in the middle of the direction arrows). Press the down arrow until you reach cell B7 then press ENTER.

3. Open a Data & Statistics page.
Press TAB to locate the label of the horizontal axis and select the variable x.
Press TAB again to locate the label of the vertical axis and select the variable y. The graph will be plotted as shown.

CASIO | THINK

1. On the Spreadsheet screen, enter the x-values into column A.
Then in cell B1, complete the entry line as:
$= 2A1 - 5$
Then press EXE.

DISPLAY/WRITE

2. Highlight cell B1 to B7.
Tap:
• Edit
• Fill
• Fill Range
• OK

3. Highlight cells A1 to B7.
Tap:
• Graph
• Scatter

4. To join the dots with a line, press:
   - MENU
   - 2: Plot Properties
   - 1: Connect Data Points

4. To join the dots with a line, tap:
   - Calc
   - Regression
   - Linear Reg

   Note that the equation is given, if required.

4.2.2 Sketching linear graphs

eles-4737

Sketching a straight line using the x- and y-intercepts

- We need a minimum of two points in order to sketch a straight-line (linear) graph.
- Since we need to label all critical points, it is most efficient to plot these graphs by determining the x- and y-intercepts.
- At the **x-intercept**, $y = 0$, so the coordinates of the x-intercept are $(x, 0)$.
- At the **y-intercept**, $x = 0$, so the coordinates of the y-intercept are $(0, y)$.

Sketching a straight-line graph

- **The x- and y-intercepts need to be labelled.**
- **The equation needs to be labelled.**

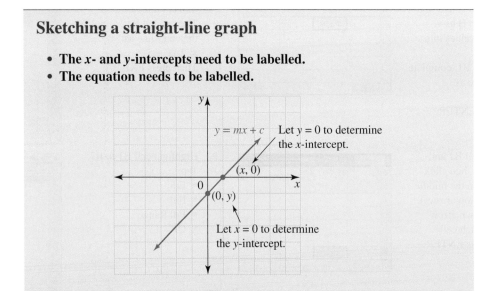

WORKED EXAMPLE 2 Sketching linear graphs

Sketch graphs of the following linear equations.

a. $2x + y = 6$

b. $y = -3x - 12$

THINK	WRITE/DRAW
a. 1. Write the equation.	a. $2x + y = 6$
2. Determine the x-intercept by substituting $y = 0$.	x-intercept: when $y = 0$, $2x + 0 = 6$ $2x = 6$ $x = 3$ The x-intercept is $(3, 0)$.

3. Determine the *y*-intercept by substituting $x = 0$.

y-intercept: when $x = 0$,
$2(0) + y = 6$
$y = 6$
The *y*-intercept is $(0, 6)$.

4. Plot both points and rule the line.

5. Label the graph.

b. 1. Write the equation.

   b. $y = -3x - 12$

2. Determine the *x*-intercept by substituting $y = 0$
   i. Add 12 to both sides of the equation.
   ii. Divide both sides of the equation by -3.

x-intercept: when $y = 0$,
$-3x - 12 = 0$
$-3x = 12$
$x = -4$
The *x*-intercept is $(-4, 0)$.

3. Determine the *y*-intercept. The equation is in the form $y = mx + c$, so compare this with our equation to determine the *y*-intercept, *c*.

$c = -12$
The *y*-intercept is $(0, -12)$.

4. Plot both points and rule the line.

5. Label the graph.

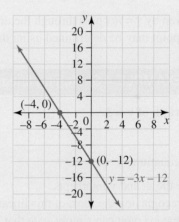

Sketching a straight line using the gradient–intercept method

- The gradient–intercept method is often used if the equation is in the form $y = mx + c$, where m represents the gradient (slope) of the straight line and c represents the *y*-intercept.
- The following steps outline how to use the gradient–intercept method to sketch a linear graph.

Step 1: Plot a point at the *y*-intercept.

Step 2: Write the gradient in the form $m = \dfrac{\text{rise}}{\text{run}}$.

Step 3: Starting from the *y*-intercept, move up the number of units suggested by the rise (move down if the gradient is negative).

Step 4: Move to the right the number of units suggested by the run and plot the second point.

Step 5: Rule a straight line through the two points.

WORKED EXAMPLE 3 Sketching more linear graphs

Sketch the graph of $y = \dfrac{2}{5}x - 3$ using the gradient–intercept method.

THINK	WRITE
1. Write the equation of the line.	$y = \dfrac{2}{5}x - 3$
2. Identify the value of c (that is, the y-intercept) and plot this point.	$c = -3$, so the y-intercept is $(0, -3)$.
3. Write the gradient, m, as a fraction.	$m = \dfrac{2}{5}$
4. $m = \dfrac{\text{rise}}{\text{run}}$; note the rise and run.	So rise $= 2$ and run $= 5$.
5. Starting from the y-intercept at $(0, -3)$, move 2 units up and 5 units to the right to find the second point $(5, -1)$. We have still not found the x-intercept.	

4.2.3 Sketching linear graphs of the form $y = mx$

eles-4738

- Graphs given by $y = mx$ pass through the origin $(0, 0)$, since $c = 0$.
- A second point may be determined using the rule $y = mx$ by substituting a value for x to determine y.

WORKED EXAMPLE 4 Sketching linear graphs of the form $y = mx$

Sketch the graph of $y = 3x$.

THINK	WRITE/DRAW
1. Write the equation.	$y = 3x$
2. Determine the x- and y-intercepts. *Note:* By recognising the form of this linear equation, $y = mx$, you can simply state that the graph passes through the origin, $(0, 0)$.	x intercept: when $y = 0$, $0 = 3x$ $x = 0$ y-intercept: $(0, 0)$ Both the x- and the y-intercept are at $(0, 0)$.
3. Determine another point to plot by calculating the y-value when $x = 1$.	When $x = 1$, $\quad y = 3 \times 1$ $\qquad\qquad\qquad = 3$ Another point on the line is $(1, 3)$.

4. Plot the two points $(0, 0)$ and $(1, 3)$ and rule a straight line through them.

5. Label the graph.

▶ 4.2.4 Sketching linear graphs of the form $y = c$ and $x = a$

eles-4739

- The line $y = c$ is parallel to the x-axis, having a gradient of zero and a y-intercept of c.
- The line $x = a$ is parallel to the y-axis and has an undefined (infinite) gradient.

Horizontal and vertical lines

- **Horizontal lines are in the form $y = c$.**
- **Vertical lines are in the form $x = a$.**

WORKED EXAMPLE 5 Sketching graphs of the form $y = c$ and $x = a$

Sketch graphs of the following linear equations.

a. $y = -3$

b. $x = 4$

THINK

a. 1. Write the equation.

   2. The y-intercept is -3. As x does not appear in the equation, the line is parallel to the x-axis, such that all points on the line have a y-coordinate equal to -3. That is, this line is the set of points $(x, -3)$ where x is an element of the set of real numbers.

   3. Sketch a horizontal line through $(0, -3)$.

   4. Label the graph.

WRITE/DRAW

a. $y = -3$

y-intercept $= -3$, $(0, -3)$

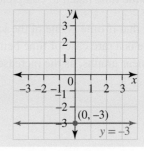

b. 1. Write the equation.

2. The x-intercept is 4. As y does not appear in the equation, the line is parallel to the y-axis, such that all points on the line have an x-coordinate equal to 4. That is, this line is the set of points $(4, y)$ where y is an element of the set of real numbers.

3. Sketch a vertical line through $(4, 0)$.

4. Label the graph.

b. $x = 4$

x-intercept $= 4, (4, 0)$

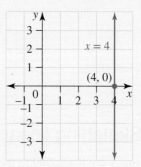

▶ 4.2.5 Using linear graphs to model real-life contexts

eles-4740

- If a real-life situation involves a constant increase or decrease at regular intervals, it can be modelled by a linear equation. Examples include water being poured from a tap into a container at a constant rate, or money being deposited into a savings account at regular intervals.
- To model a linear situation, we first need to determine which of the two given variables is the **independent variable** and which is the **dependent variable**.
- With numerical bivariate data, we often see the independent variable being referred to as the **explanatory variable**. Likewise, the dependent variable is is often referred to as the **response variable**.
- The independent variable does not depend on the value of the other variable, whereas the dependent variable takes its value depending on the value of the other variable. When plotting a graph of a linear model, the independent variable will be on the x-axis (horizontal axis) and the dependent variable will be on the y-axis (vertical axis).
- Real-life examples identifying the variables are shown in the following table.

Situation	Independent variable (explanatory variable)	Dependent variable (response variable)
Money being deposited into a savings account at regular intervals	Time	Money in account
The age of a person in years and their height in cm	Age in years	Height in cm
The temperature at a snow resort and the depth of the snow	Temperature	Depth of snow
The length of Pinocchio's nose and the number of lies he told	Number of lies Pinocchio told	Length of Pinocchio's nose
The number of workers building a house and the time taken to complete the project	Number of workers	Time

- Note that if time is one of the variables, it will usually be the independent variable. The final example above is a rare case of time being the dependent variable. Also, some of the above cases can't be modelled by linear graphs, as the increases or decreases aren't necessarily happening at constant rates.

Water is leaking from a bucket at a constant rate. After 1 minute there is 45 litres in the bucket; after 3 minutes there is 35 litres in the bucket; after 5 minutes there is 25 litres in the bucket; and after 7 minutes there is 15 litres in the bucket.
a. Define two variables to represent the given information.
b. Determine which variable is the independent variable and which is the dependent variable.
c. Represent the given information in a table of values.
d. Plot a graph to represent how the amount of water in the bucket is changing.
e. Use your graph to determine how much water was in the bucket at the start and how long it will take for the bucket to be empty.

THINK

a. Determine which two values change in the relationship given.

b. The dependent variable takes its value depending on the value of the independent variable.
In this situation the amount of water depends on the amount of time elapsed, not the other way round.

c. The independent variable should appear in the top row of the table of values, with the dependent variable appearing in the second row.

d. The values in the top row of the table represent the values on the horizontal axis, and the values in the bottom row of the table represent the values on the vertical axis. As the value for time can't be negative and there can't be a negative amount of water in the bucket, only the first quadrant needs to be drawn for the graph. Plot the 4 points and rule a straight line through them. Extend the graph to meet the vertical and horizontal axes.

e. The amount of water in the bucket at the start is the value at which the line meets the vertical axis, and the time taken for the bucket to be empty is the value at which the line meets the horizontal axis.
Note: Determining the time when the bucket will be empty is an example of *extrapolation*, as this time is determined by extending the graph beyond the known data points.

WRITE/DRAW

a. The two variables are 'time' and 'amount of water in bucket'.

b. Independent variable = time
Dependent variable = amount of water in bucket

c.

Time (minutes)	1	3	5	7
Amount of water in bucket (litres)	45	35	25	15

d.

e. There was 50 litres of water in the bucket at the start, and it will take 10 minutes for the bucket to be empty.

Exercise 4.2 Sketching linear graphs

learn on

4.2 Quick quiz on	4.2 Exercise

Individual pathways

■ PRACTISE	■ CONSOLIDATE	■ MASTER
1, 4, 7, 10, 13, 16, 21, 24	2, 5, 8, 11, 14, 17, 19, 22, 25	3, 6, 9, 12, 15, 18, 20, 23, 26

Fluency

1. **WE1** Generate a table of values and plot the linear graphs defined by the following rules for the given range of *x*-values.

Rule	***x*-values**
a. $y = 10x + 25$	$-5, -4, -3, -2, -1, 0, 1$
b. $y = 5x - 12$	$-1, 0, 1, 2, 3, 4$
c. $y = -0.5x + 10$	$-6, -4, -2, 0, 2, 4$

2. Generate a table of values and plot the linear graphs defined by the following rules for the given range of *x*-values.

Rule	***x*-values**
a. $y = 100x - 240$	$0, 1, 2, 3, 4, 5$
b. $y = -5x + 3$	$-3, -2, -1, 0, 1, 2$
c. $y = 7 - 4x$	$-3, -2, -1, 0, 1, 2$

3. Plot the linear graphs defined by the following rules for the given range of x-values.

Rule   **x-values**

a. $y = -3x + 2$

x	−6	−4	−2	0	2	4	6
y							

b. $y = -x + 3$

x	−3	−2	−1	0	1	2	3
y							

c. $y = -2x + 3$

x	−6	−4	−2	0	2	4	6
y							

WE2 For questions 4 to 6, sketch the graphs of the following linear equations by determining the x- and y-intercepts.

4. a. $5x - 3y = 10$ b. $5x + 3y = 10$ c. $-5x + 3y = 10$
   d. $-5x - 3y = 10$ e. $2x - 8y = 20$

5. a. $4x + 4y = 40$ b. $-x + 6y = 120$ c. $-2x + 8y = -20$
   d. $10x + 30y = -150$ e. $5x + 30y = -150$

6. a. $-9x + 4y = 36$ b. $6x - 4y = -24$ c. $y = 2x - 10$

   d. $y = -5x + 20$ e. $y = -\dfrac{1}{2}x - 4$

WE3 For questions 7 to 9, sketch the graphs of the following using the gradient–intercept method.

7. a. $y = 4x + 1$ b. $y = 3x - 7$ c. $y = -2x + 3$

8. a. $y = -5x - 4$ b. $y = \dfrac{1}{2}x - 2$ c. $y = -\dfrac{2}{7}x + 3$

9. a. $y = 0.6x + 0.5$ b. $y = 8x$ c. $y = x - 7$

WE4 For questions 10 to 12, sketch the graphs of the following linear equations on the same set of axes.

10. a. $y = 2x$ b. $y = \dfrac{1}{2}x$ c. $y = -2x$

11. a. $y = 5x$ b. $y = \dfrac{1}{3}x$ c. $y = -\dfrac{5}{2}x$

12. a. $y = \dfrac{2}{3}x$ b. $y = -3x$ c. $y = -\dfrac{3}{2}x$

WE5 For questions 13 to 15, sketch the graphs of the following linear equations.

13. a. $y = 10$ b. $x = -10$ c. $x = 0$

14. a. $y = -10$ b. $y = 100$ c. $x = -100$

15. a. $x = 10$ b. $y = 0$ c. $y = -12$

Understanding

For questions **16** to **18**, transpose each of the equations to standard form (that is, $y = mx + c$). State the x- and y-intercept for each.

16. a. $5(y + 2) = 4(x + 3)$ **b.** $5(y - 2) = 4(x - 3)$ **c.** $2(y + 3) = 3(x + 2)$

17. a. $10(y - 20) = 40(x - 2)$ **b.** $4(y + 2) = -4(x + 2)$ **c.** $2(y - 2) = -(x + 5)$

18. a. $-5(y + 1) = 4(x - 4)$ **b.** $5(y + 2.5) = 2(x - 3.5)$ **c.** $2.5(y - 2) = -6.5(x - 1)$

19. Determine the x- and y-intercepts of the following lines.

     a. $-y = 8 - 4x$ **b.** $6x - y + 3 = 0$ **c.** $2y - 10x = 50$

20. Explain why the gradient of a horizontal line is equal to zero and the gradient of a vertical line is undefined.

Reasoning

21. **WE6** Your friend loves to download music. She earns \$50 and spends some of it buying music online at \$1.75 per song. She saves the remainder.
Her saving is given by the function $y = 50 - 1.75x$.

     a. Determine which variable is the independent variable and which is the dependent variable.

     b. Sketch the function.

     c. Determine the number of songs your friend can buy and still save \$25.

22. Determine whether $\dfrac{x}{3} - \dfrac{y}{2} = \dfrac{7}{6}$ is the equation of a straight line by rearranging into an appropriate form and hence sketch the graph, showing all relevant features.

23. Nikita works a part-time job and is interested in sketching a graph of her weekly earnings. She knows that in a week where she does not work any hours, she will still earn \$25.00 for being 'on call'. On top of this initial payment, Nikita earns \$20.00 per hour for her regular work.
Nikita can work a maximum of 8 hours per day as her employer is unwilling to pay her overtime.

     a. Write a linear equation that represents the amount of money Nikita could earn in a week.
(*Hint:* You might want to consider the 'on call' amount as an amount of money earned for zero hours worked.)

     b. Sketch a graph of Nikita's weekly potential earnings.

     c. Determine the maximum amount of money that Nikita can earn in a single week.

Problem solving

24. The temperature in a room is rising at a constant rate. Initially (when time equals zero), the temperature of the room is 15 °C. After 1 hour, the temperature of the room has risen to 18 °C.
After 3 hours, the temperature has risen to 24 °C.

     a. Using the variables t to represent the time in hours and T to represent the temperature of the room, identify the dependent and the independent variable in this linear relationship.

     b. **i.** Construct a table of values to represent this information.

         ii. Plot this relationship on a suitable axis.

     c. If the maximum temperature of the room was recorded to be 30 °C, evaluate after how many hours was this recording taken.

25. Water is flowing from a tank at a constant rate. The equation relating the volume of water in the tank, V litres, to the time the water has been flowing from the tank, t minutes, is given by $V = 80 - 4t$, $t \geq 0$.

   a. Determine which variable is the independent variable and which is the dependent variable.
   b. Calculate how much water is in the tank initially.
   c. Explain why it is important that $t \geq 0$.
   d. Determine the rate at which the water is flowing from the tank.
   e. Determine how long it takes for the tank to empty.
   f. Sketch the graph of V versus t.

26. A straight line has a general equation defined by $y = mx + c$. This line intersects the lines defined by the rules $y = 7$ and $x = 3$. The lines $y = mx + c$ and $y = 7$ have the same y-intercept, and $y = mx + c$ and $x = 3$ have the same x-intercept.

   a. On the one set of axes, sketch all three graphs.
   b. Determine the y-axis intercept for $y = mx + c$.
   c. Determine the gradient for $y = mx + c$.
   d. **MC** The equation of the line defined by $y = mx + c$ is:

   A. $x + y = 3$
   B. $7x + 3y = 21$
   C. $3x + 7y = 21$
   D. $x + y = 7$

LESSON
4.3 Determining linear equations

LEARNING INTENTIONS

At the end of this lesson you should be able to:
   • determine the equation of a straight line when given its graph
   • determine the equation of a straight line when given the gradient and the y-intercept
   • determine the equation of a straight line passing through two points
   • formulate the equation of a straight line from a written context
   • calculate average speed and simple interest.

4.3.1 Determining a linear equation given two points

eles-4741

   • The gradient of a straight line can be calculated from the coordinates of two points (x_1, y_1) and (x_2, y_2) that lie on the line.
   • The equation of the straight line can then be found in the form $y = mx + c$, where c is the y-intercept.

Gradient of a straight line

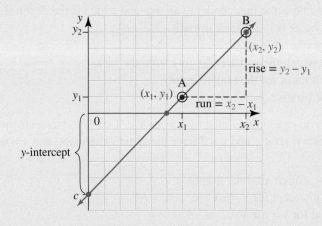

- **The equation of a straight line is given by $y = mx + c$.**
- **m is the value of the gradient and c is the value of the y-intercept.**

$$\text{Gradient} = m = \frac{\textbf{rise}}{\textbf{run}} = \frac{y_2 - y_1}{x_2 - x_1}$$

WORKED EXAMPLE 7 Determining equations with a known y-intercept

Determine the equation of the straight line shown in the graph.

THINK

1. There are two points given on the straight line: the x-intercept $(3, 0)$ and the y-intercept $(0, 6)$.

2. Calculate the gradient of the line by applying the formula $m = \dfrac{\text{rise}}{\text{run}} = \dfrac{y_2 - y_1}{x_2 - x_1}$, where $(x_1, y_1) = (3, 0)$ and $(x_2, y_2) = (0, 6)$.

3. The graph has a y-intercept of 6, so $c = 6$.
   Substitute $m = -2$ and $c = 6$ into $y = mx + c$ to determine the equation.

WRITE

$(3, 0), (0, 6)$

$m = \dfrac{\text{rise}}{\text{run}}$

$\quad = \dfrac{y_2 - y_1}{x_2 - x_1}$

$\quad = \dfrac{6 - 0}{0 - 3}$

$\quad = \dfrac{6}{-3}$

$\quad = -2$

The gradient $m = -2$.

$y = mx + c$
$y = -2x + 6$

Determine the equation of the straight line shown in the graph.

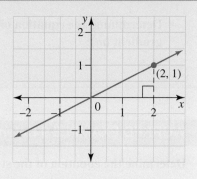

THINK

1. There are two points given on the straight line: the x- and y-intercept $(0, 0)$, and another point $(2, 1)$.

2. Calculate the gradient of the line by applying the formula $m = \dfrac{\text{rise}}{\text{run}} = \dfrac{y_2 - y_1}{x_2 - x_1}$, where $(x_1, y_1) = (0, 0)$ and $(x_2, y_2) = (2, 1)$.

3. The y-intercept is 0, so $c = 0$. Substitute $m = \dfrac{1}{2}$ and $c = 0$ into $y = mx + c$ to determine the equation.

WRITE

$(0, 0), (2, 1)$

$m = \dfrac{\text{rise}}{\text{run}}$

$= \dfrac{y_2 - y_1}{x_2 - x_1}$

$= \dfrac{1 - 0}{2 - 0}$

$= \dfrac{1}{2}$

The gradient $m = \dfrac{1}{2}$.

$y = mx + c$

$y = \dfrac{1}{2}x + 0$

$y = \dfrac{1}{2}x$

4.3.2 A straight line equation

eles-4742

- The diagram shows a line of gradient m passing through the point (x_1, y_1).
- If (x, y) is any other point on the line, then:

$$m = \dfrac{\text{rise}}{\text{run}}$$

$$m = \dfrac{y - y_1}{x - x_1}$$

$$m(x - x_1) = y - y_1$$

$$y - y_1 = m(x - x_1)$$

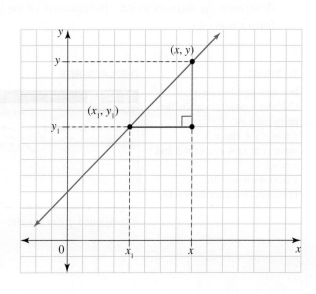

- The formula $y - y_1 = m(x - x_1)$ can be used to write down the equation of a line, given the gradient and the coordinates of one point.

The equation of a straight line

- Determining the equation of a straight line with coordinates of one point (x_1, y_1) and the gradient (m):

$$y - y_1 = m(x - x_1)$$

WORKED EXAMPLE 9 Determining the equation using the gradient and the y-intercept

Determine the equation of the straight line with a gradient of 2 and a y-intercept of -5.

THINK	WRITE
1. Write the gradient formula.	$y - y_1 = m(x - x_1)$
2. State the known variables.	$m = 2, (x_1, y_1) = (0, -5)$
3. Substitute the values into the formula.	$y - (-5) = 2(x - 0)$ $y + 5 = 2x$
4. Rearrange the formula. *Note:* You could also solve this by using the equation $y = mx + c$ and substituting directly for m and c.	$y = 2x - 5$

WORKED EXAMPLE 10 Determining the equation using the gradient and another point

Determine the equation of the straight line with a gradient of 3 and passing through the point $(5, -1)$.

THINK	WRITE
1. Write the gradient formula.	$y - y_1 = m(x - x_1)$
2. State the known variables.	$m = 3, x_1 = 5, y_1 = -1$
3. Substitute the values $m = 3, x_1 = 5, y_1 = -1$ into the formula.	$y - (-1) = 3(x - 5)$ $y + 1 = 3x - 15$
4. Rearrange the formula to state the equation of the line in the form $y = mx + c$.	$y = 3x - 16$

TI	THINK	DISPLAY/WRITE	CASIO	THINK	DISPLAY/WRITE
In a new problem on a Calculator page, complete the entry lines as: $y = m \times x + c | m = 3$ solve $(y = 3x + c, c)	x = 5$ and $y = -1$ $y = 3x + c	c = -16$ Press ENTER after each entry.	 The equation is $y = 3x - 16$.	On the Main screen, complete the entry lines as: solve $(y = 3x + c, c)$ $| x = 5 | y = -1$ Press EXE.	 The equation is $y = 3x - 16$.

WORKED EXAMPLE 11 Determining the equation of a line using two points

Determine the equation of the straight line passing through the points $(-2, 5)$ and $(1, -1)$.

THINK	WRITE
1. Write the gradient formula.	$y - y_1 = m(x - x_1)$
2. State the known variables.	$(x_1, y_1) = (-2, 5)$ $(x_2, y_2) = (1, -1)$
3. Substitute the values $(x_1, y_1) = (-2, 5)$ and $(x_2, y_2) = (1, -1)$ to calculate the gradient from the given points.	$m = \dfrac{y_2 - y_1}{x_2 - x_1}$ $m = \dfrac{-1 - 5}{1 - (-2)}$ $m = \dfrac{-6}{3}$ $= -2$
4. Substitute the values $m = -2$ and $(x_1, y_1) = (-2, 5)$ into the formula for the equation of a straight line.	$y - y_1 = m(x - x_1)$ $y - 5 = -2(x - (-2))$
5. Rearrange the formula to state the equation of the line in the form $y = mx + c$.	$y - 5 = -2(x + 2)$ $y = -2x - 4 + 5$ $y = -2x + 1$

WORKED EXAMPLE 12 Writing an equation in the form $ax + by + c = 0$

Determine the equation of the line with a gradient of -2 that passes through the point $(3, -4)$. Write the equation in general form, that is in the form $ax + by + c = 0$.

THINK	WRITE
1. Use the formula $y - y_1 = m(x - x_1)$. Write the values of x_1, y_1, and m.	$m = -2$, $x_1 = 3$, $y_1 = -4$ $y - y_1 = m(x - x_1)$
2. Substitute for x_1, y_1, and m into the equation.	$y - (-4) = -2(x - 3)$ $y + 4 = -2x + 6$
3. Transpose the equation into the form $ax + by + c = 0$.	$y + 4 + 2x - 6 = 0$ $2x + y - 2 = 0$

A printer prints pages at a constant rate. It can print 165 pages in 3 minutes and 275 pages in 5 minutes.

a. **Identify which variable is the independent variable (x) and which is the dependent variable (y).**
b. **Calculate the gradient of the equation and explain what this means in the context of the question.**
c. **Write an equation in algebraic form linking the independent and dependent variables.**
d. **Rewrite your equation in words.**
e. **Using the equation, determine how many pages can be printed in 11 minutes.**

THINK	WRITE/DRAW
a. The dependent variable takes its value depending on the value of the independent variable. In this situation the number of pages depends on the time elapsed, not the other way round.	a. Independent variable = time Dependent variable = number of pages
b. 1. Determine the two points given by the information in the question.	b. $(x_1, y_1) = (3, 165)$ $(x_2, y_2) = (5, 275)$
2. Substitute the values of these two points into the formula to calculate the gradient.	$\begin{aligned} m &= \frac{y_2 - y_1}{x_2 - x_1} \\ &= \frac{275 - 165}{5 - 3} \\ &= \frac{110}{2} \\ &= 55 \end{aligned}$
3. The gradient states how much the dependent variable increases for each increase of 1 unit in the independent variable.	In the context of the question, this means that each minute 55 pages are printed.
c. The graph travels through the origin, as the time elapsed for the printer to print 0 pages is 0 seconds. Therefore, the equation will be in the form $y = mx$. Substitute in the value of m.	c. $y = mx$ $y = 55x$
d. Replace x and y in the equation with the independent and dependent variables.	d. Number of pages $= 55 \times$ time
e. 1. Substitute $x = 11$ into the equation.	e. $\begin{aligned} y &= 55x \\ &= 55 \times 11 \\ &= 605 \end{aligned}$
2. Write the answer in words.	The printer can print 605 pages in 11 minutes.

▶ 4.3.3 Applications of linear equations

eles-6177

- There are many applications of linear equations. Here we will examine two: average speed and simple interest.

Average speed

- Average speed is a measure of how fast an object is travelling between two periods of time. We can use the gradient of the line segment between two distinct points as a measure of the rate of change of distance with respect to time to obtain a value for average speed.

Speed

- **Speed is the rate of change of distance with respect to time.**

$$\text{Speed} = \frac{\text{distance}}{\text{time}}$$

Speed = gradient (slope) of the straight line

- **The units of speed are expressed in length/time, usually metres per second (m/s) or kilometre per hour (km/h).**
- **In graphs, the sharper the slope of a line, the greater the speed.**
- **In graphs, if the slope is flat, the speed is zero.**

The graph shows the distance in metres (y-axis) over time in seconds (x-axis) of an object that moves at a constant speed. Since the speed is constant, the line is straight.

$$\text{Speed} = \frac{\text{rise}}{\text{run}} = \frac{\text{change in distance}}{\text{change in time}} = \frac{15 - 0}{3 - 0} = \frac{15}{3} = 5 \text{ m/s}$$

Note: Don't forget to write the unit of speed.

WORKED EXAMPLE 14 Calculating speed

Consider the following distance–time graph.
Calculate the speed:
a. **between 0 and 1 seconds**
b. **between 1 and 2 seconds**
c. **at $t = 3$ seconds**

▶

THINK

a. 1. To calculate the average speed between 0 and 1 seconds, calculate the gradient of the line A.

2. As the line A is a straight line, the speed is constant; that is, in the time period from $t = 0$ s to $t = 1$ s, the speed will be same. Write the answer.

b. 1. To calculate the average speed between 1 and 2 seconds, calculate the gradient of the line B.

2. As the line B is a straight line, the speed is constant; that is, in the time period from $t = 1$ s to $t = 2$ s, the speed will be same. Write the answer.

DRAW/WRITE

Distance–time graph

$$\text{Speed} = \frac{\text{rise}}{\text{run}} = \frac{\text{change in distance}}{\text{change in time}} = \frac{10 - 0}{1 - 0} = \frac{10}{1} = 10\,\text{m/s}$$

The average speed between 0 and 1 seconds is 10 m/s.

Distance–time graph

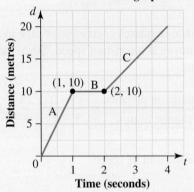

$$\text{Speed} = \frac{\text{rise}}{\text{run}} = \frac{\text{change in distance}}{\text{change in time}} = \frac{10 - 10}{2 - 1} = \frac{0}{1} = 0\,\text{m/s}$$

There is no change in the distance. This means the object did not move from $t = 1$ s to $t = 2$ s.

The average speed between 1 and 2 seconds is 0 m/s.

c. 1. To calculate the speed at $t = 3$ seconds, calculate the gradient of the line C, since the gradient at the instance of $t = 3$ will be the same as the gradient of the line C.

Distance–time graph

$$\text{Speed} = \frac{\text{rise}}{\text{run}} = \frac{\text{change in distance}}{\text{change in time}} = \frac{20 - 10}{4 - 2} = \frac{10}{2} = 5 \text{ m/s}$$

2. As the line C is a straight line, the speed is constant; that is, in the time period from $t = 2$ s to $t = 4$ s, the speed will be same. Write the answer.

The instantaneous speed at $t = 3$ seconds is 5 m/s.

Simple interest

- When you put money in a financial institution such as a bank or credit union, the amount of money you start with is called the **principal**.
- **Interest** is the fee charged for the use of someone else's money.
- The amount of interest is determined by the interest rate.
- Interest rates are quoted as a percentage over a given time period, usually a year.
- **Simple interest** is the interest paid based on the principal of an investment.
- The principal remains constant and does not change from one period to the next.
- Since the amount of interest paid for each time period is based on the principal, the amount of interest paid is constant and is therefore linear.

Formula for simple interest

$$I = Pin$$

where:
- I = the amount of interest earned or paid
- P = the principal
- i = the interest rate as a percentage per annum (yearly), written as a decimal (e.g. 2% p.a. is equal to 0.02 p.a.)
- n = the duration of the investment in years.

- The total amount of money, which combines the principal (the initial amount) and the interest earned, is known as the value of the investment.

Value of an investment or loan

$$A = P + I$$

where:
- A = the total amount of money at the end of the investment or loan period
- P = the principal (the initial amount invested or borrowed)
- I = the amount of interest earned.
  The formula could also be written as:

$$A = P(1 + in)$$

- When working with money, you need to know how to round it, either to the nearest cent, to the nearest dollar, or to a larger amount such as the tens or hundreds place.
- Two rounding techniques are commonly used when rounding money. The first is rounding to the nearest dollar. Rounding to the nearest dollar is common to use when you fill out your tax return each year.
- The second is rounding to the nearest cent. This is common when you have monetary calculations where the amounts do not come out to the exact cent.

Rounding money

- **When rounding to the nearest cent, look at the number to the right of the full cents.**
  - **If the number is 5 or more, increase the cents by 1.**
  - **If the number is 4 or less, keep the cents the same.**

WORKED EXAMPLE 15 Calculating the value of a loan at simple interest

Calculate the total amount of money at the end of the loan period for a loan of $2500 for 7 months at 5.3% p.a. simple interest. Give your answer to the nearest cent.

THINK	WRITE
1. Write the simple interest formula.	$I = Pin$
2. State the known values of the variables.	$P = \$2500$, $I = 0.053$, $n = \dfrac{7}{12}$
3. Substitute the given values into the simple interest formula and evaluate the value of interest.	$I = 2500 \times 0.053 \times \dfrac{7}{12}$ $= 77.291\,666\,666\,7$ ≈ 77.29 The value of the interest is \$77.29 (rounding to the nearest cent).
4. Substitute the values for P and I in the formula $A = P + I$ and calculate the total amount at the end of the loan period.	$A = P + I$ $= 2500 + 77.29$ $= 2577.29$
5. Write the answer in a sentence.	The total amount of the loan at the end of the loan period is \$2577.29.

Note: When using the simple interest and total amount of loan formula, we often neglect the fact that in real life, money amounts are rounded to the nearest cent. As a consequence, when the amounts are rounded, their values may differ by a few cents from the amounts given by these formulas.

DISCUSSION

What problems might you encounter when calculating the equation of a line whose graph is actually parallel to one of the axes?

 Resources

 eWorkbook   Topic 4 Workbook (worksheets, code puzzle and project) (ewbk-2095)

Video eLesson The equation of a straight line (eles-2313)

 Interactivities   Individual pathway interactivity: Determining linear equations (int-4573)
Linear graphs (int-6484)

Exercise 4.3 Determining linear equations

learn on

4.3 Quick quiz on	4.3 Exercise

Individual pathways

■ PRACTISE	■ CONSOLIDATE	■ MASTER
1, 4, 7, 10, 12, 13, 17, 20	2, 5, 8, 11, 14, 18, 21	3, 6, 9, 15, 16, 19, 22

Fluency

1. **WE7** Determine the equation of each of the straight lines shown.

a.

b.

c.

d.

2. Determine the equation of each of the straight lines shown.

a.

b.

c.

d.
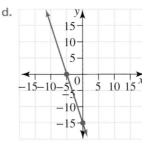

3. **WE8** Determine the equation of each of the straight lines shown.

a.

b.

c.

d.
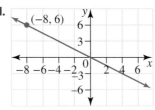

4. **WE9** Determine the linear equation given the information in each case.

a. Gradient $= 3$, y-intercept $= 3$
b. Gradient $= -3$, y-intercept $= 4$
c. Gradient $= -4$, y-intercept $= 2$
d. Gradient $= 4$, y-intercept $= 2$
e. Gradient $= -1$, y-intercept $= 4$

5. Determine the linear equation given the information in each case.

a. Gradient $= 0.5$, y-intercept $= -4$
b. Gradient $= 5$, y-intercept $= 2.5$
c. Gradient $= -6$, y-intercept $= 3$
d. Gradient $= -2.5$, y-intercept $= 1.5$
e. Gradient $= 3.5$, y-intercept $= 6.5$

6. **WE10** For each of the following, determine the equation of the straight line with the given gradient and passing through the given point.

a. Gradient $= 5$, point $= (5, 6)$
b. Gradient $= -5$, point $- (5, 6)$
c. Gradient $= -4$, point $= (-2, 7)$
d. Gradient $= 4$, point $= (8, -2)$
e. Gradient $= 3$, point $= (10, -5)$

7. For each of the following, determine the equation of the straight line with the given gradient and passing through the given point.
   a. Gradient $= -3$, point $= (3, -3)$
   b. Gradient $= -2$, point $= (20, -10)$
   c. Gradient $= 2$, point $= (2, -0.5)$
   d. Gradient $= 0.5$, point $= (6, -16)$
   e. Gradient $= -0.5$, point $= (5, 3)$

8. **WE11** Determine the equation of the straight line that passes through each of the following pairs of points.
   a. $(1, 4)$ and $(3, 6)$
   b. $(0, -1)$ and $(3, 5)$
   c. $(-1, 4)$ and $(3, 2)$

9. Determine the equation of the straight line that passes through each of the following pairs of points.
   a. $(3, 2)$ and $(-1, 0)$
   b. $(-4, 6)$ and $(2, -6)$
   c. $(-3, -5)$ and $(-1, -7)$

10. **WE14** Consider the following distance–time graph.

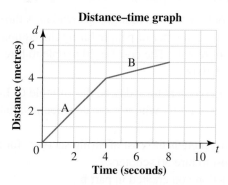

Calculate the instantaneous speed at:
   a. $t = 2$ seconds
   b. $t = 6$ seconds

11. Sam drove 60 kilometres from his home to Millgrove. He stopped and visited his friend Callum on the way.

   a. Calculate Sam's speed for the first part (line A) of his journey.
   b. Determine how much time Sam spent visiting Callum.
   c. Calculate Sam's speed for the last part (line C) of his journey.

12. **WE15** Calculate the total amount of money at the end of the loan period for a loan of $5000 for 3 years at 8% p.a simple interest. Give your answer to the nearest cent.

Understanding

13. **WE13** a. Determine which variable (time or cost) is the independent variable and which is the dependent variable in the Supa-Bowl advertisement shown.

> Save $$$ with Supa-Bowl!!!
> NEW Ten-Pin Bowling Alley
> *Shoe rental just $2 (fixed fee)*
> Rent a lane for ONLY $6/hour!

b. If t represents the time in hours and C represents cost ($), construct a table of values for 0–3 hours for the cost of playing ten-pin bowling at the new alley.

c. Use your table of values to plot a graph of time versus cost. (*Hint:* Ensure your time axis (horizontal axis) extends to 6 hours and your cost axis (vertical axis) extends to $40.)

d. i. Identify the y-intercept.

   ii. Describe what the y-intercept represents in terms of the cost.

e. Calculate the gradient and explain what this means in the context of the question.

f. Write a linear equation to describe the relationship between cost and time.

g. Use your linear equation from part f to calculate the cost of a 5-hour tournament.

h. Use your graph to check your answer to part g.

14. A local store has started renting out scooters to tour groups who pass through the city. Groups are charged based on the number of people hiring the equipment. There is a flat charge of $10.00 any time you book a day of rentals. The cost for 20 people to hire scooters is $310.00, and the cost for 40 people to hire scooters is $610.00.

a. Let the cost in dollars for hiring scooters for a day be the variable C. Let the number of people hiring scooters be the variable n. Identify which is the dependent variable and which is the independent variable.

b. Formulate a linear equation that models the cost of hiring scooters for a day.

c. Calculate how much it will cost to hire 30 scooters.

d. Sketch a graph of the cost function you created in part b.

15. If the simple interest charged on a loan of $9800 over 3 years is $2352, determine the percentage rate of interest that was charged.

16. The Robinsons' water tank sprang a leak and has been losing water at a steady rate. Four days after the leak occurred, the tank contained 552 L of water, and ten days later it held only 312 L.

a. Determine the rule linking the amount of water in the tank (w) and the number of days (t) since the leak occurred.

b. Calculate how much water was in the tank initially.

c. If water loss continues at the same rate, determine when the tank will be empty.

Reasoning

17. When using the gradient to draw a line, does it matter if you rise before you run or run before you rise? Explain your answer.

18. a. Using the graph shown, write a general formula for the gradient m in terms of x, y and c.

   b. Transpose your formula to make y the subject. Explain what you notice.

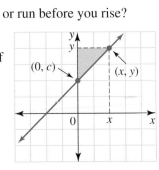

19. The points A (x_1, y_1), B (x_2, y_2) and P (x, y) all lie on the same line. P is a general point that lies anywhere on the line. Given that the gradient from A to P must be equal to the gradient from P to B, show that an equation relating these three points is given by:

$$y - y_1 = \frac{y_2 - y_1}{x_2 - x_1}(x - x_1)$$

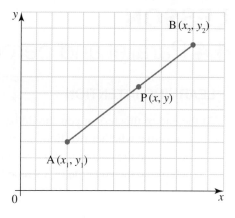

Problem solving

20. ABCD is a parallelogram with coordinates A $(2, 1)$, B $(3, 6)$ and C $(7, 10)$.

   a. Calculate the value of the gradient of the line AB.
   b. Determine the equation of the line AB.
   c. Calculate the value of the gradient of the line CD.
   d. Determine the coordinates of the point D.

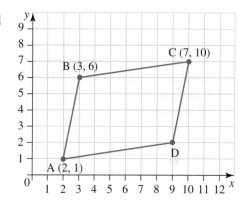

21. Show that the quadrilateral ABCD is a parallelogram.

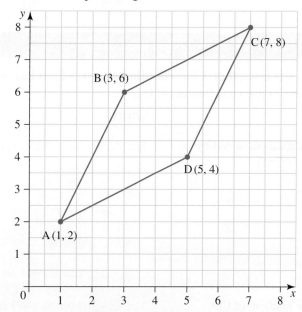

22. $2x + 3y = 5$ and $ax - 6y = b$ are the equations of two lines.

   a. If both lines have the same y-intercept, determine the value of b.
   b. If both lines have the same gradient (but a different y-intercept), determine the value of a.

LESSON
4.4 Parallel and perpendicular lines

LEARNING INTENTION

At the end of this lesson you should be able to:
- determine whether two lines are parallel, perpendicular or neither
- determine the equation of a straight line that is parallel to a given line
- determine the equation of a straight line that is perpendicular to a given line.

▶ 4.4.1 Parallel lines

eles-4743

- Lines that have the same gradient are **parallel** lines.
- The three lines (pink, green and blue) on the graph shown all have a gradient of 1 and are parallel to each other.
- Parallel lines will never intersect with on another.

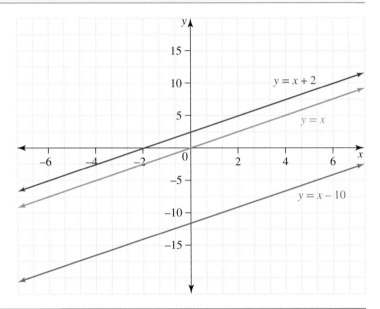

WORKED EXAMPLE 16 Proving that two lines are parallel

Show that AB is parallel to CD given that A has coordinates $(-1, -5)$, B has coordinates $(5, 7)$, C has coordinates $(-3, 1)$ and D has coordinates $(4, 15)$.

THINK	WRITE
1. Calculate the value of the gradient of AB by applying the formula $m = \dfrac{y_2 - y_1}{x_2 - x_1}$.	Let $A(-1, -5) = (x_1, y_1)$ and $B(5, 7) = (x_2, y_2)$ Since $m = \dfrac{y_2 - y_1}{x_2 - x_1}$ $m_{AB} = \dfrac{7 - (-5)}{5 - (-1)}$ $= \dfrac{12}{6}$ $= 2$
2. Calculate the value of the gradient of CD.	Let $C(-3, 1) = (x_1, y_1)$ and $D(4, 15) = (x_2, y_2)$ $m_{CD} = \dfrac{15 - 1}{4 - (-3)}$ $= \dfrac{14}{7}$ $= 2$
3. Draw a conclusion. (*Note*: ‖ means 'is parallel to'.)	Since $m_{AB} = m_{CD} = 2$, then AB ‖ CD.

⏵ 4.4.2 Perpendicular lines

eles-4744

- **Perpendicular** lines are lines that intersect at *right angles* as seen in the diagram. We can think of these lines as each having opposite gradients.
- More formally, we would state that perpendicular lines have gradients in which one gradient is the *negative reciprocal* of the other.
- Mathematically, we denote this as $m_1 \times m_2 = -1$.
- The gradient properties of parallel and perpendicular lines can be used to solve many problems.

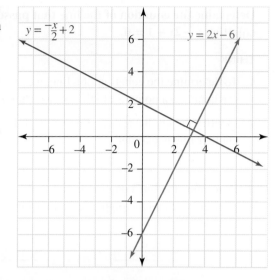

Parallel and perpendicular lines

- If we have two equations:

$$y_1 = m_1 x + c_1 \text{ and } y_2 = m_2 x + c_2$$

- the lines will be parallel if, and only if, $m_1 = m_2$
- the lines will be perpendicular if, and only if, $m_1 \times m_2 = -1$.

WORKED EXAMPLE 17 Proving two lines are perpendicular

Show that the lines $y = -5x + 2$ and $5y - x + 15 = 0$ are perpendicular.

THINK	WRITE
1. Identify the gradient of the first line $y_1 = -5x + 2$.	$y_1 = -5x + 2$ $m_1 = -5$
2. Identify the gradient of the second line $5y - x + 15 = 0$ by rearranging the equation in the form $y = mx + c$.	$5y - x + 15 = 0$ $5y = x - 15$ $y = \dfrac{x}{5} - 3$ $m_2 = \dfrac{1}{5}$
3. Test if the two points are perpendicular by checking whether the product of the two gradients is equal to -1.	$m_1 \times m_2 = -5 \times \dfrac{1}{5}$ $= -1$
4. Write the answer in a sentence.	As the product of the two gradients is equal to -1, therefore these two lines are perpendicular.

WORKED EXAMPLE 18 Determining the equation of a parallel line

Determine the equation of the line that passes through the point $(3, 1)$ and is parallel to the line with equation $y - 2x + 1 = 0$.

THINK	WRITE
1. To determine the equation of a line, we need both a gradient and a point. State the known values.	$(x_1, y_1) = (3, 1)$
2. Identify the gradient of the line $y - 2x + 1 = 0$ to which another line is parallel.	$y - 2x + 1 = 0$ $y = 2x - 1$ $m = 2$
3. Determine the line equation by using the gradient $= 2$ and the point $(3, 1)$. Rearrange and write the answer.	$y - y_1 = m(x - x_1)$ $y - 1 = 2(x - 3)$ $y - 1 = 2x - 6$ $y = 2x - 5$

WORKED EXAMPLE 19 Determining the equation of a perpendicular line

Determine the equation of the line that passes through the point $(2, 1)$ and is perpendicular to the line with a gradient of 5.

THINK	WRITE
1. To determine the equation of a line, we need both a gradient and a point. State the known values.	$(x_1, y_1) = (2, 1)$ $m_1 = 5$
2. As the lines are perpendicular so $m_1 \times m_2 = -1$. Calculate the value of the gradient of the other line using this formula.	$m_1 \times m_2 = -1$ $5 \times m_2 = -1$ $m_2 = \dfrac{-1}{5}$
3. Determine the line equation by using the gradient $= \dfrac{-1}{5}$ and the point $(2, 1)$. Rearrange and write the answer.	$y - y_1 = m(x - x_1)$ $y - 1 = \dfrac{-1}{5}(x - 2)$ $y - 1 = \dfrac{-x}{5} + \dfrac{2}{5}$ $y = \dfrac{-x}{5} + \dfrac{7}{5}$ or $5y + x = 7$

 Resources

eWorkbook   Topic 4 Workbook (worksheets, code puzzle and project) (ewbk-2095)

Interactivities Individual pathway interactivity: Parallel and perpendicular lines (int-4576)
Parallel lines (int-3841)
Perpendicular lines (int-6124)

4.4 Quick quiz on

4.4 Exercise

Individual pathways

■ PRACTISE	■ CONSOLIDATE	■ MASTER
1, 3, 7, 8, 13, 14, 18, 19, 22	2, 4, 9, 12, 17, 20, 23	5, 6, 10, 11, 15, 16, 21, 24

Fluency

WE16 For questions **1** to **4**, determine whether AB is parallel to CD given the following sets of points.

1. a. $A(4, 13)$, $B(2, 9)$, $C(0, -10)$, $D(15, 0)$
   b. $A(2, 4)$, $B(8, 1)$, $C(-6, -2)$, $D(2, -6)$
   c. $A(-3, -10)$, $B(1, 2)$, $C(1, 10)$, $D(8, 16)$

2. a. $A(1, -1)$, $B(4, 11)$, $C(2, 10)$, $D(-1, -5)$
   b. $A(1, 0)$, $B(2, 5)$, $C(3, 15)$, $D(7, 35)$
   c. $A(1, -6)$, $B(-5, 0)$, $C(0, 0)$, $D(5, -4)$

3. a. $A(1, 6)$, $B(3, 8)$, $C(4, -6)$, $D(-3, 1)$
   b. $A(2, 12)$, $B(-1, -9)$, $C(0, 2)$, $D(7, 1)$
   c. $A(1, 3)$, $B(4, 18)$, $C(-5, 4)$, $D(5, 0)$

4. a. $A(1, -5)$, $B(0, 0)$, $C(5, 11)$, $D(-10, 8)$
   b. $A(-4, 9)$, $B(2, -6)$, $C(-5, 8)$, $D(10, 14)$
   c. $A(4, 4)$, $B(-8, 5)$, $C(-6, 2)$, $D(3, 11)$

5. Determine which pairs of the following straight lines are parallel.

   a. $2x + y + 1 = 0$
   b. $y = 3x - 1$
   c. $2y - x = 3$
   d. $y = 4x + 3$
   e. $y = \dfrac{x}{2} - 1$
   f. $6x - 2y = 0$
   g. $3y = x + 4$
   h. $2y = 5 - x$

6. **WE17** Show that the lines $y = 6x - 3$ and $x + 6y - 6 = 0$ are perpendicular to one another.

7. **WE18** Determine the equation of the line that passes through the point $(4, -1)$ and is parallel to the line with equation $y = 2x - 5$.

8. **WE19** Determine the equation of the line that passes through the point $(-2, 7)$ and is perpendicular to a line with a gradient of $\dfrac{2}{3}$.

9. Determine the equations of the following lines.

   a. Gradient 3 and passing through the point $(1, 5)$
   b. Gradient -4 and passing through the point $(2, 1)$
   c. Passing through the points $(2, -1)$ and $(4, 2)$
   d. Passing through the points $(1, -3)$ and $(6, -5)$
   e. Passing through the point $(5, -2)$ and parallel to $x + 5y + 15 = 0$
   f. Passing through the point $(1, 6)$ and parallel to $x - 3y - 2 = 0$
   g. Passing through the point $(-1, -5)$ and perpendicular to $3x + y + 2 = 0$

10. Determine the equation of the line that passes through the point $(-2, 1)$ and is:

    a. parallel to the line with equation $2x - y - 3 = 0$
    b. perpendicular to the line with equation $2x - y - 3 = 0$.

11. Determine the equation of the line that contains the point $(1, 1)$ and is:

   a. parallel to the line with equation $3x - 5y = 0$

   b. perpendicular to the line with equation $3x - 5y = 0$.

Understanding

12. **MC** **a.** The vertical line passing through the point $(3, -4)$ is given by:

   A. $y = -4$ **B.** $x = 3$ **C.** $y = 3x - 4$ **D.** $y = -4x + 3$

   b. Select the point which passes through the horizontal line given by the equation $y = -5$.

   A. $(-5, 4)$ **B.** $(4, 5)$ **C.** $(3, -5)$ **D.** $(5, -4)$

   c. Select which of the following statements is true.

      A. Vertical lines have a gradient of zero.
      B. The y-coordinates of all points on a vertical line are the same.
      C. Horizontal lines have an undefined gradient.
      D. The x-coordinates of all points on a vertical line are the same.

   d. Select which of the following statements is false.

      A. Horizontal lines have a gradient of zero.
      B. The line joining the points $(1, -1)$ and $(-7, -1)$ is vertical.
      C. Vertical lines have an undefined gradient.
      D. The line joining the points $(1, 1)$ and $(-7, 1)$ is horizontal.

13. **MC** The point $(-1, 5)$ lies on a line parallel to $4x + y + 5 = 0$. Another point on the same line as $(-1, 5)$ is:

   A. $(2, 9)$ **B.** $(4, 2)$ **C.** $(4, 0)$ **D.** $(3, -11)$

14. Determine the equation of the straight line given the following conditions.

   a. Passes through the point $(-1, 3)$ and parallel to $y = -2x + 5$
   b. Passes through the point $(4, -3)$ and parallel to $3y + 2x = -3$

15. Determine which pairs of the following lines are perpendicular.

   a. $x + 3y - 5 = 0$ **b.** $y = 4x - 7$ **c.** $y = x$ **d.** $2y = x + 1$
   e. $y = 3x + 2$ **f.** $x + 4y - 9 = 0$ **g.** $2x + y = 6$ **h.** $x + y = 0$

16. Determine the equation of the straight line that cuts the x-axis at 3 and is perpendicular to the line with equation $3y - 6x = 12$.

17. Calculate the value of m for which lines with the following pairs of equations are perpendicular to each other.

   a. $2y - 5x = 7$ and $4y + 12 = mx$
   b. $5x - 6y = -27$ and $15 + mx = -3y$

18. **MC** The gradient of the line perpendicular to the line with equation $3x - 6y = 2$ is:

   A. 3   **B.** -6 **C.** 2   **D.** -2

Reasoning

19. Determine the equation of a line, in the form of $ax + by + c = 0$, that is perpendicular to the line with equation $2x - y = 3$ and passes through the point $(2, 3)$.

20. Form the equation of the line, in the form of $ax + by + c = 0$, that is perpendicular to the line with equation $-4x - 3y = 3$ and passes through the point $(-1, 4)$.

21. **MC** Triangle ABC has a right angle at B. The vertices are A($-2, 9$), B($2, 8$) and C($1, z$). The value of z is:

   A. $8\dfrac{1}{4}$ **B.** 4   **C.** 12   **D.** $7\dfrac{3}{4}$

Problem solving

22. **a.** Sketch the graph of the equation $y = 2x - 4$.
    b. On the same set of axes, sketch the graph of the line parallel to $y = 2x - 4$ that has a y-intercept of -2.
    c. Sketch the graph of the line that is perpendicular to the lines found in parts **a** and **b** that also passes through the origin.

23. Determine the value(s) of a such that there would be no point of intersection between the lines $ay + 3x = 4a$ and $2x - y = 5$.

24. A family of parallel lines has the equation $3x - 2y = k$ where k is a real number.

    a. Determine the gradient of each member of this family of lines.
    b. Show that all lines in the family contain the point (k, k).

LESSON
4.5 Graphical solution of simultaneous linear equations

4.5.1 Simultaneous linear equations and graphical solutions

eles-4763

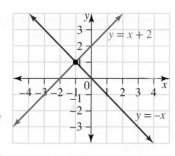

- **Simultaneous** means occurring at the same time.
- When a point lies on more than one line, the coordinates of that point are said to satisfy all equations of the lines it lies on. The equations of the lines are called **simultaneous equations**.
- A **system of equations** is a set of two or more equations with the same variables.
- Solving a system of simultaneous equations is finding the coordinates of any point or points that satisfy all equations in the system.
- Any point or points that satisfy a system of simultaneous equations is said to be the solution. For the equations shown, the solution is the point ($-1, 1$).
- Simultaneous equations can be solved by finding these points graphically or algebraically.

Graphical solution

- The solution to a pair of simultaneous equations can be found by graphing the two equations and identifying the coordinates of the point of intersection.
- The accuracy of the solution depends on having an accurate graph.

WORKED EXAMPLE 20 Solving simultaneous equations graphically

Use the graphs of the given simultaneous equations to determine the point of intersection and hence the solution of the simultaneous equations.

$$x + 2y = 4$$
$$y = 2x - 3$$

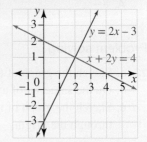

THINK	WRITE/DRAW
1. Write the equations one under the other and number them.	$x + 2y = 4$ [1] $y = 2x - 3$ [2]
2. Locate the point of intersection of the two lines. This gives the solution.	Point of intersection $(2, 1)$ Solution: $x = 2$ and $y = 1$

3. Check the solution by substituting $x = 2$ and $y = 1$ into the given equations. Comment on the results obtained.	Check equation [1]: $\begin{aligned} \text{LHS} &= x + 2y \qquad\qquad \text{RHS} = 4 \\ &= 2 + 2(1) \\ &= 4 \end{aligned}$ $\text{LHS} = \text{RHS}$ Check equation [2]: $\begin{aligned} \text{LHS} &= y \qquad\qquad \text{RHS} = 2x - 3 \\ &= 1 \qquad\qquad\qquad\quad = 2(2) - 3 \\ &\qquad\qquad\qquad\qquad\quad = 4 - 3 \\ &\qquad\qquad\qquad\qquad\quad = 1 \end{aligned}$ $\text{LHS} = \text{RHS}$
4. State the solution.	In both cases $\text{LHS} = \text{RHS}$; therefore, the solution set $(2, 1)$ is correct.

WORKED EXAMPLE 21 Verifying a solution using substitution

Verify whether the given pair of coordinates, $(5, -2)$, is the solution to the following pair of simultaneous equations.

$$3x - 2y = 19$$
$$4y + x = -3$$

THINK	WRITE
1. Write the equations and number them.	$3x - 2y = 19$ [1] $4y + x = -3$ [2]
2. Substitute $x = 5$ and $y = -2$ into equation [1].	Check equation [1]: $\begin{aligned} \text{LHS} &= 3x - 2y \qquad\qquad \text{RHS} = 19 \\ &= 3(5) - 2(-2) \\ &= 15 + 4 \\ &= 19 \end{aligned}$ $\text{LHS} = \text{RHS}$

3. Substitute $x = 5$ and $y = -2$ into equation [2].

Check equation [2]:
$$\text{LHS} = 4y + x \qquad \text{RHS} = -3$$
$$= 4(-2) + 5$$
$$= -3$$
$$\text{LHS} = \text{RHS}$$

4. State the solution.

Therefore, the solution set $(5, -2)$ is a solution to both equations.

WORKED EXAMPLE 22 Using a graphical method to solve simultaneously

Solve the following pair of simultaneous equations using a graphical method.
$$x + y = 6$$
$$2x + 4y = 20$$

THINK	WRITE/DRAW
1. Write the equations one under the other and number them.	$x + y = 6$ [1] $2x + 4y = 20$ [2]
2. Calculate the x- and y-intercepts for equation [1]. For the x-intercept, substitute $y = 0$ into equation [1].	Equation [1]: x-intercept: when $y = 0$, $x + 0 = 6$ $x = 6$ The x-intercept is $(6, 0)$.
For the y-intercept, substitute $x = 0$ into equation [1].	y-intercept: when $x = 0$, $0 + y = 6$ $y = 6$ The y-intercept is $(0, 6)$.
3. Calculate the x- and y-intercepts for equation [2]. For the x-intercept, substitute $y = 0$ into equation [2]. Divide both sides by 2.	Equation [2] x-intercept: when $y = 0$, $2x + 0 = 20$ $2x = 20$ $x = 10$ The x-intercept is $(10, 0)$.
For the y-intercept, substitute $y = 0$ into equation [2]. Divide both sides by 4.	y-intercept: when $x = 0$, $0 + 4y = 20$ $4y = 20$ $y = 5$ The y-intercept is $(0, 5)$.
4. Use graph paper to rule up a set of axes. Label the x-axis from 0 to 10 and the y-axis from 0 to 6. **5.** Plot the x- and y-intercepts for each equation. **6.** Produce a graph of each equation by ruling a straight line through its intercepts. **7.** Label each graph.	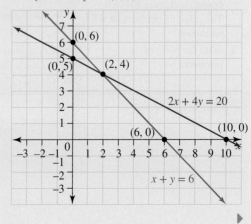

8. Locate the point of intersection of the lines.

The point of intersection is $(2, 4)$.

9. Check the solution by substituting $x = 2$ and $y = 4$ into each equation.

Check [1]:

$$\begin{aligned} \text{LHS} &= x + y & \text{RHS} &= 6 \\ &= 2 + 4 \\ &= 6 \end{aligned}$$

$\text{LHS} = \text{RHS}$

Check [2]:

$$\begin{aligned} \text{LHS} &= 2x + 4y & \text{RHS} &= 20 \\ &= 2(2) + 4(4) \\ &= 4 + 16 \\ &= 20 \end{aligned}$$

$\text{LHS} = \text{RHS}$

10. State the solution.

In both cases, $\text{LHS} = \text{RHS}$. Therefore, the solution set $(2, 4)$ is correct.
The solution is $x = 2, y = 4$.

TI	THINK	DISPLAY/WRITE	CASIO	THINK	DISPLAY/WRITE
1. In a new problem on a Graphs page, complete the function entry lines as: $f1(x) = 6 - x$ $f2(x) = 5 - \dfrac{x}{2}$ Press the down arrow between entering the functions. The graphs will be displayed.	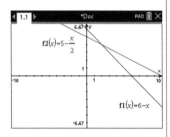	1. On a Graph & Table screen, complete the function entry lines as: $y1 = 6 - x$ $y2 = 5 - \dfrac{x}{2}$ Then tap the graphing icon. The graphs will be displayed.	
2. To locate the point of intersection, press: • MENU • 6: Analyze Graph • 4: Intersection Drag the dotted line to the left of the point of intersection (the lower bound), press ENTER, then drag the dotted line to the right of the point of intersection (the upper bound) and press ENTER. The point of intersection will be shown.		2. To locate the point of intersection, tap: • Analysis • G-Solve • Intersection The point of intersection will be shown.	
3. State the point of intersection.	The point of intersection (the solution) is $(2, 4)$. That is, $x = 2, y = 4$.	3. State the point of intersection.	The point of intersection (the solution) is $(2, 4)$. That is, $x = 2, y = 4$.

▶ 4.5.2 Solutions to coincident, parallel and perpendicular lines

eles-4764

- Two lines are **coincident** if they lie one on top of the other. For example, in the graph shown, the line in blue and the line segment in pink are coincident.

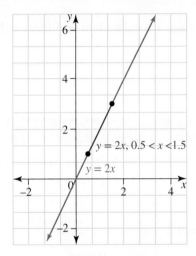

- There are an infinite number of solutions to coincident equations. Every point where the lines coincide satisfies both equations and hence is a solution to the simultaneous equations.
- Coincident equations have the same equation, although the equations may have been transposed so they look different. For example, $y = 2x + 3$ and $2y - 4x = 6$ are coincident equations.

Parallel lines

- If two lines do not intersect, there is no simultaneous solution to the equations. For example, the graph lines shown do not intersect, so there is no point that belongs to both lines.
- **Parallel** lines have the same gradient but different y-intercepts.
- For straight lines, the only situation in which the lines do not cross is if the lines are parallel *and* not coincident.

$$2x - y = 1 \qquad [1]$$
$$-y = 1 - 2x$$
$$-y = -2x - 1$$
$$y = 2x - 1$$
$$\text{Gradient } m = 2$$

$$4x - 2y = -2 \qquad [2]$$
$$-2y = -2 - 4x$$
$$-2y = -4x - 2$$
$$y = 2x + 1$$
$$\text{Gradient } m = 2$$

- Writing both equations in the form $y = mx + c$ confirms that the lines are parallel, as the gradients are equal.

Perpendicular lines

- Two lines are **perpendicular** if they intersect at right angles (90°).
- The product of the gradients of two perpendicular lines is equal to -1:

$$m_1 \times m_2 = -1 \text{ or } m_1 = -\frac{1}{m_2}$$

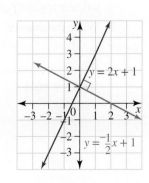

- The two lines in the graph shown are perpendicular as $m_1 \times m_2 = 2 \times -\dfrac{1}{2} = -1$.

Number of solutions for a pair of simultaneous linear equations

For two linear equations given by $y_1 = m_1 x + c_1$ and $y_2 = m_2 x + c_2$:
- If $m_1 = m_2$ and $c_1 \neq c_2$, the two lines are parallel and there will be no solutions between the two lines.
- If $m_1 = m_2$ and $c_1 = c_2$, the two lines are coincident and there will be infinite solutions between the two lines.
- If $m_1 \neq m_2$, the lines will cross once, so there will be one solution.
- If $m_1 \times m_2 = -1$, the lines are perpendicular and will intersect (once) at right angles (90°).

WORKED EXAMPLE 23 Determining the number of solutions between two lines

Determine the number of solutions between the following pairs of simultaneous equations. If there is only one solution, determine whether the lines are perpendicular.
a. $2y = 4x + 6$ and $-3y = -6x - 12$
b. $y = -3x + 2$ and $-3y = x + 15$
c. $5y = 25x - 30$ and $2y - 10x + 12 = 0$

THINK	WRITE
a. 1. Rewrite both equations in the form $y = mx + c$.	**a.** $2y = 4x + 6$ $\quad y = 2x + 3$ [1] $-3y = -6x - 12$ $\quad y = 2x + 4$ [2]
2. Determine the gradients of both lines.	$m_1 = 2$ and $m_2 = 2$
3. Check if the lines are parallel, coincident or perpendicular.	The gradients are the same and the y-intercepts different. Therefore, the two lines are parallel.
4. Write the answer.	There will be *no* solutions between this pair of simultaneous equations as the lines are parallel.
b. 1. Rewrite both equations in the form $y = mx + c$.	**b.** $y = -3x + 2$ [1] $-3y = x + 15$ $\quad y = \dfrac{x}{-3} - 5$ [2]
2. Determine the gradients of both lines.	$m_1 = -3$ and $m_2 = -\dfrac{1}{3}$
3. Check if the lines are parallel, coincident or perpendicular and comment on the number of solutions.	The gradients are different, so there will be *one* solution.
4. Determine if the lines are perpendicular by calculating the product of the gradients.	$m_1 \times m_2 = -3 \times -\dfrac{1}{3} = 1$
5. Write the answer.	The lines have one solution but they are not perpendicular.
c. 1. Rewrite both equations in the form $y = mx + c$.	**c.** $\quad 5y = 25x - 30$ $\quad\quad y = 5x - 6$ [1] $2y - 10x + 12 = 0$ $\quad\quad 2y = 10x - 12$ $\quad\quad\quad y = 5x - 6$ [2]

2. Determine the gradients of both lines.

3. Check if the lines are parallel, coincident, or perpendicular.

4. Write the answer.

$m_1 = 5$ and $m_2 = 5$

The gradients are the same and the y-intercepts are also the same. Therefore, the two lines are coincident.

The lines are coincident, so there are *infinite* solutions between the two lines.

DISCUSSION

What do you think is the major error made when solving simultaneous equations graphically?

 Resources

Exercise 4.5 Graphical solution of simultaneous linear equations

learn on

4.5 Quick quiz on	4.5 Exercise

Individual pathways

■ PRACTISE	■ CONSOLIDATE	■ MASTER
1, 4, 7, 10, 14, 15, 18	2, 5, 8, 11, 12, 16, 19	3, 6, 9, 13, 17, 20, 21

Fluency

WE20 For questions 1 to 3, use the graphs to determine the point of intersection and hence the solution of the simultaneous equations.

1. a. $x + y = 3$
       $x - y = 1$

    b. $x + y = 2$
       $3x - y = 2$

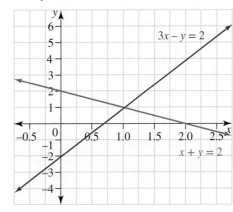

2. a. $y - x = 4$
  $3x + 2y = 8$

b. $y + 2x = 3$
  $2y + x = 0$

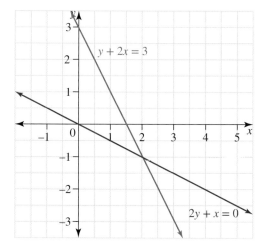

3. a. $y - 3x = 2$
  $x - y = 2$

b. $2y - 4x = 5$
  $4y + 2x = 5$

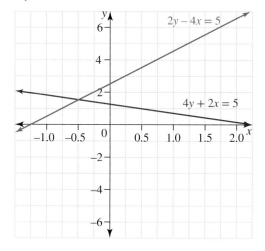

WE21 For questions **4** to **6**, use substitution to check if the given pair of coordinates is a solution.

4. a. $(7, 5)$ $\quad 3x + 2y = 31$
  $2x + 3y = 28$

b. $(3, 7)$ $\quad y - x = 4$
  $2y + x = 17$

c. $(9, 1)$ $\quad x + 3y = 12$
  $5x - 2y = 43$

d. $(2, 5)$ $\quad x - y = 7$
  $2x + 3y = 18$

5. a. $(4, -3)$ $\quad y = 3x - 15$
  $4x + 7y = -5$

b. $(6, -2)$ $\quad x - y = 7$
  $3x + y = 16$

c. $(4, -2)$ $\quad 2x + y = 6$
  $x - 3y = 8$

d. $(5, 1)$ $\quad y - 5x = -24$
  $3y + 4x = 23$

6. a. $(-2, -5)$ $\quad 3x - 2y = -4$
  $2x - 3y = 11$

b. $(-3, -1)$ $\quad y - x = 2$
  $2y - 3x = 7$

c. $\left(-\dfrac{1}{2}, 2\right)$ $\quad 6x + 4y = 5$
  $20x - 5y = 0$

d. $\left(\dfrac{3}{2}, \dfrac{5}{3}\right)$ $\quad 8x + 6x = 22$
  $10x - 9y = 0$

WE22 For questions **7** to **9**, solve each of the pairs of simultaneous equations using a graphical method.

7. **a.** $x + y = 5$
   $2x + y = 8$

   b. $x + 2y = 10$
   $3x + y = 15$

   c. $2x + 3y = 6$
   $2x - y = -10$

   d. $x - 3y = -8$
   $2x + y = -2$

8. **a.** $6x + 5y = 12$
   $5x + 3y = 10$

   b. $y + 2x = 6$
   $2y + 3x = 9$

   c. $y = 3x + 10$
   $y = 2x + 8$

   d. $y = 8$
   $3x + y = 17$

9. **a.** $4x - 2y = -5$
   $x + 3y = 4$

   b. $3x + y = 11$
   $4x - y = 3$

   c. $3x + 4y = 27$
   $x + 2y = 11$

   d. $3y + 3x = 8$
   $3y + 2x = 6$

Understanding

For questions **10** to **12**, using technology, determine which of the following pairs of simultaneous equations have no solutions. Confirm by finding the gradient of each line.

10. **a.** $y = 2x - 4$
    $3y - 6x = 10$

    b. $5x - 3y = 13$
    $4x - 2y = 10$

    c. $x + 2y = 8$
    $5x + 10y = 45$

    d. $y = 4x + 5$
    $2y - 10x = 8$

11. **a.** $3y + 2x = 9$
    $6x + 4y = 22$

    b. $y = 5 - 3x$
    $3y = -9x + 18$

    c. $4y + 3x = 7$
    $12y + 9x = 22$

    d. $2y - x = 0$
    $14y - 6x = 2$

12. **a.** $y = 3x - 4$
    $5y = 12 + 15x$

    b. $4x - 6y = 12$
    $6x - 4y = 12$

    c. $3y = 5x - 22$
    $5x = 3y + 26$

    d. $3x = 12 - 4y$
    $8y + 6x = 14$

13. Two straight lines intersect at the point $(3, -4)$. One of the lines has a y-intercept of 8. The second line is a mirror image of the first in the line $x = 3$. Determine the equation of the second line.
    (*Hint:* Draw a graph of both lines.)

Reasoning

14. At a well-known beach resort it is possible to hire a jet-ski by the hour in two different locations. On the northern beach the cost is \$20 plus \$12 per hour, and on the southern beach the cost is \$8 plus \$18 per hour. The jet-skis can be rented for up to 5 hours.
    a. Write the rules relating cost to the length of rental.
    b. On the same set of axes sketch a graph of cost (y-axis) against length of rental (x-axis) for 0–5 hours.
    c. For what rental times, if any, is the northern beach rental cheaper than the southern beach rental? Use your graph to justify your answer.
    d. For what length of rental time are the two rental schemes identical? Use the graph and your rules to justify your answer.

15. For each of the following pairs of simultaneous equations, determine whether they are the same line, parallel lines, perpendicular lines or intersecting lines. Show your working.
    a. $2x - y = -9$
    $-4x - 18 = -2y$

    b. $x - y = 7$
    $x + y = 7$

    c. $x + 6 = y$
    $2x + y = 6$

    d. $x + y = -2$
    $x + y = 7$

16. For each of the following pairs of simultaneous equations, explain if they have one solution, an infinite number of solutions or no solution.
    a. $x - y = 1$
    $2x - 3y = 2$

    b. $2x - y = 5$
    $4x - 2y = -6$

    c. $x - 2y = -8$
    $4x - 8y = -16$

17. WE23 Determine whether the following pairs of equations will have one, infinite or no solutions. If there is only one solution, determine whether the lines are perpendicular.
    a. $3x + 4y = 14$
    $4x - 3y = 2$

    b. $2x + y = 5$
    $3y + 6x = 15$

    c. $3x - 5y = -6$
    $5x - 3y = 24$

    d. $2y - 4x = 6$
    $2x - y = -10$

Problem solving

18. Use the information given to determine the value of a in each of the following equations:

    a. $y = ax + 3$, which is parallel to $y = 3x - 2$
    b. $y = ax - 2$, which is perpendicular to $y = -4x + 6$
    c. $y = ax - 4$, which intersects the line $y = 3x + 6$ when $x = 2$.

19. Line A is parallel to the line with equation $y - 3x - 3 = 0$ and passes through the point $(1, 9)$. Line B is perpendicular to the line with equation $2y - x + 6 = 0$ and passes through the point $(2, -3)$.

    a. Determine the equation of line A.
    b. Determine the equation of line B.
    c. Sketch both lines on the one set of axes to find where they intersect.

20. Solve the following system of three simultaneous equations graphically.

$$3x - y = 2$$
$$y + 3x = 4$$
$$2y - x = 1$$

21. A line with equation $4x + 5y = 4$ intersects a second line when $x = -4$. Determine the equation of the second line if it is perpendicular to the first line.

LESSON
4.6 Solving simultaneous linear equations using substitution

LEARNING INTENTIONS

At the end of this lesson you should be able to:
- identify when it is appropriate to solve using the substitution method
- solve a system of two linear simultaneous equations using the substitution method.

▶ 4.6.1 Solving simultaneous equations using the substitution method

- A variable is considered the **subject of an equation** if it is expressed in terms of the other variables. In the equation $y = 3x + 4$, the variable y is the subject.
- The **substitution method** is used when one (or both) of a pair of simultaneous equations is presented in a form where one of the two variables is the subject of the equation.
- When solving two linear simultaneous equation, the substitution method involves replacing a variable in one equation with the other equation. This produces a new third equation expressed in terms of a single variable.
- Consider the following pair of simultaneous equations:

$$y = 2x - 4$$
$$3x + 2y = 6$$

- In the first equation, y is written as the subject and is equal to $(2x - 4)$. In this case, substitution is performed by replacing y in the second equation with the expression $(2x - 4)$.

264 Jacaranda Maths Quest 10 + 10A

$$y = \underbrace{2x - 4}$$

$$3x + 2(y) = 10$$

$$3x + 2(2x - 4) = 6$$

- This produces a third equation, all in terms of x, so that the value of x can be found.
- Once a value for one variable is found, it can be substituted back into either equation to find the value of the other variable.
- It is often helpful to use brackets when substituting an expression into another equation.

WORKED EXAMPLE 24 Solving using the substitution method

Solve the simultaneous equations $y = 2x - 1$ and $3x + 4y = 29$ using the substitution method.

THINK	WRITE
1. Write the equations one under the other and number them.	$y = 2x - 1$ [1] $3x + 4y = 29$ [2]
2. y and $2x - 1$ are equal, so substitute the expression $(2x - 1)$ for y into equation [2].	Substituting $(2x - 1)$ into [2]: $3x + 4(2x - 1) = 29$
3. Solve for x.	
i. Expand the brackets on the left-hand side of the equation.	$3x + 8x - 4 = 29$
ii. Collect like terms.	$11x - 4 = 29$
iii. Add 4 to both sides of the equation.	$11x = 33$
iv. Divide both sides by 11.	$x = 3$
4. Substitute $x = 3$ into either of the equations, say [1], to find the value of y.	Substituting $x = 3$ into [1]: $y = 2(3) - 1$ $\quad = 6 - 1$ $\quad = 5$
5. Write your answer.	Solution: $x = 3, y = 5$ or $(3, 5)$
6. Check the solution by substituting $(3, 5)$ into equation [2].	Check: Substitute $(3, 5)$ into $3x + 4y = 29$. LHS $= 3(3) + 4(5)$ RHS $= 29$ $\quad = 9 + 20$ $\quad = 29$ As LHS $=$ RHS, the solution is correct.

4.6.2 Equating equations

eles-4767

- To **equate** in mathematics is to take two expressions that have the same value and make them equal to each other.
- When both linear equations are written with the same variable as the subject, we can equate the equations to solve for the other variable. Consider the following simultaneous equations:

$$y = 4x - 3$$
$$y = 2x + 9$$

- In the first equation y is equal to $(4x - 3)$ and in the second equation y is equal to $(2x + 9)$. Since both expressions are equal to the same thing (y), they must also be equal to each other. Thus, equating the equations gives:

$$y = \underbrace{4x - 3}$$

$$(y) = 2x + 9$$

$$4x - 3 = 2x + 9$$

- As can be seen above, equating equations is still a form of substitution. A third equation is produced, all in terms of x, allowing for a value of x to be solved.

WORKED EXAMPLE 25 Substitution by equating two equations

Solve the pair of simultaneous equations $y = 5x - 8$ and $y = -3x + 16$ by equating the equations.

THINK	WRITE
1. Write the equations one under the other and number them.	$y = 5x - 8$ [1] $y = -3x + 16$ [2]
2. Both equations are written with y as the subject, so equate them.	$5x - 8 = -3x + 16$

3. Solve for x.
   i. Add $3x$ to both sides of the equation.
   ii. Add 8 to both sides of the equation.
   iii. Divide both sides of the equation by 8.
	$8x - 8 = 16$
	$8x = 24$
	$x = 3$

4. Substitute the value of x into either of the original equations, say [1], and solve for y.	Substituting $x = 3$ into [1]: $y = 5(3) - 8$ $\quad = 15 - 8$ $\quad = 7$
5. Write your answer.	Solution: $x = 3, y = 7$ or $(3, 7)$
6. Check the answer by substituting the point of intersection into equation [2].	Check: Substitute into $y = -3x + 16$. LHS $= y$ $\quad = 7$ RHS $= -3x + 16$ $\quad = -3(3) + 16$ $\quad = -9 + 16$ $\quad = 7$ As LHS = RHS, the solution is correct.

DISCUSSION

When would you choose the substitution method in solving simultaneous equations?

 Resources

eWorkbook   Topic 4 Workbook (worksheets, code puzzle and a project) (ewbk-2095)

Video eLesson   Solving simultaneous equations using substitution (eles-1932)

Interactivities   Individual pathway interactivity: Solving simultaneous linear equations using substitution (int-4578)
                    Solving simultaneous equations using substitution (int-6453)

Exercise 4.6 Solving simultaneous linear equations using substitution

learn on

4.6 Quick quiz on	4.6 Exercise

Individual pathways

■ PRACTISE	■ CONSOLIDATE	■ MASTER
1, 4, 7, 8, 14, 15, 18	2, 5, 9, 12, 16, 19	3, 6, 10, 11, 13, 17, 20

Fluency

WE24 For questions **1** to **3**, solve each pair of simultaneous equations using the substitution method. Check your solutions using technology.

1. **a.** $x = -10 + 4y$
   $3x + 5y = 21$
   b. $3x + 4y = 2$
   $x = 7 + 5y$
   c. $3x + y = 7$
   $x = -3 - 3y$
   d. $3x + 2y = 33$
   $y = 41 - 5x$

2. **a.** $y = 3x - 3$
   $-5x + 3y = 3$
   b. $4x + y = 9$
   $y = 11 - 5x$
   c. $x = -5 - 2y$
   $5y + x = -11$
   d. $x = -4 - 3y$
   $-3x - 4y = 12$

3. **a.** $x = 7 + 4y$
   $2x + y = -4$
   b. $x = 14 + 4y$
   $-2x + 3y = -18$
   c. $3x + 2y = 12$
   $x = 9 - 4y$
   d. $y = 2x + 1$
   $-5x - 4y = 35$

WE25 For questions **4** to **6**, solve each pair of simultaneous equations by equating the equations. Check your solutions using technology.

4. **a.** $y = 2x - 11$ and $y = 4x + 1$
   b. $y = 3x + 8$ and $y = 7x - 12$
   c. $y = 2x - 10$ and $y = -3x$
   d. $y = x - 9$ and $y = -5x$

5. **a.** $y = -4x - 3$ and $y = x - 8$
   b. $y = -2x - 5$ and $y = 10x + 1$
   c. $y = -x - 2$ and $y = x + 1$
   d. $y = 6x + 2$ and $y = -4x$

6. **a.** $y = 0.5x$ and $y = 0.8x + 0.9$
   b. $y = 0.3x$ and $y = 0.2x + 0.1$
   c. $y = -x$ and $y = -\dfrac{2}{7}x + \dfrac{4}{7}$
   d. $y = -x$ and $y = -\dfrac{3}{4}x - \dfrac{1}{4}$

Understanding

7. A small farm has sheep and chickens. There are twice as many chickens as sheep, and there are 104 legs between the sheep and the chickens.
   Calculate the total number of chickens.

For questions **8** to **10**, use substitution to solve each pair of simultaneous equations.

8. **a.** $5x + 2y = 17$
   $y = \dfrac{3x - 7}{2}$
   b. $2x + 7y = 17$
   $x = \dfrac{1 - 3y}{4}$

9. **a.** $2x + 3y = 13$

$$y = \frac{4x - 15}{5}$$

b. $-2x - 3y = -14$

$$x = \frac{2 + 5y}{3}$$

10. **a.** $3x + 2y = 6$

$$y = 3 - \frac{5x}{3}$$

b. $-3x - 2y = -12$

$$y = \frac{5x - 20}{3}$$

11. Use substitution to solve each of the following pairs of simultaneous equations for x and y in terms of m and n.

a. $mx + y = n$
$y = mx$

b. $x + ny = m$
$y = nx$

c. $mx - ny = n$
$y = nx$

d. $mx - ny = n$
$y = x$

e. $mx - ny = -m$
$x = y - n$

f. $mx + y = m$

$$x = \frac{y + m}{n}$$

12. Determine the values of a and b so that the pair of equations $ax + by = 17$ and $2ax - by = -11$ has a unique solution of $(-2, 3)$.

13. The earliest record of magic squares is from China in about 2200 BCE. In magic squares the sums of the numbers of each row, column and diagonal are all equal to a magic number. Let z be the magic number. By creating a set of equations, solve to find the magic number and the missing values in the magic square.

m	11	7
9		
n	5	10

Reasoning

14. **a.** Consider the following pair of simultaneous equations.

$$8x - 7y = 9$$
$$x + 2y = 4$$

Identify which equation is the logical choice to make x the subject.
b. Use the substitution method to solve the system of equations. Show all your working.

15. A particular chemistry book costs \$6 less than a particular physics book. Two such chemistry books and three such physics books cost a total of \$123. Construct two simultaneous equations and solve them using the substitution method. Show your working.

16. The two shorter sides of a right triangle are 1 cm and 8 cm shorter than the hypotenuse. If the area of the triangle is $30\,\text{cm}^2$, determine the perimeter of the triangle.

17. Andrew is currently ten years older than his sister Prue. In four years' time he will be twice as old as Prue. Determine how old Andrew and Prue are now.

Problem solving

18. Use the substitution method to solve the following pair of equations.

$$2x + y - 9 = 0$$
$$4x + 5y + 3 = 0$$

19. Use the substitution method to solve the following pair of equations.

$$\frac{y - x}{2} - \frac{x + y}{3} = \frac{1}{6}$$

$$\frac{x}{5} + \frac{y}{2} = \frac{1}{2}$$

20. Consider the following pair of equations.

$$kx - \frac{y}{k} = 2$$
$$27x - 3y = 12k - 18$$

Determine the values of k when the equations have:

a. one solution   **b.** no solutions   **c.** infinite solutions.

LESSON
4.7 Solving simultaneous linear equations using elimination

LEARNING INTENTION

At the end of this lesson you should be able to:
• solve two simultaneous linear equations using the elimination method.

▶ 4.7.1 Solving simultaneous equations using the elimination method

eles-4768

- The **elimination method** is an algebraic method to solve simultaneous linear equations. It involves adding or subtracting equations in order to eliminate one of the variables.
- In order to eliminate a variable, the variable must be on the same side of the equal sign in both equations and must have the same coefficient.
- If the coefficients of the variable have the same sign, we subtract one equation from the other to eliminate the variable.
- If the coefficients of the variables have the opposite sign, we add the two equations together to eliminate the variable.

$$3x + 4y = 14 \qquad\qquad 6x - 2y = 12$$
$$5x - 4y = 2 \qquad\qquad 6x + 3y = 27$$

(add equations to eliminate y)(subtract equations to eliminate x)

Subtract

- The process of elimination is carried out by adding (or subtracting) the left-hand sides and the right-hand sides of each equation together.
- Consider the equations $2x + y = 5$ and $x + y = 3$. The process of subtracting each side of the equation from each other is visualised on the scales shown.
- To represent this process algebraically, the setting out would look like:

$$2x + y = 5$$
$$\underline{- (x + y = 3)}$$
$$x = 2$$

- Once the value of x has been found, it can be substituted into either original equation to find y.

$$2(2) + y = 5 \Rightarrow y = 1$$

Solve the following pair of simultaneous equations using the elimination method.

$$-2x - 3y = -9$$
$$2x + y = 7$$

THINK	WRITE
1. Write the equations one under the other and number them.	$-2x - 3y = -9$ [1] $2x + y = 7$ [2]
2. Look for an addition or subtraction that will eliminate either x or y. *Note:* Adding equations [1] and [2] in order will eliminate x.	$[1] + [2]$: $-2x - 3y + (2x + y) = -9 + 7$ $-2x - 3y + 2x + y = -2$ $-2y = -2$
3. Solve for y by dividing both sides of the equation by -2.	$y = 1$
4. Substitute the value of y into equation [2]. *Note:* $y = 1$ may be substituted into either equation.	Substituting $y = 1$ into [2]: $2x + 1 = 7$

5. Solve for x.
    i. Subtract 1 from both sides of the equation.
    ii. Divide both sides of the equation by 2.
	$2x = 6$
	
	 $x = 3$

6. Write the solution.	Solution: $x = 3$, $y = 1$ or $(3, 1)$
7. Check the solution by substituting $(3, 1)$ into equation [1] since equation [2] was used to find the value of x.	Check: Substitute into $-2x - 3y = -9$. LHS $= -2(3) - 3(1)$ $= -6 - 3$ $= -9$ RHS $= -9$ LHS = RHS, so the solution is correct.

4.7.2 Solving simultaneous equations by multiplying by a constant

eles-4769

- If neither variable in the two equations have the same coefficient, it is necessary to multiply one or both equations by a constant so that a variable can be eliminated.
- The equals sign in an equation acts like a balance, so as long as both sides of equation are correctly multiplied by the same value, the new statement is still a valid equation.

Double both sides and it remains balanced.

- Consider the following pairs of equations:

$$3x + 7y = 23 \quad 4x + 5y = 22$$
$$6x + 2y = 22 \quad 3x - 4y = -6$$

- For the first pair, the easiest starting point is to work towards eliminating x. This is done by first multiplying the top equation by 2 so that both equations have the same coefficient of x.

$$2(3x + 7y = 23) \Rightarrow 6x + 14y = 46$$

- For the second pair, both equations will need to be multiplied by a constant. Choosing to eliminate x would require the top equation to be multiplied by 3 and the bottom equation by 4 in order to produce two new equations with the same coefficient of x.

$$3(4x + 5y = 23) \Rightarrow 12x + 15y = 69$$
$$4(3x - 4y = -6) \Rightarrow 12x - 16y = -24$$

- Once the coefficient of one of the variables is the same, you can begin the elimination method.

WORKED EXAMPLE 27 Multiplying one equation by a constant to eliminate

Solve the following pair of simultaneous equations using the elimination method.

$$x - 5y = -17$$
$$2x + 3y = 5$$

THINK	WRITE
1. Write the equations one under the other and number them.	$x - 5y = -17$ [1] $2x + 3y = 5$ [2]
2. Look for a single multiplication that will create the same coefficient of either x or y. Multiply equation [1] by 2 and call the new equation [3].	$[1] \times 2 : 2x - 10y = -34$ [3]
3. Subtract equation [2] from [3] in order to eliminate x.	$[3] - [2]:$ $2x - 10y - (2x + 3y) = -34 - 5$ $2x - 10y - 2x - 3y = -39$ $-13y = -39$
4. Solve for y by dividing both sides of the equation by -13.	$y = 3$
5. Substitute the value of y into equation [2].	Substituting $y = 3$ into [2]: $2x + 3(3) = 5$ $2x + 9 = 5$
6. Solve for x. i. Subtract 9 from both sides of the equation.	$2x = -4$
ii. Divide both sides of the equation by 2.	$x = -2$
7. Write the solution.	Solution: $x = -2$, $y = 3$ or $(-2, 3)$
8. Check the solution by substituting into equation [1].	Check: Substitute into $x - 5y = -17$. LHS $= (-2) - 5(3)$ $= -2 - 15$ $= -17$ RHS $= -17$ LHS $=$ RHS, so the solution is correct.

Note: In this example, equation [1] could have been multiplied by -2 (instead of by 2), then the two equations added (instead of subtracted) to eliminate x.

WORKED EXAMPLE 28 Multiplying both equations by a constant to eliminate

Solve the following pair of simultaneous equations using the elimination method.

$$6x + 5y = 3$$
$$5x + 4y = 2$$

THINK	WRITE
1. Write the equations one under the other and number them.	$6x + 5y = 3$ [1] $5x + 4y = 2$ [2]
2. Decide which variable to eliminate, say y. Multiply equation [1] by 6 and call the new equation [3]. Multiply equation [2] by 5 and call the new equation [4].	Eliminate y. $[1] \times 4: \ 24x + 20y = 12$ [3] $[1] \times 5: \ 25x + 20y = 10$ [4]
3. Subtract equation [3] from [4] in order to eliminate y.	$[4] - [3]:$ $25x + 20y - (24x + 20y) = 10 - 12$ $25x + 20y - 24x - 20y = -2$ $x = -2$
4. Substitute the value of x into equation [1].	Substituting $x = -2$ into [1]: $6(-2) + 5y = 3$ $-12 + 5y = 3$

5. Solve for y.
    i.   Add 12 to both sides of the equation.
    ii.   Divide both sides of the equation by 5.
	$5y = 15$
	 $y = 3$

6. Write your answer.	Solution: $x = -2, y = 3$ or $(-2, 3)$
7. Check the answer by substituting the solution into equation [2].	Check: Substitute into $5x + 4y = 2$. $\begin{aligned} \text{LHS} &= 5(-2) + 4(3) \\ &= -10 + 12 \\ &= 2 \end{aligned}$ $\text{RHS} = 2$ $\text{LHS} = \text{RHS}$, so the solution is correct.

Note: In this example, equation [1] could have been multiplied by -4 (instead of by 4), then the two equations added (instead of subtracted) to eliminate y.

---

on Resources

Exercise 4.7 Solving simultaneous linear equations using elimination

4.7 Quick quiz on	**4.7 Exercise**

Individual pathways

■ PRACTISE	■ CONSOLIDATE	■ MASTER
1, 3, 5, 10, 13, 18	2, 6, 8, 11, 14, 15, 19	4, 7, 9, 12, 16, 17, 20

Fluency

1. **WE26** Solve the following pairs of simultaneous equations by adding equations to eliminate either x or y.

   a. $x + 2y = 5$
      $-x + 4y = 1$

   b. $5x + 4y = 2$
      $5x - 4y = -22$

   c. $-2x + y = 10$
      $2x + 3y = 14$

2. Solve the following pairs of equations by subtracting equations to eliminate either x or y.

   a. $3x + 2y = 13$
      $5x + 2y = 23$

   b. $2x - 5y = -11$
      $2x + y = 7$

   c. $-3x - y = 8$
      $-3x + 4y = 13$

3. Solve each of the following equations using the elimination method.

   a. $6x - 5y = -43$
      $6x - y = -23$

   b. $x - 4y = 27$
      $3x - 4y = 17$

   c. $-4x + y = -10$
      $4x - 3y = 14$

4. Solve each of the following equations using the elimination method.

   a. $-5x + 3y = 3$
      $-5x + y = -4$

   b. $5x - 5y = 1$
      $2x - 5y = -5$

   c. $4x - 3y - 1 = 0$
      $4x + 7y - 11 = 0$

WE27 For questions 5 to 7, solve the following pairs of simultaneous equations.

5. a. $6x + y = 9$
      $-3x + 2y = 3$

   b. $x + 3y = 14$
      $3x + y = 10$

   c. $5x + y = 27$
      $4x + 3y = 26$

6. a. $-6x + 5y = -14$
      $-2x + y = -6$

   b. $2x + 5y = 14$
      $3x + y = -5$

   c. $-3x + 2y = 6$
      $x + 4y = -9$

7. a. $3x - 5y = 7$
      $x + y = -11$

   b. $2x + 3y = 9$
      $4x + y = -7$

   c. $-x + 5y = 7$
      $5x + 5y = 19$

8. **WE28** Solve the following pairs of simultaneous equations.

   a. $-4x + 5y = -9$
      $2x + 3y = 21$

   b. $2x + 5y = -6$
      $3x + 2y = 2$

   c. $2x - 2y = -4$
      $5x + 4y = 17$

9. Solve the following pairs of simultaneous equations.

   a. $2x - 3y = 6$
      $4x - 5y = 9$

   b. $\dfrac{x}{2} + \dfrac{y}{3} = 2$
      $\dfrac{x}{4} + \dfrac{y}{3} = 4$

   c. $\dfrac{x}{3} + \dfrac{y}{2} = \dfrac{3}{2}$
      $\dfrac{x}{2} + \dfrac{y}{5} = -\dfrac{1}{2}$

Understanding

For questions **10** to **12**, solve each pair of simultaneous equations using an appropriate method. Check your answers using technology.

10. **a.** $7x + 3y = 16$
$y = 4x - 1$

    b. $2x + y = 8$
$4x + 3y = 16$

    c. $-3x + 2y = 19$
$4x + 5y = 13$

11. **a.** $-3x + 7y = 9$
$4x - 3y = 7$

    b. $-4x + 5y = -7$
$x = 23 - 3y$

    c. $y = -x$
$y = -\dfrac{2}{5}x - \dfrac{1}{5}$

12. **a.** $4x + 5y = 41$
$y = \dfrac{3x}{2} - 1$

    b. $3x - 2y = 9$
$2x + 5y = -13$

    c. $\dfrac{x}{3} + \dfrac{y}{4} = 7$
$3x - 2y = 12$

Reasoning

13. At a local bakery, a cup of coffee and a croissant costs $8.50, and an order of 5 coffees and three croissants costs $35.70. Determine the cost of one croissant.

14. Celine notices that she only has 5-cent and 10-cent coins in her coin purse. She counts up how much she has and finds that the total value of the 34 coins in the purse is $2.80.
Determine how many of each type of coin she has.

15. Abena, Bashir and Cecily wanted to weigh themselves, but the scales they had were broken and would only give readings over 100 kg. They decided to weigh themselves in pairs and calculate their weights from the results.
- Abena and Bashir weighed 119 kg.
- Bashir and Cecily weighed 112 kg.
- Cecily and Abena weighed 115 kg.

Determine the weight of each student.

16. **a.** For the general case $ax + by = e$ [1]
                        $cx + dy = f$ [2]
    y can be found by eliminating x.

    i. Multiply equation [1] by c to create equation [3].
    ii. Multiply equation [2] by a to create equation [4].
    iii. Use the elimination method to find a general solution for y.

  b. Use a similar process to that outlined above to find a general solution for x.
  c. Use the general solution for x and y to solve each of the following.

    i. $2x + 5y = 7$
$7x + 2y = 24$
    ii. $3x - 5y = 4$
$x + 3y = 5$

    Choose another method to check that your solutions are correct in each part.
  d. For y to exist, it is necessary to state that $bc - ad \neq 0$. Explain.
  e. Is there a necessary condition for x to exist? Explain.

17. A family of two parents and four children go to the movies and spend $95 on the tickets. Another family of one parent and two children go to see the same movie and spend $47.50 on the tickets. Determine if it possible to work out the costs of an adult's ticket and a child's ticket from this information.

Problem solving

18. The sum of two numbers is equal to k. The difference of the two numbers is given by $k - 20$. Determine the possible solutions for the two numbers.

19. Use the elimination method to solve the following pair of simultaneous equations.

$$\frac{x-4}{3} + y = -2$$
$$\frac{2y-1}{7} + x = 6$$

20. Use an appropriate method to solve the following simultaneous equations.

$$2x + 3y + 3z = -1$$
$$3x - 2y + z = 0$$
$$z + 2y = 0$$

LESSON
4.8 Applications of simultaneous linear equations

LEARNING INTENTIONS

At the end of this lesson you should be able to:
- define unknown quantities with appropriate variables
- form two simultaneous equations using the information presented in a problem
- choose an appropriate method to solve simultaneous equations in order to determine the solution to a problem.
- use simultaneous equations to determine the break-even point.

4.8.1 Applications of simultaneous linear equations

eles-4770

- When solving practical problems, the following steps can be useful:
  1. Define the unknown quantities using appropriate pronumerals.
  2. Use the information given in the problem to form two equations in terms of these pronumerals.
  3. Solve these equations using an appropriate method.
  4. Write the solution in words.
  5. Check the solution.

> ### Key language used in worded problems
>
> **To help set up equations from the information presented in a problem question, make sure you look out for the following key terms.**
> - **Addition: sum, altogether, add, more than, and, in total**
> - **Subtraction: difference, less than, take away, take off, fewer than**
> - **Multiplication: product, groups of, times, of, for each, double, triple**
> - **Division: quotient, split into, halve, thirds**
> - **Equals: gives, is**

WORKED EXAMPLE 29 Applying the elimination method

Ashley received better results for his Mathematics test than for his English test. If the sum of the two marks is 164 and the difference is 22, calculate the mark he received for each subject.

THINK	WRITE
1. Define the two variables.	Let x = the Mathamatics mark Let y = the English mark
2. Formulate two equations from the information given and number them. The sum of the two marks is $x + y$. The difference of the two marks is $x - y$.	$x + y = 164$ [1] $x - y = 22$ [2]
3. Use the elimination method by adding equations [1] and [2] to eliminate y.	[1] + [2]: $2x = 186$
4. Solve for x by dividing both sides of the equation by 2.	$x = 93$
5. Substitute the value of x into equation [1].	Substituting $x = 93$ into [1]: $x + y = 164$ $93 + y = 164$
6. Solve for y by subtracting 93 from both sides of the equation.	$y = 71$
7. Write the solution.	Solution: Mathematics mark $(x) = 93$ English mark $(y) = 71$
8. Check the solution by substituting $x = 93$ and $y = 71$ into equation [1].	Check: Substitute into $x + y = 164$. LHS $= 93 + 71$ RHS $= 164$ $= 164$ As LHS $=$ RHS, the solution is correct

WORKED EXAMPLE 30 Applying the substitution method

To finish a project, Genevieve bought a total of 25 nuts and bolts from a hardware store. If each nut costs 12 cents, each bolt costs 25 cents and the total purchase price is $4.30, calculate how many nuts and how many bolts Genevieve bought.

THINK	WRITE
1. Define the two variables.	Let x = the number of nuts Let y = the number of bolts
2. Formulate two equations from the information given and number them. *Note:* The total number of nuts and bolts is 25. Each nut cost 12 cents, each bolt cost 25 cents and the total cost is 430 cents ($4.30).	$x + y = 25$ [1] $12x + 25y = 430$ [2]

3. Solve simultaneously using the substitution method, since equation [1] is easy to rearrange.
Rearrange equation [1] to make x the subject by subtracting y from both sides of equation [1].

Rearrange equation [1]:
$x + y = 25$
$x = 25 - y$

4. Substitute the expression $(25 - y)$ for x into equation [2].

Substituting $(25 - y)$ into [2]:
$12(25 - y) + 25y = 430$

5. Solve for y.

$300 - 12x + 25y = 430$
$300 + 13y = 430$
$13y + 300 = 430$
$13y = 130$
$y = 10$

6. Substitute the value of y into the rearranged equation $x = 25 - y$ from step 3.

Substituting $y = 10$ into $x = 25 - y$:
$x = 25 - 10$
$x = 15$

7. Write the solution.

Solution:
The number of nuts $(x) = 15$.
The number of bolts $(y) = 10$.

8. Check the solution by substituting $x = 15$ and $y = 10$ into equation [1].

Check: Substitute into $x + y = 25$.
LHS $= 15 + 20$ RHS $= 25$
$= 25$
As LHS $=$ RHS, the solution is correct.

- It is also possible to determine solutions to worded problems using the graphical method by forming and then graphing equations.

WORKED EXAMPLE 31 Applying the graphical method

Cecilia buys 2 pairs of shorts and 3 T-shirts for \$160. Ida buys 1 pair of shorts and 2 T-shirts for \$90. Develop two equations to describe the situation and solve them graphically to determine the cost of one pair of shorts and one T-shirt.

THINK

1. Define the two variables.

2. Formulate two equations from the information given and number them.

3. Calculate the x- and y-intercepts for both graphs.

WRITE

Let $x =$ cost of a pair of shorts.
Let $y =$ cost of a T-shirt.

$2x + 3y = 160$ [1]
$x + 2y = 90$ [2]

Equation [1]:
$2x + 3y = 160$
x-intercept, $y = 0$
$2x + 3 \times 0 = 160$
$2x = 160$
$x = 80$
y-intercept, $x = 0$
$2 \times 0 + 3y = 160$
$3y = 160$
$y = 53\frac{1}{3}$

Equation [2]:
$x + 2y = 90$
x-intercept, $y = 0$
$x + 2 \times 0 = 90$
$x = 90$
y-intercept, $x = 0$
$0 + 2y = 90$
$2y = 90$
$y = 45$

4. Graph the two lines either by hand or using technology. Only the first quadrant of the graph is required, as cost cannot be negative.

5. Identify the point of intersection to solve the simultaneous equations.

The point of intersection is (50, 20).

6. Write the answer as a sentence.

The cost of one pair of shorts is $50 and the cost of one T-shirt is $20.

▶ 4.8.2 Break-even points

eles-6178

- A **break-even point** is a point where the costs of a business equal the revenue. It is also the point where there is zero profit.
- Break-even points can help businesses make decisions about their selling prices, set sales budgets and prepare their business plans.

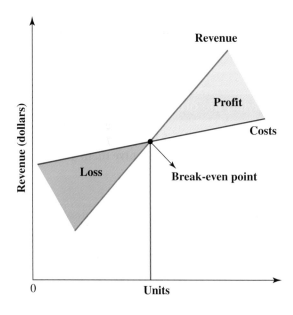

Determining the break-even point

To determine the break-even point, the equations for cost and revenue are solved simultaneously. That is, solve the equation $R = C$.

WORKED EXAMPLE 32 Using simultaneous equations to determine the break-even point

Santo sells shirts for $25. The revenue, R, for selling n shirts is represented by the equation $R = 25n$. The cost to make n shirts is represented by the equation $C = 2200 + 3n$.

a. Solve the equations simultaneously to determine the break-even point.
b. Determine the profit or loss, in dollars, for the following shirt orders.
   i. 75 shirts
   ii. 220 shirts

THINK	WRITE

a. 1. Write the two equations.

a. $C = 2200 + 3n$
$R = 25n$

2. Equate the equations ($R = C$).

$2200 + 3n = 25n$

3. Solve for the unknown.

$2200 + 3n - 3n = 25n - 3n$
$2200 = 22n$
$\dfrac{2200}{22} = n$
$n = 100$

4. Substitute back into either equation to determine the values of C and R.

$R = 25n$
$ = 25 \times 100$
$ = 2500$

5. Write the answer in the context of the problem.

The break-even point is (100, 2500). Therefore, 100 shirts need to be sold to cover the production cost, which is $2500.

b. i. 1. Write the two equations.

b. i. $C = 2200 + 3n$
$R = 25n$

2. Substitute the given value into both equations.

$n = 75$
$C = 2200 + 3 \times 75$
$ = 2425$
$R = 25 \times 75$
$ = 1875$

3. Determine the profit/loss.

Profit/loss $= R - C$
$ = 1875 - 2425$
$ = -550$

4. Write the answer in the context of the problem.

Since the answer is negative, it means that Santo lost $550 (that is, selling 75 shirts did not cover the cost to produce the shirts).

ii. 1. Write the two equations.

ii. $C = 2200 + 3n$
$R = 25n$

2. Substitute the given value into both equations.

$n = 220$
$C = 2200 + 3 \times 220$
$ = 2860$
$R = 25 \times 220$
$ = 5500$

3. Determine the profit/loss.

$$Profit/loss = R - C$$
$$= 5500 - 2860$$
$$= 2640$$

4. Write the answer in the context of the problem.

Since the answer is positive, it means that Santo made $2640 profit from selling 220 shirts.

DISCUSSION

How do you decide which method to use when solving problems using simultaneous linear equations?

 Resources

 eWorkbook　　Topic 4 Workbook (worksheets, code puzzle and a project) (ewbk-2095)

Interactivities　Individual pathway interactivity: Applications of simultaneous linear equations (int-4580)
　　　　　　　　　　Break-even points (int-6454)

Exercise 4.8 Applications of simultaneous linear equations　learn on

4.8 Quick quiz　on	4.8 Exercise

Individual pathways

■ PRACTISE	■ CONSOLIDATE	■ MASTER
1, 7, 8, 10, 13, 14, 16, 18, 23	2, 5, 9, 11, 15, 19, 20, 24	3, 4, 6, 12, 17, 21, 22, 25, 26

Fluency

1. **WE29** Rick received better results for his Maths test than for his English test. If the sum of his two marks is 163 and the difference is 31, calculate the mark recieved for each subject.

2. **WE30** Rachael buys 30 nuts and bolts to finish a project. If each nut costs 10 cents, each bolt costs 20 cents and the total purchase price is $4.20, calculate how many nuts and how many bolts she buys.

3. Eloise has a farm that raises chicken and sheep. Altogether there are 1200 animals on the farm. If the total number of legs from all the animals is 4000, calculate how many of each type of animal there is on the farm.

Understanding

4. Determine the two numbers whose difference is 5 and whose sum is 11.

5. The difference between two numbers is 2. If three times the larger number minus twice the smaller number is 13, determine the values of the two numbers.

6. One number is 9 less than three times a second number. If the first number plus twice the second number is 16, determine the values of the two numbers.

7. A rectangular house has a perimeter of 40 metres and the length is 4 metres more than the width. Calculate the dimensions of the house.

8. **WE31** Mike has 5 lemons and 3 oranges in his shopping basket. The cost of the fruit is \$3.50. Voula, with 2 lemons and 4 oranges, pays \$2.10 for her fruit.
   Develop two equations to describe the situation and solve them graphically to determine the cost of each type of fruit.

9. A surveyor measuring the dimensions of a block of land finds that the length of the block is three times the width. If the perimeter is 160 metres, calculate the dimensions of the block.

10. **WE32** Yolanda sells handmade bracelets at a market for \$12.50. The revenue, R, for selling n bracelets is represented by the equation $R = 12.50n$.
    The cost to make n bracelets is represented by the equation $C = 80 + 4.50n$.
    a. Solve the equations simultaneously to determine the break-even point.
    b. Determine the profit or loss, in dollars, if Yolanda sells:
       i. 8 bracelets
       ii. 13 bracelets.

11. Julie has \$3.10 in change in her pocket. If she has only 50-cent and 20-cent pieces and the total number of coins is 11, calculate how many coins of each type she has.

12. Mr Yang's son has a total of twenty-one \$1 and \$2 coins in his moneybox. When he counts his money, he finds that its total value is \$30. Determine how many coins of each type he has.

13. If three Magnums and two Paddlepops cost \$8.70 and the difference in price between a Magnum and a Paddlepop is 90 cents, calculate how much each type of ice-cream costs.

14. If one Red Frog and four Killer Pythons cost \$1.65, whereas two Red Frogs and three Killer Pythons cost \$1.55, calculate how much each type of lolly costs.

15. A catering firm charges a fixed cost for overheads and a price per person. It is known that a party for 20 people costs \$557, whereas a party for 35 people costs \$909.50.
    Determine the fixed cost and the cost per person charged by the company.

16. The difference between Sally's PE mark and her Science mark is 12, and the sum of the marks is 154. If the PE mark is the higher mark, calculate what mark Sally got for each subject.

17. Mozza's Cheese Supplies sells six Mozzarella cheeses and eight Swiss cheeses to Munga's deli for \$83.60, and four Mozzarella cheeses and four Swiss cheeses to Mina's deli for \$48.
    Calculate how much each type of cheese costs.

Reasoning

18. If the perimeter of the triangle in the diagram is 12 cm and the length of the rectangle is 1 cm more than the width, determine the value of x and y.

19. Mr and Mrs Waugh want to use a caterer for a birthday party for their twin sons. The manager says the cost for a family of four would be \$160. However, the sons want to invite 8 friends, making 12 people in all.

The cost for this would be $360. If the total cost in each case is made up of the same cost per person and the same fixed cost, calculate the cost per person and the fixed cost. Show your working.

20. Joel needs to buy some blank DVDs and USB sticks to back up a large amount of data that has been generated by an accounting firm. He buys 6 DVDs and 3 USB sticks for $96. He later realises these are not sufficient and so buys another 5 DVDs and 4 USB sticks for $116.
Determine how much each DVD and each USB stick cost. (Assume the same rate per item was charged for each visit.)

21. Four years ago Tim was 4 times older than his brother Matthew. In six years' time Tim will only be double his brother's age. Calculate how old the two brothers currently are.

22. A local cinema has different prices for movie tickets for children (under 12), adults and seniors (over 60). Consider the following scenarios:
    • For a senior couple (over 60) and their four grandchildren, the total cost is $80.
    • For two families with four adults and seven children, the total cost is $160.50.
    • For a son (under 12), his father and his grandfather (over 60), the total cost is $45.75.
    Determine the cost of each type of ticket.

Problem solving

23. Reika completes a biathlon (swimming and running) that has a total distance of 37 kilometres. Reika knows that it takes her 21.2 minutes to swim 1 kilometre and 4.4 minutes to run 1 kilometre.
If her total time for the race was 4 hours and 32 minutes, calculate the length of the swimming component of the race.

24. At the football, hot chips are twice as popular as meat pies and three times as popular as hot dogs. Over the period of half an hour during half time, a fast-food outlet serves 121 people who each bought one item. Determine how many serves of each of the foods were sold during this half-hour period.

25. Three jet-skis in a 300-kilometre handicap race leave at two hour intervals. Jet-ski 1 leaves first and has an average speed of 25 kilometres per hour for the entire race. Jet-ski 2 leaves two hours later and has an average speed of 30 kilometres per hour for the entire race.
Jet-ski 3 leaves last, two hours after jet-ski 2, and has an average speed of 40 kilometres per hour for the entire race.
    a. Sketch a graph to show each jet-ski's journey on the one set of axes.
    b. Determine who wins the race.
    c. Check your findings algebraically and describe what happened to each jet-ski during the course of the race.

26. Alice is competing in a cycling race on an extremely windy day. The race is an 'out and back again' course, so the wind is against Alice in one direction and assisting her in the other.
For the first half of the race the wind is blowing against Alice, slowing her down by 4 km per hour.
Given that on a normal day Alice could maintain a pace of 36 km per hour and that this race took her 4 hours and 57 minutes, calculate the total distance of the course.

LESSON
4.9 Solving simultaneous linear and non-linear equations (10A)

LEARNING INTENTIONS

At the end of this lesson you should be able to:
- determine the point or points of intersection between a linear equation and various non-linear equations using various techniques
- use digital technology to determine the points of intersection between a linear equation and a non-linear equation.

▶ 4.9.1 Solving simultaneous linear and quadratic equations

eles-4771

- The graph of a quadratic function is called a parabola.
- A parabola and a straight line may:
  - intersect at only one point

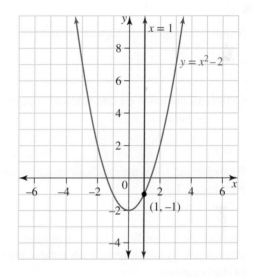

  - intersect at two points

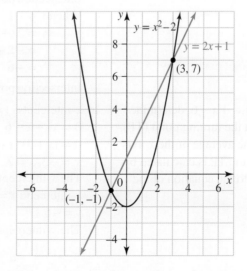

- not intersect at all.

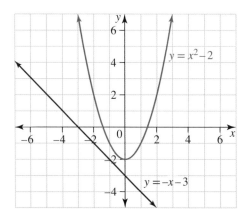

WORKED EXAMPLE 33 Solving linear and quadratic simultaneous equations

Determine the points of intersection of $y = x^2 + x - 6$ and $y = 2x - 4$:
a. algebraically
b. graphically.

THINK

WRITE/DRAW

a. 1. Number the equations. Equate [1] and [2].

a.
$$y = x^2 + x - 6 \quad [1]$$
$$y = 2x - 4 \quad [2]$$
$$x^2 + x - 6 = 2x - 4$$

2. Collect all the terms on one side and simplify.

$$x^2 + x - 6 - 2x + 4 = 2x - 4 - 2x + 4$$
$$x^2 + x - 6 - 2x + 4 = 0$$
$$x^2 - x - 2 = 0$$

3. Factorise and solve the quadratic equation using the Null Factor Law.

$$(x - 2)(x + 1) = 0$$
$$x - 2 = 0 \quad \text{or} \quad x + 1 = 0$$
$$x = 2 \qquad\qquad x = -1$$

4. Identify the y-coordinate for each point of intersection by substituting each x-value into one of the equations.

When $x = 2$,
$$y = 2(2) - 4$$
$$= 4 - 4$$
$$= 0$$
Intersection point $(2, 0)$
When $x = -1$
$$y = 2(-1) - 4$$
$$= -2 - 4$$
$$= -6$$

5. Write the solution.

Intersection point $(-1, -6)$

b. 1. To sketch the graph of $y = x^2 + x - 6$, determine the x- and y-intercepts and the turning point (TP). The x-value of the TP is the average of the x-axis intercepts. The y-value of the TP is calculated by substituting the x-value into the equation of the parabola.

b. x-intercepts: $y = 0$
$$0 = x^2 + x - 6$$
$$0 = (x + 3)(x - 2)$$
$$x = -3, x = 2$$
The x-intercepts are $(-3, 0)$ and $(2, 0)$.
y-intercept: $x = 0$
$$y = -6$$
The y-intercept is $(0, -6)$.

x-value of the turning point: $\dfrac{-3+2}{2} = -0.5$

y-value of the turning point; when $x = -0.5$:
$y = (-0.5)^2 + (-0.5) - 6$
$y = -6.25$
The turning point is $(-0.5, -6.25)$.

2. To sketch the graph of $y = 2x - 4$, determine the x- and y-intercepts.

x-intercept: $y = 0$
$0 = 2x - 4$
$x = 2$
The x-intercept is $(2, 0)$.
y-intercept: $x = 0$
$y = -4$
The y-intercept is $(0, -4)$.

3. On the same set of axes, sketch the graphs of $y = x^2 + x - 6$ and $y = 2x - 4$, labelling both.

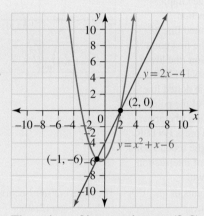

4. On the graph, locate the points of intersection and write the solutions.

The points of intersection are $(2, 0)$ and $(-1, -6)$.

| TI | THINK | DISPLAY/WRITE | CASIO | THINK | DISPLAY/WRITE |

a.

1. On a Calculator page, press:
   - MENU
   - 3: Algebra
   - 1: Solve
   Complete the entry line as:
   solve($y = x^2 + x - 6$ and $y = 2x - 4, \{x, y\}$)
   Then press ENTER.

a.

2. Write the solutions.

The points of intersection are $(-1, -6)$ and $(2, 0)$.

a.

1. On a Main screen, complete the entry line as:
   solve
   $(x^2 + x - 6 = 2x - 4, x)$
   The x-values of the solutions will be shown.
   To determine the corresponding y-values, complete the entry lines as:
   $2x - 4 | x = -1$
   $2x - 4 | x = 2$
   Press EXE after each entry.

a.

2. Write the solutions.

The points of intersection are $(-1, -6)$ and $(2, 0)$.

b.

1. On a Graphs page, complete the function entry lines as:
$f1(x) = x^2 + x - 6$
$f2(x) = 2x - 4$
Press the down arrow between entering the functions. The graphs will be displayed.

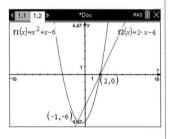

2. To locate the point of intersection, press:
   • MENU
   • 6: Analyze Graph
   • 4: Intersection
Drag the dotted line to the left of the first point of intersection (the lower bound), press ENTER, then drag the dotted line to the right of the point of intersection (the upper bound) and press ENTER. Repeat for the second point. The points of intersection will be shown.

3. State the points of intersection.

The points of intersection are $(-1, -6)$ and $(2, 0)$.

b.

1. On a Graph & Table screen, complete the function entry lines as:
$y1 = x^2 + x - 6$
$y2 = 2x - 4$
Then tap the graphing icon. The graphs will be displayed.

2. To locate the first point of intersection, press:
   • Analysis
   • G-Solve
   • Intersection
The point of intersection will be shown. Tap the right arrow for the second point.

3. State the points of intersection.

The points of intersection are $(-1, -6)$ and $(2, 0)$.

4.9.2 Solving simultaneous linear and hyperbolic equations

eles-4772

• A hyperbola and a straight line may:
  • intersect at only one point. In the first case, the line is a tangent to the curve.

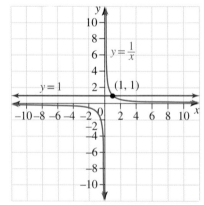

- intersect at two points

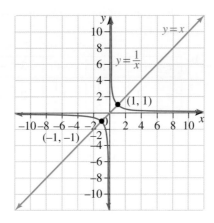

- not intersect at all.

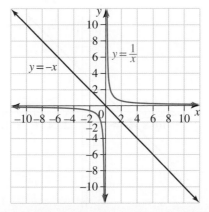

Determine the point(s) of intersection between $y = x + 5$ and $y = \dfrac{6}{x}$:

a. **algebraically**

b. **graphically.**

THINK	WRITE/DRAW
a. 1. Number the equations.	a. $y = x + 5$ [1] $y = \dfrac{6}{x}$ [2]
2. Equate [1] and [2]. Collect all terms on one side, factorise and simplify to solve for x.	$x + 5 = \dfrac{6}{x}$ $x(x + 5) = 6$ $x^2 + 5x - 6 = 0$ $(x + 6)(x - 1) = 0$ $\qquad\qquad x = -6,\ x = 1$
3. To determine the y-coordinates of the points of intersection, substitute the values of x into [1].	$x = -6 \qquad x = 1$ $y = -6 + 5 \quad y = 1 + 5$ $y = -1 \qquad\ y = 6$

4. Write the solutions.

The points of intersection are $(-6, -1)$ and $(1, 6)$.

b. 1. To sketch the graph of $y = \dfrac{6}{x}$, draw a table of values.

b.

x	-6	-5	-4	-3	-2	-1	0	1	2
y	-1	$-1\dfrac{1}{5}$	$-1\dfrac{1}{2}$	-2	-3	-6	Undef.	6	3

2. To sketch the graph of $y = x + 5$, determine the x- and y-intercepts.

x-intercept: $y = 0$
$$0 = x + 5$$
$$x = -5$$
The x-intercept is $(-5, 0)$.
y-intercept: $x = 0$
$$y = 5$$
The y-intercept is $(0, 5)$.

3. On the same set of axes, sketch the graphs of $y = x + 5$ and $y = \dfrac{6}{x}$, labelling both.

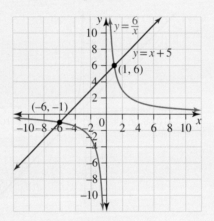

4. On the graph, locate the points of intersection and write the solutions.

The points of intersection are $(1, 6)$ and $(-6, -1)$.

TI | THINK

a.

1. In a new problem, on a Calculator page, press:
- MENU
- 1: Actions
- 1: Define

Complete the entry line as:
Define $f1(x) = x + 5$
Repeat for the second function:
Define $f2(x) = \dfrac{6}{x}$
Press ENTER after each entry.

DISLPAY/WRITE

a.

CASIO | THINK

a.

On the Main screen, tap:
- Action
- Advanced
- solve

Complete the entry lines as:
$\text{solve}\left(x + 5 = \dfrac{6}{x}, x\right)$

$x + 5 | x = -6$
$x + 5 | x = 1$
Press EXE after each entry.

DISPLAY/WRITE

a.

The points of intersection are $(-6, -1)$ and $(1, 6)$.

2. To determine the
intersection points
algebraically, press:
- MENU
- 3: Algebra
- 1: Solve

Complete the entry line
as:

solve $(f1(x) = f2(x), x)$
$f1(-6)$
$f2(1)$

The points $(-6, -1)$ and $(1, 6)$ are
the points of intersection.

b.

1. On a Graphs page,
press the up arrow ▲
to select the function
$f2(x)$, then press
ENTER. The graph
will be displayed. Now
press TAB, select the
function $f1(x)$ and press
ENTER to draw the
function.
Apply colour if you
would like to.

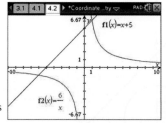

2. To determine the points
of intersection between
the two graphs, press:
- MENU
- 6: Analyze Graph
- 4: Intersection

Move the cursor to
the left of one of the
intersection points,
press ENTER, then
move the cursor
to the right of this
intersection point
and press ENTER.
The intersection point
is displayed. Repeat
for the other point of
intersection.

The points of intersection are
$(-6, -1)$ and $(1, 6)$.

b.

In the Graph & Table page,
complete the entry lines as:

$y1 = x + 5$
$y2 = \dfrac{6}{x}$

Then tap the graphing icon.
To determine the points of
intersection, tap:
- Analysis
- G-Solve
- Intersection

To determine the next point
of intersection, press the right
arrow.

b.

The points of intersection are
$(-6, -1)$ and $(1, 6)$.

Solving simultaneous linear equations and circles

- A circle and a straight line may:
  - intersect at only one point. Here, the line is a tangent to the curve.

- intersect at two points

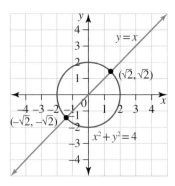

- not intersect at all.

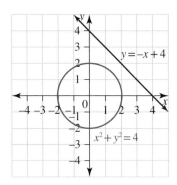

Solutions of a linear equation and a non-linear equation

Depending on the equations, a linear equation and a non-linear equation can have a different number of solutions. For a linear equation and any of the following:
- **quadratic equations**
- **hyperbolic equations**
- **circles**

the number of possible solutions (points of intersections) is $0, 1$ or 2.

DISCUSSION

What does it mean if a straight line touches a curve only once?

 Resources

 eWorkbook Topic 4 Workbook (worksheets, code puzzle and a project) (ewbk-2095)

 Interactivity Solving simultaneous linear and non-linear equations (int-6128)

4.9 Quick quiz on	4.9 Exercise

Individual pathways

■ PRACTISE	■ CONSOLIDATE	■ MASTER
1, 2, 5, 8, 9, 12, 15	3, 6, 10, 13, 16, 17	4, 7, 11, 14, 18, 19, 20

Fluency

1. Describe how a parabola and straight line may intersect. Use diagrams to illustrate your explanation.

2. **WE33** Determine the points of intersection of the following pairs of equations:
   i. algebraically
   ii. algebraically using a calculator
   iii. graphically using a calculator.
   a. $y = x^2 + 5x + 4$ and $y = -x - 1$
   b. $y = -x^2 + 2x + 3$ and $y = -2x + 7$
   c. $y = -x^2 + 2x + 3$ and $y = -6$

3. Determine the points of intersection of the following pairs of quations.
   a. $y = -x^2 + 2x + 3$ and $y = 3x - 8$
   b. $y = -(x-1)^2 + 2$ and $y = x - 1$
   c. $y = x^2 + 3x - 7$ and $y = 4x + 2$

4. Determine the points of intersection of the following pairs of equations.
   a. $y = 6 - x^2$ and $y = 4$
   b. $y = 4 + x - x^2$ and $y = \dfrac{3-x}{2}$
   c. $x = 3$ and $y = 2x^2 + 7x - 2$

5. **MC** Identify which of the following graphs shows the parabola $y = x^2 - 3x + 2$, $x \in R$, and the straight line $y = x + 3$.

A.

B.

C.

D.
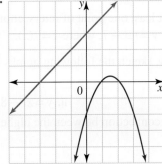

6. **MC** Identify which of the following equations are represented by the graph shown.

A. $y = 0.5(x + 1.5)^2 + 4$ and $y = -\dfrac{1}{3}x + 1$

B. $y = -0.5(x + 1.5)^2 - 4$ and $y = -\dfrac{1}{3}x + 1$

C. $y = -0.5(x - 1.5)^2 + 4$ and $y = \dfrac{1}{3}x + 1$

D. $y = 0.5(x - 1.5)^2 + 4$ and $y = -\dfrac{1}{3}x + 1$

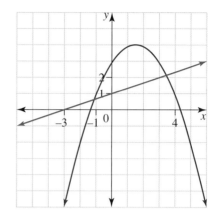

7. Determine whether the following graphs intersect.

a. $y = -x^2 + 3x + 4$ and $y = x - 4$

b. $y = -x^2 + 3x + 4$ and $y = 2x + 5$

c. $y = -(x + 1)^2 + 3$ and $y = -4x - 1$

d. $y = (x - 1)^2 + 5$ and $y = -4x - 1$

Understanding

WE34 For questions 8 to 11, determine the point(s) of intersection between each pair of equations.

8. a. $y = x$
   $y = \dfrac{1}{x}$

   b. $y = x - 2$
   $y = \dfrac{1}{x}$

   c. $y = 3x$
   $y = \dfrac{5}{x}$

   d. $y = \dfrac{6}{x}$
   $y = \dfrac{x}{2} + 2$

9. a. $y = 3x$
   $x^2 + y^2 = 10$

   b. $x^2 + y^2 = 25$
   $3x + 4y = 0$

   c. $x^2 + y^2 = 50$
   $y = 5 - 2x$

   d. $x^2 + y^2 = 9$
   $y = 2 - x$

10. a. $y = \dfrac{1}{x}$
    $y = 4x$

    b. $x^2 + y^2 = 25$
    $y = -2x + 5$

    c. $y = 2x + 3$
    $y = -4x^2 + 3$

    d. $3x + 4y = 7$
    $y = \dfrac{10}{x} - 4$

11. a. $y = x^2$
    $y = 2x - 1$

    b. $x^2 + (y + 1)^2 = 25$
    $y = 3$

    c. $y = -4x - 5$
    $y = x^2 + 2x + 3$

    d. $\dfrac{x}{3} + \dfrac{y}{4} = 7$
    $y = \dfrac{x^2}{16} + 3$

Reasoning

12. Consider the following equations: $y = \dfrac{1}{2}(x - 3)^2 + \dfrac{5}{2}$ and $y = x + k$.

    Identify for what values of k the two lines would have:

    a. no points of intersection
    b. one point of intersection
    c. two points of intersection.

13. Show that there is at least one point of intersection between the parabola $y = -2(x + 1)^2 - 5$, where $y = f(x)$, and the straight line $y = mx - 7$, where $y = f(x)$.

14. a. Using technology, sketch the following graphs and state how many ways a straight line could intersect with each graph.

    i. $y = x^3 - 4x$

    ii. $y = x^4 - 8x^2 + 16$

    iii. $y = x^5 - 8x^3 + 16x$

    b. Comment on the connection between the highest power of x and the number of possible points of intersection.

Problem solving

15. If two consecutive numbers have a product of 306, calculate the possible values for these numbers.

16. The perimeter of a rectangular paddock is 200 m and the area is 1275 m². Determine the length and width of the paddock.

17. a. Determine the point(s) of intersection between the circle $x^2 + y^2 = 50$ and the linear equation $y = 2x - 5$.
    b. Confirm your solution to part a by plotting the equation of the circle and the linear equation on the same graph.

18. The sum of two positive numbers is 21. Twice the square of the larger number minus three times the square of the smaller number is 45. Determine the value of the two numbers.

19. a. Omar is running laps around a circular park with equation $x^2 + y^2 = 32$. Chae-won is running along another track where the path is given by $y = \left(\sqrt{2} - 1 \right) x + \left(8 - 4\sqrt{2} \right)$. Determine the point(s) where the two paths intersect.
    b. Omar and Chae-won both start from the same point. If Chae-won gets between the two points in two hours, calculate the possible speeds Omar could run at along his circle in order to collide with Chae-won at the other point of intersection. Assume all distances are in kilometres and give your answer to 2 decimal places.

20. Adam and Eve are trying to model the temperature of a cup of coffee as it cools.
    • Adam's model: Temperature (°C) $= 100 - 5 \times \text{time}$
    • Eve's model: Temperature (°C) $= \dfrac{800}{10 + \text{time}} + 20$
    Time is measured in minutes.

    a. Using either model, identify the initial temperature of the cup of coffee.
    b. Determine at what times the two models predict the same temperature for the cup of coffee.
    c. Evaluate whose model is more realistic. Justify your answer.

LESSON
4.10 Solving linear inequalities

LEARNING INTENTIONS

At the end of this lesson you should be able to:
  • solve an inequality and represent the solution on a number line
  • convert a worded statement to an inequality in order to solve a problem.
  • evaluate and sketch piecewise linear graphs.

▶ 4.10.1 Inequalities between two expressions

eles-4774

  • An equation, such as $y = 2x$, is a statement of *equality* as both sides are equal to each other.
  • An **inequation**, such as $y < x + 3$, is a statement of **inequality** between two expressions.

- A linear equation such as $3x = 6$ will have a unique solution ($x = 2$), whereas an inequation such as $3x < 6$ will have an infinite number of solutions ($x = 1, 0, -1, -2 \ldots$ are all solutions).
- We use a number line to represent all possible solutions to a linear inequation. When representing an inequality on a number line, an open circle is used to represent that a value is not included, and a closed circle is used to indicate that a number is included.
- The following table shows four basic inequalities and their representations on a number line.

Mathematical statement	Worded statement	Number line diagram
$x > 2$	x is greater than 2	
$x \geq 2$	x is greater than or equal to 2	
$x < 2$	x is less than 2	
$x \leq 2$	x is less than or equal to 2	

▶ 4.10.2 Solving inequalities

eles-6179

- The following operations may be done to both sides of an inequality without affecting its truth.
  - A number can be added or subtracted from both sides of the inequality.

Adding or subtracting a number:

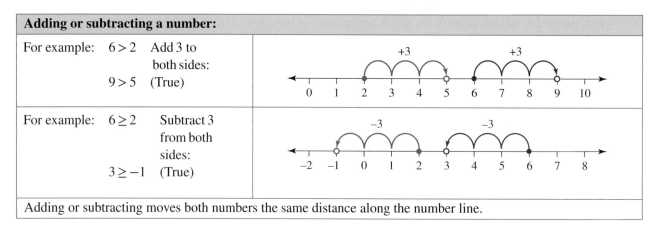

For example: $6 > 2$ Add 3 to both sides:
$9 > 5$ (True)

For example: $6 \geq 2$ Subtract 3 from both sides:
$3 \geq -1$ (True)

Adding or subtracting moves both numbers the same distance along the number line.

- A number can be multiplied or divided by a positive number.

Multiplying or dividing by a positive number:

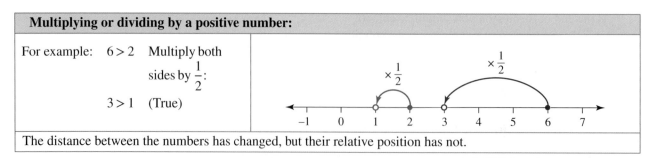

For example: $6 > 2$ Multiply both sides by $\dfrac{1}{2}$:
$3 > 1$ (True)

The distance between the numbers has changed, but their relative position has not.

- Care must be taken when multiplying or dividing by a negative number.

Multiplying or dividing by a negative number:	
For example: $6 > 2$ Multiply both sides by -1: $-6 > -2$ (False)	

Multiplying or dividing by a negative number reflects numbers about $x = 0$.
Their relative positions are reversed.

- When solving inequalities, if both sides are multiplied or divided by a negative number, then the inequality sign must be reversed.
  For example, $6 > 2$ implies that $-6 < -2$.

Solving a linear inequality

Solving a linear inequality is a similar process to solving a standard linear equation. We can perform the following inverse operations as normal:
- **a number or term can be added to or subtracted from each side of the inequality**
- **each side of an inequation can be multiplied or divided by a positive number.**

We must take care to change the direction of the inequality sign when:
- **each side of an inequation is to be multiplied or divided by a negative number.**

WORKED EXAMPLE 35 Solving linear inequalities

Solve each of the following linear inequalities and show the solution on a number line.
a. $4x - 1 < -2$
b. $6x - 7 \geq 3x + 5$

THINK

a. 1. Write the inequality.

   2. Add 1 to both sides of the inequality.

   3. Obtain x by dividing both sides of the inequality by 4.

   4. Show the solution on a number line. Use an open circle to show that the value of $-\dfrac{1}{4}$ is not included.

b. 1. Write the inequality.
   2. Subtract $3x$ from both sides of the inequality.

WRITE/DRAW

a. $4x - 1 < -2$

$4x - 1 + 1 < -2 + 1$
$$4x < -1$$
$$\frac{4x}{4} < -\frac{1}{4}$$

$$x < -\frac{1}{4}$$

b. $6x - 7 \geq 3x + 5$
$6x - 7 - 3x \geq 3x + 5 - 3x$
$$3x - 7 \geq 5$$

3. Add 7 to both sides of the inequality.

$$3x - 7 + 7 \geq 5 + 7$$
$$3x \geq 12$$

4. Obtain x by dividing both sides of the inequality by 3.

$$3x \geq 12$$
$$\frac{3x}{3} \geq \frac{12}{3}$$
$$x \geq 4$$

5. Show the solution on a number line. Use a closed circle to show that the value of 4 is included.

WORKED EXAMPLE 36 Solving complex linear inequalities

Solve each of the following linear inequalities.

a. $-3m + 5 < -7$

b. $5(x - 2) \geq 7(x + 3)$

THINK	WRITE
a. 1. Write the inequality.	**a.** $-3m + 5 < -7$
2. Subtract 5 from both sides of the inequality. (There is no change to the inequality sign.)	$-3m + 5 - 5 < -7 - 5$ $-3m < -12$
3. Obtain m by dividing both sides of the inequation by -3. Reverse the inequality sign, since we are dividing by a negative number.	$\dfrac{-3m}{-3} > \dfrac{-12}{-3}$ $m > 4$
b. 1. Write the inequality.	**b.** $5(x - 2) \geq 7(x + 3)$
2. Expand both brackets.	$5x - 10 \geq 7x + 21$
3. Subtract $7x$ from both sides of the inequality.	$5x - 10 - 7x \geq 7x + 21 - 7x$ $-2x - 10 \geq 21$
4. Add 10 to both sides of the inequation.	$-2x - 10 + 10 \geq 21 + 10$ $-2x \geq 31$
5. Divide both sides of the inequality by -2. Reverse the direction of the inequality sign as we are dividing by a negative number.	$\dfrac{-2x}{-2} \leq \dfrac{31}{-2}$ $x \leq \dfrac{-31}{2}$ $x \leq -15\dfrac{1}{2}$

▶ 4.10.3 Piecewise linear graphs

eles-6180

- A **piecewise graph** is a graph that is made up of two or more linear equations.
- Often these are used to represent a situation that involves a transition or a change in a rate.
- The piecewise graph will be continuous, and the equations of the lines will intersect at each transition point.
- Piecewise position–time graphs can be used to describe motion in real life.

Distance–time graph

COLLABORATIVE TASK: Tell me a story

Students work in pair and write a story about the following motion graph.

The piecewise graph describes the motion of a bike rider who has a goal to ride a 150-km distance in less than 10 hours.

Discuss the following questions.
1. Give a story that explains the graph.
2. State whether the rider was able to achieve their goal? Explain.
3. State the speed at each line: A, B, C, D, E, F and G.
4. The speed of the rider was zero at some intervals. State these line parts and what this tells you about the motion of the rider.

Distance–time graph

WORKED EXAMPLE 37 Evaluating and sketching a piecewise graph

The following two equations represent the distance travelled by a group of students over 5 hours. Equation 1 represents the first section of the hike, when the students are walking at a pace of 4 km/h. Equation 2 represents the second section of the hike, when the students change their walking pace.

$$\text{Equation 1: } d = 4t, \ 0 \le t \le 2$$
$$\text{Equation 2: } d = 4 + 2t, \ 2 \le t \le 5$$

The variable d is the distance in km from the campsite, and t is the time in hours.
a. Determine the time, in hours, for which the group travelled in the first section of the hike.
b. i. Determine their walking pace in the second section of their hike.
   ii. Calculate for how long, in hours, they walked at this pace.
c. Sketch a piecewise linear graph to represent the distance travelled by the group of students over the five-hour hike.

THINK	WRITE/DRAW
a. 1. Determine which equation the question applies to.	**a.** This question applies to equation 1.
2. Look at the time interval for this equation.	$0 \le t \le 2$
3. Interpret the information.	The group travelled for 2 hours.
b. i. 1. Determine which equation the question applies to.	**b. i.** This question applies to equation 2.
2. Interpret the equation. The walking pace is found by the coefficient of t, as this represents the gradient.	$d = 4 + 2t, \ 2 \le t \le 5$ The coefficient of t is 2.
3. Write the answer as a sentence.	The walking pace is 2 km/h.
ii. 1. Look at the time interval shown.	**ii.** $2 \le t \le 5$
2. Interpret the information and answer the question.	They walked at this pace for 3 hours.
c. 1. Calculate the distance travelled before the change of pace.	**c.** Change after $t = 2$ hours: $d = 4t$ $d = 4 \times 2$ $d = 8$ km
2. Using a calculator, spreadsheet or otherwise, sketch the graph $d = 4t$ between $t = 0$ and $t = 2$.	
3. Solve the simultaneous equations to determine the point of intersection.	$4t = 4 + 2t$ $4t - 2t = 4 + 2t - 2t$ $2t = 4$ Substitute $t = 2$ into $d = 4t$: $d = 4 \times 2$ $= 8$

4. Using CAS, sketch the graph of $d = 4 + 2t$ between $t = 2$ and $t = 5$.

DISCUSSION

What are the similarities and differences when solving linear inequations compared to linear equations?

 Resources

eWorkbook   Topic 4 Workbook (worksheets, code puzzle and a project) (ewbk-2095)

Interactivities   Individual pathway interactivity: Solving linear inequalities (int-4582)
Inequalities on the number line (int-6129)

Exercise 4.10 Solving linear inequalities

learn

4.10 Quick quiz	4.10 Exercise

Individual pathways

■ PRACTISE	■ CONSOLIDATE	■ MASTER
1, 4, 7, 10, 15, 18, 19, 22, 25, 29, 32	2, 5, 8, 11, 13, 16, 20, 23, 26, 27, 30, 33	3, 6, 9, 12, 14, 17, 21, 24, 28, 31, 34

Fluency

WE35a For questions **1** to **3**, solve each of the following inequalities and show the solution on a number line.

1. a. $x + 1 > 3$ b. $a + 2 > 1$ c. $y - 3 \geq 4$ d. $m - 1 \geq 3$

2. a. $p + 4 < 5$ b. $x + 2 < 9$ c. $m - 5 \leq 4$ d. $a - 2 \leq 5$

3. a. $x - 4 > -1$ b. $5 + m \geq 7$ c. $6 + q \geq 2$ d. $5 + a > -3$

For questions **4** to **6**, solve each of the following inequalities. Check your solutions by substitution.

4. **a.** $3m > 9$ **b.** $5p \le 10$ **c.** $2a < 8$ **d.** $4x \ge 20$

5. **a.** $5p > -25$ **b.** $3x \le -21$ **c.** $2m \ge -1$ **d.** $4b > -2$

6. **a.** $\dfrac{m}{3} > 6$ **b.** $\dfrac{x}{2} < 4$ **c.** $\dfrac{a}{7} \le -2$ **d.** $\dfrac{m}{5} \ge 5$

For questions **7** to **9**, solve each of the following inequalities.

7. **a.** $2m + 3 < 12$ **b.** $3x + 4 \ge 13$ **c.** $5p - 9 > 11$ **d.** $4n - 1 \le 7$

8. **a.** $2b - 6 < 4$ **b.** $8y - 2 > 14$ **c.** $10m + 4 \le -6$ **d.** $2a + 5 \ge -5$

9. **a.** $3b + 2 < -11$ **b.** $6c + 7 \le 1$ **c.** $4p - 2 > -10$ **d.** $3a - 7 \ge -28$

WE35b For questions **10** to **14**, solve each of the following linear inequalities and show the solution on a number line.

10. **a.** $2m + 1 > m + 4$ **b.** $2a - 3 \ge a - 1$
    c. $5a - 3 < a - 7$ **d.** $3a + 4 \le a - 2$

11. **a.** $5x - 2 > 40 - 2x$ **b.** $7x - 5 \le 11 - x$
    c. $7b + 5 < 2b + 25$ **d.** $2(a + 4) > a + 13$

12. **a.** $3(m - 1) < m + 1$ **b.** $5(2m - 3) \le 3m + 6$
    c. $3(5b + 2) \le -10 + 4b$ **d.** $5(3m + 1) \ge 2(m + 9)$

13. **a.** $\dfrac{x + 1}{2} \le 4$ **b.** $\dfrac{x - 2}{5} \ge -4$ **c.** $\dfrac{x + 7}{3} < -1$

14. **a.** $\dfrac{2x + 3}{4} > 6$ **b.** $\dfrac{3x - 1}{7} \ge 2$ **c.** $\dfrac{5x + 9}{6} < 0$

WE36 For questions **15** to **17**, solve each of the following inequalities.

15. **a.** $-2m > 4$ **b.** $-5p \le 15$ **c.** $-2a \ge -10$
    d. $-p - 3 \le 2$ **e.** $10 - y \ge 13$

16. **a.** $14 - x < 7$ **b.** $1 - 6p > 1$ **c.** $2 - 10a \le 0$
    d. $2(3 - x) < 12$ **e.** $-4(a + 9) \ge 8$

17. **a.** $-15 \le -3(2 + b)$ **b.** $2x - 3 > 5x + 6$ **c.** $k + 5 < 2k - 3$
    d. $3(x - 4) < 5(x + 5)$ **e.** $7(a + 4) \ge 4(2a - 3)$

18. **MC** When solving the inequality $-2x > -7$, we need to:

    A. change the sign to $\ge$ **B.** change the sign to $<$
    C. change the sign to $=$ **D.** change the sign to $\le$

For questions **19** to **24**, solve each of the following inequalities.

19. **a.** $\dfrac{2 - x}{3} > 1$ **b.** $\dfrac{5 - m}{4} \ge 2$

20. **a.** $\dfrac{-3 - x}{5} < -4$ **b.** $\dfrac{3 - 8a}{2} < -1$

21. **a.** $\dfrac{4 - 3m}{2} \le 0$ **b.** $\dfrac{-2m + 6}{10} \le 3$

22. **a.** $3k > 6$ **b.** $-a - 7 < -2$ **c.** $5 - 3m \ge 0$ **d.** $x + 4 > 9$

23. a. $10 - y \leq 3$ b. $5 + 3d < -1$ c. $\dfrac{7p}{3} \geq -2$ d. $\dfrac{1 - x}{3} \leq 2$

24. a. $\dfrac{-4 - 2m}{5} > 0$ b. $5a - 2 < 4a + 7$ c. $6p + 2 \leq 7p - 1$ d. $2(3x + 1) > 2x - 16$

Understanding

25. Write linear inequalities for the following statements, using x to represent the unknown. (Do not attempt to solve the equations.)

     a. The product of 5 and a certain number is greater than 10.
     b. When three is subtracted from a certain number the result is less than or equal to 5.
     c. The sum of seven and three times a certain number is less than 42.

26. Write linear inequalities for the following statements. Choose an appropriate letter to represent the unknown.

     a. Four more than triple a number is more than 19.
     b. Double the sum of six and a number is less than 10.
     c. Seven less the half the difference between a number and 8 is at least 9.

27. **WE37** The following two equations represent the distance travelled by a group of bike riders over 8 hours, where d is the distance in km from the starting point and t is the time in hours.
Equation 1: $d = 20t,\ 0 \leq t \leq 4$
Equation 2: $d = -40 + 30t,\ 4 \leq t \leq 8$

     a. Determine the time, in hours, for which the riders travelled in the first section of the ride.
     b. i. Determine their riding pace in the second section of their ride.
        ii. Calculate for how long, in hours, they rode at this pace.
     c. Sketch a piecewise graph to represent the distance travelled by the riders over the 8-hour ride.

28. Write linear inequalities for the following situations. Choose an appropriate letter to represent the unknown.

     a. John makes $50 profit for each television he sells. Determine how many televisions John needs to sell to make at least $650 in profit.
     b. Determine what distances a person can travel with $60 if the cost of a taxi ride is $2.50 per km with a flagfall cost of $5.

Reasoning

29. Tom is the youngest of five children. The five children were all born 1 year apart. If the sum of their ages is at most 150, set up an inequality and solve it to find the possible ages of Tom.

30. Given the positive numbers a, b, c and d and the variable x, there is the following relationship:
$$-c < ax + b < -d.$$

     a. Determine the possible range of values of x if $a = 2$, $b = 3$, $c = 10$ and $d = 1$.
     b. Rewrite the original relationship in terms of x only (x by itself between the $<$ signs), using a, b, c and d.

31. Two speed boats are racing along a section of Lake Quikalong. The speed limit along this section of the lake is 50 km/h. Ella is travelling 6 km/h faster than Steven, and the sum of the speeds at which they are travelling is greater than 100 km/h.

     a. Write an inequation and solve it to describe all possible speeds that Steven could be travelling at.
     b. At Steven's lowest possible speed, is he over the speed limit?
     c. The water police issue a warning to Ella for exceeding the speed limit on the lake.
        Show that the police were justified in issuing a warning to Ella.

Problem solving

32. Mick the painter has fixed costs (insurance, equipment etc) of $3400 per year. His running cost to travel to jobs is based on $0.75 per kilometre. Last year Mick had costs that were less than $16 000.

    a. Write an inequality and solve it to find how many kilometres Mick travelled for the year.
    b. Explain the information you have found.

33. A coffee store produces doughnuts and croissants to sell alongside its coffee. Each morning the bakery has to decide how many of each it will produce. The store has 240 minutes to produce food in the morning.
    It takes 20 minutes to make a batch of doughnuts and 10 minutes to make a batch of croissants. The store also has 36 kg of flour to use each day. A batch of doughnuts uses 2 kg of flour and a batch of croissants uses 2 kg of flour.

    a. Set up an inequality around the amount of time available to produce doughnuts and croissants.
    b. Set up an inequality around the amount of flour available to produce doughnuts and croissants.
    c. Use technology to work out the possible number of each that can be made, taking into account both inequalities.

34. I have $40 000 to invest. Part of this I intend to invest in a stable 5% simple interest account. The remainder will be invested in my friend's IT business. She has said that she will pay me 7.5% interest on any money I give to her. I am saving for a European trip so I want the best return for my money.
    Calculate the least amount of money I should invest with my friend so that I receive at least $2500 interest per year from my investments.

LESSON
4.11 Inequalities on the Cartesian plane

LEARNING INTENTIONS

At the end of this lesson you should be able to:
- sketch the graph of a half plane: the region represented by an inequality
- sketch inequalities using digital technology.

▶ 4.11.1 Inequalities on the Cartesian plane

eles-4776
- A solution to a linear inequality is any ordered pair (coordinate) that makes the inequality true.
- There is an infinite number of points that can satisfy an inequality. If we consider the inequality $x + y < 10$, the following points $(1, 7)$, $(5, 2)$ and $(4, 3)$ are all solutions, whereas $(6, 8)$ is not as it does not satisfy the inequality ($6 + 8$ is not less than 10).
- These points that satisfy an inequality are represented by a region that is found on one side of a line and is called a **half plane**.

- To indicate whether the points on a line satisfy the inequality, a specific type of **boundary line** is used.

Points on the line	Symbol	Type of boundary line used
Do not satisfy the inequality	$<$ or $>$	Dashed - - - - - - - - - - - - - - -
Satisfy the inequality	$\leq$ or $\geq$	Solid _____

- The **required region** is the region that contains the points that satisfy the inequality.
- Shading or no shading is used to indicate which side of the line is the required region, and a key is shown to indicate the region.

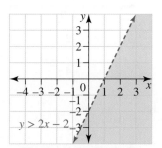

The required region is ☐.   The required region is ☐.

- Consider the line $x = 2$. It divides the **Cartesian plane** into two distinct regions or half-planes.

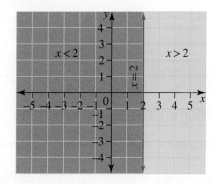

- The region on the left (shaded pink) contains all the points whose x-coordinate is less than 2, for example $(1, 3)$, so this region is given the name $x < 2$.
- The region on the right (shaded blue) contains all the points whose x-coordinate is greater than 2, for example $(3, -2)$, so this region is given the name $x > 2$.
- There are three distinct parts to the graph:
  - the boundary line, where $x = 2$
  - the pink region, where $x < 2$
  - the blue region, where $x > 2$.

Sketch a graph of each of the following regions.

a. $x \geq -1$

b. $y < 3$

THINK

a. 1. $x \geq -1$ includes the line $x = -1$ and the region $x > -1$.

2. On a Cartesian plane sketch the line $x = -1$. Because the line is required, it will be drawn as a continuous (unbroken) line.

3. Identify a point where $x > -1$, say $(2, 1)$.

4. Shade the region that includes this point. Label the region $x \geq -1$.

DRAW

a.

b. 1. The line $y = 3$ is not included.

2. Sketch the line $y = 3$. Because the line is not included, show it as a dashed (broken) line.

3. Identify a point where $y < 3$, say $(1, 2)$.

4. Shade the region where $y < 3$.

5. Label the region.

b.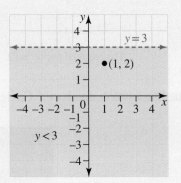

4.11.2 Determining the required region on the Cartesian plane

eles-4777

- For a more complex inequality, such as $y < 2x + 3$, first sketch the boundary line, which in this case is given by the equation $y = 2x + 3$.
  Note: The boundary line will be drawn as a solid line if it is included in the inequality ($y \leq x$) or as a broken line if it is not included ($y < x$).

- In order to determine which side of the boundary line satisfies the inequality, choose a point and test whether it satisfies the inequality. In most cases the point $(0, 0)$ is the best point to choose, but if the boundary line passes through the origin, it will be necessary to test a different point such as $(0, 1)$.
  For example:

$$\text{Inequality: } y < 2x + 3:$$
$$\text{Test } (0, 0): 0 < 2(0) + 3$$
$$0 < 3 \qquad \text{True}$$

- Since 0 is less than 3, the point $(0, 0)$ does satisfy the inequality. Thus, the half plane containing $(0, 0)$ is the required region.

WORKED EXAMPLE 39 Verifying inequalities at points on the Cartesian plane

Determine whether the points $(0, 0)$ and $(3, 4)$ satisfy either of the following inequalities.

a. $x - 2y < 3$ b. $y > 2x - 3$

THINK	WRITE
a. 1. Substitute $(0, 0)$ for x and y.	a. $x - 2y < 3$ Substitute $(0, 0)$:
2. Since the statement is true, $(0, 0)$ satisfies the inequality.	$0 - 0 < 3$ $0 < 3$ True
3. Substitute $(3, 4)$ for x and y.	$x - 2y < 3$ Substitute $(3, 4)$: $3 - 2(4) < 3$ $3 - 8 < 3$
4. Since the statement is true, $(3, 4)$ satisfies the inequality.	$-5 < 3$ True
5. Write the answer in a sentence.	The points $(0, 0)$ and $(3, 4)$ both satisfy the inequality.
b. 1. Substitute $(0, 0)$ for x and y.	b. $y > 2x - 3$ Substitute $(0, 0)$: $0 > 0 - 3$
2. Since the statement is true, $(0, 0)$ satisfies the inequality.	$0 > -3$ True
3. Substitute $(3, 4)$ for x and y.	$y > 2x - 3$ Substitute $(3, 4)$: $4 > 2(3) - 3$ $4 > 6 - 3$
4. Since the statement is true, $(3, 4)$ satisfies the inequality.	$4 > 3$ True
5. Write the answer in a sentence.	The points $(0, 0)$ and $(3, 4)$ both satisfy the inequality.

WORKED EXAMPLE 40 Sketching a linear inequality

Sketch a graph of the region $2x + 3y < 6$.

THINK	WRITE/DRAW

1. Locate the boundary line $2x + 3y < 6$ by finding the x- and y-intercepts.	$x = 0$: $0 + 3y = 6$
	         $y = 2$
	$y = 0$: $2x + 0 = 6$
	         $x = 3$

2. The line is not required due to the $<$ inequality, so rule a broken line.	

3. Test with the point $(0, 0)$. Does $(0, 0)$ satisfy $2x + 3y < 6$?

Test $(0, 0)$: $2(0) + 3(0) = 0$
As $0 < 6$, $(0, 0)$ is in the required region.

4. Shade the region that includes $(0, 0)$.

5. Label the region.

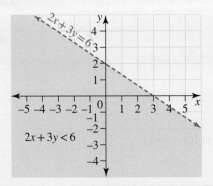

TI	THINK	DISPLAY/WRITE

Rearrange the inequality as:
$$y < 2 - \frac{2x}{3}$$
On a Graphs page at the function entry line, delete the $=$ symbol then select:

• 2: $y <$

Complete the function entry line as:
$$y < 2 - \frac{2x}{3}$$
Press ENTER. The shaded region will be displayed.

CASIO	THINK	DISPLAY/WRITE

Rearrange the inequality as:
$$y < 2 - \frac{2x}{3}$$
On a Graphs & Table screen, tap:

• Type
• Inequality
• $y <$ Type

Complete the function entry line as:
$$y < 2 - \frac{2x}{3}$$
Then tap the graphing icon. The shaded region will be displayed.

WORKED EXAMPLE 41 Modelling real-life situations

In the school holidays you have been given $160 to arrange some activities for your family. A ticket to the movies costs $10 and a ticket for the trampoline park costs $16.

a. If m represents the movie tickets and t represents the trampoline park tickets, write an inequality in terms of m and t that represents your entertainment budget.

b. Sketch the inequality from part a on the Cartesian plane.

c. Using the graph from part b explore the maximum number of movie and trampoline park tickets you can buy to use the maximum amount of your holiday budget.

THINK

a. Each movie ticket, m, costs $10, and each trampoline ticket, t, costs $16. The maximum amount you have to spend is $160.

b. 1. To draw the boundary line $10m + 16t \leq 160$, identify two points on the line. Let m be the x-axis and t be the y-axis.

WRITE

a. $10m + 16t \leq 160$

b. For the line: $10m + 16t = 160$
x-intercept: let $t = 0$.
$$10m + 16 \times 0 = 160$$
$$10m = 160$$
$$m = 16$$

The x-intercept is $(16, 0)$
y-intercept: let $m = 0$.
$$10 \times 0 + 16t = 160$$
$$16t = 160$$
$$t = 10$$
The y-intercept is $(0, 10)$.

2. Plot the two points and draw the line. As you can spend up to and including $160, the boundary line is solid.
Only the first quadrant of the graph is required, as the number of tickets cannot be negative.

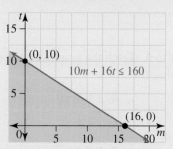

c. 1. To determine the maximum number of movie and trampoline park tickets, identify the nearest whole numbers of each to the graph line. These must be whole numbers as you cannot buy part of a ticket.

c. To spend the entire $160, only 16 movie tickets or 10 trampoline park tickets can be purchased. If less than $160 was spent you could purchase any whole-number combinations, such as 6 movie tickets and 6 trampoline park tickets for $156.

DISCUSSION

Think of some real-life situations where inequalities could be used to help solve a problem.

 Resources

 eWorkbook   Topic 4 Workbook (worksheets, code puzzle and a project) (ewbk-2095)

 Interactivities Individual pathway interactivity: Inequalities on the Cartesian plane (int-4583)
Linear inequalities in two variables (int-6488)

Exercise 4.11 Inequalities on the Cartesian plane   learn

4.11 Quick quiz on	4.11 Exercise

Individual pathways

■ PRACTISE	■ CONSOLIDATE	■ MASTER
1, 4, 7, 10, 13, 16, 19	2, 5, 8, 11, 14, 17, 20	3, 6, 9, 12, 15, 18, 21

Fluency

WE38 For questions **1** to **3**, sketch graphs of the following regions.

1. a. $x < 1$ b. $y \geq -2$ c. $x \geq 0$ d. $y < 0$

2. a. $x > 2$ b. $x \leq -6$ c. $y \geq 3$ d. $y \leq 2$

3. a. $x < \dfrac{1}{2}$ **b.** $y < \dfrac{3}{2}$ **c.** $y \geq -4$ **d.** $x \leq \dfrac{3}{2}$

 WE39 For questions **4** to **6**, determine which of the points A $(0, 0)$, B $(1, -2)$ and C $(4, 3)$ satisfy the following inequalities.

4. a. $x + y > 6$ **b.** $x - 3y < 2$

5. a. $y > 2x - 5$ **b.** $y < x + 3$

6. a. $3x + 2y < 0$ **b.** $x \geq 2y - 2$

WE40 For questions **7** to **9**, sketch the graphs for the regions given by the following inequations. Verify your solutions using technology.

7. a. $y \geq x + 1$ **b.** $y < x - 6$ **c.** $y > -x - 2$ **d.** $y < 3 - x$

8. a. $y > x - 2$ **b.** $y < 4$ **c.** $2x - y < 6$ **d.** $y \leq x - 7$

9. a. $x - y > 3$ **b.** $y < x + 7$ **c.** $x + 2y \leq 5$ **d.** $y \leq 3x$

10. **MC** The shaded region satisfying the inequality $y > 2x - 1$ is:

A.

B.

C.

D.
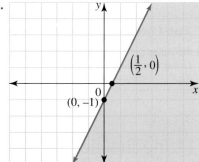

11. **MC** The shaded region satisfying the inequality $y \leq x + 4$ is:

A.

B.

C.

D.
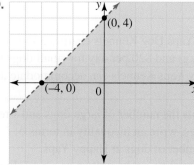

12. **MC** The region satisfying the inequality $y < -3x$ is:

A.

B.

C.

D.
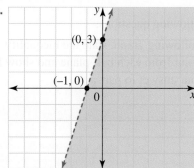

Understanding

13. a. Determine the equation of the line l shown in the diagram.
    b. Write down three inequalities that define the region R.

14. Identify all points with integer coordinates that satisfy the following inequalities:

$$x \geq 3$$
$$y > 2$$
$$3x + 2y \leq 19$$

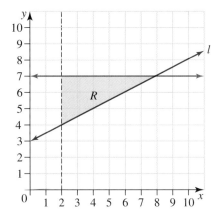

15. **WE41** Happy Yaps Dog Kennels charges \$35 per day for large dogs (dogs over 20 kg) and \$20 per day for small dogs (less than 20 kg). On any day, Happy Yaps Kennels can accommodate a maximum of 30 dogs.

    a. If l represents the number of large dogs and s represents the number of small dogs, write an inequality in terms of l and s that represents the total number of dogs at Happy Yaps.
    b. Another inequality can be written as $s \geq 12$. In the context of this problem, write down what this inequality represents.
    c. The inequality $l \leq 15$ represents the number of large dogs that Happy Yaps can accommodate on any day. Draw a graph that represents this situation.
    d. Explore the maximum numbers of small dogs and large dogs Happy Yaps Kennels can accommodate to receive the maximum amount in fees.

Reasoning

16. Use technology to sketch and then find the area of the region formed by the following inequalities.

$$y \geq -4$$
$$y < 2x - 4$$
$$2y + x \leq 2$$

17. Answer the following questions.

    a. Given the following graph, state the inequality it represents.
    b. Choose a point from each half plane and show how this point confirms your answer to part a.

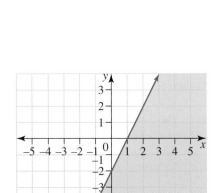

The required region is ☐.

18. Answer the following questions relating to the graph shown.

    a. Determine the equation of the line, *l*.

    b. Write an inequation to represent the unshaded region.

    c. Write an inequation to represent the shaded region.

    d. Rewrite the answer for part **b** if the line was not broken.

Problem solving

19. a. Sketch the graph of the following equation.

$$\frac{x+1}{2} - \frac{x+1}{3} = 2 - y$$

    b. Shade the region that represents the following inequation.

$$\frac{x+1}{2} - \frac{x+1}{3} \leq 2 - y$$

20. Use your knowledge about linear inequations to sketch the regions defined by:

    a. $x^2 + y^2 < 16$ **b.** $x^2 + y^2 > 36$

21. Use your knowledge about linear inequations to sketch the region defined by $y \geq x^2 + 4x + 3$.

LESSON
4.12 Review

4.12.1 Topic summary

Sketching linear graphs

- To plot linear graphs, complete a table of values to determine the points and use a rule.
- Only two points are needed in order to sketch a straight-line graph.
- The x- and y-intercept method involves calculating both axis intercepts, then drawing the line through them.
- Determine the x-intercept but substituting $y = 0$.
- Determine the y-intercept but substituting $x = 0$.
- Graphs given by $y = mx$ pass through the origin (0, 0), since $c = 0$.
- The line $y = c$ is parallel to the x-axis, having a gradient of zero and a y-intercept of c.
- The line $x = a$ is parallel to the y-axis, having a undefined (infinite) gradient and a x-intercept of a.

Equation of a straight line

- The equation of a straight line is:
$$y = mx + c$$
Where: m is the gradient and c is the y-intercept

e.g. $y = ②x + ⑤ \rightarrow$ y-intercept
$\quad\quad\;\;\downarrow$
$\quad\;$ gradient
- The rule $y = mx + c$ is called the equation of a straight line in the **gradient-intercept** form.
- The gradient of a straight line can be determined by the formulas:
$$m = \frac{\text{rise}}{\text{run}} = \frac{y_2 - y_1}{x_2 - x_1}$$

Simultaneous equations

- Solving simultaneous equations involves determining the point (or points) of intersection between two lines.
- We can determine these points by accurately sketching both equations, or using technology to find these points.

Substitution and elimination

- The substitution and elimination methods are two algebraic techniques used to solve simultaneous equations.
- We can use substitution when one (or both) of the equations have a variable as the subject.
  e.g. $y = 3x - 4$
- We use the elimination method when substitution isn't possible.
- Elimination method involves adding or subtracting equations to eliminate one of the variables.

Determining linear equations

- The formula:
$$y - y_1 = m(x - x_1)$$
can be used to write the equations of a line, given the gradient and the coordinates of one point.

Parallel and perpendicular lines

- Parallel lines will never intersect with each other.
- Two lines are parallel if they have the same gradient.
  e.g. $\quad\quad y = 3x - 6$
  $\quad\quad\quad\quad y = 3x + 1$
- Perpendicular lines are lines that intersect at *right angles.*
- Two lines are perpendicular if the product of their gradients is -1.
  e.g. $\quad\quad y = 2x + 3$
  $\quad\quad\quad\quad y = -\dfrac{x}{2} - 4$
  $m_1 \times m_2 = 2 \times -\dfrac{1}{2} = -1$

LINEAR RELATIONSHIPS

Inequations

- An equation has a = sign.
- An inequation will have one of the following:
$$>, \geq, <, \leq$$
- An inequation, such as $x < 5$, will have an infinite number of solutions: $x = \{4, 3, 2, 1, \dots\}$
- We can represent an inequality on a number line.
  e.g. $x > 2$

- An open circle mean that a value **is not included** as a solution. A closed circle means that the value **is included** in the solution.

Simultaneous linear and non-linear equations

- A system of equations which contains a linear equation and a non-linear equation can have 0, 1 or 2 solutions (points of intersection).
- The number of solutions will depend on the equations of both lines.

Number of solutions

- Parallel lines with *different* y-intercepts will never intersect.
- Parallel lines with the *same* y-intercept are called **coincident lines** and will intersect an infinite number of times.
- Perpendicular lines will intersect once and cross at right angles to each other.

Inequalities

- The graph of a linear inequality is called a **half plane** and is the region above or below a boundary line.
- If the inequality has < or > the boundary line is dotted as it is not included in the solution.
- If the inequality has ≤ or ≥ the boundary line is solid as it is included in the solution.
- When sketching a linear inequality such as $y < 3x + 4$, sketch the boundary line first, which is given by $y = 3x + 4$.
- To determine which side of the boundary line is the required region test the point (0, 0) and see if it satisfies the inequality.
  Test $(0, 0) = 0 < 3(0) + 4$
  $\quad\quad\quad\quad\quad\;\; 0 < 4$ which is true
- In this case, the region required is the region with the point (0, 0).

4.12.2 Success criteria

Tick the column to indicate that you have completed the lesson and how well you have understood it using the traffic light system.

(**Green:** I understand; **Yellow:** I can do it with help; **Red:** I do not understand)

Lesson	Success criteria	⬤	⬤	⬤
4.2	I can plot points on a graph using a rule and a table of values.			
	I can sketch linear graphs by determining the x- and y-intercepts.			
	I can sketch the graphs of horizontal and vertical lines.			
	I can model linear graphs from a worded context.			
4.3	I can determine the equation of a straight line when given its graph.			
	I can determine the equation of a straight line when given the gradient and the y-intercept.			
	I can determine the equation of a straight line passing through two points.			
	I can formulate the equation of a straight line from a written context.			
	I can calculate average speed and simple interest.			
4.4	I can determine whether two lines are parallel, perpendicular or neither.			
	I can determine the equation of a straight line that is parallel to a given line.			
	I can determine the equation of a straight line that is perpendicular to a given line.			
4.5	I can use the graph of two simultaneous equations to determine the point of intersection			
	I can determine whether two simultaneous equations will have 0, 1 or infinite solutions			
	I can determine whether two lines are parallel or perpendicular.			
4.6	I can identify when it is appropriate to solve using the substitution method.			
	I can solve a system of two linear simultaneous equations using the substitution method.			
4.7	I can solve two linear simultaneous equations using the elimination method.			
4.8	I can define unknown quantities with appropriate variables.			
	I can form two simultaneous equations using the information presented in a problem.			
	I can choose an appropriate method to solve simultaneous equations in order to find the solution to a problem.			
4.9	I can determine the point or points of intersection between a linear equation and various non-linear equations using various techniques.			
	I can use digital technology to find the points of intersection between a linear equation and a non-linear equation.			
4.10	I can solve an inequality and represent the solution on a number line.			
	I can convert a worded statement to an inequality in order to solve a problem.			
4.11	I can sketch the graph of a half plane: the region represented by an inequality.			
	I can sketch inequalities using digital technology.			

4.12.3 Project

Documenting business expenses

In business, expenses can be represented graphically so that relevant features are clearly visible. The graph compares the costs of hiring cars from two different car rental companies. It will be cheaper to use Plan A when travelling distances less than 250 kilometres, and Plan B when travelling more than 250 kilometres. Both plans cost the same when you are travelling exactly 250 kilometres.

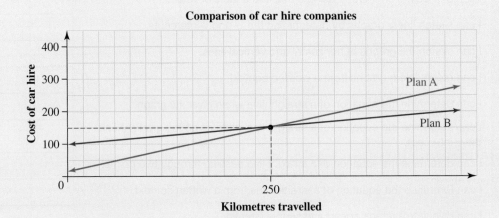

Comparison of car hire companies

Andrea works as a travelling sales representative. She needs to plan her next business trip to Port Hedland, which she anticipates will take her away from the office for 3 or 4 days. Due to other work commitments, she is not sure whether she can make the trip by the end of this month or early next month.

She plans to fly to Port Hedland and use a hire car to travel when she arrives. Andrea's boss has asked her to supply documentation detailing the anticipated costs for the hire car, based on the following quotes received.

| A1 Rentals | $35 per day plus 28c per kilometre of travel |
| Cut Price Rentals | $28 per day plus 30c per kilometre of travel |

Andrea is aware that, although the Cut Price Rentals deal looks cheaper, it could work out more expensive in the long run because of the higher cost per kilometre of travel. She intends to travel a considerable distance.

Andrea is advised by both rental companies that their daily hire charges are due to rise by $2 per day from the first day of next month.

Assuming that Andrea is able to travel this month and her trip will last 3 days, use the information given to answer questions **1** to **4**.

1. Write equations to represent the costs of hiring a car from A1 Rentals and Cut Price Rentals. Use the pronumeral C to represent the cost (in dollars) and the pronumeral d to represent the distance travelled (in kilometres).

2. Copy the following set of axes to plot the two equations from question **1** to show how the costs compare over 1500 km.

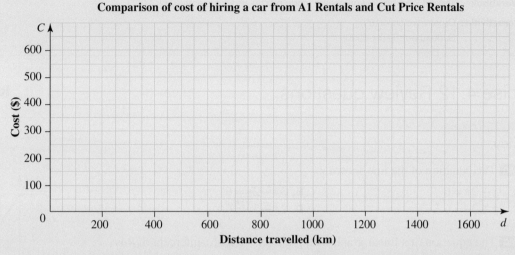

Comparison of cost of hiring a car from A1 Rentals and Cut Price Rentals

3. Use the graph to determine how many kilometres Andrea would have to travel to make the hire costs the same for both rental companies.
4. Assume Andrea's trip is extended to 4 days. Use an appropriate method to show how this changes the answer found in question **3**.

For questions **5** to **7**, assume that Andrea has delayed her trip until next month when the hire charges have increased.
5. Write equations to show the cost of hiring a car from both car rental companies for a trip lasting:
   a. 3 days
   b. 4 days.
6. Copy the following set of axes to plot the four equations from question **5** to show how the costs compare over 1500 km.

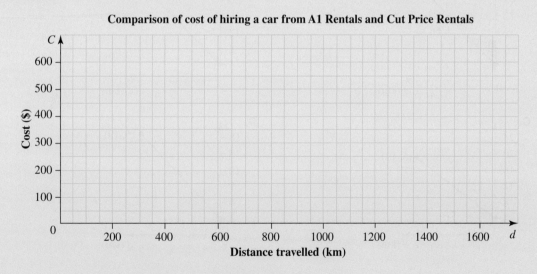

Comparison of cost of hiring a car from A1 Rentals and Cut Price Rentals

7. Comment on the results displayed in your graph.
8. Andrea needs to provide her boss with documentation of the hire car costs, catering for all options. Prepare a document for Andrea to hand to her boss.

Resources

Exercise 4.12 Review questions

learn on

Fluency

1. **MC** The equation of the line shown is:

   A. $3x + 2y = 6$ **B.** $3x - 2y = 6$

   C. $2x + 3y = 6$ **D.** $2x - 3y = 6$

2. **MC** The equation of a linear graph with gradient -3 and x-intercept of 4 is:

   A. $y = -3x - 12$ **B.** $y = -3x + 4$

   C. $y = -3x - 4$ **D.** $y = -3x + 12$

3. **MC** A music shop charges a flat rate of $5 postage for 2 CDs and $11 for 5 CDs. Identify the equation that best represents this, if C is the cost and n is the number of CDs.

   A. $C = 5n + 11$ **B.** $C = 6n + 5$ **C.** $C = n + 2$ **D.** $C = 2n + 1$

4. **MC** During a charity walk-a-thon, Sarah receives $4 plus $3 per kilometre. The graph that best represents Sarah walking up to 5 kilometres is:

   A.

   B.

   C.

   D.

5. **MC** The gradient of the line perpendicular to $3x - 4y + 7 = 0$ is:

   A. $\dfrac{3}{4}$ **B.** $\dfrac{4}{3}$ **C.** $-\dfrac{4}{3}$ **D.** 3

6. **MC** The equation of the line perpendicular to $2x + y - 1 = 0$ and passing through the point $(1, 4)$ is:

   A. $2x + y - 6 = 0$ **B.** $2x + y - 2 = 0$ **C.** $x - 2y + 7 = 0$ **D.** $x + 2y + 9 = 0$

7. **MC** Identify which of the following pairs of coordinates is the solution to the simultaneous equations given below.

$$2x + 3y = 18$$
$$5x - y = 11$$

A. $(6, 2)$ **B.** $(3, -4)$ **C.** $(3, 9)$ **D.** $(3, 4)$

8. **MC** Identify the graphical solution to the following pair of simultaneous equations.

$$y = 5 - 2x$$
$$y = 3x - 10$$

A.

B.

C.

D.

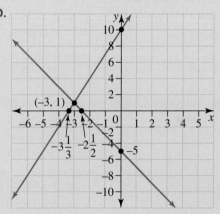

9. Sketch the half plane given by each of the following inequalities.

   a. $y \le x + 1$ **b.** $y \ge 2x + 10$ **c.** $y > 3x - 12$ **d.** $y < 5x$

   e. $x \ge 7$ **f.** $y \le \dfrac{1}{2}x + 1$ **g.** $2x + y \ge 9$ **h.** $y > -12$

10. Sketch the graph of each of the following linear equations, labelling the x- and y-intercepts.

   a. $y = 3x - 2$ **b.** $y = -5x + 15$ **c.** $y = -\dfrac{2}{3}x + 1$ **d.** $y = \dfrac{7}{5}x - 3$

11. Solve each of the following pairs of simultaneous equations using a graphical method.

   a. $4y - 2x = 8$
   $x + 2y = 0$
        b. $y = 2x - 2$
   $x - 4y = 8$
        c. $2x + 5y = 20$
   $y = 7$

12. Use the graphs showing the given simultaneous equations to write the points of intersection of the graphs and hence the solutions of the simultaneous equations.

   a. $x + 3y = 6$
      $y = 2x - 5$

   b. $3x + 2y = 12$
      $2y = 3x$

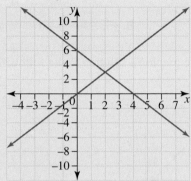

13. Solve the following pairs of simultaneous equations using the substitution method.

   a. $y = 3x + 1$
      $x + 2y = 16$

   b. $y = 2x + 7$
      $3y - 4x = 11$

   c. $2x + 5y = 6$
      $y = \dfrac{3}{2}x + 5$

   d. $y = -x$
      $y = 8x + 21$

   e. $y = 3x - 11$
      $y = 5x + 17$

   f. $y = 4x - 17$
      $y = 6x - 22$

14. Solve the following pairs of simultaneous equations using the elimination method.

   a. $3x + y = 17$
      $7x - y = 33$

   b. $4x + 3y = 1$
      $-4x + y = 11$

   c. $3x - 7y = -2$
      $-2x - 7y = 13$

   d. $4y - 3x = 9$
      $y + 3x = 6$

   e. $5x + 2y = 6$
      $4x + 3y = 2$

   f. $x - 4y = -4$
      $4x - 2y = 12$

15. Solve the following pairs of simultaneous equations using an appropriate method.

   a. $3x + 2y = 6$
      $3y + 5x = 9$

   b. $6x - 4y = -6$
      $7x + 3y = -30$

   c. $6x + 2y = 14$
      $x = -3 + 5y$

16. Sketch the following pairs of inequalities.

   a. $y \le x + 4$
      $y \ge 3$

   b. $2y - 3x \ge 12$
      $y + 3x > 0$

   c. $5x + y < 10$
      $x + 2y < 11$

17. Determine the equation of the straight line joining the point $(-2, 5)$ and the point of intersection of the straight lines with equations $y = 3x - 1$ and $y = 2x + 5$.

Problem solving

18. John has a part-time job working as a gardener and is paid $13.50 per hour.

   a. Complete the following table of values relating the amount of money received to the number of hours worked.

Number of hours	0	2	4	6	8	10
Pay ($)						

   b. Determine a linear equation relating the amount of money received to the number of hours worked.

   c. Sketch the linear equation on a Cartesian plane over a suitable domain.

   d. Using algebra, calculate the pay that John will receive if he works for $6\dfrac{3}{4}$ hours.

19. Write each of the following as a pair of simultaneous equations and solve.
    a. Determine which two numbers have a difference of 5, and their sum is 23.
    b. A rectangular house has a total perimeter of 34 metres and the width is 5 metres less than the length. Calculate the dimensions of the house.
    c. If two Chupa Chups and three Wizz Fizzes cost $2.55, but five Chupa Chups and seven Wizz Fizzes cost $6.10, determine the price of each type of lolly.

20. Laurie buys milk and bread for his family on the way home from school each day, paying with a $10 note. If he buys three cartons of milk and two loaves of bread, he receives 5 cents in change. If he buys two cartons of milk and one loaf of bread, he receives $4.15 in change. Calculate how much each item costs.

21. A paddock contains some cockatoos (2-legged) and kangaroos (4-legged). The total number of animals is 21 and they have 68 legs in total. Using simultaneous equations, determine how many cockatoos and kangaroos there are in the paddock.

22. Warwick was solving a pair of simultaneous equations using the elimination method and reached the result that $0 = -5$. Suggest a solution to the problem, giving a reason for your answer.

23. The speed of a model race car along a racetrack is given by the following graph.

Determine the speed at $t = 1$, $t = 2$, $t = 5$ and $t = 7$.

24. Charlotte has a babysitting service and charges $12.50 per hour. After Charlotte calculated her set-up and travel costs, she constructed the cost equation $C = 45 + 2.50h$, where C represents the cost in dollars per job and h represents the hours Charlotte babysits for.

    a. Write an equation that represents the revenue, R, earned by Charlotte in terms of number of hours, h.
    b. By solving the equations simultaneously, determine the number of hours Charlotte needs to babysit to cover her costs (that is the break-even point).

25. There are two sections to a concert hall. Seats in the 'Dress circle' are arranged in rows of 40 and cost $140 each. Seats in the 'Bleachers' are arranged in rows of 70 and cost $60 each. There are 10 more rows in the 'Dress circle' than in the 'Bleachers', and the capacity of the hall is 7000.

   a. If d represents the number of rows in the 'Dress circle' and b represents the number of rows in the 'Bleachers', write an equation in terms of these two variables based on the fact that there are 10 more rows in the 'Dress circle' than in the 'Bleachers'.

   b. Write an equation in terms of these two variables based on the fact that the capacity of the hall is 7000 seats.

   c. Solve the two equations from **a** and **b** simultaneously using the method of your choice to find the number of rows in each section.

   d. Now that you have the number of rows in each section, calculate the number of seats in each section.

   e. Hence, calculate the total receipts for a concert where all tickets are sold.

26. John is comparing two car rental companies, Golden Ace Rental Company and Silver Diamond Rental Company.

   Golden Ace Rental Company charges a flat rate of $38 per day and $0.20 per kilometre. Silver Diamond Rental Company charges a flat rate of $30 per day plus $0.32 per kilometre.

   a. Write an algebraic equation for the cost of renting a car for 3 days from Golden Ace Rental Company in terms of the number of kilometres travelled, k.

   b. Write an algebraic equation for the cost of renting a car for 3 days from Silver Diamond Rental Company in terms of the number of kilometres travelled, k.

   c. Determine how many kilometres John would have to travel so that the cost of hiring from each company for 3 days is the same.

   d. Write an inequation that, when solved, will tell you the number of kilometres for which it is cheaper to use Golden Ace Rental Company when renting for 3 days.

   e. Determine the number of kilometres for which it is cheaper to use Silver Diamond Rental Company for 3 days' hire.

27. Frederika has $24 000 saved for a holiday and a new stereo. Her travel expenses are $5400 and her daily expenses are $260.

   a. Write an equation for the cost of her holiday if she stays for d days.

   b. Upon her return from holidays Frederika wants to purchase a new stereo system that will cost her $2500.
   Calculate how many days can she spend on her holiday if she wishes to purchase a new stereo upon her return.

on To test your understanding and knowledge of this topic, go to your learnON title at www.jacplus.com.au and complete the **post-test**.

Answers

Topic 4 Linear relationships

4.1 Pre-test

1. 3 apps
2. $y = 3x - 3$
3. An infinite number of solutions
4. $(1, 0.2)$
5. D
6. B
7. 92
8. A
9. D
10. A, C, E, G
11. C
12. B, D
13. At two points
14. D
15. B

4.2 Sketching linear graphs

1. a.

x	y
−5	−25
−4	−15
−3	−5
−2	5
−1	15
0	25
1	35

b.

x	y
−1	−17
0	−12
1	−7
2	−2
3	3
4	8

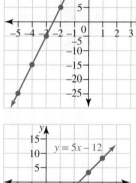

c.

x	y
−6	13
−4	12
−2	11
0	10
2	9
4	8

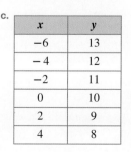

2. a.

x	y
0	−240
1	−140
2	−40
3	60
4	160
5	260

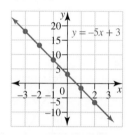

b.

x	y
−3	18
−2	13
−1	8
0	3
1	−2
2	−7

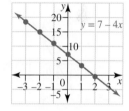

c.

x	y
−3	19
−2	15
−1	11
0	7
1	3
2	−1

3. a.

x	y
−6	20
−4	14
−2	8
0	2
2	−4
4	−10
6	−16

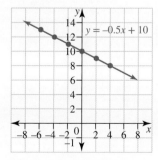

b.

x	y
−3	6
−2	5
−1	4
0	3
1	2
2	1
3	0

c.

x	y
−6	15
−4	11
−2	7
0	3
2	−1
4	−5
6	−9

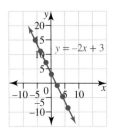

$y = -2x + 3$

4. a.

$5x − 3y = 10$

b.

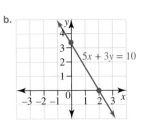

$5x + 3y = 10$

c.

$−5x + 3y = 10$

d.

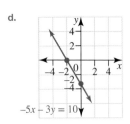

$−5x − 3y = 10$

e.

$2x − 8y = 20$

5. a.

$4x + 4y = 40$

b.

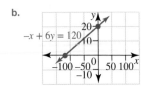

$−x + 6y = 120$

c.

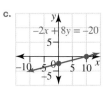

$−2x + 8y = −20$

d.

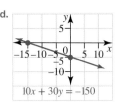

$10x + 30y = −150$

e.

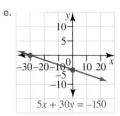

$5x + 30y = −150$

6. a.

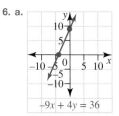

$−9x + 4y = 36$

b.

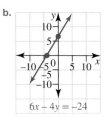

$6x − 4y = −24$

c.

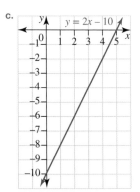

$y = 2x − 10$

d.

$y = -5x + 20$

e.

$y = -\dfrac{1}{2}x - 4$

7. a.

$y = 4x + 1$
$(1, 5)$
$(0, 1)$

b.

$y = 3x - 7$
$(1, -4)$
$(0, -7)$

c.

$(0, 3)$
$(1, 1)$
$y = -2x + 3$

8. a.

$(0, -4)$
$y = -5x - 4$
$(1, -9)$

b.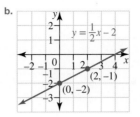

$y = \dfrac{1}{2}x - 2$
$(2, -1)$
$(0, -2)$

c.

$(0, 3)$
$y = -\dfrac{2}{7}x + 3$
$(7, 1)$

9. a.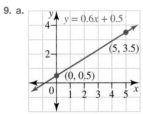

$y = 0.6x + 0.5$
$(5, 3.5)$
$(0, 0.5)$

b.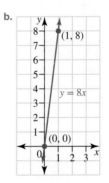

$(1, 8)$
$y = 8x$
$(0, 0)$

c.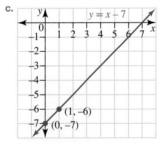

$y = x - 7$
$(1, -6)$
$(0, -7)$

10.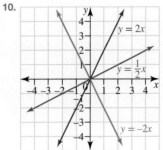

$y = 2x$
$y = \dfrac{1}{2}x$
$y = -2x$

11.

12.

13. a.   b.

c.

14. a.

b.

c.

15. a.

b.

c.

16. a. $y = \frac{4}{5}x + \frac{2}{5}$; x-intercept: -0.5; y-intercept: 0.4

b. $y = \frac{4}{5}x + \frac{2}{5}$; x-intercept: 0.5; y-intercept: -0.4

c. $y = \frac{3}{2}x$; x-intercept: 0; y-intercept: 0

17. a. $y = 4x + 12$; x-intercept: -3; y-intercept: 12

b. $y = -x - 4$; x-intercept: -4; y-intercept: -4

c. $y = -\frac{1}{2}x - \frac{1}{2}$; x-intercept: -1; y-intercept: -0.5

18. a. $y = -\frac{4}{5}x + \frac{11}{5}$; x-intercept: 2.75; y-intercept: 2.2

b. $y = \frac{2}{5}x + \frac{39}{10}$; x-intercept: 9.75; y-intercept: -3.9

c. $y = -2.6x + 4.6$; x-intercept: $\frac{23}{13} \approx 1.77$; y-intercept: 4.6

19. a. $(2, 0), (0, -8)$

b. $\left(-\frac{1}{2}, 0\right), (0, 3)$

c. $(-5, 0), (0, 25)$

20. Sample responses can be found in the worked solutions in the online resources.

21. a. Independent variable = number of songs bought, dependent variable = amount of money saved

b.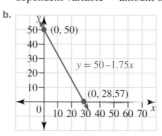

c. 14 songs

22. $y = \dfrac{2}{3}x - \dfrac{7}{3}$

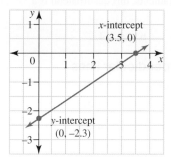

23. a. $y = 20x + 25$

b.

c. Nikita can earn a maximum of $1145.00 in a single week.

24. a. T is the dependent variable (temperature) and t is the independent variable (time).

b. i.

t	0	1	2	3	4	5
T	15	18	21	24	27	30

ii.

c. 5 hours

25. a. Independent variable = time, dependent variable = amount of water in the tank

b. Initially there are 80 litres of water.

c. Time cannot be negative.

d. 4 litres per minute

e. 20 minutes

f.

26. a.

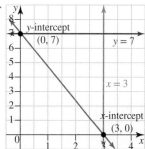

b. 7

c. $-\dfrac{7}{3}$

d. B

4.3 Determining linear equations

1. a. $y = 2x + 4$ **b.** $y = -3x + 12$
   c. $y = -x + 5$ **d.** $y = 2x - 8$

2. a. $y = \dfrac{1}{2}x + 3$ **b.** $y = -\dfrac{1}{4}x - 4$

   c. $y = 7x - 5$ **d.** $y = -3x - 15$

3. a. $y = 2x$ **b.** $y = -3x$

   c. $y = \dfrac{1}{2}x$ **d.** $y = -\dfrac{3}{4}x$

4. a. $y = 3x + 3$ **b.** $y = -3x + 4$ **c.** $y = -4x + 2$
   d. $y = 4x + 2$ **e.** $y = -x - 4$

5. a. $y = 0.5x - 4$ **b.** $y = 5x + 2.5$
   c. $y = -6x + 3$ **d.** $y = -2.5x + 1.5$
   e. $y = 3.5x + 6.5$

6. a. $y = 5x - 19$ **b.** $y = -5x + 31$
   c. $y = -4x - 1$ **d.** $y = 4x - 34$
   e. $y = 3x - 35$

7. a. $y = -3x + 6$ **b.** $y = -2x + 30$
   c. $y = 2x - 4.5$ **d.** $y = 0.5x - 19$
   e. $y = -0.5x + 5.5$

8. a. $y = x + 3$ **b.** $y = 2x - 1$ **c.** $y = -\dfrac{1}{2}x + \dfrac{7}{2}$

9. a. $y = \dfrac{1}{2}x + \dfrac{1}{2}$ **b.** $y = -2x - 2$ **c.** $y = -x - 8$

10. a. 1 m/s

   b. 0.25 m/s

11. a. 30 km/h

   b. 1 h and 30 mins

   c. 20 km/h

12. $6200

13. a. Independent variable = time (in hours), dependent variable = cost (in $)

b.

t	0	1	2	3
c	2	8	14	20

c.

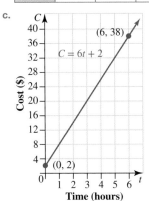

d. i. $(0, 2)$

    ii. The y-intercept represents the initial cost of bowling at the alley, which is the shoe rental.

e. $m = 6$, which represents the cost to hire a lane for an additional hour.

f. $C = 6t + 2$

g. $32

h. Sample responses can be found in the worked solutions in the online resources.

14. a. C = dependent variable, n = independent variable

b. $C = 15n + 10$

c. $460.00

d.

15. 8%

16. a. $W = -40t + 712$

    b. $712\,\text{L}$

    c. 18 days

17. It does not matter if you rise before you run or run before you rise, as long as you take into account whether the rise or run is negative.

18. a. $m = \dfrac{y - c}{x}$ **b.** $y = mx + c$

19. Sample responses can be found in the worked solutions in the online resources.

20. a. $m_{AB} = 5$ **b.** $y = 5x - 3$

    c. $m_{CD} = 5$ **d.** $D = (6, 5)$

21. $m_{AB} = m_{CD} = 2$ and $m_{BC} = m_{AD} = \dfrac{1}{2}$. As opposite sides have the same gradients, this quadrilateral is a parallelogram.

22. a. $b = -10$ **b.** $a = -4$

4.4 Parallel and perpendicular lines

1. a. No   **b.** Yes   **c.** No

2. a. No   **b.** Yes   **c.** No

3. a. No   **b.** No   **c.** No

4. a. No   **b.** No   **c.** No

5. b and f are parallel. c and e are parallel.

6. Sample responses can be found in the worked solutions in the online resources.

7. $y = 2x - 9$

8. $3x + 2y - 8 = 0$

9. a. $y = 3x + 2$ **b.** $y = -4x + 9$

    c. $y = \dfrac{3x}{2} - 4$ **d.** $y = \dfrac{-2x}{5} - \dfrac{13}{5}$

    e. $y = \dfrac{x}{5} - 1$ **f.** $y = \dfrac{x}{3} + \dfrac{17}{3}$

    g. $y = \dfrac{x}{3} - \dfrac{14}{3}$

10. a. $2x - y + 5 = 0$ **b.** $x + 2y = 0$

11. a. $3x - 5y + 2 = 0$ **b.** $5x + 3y - 8 = 0$

12. a. B   **b.** C   **c.** D   **d.** B

13. D

14. a. $y = -2x + 1$ **b.** $y = \dfrac{-2x}{3} - \dfrac{1}{3}$

15. a and e are perpendicular; b and f are perpendicular; c and h are perpendicular; d and g are perpendicular.

16. $y = \dfrac{-x}{2} + \dfrac{3}{2}$

17. a. $m = \dfrac{-8}{5}$ **b.** $m = \dfrac{18}{5}$

18. D

19. $2y + x - 8 = 0$

20. $4y - 3x + 15 = 0$

21. B

22. a.

b.

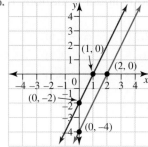

c.

23. $a = \dfrac{-3}{2}$

24. a. $\dfrac{3}{2}$

   b. Sample responses can be found in the worked solutions in the online resources.

4.5 Graphical solution of simultaneous linear equations

1. a. $(2, 1)$ **b.** $(1, 1)$

2. a. $(0, 4)$ **b.** $(2, -1)$

3. a. $(-2, -4)$ **b.** $(-0.5, 1.5)$

4. a. No   **b.** Yes   **c.** Yes   **d.** No

5. a. Yes   **b.** No   **c.** No   **d.** Yes

6. a. No   **b.** Yes   **c.** No   **d.** Yes

7. a. $(3, 2)$ **b.** $(4, 3)$
   c. $(-3, 4)$ **d.** $(-2, 2)$

8. a. $(2, 0)$ **b.** $(3, 0)$
   c. $(-2, 4)$ **d.** $(3, 8)$

9. a. $\left(-\dfrac{1}{2}, 1\dfrac{1}{2}\right)$ **b.** $(2, 5)$

   c. $(5, 3)$ **d.** $\left(2, \dfrac{2}{3}\right)$

10. a. No solution   **b.** $(2, -1)$
    c. No solution   **d.** $(1, 9)$

11. a. $(3, 1)$ **b.** No solution
    c. No solution   **d.** $(2, 1)$

12. a. No solution   **b.** $\left(\dfrac{6}{5}, -\dfrac{6}{5}\right)$

    c. No solution   **d.** No solution

13. $y = 4x - 16$

14. a. Northern beach:
    $C = 20 + 12t, 0 \le t \le 5$
    Southern beach:
    $D = 8 + 18t, 0 \le t \le 5$

b. Northern beaches in red, southern beaches in blue

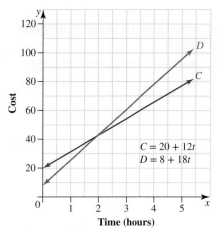

c. Time > 2 hours

d. Time = 2 hours, cost = \$44

15. a. Same line   **b.** Perpendicular
    c. Intersecting   **d.** Parallel

16. a. 1 solution
    b. No solution (parallel lines)
    c. No solution (parallel lines)

17. a. 1 solution (perpendicular lines)
    b. Infinite solutions (coincident)
    c. 1 solution
    d. No solution (parallel lines)

18. a. $a = 3$ **b.** $a = \dfrac{1}{4}$ **c.** $a = 8$

19. a. $y = 3x + 6$
    b. $y = -2x + 1$

c.

20.

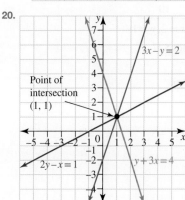

21. $y = \dfrac{5x}{4} + 9$

4.6 Solving simultaneous linear equations using substitution

1. a. $(2, 3)$ b. $(2, -1)$ c. $(3, -2)$ d. $(7, 6)$
2. a. $(3, 6)$ b. $(2, 1)$ c. $(-1, -2)$ d. $(-4, 0)$
3. a. $(-1, -2)$ b. $(6, -2)$ c. $\left(3, 1\dfrac{1}{2}\right)$ d. $(-3, -5)$
4. a. $(-6, -23)$ b. $(5, 23)$ c. $(2, -6)$ d. $\left(\dfrac{3}{2}, -\dfrac{15}{2}\right)$
5. a. $(1, -7)$ b. $\left(-\dfrac{1}{2}, -4\right)$
   c. $\left(-\dfrac{3}{2}, -\dfrac{1}{2}\right)$ d. $\left(-\dfrac{1}{5}, \dfrac{4}{5}\right)$
6. a. $(-3, -1.5)$ b. $(1, 0.3)$
   c. $\left(-\dfrac{4}{5}, \dfrac{4}{5}\right)$ d. $(1, -1)$
7. 26 chickens
8. a. $(3, 1)$ b. $(-2, 3)$
9. a. $(5, 1)$ b. $(4, 2)$
10. a. $(0, 3)$ b. $(4, 0)$
11. a. $x = \dfrac{n}{2m}, y = \dfrac{n}{2}$
    b. $x = \dfrac{m}{n^2 + 1}, y = \dfrac{mn}{n^2 + 1}$
    c. $x = \dfrac{n}{m - n^2}, y = \dfrac{n^2}{m - n^2}$
    d. $x = \dfrac{n}{m - n}, y = \dfrac{n}{m - n}$
    e. $x = \dfrac{n^2 - m}{m - n}, y = \dfrac{m(n - 1)}{m - n}$
    f. $x = \dfrac{2m}{m + n}, y = \dfrac{m(n - m)}{m + n}$
12. $a = -1, b = 5$
13. $z = 24, m = 6, n = 9$

m	11	7
9	8	7
n	5	10

14. a. $x + 2y = 4$ b. $x = 2, y = 1$
15. Chemistry \$21, physics \$27
16. 30 cm
17. Andrew is 16, Prue is 6.
18. $x = 8, y = -7$
19. $x = 0, y = 1$
20. a. $k \neq +3$ b. $k = -3$ c. $k = 3$

4.7 Solving simultaneous linear equations using elimination

1. a. $(3, 1)$ b. $(-2, 3)$ c. $(-2, 6)$
2. a. $(5, -1)$ b. $(2, 3)$ c. $(-3, 1)$
3. a. $(-3, 5)$ b. $(-5, -8)$ c. $(2, -2)$
4. a. $\left(1\dfrac{1}{2}, 3\dfrac{1}{2}\right)$ b. $\left(2, 1\dfrac{4}{5}\right)$ c. $(1, 1)$
5. a. $(1, 3)$ b. $(2, 4)$ c. $(5, 2)$
6. a. $(4, 2)$ b. $(-3, 4)$ c. $\left(-3, -1\dfrac{1}{2}\right)$
7. a. $(-6, -5)$ b. $(-3, 5)$ c. $(2, 1.8)$
8. a. $(6, 3)$ b. $(2, -2)$ c. $(1, 3)$
9. a. $(-1.5, -3)$ b. $(-8, 18)$ c. $(-3, 5)$
10. a. $(1, 3)$ b. $(4, 0)$ c. $(-3, 5)$
11. a. $(4, 3)$ b. $(8, 5)$ c. $\left(\dfrac{1}{3}, -\dfrac{1}{3}\right)$
12. a. $(4, 5)$ b. $(1, -3)$ c. $(12, 12)$
13. \$3.40 (coffee is \$5.10)
14. 12 five-cent coins and 22 ten-cent coins
15. Abena 61 kg, Bashir 58 kg, Cecily 54 kg
16. a. i. $acx + bcy = ce$ (3)
       ii. $acx + ady = af$ (4)
       iii. $y = \dfrac{ce - af}{bc - ad}$
    b. $x = \dfrac{de - bf}{ad - bc}$
    c. i. $\left(\dfrac{106}{31}, \dfrac{1}{31}\right)$ ii. $\left(\dfrac{37}{14}, \dfrac{11}{14}\right)$
    d. Because you cannot divide by 0.
    e. $ad - bc \neq 0$
17. It is not possible. When the two equations are set up it is impossible to eliminate one variable without eliminating the other.
18. $k - 10, 10$
19. $x = 7, y = -3$
20. $x = 4, y = 3, z = -6$

4.8 Applications of simultaneous linear equations

1. Maths mark $= 97$, English mark $= 66$
2. 18 nuts, 12 bolts
3. 800 sheep, 400 chickens
4. 8 and 3
5. 9 and 7
6. 6 and 5
7. Length $= 12$ m and width $= 8$ m.
8. Lemons cost 55 cents and oranges cost 25 cents.
9. Length 60 m and width 20 m

10. a. Yolanda needs to sell 10 bracelets to cover her costs.

    b. i. $16 loss

      ii. $24 profit

11. Eight 20-cent coins and three 50-cent coins

12. Twelve $1 coins and nine $2 coins

13. Paddlepops cost $1.20 and a Magnum costs $2.10.

14. Cost of the Killer Python = 35 cents and cost of the Red Frog = 25 cents.

15. Fixed costs = $87, cost per person = $23.50.

16. The PE mark is 83 and the Science mark is 71.

17. Mozzarella costs $6.20 and Swiss cheese costs $5.80.

18. $x = 3$ and $y = 4$.

19. Fixed costs = $60, cost per person = $25.

20. $4 each for DVDs and $24 each for USB sticks

21. 9 and 24 years old

22. Child $12.50, Adult $18.25, Elderly $15

23. 6.5 km

24. 66 cups of hot chips, 33 meat pies and 22 hot dogs were sold during the half-hour period.

25. a. See the graph at the bottom of the page.*

    b. Jet-ski 3 wins the race.

    c. Jet-skis 1 and 2 reach the destination at the same time, although jet-ski 2 started 2 hours after jet-ski 1. Jet-ski 3 overtakes jet-ski 1 6 hours and 40 minutes after its race begins or 10 hours and 40 minutes after jet-ski 1 starts the race. Jet-ski 3 overtakes jet-ski 2.6 hours after it starts the race or 8 hours after jet-ski 2 started the race.

26. 176 km

4.9 Solving simultaneous linear and non-linear equations (10A)

1. A parabola may intersect with a straight line twice, once or not at all.

2. a. $(-5, 4)$ and $(-1, 0)$

    b. $(2, 3)$

    c. $\left(1 - \sqrt{10}, -6\right)$ and $\left(1 + \sqrt{10}, -6\right)$

3. a. $\left(\dfrac{-1}{2} - \dfrac{3\sqrt{5}}{2}, \dfrac{-19}{2} - \dfrac{9\sqrt{5}}{2}\right)$ and
$\left(\dfrac{-1}{2} + \dfrac{3\sqrt{5}}{2}, \dfrac{-19}{2} + \dfrac{9\sqrt{5}}{2}\right)$

    b. $(-1, -2)$ and $(2, 1)$

    c. $(-2.54, -8.17)$ and $(3.54, 16.17)$

4. a. $(-1.41, 4)$ and $(1.41, 4)$

    b. $(-1, 2)$ and $\left(\dfrac{5}{2}, \dfrac{1}{4}\right)$

    c. $(3, 37)$

5. B

6. C

7. a. Yes   b. No   c. Yes   d. No

8. a. $(1, 1), (-1, -1)$

    b. $\left(1 + \sqrt{2}, -1 + \sqrt{2}\right), \left(1 - \sqrt{2}, -1 - \sqrt{2}\right)$

    c. $\left(\dfrac{-\sqrt{15}}{3}, -\sqrt{15}\right), \left(\dfrac{\sqrt{15}}{3}, \sqrt{15}\right)$

    d. $(-6, -1), (2, 3)$

9. a. $(-1, -3), (1, 3)$

    b. $(-4, 3), (4, -3)$

    c. $(-1, 7), (5, -5)$

    d. $\left(\dfrac{\left(2 - \sqrt{14}\right)}{2}, \dfrac{\left(\sqrt{14} + 2\right)}{2}\right), \left(\dfrac{\left(\sqrt{14} + 2\right)}{2}, \dfrac{\left(2 - \sqrt{14}\right)}{2}\right)$

10. a. $\left(-\dfrac{1}{2}, -2\right), \left(\dfrac{1}{2}, 2\right)$ b. $(0, 5), (4, -3)$

    c. $\left(-\dfrac{1}{2}, 2\right), (0, 3)$ d. $\left(\dfrac{8}{3}, -\dfrac{1}{4}\right), (5, -2)$

*25. a.

Jet-ski 2 $d = 30(t - 2)$

Jet-ski 3 $d = 40(t - 4)$

Jet-ski 1 $d = 25t$

Winner (11.5, 300)

Point of intersection (12, 300)

d (kilometres)

t (hours)

11. a. $(1, 1)$ **b.** $(-3, 3), (3, 3)$

   c. $(-4, 11), (-2, 3)$ **d.** $\left(-\dfrac{100}{3}, \dfrac{652}{9}\right), (12, 12)$

12. a. $k < -1$ **b.** $k = -1$ **c.** $k > -1$

13. The straight line crosses the parabola at $(0, -7)$, so no matter what value m takes, there will be at least one intersection point.

14. a.   **i.** $1, 2, 3$

      ii. $0, 1, 2, 4$

      iii. $1, 2, 3, 4, 5$

  b. The number of possible intersections between an equation and a straight line is equal to the highest power of x.

15. $17, 18$ and $-17, -18$

16. Length 15 m, width 85 m

17. a. $(5, 5), (-1, -7)$

  b.

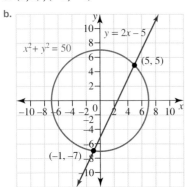

18. $9, 12$

19. a. $(4, 4), \left(-\sqrt{32}, 0\right)$

  b. 6.66 km per hour or 11.11 km per hour.

20. a. $100\,°C$

  b. 0 minutes, 6 minutes

  c. Eve's model. This model flattens out at $20\,°C$, whereas Adam's becomes negative, which would not occur.

4.10 Solving linear inequalities

1. a. $x > 2$

  b. $a > -1$

  c. $y \geq 7$

  d. $m \geq 4$

2. a. $p < 1$

  b. $x < 7$

  c. $m \leq 9$

  d. $a \leq 7$

3. a. $x > 3$

  b. $m \geq 2$

  c. $q \geq -4$

  d. $a > -8$

4. a. $m > 3$ **b.** $p \leq 2$ **c.** $a < 4$ **d.** $x \geq 5$

5. a. $p > -5$ **b.** $x \leq -7$ **c.** $m \geq -0.5$ **d.** $b > -0.5$

6. a. $m > 18$ **b.** $x < 8$ **c.** $a \leq -14$ **d.** $m \geq 25$

7. a. $m < 4.5$ **b.** $x \geq 3$ **c.** $p > 4$ **d.** $n \leq 2$

8. a. $b < 5$ **b.** $y > 2$ **c.** $m \leq -1$ **d.** $a \geq -5$

9. a. $b < -4\dfrac{1}{3}$ **b.** $c \leq -1$ **c.** $p > -2$ **d.** $a \geq -7$

10. a. $m > 3$ **b.** $a \geq 2$ **c.** $a < -1$ **d.** $a \leq -3$

11. a. $x > 6$ **b.** $x \leq 2$

   c. $b < 4$ **d.** $a > 5$

12. a. $m < 2$ **b.** $m \leq 3$

   c. $b \leq -\dfrac{16}{11}$ **d.** $m \geq 1$

13. a. $x \leq 7$ **b.** $x \geq -18$ **c.** $x < -10$

14. a. $x > 10\dfrac{1}{2}$ **b.** $x \geq 5$ **c.** $x < -1\dfrac{4}{5}$

15. a. $m < -2$ **b.** $p \geq -3$ **c.** $a \leq 5$

   d. $p \geq -5$ **e.** $y \leq -3$

16. a. $x > 7$ **b.** $p < 0$ **c.** $a \geq \dfrac{1}{5}$

  d. $x > -3$ **e.** $a \leq -11$

17. a. $b \leq 3$ **b.** $x < -3$ **c.** $k > 8$

  d. $x > -18\dfrac{1}{2}$ **e.** $a \leq 40$

18. B

19. a. $x < -1$ **b.** $m \leq -3$

20. a. $x > 17$ **b.** $a > \dfrac{5}{8}$

21. a. $m \geq 1\dfrac{1}{3}$ **b.** $m \geq -12$

22. a. $k > 2$ **b.** $a > -5$ **c.** $m \leq 1\dfrac{2}{3}$ **d.** $x > 5$

23. a. $y \geq 7$ **b.** $d < -2$ **c.** $p \geq \dfrac{-6}{7}$ **d.** $x \geq -5$

24. a. $m < -2$ **b.** $a < 9$ **c.** $p \geq 3$ **d.** $x > -4\dfrac{1}{2}$

25. a. $5x > 10$ **b.** $x - 3 \leq 5$ **c.** $7 + 3x < 42$

26. a. $4 + 3x > 19$ **b.** $2(x + 6) < 10$ **c.** $\dfrac{(x-8)}{2} - 7 \geq 9$

27. a. 4 hours

  b. i. 30 km/h **ii.** 4 hours

  c.

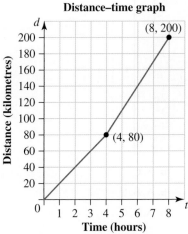

Distance–time graph

28. a. $50x \geq 650$ **b.** $2.50d + 5 \leq 60$

29. Tom could be any age from 1 to 28.

30. a. $-6.5 < x < -2$ **b.** $\dfrac{-c - b}{a} < x < \dfrac{-d - b}{a}$

31. a. $S > 47$

  b. No

  c. Sample responses can be found in the worked solutions in the online resources.

32. a. $n < 16\,800$ km

  b. Mick travelled less than 16 800 km for the year and his costs stayed below \$16 000.

33. a. $20d + 10c \leq 240$

  b. $2d + 2c \leq 36$

  c. $0 \leq d \leq 12$ and $0 \leq C \leq 18$

34. \$20 000

4.11 Inequalities on the Cartesian plane

1. a.

  b.

  c.

  d.

2. a.

b.

$x \le -6$

c.

$y \ge 3$

d.

$y \le 2$

b.

$y < \dfrac{3}{2}$

c.

$y \ge -4$

d.

$x \le \dfrac{3}{2}$

3. a.

$x < \dfrac{1}{2}$

4. a. C **b.** A, C

5. a. A, B **b.** A, B, C

6. a. B **b.** A, B, C

7. a. $y \ge x + 1$

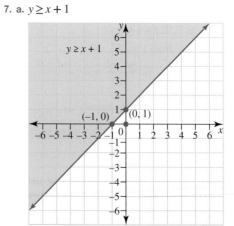

$y \ge x + 1$

$(-1, 0)$ $(0, 1)$

b. $y < x - 6$

c. $y > -x - 2$

d. $y < 3 - x$

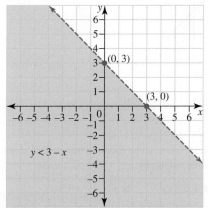

8. a. $y > x - 2$

b. $y < 4$

c. $2x - y < 6$

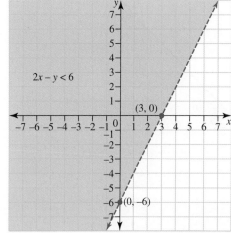

d. $y \leq x - 7$

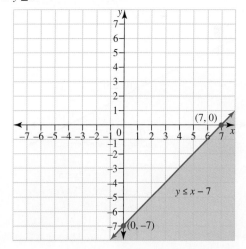

9. a. $x - y > 3$

b. $y < x + 7$

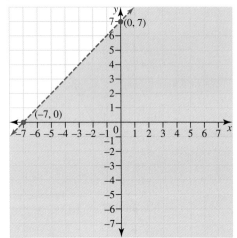

c. $x + 2y \leq 5$

d. $y \leq 3x$

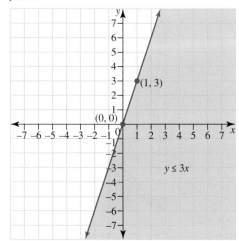

10. B

11. C

12. A

13. a. $y = \dfrac{1}{2}x + 3$

b. $y \geq \dfrac{1}{2}x + 3, x > 2, y \leq 7$

14. $(3, 3), (3, 4), (3, 5), (4, 3)$

15. a. $l + s \leq 30$

b. At least 12 small dogs

c.

d. 15 large and 15 small dogs

16. 20 units^2

17. a. $y \leq 2x - 2$

b. Sample responses can be found in the worked solutions in the online resources.

18. a. $y = -\dfrac{2}{3}x + 3$ **b.** $y > -\dfrac{2}{3}x + 3$

c. $y < -\dfrac{2}{3}x + 3$ **d.** $y \geq -\dfrac{2}{3}x + 3$

19. a. $y = \dfrac{11}{6} - \dfrac{1}{6}x$

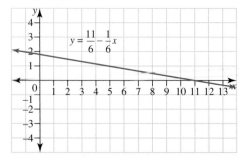

b. The unshaded region is the required region.

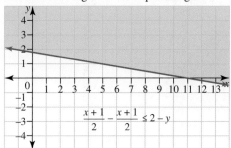

$$\frac{x+1}{2} - \frac{x+1}{2} \le 2 - y$$

20. a.

b.

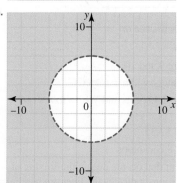

21. The unshaded region is the required region.

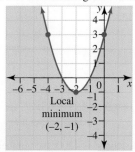

4.12 Review questions

1. A
2. D
3. D
4. C
5. C
6. C
7. D
8. A

9. *Note:* The shaded region is the region required.

a.

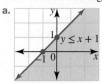

$y \le x + 1$

b.

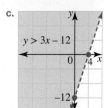

$y \ge 2x + 10$

c.

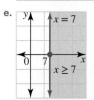

$y > 3x - 12$

d.

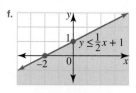

$y = 5x$

$y < 5x$

e.

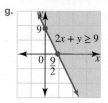

$x = 7$

$x \ge 7$

f.

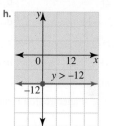

$y \le \frac{1}{2}x + 1$

g.

$2x + y \ge 9$

h.

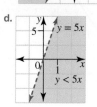

$y > -12$

10. a.

$y = 3x - 2$

$\left(\frac{2}{3}, 0\right)$ (1, 1) (0, -2)

b.

$y = -5x + 15$

(0, 15) (1, 10) (3, 0)

c.

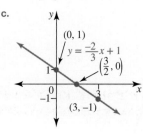

(0, 1) $y = \frac{-2}{3}x + 1$ $\left(\frac{3}{2}, 0\right)$ (3, -1)

d.

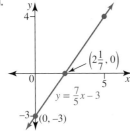

11. **a.** $(-2, 1)$ **b.** $(0, -2)$

    c. $(-7.5, 7)$

12. **a.** $(3, 1)$ **b.** $(2, 3)$

13. **a.** $(2, 7)$ **b.** $(-5, -3)$

    c. $(-2, 2)$ **d.** $\left(-\dfrac{7}{3}, \dfrac{7}{3}\right)$

    e. $(-14, -53)$ **f.** $\left(\dfrac{5}{2}, -7\right)$

14. **a.** $(5, 2)$ **b.** $(-2, 3)$

    c. $(-3, -1)$ **d.** $(1, 3)$

    e. $(2, -2)$ **f.** $(4, 2)$

15. **a.** $(0, 3)$ **b.** $(-3, -3)$

    c. $(2, 1)$

16. *Note:* The shaded region is the region required.

    a.

    b.

c.

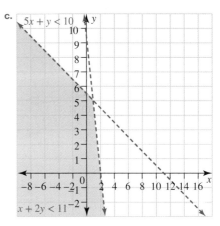

17. $3x - 2y + 16 = 0$

18. **a.** See the table at the bottom of the page.*

    b. Pay $= \$13.50 \times$ (number of hours worked)

    c.

    d. $\$91.13$

19. **a.** The numbers are 9 and 14.

    b. Length $= 11$ meters, width $= 6$ meters.

    c. Chupa-chups cost 45 cents and Whizz fizzes cost 55 cents.

20. Milk $\$1.75$, bread $\$2.35$

21. 13 kangaroos and 8 cockatoos

22. Any false statement that occurs during the solving of simultaneous equations indicates the lines are parallel and have no points of intersection.

23. At $t = 1$, $v = 2.5$ m/s.
    At $t = 2$, $v = 5$ m/s.
    At $t = 5$, $v = 10$ m/s.
    At $t = 7$, $v = 5$ m/s.

24. **a.** $R = 12.50h$ **b.** 4.5 hours

25. **a.** $d = b + 10$

    b. $7000 = 70b + 40d$

    c. $b = 60$ and $d = 70$

    d. The number of seats in 'Bleachers' is 4200; the number of seats in the 'Dress circle' is 2800.

    e. $\$644\,000$

26. **a.** $C_G = 114 + 0.2k$ **b.** $C_S = 90 + 0.32k$

    c. 200 km   **d.** $114 + 0.2k < 90 + 0.32k$

    e. $k < 200$

27. **a.** $5400 + 260d = C_H$

    b. 61 days

---

*18. **a.**

Number of hours	0	2	4	6	8	10
Pay (\$)	0	27	54	81	108	135

5 Trigonometry I

LESSON
5.1 Overview

Why learn this?

Nearly 2000 years ago Ptolemy of Alexandria published the first book of trigonometric tables, which he used to chart the heavens and plot the courses of the Moon, stars and planets. He also created geographical charts and provided instructions on how to create maps.

The word trigonometry is derived from the Greek words 'trigonon' and 'metron', meaning *triangles* and *measure* respectively. This field of mathematics has reportedly been studied across the world since the third century BCE, with major discoveries made in India, China, Greece and Persia. The works ranged from developing relationships, axioms and proofs to describing its application to everyday use and life.

Trigonometry is the branch of mathematics that makes the whole universe more easily understood. The role and use of trigonometry in navigation in the early years were crucial and its application and study grew from there. Today, it is used in architecture, surveying, astronomy and, as previously mentioned, navigation. It also provides the foundation of the study of sound and light waves, resulting in its application in the areas of music manufacturing and composition, study of tides, radiology and many other fields.

Hey students! Bring these pages to life online

▶ Watch videos

🧩 Engage with interactivities

A+ Answer questions and check solutions

Find all this and MORE in jacPLUS ▶

Reading content and rich media, including interactivities and videos for every concept

Extra learning resources

Differentiated question sets

Questions with immediate feedback, and fully worked solutions to help students get unstuck

1. Determine the value of the pronumeral w, correct to 2 decimal places.

7.3 cm

w cm

2.41 cm

2. Determine the value of x, correct to 2 decimal places.

5 cm

x cm

$2x$ cm

3. A square-based pyramid is 16 cm high. Each sloping edge is 20 cm long. Calculate the length of the sides of the base, in cm correct to 2 decimal places.

4. A cork is in the shape of a truncated cone; both the top and the base of the cork are circular.

20 cm

32 cm

30 cm

Calculate the sum of diameters of the top and the base. Give your answer in cm to 2 decimal places.

5. Evaluate $\sin(20°37')$, correct to 4 decimal places.

6. Calculate the size of the angle θ, correct to the nearest minute, given that $\cos(\theta) = 0.5712$. Give your answer in degrees and minutes.

7. Determine the size of the angle θ, correct to the nearest second.

2.4

θ

3.2

8. Calculate y, correct to 1 decimal place.

9. **MC** Tyler is standing 12 m from a flagpole and measures the angle of elevation from his eye level to the top of the pole as 62°.
The distance from Tyler's eyes to the ground is 185 cm. The height of the flagpole correct to 2 decimal places is:
   A. 20.71 metres
   B. 22.56 metres
   C. 24.42 metres
   D. 207.56 metres

10. Change each of the following compass bearings to true bearings.
    a. N20°E
    b. S47°W
    c. N33°W
    d. S17°E

11. **MC** In a right square-based pyramid, the square base has a length of 7.2 cm. If the angle between the triangular face and the base is 55°, the angle the sloping edge makes with the base is:
    A. 45.3°
    B. 55°
    C. 63.4°
    D. 47.7°

12. **MC** A boat travels 15 km from A to B on a bearing of 032°T. The bearing from B to A is:
    A. 032°T
    B. 058°T
    C. 122°T
    D. 212°T

13. A bushwalker travels N50°W for 300 m and then changes direction 220°T for 0.5 km. Determine how many metres west the bushwalker is from his starting point.
    Give your answer in km correct to 1 decimal place.

14. **MC** P and Q are two points on a horizontal line that are 120 metres apart. If the angles of elevation from P and Q to the top of the mountain are 34°5′ and 41°16′ respectively, the height of the mountain correct to 1 decimal place is:
    A. 81.2 metres
    B. 105.3 metres
    C. 120.5 metres
    D. 354.7 metres

15. Determine the value of x in the following figure, correct to 1 decimal place.

LESSON
5.2 Pythagoras' theorem

LEARNING INTENTIONS

At the end of this lesson you should be able to:
- identify similar right-angled triangles when corresponding sides are in the same ratio and corresponding angles are congruent
- apply Pythagoras' theorem to calculate the third side of a right-angled triangle when two other sides are known.

▶ 5.2.1 Similar right-angled triangles

eles-4799

- Two similar right-angled triangles have the same angles when the corresponding sides are in the same ratio.
- The **hypotenuse** is the longest side of a right-angled triangle and is always the side that is opposite the right angle.
- The corresponding sides are in the same ratio.

$$\frac{AB}{DE} = \frac{AC}{DF} = \frac{BC}{EF}$$

 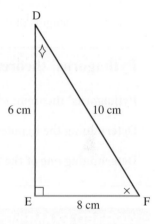

- To write this using the side lengths of the triangles gives:

$$\frac{AB}{DE} = \frac{3}{6} = \frac{1}{2}$$
$$\frac{AC}{DF} = \frac{5}{10} = \frac{1}{2}$$
$$\frac{BC}{EF} = \frac{4}{8} = \frac{1}{2}$$

This means that for right-angled triangles, when the angles are fixed, the ratios of the sides in the triangle are constant.
- We can examine this idea further by completing the following activity.
  Using a protractor and ruler, draw an angle of 70° measuring horizontal distances of 3 cm, 7 cm and 10 cm as demonstrated in the diagram below.

Note: Diagram not drawn to scale.

Measure the perpendicular heights a, b and c.

$$a \approx 8.24 \text{ cm}, \ b \approx 19.23 \text{ cm}, \ c \approx 27.47 \text{ cm}$$

- To test if the theory for right-angled triangles, that when the angles are fixed the ratios of the sides in the triangle are constant, is correct, calculate the ratios of the side lengths.

$$\frac{a}{3} \approx \frac{8.24}{3} \approx 2.75, \quad \frac{b}{7} \approx \frac{19.23}{7} \approx 2.75, \quad \frac{c}{10} \approx \frac{27.47}{10} \approx 2.75$$

The ratios are the same because the triangles are similar. This important concept forms the basis of trigonometry.

⊙ 5.2.2 Review of Pythagoras' theorem

eles-4800

- Pythagoras' theorem gives us a way of calculating the length of the third side in a right angle triangle, if we know the lengths of the two other sides.

Pythagoras' theorem

Pythagoras' theorem: $a^2 + b^2 = c^2$

Determining the hypotenuse: $c = \sqrt{a^2 + b^2}$

Determining one of the two shorter sides: $a = \sqrt{c^2 - b^2}$ or $b = \sqrt{c^2 - a^2}$

WORKED EXAMPLE 1 Calculating the hypotenuse

For the following triangle, calculate the length of the hypotenuse x, correct to 1 decimal place.

THINK	WRITE/DRAW
1. Copy the diagram and label the sides a, b and c. Remember to label the hypotenuse as c.	
2. Write Pythagoras' theorem.	$c^2 = a^2 + b^2$
3. Substitute the values of a, b and c into this rule and simplify.	$x^2 = 5^2 + 8^2$ $= 25 + 64$ $= 89$
4. Take the square root of both sides. Round the positive answer correct to 1 decimal place, since $x > 0$.	$x = \pm\sqrt{89}$ $x \approx 9.4$

WORKED EXAMPLE 2 Calculating the length of the shorter side

Calculate the length, correct to 1 decimal place, of the unmarked side of the following triangle.

THINK	WRITE/DRAW
1. Copy the diagram and label the sides a, b and c. Remember to label the hypotenuse as c; it does not matter which side is a and which side is b.	
2. Write Pythagoras' theorem.	$c^2 = a^2 + b^2$
3. Substitute the values of a, b and c into this rule and solve for a.	$14^2 = a^2 + 8^2$ $196 = a^2 + 64$ $a^2 = 196 - 64$ $\quad = 132$
4. Evaluate a by taking the square root of both sides and round to 1 decimal place ($a > 0$).	$a = \pm\sqrt{132}$ $\approx 11.5 \text{ cm}$

TI | THINK	DISPLAY/WRITE	CASIO | THINK	DISPLAY/WRITE
In a new document, on a Calculator page, to solve equations press: • MENU • 3: Algebra • 1: Solve Complete the entry line as: solve $\left(c^2 = a^2 + b^2, a\right)$ $\lvert b = 8$ and $c = 14$ and $a > 0$ Then press ENTER. Press CTRL ENTER to get a decimal approximation.	 The length of the unmarked side is $a = 2\sqrt{33} = 11.5$ correct to 1 decimal place.	On the Main screen, complete the entry line as: solve $\left(c^2 = a^2 + b^2, a\right)$ $\lvert b = 8 \rvert c = 14$ Then press EXE. To convert to decimals, highlight the answer that is greater than 0 and drag it to a new line. Change the mode to decimal then press EXE.	 The length of the unmarked side is $a = 2\sqrt{33} = 11.5$ correct to 1 decimal place.

WORKED EXAMPLE 3 Solving a practical problem using Pythagoras' theorem

A ladder that is 5.5 m long leans up against a vertical wall. The foot of the ladder is 1.5 m from the wall. Determine how far up the wall the ladder reaches. Give your answer in metres correct to 1 decimal place.

THINK	WRITE/DRAW
1. Draw a diagram and label the sides a, b and c. Remember to label the hypotenuse as c.	
2. Write Pythagoras' theorem.	$c^2 = a^2 + b^2$
3. Substitute the values of a, b and c into this rule and simplify.	$5.5^2 = a^2 + 1.5^2$ $30.25 = a^2 + 2.25$ $a^2 = 30.25 - 2.25$ $= 28$
4. Evaluate a by taking the square root of 28. Round to 1 decimal place, $a > 0$.	$a = \pm\sqrt{28}$ ≈ 5.3
5. Write the answer in a sentence.	The ladder reaches 5.3 m up the wall.

WORKED EXAMPLE 4 Determining the unknown sides

Determine the unknown side lengths of the triangle, correct to 2 decimal places.

THINK	WRITE/DRAW
1. Copy the diagram and label the sides a, b and c.	
2. Write Pythagoras' theorem.	$c^2 = a^2 + b^2$
3. Substitute the values of a, b and c into this rule and simplify.	$78^2 = (3x)^2 + (2x)^2$ $6084 = 9x^2 + 4x^2$ $6084 = 13x^2$

4. Rearrange the equation so that the pronumeral is on the left-hand side of the equation.

$13x^2 = 6084$

5. Divide both sides of the equation by 13.

$$\frac{13x^2}{13} = \frac{6084}{13}$$
$$x^2 = 468$$

6. Evaluate x by taking the square root of both sides. Round the answer correct to 2 decimal places.

$x = \pm\sqrt{468}$
≈ 21.6333

7. Substitute the value of x into $2x$ and $3x$ to determine the lengths of the unknown sides.

$2x \approx 43.27\text{m}$
$3x \approx 64.90\text{ m}$

DISCUSSION

Pythagoras' theorem was known about before the age of Pythagoras. Research which other civilisations knew about the theory and construct a timeline for its history.

 Resources

 eWorkbook   Topic 5 Workbook (worksheets, code puzzle and project) (ewbk-13288)

 Interactivities Individual pathway interactivity: Pythagoras' theorem (int-4585)
Finding the hypotenuse (int-3844)
Finding a shorter side (int-3845)

Exercise 5.2 Pythagoras' theorem

learn on

| 5.2 Quick quiz on | 5.2 Exercise |

Individual pathways

| ■ PRACTISE | ■ CONSOLIDATE | ■ MASTER |
| 1, 5, 7, 10, 11, 14, 17, 21, 22, 25 | 2, 3, 8, 12, 15, 19, 20, 23, 26 | 4, 6, 9, 13, 16, 18, 24, 27 |

Fluency

1. **WE1** For each of the following triangles, calculate the length of the hypotenuse, giving answers correct to 2 decimal places.

a.

4.7
6.3

b.

19.3
27.1

c.

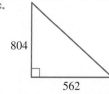

804
562

2. For each of the following triangles, calculate the length of the hypotenuse, giving answers correct to 2 decimal places.

a.

b.

c.

3. **WE2** Determine the value of the pronumeral, correct to 2 decimal places.

a.

b.

c.

4. Determine the value of the pronumeral, correct to 2 decimal places.

a.

b.

c.

5. **WE3** The diagonal of the rectangular NO SMOKING sign is 34 cm. If the height of this sign is 25 cm, calculate the width of the sign, in cm correct to 2 decimal places.

6. A right-angled triangle has a base of 4 cm and a height of 12 cm. Calculate the length of the hypotenuse in cm correct to 2 decimal places.

7. Calculate the lengths of the diagonals (in cm to 2 decimal places) of squares that have side lengths of:

   a. 10 cm   b. 17 cm   c. 3.2 cm.

8. The diagonal of a rectangle is 90 cm. One side has a length of 50 cm. Determine, correct to 2 decimal places:

   a. the length of the other side
   b. the perimeter of the rectangle
   c. the area of the rectangle.

9. **WE4** Determine the value of the pronumeral, correct to 2 decimal places for each of the following.

a.

b.

c.

Understanding

10. An isosceles triangle has a base of 25 cm and a height of 8 cm. Calculate the length of the two equal sides, in cm correct to 2 decimal places.

11. An equilateral triangle has sides of length 18 cm. Determine the height of the triangle, in cm correct to 2 decimal places.

12. A right-angled triangle has a height of 17.2 cm, and a base that is half the height. Calculate the length of the hypotenuse, in cm correct to 2 decimal places.

13. The road sign shown is based on an equilateral triangle. Determine the height of the sign and, hence, calculate its area. Round your answers to 2 decimal places.

14. A flagpole, 12 m high, is supported by three wires, attached from the top of the pole to the ground. Each wire is pegged into the ground 5 m from the pole. Determine how much wire is needed to support the pole, correct to the nearest metre.

15. Sarah goes canoeing in a large lake. She paddles 2.1 km to the north, then 3.8 km to the west. Use the triangle shown to determine how far she must then paddle to get back to her starting point in the shortest possible way, in km correct to 2 decimal places.

16. A baseball diamond is a square of side length 27 m. When a runner on first base tries to steal second base, the catcher has to throw the ball from home base to second base. Calculate the distance of the throw, in metres correct to 1 decimal place.

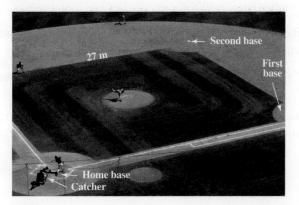

17. A rectangle measures 56 mm by 2.9 cm. Calculate the length of its diagonal in millimetres correct to 2 decimal places.

18. A rectangular envelope has a length of 24 cm and a diagonal measuring 40 cm. Calculate:
    a. the width of the envelope, correct to the nearest cm
    b. the area of the envelope, correct to the nearest cm^2.

19. A swimming pool is 50 m by 25 m. Peter is bored by his usual training routine, and decides to swim the diagonal of the pool. Determine how many diagonals he must swim to complete his normal distance of 1500 m.

20. A hiker walks 2.9 km north, then 3.7 km east. Determine how far in metres she is from her starting point. Give your answer in metres to 2 decimal places.

21. A square has a diagonal of 14 cm. Calculate the length of each side, in cm correct to 2 decimal places.

Reasoning

22. The triangles below are right-angled triangles. Two possible measurements have been suggested for the hypotenuse in each case. For each triangle, complete calculations to determine which of the lengths is correct for the hypotenuse in each case.
    Show your working.

    a.

    60 or 65, 33, 56

    b.

    185 or 195, 175, 60

    c.

    273, 305 or 308, 136

23. The square root of a number usually gives us both a positive and negative answer. Explain why we take only the positive answer when using Pythagoras' theorem.

24. Four possible side length measurements are 105, 208, 230 and 233. Three of them together produce a right-angled triangle.

    a. Explain which of the measurements could not be the hypotenuse of the triangle.
    b. Complete as few calculations as possible to calculate which combination of side lengths will produce a right-angled triangle.

Problem solving

25. The area of the rectangle MNPQ is $588\,\text{cm}^2$. Angles MRQ and NSP are right angles.

    a. Determine the integer value of x.
    b. Determine the length of MP.
    c. Calculate the value of y and hence determine the length of RS, in cm correct to 1 decimal place.

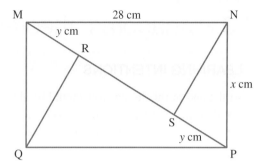

26. Triangle ABC is an equilateral triangle of side length x cm. Angles ADB and DBE are right angles. Determine the value of x in cm, correct to 2 decimal places.

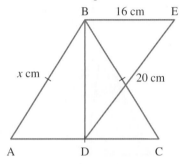

27. The distance from Earth to the Moon is approximately 385 000 km and the distance from Earth to the Sun is approximately 147 million kilometres.

    In a total eclipse of the Sun, the moon moves between the Sun and Earth, thus blocking the light of the Sun from reaching Earth and causing a total eclipse of the Sun.

    If the diameter of the Moon is approximately 3474 km, evaluate the diameter of the Sun. Express your answer to the nearest 10 000 km.

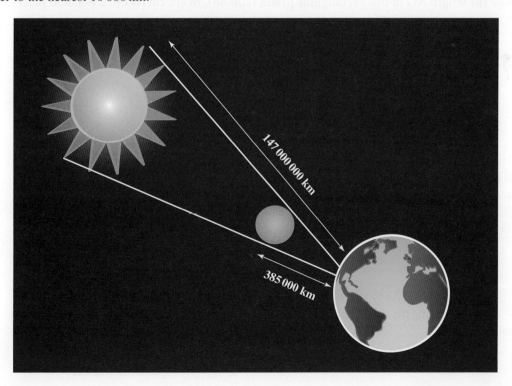

LESSON
5.3 Pythagoras' theorem in three dimensions (10A)

LEARNING INTENTIONS

At the end of this lesson you should be able to apply Pythagoras' theorem to:
* determine unknown lengths when a 3D diagram is given
* determine unknown lengths in situations by first drawing a diagram.

▶ 5.3.1 Applying Pythagoras' theorem in three dimensions

eles-4801

* Many real-life situations involve **3-dimensional** (3D) objects: objects with length, width and height. Some common 3D objects used in this section include cuboids, pyramids and right-angled wedges.

Cuboid

Pyramid

Right-angled wedge

* In diagrams of 3D objects, right angles may not look like right angles, so it is important to redraw sections of the diagram in two dimensions, where the right angles can be seen accurately.

WORKED EXAMPLE 5 Applying Pythagoras' theorem to 3D objects

Determine the length AG in this rectangular prism (cuboid), in cm correct to 2 decimal places.

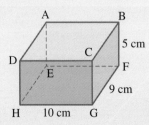

THINK	WRITE/DRAW
1. Draw the diagram in three dimensions. Draw the lines AG and EG. ∠AEG is a right angle.	
2. Draw ΔAEG, showing the right angle. Only 1 side is known, so EG must be found.	

---

3. Draw EFGH in two dimensions and label the diagonal EG as x.

4. Use Pythagoras' theorem to calculate x. $\left(c^2 = a^2 + b^2\right)$

$x^2 = 9^2 + 10^2$
$\quad = 81 + 100$
$\quad = 181$
$x = \sqrt{181}$

5. Place this information on triangle AEG. Label the side AG as y.

6. Use Pythagoras' theorem to calculate y. $\left(c^2 = a^2 + b^2\right)$

$y^2 = 5^2 + \left(\sqrt{181}\right)^2$
$\quad = 25 + 181$
$\quad = 206$
$y = \sqrt{206}$
$\quad \approx 14.35$

7. Write the answer in a sentence.

The length of AG is 14.35 cm.

WORKED EXAMPLE 6 Drawing a diagram to solve problems

A piece of cheese in the shape of a right-angled wedge sits on a table. It has a rectangular base measuring 14 cm by 8 cm, and is 4 cm high at the thickest point. An ant crawls diagonally across the sloping face.
Determine how far, to the nearest millimetre, the ant walks.

THINK

1. Draw a diagram in three dimensions and label the vertices. Mark BD, the path taken by the ant, with a dotted line. ∠BED is a right angle.

2. Draw ΔBED, showing the right angle. Only one side is known, so ED must be found.

3. Draw EFDA in two dimensions and label the diagonal ED. Label the side ED as x.

WRITE/DRAW

4. Use Pythagoras' theorem to calculate x.

$$c^2 = a^2 + b^2$$
$$x^2 = 8^2 + 14^2$$
$$= 64 + 196$$
$$= 260$$
$$x = \sqrt{260}$$

5. Place this information on triangle BED. Label the side BD as y.

6. Use Pythagoras' theorem to calculate y.

$$y^2 = 4^2 + \left(\sqrt{260}\right)^2$$
$$= 16 + 260$$
$$= 276$$
$$y = \sqrt{276}$$
$$\approx 16.61\,\text{cm}$$
$$\approx 166.1\,\text{mm}$$

7. Write the answer in a sentence.

The ant walks 166 mm, correct to the nearest millimetre.

DISCUSSION

Look around the room you are in. How many right angles you can spot in three-dimensional objects? Make a list of them and compare your list to that of another student.

---

 on Resources

 eWorkbook   Topic 5 Workbook (worksheets, code puzzle and project) (ewbk-13288)

 Video eLesson Pythagoras' theorem in three dimensions (eles-1913)

 Interactivity   Right angles in 3-dimensional objects (int-6132)

5.3 Quick quiz on	5.3 Exercise

Individual pathways

■ PRACTISE	■ CONSOLIDATE	■ MASTER
1, 4, 7, 11, 12, 15	2, 5, 8, 13, 16	3, 6, 9, 10, 14, 17

Where appropriate in this exercise, give answers correct to 2 decimal places.

Fluency

1. **WE5** Calculate the length of AG in each of the following figures.

a.

b.

c.
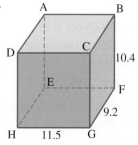

2. Consider the wedge shown. Calculate the length of CE in the wedge and, hence, obtain the length of AC .

3. If DC = 3.2 m, AC = 5.8 m, and CF = 4.5 m in the figure, calculate the lengths of AD and BF.

4. Consider the pyramid shown. Calculate the length of BD and, hence, the height of the pyramid.

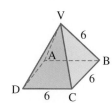

5. The pyramid ABCDE has a square base. The pyramid is 20 cm high. Each sloping edge measures 30 cm. Calculate the length of the sides of the base.

6. The sloping side of a cone is 16 cm and the height is 12 cm. Determine the length of the radius of the base.

Understanding

7. **WE6** A piece of cheese in the shape of a right-angled wedge sits on a table. It has a base measuring 20 mm by 10 mm, and is 4 mm high at the thickest point, as shown in the figure. A fly crawls diagonally across the sloping face. Determine how far, to the nearest millimetre, the fly walks.

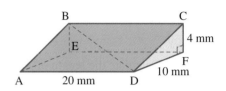

8. A 7 m high flagpole is in the corner of a rectangular park that measures 200 m by 120 m. Give your answers to the following questions correct to 2 decimal places.

   a. Calculate:
      i. the length of the diagonal of the park
      ii. the distance from A to the top of the pole
      iii. the distance from B to the top of the pole.

   b. A bird flies from the top of the pole to the centre of the park. Calculate how far it flies.

9. A candlestick is in the shape of two cones, joined at the vertices as shown. The smaller cone has a diameter and sloping side of 7 cm, and the larger one has a diameter and sloping side of 10 cm. Calculate the total height of the candlestick.

10. The total height of the shape below is 15 cm. Calculate the length of the sloping side of the pyramid.

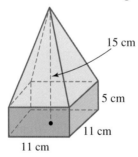

11. A sandcastle is in the shape of a truncated cone as shown. Calculate the length of the diameter of the base.

Reasoning

12. Stephano is renovating his apartment, which he accesses through two corridors. The corridors of the apartment building are 2 m wide with 2 m high ceilings, and the first corridor is at right angles to the second. Show that he can carry lengths of timber up to 6 m long to his apartment.

13. The Great Pyramid in Egypt is a square-based pyramid.
The square base has a side length of 230.35 metres and the perpendicular height is 146.71 metres.
Determine the slant height, s, of the great pyramid. Give your answer correct to 1 decimal place.

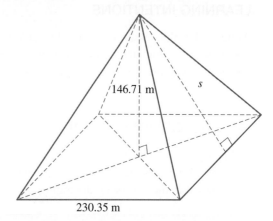

14. A tent is in the shape of a triangular prism, with a height of 140 cm as shown in the diagram. The width across the base of the door is 1 m and the tent is 2.5 m long.

    a. Calculate the length of each sloping side, in metres.
    b. Using your answer from part **a** calculate the area of fabric used in the construction of the sloping rectangles which form the sides. Show full working.

Problem solving

15. Determine the exact length of the longest steel rod that can sit inside a cuboid with dimensions $32 \text{ cm} \times 15 \text{ cm} \times 4 \text{ cm}$. Ignore the thickness of the steel rod.

16. Angles ABD, CBD and ABC are right angles. Determine the value of h, correct to 3 decimal places.

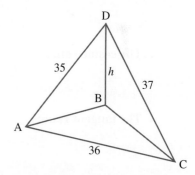

17. The roof of a squash centre is constructed to allow for maximum use of sunlight. Determine the value of h, giving your answer correct to 1 decimal place.

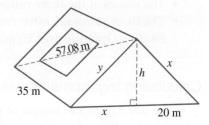

LESSON
5.4 Trigonometric ratios

LEARNING INTENTIONS

At the end of this lesson you should be able to:
- define trigonometric ratios according to the lengths of the relevant sides
- write equations for trigonometric ratios.

▶ 5.4.1 Trigonometric ratios

eles-4802
- In a right-angled triangle, the longest side is called the hypotenuse.
- If one of the two acute angles is named (for example, θ), then the other two sides can also be given names, as shown in the following diagram.

Three basic definitions

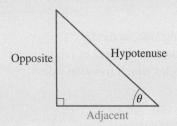

- **Using the diagram, the following three trigonometric ratios can be defined.**
  - **The sine ratio:**
  $$\text{sine}(\theta) = \frac{\text{length of Opposite side}}{\text{length of Hypotenuse}}$$
  - **The cosine ratio:**
  $$\text{cosine}(\theta) = \frac{\text{length of Adjacent side}}{\text{length of Hypotenuse}}$$
  - **The tangent ratio:**
  $$\text{tangent}(\theta) = \frac{\text{length of Opposite side}}{\text{length of Adjacent side}}$$

- **The names of the three ratios are usually shortened to $\sin(\theta)$, $\cos(\theta)$ and $\tan(\theta)$.**
- **The three ratios are often remembered using the mnemonic SOHCAHTOA, where SOH means $\text{Sin}(\theta) = \text{Opposite over Hypotenuse}$ and so on.**

Calculating trigonometric values using a calculator

- The sine, cosine and tangent of an angle have numerical values that can be found using a calculator.
- Traditionally angles were measured in **degrees**, minutes and seconds, where 60 seconds = 1 minute and 60 minutes = 1 degree. This is known as a sexagesimal system as the division are based on 60.
  For example, $50°33'48''$ means 50 degrees, 33 minutes and 48 seconds.

WORKED EXAMPLE 7 Calculating values (ratios) from angles

Calculate the value of each of the following, correct to 4 decimal places, using a calculator. (Remember to first work to 5 decimal places before rounding.)

a. $\cos(65°57')$

b. $\tan(56°45'30'')$

THINK

a. Write your answer to the required number of decimal places.

b. Write your answer to the correct number of decimal places.

WRITE

a. $\cos(65°57') \approx 0.40753$
≈ 0.4075

b. $\tan(56°45'30'') \approx 1.52573$
≈ 1.5257

TI | THINK	**DISPLAY/WRITE**	**CASIO | THINK**	**DISPLAY/WRITE**

a, b.

1. To ensure your calculator is set to degree and approximate mode, press:
   - HOME
   - 5: Settings
   - 2: Document Settings
   In the Display Digits, select Fix 4. Tab to Angle and select Degree; tab to Calculation Mode and select Approximate.
   Tab to OK and press ENTER

a, b.

2. On a Calculator page, press TRIG to access and select the appropriate trigonometric ratio. Then press ⊕ and choose the template for degrees, minutes and seconds as shown.

3. Complete the entry lines as:
   $\cos(65°57')$
   $\tan(56°45'30'')$
   Press ENTER after each entry.
   Since the Calculation Mode is set to Approximate and Fix 4, the answer are shown correct to 4 decimal places.

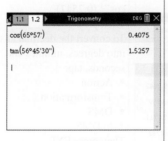

$\cos(65°57') = 0.4075$
$\tan(56°45'30'') = 1.5257$

a, b.

1. Ensure your calculator is set to degrees at the bottom right. If not, tap 'Rad' or 'Gra'. Tap unit Deg appears.
   Set to Decimal.
   Set the keyboard to Trig.

a, b.

2. Complete the entry line as:
   $\cos(\text{dms}(65, 57))$
   $\tan(\text{dms}(56, 45, 30))$
   To get dms, tap
   - Action
   - Transformation
   - DMS
   - dms
   Then press EXE after each entry.

$\cos(65°57') = 0.4075$
$\tan(56°45'30'') = 1.5257$

WORKED EXAMPLE 8 Calculating angles from ratios

Calculate the size of angle θ, correct to the nearest degree, given $\sin(\theta) = 0.7854$.

THINK	WRITE
1. Write the given equation.	$\sin(\theta) = 0.7854$
2. To calculate the size of the angle, we need to undo sine with its inverse, $\sin^{-1}$. (Ensure your calculator is in degrees mode.)	$\theta = \sin^{-1}(0.7854)$ $\approx 51.8°$
3. Write your answer to the nearest degree.	$\theta \approx 52°$

WORKED EXAMPLE 9 Expressing angles in degrees, minutes and seconds

Calculate the value of θ:
a. correct to the nearest minute, given that $\cos(\theta) = 0.2547$
b. correct to the nearest second, given that $\tan(\theta) = 2.364$.

THINK	WRITE
a. 1. Write the equation.	a. $\cos(\theta) = 0.2547$
2. Write your answer, including seconds. There are 60 seconds in 1 minute. Round to the nearest minute. (Remember $60'' = 1'$, so $39''$ is rounded up.)	$\cos^{-1}(0.2547) \approx 75°14'39''$ $\approx 75°15'$
b. 1. Write the equation.	b. $\tan(\theta) = 2.364$
2. Write the answer, rounding to the nearest second.	$\tan^{-1}(2.364) \approx 67°4'15.8''$ $\approx 67°4'16''$

TI | THINK	DISPLAY/WRITE	CASIO | THINK	DISPLAY/WRITE
a, b. On a Calculator page, press TRIG to access and select the appropriate trigonometric ratio, in the case $\cos^{-1}$. Complete the entry lines as: $\cos^{-1}(0.2547)$ To convert the decimal degree into degrees, minutes and seconds, press: • CATALOG • 1 • D Scroll and select ▶ DMS Then press ENTER. Repeat this process for $\tan^{-1}(2.364)$.	a, b. 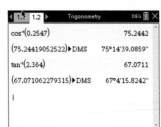 $\cos^{-1}(0.2547) = 75°15'$ rounding to the nearest minute. $\tan^{-1}(2.364) = 64°4'16''$ rounding to the nearest second.	a, b. Ensure your calculator is set to degrees at the bottom right. On the Main screen in decimal and degree mode, complete the entry line as: $\cos^{-1}(0.2547)$ Then press EXE. To convert the decimal answer into degrees, minutes and seconds, tap: • Action • Transformation • DMS • to DMS Then press EXE. Repeat this process for $\tan^{-1}(2.364)$.	a, b. $\cos^{-1}(0.2547) = 75°15'$ rounding to the nearest minute. $\tan^{-1}(2.364) = 64°4'16''$ rounding to the nearest second.

WORKED EXAMPLE 10 Expressing trigonometric ratios as equations

Write the equation that relates the two marked sides and the marked angle.

a.

b.

THINK	WRITE/DRAW
a. 1. Label the given sides of the triangle.	**a.** $8 = O$ $12 = H$ b
2. Write the ratio that contains O and H.	$\sin(\theta) = \dfrac{O}{H}$
3. Identify the values of the pronumerals.	$O = 8, H = 12$
4. Substitute the values of the pronumerals into the ratio and simplify the fraction. (Since the given angle is denoted with the letter b, replace θ with b.)	$\sin(b) = \dfrac{8}{12} = \dfrac{2}{3}$
b. 1. Label the given sides of the triangle.	**b.** $22 = A$ $x = O$ $40°$
2. Write the ratio that contains O and A.	$\tan(\theta) = \dfrac{O}{A}$
3. Identify the values of the pronumerals.	$O = x, A = 22, \theta = 40°$
4. Substitute the values of the pronumerals into the ratio.	$\tan(40°) = \dfrac{x}{22}$

DISCUSSION

Do you know of any other mnemonics that you can use to help you remember important information?

Exercise 5.4 Trigonometric ratios

5.4 Quick quiz on	5.4 Exercise

Individual pathways

■ PRACTISE	■ CONSOLIDATE	■ MASTER
1, 5, 7, 9, 11, 14, 15, 22, 25	2, 6, 12, 16, 17, 20, 23, 26, 27	3, 4, 8, 10, 13, 18, 19, 21, 24, 28

Fluency

1. Calculate each of the following, correct to 4 decimal places.
   - **a.** $\sin(30°)$
   - **b.** $\cos(45°)$
   - **c.** $\tan(25°)$
   - **d.** $\sin(57°)$
   - **e.** $\tan(83°)$
   - **f.** $\cos(44°)$

   WE7 For questions **2** to **4**, calculate each of the following, correct to 4 decimal places.

2. **a.** $\sin(40°30')$
   b. $\cos(53°57')$
   c. $\tan(27°34')$
   d. $\tan(123°40')$
   e. $\sin(92°32')$
   f. $\sin(42°8')$

3. **a.** $\cos(35°42'35'')$
   b. $\tan(27°42'50'')$
   c. $\cos(143°25'23'')$
   d. $\sin(23°58'21'')$
   e. $\cos(8°54'2'')$

4. **a.** $\sin(286)°$
   b. $\tan(420°)$
   c. $\cos(845°)$
   d. $\sin(367°35')$

5. **WE8** Calculate the size of angle θ, correct to the nearest degree, for each of the following.
   - **a.** $\sin(\theta) = 0.763$
   - **b.** $\cos(\theta) = 0.912$
   - **c.** $\tan(\theta) = 1.351$

6. Calculate the size of angle θ, correct to the nearest degree, for each of the following.
   - **a.** $\cos(\theta) = 0.321$
   - **b.** $\tan(\theta) = 12.86$
   - **c.** $\cos(\theta) = 0.756$

7. **WE9a** Calculate the size of the angle θ, correct to the nearest minute.
   - **a.** $\sin(\theta) = 0.814$
   - **b.** $\sin(\theta) = 0.110$
   - **c.** $\tan(\theta) = 0.015$

8. Calculate the size of the angle θ, correct to the nearest minute.
   - **a.** $\cos(\theta) = 0.296$
   - **b.** $\tan(\theta) = 0.993$
   - **c.** $\sin(\theta) = 0.450$

9. **WE9b** Calculate the size of the angle θ, correct to the nearest second.
   - **a.** $\tan(\theta) = 0.5$
   - **b.** $\cos(\theta) = 0.438$
   - **c.** $\sin(\theta) = 0.9047$

10. Calculate the size of the angle θ, correct to the nearest second.
    - **a.** $\tan(\theta) = 1.1141$
    - **b.** $\cos(\theta) = 0.8$
    - **c.** $\tan(\theta) = 43.76.$

For questions **11** to **13**, calculate the value of each expression, correct to 3 decimal places.

11. **a.** $3.8 \cos(42°)$
    b. $118 \sin(37°)$
    c. $2.5 \tan(83°)$
    d. $\dfrac{2}{\sin(45°)}$

12. **a.** $\dfrac{220}{\cos(14°)}$
    b. $\dfrac{2 \cos(23°)}{5 \sin(18°)}$
    c. $\dfrac{12.8}{\tan(60°32')}$
    d. $\dfrac{18.7}{\sin(35°25'42'')}$

13. **a.** $\dfrac{55.7}{\cos(89°21')}$
    b. $\dfrac{3.8 \tan(1°51'44'')}{4.5 \sin(25°45')}$
    c. $\dfrac{2.5 \sin(27°8')}{10.4 \cos(83°2')}$
    d. $\dfrac{3.2 \cos(34°52')}{0.8 \sin(12°48')}$

For questions **14** to **19**, write an expression for:

a. sine   **b.** cosine   **c.** tangent.

14.

15.

16.

17.

18.

19.

Understanding

20. **WE10** Write the equation that relates the two marked sides and the marked angle in each of the following triangles.

a.

b.

c.

21. Write the equation that relates the two marked sides and the marked angle in each of the following triangles.

a.

b.

c.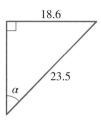

Reasoning

22. Consider the right-angled triangle shown.

    a. Label each of the sides using the letters O, A and H with respect to the 37° angle.
    b. Determine the value of each trigonometric ratio. (Where applicable, answers should be given correct to 2 decimal places.)

       i. $\sin(37°)$ ii. $\cos(37°)$ iii. $\tan(37°)$

    c. Determine the value of the unknown angle, α.

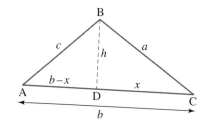

23. Consider the right-angled triangle shown in question 22.

    a. Determine the value of each of these trigonometric ratios, correct to 2 decimal places.

       i. $\sin(\alpha)$ ii. $\cos(\alpha)$ iii. $\tan(\alpha)$

    (*Hint:* First relabel the sides of the triangle with respect to angle α.)
    b. What do you notice about the relationship between $\sin(37°)$ and $\cos(\alpha)$? Explain your answer.
    c. What do you notice about the relationship between $\sin(\alpha)$ and $\cos(37°)$? Explain your answer.
    d. Make a general statement about the two angles.

24. Using a triangle labelled with a, h and o, algebraically show that $\tan(\theta) = \dfrac{\sin(\theta)}{\cos(\theta)}$.

    (*Hint:* Write all the sides in terms of the hypotenuse.)

Problem solving

25. ABC is a scalene triangle with side lengths a, b and c as shown. Angles BDA and BDC are right angles.

    a. Express h^2 in terms of a and x.
    b. Express h^2 in terms of b, c and x.
    c. Equate the two equations for h^2 to show that $c^2 = a^2 + b^2 - 2bx$.
    d. Use your knowledge of trigonometry to produce the equation $c^2 = a^2 + b^2 - 2ab \cos(C)$, which is known as the cosine rule for non-right-angled triangles.

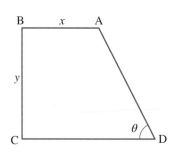

26. Determine the length of the side DC in terms of x, y and θ.

27. Explain how we determine whether to use sin, cos or tan in trigonometry questions.

28. From an observer on a boat 110 m away from a vertical cliff with height c, the angle from the base of the cliff to the top of the cliff is $\alpha°$. There is a lighthouse with height t on the cliff. From the observer, the angle from the base of the lighthouse to the top of the lighthouse is another $\theta°$ more than $\alpha°$.

Express the height of the lighthouse, t, in terms of $\theta°$ and $\alpha°$.

110 m

LESSON
5.5 Using trigonometry to calculate side lengths

LEARNING INTENTION

At the end of this lesson you should be able to:
- apply trigonometric ratios to calculate the length of an unknown side when the length of one other side and an acute angle is known.

▶ 5.5.1 Using trigonometry to calculate side lengths

eles-4804

- When one acute angle and one side length are known in a right-angled triangle, this information can be used to find all other unknown sides or angles.

WORKED EXAMPLE 11 Using trigonometry to calculate side lengths

Calculate the value of each pronumeral, giving answers correct to 3 decimal places.

a.

b.

THINK

a. 1. Label the marked sides of the triangle.

WRITE/DRAW

a.

2. Identify the appropriate trigonometric ratio to use.

$\sin(\theta) = \dfrac{O}{H}$

3. Substitute $O = a$, $H = 6$ and $\theta = 35°$.

$\sin(35°) = \dfrac{a}{6}$

4. Make a the subject of the equation.

$$6\sin(35°) = a$$
$$x = 6\sin(35°)$$

5. Calculate and round the answer, correct to 3 decimal places.

$$a \approx 3.441 \text{ cm}$$

b. 1. Label the marked sides of the triangle.

b.

2. Identify the appropriate trigonometric ratio to use.

$$\cos(\theta) = \frac{A}{H}$$

3. Substitute $A = f$, $H = 0.346$ and $\theta = 32°$.

$$\cos(32°) = \frac{f}{0.346}$$

4. Make f the subject of the equation.

$$0.346\cos(32°) = f$$
$$f = 0.346\cos(32°)$$

5. Calculate and round the answer, correct to 3 decimal places.

$$f \approx 0.293 \text{ cm}$$

5.5.2 Angle of inclination

- A **clinometer** is a tool used to measure angles of inclination. It is used in various fields, such as engineering, architecture and geology.
- An angle of inclination is the acute or obtuse angle that is formed when a non-horizontal line intersects the x-axis.
- It is measured counterclockwise from the x-axis to the line.
- In the figures shown, the angle in green is the angle of inclination since it is measured counterclockwise.

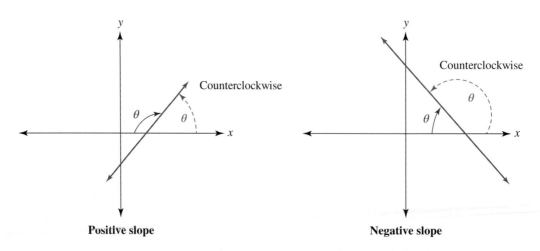

Positive slope　　　　　　　　　**Negative slope**

- When the slope is positive, the angle of inclination is an acute angle.
- When the slope is negative, the angle of inclination is an obtuse angle.

WORKED EXAMPLE 12 Using trigonometry to calculate side lengths

In the given triangle, the inclination angle is 5°. Calculate the value of the pronumeral *P*. Give the answer correct to 2 decimal places.

THINK	WRITE/DRAW
1. Label the marked sides of the triangle.	
2. Identify the appropriate trigonometric ratio to use.	$\tan(\theta) = \dfrac{O}{A}$
3. Substitute O = 120, A = *P* and $\theta = 5°$.	$\tan(5°) = \dfrac{120}{p}$

4. Make *P* the subject of the equation.
  i. Multiply both sides of the equation by *P*.
  ii. Divide both sides of the equation by tan (5°).
	$P \times \tan(5°) = 120$
	$P = \dfrac{120}{\tan(5°)}$

5. Calculate and round the answer, correct to 2 decimal places.	$P \approx 1371.61 \text{ m}$

COLLABORATIVE TASK: Clinometer

The following group activity helps students learn how to use a clinometer to measure angles of inclination.

Materials used

- Clinometers (one per group of 3 or 4 students)
- Measuring tape or ruler
- Pencils and paper
- Protractor
- Worksheets or data recording sheets
- Objects with known angles of inclination (optional)
- Access to an outdoor area or a large indoor space (with high ceilings)

Instructions

1. Explain to the students how to use the clinometer and how it is a valuable tool for accurately measuring angles of inclination.
2. Demonstrate how a clinometer works, explaining its different components (e.g. sighting mirror, scale and sighting line) and how to read the measurements.
3. Emphasise the need to hold the clinometer level and steady to ensure accurate readings.
4. Divide the students into groups of 3 or 4 and distribute the materials.

5. Have each group choose a location outside to measure the angles of inclination. This could be a hill, a building, a tree, or any other object with an inclined surface.

6. Students should measure the height of the object using the measuring tape or ruler. They should also measure the distance from the object.

7. Measure the angle of inclination of the object from their location. They should record their measurements on paper, along with the height and distance of the object.

8. Once all groups have completed their measurements, have them compare their results and discuss any discrepancies. This can be a good opportunity to discuss sources of error in the measurements and ways to improve accuracy.

9. As a class, discuss the different types of angles that the students measured (angles of elevation or angles of depression) and their practical applications.

10. Finally, have each group present their findings and measurements to the class, along with any insights they gained from the activity.

This activity can help students develop their measurement skills, spatial reasoning, and collaboration skills. It can also help them understand the practical applications of measuring angles of inclination.

Exercise 5.5 Using trigonometry to calculate side lengths   **learn on**

5.5 Quick quiz **on**	**5.5 Exercise**

Individual pathways

■ PRACTISE	■ CONSOLIDATE	■ MASTER
1, 2, 7, 9, 12	3, 5, 8, 10, 13	4, 6, 11, 14, 15

Fluency

1. **WE11** Calculate the value of the pronumeral in each of the following, correct to 2 decimal places.

a.
$71°$
m
2.3 m

b.
4.6 m
$13°$
n

c.
94 mm
$68°$
t

2. **WE12** In each of the following triangles, the inclination angle is shown. Calculate the value of the pronumeral in each of the following, correct to 3 decimal places.

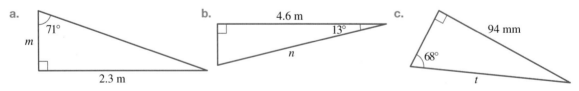

a.
10 cm
a
$60°$

b.
8
$25°$
a

c.
x
$31°$
14

3. Determine the length of the unknown side in each of the following, correct to 2 decimal places.

a.

b.

c.

4. Determine the length of the unknown side in each of the following, correct to 2 decimal places.

a.

b.

c.

5. Calculate the value of the pronumeral in each of the following, correct to 2 decimal places.

a.

b.

c.

6. Calculate the value of the pronumeral in each of the following, correct to 2 decimal places.

a.

b.

c.

Understanding

7. Given that the angle θ is 42° (the angle between the base and the hypotenuse of the right-angled triangle) and the length of the hypotenuse is 8.95 m in a right-angled triangle, calculate the length of:

   a. the opposite side
   b. the adjacent side.
   Give each answer correct to 1 decimal place.

8. A ladder rests against a wall. If the angle of inclination (between the ladder and the ground) is 35° and the foot of the ladder is 1.5 m from the wall, calculate how high up the wall the ladder reaches.
   Write your answer in metres correct to 2 decimal places.

Reasoning

9. Tran is going to construct an enclosed rectangular desktop that has an inclination angle of 15°. The diagonal length of the desktop is 50 cm. At the high end, the desktop, including top, bottom and sides, will be raised 8 cm. The desktop will be made of wood.
   The diagram below represents this information.

Side view of the desktop

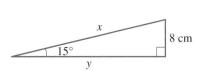

Top view of the desktop

   a. Determine the values (in centimetres) of x, y and z of the desktop. Write your answers correct to 2 decimal places.
   b. Using your answer from part a determine the minimum area of wood, in cm², Tran needs to construct his desktop including top, bottom and sides. Write your answer correct to 2 decimal places.

10. a. In a right-angled triangle, under what circumstances will the opposite side and the adjacent side have the same length?
    b. In a right-angled triangle, for what values of θ (the reference angle) will the adjacent side be longer than the opposite side?

11. In triangle ABC shown, the length of x correct to 2 decimal places is 8.41 cm. AM is perpendicular to BC. Jack found the length of x to be 5.11 cm. His working is shown below.

$$\cos(36°) = \frac{BM}{4.5}$$
$$BM = 3.641$$
$$\sin(36°) = \frac{AM}{4.5}$$
$$AM = 2.645$$
$$\tan(29°) = \frac{2.645}{MC}$$
$$MC = 1.466$$
$$BC = 3.641 + 1.466 = 5.107 \text{ cm}$$
$$= 5.11 \text{ cm (rounded to 2 decimal places)}$$

Identify his error and what he should have done instead.

Problem solving

12. A surveyor needs to determine the height of a building. She measures the angle to the top of the building from two points, A and B, which are 64 m apart. The surveyor's eye level is 195 cm above the ground.

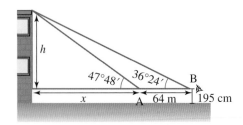

    a. Determine the expressions for the height of the building, h, in terms of x using the two angles.
    b. Solve for x by equating the two expressions obtained in part a. Give your answer to 2 decimal places.
    c. Determine the height of the building correct to 2 decimal places.

13. Building A and Building B are 110 m apart. From the base of Building A to the top of Building B, the angle is 15°.
From the top of Building A looking down to the top of Building B, the angle is 22°.
Evaluate the height of each of the two buildings correct to 1 decimal place.

14. If angles QNM, QNP and MNP are right angles, determine the length of NQ.

15. Determine how solving a trigonometric equation differs when we are calculating the length of the hypotenuse side compared to when determining the length of a shorter side.

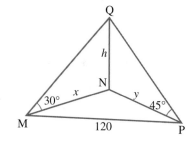

LESSON
5.6 Using trigonometry to calculate angle size

LEARNING INTENTION

At the end of this lesson you should be able to:
- apply inverse operations to calculate a known acute angle when two sides are given.

▶ 5.6.1 Using trigonometry to calculate angle size

eles-4805

- The size of any angle in a right-angled triangle can be found if the lengths of any two sides are known.
- Just as inverse operations are used to solve equations, inverse trigonometric ratios are used to solve trigonometric equations for the value of the angle.
  - Inverse sine ($\sin^{-1}$) is the inverse of sine.
  - Inverse cosine ($\cos^{-1}$) is the inverse of cosine.
  - Inverse tangent ($\tan^{-1}$) is the inverse of tangent.

Inverse operations

$$\text{If } \sin(\theta) = a, \text{ then } \sin^{-1}(a) = \theta.$$
$$\text{If } \cos(\theta) = a, \text{ then } \cos^{-1}(a) = \theta.$$
$$\text{If } \tan(\theta) = a, \text{ then } \tan^{-1}(a) = \theta.$$

For example, since $\sin(30°) = 0.5$, then $\sin^{-1}(0.5) = 30°$; this is read as 'inverse sine of 0.5 is 30 degrees'.

A calculator can be used to calculate the values of inverse trigonometric ratios.

For each of the following, calculate the size of the angle, θ, correct to the nearest degree.

a.

b.

THINK	WRITE/DRAW
a. 1. Label the given sides of the triangle.	**a.**
2. Identify the appropriate trigonometric ratio to use. We are given O and H.	$\sin(\theta) = \dfrac{O}{H}$
3. Substitute O = 3.5 and H = 5 and evaluate the expression.	$\sin(\theta) = \dfrac{3.5}{5}$ $= 0.7$
4. Make θ the subject of the equation using inverse sine.	$\theta = \sin^{-1}(0.7)$ $= 44.427\,004°$
5. Evaluate θ and round the answer, correct to the nearest degree.	$\theta \approx 44°$
b. 1. Label the given sides of the triangle.	**b.**
2. Identify the appropriate trigonometric ratio to use. We are given O and A.	$\tan(\theta) = \dfrac{O}{A}$
3. Substitute O = 5 and A = 11.	$\tan(\theta) = \dfrac{5}{11}$
4. Make θ the subject of the equation using inverse tangent.	$\theta = \tan^{-1}\left(\dfrac{5}{11}\right)$ $= 24.443\,954\,78°$
5. Evaluate θ and round the answer, correct to the nearest degree.	$\theta \approx 24°$

WORKED EXAMPLE 14 Evaluating angles in minutes and seconds

Calculate the size of angle θ:
a. **correct to the nearest second**
b. **correct to the nearest minute.**

THINK	WRITE/DRAW
a. 1. Label the given sides of the triangle.	a.
2. Identify the appropriate trigonometric ratio to use.	$\tan(\theta) = \dfrac{O}{A}$
3. Substitute O = 7.2 and A = 3.1.	$\tan(\theta) = \dfrac{7.2}{3.1}$
4. Make θ the subject of the equation using inverse tangent.	$\theta = \tan^{-1}\left(\dfrac{17.2}{3.1}\right)$
5. Evaluate θ and write the calculator display.	$\theta = 66.70543675°$
6. Use the calculator to convert the answer to degrees, minutes and seconds.	$\theta = 66°42'19.572''$
7. Round the answer to the nearest second.	$\theta \approx 66°42'20''$
b. Round the answer to the nearest minute.	b. $\theta \approx 66°42'$

| TI | THINK | DISPLAY/WRITE | CASIO | THINK | DISPLAY/WRITE |

a.
On a Calculator page, in degree mode, complete the entry line as:

$$\tan^{-1}\left(\frac{7.2}{3.1}\right)$$

To convert the decimal degree into degrees, minutes and seconds, press:
- CATALOG
- 1
- D

Scroll and select ▶ DMS. Then press ENTER.

a.

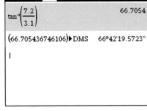

$\theta = 66°42'20''$ correct to the nearest second.

a.
On the Main screen in decimal and degree mode, complete the entry line as:

$$\tan^{-1}\left(\frac{7.2}{3.1}\right)$$

To convert the decimal answer into degrees, minutes and seconds, tap:
- Action
- Transformation
- DMS
- toDMS

Highlight the decimal answer and drag it into this line.
Then press EXE.

a.

$\theta = 66°42'20''$ correct to the nearest second.

b.
Using the same screen, round to the nearest minute.

b.
$\theta = 66°42'$ correct to the nearest minute.

b.
Using the same screen, round to the nearest minute.

b.
$\theta = 66°42'$ correct to the nearest minute.

 Resources

 eWorkbook   Topic 5 Workbook (worksheets, code puzzle and project) (ewbk-13288)

Interactivities   Individual pathway interactivity: Using trigonometry to calculate angle size (int-4589)
Finding the angle when two sides are known (int-6046)

Exercise 5.6 Using trigonometry to calculate angle size   learn on

| 5.6 Quick quiz   on | 5.6 Exercise |

Individual pathways

■ PRACTISE	■ CONSOLIDATE	■ MASTER
1, 4, 7, 9, 12	2, 5, 8, 10, 13	3, 6, 11, 14

Fluency

1. **WE13** Calculate the size of the angle, θ, in each of the following. Give your answers correct to the nearest degree.

a.

b.

c.

2. **WE14b** Calculate the size of the angle marked with the pronumeral in each of the following. Give your answers correct to the nearest minute.

a.

17

θ

12

b.

β

7.2 m

4 m

c.

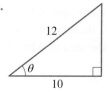

12

θ

10

3. **WE14a** Calculate the size of the angle marked with the pronumeral in each of the following. Give your answers correct to the nearest second.

a.

8

θ

2

b.

α

5 m

3 m

c.

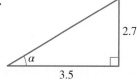

2.7

α

3.5

4. Calculate the size of the angle marked with the pronumeral in each of the following, giving your answers correct to the nearest degree.

a.

13.5

a°

15.3

b.

89.4

c°

77.3

c.

106.4

92.7

b°

5. Calculate the size of the angle marked with the pronumeral in each of the following, giving your answers correct to the nearest degree.

a.

43.7

d°

18.7

b.

12.36

13.85

e°

18.56

c.

7.3 cm

12.2 cm

9.8 cm

α°

6. Calculate the size of each of the angles in the following, giving your answers correct to the nearest minute.

a.

a°

56.3

b°

27.2

b.

d°

0.798

0.342

e°

c.

x°

5.7

2.3

y°

Understanding

7. Answer the following questions for the triangle shown.

   a. Calculate the length of the sides r, l and h. Write your answers correct to 2 decimal places.

   b. Calculate the area of ABC, correct to the nearest square centimetre.

   c. Determine the size of $\angle$BCA.

8. In the sport of air racing, small aeroplanes have to travel between two large towers (or pylons). The gap between a pair of pylons is smaller than the wingspan of the plane, so the plane has to go through on an angle with one wing higher than the other.

   The wingspan of a competition plane is 8 metres.

   a. Determine the angle, correct to 1 decimal place, that the plane has to tilt if the gap between pylons is:

      i. 7 metres   ii. 6 metres   iii. 5 metres.

   b. Because the plane has rolled away from the horizontal as it travels between the pylons it loses speed. If the plane's speed is below 96 km/h it will stall and possibly crash.

   For each degree of 'tilt', the speed of the plane is reduced by 0.98 km/h. Calculate the minimum speed the plane must go through each of the pylons in part **a**.

   Write your answer correct to 2 decimal places.

Reasoning

9. Explain how calculating the angle of a right-angled triangle is different to calculating a side length.

10. There are two important triangles commonly used in trigonometry.
    Complete the following steps and answer the questions to create these triangles.

    Triangle 1
    - Sketch an equilateral triangle with side length 2 units.
    - Calculate the size of the internal angles.
    - Bisect the triangle to form two right-angled triangles.
    - Redraw one of the triangles formed.
    - Calculate the side lengths of this right-angled triangle as exact values.
    - Fully label your diagram showing all side lengths and angles.

    Triangle 2
    - Draw a right-angled isosceles triangle.
    - Calculate the sizes of the internal angles.
    - Let the sides of equal length be 1 unit long each.
    - Calculate the length of the third side as an exact value.
    - Fully label your diagram showing all side lengths and angles.

11. a. Use the triangles formed in question **10** to calculate exact values for $\sin(30°)$, $\cos(30°)$ and $\tan(30°)$. Justify your answers.

    b. Use the exact values for $\sin(30°)$, $\cos(30°)$ and $\tan(30°)$ to show that $\tan(30°) = \dfrac{\sin(30°)}{\cos(30°)}$.

    c. Use the formulas $\sin(\theta) = \dfrac{O}{H}$ and $\cos(\theta) = \dfrac{A}{H}$ to prove that $\tan(\theta) = \dfrac{\sin(\theta)}{\cos(\theta)}$.

Problem solving

12. Determine the angle θ in degrees and minutes.

13. During a Science excursion, a class visited an underground cave to observe rock formations. They were required to walk along a series of paths and steps as shown in the diagram below.

  a. Calculate the angle of the incline (slope) required to travel down between each site. Give your answers to the nearest whole number.
  b. Determine which path would have been the most challenging; that is, which path had the steepest slope.

14. At midday, the hour hand and the minute hand on a standard clock are both pointing at the twelve.

Calculate the angles the minute hand and the hour hand have moved 24.5 minutes later. Express both answers in degrees and minutes.

LESSON
5.7 Angles of elevation and depression

LEARNING INTENTION

At the end of this lesson you should be able to:
- identify angles of elevation and depression and solve for unknown side lengths and angles.

▶ 5.7.1 Angles of elevation and depression

eles-4806

- Solving real-life problems usually involves the person measuring angles or lengths from their position using trigonometry.
- They may have to either look up at the object or look down to it; hence the terms '**angle of elevation**' and '**angle of depression**' respectively.

Angle of elevation

Consider the points A and B, where B is at a higher elevation than A.

If a horizontal line is drawn from A as shown, forming the angle θ, then θ is called the angle of elevation of B *from* A.

Angle of depression

If a horizontal line is drawn from B, forming the angle α, then α is called the angle of depression of A *from* B.

Alternate angle rule

Because the horizontal lines are parallel, θ and α have the same size (alternate angles).

WORKED EXAMPLE 15 Applying angles of elevation to solve problems

From a point P, on the ground, the angle of elevation of the top of a tree is 50°. If P is 8 metres from the tree, determine the height of the tree correct to 2 decimal places.

THINK	WRITE/DRAW
1. Let the height of the tree be h. Sketch a diagram and show the relevant information.	
2. Identify the appropriate trigonometric ratio.	$\tan(\theta) = \dfrac{O}{A}$
3. Substitute $O = h$, $A = 8$ and $\theta = 50°$.	$\tan(50°) = \dfrac{h}{8}$
4. Rearrange to make h the subject.	$h = 8 \tan(50°)$
5. Calculate and round the answer to 2 decimal places.	≈ 9.53
6. Write the answer in a sentence.	The height of the tree is 9.53 m.

WORKED EXAMPLE 16 Applying angles of depression to solve problems

The angle of depression from a helicopter, at point H, to a swimmer in distress in the water is 60°. If the helicopter is hovering 800 m above sea level, determine how far horizontally the swimmer is from the helicopter.
Write your answer in metres correct to 2 decimal places.

THINK	WRITE/DRAW
1. Let the horizontal distance between the swimmer and the helicopter be d. Sketch a diagram and show the relevant information.	
2. Identify the appropriate trigonometric ratio.	$\tan(\theta) = \dfrac{O}{A}$
3. Substitute $O = 800$, $Q = 60°$ and $A = d$.	$\tan(60°) = \dfrac{800}{d}$
4. Rearrange to make d the subject.	$d = \dfrac{800}{\tan(60°)}$
5. Calculate and round to 2 decimal places.	$d \approx 461.88$ m
6. Write the answer in a sentence.	The horizontal distance between the swimmer and the helicopter is 461.88 m.

Exercise 5.7 Angles of elevation and depression   **learn** on

5.7 Quick quiz on	**5.7 Exercise**

Individual pathways

■ PRACTISE	■ CONSOLIDATE	■ MASTER
1, 4, 6, 9, 12, 15	2, 5, 7, 10, 13, 16	3, 8, 11, 14, 17

Fluency

1. **WE15** From a point P on the ground the angle of elevation from an observer to the top of a tree is $54°22'$. If the tree is known to be 12.19 m high, determine how far P is from the tree (measured horizontally). Write your answer in metres correct to 2 decimal places.

2. **WE16** From the top of a cliff 112 m high, the angle of depression to a boat is $9°15'$. Determine how far the boat is from the foot of the cliff. Write your answer in metres correct to 1 decimal place.

3. A person on a ship observes a lighthouse on the cliff, which is 830 metres away from the ship. The angle of elevation of the top of the lighthouse is $12°$.

   a. Determine how far above sea level the top of the lighthouse is, correct to 2 decimal places.

   b. If the height of the lighthouse is 24 m, calculate the height of the cliff, correct to 2 decimal places.

4. At a certain time of the day a post that is 4 m tall casts a shadow of 1.8 m. Calculate the angle of elevation of the sun at that time. Write your answer correct to the nearest minute.

5. An observer who is standing 47 m from a building measures the angle of elevation of the top of the building as $17°$. If the observer's eye is 167 cm from the ground, determine the height of the building. Write your answer in metres correct to 2 decimal places.

Understanding

6. A surveyor needs to determine the height of a building. She measures the angle of elevation of the top of the building from two points that are 38 m apart. The surveyor's eye level is 180 cm above the ground.

   a. Determine two expressions for the height of the building, h, in terms of x using the two angles.

   b. Solve for x by equating the two expressions obtained in **a**. Write your answer in metres correct to 2 decimal places

   c. Determine the height of the building, in metres correct to 2 decimal places.

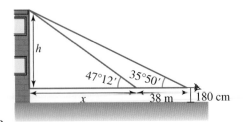

7. The height of another building needs to be determined but cannot be found directly. The surveyor decides to measure the angle of elevation of the top of the building from different sites, which are 75 m apart.
The surveyor's eye level is 189 cm above the ground.

   a. Determine two expressions for the height of the building above the surveyor's eye level, h, in terms of x using the two angles.
   b. Solve for x. Write your answer in metres correct to 2 decimal places.
   c. Determine the height of the building in metres correct to 2 decimal places.

8. A lookout tower has been erected on top of a cliff. At a distance of 5.8 km from the foot of the cliff, the angle of elevation to the base of the tower is 15.7° and to the observation deck at the top of the tower is 16° respectively, as shown in the figure below.
Determine how high from the top of the cliff the observation deck is, to the nearest metre.

9. Elena and Sonja were on a camping trip to the Grampians, where they spent their first day hiking. They first walked 1.5 km along a path inclined at an angle of 10° to the horizontal.
Then they had to follow another path, which was at an angle of 20° to the horizontal. They walked along this path for 1.3 km, which brought them to the edge of the cliff. Here Elena spotted a large gum tree 1.4 km away.
If the gum tree is 150 m high, calculate the angle of depression from the top of the cliff to the top of the gum tree. Express your answer in degrees correct to the nearest degree.

10. From a point on top of a cliff, two boats are observed. If the angles of depression are 58° and 32° and the cliff is 46 m above sea level, determine how far apart the boats are, in metres correct to 2 decimal places.

11. A 2.05-m tall man, standing in front of a street light 3.08 m high, casts a 1.5-m shadow.

   a. Calculate the angle of elevation, to the nearest degree, from the ground to the source of light.
   b. Determine how far the man is from the bottom of the light pole in metres correct to 2 decimal places.

Reasoning

12. Explain the difference between an angle of elevation and an angle of depression.

13. Joseph is asked to obtain an estimate of the height of his house using any mathematical technique. He decides to use a clinometer and basic trigonometry.
Using the clinometer, Joseph determines the angle of elevation, θ, from his eye level to the top of his house to be 42°.
The point from which Joseph measures the angle of elevation is 15 m away from his house and the distance from Joseph's eyes to the ground is 1.76 m.

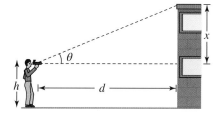

a. Determine the values for the pronumerals h, d and θ.
b. Determine the height of Joseph's house, in metres correct to 2 decimal places.

14. The angle of elevation of a vertically rising hot air balloon changes from 27° at 7:00 am to 61° at 7:03 am, according to an observer who is 300 m away from the take-off point.

a. Assuming a constant speed, calculate that speed (in m/s and km/h) at which the balloon is rising, correct to 2 decimal places.
b. The balloon then falls 120 metres. Determine the angle of elevation now.
Write your answer in degrees correct to 1 decimal place.

Problem solving

15. The competitors of a cross-country run are nearing the finish line. From a lookout 100 m above the track, the angles of depression to the two leaders, Nathan and Rachel, are 40° and 62° respectively.
Evaluate how far apart, to the nearest metre, the two competitors are.

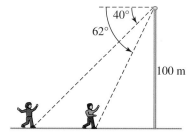

16. The angle of depression from the top of one building to the foot of another building across the same street and 45 metres horizontally away is 65°.
The angle of depression to the roof of the same building is 30°. Evaluate the height of the shorter building.
Write your answer in metres correct to 3 decimal places.

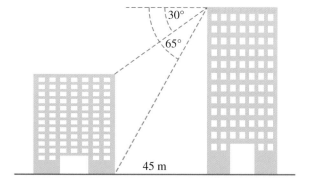

17. P and Q are two points on a horizontal line that are 120 metres apart. The angles of elevation from P and Q to the top of a mountain are 36° and 42° respectively.
Determine the height of the mountain, in metres, correct to 1 decimal place.

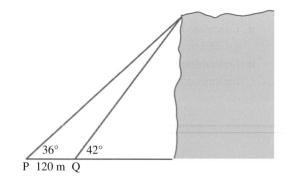

LESSON
5.8 Bearings

▶ 5.8.1 Using bearings

eles-4807

- A bearing gives the direction of travel from one point or object to another.
- The bearing of B from A tells how to get to B *from* A. A compass rose would be drawn at A.

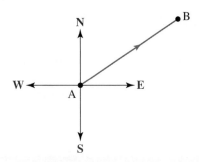

- To illustrate the bearing of A *from* B, a compass rose would be drawn at B.

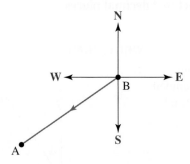

- There are two ways in which bearings are commonly written. They are compass bearings and true bearings.

Compass bearings

- A **compass bearing** (for example N40°E or S72°W) has three parts.
  - The first part is either N or S (for north or south).
  - The second part is an acute angle.
  - The third part is either E or W (for east or west).
  For example, the compass bearing S20°E means start by facing south and then turn 20° towards the east. This is the direction of travel.
  N40°W means start by facing north and then turn 40° towards the west.

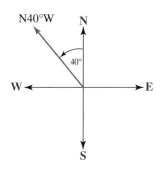

True bearings

- **True bearings** are measured from north in a clockwise direction and are expressed in 3 digits.
- The diagrams below show the bearings of 025° true and 250° true respectively.
  (These true bearings are more commonly written as 025°T and 250°T.)

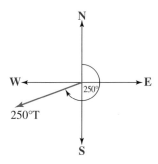

WORKED EXAMPLE 17 Solving trigonometric problems involving bearings

A boat travels a distance of 5 km from P to Q in a direction of 035°T.
a. Calculate how far east of P is Q, correct to 2 decimal places.
b. Calculate how far north of P is Q, correct to 2 decimal places.
c. Calculate the true bearing of P from Q.

THINK	WRITE/DRAW
a. 1. Draw a diagram showing the distance and bearing of Q from P. Complete a right-angled triangle travelling x km due east from P and then y km due north to Q.	a.
2. To determine how far Q is east of P, we need to determine the value of x. We are given the length of the hypotenuse (H) and need to find the length of the opposite side (O). Write the sine ratio.	$\sin(\theta) = \dfrac{O}{H}$
3. Substitute $O = x$, H = 5 and $\theta = 35°$.	$\sin(35°) = \dfrac{x}{5}$
4. Make x the subject of the equation.	$x = 5\sin(35°)$

5. Evaluate and round the answer, correct to 2 decimal places.

≈ 2.87

6. Write the answer in a sentence.

Point Q is 2.87 km east of P.

b. 1. To determine how far Q is north of P, we need to find the value of y. This can be done in several ways, namely: using the cosine ratio, the tangent ratio, or Pythagoras' theorem. Write the cosine ratio.

b. $\cos(\theta) = \dfrac{A}{H}$

2. Substitute A = y, H = 5 and $\theta = 35°$.

$\cos(35°) = \dfrac{y}{5}$

3. Make y the subject of the equation.

$y = 5 \cos(35°)$

4. Evaluate and round the answer, correct to 2 decimal places.

≈ 4.10

5. Write the answer in a sentence.

Point B is 4.10 km north of A.

c. 1. To determine the bearing of P from Q, draw a compass rose at Q. The true bearing is given by $\angle \theta$.

c.

2. The value of θ is the sum of 180° (from north to south) and 35°. Write the value of θ.

True bearing $= 180° + \alpha$
$\alpha = 35°$
True bearing $= 180° + 35°$
$= 215°$

3. Write the answer in a sentence.

The bearing of P from Q is 215°T.

5.8.2 Navigation problems involving bearings

eles-6140

- Sometimes a journey includes a change in directions. In such cases, each section of the journey should be dealt with separately.

WORKED EXAMPLE 18 Solving bearings problems with two legs

A boy walks 2 km on a true bearing of 090° and then 3 km on a true bearing of 130°.

a. Calculate how far east of the starting point the boy is at the completion of his walk, correct to 1 decimal place.

b. Calculate how far south of the starting point the boy is at the completion of his walk, correct to 1 decimal place.

c. To return directly to his starting point, calculate how far the boy must walk and on what bearing. Write your answers in km correct to 2 decimal places and in degrees and minutes correct to the nearest minute.

THINK

a. 1. Draw a diagram of the boy's journey. The first leg of the journey is due east. Label the easterly component x and the southerly component y.

2. Write the ratio to determine the value of x.

3. Substitute $O = x$, $H = 3$ and $\theta = 50°$.

4. Make x the subject of the equation.

5. Evaluate and round correct to 1 decimal place.

6. Add to this the 2 km east that was walked in the first leg of the journey and write the answer in a sentence.

b. 1. To determine the value of y (see the diagram in part **a**) we can use Pythagoras' theorem, as we know the lengths of two out of three sides in the right-angled triangle. Round the answer correct to 1 decimal place. *Note:* Alternatively, the cosine ratio could have been used.

2. Write the answer in a sentence.

c. 1. Draw a diagram of the journey and write in the results found in parts **a** and **b**. Draw a compass rose at Q.

2. Determine the value of z using Pythagoras' theorem.

3. Determine the value of α using trigonometry.

WRITE/DRAW

a.

$\sin(\theta) = \dfrac{O}{H}$

$\sin(50°) = \dfrac{x}{3}$

$x = 3\sin(50°)$

$\approx 2.3\,\text{km}$

Total distance east $= 2 + 2.3$
$= 4.3\,\text{km}$
The boy is 4.3 km east of the starting point.

b. Distance south $= y\,\text{km}$

$a^2 = c^2 - b^2$

$y^2 = 3^2 - 2.3^2$
$= 9 - 5.29$
$= 3.71$
$y = \sqrt{3.71}$
$= 1.9\,\text{km}$

The boy is 1.9 km south of the starting point.

c.

$z^2 = 1.9^2 + 4.3^2$
$= 22.1$
$z = \sqrt{22.1}$
≈ 4.70

$\tan(\alpha) = \dfrac{4.3}{1.9}$

4.	Make α the subject of the equation using the inverse tangent function.	$\alpha = \tan^{-1}\left(\dfrac{4.3}{1.9}\right)$
5.	Evaluate and round to the nearest minute.	$= 66.161\,259\,82°$ $= 66°9'40.535''$ $= 66°10'$
6.	The angle β gives the bearing.	$\beta = 360° - 66°10'$ $= 293°50'$
7.	Write the answer in a sentence.	The boy travels 4.70 km on a bearing of 293°50′ T.

DISCUSSION

Explain the difference between true bearings and compass directions.

Exercise 5.8 Bearings

learn on

5.8 Quick quiz on	5.8 Exercise

Individual pathways

■ PRACTISE	■ CONSOLIDATE	■ MASTER
1, 3, 7, 10, 14, 17	2, 5, 8, 11, 12, 15, 18	4, 6, 9, 13, 16, 19

Fluency

1. Change each of the following compass bearings to true bearings.
   a. N20°E
   b. N20°W
   c. S35°W

2. Change each of the following compass bearings to true bearings.
   a. S28°E
   b. N34°E
   c. S42°W

3. Change each of the following true bearings to compass bearings.
   a. 049°T
   b. 132°T
   c. 267°T

4. Change each of the following true bearings to compass bearings.
   a. 330°T
   b. 086°T
   c. 234°T

5. Describe the following paths using true bearings.

a.

b.

c.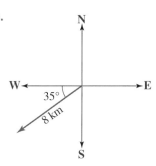

6. Describe the following paths using true bearings.

a.

b.

c.

7. Show each of the following journeys as a diagram.

a. A ship travels 040°T for 40 km and then 100°T for 30 km.

b. A plane flies for 230 km in a direction 135°T and a further 140 km in a direction 240°T.

8. Show each of the following journeys as a diagram.

a. A bushwalker travels in a direction 260°T for 0.8 km, then changes direction to 120°T for 1.3 km, and finally travels in a direction of 32° for 2.1 km.

b. A boat travels N40°W for 8 km, then changes direction to S30°W for 5 km and then S50°E for 7 km.

c. A plane travels N20°E for 320 km, N70°E for 180 km and S30°E for 220 km.

9. **WE17** A yacht travels 20 km from *A* to *B* on a bearing of 042°T.

a. Calculate how far east of A is B, in km correct to 2 decimal places.

b. Calculate how far north of A is B, in km correct to 2 decimal places.

c. Calculate the bearing of A from B.

The yacht then sails 80 km from B to C on a bearing of 130°T.

d. Show the journey using a diagram.

e. Calculate how far south of B is C, in km correct to 2 decimal places.

f. Calculate how far east of B is C, in km correct to 2 decimal places.

g. Calculate the bearing of B from C.

10. If a farmhouse is situated 220 m N35°E from a shed, calculate the true bearing of the shed from the house.

Understanding

11. A pair of hikers travel 0.7 km on a true bearing of 240° and then 1.3 km on a true bearing of 300°. Calculate how far west have they travelled from their starting point, in km correct to 3 decimal places.

12. 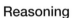 A boat travels 6 km on a true bearing of 120° and then 4 km on a true bearing of 080°.
   a. Calculate how far east the boat is from the starting point on the completion of its journey, in km correct to 3 decimal places.
   b. Calculate how far south the boat is from the starting point on the completion of its journey, in km correct to 3 decimal places.
   c. Calculate the bearing of the boat from the starting point on the completion of its journey, correct to the nearest minute.

13. A plane flies on a true bearing of 320° for 450 km. It then flies on a true bearing of 350° for 130 km and finally on a true bearing of 050° for 330 km.
   Calculate how far north of its starting point the plane is.
   Write your answer in km correct to 2 decimal places.

Reasoning

14. A bushwalker leaves her tent and walks due east for 4.12 km, then walks a further 3.31 km on a bearing of N20°E. If she wishes to return directly to her tent, determine how far she must walk and what bearing she should take.
   Write your answers in km correct to 2 decimal places and to the nearest degree.

15. A car travels due south for 3 km and then due east for 8 km. Determine the bearing of the car from its starting point, to the nearest degree. Show full working.

16. If the bearing of A from O is $\theta°$T, then (in terms of θ) determine the bearing of O from A:
   a. if $0° < \theta° < 180°$
   b. if $180° < \theta° < 360°$.

Problem solving

17. A boat sails on a compass direction of E12°S for 10 km then changes direction to S27°E for another 20 km.
   The pilot of the boat then decides to return to the starting point.
   a. Determine how far, correct to 2 decimal places, the boat is from its starting point.
   b. Determine on what bearing the boat should travel to return to its starting point. Write the angle correct to the nearest degree.

18. Samira and Tim set off early from the car park of a national park to hike for the day. Initially they walk N60°E for 12 km to see a spectacular waterfall.
   They then change direction and walk in a south-easterly direction for 6 km, then stop for lunch.
   Give all answers correct to 2 decimal places.
   a. Make a scale diagram of the hiking path they completed.
   b. Determine how far north of the car park they are at the lunch stop.
   c. Determine how far east of the car park they are at the lunch stop.
   d. Determine the bearing of the lunch stop from the car park.
   e. If Samira and Tim then walk directly back to the car park, calculate the distance they have covered after lunch.

19. Starting from their base in the national park, a group of bushwalkers travel 1.5 km at a true bearing of 030°, then 3.5 km at a true bearing of 160°, and then 6.25 km at a true bearing of 300°.

Evaluate how far, and at what true bearing, the group should walk to return to their base.

Write your answers in km correct to 2 decimal places and to the nearest degree.

LESSON
5.9 Applications

LEARNING INTENTIONS

At the end of this lesson you should be able to:
- draw well-labelled diagrams to represent information
- apply trigonometry to solve various problems involving triangles including bullet trajectory analysis..

▶ 5.9.1 Applications of trigonometry

eles-4808

- When applying trigonometry to practical situations, it is essential to draw good mathematical diagrams using points, lines and angles.
- Several diagrams may be required to show all the necessary right-angled triangles.

WORKED EXAMPLE 19 Applying trigonometry to solve problems

A ladder of length 3 m makes an angle of 32° with the wall.

a. Calculate how far the foot of the ladder is from the wall, in metres, correct to 2 decimal places.

b. Calculate how far up the wall the ladder reaches, in metres, correct to 2 decimal places.

c. Calculate the value of the angle the ladder makes with the ground.

THINK	WRITE/DRAW
Sketch a diagram and label the sides of the right-angled triangle with respect to the given angle.	

a. 1. We need to calculate the distance of the foot of the ladder from the wall (O) and are given the length of the ladder (H). Write the sine ratio.

    a. $\sin(\theta) = \dfrac{O}{H}$

   2. Substitute $O = x$, H = 3 and $\theta = 32°$.

       $\sin(32°) = \dfrac{x}{3}$

3. Make x the subject of the equation. $x = 3 \sin(32°)$

4. Evaluate and round the answer to $\approx 1.59\,\text{m}$
   2 decimal places.

5. Write the answer in a sentence.   The foot of the ladder is 1.59 m
                                             from the wall.

b. 1. We need to calculate the height the ladder   **b.** $\cos(\theta) = \dfrac{A}{H}$
      reaches up the wall (A) and are given the
      hypotenuse (H). Write the cosine ratio.

   2. Substitute A = y, H = 3 and $\theta = 32°$. $\cos(32°) = \dfrac{y}{3}$

   3. Make y the subject of the equation. $y = 3 \cos(32°)$

   4. Evaluate and round the answer to $y \approx 2.54\,\text{m}$
      2 decimal places.

   5. Write the answer in a sentence.   The ladder reaches 2.54 m up
                                                the wall.

c. 1. To calculate the angle that the ladder   **c.** $\alpha + 90° + 32° = 180°$
      makes with the ground, we could use any $\alpha + 122° = 180°$
      of the trigonometric ratios, as the lengths $\alpha = 180° - 122°$
      of all three sides are known. However, it is $\alpha = 58°$
      quicker to use the angle sum of a triangle.

   2. Write the answer in a sentence.   The ladder makes a 58° angle with
                                                the ground.

TI | THINK	DISPLAY/WRITE	CASIO | THINK	DISPLAY/WRITE
On a Calculator page, complete the entry lines as: $\text{solve}\left(\sin(32) = \dfrac{x}{3}, x\right)$ $\text{solve}\left(\cos(32) = \dfrac{y}{3}, y\right)$ $180 - (90 + 32)$ Press ENTER after each entry.	$x = 1.59$ m correct to 2 decimal places. $y = 2.54$ m correct to 2 decimal places. $\alpha = 58°$	On the Main screen, complete the entry lines as: $\text{solve}\left(\sin(32) = \dfrac{x}{3}, x\right)$ $\text{solve}\left(\cos(32) = \dfrac{y}{3}, y\right)$ $180 - (90 + 32)$ Press EXE after each entry.	$x = 1.59$ m correct to 2 decimal places. $y = 2.54$ m correct to 2 decimal places. $\alpha = 58°$

▶ 5.9.2 Bullet trajectory analysis

eles-6141

- Pythagoras' theorem and trigonometry can be used to model a crime scene.
- Trigonometry could be used to determine the trajectory of a bullet.
- Ballistics experts use these calculations to determine the impact angle and location of the shooter.

Police are investigating a shooting incident that resulted in a shot being fired that lodged in a wall. The height of the suspect is 190 cm, with a shoulder height of 160 cm. If the bullet hole is located at 250 cm from the floor and the bullet hole makes an angle of 50° with the wall (the impact angle), determine the distance of the shooter from the wall.

THINK	WRITE
1. Draw the diagram.	
2. As we know two angles in a right-angled triangle, we can calculate the value of the third angle.	Angle $c = 180° - 90° - 50°$ $\quad\quad\ = 40°$
3. Determine the length of the segment 'y'. This is the distance from the bullet hole to the shooter's shoulder height.	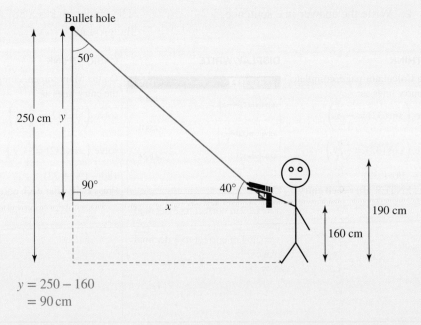 $y = 250 - 160$ $\quad = 90$ cm
4. Apply the tangent formula to determine the distance of the shooter from the wall.	$\tan 40° = \dfrac{90}{x}$ $x = \dfrac{90}{\tan 40°}$ $\quad = 107.26$ cm
5. Write the answer.	The distance of the shooter from the wall is 107.26 cm.

on Resources

eWorkbook Topic 5 Workbook (worksheets, code puzzle and project) (ewbk-13288)

Interactivity Individual pathway interactivity: Applications (int-4592)

Exercise 5.9 Applications

learn**on**

| 5.9 Quick quiz **on** | 5.9 Exercise |

Individual pathways

■ PRACTISE	■ CONSOLIDATE	■ MASTER
1, 3, 4, 7, 12, 13, 16, 19	2, 5, 8, 10, 14, 17, 20	6, 9, 11, 15, 18, 21

Fluency

1. A carpenter wants to make a roof pitched at 29°30′, as shown in the diagram. Calculate how long, in metres correct to 2 decimal places, he should cut the beam PR.

2. The mast of a boat is 7.7 m high. A guy wire from the top of the mast is fixed to the deck 4 m from the base of the mast.
   Determine the angle, correct to the nearest minute, that the wire makes with the horizontal.

Understanding

3. A steel roof truss is to be made to the following design.
   Write your answers in metres correct to 2 decimal places.

   a. Calculate how high the truss is.
   b. Determine the total length of steel required to make the truss.

4. **WE19** A ladder that is 2.7 m long is leaning against a wall at an angle of 20° as shown. If the base of the ladder is moved 50 cm further away from the wall, determine what angle the ladder will make with the wall. Write your answer correct to the nearest minute.

5. A wooden framework is built as shown.
   Bella plans to reinforce the framework by adding a strut from C to the midpoint of AB.
   Calculate the length of the strut in metres correct to 2 decimal places.

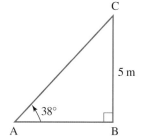

6. Atlanta is standing due south of a 20 m flagpole at a point where the angle of elevation of the top of the pole is 35°. Ginger is standing due east of the flagpole at a point where the angle of elevation of the top of the pole is 27°.
   Calculate how far, to the nearest metre, Ginger is from Atlanta.

7. **WE20** Police are investigating a shooting incident that resulted in a shot being fired that lodged in a wall. The height of the suspect is 175 cm, with a shoulder height of 155 cm. If the bullet hole is located at 350 cm from the floor and the bullet hole makes an angle of 55° with the wall (the impact angle), determine the distance of the shooter from the wall.

8. A sniper is aiming at a target that is located on top of a building. The sniper is located on the ground, 500 meters away from the building. The height of the building is 40 meters. The sniper's rifle is elevated at an angle of 30 degrees above the horizontal.
   Using Pythagoras' theorem and trigonometry, determine the height of the target above the ground.
   Also assume that the bullet travels in a straight line and does not experience any air resistance.

9. From a point at ground level, Henry measures the angle of elevation of the top of a tall building to be 41°. After walking directly towards the building, he finds the angle of elevation to be 75°. If the building is 220 m tall, determine how far Henry walked between measurements.
   Write your answer correct to the nearest metre.

10. Sailing in the direction of a mountain peak of height 893 m, Imogen measured the angle of elevation to be 14°. A short time later the angle of elevation was 27°.
    Calculate how far, in km correct to 3 decimal places, Imogen had sailed in that time.

11. A desk top of length 1.2 m and width 0.5 m rises to 10 cm.

    Calculate, correct to the nearest minute:
    a. ∠DBF
    b. ∠CBE.

12. A cuboid has a square end. If the length of the cuboid is 45 cm and its height and width are 25 cm each, calculate:

a. the length of BD, correct to 2 decimal places
b. the length of BG, correct to 2 decimal places
c. the length of BE, correct to 2 decimal places
d. the length of BH, correct to 2 decimal places
e. ∠FBG, correct to the nearest minute
f. ∠EBH, correct to the nearest minute.

If the midpoint of FG is X and the centre of the rectangle ABFE is O, calculate:

g. the length OF, correct to 2 decimal places
h. the length FX, correct to 1 decimal place
i. ∠FOX, correct to the nearest minute
j. the length OX, correct to 2 decimal places.

13. In a right square-based pyramid, the length of the side of the base is 12 cm and the height is 26 cm.

Determine:

a. the angle the triangular face makes with the base, correct to the nearest degree
b. the angle the sloping edge makes with the base, correct to the nearest minute
c. the length of the sloping edge, in cm correct to 2 decimal places.

14. In a right square-based pyramid, the length of the side of the square base is 5.7 cm.

If the angle between the triangular face and the base is 68°, calculate:

a. the height of the pyramid, in cm correct to 2 decimal places
b. the angle the sloping edge makes with the base, correct to the nearest minute
c. the length of the sloping edge, in cm correct to 2 decimal places.

15. In a right square-based pyramid, the height is 47 cm. If the angle between a triangular face and the base is 73°, calculate:

a. the length of the side of the square base, in cm correct to 2 decimal places
b. the length of the diagonal of the base, in cm correct to 2 decimal places
c. the angle the sloping edge makes with the base, correct to the nearest minute.

Reasoning

16. Explain whether sine of an acute angle can be 1 or greater.

17. A block of cheese is in the shape of a rectangular prism as shown. The cheese is to be sliced with a wide blade that can slice it in one go.

    Calculate the angle (to the vertical, correct to 2 decimal places) that the blade must be inclined if:

    a. the block is to be sliced diagonally into two identical triangular wedges

    b. the blade is to be placed in the middle of the block and sliced through to the bottom corner, as shown.

18. Aldo the carpenter is lost in a rainforest. He comes across a large river and he knows that he can not swim across it. Aldo intends to build a bridge across the river.

    He draws some plans to calculate the distance across the river as shown in the diagram below.

    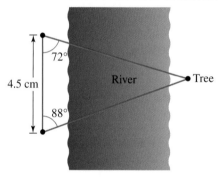

    a. Aldo used a scale of 1 cm to represent 20 m. Determine the real-life distance represented by 4.5 cm in Aldo's plans.

    b. Use the diagram below to write an equation for h in terms of d and the two angles.

    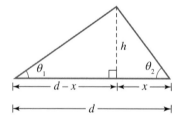

    c. Use your equation from part b to find the distance across the river, correct to the nearest metre.

Problem solving

19. A ship travels north for 7 km, then on a true bearing of 140° for another 13 km.
    a. Draw a sketch of the situation.
    b. Determine how far south the ship is from its starting point, in km correct to 2 decimal places.
    c. Evaluate the bearing, correct to the nearest degree, the ship is now from its starting point.

20. The ninth hole on a municipal golf course is 630 m from the tee. A golfer drives a ball from the tee a distance of 315 m at a 10° angle off the direct line as shown.

Determine how far the ball is from the hole and state the angle of the direct line that the ball must be hit along to go directly to the hole.
Give your answers correct to 1 decimal place.

21. A sphere of radius length 2.5 cm rests in a hollow inverted cone as shown. The height of the cone is 12.5 cm and its vertical angle is equal to 36°.

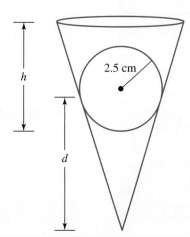

    a. Evaluate the distance, d, from the tip of the cone to the point of contact with the sphere, correct to 2 decimal places.
    b. Determine the distance, h, from the open end of the cone to the bottom of the ball, correct to 2 decimal places.

LESSON
5.10 Review

5.10.1 Topic summary

Similar triangles

- When triangles have common angles, they are said to be similar.
  e.g. Triangles OAH, OBG and ODE are similar.

- Corresponding sides of similar triangles will have the same ratio.

  e.g. $\dfrac{FC}{OC} = \dfrac{ED}{OD} = \dfrac{HA}{OA}$

Pythagoras' theorem

- When the length of two sides are known in a right-angled triangle, the third side can be found using the rule $a^2 + b^2 = c^2$.

- Length of the longest side $c = \sqrt{a^2 + b^2}$
- Length of the shorter sides $a = \sqrt{c^2 - b^2}$ or $b = \sqrt{c^2 - a^2}$

TRIGONOMETRY I

Trigonometric ratios (SOHCAHTOA)

- In a right-angled triangle, the longest side is called the hypotenuse.

- If an acute angle is known, then the trigonometric ratios can be defined as:

  $$\sin\theta = \dfrac{O}{H}, \; \cos\theta = \dfrac{A}{H}, \; \tan\theta = \dfrac{O}{A}$$

- An acute angle can be calculated when two sides are known using the inverse operation of the correct trigonometric ratio.
  e.g. Since $\sin(30°) = 0.5$, then $\sin^{-1}(0.5) = 30°$; this is read as 'inverse sine of 0.5 is 30°'.

Angles of elevation and depression

- If a horizontal line is drawn from A as shown, forming the angle θ, then θ is called the **angle of elevation** of B *from* A.

- If a horizontal line is drawn from B as shown, forming the angle α, then α is called the **angle of depression** of A *from* B.

Bearings

There are two ways in which bearings can be written:
- **Compass bearings** have 3 parts:
  - The first part is either N or S (for north or south).
  - The second part is an acute angle.
  - The third part is either E or W (for east or west).
  e.g. S20°E means start by facing south and then turn 20° towards the east.

- **True bearings** are measured from north in a clockwise direction and are expressed in 3 digits.

5.10.2 Success criteria

Tick the column to indicate that you have completed the lesson and how well you have understood it using the traffic light system.

(**Green:** I understand; **Yellow:** I can do it with help; **Red:** I do not understand)

Lesson	Success criteria	⬤	◯	⬤
5.2	I can identify similar right-angled triangles when corresponding sides are in the same ratio and corresponding angles are congruent.			
	I can apply Pythagoras' theorem to calculate the third side of a right-angled triangle when two other sides are known.			
5.3	I can apply Pythagoras' theorem to determine unknown lengths when a 3D diagram is given.			
	I can apply Pythagoras' theorem to determine unknown lengths in situations by first drawing a diagram.			
5.4	I can define trigonometric ratios according to the lengths of the relevant sides.			
	I can write equations for trigonometric ratios.			
5.5	I can apply trigonometric ratios to calculate the length of an unknown side when the length of one other side and an acute angle is known.			
5.6	I can apply inverse operations to calculate a known acute angle when two sides are given.			
5.7	I can identify angles of elevation and depression and solve for unknown side lengths and angles.			
5.8	I can draw diagrams with correct angles to represent information to help solve triangles.			
	I can apply trigonometry to solve bearing problems involving compass bearings and true bearings.			
5.9	I can draw well-labelled diagrams to represent information.			
	I can apply trigonometry to solve various problems involving triangles including bullet trajectory analysis.			

5.10.3 Project

How steep is the land?

When buying a block of land on which to build a house, the slope of the land is often not very obvious. The slab of a house built on the ground must be level, so it is frequently necessary to remove or build up soil to obtain a flat area. The gradient of the land can be determined from a contour map of the area.

Consider the building block shown. The contour lines join points having the same height above sea level. Their measurements are in metres. The plan clearly shows that the land rises from A to B. The task is to determine the angle of this slope.

1. A cross-section shows a profile of the surface of the ground. Let us look at the cross-section of the ground between A and B. The technique used is as follows.
    • Place the edge of a piece of paper on the line joining A and B.
    • Mark the edge of the paper at the points where the contour lines intersect the paper.
    • Transfer this paper edge to the horizontal scale of the profile and mark these points.
    • Choose a vertical scale within the range of the heights of the contour lines.
    • Plot the height at each point where a contour line crosses the paper.
    • Join the points with a smooth curve.

Scale 1 : 500

The cross-section has been started for you. Complete the profile of the line AB. You can now see a visual picture of the profile of the soil between A and B.

Cross-section of AB

Profile of line BA (metres)

2. We now need to determine the horizontal distance between A and B.
   a. Measure the map distance between A and B using a ruler. What is the map length?
   b. Using the scale of 1 : 500, calculate the actual horizontal distance AB (in metres).
3. The vertical difference in height between A and B is indicated by the contour lines. Calculate this vertical distance.
4. Complete the measurements on this diagram.

Vertical distance = ... m

Horizontal distance = ... m

5. The angle a represents the angle of the average slope of the land from A to B. Use the tangent ratio to calculate this angle (to the nearest minute).
6. In general terms, an angle less than 5° can be considered a gradual to moderate rise. An angle between 5° and 15° is regarded as moderate to steep while more than 15° is a steep rise. How would you describe this block of land?
7. Imagine that you are going on a bushwalk this weekend with a group of friends. A contour map of the area is shown. Starting at X, the plan is to walk directly to the hut.
   Draw a cross-section profile of the walk and calculate the average slope of the land. How would you describe the walk?

Scale 1 : 20 000

Fluency

1. **MC** The most accurate measurement for the length of the third side in the triangle is:
   A. 483 m
   B. 23.3 cm
   C. 3.94 m
   D. 4826 mm

2. **MC** The value of x in this figure is:
   A. 5.4
   B. 7.5
   C. 10.1
   D. 10.3

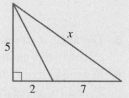

3. **MC** Select the closest length of AG of the cube.
   A. 10
   B. 30
   C. 20
   D. 17

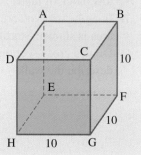

4. **MC** If $\sin(38°) = 0.6157$, identify which of the following will also give this result.
   A. $\sin(218°)$
   B. $\sin(322°)$
   C. $\sin(578°)$
   D. $\sin(142°)$

5. **MC** The angle $118°52'34''$ is also equal to:
   A. $118.5234°$

   B. $118\dfrac{52}{34}°$

   C. $118.861°$
   D. $118.876°$

6. Identify which trigonometric ratio for the triangle shown below is incorrect.

A. $\sin(\alpha) = \dfrac{b}{c}$

B. $\sin(\alpha) = \dfrac{a}{c}$

C. $\cos(\alpha) = \dfrac{a}{c}$

D. $\tan(\alpha) = \dfrac{b}{a}$

7. Identify which of the following statements is correct.
A. $\sin(55°) = \cos(55°)$
B. $\sin(45°) = \cos(35°)$
C. $\cos(15°) = \sin(85°)$
D. $\sin(42°) = \cos(48°)$

8. Identify which of the following can be used to determine the value of x
in the diagram.
A. $28.7 \sin(35°)$
B. $28.7 \cos(35°)$
C. $28.7 \tan(35°)$

D. $\dfrac{28.7}{\sin(35°)}$

9. Identify which of the following expressions can be used to determine the value of a in the triangle shown.

A. $35 \sin(75°)$

B. $\sin^{-1}\left(\dfrac{35}{75}\right)$

C. $\sin^{-1}\left(\dfrac{75}{35}\right)$

D. $\cos^{-1}\left(\dfrac{35}{75}\right)$

10. **MC** If a school is 320 m S42°W from the police station, calculate the true bearing of the police station from the school.
    A. 042°T
    B. 048°T
    C. 222°T
    D. 228°T

11. Calculate x, correct to 2 decimal places.

    a.

    b.

12. Calculate the value of the pronumeral, correct to 2 decimal places.

13. Calculate the height of this pyramid, in mm correct to 2 decimal places.

14. A person standing 23 m away from a tree observes the top of the tree at an angle of elevation of 35°. If the person's eye level is 1.5 m from the ground, calculate the height of the tree, in metres correct to 1 decimal place.

15. A man with an eye level height of 1.8 m stands at the window of a tall building. He observes his young daughter in the playground below. If the angle of depression from the man to the girl is 47° and the floor on which the man stands is 27 m above the ground, determine how far from the bottom of the building the child is, in metres correct to 2 decimal places.

16. A plane flies 780 km in a direction of 185°T. Evaluate how far west it has travelled from the starting point, in km correct to 2 decimal places.

17. A hiker travels 3.2 km on a bearing of 250°T and then 1.8 km on a bearing of 320°T. Calculate how far west she has travelled from the starting point, in km correct to 2 decimal places.

18. If a 4 m ladder is placed against a wall and the foot of the ladder is 2.6 m from the wall, determine the angle (in degrees and minutes, correct to the nearest minute) the ladder makes with the wall.

Problem solving

19. The height of a right square-based pyramid is 13 cm. If the angle the face makes with the base is 67°, determine:
    a. the length of the edge of the square base, in cm correct to 2 decimal places
    b. the length of the diagonal of the base, in cm correct to 2 decimal places
    c. the angle the slanted edge makes with the base in degrees and minutes, correct to the nearest minute.

20. A car is travelling northwards on an elevated expressway 6 m above ground at a speed of 72 km/h. At noon another car passes under the expressway, at ground level, travelling west, at a speed of 90 km/h.
    a. Determine how far apart, in metres, the two cars are 40 seconds after noon, in metres correct to 2 decimal places.
    b. At this time the first car stops, while the second car keeps going. Determine the time when they will be 3.5 km apart. Write your answer correct to the nearest tenth of a second.

21. Two towers face each other separated by a distance, d, of 20 metres. As seen from the top of the first tower, the angle of depression of the second tower's base is 59° and that of the top is 31°. Calculate the height, in metres correct to 2 decimal places, of each of the towers.

22. A piece of flat pastry is cut in the shape of a right-angled triangle. The longest side is $6b$ cm and the shortest is $2b$ cm.
    a. Determine the length of the third side. Give your answer in exact form.
    b. Determine the sizes of the angles in the triangle.
    c. Show that the area of the triangle is equal to $4\sqrt{2}b^2$ cm².

23. A yacht is anchored off an island. It is 2.3 km from the yacht club and 4.6 km from a weather station. The three points form a right-angled triangle at the yacht club.

    a. Calculate the angle at the yacht between the yacht club and the weather station.
    b. Evaluate the distance between the yacht club and the weather station, in km correct to 2 decimal places.
    The next day the yacht travels directly towards the yacht club but is prevented from reaching the club because of dense fog. The weather station notifies the yacht that it is now 4.2 km from the station.
    c. Calculate the new angle at the yacht between the yacht club and the weather station, in degrees correct to 1 decimal place.
    d. Determine how far the yacht is now from the yacht club, correct to 2 decimal places.

on To test your understanding and knowledge of this topic, go to your learnON title at www.jacplus.com.au and complete the **post-test**.

Answers

Topic 5 Trigonometry I

5.1 Pre-test

1. $w = 6.89\,\text{cm}$
2. $x = 2.24\,\text{cm}$
3. $16.97\,\text{cm}$
4. $62.28\,\text{cm}$
5. 0.3521
6. $\theta = 55°10'$
7. $\theta = 36°52'12''$
8. $y = 6.7\,\text{m}$
9. C
10. a. $020°\text{T}$ b. $227°\text{T}$ c. $327°\text{T}$ d. $163°\text{T}$
11. A
12. D
13. $551.2\,\text{m}$
14. D
15. $x = 9.0$

5.2 Pythagoras' theorem

1. a. 7.86 b. 33.27 c. 980.95
2. a. 12.68 b. 2.85 c. 175.14
3. a. 36.36 b. 1.62 c. 15.37
4. a. 0.61 b. 2133.19 c. 453.90
5. $23.04\,\text{cm}$
6. $12.65\,\text{cm}$
7. a. $14.14\,\text{cm}$ b. $24.04\,\text{cm}$ c. $4.53\,\text{cm}$
8. a. $74.83\,\text{cm}$ b. $249.67\,\text{cm}$ c. $3741.66\,\text{cm}^2$
9. a. 6.06 b. 4.24 c. 4.74
10. $14.84\,\text{cm}$
11. $15.59\,\text{cm}$
12. $19.23\,\text{cm}$
13. $72.75\,\text{cm}$; $3055.34\,\text{cm}^2$
14. $39\,\text{m}$
15. $4.34\,\text{km}$
16. $38.2\,\text{m}$
17. $63.06\,\text{mm}$
18. a. $32\,\text{cm}$ b. $768\,\text{cm}^2$
19. 26.83 diagonals, so would need to complete 27
20. $4701.06\,\text{m}$
21. $9.90\,\text{cm}$
22. a. 65 b. 185 c. 305
23. The value found using Pythagoras' theorem represents length and therefore can't be negative.
24. a. Neither 105 nor 208 can be the hypotenuse of the triangle, because they are the two smallest values. The other two values could be the hypotenuse if they enable the creation of a right-angled triangle.
    b. $105, 208, 233$
25. a. $21\,\text{cm}$
    b. $35\,\text{cm}$
    c. $y = 12.6\,\text{cm}$ and $RS = 9.8\,\text{cm}$
26. $13.86\,\text{cm}$
27. 1.33 million km

5.3 Pythagoras' theorem in three dimensions (10A)

1. a. 13.86 b. 13.93 c. 18.03
2. $12.21, 12.85$
3. $4.84\,\text{m}, 1.77\,\text{m}$
4. $8.49, 4.24$
5. $31.62\,\text{cm}$
6. $10.58\,\text{cm}$
7. $23\,\text{mm}$
8. a. i. $233.24\,\text{m}$ ii. $200.12\,\text{m}$ iii. $120.20\,\text{m}$
   b. $116.83\,\text{m}$
9. $14.72\,\text{cm}$
10. $12.67\,\text{cm}$
11. $42.27\,\text{cm}$
12. Sample responses can be found in the worked solutions in the online resources.
13. $186.5\,\text{m}$
14. a. $1.49\,\text{m}$ b. 7.43m^2
15. $\sqrt{1265}\,\text{cm}$
16. 25.475
17. $28.6\,\text{m}$

5.4 Trigonometric ratios

1. a. 0.5000 b. 0.7071 c. 0.4663
   d. 0.8387 e. 8.1443 f. 0.7193
2. a. 0.6494 b. 0.5885 c. 0.5220
   d. -1.5013 e. 0.9990 f. 0.6709
3. a. 0.8120 b. 0.5253 c. -0.8031
   d. 0.4063 e. 0.9880
4. a. -0.9613 b. 1.7321 c. -0.5736 d. 0.1320
5. a. $50°$ b. $24°$ c. $53°$
6. a. $71°$ b. $86°$ c. $41°$
7. a. $54°29'$ b. $6°19'$ c. $0°52'$
8. a. $72°47'$ b. $44°48'$ c. $26°45'$
9. a. $26°33'54''$ b. $64°1'25''$ c. $64°46'59''$
10. a. $48°5'22''$ b. $36°52'12''$ c. $88°41'27''$
11. a. 2.824 b. 71.014 c. 20.361 d. 2.828
12. a. 226.735 b. 1.192 c. 7.232 d. 32.259
13. a. 4909.913 b. 0.063 c. 0.904 d. 14.814
14. a. $\sin(\theta) = \dfrac{e}{f}$ b. $\cos(\theta) = \dfrac{d}{f}$ c. $\tan(\theta) = \dfrac{e}{d}$
15. a. $\sin(\alpha) = \dfrac{i}{g}$ b. $\cos(\alpha) = \dfrac{h}{g}$ c. $\tan(\alpha) = \dfrac{i}{h}$

16. a. $\sin(\beta) = \dfrac{1}{k}$ b. $\cos(\beta) = \dfrac{j}{k}$ c. $\tan(\beta) = \dfrac{1}{j}$

17. a. $\sin(\gamma) = \dfrac{n}{m}$ b. $\cos(\gamma) = \dfrac{o}{m}$ c. $\tan(\gamma) = \dfrac{n}{o}$

18. a. $\sin(\beta) = \dfrac{b}{c}$ b. $\cos(\beta) = \dfrac{a}{c}$ c. $\tan(\beta) = \dfrac{b}{a}$

19. a. $\sin(\gamma) = \dfrac{v}{u}$ b. $\cos(\gamma) = \dfrac{t}{u}$ c. $\tan(\gamma) = \dfrac{v}{t}$

20. a. $\sin(\theta) = \dfrac{15}{18}$ b. $\cos(\theta) = \dfrac{22}{30}$ c. $\tan(\theta) = \dfrac{7}{9}$

21. a. $\tan(\theta) = \dfrac{3.6}{p}$ b. $\sin(25°) = \dfrac{13}{t}$ c. $\sin(\alpha) = \dfrac{18.6}{23.5}$

22. a.

b. i. $\sin(37°) = 0.60$ ii. $\cos(37°) = 0.80$
    iii. $\tan(37°) = 0.75$

c. $\alpha = 53°$

23. a. i. $\sin(53°) = 0.80$ ii. $\cos(53°) = 0.60$
    iii. $\tan(53°) = 1.33$

b. They are equal.

c. They are equal.

d. The sine of an angle is equal to the cosine of its complement angle.

24. $\sin(\theta) = \dfrac{\text{opp}}{\text{hyp}}, \cos(\theta) = \dfrac{\text{adj}}{\text{hyp}} \Rightarrow \dfrac{\sin(\theta)}{\cos(\theta)} = \dfrac{\text{opp}}{\text{adj}} = \tan(\theta)$

25. a. $h^2 = a^2 - x^2$
    b. $h^2 = c^2 - b^2 + 2bx - x^2$
    c. Sample responses can be found in the worked solutions in the online resources.
    d. Sample responses can be found in the worked solutions in the online resources.

26. $DC = x + \dfrac{y}{\tan(\theta)}$

27. To determine which trigonometric ratio to apply, the sides in relation to the angle relevant in the question need to be identified and named.

28. $110\tan(\theta + \alpha) - 110\tan(\alpha)$

5.5 Using trigonometry to calculate side lengths

1. a. 0.79 b. 4.72 c. 101.38
2. a. 8.660 b. 7.250 c. 8.412
3. a. 33.45 m b. 74.89 m c. 44.82 m
4. a. 7.76 mm b. 80.82 km c. 9.04 cm
5. a. $x = 31.58$ cm b. $y = 17.67$ m c. $z = 14.87$ m
6. a. $p = 67.00$ m
   b. $p = 21.38$ km, $q = 42.29$ km
   c. $a = 0.70$ km, $b = 0.21$ km
7. a. 6.0 m b. 6.7 m
8. 1.05 m

9. a. $x = 30.91$ cm, $y = 29.86$ cm, $z = 39.30$ cm
   b. 2941.54 cm^2
10. a. In an isosceles right-angled triangle
    b. $\theta < 45°$
11. Sample responses can be found in the worked solutions in the online resources.
12. a. $h = \tan(47°48')x$ m
       $h = \tan(36°24')(x + 64)$ m
    b. 129.07 m
    c. 144.29 m
13. Building A is 73.9 m and Building B is 29.5 m.
14. 60
15. Sample responses can be found in the worked solutions in the online resources.

5.6 Using trigonometry to calculate angle size

1. a. 67° b. 47° c. 69°
2. a. 54°47′ b. 33°45′ c. 33°33′
3. a. 75°31′21″ b. 36°52′12″ c. 37°38′51″
4. a. 41° b. 30° c. 49°
5. a. 65° b. 48° c. 37°
6. a. $a = 25°47'$, $b = 64°13'$ b. $d = 25°23'$, $e = 64°37'$
   c. $x = 66°12'$, $y = 23°48'$
7. a. $r = 57.58$, $l = 34.87$, $h = 28.56$
   b. 428 cm^2
   c. 29.7°
8. a. i. 29.0° ii. 41.4° iii. 51.3°
   b. i. 124.42 km/h ii. 136.57 km/h iii. 146.27 km/h
9. To determine the size of acute angles, use inverse operations.
10. Sample responses can be found in the worked solutions in the online resources.
11. a. $\sin(30°) = \dfrac{1}{2}$, $\cos(30°) = \dfrac{\sqrt{3}}{2}$, $\tan(30°) = \dfrac{\sqrt{3}}{3}$
    b. Sample responses can be found in the worked solutions in the online resources.
    c. Sample responses can be found in the worked solutions in the online resources.
12. 58°3′
13. a. Between site 3 and site 2: 61°
       Between site 2 and site 1: 18°
       Between site 1 and bottom: 75°
    b. Between site 1 and bottom: 75° slope
14. 147°0′; 12°15′

5.7 Angles of elevation and depression

1. 8.74 m
2. 687.7 m
3. a. 176.42 m b. 152.42 m
4. 65°46′
5. 16.04 m

6. a. $h = x \tan(47°12')$ m; $h = (x + 38) \tan(35°50')$ m
   b. $x = 76.69$ m
   c. 84.62 m
7. a. $h = x \tan(43°35')$ m; $h = (x + 75) \tan(32°18')$ m
   b. 148.37 m
   c. 143.10 m
8. 0.033 km or 33 m
9. 22°
10. 44.88 m
11. a. 54°   b. 0.75 m
12. Angle of elevation is an angle measured upwards from the horizontal. Angle of depression is measured from the horizontal downwards.
13. a.

   b. 15.27 m
14. a. 2.16 m/s, 7.77 km/h   b. 54.5°
15. 66 m
16. 70.522 m
17. 451.5 m

5.8 Bearings

1. a. 020°T   b. 340°T   c. 215°T
2. a. 152°T   b. 034°T   c. 222°T
3. a. N49°E   b. S48°E   c. S87°W
4. a. N30°W   b. N86°E   c. S54°W
5. a. 3 km 325°T   b. 2.5 km 112°T   c. 8 km 235°T
6. a. 4 km 090°T, then 2.5 km 035°T
   b. 12 km 115°T, then 7 km 050°T
   c. 300 m 310°T, then 500 m 220°T
7. a.

   b.

8. a.

   b.

   c.

9. a. 13.38 km
   b. 14.86 km
   c. 222°T
   d.

   e. 51.42 km
   f. 61.28 km
   g. 310°T
10. 215°T
11. 1.732 km
12. a. 9.135 km   b. 2.305 km   c. 104°10'T
13. 684.86 km
14. 6.10 km and 239°T
15. 111°T
16. a. $(180 + \theta)$ °T   b. $(\theta - 180)$ °T
17. a. 27.42 km   b. N43°W or 317°T
18. a.

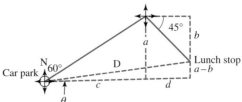

   b. 1.76 km North
   c. 14.63 km East
   d. N83.15°E
   e. D = 14.74 km
19. 3.65 km on a bearing of 108°T

5.9 Applications

1. 6.09 m
2. 62°33′
3. a. 1.82 m   b. 27.78 m
4. 31°49′
5. 5.94 m
6. 49 m
7. 278.49 cm
8. 288.68 m
9. 194 m
10. 1.829 km
11. a. 11°32′   b. 4°25′
12. a. 35.36 cm   b. 51.48 cm   c. 51.48 cm
    d. 57.23 cm   e. 29°3′   f. 25°54′
    g. 25.74 cm   h. 12.5 cm   i. 25°54′
    j. 28.61 cm
13. a. 77°   b. 71°56′   c. 27.35 cm
14. a. 7.05 cm   b. 60°15′   c. 8.12 cm
15. a. 28.74 cm   b. 40.64 cm   c. 66°37′
16. $\sin(\theta) = \dfrac{O}{H}$. Since the hypotenuse H is the longest side in the right-angled triangle, when dividing O by H the value will be between 0 and 1.
17. a. 122.97°   b. 142.37°
18. a. 90 m
    b. $h = \dfrac{d\tan(\theta_1)}{\tan(\theta_1) + \tan(\theta_2)} \times \tan(\theta_2)$
    c. 250 m
19. a.

    b. 2.96 km
    c. 110°
20. The golfer must hit the ball 324.4 m at an angle of 9.7° off the direct line.
21. a. 7.69 cm   b. 6.91 cm

Project

1.

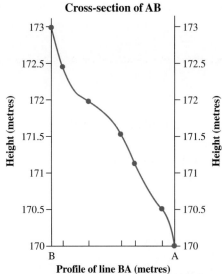

Cross-section of AB

Profile of line BA (metres)

2. a. 8 cm   b. 40 m
3. 3 m
4.

5. $a = 4°17′$
6. Gradual to moderate
7.

Cross-section X to hut

Profile of X to hut

The average slope is 11.46° — moderate to steep.

5.10 Review questions

1. D
2. D
3. D
4. D
5. D
6. B
7. D
8. B
9. B
10. A

11. a. $x = 113.06$ cm   b. $x = 83.46$ mm

12. 9.48 cm

13. 8.25 mm

14. 17.6 m

15. 26.86 m

16. 67.98 km

17. 4.16 km

18. $40°32'$

19. a. 11.04 cm   b. 15.61 cm   c. $59°1'$

20. a. 1280.64 m

    b. 2 minutes, 16.3 seconds past 12 pm

21. 33.29 m, 21.27 m

22. a. $4\sqrt{2}b$

    b. $19.5°$, $70.5°$, $90°$.

    c. Area $= \dfrac{1}{2}$ base $\times$ height

    $= \dfrac{1}{2} \times 2b \times 4\sqrt{2}b$

    $= 4\sqrt{2}b^2$ cm^2

23. a. $60°$ b. 3.98 km   c. $71.5°$ d. 1.33 km

6 Surface area and volume

LESSON
6.1 Overview

Why learn this?

People must measure! How much extra freezer space is required for the amount of ice cream on your birthday party? How many litres of water will it take to fill the new pool? These are just a few examples where measurement skills are needed.

Measuring tools have advanced significantly in their capability to measure extremely small and extremely large amounts and objects, leading to many breakthroughs in medicine, engineering, science, architecture and astronomy.

In medical research, measurement errors can occur when measuring factors such as blood pressure, heart rate, or hormone levels. This can lead to incorrect conclusions about the effectiveness of the medication. In environmental research, measurement errors can occur when measuring factors such as air pollution, water quality or temperature. For example, a study investigating the impact of air pollution on respiratory health may yield biased results if the air pollution measurements are inaccurate due to measurement error. This can lead to incorrect conclusions about the impact of air pollution on respiratory health.

Have you ever wondered why tennis balls are sold in cylindrical containers? This is an example of manufacturers wanting to minimise the amount of waste in packaging. Measurement data impacted by error can result in biased findings because the data does not accurately represent the true values of the factors being measured. This can lead to incorrect conclusions about the relationships between factors or the effectiveness of interventions.

Hey students! Bring these pages to life online

▶ Watch videos

Engage with interactivities

A⁺ Answer questions and check solutions

Find all this and MORE in jacPLUS

Reading content and rich media, including interactivities and videos for every concept

Extra learning resources

Differentiated question sets

Questions with immediate feedback, and fully worked solutions to help students get unstuck

1. Calculate the area of the shape, correct to 2 decimal places.

7.3 mm

6.1 mm

15.2 mm

2. Calculate the area of the ellipse, correct to 1 decimal place.

8 cm

5 cm

3. **MC** Select the total surface area of the rectangular prism from the following.

1.5 m

2 m

3.2 m

A. $9.6 \, m^2$ **B.** $14.2 \, m^2$ **C.** $22.0 \, m^2$ **D.** $28.4 \, m^2$

4. Calculate the total surface area of the sphere, correct to 1 decimal place.

15 cm

5. Calculate the volume of the solid.

4 cm

$A = 3 \, cm^2$

6. Calculate the area of the shape, correct to 1 decimal place.

2 cm

45°

3 cm

7. A council park is shown below.

240 m

100 m

80 m   30 m

A worker charges $30 per 1000 m² to mow the grass. Determine how much it will cost the council to have the grass mown.

8. **MC** Select the total surface area of the object shown from the following.

2.5 cm

9 cm

A. 109.96 cm²   **B.** 112.63 cm²   **C.** 151.9 cm²   **D.** 124.36 cm²

9. Determine the volume of the triangular prism.

10 mm

5 mm

3 mm

10. **MC** Select the volume of the object from the following.

A. $2748.9\,\text{cm}^3$ B. $1701.7\,\text{cm}^3$ C. $1963.5\,\text{cm}^3$ D. $7854\,\text{cm}^3$

11. **MC** The volume of the frustum of a square-based pyramid is given by:

A. $V = \dfrac{1}{3}\pi\left(x^2 + y^2\right)$ B. $V = \dfrac{1}{3}\pi\left(x^2 + xy + y^2\right)$

C. $V = \dfrac{1}{3}h\pi\left(x^2 + xy + y^2\right)$ D. $V = \dfrac{1}{3}h\left(x^2 + xy + y^2\right)$

12. The volume of a ball is given by the formula $V = \dfrac{4}{3}\pi r^3$. Evaluate the radius of a ball with a volume of $384.66\,\text{cm}^3$. Give your answer correct to 1 decimal place.

13. **MC** Determine what effect doubling the radius and halving the height of a cone will have on its volume.
    A. The volume will be the same.
    B. The volume will be halved.
    C. The volume will be doubled.
    D. The volume will be quadrupled.

14. Using Heron's formula, evaluate the area of the triangle, correct to 1 decimal place.

15. A cylindrical soft drink can has a diameter of 6.4 cm and a height of 14.3 cm.
    If the can is only half full, determine what capacity of soft drink remains, to the nearest millilitre.

LESSON
6.2 Area

LEARNING INTENTIONS

At the end of this lesson you should be able to:
- convert between units of area
- calculate the areas of plane figures using area formulas
- calculate the area of a triangle using Heron's formula.

▶ 6.2.1 Area

eles-4809

- The **area** of a figure is the amount of surface covered by the figure.
- The units used for area are mm^2, cm^2, m^2, km^2 and ha (hectares).
- One unit that is often used when measuring land is the hectare. It is equal to $10\,000\,m^2$.
- The following diagram can be used to convert between units of area.

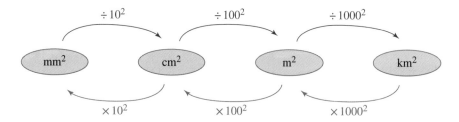

Area formulas

- The table below shows the formulas for the area of some common shapes.

Shape	Diagram	Formula
Square		$A = l^2$
Rectangle		$A = lw$
Triangle		$A = \dfrac{1}{2}bh$
Parallelogram		$A = bh$

Shape	Diagram	Formula
Trapezium		$A = \dfrac{1}{2}(a+b)h$
Kite (including rhombus)		$A = \dfrac{1}{2}xy$
Circle		$A = \pi r^2$
Sector		$A = \dfrac{\theta°}{360°} \times \pi r^2$
Ellipse		$A = \pi ab$

Heron's formula

- The area of a triangle can be calculated if the lengths of all three sides are known.

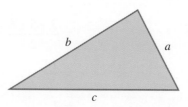

- The area, A, of a triangle given the lengths of the three sides a, b and c is:

$$A = \sqrt{s(s-a)(s-b)(s-c)}$$

where $s = \dfrac{a+b+c}{2}$, the semi-perimeter.

Digital technology

When using the number π in calculations, it is best to use a calculator. Calculators use the exact value of π, which will ensure your answer is exactly correct.

The π button on the TI-*n*spire CX CAS calculator is found near the bottom left of the calculator, as can be seen in the image.

If you do not have a calculator to hand, you can use the approximations $\pi \approx \dfrac{22}{7} \approx 3.14$; however, your answer may differ from the exact answer by a small amount.

WORKED EXAMPLE 1 Calculating areas of plane figures

Calculate the areas of the following plane figures, correct to 2 decimal places.

a.

3 cm 5 cm

6 cm

b.

2 cm

5 cm

c.

15 cm

40°

THINK	WRITE
a. 1. Three side lengths are known; apply Heron's formula.	**a.** $A = \sqrt{s(s-a)(s-b)(s-c)}$
2. Identify the values of a, b and c.	$a = 3, b = 5, c = 6$
3. Calculate the value of s, the semi-perimeter of the triangle.	$s = \dfrac{a+b+c}{2}$
	$= \dfrac{3+5+6}{2}$
	$= \dfrac{14}{2}$
	$= 7$
4. Substitute the values of a, b, c and s into Heron's formula and evaluate, correct to 2 decimal places.	$A = \sqrt{7(7-3)(7-5)(7-6)}$
	$= \sqrt{7 \times 4 \times 2 \times 1}$
	$= \sqrt{56}$
	$= 7.48 \text{ cm}^2$
b. 1. The shape shown is an ellipse. Write the appropriate area formula.	**b.** $A = \pi ab$
2. Identify the values of a and b (the semi-major and semi-minor axes).	$a = 5, b = 2$

3. Substitute the values of a and b into the formula and evaluate, correct to 2 decimal places. $(\pi = 3.14)$	$A = \pi \times 5 \times 2$ $= 31.42 \text{ cm}^2$
c. **1.** The shape shown is a sector. Write the formula for finding the area of a sector.	**c.** $A = \dfrac{\theta}{360°} \times \pi r^2$
2. Write the value of θ and r.	$\theta = 40°,\ r = 15$
3. Substitute and evaluate the expression, correct to 2 decimal places.	$A = \dfrac{40°}{360°} \times \pi \times 15^2$ $= 78.54 \text{ cm}^2$

⏵ 6.2.2 Areas of composite figures

eles-4810

- A **composite figure** is a figure made up of a combination of simple figures.
- The area of a composite figure can be calculated by:
  - calculating the sum of the areas of the simple figures that make up the composite figure
  - calculating the area of a larger shape and then subtracting the extra area involved.

WORKED EXAMPLE 2 Calculating areas of composite shapes

Calculate the area of each of the following composite shapes.

a.

AB = 8 cm
EC = 6 cm
FD = 2 cm

b.

9 cm
2 cm
5 cm
10 cm

THINK

a. **1.** ACBD is a quadrilateral that can be split into two triangles: $\triangle$ABC and $\triangle$ABD.

2. Write the formula for the area of a triangle containing base and height.

3. Identify the values of b and h for $\triangle$ABC.

4. Substitute the values of the pronumerals into the formula and calculate the area of $\triangle$ABC.

5. Identify the values of b and h for $\triangle$ABD.

WRITE

a. Area ACBD = Area $\triangle$ABC + Area $\triangle$ABD

$A_{\text{triangle}} = \dfrac{1}{2} bh$

$\triangle$ABC: $b = $ AB $= 8,\ h = $ EC $= 6$

Area of $\triangle$ABC $= \dfrac{1}{2} \times$ AB $\times$ EC

$= \dfrac{1}{2} \times 8 \times 6$

$= 24 \text{ cm}^2$

$\triangle$ABD: $b = $ AB $= 8,\ h = $ FD $= 2$

6. Calculate the area of $\triangle ABD$.

$$\text{Area of } \triangle ABD = \frac{1}{2}AB \times FD$$
$$= \frac{1}{2} \times 8 \times 2$$
$$= 8\,\text{cm}^2$$

7. Add the areas of the two triangles together to find the area of the quadrilateral ACBD.

$$\text{Area of ACBD} = 24\,\text{cm}^2 + 8\,\text{cm}^2$$
$$= 32\,\text{cm}^2$$

b. 1. One way to calculate the area of the shape shown is to find the total area of the rectangle ABGH and then subtract the area of the smaller rectangle DEFC.

b. Area = Area ABGH − Area DEFC

2. Write the formula for the area of a rectangle.

$$A_{\text{rectangle}} = l \times w$$

3. Identify the values of the pronumerals for the rectangle ABGH.

Rectangle ABGH: $l = 9 + 2 + 9$
$$= 20$$
$$w = 10$$

4. Substitute the values of the pronumerals into the formula to find the area of the rectangle ABGH.

$$\text{Area of ABGH} = 20 \times 10$$
$$= 200\,\text{cm}^2$$

5. Identify the values of the pronumerals for the rectangle DEFC.

Rectangle DEFC: $l = 5$, $w = 2$

6. Substitute the values of the pronumerals into the formula to find the area of the rectangle DEFC.

$$\text{Area of DEFC} = 5 \times 2$$
$$= 10\,\text{cm}^2$$

7. Subtract the area of the rectangle DEFC from the area of the rectangle ABGH to find the area of the given shape.

$$\text{Area} = 200 - 10$$
$$= 190\,\text{cm}^2$$

 Resources

6.2 Quick quiz on	6.2 Exercise

Individual pathways

■ PRACTISE	■ CONSOLIDATE	■ MASTER
1, 4, 7, 9, 11, 14, 15, 19, 22	2, 5, 8, 12, 16, 17, 20, 23	3, 6, 10, 13, 18, 21, 24

Unless told otherwise, where appropriate, give answers correct to 2 decimal places.

Fluency

1. Calculate the areas of the following shapes.

a.

b.

c.

2. Calculate the areas of the following shapes.

a.

b.

c.

3. Calculate the areas of the following shapes.

a.

b.

c.

4. **WE1a** Use Heron's formula to calculate the areas of the following triangles, correct to 2 decimal places.

a.

b.

5. **WE1b** Calculate the areas of the following ellipses, correct to 1 decimal place.

a.

b.

6. **WE1c** Calculate the area of each of the following shapes:

   i. stating the answer exactly, that is in terms of π
   ii. correct to 2 decimal places.

a.

b.

c.

7. **MC** A figure has an area of about $64\,\text{cm}^2$. Identify which of the following *cannot* possibly represent the figure.

   A. A triangle with base length $16\,\text{cm}$ and height $8\,\text{cm}$
   B. A circle with radius $4.51\,\text{cm}$
   C. A rectangle with dimensions $16\,\text{cm}$ and $4\,\text{cm}$
   D. A rhombus with diagonals $16\,\text{cm}$ and $4\,\text{cm}$

8. **MC** Identify from the following list all the lengths required to calculate the area of the quadrilateral shown.

   A. AB, BC, CD and AD
   B. AB, BE, AC and CD
   C. BC, BE, AD and CD
   D. AC, BE and FD

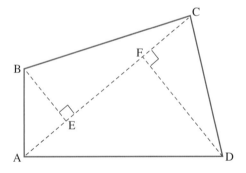

9. **WE2** Calculate the areas of the following composite shapes.

a.

b.

c.

10. Calculate the areas of the following composite shapes.

a.

2.1 m

3.8 m

b.

18 cm

5 cm

←—12 cm—→

c.

28 cm

11. Calculate the shaded area in each of the following.

a.

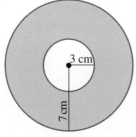

3 cm

7 cm

b.

16 m

8 m

2 m

2 m

12. Calculate the shaded area in each of the following.

a.

8 m

b.

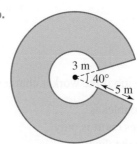

3 m

40°

5 m

13. Calculate the shaded area in each of the following.

a.

←—— 8 m ——→

3 m

2 m

13 m 7 m

b.

←——— 15 m ———→

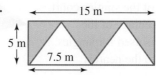

5 m

7.5 m

Understanding

14. A sheet of cardboard is 1.6 m by 0.8 m. The following shapes are cut from the cardboard:
    - a circular piece with radius 12 cm
    - a rectangular piece 20 cm by 15 cm
    - two triangular pieces with base length 30 cm and height 10 cm
    - a triangular piece with side lengths 12 cm, 10 cm and 8 cm.

    Calculate the area of the remaining piece of cardboard.

15. A rectangular block of land, 12 m by 8 m, is surrounded by a concrete path 0.5 m wide. Calculate the area of the path.

16. Concrete slabs 1 m by 0.5 m are used to cover a footpath 20 m by 1.5 m. Determine how many slabs are needed.

17. A city council builds a 0.5-m-wide concrete path around the garden as shown.

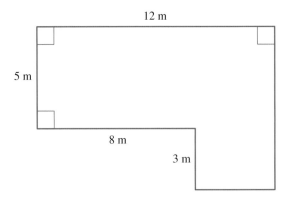

12 m

5 m

8 m

3 m

Determine the cost of the job if the worker charges $40.00 per m².

18. A tennis court used for doubles is 10.97 m wide, but a singles court is only 8.23 m wide, as shown in the diagram.

    a. Calculate the area of the doubles court.
    b. Calculate the area of the singles court.
    c. Determine the percentage of the doubles court that is used for singles. Give your answer to the nearest whole number.

8.23 m

6.40 m

10.97 m

11.89 m

Reasoning

19. Dan has purchased a country property with layout and dimensions as shown in the diagram.

    a. Show that the property has a total area of 987.5 ha.
    b. Dan wants to split the property in half (in terms of area) by building a straight-line fence running either north–south or east–west through the property. Assuming the cost of the fencing is a fixed amount per linear metre, justify where the fence should be built (that is, how many metres from the top left-hand corner and in which direction) to minimise the cost.

N

1500 m

5000 m

2000 m

1000 m

20. Ron the excavator operator has 100 metres of barricade mesh and needs to enclose an area to safely work in. He chooses to make a rectangular region with dimensions x and y. Show your working when required.

   a. Write an equation that connects x, y and the perimeter.
   b. Write y in terms of x.
   c. Write an equation for the area of the region in terms of x.
   d. Fill in the table for different values of x.

x	0	5	10	15	20	25	30	35	40	45	50
Area (m^2)											

   e. Can x have a value more than 50? Why?
   f. Sketch a graph of area against x.
   g. Determine the value of x that makes the area a maximum.
   h. Determine the value of y for maximum area.
   i. Determine the shape that encloses the maximum area.
   j. Calculate the maximum area.

   Ron decides to choose to make a circular area with the barricade mesh.

   k. Calculate the radius of this circular region.
   l. Calculate the area that is enclosed in this circular region.
   m. Determine how much extra area Ron now has compared to his rectangular region.

21. In question 20, Ron the excavator operator could choose to enclose a rectangular or circular area with 150 m of barricade mesh. In this case, the circular region resulted in a larger safe work area.

   a. Show that for 150 m of barricade mesh, a circular region again results in a larger safe work area as opposed to a rectangular region.
   b. Show that for n metres of barricade mesh, a circular region will result in a larger safe work area as opposed to a rectangular region.

Problem solving

22. A vegetable gardener is going to build four new rectangular garden beds side by side. Each garden bed measures 12.5 metres long and 3.2 metres wide. To access the garden beds, the gardener requires a path 1 metre wide between each garden bed and around the outside of the beds.

   a. Evaluate the total area the vegetable gardener would need for the garden beds and paths.
   b. The garden beds need to be mulched. A bag of mulch costing $29.50 covers an area of 25 square metres. Determine how many bags of mulch the gardener will need to purchase.
   c. The path is to be resurfaced at a cost of $39.50 per 50 square metres. Evaluate the cost of resurfacing the path.
   d. The gardener needs to spend a further $150 on plants. Determine the total cost of building these new garden beds and paths.

23. The diagram shows one smaller square drawn inside a larger square on grid paper.

   a. Determine what fraction of the area of the larger square is the area of the smaller square.

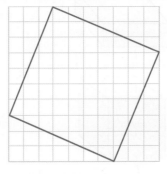

b. Another square with side lengths of 10 cm has a smaller square drawn inside. Determine the values of x and y if the smaller square is half the larger square.

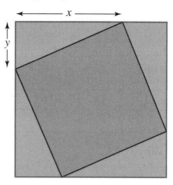

24. The shaded area in the diagram is called a segment of a circle.
A circle with a radius of 10 cm has ∠AOB equal to 90°.
A second circle, also with a radius of 10 cm, has ∠AOB equal to 120°.
Evaluate the difference in the areas of the segments of these two circles, correct to 2 decimal places.

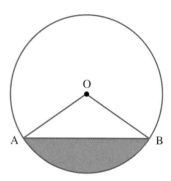

LESSON
6.3 Total surface area

LEARNING INTENTIONS

At the end of this lesson you should be able to:
- calculate the total surface areas of rectangular prisms and pyramids
- calculate the total surface areas of cylinders and spheres
- calculate the total surface areas of cones

⊳ 6.3.1 Total surface areas of solids

eles-4811
- The **total surface area (TSA)** of a solid is the sum of the areas of all the faces of that solid.

TSAs of rectangular prisms and cubes

Shape	Diagram	Formula
Rectangular prism (cuboid)		$TSA = 2\,(lh + lw + wh)$
Cube		$TSA = 6l^2$

TSAs of spheres and cylinders

Shape	Diagram	Formula
Sphere	Radius	$\text{TSA} = 4\pi r^2$
Cylinder	r h	$\text{TSA} = A_{\text{curved surface}} + A_{\text{circular ends}}$ $= 2\pi rh + 2\pi r^2$ $= 2\pi r(h + r)$

Note: Total surface area of pyramids, cones and spheres is part of the 10A course.

WORKED EXAMPLE 3 Calculating TSAs of solids

Calculate the total surface area of the solids, correct to the nearest cm².

a.
$r = 7$ cm

b.
50 cm
1.5 m

THINK	WRITE
a. 1. Write the formula for the TSA of a sphere.	**a.** $\text{TSA} = 4\pi r^2$
2. Identify the value for r.	$r = 7$
3. Substitute and evaluate.	$\text{TSA} = 4 \times \pi \times 7^2$ $\approx 615.8 \text{ cm}^2$
4. Write the answer to correct to the nearest cm².	$\approx 616 \text{ cm}^2$
b. 1. Write the formula for the TSA of a cylinder.	**b.** $\text{TSA} = 2\pi r(r + h)$
2. Identify the values for r and h. Note that the units will need to be the same.	$r = 50 \text{ cm}, h = 1.5 \text{ m}$ $= 150 \text{ cm}$
3. Substitute and evaluate.	$\text{TSA} = 2 \times \pi \times 50 \times (50 + 150)$ $= 62\,831.9 \text{ cm}^2$
4. Write the answer to correct to the nearest cm².	$\approx 62\,832 \text{ cm}^2$

▶ 6.3.2 Total surface areas of cones

eles-4812

- The total surface area of a cone can be found by considering its net, which is comprised of a small circle and a sector of a larger circle.

r = radius of the cone
l = slant height of the cone

- The sector is a fraction of the full circle of radius l with circumference $2\pi l$.
- The sector has an arc length equivalent to the circumference of the base of the cone, $2\pi r$.
- The fraction of the full circle represented by the sector can be found by writing the arc length as a fraction of the circumference of the full circle, $\dfrac{2\pi r}{2\pi l} = \dfrac{r}{l}$.

$$\text{Area of a sector} = \text{fraction of the circle} \times \pi l^2$$

$$= \frac{r}{l} \times \pi l^2$$

$$= \pi r l$$

Shape	Diagram	Formula
Cone		$\text{TSA} = A_{\text{curved surface}} + A_{\text{circular end}}$ $= \pi r^2 + \pi r l$ $= \pi r (r + l)$

WORKED EXAMPLE 4 Calculating the TSA of a cone

Calculate the total surface area of the cone shown.

15 cm
12 cm

THINK	WRITE
1. Write the formula for the TSA of a cone.	$\text{TSA} = \pi r (r + l)$
2. State the values of r and l.	$r = 12, l = 15$
3. Substitute and evaluate to obtain the answer.	$\text{TSA} = \pi \times 12 \times (12 + 15)$
	$= 1017.9 \text{ cm}^2$

TI	THINK	DISPLAY/WRITE	CASIO	THINK	DISPLAY/WRITE

On the Calculator page, complete the entry line as:
$\pi(r + s) | r = 12$ and $s = 15$
Press CTRL ENTER to get a decimal approximation.

The total surface area of the cone is 1017.9 cm² correct to 1 decimal place.

On the Main screen in decimal mode, complete the entry lines as:
$\pi r(r + s) | r = 12 | s = 15$
Then press EXE.

The total surface area of the cone is 1017.9 cm² correct to 1 decimal place.

▶ 6.3.3 Total surface areas of other solids

eles-4813

- The TSA of a solid can be found by summing the areas of each face.
- Check the total number of faces to ensure that none are left out.

WORKED EXAMPLE 5 Calculating the TSA of a pyramid

Calculate the total surface area of the square-based pyramid shown.

5 cm

6 cm

THINK

1. There are five faces: the square base and four identical triangles.

2. Calculate the area of the square base.

3. Draw and label one triangular face and write the formula for determining its area.

4. Calculate the height of the triangle, h, using Pythagoras' theorem.

WRITE/DRAW

TSA = area of square base + area of four triangular faces

Area of base = l^2, where $l = 6$
Area of base = 6^2
$\qquad = 36 \text{ cm}^2$

Area of a triangular face = $\dfrac{1}{2} bh$; $b = 6$

$a^2 = c^2 - b^2$, where $a = h$, $b = 3$, $c = 5$
$h^2 = 5^2 - 3^2$
$h^2 = 25 - 9$
$h^2 = 16$
$\quad h = 4 \text{ cm}$

5. Calculate the area of the triangular face by substituting $b = 6$ and $h = 4$.

$$\text{Area of triangular face} = \frac{1}{2} \times 6 \times 4$$
$$= 12 \text{ cm}^2$$

6. Calculate the TSA by adding the area of the square base and the area of the four identical triangular faces together.

$$\text{TSA} = 36 + 4 \times 12$$
$$= 36 + 48$$
$$= 84 \text{ cm}^2$$

Note: The area of the triangular faces can be found using Heron's formula. This method is demonstrated in the following worked example.

WORKED EXAMPLE 6 Calculating the TSA of a solid

Calculate the total surface area of the solid shown correct to 1 decimal place.

6 cm

10 cm

THINK

1. The solid shown has nine faces — five identical squares and four identical triangles.

2. Calculate the area of one square face with the side length 10 cm.

3. Draw a triangular face and label the three sides. Use Heron's formula to calculate the area.

4. State the formula for s, the semi-perimeter. Substitute the values of a, b and c and evaluate the value of s.

5. State Heron's formula for the area of one triangle. Substitute and evaluate.

WRITE/DRAW

$\text{TSA} = 5 \times \text{area of a square}$
$\qquad + 4 \times \text{area of a triangle}$

$A_{\text{square}} = l^2$, where $l = 10$
$\quad A = 10^2$
$\quad A = 100 \text{ cm}^2$

6 cm   6 cm

10 cm

$s = \dfrac{a+b+c}{2}$

$s = \dfrac{6+6+10}{2}$

$s = 11$

$A = \sqrt{s(s-a)(s-b)(s-c)}$

$A = \sqrt{11(11-6)(11-6)(11-10)}$

$A = \sqrt{275}$

$A = 16.583\,124\ldots \text{ cm}^2$

6. Determine the TSA of the solid by adding the area of the five squares and four triangles.

$$TSA = 5 \times 100 + 4 \times 16.583\,124\,\ldots$$
$$= 566.3325\,\ldots$$
$$= 566.3\,\text{cm}^2 \text{ (to 1 decimal place)}$$

Note: Rounding is not done until the final step. It is important to realise that rounding too early can affect the accuracy of results.

WORKED EXAMPLE 7 Applying surface area in worded problems

The silo shown is to be built from metal. The top portion of the silo is a cylinder of diameter 4 m and height 8 m. The bottom part of the silo is a cone of slant height 3 m. The silo has a circular opening of radius 30 cm on the top.

a. Calculate the area of metal (to the nearest m²) that is required to build the silo.
b. If it costs \$12.50 per m² to cover the surface with an anti-rust material, determine how much will it cost to cover the silo completely.

THINK	WRITE

a. 1. The surface area of the silo consists of an annulus, the curved part of the cylinder and the curved section of the cone.	**a.** TSA = area of annulus
	    + area of curved section of a cylinder
	    + area of curved section of a cone

2. To calculate the area of the annulus, subtract the area of the small circle from the area of the larger circle.
Let R = radius of small circle. Remember to convert all measurements to the same units.
	Area of annulus = $A_{\text{large circle}} - A_{\text{small circle}}$
	    $= \pi r^2 - \pi R^2$
	where $r = \dfrac{4}{2} = 2\,\text{m}$ and $R = 30\,\text{cm} = 0.3\,\text{m}$.
	Area of annulus $= \pi \times 2^2 - \pi \times 0.3^2$
	    $= 12.28\,\text{m}$

3. The middle part of the silo is the curved part of a cylinder. Determine its area. (Note that in the formula $TSA_{\text{cylinder}} = 2\pi r^2 + 2\pi rh$, the curved part is represented by $2\pi rh$.)	Area of curved section of cylinder $= 2\pi rh$ where $r = 2$, $h = 8$. Area of curved section of cylinder $= 2 \times \pi \times 2 \times 8$ $= 100.53\,\text{m}^2$
4. The bottom part of the silo is the curved section of a cone. Determine its area. (Note that in the formula $TSA_{\text{cone}} = \pi r^2 + \pi rl$, the curved part is given by πrl.)	Area of curved section of cone $= \pi rl$ where $r = 2$, $l = 3$. Area of curved section of cone $= \pi \times 2 \times 3$ $= 18.85\,\text{m}^2$

5. Calculate the total surface area of the silo by finding the sum of the surface areas calculated above.

$$TSA = 12.28 + 100.53 + 18.85$$
$$= 131.66 \, m^2$$

6. Write the answer in words.

The area of metal required is $132 \, m^2$, correct to the nearest square metre.

b. To determine the total cost, multiply the total surface area of the silo by the cost of the anti-rust material per m^2 ($12.50).

b. Cost $= 132 \times \$12.50$
$= \$1650.00$

Exercise 6.3 Total surface area

learn**on**

6.3 Quick quiz **on**	6.3 Exercise

Individual pathways

■ PRACTISE	■ CONSOLIDATE	■ MASTER
1, 5, 7, 10, 12, 17, 20	2, 6, 9, 11, 14, 15, 18, 21	3, 4, 8, 13, 16, 19, 22

Unless told otherwise, where appropriate, give answers correct to 1 decimal place.

Fluency

1. Calculate the total surface areas of the solids shown.

a.
10 cm

b.
8 cm

c.
12 cm
15 cm
20 cm

d.
2 m
1.5 m
3 m

2. **WE3** Calculate the total surface areas of the solids shown below.

a.
$r = 3$ m
r

b.
21 cm
30 cm

c.
0.5 m
2.1 m

d.
12 cm

3. **WE4** Calculate the total surface areas of the cones below.

a.

20 cm
14 cm

b.
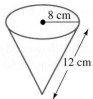
8 cm
12 cm

4. **WE5** Calculate the total surface areas of the solids below.

a.

12 cm
15 cm

b.

2.5 m
1.5 m

c.

9.1 cm
8 cm
14 cm
5.1 cm
7.2 cm

d.

6 cm
10 cm
7 cm

5. Calculate the surface areas of the following.

   a. A cube of side length 1.5 m
   b. A rectangular prism 6 m × 4 m × 2.1 m
   c. A cylinder of radius 30 cm and height 45 cm, open at one end

6. Calculate the surface areas of the following.

   a. A sphere of radius 28 mm
   b. An open cone of radius 4 cm and slant height 10 cm
   c. A square pyramid of base length 20 cm and slant edge 30 cm

7. **WE6** Calculate the total surface areas of the objects shown.

a.

8 cm
10 cm
5 cm
12 cm
5 cm
20 cm

b.

20 cm
35 cm
12 cm

c.

5 cm
3 cm

8. Calculate the total surface areas of the objects shown.

a.

2 cm
3 cm
2.5 cm

b.

3.5 cm
10 cm

c.

5 cm
20 cm
12 cm
15 cm

9. **MC** A cube has a total surface area of 384 cm². Calculate the length of the edge of the cube.

   A. 9 cm   B. 8 cm   C. 7 cm   D. 6 cm

Understanding

10. **WE7** The greenhouse shown is to be built using shade cloth. It has a wooden door of dimensions 1.2 m × 0.5 m.

    a. Calculate the total area of shade cloth needed to complete the greenhouse.
    b. Determine the cost of the shade cloth at $6.50 per m².

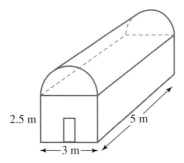

11. A cylinder is joined to a hemisphere to make a cake holder, as shown.

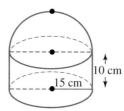

    The surface of the cake holder is to be chromed at 5.5 cents per cm².

    a. Calculate the total surface area to be chromed.
    b. Determine the cost of chroming the cake holder.

12. A steel girder is to be painted. Calculate the area of the surface to be painted.

13. Open cones are made from nets cut from a large sheet of paper 1.2 m × 1.0 m. If a cone has a radius of 6 cm and a slant height of 10 cm, determine how many cones can be made from the sheet.
    (Assume there is 5% wastage of paper.)

14. A prism of height 25 cm has a base in the shape of a rhombus with diagonals of 12 cm and 16 cm. Calculate the total surface area of the prism.

15. A hemispherical glass dome, with a diameter of 24 cm, sits on a concrete cube with sides of 50 cm. To protect the structure, all exposed sides are to be treated. The glass costs $1.50/cm² to treat and the concrete costs 5c/cm².
    Calculate the cost in treating the structure if the base of the cube is already fixed to the ground. Give your answer to the nearest dollar.

16. An inverted cone with side length 4 metres is placed on top of a sphere such that the centre of the cone's base is 0.5 metres above the centre of the sphere. The radius of the sphere is $\sqrt{2}$ metres.

a. Calculate the exact total surface area of the sphere.

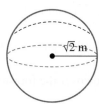

b. Calculate the radius of the cone exactly.

c. Calculate the area of the curved surface of the cone exactly.

Reasoning

17. A shower recess with dimensions 1500 mm (back wall) by 900 mm (side wall) needs to have the back and two side walls tiled to a height of 2 m.

    a. Calculate the area to be tiled in m^2.
    b. Justify that 180 tiles (including those that need to be cut) of dimension 20 cm by 20 cm will be required. Disregard the grout and assume that once a tile is cut, only one piece of the tile can be used.
    c. Evaluate the cheapest option of tiling; $1.50/tile or $39.50/box, where a box covers 1 m^2, or tiles of dimension 30 cm by 30 cm costing $3.50/tile.

18. The table shown below is to be varnished (including the base of each leg). The tabletop has a thickness of 180 mm and the cross-sectional dimensions of the legs are 50 mm by 50 mm. A friend completes the calculation without a calculator as shown. Assume there are no simple calculating errors. Analyse the working presented and justify if the TSA calculated is correct.

Tabletop (inc. leg bases)	0.96	$2 \times (0.8 \times 0.6)$
Legs	0.416	$16 \times (0.52 \times 0.05)$
Tabletop edging	0.504	$0.18 \times (2(0.8 + 0.6))$
TSA	1.88 m^2	

19. A soccer ball is made up of a number of hexagons sewn together on its surface. Each hexagon can be considered to have dimensions as shown in the diagram.

a. Calculate $\theta°$.

b. Calculate the values of x and y exactly.

c. Calculate the area of the trapezium in the diagram.

d. Hence, determine the area of the hexagon.

e. If the total surface area of the soccer ball is $192\sqrt{3}\text{ cm}^2$, determine how many hexagons are on its surface.

Problem solving

20. Tina is re-covering a footstool in the shape of a cylinder with diameter 50 cm and height 30 cm. She also intends to cover the base of the cushion. She has 1 m^2 of fabric to make this footstool.

When calculating the area of fabric required, allow an extra 20% of the total surface area to cater for seams and pattern placings.

Explain whether Tina has enough material to cover the footstool.

21. If the surface area of a sphere to that of a cylinder is in the ratio 4 : 3 and the sphere has a radius of $3a$, show that if the radius of the cylinder is equal to its height, then the radius of the cylinder is $\dfrac{3\sqrt{3}a}{2}$.

22. A frustum of a cone is a cone with the top sliced off, as shown.

When the curved side is 'opened up', it creates a shape, ABYX, as shown in the diagram.

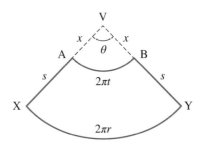

a. Write an expression for the arc length XY in terms of the angle θ. Write another expression for the arc length AB in terms of the same angle θ. Show that, in radians, $\theta = \dfrac{2\pi(r-t)}{s}$.

b. **i.** Using the above formula for θ, show that $x = \dfrac{st}{(r-t)}$.

　　ii. Use similar triangles to confirm this formula.

c. Determine the areas of sectors AVB and XVY and hence determine the area of ABYX. Add the areas of the 2 circles to the area of ABYX to determine the TSA of a frustum.

LESSON
6.4 Volume

LEARNING INTENTIONS

At the end of this lesson you should be able to:
- calculate the volumes of prisms, including cylinders
- calculate the volumes of spheres
- calculate the volumes of pyramids.

▶ 6.4.1 Volume

eles-4814

- The **volume** of a 3-dimensional object is the amount of space it takes up.
- Volume is measured in units of mm^3, cm^3 and m^3.
- The following diagram can be used to convert between units of volume.

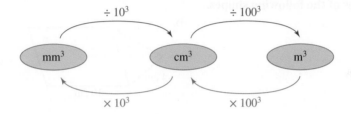

Volume of a prism

- The volume of any solid with a uniform cross-sectional area is given by the formula shown below.

> ## Volume of a solid with uniform cross-sectional area
>
> $$V = AH$$
>
> **where A is the area of the cross-section and H is the height of the solid.**

Shape	Diagram	Formula
Cube		Volume $= AH$ $=$ area of a square $\times$ height $= l^2 \times l$ $= l^3$
Rectangular prism		Volume $= AH$ $=$ area of a rectangle $\times$ height $= lwh$

(continued)

Shape	Diagram	Formula
Cylinder		Volume $= AH$ $\quad\quad\quad\;\; =$ area of a circle $\times$ height $\quad\quad\quad\;\; = \pi r^2 h$
Triangular prism		Volume $= AH$ $\quad\quad\quad\;\; =$ area of a triangle $\times$ height $\quad\quad\quad\;\; = \dfrac{1}{2}bh \times H$

WORKED EXAMPLE 8 Calculating volumes of prisms

Calculate the volumes of the following shapes.

a. 14 cm, 20 cm

b. 5 cm, 4 cm, 10 cm

THINK	WRITE
a. 1. Write the formula for the volume of the cylinder (prism).	a. $V = AH$ $\quad = \pi r^2 h$
2. Identify the value of the pronumerals.	$r = 14,\ h = 20$
3. Substitute and evaluate the answer.	$V = \pi \times 14^2 \times 20$ $\quad \approx 12\,315.04 \text{ cm}^3$
b. 1. Write the formula for the volume of a triangular prism.	b. $V = \dfrac{1}{2}bh \times H$
2. Identify the value of the pronumerals. *Note: h* is the height of the triangle and *H* is the depth of the prism.	$b = 4,\ h = 5,\ H = 10$
3. Substitute and evaluate the answer.	$V = \dfrac{1}{2} \times 4 \times 5 \times 10$ $\quad = 100 \text{ cm}^3$

WORKED EXAMPLE 9 Changing the dimensions of a prism

a. **If each of the side lengths of a cube are doubled, then determine the effect on its volume.**
b. **If the radius is halved and the height of a cylinder is doubled, then determine the effect on its volume.**

THINK	WRITE
a. 1. Write the formula for the volume of the cube.	**a.** $V = l^3$
2. Identify the value of the pronumeral. *Note:* Doubling is the same as multiplying by 2.	$l_{new} = 2l$
3. Substitute and evaluate.	$V_{new} = (2l)^3$
4. Compare the answer obtained in step 3 with the volume of the original shape.	$= 8l^3$
5. Write your answer.	Doubling each side length of a cube increases the volume by a factor of 8; that is, the new volume will be 8 times as large as the original volume.
b. 1. Write the formula for the volume of the cylinder.	**b.** $V = \pi r^2 h$
2. Identify the value of the pronumerals. *Note:* Halving is the same as dividing by 2.	$r_{new} = \dfrac{r}{2}, \ h_{new} = 2h$
3. Substitute and evaluate.	$V_{new} = \pi \left(\dfrac{r}{2}\right)^2 2h$ $= \pi \times \dfrac{r^2}{2^4} \times 2h$ $= \dfrac{\pi r^2 h}{2}$
4. Compare the answer obtained in step 3 with the volume of the original shape.	$= \dfrac{1}{2}\pi r^2 h$
5. Write your answer.	Halving the radius and doubling the height of a cylinder decreases the volume by a factor of 2; that is, the new volume will be half the original volume.

▶ 6.4.2 Volumes of common shapes

eles-4815

Volume of a sphere

- The volume of a sphere of radius r is given by the following formula.

Shape	Diagram	Formula
Sphere		$V = \dfrac{4}{3}\pi r^3$

WORKED EXAMPLE 10 Calculating the volume of a sphere

Find the volume of a sphere of radius 9 cm. Answer correct to 1 decimal place.

THINK	WRITE
1. Write the formula for the volume of a sphere.	$V = \dfrac{4}{3}\pi r^3$
2. Identify the value of r.	$r = 9$
3. Substitute and evaluate.	$V = \dfrac{4}{3} \times \pi \times 9^3$
	$= 3053.6 \text{ cm}^3$

Volume of a pyramid

- Pyramids are not prisms, as the cross-section changes from the base upwards.
- The volume of a pyramid is one-third the volume of the prism with the same base and height.

Shape	Diagram	Formula
Pyramid		$V_{\text{pyramid}} = \dfrac{1}{3}AH$

Volume of a cone

- The cone is a pyramid with a circular base.

Shape	Diagram	Formula
Cone		$V_{cone} = \dfrac{1}{3}\pi r^2 h$

Note: Volume of cones, pyramids and spheres is part of the 10A course.

WORKED EXAMPLE 11 Calculating the volume of pyramids and cones

Calculate the volume of each of the following solids.

a.

b.

THINK		WRITE
a. 1. Write the formula for the volume of a cone.		**a.** $V = \dfrac{1}{3}\pi r^2 h$
2. Identify the values of r and h.		$r = 8,\ h = 10$
3. Substitute and evaluate.		$V = \dfrac{1}{3} \times \pi \times 8^2 \times 10$
		$= 670.21\ \text{cm}^3$
b. 1. Write the formula for the volume of a pyramid.		**b.** $V = \dfrac{1}{3}AH$
2. Calculate the area of the square base.		$A = l^2$ where $l = 8$
		$A = 8^2$
		$= 64\ \text{cm}^2$
3. Identify the value of H.		$H = 12$
4. Substitute and evaluate.		$V = \dfrac{1}{3} \times 64 \times 12$
		$= 256\ \text{cm}^3$

6.4.3 Volumes of composite solids

eles-4816

- A composite solid is a combination of a number of solids.
- Calculate the volume of each solid separately.
- Sum these volumes to give the volume of the composite solid.

Calculate the volume of the composite solid shown.

THINK	WRITE
1. The given solid is a composite figure, made up of a cube and a square-based pyramid.	$V = $ Volume of cube $+$ Volume of pyramid
2. Calculate the volume of the cube.	$V_{cube} = l^3$ where $l = 3$ $V_{cube} = 3^3$ $\quad\quad = 27\,m^3$
3. Write the formula for the volume of a square-based pyramid.	$V_{square\text{-}based\ pyramid} = \dfrac{1}{3}AH$
4. Calculate the area of the square base.	$A = l^2$ $\quad = 3^2$ $\quad = 9\,m^2$
5. Identify the value of H.	$H = 1.5$
6. Substitute and evaluate the volume of the pyramid.	$V_{square\text{-}based\ pyramid} = \dfrac{1}{3} \times 9 \times 1.5$ $\quad\quad\quad\quad\quad\quad = 4.5\,m^3$
7. Calculate the total volume by adding the volume of the cube and pyramid.	$V = 27 + 4.5$ $\quad = 31.5\,m^3$

▶ 6.4.4 Capacity

eles-4817

- Some 3-dimensional objects are hollow and can be filled with liquid or some other substance.
- The amount of substance that a container can hold is called its capacity.
- **Capacity** is essentially the same as volume but is usually measured in mL, L, kL and ML (megalitres) where $1\,mL = 1\,cm^3$

$$1\,L = 1000\,cm^3$$
$$1\,kL = 1\,m^3.$$

- The following diagram can be used to convert between units of capacity.

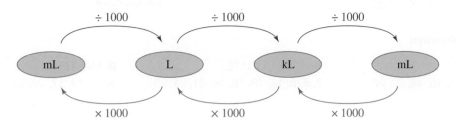

WORKED EXAMPLE 13 Calculating the capacity of a prism

Determine the capacity (in litres) of a cuboidal aquarium that is 50 cm long, 30 cm wide and 40 cm high.

THINK	WRITE
1. Write the formula for the volume of a rectangular prism.	$V = lwh$
2. Identify the values of the pronumerals.	$l = 50, w = 30, h = 40$
3. Substitute and evaluate.	$V = 50 \times 30 \times 40$ $= 60\,000 \text{ cm}^3$
4. State the capacity of the container in millilitres, using $1 \text{ cm}^3 = 1 \text{ mL}$.	$= 60\,000 \text{ mL}$
5. Since $1 \text{ L} = 1000 \text{ mL}$, to convert millilitres to litres divide by 1000.	$= 60 \text{ L}$
6. Write the answer in a sentence.	The capacity of the fish tank is 60 L.

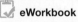 Resources

Individual pathways

■ PRACTISE	■ CONSOLIDATE	■ MASTER
1, 3, 5, 7, 9, 12, 15, 18, 23, 27	2, 6, 8, 10, 13, 16, 19, 21, 24, 28	4, 11, 14, 17, 20, 22, 25, 26, 29, 30

Fluency

1. Calculate the volumes of the following prisms.

   a.

   b.

   3 cm

   4.2 m

   c.

   12 cm

   15 cm

   20 cm

   d.

   4.2 cm

   3 cm

   7.5 cm

2. Calculate the volume of each of these solids.

   a.

   18 mm

   [Base area: 25 mm^2]

   b.

   15 cm

   [Base area: 24 cm^2]

3. **WE8** Calculate the volume of each of the following. Give each answer correct to 1 decimal place where appropriate.

   a.

   14 cm

   12 cm

   b.

   2.7 m

   1.5 m

   c.

   10 cm

   7 cm

   8 cm

4. Calculate the volume of each of the following. Give each answer correct to 1 decimal place where appropriate.

   a.

   12 mm

   8 mm

   6 mm

   b.

   6.5 m

   35°

   7.1 m

   c.

   45 cm

   18 cm

5. **WE10** Determine the volume of a sphere (correct to 1 decimal place) with a radius of:

   a. 1.2 m   b. 15 cm   c. 7 mm   d. 50 cm

6. Calculate the volume of each of these figures, correct to 2 decimal places.

   a.

   30 cm

   b.

   1.4 m

   c.

   18 mm

   d.
   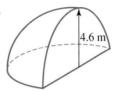
   4.6 m

7. **WE11a** Determine the volume of each of the following cones, correct to 1 decimal place.

   a.

   10 cm
   6 cm

   b.

   20 mm   22 mm

8. **WE11b** Calculate the volume of each of the following pyramids.

   a.
   12 cm

   10 cm

   b.

   42 cm
   24 cm
   30 cm

9. **WE12** Calculate the volume of each of the following composite solids correct to 2 decimal places where appropriate.

   a.

   10 cm   8 cm   5 cm   12 cm   5 cm
   20 cm

   b.

   20 cm
   35 cm
   12 cm

10. Calculate the volume of each of the following composite solids correct to 2 decimal places where appropriate.

a.

5 cm

3 cm

b.

2 cm

3 cm

2.5 cm

11. Calculate the volume of each of the following composite solids correct to 2 decimal places where appropriate.

a.

3.5 cm

10 cm

b.

5 cm

20 cm

12 cm

15 cm

Understanding

12. **WE9** Answer the following questions.

a. If the side length of a cube is tripled, then determine the effect on its volume.

b. If the side length of a cube is halved, then determine the effect on its volume.

c. If the radius is doubled and the height of a cylinder is halved, then determine the effect on its volume.

d. If the radius is doubled and the height of a cylinder is divided by four, then determine the effect on its volume.

e. If the length is doubled, the width is halved and the height of a rectangular prism is tripled, then determine the effect on its volume.

13. **MC** A hemispherical bowl has a thickness of 2 cm and an outer diameter of 25 cm.

2 cm

25 cm

If the bowl is filled with water, the capacity of the water will be closest to:

A. 1.526 L   B. 1.30833 L   C. 3.05208 L   D. 2.42452 L

14. Tennis balls of diameter 8 cm are packed in a box 40 cm × 32 cm × 10 cm, as shown.

    a. Estimate the volume of one tennis ball.
       (*Hint:* Use π as approximately equal to 3.)
    b. Estimate the volume of all the tennis balls in the box.
    c. Estimate the volume of the box.
    d. Estimate the unfilled volume of the box.
    e. Determine, correct to 2 decimal places, how much space is left unfilled.
    f. Compare your estimate of the unfilled volume to your answer in part **e**.

15. **WE13** A cylindrical water tank has a diameter of 1.5 m and a height of 2.5 m. Determine the capacity
    (in litres) of the tank, correct to 1 decimal place.

16. A monument in the shape of a rectangular pyramid (base length of 10 cm, base width of 6 cm, height of
    8 cm), a spherical glass ball (diameter of 17 cm) and conical glassware (radius of 14 cm, height of 10 cm) are
    packed in a rectangular prism of dimensions 30 cm by 25 cm by 20 cm. The extra space in the box is filled
    up by a packing material.
    Determine, correct to 2 decimal places, the volume of packing material that is required.

17. A swimming pool is being constructed so that it is the upper part of an inverted
    square-based pyramid.

    a. Calculate H.
    b. Calculate the volume of the pool.
    c. Determine how many 6-m^3 bins will be required to take the dirt away.
    d. Determine how many litres of water are required to fill this pool.
    e. Determine how deep the pool is when it is half-filled.

    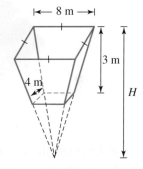

18. A soft drink manufacturer is looking to repackage cans of soft drink to minimise the cost of packaging while
    keeping the volume constant. Consider a can of soft drink with a capacity of 400 mL.

    a. If the soft drink was packaged in a spherical can:
       i. calculate the radius of the sphere, correct to 2 decimal places
       ii. determine the total surface area of this can, correct to 1 decimal place.
    b. If the soft drink was packaged in a cylindrical can with a radius
       of 3 cm:
       i. calculate the height of the cylinder, correct to 2 decimal places
       ii. determine the total surface area of this can, correct to 2 decimal places.
    c. If the soft drink was packaged in a square-based pyramid with a base side
       length of 6 cm:
       i. calculate the height of the pyramid, correct to 2 decimal places
       ii. determine the total surface area of this can, correct to 2 decimal places.
    d. Explain which can you would recommend the soft drink manufacturer use for its repackaging.

19. The volume of a cylinder is given by the formula $V = \pi r^2 h$.
   a. Transpose the formula to make h the subject.
   b. A given cylinder has a volume of $1600\,\text{cm}^3$. Calculate its height, correct to 1 decimal place, if it has a radius of:
      i. $4\,\text{cm}$ ii. $8\,\text{cm}$.
   c. Transpose the formula to make r the subject.
   d. Explain what restrictions must be placed on r.
   e. A given cylinder has a volume of $1800\,\text{cm}^3$. Determine its radius, correct to 1 decimal place, if it has a height of:
      i. $10\,\text{cm}$ ii. $15\,\text{cm}$.

20. A toy maker has enough rubber to make one super-ball of radius $30\,\text{cm}$. Determine how many balls of radius $3\,\text{cm}$ he can make from this rubber.

21. A manufacturer plans to make a cylindrical water tank to hold $2000\,\text{L}$ of water.
   a. Calculate the height, correct to 2 decimal places, if he uses a radius of $500\,\text{cm}$.
   b. Calculate the radius, correct to 2 decimal places if he uses a height of $500\,\text{cm}$.
   c. Determine the surface area of each of the two tanks. Assume the tank is a closed cylinder and give your answer in square metres correct to 2 decimal places.

22. The ancient Egyptians knew that the volume of the frustum of a square-based pyramid was given by the formula $V = \dfrac{1}{3} h \left(x^2 + xy + y^2 \right)$, although how they discovered this is unclear. (A frustum is the part of a cone or pyramid that is left when the top is cut off.)

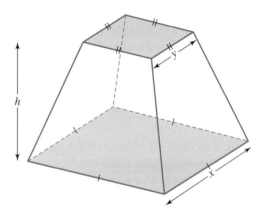

   a. Calculate the volume of the frustum below, correct to 2 decimal places.
   b. Determine the volume of the missing portion of the square-based pyramid shown, correct to 2 decimal places.

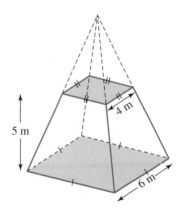

Reasoning

23. The Hastings' family house has a rectangular roof with dimensions $17\,\text{m} \times 10\,\text{m}$ providing water to three cylindrical water tanks, each with a radius of $1.25\,\text{m}$ and a height of $2.1\,\text{m}$.
Show that approximately 182 millimetres of rain must fall on the roof to fill the tanks.

24. Archimedes is considered to be one of the greatest mathematicians of all time. He discovered several of the formulas used in this topic. Inscribed on his tombstone was a diagram of his proudest discovery.
It shows a sphere inscribed (fitting exactly) into a cylinder.
Show that:

$$\frac{\text{volume of the cylinder}}{\text{volume of the sphere}} = \frac{\text{surface area of the cylinder}}{\text{surface area of the sphere}}$$

25. Marion has mixed together ingredients for a cake. The recipe requires a baking tin that is cylindrical in shape with a diameter of $20\,\text{cm}$ and a height of $5\,\text{cm}$. Marion only has a tin in the shape of a trapezoidal prism and a muffin tray consisting of 24 muffin cups. Each of the muffin cups in the tray is a portion of a cone. Both the tin and muffin cup are shown in the diagrams.
Explain whether Marion should use the tin or the muffin tray.

26. Sam is having his 16th birthday party and wants to make an ice trough to keep drinks cold. He has found a square piece of sheet metal with a side length of 2 metres.
He cuts squares of side length x metres from each corner, then bends the sides of the remaining sheet.
When four squares of the appropriate side length are cut from the corners, the capacity of the trough can be maximised at 588 litres.
Explain how Sam should proceed to maximise the capacity of the trough.

Problem solving

27. Nathaniel and Annie are going to the snow for survival camp. They plan to construct an igloo, consisting of an entrance and a hemispherical living section as shown.

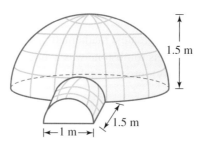

Nathaniel and Annie are asked to redraw their plans and increase the height of the liveable region (hemispherical structure) so that the total volume (including the entrance) is doubled.
Determine what must the new height of the hemisphere be to achieve this so that the total volume (including the entrance) is doubled. Write your answer in metres correct to 2 decimal places.

28. Six tennis balls are just contained in a cylinder as the balls touch the sides and the end sections of the cylinder. Each tennis ball has a radius of R cm.

   a. Express the height of the cylinder in terms of R.
   b. Evaluate the total volume of the tennis balls.
   c. Determine the volume of the cylinder in terms of R.
   d. Show that the ratio of the volume of the tennis balls to the volume of the cylinder is $2 : 3$.

29. A frustum of a square-based pyramid is a square pyramid with the top sliced off. H is the height of the full pyramid and h is the height of the frustum.

   a. Determine the volume of the large pyramid that has a square base side of X cm.
   b. Evaluate the volume of the small pyramid that has a square base side of x cm.

   c. Show that the relationship between H and h is given by $H = \dfrac{Xh}{X - x}$.

   d. Show that the volume of the frustum is given by $\dfrac{1}{3} h \left(X^2 + x^2 + Xx \right)$.

30. A large container is five-eighths full of ice-cream. After removing 27 identical scoops, it is one-quarter full. Determine how many scoops of ice-cream are left in the container.

LESSON
6.5 Review

6.5.1 Topic summary

Prisms and cylinders

- Prisms are 3D objects that have a uniform cross section and all flat surfaces.
- The surface area of a prism is calculated by adding the areas of its faces.
- The volume of a prism is $V = AH$, where A is the cross-sectional area of the prism, and H is the perpendicular height.
- A cylinder is a 3D object that has a circular cross-section.
- The curved surface area of a cylinder is $2\pi rh$, and the total surface area is $2\pi r^2 + 2\pi rh = 2\pi rh(r + h)$.
- The volume of a cylinder is $V = \pi r^2 h$.

SURFACE AREA AND VOLUME

Pyramids

- The surface area of a pyramid can be calculated by adding the surface areas of its faces.
- The volume of a pyramid is $V = \frac{1}{3}AH$, where A is the area of the base and H is the height.

Cones

- The curved surface area of a cone is $SA_{\text{curved}} = \pi rl$, where l is the slant height.
- The total surface area is $SA = \pi rl + \pi r^2 = \pi r(l + r)$.
- The volume of a cone is $V = \frac{1}{3}\pi r^2 h$.

Spheres

- The surface area of a sphere is $A = 4\pi r^2$.
- The volume of a sphere is $V = \frac{4}{3}\pi r^3$.

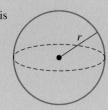

Units of area, volume and capacity

Area

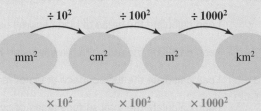

1 hectare (ha) = 10 000 m²

Volume

Capacity

$1\ \text{cm}^3 = 1\ \text{mL}$
$1\ \text{L} = 1000\ \text{cm}^3$

Area formulas

- Square: $A = l^2$
- Rectangle: $A = lw$
- Triangle: $A = \frac{1}{2}bh$
- Parallelogram: $A = bh$
- Trapezium: $A = \frac{1}{2}(a + b)h$

- Kite: $A = \frac{1}{2}xy$
- Circle: $A = \pi r^2$
- Sector: $A = \frac{\theta}{360} \times \pi r^2$
- Ellipse: $A = \pi ab$

Heron's formula

- Heron's formula is an alternate method to calculate the area of a triangle.
- For a triangle with sides of length a, b and c, the area is

$A = \sqrt{s(s-a)(s-b)(s-c)}$, where $s = \frac{a+b+c}{2}$.

6.5.2 Success criteria

Tick the column to indicate that you have completed the lesson and how well you have understood it using the traffic light system.

(**Green:** I understand; **Yellow:** I can do it with help; **Red:** I do not understand)

Lesson	Success criteria	●	○	●
6.2	I can convert between units of area.			
	I can calculate the areas of plane figures using area formulas.			
	I can calculate the area of a triangle using Heron's formula.			
6.3	I can calculate the total surface areas of rectangular prisms and pyramids.			
	I can calculate the total surface areas of cylinders and spheres.			
	I can calculate the total surface areas of cones.			
6.4	I can calculate the volumes of prisms, including cylinders.			
	I can calculate the volumes of spheres.			
	I can calculate the volumes of pyramids.			

6.5.3 Project

So close!

Humans must measure! Imagine what a chaotic world it would be if we didn't measure anything. Some of the things we measure are time, length, weight and temperature; we also use other measures derived from these such as area, volume, speed.

Accurate measurement is important. The accuracy of a measurement depends on the instrument being used to measure and the interpretation of the measurement. There is no such thing as a perfectly accurate measurement. The best we can do is learn how to make meaningful use of the numbers we read off our devices. It is also important to use appropriate units of measurement.

Measurement errors

When we measure a quantity by using a scale, the accuracy of our measurement depends on the markings on the scale. For example, the ruler shown can measure both in centimetres and millimetres.

Measurements made with this ruler would have ± 0.5 mm added to the measurement. The quantity ± 0.5 mm is called the tolerance of measurement or measurement error.

$$\text{Tolerance of measurement} = \frac{1}{2} \times \text{size of smallest marked unit}$$

For a measurement of 5.6 ± 0.5 mm, the largest possible value is 5.6 cm $+ 0.5$ mm $= 5.65$ cm, and the smallest value is 5.6 cm $- 0.5$ mm $= 5.55$ cm.

1. For the thermometer scale shown:
    a. identify the temperature
    b. state the measurement with its tolerance
    c. calculate the largest and smallest possible values.
2. Calculate the largest and smallest values for:
    a. $(56.2 \pm 0.1) - (19.07 \pm 0.05)$
    b. $(78.4 \pm 0.25) \times (34 \pm 0.1)$.

Significant figures in measurement

A significant figure is any non zero-digit, any zero appearing between two non-zero digits, any trailing zeros in a number containing a decimal point, and any digits in the decimal places. For example, the number 345.6054 has 7 significant figures, whereas 300 has 1 significant figure.

The number of significant figures is an expression of the accuracy of a measurement. The greater the number of significant figures, the more accurate the measurement. For example, a fast food chain claims it has sold 6 000 000 000 hamburgers, not 6 453 456 102. The first measurement has only 1 significant figure and is a very rough approximation of the actual number sold, which has 10 significant figures.

Reducing the number of significant figures is a process that is similar to rounding.

Rounding and measurement error in calculations

When you perform calculations, it is important to keep as many significant digits as practical and to perform any rounding as the *final* step. For example, calculating 5.34×341 by rounding to 2 significant figures *before* multiplying gives $5.30 \times 340 = 1802$, compared with 1820 if the rounding is carried out after the multiplication.

Calculations that involve numbers from measurements containing errors can result in answers with even larger errors. The smaller the tolerances, the more accurate the answers will be.

3. a. Calculate $45\,943.4503 \times 86.765\,303$ by:
   i. first rounding each number to 2 significant figures
   ii. rounding only the answer to 2 significant figures.
   b. Compare the two results.

Error in area and volume resulting from an error in a length measurement

The side length of a cube is measured and incorrectly recorded as 5 cm. The actual length is 6 cm. The effect of the length measurement error used on calculations of the surface area is shown below.

Error used in length measurement = 1 cm

Surface area calculated with incorrectly recorded value $= 5^2 \times 6 = 150\,\text{cm}^2$

Surface area calculated with actual value $= 6^2 \times 6 = 216\,\text{cm}^2$

Percentage error $= \dfrac{216 - 150}{6} \times 100\% \approx 30.5\%$

4. a. Complete a similar calculation for the volume of the cube using the incorrectly recorded length. What conclusion can you make regarding errors when the number of dimensions increase?
   b. Give three examples of a practical situation where an error in measuring or recording would have a potentially disastrous impact.

 Resources

 eWorkbook   Topic 6 workbook (worksheets, code puzzle and a project) (ewbk-13290)

 Interactivities Crossword (int-2842)
                   Sudoku puzzle (int-3593)

Exercise 6.5 Review questions   **learn**on

Unless told otherwise, where appropriate, give answers correct to 2 decimal places.

Fluency

1. **MC** If all measurements are in cm, the area of the figure is:

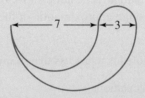

   A. $16.49\,\text{cm}^2$ B. $39.25\,\text{cm}^2$ C. $9.81\,\text{cm}^2$ D. $23.56\,\text{cm}^2$

2. **MC** If all measurements are in centimetres, the area of the figure is:
   A. 50.73 cm²
   B. 99.82 cm²
   C. 80.18 cm²
   D. 90 cm²

3. **MC** If all measurements are in centimetres, the shaded area of the figure is:

   A. 3.93 cm²   **B.** 11.52 cm²   **C.** 388.77 cm²   **D.** 129.59 cm²

4. **MC** The total surface area of the solid is:

   A. 8444.6 mm²   **B.** 9221 mm²   **C.** 14 146.5 mm²   **D.** 50 271.1 mm²

5. Calculate the areas of the following plane figures. All measurements are in cm.
   a.

   b.

   c.

6. Calculate the areas of the following plane figures. All measurements are in cm.
   a.

   b.

   c.

7. Calculate the areas of the following figures. All measurements are in cm.

a.

15

20

b.

12

10

8

c.

10

10

6

5

20

8. Calculate the blue shaded area in each of the following. All measurements are in cm.

a.

Q

$\overline{QO}$ = 15 cm
$\overline{SO}$ = 8 cm
$\overline{PR}$ = 18 cm

O

P   S   R

b.

12.5

c.

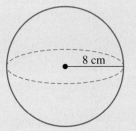

5

9. Calculate the total surface area of each of the following solids.

a.

35 cm

50 cm

b.

14 mm

20 mm

c.

8 cm

10. Calculate the total surface area of each of the following solids.

a.

14 cm

18 cm

12 cm

b.

10 mm

14 mm

10 mm

4 mm

[closed at both ends]

c.

12 cm

10 cm

10 cm

10 cm

11. Calculate the volume of each of the following.

a.

7 cm

b.

7 cm
8 cm
12 cm

c.

35 cm
40 cm

12. Determine the volume of each of the following.

a.

3.7 m
1 m

b.

10 cm
30 cm
12 cm

c.

12 cm
10 cm

13. Determine the volume of each of the following.

a.

11 cm
9 cm

b.

30 cm
20 cm
42 cm

c.

12 mm

Problem solving

14. A rectangular block of land $4\,\text{m} \times 25\,\text{m}$ is surrounded by a concrete path 1 m wide.
   a. Calculate the area of the path.
   b. Determine the cost of concreting at $45 per square metre.

15. If the radius is tripled and the height of a cylinder is divided by six, then determine the effect on its volume (in comparison with the original shape).

16. If the length is halved, the width is tripled and the height of a rectangular prism is doubled, then determine the effect on its volume (in comparison with the original shape).

17. A cylinder of radius 14 cm and height 20 cm is joined to a hemisphere of radius 14 cm to form a bread holder.
   a. Calculate the total surface area.
   b. Determine the cost of chroming the bread holder on the outside at $0.05 per cm^2.
   c. Calculate the storage volume of the bread holder.
   d. Determine how much more space is in this new bread holder than the one it is replacing, which had a quarter circle end with a radius of 18 cm and a length of 35 cm.

18. Bella Silos has two rows of silos for storing wheat.
Each row has 16 silos and all the silos are identical,
with a cylindrical base (height of 5 m, diameter of
1.5 m) and conical top (diameter of 1.5 m, height
of 1.1 m).

a. Calculate the slant height of the conical tops.
b. Determine the total surface area of all the silos.
c. Evaluate the cost of painting the silos if one litre of
paint covers $40\,\text{m}^2$ at a bulk order price of \$28.95
per litre.
d. Determine how much wheat can be stored altogether
in these silos.
e. Wheat is pumped from these silos into cartage trucks with rectangular containers 2.4 m wide, 5 m
long and 2.5 m high.
Determine how many truckloads are necessary to empty all the silos.
f. If wheat is pumped out of the silos at $2.5\,\text{m}^3/\text{min}$, determine how long it will take to fill one truck.

19. The Greek mathematician Eratosthenes developed an accurate method for calculating the circumference
of the Earth 2200 years ago!
The figure illustrates how he did this.

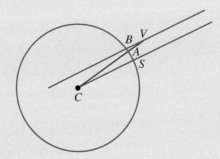

In this figure, A is the town of Alexandria and S is the town of Syene, exactly 787 km due south.
When the sun's rays (blue lines) were vertical at Syene, they formed an angle of 7.2° at Alexandria
($\angle BVA = 7.2°$), obtained by placing a stick at A and measuring the angle formed by the sun's shadow
with the stick.
a. Assuming that the sun's rays are parallel, evaluate the angle $\angle SCA$, correct to 1 decimal place.
b. Given that the arc $AS = 787$ km, determine the radius of the Earth, SC. Write your answer correct to
the nearest kilometre.
c. Given that the true radius is 6380 km, determine Eratosthenes' percentage error, correct to 1 decimal
place.

 To test your understanding and knowledge of this topic, go to your learnON title at
www.jacplus.com.au and complete the **post-test**.

Answers

Topic 6 Surface area and volume

6.1 Pre-test

1. $68.63\,\text{mm}^2$
2. $125.7\,\text{cm}^2$
3. D
4. $706.9\,\text{cm}^2$
5. $4\,\text{cm}^3$
6. $57.7\,\text{cm}^2$
7. \$864
8. B
9. $60\,\text{mm}^3$
10. C
11. D
12. $4.5\,\text{cm}$
13. C
14. $13.4\,\text{cm}^2$
15. $230\,\text{mL}$

6.2 Area

1. a. $16\,\text{cm}^2$ b. $48\,\text{cm}^2$ c. $75\,\text{cm}^2$
2. a. $120\,\text{cm}^2$ b. $706.86\,\text{cm}^2$ c. $73.5\,\text{mm}^2$
3. a. $254.47\,\text{cm}^2$ b. $21\,\text{m}^2$ c. $75\,\text{cm}^2$
4. a. $20.66\,\text{cm}^2$ b. $7.64\,\text{cm}^2$
5. a. $113.1\,\text{mm}^2$ b. $188.5\,\text{mm}^2$
6. a. i. $12\pi\,\text{cm}^2$ ii. $37.70\,\text{cm}^2$

   b. i. $\dfrac{69\pi}{2}\,\text{mm}^2$ ii. $108.38\,\text{mm}^2$

   c. i. $261\pi\,\text{cm}^2$ ii. $819.96\,\text{cm}^2$
7. D
8. D
9. a. $123.29\,\text{cm}^2$ b. $1427.88\,\text{m}^2$ c. $52\,\text{cm}^2$
10. a. $30.4\,\text{m}^2$ b. $78\,\text{cm}^2$ c. $2015.50\,\text{cm}^2$
11. a. $125.66\,\text{cm}^2$ b. $102.87\,\text{m}^2$
12. a. $13.73\,\text{m}^2$ b. $153.59\,\text{m}^2$
13. a. $27.86\,\text{m}^2$ b. $37.5\,\text{m}^2$
14. $11\,707.92\,\text{cm}^2$
15. $21\,\text{m}^2$
16. 60
17. \$840
18. a. $260.87\,\text{m}^2$ b. $195.71\,\text{m}^2$ c. 75%
19. a. Sample responses can be found in the worked solutions in the online resources.

    b. $2020.83\,\text{m}$; horizontal. For a vertical split, $987.5\,\text{m}$.

20. a. $50 = x + y$
    b. $y = 50 - x$
    c. Area $= 50x - x^2$
    d. See the table at the bottom of the page.*
    e. No, impossible to make a rectangle.
    f.
    g. $x = 25$
    h. $y = 25$
    i. Square
    j. $625\,\text{m}^2$
    k. $r = 15.92\,\text{m}$
    l. $795.77\,\text{m}^2$
    m. $170.77\,\text{m}^2$

21. a. Circular area, $1790.49\,\text{m}^2$; rectangular area, $1406.25\,\text{m}^2$

    b. Circular area, $\left(\dfrac{1}{4\pi}n^2\right)\,\text{m}^2$; rectangular (square) area, $\left(\dfrac{1}{16}n^2\right)\,\text{m}^2$. Circular area is always $\dfrac{4}{\pi}$ or 1.27 times larger.

22. a. $258.1\,\text{m}^2$ b. 7 bags c. \$79
    d. \$435.50

23. a. $\dfrac{29}{50}$ b. $x = 5, y = 5$

24. $32.88\,\text{cm}^2$

6.3 Total surface area

1. a. $600\,\text{cm}^2$ b. $384\,\text{cm}^2$ c. $1440\,\text{cm}^2$
   d. $27\,\text{m}^2$
2. a. $113.1\,\text{m}^2$ b. $6729.3\,\text{cm}^2$ c. $8.2\,\text{m}^2$
   d. $452.4\,\text{cm}^2$
3. a. $1495.4\,\text{cm}^2$ b. $502.7\,\text{cm}^2$
4. a. $506.0\,\text{cm}^2$ b. $9.4\,\text{m}^2$ c. $340.4\,\text{cm}^2$
   d. $224.1\,\text{cm}^2$
5. a. $13.5\,\text{m}^2$ b. $90\,\text{m}^2$ c. $11\,309.7\,\text{cm}^2$
6. a. $9852.0\,\text{mm}^2$ b. $125.7\,\text{cm}^2$ c. $1531.4\,\text{cm}^2$
7. a. $880\,\text{cm}^2$ b. $3072.8\,\text{cm}^2$ c. $75\,\text{cm}^2$
8. a. $70.4\,\text{cm}^2$ b. $193.5\,\text{cm}^2$ c. $1547.2\,\text{cm}^2$
9. B
10. a. $70.0\,\text{m}^2$ b. \$455
11. a. $3063.1\,\text{cm}^2$ b. \$168.47
12. $11\,216\,\text{cm}^2$

---

*20. d.

x	0	5	10	15	20	25	30	35	40	45	50
Area(m^2)	0	225	400	525	600	625	600	525	400	225	0

13. 60

14. $1592\,\text{cm}^2$

15. $1960

16. a. $8\pi\,\text{m}^2$ b. $\dfrac{\sqrt{7}}{2}\,\text{m}$ c. $4\sqrt{2}\pi\,\text{m}^2$

17. a. $6.6\,\text{m}^2$

    b. Back wall = 80 tiles
       Side wall = 50 tiles
       $80 + 50 + 50 = 180$ tiles

    c. Cheapest: 30 cm by 30 cm, $269.50; 20 cm by 20 cm
       (individually) $270; 20 cm by 20 cm (boxed) $276.50

18. The calculation is correct.

19. a. $\theta = 120°$ b. $x = 1;\, y = \sqrt{3}$
    c. $3\sqrt{3}\,\text{cm}^2$ d. $6\sqrt{3}\,\text{cm}^2$
    e. 32

20. The area of material required is $1.04\,\text{m}^2$. If Tina is careful in placing the pattern pieces, she may be able to cover the footstool.

21. $r = \dfrac{3\sqrt{3}a}{2}$

22. a. Arc length XY $= (x + s)\theta$
      Arc length AB $= x\theta$

    b. i. $x = \dfrac{2\pi t}{\theta} = \dfrac{st}{r - t}$

      ii. $\dfrac{x}{x + s} = \dfrac{t}{r}$

    c. Area of sector AVB $= \dfrac{x^2\theta}{2}$

      Area of sector XVY $= \dfrac{(s + x)^2\theta}{2}$

      Area of ABYX $= \dfrac{s\theta(s + 2x)}{2}$

      TSA of frustum $= \pi(t^2 + r^2) + \dfrac{s\theta(s + 2x)}{2}$

6.4 Volume

1. a. $27\,\text{cm}^3$ b. $74.088\,\text{m}^3$ c. $3600\,\text{cm}^3$
   d. $94.5\,\text{cm}^3$

2. a. $450\,\text{mm}^3$ b. $360\,\text{cm}^2$

3. a. $6333.5\,\text{cm}^3$ b. $19.1\,\text{m}^3$ c. $280\,\text{cm}^3$

4. a. $288\,\text{mm}^3$ b. $91.6\,\text{m}^3$ c. $21\,470.8\,\text{cm}^3$

5. a. $7.2\,\text{m}^3$ b. $14\,137.2\,\text{cm}^3$ c. $1436.8\,\text{mm}^3$
   d. $523\,598.8\,\text{cm}^3$

6. a. $113\,097.34\,\text{cm}^3$ b. $1.44\,\text{m}^3$
   c. $12\,214.51\,\text{mm}^3$ d. 101.93

7. a. $377.0\,\text{cm}^3$ b. $2303.8\,\text{mm}^3$

8. a. $400\,\text{cm}^3$ b. $10\,080\,\text{cm}^3$

9. a. $1400\,\text{cm}^3$ b. $10\,379.20\,\text{cm}^3$

10. a. $41.31\,\text{cm}^3$ b. $48.17\,\text{cm}^3$

11. a. $218.08\,\text{cm}^3$ b. $3691.37\,\text{cm}^3$

12. a. $V_{\text{new}} = 27l^3$; the volume will be 27 times as large as the original volume.

    b. $V_{\text{new}} = \dfrac{1}{8}l^2$; the volume will be $\dfrac{1}{8}$ of the original volume.

    c. $V_{\text{new}} = 2\pi r^2 h$; the volume will be twice as large as the original volume.

    d. $V_{\text{new}} = \pi r^2 h$; the volume will remain the same.

    e. $V_{\text{new}} = 3lwh$; the volume will be 3 times as large as the original value.

13. D

14. a. $256\,\text{cm}^3$ b. $5120\,\text{cm}^3$ c. $12\,000\,\text{cm}^3$
   d. $6880\,\text{cm}^3$ e. $7438.35\,\text{cm}^3$ f. $558.35\,\text{cm}^3$

15. $4417.9\,\text{L}$

16. $10\,215.05\,\text{cm}^3$

17. a. $H = 6\,\text{m}$ b. $112\,\text{m}^3$ c. 19 bins
   d. $112\,000\,\text{L}$ e. 1.95 m from floor

18. a. i. $4.57\,\text{cm}$ ii. $262.5\,\text{cm}^2$
   b. i. $14.15\,\text{cm}$ ii. $323.27\,\text{cm}^2$
   c. i. $33.33\,\text{cm}$ ii. $437.62\,\text{cm}^2$
   d. Sphere — costs less for a smaller surface area.

19. a. $h = \dfrac{V}{\pi r^2}$

   b. i. $31.8\,\text{cm}$ ii. $8.0\,\text{cm}$

   c. $\sqrt{\dfrac{V}{\pi h}}$

   d. $r \geq 0$, since r is a length.
   e. i. $7.6\,\text{cm}$ ii. $6.2\,\text{cm}$

20. 1000

21. a. $2.55\,\text{cm}$
   b. $35.68\,\text{cm}$
   c. $A_a = 157.88\,\text{m}^2$, $A_b = 12.01\,\text{m}^2$

22. a. $126.67\,\text{m}^3$ b. $53.33\,\text{m}^3$

23. Volume of water needed: $30.9\,\text{m}^3$

24. Sample responses can be found in the worked solutions in the online resources.

25. Required volume $= 1570.80\,\text{cm}^3$; tin volume $= 1500\,\text{cm}^3$; muffin tray volume $= 2814.72\,\text{cm}^3$. Marion could fill the tin and have a small amount of mixture left over, or she could almost fill 14 of the muffin cups and leave the remaining cups empty.

26. Cut squares of side length $s = 0.3\,\text{m}$ or $0.368\,\text{m}$ from the corners.

27. $1.94\,\text{m}$

28. a. $H = 12R$ b. $8\pi R^3$ c. $12\pi R^3$
   d. $8 : 12 = 2 : 3$

29. a. $\dfrac{1}{3}X^2H$ b. $\dfrac{1}{3}x^2(H - h)$

   c, d. Sample responses can be found in the worked solutions in the online resources.

30. 18 scoops

Project

1. **a.** The temperature reading is 26.5 °C.
   b. The smallest unit mark is 1°C, so the tolerance is 0.5.
   c. Largest possible value = 27 °C,
   smallest possible value = 26 °C

2. **a.** Largest value = 37.28, smallest value = 36.98
   b. Largest value = 2681.965, smallest value = 2649.285

3. **a.** **i.** 4 002 000
      ii. 4 000 000
   b. The result for i has 4 significant figures, whereas ii has only 1 significant figure after rounding. However, ii is closer to the actual value (3 986 297.386 144 940 9).

4. Volume using the incorrectly recorded value = 125 cm^3
   Volume using the actual value = 216 cm^3
   The percentage error is 42.1%, which shows that the error compounds as the number of dimensions increases.

6.5 Review questions

1. D
2. C
3. D
4. A
5. **a.** 84 cm^2 **b.** 100 cm^2 **c.** 6.50 cm^2
6. **a.** 56.52 cm^2 (when $\pi = 3.14$) or 56.55 cm^2 (when $\pi = 3.1416$)
   b. 60 cm^2
   c. 244.35 cm^2
7. **a.** 300 cm^2 **b.** 224.55 cm^2 **c.** 160 cm^2
8. **a.** 499.86 cm^2 **b.** 44.59 cm^2 **c.** 128.76 cm^2
9. **a.** 18 692.48 cm^2 **b.** 1495.40 mm^2 **c.** 804.25 cm^2
10. **a.** 871.79 cm^2 **b.** 873.36 mm^2 **c.** 760 cm^2
11. **a.** 343 cm^3 **b.** 672 cm^3 **c.** 153 938.04 cm^3
12. **a.** 1.45 m^3 **b.** 1800 cm^3 **c.** 1256.64 cm^3
13. **a.** 297 cm^3 **b.** 8400 cm^3 **c.** 7238.23 mm^3
14. **a.** 62 m^2 **b.** $2790
15. $V = \dfrac{3}{2}\pi r^2 h$; the volume will be 1.5 times as large as the original volume.
16. $V = 3lwh$; the volume will be 3 times as large as (or triple) the original volume.
17. **a.** 3606.55 cm^2 **b.** $180.33   **c.** 18 062.06 cm^3
    d. 9155.65 cm^3
18. **a.** 1.33 m
    b. 910.91 m^2
    c. $618.35 or $636.90 assuming you have to buy full litres (i.e not 0.7 of a litre)
    d. 303.48 m^3
    e. 11 trucks
    f. 12 minutes
19. **a.** 7.2°   **b.** 6263 km   **c.** 1.8% error

7 Quadratic expressions

LESSON
7.1 Overview

Why learn this?

How is your algebraic tool kit? Is there some room to expand your skills? As expressions become more complex, more power will be needed to manipulate them and to carry out basic algebraic skills such as adding, multiplying, expanding and factorising.

We have sent humans into space to live for months at a time on the International Space Station and even landed people on the moon. We have satellites circling our globe that enable us to communicate with friends and family around the world. Satellites also send out weather information, allowing meteorologists to study the patterns and changes and predict the upcoming weather conditions.

Technology connects us with others via our mobile phones and computers, often using a digital social media platform. All of these things that have become such important parts of our modern lives are only possible due to the application of mathematical techniques that we will begin to explore in this topic.

Many careers require a strong understanding of algebraic techniques. An example is a structural engineer. When designing structures, bridges, high rise buildings or domestic homes, they need to make decisions based on mathematics to ensure the structure is strong and durable, and passes all regulations. This topic will help you develop your knowledge of algebraic expressions, so that you can apply them to real-life situations.

1. Expand $-2(x+3)$.

2. Factorise $4m - 20m^2$.

3. Factorise $4a^2 + a^2b + 4ac + abc$.

4. Factorise $x^2 + 3x - 10$.

5. Expand and simplify $-(3x+1)^2$. Write your answer in descending powers.

6. **MC** $\left(\sqrt{2}+5x\right)\left(\sqrt{5}-2x\right) + \sqrt{2}x$ when expanded and simplified is:

   A. $\sqrt{10} - \sqrt{2}x + 5\sqrt{5}x - 10x^2$
   B. $\sqrt{10} - \sqrt{2}x + 5\sqrt{5}x - 10x$
   C. $10 - \sqrt{2}x + 5\sqrt{5}x - 10x^2$
   D. $10 - \sqrt{2}x + 5\sqrt{5}x - 10x$

7. **MC** $7x^2 - 700$ when factorised is:

   A. $\left(\sqrt{7}x - 100\right)\left(\sqrt{7}x + 100\right)$
   B. $\left(\sqrt{7}x - 10\right)\left(\sqrt{7}x + 10\right)$
   C. $(7x - 100)(7x + 100)$
   D. $7(x - 10)(x + 10)$

8. **MC** $4(2x+1)^2 - 9(y-4)^2$ when factorised is:

   A. $(4x + 3y - 11)(4x - 3y - 11)$
   B. $(8x - 9y + 1)(8x + 9y + 16)$
   C. $(4x + 3y - 10)(4x - 3y + 14)$
   D. $(2x + y - 3)(2x - y + 5)$

9. Complete the square to factorise $x^2 + 4x - 6$.

10. Expand $(x-2)(x+1)(2x+3)$. Write your answer in descending powers.

11. Factorise fully the algebraical expression $2x^2 - 6x - 10$.

12. **MC** $-2x^2 + 10x + 6$ when factorised is:

   A. $-2\left(x - \dfrac{5}{2} - \dfrac{\sqrt{17}}{2}\right)\left(x - \dfrac{5}{2} + \dfrac{\sqrt{17}}{2}\right)$
   B. $-2\left(x - 5 - \sqrt{37}\right)\left(x - 5 + \sqrt{37}\right)$
   C. $2\left(x + \dfrac{5}{2} - \dfrac{\sqrt{37}}{2}\right)\left(x + \dfrac{5}{2} + \dfrac{\sqrt{37}}{2}\right)$
   D. $-2\left(x - \dfrac{5}{2} - \dfrac{\sqrt{37}}{2}\right)\left(x - \dfrac{5}{2} + \dfrac{\sqrt{37}}{2}\right)$

13. Simplify $\dfrac{x^2 - 1}{x^2 + 2x - 3} \div \dfrac{x+1}{2x+6}$.

14. Factorise $4(x+6)^2 + 11(x+6) - 3$.

15. **MC** $\dfrac{6x^2 + 11x - 2}{6x^2 - x} \times \dfrac{x^2 - 4x + 4}{x^2 - 4}$ can be simplified to:

A. $\dfrac{-15x - 2}{-x}$ **B.** $\dfrac{x - 2}{x}$ **C.** -2 **D.** $\dfrac{-44x - 2}{x}$

LESSON
7.2 Expanding algebraic expressions

LEARNING INTENTIONS

At the end of this lesson you should be able to:
- expand binomial expressions using FOIL
- expand squares of binomial expressions
- expand binomial expressions using the difference of two squares (DOTS).

7.2.1 Binomial expansion

eles-4825

- Consider the rectangle of length $a + b$ and width $c + d$ shown below. Its area is equal to $(a + b)(c + d)$.

	a	$+$	b
c	ac		bc
$+$			
d	ad		bd

The diagram shows that $\underbrace{(a + b)(c + d)}_{\text{factorised form}} = \underbrace{ac + ad + bc + bd}_{\text{expanded form}}$.

- Expansion of the binomial expression $(x + 3)(x + 2)$ can be shown by this area model.

	x	$+$	3
x	$x \times x = x^2$		$3 \times x = 3x$
$+$			
2	$2 \times x = 2x$		$3 \times 2 = 6$

Expressed mathematically this is:

$$(x + 3)(x + 2) = x^2 + 2x + 3x + 6$$
$$= x^2 + 5x + 6$$

$\underbrace{\text{factorised form}} \quad \underbrace{\text{expanded form}}$

- There are several methods that can be used to expand binomial factors.

FOIL method

- The word **FOIL** provides us with an acronym for the expansion of a binomial product.
  - **First:** multiply the first terms in each bracket.

$$\overset{\text{F}}{\overparen{(x + a)(x} - b)}$$

- **Outer:** multiply the two outer terms.

$$\overset{\text{O}}{\overparen{(x + a)(x - b)}}$$

- **Inner:** multiply the two inner terms.

$$\overset{\text{I}}{(x + a\overparen{)(x} - b)}$$

- **Last:** multiply the last terms in each bracket.

$$\overset{\text{L}}{(x + a\overparen{)(x - b)}}$$

WORKED EXAMPLE 1 Expanding binomial expressions using FOIL

Expand each of the following expressions.

a. $(x + 3)(x + 2)$

b. $(x - 7)(6 - x)$

THINK	WRITE
a. 1. Write the expression.	a. $(x + 3)(x + 2)$
2. Use FOIL to expand the pair of brackets.	$(x \times x) + (x \times 2) + (3 \times x) + (3 \times 2)$
3. Simplify and then collect like terms.	$= x^2 + 2x + 3x + 6$
	$= x^2 + 5x + 6$
b. 1. Write the expression.	b. $(x - 7)(6 - x)$
2. Use FOIL to expand the pair of brackets. Remember to include the negative signs where appropriate.	$(x \times 6) + (x \times -x) + (-7 \times 6) + (-7 \times -x)$
3. Simplify and then collect like terms.	$= 6x - x^2 - 42 + 7x$
	$= -x^2 + 13x - 42$

- If there is a term outside the pair of brackets, expand the brackets and then multiply each term of the expansion by that term.

WORKED EXAMPLE 2 Expanding binomial expressions

Expand $3(x + 8)(x + 2)$.

THINK	WRITE
1. Write the expression.	$3(x + 8)(x + 2)$
2. Use FOIL to expand the pair of brackets.	$= 3(x^2 + 2x + 8x + 16)$
3. Collect like terms within the brackets.	$= 3(x^2 + 10x + 16)$
4. Multiply each of the terms inside the brackets by the term outside the brackets.	$= 3x^2 + 30x + 48$

▶ 7.2.2 The square of a binomial expression

eles-4827

- The expansion of $(a + b)^2$ can be represented by this area model.

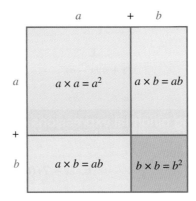

- This result provides a shortcut method for squaring binomials.

Squaring binomial expressions

$$(a + b)^2 = a^2 + 2ab + b^2$$

Similarly,

$$(a - b)^2 = a^2 - 2ab + b^2$$

- This expansion is often memorised. To determine the square of a binomial:
  - square the first term
  - multiply the two terms together and then double the product
  - square the last term.

WORKED EXAMPLE 3 Expanding and simplifying binomial expressions

Expand and simplify each of the following expressions.

a. $(2x - 5)^2$

b. $-3(2x + 7)^2$

THINK	WRITE
a. 1. Write the expression.	a. $(2x - 5)^2$
2. Expand using the rule $(a - b)^2 = a^2 - 2ab + b^2$.	$= (2x)^2 - 2 \times 2x \times 5 + (5)^2$ $= 4x^2 - 20x + 25$
b. 1. Write the expression.	b. $-3(2x + 7)^2$
2. Expand the brackets using the rule $(a + b)^2 = a^2 + 2ab + b^2$.	$= -3[(2x)^2 + 2 \times 2x \times 7 + (7)^2]$ $= -3(4x^2 + 28x + 49)$
3. Multiply every term inside the brackets by the term outside the brackets.	$\stackrel{.}{=} -12x^2 - 84x - 147$

TI	THINK	DISPLAY/WRITE	CASIO	THINK	DISPLAY/WRITE
a, b. In a new problem, on a Calculator page, press: • CATALOG • 1 • E Then scroll down to select expand(Complete the entry lines as: expand $\left((2x - 5)^2\right)$ expand $\left(-3(2x + 7)^2\right)$ Press ENTER after each entry.	a, b. $(2x - 5)^2 = 4x^2 - 20x + 25$ $-3(2x + 7)^2 = -12x^2 - 84x - 147$	a, b. On the Main screen, tap: • Action • Transformation • expand Complete the entry lines as: expand $(2x - 5)^2$ expand $\left(-3(2x + 7)^2\right)$ Press EXE after each entry.	a, b. $(2x - 5)^2 = 4x^2 - 20x + 25$ $-3(2x + 7)^2 = -12x^2 - 84x - 147$

7.2.3 The difference of two squares

eles-4828

- When $(a + b)$ is multiplied by $(a - b)$ (or vice versa),

$$(a + b)(a - b) = a^2 - ab + ab - b^2$$
$$= a^2 - b^2$$

The expression is called the difference of two squares and is often referred to as DOTS.

D Difference
O Of
T Two
S Squares

Difference of two squares

$$(a + b)(a - b) = a^2 - b^2$$

WORKED EXAMPLE 4 Expanding using DOTS

Expand and simplify each of the following expressions.

a. $(3x+1)(3x-1)$

b. $4(2x-7)(2x+7)$

THINK	WRITE
a. 1. Write the expression.	**a.** $(3x+1)(3x-1)$
2. Expand using the rule $(a+b)(a-b)=a^2-b^2$.	$=(3x)^2-(1)^2$
	$=9x^2-1$
b. 1. Write the expression.	**b.** $4(2x-7)(2x+7)$
2. Expand using the difference of two squares rule.	$=4[(2x)^2-(7)^2]$
	$=4(4x^2-49)$
3. Multiply by 4.	$=16x^2-196$

 Resources

 eWorkbook Topic 7 Workbook (worksheets, code puzzle and project) (ewbk-13292)

Video eLesson Expansion of binomial expressions (eles-1908)

Interactivities Individual pathway interactivity: Expanding algebraic expressions (int-4596)
Expanding binomial factors (int-6033)
Difference of two squares (int-6036)

Exercise 7.2 Expanding algebraic expressions

learn on

7.2 Quick quiz on	7.2 Exercise

Individual pathways

■ PRACTISE	■ CONSOLIDATE	■ MASTER
1, 4, 7, 10, 14, 16, 17, 20, 23, 28, 29, 30, 34	2, 5, 8, 11, 12, 15, 18, 21, 24, 26, 31, 35	3, 6, 9, 13, 19, 22, 25, 27, 32, 33, 36

Fluency

For questions 1–3, expand the expressions.

1. **a.** $2(x+3)$ **b.** $4(x-5)$ **c.** $3(7-x)$ **d.** $-(x+3)$

2. **a.** $x(x+2)$ **b.** $2x(x-4)$ **c.** $3x(5x-2)$ **d.** $5x(2-3x)$

3. **a.** $2x(4x+1)$ **b.** $2x^2(2x-3)$ **c.** $3x^2(2x-1)$ **d.** $5x^2(3x+4)$

For questions **4**–**6**, expand the expressions.

4. a. $(x+3)(x-4)$ b. $(x+1)(x-3)$ c. $(x-7)(x+2)$ d. $(x-1)(x-5)$

5. a. $(2-x)(x+3)$ b. $(x-4)(x-2)$ c. $(2x-3)(x-7)$ d. $(x-1)(3x+2)$

6. a. $(3x-1)(2x-5)$ b. $(3-2x)(7-x)$ c. $(5-2x)(3+4x)$ d. $(11-3x)(10+7x)$

WE2 For questions **7**–**11**, expand the expressions.

7. a. $2(x+1)(x-3)$ b. $4(2x+1)(x-4)$ c. $-2(x+1)(x-7)$

8. a. $2x(x-1)(x+1)$ b. $3x(x-5)(x+5)$ c. $6x(x-3)(x+3)$

9. a. $-2x(3-x)(x-3)$ b. $-5x(2-x)(x-4)$ c. $6x(x+5)(4-x)$

10. a. $(x-1)(x+1)(x+2)$ b. $(x-3)(x-1)(x+2)$ c. $(x-5)(x+1)(x-1)$

11. a. $(x-1)(x-2)(x-3)$ b. $(2x-1)(x+1)(x-4)$ c. $(3x+1)(2x-1)(x-1)$

For questions **12** and **13**, expand and simplify the expressions.

12. a. $(x+2)(x-1)-2x$
        b. $3x-(2x-5)(x+2)$
    c. $(2x-3)(x+1)+(3x+1)(x-2)$
        d. $(3-2x)(2x-1)+(4x-5)(x+4)$

13. a. $(x+1)(x-7)-(x+2)(x-3)$
        b. $(x-2)(x-5)-(x-1)(x-4)$
    c. $(x-3)(x+1)+\sqrt{3}x$
        d. $\left(\sqrt{2}-3x\right)\left(\sqrt{3}+2x\right)-\sqrt{5}x$

14. **MC** $(3x-1)(2x+4)$ expands to:
    A. $6x^2+10x-4$
    B. $5x^2-24x+3$
    C. $3x^2+2x-4$
    D. $6x^2-10x-4$

15. **MC** $-2x(x-1)(x+3)$ expands to:
    A. x^2+2x-3
    B. $-2x^2-4x+6$
    C. $-2x^3-4x^2+6x$
    D. $-2x^3+4x^2-6x$

16. **MC** The expression $(x-1)(x-3)(x+2)$ is *not* the same as:
    A. $(x-3)(x-1)(x+2)$
    B. $(x+3)(x-1)(x-2)$
    C. $(x-1)(x+2)(x-3)$
    D. $(x+2)(x-1)(x-3)$

WE3a For questions **17**–**19**, expand and simplify the expressions.

17. a. $(x-1)^2$ b. $(x+2)^2$ c. $(x+5)^2$ d. $(4+x)^2$

18. a. $(7-x)^2$ b. $(12-x)^2$ c. $(3x-1)^2$ d. $(12x-3)^2$

19. a. $(5x+2)^2$ b. $(2-3x)^2$ c. $(5-4x)^2$ d. $(1-5x)^2$

Understanding

WE3b For questions **20**–**22**, expand and simplify the expressions.

20. a. $2(x-3)^2$ b. $4(x-7)^2$ c. $3(x+1)^2$

21. a. $-(2x+3)^2$ b. $-(7x-1)^2$ c. $2(2x-3)^2$

22. a. $-3(2-9x)^2$ b. $-5(3-11x)^2$ c. $-4(2x+1)^2$

WE4 For questions **23**–**25**, expand and simplify the expressions.

23. a. $(x+7)(x-7)$ b. $(x+9)(x-9)$ c. $(x-5)(x+5)$

24. a. $(x-1)(x+1)$ **b.** $(2x-3)(2x+3)$ **c.** $(3x-1)(3x+1)$

25. a. $(7-x)(7+x)$ **b.** $(8+x)(8-x)$ **c.** $(3-2x)(3+2x)$

26. The length of the side of a rectangle is $(x+1)$ cm and the width is $(x-3)$ cm.

     a. Determine an expression for the area of the rectangle.
     b. Simplify the expression by expanding.
     c. If $x=5$ cm, calculate the dimensions of the rectangle and hence its area.

27. Chickens are kept in a square enclosure with sides measuring x m. The number of chickens is increasing, so the enclosure is to have 1 metre added to one side and 2 metres to the adjacent side.

     a. Draw a diagram of the original enclosure.
     b. Add to the first diagram or draw another one to show the new enclosure. Mark the lengths on each side on your diagram.
     c. Write an expression for the area of the new enclosure in factorised form.
     d. Expand and simplify the expression by removing the brackets.
     e. If the original enclosure had sides of 2 metres, calculate the area of the original square and then the area of the new enclosure.

28. Write an expression in factorised and expanded form that is:

     a. a quadratic trinomial   **b.** the square of a binomial
     c. the difference of two squares   **d.** both **a** and **b**.

Reasoning

29. Shown below are three students' attempts at expanding $(3x+4)(2x+5)$.

STUDENT A

$$(3x+4)(2x+5)$$
$$= 3x \times 2x + 3x \times 5 + 4 \times 2x + 4 \times 5$$
$$= 6x + 15x + 8x + 20$$
$$= 29x + 20$$
$$= 49x$$

STUDENT B

$$(3x+4)(2x+5)$$
$$= 3x \times 2x + 4 \times 2x + 4 \times 5$$
$$= 6x^2 + 8x + 20$$

STUDENT C

$$(3x+4)(2x+5)$$
$$= 3x \times 2x + 3x \times 5 + 4 \times 2x + 4 \times 5$$
$$= 6x^2 + 15x + 8x + 20$$
$$= 6x^2 + 23x + 20$$

     a. Identify which student's work was correct.
     b. Correct the mistakes in each wrong case as though you were the teacher of these students. Explain your answer.

30. If $a = 5$ and $b = 3$, show that $(a - b)(a + b) = a^2 - b^2$ by evaluating both expressions.

31. If $a = 5$ and $b = 3$, show that $(a + b)^2 = a^2 + 2ab + b^2$ by evaluating both expressions.

32. Show that $(a + b)(c + d) = (c + d)(a + b)$.

33. Explain the difference between 'the square of a binomial' and 'the difference between two squares'.

Problem solving

34. Determine an expanded expression for the volume of the cuboid shown.

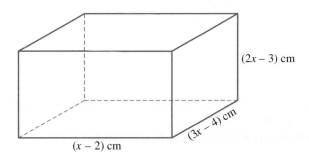

$(2x - 3)$ cm

$(3x - 4)$ cm

$(x - 2)$ cm

35. Determine an expanded expression for the total surface area of the square-based pyramid.

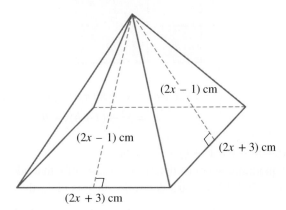

$(2x - 1)$ cm

$(2x - 1)$ cm

$(2x + 3)$ cm

$(2x + 3)$ cm

36. Expand the following expressions.

   a. $(2x + 3y - 5z)^2$

   b. $\left(\left(1 + \dfrac{1}{2x}\right) - 2x\right)^2$

LESSON
7.3 Factorising expressions with three terms

7.3.1 Factorising monic quadratic trinomials

eles-4829

- A monic quadratic expression is an expression in the form $ax^2 + bx + c$ where $a = 1$.
- To factorise a monic quadratic expression $x^2 + bx + c$, we need to determine the numbers f and h such that $x^2 + bx + c = (x + f)(x + h)$.
- Using FOIL or the area model, $(x + f)(x + h)$ can be expanded as follows:

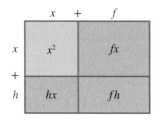

$$(x + f)(x + h) = x^2 + xh + xf + fh$$
$$= x^2 + (f + h)x + fh$$

- Equating the quadratics: $x^2 + bx + c = x^2 + (f + h)x + fh$.
- Therefore, to factorise the quadratic we must determine factors of c that add to b.
- For example, to factorise $x^2 + 7x + 12$ we must determine factors of 12 that add to 7.

Factors of 12	Sum of factors
1 and 12	13
2 and 6	8
3 and 4	7

The factors of 12 that add to 7 are 3 and 4. Therefore, $x^2 + 7x + 12 = (x + 3)(x + 4)$.
- This result can be checked by expanding using FOIL.

$$(x + 3)(x + 4) = x^2 + 4x + 3x + 12$$
$$= x^2 + 7x + 12$$

Factorising a monic quadratic

$$x^2 + bx + c = (x + f)(x + h)$$

where f and h are factors of c that sum to b.

Factorise the following quadratic expressions.

a. $x^2 + 5x + 6$　　　　　　　　　　　　　　b. $x^2 + 10x + 24$

THINK	WRITE
a. 1. List all of the factors of 6 and determine the sum of each pair of factors. Highlight the pair of factors that add to 5.	(table below)

Factors of 6	Sum of factors
1 and 6	7
2 and 3	**5**

2. Factorise the quadratic using the factor pair highlighted in step 1.

$$x^2 + 5x + 6 = (x+2)(x+3)$$

3. Check by expanding back.

$$(x+2)(x+3) = x(x+3) + 2(x+3)$$
$$= x^2 + 3x + 2x + 6$$
$$= x^2 + 5x + 6$$

b. 1. List all of the factors of 24 and determine the sum of each pair of factors. Highlight the pair of factors that add to 10.

Factors of 24	Sum of factors
1 and 24	25
2 and 12	14
3 and 8	11
4 and 6	**10**

2. Factorise the quadratic using the factor pair highlighted in step 1.

$$x^2 + 10x + 24 = (x+4)(x+6)$$

3. Check by expanding back.

$$(x+4)(x+6) = x(x+6) + 4(x+6)$$
$$= x^2 + 6x + 4x + 24$$
$$= x^2 + 10x + 24$$

▶ 7.3.2 Factorising non-monic quadratic trinomials

eles-4830

- A non-monic quadratic expression is an expression in the form $ax^2 + bx + c$, where $a \neq 1$.
- Quadratic expressions of the form $ax^2 + bx + c$ can be factorised by following the process shown below.

Steps	Process	Example: $2x^2 + 11x + 12$
1	Determine factors of ac that add up to b.	$a = 2$, $b = 11$, $c = 12$, $ac = 24$ The factors of 24 that add up to 11 are 8 and 3.
2	Rewrite the quadratic expression as $ax^2 + mx + nx + c$, where m and n are the factors of ac that add up to b (as found in step 1).	$2x^2 + 11x + 12 = 2x^2 + 8x + 3x + 12$
3	Order the terms in $ax^2 + mx + nx + c$ so that the two terms on the left share a common factor and the two terms on the right also share a common factor.	$= (2x^2 + 8x) + (3x + 12)$
4	Factorise the two left terms and two right terms independently by taking out their common factor.	$= 2x(x+4) + 3(x+4)$
5	Factorise the expression by taking out the binomial factor.	$= (x+4)(2x+3)$

- The reasoning behind step 1 is not given in the table above. The working below shows why we use factors of ac that add to b. You do not need to be able to complete this working yourself, but try to read along and understand it.

$$(dx + e)(fx + g) = dfx^2 + dgx + efx + eg$$
$$= dfx^2 + (dg + ef)x + eg$$

$$m + n = dg + ef \qquad \text{and} \qquad m \times n = dg \times ef$$
$$= b \qquad\qquad\qquad = dgef$$
$$= dfeg$$
$$= ac$$

Factorising a non-monic quadratic

To factorise a general quadratic of the form $ax^2 + bx + c$, find factors of ac that sum to b. Then rewrite the expression as four terms that can be regrouped and factorised.

$$ax^2 + bx + c = ax^2 + mx + nx + c$$

Factors of ac that sum to b

WORKED EXAMPLE 6 Factorising a non-monic quadratic trinomial

Factorise $6x^2 - 11x - 10$.

THINK	WRITE
1. Write the expression and look for common factors and special patterns. The expression is a general quadratic with $a = 6$, $b = -11$ and $c = -10$.	$6x^2 - 11x - 10$
2. List factors of -60 and determine the sum of each pair of factors. Highlight the pair of factors that add to -11.	<table><tr><th>Factors of −60 (6 × −10)</th><th>Sum of factors</th></tr><tr><td>−60, 1</td><td>−59</td></tr><tr><td>−20, 3</td><td>−17</td></tr><tr><td>−30, 2</td><td>−28</td></tr><tr><td>15, −4</td><td>11</td></tr><tr><td>−15, 4</td><td>−11</td></tr></table>
3. Rewrite the quadratic expression: $ax^2 + bx + c = ax^2 + mx + nx + c$ with $m = 4$ and $n = -15$.	$6x^2 - 11x - 10 = 6x^2 + 4x + -15x - 10$
4. Factorise using the grouping method: $6x^2 + 4x = 2x(3x + 2)$ and $-15x - 10 = -5(3x + 2)$. Write the answer.	$6x^2 - 11x - 10 = 2x(3x + 2) + -5(3x + 2)$ $= (3x + 2)(2x - 5)$

Exercise 7.3 Factorising expressions with three terms   learn on

7.3 Quick quiz on	**7.3 Exercise**

Individual pathways

■ PRACTISE	■ CONSOLIDATE	■ MASTER
1, 4, 7, 10, 12, 13, 16, 19, 22, 25	2, 5, 8, 11, 14, 17, 20, 23, 26, 27	3, 6, 9, 15, 18, 21, 24, 28, 29

Fluency

WE5 For questions **1–3**, factorise the expressions.

1. a. $x^2 + 3x + 2$
   b. $x^2 + 4x + 3$
   c. $x^2 + 10x + 16$
   d. $x^2 + 8x + 16$
   e. $x^2 - 2x - 3$

2. a. $x^2 - 3x - 4$
   b. $x^2 - 11x - 12$
   c. $x^2 - 4x - 12$
   d. $x^2 + 3x - 4$
   e. $x^2 + 4x - 5$

3. a. $x^2 + 6x - 7$
   b. $x^2 + 3x - 10$
   c. $x^2 - 4x + 3$
   d. $x^2 - 9x + 20$
   e. $x^2 + 9x - 70$

WE6 For questions **4–9**, factorise the expressions.

4. a. $-2x^2 - 20x - 18$
   b. $-3x^2 - 9x - 6$
   c. $-x^2 - 3x - 2$
   d. $-x^2 - 11x - 10$

5. a. $-x^2 - 7x - 10$
   b. $-x^2 - 13x - 12$
   c. $-x^2 - 7x - 12$
   d. $-x^2 - 8x - 12$

6. a. $2x^2 + 14x + 20$
   b. $3x^2 + 33x + 30$
   c. $5x^2 + 105x + 100$
   d. $5x^2 + 45x + 100$

7. a. $a^2 - 6a - 7$
   b. $t^2 - 6t + 8$
   c. $b^2 + 5b + 4$
   d. $m^2 + 2m - 15$

8. a. $p^2 - 13p - 48$
   b. $c^2 + 13c - 48$
   c. $k^2 + 22k + 57$
   d. $s^2 - 16s - 57$

9. a. $g^2 - g - 72$
   b. $v^2 - 28v + 75$
   c. $x^2 + 14x - 32$
   d. $x^2 - 19x + 60$

10. **MC** To factorise $-14x^2 - 49x + 21$, the first step is to:
    A. determine factors of 14 and 21 that will add to -49.
    B. take out 14 as a common factor.
    C. take out -7 as a common factor.
    D. determine factors of 14 and -49 that will add to make 21.

11. **MC** The expression $42x^2 - 9x - 6$ can be completely factorised to:

    A. $(6x - 3)(7x + 2)$ **B.** $3(2x - 1)(7x + 2)$

    C. $(2x - 1)(21x + 6)$ **D.** $3(2x + 1)(7x - 2)$

12. **MC** When factorised, $(x + 2)^2 - (y + 3)^2$ equals:

    A. $(x + y - 2)(x + y + 2)$ **B.** $(x - y - 1)(x + y - 1)$

    C. $(x - y - 1)(x + y + 5)$ **D.** $(x - y + 1)(x + y + 5)$

Understanding

For questions **13–15**, factorise the expressions using an appropriate method.

13. **a.** $2x^2 + 5x + 2$ **b.** $2x^2 - 3x + 1$ **c.** $4x^2 - 17x - 15$ **d.** $4x^2 + 4x - 3$

14. **a.** $2x^2 - 9x - 35$ **b.** $3x^2 + 10x + 3$ **c.** $6x^2 - 17x + 7$ **d.** $12x^2 - 13x - 14$

15. **a.** $10x^2 - 9x - 9$ **b.** $20x^2 + 3x - 2$ **c.** $12x^2 + 5x - 2$ **d.** $15x^2 + x - 2$

For questions **16–18**, factorise the expressions. Remember to look for a common factor first.

16. **a.** $4x^2 + 2x - 6$ **b.** $9x^2 - 60x - 21$ **c.** $72x^2 + 12x - 12$ **d.** $-18x^2 + 3x + 3$

17. **a.** $-60x^2 + 150x + 90$ **b.** $24ax^2 + 18ax - 105a$ **c.** $-8x^2 + 22x - 12$ **d.** $-10x^2 + 31x + 14$

18. **a.** $-24x^2 + 35x - 4$ **b.** $-12x^2 - 2xy + 2y^2$

    c. $-30x^2 + 85xy + 70y^2$ **d.** $-600x^2 - 780xy - 252y^2$

19. Consider the expression $(x - 1)^2 + 5(x - 1) - 6$.

    a. Substitute $w = x - 1$ in this expression.

    b. Factorise the resulting quadratic.

    c. Replace w with $x - 1$ and simplify each factor. This is the factorised form of the original expression.

For questions **20** and **21**, use the method outlined in question **19** to factorise the expressions.

20. **a.** $(x + 1)^2 + 3(x + 1) - 4$ **b.** $(x + 2)^2 + (x + 2) - 6$ **c.** $(x - 3)^2 + 4(x - 3) + 4$

21. **a.** $(x + 3)^2 + 8(x + 3) + 12$ **b.** $(x - 7)^2 - 7(x - 7) - 8$ **c.** $(x - 5)^2 - 3(x - 5) - 10$

Reasoning

22. Fabric pieces comprising yellow squares, white squares and black rectangles are sewn together to make larger squares (patches) as shown in the diagram.

    The length of each black rectangle is twice its width. These patches are then sewn together to make a patchwork quilt.

    A finished square quilt made from 100 patches has an area of $1.44\,\text{m}^2$.

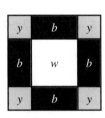

    a. Determine the size of each yellow, black and white section in one fabric piece. Show your working.

    b. Determine how much (in m^2) of each of the coloured fabrics would be needed to construct the quilt.

    (Ignore seam allowances.)

    c. Sketch a section of the finished product.

23. Each factorisation below contains an error. Identify the error in each statement.

    a. $x^2 - 7x + 12 = (x + 3)(x - 4)$ **b.** $x^2 - x - 12 = (x - 3)(x + 4)$

    c. $x^2 - x - 2 = (x - 1)(x + 2)$ **d.** $x^2 - 4x - 21 = (x - 3)(x - 7)$

24. Each factorisation below contains an error. Identify the error in each statement.

   a. $x^2 + 4x - 21 = (x + 3)(x - 7)$
   b. $x^2 - x - 30 = (x - 5)(x + 6)$
   c. $x^2 + 7x - 8 = (x + 1)(x - 8)$
   d. $x^2 - 11x + 30 = (x - 5)(x + 6)$

Problem solving

25. Cameron wants to build an in-ground 'endless' pool. Basic models have a depth of 2 metres and a length triple the width. A spa will also be attached to the end of the pool.

   a. The pool needs to be tiled. Write an expression for the surface area of the empty pool (that is, the floor and walls only).
   b. The spa needs an additional $16\,\text{m}^2$ of tiles. Write an expression for the total area of tiles needed for both the pool and the spa.
   c. Factorise this expression.
   d. Cameron decides to use tiles that are selling at a discount price, but there is only $280\,\text{m}^2$ of the tile available. Determine the maximum dimensions of the pool he can build if the width is in whole metres. Assume the spa is to be included in the tiling.
   e. Evaluate the area of tiles that is actually needed to construct the spa and pool.
   f. Determine the volume of water the pool can hold.

26. Factorise $x^2 + x - 0.75$.

27. The area of a rectangular playground is given by the general expression $\left(6x^2 + 11x + 3\right)\,\text{m}^2$, where x is a positive whole number.

   a. Determine the length and width of the playground in terms of x.
   b. Write an expression for the perimeter of the playground.
   c. If the perimeter of a particular playground is 88 metres, evaluate x.

28. Factorise:

   a. $6(3a - 1)^2 - 13(3a - 1) - 5$
   b. $3m^4 - 19m^2 - 14$
   c. $2\sin^2(x) - 3\sin(x) + 1$.

29. Students decide to make Science Day invitation cards.
The total area of each card is equal to $(x^2 - 4x - 5)\,\text{cm}^2$.

   a. Factorise the expression to determine the dimensions of the cards in terms of x.
   b. Write down the length of the shorter side in terms of x.
   c. If the shorter side of a card is $10\,\text{cm}$ in length and the longer side is $16\,\text{cm}$ in length, determine the value of x.
   d. Evaluate the area of the card proposed in part **c**.
   e. If the students want to make 3000 Science Day invitation cards, determine how much cardboard will be required. Give your answer in terms of x.

LESSON
7.4 Factorising expressions with two or four terms

LEARNING INTENTIONS

At the end of this lesson you should be able to:
- factorise expressions by taking out the highest common factor
- recognise and factorise expressions that can be written as DOTS
- factorise expressions with four terms by grouping.

7.4.1 Factorising expressions with two terms

eles-4831

- To factorise an expression with two terms, follow the following steps:

  Step 1: Look for the highest common factor (HCF) of the terms.

  Step 2: Take out the HCF.

  Step 3: Check to see if the remaining expression is a difference of two squares (DOTS).

  For example:

  $$4x^2 - 36 = 4(x^2 - 9) \quad \text{taking out the HCF of 4}$$
  $$= 4(x - 3)(x + 3) \quad \text{using DOTS}$$

  Remember, the formula for the difference of two squares is:

  $$a^2 - b^2 = (a - b)(a + b)$$

WORKED EXAMPLE 7 Factorising expressions with two terms

Factorise the following.

a. $12k^2 + 18$

b. $16a^2 - 25b^4$

THINK	WRITE
a. 1. Write the expression and look for common factors. The terms have a highest common factor of 6. Write the 6 in front of a set of brackets, then determine what must go inside the brackets: $12k^2 = 6 \times 2k^2$, $18 = 6 \times 3$. Look for patterns in the expression inside the brackets to factorise further. The expression inside the brackets cannot be factorised further.	**a.** $12k^2 + 18 = 6(2k^2 + 3)$
b. 1. Write the expression and look for common factors. The expression has no common factors.	**b.** $16a^2 - 25b^4$
2. Look for the DOTS pattern in the expression. Write the equation showing squares.	$= 4^2 a^2 - 5^2 (b^2)^2$ $= (4a)^2 - (5b^2)^2$
3. Use the pattern for DOTS to write the factors. $a^2 - b^2 = (a + b)(a - b)$	$= (4a + 5b^2)(4a - 5b^2)$

▶ 7.4.2 Factorising expressions with four terms

eles-4832

- To factorise an expression with four terms, follow the following steps.
  Step 1: Look for a common factor.
  Step 2: Group in pairs that have a common factor.
  Step 3: Factorise each pair by taking out the common factor.
  Step 4: Factorise fully by taking out the common binomial factor.
- This process is known as grouping 'two and two'.
  For example,

$$10a - 30b + 15ac - 45bc$$ **terms have a common factor of 5**
$$= 5(2a - 6b + 3ac - 9bc)$$ **grouping in pairs**
$$= 5(2(a - 3b) + 3c(a - 3b))$$ **factorising each pair**
$$= 5(a - 3b)(2 + 3c)$$ **common binomial factor**

WORKED EXAMPLE 8 Factorising expressions with four terms by grouping 'two and two'

Factorise each of the following expressions.

a. $x - 4y + mx - 4my$

b. $x^2 + 3x - y^2 + 3y$

THINK

WRITE

a. 1. Write the expression and look for a common factor. (There isn't one.)

a. $x - 4y + mx - 4my$

 2. Group the terms so that those with common factors are next to each other.

$= (x - 4y) + (mx - 4my)$

 3. Take out a common factor from each group (it may be 1).

$= 1(x - 4y) + m(x - 4y)$

 4. Factorise by taking out a common binomial factor. The factor $(x - 4y)$ is common to both groups.

$= (x - 4y)(1 + m)$

b. 1. Write the expression and look for a common factor.

b. $x^2 + 3x - y^2 + 3y$

 2. Group the terms so that those with common factors are next to each other.

$= (x^2 - y^2) + (3x + 3y)$

 3. Factorise each group.

$= (x + y)(x - y) + 3(x + y)$

 4. Factorise by taking out a common binomial factor. The factor $(x + y)$ is common to both groups.

$= (x + y)(x - y + 3)$

- Sometimes an expression containing four terms can be factorised by grouping three terms together and then factorising. This method is known as grouping 'three and one', and it is demonstrated in Worked example 9.

WORKED EXAMPLE 9 Factorising by grouping 'three and one'

Factorise the following expression: $x^2 + 12x + 36 - y^2$.

THINK	WRITE
1. Write the expression and look for a common factor.	$x^2 + 12x + 36 - y^2$
2. Group the terms so that those that can be factorised are next to each other.	$= (x^2 + 12x + 36) - y^2$
3. Factorise the quadratic trinomial. This is the form of a perfect square.	$= (x + 6)(x + 6) - y^2$ $= (x + 6)^2 - y^2$
4. Factorise the expression using $a^2 - b^2 = (a + b)(a - b)$.	$= (x + 6 + y)(x + 6 - y)$

 Resources

- **eWorkbook**   Topic 7 Workbook (worksheets, code puzzle and project) (ewbk-13292)
- **Interactivities** Individual pathway interactivity: Factorising expressions with two or four terms (int-4598)
  Factorising expressions with four terms (int-6145)

Exercise 7.4 Factorising expressions with two or four terms   learn on

7.4 Quick quiz on	7.4 Exercise

Individual pathways

■ PRACTISE	■ CONSOLIDATE	■ MASTER
1, 4, 6, 9, 12, 15, 16, 19, 22, 25, 31, 32, 35, 38	2, 5, 7, 10, 13, 17, 20, 23, 26, 27, 29, 33, 36, 39	3, 8, 11, 14, 18, 21, 24, 28, 30, 34, 37, 40

Fluency

For questions 1–3, factorise the expressions by taking out a common factor.

1. a. $x^2 + 3x$
   b. $x^2 - 4x$
   c. $3x^2 - 6x$

2. a. $4x^2 + 16x$
   b. $9x^2 - 3x$
   c. $8x - 8x^2$

3. a. $12x - 3x^2$
   b. $8x - 12x^2$
   c. $8x^2 - 11x$

For questions 4 and 5, factorise the expressions by taking out a common binomial factor.

4. a. $3x(x - 2) + 2(x - 2)$
   b. $5(x + 3) - 2x(x + 3)$
   c. $(x - 1)^2 + 6(x - 1)$

5. a. $(x + 1)^2 - 2(x + 1)$
   b. $(x + 4)(x - 4) + 2(x + 4)$
   c. $7(x - 3) - (x + 3)(x - 3)$

WE7 For questions 6–11, factorise the expressions.

6. a. $x^2 - 1$
   b. $x^2 - 9$
   c. $x^2 - 25$

7. a. $x^2 - 100$
   b. $y^2 - k^2$
   c. $4x^2 - 9y^2$

8. a. $16a^2 - 49$
   b. $25p^2 - 36q^2$
   c. $1 - 100d^2$

9. a. $4x^2 - 4$ b. $5x^2 - 80$ c. $ax^2 - 9a$

10. a. $2b^2 - 8d^2$ b. $100x^2 - 1600$ c. $3ax^2 - 147a$

11. a. $4px^2 - 256p$ b. $36x^2 - 16$ c. $108 - 3x^2$

12. **MC** If the factorised expression is $(x + 7)(x - 7)$, then the expanded expression must have been:

     A. $x^2 - 7$ **B.** $x^2 + 7$ **C.** $x^2 - 49$ **D.** $x^2 + 49$

13. **MC** If the factorised expression is $\left(\dfrac{x}{4} - \dfrac{3}{5} \right) \left(\dfrac{x}{4} + \dfrac{3}{5} \right)$, then the original expression must have been:

     A. $\dfrac{x^2}{4} - \dfrac{3}{5}$ **B.** $\dfrac{x^2}{16} - \dfrac{9}{25}$

     C. $\dfrac{x^2}{4} - \dfrac{\left(\sqrt{3} \right)^2}{\left(\sqrt{5} \right)^2}$ **D.** $\dfrac{x^2}{4} - \dfrac{9}{25}$

14. **MC** The factorised form of $64x^2 - 9y^2$ is:

     A. $(64x + 9y)(64x - 9y)$ **B.** $(8x + 3y)(8x - 3y)$

     C. $(8x - 3y)(8x - 3y)$ **D.** $(8x + 3y)(8x + 3y)$

15. **MC** Which of the following expressions would be factorised by grouping 'two and two'?

     A. $x^2 - a^2 + 12a - 36$ **B.** $x^2 - 7x - 10$

     C. $2x^2 - 6x - xy + 3y$ **D.** $(s - 5)^2 - 25(s + 3)^2$

For questions **16–18**, factorise the expressions over the set of real numbers.

16. a. $x^2 - 11$ b. $x^2 - 7$ c. $x^2 - 15$

17. a. $4x^2 - 13$ b. $9x^2 - 19$ c. $3x^2 - 66$

18. a. $5x^2 - 15$ b. $2x^2 - 4$ c. $12x^2 - 36$

Understanding

For questions **19–21**, factorise the expressions.

19. a. $(x - 1)^2 - 4$ b. $(x + 1)^2 - 25$ c. $(x - 2)^2 - 9$

20. a. $(x + 3)^2 - 16$ b. $49 - (x + 1)^2$ c. $36 - (x - 4)^2$

21. a. $(x - 1)^2 - (x - 5)^2$ b. $4(x + 2)^2 - 9(x - 1)^2$ c. $25(x - 2)^2 - 16(x + 3)^2$

WE8a For questions **22–26**, factorise the expressions.

22. a. $x - 2y + ax - 2ay$ b. $2x + ax + 2y + ay$ c. $ax - ay + bx - by$ d. $4x + 4y + xz + yz$

23. a. $ef - 2e + 3f - 6$ b. $mn - 7m + n - 7$

     c. $6rt - 3st + 6ru - 3su$ d. $7mn - 21n + 35m - 105$

24. a. $64 - 8j + 16k - 2jk$ b. $3a^2 - a^2b + 3ac - abc$

     c. $5x^2 + 10x + x^2y + 2xy$ d. $2m^2 - m^2n + 2mn - mn^2$

25. a. $xy + 7x - 2y - 14$ b. $mn + 2n - 3m - 6$ c. $pq + 5p - 3q - 15$

26. a. $s^2 + 3s - 4st - 12t$ b. $a^2b - cd - bc + a^2d$ c. $xy - z - 5z^2 + 5xyz$

WE8b For questions **27** and **28**, factorise the expressions.

27. **a.** $a^2 - b^2 + 4a - 4b$ **b.** $p^2 - q^2 - 3p + 3q$ **c.** $m^2 - n^2 + lm + ln$

28. **a.** $7x + 7y + x^2 - y^2$ **b.** $5p - 10pq + 1 - 4q^2$ **c.** $49g^2 - 36h^2 - 28g - 24h$

WE9 For questions **29** and **30**, factorise the expressions.

29. **a.** $x^2 + 14x + 49 - y^2$ **b.** $x^2 + 20x + 100 - y^2$ **c.** $a^2 - 22a + 121 - b^2$

30. **a.** $9a^2 + 12a + 4 - b^2$ **b.** $25p^2 - 40p + 16 - 9t^2$ **c.** $36t^2 - 12t + 1 - 5v^2$

31. **MC** In the expression $3(x - 2) + 4y(x - 2)$, the common binomial factor is:

     A. $3 + 4y$ **B.** $3 - 4y$ **C.** x **D.** $x - 2$

32. **MC** Identify which of the following terms is a perfect square.

     A. 9 **B.** $(x + 1)(x - 1)$ **C.** $3x^2$ **D.** $5(a + b)^2$

33. **MC** Identify which of the following expressions can be factorised using grouping.

     A. $x^2 - y^2$ **B.** $1 + 4y - 2xy + 4x^2$

     C. $3a^2 + 8a + 4$ **D.** $x^2 + x + y - y^2$

34. **MC** When factorised, $6(a + b) - x(a + b)$ equals:

     A. $6 - x(a + b)$ **B.** $(6 - x)(a + b)$ **C.** $6(a + b - x)$ **D.** $(6 + x)(a - b)$

Reasoning

35. Jack and Jill both attempt to factorise the expression $4x^2 - 12x + 8$.

Jack's solution	Jill's solution
$4x^2 - 12x + 8$	$4x^2 - 12x + 8$
$= 4(x^2 - 3x + 2)$	$= 4(x^2 - 3x + 2)$
$= 4(x - 1)(x + 3)$	$= 4(x - 2)(x - 1)$

     a. State who was correct.

     b. Explain what error was made in the incorrect solution.

36. Jack and Jill attempted to factorise another expression. Their solutions are given below. Explain why they both had the same correct answer but their working was different.

Jack's solution	Jill's solution
$6x^2 + 7x - 20$	$6x^2 + 7x - 20$
$= 6x^2 - 8x + 15x - 20$	$= 6x^2 + 15x - 8x - 20$
$= 2x(3x - 4) + 5(3x - 4)$	$= 3x(2x + 5) - 4(2x + 5)$
$= (3x - 4)(2x + 5)$	$= (2x + 5)(3x - 4)$

37. Factorise the following expressions. Explain your answer.

     a. $x^2 - 4xy + 4y^2 - a^2 + 6ab - b^2$ **b.** $12x^2 - 75y^2 - 9(4x - 3)$

Problem solving

38. The area of a rectangle is $(x^2 - 25)$ cm^2.

     a. Factorise the expression.

     b. Determine the length of the rectangle if the width is $(x + 5)$ cm.

     c. If $x = 7$ cm, calculate the dimensions of the rectangle.

     d. Hence, evaluate the area of the rectangle.

     e. If $x = 13$ cm, determine how much bigger the area of this rectangle would be.

39. A circular garden of diameter $2r$ m is to have a gravel path laid around it. The path is to be 1 m wide.

   a. Determine the area of the garden in terms of r.
   b. Determine the area of the garden and path together in terms of r, using the formula for the area of a circle.
   c. Write an expression for the area of the path in fully factorised form.
   d. If the radius of the garden is 5 m, determine the area of the path, correct to 2 decimal places.

40. A roll of material is $(x + 2)$ metres wide. Annie buys $(x + 3)$ metres of the material and Bronwyn buys 5 metres of the material.

   a. Write an expression, in terms of x, for the area of each piece of material purchased.
   b. If Annie has bought more material than Bronwyn, write an expression for how much more she has than Bronwyn.
   c. Factorise and simplify this expression.
   d. Evaluate the width of the material if Annie has 5 m^2 more than Bronwyn.
   e. Determine how much material each person has.

LESSON
7.5 Factorising by completing the square

LEARNING INTENTIONS

At the end of this lesson you should be able to:
- complete the square for a given quadratic expression
- factorise a quadratic expression by completing the square.

▶ 7.5.1 Completing the square

eles-4833

- **Completing the square** is the process of writing a general quadratic expression in turning point form.

$$\underbrace{ax^2 + bx + c}_{\text{General form}} = \underbrace{a(x-h)^2 + k}_{\text{Turning point form}}$$

- The expression $x^2 + 8x$ can be modelled as a square with a smaller square missing from the corner, as shown below.

$$x^2 + 8x \qquad = \qquad (x+4)^2 - (4)^2$$

- In 'completing the square', the general equation is written as the area of the large square minus the area of the small square.
- In general, to complete the square for $x^2 + bx$, the small square has a side length equal to half of the coefficient of x; that is, the area of the small square is $\left(\dfrac{b}{2}\right)^2$.

$$x^2 + bx \qquad = \qquad \left(x + \dfrac{b}{2}\right)^2 - \left(\dfrac{b}{2}\right)^2$$

- The process of completing the square is sometimes described as the process of adding the square of half of the coefficient of x then subtracting it, as shown in purple below.
  The result of this process is a perfect square that is then factorised, as shown in blue.

$$x^2 + bx = x^2 + bx + \left(\dfrac{b}{2}\right)^2 - \left(\dfrac{b}{2}\right)^2$$

$$= x^2 + bx + \left(\dfrac{b}{2}\right)^2 - \left(\dfrac{b}{2}\right)^2$$

$$= \left(x + \dfrac{b}{2}\right)^2 - \left(\dfrac{b}{2}\right)^2$$

WORKED EXAMPLE 10 Converting expressions into turning point form

Write the following expressions in turning point form by completing the square.

a. $x^2 + 4x$

b. $x^2 + 7x + 1$

THINK	WRITE

a.
- The square will consist of a square that has an area of x^2 and two identical rectangles with a total area of $4x$.
- The length of the large square is $(x + 2)$, so its area is $(x + 2)^2$.
- The area of the smaller square is $(2)^2$.
- Write $x^2 + 4x$ in turning point form.

a.
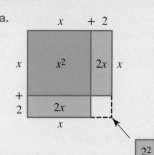

$$x^2 + 4x = (x + 2)^2 - (2)^2$$
$$= (x + 2)^2 - 4$$

b. 1.
- Complete the square with the terms containing x.
- The square will consist of a square that has an area of x^2 and two identical rectangles with a total area of $7x$.

- The length of the large square is $\left(x + \dfrac{7}{2}\right)$, so its area is $\left(x + \dfrac{7}{2}\right)^2$.

- The area of the smaller square is $\left(\dfrac{7}{2}\right)^2$.

- Write $x^2 + 7x + 1$ in turning point form.

b.

$$x^2 + 7x + 1 = \left(x + \frac{7}{2}\right)^2 - \left(\frac{7}{2}\right)^2 + 1$$

2. Simplify the last two terms.

$$= \left(x + \frac{7}{2}\right)^2 - \frac{49}{4} + \frac{4}{4}$$

$$= \left(x + \frac{7}{2}\right)^2 - \frac{45}{4}$$

▶ 7.5.2 Factorising by completing the square

eles-4834

- When an equation is written in turning point form, it can be factorised as a difference of two squares. For example, factorise $x^2 + 8x + 2$ by completing the square.

$$x^2 + 8x + \left(\frac{8}{2}\right)^2 - \left(\frac{8}{2}\right)^2 + 2$$

$$= x^2 + 8x + (4)^2 - (4)^2 + 2$$
$$= x^2 + 8x + 16 - 16 + 2$$
$$= (x + 4)^2 - 14$$
$$= (x + 4)^2 - 14$$
$$= \left(x + 4 - \sqrt{14}\right)\left(x + 4 + \sqrt{14}\right)$$

Factorise the following expressions by completing the square.

a. $x^2 + 4x + 2$

b. $x^2 - 9x + 1$

THINK

WRITE

a. 1. To complete the square, add the square of half of the coefficient of x and then subtract it.

a. $x^2 + 4x + 2 = x^2 + 4x + \left(\dfrac{4}{2}\right)^2 - \left(\dfrac{4}{2}\right)^2 + 2$

2. Write the perfect square created in its factorised form.

$= x^2 + 4x + (2)^2 - (2)^2 + 2$

$= (x+2)^2 - (2)^2 + 2$

3. Write the expression as a difference of two squares by:
 • simplifying the numerical terms
 • writing the numerical term as a square $\left(2 = \left(\sqrt{2}\right)^2\right).$

$= (x+2)^2 - 4 + 2$

$= (x+2)^2 - 2$

$= (x+2)^2 - \left(\sqrt{2}\right)^2$

4. Use the pattern for DOTS, $a^2 - b^2 = (a-b)(a+b)$, where $a = (x+2)$ and $b = \sqrt{2}$.

$= \left(x + 2 + \sqrt{2}\right)\left(x + 2 - \sqrt{2}\right)$

b. 1. To complete the square, add the square of half of the coefficient of x, then subtract it.

b. $x^2 - 9x + 1 = x^2 - 9x + \left(-\dfrac{9}{2}\right)^2 - \left(-\dfrac{9}{2}\right)^2 + 1$

2. Write the perfect square created in its factorised form.

$= x^2 - 9x + \left(-\dfrac{9}{2}\right)^2 - \left(-\dfrac{9}{2}\right)^2 + 1$

$= \left(x - \dfrac{9}{2}\right)^2 - \left(-\dfrac{9}{2}\right)^2 + 1$

3. Write the expression as a difference of two squares by:

 • simplifying the numerical terms

 • writing the numerical term as a square. $\dfrac{77}{4} = \left(\sqrt{\dfrac{77}{4}}\right)^2 = \left(\dfrac{\sqrt{77}}{2}\right)^2$

$= \left(x - \dfrac{9}{2}\right)^2 - \dfrac{81}{4} + 1$

$= \left(x - \dfrac{9}{2}\right)^2 - \dfrac{77}{4}$

$= \left(x - \dfrac{9}{2}\right)^2 - \left(\dfrac{\sqrt{77}}{2}\right)^2$

4. Use the pattern for DOTS: $a^2 - b^2 = (a-b)(a+b)$, where $a = \left(x + -\dfrac{9}{2}\right)$ and $b = \dfrac{\sqrt{77}}{2}$.

$= \left(x - \dfrac{9}{2} + \dfrac{\sqrt{77}}{2}\right)\left(x - \dfrac{9}{2} - \dfrac{\sqrt{77}}{2}\right)$

TI	THINK	DISPLAY/WRITE	CASIO	THINK	DISPLAY/WRITE

a, b.

In a new problem, on a Calculator page, press:
- CATALOG
- 1
- F

Then scroll down to select factor(

Complete the entry lines as:

factor $\left(x^2 + 4x + 2, x\right)$

factor $\left(x^2 - 9x + 1, x\right)$

Press EXE after each entry.

a, b.

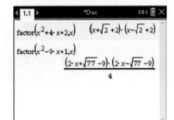

a, b.

On the Main screen, tap:
- Action
- Transformation
- factor
- rFactor

Complete the entry lines as:

rfactor $\left(x^2 + 4x + 2\right)$

rfactor $\left(x^2 - 9x + 1\right)$

Press EXE after each entry.

a, b.

$$x^2 + 4x + 2 = \left(x + \sqrt{2} + 2\right)\left(x - \sqrt{2} + 2\right)$$

$$x^2 - 9x + 1 = \frac{\left(2x + \sqrt{77} - 9\right)\left(2x - \sqrt{77} - 9\right)}{4}$$

$$x^2 + 4x + 2 = \left(x + \sqrt{2} + 2\right)\left(x - \sqrt{2} + 2\right)$$

$$x^2 - 9x + 1 = \left(x + \frac{\sqrt{77}}{2} - \frac{9}{2}\right)\left(x - \frac{\sqrt{77}}{2} - \frac{9}{2}\right)$$

- Remember that you can expand the brackets to check your answer.
- If the coefficient of $x^2 \neq 1$, factorise the expression before completing the square.

  For example, $2x^2 - 8x + 2$

  $$= 2(x^2 - 4x + 1)$$

  $$= 2\left(x^2 - 4x + \left(\frac{-4}{2}\right)^2 - \left(\frac{-4}{2}\right)^2 + 1\right)$$

  $$= 2((x - 2)^2 - 4 + 1)$$

  $$= 2((x - 2)^2 - 3)$$

  $$= 2\left(x - 2 - \sqrt{3}\right)\left(x - 2 + \sqrt{3}\right)$$

Exercise 7.5 Factorising by completing the square

7.5 Quick quiz on	7.5 Exercise

Individual pathways

■ PRACTISE	■ CONSOLIDATE	■ MASTER
1, 3, 4, 7, 10, 13, 16	2, 5, 8, 11, 14, 17	6, 9, 12, 15, 18

Fluency

WE10 For questions 1 and 2, write the expressions in turning point form by completing the square.

1. **a.** $x^2 + 10x$ **b.** $x^2 + 6x$ **c.** $x^2 - 4x$ **d.** $x^2 + 16x$ **e.** $x^2 - 20x$

2. **a.** $x^2 + 8x$ **b.** $x^2 - 14x$ **c.** $x^2 + 50x$ **d.** $x^2 + 7x$ **e.** $x^2 - x$

WE11 For questions 3–6, factorise the expressions by completing the square.

3. **a.** $x^2 - 4x - 7$ **b.** $x^2 + 2x - 2$ **c.** $x^2 - 10x + 12$ **d.** $x^2 + 6x - 10$ **e.** $x^2 + 16x - 1$

4. **a.** $x^2 - 14x + 43$ **b.** $x^2 + 8x + 9$ **c.** $x^2 - 4x - 13$ **d.** $x^2 - 12x + 25$ **e.** $x^2 - 6x + 4$

5. **a.** $x^2 - x - 1$ **b.** $x^2 - 3x - 3$ **c.** $x^2 + x - 5$ **d.** $x^2 + 3x - 1$ **e.** $x^2 + 5x + 2$

6. **a.** $x^2 + 5x - 2$ **b.** $x^2 - 7x - 1$ **c.** $x^2 - 9x + 13$ **d.** $x^2 - x - 3$ **e.** $x^2 - x - 1$

For questions 7 and 8, factorise the expressions by first looking for a common factor and then completing the square.

7. **a.** $2x^2 + 4x - 4$ **b.** $4x^2 - 8x - 20$ **c.** $5x^2 + 30x + 5$ **d.** $3x^2 - 12x - 39$ **e.** $5x^2 - 30x + 10$

8. **a.** $6x^2 + 24x - 6$ **b.** $3x^2 + 30x + 39$ **c.** $2x^2 - 8x - 14$ **d.** $6x^2 + 36x - 30$ **e.** $4x^2 - 8x - 16$

Understanding

9. From the following list, determine which method of factorising is the most appropriate for each of the expressions given.

   a. Factorising using common factors
   b. Factorising using the difference of two squares rule
   c. Factorising by grouping
   d. Factorising quadratic trinomials
   e. Completing the square

     i. $3x^2 - 8x - 3$ **ii.** $49m^2 - 16n^2$
     iii. $x^2 + 8x + 4 - y^2$ **iv.** $7x^2 - 28x$
     v. $6a - 6b + a^2 - b^2$ **vi.** $x^2 + x - 5$
     vii. $(x - 3)^2 + 3(x - 3) - 10$ **viii.** $x^2 - 7x - 1$

10. **MC** To complete the square, the term that should be added to $x^2 + 4x$ is:

   A. 16   **B.** 4   **C.** $4x$ **D.** 2

11. **MC** To factorise the expression $x^2 - 3x + 1$, the term that must be both added and subtracted is:

   A. 9   **B.** 3   **C.** $3x$ **D.** $\dfrac{9}{4}$

12. **MC** The factorised form of $x^2 - 6x + 2$ is:

   A. $\left(x + 3 - \sqrt{7}\right)\left(x + 3 + \sqrt{7}\right)$ **B.** $\left(x + 3 - \sqrt{7}\right)\left(x - 3 + \sqrt{7}\right)$

   C. $\left(x - 3 - \sqrt{7}\right)\left(x - 3 - \sqrt{7}\right)$ **D.** $\left(x - 3 + \sqrt{7}\right)\left(x - 3 - \sqrt{7}\right)$

Reasoning

13. Show that $x^2 + 4x + 6$ cannot be factorised by completing the square.

14. A square measuring x cm in side length has a cm added to its length and b cm added to its width. The resulting rectangle has an area of $\left(x^2 + 6x + 3\right)$ cm^2. Evaluate a and b, correct to 2 decimal places.

15. Show that $2x^2 + 3x + 4$ cannot be factorised by completing the square.

Problem solving

16. Students were asked to choose one quadratic expression from a given list.
    Peter chose $x^2 - 4x + 9$ and Annabelle chose $x^2 - 4x - 9$.
    Using the technique of completing the square to write the expression in turning point form, determine which student was able to factorise their expression.

17. Use the technique of completion of the square to factorise $x^2 + 2(1 - p)x + p(p - 2)$.

18. For each of the following, complete the square to factorise the expression.
    a. $2x^2 + 8x + 1$
    b. $3x^2 - 7x + 5$

LESSON
7.6 Mixed factorisation

LEARNING INTENTIONS

At the end of this lesson you should be able to:
- factorise by grouping
- factorise by completing the square
- factorise using DOTS.

▶ 7.6.1 Mixed factorisation

eles-4835

- To factorise quadratic trinomials and other expressions, remember:

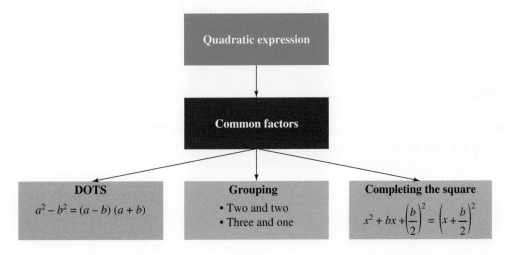

WORKED EXAMPLE 12 Applying mixed factorisation to expressions

Factorise and simplify each of the following expressions.

a. $\dfrac{6x-24}{x^2-16} \times \dfrac{5x+10}{8(x+2)}$

b. $\left(\dfrac{x^2+6x+8}{x^2+2x+1}\right) \div \left(\dfrac{x^2+5x+6}{(x+1)}\right)$

THINK	WRITE
a. 1. Write the expression.	**a.** $\dfrac{6x-24}{x^2-16} \times \dfrac{5x+10}{8(x+2)}$
2. Look for common factors.	$=\dfrac{6(x-4)}{x^2-16} \times \dfrac{5(x+2)}{8(x+2)}$
3. Look for the DOTS pattern in the expression. Use $a^2-b^2=(a+b)(a-b)$ to write the factors.	$=\dfrac{6(x-4)}{(x-4)(x+4)} \times \dfrac{5(x+2)}{8(x+2)}$
4. Cancel the common factors.	$=\dfrac{^3\cancel{6}\cancel{(x-4)}}{_1\cancel{(x-4)}(x+4)} \times \dfrac{5\cancel{(x+2)}}{_4\cancel{8}_1\cancel{(x+2)}}$
5. Simplify and write the answer.	$=\dfrac{3}{(x+4)} \times \dfrac{5}{4}$
	$=\dfrac{15}{4(x+4)}$
b. 1. Write the expression.	**b.** $\dfrac{x^2+6x+8}{x^2+2x+1} \div \dfrac{x^2+5x+6}{(x+1)}$
2. Change the division to multiplication. Flip the second expression.	$=\dfrac{x^2+6x+8}{x^2+2x+1} \times \dfrac{x+1}{x^2+5x+6}$
3. Factorise and look for common factors.	$=\dfrac{(x+2)(x+4)}{(x+1)^2} \times \dfrac{x+1}{(x+3)(x+2)}$
4. Cancel the common factors.	$=\dfrac{^1\cancel{(x+2)}(x+4)}{(x+1)^{\cancel{2}1}} \times \dfrac{\cancel{(x+1)}}{(x+3)_1\cancel{(x+2)}}$
5. Simplify and write the answer.	$=\dfrac{(x+4)}{(x+1)} \times \dfrac{1}{(x+3)}$
	$=\dfrac{(x+4)}{(x+1)(x+3)}$

on Resources

☑ **eWorkbook** Topic 7 Workbook (worksheets, code puzzle and project) (ewbk-13292)

◆ **Interactivity** Individual pathway interactivity: Mixed factorisation (int-4600)

Exercise 7.6 Mixed factorisation

7.6 Quick quiz on	7.6 Exercise

Individual pathways

■ PRACTISE	■ CONSOLIDATE	■ MASTER
1, 4, 7, 10, 11, 14, 17	2, 5, 8, 12, 15, 18	3, 6, 9, 13, 16, 19

Fluency

For questions 1–9, factorise each of the following expressions.

1. a. $3x + 9$ b. $x^2 + 4x + 4 - 9y^2$ c. $x^2 - 36$
   d. $x^2 - 49$ e. $5x^2 - 9x - 2$

2. a. $15x - 20y$ b. $5c + de + dc + 5e$ c. $5x^2 - 80$
   d. $-x^2 - 6x - 5$ e. $x^2 + x - 12$

3. a. $mn + 1 + m + n$ b. $x^2 - 7$ c. $16x^2 - 4x$
   d. $5x^2 + 60x + 100$ e. $18 + 9x - 6y - 3xy$

4. a. $x^2 - 8x + 16 - y^2$ b. $4x^2 + 8$ c. $fg + 2h + 2g + fh$
   d. $x^2 - 5$ e. $10mn - 5n + 10m - 5$

5. a. $x^2 + 6x + 5$ b. $x^2 - 10x - 11$ c. $x^2 - 4$
   d. $-5a + bc + ac - 5b$ e. $xy - 1 + x - y$

6. a. $3x^2 + 5x + 2$ b. $7x^2 - 28$ c. $-4x^2 - 28x - 24$
   d. $2p - rs + pr - 2s$ e. $3x^2 - 27$

7. a. $-3u + tv + ut - 3v$ b. $x^2 - 11$ c. $12x^2 - 7x + 1$
   d. $(x - 1)^2 - 4$ e. $(x + 2)^2 - 16$

8. a. $(2x + 3)^2 - 25$ b. $3(x + 5)^2 - 27$ c. $25 - (x - 2)^2$
   d. $4(3 - x)^2 - 16y^2$ e. $(x + 2y)^2 - (2x + y)^2$

9. a. $(x + 3)^2 - (x + 1)^2$ b. $(2x - 3y)^2 - (x - y)^2$ c. $(x + 3)^2 + 5(x + 3) + 4$
   d. $(x - 3)^2 + 3(x - 3) - 10$ e. $2(x + 1)^2 + 5(x + 1) + 2$

Understanding

10. Consider the following product of algebraic fractions.

$$\frac{x^2 + 3x - 10}{x^2 - 4} \times \frac{x^2 + 4x + 4}{x^2 - 2x - 8}$$

   a. Factorise the expression in each numerator and denominator.
   b. Cancel factors common to both the numerator and the denominator.
   c. Simplify the expression as a single fraction.

WE12 For questions 11–13, factorise and simplify each of the following expressions.
Note: You may choose to follow the procedure given in question 10.

11. a. $\dfrac{x^2 - 4x + 3}{x^2 - 4x - 12} \times \dfrac{x^2 + 5x + 6}{x^2 - 9}$ b. $\dfrac{3x^2 - 17x + 10}{6x^2 + 5x - 6} \times \dfrac{x^2 - 1}{x^2 - 6x + 5}$ c. $\dfrac{6x - 12}{x^2 - 4} \times \dfrac{3x + 6}{x(x - 5)}$

12. a. $\dfrac{6x^2 - x - 2}{2x^2 + 3x + 1} \times \dfrac{2x^2 + x - 1}{3x^2 + 10x - 8}$

b. $\dfrac{x^2 + 4x - 5}{x^2 + x - 2} \div \dfrac{x^2 + 10x + 25}{x^2 + 4x + 4}$

c. $\dfrac{x^2 - 7x + 6}{x^2 + x - 2} \div \dfrac{x^2 - x - 12}{x^2 - 2x - 8}$

13. a. $\dfrac{4ab + 8a}{(c - 3)} \div \dfrac{5ac + 5a}{c2 - 2c - 3}$

b. $\dfrac{p^2 - 7p}{p^2 - 49} \div \dfrac{p^2 + p - 6}{p^2 + 14p + 49}$

c. $\dfrac{m^2 + 4m + 4 - n^2}{4m^2 - 4m - 15} \div \dfrac{2m^2 + 4m - 2mn}{10m^2 + 15m}$

Reasoning

14. a. Determine the original expression if the factorised expression is:

   i. $\left(\dfrac{x}{4} + \dfrac{3}{5}\right)\left(\dfrac{x}{4} - \dfrac{3}{5}\right)$

   ii. $\left(\dfrac{-x}{4} + \dfrac{3}{5}\right)\left(\dfrac{-x}{4} - \dfrac{3}{5}\right)$

   b. Explain why your answers are the same.

15. Factorise the following using grouping 'three and one' and DOTS.
   a. $x^2 - 18x + 81 - y^2$
   b. $4x^2 + 12x - 16y^2 + 9$

16. **The expansion of perfect squares**
   The expansions $(a + b)^2 = a^2 + 2ab + b^2$ and
   $(a - b)^2 = a^2 - 2ab + b^2$ can be used to simplify some
   arithmetic calculations. For example:

   $$97^2 = (100 - 3)^2$$
   $$= 100^2 - 2 \times 100 \times 3 + 3^2$$
   $$= 9409$$

   Use this method to calculate the following.
   a. 103^2
   b. 62^2
   c. 997^2
   d. 1012^2
   e. 53^2
   f. 98^2

Problem solving

17. Use grouping 'two and two' and DOTS to factorise the following. Show your working.
   a. $x^2 + 3x - y^2 + 3y$
   b. $7x + 7y + x^2 - y^2$
   c. $5p - 10pq + 1 - 4q^2$

18. The expansions for the sum and difference of two cubes are given below.

   $$a^3 + b^3 = (a + b)\left(a^2 - ab + b^2\right)$$

   $$a^3 - b^3 = (a - b)\left(a^2 + ab + b^2\right)$$

   Using these expansions, simplify:

   $$\dfrac{2a^2 - 7a + 6}{a^3 + 8} \times \dfrac{5a^2 + 11a + 2}{a^3 - 8} \div \dfrac{10a^2 - 13a - 3}{a^2 - 2a + 4}$$

19. Factorise:
   a. $x^2 + 12x + 40 - 4(x^2y^2 + 1)$
   b. $225x^4y^2 - 169x^2y^6$

LESSON
7.7 Review

7.7.1 Topic summary

Expanding using FOIL

- FOIL is a process that can be applied to expand brackets.
  It is an acronym that stands for:
  First
  Outer
  Inner
  Last

  $\mathbf{F}$
  $(x + a)(x - b)$

  $\mathbf{O}$
  $(x + a)(x - b)$

  $\mathbf{I}$
  $(x + a)(x - b)$

  $\mathbf{L}$
  $(x + a)(x - b)$

  e.g.
  $(x + 2)(x - 5)$
  $= (x \times x) + (x \times -5) + (2 \times x) + (2 \times -5)$
  $= x^2 - 5x + 2x - 10$
  $= x^2 - 3x - 10$

Factorising monic quadratics

- To factorise monic quadratics, $x^2 + bx + c$, determine factors of c that add to b.
- $x^2 + bx + c = (x + m)(x + n)$ where m and n are the factors of c that add to b.

  e.g.
  $x^2 - 2x - 15$
  The factors of -15 that add to -2 are -5 and 3.
  $x^2 - 2x - 15 = (x - 5)(x + 3)$.

Factorising non-monic quadratics

- To factorise non-monic quadratics, $ax^2 + bx + c$, determine factors of ac that add to b.
- Split the bx term into two terms based on the factors of ac that add to b.
- Pair the four terms so that the terms on the left share a common factor and the terms on the right share a common factor.
- Take the common factors out of the pairs of terms.
- Take out the binomial factor.

  e.g.
  $5x^2 + 8x + 3$
  The factors of 15 that add to 8 are 5 and 3.
  $5x^2 + 8x + 3$
  $= 5x^2 + 5x + 3x + 3$
  $= 5x(x + 1) + 3(x + 1)$
  $= (x + 1)(5x + 3)$

QUADRATIC EXPRESSIONS

Binomial expansion (perfect squares)

- When a binomial expression is squared, the expansion can be determined using the following formulas.
  - $(a + b)^2 = a^2 + 2ab + b^2$
  - $(a - b)^2 = a^2 - 2ab + b^2$

  e.g.
  $(3x - 5)^2$
  $= (3x)^2 - 2 \times 3x \times 5 + (5)^2$
  $= 9x^2 - 30x + 25$

Difference of two squares (DOTS)

- $a^2 - b^2 = (a + b)(a - b)$
- This formula can be used to expand or factorise quadratic expressions.

  e.g.
  $4x^2 - 9$
  $= (2x)^2 - (3)^2$
  $= (2x + 3)(2x - 3)$

Completing the square

- Completing the square is a method to convert a quadratic expression into turning point form.

- $x^2 + bx + \left(\dfrac{b}{2}\right)^2 = \left(x + \dfrac{b}{2}\right)^2$

- By adding $\left(\dfrac{b}{2}\right)^2$ to the expression, a perfect square is formed. This part of the expression can then be factorised.

  e.g.
  $x^2 + 6x = x^2 + 6x + \left(\dfrac{6}{2}\right)^2 - \left(\dfrac{6}{2}\right)^2$
  $= x^2 + 6x + 9 - 9$
  $= (x + 3)^2 - 9$

- Once the square has been completed, the quadratic can often be factorised further using DOTS.

  e.g.
  $(x + 3)^2 - 9 = (x + 3)^2 - (3)^2$
  $= (x + 3 - 3)(x + 3 + 3)$
  $= x(x + 6)$

7.7.2 Success criteria

Tick the column to indicate that you have completed the lesson and how well you have understood it using the traffic light system.

(**Green:** I understand; **Yellow:** I can do it with help; **Red:** I do not understand)

Lesson	Success criteria	⬤	◯	⬤
7.2	I can expand binomial expressions using FOIL.			
	I can expand squares of binomial expressions.			
	I can expand binomial expressions using the difference of two squares (DOTS).			
7.3	I can recognise monic and non-monic quadratic expressions.			
	I can factorise monic quadratic trinomials.			
	I can factorise non-monic quadratic trinomials.			
7.4	I can factorise expressions by taking out the highest common factor.			
	I can recognise and factorise expressions that can be written as DOTS.			
	I can factorise expressions with four terms by grouping.			
7.5	I can complete the square for a given quadratic expression.			
	I can factorise a quadratic expression by completing the square.			
7.6	I can factorise by grouping.			
	I can factorise by completing the square.			
	I can factorise using DOTS.			

7.7.3 Project

Celebrity squares and doubles

In small groups or as a class, use the process of elimination to find your 'square and double pair' by playing 'Celebrity squares and doubles' as outlined below.

Equipment: roll of calculator paper, scissors, sticky tape, marker pen

1. **Set-up**
   - Make a class set of headbands. Each headband will be part of a matching pair made by a number being squared and that same original number being doubled (16 and 8 would be a pair, because $4^2 = 16$ and $4 \times 2 = 8$). Your teacher will direct the class as to what number should be written on each headband.
   - Place the headbands randomly on a table.

2. **Beginning the game**
   - There is to be no communication between players at this time.
   - Your teacher will randomly allocate a headband to each player by placing a headband on their head without the player seeing the number on their headband.

3. **Playing the game**

   The object of the game
   - The object of the game is to use the process of elimination for you to find your pair.

   Starting the game
   - Once all headbands have been allocated, stand in a circle or walk around freely.
   - Without speaking, determine who is a match; then, by a process of elimination, determine who might be your match.

   Making a match
   - When you think you have found your match, approach that person and say 'I think I am your match.'
   - The other player should now check to see if you have a match elsewhere and can reply by saying one of two things: 'Yes, I think I am your match,' or 'I know your match is still out there.'
   - If a match is agreed upon, the players should sit out for the remainder of the game. If a match is not agreed upon, players should continue looking.

   Ending the game
   - The class should continue until everyone is in a pair, at which time the class can check their results.
   - The class should now discuss the different trains of thought they used to find their pair and how this relates to factorising quadratic trinomials.

---

on Resources

eWorkbook Topic 7 Workbook (worksheets, code puzzle and project) (ewbk-13292)

Interactivities Crossword (int-2845)

Sudoku puzzle (int-3594)

Fluency

1. **MC** When expanded, $-3x(x+4)(5-x)$ becomes:

   A. $-3x^3 - 3x^2 - 27x$ **B.** $-3x^3 + 3x - 27x$

   C. $3x^3 + 3x^2 - 60x$ **D.** $3x^3 - 3x^2 - 60x$

2. **MC** When expanded, $(3x+7)^2$ becomes:

   A. $9x^2 + 49$ **B.** $3x^2 + 49$

   C. $3x^2 + 21x + 49$ **D.** $9x^2 + 42x + 49$

3. **MC** The factorised form of $-3d^2 - 9d + 30$ is:

   A. $-3(d-5)(d-2)$ **B.** $-3(d+5)(d-6)$

   C. $-(3d+5)(d-2)$ **D.** $-3(d+5)(d-2)$

4. **MC** If the factorised expression is $(2x-5)(2x+5)$, then the original expression must have been:

   A. $2x^2 - 5$ **B.** $4x^2 - 5$

   C. $4x^2 - 25$ **D.** $4x^2 - 20x + 25$

5. **MC** To factorise $-5x^2 - 45x + 100$, the first step is to:

   A. determine factors of 5 and 100 that sum to -45.

   B. take out 5 as a common factor.

   C. take out -5 as a common factor.

   D. determine factors of 5 and -45 that will add to make 100.

6. **MC** To complete the square, the term that should be added to $x^2 - 12x$ is:

   A. 36   **B.** -12 **C.** $-12x$ **D.** -6

7. **MC** Identify which of the following is equivalent to $5x^2 - 20x - 5$.

   A. $5(x-2)^2$ **B.** $5(x-2)^2 - 3$

   C. $5(x-2)^2 - 15$ **D.** $5(x-2)^2 - 25$

8. **MC** In the expanded form of $(x-3)(x+5)$, determine which of the following is incorrect.

   A. The value of the constant is -15.

   B. The coefficient of the x term is 2.

   C. The coefficient of the x term is -8.

   D. The coefficient of the x^2 term is 1.

For questions **9** and **10**, expand each of the following and simplify where necessary.

9. **a.** $3x(x-4)$ **b.** $-7x(3x+1)$ **c.** $(x-7)(x+1)$

   d. $(2x-5)(x-3)$ **e.** $(4x-1)(3x-5)$

10. **a.** $3(x-4)(2x+7)$ **b.** $(2x-5)(x+3)(x+7)$

    c. $(x+5)(x+7) + (2x-5)(x-6)$ **d.** $(x+3)(5x-1) - 2x$

For questions **11–13**, expand and simplify each of the following.

11. a. $(x-7)^2$ **b.** $(2-x)^2$ **c.** $(3x+1)^2$

12. a. $-2(3x-2)^2$ **b.** $-7(2x+5)^2$ **c.** $-10(4x-5)^2$

13. a. $(x+9)(x-9)$ **b.** $(3x-1)(3x+1)$ **c.** $(5+2x)(5-2x)$

For questions **14–17**, factorise each of the following.

14. a. $2x^2-8x$ **b.** $-4x^2+12x$ **c.** $3ax-2ax^2$

15. a. $(x+1)^2+(x+1)$ **b.** $3(2x-5)-(2x-5)^2$ **c.** $(x-4)(x+2)-(x-4)$

16. a. x^2-16 **b.** x^2-25 **c.** $2x^2-72$

17. a. $3x^2-27y^2$ **b.** $4ax^2-16ay^2$ **c.** $(x-4)^2-9$

For questions **18–21**, factorise each of the following by grouping.

18. a. $ax-ay+bx-by$ **b.** $7x+ay+ax+7y$ **c.** $xy+2y+5x+10$

19. a. $mn-q-2q^2+2mnq$ **b.** $pq-5r^2-r+5pqr$ **c.** $uv-u+9v-9$

20. a. $a^2-b^2+5a-5b$ **b.** $d^2-4c^2-3d+6c$ **c.** $2+2m+1-m^2$

21. a. $4x^2+12x+9-y^2$ **b.** $49a^2-28a+4-4b^2$ **c.** $64s^2-16s+1-3t$

For questions **22–24**, factorise each of the following.

22. a. $x^2+10x+9$ **b.** $x^2-11x+18$ **c.** $x^2-4x-21$
     d. $x^2+3x-28$ **e.** $-x^2+6x-9$

23. a. $3x^2+33x-78$ **b.** $-2x^2+8x+10$ **c.** $-3x^2+24x-36$
     d. $8x^2+2x-1$ **e.** $6x^2+x-1$

24. a. $8x^2+4x-12$ **b.** $105x^2-10x-15$ **c.** $-12x^2+62x-70$
     d. $-45x^2-3x+6$ **e.** $-60x^2-270x-270$

For questions **25** and **26**, factorise each of the following by completing the square.

25. a. x^2+6x+1 **b.** $x^2-10x-3$ **c.** x^2+4x-2

26. a. x^2-5x+2 **b.** x^2+7x-1 **c.** $2x^2+18x-2$

For questions **27** and **28**, factorise each of the following using the most appropriate method.

27. a. $3x^2-12x$ **b.** x^2+6x+2 **c.** $4x^2-25$

28. a. $2x^2+9x+10$ **b.** $2ax+4x+3a+6$ **c.** $-3x^2-3x+18$

29. First factorise then simplify each of the following.

     a. $\dfrac{x+4}{5x-30}\times\dfrac{2x-12}{x+1}$ **b.** $\dfrac{3x+6}{4x-24}\times\dfrac{7x-42}{6x+12}$ **c.** $\dfrac{x^2-4}{x^2+5x}\times\dfrac{x^2+4x-5}{x^2-2x-8}$

Problem solving

30. A large storage box has a square base with sides measuring $(x + 2)$ cm and is 32 cm high.

    a. Write an expression for the area of the base of the box.

    b. Write an expression for the volume of the box (V = area of base × height).

    c. Simplify the expression by expanding the brackets.

    d. If $x = 30$ cm, calculate the volume of the box in cm^3.

31. A section of garden is to have a circular pond of radius $2r$ with a 2-metre path around its edge.

    a. State the diameter of the pond.

    b. State the radius of the pond and path.

    c. Calculate the area of the pond.

    d. Calculate the area of the pond and path.

    e. Write an expression to determine the area of the path only and write it in factorised form.

    f. If the radius of the pond is 3 metres, calculate the area of the path.

32. In order to make the most of the space available for headlines and stories, the front page of a newspaper is given an area of $(x^2 - 5x - 14)$ cm^2.

    a. If the length is $(x + 2)$ cm, calculate the width.

    b. Write down the length of the shorter side in terms of x.

    c. If the shorter side of the front page is 28 cm, determine the value of x.

    d. Evaluate the area of this particular paper.

33. Here is a well-known puzzle. Let $a = b = 1$.

Step 1: Write $a = b$.	$a = b$
Step 2: Multiply both sides by a.	$a^2 = ab$
Step 3: Subtract b^2 from both sides.	$a^2 - b^2 = ab - b^2$
Step 4: Factorise.	$(a + b)(a - b) = b(a - b)$
Step 5: Simplify by dividing by $(a - b)$.	$(a + b) = b$
Step 6: Substitute $a = b = 1$.	$1 + 1 = 1$

Explain where the error is. Show your thinking.

on To test your understanding and knowledge of this topic, go to your learnON title at www.jacplus.com.au and complete the **post-test**.

Answers

Topic 7 Quadratic expressions

7.1 Pre-test

1. $-2x - 6$
2. $4m(1 - 5m)$
3. $a(a + c)(4 + b)$
4. $(x + 5)(x - 2)$
5. $-9x^2 - 6x - 1$
6. A
7. D
8. C
9. $\left(x + 2 - \sqrt{10}\right)\left(x + 2 + \sqrt{10}\right)$
10. $2x^3 + x^2 - 7x - 6$
11. $2\left(x - \dfrac{3}{2} - \dfrac{\sqrt{29}}{2}\right)\left(x - \dfrac{3}{2} + \dfrac{\sqrt{29}}{2}\right)$
12. D
13. 2
14. $(4x + 23)(x + 9)$
15. B

7.2 Expanding algebraic expressions

1. a. $2x + 6$ b. $4x - 20$
   c. $21 - 3x$ d. $-x - 3$
2. a. $x^2 + 2x$ b. $2x^2 - 8x$
   c. $15x^2 - 6x$ d. $10x - 15x^2$
3. a. $8x^2 + 2x$ b. $4x^3 - 6x^2$
   c. $6x^3 - 3x^2$ d. $15x^3 + 20x^2$
4. a. $x^2 - x - 12$ b. $x^2 - 2x - 3$
   c. $x^2 - 5x - 14$ d. $x^2 - 6x + 5$
5. a. $-x^2 - x + 6$ b. $x^2 - 6x + 8$
   c. $2x^2 - 17x + 21$ d. $3x^2 - x - 2$
6. a. $6x^2 - 17x + 5$ b. $21 - 17x + 2x^2$
   c. $15 + 14x - 8x^2$ d. $110 + 47x - 21x^2$
7. a. $2x^2 - 4x - 6$ b. $8x^2 - 28x - 16$
   c. $-2x^2 + 12x + 14$
8. a. $2x^3 - 2x$ b. $3x^3 - 75x$ c. $6x^3 - 54x$
9. a. $2x^3 - 12x^2 + 18x$ b. $5x^3 - 30x^2 + 40x$
   c. $-6x^3 - 6x^2 + 120x$
10. a. $x^3 + 2x^2 - x - 2$ b. $x^3 - 2x^2 - 5x + 6$
    c. $x^3 - 5x^2 - x + 5$
11. a. $x^3 - 6x^2 + 11x - 6$ b. $2x^3 - 7x^2 - 5x + 4$
    c. $6x^3 - 7x^2 + 1$
12. a. $x^2 - x - 2$ b. $-2x^2 + 4x + 10$
    c. $5x^2 - 6x - 5$ d. $19x - 23$
13. a. $-5x - 1$
    b. $-2x + 6$
    c. $x^2 - 2x - 3 + \sqrt{3}x$
    d. $\sqrt{6} + 2\sqrt{2}x - 3\sqrt{3}x - 6x^2 - \sqrt{5}x$

14. A
15. C
16. B
17. a. $x^2 - 2x + 1$ b. $x^2 + 4x + 4$
    c. $x^2 + 10x + 25$ d. $16 + 8x + x^2$
18. a. $49 - 14x + x^2$ b. $144 - 24x + x^2$
    c. $9x^2 - 6x + 1$ d. $144x^2 - 72x + 9$
19. a. $25x^2 + 20x + 4$ b. $4 - 12x + 9x^2$
    c. $25 - 40x + 16x^2$ d. $1 - 10x + 25x^2$
20. a. $2x^2 - 12x + 18$ b. $4x^2 - 56x + 196$
    c. $3x^2 + 6x + 3$
21. a. $-4x^2 - 12x - 9$ b. $-49x^2 + 14x - 1$
    c. $8x^2 - 24x + 18$
22. a. $-12 + 108x - 243x^2$ b. $-45 + 330x - 605x^2$
    c. $-16x^2 - 16x - 4$
23. a. $x^2 - 49$ b. $x^2 - 81$ c. $x^2 - 25$
24. a. $x^2 - 1$ b. $4x^2 - 9$ c. $9x^2 - 1$
25. a. $49 - x^2$ b. $64 - x^2$ c. $9 - 4x^2$
26. a. $(x + 1)(x - 3)$ b. $x^2 - 2x - 3$
    c. $6\,\text{cm}, 2\,\text{cm}, 12\,\text{cm}^2$
27. a.

x m

b.

$(x + 1)$ m

$(x + 2)$ m

c. $(x + 1)(x + 2)$
d. $x^2 + 3x + 2$
e. $4\,\text{m}^2, 12\,\text{m}^2$

28. Sample responses can be found in the worked solutions in the online resources. Examples are shown.
    a. $(x + 4)(x + 3) = x^2 + 7x + 12$
    b. $(x + 4)^2 = x^2 + 8x + 16$
    c. $(x + 4)(x - 4) = x^2 - 16$
    d. $(x + 4)^2 = x^2 + 8x + 16$

29. a. Student C
    b. Student B
$$(3x + 4)(2x + 5)$$
$$= 3x \times 2x + 3x \times 5 + 4 \times 2x + 4 \times 5$$
$$= 6x^2 + 23x + 20$$
    Student A
$$(3x + 4)(2x + 5)$$
$$= 3x \times 2x + 3x \times 5 + 4 \times 2x + 4 \times 5$$
$$= 6x^2 + 15x + 8x + 20$$
$$= 6x^2 + 23x + 20$$

30. Sample responses can be found in the worked solutions in the online resources.

31. Sample responses can be found in the worked solutions in the online resources.

32. Sample responses can be found in the worked solutions in the online resources.

33. The square of a binomial is a trinomial; the difference of two squares has two terms.

34. $V = 6x^3 - 29x^2 + 46x - 24$

35. $TSA = 12x^2 + 20x + 3$

36. a. $4x^2 + 12xy - 20xz + 9y^2 - 30yz + 25z^2$

   b. $\dfrac{1}{x} + \dfrac{1}{4x^2} - 4x + 4x^2 - 1$

7.3 Factorising expressions with three terms

1. a. $(x+2)(x+1)$ b. $(x+3)(x+1)$
   c. $(x+8)(x+2)$ d. $(x+4)^2$
   e. $(x-3)(x+1)$

2. a. $(x-4)(x+1)$ b. $(x-12)(x+1)$
   c. $(x-6)(x+2)$ d. $(x+4)(x-1)$
   e. $(x+5)(x-1)$

3. a. $(x+7)(x-1)$ b. $(x+5)(x-2)$
   c. $(x-3)(x-1)$ d. $(x-4)(x-5)$
   e. $(x+14)(x-5)$

4. a. $-2(x+9)(x+1)$ b. $-3(x+2)(x+1)$
   c. $-(x+2)(x+1)$ d. $-(x+10)(x+1)$

5. a. $-(x+2)(x+5)$ b. $-(x+12)(x+1)$
   c. $-(x+3)(x+4)$ d. $-(x+2)(x+6)$

6. a. $2(x+2)(x+5)$ b. $3(x+1)(x+10)$
   c. $5(x+20)(x+1)$ d. $5(x+4)(x+5)$

7. a. $(a-7)(a+1)$ b. $(t-4)(t-2)$
   c. $(b+4)(b+1)$ d. $(m+5)(m-3)$

8. a. $(p-16)(p+3)$ b. $(c+16)(c-3)$
   c. $(k+19)(k+3)$ d. $(s-19)(s+3)$

9. a. $(g+8)(g-9)$ b. $(v-25)(v-3)$
   c. $(x+16)(x-2)$ d. $(x-15)(x-4)$

10. C

11. B

12. C

13. a. $(2x+1)(x+2)$ b. $(2x-1)(x-1)$
    c. $(4x+3)(x-5)$ d. $(2x-1)(2x+3)$

14. a. $(x-7)(2x+5)$ b. $(3x+1)(x+3)$
    c. $(3x-7)(2x-1)$ d. $(4x-7)(3x+2)$

15. a. $(5x+3)(2x-3)$ b. $(4x-1)(5x+2)$
    c. $(3x+2)(4x-1)$ d. $(3x-1)(5x+2)$

16. a. $2(x-1)(2x+3)$ b. $3(3x+1)(x-7)$
    c. $12(2x+1)(3x-1)$ d. $-3(3x+1)(2x-1)$

17. a. $-30(2x+1)(x-3)$ b. $3a(4x-7)(2x+5)$
    c. $-2(4x-3)(x-2)$ d. $-(2x-7)(5x+2)$

18. a. $-(8x-1)(3x-4)$
    b. $-2(3x-y)(2x+y)$
    c. $-5(2x-7y)(3x+2y)$
    d. $-12(5x+3y)(10x+7y)$

19. a. $w^2 + 5w - 6$ b. $(w+6)(w-1)$
    c. $(x+5)(x-2)$

20. a. $x(x+5)$
    b. $x(x+5)$
    c. $(x-1)^2$

21. a. $(x+9)(x+5)$ b. $(x-15)(x-6)$
    c. $(x-10)(x-3)$

22. a. Yellow $= 3\,\text{cm} \times 3\,\text{cm}$
    Black $= 3\,\text{cm} \times 6\,\text{cm}$
    White $= 6\,\text{cm} \times 6\,\text{cm}$
    b. Yellow $= 0.36\,\text{m}^2$
    Black $= 0.72\,\text{m}^2$
    White $= 0.36\,\text{m}^2$
    c.

23. a. $x^2 - 7x + 12 = (x-3)(x-4)$
    b. $x^2 - x - 12 = (x-4)(x+3)$
    c. $x^2 - x - 2 = (x-2)(x+1)$
    d. $x^2 - 4x - 21 = (x+3)(x-7)$

24. a. $x^2 + 4x - 21 = (x-3)(x+7)$
    b. $x^2 - x - 30 = (x+5)(x-6)$
    c. $x^2 + 7x - 8 = (x-1)(x+8)$
    d. $x^2 - 11x + 30 = (x-5)(x-6)$

25. a. $SA = 3x^2 + 16x$
    b. Total area $= 3x^2 + 16x + 16$
    c. $(3x+4)(x+4)$
    d. $l = 21\,\text{m}$; $w = 7\,\text{m}$; $d = 2\,\text{m}$
    e. $275\,\text{m}^2$
    f. $294\,\text{m}^3$

26. $(x-0.5)(x+1.5)$

27. a. $(2x+3)(3x+1)$
    b. $P = 10x + 8$
    c. $x = 8$ metres

28. a. $(9a-2)(6a-7)$
    b. $(3m^2+2)(m-\sqrt{7})(m+\sqrt{7})$
    c. $(2\sin(x)-1)(\sin(x)-1)$

29. a. $(x-5)(x+1)$
    b. $(x-5)$ cm
    c. $x = 15\,\text{m}$
    d. $160\,\text{cm}^2$
    e. $3000(x-5)(x+1)\,\text{cm}^2$ or $(3000x^2 - 12\,000x - 15\,000)\,\text{cm}^2$

7.4 Factorising expressions with two or four terms

1. a. $x(x+3)$ b. $x(x-4)$ c. $3x(x-2)$

2. a. $4x(x+4)$ b. $3x(3x-1)$ c. $8x(1-x)$

3. a. $3x(4-x)$ b. $4x(2-3x)$ c. $x(8x-11)$

4. a. $(x-2)(3x+2)$ b. $(x+3)(5-2x)$
   c. $(x-1)(x+5)$

5. a. $(x+1)(x-1)$ b. $(x+4)(x-2)$
    c. $(x-3)(4-x)$

6. a. $(x+1)(x-1)$ b. $(x+3)(x-3)$
    c. $(x+5)(x-5)$

7. a. $(x+10)(x-10)$ b. $(y+k)(y-k)$
    c. $(2x+3y)(2x-3y)$

8. a. $(4a+7)(4a-7)$ b. $(5p+6q)(5p-6q)$
    c. $(1+10d)(1-10d)$

9. a. $4(x+1)(x-1)$ b. $5(x+4)(x-4)$
    c. $a(x+3)(x-3)$

10. a. $2(b+2d)(b-2d)$ b. $100(x+4)(x-4)$
    c. $3a(x+7)(x-7)$

11. a. $4p(x+8)(x-8)$ b. $4(3x+2)(3x-2)$
    c. $3(6+x)(6-x)$

12. C

13. B

14. B

15. C

16. a. $\left(x+\sqrt{11}\right)\left(x-\sqrt{11}\right)$

    b. $\left(x+\sqrt{7}\right)\left(x-\sqrt{7}\right)$

    c. $\left(x+\sqrt{15}\right)\left(x-\sqrt{15}\right)$

17. a. $\left(2x+\sqrt{13}\right)\left(2x-\sqrt{13}\right)$

    b. $\left(3x+\sqrt{19}\right)\left(3x-\sqrt{19}\right)$

    c. $3\left(x+\sqrt{22}\right)\left(x-\sqrt{22}\right)$

18. a. $5\left(x+\sqrt{3}\right)\left(x-\sqrt{3}\right)$

    b. $2\left(x+\sqrt{2}\right)\left(x-\sqrt{2}\right)$

    c. $12\left(x+\sqrt{3}\right)\left(x-\sqrt{3}\right)$

19. a. $(x-3)(x+1)$ b. $(x-4)(x+6)$
    c. $(x-5)(x+1)$

20. a. $(x-1)(x+7)$ b. $(6-x)(x+8)$
    c. $(10-x)(x+2)$

21. a. $8(x-3)$ b. $(7-x)(5x+1)$
    c. $(x-22)(9x+2)$

22. a. $(x-2y)(1+a)$ b. $(x+y)(2+a)$
    c. $(x-y)(a+b)$ d. $(x+y)(4+z)$

23. a. $(f-2)(e+3)$ b. $(n-7)(m+1)$
    c. $3(2r-s)(t+u)$ d. $7(m-3)(n+5)$

24. a. $2(8-j)(4+k)$ b. $a(3-b)(a+c)$
    c. $x(5+y)(x+2)$ d. $m(m+n)(2-n)$

25. a. $(y+7)(x-2)$ b. $(m+2)(n-3)$
    c. $(q+5)(p-3)$

26. a. $(s+3)(s-4t)$ b. $(b+d)(a^2-c)$
    c. $(1+5z)(xy-z)$

27. a. $(a-b)(a+b+4)$
    b. $(p-q)(p+q-3)$
    c. $(m+n)(m-n+l)$

28. a. $(x+y)(7+x-y)$
    b. $(1-2q)(5p+1+2q)$
    c. $(7g+6h)(7g-6h-4)$

29. a. $(x+7+y)(x+7-y)$
    b. $(x+10+y)(x+10-y)$
    c. $(a-11+b)(a-11-b)$

30. a. $(3a+2+b)(3a+2-b)$
    b. $(5p-4+3t)(5p-4-3t)$
    c. $\left(6t-1+\sqrt{5}v\right)\left(6t-1-\sqrt{5}v\right)$

31. D

32. A

33. D

34. B

35. a. Jill
    b. The factors have to add to the -3 and multiple to 2. Jack's pair added to 2 and multiplied to -3, the wrong way around.

36. When splitting the middle term, the order is not important. The order will change the first binomial factor but gives the same answer.

37. a. $(x-2y-a+3b)(x-2y+a-3b)$
    b. $3(2x-5y-3)(2x+5y-3)$

38. a. $(x-5)(x+5)$
    b. $(x-5)$ cm, $(x+5)$ cm
    c. 2 cm, 12 cm
    d. $24\,\text{cm}^2$
    e. $120\,\text{cm}^2$ or 6 times bigger

39. a. $A_1 = \pi r^2\,\text{m}^2$
    b. $A_2 = \pi(r+1)^2\,\text{m}^2$
    c. $A = \pi(r+1)^2 - \pi r^2 = \pi(2r+1)\,\text{m}^2$
    d. $34.56\,\text{m}^2$

40. a. Annie $= (x+3)(x+2)\,\text{m}^2$ Bronwyn $= 5(x+2)\,\text{m}^2$
    b. $(x+3)(x+2) - 5(x+2)$
    c. $(x+2)(x-2) = x^2 - 4$
    d. Width $= 5\,\text{m}$
    e. Annie has $30\,\text{m}^2$ and Bronwyn has $25\,\text{m}^2$.

7.5 Factorising by completing the square

1. a. 25   b. 9   c. 4
    d. 64   e. 100

2. a. 16   b. 49   c. 625
    d. $\dfrac{49}{4}$ e. $\dfrac{1}{4}$

3. a. $\left(x-2+\sqrt{11}\right)\left(x-2-\sqrt{11}\right)$

    b. $\left(x+1+\sqrt{3}\right)\left(x+1-\sqrt{3}\right)$

    c. $\left(x-5+\sqrt{13}\right)\left(x-5-\sqrt{13}\right)$

d. $\left(x+3+\sqrt{19}\right)\left(x+3-\sqrt{19}\right)$

e. $\left(x+8+\sqrt{65}\right)\left(x+8-\sqrt{65}\right)$

4. a. $\left(x-7+\sqrt{6}\right)\left(x-7-\sqrt{6}\right)$

b. $\left(x+4+\sqrt{7}\right)\left(x+4-\sqrt{7}\right)$

c. $\left(x-2+\sqrt{17}\right)\left(x-2-\sqrt{17}\right)$

d. $\left(x-6+\sqrt{11}\right)\left(x-6-\sqrt{11}\right)$

e. $\left(x-3-\sqrt{5}\right)\left(x-3+\sqrt{5}\right)$

5. a. $\left(x-\dfrac{1}{2}+\dfrac{\sqrt{5}}{2}\right)\left(x-\dfrac{1}{2}-\dfrac{\sqrt{5}}{2}\right)$

b. $\left(x-\dfrac{3}{2}+\dfrac{\sqrt{21}}{2}\right)\left(x-\dfrac{3}{2}-\dfrac{\sqrt{21}}{2}\right)$

c. $\left(x+\dfrac{1}{2}+\dfrac{\sqrt{21}}{2}\right)\left(x+\dfrac{1}{2}-\dfrac{\sqrt{21}}{2}\right)$

d. $\left(x+\dfrac{3}{2}+\dfrac{\sqrt{13}}{2}\right)\left(x+\dfrac{3}{2}-\dfrac{\sqrt{13}}{2}\right)$

e. $\left(x+\dfrac{5}{2}+\dfrac{\sqrt{17}}{2}\right)\left(x+\dfrac{5}{2}-\dfrac{\sqrt{17}}{2}\right)$

6. a. $\left(x+\dfrac{5}{2}+\dfrac{\sqrt{33}}{2}\right)\left(x+\dfrac{5}{2}-\dfrac{\sqrt{33}}{2}\right)$

b. $\left(x-\dfrac{7}{2}+\dfrac{\sqrt{53}}{2}\right)\left(x-\dfrac{7}{2}-\dfrac{\sqrt{53}}{2}\right)$

c. $\left(x-\dfrac{9}{2}+\dfrac{\sqrt{29}}{2}\right)\left(x-\dfrac{9}{2}-\dfrac{\sqrt{29}}{2}\right)$

d. $\left(x-\dfrac{1}{2}+\dfrac{\sqrt{13}}{2}\right)\left(x-\dfrac{1}{2}-\dfrac{\sqrt{13}}{2}\right)$

e. $\left(x-\dfrac{1}{2}-\dfrac{\sqrt{5}}{2}\right)\left(x-\dfrac{1}{2}+\dfrac{\sqrt{5}}{2}\right)$

7. a. $2\left(x+1+\sqrt{3}\right)\left(x+1-\sqrt{3}\right)$

b. $4\left(x-1+\sqrt{6}\right)\left(x-1-\sqrt{6}\right)$

c. $5\left(x+3+2\sqrt{2}\right)\left(x+3-2\sqrt{2}\right)$

d. $3\left(x-2+\sqrt{17}\right)\left(x-2-\sqrt{17}\right)$

e. $5\left(x-3+\sqrt{7}\right)\left(x-3-\sqrt{7}\right)$

8. a. $6\left(x+2+\sqrt{5}\right)\left(x+2-\sqrt{5}\right)$

b. $3\left(x+5+2\sqrt{3}\right)\left(x+5-2\sqrt{3}\right)$

c. $2\left(x-2+\sqrt{11}\right)\left(x-2-\sqrt{11}\right)$

d. $6\left(x+3+\sqrt{14}\right)\left(x+3-\sqrt{14}\right)$

e. $4\left(x-1-\sqrt{5}\right)\left(x-1+\sqrt{5}\right)$

9. i. d   ii. b   iii. c   iv. a
   v. c   vi. d   vii. d   viii.e

10. B

11. D

12. D

13. This expression cannot be factorised as there is no difference of two squares after completing the square.

14. $a = 0.55$; $b = 5.45$

15. This expression cannot be factorised as it does not become a difference of two squares.

16. Annabelle: $\left(x-2-\sqrt{13}\right)\left(x-2+\sqrt{13}\right)$

17. $(x-p)(x-p+2)$

18. a. $2\left(x+2-\dfrac{\sqrt{14}}{2}\right)\left(x+2+\dfrac{\sqrt{14}}{2}\right)$

b. This expression cannot be factorised as there is no difference of two squares.

7.6 Mixed factorisation

1. a. $3(x+3)$
   b. $(x+2+3y)(x+2-3y)$
   c. $(x+6)(x-6)$
   d. $(x+7)(x-7)$
   e. $(5x+1)(x-2)$

2. a. $5(3x-4y)$
   b. $(c+e)(5+d)$
   c. $5(x+4)(x-4)$
   d. $-(x+5)(x+1)$
   e. $(x+4)(x-3)$

3. a. $(m+1)(n+1)$
   b. $\left(x+\sqrt{7}\right)\left(x-\sqrt{7}\right)$
   c. $4x(4x-1)$
   d. $5(x+10)(x+2)$
   e. $3(3-y)(x+2)$

4. a. $(x-4+y)(x-4-y)$
   b. $4(x^2+2)$
   c. $(g+h)(f+2)$
   d. $\left(x+\sqrt{5}\right)\left(x-\sqrt{5}\right)$
   e. $5(n+1)(2m-1)$

5. a. $(x+5)(x+1)$
   b. $(x+1)(x-11)$
   c. $(x+2)(x-2)$
   d. $(a+b)(c-5)$
   e. $(y+1)(x-1)$

6. a. $(3x+2)(x+1)$
   b. $7(x+2)(x-2)$
   c. $-4(x+6)(x+1)$
   d. $(2+r)(p-s)$
   e. $3(x+3)(x-3)$

7. a. $(u+v)(t-3)$

b. $\left(x+\sqrt{11}\right)\left(x-\sqrt{11}\right)$

c. $(4x-1)(3x-1)$

d. $(x+1)(x-3)$

e. $(x+6)(x-2)$

8. a. $4(x-1)(x+4)$

b. $3(x+2)(x+8)$

c. $(3+x)(7-x)$

d. $4(3-x+2y)(3-x-2y)$

e. $3(y+x)(y-x)$

9. a. $4(x+2)$

b. $(3x-4y)(x-2y)$

c. $(x+7)(x+4)$

d. $(x+2)(x-5)$

e. $(2x+3)(x+3)$

10. a. $\dfrac{(x+5)(x-2)}{(x+2)(x-2)} \times \dfrac{(x+2)(x+2)}{(x-4)(x+2)}$

b. $\dfrac{(x+5)\,\cancel{(x-2)}}{\cancel{(x+2)}\,\cancel{(x-2)}} \times \dfrac{\cancel{(x+2)}\,\cancel{(x+2)}}{(x-4)\,\cancel{(x+2)}}$

c. $\dfrac{x+5}{x-4}$

11. a. $\dfrac{x-1}{x-6}$ **b.** $\dfrac{x+1}{2x+3}$ **c.** $\dfrac{18}{x(x-5)}$

12. a. $\dfrac{2x-1}{x+4}$ **b.** $\dfrac{x+2}{x+5}$ **c.** $\dfrac{x-6}{x+3}$

13. a. $\dfrac{4(b+2)}{5}$

b. $\dfrac{p(p+7)}{(p+3)(p-2)}$

c. $\dfrac{5(m+2+n)}{2(2m-5)}$

14. a. i. $\dfrac{x^2}{16} - \dfrac{9}{25}$

ii. $\dfrac{x^2}{16} - \dfrac{9}{25}$

b. $\left(\dfrac{-x}{4}\right)^2 = \left(\dfrac{x}{4}\right)^2 = \dfrac{x^2}{16}$

15. a. $(x-9-y)(x-9+y)$

b. $(2x+3-4y)(2x+3+4y)$

16. a. $10\,609$ **b.** 3844 **c.** $99\,409$

d. $1\,024\,144$ **e.** 2809 **f.** 9604

17. a. $(x+y)(x-y+3)$

b. $(x+y)(7+x-y)$

c. $(1-2q)(5p+1+2q)$

18. $\dfrac{1}{a^2+2a+4}$

19. a. $(x+6-2xy)(x+6+2xy)$

b. $x^2y^2(15x-13y^2)(15x+13y^2)$ or
$x^2y^2(-15x-13y^2)(-15x+13y^2)$

Project

Students will apply their skills of elimination to find 'square and double pair' by playing the game with the given instructions.

7.7 Review questions

1. D
2. D
3. D
4. C
5. C
6. A
7. D
8. C

9. a. $3x^2 - 12x$ **b.** $-21x^2 - 7x$

c. $x^2 - 6x - 7$ **d.** $2x^2 - 11x + 15$

e. $12x^2 - 23x + 5$

10. a. $6x^2 - 3x - 84$ **b.** $2x^3 + 15x^2 - 8x - 105$

c. $3x^2 - 5x + 65$ **d.** $5x^2 + 12x - 3$

11. a. $x^2 - 14x + 49$ **b.** $4 - 4x + x^2$ **c.** $9x^2 + 6x + 1$

12. a. $-18x^2 + 24x - 8$

b. $-28x^2 - 140x - 175$

c. $-160x^2 + 400x - 250$

13. a. $x^2 - 81$ **b.** $9x^2 - 1$ **c.** $25 - 4x^2$

14. a. $2x(x-4)$ **b.** $-4x(x-3)$ **c.** $ax(3-2x)$

15. a. $(x+1)(x+2)$

b. $2(2x-5)(4-x)$

c. $(x-4)(x+1)$

16. a. $(x+4)(x-4)$

b. $(x+5)(x-5)$

c. $2(x+6)(x-6)$

17. a. $3(x+3y)(x-3y)$

b. $4a(x+2y)(x-2y)$

c. $(x-1)(x-7)$

18. a. $(x-y)(a+b)$ **b.** $(x+y)(7+a)$

c. $(x+2)(y+5)$

19. a. $(1+2q)(mn-q)$ **b.** $(5r+1)(pq-r)$

c. $(v-1)(u+9)$

20. a. $(a-b)(a+b+5)$ **b.** $(d-2c)(d+2c-3)$

c. $(1+m)(3-m)$

21. a. $(2x+3+y)(2x+3-y)$

b. $(7a-2+2b)(7a-2-2b)$

c. $\left(8s-1+\sqrt{3}t\right)\left(8s-1-\sqrt{3}t\right)$

22. a. $(x+9)(x+1)$ **b.** $(x-9)(x-2)$

c. $(x-7)(x+3)$ **d.** $(x+7)(x-4)$

e. $-(x-3)^2$

23. a. $3(x+13)(x-2)$ **b.** $-2(x-5)(x+1)$

c. $-3(x-6)(x-2)$ **d.** $(4x-1)(2x+1)$

e. $(3x-1)(2x+1)$

24. a. $4(2x+3)(x-1)$ **b.** $5(7x-3)(3x+1)$

c. $-2(3x-5)(2x-7)$ **d.** $-3(3x-1)(5x+2)$

e. $-30(2x+3)(x+3)$

25. a. $\left(x + 3 + 2\sqrt{2}\right)\left(x + 3 - 2\sqrt{2}\right)$

b. $\left(x - 5 + 2\sqrt{7}\right)\left(x - 5 - 2\sqrt{7}\right)$

c. $\left(x + 2 + \sqrt{6}\right)\left(x + 2 - \sqrt{6}\right)$

26. a. $\left(x - \dfrac{5}{2} + \dfrac{\sqrt{17}}{2}\right)\left(x - \dfrac{5}{2} - \dfrac{\sqrt{17}}{2}\right)$

b. $\left(x + \dfrac{7}{2} + \dfrac{\sqrt{53}}{2}\right)\left(x + \dfrac{7}{2} - \dfrac{\sqrt{53}}{2}\right)$

c. $2\left(x + \dfrac{9}{2} + \dfrac{\sqrt{85}}{2}\right)\left(x + \dfrac{9}{2} - \dfrac{\sqrt{85}}{2}\right)$

27. a. $3x(x - 4)$ c. $(2x + 5)(2x - 5)$

b.
$\left(x + 3 + \sqrt{7}\right)\left(x + 3 - \sqrt{7}\right)$

28. a. $(2x + 5)(x + 2)$ b. $(a + 2)(2x + 3)$

c. $-3(x - 2)(x + 3)$

29. a. $\dfrac{2(x + 4)}{5(x + 1)}$ b. $\dfrac{7}{8}$

c. $\dfrac{(x - 2)(x - 1)}{x(x - 4)}$

30. a. $(x + 2)^2$ b. $32(x + 2)^2$

c. $32x^2 + 128x + 128$ d. $32768 \, \text{cm}^3$

31. a. $4r$ b. $2r + 2$

c. $4\pi r^2$ d. $(4r^2 + 8r + 4)\pi$

e. $4\pi(2r + 1)$ f. $28\pi \, \text{m}^2$

32. a. $(x - 7)$ b. $(x - 7) \, \text{cm}$

c. 35 d. $1036 \, \text{cm}^2$

33. Dividing by zero in step 5

8 Quadratic equations

LESSON SEQUENCE

LESSON
8.1 Overview

Why learn this?

Have you ever thought about the shape a ball makes as it flies through the air? Or have you noticed the shape that the stream of water from a drinking fountain makes? These are both examples of quadratic equations in the real world. When you learn about quadratic equations, you learn about the mathematics of these real-world shapes, but quadratic equations are so much more than interesting shapes.

Being able to use and understand quadratic equations lets you unlock incredible problem-solving skills. It allows you to quickly understand situations and solve complicated problems that would be close to impossible without these skills. Many professionals rely on their understanding of quadratic equations to make important decisions, designs and discoveries. Some examples include: sports people finding the perfect spot to intercept a ball, architects designing complex buildings such as the Sydney Opera House and scientists listening to sound waves coming from the furthest regions of space using satellite dishes. All of these professionals use the principles of quadratic equations.

By learning about quadratic equations you are also embarking on your first step on the path to learning about non-linear equations. When you learn about non-linear equations, a whole world of different shaped curves and different problem-solving skills will open up.

Hey students! Bring these pages to life online

- ▶ **Watch videos**
- 🧩 **Engage with interactivities**
- A⁺ **Answer questions and check solutions**

Find all this and MORE in jacPLUS ▶

Reading content and rich media, including interactivities and videos for every concept

Extra learning resources

Differentiated question sets

Questions with immediate feedback, and fully worked solutions to help students get unstuck

1. State whether the following statement is true or false.
   The Null Factor Law states that if the product of two numbers is zero then one or both of the numbers must equal zero.

2. Calculate the two solutions for the equation $(2x - 1)(x + 5) = 0$.

3. Solve $16x^2 - 9 = 0$ for x.

4. **MC** The solutions for $x(x - 3)(x + 2) = 0$ are:
   A. $x = 3$ or $x = -2$
   C. $x = 0$ or $x = -3$ or $x = 2$
   B. $x = 3$ or $x = -2$
   D. $x = 0$ or $x = 3$ or $x = -2$

5. **MC** The solutions to the equation $x^2 + x - 6 = 0$ are:
   A. $x = 3$ or $x = -2$
   C. $x = 3$ or $x = 2$
   B. $x = -3$ or $x = 2$
   D. $x = -6$ or $x = 1$

6. **MC** Four times a number is subtracted from 3 times its square. If the result is 4, the possible numbers are:
   A. $x = -4$
   C. $x = 0$ or $x = 4$
   B. $x = 0$ or $x = -4$
   D. $x = -\dfrac{2}{3}$ or $x = 2$

7. **MC** An exact solution to the equation $x^2 - 3x - 1 = 0$ is $x =$
   A. $\dfrac{3 \pm \sqrt{13}}{2}$
   B. $\dfrac{-3 \pm \sqrt{13}}{2}$
   C. $\dfrac{3 \pm 2\sqrt{2}}{2}$
   D. $\dfrac{-3 \pm 2\sqrt{2}}{2}$

8. Calculate the discriminant for $5x^2 - 8x + 2 = 0$.

9. Match the discriminant value to the number of solutions for a quadratic equation.

Discriminant value	Number of solutions
a. $\Delta = 0$	No real solutions
b. $\Delta < 0$	Two solutions
c. $\Delta > 0$	Only one solution

10. **MC** If $\Delta = 0$, the graph of a quadratic:
    A. does not cross or touch the x-axis
    B. touches the x-axis
    C. intersects the x-axis twice
    D. intersects the x-axis three times

▶

11. **MC** The equation $3x^2 - 27 = 0$ has:
    A. two rational solutions
    B. two irrational solutions
    C. one solution
    D. one rational and one irrational solution.

12. A rectangle's length is 4 cm more than its width. Determine the dimensions of the rectangle if its area is $(4x + 9)$ cm².

13. Calculate the value of m for which $mx^2 - 12x + 9 = 0$ has one solution.

14. **MC** Identify for what values of a the straight line $y = ax - 12$ intersects once with the parabola $y = x^2 - 2x - 8$.
    A. $a = -6$ or $a = 2$
    B. $a = 6$ or $a = -2$
    C. $a = 4$ or $a = -2$
    D. $a = -4$ or $a = 2$

15. **MC** Solve for x, $\dfrac{x}{x-2} - \dfrac{x+1}{x+4} = 1$.

    A. $x = -1$ or $x = -4$
    B. $x = 0$ or $x = 2$
    C. $x = 0$ or $x = -1$
    D. $x = 5$ or $x = -2$

LESSON
8.2 Solving quadratic equations algebraically

LEARNING INTENTIONS

At the end of this lesson you should be able to:
- use the Null Factor Law to solve quadratic equations
- use the completing the square technique to solve quadratic equations
- solve worded questions algebraically.

▶ 8.2.1 Quadratic equations

eles-4843

- **Quadratic equations** are equations in the form:
  $$ax^2 + bx + c = 0, \text{ where } a, b \text{ and } c \text{ are numbers.}$$
- For example, the equation $3x^2 + 5x - 7 = 0$ has $a = 3, b = 5$ and $c = -7$.

The Null Factor Law

- We know that any number multiplied by zero will equal zero.

> **Null Factor Law**
>
> - **If the product of two numbers is zero, then one or both numbers must be zero.**
>
>   **a number × another number = 0**
> - **In the above equation, we know that a number = 0 and/or another number = 0.**

- To apply the Null Factor Law, quadratic equations must be in a factorised form.
- To review factorising quadratic expressions, see topic 7.

WORKED EXAMPLE 1 Applying the Null Factor Law to solve quadratic equations

Solve the equation $(x - 7)(x + 11) = 0$.

THINK	WRITE
1. Write the equation and check that the right-hand side equals zero. (The product of the two numbers is zero.)	$(x - 7)(x + 11) = 0$
2. The left-hand side is factorised, so apply the Null Factor Law.	$x - 7 = 0$ or $x + 11 = 0$
3. Solve for x.	$x = 7$ or $x = -11$

WORKED EXAMPLE 2 Factorising then applying the Null Factor Law

Solve each of the following equations.
a. $x^2 - 3x = 0$
b. $3x^2 - 27 = 0$
c. $x^2 - 13x + 42 = 0$
d. $36x^2 - 21x = 2$

THINK	WRITE
a. 1. Write the equation. Check that the right-hand side equals zero.	a. $x^2 - 3x = 0$
2. Factorise by taking out the common factor of x^2 and $3x$, which is x.	$x(x - 3) = 0$
3. Apply the Null Factor Law.	$x = 0$ or $x - 3 = 0$
4. Solve for x.	$x = 0$ or $x = 3$
b. 1. Write the equation. Check that the right-hand side equals zero.	b. $3x^2 - 27 = 0$
2. Factorise by taking out the common factor of $3x^2$ and 27, which is 3.	$3(x^2 - 9) = 0$
3. Factorise using the difference of two squares rule.	$3(x^2 - 3^2) = 0$ $3(x + 3)(x - 3) = 0$
4. Apply the Null Factor Law.	$x + 3 = 0$ or $x - 3 = 0$
5. Solve for x.	$x = -3$ or $x = 3$ (Alternatively, $x = \pm 3$.)
c. 1. Write the equation. Check that the right-hand side equals zero.	c. $x^2 - 13x + 42 = 0$
2. Factorise by identifying a factor pair of 42 that adds to -13.	<table><tr><th>Factors of 42</th><th>Sum of factors</th></tr><tr><td>−6 and −7</td><td>−13</td></tr></table>
3. Use the Null Factor Law to write two linear equations.	$(x - 6)(x - 7) = 0$ $x - 6 = 0$ or $x - 7 = 0$
4. Solve for x.	$x = 6$ or $x = 7$

d. 1. Write the equation. Check that the right-hand side equals zero. (It does not.)

d. $36x^2 - 21x = 2$

2. Rearrange the equation so the right-hand side of the equation equals zero as a quadratic trinomial in the form $ax^2 + bx + c = 0$.

$36x^2 - 21x - 2 = 0$

3. Recognise that the expression to factorise is a quadratic trinomial.
Calculate $ac = 36 \times -2 = -72$.
Factorise by identifying a factor pair of -72 that adds to -21.

Factors of -72	Sum of factors
3 and -24	-21

4. Factorise the expression.

$36x^2 - 24x + 3x - 2 = 0$

$12x(3x - 2) + (3x - 2) = 0$

$(3x - 2)(12x + 1) = 0$

5. Use the Null Factor Law to write two linear equations.

$3x - 2 = 0$ or $12x + 1 = 0$

$3x = 2$ or $12x = -1$

6. Solve for x.

$x = \dfrac{2}{3}$ or $x = -\dfrac{1}{12}$

⊙ 8.2.2 Solving quadratic equations by completing the square

eles-4844

- Completing the square is another technique that can be used for factorising quadratics.
- Use completing the square when other techniques will not work.
- To complete the square, the value of a must be 1. If a is not 1, divide everything by the coefficient of x^2 so that $a = 1$.
- The completing the square technique was introduced in topic 9.

Steps to complete the square

Factorise $x^2 + 6x + 2 = 0$.

Step 1: $ax^2 + bx + c = 0$. Since $a = 1$ in this example, we can complete the square.

Step 2: Add and subtract $\left(\dfrac{1}{2}b\right)^2$. In this example, $b = 6$.

$$x^2 + 6x + \left(\dfrac{1}{2} \times 6\right)^2 + 2 - \left(\dfrac{1}{2} \times 6\right)^2 = 0$$

$$x^2 + 6x + 9 + 2 - 9 = 0$$

Step 3: Factorise the first three terms.

$$x^2 + 6x + 9 - 7 = 0$$
$$(x + 3)^2 - 7 = 0$$

Step 4: Factorise the quadratic by using difference of two squares

$$(x + 3)^2 - \left(\sqrt{7}\right)^2 = 0 \text{ because } \left(\sqrt{7}\right)^2 = 7$$

$$\left(x + 3 + \sqrt{7}\right)\left(x + 3 - \sqrt{7}\right) = 0$$

When $a \neq 1$:
- factorise the expression if possible before completing the square
- if a is not a factor of each term, divide each term by a to ensure the coefficient of x^2 is 1.

Solve the equation $x^2 + 2x - 4 = 0$ by completing the square. Give exact answers.

THINK	WRITE
1. Write the equation.	$x^2 + 2x - 4 = 0$
2. Identify the coefficient of x, halve it and square the result.	$\left(\dfrac{1}{2} \times 2\right)^2$
3. Add the result of step **2** to the equation, placing it after the x-term. To balance the equation, we need to subtract the same amount as we have added.	$x^2 + 2x + \left(\dfrac{1}{2} \times 2\right)^2 - 4 - \left(\dfrac{1}{2} \times 2\right)^2 = 0$ $x^2 + 2x + (1)^2 - 4 - (1)^2 = 0$ $x^2 + 2x + 1 - 4 - 1 = 0$
4. Insert brackets around the first three terms to group them and then simplify the remaining terms.	$(x^2 + 2x + 1) - 5 = 0$
5. Factorise the first three terms to produce a perfect square.	$(x + 1)^2 - 5 = 0$
6. Express as the difference of two squares and then factorise.	$(x + 1)^2 - \left(\sqrt{5}\right)^2 = 0$ $\left(x + 1 + \sqrt{5}\right)\left(x + 1 - \sqrt{5}\right) = 0$
7. Apply the Null Factor Law to identify linear equations.	$x + 1 + \sqrt{5} = 0$ or $x + 1 - \sqrt{5} = 0$
8. Solve for x. Keep the answer in surd form to provide an exact answer.	$x = -1 - \sqrt{5}$ or $x = -1 + \sqrt{5}$ (Alternatively, $x = -1 \pm \sqrt{5}$.)

TI | THINK	DISPLAY/WRITE	CASIO | THINK	DISPLAY/WRITE
In a new document, on a Calculator page, press: • Menu • 3: Algebra • 1: Solve Complete the entry line as: solve $(x^2 + 2x - 4 = 0, x)$ Then press ENTER.	 $x^2 + 2x - 4 = 0$ $\Rightarrow x = -1 + \sqrt{5}$ or $-1 - \sqrt{5}$	On the Main screen in standard, tap: • Action • Advanced • solve Then press EXE.	 $x^2 + 2x - 4 = 0$ $\Rightarrow x = -1 + \sqrt{5}$ or $-1 - \sqrt{5}$

8.2.3 Solving worded questions

eles-4845

- For worded questions:
  - start by identifying the unknowns
  - then write the equation and solve it
  - give answers in a full sentence and include units if required.

WORKED EXAMPLE 4 Solving worded questions

When two consecutive numbers are multiplied together, the result is 20. Determine the numbers.

THINK	WRITE
1. Define the unknowns. First number $= x$, second number $= x + 1$.	Let the two numbers be x and $(x + 1)$.
2. Write an equation using the information given in the question.	$x(x + 1) = 20$
3. Transpose the equation so that the right-hand side equals zero.	$x(x + 1) - 20 = 0$
4. Expand to remove the brackets.	$x^2 + x - 20 = 0$
5. Factorise.	$(x + 5)(x - 4) = 0$
6. Apply the Null Factor Law to solve for x.	$x + 5 = 0$ or $x - 4 = 0$ $x = -5$ or $x = 4$
7. Use the answer to determine the second number.	If $x = -5, x + 1 = -4$. If $x = 4, x + 1 = 5$.
8. Check the solutions.	Check: $4 \times 5 = 20 \quad (-5) \times (-4) = 20$
9. Write the answer in a sentence.	The numbers are 4 and 5 or -5 and -4.

WORKED EXAMPLE 5 Solving worded application questions

The height of a football after being kicked is determined by the formula $h = -0.1d^2 + 3d$, where d is the horizontal distance from the player in metres.

a. Calculate how far the ball is from the player when it hits the ground.

b. Calculate the horizontal distance the ball has travelled when it first reaches a height of 20 m.

THINK	WRITE
a. 1. Write the formula.	a. $h = -0.1d^2 + 3d$
2. The ball hits the ground when $h = 0$. Substitute $h = 0$ into the formula.	$-0.1d^2 + 3d = 0$
3. Factorise.	$-0.1d^2 + 3d = 0$ $d(-0.1d + 3) = 0$

---

4. Apply the Null Factor Law and simplify.

$$d = 0 \text{ or } -0.1d + 3 = 0$$
$$-0.1d = -3$$
$$d = \frac{-3}{-0.1}$$
$$= 30$$

5. Interpret the solutions.

$d = 0$ is the origin of the kick.
$d = 30$ is the distance from the origin that the ball has travelled when it lands.

6. Write the answer in a sentence.

The ball is 30 m from the player when it hits the ground.

b. 1. The height of the ball is 20 m, so, substitute $h = 20$ into the formula.

b. $h = -0.1d^2 + 3d$
$20 = -0.1d^2 + 3d$

2. Transpose the equation so that zero is on the right-hand side.

$0.1d^2 - 3d + 20 = 0$

3. Multiply both sides of the equation by 10 to remove the decimal from the coefficient.

$d^2 - 30d + 200 = 0$

4. Factorise.

$(d - 20)(d - 10) = 0$

5. Apply the Null Factor Law.

$d - 20 = 0 \quad \text{or} \quad d - 10 = 0$

6. Solve.

$d = 20 \quad \text{or} \quad d = 10$

7. Interpret the solution. The ball reaches a height of 20 m on the way up and on the way down. The *first* time the ball reaches a height of 20 m is the smaller value of d. Write the answer in a sentence.

The ball first reaches a height of 20 m after it has travelled a distance of 10 m.

DISCUSSION

What does the Null Factor Law mean?

 Resources

 eWorkbook   Topic 8 Workbook (worksheets, code puzzle and a project) (ewbk-2034)

 Video eLesson The Null Factor Law (eles-2312)

 Interactivities Individual pathway interactivity: Solving quadratic equations algebraically (int-4601)
                     The Null Factor Law (int-6095)

Exercise 8.2 Solving quadratic equations algebraically

8.2 Quick quiz $\quad$ on	8.2 Exercise

Individual pathways

■ PRACTISE	■ CONSOLIDATE	■ MASTER
1, 4, 7, 10, 13, 16, 17, 18, 21, 27, 30, 31, 35, 39, 44	2, 5, 8, 11, 14, 19, 22, 24, 25, 28, 32, 33, 36, 40, 41, 45	3, 6, 9, 12, 15, 20, 23, 26, 29, 34, 37, 38, 42, 43, 46

Fluency

WE1 For questions 1–6 solve each of the following equations.

1. a. $(x+7)(x-9)=0$
   b. $(x-3)(x+2)=0$
   c. $(x-2)(x-3)=0$
   d. $x(x-3)=0$

2. a. $x(x-1)=0$
   b. $x(x+5)=0$
   c. $2x(x-3)=0$
   d. $9x(x+2)=0$

3. a. $\left(x-\dfrac{1}{2}\right)\left(x+\dfrac{1}{2}\right)=0$
   b. $-(x+1.2)(x+0.5)=0$
   c. $2(x-0.1)(2x-1.5)=0$
   d. $\left(x+\sqrt{2}\right)\left(x-\sqrt{3}\right)=0$

4. a. $(2x-1)(x-1)=0$
   b. $(3x+2)(x+2)=0$
   c. $(4x-1)(x-7)=0$

5. a. $(7x+6)(2x-3)=0$
   b. $(5x-3)(3x-2)=0$
   c. $(8x+5)(3x-2)=0$

6. a. $x(x-3)(2x-1)=0$
   b. $x(2x-1)(5x+2)=0$
   c. $x(x+3)(5x-2)=0$

WE2a For questions 7–9 solve each of the following equations.

7. a. $x^2-2x=0$
   b. $x^2+5x=0$
   c. $x^2=7x$

8. a. $3x^2=-2x$
   b. $4x^2-6x=0$
   c. $6x^2-2x=0$

9. a. $4x^2-2\sqrt{7}x=0$
   b. $3x^2+\sqrt{3}x=0$
   c. $15x-12x^2=0$

WE2b For questions 10–12 solve each of the following equations.

10. a. $x^2-4=0$
    b. $x^2-25=0$
    c. $3x^2-12=0$
    d. $4x^2-196=0$

11. a. $9x^2-16=0$
    b. $4x^2-25=0$
    c. $9x^2=4$
    d. $36x^2=9$

12. a. $x^2-\dfrac{1}{25}=0$
    b. $\dfrac{1}{36}x^2-\dfrac{4}{9}=0$
    c. $x^2-5=0$
    d. $9x^2-11=0$

WE2c For questions 13–15 solve each of the following equations.

13. a. $x^2-x-6=0$
    b. $x^2+6x+8=0$
    c. $x^2-6x-7=0$
    d. $x^2-8x+15=0$

14. a. $x^2 - 3x - 4 = 0$
    b. $x^2 - 10x + 25 = 0$
    c. $x^2 - 3x - 10 = 0$
    d. $x^2 - 8x + 12 = 0$
    e. $x^2 - 4x - 21 = 0$

15. a. $x^2 - x - 30 = 0$
    b. $x^2 - 7x + 12 = 0$
    c. $x^2 - 8x + 16 = 0$
    d. $x^2 + 10x + 25 = 0$
    e. $x^2 - 20x + 100 = 0$

16. **MC** The solutions to the equation $x^2 + 9x - 10 = 0$ are:

    A. $x = 1$ and $x = 10$
    B. $x = 1$ and $x = -10$
    C. $x = -1$ and $x = 10$
    D. $x = -1$ and $x = -10$

17. **MC** The solutions to the equation $x^2 - 100 = 0$ are:

    A. $x = 0$ and $x = 10$
    B. $x = 0$ and $x = -10$
    C. $x = -10$ and $x = 10$
    D. $x = 0$ and $x = 100$

WE2d For questions **18–20** solve each of the following equations.

18. a. $2x^2 - 5x = 3$
    b. $3x^2 + x - 2 = 0$
    c. $5x^2 + 9x = 2$
    d. $6x^2 - 11x + 3 = 0$

19. a. $14x^2 - 11x = 3$
    b. $12x^2 - 7x + 1 = 0$
    c. $6x^2 - 7x = 20$
    d. $12x^2 + 37x + 28 = 0$

20. a. $10x^2 - x = 2$
    b. $6x^2 - 25x + 24 = 0$
    c. $30x^2 + 7x - 2 = 0$
    d. $3x^2 - 21x = -36$

WE3 For questions **21–26** solve the following equations by completing the square. Give exact answers.

21. a. $x^2 - 4x + 2 = 0$
    b. $x^2 + 2x - 2 = 0$
    c. $x^2 + 6x - 1 = 0$

22. a. $x^2 - 8x + 4 = 0$
    b. $x^2 - 10x + 1 = 0$
    c. $x^2 - 2x - 2 = 0$

23. a. $x^2 + 2x - 5 = 0$
    b. $x^2 + 4x - 6 = 0$
    c. $x^2 + 4x - 11 = 0$

24. a. $x^2 - 3x + 1 = 0$
    b. $x^2 + 5x - 1 = 0$
    c. $x^2 - 7x + 4 = 0$

25. a. $x^2 - 5 = x$
    b. $x^2 - 11x + 1 = 0$
    c. $x^2 + x = 1$

26. a. $x^2 + 3x - 7 = 0$
    b. $x^2 - 3 = 5x$
    c. $x^2 - 9x + 4 = 0$

For questions **27–29** solve each of the following equations, rounding answers to 2 decimal places.

27. a. $2x^2 + 4x - 6 = 0$
    b. $3x^2 + 12x - 3 = 0$
    c. $5x^2 - 10x - 15 = 0$

28. a. $4x^2 - 8x - 8 = 0$
    b. $2x^2 - 6x + 2 = 0$
    c. $3x^2 - 9x - 3 = 0$

29. a. $5x^2 - 15x - 25 = 0$
    b. $7x^2 + 7x - 21 = 0$
    c. $4x^2 + 8x - 2 = 0$

Understanding

30. **WE4** When two consecutive numbers are multiplied, the result is 72. Determine the numbers.

31. When two consecutive even numbers are multiplied, the result is 48. Determine the numbers.

32. When a number is added to its square the result is 90. Determine the number.

33. Twice a number is added to three times its square. If the result is 16, determine the number.

34. Five times a number is added to two times its square. If the result is 168, determine the number.

35. **WE5** A soccer ball is kicked. The height, h, in metres, of the soccer ball t seconds after it is kicked can be represented by the equation $h = -t(t - 6)$. Calculate how long it takes for the soccer ball to hit the ground again.

36. The length of an Australian flag is twice its width and the diagonal length is 45 cm.
    a. If x cm is the width of the flag, express its length in terms of x.
    b. Draw a diagram of the flag marking in the diagonal. Label the length and the width in terms of x.
    c. Use Pythagoras' theorem to write an equation relating the lengths of the sides to the length of the diagonal.
    d. Solve the equation to calculate the dimensions of the Australian flag. Round your answer to the nearest cm.

37. If the length of a paddock is 2 m more than its width and the area is 48 m², calculate the length and width of the paddock.

38. Solve for x.
    a. $x + 5 = \dfrac{6}{x}$
    b. $x = \dfrac{24}{x - 5}$
    c. $x = \dfrac{1}{x}$

Reasoning

39. The sum of the first n numbers $1, 2, 3, 4 \dots n$ is given by the formula $S = \dfrac{n(n+1)}{2}$.
    a. Use the formula to calculate the sum of the first 6 counting numbers.
    b. Determine how many numbers are added to give a sum of 153.

40. If these two rectangles have the same area, determine the value of x.

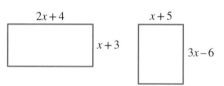

41. Henrietta is a pet rabbit who lives in an enclosure that is 2 m wide and 4 m long. Her human family has decided to purchase some more rabbits to keep her company and so the size of the enclosure must be increased.
    a. Draw a diagram of Henrietta's enclosure, clearly marking the lengths of the sides.
    b. If the length and width of the enclosure are increased by x m, express the new dimensions in terms of x.
    c. If the new area is to be 24 m², write an equation relating the sides and the area of the enclosure (area = length × width).
    d. Use the equation to calculate the value of x and, hence, the length of the sides of the new enclosure. Justify your answer.

42. The cost per hour, C, in thousands of dollars of running two cruise ships, *Annabel* and *Betty*, travelling at a speed of s knots is given by the following relationships.
    $C_{Annabel} = 0.3s^2 + 4.2s + 12$ and $C_{Betty} = 0.4s^2 + 3.6s + 8$
    a. Determine the cost per hour for each ship if they are both travelling at 28 knots.
    b. Calculate the speed in knots at which both ships must travel for them to have the same cost.
    c. Explain why only one of the solutions obtained in your working for part **b** is valid.

43. Explain why the equation $x^2 + 4x + 10 = 0$ has no real solutions.

Problem solving

44. Solve $(x^2 - x)^2 - 32(x^2 - x) + 240 = 0$ for x.

45. Solve $\dfrac{3z^2 - 35}{16} - z = 0$ for z.

46. A garden measuring 12 metres by 16 metres is to have a pedestrian pathway installed all around it, increasing the total area to 285 square metres. Determine the width of the pathway.

LESSON
8.3 The quadratic formula

LEARNING INTENTIONS

At the end of this lesson you should be able to:
- use the quadratic formula to solve equations
- use the quadratic formula to solve worded questions.

▶ 8.3.1 Using the quadratic formula

eles-4846

- The **quadratic formula** can be used to solve any equations in the form:

  $$ax^2 + bx + c = 0, \text{ where } a, b \text{ and } c \text{ are numbers.}$$

- To use the quadratic formula:
  - identify the values for a, b and c
  - substitute the values into the quadratic formula.
- Note that the derivation of this formula is beyond the scope of this course.

The quadratic formula

The quadratic formula is:

$$x = \frac{-b \pm \sqrt{b^2 - 4ac}}{2a}$$

Note that the $\pm$ symbol is called the plus–minus sign. To use it, identify the first solution by treating it as a plus, then identify the second solution by treating it as a minus.

- There will be no real solutions if the value in the square root is negative; that is, there will be no solutions if $b^2 - 4ac < 0$.

WORKED EXAMPLE 6 Using the quadratic formula to solve equations

Use the quadratic formula to solve each of the following equations.

a. $3x^2 + 4x + 1 = 0$ (exact answer)

b. $-3x^2 - 6x - 1 = 0$ (round to 2 decimal places)

THINK	WRITE
a. 1. Write the equation.	a. $3x^2 + 4x + 1 = 0$
2. Write the quadratic formula.	$x = \dfrac{-b \pm \sqrt{b^2 - 4ac}}{2a}$
3. State the values for a, b and c.	where $a = 3, b = 4, c = 1$
4. Substitute the values into the formula.	$x = \dfrac{-4 \pm \sqrt{4^2 - 4 \times 3 \times 1}}{2 \times 3}$
5. Simplify and solve for x.	$= \dfrac{-4 \pm \sqrt{4}}{6}$
	$= \dfrac{-4 \pm 2}{6}$
	$x = \dfrac{-4 + 2}{6}$ or $x = \dfrac{-4 - 2}{6}$
6. Write the two solutions.	$x = -\dfrac{1}{3}$ or $x = -1$
b. 1. Write the equation.	b. $-3x^2 - 6x - 1 = 0$
2. Write the quadratic formula.	$x = \dfrac{-b \pm \sqrt{b^2 - 4ac}}{2a}$
3. State the values for a, b and c.	where $a = -3, b = -6, c = -1$
4. Substitute the values into the formula.	$x = \dfrac{-(-6) \pm \sqrt{(-6)^2 - 4 \times (-3) \times (-1)}}{2 \times -3}$
5. Simplify the fraction.	$= \dfrac{6 \pm \sqrt{24}}{-6}$
	$= \dfrac{6 \pm 2\sqrt{6}}{-6}$
	$= \dfrac{3 \pm \sqrt{6}}{-3}$
	$x = \dfrac{3 + \sqrt{6}}{-3}$ or $\dfrac{3 - \sqrt{6}}{-3}$
6. Write the two solutions correct to 2 decimal places.	$x \approx -1.82$ or $x \approx -0.18$

Note: When asked to give an answer in exact form, you should simplify any surds as necessary.

| **TI | THINK** | **DISPLAY/WRITE** | **CASIO | THINK** | **DISPLAY/WRITE** |

TI | THINK

a-b.

In a new problem, on a Calculator page, complete the entry lines as:
solve $(3x^2 + 4x + 1 = 0)$
solve $(-3x^2 - 6x - 1 = 0)$
Then press ENTER after each entry.
Press CTRL ENTER to get a decimal approximation for **b.**

DISPLAY/WRITE

a-b.

$$3x^2 + 4x + 1 = 0$$
$$\Rightarrow x = -1 \quad \text{or} \quad -\frac{1}{3}$$
$$-3x^2 - 6x - 1 = 0$$
$$\Rightarrow x = \frac{-\left(\sqrt{6} + 3\right)}{3} \quad \text{or} \quad \frac{\sqrt{6} - 3}{3}$$
$x = -1.82$ or -0.18 rounding to 2 decimal places.

CASIO | THINK

a-b.

On the Main screen in standard, complete the entry lines as:
solve $(3x^2 + 4x + 1 = 0)$
solve $(-3x^2 - 6x - 1 = 0)$
Press EXE after each entry.
To change the final answer to a decimal, change from standard to decimal mode and press EXE again.

DISPLAY/WRITE

a-b.

$$3x^2 + 4x + 1 = 0$$
$$\Rightarrow x = -1 \quad \text{or} \quad -\frac{1}{3}$$
$$-3x^2 - 6x - 1 = 0$$
$$\Rightarrow x = \frac{-\left(\sqrt{6} + 3\right)}{3} \quad \text{or} \quad \frac{\sqrt{6} - 3}{3}$$
$x = -1.82$ or -0.18 rounding to 2 decimal places.

DISCUSSION

What kind of answer will you get if the value inside the square root sign in the quadratic formula is zero?

 Resources

 eWorkbook   Topic 8 Workbook (worksheets, code puzzle and a project) (ewbk-2034)

Video eLesson The quadratic formula (eles-2314)

Interactivities Individual pathway interactivity: The quadratic formula (int-4602)

The quadratic formula (int-2561)

Exercise 8.3 The quadratic formula

learn

| **8.3 Quick quiz** on | **8.3 Exercise** |

Individual pathways

■ PRACTISE	■ CONSOLIDATE	■ MASTER
1, 3, 6, 9, 13, 18, 19, 22	2, 4, 7, 10, 11, 14, 17, 20, 23, 24	5, 8, 12, 15, 16, 21, 25

Fluency

1. State the values for a, b and c in each of the following equations of the form $ax^2 + bx + c = 0$.

   a. $3x^2 - 4x + 1 = 0$ **b.** $7x^2 - 12x + 2 = 0$

   c. $8x^2 - x - 3 = 0$ **d.** $x^2 - 5x + 7 = 0$

2. State the values for a, b and c in each of the following equations of the form $ax^2 + bx + c = 0$.

a. $5x^2 - 5x - 1 = 0$
b. $4x^2 - 9x - 3 = 0$
c. $12x^2 - 29x + 103 = 0$
d. $43x^2 - 81x - 24 = 0$

WE6a For questions 3–5 use the quadratic formula to solve each of the following equations. Give exact answers.

3. a. $x^2 + 5x + 1 = 0$
   b. $x^2 + 3x - 1 = 0$
   c. $x^2 - 5x + 2 = 0$
   d. $x^2 - 4x - 9 = 0$

4. a. $x^2 + 2x - 11 = 0$
   b. $x^2 - 7x + 1 = 0$
   c. $x^2 - 9x + 2 = 0$
   d. $x^2 - 6x - 3 = 0$

5. a. $x^2 + 8x - 15 = 0$
   b. $-x^2 + x + 5 = 0$
   c. $-x^2 + 5x + 2 = 0$
   d. $-x^2 - 2x + 7 = 0$

WE6b For questions 6–8 use the quadratic formula to solve each of the following equations. Give approximate answers rounded to 2 decimal places.

6. a. $3x^2 - 4x - 3 = 0$
   b. $4x^2 - x - 7 = 0$
   c. $2x^2 + 7x - 5 = 0$
   d. $7x^2 + x - 2 = 0$
   e. $5x^2 - 8x + 1 = 0$

7. a. $2x^2 - 13x + 2 = 0$
   b. $-3x^2 + 2x + 7 = 0$
   c. $-7x^2 + x + 8 = 0$
   d. $-12x^2 + x + 9 = 0$
   e. $-6x^2 + 4x + 5 = 0$

8. a. $-11x^2 - x + 1 = 0$
   b. $-4x^2 - x + 7 = 0$
   c. $-2x^2 + 12x - 1 = 0$
   d. $-5x^2 + x + 3 = 0$

9. **MC** The solutions of the equation $3x^2 - 7x - 2 = 0$ are:

A. $1, 2$
B. $1, -2$
C. $-0.257, 2.59$
D. $-0.772, 7.772$

10. **MC** In the expansion of $(6x - 5)(3x + 4)$, the coefficient of x is:

A. 18
B. -15
C. 9
D. 6

11. **MC** In the expanded form of $(x - 2)(x + 4)$, identify which of the following statements is incorrect.

A. The value of the constant is -8.
B. The coefficient of the x term is -6.
C. The coefficient of the x term is 2.
D. The coefficient of the x^2 term is 1.

12. **MC** Identify an exact solution to the equation $x^2 + 2x - 5 = 0$.

A. -3.449
B. $-1 + \sqrt{24}$
C. $-1 + \sqrt{6}$
D. $\dfrac{2 + \sqrt{-16}}{2}$

Understanding

For questions 13–15 solve each of the following equations using any suitable method. Round to 3 decimal places where appropriate.

13. a. $2x^2 - 7x + 3 = 0$
    b. $x^2 - 5x = 0$
    c. $x^2 - 2x - 3 = 0$
    d. $x^2 - 3x + 1 = 0$

14. a. $x^2 - 6x + 8 = 0$
    b. $x^2 - 5x + 8 = 0$
    c. $x^2 - 7x - 8 = 0$
    d. $x^2 + 2x - 9 = 0$

15. a. $2x^2 + 11x - 21 - 0$
    b. $7x^2 - 2x + 1 = 0$
    c. $-x^2 + 9x - 14 - 0$
    d. $-6x^2 - x + 1 = 0$

16. The surface area of a closed cylinder is given by the formula $SA = 2\pi r(r + h)$, where r cm is the radius of the can and h cm is the height.
The height of a can of wood finish is 7 cm and its surface area is 231 cm^2.

a. Substitute values into the formula to form a quadratic equation using the pronumeral, r.
b. Use the quadratic formula to solve the equation and, hence, determine the radius of the can correct to 1 decimal place.
c. Calculate the area of the curved surface of the can, correct to the nearest square centimetre.

17. To satisfy lighting requirements, the window shown must have an area of 1500 cm^2.

a. Write an expression for the area of the window in terms of x.
b. Write an equation so that the window satisfies the lighting requirements.
c. Use the quadratic formula to solve the equation and calculate x to the nearest mm.

18. When using the quadratic formula, you are required to calculate b^2 (inside the square root sign). When $b = -3$, Breanne says that $b^2 = -(3)^2$ but Kelly says that $b^2 = (-3)^2$.

a. What answer did Breanne calculate for b^2?
b. What answer did Kelly calculate for b^2?
c. Identify who is correct and explain why.

Reasoning

19. There is one solution to the quadratic equation if $b^2 - 4ac = 0$. The following have one solution only. Determine the missing value.

a. $a = 1$ and $b = 4$, $c = ?$
b. $a = 2$ and $c = 8$, $b = ?$
c. $2x^2 + 12x + c = 0$, $c = ?$

20. Two competitive neighbours build rectangular pools that cover the same area but are different shapes. Pool A has a width of $(x + 3)$ m and a length that is 3 m longer than its width. Pool B has a length that is double the width of Pool A. The width of Pool B is 4 m shorter than its length.

a. Determine the exact dimensions of each pool if their areas are the same.
b. Verify that the areas are the same.

21. A block of land is in the shape of a right-angled triangle with a perimeter of 150 m and a hypotenuse of 65 m. Determine the lengths of the other two sides. Show your working.

Problem solving

22. Solve $\left(x + \dfrac{1}{x}\right)^2 - 14\left(x + \dfrac{1}{x}\right) = 72$ for x.

23. Gunoor's tennis serve can be modelled using $y = -\dfrac{1}{16}x^2 + \dfrac{7}{8}x + 2$. The landing position of his serve will be one of the solutions to this equation when $y = 0$.

a. For $-\dfrac{1}{16}x^2 + \dfrac{7}{8}x + 2 = 0$, determine the values of a, b and c in $ax^2 + bx + c = 0$.

b. Using a calculator, calculate the solutions to this equation.
c. Gunoor's serve will be 'in' if one solution is between 12 and 18, and a 'fault' if no solutions are between those values. Interpret whether Gunoor's serve was 'in' or if it was a 'fault'.

24. Triangle MNP is an isosceles triangle with sides $MN = MP = 3$ cm. Angle MPN is equal to $72°$. The line NQ bisects the angle MNP.

    a. Prove that triangles MNP and NPQ are similar.
    b. If $NP = m$ cm and $PQ = 3 - m$ cm, show that $m^2 + 3m - 9 = 0$.
    c. Solve the equation $m^2 + 3m - 9 = 0$ and determine the side length of NP, giving your answer correct to 2 decimal places.

25. The equation $ax^4 + bx^2 + c = 0$ can be solved by applying substitution and the rules used to solve quadratics.
    For example, $x^4 - 5x^2 + 4 = 0$ is solved for x as follows.
    Notice that $x^4 - 5x^2 + 4 = (x^2)^2 - 5(x)^2 + 4$. Now let $x^2 = u$ and substitute.

    $$(x^2)^2 - 5(x)^2 + 4 = u^2 - 5u + 4$$

    Solve for u. That is,

    $$u^2 - 5u + 4 = 0$$
    $$(u - 4)(u - 1) = 0$$
    $$u - 4 = 0 \text{ or } u - 1 = 0$$
    $$u = 4 \text{ or } u = 1$$

    Since $x^2 = u$, that implies that
    $$x^2 = 4 \text{ or } x^2 = 1$$
    $$x = \pm 2 \text{ or } x = \pm 1$$

    Using this or another method, solve the following for x.

    a. $x^4 - 13x^2 + 36 = 0$ b. $4x^4 - 17x^2 = -4$

LESSON
8.4 Solving quadratic equations graphically

LEARNING INTENTIONS

At the end of this lesson you should be able to:
- identify the solutions to $ax^2 + bx + c = 0$ by inspecting the graph of $y = ax^2 + bx + c$
- recognise whether $ax^2 + bx + c = 0$ has two, one or no solutions by looking at the graph.

▶ 8.4.1 Solving quadratic equations graphically

eles-4847

- To solve quadratic equations, look at the graph of $y = ax^2 + bx + c$.
  - The solutions to $ax^2 + bx + c = 0$ are the x-axis intercepts, where the graph 'cuts' the x-axis.
- The number of x-axis intercepts will indicate the number of solutions.

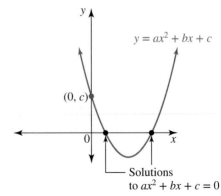

Number of solutions

Two solutions	One solution	No real solutions
The graph 'cuts' the *x*-axis twice. For example:	The graph 'touches' the *x*-axis. For example:	The graph does not reach the *x*-axis. For example:
		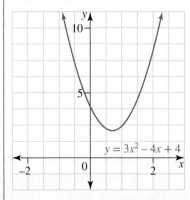

WORKED EXAMPLE 7 Determining solutions from the graph

Determine the solutions of each of the following quadratic equations by inspecting their corresponding graphs. Give answers to 1 decimal place where appropriate.

a. $x^2 + x - 2 = 0$

b. $2x^2 + 4x - 5 = 0$

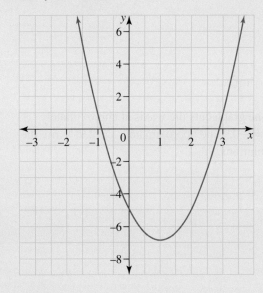

THINK

WRITE/DRAW

a. 1. Examine the graph of $y = x^2 + x - 2$ and locate the points where $y = 0$, that is, where the graph intersects the x-axis.

a.

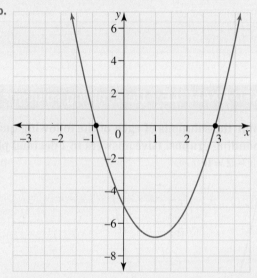

2. The graph cuts the x-axis ($y = 0$) at $x = 1$ and $x = -2$. Write the solutions.

From the graph, the solutions are $x = 1$ and $x = -2$.

b. 1. The graph of $y = 2x^2 - 4x - 5$ is equal to zero when $y = 0$. Look at the graph to see where $y = 0$, that is, where it intersects the x-axis. By sight, we can only give estimates of the solutions.

b.

2. The graph cuts the x-axis at approximately $x = -0.9$ and approximately $x = 2.9$. Write the solutions.

From the graph, the solutions are $x \approx -0.9$ and $x \approx 2.9$.

TI	THINK	DISPLAY/WRITE	CASIO	THINK	DISPLAY/WRITE

a.

In a new problem, on a Graphs page, complete the function entry line as:
$f1(x) = x^2 + x - 2$
Then press ENTER. The graph will be displayed.
To locate the x-intercepts, press:
- MENU
- 6: Analyze Graph
- 1: Zero

Move the cursor to the left of the zero, press ENTER, then move the cursor to the right of the zero and press ENTER. The coordinates of the x-intercept are displayed. Repeat for the other x-intercept.

a.

$x^2 + x - 2 = 0$
$\Rightarrow x = 1 \text{ or } -2$

a.

On the Graph & Table screen, complete the function entry line as:
$y1 = x^2 + x - 2$
Press EXE.
Then tap the graphing icon. The graph will be displayed. To locate the x-intercepts, tap the Y = 0 icon.
To locate the second root, tap the right arrow.

a.

$x^2 + x - 2 = 0$
$\Rightarrow x = 1 \text{ or } -2$

b.

On a Graphs page, complete the function entry line as:
$f1(x) = 2x^2 - 4x - 5$
Then press ENTER. The graph will be displayed.
To locate the x-intercepts, press:
- MENU
- 6: Analyze Graph
- 1: Zero

Move the cursor to the left of the zero, press ENTER, then move the cursor to the right of the zero and press ENTER. The coordinates of the x-intercept are displayed. Repeat for the other x-intercept.

b.

$2x^2 - 4x - 5 = 0$
$\Rightarrow x \approx -0.9 \text{ or } 2.91$
correct to 1 decimal place.

b.

On the Graph & Table screen, complete the function entry line as:
$y1 = 2x^2 - 4x - 5$
Press EXE.
Then tap the graphing icon. The graph will be displayed. To locate the x-intercepts, tap the Y = 0 icon.
To locate the second root, tap the right arrow.

b.

$2x^2 - 4x - 5 = 0$
$\Rightarrow x \approx -0.9 \text{ or } 2.9$
correct to 1 decimal place.

▶ 8.4.2 Confirming solutions of quadratic equations

eles-4848

- To confirm a solution is correct:
  - substitute one solution for x into the quadratic equation
  - the solution is confirmed if the left-hand side (LHS) equals the right-hand side (RHS)
  - repeat for the second solution if applicable.

WORKED EXAMPLE 8 Confirming solutions by substitution

Confirm, by substitution, the solutions obtained in Worked example 7a:
$x^2 + x - 2 = 0$; solutions: $x = 1$ and $x = -2$.

THINK

1. Write the left-hand side of the equation and substitute $x = 1$ into the expression.

WRITE

When $x = 1$,
LHS: $x^2 + x - 2 = 1^2 + 1 - 2$
$= 0$

▶

2. Write the right-hand side.

$RHS = 0$

3. Confirm the solution.

$LHS = RHS \Rightarrow$ Solution is confirmed.

4. Write the left-hand side and substitute $x = -2$.

When $x = -2$,

$$LHS: x^2 + x - 2 = (-2)^2 + -2 - 2$$
$$= 4 - 2 - 2$$
$$= 0$$

5. Write the right-hand side.

$RHS = 0$

6. Confirm the solution.

$LHS = RHS \Rightarrow$ Solution is confirmed

WORKED EXAMPLE 9 Solving application questions using a graph

A golf ball hit along a fairway follows the path shown in the following graph. The height, h metres after it has travelled x metres horizontally, follows the rule $h = \dfrac{1}{270}\left(x^2 - 180x\right)$.

Use the graph to identify how far the ball landed from the golfer.

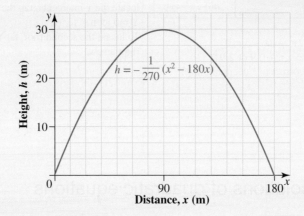

THINK

On the graph, the ground is represented by the x-axis since this is where $h = 0$. The golf ball lands when the graph intersects the x-axis.

WRITE

The golf ball lands 180 m from the golfer.

DISCUSSION

What does 'the solution of a graph' mean?

Exercise 8.4 Solving quadratic equations graphically

learn on

8.4 Quick quiz on	8.4 Exercise

Individual pathways

■ PRACTISE	■ CONSOLIDATE	■ MASTER
1, 4, 7, 10	2, 5, 8, 11	3, 6, 9, 12

Fluency

1. a. **WE7** Determine the solutions of each of the following quadratic equations by inspecting the corresponding graphs.

   i. $x^2 - x - 6 = 0$

   ii. $x^2 - 11x + 10 = 0$

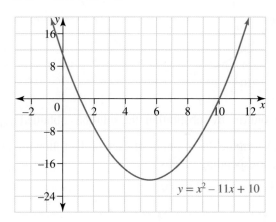

   iii. $-x^2 + 25 = 0$

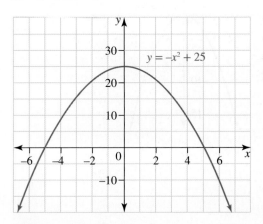

   b. **WE8** Confirm, by substitution, the solutions obtained in part **a**.

2. a. Determine the solutions of each of the following quadratic equations by inspecting the corresponding graphs. Give answers correct to 1 decimal place where appropriate.

   i. $2x^2 - 8x + 8 = 0$

   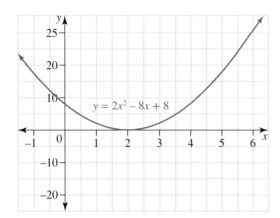

   ii. $x^2 - 3x - 4 = 0$

   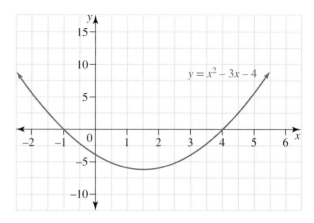

   iii. $x^2 - 3x - 6 = 0$

   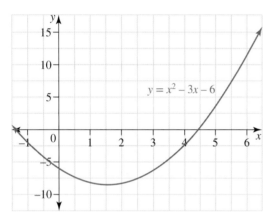

   b. Confirm, by substitution, the solutions obtained in part a.

3. a. Determine the solutions of each of the following quadratic equations by inspecting the corresponding graphs.

   i. $x^2 + 15x - 250 = 0$

   ii. $-x^2 = 0$

   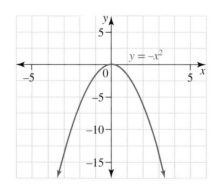

iii. $x^2 + x - 3 = 0$

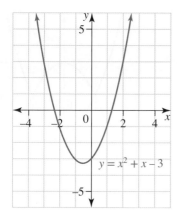

iv. $2x^2 + x - 3 = 0$

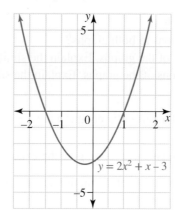

b. Confirm, by substitution, the solutions obtained in part **a**.

Understanding

4. **WE9** A golf ball hit along a fairway follows the path shown in the graph.

   The height, h metres after it has travelled x metres horizontally, follows the rule $h = -\dfrac{1}{200}(x^2 - 150x)$. Use the graph to identify how far the ball lands from the golfer.

5. **MC** Use the graph to determine how many solutions the equation $x^2 + 3 = 0$ has.

   A. No real solutions
   B. One solution
   C. Two solutions
   D. Three solutions

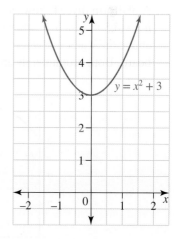

6. **MC** For the equation $x^2 + c = 0$, what values of c will give two solutions, one solution and no real solutions?

   A. Two solutions $c > 0$; one solution $c = 0$; no solutions $c < 0$
   B. Two solutions $c = 2$; one solution $c = 0$; no solutions $c = -2$
   C. Two solutions $c < 0$; one solution $c = 0$; no solutions $c > 0$
   D. Two solutions $c = 2$; one solution $c = 1$; no solutions $c = 0$

Reasoning

7. Two graphs are shown.

   a. What are the solutions to both graphs?

   b. Look at the shapes of the graphs. Explain the similarities and differences.

   c. Look at the equations of the graphs. Explain the similarities and differences.

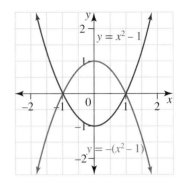

8. **a.** The x-intercepts of a particular equation are $x = 2$ and $x = 5$. Suggest a possible equation.

   b. If the y-intercept in part **a** is $(0, 4)$, give the exact equation.

9. **a.** The x-intercepts of a particular equation are $x = p$ and $x = q$. Suggest a possible equation.

   b. If the y-intercept in part **a** is $(0, r)$, give the exact equation.

Problem solving

10. A ball is thrown upwards from a building and follows the path shown in the graph until it lands on the ground.

    The ball is h metres above the ground when it is a horizontal distance of x metres from the building.

    The path of the ball follows the rule $h = -x^2 + 4x + 21$.

    a. Use the graph to identify how far from the building the ball lands.

    b. Use the graph to determine the height the ball was thrown from.

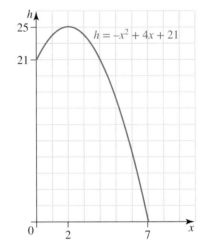

11. A platform diver follows a path determined by the equation $h = -0.5d^2 + 2d + 6$, where h represents the height of the diver above the water and d represents the distance from the diving board. Both pronumerals are measured in metres.

Use the graph to determine:

   a. how far the diver landed from the edge of the diving board

   b. how high the diving board is above the water

   c. the maximum height reached by the diver when they are in the air.

12. Determine the equation of the given parabola. Give your answer in the form $y = ax^2 + bx + c$.

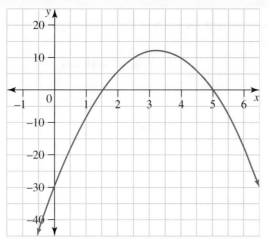

LESSON
8.5 The discriminant

LEARNING INTENTIONS

At the end of this lesson you should be able to:
- calculate the value of the discriminant
- use the value of the discriminant to determine the number of solutions and type of solutions
- calculate the discriminant for simultaneous equations to determine whether two graphs intersect.

⊙ 8.5.1 Using the discriminant

- **The discriminant** is the value inside the square root sign in the quadratic formula.
- The discriminant can be found for equations in the form $ax^2 + bx + c = 0$.

The discriminant

$$\Delta = b^2 - 4ac$$

- The symbol Δ represents the *discriminant*. This symbol is the Greek capital letter delta.
- The discriminant is found from:

$$x = \frac{-b \pm \sqrt{b^2 - 4ac}}{2a}$$

$$x = \frac{-b \pm \sqrt{\Delta}}{2a}$$

- The value of the discriminant indicates the number of solutions for a quadratic equation.
  - $\Delta < 0$: if the discriminant is negative there are *no real solutions*.
  - $\Delta = 0$: if the discriminant is zero there is *one solution*.
  - $\Delta > 0$: if the discriminant is positive there are *two solutions*.
    – A positive discriminant that is a perfect square (e.g. 16, 25, 144) gives two *rational* solutions.
    – A positive discriminant that is not a perfect square (e.g. 7, 11, 15) gives two *irrational (surd)* solutions.

Calculate the value of the discriminant for each of the following and use it to determine how many solutions the equation will have.

a. $2x^2 + 9x - 5 = 0$

b. $x^2 + 10 = 0$

THINK	WRITE
a. 1. Write the expression and determine the values of a, b and c given $ax^2 + bx + c = 0$.	a. $2x^2 + 9x - 5 = 0$ $2x^2 + 9x + -5 = 0$ $a = 2, b = 9, c = -5$
2. Write the formula for the discriminant and substitute values of a, b and c.	$\Delta = b^2 - 4ac$ $= 9^2 - 4 \times 2 \times -5$
3. Simplify the equation and solve.	$= 81 - (-40)$ $= 121$
4. State the number of solutions. In this case $\Delta > 0$, which means there are two solutions.	$\Delta > 0$, so there will be two solutions to the equation $2x^2 + 9x - 5 = 0$.
b. 1. Write the expression and determine the values of a, b and c given $ax^2 + bx + c = 0$.	b. $x^2 + 10 = 0$ $1x^2 + 0x + 10 = 0$ $a = 1, b = 0, c = 10$
2. Write the formula for the discriminant and substitute the values of a, b and c.	$\Delta = b^2 - 4ac$ $= 0^2 - 4 \times 1 \times 10$ $= 0 - 40$ $= -40$
3. State the number of solutions. In this case $\Delta < 0$, which means there are no solutions.	$\Delta < 0$, so there will be no solutions to the equation $x^2 + 10 = 0$.

8.5.2 Using the discriminant to determine if graphs intersect

eles-4850

- The table below summarises the number of points of intersection by the graph, indicated by the discriminant.

	$\Delta < 0$ (negative)	$\Delta = 0$ (zero)	$\Delta > 0$ (positive)	
			Perfect square	**Not a perfect square**
Number of solutions	No real solutions	1 rational solution	2 rational solutions	2 irrational (surd) solutions
Description	Graph does not cross or touch the x-axis	Graph touches the x-axis	Graph intersects the x-axis twice	
Graph				

WORKED EXAMPLE 11 Using the discriminant to determine the number and type of solutions

By using the discriminant, determine whether the following equations have:

  i. **two rational solutions**   ii. **two irrational solutions**
iii. **one solution**   iv. **no real solutions.**

a. $x^2 - 9x - 10 = 0$ b. $x^2 - 2x - 14 = 0$
c. $x^2 - 2x + 14 = 0$ d. $x^2 - 14x = -49$

THINK	WRITE
a. 1. Write the equation.	**a.** $x^2 - 9x - 10 = 0$
2. Identify the coefficients a, b and c.	$a = 1, b = -9, c = -10$
3. Calculate the discriminant.	$\Delta = b^2 - 4ac$ $= (-9)^2 - 4 \times 1 \times (-10)$ $= 121$
4. Identify the number and type of solutions when $\Delta > 0$ and is a perfect square.	The equation has two rational solutions.
b. 1. Write the equation.	**b.** $x^2 - 2x - 14 = 0$
2. Identify the coefficients a, b and c.	$a = 1, b = -2, c = -14$
3. Calculate the discriminant.	$\Delta = b^2 - 4ac$ $= (-2)^2 - 4 \times 1 \times (-14)$ $= 60$
4. Identify the number and type of solutions when $\Delta > 0$ but not a perfect square.	The equation has two irrational solutions.
c. 1. Write the equation.	**c.** $x^2 - 2x + 14 = 0$
2. Identify the coefficients a, b and c.	$a = 1, b = -2, c = 14$
3. Calculate the discriminant.	$\Delta = b^2 - 4ac$ $= (-2)^2 - 4 \times 1 \times 14$ $= -52$
4. Identify the number and type of solutions when $\Delta < 0$.	The equation has no real solutions.
d. 1. Write the equation, then rewrite it so the right side equals zero.	**d.** $\quad\quad x^2 + 14x = -49$ $x^2 + 14x + 49 = 0$
2. Identify the coefficients a, b and c.	$a = 1, b = 14, c = 49$
3. Calculate the discriminant.	$\Delta = b^2 - 4ac$ $= 14^2 - 4 \times 1 \times 49$ $= 0$
4. Identify the number and types of solutions when $\Delta = 0$.	The equation has one solution.

| TI | THINK | DISPLAY/WRITE | CASIO | THINK | DISPLAY/WRITE |

TI | THINK

a-d.

On a Calculator page, complete the entry lines as:

$b^2 - 4ac \,|\, a = 1$ and $b = -9$ and $c = -10$

$b^2 - 4ac \,|\, a = 1$ and $b = -2$ and $c = -14$

$b^2 - 4ac \,|\, a = 1$ and $b = -2$ and $c = 14$

$b^2 - 4ac \,|\, a = 1$ and $b = 14$ and $c = 49$

Press ENTER after each entry.

DISPLAY/WRITE

a-d.

If $\Delta = 121$, the equation has two rational solutions.
If $\Delta = 60$, the equation has two irrational solutions.
If $\Delta = -52$, the equation has no real solutions.
If $\Delta = 0$, the equation has one solution.

CASIO | THINK

a-d.

On the Main screen, complete the entry lines as:

$b^2 - 4ac \,|\, a = 1 \,|\, b = -9 \,|$ $c = -10$

$b^2 - 4ac \,|\, a = 1 \,|\, b = -2 \,|$ $c = -14$

$b^2 - 4ac \,|\, a = 1 \,|\, b = -2 \,|$ $c = 14$

$b^2 - 4ac \,|\, a = 1 \,|\, b = 14 \,|\, c = 49$

Press EXE after each entry.

DISPLAY/WRITE

a-d.

If $\Delta = 121$, the equation has two rational solutions.
If $\Delta = 60$, the equation has two irrational solutions.
If $\Delta = -52$, the equation has no real solutions.
If $\Delta = 0$, the equation has one solution.

- In topic 4 we saw that simultaneous equations can be solved graphically, where the intersection of the two graphs is the solution.
- The discriminant can be used to determine whether a solution exists for two equations and, hence, whether the graphs intersect.

WORKED EXAMPLE 12 Using the discriminant to determine if graphs intersect

Determine whether the parabola $y = x^2 - 2$ and the line $y = x - 3$ intersect.

THINK	WRITE
1. If the parabola and the line intersect, there will be at least one solution to the simultaneous equations: let $y_1 = y_2$.	$y_1 = x^2 - 2$ $y_2 = x - 3$ $y_1 = y_2$ $x^2 - 2 = x - 3$
2. Collect all terms on one side and simplify.	$x^2 - 2 - x + 3 = x - 3 - x + 3$ $x^2 - 2 - x + 3 = 0$ $x^2 - x + 1 = 0$
3. Use the discriminant to check if any solutions exist. If $\Delta < 0$, then no solutions exist.	$\Delta = b^2 - 4ac$ $a = 1, b = -1, c = 1$ $\Delta = (-1)^2 - 4 \times 1 \times 1$ $= 1 - 4$ $= -3$ $\Delta < 0, \therefore$ no solutions exist
4. Write the answer in a sentence.	The parabola and the line do not intersect.

Resources

eWorkbook Topic 8 Workbook (worksheets, code puzzle and a project) (ewbk-2034)

Video eLesson The discriminant (eles-1946)

Interactivities Individual pathway interactivity: The discriminant (int-4604)

The discriminant (int-2560)

Exercise 8.5 The discriminant

learn

| **8.5 Quick quiz** | **8.5 Exercise** |

Individual pathways

■ PRACTISE	■ CONSOLIDATE	■ MASTER
1, 4, 9, 11, 15, 19, 22	2, 5, 7, 12, 14, 18, 20, 23	3, 6, 8, 10, 13, 16, 17, 21, 24

Fluency

WE10 For questions **1–3** calculate the value of the discriminant for each of the following and use it to determine how many solutions the equation will have.

1. **a.** $6x^2 + 13x - 5 = 0$ **b.** $x^2 + 9x - 90 = 0$ **c.** $x^2 + 4x - 2 = 0$
   d. $36x^2 - 1 = 0$ **e.** $x^2 + 2x + 8 = 0$

2. **a.** $x^2 - 5x - 14 = 0$ **b.** $36x^2 + 24x + 4 = 0$ **c.** $x^2 - 19x + 88 = 0$
   d. $x^2 - 10x + 17 = 0$ **e.** $30x^2 + 17x - 21 = 0$

3. **a.** $x^2 + 16x + 62 = 0$ **b.** $9x^2 - 36x + 36 = 0$ **c.** $2x^2 - 16x = 0$ **d.** $x^2 - 64 = 0$

4. **WE11** By using the discriminant, determine whether the equations below have:

   i. two rational solutions **ii.** two irrational solutions
   iii. one solution **iv.** no real solutions.

   a. $x^2 - 3x + 5$ **b.** $4x^2 - 20x + 25 = 0$
   c. $x^2 + 9x - 22 = 0$ **d.** $9x^2 + 12x + 4$

5. By using the discriminant, determine whether the equations below have:

   i. two rational solutions **ii.** two irrational solutions
   iii. one solution **iv.** no real solutions.

   a. $x^2 + 3x - 7 = 0$ **b.** $25x^2 - 10x + 1 = 0$
   c. $3x^2 - 2x - 4 = 0$ **d.** $2x^2 - 5x + 4 = 0$

6. By using the discriminant, determine whether the equations below have:

   i. two rational solutions **ii.** two irrational solutions
   iii. one solution **iv.** no real solutions.

   a. $x^2 - 10x + 26 = 0$ **b.** $3x^2 + 5x - 7 = 0$
   c. $2x^2 + 7x - 10 = 0$ **d.** $x^2 - 11x + 30 = 0$

7. **WE12** Determine whether the following graphs intersect.

    a. $y = -x^2 + 3x + 4$ and $y = x - 4$ b. $y = -x^2 + 3x + 4$ and $y = 2x + 5$

    c. $y = -(x + 1)^2 + 3$ and $y = -4x - 1$ d. $y = (x - 1)^2 + 5$ and $y = -4x - 1$

8. Consider the equation $3x^2 + 2x + 7 = 0$.

    a. Identify the values of a, b and c.

    b. Calculate the value of $b^2 - 4ac$.

    c. State how many real solutions, and hence x-intercepts, are there for this equation.

9. Consider the equation $-6x^2 + x + 3 = 0$.

    a. Identify the values of a, b and c.

    b. Calculate the value of $b^2 - 4ac$.

    c. State how many real solutions, and hence x-intercepts, there are for this equation.

10. Consider the equation $-6x^2 + x + 3 = 0$. With the information gained from the discriminant, use the most efficient method to solve the equation. Give an exact answer.

11. **MC** Identify the discriminant of the equation $x^2 - 4x - 5 = 0$.

    A. 36   **B.** 11   **C.** 4   **D.** 0

12. **MC** Identify which of the following quadratic equations has two irrational solutions.

    A. $x^2 - 8x + 16 = 0$ **B.** $2x^2 - 7x = 0$ **C.** $x^2 + 8x + 9 = 0$ **D.** $x^2 - 4 = 0$

13. **MC** The equation $x^2 = 2x - 3$ has:

    A. two rational solutions   **B.** exactly one solution

    C. no solutions   **D.** two irrational solutions.

Understanding

14. Determine the value of k if $x^2 - 2x - k = 0$ has one solution.

15. Determine the value of m for which $mx^2 - 6x + 5 = 0$ has one solution.

16. Determine the values of n when $x^2 - 3x - n = 0$ has two solutions.

17. The path of a dolphin as it leaps out of the water can be modelled by the equation $h = -0.4d^2 + d$, where h is the dolphin's height above water and d is the horizontal distance from its starting point. Both h and d are in metres.

    a. Calculate how high above the water the dolphin is when it has travelled 2 m horizontally from its starting point.

    b. Determine the horizontal distance the dolphin has covered when it first reaches a height of 25 cm.

    c. Determine the horizontal distance the dolphin has covered when it next reaches a height of 25 cm. Explain your answer.

    d. Determine the horizontal distance the dolphin covers in one leap. (*Hint:* What is the value of h when the dolphin has completed its leap?)

    e. During a leap, determine whether this dolphin reaches a height of:

        i. 0.5 m   ii. 1 m.

        Explain how you can determine this without actually solving the equation.

    f. Determine the greatest height the dolphin reaches during a leap.

18. **a.** Determine how many times the parabolas $y = x^2 - 4$ and $y = 4 - x^2$ intersect.
    b. Determine the coordinates of their points of intersection.

Reasoning

19. Show that $3x^2 + px - 2 = 0$ will have real solutions for all values of p.

20. Answer the following questions.
    a. Determine the values of a for which the straight line $y = ax + 1$ will have one intersection with the parabola $y = -x^2 - x - 8$.
    b. Determine the values of b for which the straight line $y = 2x + b$ will *not* intersect with the parabola $y = x^2 + 3x - 5$.

21. Answer the following questions.
    a. Identify how many points of intersection exist between the parabola $y = -2(x + 1)^2 - 5$, where $y = f(x), x \in R$, and the straight line $y = mx - 7$, where $y = f(x), x \in R$.
    b. Determine the value of m (where $m < 0$) such that $y = mx - 7$ has one intersection point with $y = -m(x + 1)^2 - 5$.

Problem solving

22. Answer the following questions.
    a. If $\Delta = 9$, $a = 1$ and $b = 5$, use the quadratic formula to determine the two solutions of x.
    b. If $\Delta = 9$, $a = 1$ and $b = 5$, calculate the value of c.

23. The parabola with the general equation $y = ax^2 + bx + 9$ where $0 < a < 0$ and $0 < b < 20$ touches the x-axis at one point only. The graph passes through the point $(1, 25)$. Determine the values of a and b.

24. The line with equation $kx + y = 3$ is a tangent to the curve with equation $y = kx^2 + kx - 1$. Determine the value of k.

LESSON
8.6 Review

8.6.1 Topic summary

Standard quadratic equation

- Quadratic equations are in the form:
  $ax^2 + bx + c = 0$
  where a, b and c are numbers.
  e.g. In the equation $-5x^2 + 2x - 4 = 0$,
  $a = -5$, $b = 2$ and $c = -4$.

Solving quadratic equations graphically

- Inspect the graph of $y = ax^2 + bx + c$.
- The solutions to $ax^2 + bx + c = 0$ are the x-axis intercepts.

The discriminant

- The discriminant indicates the number and type of solutions.

  $$\Delta = b^2 - 4ac$$

- $\Delta < 0$: no real solutions
- $\Delta = 0$: one solution
- $\Delta > 0$: two solutions (if perfect square then two rational solutions, if not a perfect square then two irrational solutions)
- A positive discriminant that is a perfect square (e.g. 16, 25, 144) gives two *rational* solutions.
- A positive discriminant that is not a perfect square (e.g. 7, 11, 15) gives two *irrational* (surd) solutions.

QUADRATIC EQUATIONS

Solving quadratic equations algebraically

- Use algebra to determine the x-value solutions to $ax^2 + bc + c = 0$.
- There will be two, one or no solutions.

Factorising

Factorise using:
- common factor
- difference of two squares
- factor pairs
- completing the square.

Quadratic formula

To use the quadratic formula:
- identify the values for a, b and c
- substitute the values into the quadratic formula.

  $$x = \frac{-b \pm \sqrt{b^2 - 4ac}}{2a}$$

Note: The $\pm$ symbol is called the plus–minus sign.

Null Factor Law

- Once an equation is factorised, then apply the Null Factor Law.
  - Set the products equal to zero.
  - Solve for x.
- If the product of two numbers is zero, then one or both numbers must be zero.

8.6.2 Success criteria

Tick the column to indicate that you have completed the lesson and how well you have understood it using the traffic light system.

(**Green:** I understand; **Yellow:** I can do it with help; **Red:** I do not understand)

Lesson	Success criteria	⬤	◯	⬤
8.2	I can use the Null Factor Law to solve quadratic equations.			
	I can use the completing the square technique to solve quadratic equations.			
	I can solve worded questions algebraically.			
8.3	I can use the quadratic formula to solve equations.			
	I can use the quadratic formula to solve worded questions.			
8.4	I can identify the solutions to $ax^2 + bx + c = 0$ by inspecting the graph of $y = ax^2 + bx + c$.			
	I can recognise whether $ax^2 + bx + c = 0$ has two, one or no solutions by looking at the graph.			
8.5	I can calculate the value of the discriminant.			
	I can use the value of the discriminant to determine the number of solutions and type of solutions.			
	I can calculate the discriminant for simultaneous equations to determine whether two graphs intersect.			

8.6.3 Project

Weaving

Many articles of clothing are sewn from materials that show designs and patterns made by weaving together threads of different colours. Intricate and complex designs can result. Let's investigate some very simple repetitive patterns. Knowledge of quadratic equations and the quadratic formula is helpful in creating these designs.

We need to understand the process of weaving. Weaving machines have parts called *warps*. Each warp is divided into a number of *blocks*. Consider a pattern that is made up of a series of blocks, where the first block is all one colour except for the last thread, which is a different colour.

Let's say our pattern is red and blue. The first block contains all red threads, except for the last one, which is blue. The next block has all red threads, except for the last two threads, which are blue. The pattern continues in this manner. The last block has the first thread as red and the remainder as blue. The warp consists of a particular number of threads, let's say 42 threads.

How many blocks and threads per block would be necessary to create a pattern of this type?

▶

To produce this pattern, we need to divide the warp into equally sized blocks, if possible. What size block and how many threads per block would give us the 42-thread warp? We will need to look for a mathematical pattern. Look at the table (below), where we consider the smallest block consisting of 2 threads through to a block consisting of 7 threads.

Block 1

Block 2

Block n

Pattern	Number of threads per block	Number of blocks	Total threads in warp
RB	2	1	2
RRB RBB	3	2	6
RRRB RRBB RBBB	4		
	5		
	6		
	7		

1. Complete the entries in the table.
2. Consider a block consisting of n threads.
   a. How many blocks would be needed?
   b. What would be the total number of threads in the warp?

The 42-thread warp was chosen as a simple example to show the procedure involved in determining the number of blocks required and the number of threads per block. In this particular case, 6 blocks of 7 threads per block would give us our design for a 42-thread warp. In practice, you would not approach the problem by drawing up a table to determine the number of blocks and the size of each block.

3. Take your expression in question 2b and let it equal 42. This should form a quadratic equation. Solve this equation to verify that you would need 6 blocks with 7 threads per block to fulfil the size of a 42-thread warp.
4. In reality, the size of each block is not always clearly defined. Also, the thread warp sizes are generally much larger, about 250. Let's determine the number of threads per block and the number of blocks required for a 250-thread warp.
   a. Form your quadratic equation with the thread warp size equal to 250.
   b. A solution to this equation can be found using the quadratic formula. Use the quadratic formula to determine a solution.
   c. The number of threads per block is represented by n and this obviously must be a whole number. Round your solution down to the nearest whole number.
   d. How many whole blocks are needed?
   e. Use your solutions to c and d to determine the total number of threads used for the pattern.
   f. How many more threads do you need to make the warp size equal to 250 threads?
   g. Distribute these threads by including them at the beginning of the first block and the end of the last block. Describe your overall pattern.
5. Investigate the number of blocks required and threads per block required for a 400-thread warp.
6. Investigate changing the pattern. Let the first block be all red. In the next block, change the colour of the first and last threads to blue. With each progressive block, change the colour of an extra thread at the top and bottom to blue until the last block is all blue. On a separate sheet of paper, draw a table to determine the thread warp size for a block size of n threads. Draw the pattern and describe the result for a particular warp size.

Exercise 8.6 Review questions

Fluency

1. **MC** Identify the solutions to the equation $x^2 + 10x - 11 = 0$.
    - **A.** $x = 1$ and $x = 11$
    - **B.** $x = 1$ and $x = -11$
    - **C.** $x = -1$ and $x = 11$
    - **D.** $x = -1$ and $x = -11$

2. **MC** Identify the solutions to the equation $-5x^2 + x + 3 = 0$.
    - **A.** $x = 1$ and $x = \dfrac{3}{5}$
    - **B.** $x = -0.68$ and $x = 0.88$
    - **C.** $x = 3$ and $x = -5$
    - **D.** $x = 0.68$ and $x = -0.88$

3. **MC** Identify the discriminant of the equation $x^2 - 11x + 30 = 0$.
    - **A.** 1
    - **B.** 241
    - **C.** 91
    - **D.** 19

4. **MC** Choose from the following equations which has two irrational solutions.
    - **A.** $x^2 - 6x + 9 = 0$
    - **B.** $4x^2 - 11x = 0$
    - **C.** $x^2 - 25 = 0$
    - **D.** $x^2 + 8x + 2 = 0$

5. The area of a pool is $(6x^2 + 11x + 4)\,\text{m}^2$. Determine the length of the rectangular pool if its width is $(2x + 1)\,\text{m}$.

6. Determine the solutions of the following equations, by first factorising the left-hand side.

   a. $x^2 + 8x + 15 = 0$ b. $x^2 + 7x + 6 = 0$ c. $x^2 + 11x + 24 = 0$
   d. $x^2 + 4x - 12 = 0$ e. $x^2 - 3x - 10 = 0$

7. Determine the solutions of the following equations, by first factorising the left-hand side.

   a. $x^2 + 3x - 28 = 0$ b. $x^2 - 4x + 3 = 0$
   c. $x^2 - 11x + 30 = 0$ d. $x^2 - 2x - 35 = 0$

8. Determine the solutions of the following equations, by first factorising the left-hand side.

   a. $2x^2 + 16x + 24 = 0$ b. $3x^2 + 9x + 6 = 0$ c. $4x^2 + 10x - 6 = 0$
   d. $5x^2 + 25x - 70 = 0$ e. $2x^2 - 7x - 4 = 0$

9. Determine the solutions of the following equations, by first factorising the left-hand side.

   a. $6x^2 - 8x - 8 = 0$ b. $2x^2 - 6x + 4 = 0$
   c. $6x^2 - 25x + 25 = 0$ d. $2x^2 + 13x - 7 = 0$

10. Determine the solutions to the following equations by completing the square.

    a. $x^2 + 8x - 1 = 0$ b. $3x^2 + 6x - 15 = 0$ c. $-4x^2 - 3x + 1 = 0$

11. Ten times an integer is added to seven times its square. If the result is 152, calculate the original number.

12. By using the quadratic formula, determine the solutions to the following equations. Give your answers correct to 3 decimal places.

    a. $4x^2 - 2x - 3 = 0$ b. $7x^2 + 4x - 1 = 0$ c. $-8x^2 - x + 2 = 0$

13. By using the quadratic formula, determine the solutions to the following equations. Give your answers correct to 3 decimal places.

    a. $18x^2 - 2x - 7 = 0$ b. $29x^2 - 105x - 24 = 0$ c. $-5x^2 + 2 = 0$

14. The graph of $y = x^2 - 4x - 21$ is shown.

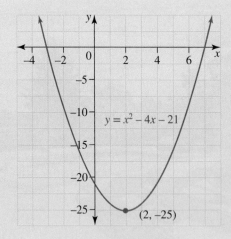

Use the graph to identify the solutions to the quadratic equation $x^2 - 4x - 21 = 0$.

15. Determine the solutions to the equation $-2x^2 - 4x + 6 = 0$.

16. By using the discriminant, determine the number and nature of the solutions for the following equations.

   a. $x^2 + 11x + 9 = 0$ **b.** $3x^2 + 2x - 5 = 0$ **c.** $x^2 - 3x + 4 = 0$

17. What are the solutions to the pair of simultaneous equations shown below?

$$y = x^2 + 4x - 10$$
$$y = 6 - 2x$$

18. Solve the following pair of simultaneous equations to determine the point(s) of intersection.

$$y = x^2 - 7x + 20$$
$$y = 3x - 5$$

19. Determine the solutions to the pair of simultaneous equations shown to determine the point(s) of intersection.

$$y = x^2 + 7x + 11$$
$$y = x$$

20. For each of the following pairs of equations:

   i. solve simultaneously to determine the points of intersection
   ii. illustrate the solution (or lack of solution) by sketching a graph.
   a. $y = x^2 + 6x + 5$ and $y = 11x - 1$
   b. $y = x^2 + 5x - 6$ and $y = 8x - 8$
   c. $y = x^2 + 9x + 14$ and $y = 3x + 5$

21. For each of the following pairs of equations:

   i. solve simultaneously to determine the points of intersection
   ii. illustrate the solution (or lack of solution) by sketching a graph.
   a. $y = x^2 - 7x + 10$ and $y = -11x + 6$
   b. $y = -x^2 + 14x - 48$ and $y = 13x - 54$
   c. $y = -x^2 + 4x + 12$ and $y = 9x + 16$

Problem solving

22. When a number is added to its square, the result is 56. Determine the number.

23. Leroy measures his bedroom and finds that its length is 3 metres more than its width. If the area of the bedroom is $18\,\text{m}^2$, calculate the length and width of the room.

24. The surface area of a cylinder is given by the formula $SA = 2\pi r\,(r + h)$, where $r\,\text{cm}$ is the radius of the cylinder and $h\,\text{cm}$ is the height.
    The height of a can of soft drink is $10\,\text{cm}$ and its surface area is $245\,\text{cm}^2$.

   a. Substitute values into the formula to form a quadratic equation using the pronumeral r.
   b. Use the quadratic formula to solve the equation and, hence, determine the radius of the can. Round your answer to 1 decimal place.
   c. Calculate the area of the label on the can. The label covers the entire curved surface. Round the answer to the nearest square centimetre.

25. Determine the value of d when $2x^2 - 5x - d = 0$ has one solution.

26. Determine the values of k where $(k - 1)\,x^2 - (k - 1)\,x + 2 = 0$ has two distinct solutions.

27. Let m and n be the solutions to the quadratic equation $x^2 - 2\sqrt{5}x - 2 = 0$. Determine the value of $m^2 + n^2$.

28. Although it requires a minimum of two points to determine the graph of a line, it requires a minimum of three points to determine the shape of a parabola. The general equation of a parabola is $y = ax^2 + bx + c$, where a, b and c are the constants to be determined.
   a. Determine the equation of the parabola that has a y-intercept of $(0, -2)$, and passes though the points $(1, -5)$ and $(-2, 16)$.
   b. Determine the equation of a parabola that goes through the points $(0, 0)$, $(2, 2)$ and $(5, 5)$. Show full working to justify your answer.

29. When the radius of a circle increases by $6\,\text{cm}$, its area increases by 25%. Use the quadratic formula to calculate the exact radius of the original circle.

30. A football player received a hand pass and ran directly towards goal. Right on the 50-metre line he kicked the ball and scored a goal. The graph shown represents the path of the ball. Using the graph, answer the following questions.

a. State the height of the ball from the ground when it was kicked.

b. Identify the greatest height the ball reached.

c. Identify the length of the kick.

d. If there were defenders in the goal square, explain if it would have been possible for one of them to mark the ball right on the goal line to prevent a goal. (*Hint:* What was the height of the ball when it crossed the goal line?)

e. As the footballer kicked the ball, a defender rushed at him to try to smother the kick. If the defender can reach a height of 3 m when he jumps, determine how close to the player kicking the ball he must be to just touch the football as it passes over his outstretched hands.

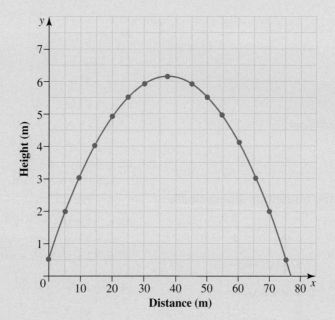

31. The quadratic formula is $x = \dfrac{-b \pm \sqrt{b^2 - 4ac}}{2a}$

An alternative form of the quadratic formula is $x = \dfrac{2c}{-b \pm \sqrt{b^2 - 4ac}}$.

Choose a quadratic equation and show that the two formulas give the same answers.

on To test your understanding and knowledge of this topic, go to your learnON title at www.jacplus.com.au and complete the **post-test**.

Answers

Topic 8 Quadratic equations

8.1 Pre-test

1. True
2. $x = -5$ or $x = \dfrac{1}{2}$
3. $x = \pm \dfrac{3}{4}$
4. D
5. B
6. D
7. A
8. 24
9. a. Only one solution
   b. No real solutions
   c. Two solutions
10. B
11. A
12. Width $= 3\,\text{cm}$, length $= 7\,\text{cm}$
13. $m = 4$
14. A
15. D

8.2 Solving quadratic equations algebraically

1. a. $-7, 9$ b. $-2, 3$ c. $2, 3$ d. $0, 3$
2. a. $0, 1$ b. $-5, 0$ c. $0, 3$ d. $-2, 0$
3. a. $-\dfrac{1}{2}, \dfrac{1}{2}$ b. $-1.2, -0.5$
   c. $0.1, 0.75$ d. $-\sqrt{2}, \sqrt{3}$
4. a. $\dfrac{1}{2}, 1$ b. $-2, -\dfrac{2}{3}$ c. $\dfrac{1}{4}, 7$
5. a. $-\dfrac{6}{7}, 1\dfrac{1}{2}$ b. $\dfrac{3}{5}, \dfrac{2}{3}$ c. $-\dfrac{5}{8}, \dfrac{2}{3}$
6. a. $0, \dfrac{1}{2}, 3$ b. $0, \dfrac{1}{2}, -\dfrac{2}{5}$ c. $0, -3, \dfrac{2}{5}$
7. a. $0, 2$ b. $-5, 0$ c. $0, 7$
8. a. $-\dfrac{2}{3}, 0$ b. $0, 1\dfrac{1}{2}$ c. $0, \dfrac{1}{3}$
9. a. $0, \dfrac{\sqrt{7}}{2}$ b. $-\dfrac{\sqrt{3}}{3}, 0$ c. $0, 1\dfrac{1}{4}$
10. a. $-2, 2$ b. $-5, 5$
    c. $-2, 2$ d. $-7, 7$
11. a. $-1\dfrac{1}{3}, 1\dfrac{1}{3}$ b. $-2\dfrac{1}{2}, 2\dfrac{1}{2}$
    c. $-\dfrac{2}{3}, \dfrac{2}{3}$ d. $-\dfrac{1}{2}, \dfrac{1}{2}$
12. a. $-\dfrac{1}{5}, \dfrac{1}{5}$ b. $-4, 4$
    c. $-\sqrt{5}, \sqrt{5}$ d. $-\dfrac{\sqrt{11}}{3}, \dfrac{\sqrt{11}}{3}$

13. a. $-2, 3$ b. $-4, -2$ c. $-1, 7$ d. $3, 5$
14. a. $-1, 4$ b. 5 c. $-2, 5$ d. $2, 6$ e. $-3, 7$
15. a. $-5, 6$ b. $3, 4$ c. 4 d. -5 e. 10
16. B
17. C
18. a. $-\dfrac{1}{2}, 3$ b. $\dfrac{2}{3}, -1$
    c. $-2, \dfrac{1}{5}$ d. $\dfrac{1}{3}, 1\dfrac{1}{2}$
19. a. $-\dfrac{3}{14}, 1$ b. $\dfrac{1}{4}, \dfrac{1}{3}$
    c. $-1\dfrac{1}{3}, 2\dfrac{1}{2}$ d. $-1\dfrac{3}{4}, -1\dfrac{1}{3}$
20. a. $-\dfrac{2}{5}, \dfrac{1}{2}$ b. $1\dfrac{1}{2}, 2\dfrac{2}{3}$
    c. $-\dfrac{2}{5}, \dfrac{1}{6}$ d. $3, 4$
21. a. $2 + \sqrt{2}, 2 - \sqrt{2}$
    b. $-1 + \sqrt{3}, -1 - \sqrt{3}$
    c. $-3 + \sqrt{10}, -3 - \sqrt{10}$
22. a. $4 + 2\sqrt{3}, 4 - 2\sqrt{3}$
    b. $5 + 2\sqrt{6}, 5 - 2\sqrt{6}$
    c. $1 + \sqrt{3}, 1 - \sqrt{3}$
23. a. $-1 + \sqrt{6}, -1 - \sqrt{6}$
    b. $-2 + \sqrt{10}, -2 - \sqrt{10}$
    c. $-2 + \sqrt{15}, -2 - \sqrt{15}$
24. a. $\dfrac{2}{3} + \dfrac{\sqrt{5}}{2}, \dfrac{3}{2} - \dfrac{\sqrt{5}}{2}$
    b. $-\dfrac{5}{2} + \dfrac{\sqrt{29}}{2}, -\dfrac{5}{2} - \dfrac{\sqrt{29}}{2}$
    c. $\dfrac{7}{2} + \dfrac{\sqrt{33}}{2}, \dfrac{7}{2} - \dfrac{\sqrt{33}}{2}$
25. a. $\dfrac{1}{2} + \dfrac{\sqrt{21}}{2}, \dfrac{1}{2} - \dfrac{\sqrt{21}}{2}$
    b. $\dfrac{11}{2} + \dfrac{\sqrt{117}}{2}, \dfrac{11}{2} - \dfrac{\sqrt{117}}{2}$
    c. $-\dfrac{1}{2} + \dfrac{\sqrt{5}}{2}, -\dfrac{1}{2} - \dfrac{\sqrt{5}}{2}$
26. a. $\dfrac{3}{2} + \dfrac{\sqrt{37}}{2}, -\dfrac{3}{2} - \dfrac{\sqrt{37}}{2}$
    b. $\dfrac{5}{2} + \dfrac{\sqrt{37}}{2}, \dfrac{5}{2} - \dfrac{\sqrt{37}}{2}$
    c. $\dfrac{9}{2} + \dfrac{\sqrt{65}}{2}, \dfrac{9}{2} - \dfrac{\sqrt{65}}{2}$
27. a. $-3, 1$ b. $-4.24, 0.24$ c. $-1, 3$
28. a. $-0.73, 2.73$ b. $0.38, 2.62$ c. $-0.30, 3.30$
29. a. $-1.19, 4.19$ b. $-2.30, 1.30$ c. $-2.22, 0.22$
30. 8 and 9 or -8 and -9
31. 6 and 8, -6 and -8

32. 9 or -10

33. 2 or $-2\dfrac{2}{3}$

34. 8 or $-10\dfrac{1}{2}$

35. 6 seconds

36. a. $l = 2x$

b.

c. $x^2 + (2x)^2 = 45^2, 5x^2 = 2025$

d. Length 40 cm, width 20 cm

37. 8 m, 6 m

38. a. $-6, 1$ b. $8, -3$ c. $x = \pm 1$

39. a. 21   b. 17

40. a. 7

41. a.

2 m

4 m

b. $(2 + x)\,\text{m}, (4 + x)\,\text{m}$

c. $(2 + x)(4 + x) = 24$ d. $x = 2$, 4 m wide, 6 m long

42. a. $C_{\text{Annabel}}(28) = \$364\,800, C_{\text{Betty}}(28) = \$422\,400$

b. 10 knots

c. Speed can only be a positive quantity, so the negative solution is not valid.

43. No real solutions — when we complete the square we get the sum of two squares, not the difference of two squares and we cannot factorise the expression.

44. $x = 5, -4, 4, -3$

45. $z = -\dfrac{5}{3}, 7$

46. The width of the pathway is 1.5 m.

8.3 The quadratic formula

1. a. $a = 3, b = -4, c = 1$ b. $a = 7, b = -12, c = 2$
   c. $a = 8, b = -1, c = -3$ d. $a = 1, b = -5, c = 7$

2. a. $a = 5, b = -5, c = -1$
   b. $a = 4, b = -9, c = -3$
   c. $a = 12, b = -29, c = 103$
   d. $a = 43, b = -81, c = -24$

3. a. $\dfrac{-5 \pm \sqrt{21}}{2}$ b. $\dfrac{-3 \pm \sqrt{13}}{2}$
   c. $\dfrac{5 \pm \sqrt{17}}{2}$ d. $2 \pm \sqrt{13}$

4. a. $-1 \pm 2\sqrt{3}$ b. $\dfrac{7 \pm 3\sqrt{5}}{2}$
   c. $\dfrac{9 \pm \sqrt{73}}{2}$ d. $3 \pm 2\sqrt{3}$

5. a. $-4 \pm \sqrt{31}$ b. $\dfrac{1 \pm \sqrt{21}}{2}$
   c. $\dfrac{5 \pm \sqrt{33}}{2}$ d. $-1 \pm 2\sqrt{2}$

6. a. $-0.54, 1.87$ b. $-1.20, 1.45$
   c. $-4.11, 0.61$ d. $-0.61, 0.47$
   e. $0.14, 1.46$

7. a. $0.16, 6.34$ b. $-1.23, 1.90$ c. $-1.00, 1.14$
   d. $-0.83, 0.91$ e. $-0.64, 1.31$

8. a. $-0.35, 0.26$ b. $-1.45, 1.20$
   c. $0.08, 5.92$ d. $-0.68, 0.88$

9. C

10. C

11. B

12. C

13. a. $0.5, 3$ b. $0, 5$
    c. $-1, 3$ d. $0.382, 2.618$

14. a. $2, 4$ b. No real solution
    c. $-1, 8$ d. $-4.162, 2.162$

15. a. $-7, 1.5$ b. No real solution
    c. $2, 7$ d. $-\dfrac{1}{2}, \dfrac{1}{3}$

16. a. $2\pi r^2 + 14\pi r - 231 = 0$
    b. 3.5 cm
    c. $154\,\text{cm}^2$

17. a. $x(x + 30)$
    b. $x(x + 30) = 1500$
    c. 265 mm

18. a. -9 b. 9   c. Kelly

19. a. 4   b. 8   c. 18

20. a. Pool A: $3\dfrac{2}{3}$ m by $6\dfrac{2}{3}$ m; Pool B: $3\dfrac{1}{3}$ m by $7\dfrac{1}{3}$ m
    b. The area of each is $24\dfrac{4}{9}\,\text{m}^2$.

21. 25 m, 60 m

22. a. $a = -\dfrac{1}{16}$ $b = \dfrac{7}{8}$ $c = 2$
    b. $x = -2$ and $x = 16$
    c. One solution is between 12 and 18 ($x = 16$), which means his serve was 'in'.

23. $-2 \pm 3, 9 \pm 45$

24. a. Sample responses can be found in the worked solutions in the online resources.
    b. Sample responses can be found in the worked solutions in the online resources.
    c. $m = 1.85$ so NP is 1.85 cm.

25. a. $x = \pm 2$ or $x = \pm 3$
    b. $x = \pm \dfrac{1}{2}$ or $x = \pm 2$

8.4 Solving quadratic equations graphically

1. a. i. $x = -2, x = 3$ ii. $x = 1, x = 10$
      iii. $x = -5, x = 5$
   b. Sample responses can be found in the worked solutions in the online resources.

2. a. i. $x = 2$ ii. $x = -1, x = 4$
      iii. $x \approx -1.4, x \approx 4.4$
   b. Sample responses can be found in the worked solutions in the online resources.

3. **a.** i. $x = -25, x = 10$ ii. $x = 0$
   iii. $x \approx -2.3, x \approx 1.3$ iv. $x \approx -1.5, x = 1$

   b. Sample responses can be found in the worked solutions in the online resources.

4. 150 m

5. A

6. C

7. **a.** $x = -1$ and $x = 1$

   b. Similarity: shapes; Difference: one is inverted

   c. Similarity: both have $x^2 - 1$; Difference: one has a negative and brackets

8. **a.** $y = a(x - 2)(x - 5)$

   b. $y = \dfrac{2}{5}(x - 2)(x - 5)$

9. **a.** $y = a(x - p)(x - q)$

   b. $y = \dfrac{r}{pq}(x - p)(x - q)$

10. **a.** 7 m   **b.** 21 m

11. **a.** 6 m   **b.** 6 m   **c.** 8 m

12. $y = -4x^2 + 26x - 30$

8.5 The discriminant

1. **a.** $\Delta = 289$, 2 solutions   **b.** $\Delta = 441$, 2 solutions
   c. $\Delta = 24$, 2 solutions   **d.** $\Delta = 144$, 2 solutions
   e. $\Delta = -28$, 0 solutions

2. **a.** $\Delta = 81$, 2 solutions   **b.** $\Delta = 0$, 1 solution
   c. $\Delta = 9$, 2 solutions   **d.** $\Delta = 32$, 2 solutions
   e. $\Delta = 2809$, 2 solutions

3. **a.** $\Delta = 8$, 2 solutions   **b.** $\Delta = 0$, 1 solution
   c. $\Delta = 256$, 2 solutions   **d.** $\Delta = 256$, 2 solutions

4. **a.** No real solutions   **b.** 1 solution
   c. 2 rational solutions   **d.** 1 solution

5. **a.** 2 irrational solutions   **b.** 1 solution
   c. 2 irrational solutions   **d.** No real solutions

6. **a.** No real solutions   **b.** 2 irrational solutions
   c. 2 irrational solutions   **d.** 2 rational solutions

7. **a.** Yes   **b.** No   **c.** Yes   **d.** No

8. **a.** $a = 3, b = 2, c = 7$ **b.** -80
   c. No real solutions

9. **a.** $a = -6, b = 1, c = 3$ **b.** 73
   c. 2 real solutions

10. $\dfrac{1 \pm \sqrt{73}}{12}$

11. A

12. C

13. C

14. $k = -1$

15. $m = 1.8$

16. $n > -\dfrac{9}{4}$

17. **a.** 0.4 m

   b. 0.28 m

   c. 2.22 m

   d. 2.5 m

   e. i. Yes

   ii. No

   Identify the halfway point between the beginning and the end of the leap, and substitute this value into the equation to determine the maximum height.

   f. 0.625 m

18. **a.** Two times   **b.** $(-2, 0), (2, 0)$

19. p^2 can only give a positive number, which, when added to 24, is always a positive solution.

20. **a.** $a = -7$ or 5 will give one intersection point.

   b. For values of $< -\dfrac{21}{4}$, there will be no intersection points.

21. **a.** The straight line crosses the parabola at $(0, -7)$, so no matter what value m takes, there will be at least one intersection point and a maximum of two.

   b. $m = -\dfrac{8}{5}$

22. **a.** $x = -4$ and $x = -1$
   b. 4

23. $a = 4, b = 12$

24. $k = -4$

Project

1. See table at the bottom of the page.*

---

*1.

Pattern	Number of threads per block	Number of blocks	Total threads in warp
RB	2	1	2
RRB RRB	3	2	6
RRRB RRBB RBBB	4	3	12
RRRRB RRRBB RRBBB RBBBB	5	4	20
RRRRRB RRRRBB RRRBBB RRBBBB RBBBBB	6	5	30
RRRRRRB RRRRRBB RRRRBBB RRRBBBB RRBBBBB RBBBBBB	7	6	42

2. a. $n - 1$ b. $n^2 - n$

3. Answers will vary. Students should form a quadratic equation and let it equal 42. Then solve this equation.

4. a. $n^2 - n = 250$

   b. $n = \dfrac{\sqrt{1001} + 1}{2}$

   c. $n = 16$

   d. 15

   e. 240

   f. 10

   g. Answers will vary. Students should show overall pattern.

5. Answers will vary. Students should investigate the number of blocks required and threads per block required for a 400-thread warp.

6. Answers will vary. Students should investigate changing the pattern. Students should draw a table to determine the thread warp size for a block size of n threads and also draw the pattern and describe the result for a particular warp size.

8.6 Review questions

1. B

2. B

3. A

4. D

5. $(3x + 4)$ m

6. a. $-5, -3$ b. $-6, -1$ c. $-8, -3$
   d. $2, -6$ e. $5, -2$

7. a. $4, -7$ b. $3, 1$ c. $5, 6$
   d. $7, -5$

8. a. $-2, -6$ b. $-2, -1$ c. $\dfrac{1}{2}, -3$
   d. $2, -7$ e. $-\dfrac{1}{2}, 4$

9. a. $-\dfrac{2}{3}, 2$ b. $2, 1$ c. $\dfrac{5}{3}, \dfrac{5}{2}$
   d. $-7, \dfrac{1}{2}$

10. a. $-4 \pm \sqrt{17}$ b. $-1 \pm \sqrt{6}$ c. $-1, \dfrac{1}{4}$

11. 4

12. a. $-0.651, 1.151$ b. $-0.760, 0.188$
    c. $0.441, -0.566$

13. a. $-0.571, 0.682$ b. $-0.216, 3.836$
    c. $-0.632, 0.632$

14. $-3, 7$

15. $-3, 1$

16. a. 2 irrational solutions
    b. 2 rational solutions
    c. No real solutions

17. a. $(-8, 22)$ and $(2, 2)$

18. $(5, 10)$

19. No solution

20. a.

    b.

    c.

21. a.

    b.

c.

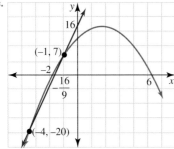

22. -8 and 7

23. Length $= 6\,\text{m}$, width $= 3\,\text{m}$

24. a. $2\pi r(r + 10) = 245$

    b. $3.0\,\text{cm}$

    c. $188\,\text{cm}^2$

25. $-\dfrac{25}{8}$

26. $k > 9$ and $k < 1$

27. 24

28. a. $y = 2x^2 - 5x - 2$

    b. No parabola is possible. The points are on the same straight line.

29. $12\left(\sqrt{5} + 2\right)\,\text{cm}$

30. a. $0.5\,\text{m}$

    b. $6.1\,\text{m}$

    c. $76.5\,\text{m}$

    d. No, the ball is $5.5\,\text{m}$ off the ground and nobody can reach it.

    e. $9.5\,\text{m}$ away

31. Sample responses can be found in the worked solutions in the online resources.

Semester review 1

The learnON platform is a powerful tool that enables students to complete revision independently and allows teachers to set mixed and spaced practice with ease.

Student self-study

Review the **Course Content** to determine which topics and subtopics you studied throughout the year. Notice the green bubbles showing which elements were covered.

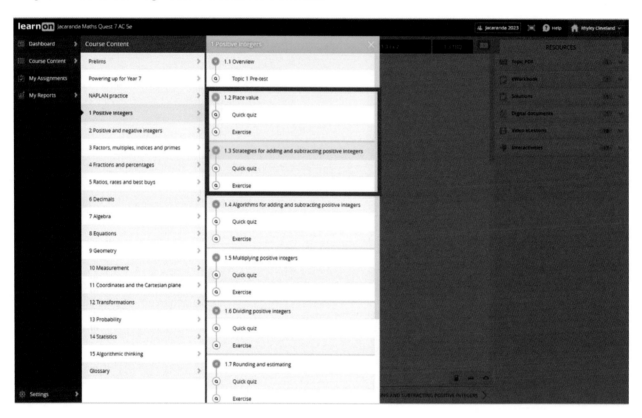

Review your results in **My Reports** and highlight the areas where you may need additional practice.

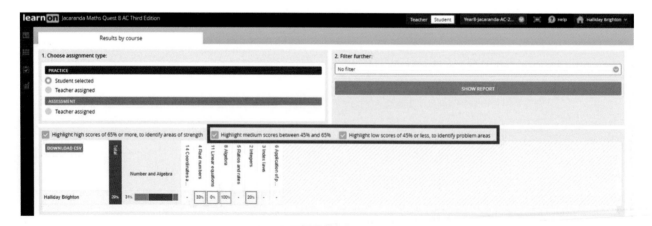

Use these and other tools to help identify areas of strengths and weakness and target those areas for improvement.

Teachers

It is possible to set questions that span multiple topics. These assignments can be given to individual students, to groups or to the whole class in a few easy steps.

Go to **Menu** and select **Assignments** and then **Create Assignment**. You can select questions from one or many topics simply by ticking the boxes as shown below.

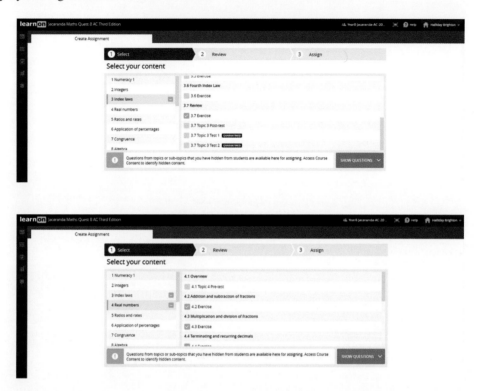

Once your selections are made, you can assign to your whole class or subsets of your class, with individualised start and finish times. You can also share with other teachers.

More instructions and helpful hints are available at www.jacplus.com.au.

9 Non-linear relationships

LESSON SEQUENCE

LESSON
9.1 Overview

Why learn this?

So far, throughout high school, much of the focus of algebraic sketching has been on linear graphs. A linear graph is the graph of a straight line; therefore, a non-linear graph is any graph of a curve that is not straight. This means that non-linear graphing is a huge field of mathematics, encompassing many topics and areas of study.

If we think of the purpose of sketching graphs, to some extent it is to model relationships between real-life variables. Yet how often do we actually come up across a linear relationship? Throwing a ball through the air, the speed of a car as it accelerates from rest, the path of a runner around a track, the temperature of coffee as it cools — none of these relationships are linear. Even the path of light from a star in the night sky may not be a straight line through space to Earth, due to the curvature of space itself based on gravity!

Non-linear graphs, whether they be parabolas, hyperbolas, exponentials or circles, are some of the more common types of graphs that are used to model phenomena in everyday life. Thus, it is important to study these relationships and their graphs so we can use them to help us model concepts such as exponential growth of a colony of bacteria, or an inversely proportional relationship such as the decay of radioactive material over time.

1. Complete the table of values for the equation $y = \dfrac{1}{2}(x-2)^2 - 3$.

x	-3	-2	-1	0
y				

2. For the graph of $y = -3(x+1)^2 - 2$, state the equation of the axis of symmetry.

3. **MC** For the graph of $y = 2(x+5)^2 + 8$, the turning point is:
   A. $(-5, -8)$
   B. $(-5, 8)$
   C. $(5, 8)$
   D. $(8, 5)$

4. **MC** The graph of $y = -(x-1)^2 - 3$ has:
   A. a maximum turning point
   B. a minimum turning point
   C. no turning point
   D. two maximum turning points.

5. **MC** The x-intercepts of $y = 2(x-2)^2 - 8$ are:
   A. $x = 2$ or $x = 8$
   B. $x = 0$ or $x = 8$
   C. $x = 10$ or $x = -6$
   D. $x = 0$ or $x = 4$

6. Calculate the coordinates of the turning point for the graph of $y = (x+3)(x+5)$. Write your answer in the form (a, b).

7. The equation $y = 2x^2 + bx - 1200$ has x-intercepts of $(-30, 0)$ and $(20, 0)$. Determine the value of b.

8. **MC** The radius r, of the circle $4x^2 + 4y^2 = 16$ is:
   A. $r = 16$
   B. $r = 8$
   C. $r = 4$
   D. $r = 2$

9. **MC** The center of the circle with equation $x^2 + 4x + y^2 - 6y + 9 = 0$ is:
   A. $(-4, -3)$
   B. $(4, -6)$
   C. $(-4, 6)$
   D. $(-2, 3)$

10. Calculate the points of intersection between the parabola $y = x^2$ and the circle $x^2 + y^2 = 1$, correct to two decimal places.

11. **MC** The horizontal asymptote for $y = 3^{-x} + 1$ is:
    A. $y = 0$
    B. $y = 1$
    C. $x = -1$
    D. $x = 0$

12. **MC** For the equation of a hyperbola $y = \dfrac{2}{x+3} - 1$, the vertical and horizontal asymptotes are:

   A. $x = \dfrac{2}{3}$ and $y = -1$
   B. $x = 3$ and $y = -1$

   C. $x = -3$ and $y = -1$
   D. $x = -1$ and $y = -3$

13. Match each graph with its correct equation.

Equation	Graph
a. $y = 2^x$	**1.**
b. -2^x	**2.**

14. MC From the graph of the hyperbola, the equation is:

A. $y = \dfrac{1}{x+2} - 1$

B. $y = \dfrac{-1}{x+2} - 1$

C. $y = \dfrac{2}{x+2} - 1$

D. $y = \dfrac{-2}{x+2} - 1$

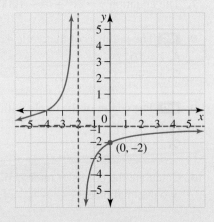

15. **MC** The graph of the truncus with equation $y = \dfrac{1}{x^2}$ reflected in the x-axis looks like:

A.

B.

C.

D.

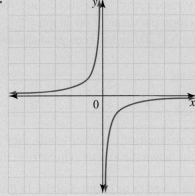

LESSON
9.2 Plotting parabolas from a table of values

LEARNING INTENTIONS

At the end of this lesson you should be able to:
- create a table of values and use this to sketch the graph of a parabola
- identify the axis of symmetry, turning point and y-intercept of a parabola.

9.2.1 Plotting parabolas

eles-4864

- The graphs of all quadratic relationships are called **parabolas**.
- If the equation of the parabola is given, a table of values can be produced by substituting x-values into the equation to obtain the corresponding y-values. These x-and y-values provide the coordinates for points that can be plotted and joined to form the shape of the graph.
- The graph of $y = x^2$ shown has been produced by generating a table of values.

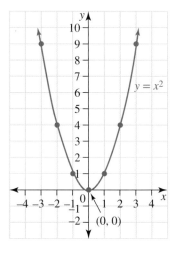

x	-3	-2	-1	0	1	2	3
y	9	4	1	0	1	4	9

- Parabolas are symmetrical; in other words, they have an **axis of symmetry**. In the parabola shown the axis of symmetry is the y-axis, also called the line $x = 0$.
- A parabola has a **vertex** or **turning point**. In this case the vertex is at the origin and is called a 'minimum turning point'.
- The **y-intercept** of a quadratic is the coordinate where the parabola cuts the y-axis. This can be found from a table of value by looking for the point where $x = 0$.
- The **x-intercept** of a quadratic is the coordinate where the parabola cuts the x-axis. This can be found from a table of value by looking for the point where $y = 0$.
- Consider the key features of the equation $y = x^2 + 2x + 8$.

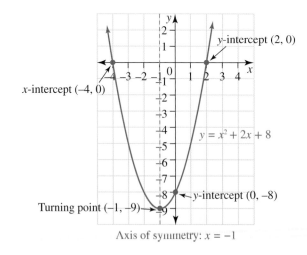

Shapes of parabolas

- **Parabolas with the shape ∪ are said to be 'concave up' and have a minimum turning point.**
- **Parabolas with the shape ∩ are said to be 'concave down' and have a maximum turning point.**

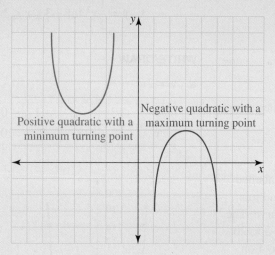

Positive quadratic with a minimum turning point

Negative quadratic with a maximum turning point

Parabolas in the world around us

- Parabolas abound in the world around us. Here are some examples.

Satellite dishes

Water droplets from a hose

The cables from a suspension bridge

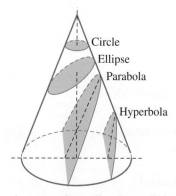

A cone when sliced parallel to its edge reveals a parabola.

Plot the graph of each of the following equations. In each case, use the values of x shown as the values in your table. State the equation of the axis of symmetry and the coordinates of the turning point.

a. $y = 2x^2$ for $-3 \leq x \leq 3$

b. $y = \dfrac{1}{2}x^2$ for $-3 \leq x \leq 3$

THINK

WRITE/DRAW

a. 1. Write the equation.

a. $y = 2x^2$

2. Produce a table of values using x-values from -3 to 3.

x	-3	-2	-1	0	1	2	3
y	18	8	2	0	2	8	18

3. Draw a set of clearly labelled axes, plot the points and join them with a smooth curve. The scale would be from -2 to 20 on the y-axis and -4 to 4 on the x-axis.

4. Label the graph.

5. Write the equation of the axis of symmetry that divides the parabola exactly in half.

The equation of the axis of symmetry is $x = 0$.

6. Write the coordinates of the turning point.

The turning point is $(0, 0)$.

b. 1. Write the equation.

b. $y = \dfrac{1}{2}x^2$

2. Produce a table of values using x-values from -3 to 3.

x	-3	-2	-1	0	1	2	3
y	4.5	2	0.5	0	0.5	2	4.5

3. Draw a set of clearly labelled axes, plot the points and join them with a smooth curve. The scale would be from -2 to 6 on the y-axis and -4 to 4 on the x-axis.

4. Label the graph.

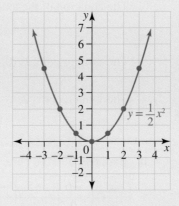

5. Write the equation of the line that divides the parabola exactly in half.

The equation of the axis of symmetry is $x = 0$.

6. Write the coordinates of the turning point.

The turning point is $(0, 0)$.

WORKED EXAMPLE 2 Determining the key features of a quadratic equation

Plot the graph of each of the following equations. In each case, use the values of x shown as the values in your table. State the equation of the axis of symmetry, the coordinates of the turning point and the y-intercept for each one.

a. $y = x^2 + 2$ for $-3 \leq x \leq 3$

b. $y = (x + 3)^2$ for $-6 \leq x \leq 0$

c. $y = -x^2$ for $-3 \leq x \leq 3$

THINK

a. 1. Write the equation.

   2. Produce a table of values.

   3. Draw a set of clearly labelled axes, plot the points and join them with a smooth curve. The scale on the y-axis would be from -2 to 12 and -4 to 4 on the x-axis.

   4. Label the graph.

WRITE/DRAW

a. $y = x^2 + 2$

x	-3	-2	-1	0	1	2	3
y	11	6	3	2	3	6	11

5. Write the equation of the line that divides the parabola exactly in half.

The equation of the axis of symmetry is $x = 0$.

6. Write the coordinates of the turning point.

The turning point is $(0, 2)$.

7. Determine the y-coordinate of the point where the graph crosses the y-axis.

The y-intercept is 2.

b. 1. Write the equation.

b. $y = (x + 3)^2$

2. Produce a table of values.

x	-6	-5	-4	-3	-2	-1	0
y	9	4	1	0	1	4	9

3. Draw a set of clearly labelled axes, plot the points and join them with a smooth curve. The scale on the y-axis would be from -2 to 10 and -7 to 1 on the x-axis.

4. Label the graph.

5. Write the equation of the line that divides the parabola exactly in half.

The equation of the axis of symmetry is $x = -3$.

6. Write the coordinates of the turning point.

The turning point is $(-3, 0)$.

7. Determine the y-coordinate of the point where the graph crosses the y-axis.

The y-intercept is 9.

c. 1. Write the equation.

c. $y = -x^2$

2. Produce a table of values.

x	-3	-2	-1	0	1	2	3
y	-9	-4	-1	0	-1	-4	-9

3. Draw a set of clearly labelled axes, plot the points and join them with a smooth curve. The scale on the y-axis would be from -10 to 1 and from -4 to 4 on the x-axis.

4. Label the graph.

$y = -x^2$

5. Write the equation of the line that divides the parabola exactly in half.

The equation of the axis of symmetry is $x = 0$.

6. Write the coordinates of the turning point.

The turning point is $(0, 0)$.

7. Determine the y-coordinate of the point where the graph crosses the y-axis.

The y-intercept is 0.

DISCUSSION

What x-values can a parabola have? What y-values can a parabola have?

 Resources

 eWorkbook   Topic 9 Workbook (worksheets, code puzzle and a project) (ewbk-2035)

 Interactivities Individual pathway interactivity: Plotting parabolas (int-4605)
                 Plotting quadratic graphs (int-6150)
                 Parabolas in the world around us (int-7539)

9.2 Quick quiz on	9.2 Exercise

Individual pathways

■ PRACTISE	■ CONSOLIDATE	■ MASTER
1, 3, 5, 11, 13, 15, 19, 22, 26, 29	2, 6, 8, 10, 16, 17, 20, 23, 27, 30	4, 7, 9, 12, 14, 18, 21, 24, 25, 28, 31

Fluency

1. **WE1** Plot the graph of each of the following equations. In each case, use the values of x shown as the values in your table. State the equation of the axis of symmetry and the coordinates of the turning point.

   a. $y = 3x^2$ for $-3 \leq x \leq 3$

   b. $y = \dfrac{1}{4}x^2$ for $-3 \leq x \leq 3$

2. Compare the graphs drawn for question 1 with that of $y = x^2$. Explain how placing a number in front of x^2 affects the graph obtained.

3. **WE2a** Plot the graph of each of the following for values of x between -3 and 3. State the equation of the axis of symmetry, the coordinates of the turning point and the y-intercept for each one.

   a. $y = x^2 + 1$
   b. $y = x^2 + 3$
   c. $y = x^2 - 3$
   d. $y = x^2 - 1$

4. Compare the graphs drawn for question 3 with the graph of $y = x^2$. Explain how adding to or subtracting from x^2 affects the graph obtained.

WE2b For questions 5 to 8, plot the graph of each of the following equations. In each case, use the values of x shown as the values in your table. State the equation of the axis of symmetry and the coordinates of the turning point and the y-intercept for each one.

5. $y = (x + 1)^2$ $-5 \leq x \leq 3$

6. $y = (x - 2)^2$ $-1 \leq x \leq 5$

7. $y = (x - 1)^2$ $-2 \leq x \leq 4$

8. $y = (x + 2)^2$ $-6 \leq x \leq 2$

9. Compare the graphs drawn for questions 5 to 8 with that for $y = x^2$. Explain how adding to or subtracting from x before squaring affects the graph obtained.

WE2c For questions 10 to 13, plot the graph of each of the following equations. In each case, use the values of x shown as the values in your table. State the equation of the axis of symmetry, the coordinates of the turning point and the y-intercept for each one.

10. $y = -x^2 + 1$ $-3 \leq x \leq 3$

11. $y = -(x + 2)^2$ $-5 \leq x \leq 1$

12. $y = -x^2 - 3$ $-3 \leq x \leq 3$

13. $y = -(x - 1)^2$ $-2 \leq x \leq 4$

14. Compare the graphs drawn for questions 10 to 13 with that for $y = x^2$. Explain how a negative sign in front of x^2 affects the graph obtained. Also compare the graphs obtained in questions 10 to 13 with those in questions 3 and 5 to 8. State which graphs have the same turning point. Describe how are they different.

Understanding

For questions **15** to **20**:

a. plot the graph

b. state the equation of the axis of symmetry

c. state the coordinates of the turning point and whether it is a maximum or a minimum

d. state the y-intercept.

15. $y = (x-5)^2 + 1 \qquad 0 \le x \le 6$

16. $y = 2(x+2)^2 - 3 \qquad -5 \le x \le 1$

17. $y = -(x-3)^2 + 4 \qquad 0 \le x \le 6$

18. $y = -3(x-1)^2 + 2 \qquad -2 \le x \le 4$

19. $y = x^2 + 4x - 5 \qquad -6 \le x \le 2$

20. $y = -3x^2 - 6x + 24 \qquad -5 \le x \le 3$

21. Use the equation $y = a(x-b)^2 + c$ to answer the following.

   a. Explain how you can determine whether a parabola has a minimum or maximum turning point by looking only at its equation.

   b. Explain how you can determine the coordinates of the turning point of a parabola by looking only at the equation.

   c. Explain how you can obtain the equation of the axis of symmetry by looking only at the equation of the parabola.

22. **MC** For the graph of $y = (x-2)^2 + 5$, the turning point is:

   A. $(5, 2)$ **B.** $(2, -5)$ **C.** $(2, 5)$ **D.** $(-2, -5)$

23. **MC** For the graph of $y = 3(x-1)^2 + 12$, the turning point is:

   A. $(3, 12)$ **B.** $(1, 12)$ **C.** $(-1, 12)$ **D.** $(-3, 12)$

24. **MC** For the graph of $y = (x+2)^2 - 7$, the y-intercept is:

   A. -2 **B.** -7 **C.** -3 **D.** -11

25. **MC** Select which of the following is true for the graph of $y = -(x-3)^2 + 4$.

   A. Turning point $(3, 4)$, y-intercept -5
   B. Turning point $(3, 4)$, y-intercept 5
   C. Turning point $(-3, 4)$, y-intercept -5
   D. Turning point $(-3, 4)$, y-intercept 5

Reasoning

26. A ball is thrown into the air. The height, h metres, of the ball at any time, t seconds, can be found by using the equation $h = -(t-4)^2 + 16$.

   a. Plot the graph for values of t between 0 and 8.
   b. Use the graph to determine:
      i. the maximum height of the ball
      ii. how long it takes for the ball to fall back to the ground from the moment it is thrown.

27. From a crouching position in a ditch, an archer wants to fire an arrow over a horizontal tree branch, which is 15 metres above the ground. The height, in metres (h), of the arrow t seconds after it has been fired is given by the equation $h = -8t(t - 3)$.

   a. Plot the graph for $t = 0, 1, 1.5, 2, 3$.
   b. From the graph, determine:
      i. the maximum height the arrow reaches
      ii. whether the arrow clears the branch and the distance by which it clears or falls short of the branch
      iii. the time it takes to reach maximum height
      iv. how long it takes for the arrow to hit the ground after it has been fired.

28. There are $0, 1, 2$ and infinite possible points of intersection for two parabolas.

   a. Illustrate these on separate graphs.
   b. Explain why infinite points of intersection are possible. Give an example.
   c. Determine how many points of intersection are possible for a parabola and a straight line. Illustrate these.

Problem solving

29. The area of a rectangle in cm^2 is given by the equation $A = \dfrac{-8w}{5}(w - 6)$, where w is the width of the rectangle in centimetres.

   a. Complete a table of values for $-1 \le w \le 7$.
   b. Explain which of the values for w from part a should be discarded and why.
   c. Sketch the graph of A for suitable values of w.
   d. Evaluate the maximum possible area of the rectangle. Show your working.
   e. Determine the dimensions of the rectangle that produce the maximum area found in part d.

30. The path taken by a netball thrown by a rising Australian player is given by the quadratic equation $y = -x^2 + 3.2x + 1.8$, where y is the height of the ball and x is the horizontal distance from the player's upstretched hand.

   a. Complete a table of values for $-1 \le x \le 4$.
   b. Plot the graph.
   c. Explain what values of x are 'not reasonable'.
   d. Evaluate the maximum height reached by the netball.
   e. Assuming that nothing hits the netball, determine how far away from the player the netball will strike the ground.

31. The values of a, b and c in the equation $y = ax^2 + bx + c$ can be calculated using three points that lie on the parabola. This requires solving triple simultaneous equations by algebra. This can also be done using a CAS calculator. If the points $(0, 1)$, $(1, 0)$ and $(2, 3)$ all lie on one parabola, determine the equation of the parabola.

LESSON
9.3 Sketching parabolas using transformations

LEARNING INTENTIONS

At the end of this lesson you should be able to:
- be familiar with the key features of the basic graph of a quadratic, $y = x^2$
- determine whether a dilation has made the graph of quadratic narrower or wider
- determine whether a translation has moved the graph of a quadratic function left/right or up/down
- sketch the graph of a quadratic equation that has undergone any of the following transformations: dilations, a reflection, or translations.

▶ 9.3.1 Sketching parabolas

eles-4865

- A sketch graph of a parabola does not show a series of plotted points, but it does accurately locate important features such as x- and y-intercepts and turning points.
- The basic quadratic graph has the equation $y = x^2$. Transformations or changes in the features of the graph can be observed when the equation changes. These transformations include:
  - dilation
  - translation
  - reflection.

Dilation

- A dilation stretches a graph away from an axis. A dilation of factor 3 from the x-axis triples the distance of each point from the x-axis. This means the point $(2, 4)$ would become $(2, 12)$.
- Compare the graph of $y = 2x^2$ with that of $y = x^2$. This graph is thinner or closer to the y-axis and has a dilation factor of 2 from the x-axis. As the magnitude (or size) of the coefficient of x^2 increases, the graph becomes **narrower** and closer to the y-axis.

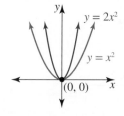

- The turning point has not changed under the transformation and is still $(0, 0)$.
- Compare the graph $y = \frac{1}{4}x^2$ with that of $y = x^2$.

The graph is **wider** or closer to the x-axis and has a dilation factor of factor $\frac{1}{4}$.

The turning point has not changed and is still $(0, 0)$. As the coefficient of x^2 decreases (but remains positive), the graph becomes wider or closer to the x-axis.

WORKED EXAMPLE 3 Determining dilation and the turning point

State whether each of the following graphs is wider or narrower than the graph of $y = x^2$ and state the coordinates of the turning point of each one.

a. $y = \dfrac{1}{5}x^2$

b. $y = 4x^2$

THINK	WRITE
a. 1. Write the equation.	a. $y = \dfrac{1}{5}x^2$
2. Look at the coefficient of x^2 and decide whether it is greater than or less than 1.	$\dfrac{1}{5} < 1$, so the graph is wider than that of $y = x^2$.
3. The dilation doesn't change the turning point.	The turning point is $(0, 0)$.
b. 1. Write the equation.	b. $y = 4x^2$
2. Look at the coefficient of x^2 and decide whether it is greater than or less than 1.	$4 > 1$, so the graph is narrower than that of $y = x^2$.
3. The dilation doesn't change the turning point.	The turning point is $(0, 0)$.

9.3.2 Vertical translation

eles-4866

- Compare the graph of $y = x^2 + 2$ with that of $y = x^2$.
  The whole graph has been moved or translated 2 units upwards. The turning point has become $(0, 2)$.

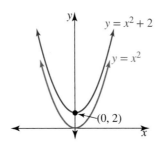

- Compare the graph of $y = x^2 - 3$ with that of $y = x^2$.
  The whole graph has been moved or translated 3 units downwards. The turning point has become $(0, -3)$.

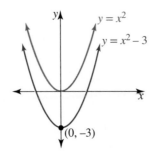

State the vertical translation and the coordinates of the turning point for the graphs of the following equations when compared to the graph of $y = x^2$.

a. $y = x^2 + 5$

b. $y = x^2 - 4$

THINK	WRITE
a. 1. Write the equation.	**a.** $y = x^2 + 5$
2. $+5$ means the graph is translated upwards 5 units.	Vertical translation of 5 units up.
3. Translate the turning point of $y = x^2$ which is $(0, 0)$. The x-coordinate of the turning point remains 0, and the y-coordinate has 5 added to it.	The turning point becomes $(0, 5)$.
b. 1. Write the equation.	**b.** $y = x^2 - 4$
2. -4 means the graph is translated downwards 4 units.	Vertical translation of 4 units down.
3. Translate the turning point of $y = x^2$, which is $(0, 0)$. The x-coordinate of the turning point remains 0, and the y-coordinate has 4 subtracted from it.	The turning point becomes $(0, -4)$.

9.3.3 Horizontal translation

eles-4867

- Compare the graph of $y = (x - 2)^2$ with that of $y = x^2$.

  The whole graph has been moved or translated 2 units to the right. The turning point has become $(2, 0)$.

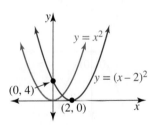

- Compare the graph of $y = (x + 1)^2$ with that of $y = x^2$.

  The whole graph has been moved or translated 1 unit left. The turning point has become $(-1, 0)$.

- *Note:* Horizontal translations appear to cause the graph to move in the opposite direction to the sign inside the brackets.

State the horizontal translation and the coordinates of the turning point for the graphs of the following equations when compared to the graph of $y = x^2$.

a. $y = (x - 3)^2$ b. $y = (x + 2)^2$

THINK	WRITE
a. 1. Write the equation.	a. $y = (x - 3)^2$
2. -3 means the graph is translated to the right 3 units.	Horizontal translation of 3 units to the right
3. Translate the turning point of $y = x^2$, which is $(0, 0)$. The y-coordinate of the turning point remains 0, and the x-coordinate has 3 added to it.	The turning point becomes $(3, 0)$.
b. 1. Write the equation.	b. $y = (x + 2)^2$
2. $+2$ means the graph is translated to the left 2 units.	Horizontal translation of 2 units to the left
3. Translate the turning point of $y = x^2$, which is $(0, 0)$. The y-coordinate of the turning point remains 0, and the x-coordinate has 2 subtracted from it.	The turning point becomes $(-2, 0)$.

⏵ 9.3.4 Reflection

eles-4868

- Compare the graph of $y = -x^2$ with that of $y = x^2$.

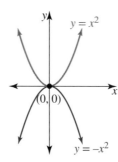

In each case the axis of symmetry is the line $x = 0$ and the turning point is $(0, 0)$. The only difference between the equations is the negative sign in $y = -x^2$, and the difference between the graphs is that $y = x^2$ 'sits' on the x-axis and $y = -x^2$ 'hangs' from the x-axis. (One is a reflection or mirror image of the other.) The graph of $y = x^2$ has a minimum turning point, and the graph of $y = -x^2$ has a maximum turning point.

Shapes of quadratic graphs

Any quadratic graph where x^2 is positive has a $\cup$ shape and is said to be upright. Conversely, if x^2 is negative the graph has a $\cap$ shape and is said to be inverted.

For each of the following graphs, identify the coordinates of the turning point and state whether it is a maximum or a minimum.

a. $y = -(x-7)^2$

b. $y = 5 - x^2$

THINK	WRITE
a. 1. Write the equation.	**a.** $y = -(x-7)^2$
2. It is a horizontal translation of 7 units to the right, so 7 units is added to the x-coordinate of $(0, 0)$.	The turning point is $(7, 0)$.
3. The sign in front of the x^2 term is negative, so it is inverted.	Maximum turning point.
b. 1. Write the equation.	**b.** $y = 5 - x^2$
2. Rewrite the equation so that the x^2 term is first.	$y = -x^2 + 5$
3. The vertical translation is 5 units up, so 5 units is added to the y-coordinate of $(0, 0)$.	The turning point is $(0, 5)$.
4. The sign in front of the x^2 term is negative, so the graph is inverted.	Maximum turning point.

TI	THINK	DISPLAY/WRITE	CASIO	THINK	DISPLAY/WRITE
a.	**a.**	**a.**	**a.**
In a new problem, on a Graphs page, complete the function entry line as: $f1(x) = -(x-7)^2$ Then press ENTER. To locate the turning point, press: • MENU • 6: Analyze Graph • 3: Maximum Drag the dotted line to the left of the turning point (the lower bound), click ENTER and then drag the dotted line to the right of the turning point (the upper bound) and press ENTER. The turning point will be shown.	 The turning point $(7, 0)$ is a maximum.	On the Graph & Table screen, complete the function entry line as: $y1 = -(x-7)^2$ Tap the graphing icon and the graph will be displayed. To locate the turning point, tap: • Analysis • G-Solve • Max The turning point will be shown.	 The turning point $(7, 0)$ is a maximum.

b.

In a new problem, on a new Graphs page, complete the function entry line as:
$f1(x) = 5 - x^2$
Then press ENTER.
To locate the turning point, press:
• MENU
• 6: Analyze Graph
• 3: Maximum
Drag the dotted line to the left of the turning point (the lower bound), click ENTER and then drag the dotted line to the right of the turning point (the upper bound) and press ENTER.
The turning point will be shown.

b.

The turning point $(0, 5)$ is a maximum.

b.

On the Graph & Table screen, complete the function entry line as:
$y1 = 5 - x^2$
Tap the graphing icon and the graph will be displayed.
To locate the turning point, press:
• Analysis
• G-Solve
• Max
The turning point will be shown.

b.

The turning point $(0, 5)$ is a maximum.

9.3.5 Combining transformations

eles-4869

- Often, multiple transformations will be applied to the equation $y = x^2$ to produce a new graph.
- We can determine the transformations applied by looking at the equation of the resulting quadratic.

Combining transformations

A quadratic of the form $y = a(x - h)^2 + k$ has been:
- **dilated by a factor of $|a|$ from the x-axis, where $|a|$ is the magnitiude (or size) of a.**
- **reflected in the x-axis if $a < 0$**
- **translated h units horizontally:**
  - if $h > 0$, the graph is translated to the *right*
  - if $h < 0$, the graph is translated to the *left*
- **translated k units vertically:**
  - if $k > 0$, the graph is translated *upwards*
  - if $k < 0$, the graph is translated *downwards*.

WORKED EXAMPLE 7 Determining transformations

For each of the following quadratic equations:
   i. state the appropriate dilation, reflection and translation of the graph of $y = x^2$ needed to obtain the graph
   ii. state the coordinates of the turning point
   iii. hence, sketch the graph.

a. $y = (x + 3)^2$

b. $y = -2x^2$

THINK	WRITE/DRAW
a. 1. Write the quadratic equation.	**a.** $y = (x + 3)^2$
2. Identify the transformation needed — horizontal translation only, no dilation or reflection.	**i.** Horizontal translation of 3 units to the left

3. State the turning point.

4. Sketch the graph of $y = (x+3)^2$. You may find it helpful to lightly sketch the graph of $y = x^2$ on the same set of axes first.

ii. The turning point is $(-3, 0)$.

iii.

b. 1. Write the quadratic equation.

2. Identify the transformations needed — dilation (2 in front of x^2) and reflection (negative in front of x^2), no translation.

3. The turning point remains the same as there is no translation.

4. Sketch the graph of $y = -2x^2$. You may find it helpful to lightly sketch the graph of $y = x^2$ on the same set of axes first.

b. $y = -2x^2$

i. This is a reflection, so the graph is inverted. As $2 > 1$, the graph is narrower than that of $y = x^2$.

ii. The turning point is $(0, 0)$.

iii.

TI | THINK

a.
1. On a Graphs page, complete the function entry lines as:

$f1(x) = x^2$

$f2(x) = (x+3)^2$

The graphs will be displayed.

DISPLAY/WRITE

a.

CASIO | THINK

a.
1. On a Graph & Table screen, complete the function entry lines as:

$y1 = x^2$

$y2 = (x+3)^2$

Tap the graphing icon and the graphs will be displayed.

DISPLAY/WRITE

a.

2. To locate the turning point, press:
• MENU
• 6: Analyze Graph
• 2: Minimum
Drag the dotted line to the left of the turning point (the lower bound), click ENTER and then drag the dotted line to the right of the turning point (the upper bound) and press ENTER.
The turning point will be shown.

2. To locate the turning point, press:
• Analysis
• G-Solve
• Min
Select the graph you want using the up and down arrow keys and then press EXE. The turning point will be shown.

b.
1. On a Graphs page, complete the function entry lines as:

$f1(x) = x^2$

$f2(x) = -2x^2$

The graphs will be displayed.

b.
1. On a Graph & Table screen, complete the function entry lines as:

$y1 = x^2$

$y2 = -2x^2$

Tap the graphing icon and the graphs will be displayed.

2. To locate the turning point, press:
- MENU
- 6: Analyze Graph
- 2: Minimum

Drag the dotted line to the left of the turning point (the lower bound), click ENTER and then drag the dotted line to the right of the turning point (the upper bound) and press ENTER.
The turning point will be shown.

2. To locate the point of intersection, press:
- Analysis
- G-Solve
- , Max

Select the graph you want using the up and down arrow keys and then press EXE. The turning point will be shown.

DISCUSSION

Determine the turning points of the graphs $y = x^2 + k$ and $y = (x-h)^2$.

  Resources

eWorkbook   Topic 9 Workbook (worksheets, code puzzle and a project) (ewbk-2035)

Interactivities Individual pathway interactivity: Sketching parabolas (int-4606)
        Horizontal translations of parabolas (int-6054)
        Vertical translations of parabolas (int-6055)
        Dilation of parabolas (int-6096)
        Reflection of parabolas (int-6151)

Exercise 9.3 Sketching parabolas using transformations

9.3 Quick quiz on	9.3 Exercise

Individual pathways

■ PRACTISE	■ CONSOLIDATE	■ MASTER
1, 4, 7, 10, 13, 16, 17, 22, 25	2, 5, 8, 11, 14, 18, 19, 23, 26	3, 6, 9, 12, 15, 20, 21, 24, 27, 28

Fluency

WE3 For questions **1** to **3**, state whether each of the following graphs is wider or narrower than the graph of $y = x^2$ and state the coordinates of the turning point of each one.

1. a. $y = 5x^2$ b. $y = \dfrac{1}{3}x^2$

2. a. $y = 7x^2$ b. $y = 10x^2$ c. $y = \dfrac{2}{5}x^2$

3. a. $y = 0.25x^2$ b. $y = 1.3x^2$ c. $y = \sqrt{3}x^2$

WE4 For questions **4** to **6**, state the vertical translation and the coordinates of the turning point for the graphs of each of the following equations when compared to the graph of $y = x^2$.

4. a. $y = x^2 + 3$ b. $y = x^2 - 1$

5. a. $y = x^2 - 7$ b. $y = x^2 + \dfrac{1}{4}$ c. $y = x^2 - \dfrac{1}{2}$

6. a. $y = x^2 - 0.14$ b. $y = x^2 + 2.37$ c. $y = x^2 + \sqrt{3}$

WE5 For questions **7** to **9**, state the horizontal translation and the coordinates of the turning point for the graphs of the following equations when compared to the graph of $y = x^2$.

7. a. $y = (x - 1)^2$ b. $y = (x - 2)^2$

8. a. $y = (x + 10)^2$ b. $y = (x + 4)^2$ c. $y = \left(x - \dfrac{1}{2}\right)^2$

9. a. $y = \left(x + \dfrac{1}{5}\right)^2$ b. $y = (x + 0.25)^2$ c. $y = (x + \sqrt{3})^2$

WE6 For questions **10** to **12**, for each of the following graphs identify the coordinates of the turning point and state whether it is a maximum or a minimum.

10. a. $y = -x^2 + 1$ b. $y = x^2 - 3$

11. a. $y = -(x + 2)^2$ b. $y = 3x^2$ c. $y = 4 - x^2$

12. a. $y = -2x^2$ b. $y = (x - 5)^2$ c. $y = 1 + x^2$

For questions **13** to **15**, in each of the following state whether the graph is wider or narrower than $y = x^2$ and whether it has a maximum or a minimum turning point.

13. **a.** $y = 3x^2$ **b.** $y = -3x^2$

14. **a.** $y = \dfrac{1}{2}x^2$ **b.** $y = -\dfrac{1}{5}x^2$ **c.** $y = -\dfrac{4}{3}x^2$

15. **a.** $y = 0.25x^2$ **b.** $y = \sqrt{3}x^2$ **c.** $y = -0.16x^2$

Understanding

WE7 For questions **16** to **21**:
 i. state the appropriate dilation, reflection and translation of the graph of $y = x^2$ needed to obtain the graph
 ii. state the coordinates of the turning point
 iii. hence, sketch the graph.

16. **a.** $y = (x + 1)^2$ **b.** $y = -3x^2$ **c.** $y = x^2 + 1$

17. **a.** $y = \dfrac{1}{3}x^2$ **b.** $y = x^2 - 3$

18. **a.** $y = (x - 4)^2$ **b.** $y = -\dfrac{2}{5}x^2$ **c.** $y = 5x^2$

19. **a.** $y = -x^2 + 2$ **b.** $y = -(x - 6)^2$

20. **a.** $y = -x^2 - 4$ **b.** $y = 2(x + 1)^2 - 4$ **c.** $y = \dfrac{1}{2}(x - 3)^2 + 2$

21. **a.** $y = -\dfrac{1}{3}(x + 2)^2 + \dfrac{1}{4}$ **b.** $y = -\dfrac{7}{4}(x - 1)^2 - \dfrac{3}{2}$

Reasoning

22. A vase 25 cm tall is positioned on a bench near a wall as shown. The shape of the vase follows the curve $y = (x - 10)^2$, where y cm is the height of the vase and x cm is the distance of the vase from the wall.

    a. Identify how far the base of the vase is from the wall.
    b. Determine the shortest distance from the top of the vase to the wall.
    c. If the vase is moved so that the top just touches the wall, determine the new distance from the wall to the base.
    d. Determine the new equation that follows the shape of the vase.

23. Tom is standing at the start of a footpath at $(0, 0)$ that leads to the base of a hill. The height of the hill is modelled by the equation $h = -\dfrac{1}{10}(d - 25)^2 + 40$, where h is the height of the hill in metres and d is the horizontal distance from the start of the path.

    a. Calculate how tall the hill is.
    b. Determine how far the base of the hill is from the beginning of the footpath.
    c. If the footpath is to be extended so the lead in to the hill is 50 m, determine the new equation that models the height of the hill.
    d. The height of the hill has been incorrectly measured and should actually be 120 m. Adjust the equation from part **c** to correct this error and state the transformation applied.

24. A ball is thrown vertically upwards. Its height in metres after t seconds is given by $h = 7t - t^2$.

  a. Sketch the graph of the height of the ball against time.

  b. Evaluate the highest point reached by the ball. Show your full working.

A second ball is thrown vertically upwards. Its height in metres after t seconds is given by $h = 10t - t^2$.

  c. On the same set of axes used for part **a**, sketch the graph of the height of the second ball against time.

  d. State the difference in the highest point reached by the two balls.

Problem solving

25. Consider the quadratic equation $y = x^2 - 4x + 7$.

  a. Determine the equivalent inverted equation of the quadratic that just touches the one above at the turning point.

  b. Confirm your result graphically.

26. Consider the equation $y = 3(x - 2)^2 - 7$.

  a. State the coordinates of the turning point and y-intercept.

  b. State a sequence of transformations that when applied to the graph of $y = \dfrac{-3}{2}(x - 2)^2 - 7$ will produce the graph of $y = x^2$.

27. A parabola has the equation $y = -\dfrac{1}{2}(x - 3)^2 + 4$. A second parabola has an equation defined by $Y = 2(y - 1) - 3$.

  a. Determine the equation relating Y to x.

  b. State the appropriate dilation, reflection and translation of the graph of $Y = x^2$ required to obtain the graph of $Y = 2(y - 1) - 3$.

  c. State the coordinates of the turning point $Y = 2(y - 1) - 3$.

  d. Sketch the graph of $Y = 2(y - 1) - 3$.

28. A ball shot at a certain angle to the horizontal follows a parabolic path. It reaches a maximum height of 200 m when its horizontal distance from its starting point is 10 m. When the ball's horizontal distance from the starting point was 1 m, the ball had reached a height of 38 m.
Determine an equation to model the ball's flight, clearly defining your chosen pronumerals.

LESSON
9.4 Sketching parabolas using turning point form

LEARNING INTENTIONS

At the end of this lesson you should be able to:
- determine the axis of symmetry and turning point of a quadratic in turning point form
- calculate the y-intercept and any x-intercepts of a quadratic in turning point form.

9.4.1 Turning point form

eles-4870

- When a quadratic equation is expressed in the form $y = a(x-h)^2 + k$:
  - the turning point is the point (h, k)
  - the axis of symmetry is $x = h$
  - the x-intercepts are calculated by solving $a(x-h)^2 + k = 0$.
- Changing the values of a, h and k in the equation transforms the shape and position of the parabola when compared with the parabola $y = x^2$.

$$y = a(x-h)^2 + k$$

Reflects and dilates   Translates left and right   Translates up and down

---

Turning point form

A quadratic of the form $y = a(x-h)^2 + k$ has:
- a turning point at the coordinate (h, k)
  - the turning point will be a minimum if $a > 0$
  - the turning point will be a maximum if $a < 0$
- an axis of symmetry of $x = h$
- a y-intercept of $(0, ah^2 + k)$.

The number of x-intercepts depends on the values of a, h and k. Changing the value of a does not change the position of the turning point, only h and k.

WORKED EXAMPLE 8 Determining the turning point from turning point form

For each of the following equations, state the coordinates of the turning point of the graph and whether it is a maximum or a minimum.

a. $y = (x-6)^2 - 4$ b. $y = -(x+3)^2 + 2$

THINK	WRITE
a. 1. Write the equation.	a. $y = (x-6)^2 - 4$
2. Identify the transformations — horizontal translation of 6 units to the right and a vertical translation of 4 units down. State the turning point.	The turning point is $(6, -4)$.
3. As a is positive ($a = 1$), the graph is upright with a minimum turning point.	Minimum turning point.
b. 1. Write the equation	b. $y = -(x+3)^2 + 2$
2. Identify the transformations — horizontal translation of 3 units to the left and a vertical translation of 2 units up. State the turning point.	The turning point is $(-3, 2)$.
3. As a is negative ($a = -1$), the graph is inverted with a maximum turning point.	Maximum turning point.

▶ 9.4.2 x- and y-intercepts of quadratic graphs

eles-4871
- Other key features such as the x- and y-intercepts can also be determined from the equation of a parabola.
- The point(s) where the graph cuts or touches the x-axis are called the x-intercept(s). At these points, $y = 0$.
- The point where the graph cuts the x-axis is called the y-intercept. At this point, $x = 0$.

WORKED EXAMPLE 9 Determining the axial intercepts from turning point form

For the parabolas with the following equations:
i. determine the y-intercept
ii. determine the x-intercepts (where they exist).

a. $y = (x + 3)^2 - 4$ **b.** $y = 2(x - 1)^2$ **c.** $y = -(x + 2)^2 - 1$

THINK	WRITE
a. 1. Write the equation.	**a.** $y = (x + 3)^2 - 4$
2. Calculate the y-intercept by substituting $x = 0$ into the equation.	y-intercept: when $x = 0$, $y = (0 + 3)^2 - 4$ $= 9 - 4$ $= 5$ The y-intercept is 5.
3. Calculate the x-intercepts by substituting $y = 0$ into the equation and solving for x. Add 4 to both sides of the equation. Take the square root of both sides of the equation. Subtract 3 from both sides of the equation. Solve for x.	x-intercepts: when $y = 0$, $(x + 3)^2 - 4 = 0$ $(x + 3)^2 = 4$ $(x + 3) = +2 \text{ or } -2$ $x = 2 - 3 \text{ or } x = -2 - 3$ $x = -1 \qquad x = -5$ The x-intercepts are -5 and -1.
b. 1. Write the equation.	**b.** $y = 2(x - 1)^2$
2. Calculate the y-intercept by substituting $x = 0$ into the equation.	y-intercept: when $x = 0$, $y = 2(0 - 1)^2$ $= 2 \times 1$ $= 2$ The y-intercept is 2.
3. Calculate the x-intercepts by substituting $y = 0$ into the equation and solving for x. Note that there is only one solution for x and so there is only one x-intercept. (The graph touches the x-axis.)	x-intercepts: when, $y = 0$, $2(x - 1)^2 = 0$ $(x - 1)^2 = 0$ $x - 1 = 0$ $x = 0 + 1$ $x = 1$ The x-intercept is 1.
c. 1. Write the equation.	**c.** $y = -(x + 2)^2 - 1$
2. Calculate the y-intercept by substituting $x = 0$ into the equation.	y-intercept: when $x = 0$, $y = -(0 + 2)^2 - 1$ $= -4 - 1$ $= -5$ The y-intercept is -5.

▶

3. Calculate the x-intercepts by substituting $y = 0$ into the equation and solving for x. We cannot take the square root of -1 to obtain real solutions; therefore, there are no x-intercepts.

x-intercepts: when $y = 0$,
$-(x + 2)^2 - 1 = 0$
$$(x + 2)^2 = -1$$
There are no real solutions, so there are no x-intercepts.

WORKED EXAMPLE 10 Sketching a quadratic in turning point form

For each of the following:
 i. write the coordinates of the turning point
 ii. state whether the graph has a maximum or a minimum turning point
 iii. state whether the graph is wider, narrower or the same width as the graph of $y = x^2$
 iv. calculate the y-intercept
 v. calculate the x-intercepts
 vi. sketch the graph.

a. $y = (x - 2)^2 + 3$

b. $y = -2(x + 1)^2 + 6$

THINK

WRITE/DRAW

a. 1. Write the equation.

a. $y = (x - 2)^2 + 3$

2. State the coordinates of the turning point from the equation. Use (h, k) as the equation is in the turning point form of $y = a(x - h)^2 + k$ where $a = 1, h = 2$ and $k = 3$.

The turning point is $(2, 3)$.

3. State the nature of the turning point by considering the sign of a.

The graph has a minimum turning point as the sign of a is positive.

4. Specify the width of the graph by considering the magnitude of a.

The graph has the same width as $y = x^2$ since $a = 1$.

5. Calculate the y-intercept by substituting $x = 0$ into the equation.

y-intercept: when $x = 0$,
$y = (0 - 2)^2 + 3$
$\quad = 4 + 3$
$\quad = 7$
y-intercept is 7.

6. Calculate the x-intercepts by substituting $y = 0$ into the equation and solving for x. As we have to take the square root of a negative number, we cannot solve for x.

x-intercepts: when $y = 0$,
$(x - 2)^2 + 3 = 0$
$$(x - 2)^2 = -3$$
There are no real solutions, and hence no x-intercepts.

7. Sketch the graph, clearly showing the turning point and the y-intercept.

8. Label the graph.

b. 1. Write the equation.

b. $y = -2(x+1)^2 + 6$

2. State the coordinates of the turning point from the equation. Use (h, k) as the equation is in the turning point form of $y = a(x-h)^2 + k$ where $a = -2, h = -1$ and $k = 6$.

The turning point is $(-1, 6)$.

3. State the nature of the turning point by considering the sign of a.

The graph has a maximum turning point as the sign of a is negative.

4. Specify the width of the graph by considering the magnitude of a.

The graph is narrower than $y = x^2$ since $|a| > 1$.

5. Calculate the y-intercept by substituting $x = 0$ into the equation.

y-intercept: when $x = 0$,
$$y = -2(0 + 1)^2 + 6$$
$$= -2 \times 1 + 6$$
$$= 4$$
The y-intercept is 4.

6. Calculate the x-intercepts by substituting $y = 0$ into the equation and solving for x.

x-intercepts: when $y = 0$,
$$-2(x+1)^2 + 6 = 0$$
$$2(x+1)^2 = 6$$
$$(x+1)^2 = 3$$
$$x + 1 = \sqrt{3} \text{ or } x + 1 = -\sqrt{3}$$
$$x = -1 + \sqrt{3} \text{ or } x = -1 - \sqrt{3}$$
The x-intercepts are $-1 - \sqrt{3}$ and $-1 + \sqrt{3}$ (or approximately -2.73 and 0.73).

7. Sketch the graph, clearly showing the turning point and the x- and y-intercepts.

8. Label the graph.

Note: **Unless otherwise stated, exact values for the intercepts should be shown on sketch graphs.**

DISCUSSION

Does a in the equation $y = a(x-h)^2 + k$ have any impact on the turning point?

Exercise 9.4 Sketching parabolas using turning point form learn

9.4 Quick quiz on

9.4 Exercise

Individual pathways

■ PRACTISE	■ CONSOLIDATE	■ MASTER
1, 4, 7, 9, 12, 14, 17, 20, 23, 27	2, 5, 8, 10, 13, 15, 18, 21, 24, 28	3, 6, 11, 16, 19, 22, 25, 26, 29

Fluency

WE8 For questions **1** to **3**, for each of the following equations, state the coordinates of the turning point of the graph and whether it is a maximum or a minimum.

1. **a.** $y = (x-1)^2 + 2$ **b.** $y = (x+2)^2 - 1$ **c.** $y = (x+1)^2 + 1$

2. **a.** $y = -(x-2)^2 + 3$ **b.** $y = -(x-5)^2 + 3$ **c.** $y = (x+2)^2 - 6$

3. **a.** $y = \left(x - \dfrac{1}{2}\right)^2 - \dfrac{3}{4}$ **b.** $y = \left(x - \dfrac{1}{3}\right)^2 + \dfrac{2}{3}$ **c.** $y = (x+0.3)^2 - 0.4$

4. For each of the following, state:
   i. the coordinates of the turning point
   ii. whether the graph has a maximum or a minimum turning point
   iii. whether the graph is wider, narrower or the same width as that of $y = x^2$.

   a. $y = 2(x+3)^2 - 5$ **b.** $y = -(x-1)^2 + 1$

5. For each of the following, state:
   i. the coordinates of the turning point
   ii. whether the graph has a maximum or a minimum turning point
   iii. whether the graph is wider, narrower or the same width as that of $y = x^2$.

   a. $y = -5(x+2)^2 - 4$ **b.** $y = \dfrac{1}{4}(x-3)^2 + 2$

6. For each of the following, state:
   i. the coordinates of the turning point
   ii. whether the graph has a maximum or a minimum turning point
   iii. whether the graph is wider, narrower or the same width as that of $y = x^2$.

   a. $y = -\dfrac{1}{2}(x+1)^2 + 7$ **b.** $y = 0.2\left(x - \dfrac{1}{5}\right)^2 - \dfrac{1}{2}$

7. Select the equation that best suits each of the following graphs.

i.

ii.

iii.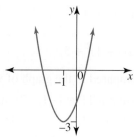

   a. $y = (x+1)^2 - 3$
   b. $y = -(x-2)^2 + 3$
   c. $y = -x^2 + 1$

8. Select the equation that best suits each of the following graphs.

i.

ii.

iii.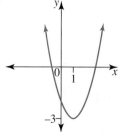

   a. $y = (x-1)^2 - 3$
   b. $y = -(x+2)^2 + 3$
   c. $y = x^2 - 1$

9. **MC** The translations required to change $y = x^2$ into $y = \left(x - \dfrac{1}{2}\right)^2 + \dfrac{1}{3}$ are:

   A. right $\dfrac{1}{2}$, up $\dfrac{1}{3}$

   B. left $\dfrac{1}{2}$, down $\dfrac{1}{3}$

   C. right $\dfrac{1}{2}$, down $\dfrac{1}{3}$

   D. left $\dfrac{1}{2}$, up $\dfrac{1}{3}$

10. **MC** For the graph $\dfrac{1}{4}\left(x - \dfrac{1}{2}\right)^2 + \dfrac{1}{3}$, the effect of the $\dfrac{1}{4}$ on the graph is:

   A. no effect
   B. to make the graph narrower
   C. to make the graph wider
   D. to invert the graph

11. **MC** Compared to the graph of $y = x^2$, $y = -2(x+1)^2 - 4$ is:
   A. inverted and wider
   B. inverted and narrower
   C. upright and wider
   D. upright and narrower

12. **MC** A graph that has a minimum turning point $(1, 5)$ and that is narrower than the graph of $y = x^2$ is:

A. $y = (x-1)^2 + 5$

B. $y = \dfrac{1}{2}(x+1)^2 + 5$

C. $y = 2(x-1)^2 + 5$

D. $y = 2(x+1)^2 + 5$

13. **MC** Compared to the graph of $y = x^2$, the graph of $y = -3(x-1)^2 - 2$ has the following features.

A. Maximum TP at $(-1, -2)$, narrower

B. Maximum TP at $(1, -2)$, narrower

C. Maximum TP at $(1, 2)$, wider

D. Minimum TP at $(1, -2)$, narrower

WE9 For questions **14** to **16**, for the parabolas with the following equations:

i. determine the y-intercept

ii. determine the x-intercepts (where they exist).

14. a. $y = (x+1)^2 - 4$

b. $y = 3(x-2)^2$

15. a. $y = -(x+4)^2 - 2$

b. $y = (x-2)^2 - 9$

16. a. $y = 2x^2 + 4$

b. $y = (x+3)^2 - 5$

Understanding

17. **WE10** For each of the following:

i. write the coordinates of the turning point

ii. state whether the graph has a maximum or a minimum turning point

iii. state whether the graph is wider, narrower or the same width as the graph of $y = x^2$

iv. calculate the y-intercept

v. calculate the x-intercepts

vi. sketch the graph.

a. $y = (x-4)^2 + 2$

b. $y = (x-3)^2 - 4$

c. $y = (x+1)^2 + 2$

18. For each of the following:

i. write the coordinates of the turning point

ii. state whether the graph has a maximum or a minimum turning point

iii. state whether the graph is wider, narrower or the same width as the graph of $y = x^2$

iv. calculate the y-intercept

v. calculate the x-intercepts

vi. sketch the graph.

a. $y = (x+5)^2 - 3$

b. $y = -(x-1)^2 + 2$

c. $y = -(x+2)^2 - 3$

19. For each of the following:

i. write the coordinates of the turning point

ii. state whether the graph has a maximum or a minimum turning point

iii. state whether the graph is wider, narrower or the same width as the graph of $y = x^2$

iv. calculate the y-intercept

v. calculate the x-intercepts

vi. sketch the graph.

a. $y = -(x+3)^2 - 2$

b. $y = 2(x-1)^2 + 3$

c. $y = -3(x+2)^2 + 1$

20. Consider the equation $2x^2 - 3x - 8 = 0$.

   a. Complete the square.
   b. Use the result to determine the exact solutions to the original equation.
   c. Determine the turning point of $y = 2x^2 - 3x - 8$ and indicate its type.

21. Answer the following questions.

   a. Determine the equation of a quadratic that has a turning point of $(-4, 6)$ and has an x-intercept at $(-1, 0)$.
   b. State the other x-intercept (if any).

22. Write the new equation for the parabola $y = x^2$ that has been:

   a. reflected in the x-axis
   b. dilated by a factor of 7 away from the x-axis
   c. translated 3 units in the negative direction of the x-axis
   d. translated 6 units in the positive direction of the y-axis
   e. dilated by a factor of $\dfrac{1}{4}$ from the x-axis, reflected in the x-axis, and translated 5 units in the positive direction of the x-axis and 3 units in the negative direction of the y-axis.

Reasoning

23. The price of shares in fledgling company 'Lollies'r'us' plunged dramatically one afternoon, following the breakout of a small fire on the premises. However, Ms Sarah Sayva of Lollies Anonymous agreed to back the company, and share prices began to rise.
   Sarah noted at the close of trade that afternoon that the company's share price followed the curve:
   $P = 0.1(t - 3)^2 + 1$ where P is the price of shares t hours after noon.

   a. Sketch a graph of the relationship between time and share price to represent the situation.
   b. Determine the initial share price.
   c. Determine the lowest price of shares that afternoon.
   d. Evaluate the time when the price was at its lowest.
   e. Determine the final price of 'Lollies'r'us' shares as trade closed at 5 pm.

24. Rocky is practising for a football kicking competition. After being kicked, the path that the ball follows can be modelled by the quadratic relationship:

$$h = -\frac{1}{30}(d - 15)^2 + 8$$

   where h is the vertical distance the ball reaches (in metres), and d is the horizontal distance (in metres).

   a. Determine the initial vertical height of the ball.
   b. Determine the exact maximum horizontal distance the ball travels.
   c. Write down both the maximum height and the horizontal distance when the maximum height is reached.

25. Answer the following questions.

   a. If the turning point of a particular parabola is $(2, 6)$, suggest a possible equation for the parabola.
   b. If the y-intercept in part a is $(0, 4)$, determine the exact equation for the parabola.

26. Answer the following questions.

   a. If the turning point of a particular parabola is (p, q), suggest a possible equation for the parabola.
   b. If the y-intercept in part a is $(0, r)$, determine the exact equation for the parabola.

Problem solving

27. Use the completing the square method to write each of the following in turning point form and sketch the parabola for each part.

   a. $y = x^2 - 8x + 1$ b. $y = x^2 + 4x - 5$ c. $y = x^2 + 3x + 2$

28. Use the information given in the graph shown to answer the following questions.

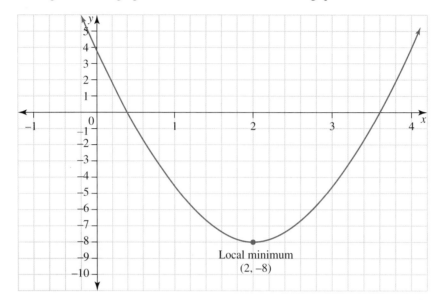

   a. Determine the equation of the parabola shown.
   b. State the dilation and translation transformations that have been applied to $y = x^2$ to achieve this parabola.
   c. This graph is reflected in the x-axis. Determine the equation of the reflected graph.
   d. Sketch the graph of the reflected parabola.

29. The graph of a quadratic equation has a turning point at $(-3, 8)$ and passes through the point $(-1, 6)$.

   a. Determine the equation of this parabola.
   b. State the transformations that have been applied to $y = x^2$ to produce this parabola.
   c. Calculate the x and y-intercepts of this parabola.
   d. The graph is reflected in the x-axis, dilated by a factor of 2 from the x-axis, and then reflected in the y-axis. Sketch the graph of this new parabola.

LESSON
9.5 Sketching parabolas in expanded form

LEARNING INTENTIONS

At the end of this lesson you should be able to:
- determine the x-intercept/s, y-intercept and turning point of a quadratic equation and sketch its graph by first factorising the equation
- determine the x-intercept/s, y-intercept and turning point of a quadratic equation and sketch its graph using the quadratic formula.

9.5.1 Parabolas of the form $y = ax^2 + bx + c$

- The general form of a quadratic equation is $y = ax^2 + bx + c$ where a, b and c are constants.
- The x-intercepts can be found by letting $y = 0$, factorising and using the Null Factor Law to solve for x.
- The y-intercept can be found by letting $x = 0$ and solving for y.
- The x-coordinate of the turning point lies midway between the x-intercepts.
- Once the midpoint of the x-intercepts is found, this value can be substituted into the original equation to find the y-coordinate of the turning point.
- Once these intercepts and turning point have been found, it is possible to sketch the parabola.
- If an equation is not written in turning point form, and cannot be readily factorised, then we will need to use the quadratic formula to help find all key points.

Quadratic formula

A quadratic of the form $y = ax^2 + bx + c$ has:
- **a y-intercept at the coordinate $(0, c)$.**
- **x-intercepts that can be found using the quadratic formula; that is, when the equation $ax^2 + bx + c = 0$**

$$x = \frac{-b \pm \sqrt{b^2 - 4ac}}{2a}$$

- **a turning point at the coordinate $\left(-\dfrac{b}{2a}, c - \dfrac{b^2}{4a}\right)$.**

For example, given the equation $y = x^2 + 4x - 6$:
- In this example we have $a = 1, b = 4$ and $c = -6$.
- The y-intercept is $(0, c) = (0, -6)$
- The x-intercepts are given by:

$$x = \frac{-b \pm \sqrt{b^2 - 4ac}}{2a} = \frac{-(4) \pm \sqrt{(4)^2 - 4 \times (1) \times (-6)}}{2 \times (1)} = \frac{-4 \pm \sqrt{40}}{2} = \frac{-4 \pm 2\sqrt{10}}{2} = -2 \pm \sqrt{10}$$

- Therefore, the x-intercepts are $\left(-2 + \sqrt{10}, 0\right)$ and $\left(-2 - \sqrt{10}, 0\right)$.

- The turning point is given by $\left(-\dfrac{b}{2a}, c - \dfrac{b^2}{4a}\right) = \left(-\dfrac{(4)}{2 \times (1)}, -6 - \dfrac{(4)^2}{4 \times (1)}\right) = (-2, -10)$.

Note: Do not convert answer to decimals unless specified by the question. It is always best practice to leave coordinates in exact form.

WORKED EXAMPLE 11 Sketching a factorised quadratic equation

Sketch the graph of $y = (x - 3)(x + 2)$.

THINK	WRITE/DRAW
1. The equation is in factorised form. To calculate the x-intercepts, let $y = 0$ and use the Null Factor Law.	$y = (x - 3)(x + 2)$ $0 = (x - 3)(x + 2)$ $x - 3 = 0$ or $x + 2 = 0$ (NFL) $\quad x = 3$ or $\quad\quad x = -2$ x-intercepts: $(3, 0)\ (-2, 0)$
2. The x-coordinate of the turning point is midway between the x-intercepts. Calculate the average of the two x-intercepts to determine the midpoint between them.	$x_{TP} = \dfrac{3 + (-2)}{2}$ $\quad\ = 0.5$
3. • To calculate the y-coordinate of the turning point, substitute x_{TP} into the equation. • State the turning point.	$y = (x - 3)(x + 2)$ $y_{TP} = (0.5 - 3)(0.5 + 2)$ $\quad\ = -6.25$ Turning point: $(0.5, -6.25)$
4. • To calculate the y-intercept, let $x = 0$ and substitute. • State the y-intercept.	$y = (0 - 3)(0 + 2)$ $\quad\ = -6$ y-intercept: $(0, -6)$
5. • Sketch the graph, showing all the important features. • Label the graph.	

WORKED EXAMPLE 12 Sketching a quadratic in expanded form

Sketch the graph of $y = 2x^2 - 6x - 6$.

THINK	WRITE/DRAW
1. The equation is not in factorised form, but there is a common factor of 2. Take out the common factor of 2.	$y = 2x^2 - 6x - 6$ $\quad = 2(x^2 - 3x - 3)$

---

2. The equation cannot be factorised (no factors of -3 add to -3), so use completing the square to write the equation in turning point form.
   - Halve and then square the coefficient of x.
   - Add this and then subtract it from the equation.
   - Collect the terms for and write the perfect square.
   - Simplify the brackets to write the equation in turning point form.
   - Identify the coordinates of the turning point (h, k).

$$y = 2\left(x^2 - 3x + \left(\frac{3}{2}\right)^2 - \left(\frac{3}{2}\right)^2 - 3\right)$$

$$= 2\left(\left[x - \frac{3}{2}\right]^2 - \left(\frac{3}{2}\right)^2 - 3\right)$$

$$= 2\left(\left[x - \frac{3}{2}\right]^2 - \frac{9}{4} - 3\right)$$

$$= 2\left(\left[x - \frac{3}{2}\right]^2 - \frac{21}{4}\right)$$

$$= 2\left(x - \frac{3}{2}\right)^2 - 2 \times \frac{21}{4}$$

$$= 2\left(x - \frac{3}{2}\right)^2 - \frac{21}{2}$$

Turning point : $\left(\dfrac{3}{2}, \dfrac{-21}{2}\right)$

3.
   - To calculate the x-intercepts, let $y = 0$. No factors of -3 add to -3, so use the quadratic formula to calculate the x-intercepts.

x-intercepts: let $y = 0$.
$$0 = 2x^2 - 6x - 6$$

$$x = \frac{-b \pm \sqrt{b^2 - 4ac}}{2a}$$

where $a = 2, b = -6, c = -6$

$$x = \frac{-(-6) \pm \sqrt{(-6)^2 - 4 \times 2 \times (-6)}}{2(2)}$$

$$= \frac{6 \pm \sqrt{36 + 48}}{4}$$

$$= \frac{6 \pm \sqrt{84}}{4} = \frac{6 \pm 2\sqrt{21}}{4}$$

   - State the x-intercepts.

The x-intercepts are:
$$x = \frac{3 + \sqrt{21}}{2} \quad \text{and} \quad x = \frac{3 - \sqrt{21}}{2}$$

$x \approx 3.79 \qquad\qquad x \approx -0.79$

4.
   - To calculate the y-intercepts, let $x = 0$ and substitute.
   - State the y-intercept.

$$y = 2x^2 - 6x - 6$$
$$y = 2(0)^2 - 6(0) - 6$$
$$= -6$$
y-intercept: $(0, -6)$

5.
- Sketch the graph, showing all the important features.
- Label the graph and show the exact values of the x-intercepts.

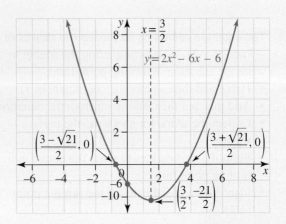

TI | THINK

1. On a Graphs page, complete the function entry line as:
$$f1(x) = 2x^2 - 6x - 6$$
The graph will be displayed.

DISPLAY/WRITE

2. To locate the turning point, press:
- MENU
- 6: Analyze Graph
- 2: Minimum

Drag the dotted line to the left of the turning point (the lower bound), click ENTER and then drag the dotted line to the right of the turning point (the upper bound) and press ENTER.

To locate the intercepts, press:
- MENU
- 6: Analyze Graph
- 1: Zero

Locate the points as described above.

The points of interest will be shown.

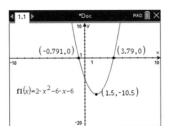

The graph is shown, along with the critical points.

CASIO | THINK

1. On a Graph & Table screen, complete the function entry line as:
$$y1 = 2x^2 - 6x - 6$$
Tap the graphing icon and the graph will be displayed.

DISPLAY/WRITE

2. To locate the turning point, press:
- Analysis
- G-Solve
- Min

To locate the turning point, press:
- Analysis
- G-Solve
- Root

To locate the second root, tap the right arrow.
The points of interest will be shown.

The turning point is $(1.5, 10.5)$.
The x-intercepts are $(-0.791, 0)$ and $(3.791, 0)$.

DISCUSSION

What strategy can you use to remember all of the information necessary to sketch a parabola?

Exercise 9.5 Sketching parabolas in expanded form   learn

9.5 Quick quiz on	9.5 Exercise

Individual pathways

■ PRACTISE	■ CONSOLIDATE	■ MASTER
1, 2, 5, 8, 11, 14	3, 6, 9, 12, 15	4, 7, 10, 13, 16

Fluency

1. What information is necessary to be able to sketch a parabola?

WE11 For questions **2** to **4**, sketch the graph of each of the following.

2. **a.** $y = (x - 5)(x - 2)$ **b.** $y = (x + 4)(x - 7)$

3. **a.** $y = (x + 3)(x + 5)$ **b.** $y = (2x + 3)(x + 5)$

4. **a.** $y = (4 - x)(x + 2)$ **b.** $y = \left(\dfrac{x}{2} + 3\right)(5 - x)$

WE12 For questions **5** to **7**, sketch the graph of each of the following.

5. **a.** $y = x^2 + 4x + 2$ **b.** $y = x^2 - 4x - 5$ **c.** $y = 2x^2 - 4x - 3$

6. **a.** $y = -2x^2 + 11x + 5$ **b.** $y = -2x^2 + 12x$ **c.** $y = 3x^2 + 6x + 1$

7. **a.** $y = -3x^2 - 5x + 2$ **b.** $y = 2x^2 + 8x - 10$ **c.** $y = -3x^2 + 7x + 3$

Understanding

8. The path of a soccer ball kicked by the goal keeper can be modelled by the equation $y = -\dfrac{1}{144}(x^2 - 24x)$ where y is the height of the soccer ball and x is the horizontal distance from the goalie, both in metres.

   a. Sketch the graph.
   b. Calculate how far away from the player does the ball first bounce.
   c. Calculate the maximum height of the ball.

9. The monthly profit or loss, p, (in thousands of dollars) for a new brand of chicken loaf is given by $p = 3x^2 - 15x - 18$, where x is the number of months after its introduction (when $x = 0$).

   a. Sketch the graph.
   b. Determine during which month a profit was first made.
   c. Calculate the month in which the profit is $54\,000$.

10. The height, h metres, of a model rocket above the ground t seconds after launch is given by the equation $h = 4t(50 - t)$, where $0 \leq t \leq 50$.

    a. Sketch the graph of the rocket's flight.
    b. State the height of the rocket above the ground when it is launched.
    c. Calculate the greatest height reached by the rocket.
    d. Determine how long the rocket takes to reach its greatest height.
    e. Determine how long the rocket is in the air.

Reasoning

11. The equation $y = x^2 + bx + 7500$ has x-intercepts of $(-150, 0)$ and $(-50, 0)$. Determine the value of b in the equation. Justify your answer.

12. The equation $y = x^2 + bx + c$ has x-intercepts of m and n. Determine the value of b in the equation. Justify your answer.

13. A ball thrown from a cliff follows a parabolic path of the form $y = ax^2 + bx + c$. The ball is released at the point $(0, 9)$, reaches a maximum height at $(2, 11)$ and passes through the point $(6, 3)$ on its descent.
    Determine the equation of the ball's path. Show full working.

Problem solving

14. A ball is thrown upwards from a building and follows the path given by the formula $h = -x^2 + 4x + 21$. The ball is h metres above the ground when it is a horizontal distance of x metres from the building.

    a. Sketch the graph of the path of the ball.
    b. Determine the maximum height of the ball.
    c. Determine how far the ball is from the wall when it reaches the maximum height.
    d. Determine how far from the building the ball lands.

15. During an 8-hour period, an experiment is done in which the temperature of a room follows the relationship $T = h^2 - 8h + 21$, where T is the temperature in degrees Celsius h hours after starting the experiment.

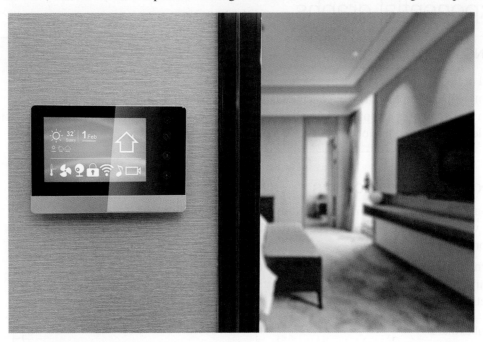

   a. Sketch the graph of this quadratic.
   b. Identify the initial temperature.
   c. Determine if the temperature is increasing or decreasing after 3 hours.
   d. Determine if the temperature is increasing or decreasing after 5 hours.
   e. Determine the minimum temperature and when it occurred.
   f. Determine the temperature after 8 hours.

16. A ball is thrown out of a window, passing through the points $(0, 7), (6, 18)$ and $(18, 28)$. A rule for the height of the ball in metres is given by $h = ax^2 + bx + c$ where x is the horizontal distance, in metres, covered by the ball.

   a. Determine the values of a, b and c in the rule for the height of the ball.
   b. Calculate the height that the ball was thrown from.
   c. Evaluate the maximum height reached by the ball.
   d. Determine horizontal distance covered by the ball when it hits the ground.
   e. Sketch the flight path of the ball, making sure you show all key points.

LESSON
9.6 Exponential graphs

▶ 9.6.1 Exponential functions

eles-4873

- Relationships of the form $y = a^x$ are called **exponential functions** with base a, where a is a real number not equal to 1, and x is the index power or exponent.

- The term 'exponential' is used, as x is an exponent (or index).
  For example, the graph of the exponential function $y = 2^x$ can be plotted by completing a table of values.

  Remember that $2^{-3} = \dfrac{1}{2^3}$

  $= \dfrac{1}{8}$, and so on.

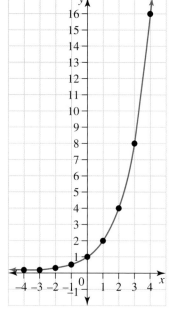

x	-4	-3	-2	-1	0	1	2	3	4
y	$\dfrac{1}{16}$	$\dfrac{1}{8}$	$\dfrac{1}{4}$	$\dfrac{1}{2}$	1	2	4	8	16

- For an exponential function, the ratio of consecutive y-values is always constant. In the table above the ratio of consecutive y-values is 2.
- The graph has many significant features.
  - The y-intercept is 1.
  - The value of y is always greater than zero.
  - As x decreases, y gets closer to but never reaches zero. So the graph gets closer to but never reaches the x-axis. The x-axis (or the line $y = 0$) is called an **asymptote.**
  - As x increases, y becomes very large.

Comparing exponential graphs

- The diagram at right shows the graphs of $y = 2^x$ and $y = 3^x$.
- The graphs both pass through the point $(0, 1)$.
- The graph of $y = 3^x$ climbs more steeply than the graph of $y = 2^x$.
- $y = 0$ is an asymptote for both graphs.

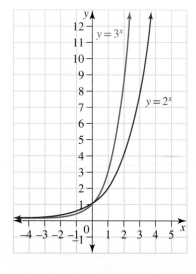

Translation and reflection

Vertical translation	Reflection about the *x*-axis	Reflection about the *y*-axis
		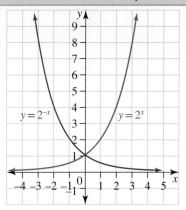

Vertical translation

- The diagram shows the graphs of $y = 2^x$ and $y = 2^x + 3$.
- The graphs have identical shape.
- Although they appear to get closer to each other, the graphs are constantly 3 vertical units apart.
- As x becomes very small, the graph of $y = 2^x + 3$ approaches but never reaches the line $y = 3$, so $y = 3$ is the horizontal asymptote.
- When the graph of $y = 2^x$ is translated 3 units upward, it becomes the graph of $y = 2^x + 3$.

Reflection about the *x*-axis

- The diagram shows the graphs of $y = 2^x$ and $y = -2^x$.
- The graphs have identical shape.
- The graph of $y = -2^x$ is a reflection about the *x*-axis of the graph of $y = 2^x$.
- The *x*-axis ($y = 0$) is an asymptote for both graphs.
- In general, the graph of $y = -a^x$ is a reflection about the *x*-axis of the graph of $y = a^x$.

Reflection about the *y*-axis

- The diagram shows the graphs of $y = 2^x$ and $y = 2^{-x}$.
- The graphs have identical shape.
- The graph of $y = 2^{-x}$ is a reflection about the *y*-axis of the graph of $y = 2^x$.
- Both graphs pass through the point (0, 1).
- The *x*-axis ($y = 0$) is an asymptote for both graphs.
- In general, the graph of $y = a^{-x}$ is a reflection about the *y*-axis of the graph of $y = a^x$.

WORKED EXAMPLE 13 Sketching exponential graphs

Given the graph of $y = 4^x$, sketch on the same axes the graphs of:

a. $y = 4^x - 2$

b. $y = -4^x$

c. $y = 4^{-x}$.

THINK

a. The graph of $y = 4^x$ has already been drawn. It has a y-intercept of 1 and a horizontal asymptote at $y = 0$. The graph of $y = 4^x - 2$ has the same shape as $y = 4^x$ but is translated 2 units vertically down. It has a y-intercept of -1 and a horizontal asymptote at $y = -2$.

b. $y = -4^x$ has the same shape as $y = 4^x$ but is reflected about the x-axis. It has a y-intercept of -1 and a horizontal asymptote at $y = 0$.

c. $y = 4^{-x}$ has the same shape as $y = 4^x$ but is reflected about the y-axis. The graphs have the same y-intercept and the same horizontal asymptote ($y = 0$).

DRAW

a.

b.

c.

▶ 9.6.2 Combining transformations of exponential graphs

eles-4874

* It is possible to combine translations, dilations and reflections in one graph.

WORKED EXAMPLE 14 Sketching exponentials with multiple transformations

By considering transformations to the graph of $y = 2^x$, sketch the graph of $y = -2^x + 1$.

THINK

1. Start by sketching $y = 2^x$.
   It has a y-intercept of 1 and a horizontal asymptote at $y = 0$.
2. Sketch $y = -2^x$ by reflecting $y = 2^x$ about the x-axis.
   It has a y-intercept of -1 and a horizontal asymptote at $y = 0$.
3. Sketch $y = -2^x + 1$ by translating $y = -2^x$ upwards by 1 unit.
   The graph has a y-intercept of 0 and a horizontal asymptote at $y = 1$.

DRAW

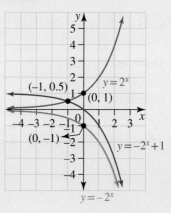

TI | THINK

1. In a new problem, on a Graphs page, complete the function entry line as:
   $f1(x) = 2^x$
   Then press ENTER.

DISPLAY/WRITE

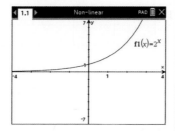

CASIO | THINK

Open the Graph & Table screen and complete the function entry line as:
$y1 = 2^x$
$y2 = -2^x$
$y3 = -2^x + 1$
Then tap the graphing icon and the graphs will be displayed.

DISPLAY/WRITE

2. Press TAB and complete the function entry line as:
   $f2(x) = -f1(x)$
   Then press ENTER.
   Press TAB and complete the function entry line as:
   $f3(x) = f2(x) + 1$
   Then press ENTER.

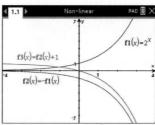

The graph of $y = -2^x$ is the reflection of the graph of $y = 2^x$ in the x-axis.
The graph of $y = -2^x + 1$ is the graph of $y = -2^x$ translated upwards by 1 unit.
The graph of $y = -2^x + 1$ passes through the origin and has a horizontal asymptote at $y = 1$.

From $y1$ to $y2$ the graph has undergone a reflection about the x-axis. From $y2$ to $y3$ the graph has undergone a translation upwards by 1 unit.

DISCUSSION

Explain whether the graph of an exponential function will always have a horizontal asymptote.

Exercise 9.6 Exponential graphs   **learn**on

9.6 Quick quiz on	9.6 Exercise

Individual pathways

■ PRACTISE	■ CONSOLIDATE	■ MASTER
1, 6, 8, 10, 13, 15, 18	2, 4, 7, 11, 14, 16, 19	3, 5, 9, 12, 17, 20

Fluency

1. Complete the following table and use it to plot the graph of $y = 3^x$, for $-3 \le x \le 3$.

x	-3	-2	-1	0	1	2	3
y							

2. If $x = 1$, calculate the value of y when:

   a. $y = 2^x$ b. $y = 3^x$ c. $y = 4^x$

3. If $x = 1$, calculate the value of y when:

   a. $y = 10^x$ b. $y = a^x$.

4. Using a calculator or graphing program, sketch the graphs of $y = 2^x$, $y = 3^x$ and $y = 4^x$ on the same set of axes.

   a. Describe the common features among the graphs.
   b. Describe how the value of the base $(2, 3, 4)$ affects the graph.
   c. Predict where the graph $y = 8^x$ would lie and sketch it in.

5. Using graphing technology, sketch the following graphs on one set of axes.
   $y = 3^x$, $y = 3^x + 2$, $y = 3^x + 5$ and $y = 3^x - 3$

   a. State what remains the same in all of these graphs.
   b. State what is changed.
   c. For the graph of $y = 3^x + 10$, write down:
      i. the y-intercept
      ii. the equation of the horizontal asymptote.

6. Using graphing technology, sketch the graphs of:

   a. $y = 2^x$ and $y = -2^x$
   b. $y = 3^x$ and $y = -3^x$
   c. $y = 6^x$ and $y = -6^x$.
   d. State the relationship between these pairs of graphs.

7. Using graphing technology, sketch the graphs of:
   a. $y = 2^x$ and $y = 2^{-x}$
   b. $y = 3^x$ and $y = 3^{-x}$
   c. $y = 6^x$ and $y = 6^{-x}$
   d. State the relationship between these pairs of graphs.

8. **WE13** Given the graph of $y = 2^x$, sketch on the same axes the graphs of:
   a. $y = 2^x + 6$
   b. $y = -2^x$
   c. $y = 2^{-x}$.

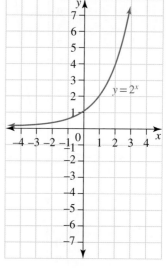

9. Given the graph of $y = 3^x$, sketch on the same axes the graphs of:
   a. $y = 3^x + 2$
   b. $y = -3^x$.

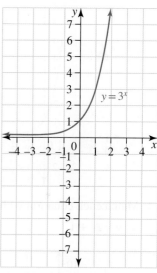

10. Given the graph of $y = 4^x$, sketch on the same axes the graphs of:
    a. $y = 4^x - 3$
    b. $y = 4^{-x}$.

Understanding

11. **WE14** By considering transformations of the graph of $y = 2^x$, sketch the following graphs on the same set of axes.
    a. $y = 2^{-x} + 2$
    b. $y = -2^x + 3$

12. By considering transformations of the graph of $y = 5^x$, sketch the following graphs on the same set of axes.
    a. $y = -5^x + 10$
    b. $y = 5^{-x} + 10$

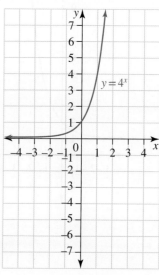

13. Match each equation with its correct graph.

A. $y = 2^x$ B. $y = 3^x$ C. $y = -4^x$ D. $y = 5^{-x}$

a.

b.

c.

d.

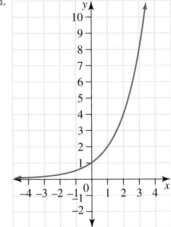

14. Match each equation with its correct graph. Explain your answer.

A. $y = 2^x + 1$ B. $y = 3^x + 1$ C. $y = -2^x + 1$ D. $y = 2^{-x} + 1$

a.

b.

c.

d.

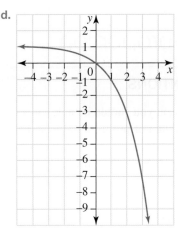

Reasoning

15. By considering transformations of the graph of $y = 3^x$, sketch the graph of $y = -3^{-x} - 3$.

16. The graph of $f(x) = 16^x$ can be used to solve for x in the exponential equation $16^x = 32$. Sketch a graph of $f(x) = 16^x$ and use it to solve $16^x = 32$.

17. The graph of $f(x) = 6^{x-1}$ can be used to solve for x in the exponential equation $6^{x-1} = 36\sqrt{6}$. Sketch a graph of $f(x) = 6^{x-1}$ and use it to solve $6^{x-1} = 36\sqrt{6}$.

Problem solving

18. Sketch the graph of $y = -2^{-x} + 2$.

19. The number of bacteria, N, in a certain culture is reduced by a third every hour so

$$N = N_0 \times \left(\frac{1}{3}\right)^t$$

where t is the time in hours after 12 noon on a particular day. Initially there are 10 000 bacteria present.
a. State the value of N_0.
b. Calculate the number of bacteria, correct to the nearest whole number, in the culture when:

   i. $t = 2$ ii. $t = 5$ iii. $t = 10$.

20. a. The table shows the population of a city between 1850 and 1930. Explain if the population growth is exponential.

Year	1850	1860	1870	1880	1890	1900	1910	1920	1930
Population (million)	1.0	1.3	1.69	2.197	2.856	3.713	4.827	6.275	8.157

b. Determine the common ratio in part **a**.
c. Evaluate the percentage increase every ten years.
d. Estimate the population in 1895.
e. Estimate the population in 1980.

LESSON
9.7 Inverse proportion

▶ 9.7.1 Inverse proportion

eles-4875

- If 24 sweets are shared between 4 children, then each child will receive 6 sweets. If the sweets are shared by 3 children, then each will receive 8 sweets.
- The relationship between the number of children (C) and the number sweets for each child (n) can be given in a table.

C	1	2	3	4	6	8	12
n	24	12	8	6	4	3	2

- As the number of children (C) increases, the number of sweets for each child (n) decreases. This is an example of **inverse proportion** or inverse variation.
- We say that 'n is inversely proportional to C' or 'n varies inversely as C'.
- This is written as $C \propto \dfrac{1}{n}$, or $C = \dfrac{k}{n}$, where k is a constant (the **constant of proportionality**). This formula can be rearranged to $Cn = k$. Note that multiplying any pair of values in the table $(3 \times 8, 12 \times 2)$ gives the same result.
- The relationship has some important characteristics:
  - As C increases, n decreases, and vice versa.
  - The graph of the relationship is a **hyperbola**.

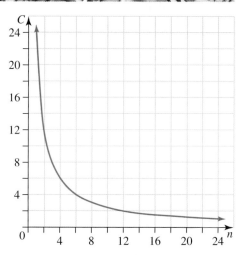

WORKED EXAMPLE 15 Calculating the constant of proportionality

y is inversely proportional to x and $y = 10$ when $x = 2$.
a. **Calculate the constant of proportionality, k, and hence the rule relating x and y.**
b. **Plot a graph of the relationship between x and y, for values of x from 2 to 10.**

THINK	WRITE/DRAW
a. 1. Write the relationship between the variables.	a. $y \propto \dfrac{1}{x}$
2. Rewrite as an equation using k, the constant of proportionality	$y = \dfrac{k}{x}$

---

3. Substitute $y = 10, x = 2$, into $y = \dfrac{k}{x}$ and solve for k

$$10 = \dfrac{k}{2}$$

$$k = 20$$

4. Write the rule by substituting $k = 20$ into $y = \dfrac{k}{x}$.

$$y = \dfrac{20}{x}$$

b. 1. Use the rule $y = \dfrac{20}{x}$ to set up a table of values for x and y, taking values for x which are positive factors of k so that only whole number values of y are obtained. For example, $x = 4, y = \dfrac{20}{4} = 5$.

b.

x	2	4	6	8	10
y	10	5	3.3	2.5	2

2. Plot the points on a clearly labelled set of axes and join the points with a smooth curve. Label the graph.

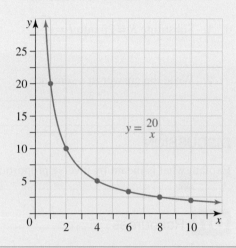

WORKED EXAMPLE 16 Determining the rule for an inversly proportional relationship

When a wire is connected to a power source, the amount of electrical current (I) passing through the wire is inversely proportional to the resistance (R) of the wire. If a current of 0.2 amperes flows through a wire of resistance 60 ohms:

a. calculate the constant of proportionality
b. determine the rule relating R and I
c. calculate the resistance if the current equals 5 amperes
d. determine the current that will flow through a wire of resistance 20 ohms.

THINK

Summarise the information in a table.

$I \propto \dfrac{1}{R}$

Write the rule.

a. 1. Substitute $R = 60, I = 0.2$, into $I = \dfrac{k}{R}$.

2. Solve for k.

WRITE

R	60		20
I	0.2	5	

$$I = \dfrac{k}{R}$$

a. $$0.2 = \dfrac{k}{60}$$

$$0.2 \times 60 = k$$

$$k = 12$$

b. Write the rule using $k = 12$.

b. $I = \dfrac{12}{R}$

c. 1. Substitute $I = 5$ into $I = \dfrac{12}{R}$.

c. $5 = \dfrac{12}{R}$

$5R = 12$

$R = \dfrac{12}{5}$

2. Solve for R.

$= 2.4$

3. Write the answer in a sentence.

The resistance equals 2.4 ohms.

d. 1. Substitute $R = 20$ into $I = \dfrac{12}{R}$.

d. $I = \dfrac{12}{20}$

$= 0.6$

2. Write the answer in a sentence.

The current will be 0.6 amperes.

DISCUSSION

Explain what is meant by inverse proportion.

 Resources

 eWorkbook   Topic 9 Workbook (worksheets, code puzzle and a project) (ewbk-2035)

 Interactivities Individual pathway interactivity: Inverse proportion (int-4512)

Inverse proportion (int-6058)

Exercise 9.7 Inverse proportion

learnon

9.7 Quick quiz on	9.7 Exercise

Individual pathways

■ PRACTISE	■ CONSOLIDATE	■ MASTER
1, 3, 6, 10, 13	2, 4, 7, 11, 14	5, 8, 9, 12, 15

Fluency

1. Decide whether inverse proportion exists between each pair of variables. If it does exist, write an equation to describe the relationship.

   a. The speed of a car (s) and the time (t) it takes to complete one lap of a race circuit.
   b. The amount of money (D) that I have and the number (n) of cards that I can buy.
   c. The time (t) that it takes to make a pair of jeans and the number of pairs (p) that can be made in one day.
   d. The price (P) of petrol and the amount (L) that can be bought for \$80.
   e. The price (P) of petrol and the cost (C) of buying 80 L.
   f. The number of questions (n) in a test and the amount of time (t) available to answer each one.

2. List three examples of inverse proportion.

3. **WE15** y varies inversely as x and $y = 100$ when $x = 10$.

   a. Calculate the constant of proportionality, k, and hence the rule relating x and y.
   b. Plot a graph of the relationship between x and y, for values of x that are positive factors of k less than 21.

4. p is inversely proportional to q and $p = 12$ when $q = 4$.

   a. Calculate the constant of proportionality, k, and hence the rule relating p and q.
   b. Plot a graph of the relationship between q and p, for values of q that are positive factors of k less than 11.

5. y varies inversely as x and $y = 42$ when $x = 1$.

   a. Calculate the constant of proportionality, k, and hence the rule relating x and y.
   b. Plot a graph of the relationship between x and y, for values of x from 1 to 10.

Understanding

6. **WE16** When a constant force is applied to an object, its acceleration is inversely proportional to its mass. When the acceleration of an object is $40 \, \text{m/s}^2$, the corresponding mass is $100 \, \text{kg}$.

   a. Calculate the constant of proportionality.
   b. Determine the rule relating mass and acceleration.
   c. Determine the acceleration of a $200 \, \text{kg}$ object.
   d. Determine the acceleration of a $1000 \, \text{kg}$ object.

7. The number of colouring pencils sold is inversely proportional to the price of each pencil.
   Two thousand pencils are sold when the price is $0.25 each.

   a. Calculate the constant of proportionality.
   b. Determine the number of pencils that could be sold for $0.20 each.
   c. Determine the number of pencils that could be sold for $0.50 each.

8. The time taken to complete a journey is inversely proportional to the speed travelled. A trip is completed in 4.5 hours travelling at 75 km per hour.

   a. Calculate the constant of proportionality.
   b. Determine how long, to the nearest minute, the trip would take if the speed was 85 km per hour.
   c. Determine the speed required to complete the journey in 3.5 hours, correct to 1 decimal place.
   d. Determine the distance travelled in each case.

9. The cost per person travelling in a charter plane is inversely proportional to the number of people in the charter group. It costs $350 per person when 50 people are travelling.

   a. Calculate the constant of variation.
   b. Determine the cost per person, to the nearest cent, if there are 75 people travelling.
   c. Determine how many people are required to reduce the cost to $250 per person.
   d. Determine the total cost of hiring the charter plane.

Reasoning

10. The electrical current in a wire is inversely proportional to the resistance of the wire to that current. There is a current of 10 amperes when the resistance of the wire is 20 ohms.

    a. Calculate the constant of proportionality.
    b. Determine the current possible when the resistance is 200 ohms.
    c. Determine the resistance of the wire when the current is 15 amperes.
    d. Justify your answer to parts b and c using a graph.

11. The pressure of an ideal gas is inversely proportional to the volume taken up by the gas. A balloon is filled with air so it takes up 3 L at a pressure of 5 atmospheres.

    a. Calculate the constant of proportionality.
    b. Determine the new volume of the balloon if the pressure was dropped to 0.75 atmospheres.
    c. Determine the pressure if the same amount of air took up a volume of 6 L.

12. Two equations relating the time of a trip, T, and the speed at which they travel, S, are given. For both cases the time is inversely proportional to the speed: $T_1 = \dfrac{5}{S_1}$ and $T_2 = \dfrac{7}{S_2}$. Explain what impact the different constants of proportionality have on the time of the trip.

Problem solving

13. The time it takes to pick a field of strawberries is inversely proportional to the number of pickers. It takes 2 people 5 hours to pick all of the strawberries in a field.

    a. Calculate the constant of proportionality.
    b. Determine the rule relating time (T) and the number of pickers (P).
    c. Determine the time spent if there are 6 pickers.

14. For a constant distance covered by a sprinter, the sprinter's speed is inversely proportional to their time. If a sprinter runs at a speed of 10.4 m/s, the corresponding time is 9.62 seconds.

    a. Calculate the constant of variation.
    b. Determine the rule relating speed (V) and time (T).
    c. Determine the time, correct to 2 decimal places, if they ran at a speed of 10.44 m/s.
    d. Determine the time, correct to 2 decimal places, if they ran at a speed of 6.67 m/s.

15. A holiday hostel is built to accommodate group bookings of up to 45 people. It is known that it would cost each of the individuals in a group of 20 people $67.50 per night to rent this venue. The cost of the venue will remain the same no matter how many people are part of the group booking.

    a. Write the rule for the cost per person (C) and the number of people in a group booking (n).
    b. If 6 people from the original group are no longer able to attend, determine the new cost per person.
    c. Calculate the cheapest possible cost per person.

LESSON
9.8 Sketching the graph of a reciprocal

▶ 9.8.1 Hyperbolas

eles-4876

- The graph of a reciprocal function is called a hyperbola.
- A hyperbola is a function of the form $xy = k$ or $y = \dfrac{k}{x}$; that is, the product of the two variables is a fixed constant.

WORKED EXAMPLE 17 Creating a table of values in order to graph a hyperbola

Complete the table of values below and use it to plot the graph of $y = \dfrac{1}{x}$.

x	-3	-2	-1	$-\dfrac{1}{2}$	0	$\dfrac{1}{2}$	1	2	3
y									

THINK

1. Substitute each x-value into the function $y = \dfrac{1}{x}$ to obtain the corresponding y-value.

WRITE/DRAW

x	-3	-2	-1	$-\dfrac{1}{2}$	0	$\dfrac{1}{2}$	1	2	3
y	$-\dfrac{1}{3}$	$-\dfrac{1}{2}$	-1	-2	Undef.	2	1	$\dfrac{1}{2}$	$\dfrac{1}{3}$

2. Draw a set of axes and plot the points from the table. Join them with a smooth curve.

- The graph in Worked example 17 has several important features.
  1. There is no function value (y-value) when $x = 0$. At this point the hyperbola is undefined. When this occurs, the line that the graph approaches ($x = 0$) is called a vertical asymptote.
  2. As x becomes larger and larger, the graph gets very close to but will never touch the x-axis. The same is true as x becomes smaller and smaller. The hyperbola also has a horizontal asymptote at $y = 0$.
  3. The hyperbola has two separate branches. It cannot be drawn without lifting your pen from the page and is an example of a discontinuous graph.
- Graphs of the form $y = \dfrac{k}{x}$ are the same basic shape as $y = \dfrac{1}{x}$ with y-values dilated by a factor of k.

General form of a hyperbola

The general form of a hyperbola with multiple transformations is given by the equation

$$y = \frac{a}{x-h} + k.$$

- This graph has a horizontal asymptote with equation $y = k$. This occurs because the fraction $\dfrac{a}{x-h}$ cannot be made to equal 0, which means y can never be equal to k.
- The graph has a vertical asymptote with equation $x = h$. This is because the denominator of a fraction cannot be equal to 0, so x can never be equal to h.
- The graph will be reflected in the x-axis if $a < 0$.

Note: Not all exponential graphs will have an x- or y-intercept. This will depend on the equation of the asymptotes.

WORKED EXAMPLE 18 Identifying the asymptotes of a hyperbola

a. Plot the graph of $y = \dfrac{4}{x}$ for $-2 \le x \le 2$.

b. Write down the equation of each asymptote.

THINK

a. 1. Prepare a table of values taking x-values from -2 to 2. Fill in the table by substituting each x-value into the given equation to find the corresponding y-value.

2. Draw a set of axes and plot the points from the table. Join them with a smooth curve.

b. Consider any lines that the curve approaches but does not cross.

WRITE/DRAW

a.

x	-2	-1	$-\dfrac{1}{2}$	0	$\dfrac{1}{2}$	1	2
y	-2	-4	-8	Undef.	8	4	2

b. Vertical asymptote is $x = 0$.
Horizontal asymptote is $y = 0$.

WORKED EXAMPLE 19 Sketching a reflected hyperbola

Plot the graph of $y = \dfrac{-3}{x}$ for $-3 \le x \le 3$.

THINK

1. Draw a table of values and substitute each x-value into the given equation to find the corresponding y-value.

WRITE/DRAW

x	-3	-2	-1	$-\dfrac{1}{2}$	0	$\dfrac{1}{2}$	1	2	3
y	1	1.5	3	6	Undef.	-6	-3	-1.5	-1

2. Draw a set of axes and plot the points from the table. Join them with a smooth curve.

$$y = -\frac{3}{x}$$

TI	THINK	DISPLAY/WRITE

TI | THINK

1. In a new problem, open a Graphs page to sketch the graph, complete the entry line as:
$$f1(x) = \frac{-3}{x}$$
Then press ENTER.

DISPLAY/WRITE

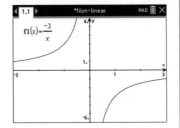

2. To see the table of values (for x and y values), press:
   • Menu
   • 7: Table
   • 1: Split-screen Table (Ctrl+T)
   Scroll up to see the negative x-values.

CASIO | THINK

1. Open the Graph & Table screen and complete the function entry line as:
$$f1(x) = \frac{-3}{x}$$
Then tap the graphing icon and the graph will be displayed.

DISPLAY/WRITE

2. To see the table of values (for x and y values), tap on the table of values icon. The values displayed can be changed by editing the Table Input.

WORKED EXAMPLE 20 Determining the equation of the asymptotes of a hyperbola

Write the equation of the asymptotes of the following hyperbolas:

a. $y = \dfrac{3}{x+2}$

b. $y = -\dfrac{1}{x} + 4$

c. $y = \dfrac{5}{10-x} - 1$

THINK

a. Consider what value of x would make the denominator equal to 0, as well as any value added to the fraction.

b. Consider what value of x would make the denominator equal to 0, as well as any value added to the fraction.

c. Consider what value of x would make the denominator equal to 0, as well as any value added to the fraction.

WRITE

a. Vertical asymptote is $x = -2$
   Horizontal asymptote is $y = 0$

b. Vertical asymptote is $x = 0$
   Horizontal asymptote is $y = 4$

c. Vertical asymptote is $x = 10$
   Horizontal asymptote is $y = -1$

Plot the graph of $y = -\dfrac{2}{x+1} + 3.$

THINK

1. Consider what value of x would make the denominator equal to 0, as well as any value added to the fraction.

2. Calculate the y-intercept (when $x = 0$).

3. Calculate the x-intercept (when $y = 0$).

4. Sketch the graph with coordinates of both intercepts and both asymptotes labelled.

WRITE

Vertical asymptote is $x = -1$
Horizontal asymptote is $y = 3$

$y = -\dfrac{2}{0+1} + 3 = 1$
y-intercept is $(0, 1)$

$0 = -\dfrac{2}{x+1} + 3$

$-3(x+1) = -2$

$x + 1 = \dfrac{-2}{-3}$

$x = \dfrac{2}{3} - 1 = -\dfrac{1}{3}$

The x-intercept is $\left(-\dfrac{1}{3}, 0\right)$ $x = -1$

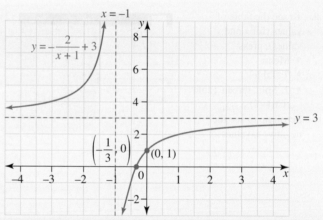

DISCUSSION

How could you summarise the effect of the transformations dealt with in this lesson on the shape of the basic hyperbola $y = \dfrac{1}{x}$?

| **9.8 Quick quiz** on | **9.8 Exercise** |

Individual pathways

■ PRACTISE	■ CONSOLIDATE	■ MASTER
1, 4, 7, 10, 14, 17, 20	2, 5, 8, 11, 15, 18, 21	3, 6, 9, 12, 13, 16, 19, 22, 23

Fluency

1. **WE17** Complete the table of values below and use it to plot the graph of $y = \dfrac{10}{x}$.

x	-5	-4	-3	-2	-1	0	1	2	3	4	5
y											

WE18 For questions 2 to 4, answer the following.
a. Plot the graph of each hyperbola.
b. Write down the equation of each asymptote.

2. $y = \dfrac{5}{x}$

3. $y = \dfrac{20}{x}$

4. $y = \dfrac{100}{x}$

5. Draw the graphs of $y = \dfrac{2}{x}$, $y = \dfrac{3}{x}$ and $y = \dfrac{4}{x}$, on the same set of axes.

6. Describe the effect of increasing the value of k on the graph of $y = \dfrac{k}{x}$.

7. **WE19** Plot the graph of $y = \dfrac{-10}{x}$ for $-5 \le x \le 5$.

8. Draw the graphs of $y = \dfrac{6}{x}$ and $y = \dfrac{-6}{x}$, on the same set of axes.

9. Describe the effect of the negative in $y = \dfrac{-k}{x}$.

10. Complete the table of values below and use the points to plot $y = \dfrac{1}{x-1}$. State the equation of the vertical asymptote.

x	-3	-2	-1	0	1	2	3	4
y								

Understanding

11. Plot the graph of each hyperbola and label the vertical asymptote.

a. $y = \dfrac{1}{x-2}$

b. $y = \dfrac{1}{x-3}$

12. Plot the graph of $y = \dfrac{1}{x+1}$ and label the vertical asymptote.

13. Describe the effect of a in $y = \dfrac{1}{x-a}$.

For questions **14** to **16**:
  i. **WE20** write the equations of the asymptotes of the following hyperbolas
  ii. sketch the graphs of the following hyperbolas.

14. a. $y = \dfrac{-4}{x+1}$ **b.** $y = \dfrac{2}{x-1}$

15. a. $y = \dfrac{5}{x+2}$ **b.** $y = \dfrac{3}{x+2} - 2$

16. a. $y = -\dfrac{4}{x+2} + 1$ **b.** $y = \dfrac{7}{3-x} + 5$

Reasoning

17. Give an example of the equation of a hyperbola that has a vertical asymptote of:

     a. $x = 3$ **b.** $x = -10$.

18. Give an example of a hyperbola that has the following key features.

     a. Asymptotes of $x = 2$ and $y = 3$
     b. Asymptotes of $x = -2$ and $y = 4$ and a y-intercept of -3

19. The graph of $y = \dfrac{1}{x}$ is reflected in the x-axis, dilated by a factor of 2 parallel to the y-axis or from the x-axis and translated 3 units to the left and down 1 unit. Determine the equation of the resulting hyperbola and give the equations of any asymptotes.

Problem solving

20. a. Complete the following table in order to graph the hyperbola defined by $y = \dfrac{1}{x^2}$.

x	-2	-1	$-\dfrac{1}{2}$	$\dfrac{1}{2}$	1	2
y						

     b. This hyperbola is also known as a truncus. Give the equations of any asymptotes.

     c. Determine the equation of the truncus which results when $y = \dfrac{1}{x^2}$ is reflected in the x-axis.

     d. Determine the equation of the truncus which results when $y = \dfrac{1}{x^2}$ is reflected in the y-axis.

21. The temperature of a cup of coffee as it cools is modelled by the equation $T = \dfrac{780}{t+10} + 22$, where T represents the temperature in °C and t is the time in minutes since the coffee was first made.

     a. State the initial temperature of the cup of coffee.
     b. Calculate the temperature, to 1 decimal place, of the coffee after it has been left to cool for an hour.
     c. A coffee will be too hot to drink unless its temperature has dropped below 50 °C. Determine how long someone would have to wait, to the nearest minute, before drinking the coffee.
     d. Explain whether the coffee will ever cool to 0 °C. Justify your answer.

22. Consider the equation $y = \dfrac{6}{x^2 - 4} + 2$.

    a. Calculate the x- and y-intercepts of the graph of the equation.

    b. Use your knowledge of quadratics and hyperbolas to state the equation of any asymptotes of the above equation.

    c. Using a table of values, or otherwise, sketch the graph of the equation.

23. Consider the truncus defined by $y = \dfrac{1}{x^2}$. This hyperbola is reflected in the x-axis, dilated by a factor of 3 parallel to the y-axis or from the x-axis and translated 1 unit to the left and up 2 units. Determine the equation of the resulting hyperbola and give the equations of any asymptotes.

LESSON
9.9 Sketching the circle

▶ 9.9.1 Circles

eles-5352

- A circle is the path traced out by a point at a constant distance (the radius) from a fixed point (the centre).
- Consider the circles shown. The first circle has its centre at the origin and radius r. Let $P(x, y)$ be a point on the circle.
  By Pythagoras: $x^2 + y^2 = r^2$.
  This relationship is true for all points, P, on the circle.

  The equation of a circle, with centre $(0, 0)$ and radius r is:

  $$x^2 + y^2 = r^2$$

- If the circle is translated h units to the right, parallel to the x-axis, and k units upwards, parallel to the y-axis, then the centre of the circle will become (h, k). The radius will remain unchanged.

Equation of a circle

- **The equation of a circle, with centre $(0, 0)$ and radius r, is:**

  $$x^2 + y^2 = r^2$$

- **The equation of a circle, with centre (h, k) and radius r, is:**

  $$(x - h)^2 + (y - k)^2 = r^2$$

Note: **We can produce an ellipse by dilating a circle from one or both of the axes.**

WORKED EXAMPLE 22 Identifying the centre and radius of a circle from its equation

Sketch the graph of $4x^2 + 4y^2 = 25$, stating the centre and radius.

THINK	WRITE/DRAW
1. Express the equation in general form by dividing both sides by 4.	$x^2 + y^2 = r^2$ $4x^2 + 4y^2 = 25$ $x^2 + y^2 = \dfrac{25}{4}$
2. State the coordinates of the centre.	Centre $(0, 0)$
3. Calculate the length of the radius by taking the square root of both sides. (Ignore the negative results.)	$r^2 = \dfrac{25}{4}$ $r = \dfrac{5}{2}$ Radius $= 2.5$ units
4. Sketch the graph.	

WORKED EXAMPLE 23 Sketching a circle from its equation

Sketch the graph of $(x - 2)^2 + (y + 3)^2 = 16$, clearly showing the centre and radius.

THINK	WRITE/DRAW
1. Express the equation in general form.	$(x - h)^2 + (y - k)^2 = r^2$ $(x - 2)^2 + (y + 3)^2 = 16$
2. State the coordinates of the centre.	Centre $(2, -3)$
3. State the length of the radius.	$r^2 = 16$ $r = 4$ Radius $= 4$ units
4. Sketch the graph.	

WORKED EXAMPLE 24 Completing the square to determine the equation of a circle

Sketch the graph of the circle $x^2 + 2x + y^2 - 6y + 6 = 0$.

THINK	WRITE/DRAW
1. Express the equation in general form by completing the square on the x terms and again on the y terms.	$(x-h)^2 + (y-k)^2 = r^2$ $x^2 + 2x + y^2 - 6y + 6 = 0$ $(x^2 + 2x + 1) - 1 + (y^2 - 6y + 9) - 9 + 6 = 0$ $(x+1)^2 + (y-3)^2 - 4 = 0$ $(x+1)^2 + (y-3)^2 = 4$
2. State the coordinates of the centre.	Centre $(-1, 3)$
3. State the length of the radius.	$r^2 = 4$ $r = 2$ Radius $= 2$ units
4. Sketch the graph.	

WORKED EXAMPLE 25 Sketching a circle including the intercepts

Sketch the graph of the circle $(x + 3)^2 + (y - 2)^2 = 25$. Make sure to show all axial intercepts.

THINK	WRITE
1. State the coordinate of the centre.	Centre $(-3, 2)$
2. State the length of the radius.	$r^2 = 25$ $r = 5$ Radius $= 5$ units
3. To determine the y-intercepts let $x = 0$.	$(0+3)^2 + (y-2)^2 = 25$ $(y-2)^2 = 25 - 9$ $(y-2)^2 = 16$ $y = \pm 4$ $y = 4 + 2$ or $4 + 2$ The y-intercepts are $(0, 6)$ and $(0, -2)$
4. To determine the x-intercepts let $y = 0$.	$(x+3)^2 + (0-2)^2 = 25$ $(x+3)^2 = 25 - 4$ $x + 3 = \pm\sqrt{21}$ $x = -3 \pm \sqrt{21}$ The x-intercepts are $\left(-3 + \sqrt{21}, 0\right)$ and $\left(-3 - \sqrt{21}, 0\right)$.

5. Sketch the graph.

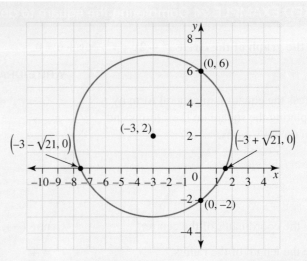

| TI | THINK | DISPLAY/WRITE | CASIO | THINK | DISPLAY/WRITE |

TI | THINK

On a Graphing page, press:
- MENU
- 3: Graph Entry/Edit
- 2: Relation

Then type:

$(x + 3)^2 + (y - 2)^2 = 25$

Press ENTER and the graph will be displayed.

DISPLAY/WRITE

CASIO | THINK

On a Conics screen, type:

$(x + 3)^2 + (y - 2)^2 = 25$

Press the graphing icon and the graph will be displayed.

DISPLAY/WRITE

DISCUSSION

How could you write equations representing a set of concentric circles (circles with the same centre, but different radii)?

 Resources

 eWorkbook   Topic 9 Workbook (worksheets, code puzzle and a project) (ewbk-2035)

 Interactivities   Individual pathway interactivity: The circle (int-4611)

Graphs of circles (int-6156)

9.9 Quick quiz on	9.9 Exercise

Individual pathways

■ PRACTISE	■ CONSOLIDATE	■ MASTER
1, 4, 7, 10, 13, 17	2, 5, 8, 11, 14, 15, 18	3, 6, 9, 12, 16, 19, 20

Fluency

WE22 For questions **1** to **3**, sketch the graphs of the following, stating the centre and radius of each.

1. a. $x^2 + y^2 = 49$
   b. $x^2 + y^2 = 4^2$

2. a. $x^2 + y^2 = 36$
   b. $x^2 + y^2 = 81$

3. a. $2x^2 + 2y^2 = 50$
   b. $9x^2 + 9y^2 = 100$

WE23 For questions **4** to **6**, sketch the graphs of the following, clearly showing the centre and the radius.

4. a. $(x-1)^2 + (y-2)^2 = 5^2$
   b. $(x+2)^2 + (y+3)^2 = 6^2$

5. a. $(x+3)^2 + (y-1)^2 = 49$
   b. $(x-4)^2 + (y+5)^2 = 64$

6. a. $x^2 + (y+3)^2 = 4$
   b. $(x-5)^2 + y^2 = 100$

WE24 For questions **7** to **9**, sketch the graphs of the following circles.

7. a. $x^2 + 4x + y^2 + 8y + 16 = 0$
   b. $x^2 - 10x + y^2 - 2y + 10 = 0$

8. a. $x^2 - 14x + y^2 + 6y + 9 = 0$
   b. $x^2 + 8x + y^2 - 12y - 12 = 0$

9. a. $x^2 + y^2 - 18y - 19 = 0$
   b. $2x^2 - 4x + 2y^2 + 8y - 8 = 0$

Understanding

10. **MC** The graph of $(x-2)^2 + (y+5)^2 = 4$ is:

A.

B.

C.

D.
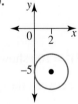

11. **MC** The centre and radius of the circle $(x+1)^2 + (y-3)^2 = 4$ is:
    A. $(1, -3), 4$
    B. $(-1, 3), 2$
    C. $(3, -1), 4$
    D. $(1, -3), 2$

12. **MC** The centre and radius of the circle with equation $x^2 + y^2 + 8x - 10y = 0$ is:
    A. $(4, 5), \sqrt{41}$
    B. $(-4, 5), 9$
    C. $(4, -5), 3$
    D. $(-4, 5), \sqrt{41}$

Reasoning

13. Determine the equation representing the outer edge of the galaxy as shown in the photo below, using the astronomical units provided.

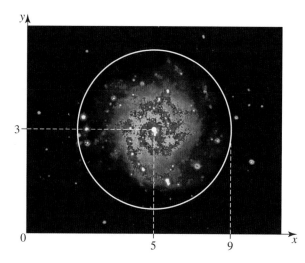

14. Circular ripples are formed when a water drop hits the surface of a pond.
    If one ripple is represented by the equation $x^2 + y^2 = 4$ and then 3 seconds later by $x^2 + y^2 = 190$, where the length of measurements are in centimetres:

    a. identify the radius (in cm) of the ripple in each case
    b. calculate how fast the ripple is moving outwards.
       (State your answers to 1 decimal place.)

15. Two circles with equations $x^2 + y^2 = 4$ and $(x-1)^2 + y^2 = 9$ intersect. Determine the point(s) of intersection. Show your working.

16. a. Graph the line $y = x$, the parabola $y = x^2$ and the circle $x^2 + y^2 = 1$ on the one set of axes.
    b. Evaluate algebraically the points of intersection of:
       i. the line and the circle
       ii. the line and the parabola
       iii. the parabola and the circle.

Problem solving

17. Sketch the graph of $(x+6)^2 + (y-3)^2 = 100$ showing all axial intercepts.

18. Determine the points of intersection between the quadratic equation $y = x^2 - 5$ and the circle given by $x^2 + y^2 = 25$.

19. Determine the point(s) of intersection of the circles $x^2 + y^2 - 2x - 2y - 2 = 0$ and $x^2 + y^2 - 8x - 2y + 16 = 0$ both algebraically and graphically.

20. The general equation of a circle is given by $x^2 + y^2 + ax + by + c = 0$. Determine the equation of the circle which passes through the points $(4, 5)$, $(2, 3)$ and $(0, 5)$. State the centre of the circle and its radius.

LESSON
9.10 Review

9.10.1 Topic summary

Transformations of the parabola

- The parabola with the equation $y = x^2$ has a turning point at $(0,0)$.
- A **dilation** from the x-axis stretches the parabola.
- A **dilation** of factor a from the x-axis produces the equation $y = ax^2$ and:
  - will produce a narrow graph for $a > 0$
  - will produce a wider graph for $0 < a < 1$
  - For $a < 0$ (a is negative), the graph is **reflected** in the x-axis.

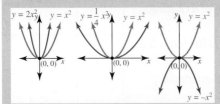

- The graph of $y = (x - h)^2$ has been translated left/right h units.
- The graph of $y = x^2 + k$ has been translated up/down k units.

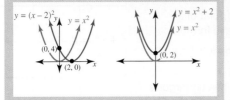

The parabola

- The graph of a quadratic equation is called a parabola.

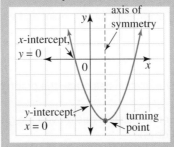

NON-LINEAR RELATIONSHIPS

Exponential graphs

- Equations $y = a^x$ are called exponential functions.
- The general form $y = a^{(x - h)} + k$ has:
  - a horizontal asymptote at $y = k$
  - a y-intercept at $(0, a^{-h} + k)$
- A negative in front of a results in a reflection in the x-axis.
- A negative in front of x results in a reflection in the y-axis.

Turning point form

- Completing the square allows us to write a quadratic in turning point form.
- Turning point form:
  $$y = a(x - h)^2 + k$$
- Turning point at (h, k)
- The y-intercept is determined by setting $x = 0$ and solving for y.
- x-intercept/s is determined by setting $y = 0$ and solving for x.

Expanded form

- The expanded form of a quadratic is $y = ax^2 + bx + c$.
- There are two methods for sketching: factorising or using the quadratic formula.
- e.g. Consider $y = x^2 - 2x - 3$.
- To sketch, first factorise:
  $y = (x - 3)(x + 1)$
- Use the null-factor law to determine the x-intercepts: when $y = 0$, $0 = (x - 3)(x + 1)$ so $x = 3$ or -1.
- Axis of symmetry: $x = \dfrac{3 + (-1)}{2} = 1$
- y-coordinate of the turning point: $x = 1$, $y = (1 - 3)(1 + 1) = -4$. Turning point is $(1, -4)$. The y-intercept when $x = 0$ is $(0, -3)$.

Inverse proportion and reciprocals

- The graph of a reciprocal function is called a hyperbola.
- If y and x are inversely proportional then $y = \dfrac{1}{x}$ or $y = \dfrac{k}{x}$, where k is the constant of proportionality.
- The graph of an equation in the form $y = \dfrac{k}{x}$ is called a hyperbola.
- The general form of a hyperbola is $y = \dfrac{a}{x - h} + k$ and has:
  - vertical asymptote at $x = h$
  - horizontal asymptote at $y = k$
  - y-intercept at $\left(0, \dfrac{-a}{h + k}\right)$
  - x-intercept at $\left(\dfrac{-a}{k + h}, 0\right)$.
- If $a < 0$ the graph is reflected in the x-axis.

Circles

- $x^2 + y^2 = r^2$ is a circle with centre $(0, 0)$ and radius r.
- $(x - h)^2 + (y - k)^2 = r^2$ is a circle with centre (h, k) and radius r.
- Use completing the square to determine the equation of a circle:
  $x^2 + 6x + y^2 + 8x = 0$
  $x^2 + 6x + 9 + y^2 + 8x + 16 = 9 + 16$
  $(x + 3)^2 + (y + 4)^2 = 25$

Quadratic formula

- For $y = ax^2 + bx + c$
  - axis of symmetry: $x = -\dfrac{b}{2a}$
  - turning point: $\left(-\dfrac{b}{2a}, c - \dfrac{b^2}{4a}\right)$
- x-intercepts can be calculated as: $x = \dfrac{-b \pm \sqrt{b^2 - 4ac}}{2a}$
- Discriminant: $\Delta = b^2 - 4ac$ is used to determine the number of x-intercepts.
  - $\Delta > 0$, two x-intercepts
  - $\Delta = 0$, one x-intercept
  - $\Delta < 0$, no x-intercept

9.10.2 Success criteria

Tick the column to indicate that you have completed the lesson and how well you have understood it using the traffic light system.

(**Green:** I understand; **Yellow:** I can do it with help; **Red:** I do not understand)

Lesson	Success criteria	⬤	⬤	⬤
9.2	I can create a table of values and use this to sketch the graph of a parabola.			
	I can identify the axis of symmetry, turning point and y-intercept of a parabola.			
9.3	I am familiar with the key features of the basic quadratic $y = x^2$.			
	I can determine whether a dilation has made the graph of quadratic narrower or wider			
	I can determine whether a translation has moved the graph of a quadratic function left/right or up/down			
	I can sketch a quadratic equation that has undergone any of the following transformations: dilations, reflections or translations.			
9.4	I can determine the axis of symmetry and turning point of a quadratic expressed in turning point form.			
	I can calculate the y-intercept and any x-intercepts of a quadratic in turning point form.			
9.5	I can determine the x-intercept/s, y-intercept and turning point of a quadratic equation and sketch its graph by first factorising the equation.			
	I can determine the x-intercept/s, y-intercept and turning point of a quadratic equation and sketch its graph using the quadratic formula.			
9.6	I can determine the equation of the asymptote of an exponential equation and sketch a graph of the equation.			
	I can sketch the graph of an exponential equation after it has undergone any number of transformations that may include dilations, a reflection in the x- or y-axis, or translations.			
9.7	I can calculate the constant of proportionality, k, between two variables that are inversely proportional to each other.			
	I can determine the rule between two inversely proportional variables and plot the graph of their relationship.			
9.8	I can determine the equation of the vertical and horizontal asymptote of a hyperbola.			
	I can sketch the graph of a hyperbola, and understand the effect dilations and reflections have on the shape of the graph.			
9.9	I can identify the centre and radius of the graph of a circle from its equation and then sketch its graph.			
	I can use completing the square to turn an equation in expanded form into the form of an equation of a circle.			

9.10.3 Project

Parametric equations

You are familiar with the quadratic equation $y = x^2$ and its resulting graph. Let us consider an application of this equation by forming a relationship between x and y through a third variable, say, t.

$$x = t \quad \text{and} \quad y = t^2$$

It is obvious that these two equations are equivalent to the equation $y = x^2$. This third variable t is known as a parameter, and the two equations are now called parametric equations. We cannot automatically assume that the resulting graph of these two parametric equations is the same as that of $y = x^2$ for all real values of x. It is dependent on the range of values of t.

Consider the parametric equations $x = t$ and $y = t^2$ for values of the parameter $t \geq 0$ for questions **1** to **3**.

1. Complete the following table by calculating x- and y-values from the corresponding t-value.

t	x	y
0		
1		
2		
3		
4		
5		

2. Graph the x-values and corresponding y-values on this Cartesian plane. Join the points with a smooth curve and place an arrow on the curve to indicate the direction of increasing t-values.
3. Is there any difference between this graph and that of $y = x^2$? Explain your answer.
4. Consider now the parametric equations $x = 1 - t$ and $y = (1 - t)^2$. These clearly are also equivalent to the equation $y = x^2$. Complete the table and draw the graph of these two equations for values of the parameter $t \geq 0$. Draw an arrow on the curve in the direction of increasing t-values.

Parametric equations

t	x	y
0		
1		
2		
3		
4		
5		

Parametric equations

Describe the shape of your resulting graph. What values of the parameter t would produce the same curve as that obtained in question **2**?

5. The graph of $y = -x^2$ is a reflection of $y = x^2$ in the x-axis. Construct a table and draw the graph of the parametric equations $x = t$ and $y = -t^2$ for parameter values $t \geq 0$. Remember to place an arrow on the curve in the direction of increasing t-values.

Parametric equations
$x = t$ and $y = -t^2$

6. Without constructing a table, predict the shape of the graph of the parametric equations $x = 1 - t$ and $y = -(1 - t)^2$ for parameter values $t \geq 0$. Draw a sketch of the shape.

Exercise 9.10 Review questions

learnon

Fluency

1. **MC** The turning point for the graph $y = 3x^2 - 4x + 9$ is:

   A. $\left(\dfrac{1}{3}, 1\dfrac{2}{3}\right)$

   B. $\left(\dfrac{1}{3}, \dfrac{2}{3}\right)$

   C. $\left(\dfrac{1}{6}, 1\dfrac{1}{6}\right)$

   D. $\left(\dfrac{2}{3}, 7\dfrac{2}{3}\right)$

2. **MC** Select which graph of the following equations has the x-intercepts closest together.

   A. $y = x^2 + 3x + 2$

   B. $y = x^2 + x - 2$

   C. $y = 2x^2 + x - 15$

   D. $y = 4x^2 + 27x - 7$

3. **MC** Select which graph of the equations below has the largest y-intercept.

   A. $y = 3(x - 2)^2 + 9$

   B. $y = 5(x - 1)^2 + 8$

   C. $y = 2(x - 1)^2 + 19$

   D. $y = 2(x - 5)^2 + 4$

4. **MC** The translation required to change $y = x^2$ into $y = (x - 3)^2 + \dfrac{1}{4}$ is:

   A. right 3, up $\dfrac{1}{4}$

   B. right 3, down $\dfrac{1}{4}$

   C. left 3, down $\dfrac{1}{4}$

   D. left 3, up $\dfrac{1}{4}$

5. **MC** If y is inversely proportional to x, then select which of the following statements is true.

   A. $x + y$ is a constant value.

   B. $y \div x$ is a constant value.

   C. $y \times x$ is a constant value.

   D. $y - x$ is a constant value.

6. The number of calculators a company sells is inversely proportional to the selling price. If a company can sell 1000 calculators when the price is $22, determine how many they could sell if they reduced the price to $16.

    A. 2000   **B.** 727   **C.** 6000   **D.** 1375

7. **MC** The graph of $y = -3 \times 2^x$ is best represented by:

    A.

    B.

    C.

    D.

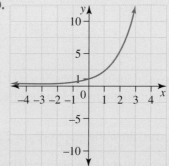

8. Use the completing the square method to determine the turning point for each of the following graphs.

    a. $y = x^2 - 8x + 1$

    b. $y = x^2 + 4x - 5$

9. For the graph of the equation $y = x^2 + 8x + 7$, produce a table of values for the x-values between -9 and 1, and then plot the graph. Show the y-intercept and turning point. From your graph, state the x-intercepts.

10. For each of the following, determine the coordinates of the turning point and the x- and y-intercepts and sketch the graph.

    a. $y = (x - 3)^2 + 1$

    b. $y = 2(x + 1)^2 - 5$

11. For the equation $y = -x^2 - 2x + 15$, sketch the graph and determine the x- and y-intercepts and the coordinates of the turning point.

12. Draw the graph of $y = 10 \times 3^x$ for $-2 \leq x \leq 4$.

13. Draw the graph of $y = 10^{-x}$ for $-2 \leq x \leq 2$.

14. For the exponential function $y = 5^x$:
    a. complete the table of values

x	y
-3	
-2	
-1	
0	
1	
2	
3	

    b. plot the graph.

15. a. On the same axes, draw the graphs of $y = (1.2)^x$ and $y = (1.5)^x$.
    b. Use your answer to part a to explain the effect of changing the value of a in the equation of $y = a^x$.

16. a. On one set of axes, draw the graphs of $y = 2 \times 3^x$, $y = 5 \times 3^x$ and $y = \dfrac{1}{2} \times 3^x$
    b. Use your answer to part a to explain the effect of changing the value of k in the equation of $y = ka^x$.

17. a. On the same set of axes, sketch the graphs of $y = (2.5)^x$ and $y = (2.5)^{-x}$.
    b. Use your answer to part a to explain the effect of a negative index on the equation $y = a^x$.

18. Sketch each of the following.
    a. $y = \dfrac{4}{x}$

    b. $y = -\dfrac{2}{x}$

19. Sketch $y = \dfrac{-3}{x-2}$.

20. Give an example of an equation of a hyperbola that has a vertical asymptote at $x = -3$.

21. Sketch each of these circles. Clearly show the centre and the radius.
    a. $x^2 + y^2 = 16$
    b. $(x-5)^2 + (y+3)^2 = 64$

22. Sketch the following circles. Remember to first complete the square.
    a. $x^2 + 4x + y^2 - 2y = 4$
    b. $x^2 + 8x + y^2 + 8y = 32$

23. Determine the equation of this circle.

Problem solving

24. The height, h, in metres of a golf ball t seconds after it is hit is given by the formula $h = 4t - t^2$.
    a. Sketch the graph of the path of the ball.
    b. Calculate the maximum height the golf ball reaches.
    c. Determine how long it takes for the ball to reach the maximum height.
    d. Determine how long is it before the ball lands on the ground after it has been hit.

25. A soccer ball is kicked upwards in the air. The height, h, in metres, t seconds after the kick is modelled by the quadratic equation $h = -5t^2 + 20t$.
    a. Sketch the graph of this relationship.
    b. Determine how many seconds the ball is in the air for.
    c. Determine how many seconds the ball is above a height of 15 m. That is, solve the quadratic inequation $-5t^2 + 20t > 15$.
    d. Calculate how many seconds the ball is above a height of 20 m.

26. The height of the water level in a cave is determined by the tides. At any time, t, in hours after 9 am, the height, $h(t)$, in metres, can be modelled by the function $h(t) = t^2 - 12t + 32, 0 \le t \le 12$.
    a. Determine the values of t for which the model is valid. Write your answer in interval notation.
    b. State the initial height of the water.
    c. Bertha has dropped her keys onto a ledge which is 7 metres from the bottom of the cave. By using a graphics calculator, determine the times in which she would be able to climb down to retrieve her keys. Write your answers correct to the nearest minute.

27. When a drop of water hits the flat surface of a pool, circular ripples are made. One ripple is represented by the equation $x^2 + y^2 = 9$ and 5 seconds later, the ripple is represented by the equation $x^2 + y^2 = 225$, where the lengths of the radii are in cm.
    a. State the radius of each of the ripples.
    b. Sketch these graphs.
    c. Evaluate how fast the ripple are moving outwards.
    d. If the ripple continues to move at the same rate, determine when it will hit the edge of the pool which is 2 m from its centre.

28. Seventy grams of ammonium sulfate crystals are dissolved in 0.5 L of water.
    a. Determine the concentration of the solution in g/mL.
    b. Another 500 mL of water is added. Determine the concentration of the solution now.

29. Use technology to graph the equations $y = x^2$ and $y = 2^x$ on the same axes. Using the graphs, determine the solution to $x^2 = 2^x$. Give your answers correct to 2 decimal places. How can you be sure you have found all the solutions?

▶

30. A grassed area is planted in a courtyard that has a width of 5 metres and length of 7 metres. The shape of the grassed area is described by the function $P = -x^2 + 5x$, where P is the distance, in metres, from the house and x is the distance, in metres from the side wall. The diagram represents this information on a Cartesian plane.

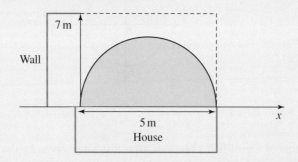

a. In terms of P, write down an inequality that describes the region where the grass has been planted.

b. Determine the maximum distance the grass area is planted from the house.

c. The owners of the house have decided that they would prefer all of the grass to be within a maximum distance of 3.5 metres from the house. The shape of the lawn following this design can be described by the equation $N(x) = ax^2 + bx + c$

i. Using algebra, show that this new design can be described by the function $N(x) = -0.56x(x - 5)$.

ii. Describe the transformation that maps $P(x)$ to $N(x)$.

d. If the owners decide on the first design, $P(x)$, the percentage of area within the courtyard without grass is 40.5%. By using any method, determine the approximate percentage of area of courtyard without lawn with the new design, $N(x)$.

31. A stone arch bridge has a span of 50 metres. The shape of the curve AB can be modelled using a quadratic equation.

a. Taking A as the origin $(0, 0)$ and given that the maximum height of the arch above the water level is 4.5 metres, show using algebra, that the shape of the arch can be modelled using the equation $b(x) = -0.0072x^2 + 0.36x$, where $b(x)$ is the vertical height of the bridge, in metres, and x is the horizontal distance, in metres.

b. A floating platform p metres high is towed under the bridge. Given that the platform needs to have a clearance of at least 30 centimetres on each side, explain why the maximum value of p is 10.7 centimetres.

on To test your understanding and knowledge of this topic, go to your learnON title at www.jacplus.com.au and complete the **post-test**.

Answers

Topic 9 Non-linear relationships

9.1 Pre-test

1.

x	-3	-2	-1	0
y	9.5	5	1.5	-1

2. $x = -1$
3. B
4. A
5. D
6. $(-4, -1)$
7. $b = 20$
8. D
9. D
10. $(-0.79, 0.62)$ and $(0.79, 0.62)$
11. B
12. C
13. a. 2   b. 1
14. D
15. B

9.2 Plotting parabolas from a table of values

1. a.

$x = 0, (0, 0)$

b.

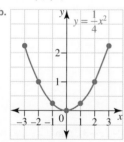

$x = 0, (0, 0)$

2. Placing a number greater than 1 in front of x^2 makes the graph thinner. Placing a number greater than 0 but less than 1 in front of x^2 makes the graph wider.

3. a.

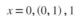

$x = 0, (0, 1), 1$

b.

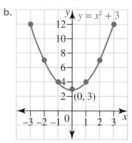

$x = 0, (0, 3), 3$

c.

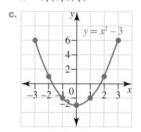

$x = 0, (0, -3), -3$

d.

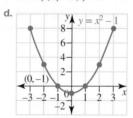

$x = 0, (0, -1), -1$

4. Adding a number raises the graph of $y = x^2$ vertically that number of units. Subtracting a number lowers the graph of $y = x^2$ vertically that number of units.

5.

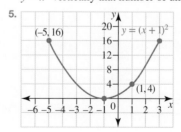

$x = -1, (-1, 0), 1$

6.

$x = 2, (2, 0), 4$

7.

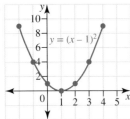

$x = 1, (1, 0), 1$

8.

$x = -2, (-2, 0), 4$

9. Adding a number moves the graph of $y = x^2$ horizontally to the left by that number of units. Subtracting a number moves the graph of $y = x^2$ horizontally to the right by that number of units.

10.

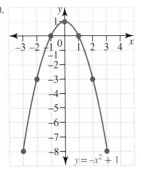

$x = 0, (0, 1), 1$

11.

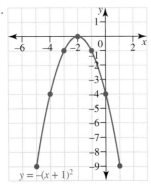

$x = -2, (-2, 0), -4$

12.

$x = 0, (0, -3), -3$

13.

$x = 1, (1, 0), -1$

14. The negative sign inverts the graph of $y = x^2$. The graphs with the same turning points are: $y = x^2 + 1$ and $y = -x^2 + 1$; $y = (x - 1)^2$ and $y = -(x - 1)^2$; $y = (x + 2)$ and $y = -(x + 2)^2$; $y = x^2 - 3$ and $y = -x^2 - 3$. They differ in that the first graph is upright while the second graph is inverted.

15. a.

b. $x = 5$ **c.** $(5, 1)$, min   **d.** 26

16. a.

b. $x = -2$ **c.** $(-2, -3)$, min

d. 5

17. a.

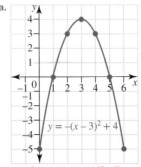

b. $x = 3$ **c.** $(3, 4)$, max   **d.** -5

18. a.

$y = -3(x - 1)^2 + 2$

b. $x = 1$ **c.** $(1, 2)$, max   **d.** -1

19. a.

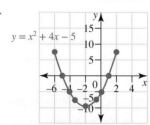

$y = x^2 + 4x - 5$

b. $x = -2$ **c.** $(-2, -9)$, min

d. -5

20. a. $y = -3x^2 - 6x + 24$

b. $x = -1$ **c.** $(-1, 27)$, max

d. 24

21. a. If the x^2 term is positive, the parabola has a minimum turning point. If the x^2 term is negative, the parabola has a maximum turning point.

b. If the equation is of the form $y = a(x - b)^2 + c$, the turning point has coordinates (b, c).

c. The equation of the axis of symmetry can be found from the x-coordinate of the turning point. That is, $x = b$.

22. C

23. B

24. C

25. A

26. a.

$h = -(t - 4)^2 + 16$

b. i. 16 m   **ii.** 8 s

27. a.

b. i. 18 m   **ii.** Yes, by 3 m

iii. 1.5 s   **iv.** 3 s

28. a.

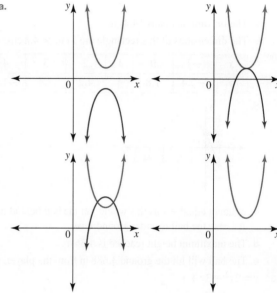

b. An infinite number of points of intersection occur when the two equations represent the same parabola, with the effect that the two parabolas superimpose. For example $y = x^2 + 4x + 3$ and $2y = 2x^2 + 8x + 6$.

c. It is possible to have 0, 1 or 2 points of intersection.

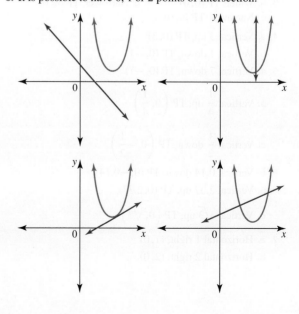

29. a.

w cm	−1	0	1	2	3	4	5	6	7
A cm²	−11.2	0	8	12.8	14.4	12.8	8	0	−11.2

b. The length of a rectangle must be positive, and the area must also be positive, so we can discard $w = -1, 0, 6$ and 7.

c.

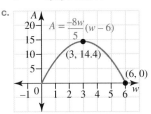

d. The maximum area is $14.4\,\text{cm}^2$.

e. The dimensions of this rectangle are $3\,\text{cm} \times 4.8\,\text{cm}$.

30. a.

x	−1	0	1	2	3	4
y	−2.4	1.8	4	4.2	2.4	−1.4

b.

c. x cannot equal -1 as this would put the ball behind her; at $x = 4$, the ball is under ground level.

d. The maximum height reached is $4.36\,\text{m}$.

e. The ball will hit the ground $3.688\,\text{m}$ from the player.

31. $y = 2x^2 - 3x + 1$

9.3 Sketching parabolas using transformations

1. a. Narrower, TP $(0, 0)$ **b.** Wider, $(0, 0)$

2. a. Narrower, TP $(0, 0)$ **b.** Narrower, TP $(0, 0)$
   c. Wider, TP $(0, 0)$

3. a. Wider, TP $(0, 0)$ **b.** Narrower, TP $(0, 0)$
   c. Narrower, TP $(0, 0)$

4. a. Vertical 3 up, TP $(0, 3)$
   b. Vertical 1 down, TP $(0, -1)$

5. a. Vertical 7 down, TP $(0, -7)$
   b. Vertical $\frac{1}{4}$ up, TP $\left(0, \frac{1}{4}\right)$
   c. Vertical $\frac{1}{2}$ down, TP $\left(0, -\frac{1}{2}\right)$

6. a. Vertical 0.14 down, TP $(0, -0.14)$
   b. Vertical 2.37 up, TP $(0, 2.37)$
   c. Vertical $\sqrt{3}$ up, TP $\left(0, \sqrt{3}\right)$

7. a. Horizontal 1 right, $(1, 0)$
   b. Horizontal 2 right, $(2, 0)$

8. a. Horizontal 10 left, $(-10, 0)$
   b. Horizontal 4 left, $(-4, 0)$
   c. Horizontal $\frac{1}{2}$ right, $\left(\frac{1}{2}, 0\right)$

9. a. Horizontal $\frac{1}{5}$ left, $\left(-\frac{1}{5}, 0\right)$
   b. Horizontal 0.25 left, $(-0.25, 0)$
   c. Horizontal $\sqrt{3}$ left, $\left(-\sqrt{3}, 0\right)$

10. a. $(0, 1)$, max   **b.** $(0, -3)$, min

11. a $(-2, 0)$, max   **b** $(0, 0)$, min   **c** $(0, 4)$, max

12. a. $(0, 0)$, max   **b.** $(5, 0)$, min   **c.** $(0, 1)$, min

13. a. Narrower, min   **b.** Narrower, max

14. a. Wider, min   **b.** Wider, max   **c.** Narrower, max

15. a. Wider, min   **b.** Narrower, min
   c. Wider, max

16. a. i. Horizontal translation 1 left
   ii. $(-1, 0)$
   iii.

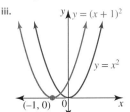

b. i. Reflected, narrower (dilation)
   ii. $(0, 0)$
   iii.

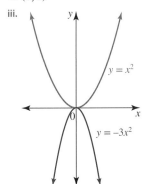

c. i. Vertical translation 1 up
   ii. $(0, 1)$
   iii.

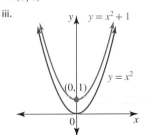

17. a. i. Wider (dilation)
   ii. $(0, 0)$

iii.

b. i. Vertical translation 3 down
   ii. $(0, -3)$
   iii.

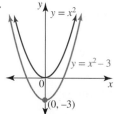

18. a. i. Horizontal translation 4 right
    ii. $(4, 0)$
    iii.

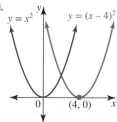

b. i. Reflected, wider (dilation)
   ii. $(0, 0)$
   iii.

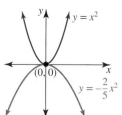

c. i. Narrower (dilation)
   ii. $(0, 0)$
   iii.

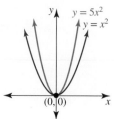

19. a. i. Reflected, vertical translation 2 up
    ii. $(0, 2)$

iii.

b. i. Reflected, horizontal translation 6 right
   ii. $(6, 0)$
   iii.

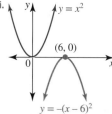

20. a. i. Reflected, vertical translation 4 down
    ii. $(0, -4)$
    iii.

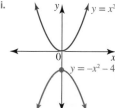

b. i. Narrower (dilation), horizontal translation 1 left,
      vertical translation 4 down
   ii. $(-1, -4)$
   iii.

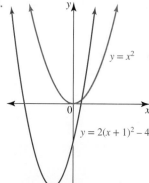

c. i. Wider (dilation), horizontal translation 3 right, vertical
      translation 2 up
   ii. $(3, 2)$
   iii.

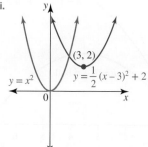

21. a. **i.** Wider (dilation), reflected, horizontal translation 2 left, vertical translation $\dfrac{1}{4}$ up

ii. $\left(-2, \dfrac{1}{4}\right)$

iii.

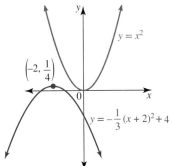

b. **i.** Narrower (dilation), reflected, horizontal translation 1 right, vertical translation $\dfrac{3}{2}$ down

ii. $\left(1, -\dfrac{3}{2}\right)$

iii.

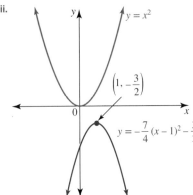

22. a. 10 cm   **b.** 5 cm   **c.** 5 cm

d. $y = (x - 5)^2$

23. a. 40 m

b. 5 m

c. $h = -\dfrac{1}{10}(d - 70)^2 + 40$

d. $h = -\dfrac{3}{10}(d - 70)^2 + 120$, this is a dilation by a factor of 3 from the d axis.

24. a. and **c.**
See figure at the bottom of the page.*
b. 12.25 m   **d.** 12.75 m

25. a. $y = -(x - 2)^2 + 3 = -x^2 + 4x - 1$

b.

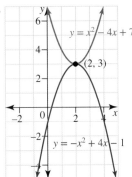

26. a. $(2, -7)$ and $(0, 5)$

b. In the following order: translate 2 units left, translate 7 units up, reflect in the x-axis then dilate by a factor of $\dfrac{2}{3}$ from the x-axis.

*24. a. and c.

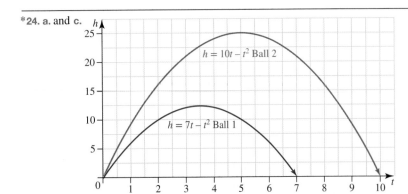

27. a. $Y = -(x - 3)^2 + 3$

b. Reflected in x-axis, translated 3 units to the right and up 3 units. No dilation.

c. $(3, 3)$

d. See figure at the bottom of the page.*

28. $v = -2h^2 + 40h$, where v is the vertical distance h is the horizontal distance.

9.4 Sketching parabolas using turning point form

1. a. $(1, 2)$, min

b. $(-2, -1)$, min

c. $(-1, 1)$, min

2. a. $(2, 3)$, max   **b.** $(5, 3)$, max   **c.** $(-2, -6)$, min

3. a. $\left(\dfrac{1}{2}, -\dfrac{3}{4}\right)$, min   **b.** $\left(\dfrac{1}{3}, \dfrac{2}{3}\right)$, min

c. $(-0.3, -0.4)$, min

4. a. i. $(-3, -5)$ **ii.** Min   **iii.** Narrower

b. i. $(1, 1)$ **ii.** Max   **iii.** Same

5. a. i. $(-2, -4)$ **ii.** Max   **iii.** Narrower

b. i. $(3, 2)$ **ii.** Min   **iii.** Wider

6. a. i. $(-1, 7)$ **ii.** Max   **iii.** Wider

b. i. $\left(\dfrac{1}{5}, -\dfrac{1}{2}\right)$ **ii.** Min   **iii.** Wider

7. a. $y = (x + 1)^2 - 3$, graph iii

b. $y = -(x - 2)^2 + 3$, graph i

c. $y = -x^2 + 1$, graph ii

8. a. $y = (x - 1)^2 - 3$, graph iii

b. $y = -(x + 2)^2 + 3$, graph i

c. $y = x^2 - 1$, graph ii

9. A

10. C

11. B

12. C

13. B

14. a. i. -3 **ii.** $-3, 1$

b. i. 12 **ii.** 2

15. a. i. -18 **ii.** No x-intercepts

b. i. -5 **ii.** $-1, 5$

16. a. i. 4 **ii.** No x-intercepts

b. i. 4

ii. $-3 - \sqrt{5}, -3 + \sqrt{5}$ (approx. $-5.24, -0.76$)

17. a. i. $(4, 2)$

ii. Min

iii. Same width

iv. 18

v. No x-intercepts

vi.

b. i. $(3, -4)$

ii. Min

iii. Same width

iv. 5

v. $1, 5$

vi.

c. i. $(-1, 2)$

ii. Min

iii. Same width

iv. 3

v. No x-intercepts

*27. d.

vi.

$y = (x + 1)^2 + 2$

(−1, 2)   (0, 3)

18. a.
   i. $(−5, −3)$
   ii. Min
   iii. Same width
   iv. 22
   v. $−5 − \sqrt{3}, −5 + \sqrt{3}$ (approx. $−6.73, −3.27$)
   vi.

$y = (x + 5)^2 − 3$

22 (0, 22)

$−5 + \sqrt{3}$

$−5 − \sqrt{3}$ $(−5, −3)$

b.
   i. $(1, 2)$
   ii. Max
   iii. Same width
   iv. 1
   v. $1 − \sqrt{2}, 1 + \sqrt{2}$ (approx. $−0.41, 2.41$)
   vi.

$(1, 2)$

$1 − \sqrt{2}$ $1 + \sqrt{2}$

$y = −(x − 1)2 + 2$

c.
   i. $(−2, −3)$
   ii. Max
   iii. Same width
   iv. $−7$
   v. No x-intercepts
   vi.

$(−2, −3)$

$(0, −7)$

$y = −(x + 2)^2 − 3$

19. a.
   i. $(−3, −2)$
   ii. Max
   iii. Same width
   iv. $−11$
   v. No x-intercepts

vi.

$(−3, −2)$

$(0, −11)$

$y = −(x + 3)^2 − 2$

b.
   i. $(1, 3)$
   ii. Min
   iii. Narrower
   iv. 5
   v. No x-intercepts
   vi.

$y = 2(x − 1)^2 + 3$

5

$(1, 3)$

c.
   i. $(−2, 1)$
   ii. Max
   iii. Narrower
   iv. $−11$
   v. $−2 − \dfrac{1}{\sqrt{3}}, −2 + \dfrac{1}{\sqrt{3}}$ (approx. $−2.58, −1.42$)
   vi.

$(−2, 1)$

$−2 − \dfrac{1}{\sqrt{3}}$ $−2 + \dfrac{1}{\sqrt{3}}$

$−11$ $y = −3(x + 2)^2 + 1$

20. a. $2\left(x − \dfrac{3}{4}\right)^2 − \dfrac{73}{8} = 0$

b. $x = \dfrac{3}{4} \pm \dfrac{\sqrt{73}}{4}$

c. $\left(\dfrac{3}{4}, −\dfrac{73}{8}\right)$, minimum

21. a. $y = −\dfrac{2}{3}(x + 4)^2 + 6$

b. $(−7, 0)$

22. a. $y = −x^2$

b. $y = 7x^2$

c. $y = (x + 3)^2$

d. $y = x^2 + 6$

e. $y = −\dfrac{1}{4}(x − 5)^2 − 3$

23. a.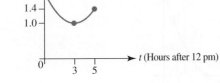

b. $1.90 **c.** $1 **d.** 3 pm **e.** $1.40

24. a. 0.5 m

b. $\left(15 + 4\sqrt{15}\right)$ m

c. Maximum height is 8 metres when horizontal distance is 15 metres.

25. a. Sample responses can be found in the worked solutions in the online resources.
An example is $y = (x-2)^2 + 6$.

b. $y = -\dfrac{1}{2}(x-2)^2 + 6$

26. a. Sample responses can be found in the worked solutions in the online resources.
An example is $y = (x-p)^2 + q$.

b. $y = \left(\dfrac{r-q}{p^2}\right)(x-p)^2 + q$

27. a. $y = (x-4)^2 - 15$

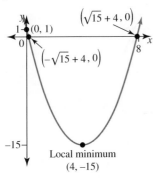

b. $y = (x+2)^2 - 9$

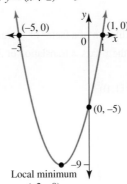

c. $y = \left(x + \dfrac{3}{2}\right)^2 - \dfrac{1}{4}$

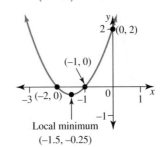

28. a. $y = 3(x-2)^2 - 8$

b. Dilated by a factor of 3 parallel to the y-axis or from the x-axis as well as being translated 2 units to the right and down 8 units.

c. $y = -3(x-2)^2 + 8$

d. See figure at the bottom of the page.*

---

*28. d.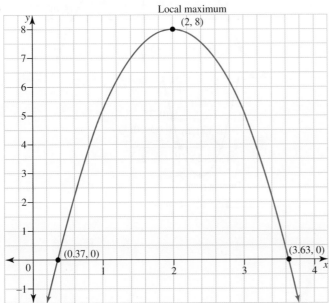

29. a. $y = -\dfrac{1}{2}(x+3)^2 + 8$

b. In the following order: a dilation of a factor of $\dfrac{1}{2}$ from the x-axis, a reflection in the x-axis, a translation of 3 units left and 8 units up.

c. x- intercepts: $(-7, 0)$ and $(1, 0)$

y- intercept: $\left(0, \dfrac{7}{2}\right)$

d. The new rule is $y = (x-3)^2 - 16$

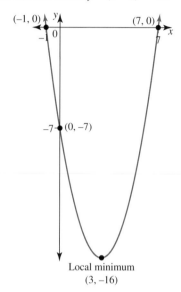

9.5 Sketching parabolas in expanded form

1. You need the x-intercepts, the y-intercept and the turning point to sketch a parabola.

2. a.

b.

3. a.

b.

4. a.

b.

5. a.

b.

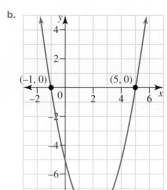

(−1, 0) (5, 0)

Turning point
(2, −9)

c.

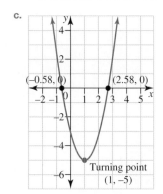

(−0.58, 0) (2.58, 0)

Turning point
(1, −5)

6. a.

Turning point
(2.75, 20.1)

(0, 5)

b.

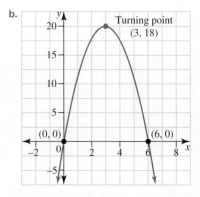

Turning point
(3, 18)

(0, 0) (6, 0)

c.

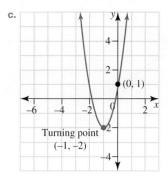

(0, 1)

Turning point
(−1, −2)

7. a.

Turning point
$\left(\dfrac{-5}{6}, \dfrac{49}{12}\right)$

(0, 2)

b.

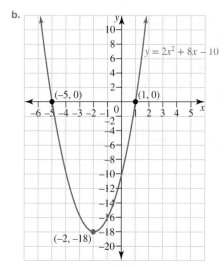

$y = 2x^2 + 8x - 10$

(−5, 0) (1, 0)

(−2, −18)

c.

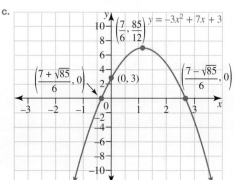

$\left(\dfrac{7}{6}, \dfrac{85}{12}\right)$ $y = -3x^2 + 7x + 3$

$\left(\dfrac{7 + \sqrt{85}}{6}, 0\right)$ (0, 3) $\left(\dfrac{7 - \sqrt{85}}{6}, 0\right)$

8. a.

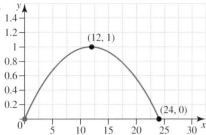

b. 24 m

c. 1 m

9. a.

b. 6th month

c. The company breaks even at the end of the 6th month. The first month in which a profit is made is the 7th month.

10. a.

b. 0

c. 2500 m

d. 25 seconds

e. 50 seconds

11. 200

12. $-(m + n)$

13. $y = -\dfrac{1}{2}x^2 + 2x + 9$

14. a.

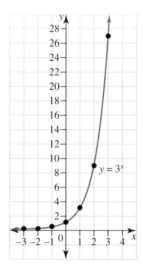

b. 25 m

c. 2 m

d. 7 m

15. a.

b. 21 °C

c. Decreasing

d. Increasing

e. 5 °C after 4 hours

f. 21 °C

16. a. $a = -\dfrac{1}{18}, b = \dfrac{13}{6}, c = 7$

b. 7 metres

c. $\dfrac{225}{8}$ metres

d. 42 metres

e.

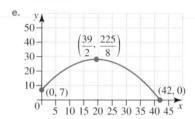

9.6 Exponential graphs

1.

x	-3	-2	-1	0	1	2	3
y	$\dfrac{1}{27}$	$\dfrac{1}{9}$	$\dfrac{1}{3}$	1	3	9	27

2. a. 2   **b.** 3   **c.** 4

3. a. 10   **b.** a

4.

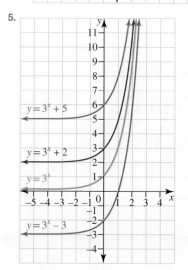

a. The graphs all pass through $(0, 1)$. The graphs have the same horizontal asymptote, $(y = 0)$. The graphs are all very steep.

b. As the base grows larger, the graphs become steeper.

c.

5.

a. The shape of each graph is the same.

b. Each graph has a different y-intercept and a different horizontal asymptote.

c. i. $(0, 11)$

ii. $y = 10$

6. a.

b.

c.

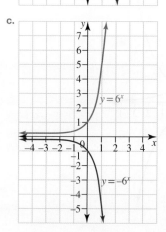

d. In each case the graphs are symmetric about the x-axis.

7. a.

b.

c.

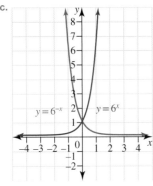

d. In each case the graphs are symmetric about the *y*-axis.

8. a–c.

9. a, b

10. a, b

11. a, b

12. a, b

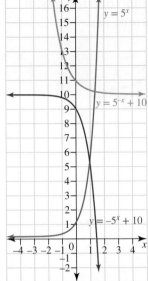

13. a. B **b.** C
  c. D **d.** A

14. a. B **b.** D
  c. A **d.** C

15.

16.

$x = 1.25$

17.

$x = 3.5$

18.

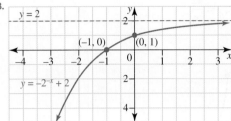

19. a. 10 000
  b. i. 1111
    ii. 41
    iii. 0

20. a. Yes
  b. There is a constant ratio of 1.3.
  c. 30%
  d. 3.26 million
  e. 30.26 million

9.7 Inverse proportion

1. a. $s = \dfrac{k}{t}$ or $t = \dfrac{k}{s}$ **b.** No

  c. $t = \dfrac{k}{p}$ or $p = \dfrac{k}{t}$ **d.** $L = \dfrac{k}{p}$ or $p = \dfrac{k}{L}$

  e. No **f.** $t = \dfrac{k}{n}$ or $n = \dfrac{k}{t}$

2. Sample responses can be found in the worked solutions in the online resources.

3. a. $k = 1000, y = \dfrac{1000}{x}$

  b.

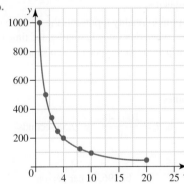

4. a. $k = 48, p = \dfrac{48}{q}$

  b.

5. a. $k = 42, y = \dfrac{42}{x}$

b.

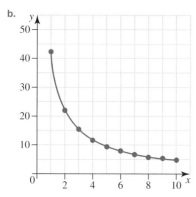

6. a. 4000 **b.** $a = \dfrac{4000}{m}$

c. $20 \, \text{m/s}^2$ **d.** $4 \, \text{m/s}^2$

7. a. 500 **b.** 2500 pencils

c. 1000 pencils

8. a. 337.5 **b.** 3.97 hours $= 3\,\text{h}\,58\,\text{min}$

c. 96.4 km/h **d.** 337.5 km

9. a. 17 500 **b.** \$233.33

c. 70 people **d.** \$17 500

10. a. 200

b. 1 ampere

c. 13.3 ohms

d. Sample responses can be found in the worked solutions in the online resources.

11. a. $k = 15$ **b.** $V = 20\,L$

c. $P = 2.5$ atmospheres

12. The constant of proportionality represents the distance of the trip; therefore, when this value is smaller, the time taken to complete the trip at the same speed is also smaller.

13. a. 10 **b.** $T = \dfrac{10}{P}$

c. 1 hour 40 minutes

14. a. 100.048 **b.** $T = \dfrac{100.048}{V}$ **c.** 9.58 seconds

d. 15.00 seconds

15. a. $C = \dfrac{1350}{n}$ **b.** \$96.43 per person

c. \$30 per person

9.8 Sketching the graph of a reciprocal

1. See table at the bottom of the page.*

2. a.

b. $x = 0, y = 0$

3. a.

b. $x = 0, y = 0$

4. a.

b. $x = 0, y = 0$

5.

6. It increases the y-values by a factor of k and hence dilates the curve by a factor of k.

*1.

x	−5	−4	−3	−2	−1	0	1	2	3	4	5
y	−2	−2.5	−3.3	−5	−10	Undef.	10	5	3.3	2.5	2

7.

$$y = \frac{-10}{x}$$

8.

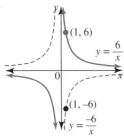

(1, 6)

$$y = \frac{6}{x}$$

(1, −6)

$$y = \frac{-6}{x}$$

9. The negative reflects the curve $y = \dfrac{k}{x}$ in the x-axis.

10. See table bottom of the page.*

$$y = \frac{1}{x - 1}$$

$x = 1$

Equation of vertical asymptote is $x = 1$.

11. a.

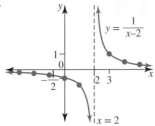

$$y = \frac{1}{x - 2}$$

$x = 2$

b.

$$y = \frac{1}{x - 3}$$

$x = 3$

12.

$$y = \frac{1}{x + 1}$$

$x = -1$

13. The a translates the graph left or right, and $x = a$ becomes the vertical asymptote.

14. a. i. $x = -1, y = 0$

ii.

(−2, 4)

$$y = \frac{-4}{x + 1}$$

−4

b. i. $x = 1, y = 0$

ii.

$$y = \frac{2}{x - 1}$$

(2, 2)

−2

15. a. i. $x = -2, y = 0$

ii.

$2\frac{1}{2}$

$$y = \frac{5}{x + 2}$$

(−3, −5)

b. i. $x = -2, y = -2$

ii.

$$y = \frac{3}{x + 2} - 2$$

(−0.5, 0)

(0, −0.5)

*10. a.

x	−3	−2	−1	0	1	2	3	4
y	−0.25	−0.33	−0.5	−1	Undef.	1	0.5	0.33

16. a. i. $x = -2, y = 1$

ii.

b. i. $x = 3, y = 5$

ii.

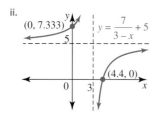

17. Sample responses can be found in the worked solutions in the online resources. Possible answers:

a. $y = \dfrac{1}{x - 3}$

b. $y = \dfrac{1}{x + 10}$

18. a. Any equation of the form $y = \dfrac{a}{x - 2} + 3$.

b. $y = \dfrac{-14}{x + 2} + 4$

19. $y = \dfrac{-2}{x + 3} - 1, x = -3, y = -1$

20. a.

x	-2	-1	$-\dfrac{1}{2}$	$\dfrac{1}{2}$	1	2
y	$\dfrac{1}{4}$	1	4	4	1	$\dfrac{1}{4}$

See figure at the bottom of the page.*

b. $x = 0$ **c.** $y = -\dfrac{1}{x^2}$ **d.** $y = \dfrac{1}{x^2}$

21. a. $100\,°C$

b. $33.1\,°C$

c. 17.85 minutes ≈ 18 minutes

d. No. The horizontal asymptote is $22\,°C$, so the temperature will never drop below this value.

22. a. y-intercept: $\left(0, \dfrac{1}{2}\right)$

x-intercepts: $(1, 0)$ and $(-1, 0)$

b. $x = 2, x = -2$ and $y = 2$

c.

***20. a.**

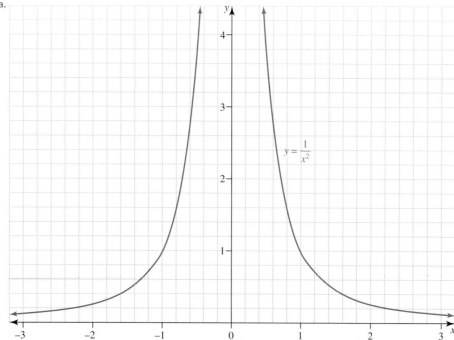

$x = 0, y = 0$

23. $y = -\dfrac{3}{(x+1)^2} + 2, x = -1, y = 2$

9.9 Sketching the circle

1. a.

Centre $(0, 0)$, radius 7

b.

Centre $(0, 0)$, radius 4

2. a.

Centre $(0, 0)$, radius 6

b.

Centre $(0, 0)$, radius 9

3. a.

Centre $(0, 0)$, radius 5

b.

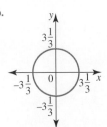

Centre $(0, 0)$, radius $\dfrac{10}{3}$

4. a.

b.

5. a.

b.

6. a.

b.

7. a. $(x+2)^2 + (y+4)^2 = 2^2$

b. $(x-5)^2 + (y-1)^2 = 4^2$

8. a. $(x-7)^2 + (y+3)^2 = 7^2$

b. $(x+4)^2 + (y-6)^2 = 8^2$

9. a. $x^2 + (y-9)^2 = 10^2$

b. $(x-1)^2 + (y+2)^2 = 3^2$

10. D

11. B

12. D

13. $(x-5)^2 + (y-3)^2 = 16$

14. a. 2 cm, 13.8 cm

 b. 3.9 cm/s

15. $(-2, 0)$

16. a. See the figure at the bottom of the page.*

 b. **i.** $(0.707, 0.707)$ and $(-0.707, -0.707)$

   ii. $(0, 0)$ and $(1, 1)$

   iii. $(0.786, 0.618)$ and $(-0.786, 0.618)$

17.

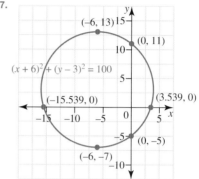

18. $(-3, 4), (3, 4), (0, 5)$ and $(0, -5)$

---

*16. a.

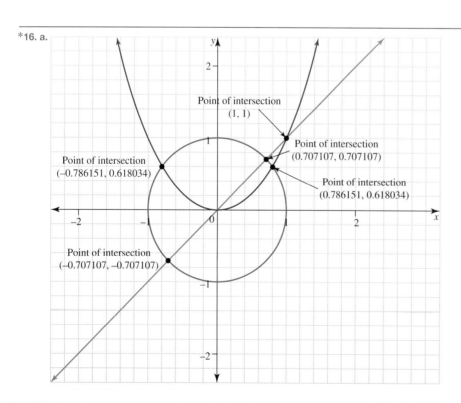

19. $(x-1)^2 + (y-1)^2 = 4$ centre at $(1, 1)$ and radius of 2 units.
$(x-4)^2 + (y-1)^2 = 1$ centre at $(4, 1)$ and radius of 1 unit.
The circles intersect (touch) at $(3, 1)$.
See figure at the bottom of the page.*

20. $(x-2)^2 + (y-5)^2 = 4$ centre at $(2, 5)$ and radius of 2 units.

Project

1.

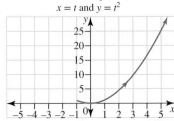

t	x	y
0	0	0
1	1	1
2	2	4
3	3	9
4	4	16
5	5	25

2.

Parametric equations
$x = t$ and $y = t^2$

3. Sample response: Yes the graph is different. $Y = x^2$ is a parabola and therefore, the graph appears in the I and the II quadrants. Where as in Q2 the graph only appears in the I quadrant. Other sample responses can be found in the worked solutions in the online resources.

4.

t	x	y
0	1	1
1	0	0
2	−1	1
3	−2	4
4	−3	9
5	−4	16

Sample responses can be found in the worked solutions in the online resources.
$t = 1, 0, -1, -2, -3, -4.$

Parametric equations
$x = 1 - t$ and $y = (1 - t)^2$

5.

t	x	y
0	0	0
1	1	−1
2	2	−4
3	3	−9
4	4	−16
5	5	−25

*19.

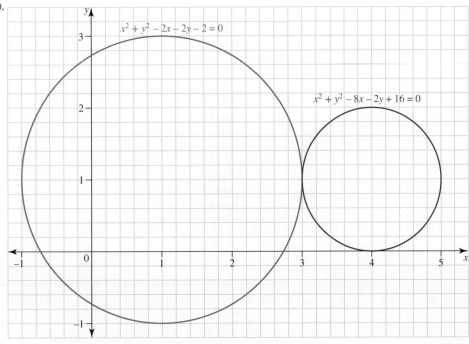

$x^2 + y^2 - 2x - 2y - 2 = 0$

$x^2 + y^2 - 8x - 2y + 16 = 0$

Parametric equations
$x = t$ and $y = -t^2$

6. Sample response: The shape of the graph will be the same as shown in Q5 but it will be in quadrant III (reflection of the graph in Q5). Other sample responses can be found in the worked solutions in the online resources.

Parametric equations
$x = t$ and $y = t^2$

9.10 Review questions

1. D
2. A
3. D
4. A
5. C
6. D
7. B
8. a. $(4, -15)$ b. $(-2, -9)$
9. See table at the bottom of the page.*

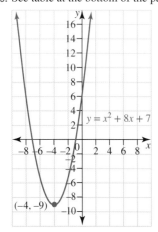

TP $(-4, -9)$; x-intercepts: -7 and -1

10. a. TP $(3, 1)$; no x-intercepts; y-intercept: $(0, 10)$

b. TP $(-1, -5)$; x-intercepts: $-1 - \sqrt{\dfrac{5}{2}}, -1 + \sqrt{\dfrac{5}{2}}$; y-intercept: $(0, -3)$

11. TP $(-1, 16)$; x-intercepts: -5 and 3; y-intercept: $(0, 15)$

12.

13.

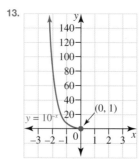

*9.

x	-9	-8	-7	-6	-5	-4	-3	-2	-1	0	1
y	16	7	0	-5	-8	-9	-8	-5	0	7	16

14. a. See table at the bottom of the page.*

b.

15. a.

b. Increasing the value of a makes the graph steeper for positive x-values and flatter for negative x-values.

16. a.

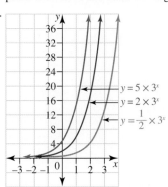

b. Increasing the value of k makes the graph steeper.

17. a.

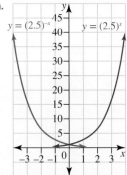

b. Changing the sign of the index reflects the graph in the y-axis.

18. a.

b.

19.

20. Sample responses can be found in the worked solutions in the online resources. Possible answer is $y = \dfrac{1}{x + 3}$.

21. a.

b.

22. a.

b.

---

*14. a.

x	-3	-2	-1	0	1	2	3
y	0.008	0.04	0.2	1	5	25	125

23. $x^2 + y^2 = 36$

24. a.

b. 4 m

c. 2 seconds

d. 4 seconds

25. a.

b. 4 seconds

c. 2 seconds $(1 < t < 3)$

d. The ball is never above a height of 20 m.

26. a. $\left[0, 12\right]$

b. 32 m

c. 11:41 am to 6:19 pm.

27. a. First ripple's radius is 3 cm, second ripple's radius is 15 cm.

b.

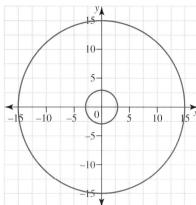

c. 2.4 cm/s

d. 1 minute 22.1 seconds after it is dropped.

28. a. 0.14 g/ml   **b.** 0.7 g/ml

29. $(-0.77, 0.59), (2, 4), (4, 16)$

By using the zoom function, the graph of $y = 2^x$ rises more steeply than the graph of $y = x^2$, so there are no further points of intersection.

30. a. $P \leq -x^2 + 5x, 0 \leq x \leq 5$

b. 6.25 m

c. i. Sample responses can be found in the worked solutions in the online resources.

ii. Dilation by a factor of 0.56 parallel to the y-axis.

d. 66.7%

31. a. Sample responses can be found in the worked solutions in the online resources.

b. When $x = 0.3$, $b = 10.7$. Therefore if p is greater than 10.7 cm the platform would hit the bridge.

10 Deductive geometry

LESSON
10.1 Overview

Why learn this?

Geometry is an area of mathematics that has an abundance of real-life applications. The first important skill that geometry teaches is the ability to reason deductively and prove logically that certain mathematical statements are true.

Euclid (c. 300 BCE) was the mathematician who developed a systematic approach to geometry, now referred to as Euclidean geometry, that relied on mathematical proofs. Mathematicians and research scientists today spend a large part of their time trying to prove new theories, and they rely heavily on all the proofs that have gone before.

Geometry is used extensively in professions such as navigation and surveying. Planes circle our world, land needs to be surveyed before any construction can commence, and architects, designers and engineers all use geometry in their drawings and plans. Geometry is also used extensively in software and computing. Computer-aided design programs, computer imaging, animations, video games and 3D printers all rely greatly on built-in geometry packages.

Just about every sport involves geometry heavily. In cricket alone there are many examples: bowlers adjust the angle at which they release the ball to make the ball bounce towards the batsmen at different heights; fielders are positioned so they cover as much of the ground as efficiently as possible; and batsmen angle their bat as they hit the ball to ensure the ball rolls along the ground instead of in the air. Netballers must consider the angle at which they shoot the ball to ensure it arcs into the ring, and cyclists must consider the curved path of their turns that will allow them to corner in the quickest and most efficient way.

Hey students! Bring these pages to life online

- **Watch videos**
- **Engage with interactivities**
- **Answer questions and check solutions**

Find all this and MORE in jacPLUS

Reading content and rich media, including interactivities and videos for every concept

Extra learning resources

Differentiated question sets

Questions with immediate feedback, and fully worked solutions to help students get unstuck

1. Calculate the value of the pronumeral x.

2. Calculate the value of the pronumeral x.

3. **MC** If the scale factor of a diagram is 1 : 250, a length of 7.5 cm on the diagram represents an actual length of:

   A. 33.33 m   **B.** 1875 cm   **C.** 7500 cm   **D.** 1875 m

4. Determine the values of the pronumerals x and y.

5. Triangles ABC and DEF are congruent; calculate the values of the pronumerals x, y and z.

6. **MC** Choose which congruency test will prove these triangles are congruent.

   A. SAS   **B.** SSS   **C.** AAS   **D.** RHS

7. **MC** A front fence is 24 m long and 1.2 m high. It is to be made with bricks 40 cm long and 12 cm high. An estimate of the number of bricks required to build the fence is:

   A. 240   **B.** 300   **C.** 400   **D.** 600

8. Calculate the value of the pronumeral x in the quadrilateral shown.

9. Determine the value of the pronumeral x in the quadrilateral shown.

10. Determine the exterior angle of a regular pentagon.

11. Evaluate the sum of the interior angles of an octagon.

12. **MC** Select the correct values of the pronumerals x and y.

A. $x = \dfrac{1}{3}$, $y = 7\sqrt{2}$

B. $x = \dfrac{2}{3}$, $y = 5\sqrt{2}$

C. $x = \dfrac{2}{3}$, $y = 7\sqrt{2}$

D. $x = 1\dfrac{1}{3}$, $y = 7\sqrt{2}$

13. Calculate the value of the pronumeral x.

14. **MC** Choose the correct values for the pronumeral x.

A. $x = 6$ or $x = 5$

B. $x = 4$ or $x = 4$

C. $x = 5$ or $x = 4$

D. $x = 6$ or $x = 4$

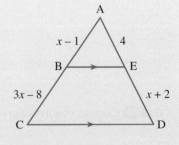

15. **MC** Select the correct values of the pronumerals x and y.

A. $x = 1.25$ and $y = 6.5$

B. $x = 0.8$ and $y = 5.6$

C. $x = 0.25$ and $y = 1.125$

D. $x = 5$ and $y = 3.5$

LESSON
10.2 Angles, triangles and congruence

LEARNING INTENTIONS

At the end of this lesson you should be able to:
- apply properties of straight lines and triangles to determine the value of an unknown angle
- construct simple geometric proofs for angles in triangles or around intersecting lines
- prove that triangles are congruent by applying the appropriate congruency test.

▶ 10.2.1 Proofs and theorems of angles

eles-4892

- Euclid (c. 300 BCE) was the mathematician who developed a systematic approach to geometry, now referred to as Euclidean geometry, that relied on mathematical proofs.
- A **proof** is an argument that shows why a statement is true.
- A **theorem** is a statement that can be demonstrated to be true. It is conventional to use the following structure when setting out a theorem.
  - **Given:** a summary of the information given
  - **To prove:** a statement that needs to be proven
  - **Construction:** a description of any additions to the diagram given
  - **Proof:** a sequence of steps that can be justified and form part of a formal mathematical proof.

Sums of angles

Angles at a point

- The sum of the angles at a point is 360°.

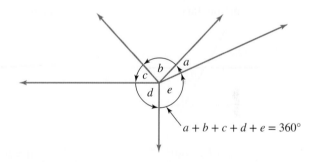

$$a + b + c + d + e = 360°$$

Supplementary angles

- The sum of the angles on a straight line is 180°.
- Angles that add up to 180° are called **supplementary angles**.
- In the diagram angles a, b and c are supplementary.

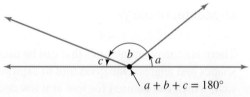

$$a + b + c = 180°$$

Complementary angles

- The sum of the angles in a right angle is 90°.
- Angles that add up to 90° are called **complementary angles**.
- In the diagram angles a, b and c are complementary.

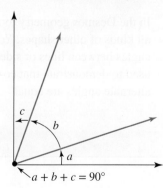

$$a + b + c = 90°$$

Vertically opposite angles

Theorem 1

- **Vertically opposite angles** are equal.

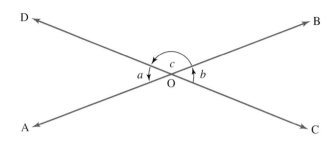

Given:	Straight lines AB and CD, which intersect at O
To prove:	$\angle AOD = \angle BOC$ and $\angle BOD = \angle AOC$
Construction:	Label $\angle AOD$ as a, $\angle BOC$ as b and $\angle BOD$ as c.
Proof:	Let $\angle AOD = a°$, $\angle BOC = b°$ and $\angle BOD = c°$.

$$a + c = 180° \qquad \text{(supplementary angles)}$$
$$b + c = 180° \qquad \text{(supplementary angles)}$$
$$\therefore a + c = b + c$$
$$\therefore a = b$$

So, $\angle AOD = \angle BOC$.
Similarly, $\angle BOD = \angle AOC$.

Parallel lines

- If two lines are parallel and cut by a **transversal**, then:
  - co-interior angles are supplementary
  - corresponding angles are equal
  - alternate angles are equal.

$$a + b = 180°$$

$$a = b$$

$$a = b$$

Digital technology

There are many online tools that can be used to play around with lines, shapes and angles. One good tool to explore is the Desmos geometry tool, which can be used for free at www.desmos.com/geometry.

In the Desmos geometry tool, you can draw lines, circles, polygons and all kinds of other shapes. You can then use the angle tool to explore the angles between lines or sides. The figure shows the angle tool being used to demonstrate that co-interior angles are supplementary and alternate angles are equal.

10.2.2 Angle properties of triangles

eles-5353

Theorem 2

- The sum of the interior angles of a triangle is 180°.

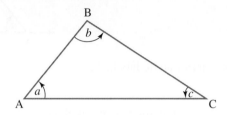

Given:	$\triangle ABC$ with interior angles a, b and c
To prove:	$a + b + c = 180°$
Construction:	Draw a line parallel to AC passing through B and label it DE as shown. Label $\angle ABD$ as x and $\angle CBE$ as y.

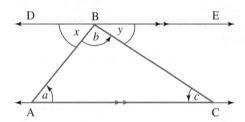

Proof:

$$a = x \qquad \text{(alternate angles)}$$
$$c = y \qquad \text{(alternate angles)}$$
$$x + b + y = 180° \qquad \text{(supplementary angles)}$$
$$\therefore a + b + c = 180°$$

Equilateral triangles

- It follows from Theorem 2 that each interior angle of an **equilateral triangle** is 60°, and conversely, if the three angles of a triangle are equal, the triangle is equiangular.

$$a + a + a = 180° \qquad \text{(sum of interior angles in a triangle is 180°)}$$
$$3a = 180°$$
$$a = 60°$$

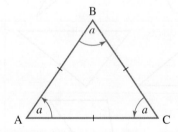

Theorem 3

- The exterior angle of a triangle is equal to the sum of the opposite interior angles.

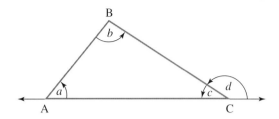

Given:	$\triangle ABC$ with the exterior angle labelled d
To prove:	$d = a + b$
Proof:	$c + d = 180°$ (supplementary angles)
	$a + b + c = 180°$ (sum of interior angles in a triangle is $180°$)
	$\therefore d = a + b$

▶ 10.2.3 Congruent triangles

eles-4897

- **Congruent triangles** have the same size and the same shape; that is, they are identical in all respects.
- The symbol used for congruency is $\cong$.
  For example, $\triangle ABC$ in the diagram below is congruent to $\triangle PQR$.
  This is written as $\triangle ABC \cong \triangle PQR$.
  This can be proven by adjusting the slope so $\triangle ABC$ sits exactly on $\triangle PQR$.

 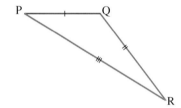

- Note that the vertices of the two triangles are written in corresponding order.
- There are four tests designed to check whether triangles are congruent. The tests are summarised in the table below.

Congruence test	Example	Description
Side-side-side (SSS)		The three corresponding sides are the same lengths.
Side-angle-side (SAS)		Two corresponding sides are the same length and the angle in between these sides is equal.
Angle-angle-side (AAS)		Two corresponding angles are equal and the side in between these angles is the same length.

Congruence test	Example	Description
		A pair of corresponding angles and a non-contained side are equal.
Right angle-hypotenuse-side (RHS)		The triangles are right-angled, and the hypotenuse and one other side of one triangle are equal to the hypotenuse and a side of the other triangle.

- In each of the tests we need to show three equal measurements about a pair of triangles in order to show they are congruent.

WORKED EXAMPLE 1 Determining pairs of congruent triangles

Select a pair of congruent triangles from the diagrams shown, giving a reason for your answer.

THINK	WRITE
1. In each triangle the length of the side opposite the 95° angle is given. If triangles are to be congruent, the sides opposite the angles of equal size must be equal in length. Draw your conclusion.	All three triangles have equal angles, but the sides opposite the angle 95° are not equal. AC = PR = 15 cm and LN = 18 cm
2. To test whether △ABC is congruent to △PQR, first evaluate the angle C.	△ABC: ∠A = 50°, ∠B = 95°, ∠C = 180° − 50° − 95° = 35°
3. Apply a test for congruence. Triangles ABC and PQR have a pair of corresponding sides equal in length and 2 pairs of angles the same, so draw your conclusion.	A pair of corresponding angles (∠B = ∠Q and ∠C = ∠R) and a corresponding side (AC = PR) are equal. △ABC ≅ △PQR (AAS)

▶ ## 10.2.4 Isosceles triangles

eles-4898

- A triangle is isosceles if the lengths of two sides are equal but the third side is not equal.

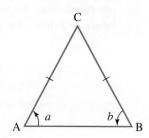

Theorem 4

- The angles at the base of an **isosceles triangle** are equal.

Given: $AC = CB$

To prove: $\angle BAC = \angle CBA$

Construction: Draw a line from the vertex C to the midpoint of the base AB and label the midpoint D. CD is the bisector of $\angle ACB$.

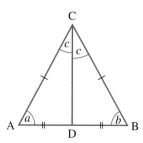

Proof: In $\triangle ACD$ and $\triangle BCD$,

$CD = CD$	(common side)
$AD = DB$	(construction, D is the midpoint of AB)
$AC = CB$	(given)
$\Rightarrow \triangle ACD \cong \triangle BCD$	(SSS)
$\therefore \angle BAC = \angle CBA$	

- Conversely, if two angles of a triangle are equal, then the sides opposite those angles are equal.

WORKED EXAMPLE 2 Determining values in congruent triangles

Given that $\triangle ABD \cong \triangle CBD$, determine the values of the pronumerals in the figure shown.

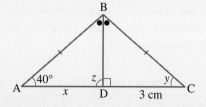

THINK	WRITE
1. In congruent triangles, corresponding sides are equal in length. Side AD (marked x) corresponds to side DC, so state the value of x.	$\triangle ABD \cong \triangle CBD$ $AD = CD$, $AD = x$, $CD = 3$ So $x = 3$ cm.
2. Since the triangles are congruent, corresponding angles are equal. State the angles corresponding to y and z and hence determine the values of these pronumerals.	$\angle BAD = \angle BCD$ $\angle BAD = 40°$, $\angle BCD = y$ So $y = 40°$. $\angle BDA = \angle BDC$ $\angle BDA = z$, $\angle BDC = 90°$ So $z = 90°$.

Prove that ΔPQS is congruent to ΔRSQ.

THINK

1. Write the information given.
2. Write what needs to be proved.

3. Select the appropriate congruency test for proof. (In this case, it is RHS because the triangles have an equal side, a right angle and a common hypotenuse.)

WRITE

Given: Rectangle PQRS with diagonal QS

To prove: ΔPQS is congruent to ΔRSQ.
QP = SR (opposite sides of a rectangle)
∠SPQ = ∠SRQ = 90° (given)
QS is common.
So ΔPQS ≅ ΔRSQ (RHS).

Digital technology: Drawing an isosceles triangle and measuring the angles using GeoGebra

Steps	
Open GeoGebra and select Geometry from the Perspective menu on the side bar.	≡ GeoGebra Geometry **Basic Tools** Move · Point · Segment Line · Polygon · Circle with Center **Edit** Select Objects · Show / Hide Label · Show / Hide Object Delete **Construct** Midpoint or Center · Perpendicular Line · Perpendicular Bisector Parallel Line · Angle Bisector · Tangents **Measure** Angle · Angle with Given Size · Distance or Length
Use the **Circle with center through point** ⊙ tool to construct a circle with centre A that passes through B.	(circle with centre A and point B on circle)

Select the **Point on object** ✐ tool. With this tool selected, touch the circle in 2 different spots to plot two different points, C and D, on the circle itself.	
Use the **Polygon** ▷ tool to construct the triangle ACD.	
Select the **Move** ⇖ tool. Click on each side of the triangle. Select the **AA** icon to show the value. Select **Show Value** to show the length of this segment. Repeat this action for the other two sides.	
Use the **Angle** ∡ tool to measure and display all 3 angles of this triangle.	

Resources

10.2 Quick quiz on	10.2 Exercise

Individual pathways

■ PRACTISE	■ CONSOLIDATE	■ MASTER
1, 4, 7, 8, 9, 16	2, 5, 10, 11, 12, 17	3, 6, 13, 14, 15, 18

Fluency

1. Determine the value of the unknown in each of the following diagrams.

a.

b.

c.

d.

e.
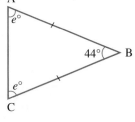

2. Determine the values of the pronumerals in the following diagrams.

a.

b.

c.

d.

3. **WE1** Select a pair of congruent triangles in each of the following, giving a reason for your answer. All side lengths are in cm.

a.

b.

c.

d.

Understanding

4. Determine the values of x and y in each of the following diagrams. Give reasons for your answers.

a.

b.

c.

d.

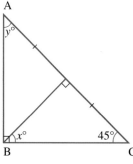

5. Determine the values of the pronumerals. Give reasons for your answers.

a.

b.

6. **WE2** Determine the value of the pronumeral in each of the following pairs of congruent triangles. All side lengths are in cm.

a.

b.

c.

d.

e.

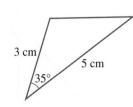

7. **MC** Choose which of the following is congruent to the triangle shown.
Note: There may be more than one correct answer.

A.

B.

C.

D.

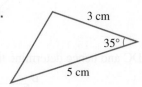

Reasoning

8. Prove that $\triangle ABC \cong \triangle ADC$ and hence determine the values of the pronumerals.

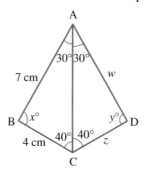

9. If $DA = DB = DC$, prove that $\angle ABC$ is a right angle.

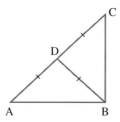

10. **WE3** Prove that each of the following pairs of triangles are congruent.

a.

b.

c.

d.

e.

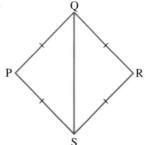

11. Prove that $\triangle ABC \cong \triangle ADC$ and hence determine the value of x.

12. Explain why the triangles shown are not necessarily congruent.

 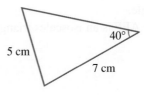

13. Prove that $\triangle ABC \cong \triangle ADC$ and hence determine the values of the pronumerals.

14. Explain why the triangles shown are not congruent.

15. If $AC = CB$ and $DC = CE$ in the diagram shown, prove that $AB\|DE$.

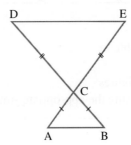

Problem solving

16. Show that $\triangle ABO \cong \triangle ACO$, if O is the centre of the circle.

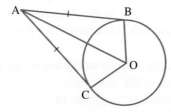

17. Triangles ABC and DEF are congruent.

 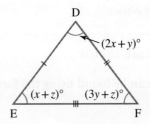

Determine the values of the pronumerals x, y and z.

18. ABC is an isosceles triangle in which AB and AC are equal in length. BDF is a right-angled triangle.
Show that triangle AEF is an isosceles triangle.

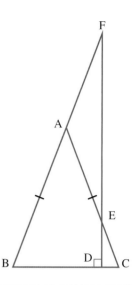

LESSON
10.3 Scale factor and similar triangles

eles-4899

LEARNING INTENTIONS

At the end of this lesson you should be able to:
- identify similar figures
- calculate the scale factor in similar figures
- show that two triangles are similar using the appropriate similarity test.

▶ 10.3.1 Similar figures

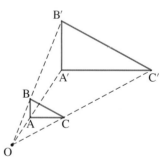

- Two geometric shapes are **similar** when one is an **enlargement** or reduction of the other shape.
  - An enlargement increases the length of each side of a figure in all directions by the same factor. For example, in the diagram shown, triangle A′B′C′ is an enlargement of triangle ABC by a factor of 3 from its **centre of enlargement** at O.
- Similar figures have the same shape. The corresponding angles are the same, and each pair of corresponding sides is in the same ratio.
- The symbol for similarity is ~ and is read as 'is similar to'.
- The **image** of the original object is the enlarged or reduced shape.
- To create a similar shape, use a **scale factor** to enlarge or reduce the original shape called the object.

Calculating scale factor

The scale factor can be found using the formula below and the lengths of a pair of corresponding sides.

$$\text{Scale factor} = \frac{\text{image side length}}{\text{object side length}}$$

- If the scale factor is less than 1, the image is a reduced version of the original shape. If the scale factor is greater than 1, the image is an enlarged version of the original shape.

Similar triangles

- Two triangles are similar if:
  - the angles are equal, or
  - the corresponding sides are proportional.
- Consider the pair of **similar triangles** below.

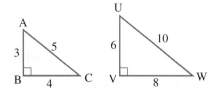

- The following statements are true for these triangles.
  - Triangle UVW is similar to triangle ABC, or using symbols, ΔUVW ~ ΔABC.
  - The corresponding angles of the two triangles are equal in size:

$$\angle CAB = \angle WUV, \angle ABC = \angle UVW \text{ and } \angle ACB = \angle UWV$$

  - The corresponding sides of the two triangles are in the same ratio. $\dfrac{UV}{AB} = \dfrac{VW}{BC} = \dfrac{UW}{AC} = 2$; that is,

    ΔUVW has each of its sides twice as long as the corresponding sides in ΔABC.
  - The scale factor is 2.

▶ 10.3.2 Testing triangles for similarity

eles-4900

- Triangles can be checked for similarity using one of the tests described in the table below.

Similarity test	Example	Description
Angle-angle-angle (AAA)		The three corresponding angles are equal.
Side-side-side (SSS)	a b c ka kc kb	The three sides of one triangle are proportional to the three sides of the other triangle.
Side-angle-side (SAS)	a c ka kc	Two sides of one triangle are proportional to two sides of the other triangle, and the included angle is equal.

Note: When using the equiangular test, only two corresponding angles have to be checked. Since the sum of the interior angles in any triangle is a constant number (180°), the third pair of corresponding angles will automatically be equal, provided that the first two pairs match exactly.

Determine a pair of similar triangles among those shown. Give a reason for your answer.

a.
b.
c.

THINK

1. In each triangle the lengths of two sides and the included angle are known, so the SAS test can be applied. Since all included angles are equal (140°), we need to the calculate ratios of corresponding sides, taking two triangles at a time.

2. Only triangles **a** and **b** have corresponding sides in the same ratio (and the included angle is equal). State your conclusion, specifying the similarity test you used.

WRITE

For triangles **a** and **b**: $\frac{6}{3} = \frac{4}{2} = 2$

For triangles **a** and **c**: $\frac{5}{3} = 1.67$, $\frac{3}{2} = 1.5$

For triangles **b** and **c**: $\frac{5}{6} = 0.83$, $\frac{3}{4} = 0.75$

Triangle **a** ~ triangle **b** (SAS)

Prove that △ABC is similar to △EDC.

THINK

1. Write the information given. AB is parallel to DE. Transversal BD forms two alternate angles: ∠ABC and ∠EDC.

2. Write what is to be proved.

3. Write the proof.

WRITE

Given:
△ABC and △DCE
AB || DE
C is common.

To prove: △ABC ~ △EDC

Proof:
∠ABC = ∠EDC (alternate angles)
∠BAC = ∠DEC (alternate angles)
∠BCA = ∠DCE (vertically opposite angles)
∴ △ABC ~ △EDC (equiangular, AAA)

Engineers and designers are working on a new car. To demonstrate what the new car looks like, they have produced a scaled-down model of the actual car. The model car is made with a length of 233 mm and a width of 95 mm.

The model car was designed with a scale factor of 20 relative to the real car.

a. Calculate the length of the real car.
b. Calculate the width of the new car.
c. If the model was not made accurately and instead the model's measurements were rounded to 230 mm by 100 mm, using the same scale factor, calculate the new dimensions of the car.
d. Calculate the percentage error in the length and width of the actual car from the two model cars.
e. Comment on the difference to the actual car's dimensions.

THINK	WRITE
a. 1. With a scale factor of 20 the length needs to be multiplied by 20.	a. $\text{Length} = 233 \times 20$ $\qquad\quad = 4660 \text{ mm}$
2. Write the answer.	The length of the actual car is 4660 mm.
b. 1. With a scale factor of 20 the width needs to be multiplied by 20.	b. $\text{Width} = 95 \times 20$ $\qquad\quad = 1900 \text{ mm}$
2. Write the answer.	The width of the actual car is 1900 mm.
c. 1. With a scale factor of 20 the dimensions need to be multiplied by 20.	c. $\text{Length} = 230 \times 20$ $\qquad\qquad = 4600 \text{ mm}$ $\text{Width} = 100 \times 20$ $\qquad\quad = 2000 \text{ mm}$
2. Write the answer.	The length and width of the actual car using the rounded model are 4600 mm and 2000 mm respectively.
d. 1. To calculate the percentage error: $\% \text{Error} = \dfrac{\text{difference in values}}{\text{actual value}} \times \dfrac{100}{1}$	d. $\% \text{Error(Length)} = \dfrac{4660 - 4600}{4660} \times \dfrac{100}{1}$ $\qquad\qquad\qquad\quad = 1.29\%$ $\% \text{Error(Width)} = \dfrac{2000 - 1900}{1900} \times \dfrac{100}{1}$ $\qquad\qquad\qquad\quad = 5.26\%$
2. Write the answer.	The percentage error for the length and width is 1.29% and 5.26% respectively.
e. Write a comment on the difference to the actual car's dimensions.	e. This shows the importance of getting the dimension correct on the model, since even though the percentage error is the same, the actual error is magnified when scaled up by the factor of 20. With the rounded model, the actual car is 60 mm shorter in length and 100 mm wider. This would affect the amount of material required to build the car and therefore the cost of building the car.

Exercise 10.3 Scale factor and similar triangles   **learn** on

10.3 Quick quiz on	10.3 Exercise

Individual pathways

■ PRACTISE	■ CONSOLIDATE	■ MASTER
1, 3, 6, 11, 14, 15	2, 4, 7, 9, 12, 16, 17	5, 8, 10, 13, 18, 19, 20

Fluency

1. **WE4** Select a pair of similar triangles among those shown in each part. Give a reason for your answer.

a.   i.   ii.   iii.

b.   i.   ii.   iii.

c.   i.   ii.   iii.

d.   i.   ii.   iii.

e.   i.   ii.   iii.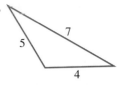

2. Name two similar triangles in each of the following diagrams.

a.

b.

c.

d.

e.

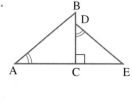

3. Consider the following diagram.

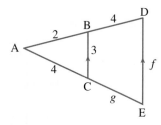

a. Use the given diagram to complete this statement: $\dfrac{AB}{AD} = \dfrac{BC}{} = \dfrac{}{AE}$.

b. Determine the values of the pronumerals in the diagram.

4. Determine the value of the pronumeral in the diagram shown.

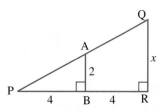

5. The triangles shown are similar.

Determine the values of the pronumerals x and y.

Understanding

6. **a.** State why the two triangles shown are similar.

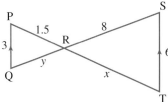

b. Determine the values of the pronumerals x and y in the diagram.

7. Calculate the values of the pronumerals in the following diagrams.

a.

b.

c.

d.

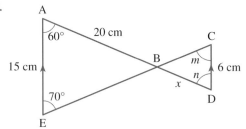

8. Calculate the values of the pronumerals in the following diagrams.

a.

b.

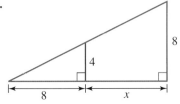

9. Determine the value of x in the diagram.

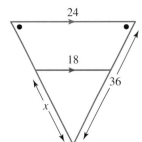

10. Calculate the values of the pronumerals in the following diagrams.

a.

b.

Reasoning

11. WE5 Prove that $\triangle ABC$ is similar to $\triangle EDC$ in each of the following diagrams.

a.

b.

c.

d.

12. $\triangle ABC$ is a right-angled triangle. A line is drawn from A to D as shown so that $AD \perp BC$.

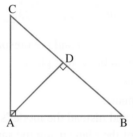

Prove that:

a. $\triangle ABD \sim \triangle ACB$

b. $\triangle ACD \sim \triangle ACB$.

13. Explain why the AAA test cannot be used to prove congruence but can be used to prove similarity.

Problem solving

14. A student casts a shadow that is 2.8 m long. Another student, who is taller, stands in the same spot at the same time of day. If the diagrams are to the same scale, determine the length of the shadow cast by the taller student.

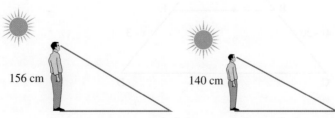

156 cm   140 cm

15. WE6 A 3D laser printer was used to produce a scaled-down prototype of a boat with a length of 15 cm and a width of 2.5 cm. If the prototype is made with a scale factor of 40, calculate the dimensions of the actual boat.

16. A waterslide is 4.2 m high and has a support 2.4 m tall. If a student reaches this support when she is 3.1 m down the slide, evaluate the length of the slide.

17. Prove that $\Delta EFO \sim \Delta GHO$ in the diagram shown.

18. A model plane was made with a length of 345 mm and a wingspan of 260 mm. The model plane was designed with a scale factor of 60 relative to the actual plane.

    a. Calculate the length of the actual plane.
    b. Calculate the wingspan of the actual plane.
    c. If the model was not made accurately and instead the model's measurements were rounded to 35 cm by 25 cm, calculate the new dimensions of the plane using the same scale factor.
    d. Comment on the difference to the actual plane's dimensions.

19. A storage tank is made of a 4-m-tall cylinder joined by a 3-m-tall cone, as shown in the diagram. If the diameter of the cylinder is 5 m, evaluate the radius of the end of the cone if 0.75 m has been cut off the tip.

20. Determine the value of x in the diagram shown.

LESSON
10.4 Calculating measurements from scale drawings

- All diagrams, maps or plans are drawn using a given scale. This scale is used to convert the lengths in the diagram to the actual lengths.

▶ 10.4.1 Perimeters

eles-6185

- Calculating the actual perimeters of shapes from scaled drawings involves first calculating the lengths required.
- Once the lengths of the sides of the shape are known, the perimeter can be calculated by adding up all the sides of the shape.

WORKED EXAMPLE 7 Using scale to calculate lengths and perimeters

Given a scale of 1 : 300 for the diagram of the backyard shown, calculate:

a. the actual length and width of the pool, given the diagram measurements are 2.4 cm by 1.2 cm

b. the actual perimeter of the pool.

THINK	WRITE
a. 1. Write the length and width of the pool on the diagram.	a. Length = 2.4 cm Width = 1.2 cm
2. Write the scale of the drawing and state the scale factor.	1 : 300 This means that every 1 cm on the drawing represents 300 cm of the actual dimension. The scale factor is $\dfrac{1}{300}$.

▶

3. Divide both dimensions by the scale factor.

$$\text{Length of the pool} = 2.4 \div \frac{1}{300}$$
$$= 2.4 \times 300$$
$$= 720\,\text{cm}$$
$$= 7.2\,\text{m}$$

$$\text{Width of the pool} = 1.2 \div \frac{1}{300}$$
$$= 1.2 \times 300$$
$$= 360\,\text{cm}$$
$$= 3.6\,\text{m}$$

b. Calculate the perimeter of the pool.

b. $\text{Perimeter of the pool} = 2 \times \text{length} + 2 \times \text{width}$
$$= 2 \times 7.2 + 2 \times 3.6$$
$$= 14.4 + 7.2$$
$$= 21.6\,\text{m}$$

▶ 10.4.2 Areas

eles-6186

- Areas of surfaces from scaled drawings can be calculated in a similar way to the perimeter. The dimensions of the shape have to be calculated first.
- Once the lengths of the shape are known, the area can be calculated using an appropriate formula.

WORKED EXAMPLE 8 Calculating actual areas from scale drawings

Given a scale of 1 : 15 for the diagram of the tile shown, calculate:
a. **the actual length and width of the tile, given the diagram measurements are 2 cm by 2 cm**
b. **the actual area of the tile**
c. **the actual area covered by 20 tiles.**

THINK

a. 1. Write the length and the width of the shape.

2. Write the scale of the drawing and state the scale factor.

3. Since both dimensions are the same, divide the dimension by the scale factor.

WRITE

a. Length = 2 cm
Width = 2 cm

1 : 15
This means that 1 cm on the drawing represents 15 cm of the actual dimension.

The scale factor is $\dfrac{1}{15}$.

$\text{Length/width of the tile} = 2 \div \dfrac{1}{15}$
$$= 2 \times 15$$
$$= 30\,\text{cm}$$
The length and width of the tile are 30 cm by 30 cm.

b. Calculate the area of the tile using the formula for the area of a square. Use $A = l^2$.

b. Area of one tile $= \text{length}^2$
$$= 30^2$$
$$= 900 \text{ cm}^2$$

c. Calculate the area required by multiplying the area of one tile by 20.
Note: Recall the conversion from cm^2 is converted to m^2 by dividing by 100^2.

c. Total area $=$ area of one tile $\times$ 20
$$= 900 \times 20$$
$$= 18\,000 \text{ cm}^2$$
$$= 18\,000 \div 100^2$$
$$= 1.8 \text{ m}^2$$

⊙ 10.4.3 Problem-solving with scale drawings

eles-6187

- Scaled diagrams are used in many areas of work in order to estimate elements such as production costs and quantities of materials required.

Packaging

- There are many products that are usually placed in a box when purchased or delivered. Boxes are cut out of cardboard or other materials using a template.
- The dimensions of the package are determined by using a net of the three-dimensional shape, as shown in the diagram.

The shape that encloses the whole net

$200 \times 240 \times 100$

Width

Length

- The required area can be estimated by calculating the area of the rectangle that encloses the whole net.

Costs

- To calculate the cost of the materials used, we use the cost per m or m^2, known as the cost per unit.

Cost of materials

$$\text{Cost required} = \frac{\text{area of material}}{\text{area per squared unit}} \times \text{cost per unit}$$

- For example, if the area of the packaging is 520 cm² and the cost of the material is $1.20 per square metre, then the cost for packaging is as follows.

$$\text{Cost required} = \frac{\text{area of material}}{\text{area per squared unit}} \times \text{cost per unit}$$

$$= \frac{520 \text{ cm}^2}{1 \text{ m}^2} \times 1.20$$

(*Note:* The two area units have to be the same: $1 \text{ m}^2 = 10\,000 \text{ cm}^2$.)

$$= \frac{520}{10\,000} \times 1.20$$

$$= 0.624\ldots$$

$$\approx \$0.62 \text{ per package}$$

WORKED EXAMPLE 9 Solving problems using scales

Consider the gift box shown and the template required to make it. The scale of the diagram is 1 : 15, and the template measurements are 6 cm by 4 cm.
a. Calculate the width and the height of the packaging template.
b. Calculate the total area of material required.
c. If the cost of materials is $0.25 per square metre, calculate how much it would cost to make the template for the gift box.

Gift box Template.
Print, cut, fold it.

THINK

a. 1. State the width and the height of the template.

2. Write the scale of the drawing and state the scale factor.

3. Divide both dimensions by the scale factor.

WRITE

a. Width $= 6$ cm
Height $= 4$ cm

1 : 15
This means that every 1 cm on the drawing represents 15 cm of the actual dimension.
The scale factor is $\dfrac{1}{15}$.

Width of template $= 6 \div \dfrac{1}{15}$

$= 6 \times 15$

$= 90$ cm

Height of template $= 4 \div \dfrac{1}{15}$

$= 4 \times 15$

$= 60$ cm

b. Calculate the area of the box template using the formula for the area of a rectangle.

b. Area of template = width × height
$$= 90 \times 60$$
$$= 5400 \, \text{cm}^2$$

c. 1. Calculate the cost of the material.

c. Cost is $0.25 per m² or 10 000 cm².

$$\text{Cost for the area required} = \frac{\text{area of material}}{\text{area per squared unit}} \times \text{cost per unit}$$
$$= \frac{5400}{10\,000} \times 0.25$$
$$= \$0.135$$
$$\approx 0.14$$

2. State the answer.

It will cost $0.14 or 14 cents to make the template for the gift box.

Painting

- Painting a wall requires three pieces of information to calculate the amount of paint needed:
  - the area of the surface to be painted
  - the amount of paint per square metre
  - the cost of paint per litre.

Painting a wall

Amount of paint required = area to be painted × amount of paint per square metre
Total cost = amount of paint required × cost of paint per litre

---

WORKED EXAMPLE 10 Solving problems involving area

Indra is going to paint a feature wall in her lounge room. She has a diagram of the room drawn at a scale of 1 : 100, with a length of 4 cm and height of 2.5 cm.
Calculate:
a. the length and the height of the wall
b. the area of the wall
c. the amount of paint Indra has to buy if the amount of paint needed is 3 L per m²
d. the total cost of the paint if the price is $7.60 per litre.

THINK

a. 1. State the length and the height of the shape.

2. Write the scale of the drawing and state the scale factor.

WRITE

a. Length = 4 cm
Height = 2.5 cm

1 : 100
This means that every 1 cm on the drawing represents 100 cm of the actual dimension.
The scale factor is $\frac{1}{100}$.

3. Divide both dimensions by the scale factor.

$$\text{Length of the wall} = 4 \div \frac{1}{100}$$
$$= 4 \times 100$$
$$= 400\,\text{cm}$$
$$= 4\,\text{m}$$

$$\text{Height of the wall} = 2.5 \div \frac{1}{100}$$
$$= 2.5 \times 100$$
$$= 250\,\text{cm}$$
$$= 2.5\,\text{m}$$

b. Calculate the area of the wall using the formula for the area of a rectangle.

b. Area wall = length × width (or height)
$$= 4 \times 2.5$$
$$= 10\,\text{m}^2$$

c. Calculate the amount of paint required.

c. Amount of paint required
= area to be painted × amount of paint per square metre
$$= 10 \times 3\,\text{L}$$
$$= 30\,\text{L}$$

d. Calculate the cost of the paint.

d. Total cost = amount of paint required × cost of paint per litre
$$= 30 \times 7.60$$
$$= \$228$$

Bricklaying

eles-6188

- Bricks are frequently used as a basic product in the construction of houses or fences.
- The dimensions of a standard brick are shown in the diagram using a scale of 1 : 10.

- Bricklayers have to calculate the number of bricks required to build a wall by considering the length, width and height of the brick, and the dimensions of the wall.
- To give the wall strength, bricks have to be laid so that their ends sit in the middle of the layer underneath, as shown in the diagram.

Calculating the number of bricks

$$\text{Number of bricks} = \frac{\text{area of the wall}}{\text{area of the exposed side of the brick}}$$

- Imagine that you want to calculate the number of bricks required to build a 2-m^2 wall.
The area of the brick is the area of the side that shows on the wall. In the diagram of the wall, using the standard brick dimensions, the exposed side of the brick has length 230 mm and height 76 mm.
The area of the exposed side of the brick is $230 \times 76 = 17\,480\,\text{mm}^2$.

The two areas have to be written in the same unit.

$$\text{Area of the wall} = 2\,\text{m}^2$$
$$= 2\,000\,000\,\text{mm}^2$$
$$\text{Number of bricks} = \frac{\text{area of the wall}}{\text{area of the exposed side of the brick}}$$
$$= \frac{2\,000\,000}{17\,480}$$
$$= 114.4\,\text{bricks}$$
$$= 115\,\text{bricks}$$

Note: Notice that although the answer is 114.4 bricks, we round up because we need 115 bricks so we can cut 0.4 of a brick to finish the whole wall.

• This answer is an estimated value, as we did not take into account the mortar that has to be added to the structure of the wall.

WORKED EXAMPLE 11 Solving problems using scales 2

Francis is going to build a brick fence with the dimensions shown in the diagram. The scale factor of the drawing is $\frac{1}{50}$.
Calculate:
a. the length and the height of the fence, given that it measures 7 cm by 1 cm on the diagram
b. the area of the front side of the fence
c. the number of bricks required if the dimensions of the exposed side of one brick are 20 cm and 10 cm
d. the cost of building the fence if the price of one brick is $1.20.

THINK	WRITE
a. 1. Write the length and the height of the fence.	a. Length $= 7\,\text{cm}$ Height $= 1\,\text{cm}$
2. Write the scale factor.	The scale factor is $\frac{1}{50}$.
3. Divide both dimensions by the scale factor.	Length of the fence $= 7 \div \frac{1}{50}$ $= 7 \times 50$ $= 350\,\text{cm}$ Height of the fence $= 1 \div \frac{1}{50}$ $= 1 \times 50$ $= 50\,\text{cm}$

b. Calculate the area of the front face of the fence using the formula for the area of a rectangle.

b. Area of the front face of the fence = length × width (or height)
= 350 × 50
= 17 500 cm²

c. 1. Calculate the area of the exposed face of one brick.

c. Area of the exposed face of the brick = 20 × 10
= 200 cm²

2. Calculate the number of bricks required.

Number of bricks = $\dfrac{17\,500}{200}$
= 87.5 bricks
= 88 bricks

d. Calculate the cost.

d. Cost = number of bricks × cost of one brick
= 88 × 1.20
= $105.60

Exercise 10.4 Calculating measurements from scale drawings **learn** on

10.4 Quick quiz on	10.4 Exercise

Individual pathways

■ PRACTISE	■ CONSOLIDATE	■ MASTER
1, 3, 7, 10, 13	2, 5, 9, 11, 14	4, 6, 8, 12, 15

Fluency

1. **WE7** The floor plan of the apartment shown has a scale of 1 : 250.

Calculate:

a. the actual length and width of the apartment, given the diagram measurements are 5 cm by 3 cm
b. the actual perimeter of the apartment.

2. If the scale of the diagram shown is 1 : 800, calculate, to the nearest metre, the radius and the circumference of the Ferris wheel shown, given the radius measures 1.5 cm on the diagram. (*Note:* Circumference $= 2\pi r$)

3. **WE8** Given a scale of 1 : 40 for the window shown, calculate:
   a. the actual width and height of each piece of glass needed to cover the window, given that the large piece of glass measures 1.5 cm by 2.5 cm on the diagram, and the width of the small piece of glass measures 0.75 cm
   b. the actual areas of the two glass pieces
   c. the total area covered by glass.

4. If the scale of the diagram shown is 1 : 20, calculate, to the nearest centimetre, the radius and the area of the clock face shown, given the radius measures 1.5 cm on the diagram. (*Note:* Area $= \pi r^2$)

5. **WE9** Consider the DVD mailing box shown and the template required to make it. The scale of the diagram is 1 : 8, and the diagram measurements are 3.5 cm by 1.5 cm.
   a. Calculate the length and the width of the packaging template.
   b. Estimate the total area of material required.
   c. If the cost of materials is $0.85 per square metre, calculate how much it would cost to make this DVD mailing box.

6. Jonathan wants to paint a wall that is 5.6 m long and 2.4 m high.
   a. Calculate how many litres of paint are required if 1.5 litres of paint cover 1 m^2.
   b. Calculate the cost of the paint if the price per litre is $15.70.

7. **WE10** Ilia was asked by a friend to paint a bedroom wall. The friend gave Ilia a diagram of the bedroom drawn at a scale of 1 : 100. If the length and height of the wall are 1.5 cm and 2.5 cm respectively, calculate:

a. the length and the height of the wall
b. the area of the wall
c. the amount of paint Ilia has to buy if the amount of paint needed is 2.6 L per m²
d. the total cost of the paint if the price is $12.80 per litre.

8. The three identical wooden cubes shown are pictured at a scale of 1 : 20. All three cubes need to be painted. Calculate:

a. the length of their sides, given each side measures 1.3 cm on the diagram
b. the area of each face
c. the total area of the three cubes to be painted (state the answer in m² correct to 1 decimal place)
d. the amount of paint required if 2 L of paint are needed per m²
e. the total cost of the paint if the price is $9.25 per litre.

9. **WE11** Consider a brick fence with the dimensions shown in the diagram. The scale factor of the brick wall shown is $\frac{1}{100}$. Calculate:

a. the length and the height of the fence, given that it measures 6 cm by 2.5 cm on the diagram
b. the area of the front side of the fence
c. the number of bricks required if the dimensions of the exposed side of one brick are 20 cm by 5 cm
d. the cost of the fence if the price of one brick is $0.75.

Reasoning

10. Consider the milk box shown and the template required to make it. The scale of the diagram is 1 : 7.

a. Calculate the length and the width of the packaging template, given the dimensions measure 3 cm by 3 cm on the diagram.
b. Estimate the total area of material required.
c. If the cost of materials is $0.45 per square metre, calculate how much it would cost to make this milk box.

11. Sky wants to pave a garden path with bricks. The garden path is shown at a scale of 1 : 150. Calculate:

a. the actual dimensions of the garden path, given the diagram measures 8.5 cm by 0.75 cm
b. the area of the garden path
c. the number of bricks required if the bricks used are squares with side length 24 cm
d. the cost of paving the garden path if the price of one brick is $1.65.

12. The scale of the house plan shown is 1 : 200. Calculate:

   a. the length of the bathroom, given that the length and width measure 1.2 cm and 1 cm respectively on the plan

   b. the width of the hallway, given that it is 0.6 cm wide on the plan

   c. the perimeter of the kitchen, given that it measures 1 cm by 2 cm on the plan

   d. the floor area of the house, given that it measures 4.5 cm by 3.5 cm on the plan.

Problem solving

13. Calculate the radius, the circumference and the area of the circular chopping board shown, given the radius measures 1.5 cm on the diagram and the scale is 1 : 12.

14. A brick has length 215 mm and height 65 mm. A bricklayer is building a wall 6 m long and 7.5 m high. Calculate:

   a. the number of bricks required

   b. the total cost of building the wall if one brick costs $1.95.

15. Consider the gift box shown and the template required to make it. The scale of the diagram is 1 : 8.

   a. Calculate the actual length and the width of the packaging template, given that it measures 8 cm by 5 cm on the diagram.

   b. Estimate the total area of material required.

   c. If the cost of materials is $2.15 per square metre, calculate how much it would cost to make this gift box.

LESSON
10.5 Creating scale drawings

LEARNING INTENTION

At the end of this lesson you should be able to:
- calculate the dimensions for a scale drawing
- construct and label scale drawings.

▶ 10.5.1 Calculating the dimensions of a drawing

eles-6189

- **Scale drawings** are drawings on paper of real-life objects. All dimensions of the drawn object are kept in the same ratio to the actual dimensions of the object using a scale.
- Calculating the dimensions of a drawing always requires a scale or a scale factor.

Drawing dimensions

The formula for the scale is used to calculate the dimensions of the drawing.

$$\text{If scale factor} = \frac{\text{dimension on the drawing}}{\text{actual dimension}},$$

then:

dimension on the drawing = scale factor × actual dimension.

WORKED EXAMPLE 12 Calculating dimensions

Calculate the dimensions required to create a scale drawing of a shipping container 12 m long, 2.5 m wide and 3 m high using a scale of 1 : 100.

THINK	WRITE
1. Convert all dimensions to an appropriate drawing unit.	The dimensions of the diagram are going to be smaller than the actual dimensions of the shipping container. Convert m to cm. 12 m = 1200 cm 2.5 m = 250 cm 3 m = 300 cm
2. Write the scale as a scale factor.	$1 : 100 = \dfrac{1}{100}$
3. Calculate the length of the object on the scale drawing.	$\dfrac{\text{dimension on the drawing}}{\text{actual dimension}} = \dfrac{1}{100}$ Length on the drawing = scale factor × actual dimension $= \dfrac{1}{100} \times 1200$ $= \dfrac{1200}{100}$ $= 12 \text{ cm}$

4. Calculate the width of the object on the scale drawing.

Width on the drawing = scale factor × actual dimension

$$= \frac{1}{100} \times 250$$

$$= \frac{250}{100}$$

$$= 2.5 \text{ cm}$$

5. Calculate the height of the object on the scale drawing.

Height on the drawing = scale factor × actual dimension

$$= \frac{1}{100} \times 300$$

$$= \frac{300}{100}$$

$$= 3 \text{ cm}$$

▶ 10.5.2 Constructing scale drawings

eles-6190

- Scale diagrams are drawn on graph paper or plain paper using a pencil and a ruler.
- Accuracy is very important when constructing these drawings.

WORKED EXAMPLE 13 Constructing scale drawings

Construct a plan-view, front-view and side-view scale drawing of a shipping container with length 12 cm, width 2.5 cm and height 3 cm.

THINK

1. Write the drawing dimensions of the plan view of the object.

2. Draw the plan view of the object. Recall that a plan view is a diagram of the object from above.

WRITE/DRAW

Drawing dimensions of the plan view of the shipping container are length = 12 cm and width = 2.5 cm.

To draw the length of the object, construct a horizontal line 12 cm long.

Construct two 90° angles on both sides of the line segment. Ensure the two vertical lines are 2.5 cm long.

Connect the bottom ends of the vertical lines to complete the rectangle.

12 cm

2.5 cm

3. State the drawing dimensions of the front view of the object.

The drawing dimensions of the front view of the shipping container are length = 12 cm and height = 3 cm.

4. Draw the front view of the object. Recall that the front view is a diagram of the object looking straight at the object from the front.

To draw the length of the object, construct a horizontal line 12 cm long.

Construct two 90° angles on both sides of the line segment. Ensure the two vertical lines are 3 cm long.

Connect the bottom ends of the vertical lines to complete the rectangle.

5. State the drawing dimensions of the side view of the object.

The drawing dimensions of the side view of the shipping container are width = 2.5 cm and height = 3 cm.

6. Draw the side view of the object. Recall that the side view is a diagram of the object looking at the object from one side.

Draw the length of the object, which is the width of the shipping container. Draw a horizontal line 2.5 cm long.

Construct two 90° angles on both sides of the line segment. Ensure the two vertical lines are 3 cm long.

Connect the bottom ends of the vertical lines to complete the rectangle.

10.5.3 Labelling scale drawings

- Scale diagrams are labelled using the actual dimensions of the object.
- The labels of a scale drawing have to be clear and easily read.

Conventions and line styles

- The line style and its thickness indicate different details of the drawing.
  - A *thick continuous line* represents a visible line — a contour of a shape or an object.
  - A *thick dashed line* represents a hidden line — a hidden contour of a shape or an object.
  - *Thin continuous lines* are used to mark the dimensions of the shape or object.
- **Dimension lines** are thin continuous lines showing the dimension of a line. They have arrowheads at both ends.
- **Projection lines** are thin continuous lines drawn perpendicular to the measurement shown. These lines do not touch the object.

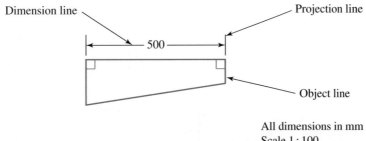

All dimensions in mm
Scale 1 : 100

- All dimensions on a drawing must be given in millimetres; however, 'mm' is not to be written on the drawing. This can be written near the scale, as shown in the diagram above.
- Dimensions must be written in the middle of the dimension line.

WORKED EXAMPLE 14 Labelling scale drawings

Label the three scale drawings of the shipping container from Worked example 13.

THINK	WRITE/DRAW
1. Draw the projection lines of the object in the plan-view drawing using thin continuous lines.	
2. Draw the dimension lines of the object in the plan-view drawing using thin continuous lines with arrowheads at both ends.	
3. Label the drawing.	The actual dimensions of the plan view of the shipping container are length 12 m and width 2.5 m. *Note:* The dimensions have to be written in mm. 12 m = 12 000 mm and 2.5 m = 2500 mm 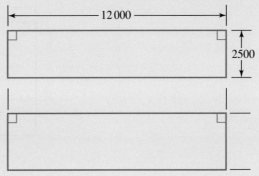
4. Draw the projection lines of the object in the front-view drawing using thin continuous lines.	

5. Draw the dimension lines of the object in the front-view drawing using thin continuous lines with arrowheads at both ends.

Note: The arrowheads of the dimension line for the height are directed from the outside of the projection line. This happens because the space between the projection lines is too small.

6. Label the drawing.

The actual dimensions of the front view of the shipping container are length 12 m and height 3 m.
Note: The dimensions have to be written in mm.
12 m = 12 000 mm and 3 m = 3000 mm

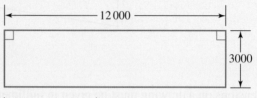

7. Draw the projection lines of the object in the side-view drawing using thin continuous lines.

8. Draw the dimension lines of the object in the side-view drawing using thin continuous lines with arrowheads at both ends.

9. Label the drawing.

The actual dimensions of the side view of the shipping container are 2.5 m wide and 3 m high.
Note: The dimensions have to be written in mm.
2.5 m = 2500 mm and 3 m = 3000 mm

Labelling circles

- A circle on a scale diagram is labelled using either the Greek letter Φ (phi) for diameter or *R* for radius.
- The diagram shows three correct notations for a 2-cm diameter.

Note: The dimensions must be written in millimetres.

WORKED EXAMPLE 15 Labelling circle scale drawings

The food can shown in the diagram has a diameter of 10 cm. Construct a scale drawing of the plan view of the food can using a scale of 1 : 4.

THINK	WRITE/DRAW
1. State the scale factor and calculate the dimensions of the object on the drawing.	Scale factor $= \dfrac{1}{4}$ Diameter on the diagram $= \dfrac{1}{4} \times 10$ $= \dfrac{10}{4}$ $= 2.5 \text{ cm}$ $= 25 \text{ mm}$
2. Draw the plan view of the object.	
3. Label the drawing.	 φ25

Note: The dimensions are written in mm.

Exercise 10.5 Creating scale drawings

learnon

| **10.5 Quick quiz** on | **10.5 Exercise** |

Individual pathways

■ PRACTISE	■ CONSOLIDATE	■ MASTER
1, 2, 5, 7, 11, 14	3, 8, 10, 12, 15	4, 6, 9, 13, 16

Fluency

1. **MC** A projection line is drawn using:
   A. a thin continuous line with two arrowheads.　　**B.** a thick dashed line.
   C. a thin continuous line.　　**D.** a thick continuous line.

2. **WE12** Calculate the dimensions required to create a scale drawing of a classroom 17 m long and 12 m wide to a scale of 1 : 100.

3. **a.** Calculate the dimensions required to create a scale drawing of a door 1.5 m wide, 2.4 m tall and 9 cm thick to a scale of 1 : 30.
   b. State the measurements that are to be written on a scale drawing.
   c. State whether the projection lines are parallel or perpendicular to the measurement shown.
   d. Write the symbol used to label the diameter of a circle on a drawing.

4. The drawings described are constructed to a scale of 1 : 50.
   a. Calculate the height, on the drawing, of a tree 3.6 m tall.
   b. Calculate the dimensions, on the drawing, of a carpet 2.8 m by 1.4 m.
   c. Calculate the height, on the drawing, of a student 1.8 m tall.

5. **a.** **WE13** Construct a plan-view, front-view and side-view scale drawing of a skip bin with drawing dimensions of length 5 cm, width 2 cm and depth 1.5 cm.
   b. **WE14** Label the three scale drawings of the skip from part **a**, given that the actual dimensions of the skip are length 2.5 m, width 1 m and depth 0.75 m.

6. Construct a plan-view, front-view and side-view scale drawing of a rectangular lunch box with drawing dimensions of length 5.3 cm, width 2.7 cm and depth 0.9 cm.

7. **WE15** The pipe shown has a diameter of 28 cm. Construct a scale drawing of the plan view of the pipe drawn to a scale of 1 : 10.

8. Calculate the dimensions to construct a scale drawing of the following shapes to the scale stated in brackets.
   a. A square of side length 1.7 cm (2 : 1)
   b. A rectangular pool table with dimensions 4.8 m by 4.0 m (1 : 80)
   c. A circular table with diameter 1.4 m (1 : 200)

9. Draw a house floor plan with actual dimensions 27.6 m by 18.3 m drawn to a scale of 1 : 600.

10. **MC** The dimension on a drawing is calculated using the formula:

    A. scale factor × actual length
    B. scale factor + actual length
    C. $\dfrac{\text{scale factor}}{\text{actual length}}$
    D. actual length − scale factor

11. Construct and label a scale drawing of the floor of a bedroom 3.5 m long and 2.5 m wide drawn to a scale of 1 : 100.

12. Construct and label a scale drawing of a window frame 1.8 m wide and 1.4 m tall drawn to a scale of 1 : 40.

13. Draw and label the projection and dimension lines on the following diagrams.
    a. Square with side length 6.1 m

    b. Rectangle 430 cm long and 210 cm wide

c. Equilateral triangle with side length 49.7 cm and height 43.0 cm

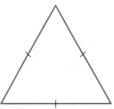

d. Circle with diameter 7.6 m

Problem solving

14. The hat box shown in the diagram has a diameter of 36 cm. Construct a scale drawing of the plan view of the box drawn to a scale of 1 : 6.

15. A sculptor plans a sculpture by first creating a scale drawing. Determine the height of the sculpture on a drawing if its actual height is intended to be 5.2 m and the scale factor is $\frac{1}{26}$.

16. Measure the dimensions of your bedroom, bed and desk and construct a scale drawing using these measurements and an appropriate scale.

LESSON
10.6 Quadrilaterals

▶ 10.6.1 Quadrilaterals

eles-4901

- Quadrilaterals are four-sided plane shapes whose interior angles sum to 360°.

Theorem 5

- The sum of the interior angles in a quadrilateral is 360°.

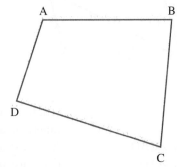

Given:	A quadrilateral ABCD
To prove:	$\angle ABC + \angle BCD + \angle ADC + \angle BAD = 360°$
Construction:	Draw a line joining vertex A to vertex C. Label the interior angles of the triangles formed.

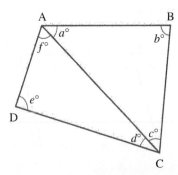

Proof:	$a + b + c = 180°$	(sum of interior angles in a triangle is 180°)
	$d + e + f = 180°$	(sum of interior angles in a triangle is 180°)
	$\Rightarrow a + b + c + d + e + f = 360°$	
	$\therefore \angle ABC + \angle BCD + \angle ADC + \angle BAD = 360°$	

▶ 10.6.2 Parallelograms

eles-5354

- A **parallelogram** is a quadrilateral with two pairs of parallel sides.

Theorem 6

- Opposite angles of a parallelogram are equal.

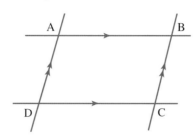

Given:	AB ∥ DC and AD ∥ BC
To prove:	∠ABC = ∠ADC
Construction:	Draw a diagonal from B to D.

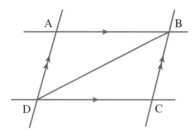

Proof:	∠ABD = ∠BDC	(alternate angles)
	∠ADB = ∠CBD	(alternate angles)
	∠ABC = ∠ABD + ∠CBD	(by construction)
	∠ADC = ∠BDC + ∠ADB	(by construction)
	∴ ∠ABC = ∠ADC	

- Conversely, if each pair of opposite angles of a quadrilateral is equal, then it is a parallelogram.

Theorem 7

- Opposite sides of a parallelogram are equal.

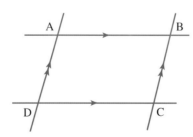

Given:	AB ∥ DC and AD ∥ BC
To prove:	AB = DC
Construction:	Draw a diagonal from B to D.

Proof:
∠ABD = ∠BDC   (alternate angles)
∠ADB = ∠CBD   (alternate angles)
BD is common to ΔABD and ΔBCD.
⇒ ΔABD ≅ ΔBCD (ASA)
∴ AB = DC

- Conversely, if each pair of opposite sides of a quadrilateral is equal, then it is a parallelogram.

Theorem 8

- The diagonals of a parallelogram bisect each other.

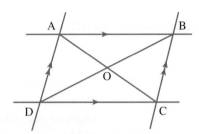

Given:	AB ‖ DC and AD ‖ BC with diagonals AC and BD
To prove:	AO = OC and BO = OD
Proof:	In ΔAOB and ΔCOD,

∠OAB = ∠OCD   (alternate angles)
∠OBA = ∠ODC   (alternate angles)
AB = CD   (opposite sides of a parallelogram)
⇒ ΔAOB ≅ ΔCOD   (ASA)
⇒ AO = OC   (corresponding sides in congruent triangles)
and BO = OD.   (corresponding sides in congruent triangles)

 ▶ Rectangles

eles-5355

- A rectangle is a parallelogram with four right angles.

Theorem 9

- A parallelogram with a right angle is a rectangle.

Given:	Parallelogram ABCD with ∠BAD = 90°
To prove:	∠BAD = ∠ABC = ∠BCD = ∠ADC = 90°
Proof:	AB ‖ CD (properties of a parallelogram)

⇒ ∠BAD + ∠ADC = 180°   (co-interior angles)
But ∠BAD = 90°.   (given)
⇒ ∠ADC = 90°
Similarly, ∠BCD = ∠ADC = 90°.
∴ ∠BAD = ∠ABC = ∠BCD = ∠ADC = 90°

Theorem 10

- The diagonals of a rectangle are equal.

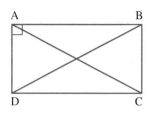

Given:	Rectangle ABCD with diagonals AC and BD
To prove:	AC = BD
Proof:	In $\triangle$ADC and $\triangle$BCD,

AD = BC	(opposite sides equal in a rectangle)
DC = CD	(common)
$\angle$ADC = $\angle$BCD = 90°	(right angles in a rectangle)
$\Rightarrow \triangle$ADC $\cong \triangle$BCD	(SAS)
$\therefore$ AC = BD	

▶ Rhombuses

eles-5356

- A **rhombus** is a parallelogram with four equal sides.

Theorem 11

- The diagonals of a rhombus are perpendicular.

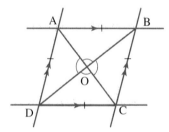

Given:	Rhombus ABCD with diagonals AC and BD
To prove:	AC $\perp$ BD
Proof:	In $\triangle$AOB and $\triangle$BOC,

AO = OC	(property of parallelogram)
AB = BC	(property of rhombus)
BO = OB	(common)
$\Rightarrow \triangle$AOB $\cong \triangle$BOC	(SSS)
$\Rightarrow \angle$AOB = $\angle$BOC	
But $\angle$AOB + $\angle$BOC = 180°.	(supplementary angles)
$\Rightarrow \angle$AOB = $\angle$BOC = 90°	
Similarly, $\angle$AOD = $\angle$DOC = 90°.	
Hence, AC$\perp$BD.	

a.

b.

THINK

a. 1. Identify the shape.

2. To determine the values of $x°$, apply theorem 6, which states that opposite angles of a parallelogram are equal.

b. 1. Identify the shape.

2. To determine the values of $y°$, apply theorem 5, which states the sum of interior angles in a quadrilateral is 360°.

3. Simplify and solve for $y°$.

WRITE

a. The shape is a parallelogram as the shape has two pairs of parallel sides.

$x° = 120°$

b. The shape is a trapezium, as one pair of opposite sides is parallel but not equal in length.

The sum of all the angles = 360°

$y° + 110° + 80° + 62° = 360°$

$$y° + 110° + 80° + 62° = 360°$$
$$y° + 252° = 360°$$
$$y° = 360° - 252°$$
$$y° = 108°$$

- A summary of the definitions and properties of quadrilaterals is shown in the following table.

Shape	Definition	Properties
Trapezium	A trapezium is a quadrilateral with one pair of opposite sides parallel.	• One pair of opposite sides is parallel but not equal in length.
Parallelogram	A parallelogram is a quadrilateral with both pairs of opposite sides parallel.	• Opposite angles are equal. • Opposite sides are equal. • Diagonals bisect each other.
Rhombus	A rhombus is a parallelogram with four equal sides.	• Diagonals bisect each other at right angles. • Diagonals bisect the angles at the vertex through which they pass.

Shape	Definition	Properties
Rectangle	A rectangle is a parallelogram whose interior angles are right angles.	• Diagonals are equal. • Diagonals bisect each other.
Square	A square is a parallelogram whose interior angles are right angles with four equal sides.	• All angles are right angles. • All side lengths are equal. • Diagonals are equal in length and bisect each other at right angles. • Diagonals bisect the vertex through which they pass (45°).

10.6.3 The midpoint theorem

eles-4905

• Now that the properties of quadrilaterals have been explored, the midpoint theorem can be tackled.

Theorem 12

• The interval joining the midpoints of two sides of a triangle is parallel to the third side and half its length.

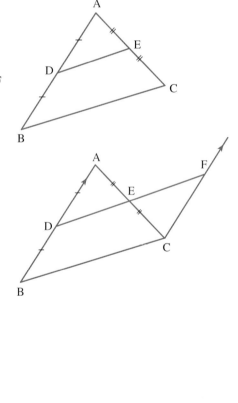

Given: $\triangle ABC$ in which $AD = DB$ and $AE = EC$

To prove: $DE \parallel BC$ and $DE = \dfrac{1}{2}BC$

Construction: Draw a line through C parallel to AB. Extend DE to F on the parallel line.

Proof:
In $\triangle ADE$ and $\triangle CEF$,
$AE = EC$ (E is the midpoint of AC, given)
$\angle AED = \angle CEF$ (vertically opposite angles)
$\angle EAD = \angle ECF$ (alternate angles)
$\Rightarrow \triangle ADE \cong \triangle CEF$ (ASA)
$\therefore AD = CF$ and $DE = EF$ (corresponding sides in congruent triangles)
So, $AD = DB = CF$.
We have $AB \parallel CF$.
So BDFC is a parallelogram.
$\Rightarrow DE \parallel BC$ (by construction)
Also, $BC = DF$.
But $DE = EF = \dfrac{1}{2}DF$, (opposite sides in parallelogram)
$\Rightarrow DE = \dfrac{1}{2}BC$ (sides in congruent triangles)
Therefore, $DE \parallel BC$
and $DE = \dfrac{1}{2}BC$.

• Conversely, if a line interval is drawn parallel to a side of a triangle and half the length of that side, then the line interval bisects each of the other two sides of the triangle.

WORKED EXAMPLE 17 Applying the midpoint theorem to determine the unknown length

In triangle ABC, the midpoints of AB and AC are D and E respectively. Determine the value of DE if
BC = 18 cm.

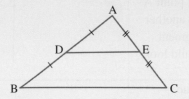

THINK	WRITE
1. Determine the midpoints on the line AB and AC.	D is the midpoint of AB and E is the midpoint of AC.
2. Apply the midpoint theorem to determine the length of DE.	$DE = \dfrac{1}{2} BC$
3. Substitute the value of BC = 18 cm into the formula.	$DE = \dfrac{1}{2} \times 18$
4. Simplify and determine the length of DE.	$DE = 9 \text{ cm}$

Digital technology: Using GeoGebra to construct a rectangle

Steps	
Open GeoGebra and select Geometry from the Perspective menu on the sidebar.	
To construct a rectangle, select the **Segment between Two Points** tool and click two distinct places on the Graphics view to construct segment AB.	

We construct two lines that are both perpendicular to **AB**, one passing through **A** and the other through **B**. To do this, click the **Perpendicular Line** tool ⊥, click point A and click the segment. To construct another line, click point B then click the segment. After these steps, your drawing should look like the one shown.	
Next, we construct a point on the line passing through point B. To do this, click the **Point** tool •A, and then click on the line passing through B. This point will be point C.	
Now, to create a line parallel to AB passing through point C, click the **Parallel line** tool ⇌, click on point C then click on the segment. Your drawing should look like the one shown.	
Using the intersect tool, we construct the intersection of the line passing C (parallel to AB) and the line passing through A. To do this, click the **Intersect** tool ✕ and click the two lines mentioned.	
Now, we hide the three lines. To do this, click the **Show/Hide Object** tool •○ and click the three lines. Notice that the lines are highlighted. Click the **Move** tool ⤢ to hide the lines.	
Use the **Segment between Two Points** tool ⤢ to connect the points to construct the rectangle.	

Click on the sides to show the value of the side lengths on the diagram. You could also use the **Distance or Length** tool ⌐̸ to show the side lengths. This tool can be used to calculate the distance between two points.	

Geogebra can be used to investigate the shortest path that touches three sides of a rectangle. Here is a link to a class activity designed to develop conceptual understandings of geometric proofs and consolidate understandings of Pythagoras' theorem and similar triangles:

https://www.resolve.edu.au/geometry-lunch-lap-trial

Students participate in an investigation to calculate the length of a path that touches three sides of a rectangle, starting and finishing at the same point on the fourth side. Students could also prove that the path forms a parallelogram.

Students will be able to apply Pythagoras' theorem and complete routine calculations to calculate the length of the hypotenuse or one of the shorter sides.

on Resources

eWorkbook   Topic 10 Workbook (worksheets, code puzzle and project) (ewbk-13298)

Interactivities Individual pathway interactivity: Quadrilaterals (int-4614)
Quadrilateral definitions (int-2786)
Quadrilaterals (int-3756)
Angles in a quadrilateral (int-3967)
Opposite angles of a parallelogram (int-6160)
Opposite sides of a parallelogram (int-6161)
Diagonals of a parallelogram (int-6162)
Diagonals of a rectangle (int-6163)
Diagonals of a rhombus (int-6164)
The midpoint theorem (int-6165)

Exercise 10.6 Quadrilaterals   learn on

10.6 Quick quiz on	10.6 Exercise

Individual pathways

■ PRACTISE	■ CONSOLIDATE	■ MASTER
1, 3, 7, 9, 12, 13, 18	2, 4, 8, 10, 14, 15, 19, 20	5, 6, 11, 16, 17, 21, 22

Fluency

1. Use the definitions of the five special quadrilaterals to decide if the following statements are true or false.

   a. A square is a rectangle.
   c. A square is a rhombus.
   b. A rhombus is a parallelogram.
   d. A rhombus is a square.

2. Use the definitions of the five special quadrilaterals to decide if the following statements are true or false.

   a. A square is a trapezium.
   c. A trapezium is a rhombus.
   b. A parallelogram is a rectangle.
   d. A rectangle is a square.

3. **WE16** Determine the value of the pronumeral in each of the following quadrilaterals.

a.

b.

c.

d.

4. Determine the values of the pronumerals in each of the following diagrams.

a.

b.

c.

d.

5. Determine the values of the pronumerals in each of the following diagrams.

a.

b.

6. Determine the values of x and y in each of the following diagrams.

a.

b.

c.

d.

Understanding

7. Draw three different trapeziums. Using your ruler, compass and protractor, decide which of the following properties are true in a trapezium.

   a. Opposite sides are equal.
   b. All sides are equal.
   c. Opposite angles are equal.
   d. All angles are equal.
   e. Diagonals are equal in length.
   f. Diagonals bisect each other.
   g. Diagonals are perpendicular.
   h. Diagonals bisect the angles they pass through.

8. Draw three different parallelograms. Using your ruler and protractor to measure, decide which of the following properties are true in a parallelogram.

   a. Opposite sides are equal.
   b. All sides are equal.
   c. Opposite angles are equal.
   d. All angles are equal.
   e. Diagonals are equal in length.
   f. Diagonals bisect each other.
   g. Diagonals are perpendicular.
   h. Diagonals bisect the angles they pass through.

9. Choose which of the following properties are true in a rectangle.

   a. Opposite sides are equal.
   b. All sides are equal.
   c. Opposite angles are equal.
   d. All angles are equal.
   e. Diagonals are equal in length.
   f. Diagonals bisect each other.
   g. Diagonals are perpendicular.
   h. Diagonals bisect the angles they pass through.

10. Name four quadrilaterals that have at least one pair of opposite sides that are parallel and equal.

11. Name a quadrilateral that has equal diagonals that bisect each other and bisect the angles they pass through.

Reasoning

12. Prove that the diagonals of a rhombus bisect each other.

13. Give reasons why a square is a rhombus, but a rhombus is not necessarily a square.

14. **WE17** ABCD is a parallelogram. X is the midpoint of AB and Y is the midpoint of DC. Prove that AXYD is also a parallelogram.

15. The diagonals of a parallelogram meet at right angles. Prove that the parallelogram is a rhombus.

16. ABCD is a parallelogram. P, Q, R and S are all midpoints of their respective sides of ABCD.

    a. Prove △PAS ≅ △RCQ.
    b. Prove △SDR ≅ △PBQ.
    c. Hence, prove that PQRS is also a parallelogram.

17. Two congruent right-angled triangles are arranged as shown. Show that PQRS is a parallelogram.

Problem solving

18. ABCD is a trapezium.

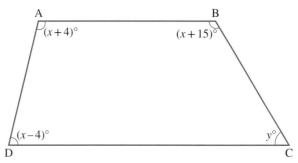

    a. Describe a fact about a trapezium.

    b. Determine the values of x and y.

19. ABCD is a kite where $AC = 8$ cm, $BE = 5$ cm and $ED = 9$ cm

    a. Determine the exact values of:

        i. x

        ii. y.

    b. Evaluate angle BAD and hence angle BCD. Write your answer in degrees and minutes, correct to the nearest minute.

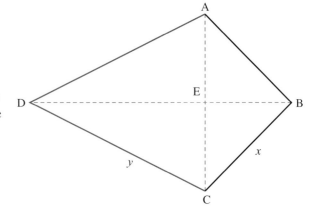

20. Pool is played on a rectangular table. Balls are hit with a cue and bounce off the sides of the table until they land in one of the holes or pockets.

    a. Draw a rectangular pool table measuring 5 cm by 3 cm on graph paper. Mark the four holes, one in each corner.

    b. A ball starts at A. It is hit so that it travels at a 45° diagonal across the grid. When it hits the side of the table, it bounces off at a 45° diagonal as well. Determine how many sides the ball bounces off before it goes in a hole.

    c. A different size table is 7 cm by 2 cm. Determine how many sides a ball bounces off before it goes in a hole when hit from A in the same way.

    d. Complete the following table.

Table size	Number of sides hit
5 cm × 3 cm	
7 cm × 2 cm	
4 cm × 3 cm	
4 cm × 2 cm	
6 cm × 3 cm	
9 cm × 3 cm	
12 cm × 4 cm	

e. Can you see a pattern? Determine how many sides a ball would bounce off before going in a hole when hit from A on an $m \times n$ table.

f. The ball is now hit from B on a 5 cm × 3 cm pool table. Determine how many *different* paths a ball can take when hit along 45° diagonals.
Do these paths all hit the same number of sides before going in a hole? Does the ball end up in the same hole each time?
Justify your answer.

B

g. The ball is now hit from C along the path shown. Determine what type of triangles and quadrilaterals are formed by the path of the ball with itself and the sides of the table. Determine if any of the triangles are congruent.

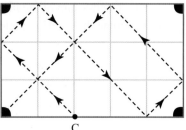

C

21. ABCD is called a cyclic quadrilateral because it is inscribed inside a circle.

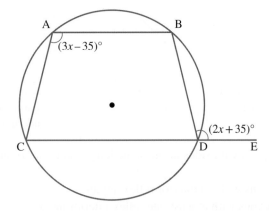

A characteristic of a cyclic quadrilateral is that the opposite angles are supplementary. Determine the value of x.

22. The perimeter of this kite is 80 cm. Determine the exact value of x.

LESSON
10.7 Polygons

▶ 10.7.1 Polygons

eles-4906

- **Polygons** are closed shapes that have three or more straight sides.

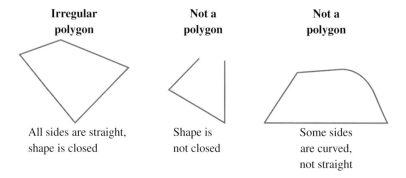

Irregular polygon	Not a polygon	Not a polygon
All sides are straight, shape is closed	Shape is not closed	Some sides are curved, not straight

- **Regular polygons** are polygons with sides of the same length and interior angles of the same size, like the hexagon shown below.
- **Convex polygons** are polygons with no interior reflex angles.
- **Concave polygons** are polygons with at least one reflex interior angle.

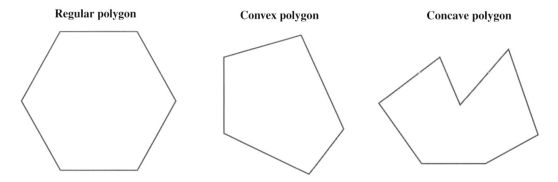

Regular polygon   Convex polygon   Concave polygon

Interior angles of a polygon

- The sum of the interior angles of a polygon is given by the following formula.

> **Sum of interior angles of a polygon**
>
> **Sum of interior angles $= 180° \times (n - 2)$**
>
> **where $n =$ the number of sides of the polygon.**

Calculate the value of the pronumerals in the figure shown.

THINK	WRITE
1. Angles a and $110°$ form a straight line, so they are supplementary (add to $180°$).	$a + 110° = 180°$ $a + 110° - 110° = 180° - 110°$ $a = 70°$
2. The interior angles of a triangle sum to $180°$.	$b + a + 80° = 180$
3. Substitute $70°$ for a and solve for b.	$b + 70° + 80° = 180°$ $b + 150° = 180°$ $b = 30°$
4. Write the value of the pronumerals.	$a = 70°$, $b = 30°$

Exterior angles of a polygon

- The exterior angles of a polygon are formed by the side of the polygon and an extension of its adjacent side.

  For example, x, y and z are exterior angles for the polygon (triangle) below, and q, r, s and t are the exterior angles of the quadrilateral.

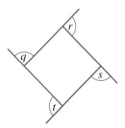

- The exterior angle and interior angle at that vertex are supplementary (add to $180°$). For example, in the triangle above, $x + a = 180°$.
- Exterior angles of polygons can be measured in a clockwise or anticlockwise direction.
- In a regular polygon, the size of the exterior angle can be found by dividing $360°$ by the number of sides.

Exterior angles of a regular polygon

$$\text{Exterior angles of a regular polygon} = \frac{360°}{n}$$

where n = the number of sides of the regular polygon.

- The sum of the exterior angles of a polygon equals $360°$.
- The exterior angle of a triangle is equal to the sum of the opposite interior angles.

Resources

> **eWorkbook**   Topic 10 Workbook (worksheets, code puzzle and project) (ewbk-13298)
>
> **Interactivities**   Individual pathway interactivity: Polygons (int-4615)
>   Interior angles of a polygon (int-6166)
>   Exterior angles of a polygon (int-6167)

Exercise 10.7 Polygons   **learn**

10.7 Quick quiz	**10.7 Exercise**

Individual pathways

■ **PRACTISE**	■ **CONSOLIDATE**	■ **MASTER**
1, 3, 6, 7, 11, 14	2, 4, 8, 12, 15, 16	5, 9, 10, 13, 17, 18

Fluency

1. **WE18** Calculate the values of the pronumerals in the diagrams shown.

    a.

    b.

    c.

    d.

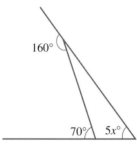

2. Determine the value of the pronumeral in each of the following diagrams.

    a.

    b.

    c.

    d.

3. For the triangles shown, evaluate the pronumerals and determine the sizes of the interior angles.

a.

b.

c.

d.

e.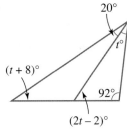

4. For each of the five quadrilaterals shown:

   i. label the quadrilateral as regular or irregular

   ii. determine the value of the pronumeral.

a.

b.

c.

d.

e.

5. For each of the four polygons shown:

   i. calculate the sum of the interior angles

   ii. determine the value of the pronumeral.

a.

b.

c.

d.

6. Explain how the interior and exterior angles of a polygon are related to the number of sides in a polygon.

Understanding

7. The photograph shows a house built on the side of a hill. Use your knowledge of angles to determine the values of the pronumerals. Show full working.

8. Determine the values of the four interior angles of the front face of the building in the photograph shown. Show full working.

9. Determine the values of the pronumerals for the irregular polygons shown. Show full working.

a.

b.

10. Calculate the size of the exterior angle of a regular hexagon (6 sides).

Reasoning

11. State whether the following polygons are regular or irregular. Give a reason for your answer.

a.

b.

c.

d.

e.

f.

12. A diagonal of a polygon joins two vertices.

a. Calculate the number of diagonals in a regular polygon with:

   i. 4 sides   ii. 5 sides   iii. 6 sides   iv. 7 sides.

b. Using your results from part a, show that the number of diagonals for an n-sided polygon is $\frac{1}{2}n(n-3)$.

13. The exterior angle of a polygon can be calculated using the formula:

$$\text{Exterior angle} = \frac{360°}{n}$$

Use the relationship between interior and exterior angles of a polygon to write a formula for the internal angle of a regular polygon.

Problem solving

14. a. Name the polygon that best describes the road sign shown.
   b. Determine the value of the pronumeral m.

15. The diagram shows a regular octagon with centre O.

    a. Calculate the size of ∠CBD.
    b. Calculate the size of ∠CBO.
    c. Calculate the size of the exterior angle of the octagon, ∠ABD.
    d. Calculate the size of ∠BOC.

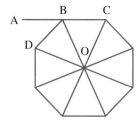

16. Answer the following questions for the given shape.

    a. Evaluate the sum of the interior angles of this shape.
    b. Determine the value of the pronumeral x.

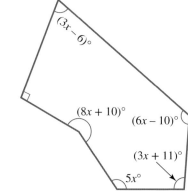

17. ABCDEFGH is an eight-sided polygon.

    a. Evaluate the sum of the interior angles of an eight-sided
       polygon.
    b. Determine the value of the pronumeral x.

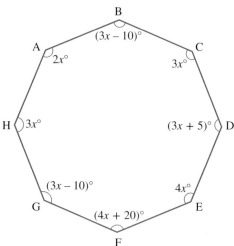

18. Answer the following questions for the given shape.

    a. Evaluate the sum of the interior angles of this shape.
    b. Determine the value of the pronumeral x.

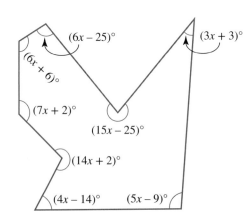

LESSON
10.8 Review

10.8.1 Topic summary

Polygons

- Polygons are closed shapes with straight sides.
- The number of sides a polygon has is denoted n.
- The sum of the interior angles is given by the formula:
  Interior angle sum = $180°(n-2)$
- **Regular polygons** have all sides the same length and all interior angles equal.
- **Convex polygons** have no internal reflex angles.
- **Concave polygons** have at least one internal reflex angle.
- The exterior angles of a regular polygon arc given by the formula:
  Exterior angles $= \dfrac{360°}{n}$

DEDUCTIVE GEOMETRY

Quadrilaterals

- Quadrilaterals are four-sided polygons.
- The interior angles sum to 360°.
- There are many types of quadrilaterals:
  - **Trapeziums** have 1 pair of parallel sides.
  - **Parallelograms** have 2 pairs of parallel sides.
  - **Rhombuses** are parallelograms that have 4 equal sides.
  - A **rectangle** is a parallelogram whose interior angles are right angles.
  - A **square** is a rectangle with 4 equal sides.

Calculating measurements from scale drawings

- Areas and costs:

  Cost required $= \dfrac{\text{area of material}}{\text{area per squared unit}} \times$ cost per unit

- Amount of material and cost:

  Amount of paint required = area to be painted × amount of paint per squre metre

  Total cost = amount of paint required × cost of paint per litre

- Number of bricks:

  Number of bricks $= \dfrac{\text{area of the wall}}{\text{area of the expected side of the brick}}$

Angles and parallel lines

- Supplementary angles are angles that add up to 180°.
- Complementary angles are angles that add up to 90°.

| Alternate angles a and b are equal. | Corresponding angles a and b are equal. | Co-interior angles are supplementary: $a + b = 180°$ |

Congruent and similar figures

- Congruent figures have the same size and shape.
- The symbol for congruence is ≅.
- The following tests can be used to determine whether two triangles are congruent:
  - Side-side-side (SSS)
  - Side-angle-side (SAS)
  - Angle-angle-side (AAS)
  - Right angle-hypotenuse-side (RHS)
- Similar figures have the same shape but different size.
- The symbol for similarity is ~.
- Scale factor $= \dfrac{\text{image side length}}{\text{object side length}}$
- The following tests can be used to determine whether two triangles are similar:

Creating scale drawings

- Scale drawings are drawings on paper of real-life objects.
- All dimensions of the drawn object are kept in the same ratio to the actual dimensions of the object using a scale.
- The dimension on a drawing can be calculated using the formula:
  Dimension on the drawing = scale factor × actual dimension
- Conventions and line styles:
  - A *thick continuous line* represents a visible line or contour of a shape or object.
  - A *thick dashed line* represents a hidden line or contour of a shape or object.
  - *Thin continuous lines* are used to mark the dimensions of a shape or object.
- Dimension lines are thin continuous lines showing the dimension of a line. They have arrowheads at both ends.
- Projection lines are thin continuous lines drawn perpendicular to the measurement shown. These lines do not touch the object.

All dimensions in mm
Scale 1 : 100

- A circle on a scale diagram is labelled using either the Greek letter φ (phi) for diameter or R for radius. The diameter should be stated in millimetres.

10.8.2 Success criteria

Tick the column to indicate that you have completed the lesson and how well you have understood it using the traffic light system.

(**Green:** I understand; **Yellow:** I can do it with help; **Red:** I do not understand)

Lesson	Success criteria	🟢	⚪	⚫
10.2	I can apply properties of straight lines and triangles to determine the value of an unknown angle.			
	I can construct simple geometric proofs for angles in triangles or around intersecting lines.			
	I can prove that triangles are congruent by applying the appropriate congruency test.			
10.3	I can identify similar figures.			
	I can calculate the scale factor in similar figures.			
	I can show that two triangles are similar using the appropriate similarity test.			
10.4	I can use scale drawings to calculate lengths, perimeters and areas.			
	I can use scale drawings to solve real-world problems.			
10.5	I can calculate the dimensions for a scale drawing.			
	I can construct and label scale drawings.			
10.6	I can identify the different types of quadrilaterals.			
	I can construct simple geometric proofs for angles, sides and diagonals in quadrilaterals.			
10.7	I can identify regular and irregular polygons.			
	I can calculate the sum of the interior angles of a polygon.			
	I can determine the exterior angles of a regular polygon.			

10.8.3 Project

Enlargement activity

Enlargement is the construction of a bigger picture from a small one. The new picture is identical to the old one except that it is bigger. The new picture is often called the image. This can also be called creating a similar figure.

The geometrical properties shared by a shape and its image under enlargement can be listed as:
- lines are enlarged as lines
- sides are enlarged to corresponding sides by the same factor
- matching angles on the two shapes are equal.

In this activity, we will start with a small cartoon character and 'blow it up' to almost life-size.

Equipment: ruler, pencil, cartoon print, butcher's paper or some other large piece of paper.

1. Do some research on the internet and select a cartoon character or any character of your choice.
2. Draw a grid of 2-cm squares over the small cartoon character. Example: The cat is 9 squares wide and 7 squares tall.
3. Label the grids with letters across the top row and numbers down the first column.
4. Get a large piece of paper and draw the same number of squares. You will have to work out the ratio of similitude (e.g. 2 cm : 8 cm).
5. If your small cartoon character stretches from one side of the 'small' paper (the paper the image is printed on) to the other, your 'large' cat must stretch from one side of the 'big' paper to the other. Your large grid squares may have to be 8 cm by 8 cm or larger, depending on the paper size.
6. Draw this enlarged grid on your large paper. Use a metre ruler or some other long straight-edged tool. Be sure to keep all of your squares the same size.
   - At this point, you are ready to draw. Remember, you do NOT have to be an artist to produce an impressive enlargement.
   - All you do is draw EXACTLY what you see in each small cell into its corresponding large cell.
   - For example, in cell B3 of the cat enlargement, you see the tip of his ear, so draw this in the big grid.
   - If you take your time and are very careful, you will produce an extremely impressive enlargement.
   - What you have used is called a 'ratio of similitude'. This ratio controls how large the new picture will be.

A 2 : 5 ratio will give you a smaller enlargement than a 2 : 7 ratio, because for every 2 units on the original you are generating only 5 units of enlargement instead of 7.

If the cat ratio is 1 : 4, it produces a figure that has a linear measure that is four times bigger.

The big cat's overall **area**, however, will be **16 times larger** than the small cat's. This is because area is found by taking length times width.

The length is 4 times longer and the width is 4 times longer. Thus, the **area** is $4 \times 4 = 16$ times **larger** than the original cat.

The overall **volume** would be $4 \times 4 \times 4$ or 64 **times larger!** This means that the big cat would weigh 64 times more than the small cat.

Resources

 eWorkbook   Topic 10 Workbook (worksheets, code puzzle and project) (ewbk-13298)

 Interactivities Crossword (int-9178)

Sudoku puzzle (int-3597)

Fluency

1. Select a pair of congruent triangles in each of the following sets of triangles, giving a reason for your answer. All angles are in degrees and side lengths in cm. (The figures are not drawn to scale.)

   a.

   b.

2. Determine the value of the pronumeral in each pair of congruent triangles. All angles are given in degrees and side lengths in cm.

   a.

   b.

   c.

3. a. Prove that the two triangles shown in the diagram are congruent.

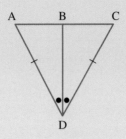

b. Prove that ΔPQR is congruent to ΔQPS.

4. Test whether the following pairs of triangles are similar. For similar triangles, determine the scale factor. All angles are in degrees and side lengths in cm.

a.

b.

c.

5. Determine the value of the pronumeral in each pair of similar triangles. All angles are given in degrees and side lengths in cm.

a.

b.

c.
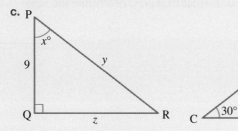

6. Prove that △ABC ~ △EDC.

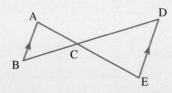

7. Prove that △PST ~ △PRQ.

8. State the definition of a rhombus.

9. **MC** Two corresponding sides in a pair of similar octagons have lengths of 4 cm and 60 mm. The respective scale factor in length is:

A. 1 : 15   B. 3 : 20   C. 2 : 3   D. 3 : 2

10. A regular nonagon has side length x cm. Use a scale factor of $\dfrac{x+1}{x}$ to calculate the side length of a similar nonagon.

11. **MC** Kirrily wants to paint a feature wall in her lounge room. The wall is 6.2 m by 2.8 m. The paint she wants to use requires 2 L per m² and costs \$9.85 per litre. The cost to put two coats of paint on the wall is closest to:

    A. \$683.98   **B.** \$341.99   **C.** \$586.58   **D.** \$872.50

12. Given a scale of 1 : 300 for the floor plan of the apartment shown, calculate:
    a. the length and the width of the apartment, given that it measures 6 cm by 4 cm on the diagram
    b. the perimeter of the apartment.

Problem solving

13. ABC is a triangle. D is the midpoint of AB, E is the midpoint of AC and F is the midpoint of BC. DG ⊥ AB, EG ⊥ AC and FG ⊥ BC.
    a. Prove that $\triangle GDA \cong \triangle GDB$.
    b. Prove that $\triangle GAE \cong \triangle GCE$.
    c. Prove that $\triangle GBF \cong \triangle GCF$.
    d. State what this means about AG, BG and CG.
    e. A circle centred at G is drawn through A. Determine what other points it must pass through.

14. PR is the perpendicular bisector of QS. Prove that $\triangle PQS$ is isosceles.

15. A sculptor plans a sculpture by first creating a scale drawing. Determine the height of the sculpture on a drawing if its actual height is intended to be 6.3 m and the scale factor is $\frac{1}{20}$.

16. A brick has length 210 mm and height 55 mm. A bricklayer is building a wall 10 m long and 8 m high. Calculate:
    a. the number of bricks required
    b. the total cost of building the wall if one brick costs $2.15.

17. Consider the juice container shown and the template required to make it. The scale of the diagram is 1 : 5.

    a. Calculate the length and the width of the packaging template, given they both measure 4 cm on the diagram.
    b. Estimate the total area of material required.
    c. If the cost of materials is $0.55 per square metre, calculate how much it would cost to make this juice container, to the nearest cent.

18. The objects in the diagram shown are drawing instruments.

    a. Calculate the scale factor of the drawing, given the length that is 32 mm measures 24 mm on the scale drawing.
    b. Calculate the actual measurements of the unknown lengths x, y and z, given they measure 6 mm, 21 mm and 51 mm, respectively, on a scale drawing.
    c. Determine the lengths of x, y and z on a drawing with a scale of 5 : 4.
    d. Construct a scale drawing using the new drawing dimensions from part c. Clearly label all measurements.

19. Name any quadrilaterals that have diagonals that bisect the angles they pass through.

20. State three tests that can be used to show that a quadrilateral is a rhombus.

21. Prove that WXYZ is a parallelogram.

22. Prove that the diagonals in a rhombus bisect the angles they pass through.

23. Explain why the triangles shown are not congruent.

24. Prove that the angles opposite the equal sides in an isosceles triangle are equal.

25. Name any quadrilaterals that have equal diagonals.

26. This 8 cm by 12 cm rectangle is cut into two sections as shown.

   a. Draw labelled diagrams to show how the two sections can be rearranged to form:
      i. a parallelogram
      ii. a right-angled triangle
      iii. a trapezium.
   b. Comment on the perimeters of the figures.

on To test your understanding and knowledge of this topic, go to your learnON title at www.jacplus.com.au and complete the **post-test**.

Answers

Topic 10 Deductive geometry

10.1 Pre-test

1. $17.5°$
2. $69°$
3. B
4. $x = 12$, $y = 11$
5. $x = 58$, $y = -76$, $z = 308$
6. A
7. D
8. $95°$
9. $45°$
10. $72°$
11. $1080°$
12. C
13. $151°$
14. A
15. A

10.2 Angles, triangles and congruence

1. a. $a = 56°$ b. $b = 30°$ c. $c = 60°$
   d. $d = 120°$ e. $e = 68°$
2. a. $x = 115°$
   b. $y = 80°$
   c. $a = 120°$, $b = 60°$, $c = 120°$, $d = 60°$
   d. $x = 30°$
3. a. I and III, SAS   b. I and II, AAS
   c. II and III, RHS   d. I and II, SSS
4. a. $x = 6$, $y = 60°$ b. $x = 80°$, $y = 50°$
   c. $x = 32°$, $y = 67°$ d. $x = 45°$, $y = 45°$
5. a. $b = 65°$, $c = 10°$, $d = 50°$, $e = 130°$
   b. $f = 60°$, $g = 60°$, $h = 20°$, $i = 36°$
6. a. $x = 3$ cm
   b. $x = 85°$
   c. $x = 80°$, $y = 30°$, $z = 70°$
   d. $x = 30°$, $y = 7$ cm
   e. $x = 40°$, $y = 50°$, $z = 50°$, $m = 90°$, $n = 90°$
7. C, D
8. $x = 110°$, $y = 110°$, $z = 4$ cm, $w = 7$ cm
9. Sample responses can be found in the worked solutions in the online resources.
10. Sample responses can be found in the worked solutions in the online resources.
    a. Use SAS.   b. Use SAS.   c. Use ASA.
    d. Use ASA.   e. Use SSS.
11. Sample responses can be found in the worked solutions in the online resources.
12. The third sides are not necessarily the same.
13. Sample responses can be found in the worked solutions in the online resources.
14. Corresponding sides are not the same.
15. Sample responses can be found in the worked solutions in the online resources.
16. Use SSS. Sample responses can be found in the worked solutions in the online resources.
17. $x = 20°$, $y = 10°$, $z = 40°$
18. Sample responses can be found in the worked solutions in the online resources.

10.3 Scale factor and similar triangles

1. a. i and iii, RHS   b. i and ii, SAS
   c. i and iii, SSS   d. i and iii, AAA
   e. i and ii, SSS
2. a. Triangles PQR and ABC
   b. Triangles ADB and ADC.
   c. Triangles PQR and TSR.
   d. Triangles ABC and DEC.
   e. Triangles ABC and DEC.
3. a. $\dfrac{AB}{AD} = \dfrac{BC}{DE} = \dfrac{AC}{AE}$ b. $f = 9$, $g = 8$
4. $x = 4$
5. $x = 20°$, $y = 2\dfrac{1}{4}$
6. a. AAA   b. $x = 3$, $y = 4$
7. a. $x = 12$
   b. $x = 4$ cm
   c. $x = 16$
   d. $x = 8$ cm, $n = 60°$, $m = 70°$
8. a. $x = 7.5$, $y = 6.4$ b. $x = 8$
9. $x = 27$
10. a. $x = 1$, $y = 7\sqrt{2}$ b. $x = 2.5$, $y = 3.91$
11. Sample responses can be found in the worked solutions in the online resources.
12. a. $\angle ABD = \angle ABC$ (common angle)
    $\angle ADB = \angle BAC = 90°$
    $\Delta ABD \sim \Delta ACB$ (AAA)
    b. $\angle ACD = \angle BCA$ (common angle)
    $\angle ADC = \angle CAB = 90°$
    $\Delta ACD \sim \Delta ACB$ (AAA)
13. Congruent triangles must be identical; that is, the angles must be equal and the side lengths must be equal. Therefore, it is not enough just to prove that the angles are equal.
14. The taller student's shadow is 3.12 m long.
15. Length $= 6$ m and width $= 1$ m
16. The slide is 7.23 m long.
17. $\angle FEO = \angle OGH$ (alternate angles equal as EF ∥ HG)
    $\angle EFO = \angle OHG$ (alternate angles equal as EF ∥ HG)
    $\angle EOF = \angle HOG$ (vertically opposite angles equal)
    $\therefore \Delta EFO \sim \Delta GHO$ (equiangular)
18. a. 20.7 m
    b. 15.6 m
    c. Length $= 21$ m
    Wingspan $= 15$ m
    d. Due to rounding up, the length has gone from 20.7 m to 21 m. This is an increase of 30 cm. Due to rounding down, the wingspan has gone from 15.6 m to 15 m. This is a reduction of 60 cm.
19. Radius $= 0.625$ m
20. $x = 6$ or $x = 11$

10.4 Calculating measurements from scale drawings

1. a. Width = 7.5 m
      Length = 12.5 m
   b. 40 m

2. Radius = 12 m
   Circumference = 75 m

3. a. Height = 1 m
      Big glass width = 0.6 m
      Small glass width = 0.3 m
   b. Area of big glass = 0.6 m^2
      Area of small glass = 0.3 m^2
   c. 0.9 m^2

4. Radius = 30 cm and area = 2827 cm^2

5. a. Length = 28 cm
      Width = 12 cm
   b. 336 cm^2
   c. 3 cents each

6. a. 20.16 L
   b. $316.51

7. a. Length = 1.5 m
      Height = 2.5 m
   b. 3.75 m^2
   c. 9.75 L
   d. $124.80

8. a. Width = length = 0.26 m
   b. 0.0676 m^2
   c. 1.2 m^2
   d. 2.4 L
   e. $22.20

9. a. 6 m, 2.5 m   b. 15 m^2 c. 1500 bricks
   d. $1125

10. a. Length = width = 21 cm
    b. 441 cm^2
    c. 2 cents each

11. a. 12.75 m, 1.125 m   b. 14.34 m^2
    c. 250 bricks   d. $412.50

12. a. Length = 2.4 m   b. Width = 1.2 m
    c. Perimeter = 12 m   d. Area = 63 m^2

13. Radius = 18 cm
    Circumference = 113.10 cm
    Area = 1017.88 cm^2

14. a. 3221 bricks   b. $6280.95

15. a. Length = 64 cm
       Width = 40 cm
    b. 0.256 m^2
    c. $0.55

10.5 Creating scale drawings

1. C

2. Length = 17 cm
   Width = 12 cm

3. a. Width = 5 cm; height = 8 cm; depth = 0.3 cm
   b. Actual measurements

c. Perpendicular to the measurement shown
d. Φ (the Greek letter phi)

4. a. Height = 7.2 cm   b. 5.6 cm; 2.8 cm
   c. Height = 3.6 cm

5. a.
   Plan view

   Front view   Side view

   b.
   Plan view

   Front view   Side view

6.
   Plan view

   Front view   Side view

7.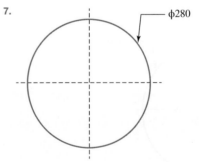

8. a. 3.4 cm   b. 6 cm; 5 cm   c. 0.7 cm

9.

10. A

11.

12.

13. a.

b.

c.

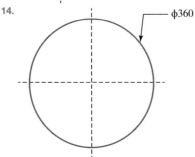

d. $\phi 7600$

14. $\phi 360$

15. 20 cm

16. Sample responses can be found in the worked solutions in the online resources.

10.6 Quadrilaterals

1. a. True   **b.** True
   c. True   **d.** False

2. a. False   **b.** False
   c. False   **d.** False

3. a. $x = 145°$
   b. $t = 174°$
   c. $m = 66°$
   d. $q = 88°$

4. a. $a = 35°$, $b = 65°$
   b. $c = 62°$, $d = 28°$, $e = 90°$
   c. $f = 40°$, $g = 140°$
   d. $h = 75°$, $i = 75°$

5. a. $x = 69°$
   b. $x = 26°$, $y = 128°$

6. a. $x = 36°$, $y = 62°$
   b. $x = 5\,\text{cm}$, $y = 90°$
   c. $x = 10°$, $y = 70°$
   d. $x = 40°$, $y = 60°$

7. None are true, unless the trapezium is a regular trapezium, then e is true.

8. a, c, f

9. a, c, d, e, f

10. Parallelogram, rhombus, rectangle, square

11. Square

12. Use AAS. Sample responses can be found in the worked solutions in the online resources.

13. All the sides of a square are equal, so a square is a special rhombus. But the angles of a rhombus are not equal, so a rhombus can't be a square.

14. AX ∥ DY because ABCD is a parallelogram.
AX = DY (given)
∴ AXYD is a parallelogram since opposite sides are equal and parallel.

15. Sample responses can be found in the worked solutions in the online resources.

16. a. Use SAS. Sample responses can be found in the worked solutions in the online resources.
   b. Use SAS. Sample responses can be found in the worked solutions in the online resources.
   c. Opposite sides are equal. Sample responses can be found in the worked solutions in the online resources.

17. PS = QR (corresponding sides in congruent triangles are equal)
PS ∥ QR (alternate angles are equal)
∴ PQRS is a parallelogram since one pair of opposite sides are parallel and equal.

18. a. One pair of opposite sides are parallel.
   b. $x = 90°$, $y = 75°$

19. a. i. $x = \sqrt{41}$ **ii.** $y = \sqrt{97}$
   b. $\angle BAD = \angle BCD = 117°23'$

20. a.

   b. 6
   c. 7

Table size	Number of sides hit
$5\,cm \times 3\,cm$	6
$7\,cm \times 2\,cm$	7
$4\,cm \times 3\,cm$	5
$4\,cm \times 2\,cm$	1
$6\,cm \times 3\,cm$	1
$9\,cm \times 3\,cm$	2
$12\,cm \times 4\,cm$	2

   e. If the ratio of the sides is written in simplest form, the pattern is $m + n - 2$.

   f. There are two routes for the ball when hit from B. Either 2 or 3 sides are hit. The ball does not end up in the same hole each time.
A suitable justification would be a diagram — student to draw.

   g. Isosceles triangles, a square, a rectangle and a trapezium. There are two sets of congruent triangles.

21. $70°$

22. $x = \sqrt{10}\,cm$

10.7 Polygons

1. a. $m = 60°$ b. $a = 45°$, $b = 45°$
    c. $t = 35°$ d. $x = 10°$

2. a. $a = 85°$, $b = 50°$, $c = 45°$
    b. $d = 140°$, $e = 110°$, $f = 110°$
    c. $g = 90°$, $h = 110°$, $i = 70°$
    d. $j = 100°$, $k = 100°$, $l = 130°$, $m = 130°$

3. a. $y = 35°$ b. $l = 5°$ c. $n = 81°$
    d. $x = 15°$ e. $t = 30°$

4. a. i. Irregular   ii. $x = 95°$
    b. i. Irregular   ii. $p = 135°$
    c. i. Irregular   ii. $t = 36°$
    d. i. Irregular   ii. $y = 70°$
    e. i. Irregular   ii. $p = 36°$

5. a. i. $540°$ ii. $b = 110°$
    b. i. $720°$ ii. $c = 134°$
    c. i. $900°$ ii. $d = 24°$
    d. i. $720°$ ii. $h = 85°$

6. The sum of the interior angles is based on the number of sides of the polygon. The size of the exterior angle can be found by dividing $360°$ by the number of sides.

7. $w = 75°$, $x = 105°$, $y = 94°$, $z = 133°$

8. $82.5°, 82.5°, 97.5°, 97.5°$

9. a. $a = 120°$, $b = 120°$, $c = 60°$, $d = 60°$, $e = 120°$, $f = 240°$
    b. $m = 10°$, $n = 270°$, $o = 50°$

10. $60°$

11. a. Regular: all sides and interior angles are equal.
    b. Irregular: neither the sides nor the internal angle are equal.
    c. Regular: all sides and interior angles are equal.
    d. Regular: all sides and interior angles are equal.

    e. Irregular: the sides are all equal, but the interior angles are not equal.
    f. Regular: all sides and interior angles are equal.

12. a. i. 2   ii. 5   iii. 9   iv. 14
    b. Sample responses can be found in the worked solutions in the online resources.

13. Internal angle $= 180° - \dfrac{360°}{n}$

14. a. Equilateral triangle
    b. $m = 150°$

15. a. $135°$ b. $67.5°$
    c. $45°$ d. $45°$

16. a. $720°$ b. $x = 25°$

17. a. $1080°$ b. $43°$

18. a. $1080°$ b. $x = 17°$

Project

Students will apply the knowledge of deductive geometry to enlarge a cartoon character to almost life-size. Sample responses can be found in the worked solutions in the online resources.

10.8 Review questions

1. a. I and III, ASA or SAS
    b. I and II, RHS

2. a. $x = 8\,cm$
    b. $x = 70°$
    c. $x = 30°$, $y = 60°$, $z = 90°$

3. Sample responses can be found in the worked solutions in the online resources.
    a. Use SAS.   b. Use ASA.

4. a. Similar, scale factor $= 1.5$
    b. Not similar
    c. Similar, scale factor $= 2$

5. a. $x = 48°$, $y = 4.5\,cm$
    b. $x = 86°$, $y = 50°$, $z = 12\,cm$
    c. $x = 60°$, $y = 15\,cm$, $z = 12\,cm$

6. Use the equiangular test. Sample responses can be found in the worked solutions in the online resources.

7. Use the equiangular test. Sample responses can be found in the worked solutions in the online resources.

8. A rhombus is a parallelogram with four equal sides.

9. C

10. $x + 1$

11. A

12. a. $18\,m$; $12\,m$
    b. $60\,m$

13. a. Use SAS. Sample responses can be found in the worked solutions in the online resources.
    b. Use SAS. Sample responses can be found in the worked solutions in the online resources.
    c. Use SAS. Sample responses can be found in the worked solutions in the online resources.
    d. They are all the same length.
    e. B and C

14. Use SAS.
PQ = PS (corresponding sides in congruent triangles are equal)
Sample responses can be found in the worked solutions in the online resources.

15. 31.5 m

16. a. 6927 bricks   b. $14 893.05

17. a. Length = width = 20 cm
b. $400 \, \text{cm}^2$
c. 2 cents each

18. a. 3 : 4
b. $x = 8$ mm; $y = 28$ mm; $z = 68$ mm
c. $x = 10$ mm; $y = 35$ mm; $z = 85$ mm
d.

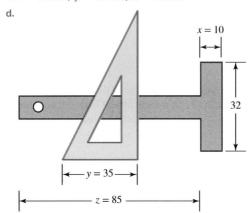

19. Rhombus, square

20. A quadrilateral is a rhombus if:
1. all sides are equal
2. the diagonals bisect each other at right angles
3. the diagonals bisect the angles they pass through.

21. WZ ∥ XY (co-interior angles are supplementary) and WZ = XY (given)
∴ WXYZ is a parallelogram since one pair of sides is parallel and equal.

22. Sample responses can be found in the worked solutions in the online resources.

23. Corresponding sides are not the same.

24. Sample responses can be found in the worked solutions in the online resources.

25. Rectangle, square

26. a. i.

ii.

iii.

b. Perimeter of rectangle = 40 cm
Perimeter of parallelogram = 44 cm
Perimeter of triangle = 48 cm
Perimeter of trapezium = 44 cm
The triangle has the largest perimeter and the rectangle has the smallest.

11 Probability

LESSON SEQUENCE

LESSON
11.1 Overview

Why learn this?

Probability is a broad and interesting area of mathematics that affects our day-to-day lives far more than we can imagine. Here is a fun fact: did you know there are so many possible arrangements of the 52 cards in a deck $\left(52! = 8.0658 \times 10^{67}\right)$ that the probability of ever getting the same arrangement after shuffling is virtually zero? This means every time you shuffle a deck of cards, you are almost certainly producing an arrangement that has never been seen before. Probability is also a big part of computer and board games; for example, the letters X and Q in Scrabble are worth more points because you are less likely to be able to form a word using those letters. It goes without saying that probability is a big part of any casino game and of the odds and payouts when gambling on the outcome of racing or sports.

Although it is handy to know probability factoids and understand gambling, this isn't the reason we spend time learning probability. Probability helps us build critical thinking skills, which are required for success in almost any career and even just for navigating our own lives. For example, if you were told your chance of catching a rare disease had doubled, you probably wouldn't need to worry, as a 1-in-a-million-chance becoming a 2-in-a-million chance isn't a significant increase in the probability of you developing the disease. On the other hand, if a disease has a 1% mortality rate, that may seem fairly low, but it means that if a billion people developed that disease, then 10 million would die. Using probability to understand risk helps us steer clear of manipulation by advertising, politicians and the media. Building on this understanding helps us as individuals to make wise decisions in our day-to-day lives, whether it be investing in the stock market, avoiding habits that increase our risk of sickness, or building our career.

1. Calculate $\Pr(A \cap B)$ if $\Pr(A = 0.4)$, $\Pr(B = 0.3)$, and $\Pr(A \cup B) = 0.5$

2. If events A and B are mutually exclusive, and $\Pr(B) = 0.38$ and $\Pr(A \cup B) = 0.89$, calculate $\Pr(A)$.

3. State whether the events $A = \{$drawing a red marble from a bag$\}$ and $B = \{$rolling a 1 on a die$\}$ are independent or dependent.

4. The Venn diagram shows the number of university students in a group of 25 who own a computer and/or tablet.
   In simplest form, calculate the probability that a university student selected at random will own only a tablet.

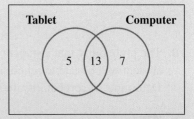

5. **MC** Two unbiased four-sided dice are rolled. Determine the probability that the total sum of two face-down numbers obtained is 6.

   A. $\dfrac{1}{8}$ **B.** $\dfrac{1}{16}$ **C.** $\dfrac{3}{16}$ **D.** $\dfrac{1}{4}$

6. **MC** Identify which Venn diagram best illustrates $\Pr(A \cup B)'$.

   A.

   B.

   C.

   D.

   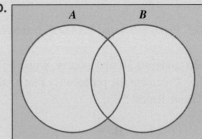

7. **MC** From a group of 25 people, 12 use Facebook (F), 14 use Snapchat (S), and 6 use both Facebook and Snapchat applications. Determine the probability that a person selected will use neither application.

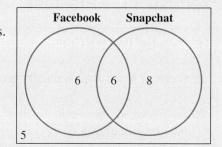

A. $\dfrac{1}{5}$ B. $\dfrac{6}{25}$

C. $\dfrac{8}{25}$ D. $\dfrac{12}{25}$

8. The probability that a student will catch a bus to school is 0.7 and the independent probability that a student will be late to school is 0.2.
Determine the probability, in simplest form, that a student catches a bus and is not late to school.

9. **MC** From events A and B in the Venn diagram, calculate $Pr(A|B)$.

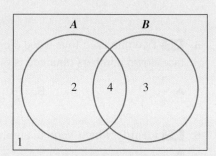

A. $\dfrac{4}{7}$ B. $\dfrac{4}{9}$

C. $\dfrac{2}{5}$ D. $\dfrac{1}{3}$

10. In the Venn diagram, events A and B are independent.
State whether this statement is true or false.

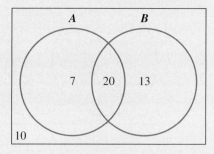

11. If $Pr(A) = 0.5$, $Pr(B) = 0.4$ and $Pr(A \cup B) = 0.8$, calculate $Pr(B|A)$, correct to 1 decimal place.

12. **MC** From 20 students, 10 play soccer, while 15 play Aussie Rules and 8 play both soccer and Aussie Rules. Calculate the probability that a student randomly selected plays soccer given that he or she plays Aussie Rules.

A. $\dfrac{8}{15}$ B. $\dfrac{2}{5}$ C. $\dfrac{2}{3}$ D. $\dfrac{10}{23}$

13. **MC** Two cards are drawn successively without replacement from a pack of playing cards. Calculate the probability of drawing 2 spades.

A. $\dfrac{1}{17}$ B. $\dfrac{2}{2652}$ C. $\dfrac{1}{2652}$ D. $\dfrac{1}{2704}$

14. A survey of a school of 800 students found that 100 used a bus (B) to get to school, 75 used a train (T) and 650 used neither.
In simplest form, determine the probability that a student uses both a bus and a train to get to school.

15. **MC** On the first day at school, students are asked to tell the class about their holidays. There are 30 students in the class and all have spent part or all of their holidays at one of the following: a coastal resort, interstate, or overseas.

The teacher finds that:
- 5 students went to a coastal resort only
- 2 students went interstate only
- 2 students holidayed in all three ways
- 8 students went to a coastal resort and travelled overseas only
- 20 students went to a coastal resort
- no less than 4 students went overseas only
- no less than 13 students travelled interstate.

Determine the probability that a student travelled overseas and interstate only.

A. $\dfrac{2}{15}$ B. $\dfrac{5}{6}$ C. $\dfrac{3}{15}$ D. $\dfrac{1}{3}$

LESSON
11.2 Review of probability

LEARNING INTENTIONS

At the end of this lesson you should be able to:
- use key probability terminology such as trials, frequency, sample space, and likely and unlikely events
- use two-way tables to represent sample spaces
- calculate experimental and theoretical probabilities
- determine the probability of complementary events and mutually exclusive events
- use the addition rule to calculate the probability of event 'A or B'.

11.2.1 The language of probability

eles-4922

- **Probability** measures the chance of an event taking place and ranges from 0 for an impossible event to 1 for a certain event.

- The **experimental probability** of an event is based on the outcomes of experiments, simulations or surveys.
- A **trial** is a single experiment, for example, a single flip of a coin.

Experimental probability

$$\text{Experimental probability} = \frac{\text{number of successful trials}}{\text{total number of trials}}$$

- The experimental probability of an event is also known as the **relative frequency**.
- The list of all possible outcomes of an experiment is known as the **event space** or **sample space**. For example, when flipping a coin there are two possible outcomes: Heads or Tails. The event space can be written using set notation as {H, T}.

WORKED EXAMPLE 1 Sample space and calculating relative frequency

The spinner shown here is made up of 4 equal-sized segments. It is known that the probability that the spinner will land on any one of the four segments from one spin is $\frac{1}{4}$. To test if the spinner shown here is fair, a student spun the spinner 20 times and each time recorded the segment in which the spinner stopped. The spinner landed as follows.

Segment	I	II	III	IV
Tally	5	4	8	3

a. List the sample space.
b. Given the experimental results, determine the experimental probability of each segment.
c. Compare the experimental probabilities with the known probabilities and suggest how the experiment could be changed to ensure that the results give a better estimate of the true probability.

THINK	WRITE
a. The sample space lists all possible outcomes from one spin of the spinner. There are four possible outcomes.	a. Sample space = {I, II, III, IV}
b. 1. For segment I there were 5 successful trials out of the 20. Substitute these values into the experimental probability formula.	b. $\text{Experimental probability}_I = \dfrac{\text{number of successful trials}}{\text{total number of trials}}$ $= \dfrac{5}{20}$ $= 0.25$
2. Repeat for segments: • II (4 successes) • III (8 successes) • IV (3 successes).	$\text{Experimental probability}_{II} = \dfrac{4}{20}$ $= 0.2$ $\text{Experimental probability}_{III} = \dfrac{8}{20}$ $= 0.4$ $\text{Experimental probability}_{IV} = \dfrac{3}{20}$ $= 0.15$

c. Compare the experimental frequency values with the known value of $\frac{1}{4}$ (0.25). Answer the question.	c. The experimental probability of segment I was the only segment that mirrored the known value. To ensure that experimental probability gives a better estimate of the true probability, the spinner should be spun many more times.

Simulations

- Sometimes it is not possible to conduct trials of a real experiment because it is too expensive, too difficult or impractical.
- Instead, outcomes of events can be modelled using spinners, dice, coins, or digital tools to simulate what happens in real life.
- These simulated experiments are called **simulations** and can be used to model real-life situations such as the number of people likely to be infected with a virus like COVID-19, or for decision-making such as insurance risk or supply and demand of products.

WORKED EXAMPLE 2 Applying simulation to model real-life situations

Six runners are to compete in a race.
a. Explain how simulation could be used to determine the winner.
b. State any limitations in your answer.

THINK	WRITE
a. There are six runners, so six numbers are needed to represent them. A die is the obvious choice. Assign numbers 1 to 6 to each runner.	Each runner could be assigned a number on a die, which could be rolled to determine the winner of the race. When the die is rolled, it simulates a race. The number that lands uppermost is the winner of the race.
b. In this simulation, each runner had an equally likely chance of winning each game.	The answer depends on each runner having an equally likely chance of winning each game. This is rarely the case.

Two-way tables

- A sample space can be displayed using a **two-way table**.
- A two-way table represents two of the outcomes of events in a two-dimensional table.
  A two-way table for the experiment of tossing a coin and rolling a die simultaneously is shown below.

		Die outcome					
		1	**2**	**3**	**4**	**5**	**6**
Coin outcome	**H**	H, 1	H, 2	H, 3	H, 4	H, 5	H, 6
	T	T, 1	T, 2	T, 3	T, 4	T, 5	T, 6

Two dice are rolled and the values on the two uppermost faces are multiplied together. Draw a diagram to illustrate the sample space.

THINK

The sample space for rolling 1 die is {1, 2, 3, 4, 5, 6}. When two dice are rolled and the two uppermost faces are multiplied, the sample space is made up of 36 products. This is best represented with the use of a two-way table.

- Draw a 7×7 grid.
- In the first row and column, list the outcomes of each die.
- At the intersection of a column and row, write the product of the relevant die outcomes.

WRITE

			First die				
	×	1	2	3	4	5	6
	1	1	2	3	4	5	6
	2	2	4	6	8	10	12
Second die	3	3	6	9	12	15	18
	4	4	8	12	16	20	24
	5	5	10	15	20	25	30
	6	6	12	18	24	30	36

Theoretical probability

- **Theoretical probability** is the probability of an event occurring, based on the number of possible favourable outcomes, $n(E)$, and the total number of possible outcomes, $n(\xi)$.

Theoretical probability

When all outcomes are equally likely, the theoretical probability of an event can be calculated using the formula:

$$\text{Pr (event)} = \frac{\text{number of favourable outcomes}}{\text{total number of possible outcomes}} \quad \text{or} \quad \text{Pr (event)} = \frac{n(E)}{n(\xi)}$$

where $n(E)$ is the number of favourable events and $n(\xi)$ is the total number of possible outcomes.

A fair die is rolled and the value of the uppermost side is recorded. Calculate the theoretical probability that a 4 is uppermost.

THINK

1. Write the number of favourable outcomes and the total number of possible outcomes. The number of 4s on a fair die is 1. There are 6 possible outcomes.

2. Substitute the values found in part 1 to calculate the probability of the event that a 4 is uppermost when a die is rolled.

3. Write the answer in a sentence.

WRITE

$n(E) = 1$
$n(\xi) = 6$

$\text{Pr(a 4)} = \dfrac{n(E)}{n(\xi)}$

$= \dfrac{1}{6}$

The probability that a 4 is uppermost when a fair die is rolled is $\dfrac{1}{6}$.

▶ 11.2.2 Properties of probability events

eles-6142

Complementary events

- The **complement** of the set A is the set of all elements that belong to the universal set (ξ) but that do *not* belong to A.
- The complement of A is written as A' and is read as 'A dashed', 'A prime' or 'not A'.
- On a Venn diagram, **complementary events** appear as separate regions that together occupy the whole universal set.

> ### Complementary events
>
> **Since complementary events fill the entire sample space:**
>
> $$\Pr(A) + \Pr(A') = 1$$

- As an example, the complement of {drawing a diamond} from a deck of cards is {not drawing a diamond}, which can also be described as {drawing a heart, spade or club}.
  This is shown in the Venn diagram.

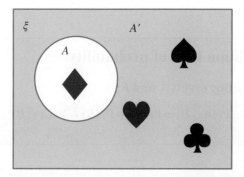

WORKED EXAMPLE 5 Determining complementary events

A player is chosen from a cricket team. Are the events 'selecting a batter' and 'selecting a bowler' complementary events if a player can have more than one role?
Give a reason for your answer.

THINK	WRITE
Explain the composition of a cricket team. Players who can bat or bowl are not necessarily the only players in a cricket team. There is a wicket-keeper as well. Some players can both bat and bowl.	No, the events 'selecting a batter' and 'selecting a bowler' are not complementary events. These events may have common elements, that is the players in the team who can bat and bowl. The cricket team also includes a wicket-keeper.

The intersection and union of *A* and *B*

- The intersection of two events *A* and *B* is written $A \cap B$. These are the outcomes that are in *A* 'and' in *B*, so the intersection is often referred to as the event '*A* and *B*'.
- The union of two events *A* and *B* is written $A \cup B$. These are the outcomes that are in *A* 'or' in *B*, so the union is often referred to as the event '*A* or *B*'.

 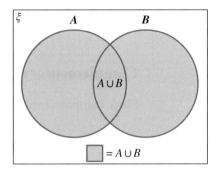

- When calculating the probability of $A \cup B$, we cannot simply add the probabilities of *A* and *B*, as $A \cap B$ would be counted twice.
- The formula for the probability of $A \cup B$ is therefore given by the following equation, which is known as the **Addition Law of probability**.

The Addition Law of probability

For intersecting events *A* and *B*:

$$Pr(A \cup B) = Pr(A) + Pr(B) - Pr(A \cap B)$$

Mutually exclusive events

- Two events are **mutually exclusive** if one event happening excludes the other from happening. These events may not encompass all possible events.

  For example, when selecting a card from a deck of cards, selecting a black card excludes the possibility that the card is a heart.
- On a Venn diagram, mutually exclusive events appear as disjointed sets within the universal set.
- For mutually exclusive events *A* and *B*, $Pr(A \cap B) = 0$.

  Therefore, the formula for $Pr(A \cup B)$ is simplified to the following.

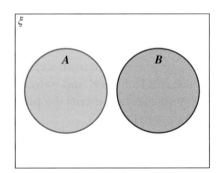

Mutually exclusive probabilities

For mutually exclusive events *A* and *B*:

$$Pr(A \cup B) = Pr(A) + Pr(B)$$

A card is drawn from a pack of 52 playing cards. Determine the probability that the card is a heart or a club.

THINK	WRITE
1. Determine whether the given events are mutually exclusive.	The two events are mutually exclusive as they have no common elements.
2. Determine the probability of drawing a heart and of drawing a club.	$Pr(\text{heart}) = \dfrac{13}{52}$ $\qquad$ $Pr(\text{club}) = \dfrac{13}{52}$ $\qquad\qquad\ = \dfrac{1}{4}$ $\qquad\qquad\qquad = \dfrac{1}{4}$
3. Write the Addition Law for two mutually exclusive events.	$Pr(A \text{ or } B) = Pr(A) + Pr(B)$ where A = drawing a heart and B = drawing a club
4. Substitute the known values into the rule.	$Pr(\text{heart or club}) = Pr(\text{heart}) + Pr(\text{club})$ $= \dfrac{1}{4} + \dfrac{1}{4}$ $= \dfrac{2}{4}$
5. Evaluate and simplify.	$= \dfrac{1}{2}$
6. Write your answer.	The probability of drawing a heart or a club is $\dfrac{1}{2}$.
Note: Alternatively, we can use the formula for theoretical probability.	$Pr(\text{heart or club}) = \dfrac{n(\text{heart or club})}{n(\xi)}$ $= \dfrac{26}{52}$ $= \dfrac{1}{2}$

Given $Pr(A) = 0.6$, $Pr(B) = 0.4$ and $Pr(A \cup B) = 0.9$:
a. **use the Addition Law of probability to calculate the value of $Pr(A \cap B)$**
b. **draw a Venn diagram to represent the universal set**
c. **calculate $Pr(A \cap B')$.**

THINK	WRITE
a. 1. Write the Addition Law of probability and substitute given values.	a. $Pr(A \cup B) = Pr(A) + Pr(B) - Pr(A \cap B)$ $0.9 = 0.6 + 0.4 - Pr(A \cap B)$
2. Collect like terms and rearrange to make $Pr(A \cap B)$ the subject. Solve the equation.	$0.9 = 1.0 - Pr(A \cap B)$ $Pr(A \cap B) = 1.0 - 0.9$ $= 0.1$

▶

b. **1.** Draw intersecting sets A and B within the universal set and write $\Pr(A \cap B) = 0.1$ inside the overlapping section, as shown.

b.

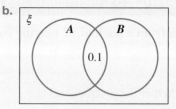

2. • As $\Pr(A) = 0.6$, 0.1 of this belongs in the overlap; the remainder of set A is 0.5 $(0.6 - 0.1)$.
• Since $\Pr(B) = 0.4$, 0.1 of this belongs in the overlap; the remainder of set B is 0.3 $(0.4 - 0.1)$.

3. The total probability for the universal set is 1. That means $\Pr(A \cup B)' = 0.1$.
Write 0.1 outside sets A and B to form the remainder of the universal set.

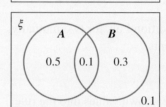

c. $\Pr(A \cap B')$ is the overlapping region of $\Pr(A)$ and $\Pr(B')$. Shade the region and write down the corresponding probability value for this area.

c.

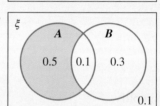

$\Pr(A \cap B') = 0.5$

WORKED EXAMPLE 8 Using a Venn diagram to represent sets and find probabilities

a. Draw a Venn diagram representing the relationship between the following sets. Show the position of all the elements in the Venn diagram.

$\xi = \{1, 2, 3, 4, 5, 6, 7, 8, 9, 10, 11, 12, 13, 14, 15, 16, 17, 18, 19, 20\}$
$A = \{3, 6, 9, 12, 15, 18\}$
$B = \{2, 4, 6, 8, 10, 12, 14, 16, 18, 20\}$

b. Determine:

   i. $\Pr(A)$ **ii. $\Pr(B)$**
 iii. $\Pr(A \cap B)$ **iv. $\Pr(A \cup B)$**
   v. $\Pr(A' \cap B')$

THINK

a. **1.** Draw a rectangle with two partly intersecting circles labelled A and B.

2. Analyse sets A and B and place any common elements in the central overlap.

3. Place the remaining elements of set A in circle A.

4. Place the remaining elements of set B in circle B.

5. Place the remaining elements of the universal set ξ in the rectangle.

WRITE/DRAW

a.

$n(\xi) = 20$

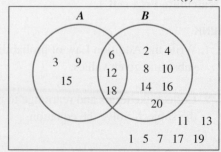

b. **i.** **1.** Write the number of elements that belong to set A and the total number of elements.

2. Write the rule for probability.

3. Substitute the known values into the rule.

4. Evaluate and simplify.

ii. **1.** Write the number of elements that belong to set B and the total number of elements.

2. Repeat steps **2** to **4** of part **b i.**

iii. **1.** Write the number of elements that belong to set $(A \cap B)$ and the total number of elements.

2. Repeat steps **2** to **4** of part **b i.**

iv. **1.** Write the number of elements that belong to set $(A \cup B)$ and the total number of elements.

2. Repeat steps **2** to **4** of part **b i.**

v. **1.** Write the number of elements that belong to set $(A' \cap B')$ and the total number of elements.

2. Repeat steps **2** to **4** of part **b i.**

b. **i.** $n(A) = 6, n(\xi) = 20$

$\Pr(A) = \dfrac{n(A)}{n(\xi)}$

$\Pr(A) = \dfrac{6}{20}$

$= \dfrac{3}{10}$

ii. $n(B) = 10, n(\xi) = 20$

$\Pr(B) = \dfrac{n(B)}{n(\xi)}$

$\Pr(B) = \dfrac{10}{20}$

$= \dfrac{1}{2}$

iii. $n(A \cap B) = 3, n(\xi) = 20$

$\Pr(A \cap B) = \dfrac{n(A \cap B)}{n(\xi)}$

$\Pr(A \cap B) = \dfrac{3}{20}$

iv. $n(A \cup B) = 13, n(\xi) = 20$

$n(A \cup B) = \dfrac{n(A \cup B)}{n(\xi)}$

$n(A \cup B) = \dfrac{13}{20}$

v. $n(A' \cap B') = 7, n(\xi) = 20$

$\Pr(A' \cap B') = \dfrac{n(A' \cap B')}{n(\xi)}$

$\Pr(A' \cap B') = \dfrac{7}{20}$

In a class of 35 students, 6 students like all three subjects: PE, Science and Music.
Eight of the students like PE and Science, 10 students like PE and Music, and 12 students like Science and Music. Also, 22 students like PE, 18 students like Science and 17 like Music.
Two students don't like any of the subjects.
a. Display this information on a Venn diagram.
b. Determine the probability of selecting a student who:
   i. likes PE only
   ii. does not like Music.
c. Determine $\Pr[(\text{Science} \cup \text{Music}) \cap \text{PE}']$.

THINK

a. 1. Draw a rectangle with three partly intersecting circles, labelled PE, Science and Music.

2. Extract the information relating to students liking all three subjects.
Note: The central overlap is the key to solving these problems. Six students like all three subjects, so place the number 6 into the section corresponding to the intersection of the three circles.

3. Extract the relevant information from the second sentence and place it into the appropriate position.
Note: Eight students like PE and Science; however, 6 of these students have already been accounted for in step 2.
Therefore, 2 will fill the intersection of only PE and Science. Similarly, 4 of the 10 who like and Music will fill the intersection of only PE and Music, and 6 of the 12 students will fill the intersection of only Science and Music.

WRITE/DRAW

a.

$n(\xi) = 35$

$n(\xi) = 35$

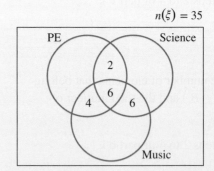

$n(\xi) = 35$

4. Extract the relevant information from the third sentence and place it into the appropriate position.
Note: Twenty-two students like PE and 12 have already been accounted for in the set. Therefore, 10 students are needed to fill the circle corresponding to PE only. Similarly, 4 students are needed to fill the circle corresponding to Science only to make a total of 18 for Science. One student is needed to fill the circle corresponding to Music only to make a total of 17 for Music.

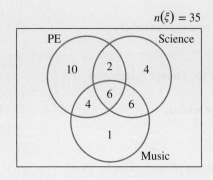

5. Extract the relevant information from the final sentence and place it into the appropriate position.
Note: Two students do not like any of the subjects, so they are placed in the rectangle outside the three circles.

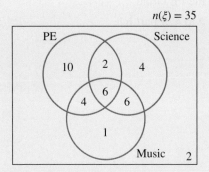

6. Check that the total number in all positions is equal to the number in the universal set.

$$10 + 2 + 4 + 4 + 6 + 6 + 1 + 2 = 35$$

b. i. 1. Write the number of students who like PE only and the total number of students in the class.

b. i. $n(\text{students who like PE only}) = 10$
$$n(\xi) = 35$$

2. Write the rule for probability.

$$\text{Pr(likes PE only)} = \frac{n(\text{likes PE only})}{n(\xi)}$$

3. Substitute the known values into the rule.

$$\text{Pr(likes PE only)} = \frac{10}{35}$$

4. Evaluate and simplify.

$$= \frac{2}{7}$$

5. Write your answer.

The probability of selecting a student who likes PE only is $\frac{2}{7}$.

ii. 1. Write the number of students who do not like Music and the total number of students in the class.
Note: Add all the values that do not appear in the Music circle as well as the two that sit in the rectangle outside the circles.

ii. $n(\text{students who do not like Music}) = 18$
$$n(\xi) = 35$$

2. Write the rule for probability.

$$\text{Pr(does not like Music)} = \frac{n(\text{does not like Music})}{n(\xi)}$$

3. Substitute the known values into the rule.

$$\text{Pr(does not like Music)} = \frac{18}{35}$$

4. Write your answer.

The probability of selecting a student who does not like Music is $\frac{18}{35}$.

c. 1. Write the number of students who like Science and Music but not PE.
Note: Add the values that appear in the Science and Music circles but do not overlap with the PE circle.

c. $n[(\text{Science} \cup \text{Music}) \cap \text{PE}'] = 11$
$n(\xi) = 35$

2. Repeat steps 2 to 4 of part b ii.

$$\text{Pr}[(\text{Science} \cup \text{Music}) \cap \text{PE}']$$
$$= \frac{n[(\text{Science} \cup \text{Music}) \cap \text{PE}']}{n(\xi)}$$
$$\text{Pr}[(\text{Science} \cup \text{Music}) \cap \text{PE}'] = \frac{11}{35}.$$

The probability of selecting a student who likes Science or Music but not PE is $\frac{11}{35}$.

COLLABORATIVE TASK: Probability of matching birthdays

Use simulation to calculate the probability that two people were born on the same day of the year, clearly explaining the method you used. Then research the simulation of the birthday problem and report on your findings.

---

on Resources

Exercise 11.2 Review of probability

11.2 Quick quiz on	11.2 Exercise

Individual pathways

■ PRACTISE	■ CONSOLIDATE	■ MASTER
1, 3, 4, 5, 8, 13, 17, 20	2, 6, 9, 10, 14, 18, 21	7, 11, 12, 15, 16, 19, 22, 23

Fluency

1. **WE1** The spinner shown was spun 50 times and the outcome each time was recorded in the table.

Segment	I	II	III	IV	V	VI
Tally	10	6	8	7	12	7

   a. List the event space.
   b. Given the experimental results, determine the relative frequency for each segment.

   c. The theoretical probability of the spinner landing on any particular segment with one spin is $\frac{1}{6}$. State how the experiment could be changed to give a better estimate of the true probabilities.

2. A laptop company conducted a survey to see what were the most appealing colours for laptop computers among 15–18-year-old students. The results were as follows.

Colour	Black Black	Sizzling Silver	Power Pink	Blazing Blue	Gooey Green	Glamour Gold
Number	102	80	52	140	56	70

   a. Calculate the number of students who were surveyed.
   b. Calculate the relative frequency of students who found silver the most appealing laptop colour.
   c. Calculate the relative frequency of students who found black and green to be their most appealing colours.
   d. State which colour was found to be most appealing.

3. **WE2** State the device you would use to simulate the following experiments.

   a. Collecting toys from a particular brand of cereal where there are six in the set
   b. Whether employees are satisfied with their jobs
   c. Whether two people who meet will be born in the same month

4. **WE3** Two dice are rolled and the values on the two uppermost faces are added together.

   a. Construct a table to illustrate the sample space.
   b. Calculate the most likely outcome.
   c. Calculate the least likely outcome.

5. **WE4** A die is rolled. Calculate the probability that the outcome is an even number or a 5.

6. A card is drawn from a well-shuffled pack of 52 playing cards. Calculate:

   a. Pr(a king is drawn)
   b. Pr(a heart is drawn)
   c. Pr(a king or a heart is drawn).

7. **WE5** For each of the following pairs of events:

   i. state, giving justification, if the pair are complementary events
   ii. alter the statements, where applicable, so that the events become complementary events.

   a. Having Weet Bix or having Strawberry Pops for breakfast
   b. Walking to a friend's place or driving there
   c. Watching TV or reading as a leisure activity
   d. Rolling a number less than 5 or rolling a number greater than 5 with a ten-sided die with faces numbered 1 to 10
   e. Passing a maths test or failing a maths test

8. **WE6** A card is drawn from a shuffled pack of 52 cards. Calculate the probability that the card drawn is:

   a. an ace
   b. a club
   c. a red card
   d. not a jack
   e. a green card
   f. not a red card.

9. A bag contains 4 blue marbles, 7 red marbles and 9 yellow marbles. All marbles are the same size. A marble is selected at random. Calculate the probability that the marble is:

   a. blue
   b. red
   c. not yellow
   d. black.

10. **WE7** Given $\Pr(A) = 0.5$, $\Pr(B) = 0.4$ and $\Pr(A \cup B) = 0.8$:

    a. use the Addition Law of probability to calculate the value of $\Pr(A \cap B)$
    b. draw a Venn diagram to represent the universal set
    c. calculate $\Pr(A \cap B')$.

11. Let $\Pr(A) = 0.25$, $\Pr(B) = 0.65$ and $\Pr(A \cap B) = 0.05$.

    a. Calculate:
       i. $\Pr(A \cup B)$
       ii. $\Pr(A \cap B)'$.

    b. **MC** Choose which Venn diagram below best illustrates $\Pr(A \cap B)'$.

    A.

    B.

    C.

    D.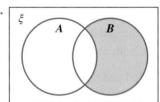

12. a. **WE8** Draw a Venn diagram representing the relationship between the following sets. Show the position of all the elements in the Venn diagram.

    $$\xi = \{1, 2, 3, 4, 5, 6, 7, 8, 9, 10, 11, 12, 13, 14, 15, 16, 17, 18, 19, 20\}$$
    $$A = \{1, 3, 5, 7, 9, 11, 13, 15, 17, 19\}$$
    $$B = \{1, 4, 9, 16\}$$

    b. Calculate:
       i. $\Pr(A)$
       ii. $\Pr(B)$
       iii. $\Pr(A \cap B)$
       iv. $\Pr(A \cup B)$
       v. $\Pr(A' \cap B')$.

Understanding

13. You and a friend are playing a dice game. You have an eight-sided die (with faces numbered 1 to 8 inclusive) and your friend has a six-sided die (with faces numbered 1 to 6 inclusive). You each roll your own die.

    a. The person who rolls the number 4 wins. Determine if this game is fair.
    b. The person who rolls an odd number wins. Determine if this game is fair.

14. A six-sided die has three faces numbered 5; the other faces are numbered 6. Determine if the events 'rolling a 5' and 'rolling a 6' are equally likely.

15. (*Note:* You will need a six-sided die for this question.)
    The latest promotion by a chocolate bar manufacturer claims 'Every third one is free'. When you get the special message inside the wrapper, you get a refund on the purchase price.
    The manufacturer has put the special message in every third wrapper. When the chocolate bars are packaged into boxes, they are mixed randomly.
    Use the numbers 3 and 6 on a die to represent getting a free chocolate bar.

    a. Carry out 50 trial purchases of chocolate bars.
    b. Calculate the relative frequency of a free chocolate bar.
    c. Determine how many free chocolate bars you would expect to receive in 100 trials.

16. **WE9** Thirty students were asked which lunchtime sports they enjoyed — volleyball, soccer or tennis. Five students chose all three sports. Six students chose volleyball and soccer, 7 students chose volleyball and tennis, and 9 chose soccer and tennis.
    Fifteen students chose volleyball, 14 students chose soccer and 18 students chose tennis.

    a. Copy the Venn diagram shown and enter the given information.
    b. If a student is selected at random, determine the probability of selecting a student who:
       i. chose volleyball
       ii. chose all three sports
       iii. chose both volleyball and soccer but not tennis
       iv. did not choose tennis
       v. chose soccer.
    c. Determine:
       i. $\Pr[(\text{soccer} \cup \text{tennis}) \cap \text{volleyball}']$
       ii. $\Pr[(\text{volleyball} \cup \text{tennis}) \cap \text{soccer}']$.

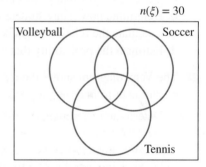

Reasoning

17. A six-sided die has three faces numbered 1 and the other three faces numbered 2. Determine if the events 'rolling a 1' and 'rolling a 2' are equally likely.

18. With the use of diagrams, show that $\Pr(A' \cap B') = \Pr(A \cup B)'$.

19. A drawer contains purple socks and red socks. The chance of obtaining a red sock is 2 in 9. There are 10 red socks in the drawer.
    Determine the smallest number of socks that need to be added to the drawer so that the probability of drawing a red sock increases to 3 in 7.

Problem solving

20. Ninety students were asked which lunchtime sports on offer, basketball, netball and soccer, they had participated in on at least one occasion in the last week.
The results are shown in the following table.

Sport	Basketball	Netball	Soccer	Basketball and netball	Basketball and soccer	Netball and soccer	All three
Number of students	35	25	39	5	18	8	3

a. Copy and complete the Venn diagram shown below to illustrate the sample space.

b. Determine how many students did not play basketball, netball or soccer at lunchtime.
c. Determine how many students played basketball and/or netball but not soccer.
d. Determine how many students are represented by the region (basketball ∩ not netball ∩ soccer).
e. Calculate the relative frequency of the region described in part d above.
f. Estimate the probability that a student will play three of the sports offered.

21. The Venn diagram shows the results of a survey completed by a Chinese restaurateur to find out the food preferences of his regular customers.

a. Determine the number of customers:
   i. surveyed
   ii. showing a preference for fried rice only
   iii. showing a preference for fried rice
   iv. showing a preference for chicken wings and dim sims.

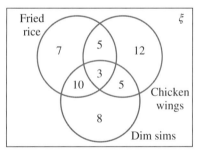

b. A customer from this group won the draw for a lucky door prize. Determine the probability that this customer:
   i. likes fried rice
   ii. likes all three — fried rice, chicken wings and dim sims
   iii. prefers chicken wings only.

c. A similar survey was conducted a month later with another group of 50 customers. This survey yielded the following results:
2 customers liked all three foods; 6 preferred fried rice and chicken wings only;
7 preferred chicken wings and dim sims only; 8 preferred fried rice and dim sims only; 22 preferred fried rice; 23 preferred chicken wings; and 24 preferred dim sims.
   i. Display this information on a Venn diagram.
   ii. Determine the probability of selecting a customer who prefers all three foods, if a random selection is made.

22. A pair of dice is rolled and the sum of the numbers shown is noted.

   a. Show the sample space in a two-way table.
   b. Determine how many different ways the sum of 7 can be obtained.
   c. Determine if all outcomes are equally likely.
   d. Complete the given table.

Sum	2	3	4	5	6	7	8	9	11	12
Frequency										

   e. Determine the relative frequencies of the following sums.
      i. 2   ii. 7   iii. 11
   f. Determine the probability of obtaining the following sums.
      i. 2   ii. 7   iii. 11
   g. If a pair of dice is rolled 300 times, calculate how many times you would expect the sum of 7.

23. The mouse in the maze is trying to reach the cheese. Use a coin to help the mouse to decide to go left or right when a new choice is available. If you come to a dead end, turn around and go back to where you have a choice to go left or right.

   There are two possible outcomes for the mouse:
   - escaping the maze
   - reaching the cheese.

   Use simulation to determine the relative frequencies of each of the outcomes.

LESSON
11.3 Tree diagrams

LEARNING INTENTIONS

At the end of this lesson you should be able to:
- create tree diagrams to represent two- and three-step chance experiments
- use tree diagrams to solve probability problems involving two or more trials or events.

▶ 11.3.1 Two-step chance experiments

eles-4924

- In **two-step chance experiments** the result is obtained after performing two trials. Two-step chance experiments are often represented using tree diagrams.
- **Tree diagrams** are used to list all possible outcomes of two or more events that are not necessarily equally likely.
- The probability of obtaining the result for a particular event is listed on the branches.
- The probability for each outcome in the sample space is the product of the probabilities associated with the respective branches. For example, the tree diagram shown here represents the sample space for flipping a coin and then choosing a marble from a bag containing three red marbles and one black marble.

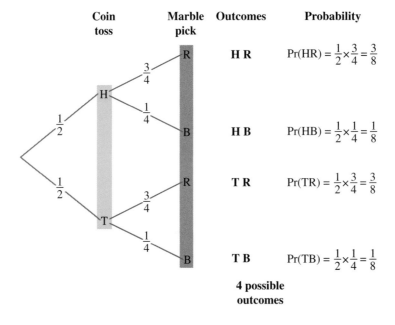

- When added together, all the probabilities for the outcomes should sum to 1. They are complementary events.
  For example,

$$\Pr(HR) + \Pr(HB) + \Pr(TR) + \Pr(TB) = \frac{3}{8} + \frac{1}{8} + \frac{3}{8} + \frac{1}{8}$$
$$= 1$$

- Other probabilities can also be calculated from the tree diagram.
  For example, the probability of getting an outcome that contains a red marble can be calculated by summing the probabilities of each of the possible outcomes that include a red marble.

Outcomes that contain a red marble are HR and TR. Therefore:

$$Pr(\text{red marble}) = Pr(HR) + Pr(TR)$$

$$= \frac{3}{8} + \frac{3}{8}$$

$$= \frac{6}{8}$$

$$= \frac{3}{4}$$

WORKED EXAMPLE 10 Using a tree diagram for a two-step chance experiment

A three-sided die is rolled and an object is picked out of a hat that contains 3 balls and 7 cubes.
a. **Construct a tree diagram to display the sample space.**
b. **Calculate the probability of:**
   i. **rolling a 3, then choosing a cube**
   ii. **choosing a cube after rolling an odd number.**

THINK

WRITE

a. 1. Draw 3 branches from the starting point to show the 3 possible outcomes of rolling a three-sided die (shown in blue) and then draw 2 branches off each of these to show the 2 possible outcomes of choosing an object (shown in red).

   2. Write probabilities on the branches to show the individual probabilities of rolling a 1, 2 or 3 on a three-sided die. As these are equally likely outcomes,
   $Pr(1) = Pr(2) = Pr(3) = \dfrac{1}{3}$.

   3. Write probabilities on the branches to show the individual probabilities of choosing an object. Since there are 3 balls and 7 cubes in the hat,
   $Pr(B) = \dfrac{3}{10}$ and $Pr(C) = \dfrac{7}{10}$.

a.

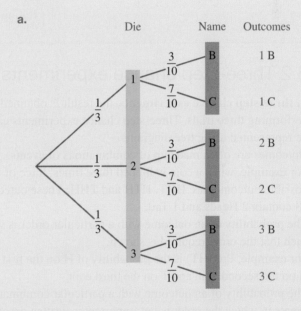

b. i. 1. Follow the pathway of rolling a 3 $\left[Pr(3) = \dfrac{1}{3}\right]$ and choosing a cube $\left[Pr(C) = \dfrac{7}{10}\right]$, and multiply the probabilities.

b. i. $Pr(3C) = Pr(3) \times Pr(C)$

$$= \frac{1}{3} \times \frac{7}{10}$$

$$= \frac{7}{30}$$

2. Write the answer.

The probability of rolling a 3 and then choosing a cube is $\dfrac{7}{30}$.

ii. 1. To roll an odd number (1 or 3) and then choose a cube:
 • roll a 1 and then choose a cube,
   or
 • roll a 3 and then choose a cube.
 Calculate the probability of each of these and add them together to calculate the total probability. Simplify the result if possible.

ii. $\Pr(\text{odd } C) = \Pr(1C) + \Pr(3C)$

$= \Pr(1) \times \Pr(C) + \Pr(3) \times \Pr(C)$

$= \dfrac{1}{3} \times \dfrac{7}{10} + \dfrac{1}{3} \times \dfrac{7}{10}$

$= \dfrac{7}{30} + \dfrac{7}{30}$

$= \dfrac{14}{30}$

$= \dfrac{7}{15}$

2. Write the answer.

The probability of choosing a cube after rolling an odd number is $\dfrac{7}{15}$.

▶ 11.3.2 Three-step chance experiments

eles-4925

- In **three-step chance experiments**, the result is obtained after performing three trials. Three-step chance experiments can also be represented using tree diagrams.
- Outcomes are often made up of combinations of events.
  For example, when a coin is flipped three times, three of the possible outcomes are HHT, HTH and THH. These outcomes all contain 2 Heads and 1 Tail.
- The probability of an outcome with a particular order is written such that the order required is shown.
  For example, Pr(HHT) is the probability of H on the first coin, H on the second coin and T on the third coin.
- The probability of an outcome with a particular combination of events in which the order is not important is written describing the particular combination required, for example Pr(2 Heads and 1 Tail).

WORKED EXAMPLE 11 Using a tree diagram for a three-step chance experiment

A coin is biased so that the chance of it falling as a Head when flipped is 0.75.
a. **Construct a tree diagram to represent the coin being flipped three times.**
b. **Calculate the following probabilities, correct to 3 decimal places:**

 i. **Pr(HTT)**
 ii. **Pr(1H and 2T)**
 iii. **Pr(at least 2 Tails).**

a. 1. Tossing a coin has two outcomes. Draw 2 branches from the starting point to show the first toss, 2 branches off each of these to show the second toss, and 2 branches off each of these to show the third toss.

2. Write probabilities on the branches to show the individual probabilities of tossing a Head (0.75) and a Tail. Because tossing a Head and tossing a Tail are mutually exclusive, $\text{Pr}(T) = 1 - \text{Pr}(H) = 0.25$.

a.

	1st toss	2nd toss	3rd toss	Outcomes

b. i. 1. Pr(HTT) implies the order H(0.75), T(0.25) and T(0.25).

2. Multiply the probabilities and round.

b. i. $\text{Pr}(HTT) = \text{Pr}(H) \times \text{Pr}(T) \times \text{Pr}(T)$

$= (0.75) \times (0.25)^2$

$= 0.046875$

≈ 0.047

ii. 1. Pr(1H and 2T) implies Pr(HTT), Pr(THT) and Pr(TTH).

2. Multiply the probabilities and round.

ii. $\text{Pr}(1H \text{ and } 2T) = \text{Pr}(HTT) + \text{Pr}(THT) + \text{Pr}(TTH)$

$= 3(0.75 \times 0.25^2)$

$= 0.140625$

≈ 0.141

iii. 1. Pr(at least 2 Tails) implies Pr(HTT), Pr(THT), Pr(TTH) and Pr(TTT).

2. Add these probabilities and round.

iii. $\text{Pr}(\text{at least } 2T) = \text{Pr}(HTT) + \text{Pr}(THT) + \text{Pr}(TTH) + \text{Pr}(TTT)$

$= 3(0.75 \times 0.25^2) + 0.25^3$

$= 0.15625$

≈ 0.156

TI | THINK	DISPLAY/WRITE	CASIO | THINK	DISPLAY/WRITE

a. i.

In a new document on a Calculator page, to calculate the probability of an exact number of successes from a set number of trials. press:
- MENU
- 5: Probability
- 5: Distributions
- A: BinomialPdf

Enter the values:
- Num Trials, n: 3
- Prob Success, p: 0.75
- X Value: 1

Then tap OK.

a. i.

Pr(1H and 2T) = 0.141 (correct to 3 decimal places)

a. i.

On the Main screen, to calculate the probability of an exact number of successes from a set number of trials, tap:
- Interactive
- Distribution/Inv. Dist
- Discrete
- binomialPDF

Enter the values:
- X: 1
- Numtrial: 3
- pos: 0.75

Then tap OK.

a. i.

Pr(1H and 2T) = 0.141
(correct to 3 decimal places)

ii.

In a new document on a Calculator page, to calculate the probability of at least *x* successes from a set number of trials, press:
- MENU
- 5: Probability
- 5: Distributions
- B: BinomialCdf

Enter the values:
- Num Trials, n: 3
- Prob Success, p: 0.25
- Lower Bound: 2
- Upper Bound: 3

Then tap OK.

Note: In situations with repeated trials of independent events, such as flipping a coin, it becomes unfeasible to draw a tree diagram when the number of trials is 4 or more. BinomialPdf and BinomialCdf are very helpful in these cases.

ii.

Pr(at least 2T) = 0.156 (correct to 3 decimal places)

ii.

On the Main screen, to calculate the probability of at least *x* successes from a set number of trials, tap:
- Interactive
- Distribution/Inv. Dist
- Discrete
- binomialCDF

Enter the values:
- Lower: 2
- Upper: 3
- Numtrial: 3
- pos: 0.25

Then tap OK.

Note: In situations with repeated trials of independent events, such as flipping a coin, it becomes unfeasible to draw a tree diagram when the number of trials is 4 or more. BinomialPDF and BinomialCDF are very helpful in these cases.

ii.

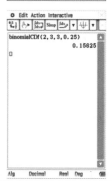

Pr(at least 2T) = 0.156
(correct to 3 decimal places)

Exercise 11.3 Tree diagrams

learn on

11.3 Quick quiz on	11.3 Exercise

Individual pathways

■ PRACTISE	■ CONSOLIDATE	■ MASTER
1, 2, 6, 9, 12	3, 5, 7, 10, 13	4, 8, 11, 14

Fluency

1. Explain how a tree diagram can be used to calculate probabilities of events that are not equally likely.

2. Use the tree diagram shown to answer the following questions.

   a. Identify how many different outcomes there are.
   b. Explain whether all outcomes are equally likely.
   c. State whether getting a red fish is more, less or equally likely than getting a green elephant.
   d. Determine the most likely outcome.
   e. Calculate the following probabilities.

      i. Pr(blue elephant)
      ii. Pr(indigo elephant)
      iii. Pr(donkey)

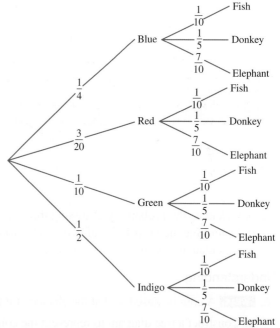

3. **WE10** The spinner shown is divided into 3 equal-sized wedges labelled 1, 2 and 3. The spinner is spun three times, and it is noted whether the spinner lands on a prime number, $Pr = \{2, 3\}$ = 'prime', or not a prime number, $Pr' = \{1\}$ = 'not prime'.

   a. Construct a labelled tree diagram for 3 spins of the spinner, showing probabilities on the branches and all possible outcomes.
   b. Calculate the following probabilities.

      i. Pr(3 prime numbers)
      ii. Pr(PPP′ in this order)
      iii. Pr(PPP′ in any order)

4. This question relates to rolling a fair die.

  a. Calculate the probability of each of the following outcomes from one roll of a die.

    i. Pr(rolling number < 4)
    ii. Pr(rolling a 4)
    iii. Pr(rolling a number other than a 6)

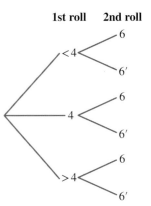

  b. The tree diagram shown has been condensed to depict rolling a die twice, noting the number relative to 4 on the first roll and 6 on the second. Complete a labelled tree diagram, showing probabilities on the branches and all outcomes, similar to that shown.

  c. Determine the probability of rolling the following with 2 rolls of the die.

    i. Pr(a 4 then a 6)
    ii. Pr(a number less than 4 then a 6)
    iii. Pr(a 4 then 6′)
    iv. Pr(a number > 4 then a number < 6)

5. a. Copy the tree diagram shown and complete the labelling for tossing a biased coin three times when the chance of tossing one Head in one toss is 0.7.

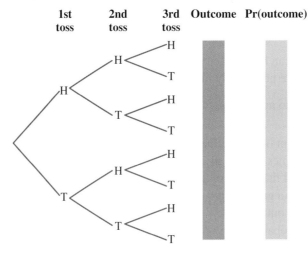

  b. Calculate the probability of tossing three Heads.
  c. Determine the probability of getting at least one Tail.
  d. Calculate the probability of getting exactly two Tails.

Understanding

6. **WE11** A coin is biased so that the chance of it falling as a Tail when tossed is 0.2.

  a. Construct a tree diagram to represent the coin being tossed 3 times.
  b. Determine the probability of getting the same outcome on each toss.

7. A die is tossed twice and each time it is recorded whether or not the number is a multiple of 3.
If $M =$ the event of getting a multiple of 3 on any one toss and $M' =$ the event of not getting a multiple of 3 on any one toss:

  a. construct a tree diagram to represent the 2 tosses
  b. calculate the probability of getting two multiples of 3.

8. The biased spinner illustrated is spun three times.

  a. Construct a completely labelled tree diagram for 3 spins of the spinner, showing probabilities on the branches and all possible outcomes and associated probabilities.
  b. Calculate the probability of getting exactly two 1s.
  c. Calculate the probability of getting at most two 1s.

Reasoning

9. Each morning when Ahmed gets dressed for work he has the following choices:
   - three suits that are grey, blue and white
   - four shirts that are white, blue, pink and grey
   - two ties that are grey and blue.

   a. Construct a fully labelled tree diagram showing all possible clothing choices.
   b. Calculate the probability of picking a grey suit, a pink shirt and a blue tie.
   c. Calculate the probability of picking the same colour for all three options.
   d. Ahmed works five days a week for 48 weeks of the year. Determine how many times each combination would get repeated over the course of one year at work.
   e. Determine how many more combinations of clothing he would have if he bought another tie and another shirt.

10. A restaurant offers its customers a three-course dinner, where they choose between two entrées, three main meals and two desserts. The managers find that 30% choose soup and 70% choose prawn cocktail for the entrée; 20% choose vegetarian, 50% chicken, and the rest have beef for their main meal; and 75% have sticky date pudding while the rest have apple crumble for dessert.

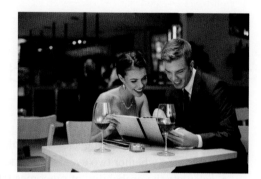

   a. Construct a fully labelled tree diagram showing all possible choices.
   b. Determine the probability that a customer will choose soup, chicken and sticky date pudding.
   c. If there are 210 people booked for the following week at the restaurant, determine how many you would expect to have the meal combination referred to in part b.

11. A bag contains 7 red and 3 white balls. A ball is taken at random, its colour is noted, and it is then placed back in the bag before a second ball is chosen at random and its colour is noted.

   a. i. Show the possible outcomes with a fully labelled tree diagram.
      ii. As the first ball was chosen, determine how many balls were in the bag.
      iii. As the second ball was chosen, determine how many balls were in the bag.
      iv. Explain whether the probability of choosing a red or white ball changes from the first selection to the second.
      v. Calculate the probability of choosing a red ball twice.
   b. Suppose that after the first ball had been chosen it was not placed back in the bag.
      i. As the second ball is chosen, determine how many balls are in the bag.
      ii. Explain if the probability of choosing a red or white ball changes from the first selection to the second.
      iii. Construct a fully labelled tree diagram to show all possible outcomes.
      iv. Evaluate the probability of choosing two red balls.

Problem solving

12. An eight-sided die is rolled three times to see whether 5 occurs.
   a. Construct a tree diagram to show the sample space.
   b. Calculate:
      i. Pr(three 5s)
      ii. Pr(no 5s)
      iii. Pr(two 5s)
      iv. Pr(at least two 5s).

13. A tetrahedral die (four faces labelled 1, 2, 3 and 4) is rolled and a coin is tossed simultaneously.

    a. Show all the outcomes on a two-way table.
    b. Construct a tree diagram and list all outcomes and their respective probabilities.
    c. Determine the probability of getting a Head on the coin and an even number on the die.

14. A biased coin that has an 80% chance of getting a Head is flipped four times. Use a tree diagram to answer to the following.

    a. Calculate the probability of getting 4 Heads.
    b. Determine the probability of getting 2 Heads then 2 Tails in that order.
    c. Calculate the probability of getting 2 Heads and 2 Tails in any order.
    d. Determine the probability of getting more Tails than Heads.

LESSON
11.4 Independent and dependent events

LEARNING INTENTIONS

At the end of this lesson you should be able to:
- determine whether two events are dependent or independent
- calculate the probability of outcomes from two-step chance experiments involving independent or dependent events.

▶ 11.4.1 Independent events

eles-4926

- **Independent events** are events that have no effect on each other. The outcome of the first event does not influence the outcome of the second.
- An example of independent events is successive coin tosses. The outcome of the first toss has no effect on the second coin toss. Similarly, the outcome of the roll on a die will not affect the outcome of the next roll.
- If events A and B are independent, then the **Multiplication Law of probability** states that:
  $\Pr(A \text{ and } B) = \Pr(A) \times \Pr(B)$ or $\Pr(A \cap B) = \Pr(A) \times \Pr(B)$
- The reverse is also true. If:
  $\Pr(A \text{ and } B) = \Pr(A) \times \Pr(B)$ or $\Pr(A \cap B) = \Pr(A) \times \Pr(B)$
  then event A and event B are independent events.

Independent events

For independent events A and B:
$$\Pr(A \cap B) = \Pr(A) \times \Pr(B)$$

WORKED EXAMPLE 12 Calculating probability if two events are independent

If A and B are independent events and $\Pr(A) = 0.6$ and $\Pr(B) = 0.3$, calculate:

a. $\Pr(A \cap B)$ **b.** $\Pr(A' \cap B)$ **c.** $\Pr(A \cap B')$ **d.** $\Pr(A' \cap B')$

THINK	WRITE
a. The events are independent, so the formula $\Pr(A \cap B) = \Pr(A) \times \Pr(B)$ can be applied.	$\begin{aligned} \Pr(A \cap B) &= \Pr(A) \times \Pr(B) \\ &= 0.6 \times 0.3 \\ &= 0.18 \end{aligned}$

b. **1.** A' is the complement of A, so $\Pr(A')$ can be calculated using $\Pr(A') = 1 - \Pr(A)$.

$$\Pr(A') = 1 - \Pr(A)$$
$$= 1 - 0.6$$
$$= 0.4$$

2. Then apply the formula for independent events.

$$\Pr(A' \cap B) = \Pr(A') \times \Pr(B)$$
$$= 0.4 \times 0.3$$
$$= 0.12$$

c. B' is the complement of B, so $\Pr(B')$ can be calculated using $\Pr(B') = 1 - \Pr(B)$.

$$\Pr(B') = 1 - \Pr(B)$$
$$= 1 - 0.3$$
$$= 0.7$$
$$\Pr(A \cap B') = \Pr(A) \times \Pr(B')$$
$$= 0.6 \times 0.7$$
$$= 0.42$$

d. Use the formula for independent events.

$$\Pr(A' \cap B') = \Pr(A') \times \Pr(B')$$
$$= 0.4 \times 0.7$$
$$= 0.28$$

▶ 11.4.2 Dependent events

eles-4927

- **Dependent events** are events where the outcome of one event affects the outcome of the other event.
- An example of dependent events is drawing a card from a deck of playing cards. The probability that the first card drawn is a heart, $\Pr(\text{heart})$, is $\dfrac{13}{52}$ (or $\dfrac{1}{4}$).

  If this card is a heart and is not replaced, then this will affect the probability of subsequent draws. The probability that the second card drawn is a heart will be $\dfrac{12}{51}$, and the probability that the second card is not a heart will be $\dfrac{39}{51}$.

- If two events are dependent, then the probability of the occurrence of one event affects that of the subsequent event.

WORKED EXAMPLE 13 Calculating the probability of outcomes of dependent events

A bag contains 5 blue, 6 green and 4 yellow marbles. The marbles are identical in all respects except their colours. Two marbles are picked in succession without replacement.
Calculate the probability of picking 2 blue marbles.

THINK

1. Determine the probability of picking the first blue marble.

WRITE

$$\Pr(\text{picking a blue marble}) = \frac{n(B)}{n(\xi)}$$

$$\Pr(\text{picking a blue marble}) = \frac{5}{15}$$

$$= \frac{1}{3}$$

2. **Determine the probability of picking the second blue marble.**
   Note: The two events are dependent since marbles are not being replaced. Since we have picked a blue marble, this leaves 4 blue marbles remaining out of a total of 14 marbles.

$$\text{Pr(picking second blue marble)} = \frac{n(B)}{n(\xi)}$$

$$\text{Pr(picking second blue marble)} = \frac{4}{14}$$

$$= \frac{2}{7}$$

3. **Calculate the probability of obtaining 2 blue marbles.**

$$\text{Pr(2 blue marbles)} = \text{Pr(1st blue)} \times \text{Pr(2nd blue)}$$

$$= \frac{1}{3} \times \frac{2}{7}$$

$$= \frac{2}{21}$$

4. **Write your answer.**

The probability of obtaining 2 blue marbles is $\frac{2}{21}$.

Note: Alternatively, a tree diagram could be used to solve this question.
The probability of selecting 2 blue marbles successively can be read directly from the first branch of the tree diagram.

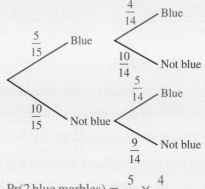

$$\text{Pr(2 blue marbles)} = \frac{5}{15} \times \frac{4}{14}$$

$$= \frac{1}{3} \times \frac{2}{7}$$

$$= \frac{2}{21}$$

---

on Resources

 eWorkbook　　Topic 11 Workbook (code puzzle and project) (ewbk-13300)

 Interactivities　Individual pathway interactivity: Independent and dependent events (int-4618)
　　　　　　　　　Independent and dependent events (int-2787)
　　　　　　　　　Multiplication Law of probability (int-6172)
　　　　　　　　　Dependent events (int-6173)

Exercise 11.4 Independent and dependent events

| 11.4 Quick quiz on | 11.4 Exercise |

Individual pathways

■ PRACTISE	■ CONSOLIDATE	■ MASTER
1, 4, 8, 11, 14, 16, 20	2, 5, 9, 12, 15, 17, 21	3, 6, 7, 10, 13, 18, 19, 22

Fluency

1. **WE12** If A and B are independent events and $Pr(A) = 0.7$ and $Pr(B) = 0.4$, calculate:
   a. $Pr(A \text{ and } B)$
   b. $Pr(A' \text{ and } B)$ where A' is the complement of A
   c. $Pr(A \text{ and } B')$ where B' is the complement of B
   d. $Pr(A' \text{ and } B')$.

2. Determine whether two events A and B with $Pr(A) = 0.6$, $Pr(B') = 0.84$ and $Pr(A \cap B) = 0.96$ are independent or dependent.

3. Determine whether two events A and B with $Pr(A) = 0.25$, $Pr(B) = 0.72$ and $Pr(A \cup B) = 0.79$ are independent or dependent.

Understanding

4. A die is rolled and a coin is tossed.
   a. Explain whether the outcomes are independent.
   b. Calculate:
      i. $Pr(\text{Head})$ on the coin
      ii. $Pr(6)$ on the die.
   c. Calculate $Pr(6 \text{ on the die and Head on the coin})$.

5. A tetrahedral (4-faced) die and a 10-sided die are rolled simultaneously. Calculate the probability of getting a 3 on the tetrahedral die and an 8 on the 10-sided die.

6. A blue die and a green die are rolled. Calculate the probability of getting a 5 on the blue die and not a 5 on the green die.

7. Dean is an archer. The experimental probability that Dean will hit the target is $\frac{4}{5}$.

   a. Calculate the probability that Dean will hit the target on two successive attempts.
   b. Calculate the probability that Dean will hit the target on three successive attempts.
   c. Calculate the probability that Dean will not hit the target on two successive attempts.
   d. Calculate the probability that Dean will hit the target on the first attempt but miss on the second attempt.

8. **MC** A bag contains 20 apples, of which 5 are bruised. Peter picks an apple and realises that it is bruised. He puts the apple back in the bag and picks another one.

   a. The probability that Peter picks 2 bruised apples is:

     A. $\dfrac{1}{4}$ **B.** $\dfrac{1}{2}$ **C.** $\dfrac{1}{16}$ **D.** $\dfrac{3}{4}$

   b. The probability that Peter picks a bruised apple first but a good one on his second attempt is:

     A. $\dfrac{1}{4}$ **B.** $\dfrac{1}{2}$ **C.** $\dfrac{3}{4}$ **D.** $\dfrac{3}{16}$

9. The probability that John will be late for a meeting is $\dfrac{1}{7}$ and the probability that Phil will be late for a meeting is $\dfrac{3}{11}$.

   Calculate the probability that:

   a. John and Phil are both late
   b. neither of them is late
   c. John is late but Phil is not late
   d. Phil is late but John is not late.

10. On the roulette wheel at the casino there are 37 numbers, 0 to 36 inclusive. Bidesi puts his chip on number 8 in game 20 and on number 13 in game 21.

   a. Calculate the probability that he will win in game 20.
   b. Calculate the probability that he will win in both games.
   c. Calculate the probability that he wins at least one of the games.

11. Based on her progress through the year, Karen was given a probability of 0.8 of passing the Physics exam. If the probability of passing both Maths and Physics is 0.72, determine her probability of passing the Maths exam.

12. Suresh found that, on average, he is delayed 2 times out of 7 at Melbourne airport. Rakesh made similar observations at Brisbane airport but found that he was delayed 1 out of every 4 times.
Determine the probability that both Suresh and Rakesh will be delayed if they are flying out of their respective airports.

13. Bronwyn has 3 pairs of Reebok and 2 pairs of Adidas running shoes. She has 2 pairs of Reebok, 3 pairs of Rio and 1 pair of Red Robin socks.
Preparing for an early morning run, she grabs at random for a pair of socks and a pair of shoes.
Calculate the probability that she chooses:

   a. Reebok shoes and Reebok socks
   b. Rio socks and Adidas shoes
   c. Reebok shoes and Red Robin socks
   d. Adidas shoes and socks that are not Red Robin.

14. **WE13** Two cards are drawn successively and without replacement from a pack of playing cards. Calculate the probability of drawing:

 a. 2 hearts
 b. 2 kings
 c. 2 red cards.

15. In a class of 30 students there are 17 students who study Music. Two students are picked randomly to represent the class in the Student Representative Council. Calculate the probability that:

 a. both students don't study Music
 b. both students do study Music
 c. one of the students doesn't study Music.

Reasoning

16. Greg has tossed a Tail on each of 9 successive coin tosses. He believes that his chances of tossing a Head on his next toss must be very high. Explain whether Greg is correct.

17. The Multiplication Law of probability relates to *independent events*. Tree diagrams can illustrate the sample space of successive *dependent events*, and the probability of any one combination of events can be calculated by *multiplying* the stated probabilities along the branches.
 Explain whether this a contradiction to the Multiplication Law of probability.

18. Explain whether it is possible for two events, *A* and *B*, to be mutually exclusive and independent.

19. Consider the following sets:

$$\xi = \{1, 2, 3, 4, 5, 6, 7, 8, 9, 10, 11, 12, 13, 14, 15, 16, 17, 18, 19, 20\}$$

$$A = \{\text{evens in } \xi\}$$

$$B = \{\text{multiples of 3 in } \xi\}$$

 a. Explain whether the events *A* and *B* are independent.
 b. If ξ is changed to the integers from 1 to 10 only, explain whether the result from part **a** changes.

Problem solving

20. There are three coins in a box. One coin is a fair coin, one coin is biased with an 80% chance of landing Heads, and the third is a biased coin with a 40% chance of landing Heads. A coin is selected at random and flipped.
 If the result is a Head, determine the probability that the fair coin was selected.

21. A game at a carnival requires blindfolded contestants to throw balls at numbered ducks sitting on 3 shelves. The game ends when 3 ducks have been knocked off the shelves.
 Assume that the probability of hitting each duck is equal.

 a. Explain whether the events described in the game are dependent or independent.
 b. Determine the initial probabilities of hitting each number.
 c. Draw a labelled tree diagram to show the possible outcomes for a contestant.
 d. Calculate the probabilities of hitting the following.

 i. Pr(1, 1, 1)
 ii. Pr(2, 2, 2)
 iii. Pr(3, 3, 3)
 iv. Pr(at least one 3)

22. Question **21** described a game at a carnival. A contestant pays $3 to play and must make 3 direct hits to be eligible for a prize. The numbers on the ducks hit are then summed and the contestant wins a prize according to the winners' table.

Winners' table	
Total score	**Prize**
9	Major prize ($30 value)
7 or 8	Minor prize ($10 value)
5 or 6	$2 prize
3 or 4	No prize

a. Determine the probability of winning each prize listed.
b. Suppose 1000 games are played on an average show day. Evaluate the profit that could be expected to be made by the sideshow owner on any average show day.

LESSON
11.5 Conditional probability

LEARNING INTENTIONS

At the end of this lesson you should be able to:
• determine from the language in a question whether it involves conditional probability
• use the rule for conditional probability to calculate other probabilities.

▶ 11.5.1 Recognising conditional probability

eles-4928

• **Conditional probability** is when the probability of an event is conditional on (depends on) another event occurring first.
• The effect of conditional probability is to reduce the sample space and thus increase the probability of the desired outcome.
• For two events, A and B, the conditional probability of event B, given that event A occurs, is denoted by $Pr(B|A)$ and can be calculated using the following formula.

Probability of B given A

$$Pr(B|A) = \frac{Pr(A \cap B)}{Pr(A)}, \ Pr(A) \neq 0$$

• Conditional probability can be expressed using a variety of language. Some examples of conditional probability statements follow. The key words to look for in a conditional probability statement have been highlighted in each instance.
  • **If** a student receives a B+ or better in their first Maths test, **then** the chance of them receiving a B+ or better in their second Maths test is 75%.
  • **Given that** a red marble was picked out of the bag with the first pick, the probability of a blue marble being picked out with the second pick is 0.35.
  • **Knowing that** the favourite food of a student is hot chips, the probability of their favourite condiment being tomato sauce is 68%.

WORKED EXAMPLE 14 Using a Venn diagram to calculate conditional probabilities

A group of students was asked whether they like spaghetti (*S*) or lasagne (*L*). The results are illustrated in the Venn diagram shown. Use the Venn diagram to calculate the following probabilities relating to a student's preferred food.

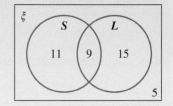

a. Calculate the probability that a randomly selected student likes spaghetti.

b. Determine the probability that a randomly selected student likes lasagne given that they also like spaghetti.

THINK	WRITE/DRAW
a. 1. Determine how many students were surveyed to identify the total number of possible outcomes. Add each of the numbers shown on the Venn diagram.	**a.** Total number of students $= 11 + 9 + 15 + 5 = 40$

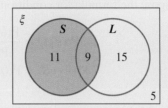

a. 2. There are 20 students who like 'spaghetti' or 'spaghetti and lasagne', as shown in pink.

a. 3. The probability that a randomly selected student likes spaghetti is found by substituting these values into the probability formula.

$$\Pr(\text{event}) = \frac{\text{number of favourable outcomes}}{\text{total number of possible outcomes}}$$

$$\Pr(\text{spaghetti}) = \frac{20}{40}$$

$$= \frac{1}{2}$$

b. 1. The condition imposed 'given they also like spaghetti' alters the sample space to the 20 students described in part **a**, as shaded in blue. Of these 20 students, 9 stated that they liked lasagne and spaghetti, as shown in pink.

b.

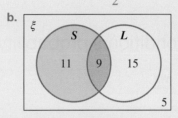

b. 2. The probability that a randomly selected student likes lasagne, given that they like spaghetti, is found by substituting these values into the probability formula for conditional probability.

$$\Pr(B|A) = \frac{\Pr(A \cap B)}{\Pr(A)}$$

$$\Pr(L|S) = \frac{\frac{9}{40}}{\frac{1}{2}}$$

$$= \frac{9}{20}$$

WORKED EXAMPLE 15 Using the rule for conditional probability

If $\Pr(A) = 0.3$, $\Pr(B) = 0.5$ and $\Pr(A \cup B) = 0.6$, calculate:

a. $\Pr(A \cap B)$

b. $\Pr(B|A)$

THINK	WRITE
a. 1. State the Addition Law for probability.	**a.** $\Pr(A \cup B) = \Pr(A) + \Pr(B) - \Pr(A \cap B)$

▶

2. Substitute the values given in the question into this formula and simplify.

$$0.6 = 0.3 + 0.5 - \Pr(A \cap B)$$
$$\Pr(A \cap B) = 0.3 + 0.5 - 0.6$$
$$= 0.2$$

b. 1. State the formula for conditional probability.

b. $\Pr(B|A) = \dfrac{\Pr(A \cap B)}{\Pr(A)}, \Pr(A) \neq 0$

2. Substitute the values given in the question into this formula and simplify.

$$\Pr(B|A) = \dfrac{0.2}{0.3}$$
$$= \dfrac{2}{3}$$

- It is possible to transpose the formula for conditional probability to calculate $\Pr(A \cap B)$:

$$\Pr(B|A) = \dfrac{\Pr(A \cap B)}{\Pr(A)}, \Pr(A) \neq 0$$
$$\Pr(A \cap B) = \Pr(A) \times \Pr(B|A)$$

 Resources

 eWorkbook Topic 11 Workbook (worksheets, code puzzle and project) (ewbk-13300)

 Video eLesson Conditional probability (eles-1928)

 Interactivities Individual pathway interactivity: Conditional probability (int-4619)
Conditional probability (int-6085)

Exercise 11.5 Conditional probability

learnon

| 11.5 Quick quiz on | 11.5 Exercise |

Individual pathways

■ PRACTISE	■ CONSOLIDATE	■ MASTER
1, 3, 5, 10, 12, 15	2, 6, 9, 13, 16	4, 7, 8, 11, 14, 17

Fluency

1. **WE14** A group of students was asked whether they liked the following forms of dance: hip hop (H) or jazz (J). The results are illustrated in the Venn diagram.
Use the Venn diagram to calculate the following probabilities relating to a student's favourite form of dance.

   a. Calculate the probability that a randomly selected student likes jazz.
   b. Determine the probability that a randomly selected student likes hip hop, given that they like jazz.

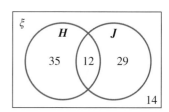

2. A group of students was asked which seats they liked: the seats in the computer lab or the seats in the science lab. The results are illustrated in the Venn diagram.

Use the Venn diagram to calculate the following probabilities relating to the most comfortable seats.

a. Calculate the probability that a randomly selected student likes the seats in the science lab.

b. Determine the probability that a randomly selected student likes the seats in the science lab, given that they like the seats in the computer lab or the science lab.

3. **WE15** If $Pr(A) = 0.7$, $Pr(B) = 0.5$ and $Pr(A \cup B) = 0.9$, calculate:

   a. $Pr(A \cap B)$
   b. $Pr(B|A)$.

4. If $Pr(A) = 0.65$, $Pr(B) = 0.75$ and $Pr(A \cap B) = 0.45$, calculate:

   a. $Pr(B|A)$
   b. $Pr(A|B)$.

Understanding

5. A medical degree requires applicants to participate in two tests: an aptitude test and an emotional maturity test. This year 52% passed the aptitude test and 30% passed both tests. Use the conditional probability formula to calculate the probability that a student who passed the aptitude test also passed the emotional maturity test.

6. At a school classified as a 'Music school for excellence', the probability that a student elects to study Music and Physics is 0.2. The probability that a student takes Music is 0.92.

   Determine the probability that a student takes Physics, given that the student is taking Music.

7. The probability that a student is well and misses a work shift the night before an exam is 0.045, and the probability that a student misses a work shift is 0.05. Determine the probability that a student is well, given they miss a work shift the night before an exam.

8. Two marbles are chosen, without replacement, from a jar containing only red and green marbles. The probability of selecting a green marble and then a red marble is 0.67. The probability of selecting a green marble on the first draw is 0.8.

   Determine the probability of selecting a red marble on the second draw, given the first marble drawn was green.

9. **MC** A group of 80 schoolgirls consists of 54 dancers and 35 singers. Each member of the group is either a dancer, a singer, or both. The probability that a randomly selected student is a singer given that she is a dancer is:

   A. 0.17   **B.** 0.44   **C.** 0.68   **D.** 0.11

10. Consider rolling a red and a black die and the probabilities of the following events:

    Event A the red die lands on 5
    Event B the black die lands on 2
    Event C the sum of the dice is 10.

    a. **MC** The initial probability of each event described is:

    A. $Pr(A) = \dfrac{1}{6}$ **B.** $Pr(A) = \dfrac{5}{6}$ **C.** $Pr(A) = \dfrac{5}{6}$ **D.** $Pr(A) = \dfrac{1}{6}$

    $Pr(B) = \dfrac{1}{6}$ $Pr(B) = \dfrac{2}{6}$ $Pr(B) = \dfrac{2}{6}$ $Pr(B) = \dfrac{1}{6}$

    $Pr(C) = \dfrac{1}{6}$ $Pr(C) = \dfrac{7}{36}$ $Pr(C) = \dfrac{5}{18}$ $Pr(C) = \dfrac{1}{12}$

    b. Calculate the following probabilities.

       i. $Pr(A|B)$ ii. $Pr(B|A)$ iii. $Pr(C|A)$ iv. $Pr(C|B)$

11. The following is the blood pressure data from 232 adult patients admitted to a hospital over a week. The results are displayed in a two-way frequency table.

Age	Blood pressure			Total
	Low	Medium	High	
Under 60 years	92	44	10	146
60 years or above	17	46	23	86
Total	109	90	33	**232**

    a. Calculate the probability that a randomly chosen patient has low blood pressure.
    b. Determine the probability that a randomly chosen patient is under 60 years of age.
    c. Calculate the probability that a randomly chosen patient has high blood pressure, given they are aged 60 years or above.
    d. Determine the probability that a randomly chosen patient is under the age of 60, given they have medium blood pressure.

Reasoning

12. Explain how imposing a condition alters probability calculations.

13. At a school, 65% of the students are male and 35% are female. Of the male students, 10% report that dancing is their favourite activity; of the female students, 25% report that dancing is their favourite activity.
    Determine the probability that:
    a. a student selected at random prefers dancing and is female
    b. a student selected at random prefers dancing and is male.
    c. Construct a tree diagram to present the information given and use it to calculate:
       i. the probability that a student is male and does not prefer dancing
       ii. the overall percentage of students who prefer dancing.

14. Consider the following sets of numbers.

$$\xi = \{1, 2, 3, 4, 5, 6, 7, 8, 9, 10, 11, 12, 13, 14, 15, 16, 17, 18, 19, 20\}$$

$$A = \{1, 2, 3, 4, 5, 6, 7, 8, 9, 10\}$$

$$B = \{2, 3, 5, 7, 11, 13, 17, 19\}$$

a. Calculate the following probabilities.
   i. $\Pr(A)$
   ii. $\Pr(B)$
   iii. $\Pr(A \cap B)$
   iv. $\Pr(A|B)$
   v. $\Pr(B|A)$

b. Explain whether the events A and B are independent.
c. Write a statement that connects $\Pr(A), \Pr(A \cap B), \Pr(A|B)$ and independent events.

Problem solving

15. The rapid test used to determine whether a person is infected with COVID-19 is not perfect. For one type of rapid test, the probability of a person with the disease returning a positive result is 0.98, while the probability of a person without the disease returning a positive result is 0.04.
   At its peak in a certain country, the probability that a randomly selected person has COVID-19 is 0.05.
   Determine the probability that a randomly selected person will return a positive result.

16. Two marbles are chosen, without replacement, from a jar containing only red and green marbles. The probability of selecting a green marble and then a red marble is 0.72. The probability of selecting a green marble on the first draw is 0.85.
   Determine the probability of selecting a red marble on the second draw if the first marble drawn was green.

17. When walking home from school during the summer months, Harold buys either an ice-cream or a drink from the corner shop.
   If Harold bought an ice-cream the previous day, there is a 30% chance that he will buy a drink the next day. If he bought a drink the previous day, there is a 40% chance that he will buy an ice-cream the next day. On Monday, Harold bought an ice-cream. Determine the probability that he buys an ice-cream on Wednesday.

LESSON
11.6 Counting principles and factorial notation (10A)

▶ 11.6.1 Counting techniques

eles-6143

- Counting techniques allow us to determine the number of ways an activity can occur. This in turn allows us to calculate the probability of an event.
- When considering a set with n elements, the number of different subsets formed can be found by considering and systematically listing each element for inclusion.
  For example, the set $\{a, b, c\}$ has subsets $\{\emptyset, \{a\}, \{b\}, \{c\}, \{a, b\}, \{a, c\}, \{b, c\}, \{a, b, c\}\}$.
- In the above example there are 8 subsets. This could have also been calculated by $2^n = 2^3$, where n is the number of elements in the set.
- Tree diagrams can be used to determine the number of solutions to simple counting problems.
  For example, a tree diagram can be made to answer the following question:

If driving from Port Douglas to Cairns I can take any one of 4 different roads and from Cairns to Townsville I can take 3 different roads, determine how many different routes I can take if I want to drive from Port Douglas to Townsville.

- There are four ways of getting from Port Douglas to Cairns and three ways of getting from Cairns to Townsville, so there are $4 \times 3 = 12$ ways of getting from Port Douglas to Townsville.
- This is known as the **multiplication principle** and should be used when there are events where one event is followed by the other — that is, when order is important.

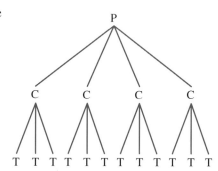

The multiplication principle

If there are n ways of performing operation A and m ways of performing operation B, then there are $n \times m$ ways of performing A *and* B in the order AB.

- A useful technique for solving problems based on the multiplication principle is to use boxes. In the example above we would write the following.

1st	2nd
4	3

- The value in the '1st' box represents the number of ways the first operation can be performed.
- The value in the '2nd' box stands for the number of ways the second operation can be performed.
- To apply the multiplication principle you multiply the numbers in the boxes.

WORKED EXAMPLE 16 Applying the multiplication principle

Two letters are to be chosen from the set of 5 letters A, B, C, D and E, where order is important.
a. List all the different ways that this may be done.
b. Use the multiplication principle to calculate the number of ways that this may be done.
c. Determine the probability the first letter will be a C.
d. Use the multiplication principle to calculate the number of ways the two letters can be chosen if the first letter is replaced after selection.

THINK

a. 1. Begin with A in first place and make a list of each of the possible pairs.

2. Make a list of each of the possible pairs with B in the first position.

3. Make a list of each of the possible pairs with C in the first position.

4. Make a list of each of the possible pairs with D in the first position.

5. Make a list of each of the possible pairs with E in the first position.
Note: AB and BA need to be listed separately as order is important.

b. The multiplication principle could have been used to determine the number of ordered pairs.

1. Rule up two boxes that represent the pair.

2. Write down the number of letters that may be selected for the first box. That is, in first place any of the 5 letters may be used.

3. Write down the number of letters that may be selected for the second box. That is, in second place any of the 4 letters may be used.
Note: One less letter is used to avoid repetition.

4. Evaluate.

5. Write the answer.

c. 1. Recall the probability formula. The total number in the set $n(\xi)$ was determined in part **b**.

2. Let A be the event that the pair starts with a C. Draw a table showing the requirement imposed by the first letter to be C.

WRITE

a. AB   AC   AD   AE

   BA   BC   BD   BE

   CA   CB   CD   CE

   DA   DB   DC   DE

   EA   EB   EC   ED

b.

5	4

$5 \times 4 = 20$ ways

There are 20 ways in which 2 letters may be selected from a group of 5 where order is important.

c. $\Pr(A) = \dfrac{n(A)}{n(\xi)}$

$n(\xi) = 20$

1	

3. Complete the table. Once the first letter has been completed, there are 4 choices for the second letter. Use the multiplication principle to determine the number of combinations starting with C.

There are $1 \times 4 = 4$ possible combinations beginning with C.
So, $n(A) = 4$.

4. Use the probability formula to determine the probability the first letter will be a C.

$$Pr(A) = \frac{n(A)}{n(\xi)}$$
$$= \frac{4}{20}$$
$$= \frac{1}{5}$$

This is confirmed by examining the answer to part **a**.

d. 1. Rule up two boxes that represent the pair.

d.
5	5

2. Write down the number of letters that may be selected for the first box. That is, in first place any of the 5 letters may be used.

3. Write down the number of letters that may be selected for the second box. That is, in second place, any of the 5 letters may be used.
Note: The first letter has been replaced so there are 5 to choose from again.

4. Evaluate.

$5 \times 5 = 25$ ways

5. Write the answer.

There are 25 ways in which 2 letters may be selected from a group of 5 with replacement.

WORKED EXAMPLE 17 Applying the multiplication principle to determine the number of ways and probability of an activity

a. **Use the multiplication principle to calculate how many ways an arrangement of 5 numbers can be chosen from {1, 2, 3, 4, 5, 6}.**

b. **Determine the probability of the number ending with 4.**

THINK

a. 1. Instead of listing all possibilities, draw 5 boxes to represent the 5 numbers chosen.
Label each box on the top row as 1st, 2nd, 3rd, 4th and 5th.
Note: The word 'arrangement' implies order is important.

WRITE

a.

2. Fill in each of the boxes showing the number of ways a number may be chosen.
   In the 1st box there are 6 choices for the first number. In the 2nd box there are 5 choices for the second number as 1 number has already been used. In the 3rd box there are 4 choices for the third number as 2 numbers have already been used. Continue this process until each of the 5 boxes is filled.

1st	2nd	3rd	4th	5th
6	5	4	3	2

3. Use the multiplication principle as this is an '*and*' situation.

$$\text{Number of ways} = 6 \times 5 \times 4 \times 3 \times 2$$
$$= 720$$

4. Write the answer.

An arrangement of 5 numbers may be chosen 720 ways.

b. 1. Recall the probability formula.
   The total number of arrangements, $n(\xi)$, was determined in part **a**.

b. $Pr(A) = \dfrac{n(A)}{n(\xi)}$

$n(\xi) = 720$

2. Let A be the event that the number ends with 4. Draw a table showing the requirement imposed by the last letter to be 4.

1st	2nd	3rd	4th	5th
				1

3. Complete the table. Once the last number has been completed, there are 5 choices for the number in the first position and 4 choices for the next number. Continue this process until each of the 5 columns has been filled. Use the multiplication principle to determine the number of combinations ending with 4.

1st	2nd	3rd	4th	5th
5	4	3	2	1

There are $5 \times 4 \times 3 \times 2 \times 1 = 120$ possible combinations ending with 4.
So, $n(A) = 120$.

4. Use the probability formula to determine the probability of the number ending with 4 .

$Pr(A) = \dfrac{n(A)}{n(\xi)}$

$= \dfrac{120}{720}$

$= \dfrac{1}{6}$

▶ 11.6.2 Factorials

- The multiplication principle can be used to determine the number of ways in which five basketball championship trophies could be displayed on a shelf in the school foyer.
- The display may be done in the following way:

Position 1	Position 2	Position 3	Position 4	Position 5
5	4	3	2	1

That is, there are $5 \times 4 \times 3 \times 2 \times 1 = 120$ ways.

- If there were a large number of items to arrange, this method could be time consuming, but there is a more efficient method using **factorials**.
- In general when we need to multiply each of the integers from a particular number, n, down to 1, we write $n!$, which is read as 'n factorial'.
  Hence:

$$6! = 6 \times 5 \times 4 \times 3 \times 2 \times 1$$
$$= 720$$
$$8! = 8 \times 7 \times 6 \times 5 \times 4 \times 3 \times 2 \times 1$$
$$= 40\,320$$
$$n! = n \times (n-1) \times (n-2) \times (n-3) \times \ldots \times 3 \times 2 \times 1$$

- The number of ways n distinct objects may be arranged is $n!$ (n factorial).

Factorials

$$n! = n \times (n-1) \times (n-2) \times (n-3) \times \ldots \times 3 \times 2 \times 1$$

That is, $n!$ is the product of each of the integers from n down to 1.

A special case of the factorial function is $0! = 1$.

WORKED EXAMPLE 18 Evaluating factorials

Evaluate the following factorials.

a. $4!$ b. $7!$ c. $\dfrac{8!}{5!}$

THINK	WRITE
a. 1. Write 4! in its expanded form and evaluate.	a. $4! = 4 \times 3 \times 2 \times 1$ $= 24$
2. Verify the answer obtained using the factorial function on a calculator.	
b. 1. Write 7! in its expanded form and evaluate.	b. $7! = 7 \times 6 \times 5 \times 4 \times 3 \times 2 \times 1$ $= 5040$
2. Verify the answer obtained using the factorial function on a calculator.	
c. 1. Write each factorial term in its expanded form.	c. $\dfrac{8!}{5!} = \dfrac{8 \times 7 \times 6 \times 5 \times 4 \times 3 \times 2 \times 1}{5 \times 4 \times 3 \times 2 \times 1}$
2. Cancel down like terms.	$= 8 \times 7 \times 6$
3. Evaluate.	$= 336$
4. Verify the answer obtained using the factorial function on a calculator.	

TI	THINK	DISPLAY/WRITE	CASIO	THINK	DISPLAY/WRITE
c.	On a Calculator page, complete the entry line as shown.		c. From the Advance palette on the Keyboard, complete the entry line as shown by selecting !.	

11.6.3 Permutations

eles-6145

- When objects are arranged in a particular order, it is called an arrangement or **permutation**.
- Previously, we learned that if you select 3 letters from 7 where the *order is important*, the number of possible arrangements can be found using the multiplication principle.

1st	2nd	3rd
7	6	5

The number of arrangements $= 7 \times 6 \times 5$
$$= 210$$

- This could also be expressed in factorial form:

$$7 \times 6 \times 5 = \frac{7 \times 6 \times 5 \times 4 \times 3 \times 2 \times 1}{4 \times 3 \times 2 \times 1}$$
$$= \frac{7!}{4!}$$
$$= \frac{7!}{(7-3)!}$$

- Using more formal terminology we say that in choosing 3 things from 7 things where order is important, the number of permutations is $^7P_3 = 7 \times 6 \times 5$. The letter P is used to remind us that we are finding permutations.
- The number of ways of choosing r things from n distinct things is given by the following rule.

> **Permutations**
>
> $$^nP_r = \frac{n!}{(n-r)!}$$

WORKED EXAMPLE 19 Determining the number of ways an activity can be arranged

Courtney has purchased 9 new books but only has room for 3 of them on her bookshelf. Determine how many ways the books can be arranged.

THINK	WRITE
1. Recall the rule for permutations.	$^nP_r = \dfrac{n!}{(n-r)!}$

2. Substitute the given values of n and r into the permutation formula.

$$^9P_3 = \frac{9!}{(9-3)!}$$

$$= \frac{9!}{6!}$$

3. Use a calculator to evaluate 9! and 6!.

$$= 504$$

4. Write the answer.

There are 504 ways the books can be arranged.

WORKED EXAMPLE 20 Using permutations to calculate probabilities

The netball club needs to appoint a president, secretary and treasurer. From the committee 7 people have volunteered for these positions. Each of the 7 nominees is happy to fill any one of the 3 positions.
a. Determine how many different ways these positions can be filled.
b. For three years, the same 7 people volunteer for these positions. Determine the probability one of them is president 3 years in a row.

THINK

a. 1. Recall the rule for permutations.
Note: Order is important, so use permutations.

2. Substitute the given values of n and r into the permutation formula.

3. Use a calculator to evaluate 7! and 4!.

4. Evaluate.

5. Write the answer.

b. 1. In three years of the president's position, each year this could be awarded to one of the 7 people.

2. Let A be the event that the same person fills the position three years in a row.

3. Calculate the probability that the same person fills the president's position three years in a row.

WRITE

a. $^nP_r = \frac{n!}{(n-r)!}$

$^7P_3 = \frac{7!}{(7-3)!}$

$= \frac{7!}{4!}$

$= \frac{5040}{24}$

$= 210$

There are 210 different ways of filling the positions of president, secretary and treasurer.

b.
7	7	7

The total number of ways the president's position could be filled is $7 \times 7 \times 7$.
$n(\xi) = 7 \times 7 \times 7$
$ = 343$

There are 7 choices of the same person to fill the positions three years in a row.
$n(A) = 7$

$P(A) = \frac{n(A)}{n(\xi)}$

$= \frac{7}{7 \times 7 \times 7}$

$= \frac{1}{49}$

The probability that one of the 7 volunteers fills the president's position three years in a row is $\frac{1}{49}$.

11.6.4 Permutations with like objects

eles-6146

- If some objects are repeated or are like objects, it will decrease the number of different arrangements. For example, a 5-letter word made from 5 different letters would have $5! = 120$ arrangements, but if the letters were 3 A's and 2 B's the number of arrangements would be less.
- To analyse the situation let us imagine that we can distinguish one A from another. We will write A_1, A_2, A_3, B_1 and B_2 to represent the 5 letters.
- As we list some of the possible arrangements we notice that some are actually the same, as shown in the table.

$A_1A_2B_1A_3B_2$	$A_1A_2B_2A_3B_1$	Each of these 12 arrangements is the same — AABAB — if $A_1 = A_2 = A_3$ and $B_1 = B_2$.
$A_1A_3B_1A_2B_2$	$A_1A_3B_2A_2B_1$	
$A_2A_1B_1A_3B_2$	$A_2A_1B_2A_3B_1$	
$A_2A_3B_1A_1B_2$	$A_2A_3B_2A_1B_1$	
$A_3A_1B_1A_2B_2$	$A_3A_1B_2A_2B_1$	
$A_3A_2B_1A_1B_2$	$A_3A_2B_2A_1B_1$	
$B_2A_1A_2B_1A_3$	$B_1A_1A_2B_2A_3$	Each of these 12 arrangements is the same — BAABA — if $A_1 = A_2 = A_3$ and $B_1 = B_2$.
$B_2A_1A_3B_1A_2$	$B_1A_1A_3B_2A_2$	
$B_2A_2A_1B_1A_3$	$B_1A_2A_1B_2A_3$	
$B_2A_2A_3B_1A_1$	$B_1A_2A_3B_2A_1$	
$B_2A_3A_1B_1A_2$	$B_1A_3A_1B_2A_2$	
$B_2A_3A_2B_1A_1$	$B_1A_3A_2B_2A_1$	

- The number of repetitions is $3!$ for the As and $2!$ for the Bs. Thus, the number of different arrangements in choosing 5 letters from 3 As and 2 Bs is $\dfrac{5!}{3! \times 2!} = 10$.

Permutations with like objects

The number of different ways of arranging n objects made up of groups of repeated (identical) objects, n_1 in the first group, n_2 in the second group and so on, is:

$$\frac{n!}{n_1! \, n_2! \, n_3! \ldots n_r!}.$$

WORKED EXAMPLE 21 Calculating permutations with like objects

Determine how many different permutations of 7 counters can be made from 4 black and 3 white counters.

THINK	WRITE
1. Write down the total number of counters.	There are 7 counters in all; therefore, $n = 7$.
2. Write down the number of times any of the coloured counters are repeated.	There are 3 white counters; therefore, $n_1 = 3$. There are 4 black counters; therefore, $n_2 = 4$.
3. Write down the formula for arranging groups of like things.	$\dfrac{n!}{n_1! \, n_2! \, n_3! \ldots n_r!}$
4. Substitute the values of n, n_1 and n_2 into the rule.	$= \dfrac{7!}{3! \times 4!}$

5. Expand each of the factorials.	$= \dfrac{7 \times 6 \times 5 \times 4 \times 3 \times 2 \times 1}{3 \times 2 \times 1 \times 4 \times 3 \times 2 \times 1}$
6. Simplify the fraction.	$= \dfrac{7 \times 6 \times 5}{6}$
7. Evaluate.	$= 35$
8. Write the answer.	Thirty-five different arrangements can be made from 7 counters, of which 3 are white and 4 are black.

11.6.5 Combinations

eles-6147

- A group of things chosen from a larger group where order is not important is called a **combination**.
- Suppose 5 people have nominated for a committee consisting of 3 members. It does not matter in what order the candidates are placed on the committee, it matters only whether they are there or not.
- If order was important we know there would be $^5P_3$, or 60, ways in which this could be done. Here are the possibilities:

ABC	ACB	BAC	BCA	CAB	CBA
ABD	ADB	BAD	BDA	DAB	DBA
ABE	AEB	BAE	BEA	EAB	ABA
ACE	AEC	CAE	CEA	EAC	ECA
ACD	ADC	CAD	CDA	DAC	DCA
ADE	AED	DAE	DEA	EAD	EDA
BCD	BDC	CBD	CDB	DBC	DCB
BCE	BEC	CBE	CEB	EBC	ECB
BDE	BED	DBE	DEB	EBD	EDB
CDE	CED	DCE	DEC	ECD	EDC

- The 60 arrangements are different only if we take order into account; that is, ABC is different from CAB and so on. You will notice in this table that there are 10 distinct committees corresponding to the 10 distinct rows. Each row merely repeats, in a different order, the committee in the first column. This result (10 distinct committees) can be arrived at logically:
  1. There are $^5P_3$ ways of choosing or selecting 3 from 5 in order.
  2. Each choice of 3 is repeated 3! times.
  3. The number of distinct selections or combinations is $^5P_3 \div 3! = 10$.
- This leads to the general rule of selecting r objects from n objects.

> **Combinations**
>
> **The number of ways of choosing or selecting r objects from n distinct objects, where order is not important, is given by ^nC_r:**
>
> $$^nC_r = \dfrac{^nP_r}{r!}$$
>
> **C is used to represent combinations.**

Write these combinations as statements involving permutations and then calculate them.

a. $^7C_2$ b. $^{20}C_3$

THINK	WRITE
a. 1. Recall the rule for ^nC_r.	a. $^nC_r = \dfrac{^nP_r}{r!}$
2. Substitute the given values of n and r into the combination formula.	$^7C_2 = \dfrac{^7P_2}{2!}$
3. Simplify the fraction.	$= \dfrac{\left(\frac{7!}{5!}\right)}{2!}$
	$= \dfrac{7!}{5!} \div 2!$
	$= \dfrac{7!}{5!} \times \dfrac{1}{2!}$
	$= \dfrac{7 \times 6 \times 5!}{5! \times 2 \times 1}$
4. Evaluate.	$= \dfrac{7 \times 6}{2 \times 1}$
	$= \dfrac{42}{2}$
	$= 21$
b. 1. Write down the rule for ^nC_r.	b. $^nC_r = \dfrac{^nP_r}{r!}$
2. Substitute the values of n and r into the formula.	$^{20}C_3 = \dfrac{^{20}P_3}{3!}$
3. Simplify the fraction.	$= \dfrac{\left(\frac{20!}{17!}\right)}{3!}$
	$= \dfrac{20!}{17!} \div 3!$
	$= \dfrac{20!}{17!} \times \dfrac{1}{3!}$
	$= \dfrac{20 \times 19 \times 18 \times 17!}{17! \times 3 \times 2 \times 1}$
4. Evaluate.	$= \dfrac{20 \times 19 \times 18}{3 \times 2 \times 1}$
	$= \dfrac{6840}{6}$
	$= 1140$

WORKED EXAMPLE 23 Applying combinations

Apply the concept of ^nC_r to calculate the number ways a basketball team of 5 players be selected from a squad of 9 if the order in which they are selected does not matter.

THINK	WRITE
1. Recall the rule for ^nC_r. *Note:* Since order does not matter, use the ^nC_r rule.	$^nC_r = \dfrac{^nP_r}{r!}$
2. Substitute the values of n and r into the formula.	$^9C_5 = \dfrac{^9P_5}{5!}$
3. Simplify the fraction.	$= \dfrac{\left(\frac{9!}{4!}\right)}{5!}$ $= \dfrac{9!}{4!} \div 5!$ $= \dfrac{9!}{4!} \times \dfrac{1}{5!}$ $= \dfrac{9!}{4!\ 5!}$ $= \dfrac{9 \times 8 \times 7 \times 6 \times 5!}{4 \times 3 \times 2 \times 1 \times 5!}$
4. Evaluate.	$= \dfrac{9 \times 8 \times 7 \times 6}{4 \times 3 \times 2 \times 1}$ $= \dfrac{3024}{24}$ $= 126$

- To calculate probabilities using combinations, express the numerator and denominator in terms of the appropriate coefficients and then carry out the calculations.

WORKED EXAMPLE 24 Calculating probabilities using combinations

A committee of 5 students is to be chosen from 7 boys and 4 girls. Use ^nC_r and the multiplication and addition principles to answer the following.
a. **Calculate how many committees can be formed.**
b. **Calculate how many of the committees contain exactly 2 boys and 3 girls.**

THINK	WRITE
a. 1. As there is no restriction, choose the committee from the total number of students.	**a.** There are 11 students in total from whom 5 students are to be chosen. This can be done in $^{11}C_5$ ways.
2. Use the formula $^nC_r = \dfrac{n!}{r! \times (n-r)!}$ to calculate the answer.	$^{11}C_5 = \dfrac{11!}{5! \times (11-5)!}$
	$= \dfrac{11!}{5! \times 6!}$
	$= \dfrac{11 \times 10 \times 9 \times 8 \times 7 \times \cancel{6!}}{5! \times \cancel{6!}}$
	$= \dfrac{11 \times 10 \times 9 \times 8 \times 7}{5 \times 4 \times 3 \times 2 \times 1}$
	$= 462$
	There are 462 possible committees.
b. 1. Select the committee to satisfy the given restriction.	**b.** The 2 boys can be chosen from the 7 boys available in $^7C_2$ ways. The 3 girls can be chosen from the 4 girls available in $^4C_3$ ways.
2. Use the multiplication principle to form the total number of committees. *Note:* The upper numbers on the combinatoric coefficients sum to the total available, $7 + 4 = 11$, and the lower numbers sum to the number that must be on the committee, $2 + 3 = 5$.	The total number of committees that contain two boys and three girls is $^7C_2 \times {}^4C_3$.
3. Calculate the answer.	$^7C_2 \times {}^4C_3 = \dfrac{7!}{2! \times 5!} \times 4$
	$= \dfrac{7 \times 6}{2!} \times 4$
	$= 21 \times 4$
	$= 84$
	There are 84 committees possible with the given restriction.

 Resources

eWorkbook Topic 11 Workbook (worksheets, code puzzle and project) (ewbk-13300)

11.6 Quick quiz on	11.6 Exercise

Individual pathways

■ PRACTISE	■ CONSOLIDATE	■ MASTER
1, 4, 7, 12, 14, 15, 17, 22, 27	2, 5, 8, 10, 13, 16, 18, 20, 23, 24, 28, 29	3, 6, 9, 11, 19, 21, 25, 26, 30, 31, 32

Fluency

1. **WE16** Two letters are to be chosen from A, B and C, where order is important.
   a. List all the different ways that this may be done.
   b. Use the multiplication principle to calculate the number of ways that this may be done.
   c. Determine the probability the last letter will be a B.
   d. Use the multiplication principle to calculate the number of ways the two letters can be chosen if the first letter is replaced after selection.

2. List all the different arrangements possible for a group of 2 colours to be chosen from B (blue), G (green), Y (yellow) and R (red).

3. List all the different arrangements possible for a group of 3 letters to be chosen from A, B and C.

4. a. **WE17** Use the multiplication principle to calculate how many ways an arrangement of 2 letters can be chosen from A, B, C, D, E, F and G.
   b. Calculate how many ways an arrangement of 3 letters can be chosen from 7 different letters.
   c. Calclate how many ways an arrangement of 4 letters can be chosen from 7 different letters.
   d. Calculate how many different arrangements of 5 letters can be made from 7 letters.
   e. Determine how many possible combinations there are beginning with an E.

5. Determine the number of different subsets formed from a set with the following numbers of elements.
   a. 3
   b. 5
   c. 9

6. Recall the definition of $n!$ and write each of the following in expanded form.
   a. 4!
   b. 5!
   c. 6!
   d. 7!

7. **WE18** Evaluate the following factorials without using a calculator.
   a. $\dfrac{9!}{5!}$
   b. $\dfrac{10!}{4!}$
   c. $\dfrac{7!}{3!}$
   d. $\dfrac{6!}{0!}$

8. Evaluate the following factorials.
   a. 10!
   b. 14!
   c. 9!
   d. 16!

9. Calculate each of the following by expressing it in expanded form.
   a. $^8P_2$
   b. $^7P_5$
   c. $^8P_7$

10. Write each of the following as a quotient of factorials and hence evaluate.
    a. $^9P_6$
    b. $^5P_2$
    c. $^{18}P_5$

11. Use your calculator to determine the value of each of the following.
    a. $^{20}P_6$
    b. $^{800}P_2$
    c. $^{18}P_5$

12. **WE 21** Determine how many different arrangements of 5 counters can be made using 3 red and 2 blue counters.

13. Determine how many different arrangements of 9 counters can be made using 4 black, 3 red and 2 blue counters.

14. **WE 22** Write each of the following as statements in terms of permutations.

   a. $^8C_3$
   b. $^{19}C_2$
   c. $^1C_1$
   d. $^5C_0$

Understanding

15. a. A teddy bear's wardrobe consists of 3 different hats, 4 different shirts and 2 different pairs of pants. Determine how many different outfits the teddy bear can wear.
   b. A surfboard is to have 1 colour on its top and a different colour on its bottom. The 3 possible colours are red, blue and green.
      Calculate how many different ways the surfboard can be coloured.
   c. A new phone comes with a choice of 3 cases, 2 different-sized screens and 2 different storage capacities. With these choices, determine how many different arrangements are possible.
   d. Messages can be sent by placing 3 different coloured flags in order on a pole. If the flags come in 4 colours, determine how many different messages can be sent.

16. Hani and Mary's restaurant offers its patrons a choice of 4 entrees, 10 main courses and 5 desserts.
   a. Determine how many choices of 3-course meals (entree, main, dessert) are available.
   b. Determine how many choices of entree and main course are offered.
   c. Determine how many choices of meals comprising a main course and dessert are offered.

17. **WE20** A soccer club will appoint a president and a vice-president. Eight people have volunteered for either of the two positions.
   a. Determine how many different ways these positions can be filled.
   b. For two years the same 8 people volunteered for these positions.
      Determine the probability that one of them is president for both years.

18. Determine how many different permutations can be made using the 11 letters of the word ABRACADABRA.

19. A collection of 12 books is to be arranged on a shelf.

   The books consist of 3 copies of *Great Expectations*, 5 copies of *Catcher in the Rye* and 4 copies of *Huntin', Fishin' and Shootin'*. Determine how many different arrangements of these books are possible.

20. **WE23** Apply the concept of ^nC_r to calculate the number of ways three types of ice-cream can be chosen in any order from a supermarket freezer if the freezer contains the following numbers of types of ice-cream.

   a. 3 types
   b. 6 types
   c. 10 types
   d. 12 types

21. A mixed netball team must have 3 women and 4 men in the side. If the squad has 6 women and 5 men wanting to play, determine how many different teams are possible.

Reasoning

22. Jasmin has a phone that has a 4-digit security code. She remembers that the first number in the code was 9 and that the others were 3, 4 and 7 but forgets the order of the last 3 digits.
Determine how many different trials she must make to be sure of unlocking the phone.

23. Determine how many different 4-digit numbers can be made from the numbers 1, 3, 5 and 7 if the numbers can be repeated (that is 3355 and 7777 are valid).

24. There are 26 players in an online game. Determine how many different results for 1st, 2nd, 3rd and 4th can occur.

25. A rowing crew consists of 4 rowers who sit in a definite order. Determine how many different crews are possible if 5 people try out for selection.

26. The school musical needs a producer, director, musical director and script coach. Nine people have volunteered for any of these positions.

Determine how many different ways the positions can be filled. (*Note:* One person cannot take on more than 1 position.)

Problem solving

27. A shelf holding 24 cans of dog food is to be stacked using 9 cans of Yummy and 15 cans of Ruff for Dogs. Determine how many different ways the shelf can be stocked.

28. **WE24** A committee of 5 students is to be chosen from 6 boys and 8 girls. Use ^nC_r and the multiplication principle to answer the following questions.

    a. Calculate how many committees can be formed.
    b. Calculate how many of the committees contain exactly 2 boys and 3 girls.

29. a. Calculate the values of:

    i. $^{12}C_3$ and $^{12}C_9$
    iii. $^{10}C_1$ and $^{10}C_9$

    ii. $^{15}C_8$ and $^{15}C_7$
    iv. $^8C_3$ and $^8C_5$

    b. Describe what you notice in your results for part a.
    Give your answer as a general statement such as 'The value of ^nC_r is...'.

30. Lotto is a game played by choosing 6 numbers from 45. Players try to match their choice with those numbers chosen at the official draw. No number can be drawn more than once, and the order in which the numbers are selected does not matter.

    a. Calculate how many different selections of 6 numbers can be made from 45.
    b. Suppose the first numbers drawn at the official draw are 12, 3 and 18.
    Determine how many selections of 6 numbers will contain these 3 numbers.

31. A bag contains 7 red and 3 white balls. A ball is taken at random, its colour is noted, and it is then replaced before a second ball is chosen at random and its colour is noted.

a. Show the possible outcomes with a fully labelled tree diagram.

b. As the first ball was chosen, how many balls were in the bag?

c. As the second ball was chosen, how many balls were in the bag?

d. Does the probability of choosing a red or white ball change from the first selection to the second? Explain.

e. Calculate the probability of choosing a red ball twice.

The process is repeated. However, this time the first selection is not replaced.

f. As the second ball is chosen, how many balls are in the bag?

g. Does the probability of choosing a red or white ball change from the first selection to the second? Explain.

h. Construct a fully labelled tree diagram to show all possible outcomes.

i. Calculate the probability of choosing two red balls.

32. A combination lock has 3 digits, each from 0 to 9.

a. Determine how many combinations are possible.

The lock mechanism becomes loose and will open if the digits are within 1 either side of the correct digit. For example, if the true combination is 382, then the lock will open on 271, 272, 371, 493 and so on.

b. Determine how many combinations would unlock the safe.

c. List the possible combinations that would open the lock if the true combination is 382.

LESSON
11.7 Review

11.7.1 Topic summary

Review of probability

- The sample space is the set of all possible outcomes. It is denoted ξ. For example, the sample space of rolling a die is $\xi = \{1, 2, 3, 4, 5, 6\}$.
- An event is a set of favourable outcomes. For example, if A is the event {rolling an even number on a die}, $A = \{2, 4, 6\}$.
- The probability of an outcome or event is always between 0 and 1.
- The sum of all probabilities in a sample space is 1.
- An event that is **certain** has a probability of 1.
- An event that is **impossible** has a probability of 0.
- When all outcomes are equally likely the theoretical probability of an event is given by the following formula:

$$\text{Pr(event)} = \frac{\text{number of favourable outcomes}}{\text{total number of outcomes}} = \frac{n(E)}{n(\xi)}$$

- The experimental probability is given by the following formula:

$$\text{Experimental Pr} = \frac{n(\text{successful trials})}{n(\text{trials})}$$

- The complement of an event A is denoted A' and is the set of outcomes that are not in A.
  - $n(A) + n(A') = n(\xi)$ • $\text{Pr}(A) + \text{Pr}(A') = 1$

PROBABILITY

Intersection and union

- The intersection of two events A and B is written $A \cap B$. These are the outcomes that are in A '**and**' in B.

- The union of two events A and B is written $A \cup B$. These are the outcomes that are in A '**or**' in B.

- A Venn diagram can be split into four distinct sections as shown at right.

- The Addition Law of probability states that:
$$\text{Pr}(A \cup B) = \text{Pr}(A) + \text{Pr}(B) - \text{Pr}(A \cap B)$$

Counting principles and factorial notation

- Counting techniques depend on whether order is important.
- The multiplication principle should be used when order is important.
- $n!$: the number of ways n distinct objects can be arranged
$$n! = n \times (n-1) \times (n-2) \times (n-3) \times ... \times 3 \times 2 \times 1$$
- Permutations are used to count when order is important.
$$^nP_r = \frac{n!}{(n-r)!}$$
- Combinations are used to count when order is *not* important.
$$^nC_r = \frac{^nP_r}{r!}$$

Independent, dependent events and conditional probability

- Two events are **independent** if the outcome of one event does not affect the outcome of the other event.
- If A and B are independent, then:
$$\text{Pr}(A \cap B) = \text{Pr}(A) \times \text{Pr}(B)$$
- Two events are **dependent** if the outcome of one event affects the outcome of the other event.
- Conditional probability applies to dependent events.
- The probability of A given that B has already occurred is given by:
$$\text{Pr}(A \mid B) = \frac{\text{Pr}(A \cap B)}{\text{Pr}(B)}$$
- This can be rearranged into:
$$\text{Pr}(A \cap B) = \text{Pr}(A \mid B) \times \text{Pr}(B)$$

Mutually exclusive events

- Mutually exclusive events have no elements in common.
- Venn diagrams show mutually exclusive events having no intersection. In the Venn diagram below A and B are mutually exclusive.

- For mutually exclusive events A and B:
  - $\text{Pr}(A \cap B) = 0$
  - $\text{Pr}(A \cup B) = \text{Pr}(A) + \text{Pr}(B)$

Multiple-step chance experiments

- Some chance experiments involve multiple trials performed separately.
- Two-way tables can be used to show the sample space of two-step experiments.
- Tree diagrams can be used to show the sample space and probabilities of multi-step experiments.

		Coin 1	
		H	**T**
Coin 2	**H**	HH	TH
	T	HT	TT

First selection	Second selection	Sample space
B	R	BR
	G	BG
	Y	BY
W	R	WR
	G	WG
	Y	WY

11.7.2 Success criteria

Tick the column to indicate that you have completed the lesson and how well you have understood it using the traffic light system.

(**Green:** I understand; **Yellow:** I can do it with help; **Red:** I do not understand)

Lesson	Success criteria	⬤	⬤	⬤
11.2	I can use key probability terminology such as trials, frequency, sample space, and likely and unlikely events.			
	I can use two-way tables to represent sample spaces.			
	I can calculate experimental and theoretical probabilities.			
	I can determine the probability of complimentary events and mutually exclusive events.			
	I can use the addition rule to calculate the probability of event 'A or B'.			
11.3	I can create tree diagrams to represent two- and three-step chance experiments.			
	I can use tree diagrams to solve probability problems involving two or more trials or events.			
11.4	I can determine whether two events are dependent or independent.			
	I can calculate the probability of outcomes from two-step chance experiments involving dependent or independent events.			
11.5	I can determine from the language in a question whether it involves conditional probability.			
	I can use the rule for conditional probability to calculate other probabilities.			
11.6	I can apply the multiplication principle as a counting technique.			
	I can calculate permutations using factorials.			
	I can calculate permutations for situations involving like objects.			
	I can calculate the number of possible combinations in situations where order is not important.			

11.7.3 Project

Tricky dice

Dice games have been played throughout the world for many years. Professional gamblers resort to all types of devious measures in order to win. Often the other players are unaware of the tricks employed.

Imagine you are playing a game that involves rolling two dice. Instead of having each die marked with the numbers 1 to 6, let the first die have only the numbers 1, 2 and 3 (two of each) and the second die the numbers 4, 5 and 6 (two of each). If you were an observer to this game, you would see the numbers 1 to 6 occurring and probably not realise that the dice were not the regular type.

1. Complete the grid below to show the sample space on rolling these two dice.
2. Identify how many different outcomes there are. Compare this with the number of different outcomes using two regular dice.

Die 1

Die 2	1	2	3	1	2	3
4						
5						
6						
4						
5						
6						

3. Calculate the chance of rolling a double using these dice.
4. The numbers on the two dice are added after rolling. Complete the table below to show the totals possible.

Die 1

Die 2	1	2	3	1	2	3
4						
5						
6						
4						
5						
6						

5. Identify how many different totals are possible and list them.
6. State which total you have the greatest chance of rolling. State which total you have the least chance of rolling.
7. If you played a game in which you had to bet on rolling a total of less than 7, equal to 7 or greater than 7, explain which option would you be best to take.
8. If you had to bet on an even-number outcome or an odd-number outcome, explain which would be the better option.
9. The rules are changed to subtracting the numbers on the two dice instead of adding them. Complete the following table to show the outcomes possible.

Die 1

Die 2	1	2	3	1	2	3
4						
5						
6						
4						
5						
6						

10. Identify how many different outcomes are possible in this case and list them.
11. State the most frequently occurring outcome and how many times it occurs.
12. Devise a game of your own using these dice. On a separate sheet of paper, write out rules for your game and provide a solution, indicating the best options for winning.

Exercise 11.7 Review questions

learnon

Fluency

1. **MC** Choose which of the following is always true for an event, M, and its complementary event, M'.
   A. $\Pr(M) + \Pr(M') = 1$
   B. $\Pr(M) - \Pr(M') = 1$
   C. $\Pr(M) + \Pr(M') = 0$
   D. $\Pr(M) - \Pr(M') = 0$

2. **MC** A number is chosen from the set $\{0, 1, 2, 3, 4, 5, 6, 7, 8, 9, 10\}$. Select which of the following pairs of events is mutually exclusive.
   A. $\{2, 4, 6\}$ and $\{4, 6, 7, 8\}$
   B. $\{1, 2, 3, 5\}$ and $\{4, 6, 7, 8\}$
   C. $\{0, 1, 2, 3\}$ and $\{3, 4, 5, 6\}$
   D. $\{$multiples of $2\}$ and $\{$factors of $8\}$

3. **MC** Choose which of the following states the Multiplication Law of probability correctly.
   A. $\Pr(A \cap B) = \Pr(A) + \Pr(B)$
   B. $\Pr(A \cap B) = \Pr(A) \times \Pr(B)$
   C. $\Pr(A \cup B) = \Pr(A) \times \Pr(B)$
   D. $\Pr(A \cup B) = \Pr(A) + \Pr(B)$

4. **MC** Given $\xi = \{1, 2, 3, 4, 5, 6, 7, 8, 9, 10\}$, $A = \{2, 3, 4\}$ and $B = \{3, 4, 5, 8\}$, $A \cap B$ is:
   A. $\{2, 3, 3, 4, 4, 5, 8\}$
   B. $\{3, 4\}$
   C. $\{2, 3, 4\}$
   D. $\{2, 3, 4, 5, 8\}$

5. **MC** Given $\xi = \{1, 2, 3, 4, 5, 6, 7, 8, 9, 10\}$, $A = \{2, 3, 4\}$ and $B = \{3, 4, 5, 8\}$, $A \cap B'$ is:
   A. $\{3, 4\}$
   B. $\{2\}$
   C. $\{2, 3, 4, 5, 8\}$
   D. $\{2, 3, 4\}$

6. **MC** There are 12 people on the committee at the local football club. Determine how many ways can a president and a secretary be chosen from this committee.
   A. 2
   B. 23
   C. 132
   D. 144

7. Shade the region stated for each of the following Venn diagrams.
   a. $A' \cup B$
   b. $A' \cap B'$
   c. $A' \cap B' \cap C$

8. a. State the definition of ^nP_r.
   b. Without a calculator, compute the value of $^{10}P_3$.
   c. Prove that $^nP_r = \dfrac{n!}{(n-r)!}$.

Problem solving

9. **MC** From past experience, it is concluded that there is a 99% probability that July will be a wet month in Launceston (it has an average rainfall of approximately 80 mm). The probability that July will not be a wet month next year in Launceston is:

   A. 99% B. 0.99 C. $\dfrac{1}{100}$ D. 1

10. **MC** A card is drawn from a well-shuffled deck of 52 cards. Select the theoretical probability of not selecting a red card.

    A. $\dfrac{3}{4}$ B. $\dfrac{1}{4}$ C. $\dfrac{1}{13}$ D. $\dfrac{1}{2}$

11. **MC** Choose which of the following events is not equally likely.
    A. Obtaining a 5 or obtaining a 1 when a die is rolled
    B. Obtaining a club or obtaining a diamond when a card is drawn from a pack of cards
    C. Obtaining 2 Heads or obtaining 2 Tails when a coin is tossed twice
    D. Obtaining 2 Heads or obtaining 1 Head when a coin is tossed twice

12. **MC** The Australian cricket team has won 12 of the last 15 Test matches. Select the experimental probability of Australia not winning its next Test match.

    A. $\dfrac{4}{5}$ B. $\dfrac{1}{5}$ C. $\dfrac{1}{4}$ D. $\dfrac{3}{4}$

13. A card is drawn from a well-shuffled pack of 52 cards. Calculate the theoretical probability of drawing:
    a. an ace b. a spade c. a queen or a king d. not a heart.

14. A die is rolled five times.
    a. Calculate the probability of rolling five 6s. b. Calculate the probability of not rolling five 6s.

15. Alan and Mary own 3 of the 8 dogs in a race. Evaluate the probability that:
    a. one of Alan's or Mary's dogs will win b. none of Alan's or Mary's dogs will win.

16. A die is rolled. Event A is obtaining an even number. Event B is obtaining a 3.
    a. Explain if events A and B are mutually exclusive.
    b. Calculate $\Pr(A)$ and $\Pr(B)$.
    c. Calculate $\Pr(A \cup B)$.

17. A card is drawn from a shuffled pack of 52 playing cards. Event A is drawing a club and event B is drawing an ace.
    a. Explain if events A and B are mutually exclusive.
    b. Calculate $\Pr(A)$, $\Pr(B)$ and $\Pr(A \cap B)$.
    c. Calculate $\Pr(A \cup B)$.

18. A tetrahedral die is numbered $0, 1, 2$ and 3. Two of these dice are rolled and the sum of the numbers (the number on the face that the die sits on) is taken.
    a. Show the possible outcomes in a two-way table.
    b. Determine if all the outcomes are equally likely.
    c. Determine which total has the least chance of being rolled.
    d. Determine which total has the best chance of being rolled.
    e. Determine which sums have the same chance of being rolled.

19. A bag contains 20 pears, of which 5 are bad. Cathy picks 2 pears (without replacement) from the bag. Evaluate the probability that:
    a. both pears are bad
    b. both pears are good
    c. one of the two pears is good.

20. Determine the probability of drawing 2 aces from a pack of cards if:
    a. the first card is replaced before the second one is drawn
    b. the first card drawn is not replaced.

21. A free-style snowboard competition has 15 entrants. Determine how many ways the first, second and third places can be filled. You may wish to use technology to answer this question.

22. The main cricket ground in Brisbane is called the Gabba, which is short for Woolloongabba. Determine how many different arrangements of letters can be made from the word WOOLLOONGABBA. You may wish to use technology to answer the question.

23. A committee of 5 men and 5 women is to be chosen from 8 men and 9 women. Determine how many ways this can be done.

Questions 24–26 refer to the following information.

The Maryborough Tennis Championships involve 16 players. The organisers plan to use 3 courts and assume that each match will last on average 2 hours and that no more than 4 matches will be played on any court per day.

24. In a 'round robin' each player plays every other player once.
    a. If the organisers use a round robin format, determine how many matches will be played in all.
    b. Determine how long the tournament lasts.

25. The organisers split the 16 players into two pools of 8 players each. After a 'round robin' within each pool, a final is played between the winners of each pool.
    a. Determine how many matches are played in the tournament.
    b. Determine how long the tournament lasts.

26. A 'knock out' format is one in which the loser of every match drops out and the winners proceed to the next round until there is only one winner left.
    a. If the game starts with 16 players, determine how many matches are needed before a winner is obtained.
    b. Determine how long the tournament lasts.

27. On grandparents day at a school, a group of grandparents was asked where they most like to take their grandchildren — the beach (*B*) or shopping (*S*). The results are illustrated in the Venn diagram.
    Use the Venn diagram to calculate the following probabilities relating to the place grandparents most like to take their grandchildren.
    a. Determine the probability that a randomly selected grandparent preferred to take their grandchildren to the beach or shopping.
    b. Determine the probability that a randomly selected grandparent liked to take their grandchildren to the beach, given that they liked to take their grandchildren shopping.

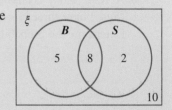

28. When all of Saphron's team players turn up for their twice weekly netball training, the chance that they then win their Saturday game is 0.65. If not all players are at the training session, then the chance of winning their Saturday game is 0.40.

    Over a four-week period, Saphron's players all turn up for training three times.

    a. Using a tree diagram, with T to represent all players training and W to represent a win, represent the winning chance of Saphron's netball team.

    b. Using the tree diagram constructed in part **a**, determine the probability of Saphron's team winning their Saturday game.

    Write your answer correct to 4 decimal places.

    c. Determine the exact probability that Saphron's team did not train given that they won their Saturday game.

29. Andrew does not know the answer to two questions on a multiple-choice exam. The first question has four choices and the second question he does not know has five choices.

    a. Determine the probability that he will get both questions wrong.

    b. If he is certain that one of the choices cannot be the answer in the first question, determine how this will change the probability that he will get both questions wrong.

30. Mariah the Mathematics teacher wanted to give her students a chance to win a reward at the end of the term. She placed 20 cards into a box and wrote the word ON on 16 cards and OFF on 4 cards.

    After a student chooses a card, that card is replaced into the box for the next student to draw. If a student chooses an OFF card, then they do not have to attend school on a specified day.

    If they choose an ON card, then they do not receive a day off.

    a. Mick, a student, chose a random card from the box. Calculate the probability he received a day off.

    b. Juanita, a student, chose a random card from the box after Mick.

    Calculate the probability that she did not receive a day off.

    c. Determine the probability that Mick and Juanita both received a day off.

31. In the game of draw poker, a player is dealt 5 cards from a deck of 52. To obtain a flush, all 5 cards must be of the same suit.

    a. Determine the probability of getting a diamond flush.

    b. Determine the probability of getting any flush.

32. a. A Year 10 boy is talking with a Year 10 girl and asks her if she has any brothers or sisters. She says, 'Yes, I have one.'

    Determine the probability that she has at least one sister.

    b. A Year 10 boy is talking with a Year 10 girl and asks her if she has any brothers or sisters. She says, 'Yes, I have three.'

    Determine the probability that she has at least one sister.

on To test your understanding and knowledge of this topic, go to your learnON title at www.jacplus.com.au and complete the **post-test**.

Answers

Topic 11 Probability

11.1 Pre-test

1. $Pr(A \cap B) = 0.2$
2. $Pr(A) = 0.51$
3. Independent
4. $\dfrac{1}{5}$
5. C
6. D
7. A
8. 0.56
9. A
10. False
11. $Pr(B|A) = 0.2$
12. A
13. A
14. 0.031 25
15. A

11.2 Review of probability

1. a. $\{I, II, III, IV, V, VI\}$
   b. Relative frequency for $I = 0.2$
      Relative frequency for $II = 0.12$
      Relative frequency for $III = 0.16$
      Relative frequency for $IV = 0.14$
      Relative frequency for $V = 0.24$
      Relative frequency for $VI = 0.14$
   c. The spinner should be spun a larger number of times.
2. a. 500 students
   b. Frequency for silver $= 0.16$
   c. Frequency for black and green $= 0.316$
   d. Blazing Blue
3. Sample responses can be found in the worked solutions in the online resources.
4. a.

+	1	2	3	4	5	6
1	2	3	4	5	6	7
2	3	4	5	6	7	8
3	4	5	6	7	8	9
4	5	6	7	8	9	10
5	6	7	8	9	10	11
6	7	8	9	10	11	12

   b. The sum of 7
   c. The sum of 2 or 12
5. $\dfrac{2}{3}$
6. a. $\dfrac{1}{13}$ b. $\dfrac{1}{4}$ c. $\dfrac{4}{13}$

7. Sample responses are given for part ii.
   a. i. No. There are many others foods one could have.
      ii. Having Weet Bix and not having Weet Bix
   b. i. No. There are other means of transport, for example catching a bus.
      ii. Walking to a friend's place and not walking to a friend's place
   c. i. No. There are other possible leisure activities.
      ii. Watching TV and not watching TV
   d. i. No. The number 5 can be rolled too.
      ii. Rolling a number less than 5 and rolling a number 5 or greater
   e. i. Yes. There are only two possible outcomes: passing or failing.
      ii. No change is required to make these events complementary.

8. a. $\dfrac{1}{13}$ b. $\dfrac{1}{4}$ c. $\dfrac{1}{2}$ d. $\dfrac{12}{13}$
   e. 0   f. $\dfrac{1}{2}$

9. a. $\dfrac{1}{5}$ b. $\dfrac{7}{20}$ c. $\dfrac{11}{20}$ d. 0

10. a. $Pr(A \cap B) = 0.1$
    b.

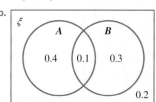

    c. $Pr(A \cap B') = 0.4$

11. a. i. $Pr(A \cup B) = 0.85$ ii. $Pr(A \cap B)' = 0.95$
    b. C

12. a.

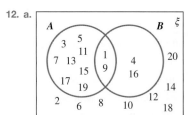

    b. i. $\dfrac{10}{20} = \dfrac{1}{2}$ ii. $\dfrac{4}{20} = \dfrac{1}{5}$ iii. $\dfrac{2}{20} = \dfrac{1}{10}$
       iv. $\dfrac{12}{20} = \dfrac{3}{5}$ v. $\dfrac{8}{20} = \dfrac{2}{5}$

13. a. No. For a 6-sided die, $Pr(4) = \dfrac{1}{6}$; for an 8-sided die, $Pr(4) = \dfrac{1}{8}$.
    b. Yes; $Pr(odd) = \dfrac{1}{2}$ for both 6-sided and 8-sided dice

14. Yes; $Pr(5) = \dfrac{1}{2}, Pr(6) = \dfrac{1}{2}$.

15. Sample responses can be found in the worked solutions in the online resources.

16. a.

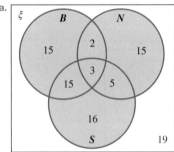

$\xi = 30$

Volleyball — Soccer — Tennis

7, 1, 4, 5, 2, 4, 7

b. i. $\dfrac{1}{2}$ **ii.** $\dfrac{1}{6}$ **iii.** $\dfrac{1}{30}$ **iv.** $\dfrac{2}{5}$ **v.** $\dfrac{7}{15}$

c. i. $\dfrac{1}{2}$ **ii.** $\dfrac{8}{15}$

17. Yes. Both have a probability of $\dfrac{1}{2}$.

18. Sample responses can be found in the worked solutions in the online resources.

19. 17 red and 1 purple, i.e. 18 more socks

20. a.

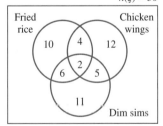

ξ

B, N, S

15, 2, 15, 3, 15, 5, 16, 19

b. 19 students

c. 32 students

d. 15 students

e. Frequency $= \dfrac{15}{90}$ or $\dfrac{1}{6}$

f. Probability $= \dfrac{3}{90} = \dfrac{1}{30}$

21. a. i. 50 **ii.** 7 **iii.** 25 **iv.** 8

b. i. $\dfrac{1}{2}$ **ii.** $\dfrac{3}{50}$ **iii.** $\dfrac{6}{25}$

c. i.

$n(\xi) = 50$

Fried rice — Chicken wings — Dim sims

10, 4, 12, 6, 2, 5, 11

ii. $\dfrac{1}{25}$

22. a.

Die 2 outcomes

	1	2	3	4	5	6
1	2	3	4	5	6	7
2	3	4	5	6	7	8
3	4	5	6	7	8	9
4	5	6	7	8	9	10
5	6	7	8	9	10	11
6	7	8	9	10	11	12

Die 1 outcomes

b. 6

c. No. The frequency of the numbers is different.

d.

Sum	2	3	4	5	6	7	8	9	10	11	12
Frequency	1	2	3	4	5	6	5	4	3	2	1

e. i. $\dfrac{1}{36}$ **ii.** $\dfrac{1}{6}$ **iii.** $\dfrac{1}{18}$

f. i. $\dfrac{1}{36}$ **ii.** $\dfrac{1}{6}$ **iii.** $\dfrac{1}{6}$

g. 50

23. Sample responses can be found in the worked solutions in the online resources.

11.3 Tree diagrams

1. If the probabilities of two events are different, the first column of branches indicates the probabilities for the first event and the second column of branches indicates the probabilities for the second event. The product of each branch gives the probability. All probabilities add to 1.

2. a. 12 different outcomes

b. No. Each branch is a product of different probabilities.

c. Less likely

d. Indigo elephant

e. i. $\Pr(\text{Blue elephant}) = \dfrac{7}{40}$

ii. $\Pr(\text{Indigo elephant}) = \dfrac{7}{20}$

iii. $\Pr(\text{Donkey}) = \dfrac{1}{5}$

3. a.

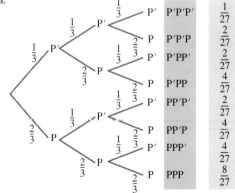

P'P'P'	$\dfrac{1}{27}$
P'P'P	$\dfrac{2}{27}$
P'PP'	$\dfrac{2}{27}$
P'PP	$\dfrac{4}{27}$
PP'P'	$\dfrac{2}{27}$
PP'P	$\dfrac{4}{27}$
PPP'	$\dfrac{4}{27}$
PPP	$\dfrac{8}{27}$

b. i. $\dfrac{8}{27}$ ii. $\dfrac{4}{27}$ iii. $\dfrac{12}{27}$

4. a. i. $\dfrac{1}{2}$ ii. $\dfrac{1}{6}$ iii. $\dfrac{5}{6}$

b.

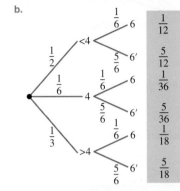

c. i. $\dfrac{1}{36}$ ii. $\dfrac{1}{12}$ iii. $\dfrac{5}{36}$ iv. $\dfrac{5}{18}$

5. a.

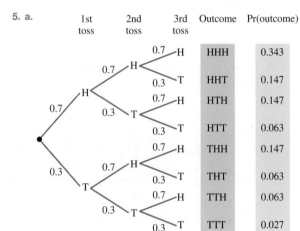

b. Pr(HHH) = 0.343

c. Pr(at least 1 Tail) = 0.657

d. Pr(exactly 2 Tails) = 0.189

6. a.

b. 0.520

7. a.

b. $\dfrac{1}{9}$

8. a.

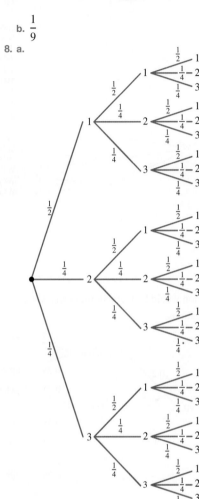

b. $\dfrac{3}{8}$

c. $\dfrac{7}{8}$

9. a.

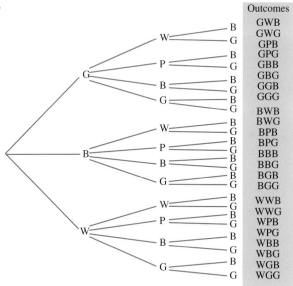

b. $\dfrac{1}{24}$

c. $\dfrac{1}{12}$

d. Each combination would be worn 10 times in a year.

e. 21 new combinations.

10. a.

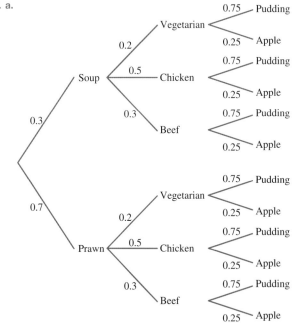

b. 0.1125

c. 24 people

11. a. i.

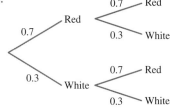

ii. 10 balls

iii. 10 balls

iv. No; the events are independent.

v. $\Pr(RR) = 0.49$

b. i. 9 balls

ii. Yes. One ball has been removed from the bag.

iii.

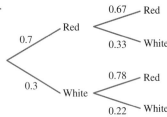

iv. $\Pr(RR) = \dfrac{7}{15}$ or 0.469 using the rounded values from iii.

12. a.

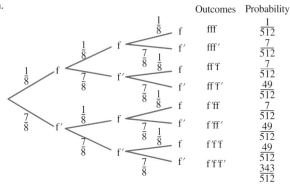

$f = $ outcome of 5

b. i. $\dfrac{1}{512}$ **ii.** $\dfrac{343}{512}$ **iii.** $\dfrac{21}{512}$ **iv.** $\dfrac{11}{256}$

13. a.

Die outcomes

Coin outcomes	1	2	3	4
H	(H, 1)	(H, 2)	(H, 3)	(H, 4)
T	(T, 1)	(T, 2)	(T, 3)	(T, 4)

b.

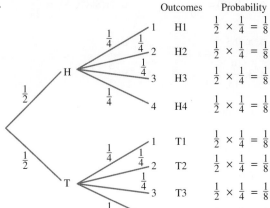

Outcomes	Probability
H1	$\frac{1}{2} \times \frac{1}{4} = \frac{1}{8}$
H2	$\frac{1}{2} \times \frac{1}{4} = \frac{1}{8}$
H3	$\frac{1}{2} \times \frac{1}{4} = \frac{1}{8}$
H4	$\frac{1}{2} \times \frac{1}{4} = \frac{1}{8}$
T1	$\frac{1}{2} \times \frac{1}{4} = \frac{1}{8}$
T2	$\frac{1}{2} \times \frac{1}{4} = \frac{1}{8}$
T3	$\frac{1}{2} \times \frac{1}{4} = \frac{1}{8}$
T4	$\frac{1}{2} \times \frac{1}{4} = \frac{1}{8}$
	$\overline{1}$

c. $\frac{1}{4}$

14. a. $\frac{256}{625}$ **b.** $\frac{16}{625}$

    c. $\frac{96}{256}$ **d.** $\frac{17}{625}$

11.4 Independent and dependent events

1. a. 0.28   **b.** 0.12   **c.** 0.42   **d.** 0.18

2. Dependent

3. Independent

4. a. Yes, the outcomes are independent.

    b. i. $\frac{1}{2}$ **ii.** $\frac{1}{6}$

    c. $\frac{1}{12}$

5. $\frac{1}{40}$

6. $\frac{5}{36}$

7. a. $\frac{16}{25}$ **b.** $\frac{64}{125}$

    c. $\frac{1}{25}$ **d.** $\frac{4}{25}$

8. a. C   **b.** D

9. a. $\frac{3}{77}$ **b.** $\frac{48}{77}$

    c. $\frac{8}{77}$ **d.** $\frac{18}{77}$

10. a. $\frac{1}{37}$ **b.** $\frac{1}{1369}$ **c.** $\frac{73}{1369}$

11. 0.9

12. $\frac{1}{14}$

13. a. $\frac{1}{5}$ **b.** $\frac{1}{5}$

    c. $\frac{1}{10}$ **d.** $\frac{1}{3}$

14. a. $\frac{1}{17}$ **b.** $\frac{1}{221}$ **c.** $\frac{25}{102}$

15. a. $\frac{26}{145}$ **b.** $\frac{136}{435}$ **c.** $\frac{221}{435}$

16. No. Coin tosses are independent events. No one toss affects the outcome of the next. The probability of a Head or Tail on a fair coin is always 0.5. Greg has a 50% chance of tossing a Head on the next coin toss as was the chance in each of the previous 9 tosses.

17. No. As events are illustrated on a tree diagram, the individual probability of each outcome is recorded. The probability of a dependent event is calculated (altered according to the previous event) and can be considered as if it was an independent event. As such, the Multiplication Law of probability can be applied along the branches to calculate the probability of successive events.

18. Only if $\Pr(A) = 0$ or $\Pr(B) = 0$ can two events be independent and mutually exclusive. For an event to have a probability of 0 means that it is impossible, so it is a trivial scenario.

19. a. A and B are independent.

    b. A and B are not independent in this situation.

20. $\frac{5}{17}$

21. a. Dependent

    b. $\Pr(1) = \frac{7}{15}$

        $\Pr(2) = \frac{5}{15} = \frac{1}{3}$

        $\Pr(3) = \frac{3}{15} = \frac{1}{5}$

c.

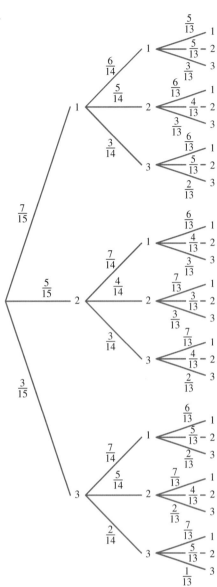

d. i. $\Pr(1, 1, 1) = \dfrac{1}{13}$ ii. $\Pr(2, 2, 2) = \dfrac{2}{91}$

iii. $\Pr(3, 3, 3) = \dfrac{1}{455}$ iv. $\Pr(\text{at least one } 3) = \dfrac{47}{91}$

22. a. $\Pr(9) = \dfrac{1}{455}$

      $\Pr(7 \text{ or } 8) = \dfrac{66}{455}$

      $\Pr(5 \text{ or } 6) = \dfrac{248}{455}$

      $\Pr(3 \text{ or } 4) = \dfrac{4}{13}$

   b. $393.41

11.5 Conditional probability

1. a. $\Pr(J) = \dfrac{41}{90}$ b. $\Pr(z|J) = \dfrac{12}{41}$

2. a. $\Pr(S) = \dfrac{13}{30}$ b. $\Pr(S|(C \cup S)) = \dfrac{13}{28}$

3. a. 0.3   b. $\dfrac{3}{7}$

4. a. $\dfrac{9}{13}$ b. $\dfrac{3}{5}$

5. 0.58 or $\dfrac{15}{26}$

6. 0.22 or $\dfrac{5}{23}$

7. 0.9

8. 0.8375

9. A

10. a. D

   b. i. $\Pr(A|B) = \dfrac{1}{6}$ ii. $\Pr(B|A) = \dfrac{1}{6}$

     iii. $\Pr(C|A) = \dfrac{1}{6}$ iv. $\Pr(C|B) = 0$

11. a. $\dfrac{109}{232}$ b. $\dfrac{73}{116}$ c. $\dfrac{23}{86}$ d. $\dfrac{22}{45}$

12. Conditional probability is when the probability of one event depends on the outcome of another event.

13. a. 0.0875

   b. 0.065

   c. i. 0.585   ii. 15.25%

14. a. i. $\dfrac{1}{2}$ ii. $\dfrac{2}{5}$ iii. $\dfrac{1}{5}$ iv. $\dfrac{1}{2}$ v. $\dfrac{2}{5}$

   b. Yes, A and B are independent.

   c. If $\Pr(A|B) = \dfrac{\Pr(A \cap B)}{\Pr(B)} = \Pr(A)$, then A and B will be independent events. This is because the probability of A given B occurs is the same as the probability of A, meaning the probability of A is independent of B occurring.

15. 8.7% chance

16. 0.847

17. 0.61

11.6 Counting principles and factorial notation (10A)

1. a. AB   AC
     BA   BC
     CA   CB

   b. 6

   c. $\dfrac{1}{3}$

   d. 9

2. BG   BY   BR
   GB   GY   GR
   YB   YG   YR
   RB   RG   RY

3. ABC   ACB   BAC
   BCA   CAB   CBA

4. a. 42   b. 210   c. 840   d. 2520   e. 360

5. a. 8   b. 32   c. 512

6. a. $4! = 4 \times 3 \times 2 \times 1$

   b. $5! = 5 \times 4 \times 3 \times 2 \times 1$

   c. $6! = 6 \times 5 \times 4 \times 3 \times 2 \times 1$

   d. $7! = 7 \times 6 \times 5 \times 4 \times 3 \times 2 \times 1$

7. a. 3024

   b. 151 200

   c. 840

   d. 720

8. a. 3 628 800

   b. 87 178 291 200

   c. 362 880

   d. 20 922 789 888 000

9. a. $^8P_2 = 8 \times 7 = 56$

   b. $^7P_5 = 7 \times 6 \times 5 \times 4 \times 3 = 2520$

   c. $^8P_7 = 8 \times 7 \times 6 \times 5 \times 4 \times 3 \times 2 = 40\,320$

10. a. $^9P_6 = \dfrac{9!}{(9-6)!}$

    $= \dfrac{9!}{3!}$

    $= \dfrac{3\,62\,880}{6}$

    $= 60\,480$

    b. $^5P_2 = \dfrac{5!}{(5-2)!}$

    $= \dfrac{5!}{3!}$

    $= \dfrac{120}{6}$

    $= 20$

    c. $^{18}P_5 = \dfrac{18!}{(18-5)!}$

    $= \dfrac{18!}{13!}$

    $= \dfrac{6\,402\,373\,705\,728\,000}{6\,227\,020\,800}$

    $= 1\,028\,160$

11. a. 27 907 200   b. 639 200   c. 1 028 160

12. 10

13. 1260

14. a. $^8C_3 = \dfrac{^8P_3}{3!}$ b. $^{19}C_2 = \dfrac{^{19}P_2}{2!}$

    c. $^1C_1 = \dfrac{^1P_1}{1!}$ d. $^5C_0 = \dfrac{^5P_0}{0!}$

15. a. 24   b. 6
    c. 12   d. 24

16. a. 200   b. 40   c. 50

17. a. 56

    b. $\dfrac{1}{8}$

18. 83 160

19. 27 720

20. a. 1   b. 20   c. 120   d. 220

21. 100

22. 6

23. 256

24. 358 800

25. 120

26. 3024

27. 1 307 504

28. a. 2002   b. 840

29. a. i. $^{12}C_3 = 220, {}^{12}C_9 = 220$

    ii. $^{15}C_8 = 6435, {}^{15}C_7 = 6435$

    iii. $^{10}C_1 = 10, {}^{10}C_9 = 10$

    iv. $^8C_3 = 56, {}^8C_5 = 56$

    b. The value of ^nC_r is the same as $^nC_{n-r}$.

30. a. 8 145 060   b. 11 480

31. a.

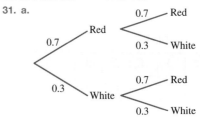

    b. 10 balls

    c. 10 balls

    d. No; the events are independent.

    e. P(RR) = 0.49

    f. 9 balls

    g. Yes. One ball has been removed from the bag.

    h.

    i. $P(RR) = \dfrac{7}{15} = 0.4\dot{6}$

32. a. 1000   b. 27

    c.
    | 271 | 371 | 471 |
    | 272 | 372 | 472 |
    | 273 | 373 | 473 |
    | 281 | 381 | 481 |
    | 282 | 382 | 482 |
    | 283 | 383 | 483 |
    | 291 | 391 | 491 |
    | 292 | 392 | 492 |
    | 293 | 393 | 493 |

Project

1.

Die 1

		1	2	3	1	2	3
Die 2	**4**	(1, 4)	(2, 4)	(3, 4)	(1, 4)	(2, 4)	(3, 4)
	5	(1, 5)	(2, 5)	(3, 5)	(1, 5)	(2, 5)	(3, 5)
	6	(1, 6)	(2, 6)	(3, 6)	(1, 6)	(2, 6)	(3, 6)
	4	(1, 4)	(2, 4)	(3, 4)	(1, 4)	(2, 4)	(3, 4)
	5	(1, 5)	(2, 5)	(3, 5)	(1, 5)	(2, 5)	(3, 5)
	6	(1, 6)	(2, 6)	(3, 6)	(1, 6)	(2, 6)	(3, 6)

2. 9

3. 0

4.

Die 1

		1	2	3	1	2	3
Die 2	**4**	5	6	7	5	6	7
	5	6	7	8	6	7	8
	6	7	8	9	7	8	9
	4	5	6	7	5	6	7
	5	6	7	8	6	7	8
	6	7	8	9	7	8	9

5. 5; 5, 6, 7, 8, 9

6. 7; 5, 9

7. Equal to 7; probability is the highest.

8. Odd-number outcome; probability is higher.

9.

Die 1

		1	2	3	1	2	3
Die 2	**4**	3	2	1	3	2	1
	5	4	3	2	4	3	2
	6	5	4	3	5	4	3
	4	3	2	1	3	2	1
	5	4	3	2	4	3	2
	6	5	4	3	5	4	3

10. 5; 1, 2, 3, 4, 5,

11. 3; 12

12. Students need to apply the knowledge of probability and create a new game using dice given. They also need to provide rules for the game and solution, indicating the best options for winning.

11.7 Review questions

1. A
2. B
3. B
4. B
5. B
6. C
7. a.
   b.
   c.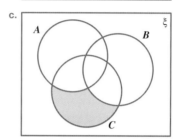

8. a. ^nP_r is the number of ways of choosing r objects from n distinct things when order is important. $^nP_r = \dfrac{n!}{(n-r)!}$

   b. $^{10}P_3 = 720$

   c. Sample responses can be found in the worked solutions in the online resources.

9. C

10. D

11. D

12. B

13. a. $\dfrac{1}{13}$ b. $\dfrac{1}{4}$

    c. $\dfrac{2}{13}$ d. $\dfrac{3}{4}$

14. a. $\dfrac{1}{7776}$

    b. $\dfrac{7775}{7776}$

15. a. $\dfrac{3}{8}$

    b. $\dfrac{5}{8}$

16. a. Yes. It is not possible to roll an even number and for that number to be a 3.

    b. $Pr(A) = \dfrac{1}{2}$ and $Pr(B) = \dfrac{1}{6}$

    c. $\dfrac{2}{3}$

17. a. No. It is possible to draw a card that is a club and an ace.

b. $\Pr(A) = \dfrac{1}{4}$ and $\Pr(B) = \dfrac{1}{13}$, $\Pr(A \cap B) = \dfrac{1}{52}$

c. $\dfrac{4}{13}$

18. a.

Die 2 outcomes

	0	**1**	**2**	**3**
0	0	1	2	3
1	1	2	3	4
2	2	3	4	5
3	3	4	5	6

Die 1 outcomes

b. No

c. 0 and 6

d. 3

e. 0 and 6, 1 and 5, 2 and 4

19. a. $\dfrac{1}{19}$

b. $\dfrac{21}{38}$

c. $\dfrac{15}{38}$

20. a. $\dfrac{1}{169}$

b. $\dfrac{1}{221}$

21. 2730

22. 32 432 400

23. 7056

24. a. 120   **b.** 10 days

25. a. 57   **b.** 4 days 6 hours

26. a. 15   **b.** 1 day 4 hours

27. a. $\dfrac{15}{25} = \dfrac{3}{5}$

b. $\dfrac{8}{10} = \dfrac{4}{5}$

28. a.

b. 0.5875

c. $\dfrac{8}{47}$

29. a. $\dfrac{3}{5}$

b. $\dfrac{8}{15}$

30. a. $\dfrac{1}{5}$

b. $\dfrac{4}{5}$

c. $\dfrac{1}{25}$

31. a. 0.000 495

b. 0.001 981

32. a. $\dfrac{1}{2}$

b. $\dfrac{7}{8}$

12 Univariate data

LESSON SEQUENCE

LESSON
12.1 Overview

Why learn this?

According to the novelist Mark Twain, 'There are three kinds of lies: lies, damned lies and statistics.' Statistics can easily be used to manipulate people unless they have an understanding of the basic concepts involved.

Statistics, when used properly, can be an invaluable aid to good decision-making. However, deliberate distortion of the data or meaningless pictures can be used to support almost any claim or point of view. Whenever you read an advertisement, hear a news report or are given some data by a friend, you need to have a healthy degree of scepticism about the reliability of the source and nature of the data presented. A solid understanding of statistics is crucially important, as it is very easy to fall prey to statistics that are designed to confuse and mislead.

In 2020 when the COVID-19 pandemic hit, news and all forms of media were flooded with statistics. These statistics were used to inform governments worldwide about infection rates, recovery rates and all sorts of other important information. These statistics guided the decision-making process in determining the restrictions that were imposed or relaxed to maintain a safe community.

Statistics are also used to provide more information about a population in order to inform government policies. For example, the results of a census might indicate that the people in a particular city are fed up with traffic congestion. With this information now known, the government might prioritise works on public roads, or increase funding of public transport to try to create a more viable alternative to driving.

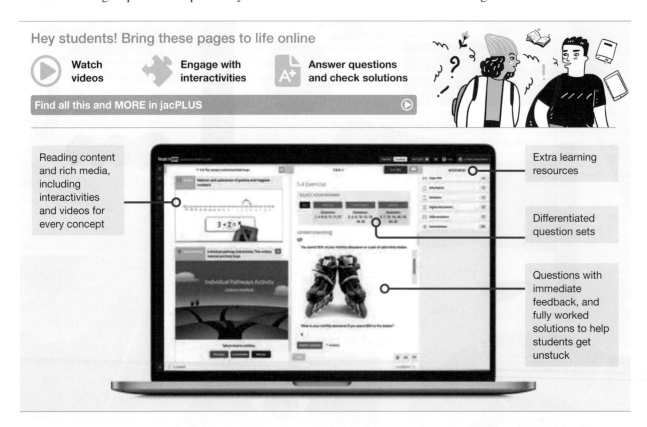

Hey students! Bring these pages to life online

▶ Watch videos

🧩 Engage with interactivities

A⁺ Answer questions and check solutions

Find all this and MORE in jacPLUS ▶

Reading content and rich media, including interactivities and videos for every concept

Extra learning resources

Differentiated question sets

Questions with immediate feedback, and fully worked solutions to help students get unstuck

1. The following data show the number of cars in each of the 12 houses along a street.

$$2, 3, 3, 2, 2, 3, 2, 4, 3, 1, 1, 0$$

   Calculate the median number of cars.

2. Calculate the range of the following data set: $5, 15, 23, 6, 31, 24, 26, 14, 12, 34, 18, 9, 17, 32$.

3. The frequency table shows the scores obtained by 102 professional golfers in the final round of a tournament.

Score	Frequency
67	2
68	6
69	7
70	11
71	16
72	23
73	17
74	11
75	9

   Identify the modal score.

4. A sample of 15 people was selected at random from those attending a local swimming pool. Their ages (in years) were recorded as follows.

$$19, 7, 83, 41, 17, 23, 62, 55, 15, 25, 32, 29, 11, 18, 10$$

   Calculate the mean age of people attending the swimming pool, correct to 1 decimal place.

5. Complete the following sentence.
   A sample is a _____ of a population.

6. At Einstein Secondary School a Year 10 mathematics class has 22 students. The following were the test scores for the class.

$$34, \ 47, \ 54, \ 59, \ 60, \ 63, \ 66, \ 69, \ 73, \ 77, \ 78$$

$$78, \ 79, \ 80, \ 82, \ 83, \ 85, \ 86, \ 88, \ 89, \ 90, \ 91$$

   Calculate the interquartile range (IQR).

7. The mean of a set of five scores is 11.8. If four of the scores are $17, 9, 14$ and 6, calculate the fifth score.

8. The box plot below shows the price in dollars of a meal for one person from ten fast-food shops.

State whether the data is negatively skewed, positively skewed or symmetrical.

9. A frequency table for the time taken by 20 people to put together an item of flat-pack furniture is shown.

Time taken (min)	Frequency
0–4	1
5–9	3
10–14	5
15–19	2
20–24	4
25–29	2
30–34	2
35–39	1

Calculate the cumulative frequency to put together an item of flat-pack furniture in less than 20 minutes.

10. **MC** The frequency table below shows the scores obtained by 100 professional golfers in the final round of a tournament.

Score	Frequency
67	2
68	5
69	8
70	11
71	16
72	22
73	14
74	13
75	9

Select the median score.

A. 71   B. 71.75   C. 72   D. 72.5

11. The heights of six basketball players (in cm) are:

$$178.1 \quad 185.6 \quad 173.3 \quad 193.4 \quad 183.1 \quad 193.0$$

Calculate the mean and standard deviation.

12. MC A group of 22 people recorded how many cans of soft drink they drank in a day. The table shows the number of cans drunk by each person.

0	2	2	2	1	1	3	4	4	2	1
2	4	1	6	3	3	5	4	1	2	5

Select the statement that is not true.
A. The maximum number of soft drinks cans drank is 6.
B. The minimum number of soft drink cans drank is 0.
C. The interquartile range is 3.
D. The median number of soft drink cans is 2.5.

13. MC Select the approximate median in the cumulative frequency percentage graph shown.

A. 30   **B.** 32   **C.** 40   **D.** 50

14. The following back-to-back stem-and-leaf plot shows the typing speed in words per minute (wpm) of 30 Year 8 and Year 10 students.

Key: 2|6 = 26 wpm

Leaf: Year 8	Stem	Leaf: Year 10
9 2 2	0	
9 8 6 5 4 2	1	4 9
8 8 8 6 4 2 1 0 0	2	2 3 6 8 9
9 7 7 6 4 1 0	3	0 3 4 5 5 7 8 8
7 6 5 2 0	4	1 2 5 8 8 9 9
	5	0 3 5 7 8
	6	0 0 1

Calculate the mean typing speed and interquartile range for Year 8 and Year 10. Comment on your answers.

15. **MC** A survey was conducted on the favourite take-away foods for university students and the results were graphed using two bar charts.

Graph 1

Graph 2

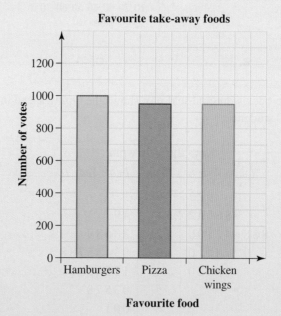

Select the statement that best describes these graphs.

A. Graph 1 is misleading as it suggests that students like hamburgers 7 times more than chicken wings.

B. Graph 1 is misleading as it suggests that students like pizza 3 times more than hamburgers.

C. Graph 1 is misleading as it suggests that students like pizza four times more than chicken wings.

D. Graph 1 is misleading as the vertical axis does not start at zero.

LESSON
12.2 Measures of central tendency

LEARNING INTENTION

At the end of this lesson you should be able to:
- calculate the mean, median and mode of data presented as ungrouped data, frequency distribution tables and grouped data.

▶ 12.2.1 Mean, median and mode of univariate data

eles-4949

- **Univariate data** are **data** with one variable, for example the heights of Year 10 students.
- **Measures of central tendency** are summary statistics that measure the centre of the data. These are known as the mean, median and mode.
  - The **mean** is the average of all observations in a set of data.
  - The **median** is the middle observation in an ordered set of data.
  - The **mode** is the most frequent observation in a data set.

The mean

- The *mean* of a set of data is what is referred to in everyday language as the *average*.
- The mean of a set of values is the sum of all the values divided by the number of values.
- The symbol we use to represent the mean is $\bar{x}$; that is, a lower-case x with a bar on top.

Calculating the mean

$$\text{Mean} = \frac{\text{sum of data values}}{\text{total number of data values}}$$

Using mathematical notation, this is written as:

$$\bar{x} = \frac{\sum x}{n}$$

The median

- The median represents the *middle* score when the data values are in ascending order such that an equal number of data values will lie below the median and above it.

Calculating the median

When calculating the median, use the following steps:
1. Arrange the data values in order.
2. The *position* of the median is the $\left(\dfrac{n+1}{2}\right)$th data value, where n is the total number of data values.

Note: **If there are an even number of data values, there will be two middle values. In this case the median is the average of those data values.**

- When there are an odd number of data values, the median is the middle value.

$$1 \qquad 1 \qquad 3 \qquad \boxed{4} \qquad 6 \qquad 7 \qquad 8$$

$$\text{median} = 4$$

- When there are an even number of data values, the median is the average of the two middle values.

$$2 \qquad 3 \qquad 3 \qquad \boxed{5} \quad \boxed{6} \qquad 6 \qquad 7 \qquad 9$$

$$\text{median} = \frac{5+6}{2} = 5.5$$

The mode

- The mode is the observation that occurs most often.
- The data set can have no modes, one mode, two modes (bimodal) or more than two modes (multimodal).

Identifying the mode

When identifying the mode, look for the number that occurs most often (has the highest frequency).

- If no value in a data set appears more than once then there is no mode.
- If a data set has multiple values that appear the most then it has multiple modes. All values that appear the most are modes.

For example, the set 1, 2, 2, 4, 5, 5, 7 has two modes, 2 and 5.

WORKED EXAMPLE 1 Calculating mean, median and mode

For the data 6, 2, 4, 3, 4, 5, 4, 5, calculate:
a. the mean   **b. the median**   **c. the mode.**

THINK	WRITE
a. 1. Calculate the sum of the scores; that is, $\sum x$.	a. $\sum x = 6 + 2 + 4 + 3 + 4 + 5 + 4 + 5$ $= 33$
2. Count the number of scores; that is, n.	$n = 8$
3. Write the rule for the mean.	$\bar{x} = \dfrac{\sum x}{n}$
4. Substitute the known values into the rule.	$= \dfrac{33}{8}$
5. Evaluate.	$= 4.125$
6. Write the answer.	The mean is 4.125.

b. 1. Write the scores in ascending numerical order.

b. $2, 3, 4, 4, 4, 5, 5, 6$

$$\text{Median} = \frac{n+1}{2}\text{th score}$$

2. Locate the position of the median using the rule $\frac{n+1}{2}$, where $n = 8$. This places the median as the 4.5th score; that is, between the 4th and 5th score.

$$= \frac{8+1}{2}\text{th score}$$

$$= 4.5\text{th score}$$

$$2, 3, 4, \mathbf{4}, \mathbf{4}, 5, 5, 6$$

3. Obtain the average of the two middle scores.

$$\text{Median} = \frac{4+4}{2}$$

$$= \frac{8}{2}$$

$$= 4$$

4. Write the answer

The median is 4.

c. 1. Systematically work through the set and make note of any repeated values (scores). Identify the most frequently occurring observation.

c.

$$2 \quad 3 \quad 4 \quad 4 \quad 4 \quad 5 \quad 5 \quad 6$$

2. Write the answer.

The mode is 4.

TI | THINK	**DISPLAY/WRITE**	**CASIO | THINK**	**DISPLAY/WRITE**

1. In a new document, on a Lists & Spreadsheet page, label column A as 'xvalues', and enter the values in the data set. Press ENTER after entering each value.

1. On the Statistics screen, label list1 as 'x' and enter the values as shown in the table.
Press EXE after entering each value.

2. Although you can find many summary statistics, to find the mean only, open a Calculator page and press:

- MENU
- 6: Statistics
- 3: List Math
- 3: Mean

Press VAR and select 'xvalues', then press ENTER. To calculate the median only, press:

- MENU
- 6: Statistics
- 3: List Math
- 4: Median

Press VAR and select 'xvalues', then press ENTER.

The mean is 4.125 and the median is 4.
The mode is 4.

2. To calculate the mean, median and mode, tap:

- Calc
- One-Variable

Set values as:

- XList: main\x
- Freq: 1

Tap OK.
The mean is at the top $(\bar{x})$.
Scroll down to find the median and mode.

The mean is 4.125 and the median is 4.
The mode is 4.

▶ 12.2.2 Calculating mean, median and mode from a frequency distribution table

- If data is provided in the form of a frequency distribution table we can determine the mean, median and mode using slightly different methods.
- The mode is the score with the highest **frequency**.
- To calculate the median, add a cumulative frequency column to the table and use it to determine the score that is the $\left(\dfrac{n+1}{2}\right)$ th data value.
- To calculate the mean, add a column that is the score multiplied by its frequency $f \times x$. The following formula can then be used to calculate the mean, where $\sum (f \times x)$ is the sum of the $(f \times x)$ column. $\sum$ is the uppercase Greek letter sigma.

> **Calculating the mean from a frequency table**
>
> $$\bar{x} = \frac{\sum (f \times x)}{n}$$

WORKED EXAMPLE 2 Calculations from a frequency distribution table

Using the frequency distribution table, calculate:
a. the mean   b. the median   c. the mode.

Score (x)	Frequency (f)
4	1
5	2
6	5
7	4
8	3
Total	15

THINK

1. Rule up a table with four columns titled Score (x), Frequency (f), Frequency × score (f×x) and Cumulative frequency (cf).

2. Enter the data and complete both the f×x and cumulative frequency columns.

WRITE

Score (x)	Frequency (f)	Frequency × score (f×x)	Cumulative frequency (cf)
4	1	4	1
5	2	10	$1+2=3$
6	5	30	$3+5=8$
7	4	28	$8+4=12$
8	3	24	$12+3=15$
	$n=15$	$\sum(f \times x)=96$	

a. 1. Write the rule for the mean.

a. $\bar{x} = \dfrac{\sum (f \times x)}{n}$

2. Substitute the known values into the rule and evaluate.

$\bar{x} = \dfrac{96}{15}$

$= 6.4$

3. Write the answer.

The mean of the data set is 6.4.

b. 1. Locate the position of the median using the rule $\dfrac{n+1}{2}$, where $n = 15$.
This places the median as the 8th score.

b. The median is the $\left(\dfrac{15+1}{2}\right)$th or 8th score.

2. Use the cumulative frequency column to find the 8th score and write the answer.

The median of the data set is 6.

c. 1. The mode is the score with the highest frequency.

c. The score with the highest frequency is 6.

2. Write the answer.

The mode of the data set is 6.

TI | THINK

1. In a new problem, on a Lists & Spreadsheet page, label column A as 'score' and column B as 'f'. Enter the values as shown in the table and press ENTER after entering each value.

DISPLAY/WRITE

CASIO | THINK

1. On the Statistics screen, label list1 as 'score' and list2 as 'f', and enter the values as shown in the table.
Press EXE after entering each value.

DISPLAY/WRITE

2. To find the summary statistics, press:
- MENU
- 4: Statistics
- 1: Stat Calculations
- 1: One-Variable Statistics...

Select 1 as the number of lists. Then on the One-Variable Statistics page select 'score' as the X1 List and 'f' as the Frequency List. Leave the next fields empty, TAB to OK and press ENTER.

2. To find the summary statistics, tap:
- Calc
- One-Variable

Set values as:
- XList: main\score
- Freq: main\f

3. The results are displayed.
   The mean $\bar{x} = 6.4$ and the
   median is 6.
   The mode is the data set
   with the highest frequency
   value, which in this case
   is 6.

The mean is $\bar{x} = 6.4$ and the
median is 6. The mode is the data
set with the highest frequency
value, which is 6.

3. Tap OK.
   The mean is at the top $\bar{x}$.
   Scroll down to find the
   median and mode.

The mean is $\bar{x} = 6.4$ and
the median is 6. The mode
is the data set with the
highest frequency value,
which is 6.

Mean, median and mode of grouped data

- When the data are grouped into class intervals, the actual values (or data) are lost. In such cases we have to approximate the real values with the midpoints of the intervals into which these values fall.
  For example, if in a grouped frequency table showing the heights of different students, 4 students had a height between 180 and 185 cm, we assume that each of those 4 students is 182.5 cm tall.

Mean

- The formula for calculating the mean is the same as the formula used when the data is displayed in a frequency distribution table:

$$\bar{x} = \frac{\sum(f \times x)}{n}$$

Here, x represents the midpoint (or class centre) of each class interval, f is the corresponding frequency and n is the total number of observations in a set.

Median

- The median is found by drawing a **cumulative frequency** curve (ogive) of the data and estimating the median from the 50th percentile (see section 12.2.3).

Modal class

- The modal class is the class interval that has the highest frequency.

▶ 12.2.3 Cumulative frequency curves (ogives)

eles-4951

Ogives

- Data from a cumulative frequency table can be plotted to form a **cumulative frequency curve** (sometimes referred to as cumulative frequency polygon), which is also called an **ogive** (pronounced '*oh-jive*').
- To plot an ogive for data that is in class intervals, the maximum value for the class interval is used as the value against which the cumulative frequency is plotted.

Quantiles

- An ogive can be used to divide the data into any given number of equal parts called **quantiles**.
- Quantiles are named after the number of parts that the data are divided into.
  - **Percentiles** divide the data into 100 equal-sized parts.
  - **Quartiles** divide the data into 4 equal-sized parts. For example, 25% of the data values lie at or below the first quartile.

Percentile	Quartile and symbol	Common name
25th percentile	First quartile, Q_1	Lower quartile
50th percentile	Second quartile, Q_2	Median
75th percentile	Third quartile, Q_3	Upper quartile
100th percentile	Fourth quartile, Q_4	Maximum

- A percentile is named after the percentage of data that lies at or below that value.
  For example, 60% of the data values lie at or below the 60th percentile.
- Percentiles can be read off a percentage cumulative frequency curve.
- A percentage cumulative frequency curve is created by:
  - writing the cumulative frequencies as a percentage of the total number of data values
  - plotting the percentage cumulative frequencies against the maximum value for each interval.
- For example, the following table and graph show the mass of cartons of eggs ranging from 55 g to 65 g.

Mass (g)	Frequency (f)	Cumulative frequency (cf)	Percentage cumulative frequency ($\%cf$)
$55-<57$	2	2	6%
$57-<59$	6	$2+6=8$	22%
$59-<61$	12	$8+12=20$	56%
$61-<63$	11	$20+11=31$	86%
$63-<65$	5	$31+5=36$	100%

For the given data:

Class interval	Frequency
$60 - <70$	5
$70 - <80$	7
$80 - <90$	10
$90 - <100$	12
$100 - <110$	8
$110 - <120$	3
Total	45

a. estimate the mean   b. estimate the median   c. determine the modal class.

THINK

1. Draw up a table with 5 columns headed Class interval, Class centre (x), Frequency (f), Frequency × class centre ($f \times x$) and Cumulative frequency (cf).

2. Complete the x, f, $f \times x$ and cf columns.

WRITE

Class interval	Class centre (x)	Freq. (f)	Frequency × class centre ($f \times x$)	Cumulative frequency (cf)
$60 - <70$	65	5	325	5
$70 - <80$	75	7	525	12
$80 - <90$	85	10	850	22
$90 - <100$	95	12	1140	34
$100 - <110$	105	8	840	42
$110 - <120$	115	3	345	45
		$n = 45$	$\sum(f \times x) = 4025$	

a. 1. Write the rule for the mean.

   2. Substitute the known values into the rule and evaluate.

   3. Write the answer.

a. $\bar{x} = \dfrac{\sum(f \times x)}{n}$

$\bar{x} = \dfrac{4025}{45}$

$\simeq 89.4$

The mean for the given data is approximately 89.4.

b. 1. Draw a combined cumulative frequency histogram and ogive, labelling class centres on the horizontal axis and cumulative frequency on the vertical axis. Join the end points of each class interval with a straight line to form the ogive.

b.

2. Locate the middle of the cumulative frequency axis, which is 22.5.

3. Draw a horizontal line from this point to the ogive and a vertical line to the horizontal axis.

4. Read off the value of the median from the *x*-axis and write the answer.

The median for the given data is approximately 90.

c. 1. The modal class is the class interval with the highest frequency.

c. The class internal 90– < 100 occurs twelve times, which is the highest frequency.

2. Write the answer.

The modal class is the 90– < 100 class interval.

 Resources

Exercise 12.2 Measures of central tendency

12.2 Quick quiz on	12.2 Exercise

Individual pathways

■ PRACTISE	■ CONSOLIDATE	■ MASTER
1, 2, 7, 10, 14, 17, 18, 23	3, 4, 6, 8, 11, 15, 19, 20, 24	5, 9, 12, 13, 16, 21, 22, 25

Fluency

WE1 For questions **1** to **5**, calculate:

a. the mean   **b.** the median   **c.** the mode.

1. 3, 5, 6, 8, 8, 9, 10

2. 4, 6, 7, 4, 8, 9, 7, 10

3. 17, 15, 48, 23, 41, 56, 61, 52

4. 4.5, 4.7, 4.8, 4.8, 4.9, 5.0, 5.3

5. $7\frac{1}{2}$, $10\frac{1}{4}$, 12, $12\frac{1}{4}$, 13, $13\frac{1}{2}$, $13\frac{1}{2}$, 14

6. The back-to-back stem-and-leaf plot shows the test results of 25 Year 10 students in Mathematics and Science. Calculate the mean, median and mode for each of the two subjects.

Key: 3 | 2 = 32

Leaf: Science	Stem	Leaf: Mathematics
8 7 3	3	2 9
9 6 2 2 1	4	0 6 8
8 7 6 1 1 0	5	1 3 5
9 7 4 3 2	6	2 6 7 9
8 5 1 0	7	3 6 7 8
7 3	8	0 4 4 6 8 9
	9	2 5 8

7. **WE2** Use the frequency distribution table shown to calculate:

    a. the mean   **b.** the median   **c.** the mode.

Score (x)	Frequency (f)
4	3
5	6
6	9
7	4
8	2
Total	24

8. Use the frequency distribution table shown to calculate:

a. the mean   b. the median   c. the mode.

Score (x)	Frequency (f)
12	4
13	5
14	10
15	12
16	9
Total	40

9. The following data show the number of bedrooms in each of the 10 houses in a particular neighbourhood:

2, 1, 3, 4, 2, 3, 2, 2, 3, 3.

a. Calculate the mean and median number of bedrooms.

b. A local motel contains 20 rooms. Add this observation to the set of data and recalculate the values of the mean and median.

c. Compare the answers obtained in parts **a** and **b** and complete the following statement:
When the set of data contains an unusually large value(s), called an outlier, the _____ (mean/median) is the better measure of central tendency, as it is less affected by this extreme value.

10. **WE3** For the given data:

a. estimate the mean   b. estimate the median   c. determine the modal class.

Class interval	Frequency
$40 - <50$	4
$50 - <60$	4
$60 - <70$	6
$70 - <80$	9
$80 - <90$	5
$90 - <100$	4
Total	30

11. Calculate the mean of the grouped data shown in the table below.

Class interval	Frequency
100–109	3
110–119	7
120–129	10
130–139	6
140–149	4
Total	30

12. Determine the modal class of the data shown in the table below.

Class interval	Frequency
$50 - <55$	1
$55 - <60$	3
$60 - <65$	4
$65 - <70$	5
$70 - <75$	3
$75 - <80$	2
Total	18

13. The number of textbooks sold by various bookshops during the second week of December was recorded. The results are summarised in the table.

Number of books sold	Frequency
220–229	2
230–239	2
240–249	3
250–259	5
260–269	4
270–279	4
Total	20

a. **MC** The modal class of the data is given by the class interval(s):

A. 220–229 and 230–239   **B.** 250–259
C. 260–269 and 270–279   **D.** of both A and C

b. **MC** The class centre of the first class interval is:

A. 224   **B.** 224.5
C. 224.75   **D.** 225

c. **MC** The median of the data is in the interval:

A. 230–239   **B.** 240–249
C. 250–259   **D.** 260–269

d. **MC** The estimated mean of the data is:

A. 251   **B.** 252
C. 253   **D.** 254

Understanding

14. A random sample was taken, composed of 30 people shopping at a supermarket on a Tuesday night. The amount of money (to the nearest dollar) spent by each person was recorded as follows.

6	32	66	17	45	1	19	52	36	23	28	20	7	47	39
6	68	28	54	9	10	58	40	12	25	49	74	63	41	13

a. Calculate the mean and median amount of money spent at the checkout by the people in this sample.
b. Group the data into class intervals of 10 and complete the frequency distribution table. Use this table to estimate the mean amount of money spent.
c. Add the cumulative frequency column to your table and fill it in. Hence, construct the ogive. Use the ogive to estimate the median.
d. Compare the mean and the median of the original data from part **a** with the mean and the median obtained for grouped data in parts **b** and **c**.
Explain if the estimates obtained in parts **b** and **c** were good enough.

15. Answer the following question and show your working.

a. Add one more number to the set of data 3, 4, 4, 6 so that the mean of a new set is equal to its median.
b. Design a set of five numbers so that mean = median = mode = 5.
c. In the set of numbers 2, 5, 8, 10, 15, change one number so that the median remains unchanged while the mean increases by 1.

16. Thirty men were asked to reveal the number of hours they spent doing housework each week. The results are detailed below.

$$1, \quad 5, \quad 2, \quad 12, \quad 2, \quad 6, \quad 2, \quad 8, \quad 14, \quad 18,$$
$$0, \quad 1, \quad 1, \quad 8, \quad 20, \quad 25, \quad 3, \quad 0, \quad 1, \quad 2,$$
$$7, \quad 10, \quad 12, \quad 1, \quad 5, \quad 1, \quad 18, \quad 0, \quad 2, \quad 2$$

a. Present the data in a frequency distribution table. (Use class intervals of 0–4, 5–9 etc.)
b. Use your table to estimate the mean number of hours that the men spent doing housework.
c. Determine the median class for hours spent by the men at housework.
d. Identify the modal class for hours spent by the men at housework.

Reasoning

17. The data shown give the age of 25 patients admitted to the emergency ward of a hospital.

$$18, \quad 16, \quad 6, \quad 75, \quad 24,$$
$$23, \quad 82, \quad 75, \quad 25, \quad 21,$$
$$43, \quad 19, \quad 84, \quad 76, \quad 31,$$
$$78, \quad 24, \quad 20, \quad 63, \quad 79,$$
$$80, \quad 20, \quad 23, \quad 17, \quad 19$$

a. Present the data in a frequency distribution table. (Use class intervals of $0-<15$, $15-<30$ and so on.)
b. Draw a histogram of the data.
c. Suggest a word to describe the pattern of the data in this distribution.
d. Use your table to estimate the mean age of patients admitted.
e. Determine the median class for age of patients admitted.
f. Identify the modal class for age of patients admitted.
g. Draw an ogive of the data.
h. Use the ogive to determine the median age.
i. Explain if any of your statistics (mean, median or mode) give a clear representation of the typical age of an emergency ward patient.
j. Give some reasons which could explain the pattern of the distribution of data in this question.

18. **MC** In a set of data there is one score that is extremely small when compared to all the others. This outlying value is most likely to:

A. have greatest effect upon the mean of the data
B. have greatest effect upon the median of the data
C. have greatest effect upon the mode of the data
D. have very little effect on any of the statistics as we are told that the number is extremely small

19. The batting scores for two cricket players over 6 innings are as follows.

Player A   31, 34, 42, 28, 30, 41
Player B   0, 0, 1, 0, 250, 0

a. Calculate the mean score for each player.
b. State which player appears to be better, based upon mean result. Justify your answer.
c. Determine the median score for each player.
d. State which player appears to be better when the decision is based on the median result. Justify your answer.
e. State which player do you think would be the most useful to have in a cricket team. Justify your answer. Explain how can the mean result sometimes lead to a misleading conclusion.

20. The following frequency table gives the number of employees in different salary brackets for a small manufacturing plant.

Position	Salary ($)	Number of employees
Machine operator	18 000	50
Machine mechanic	20 000	15
Floor steward	24 000	10
Manager	62 000	4
Chief executive officer	80 000	1

a. Workers are arguing for a pay rise but the management of the factory claims that workers are well paid because the mean salary of the factory is $22 100.
Explain whether the management is being honest.
b. Suppose that you were representing the factory workers and had to write a short submission in support of the pay rise.
How could you explain the management's claim? Quote some other statistics in favour of your case.

21. The resting pulse rate of 20 female athletes was measured. The results are detailed below.

50, 52, 48, 52, 71, 61, 30, 45, 42, 48,
43, 47, 51, 62, 34, 61, 44, 54, 38, 40

a. Construct a frequency distribution table. (Use class sizes of $1 -< 10$, $10 -< 20$ and so on.)
b. Use your table to estimate the mean of the data.
c. Determine the median class of the data.
d. Identify the modal class of the data.
e. Draw an ogive of the data. (You may like to use a graphics calculator for this.)
f. Use the ogive to determine the median pulse rate.

22. Design a set of five numbers with:
a. mean = median = mode
b. mean > median > mode
c. mean < median = mode.

Problem solving

23. The numbers 15, a, 17, b, 22, c, 10 and d have a mean of 14. Calculate the mean of a, b, c and d.

24. The numbers m, n, p, q, r, and s have a mean of a while x, y and z have a mean of b. Calculate the mean of all nine numbers.

25. The mean and median of six two-digit prime numbers is 39 and the mode is 31. The smallest number is 13. Determine the six numbers.

LESSON
12.3 Measures of spread

LEARNING INTENTION

At the end of this lesson you should be able to:
- calculate the range and interquartile range of a data set.

▶ 12.3.1 Measures of spread
eles-4952

- **Measures of spread** describe how far data values are spread from the centre or from each other.
- A shoe store proprietor has stores in Newcastle and Wollongong. The number of pairs of shoes sold each day over one week is recorded below.

$$\text{Newcastle:} \quad 45, 60, 50, 55, 48, 40, 52$$
$$\text{Wollongong:} \quad 20, 85, 50, 15, 30, 60, 90$$

In each of these data sets consider the measures of central tendency.

Newcastle:	Mean $= 50$	Wollongong:	Mean $= 50$
	Median $= 50$		Median $= 50$
	No mode		No mode

With these measures being the same for both data sets we could come to the conclusion that both data sets are very similar; however, if we look at the data sets, they are very different. We can see that the data for Newcastle are very clustered around the mean, whereas the Wollongong data are spread out more.
The data from Newcastle are between 40 and 60, whereas the Wollongong data are between 15 and 90.
- **Range** and **interquartile range (IQR)** are both measures of spread.

Range

- The most basic measure of spread is the range.
- The range is defined as the difference between the highest and the lowest values in the set of data.

Calculating the range of a data set
$$\textbf{Range} = \textbf{highest score} - \textbf{lowest score}$$
$$= X_{\text{max}} - X_{\text{min}}$$

WORKED EXAMPLE 4 Calculating the range of a data set

Calculate the range of the given data set: **2.1, 3.5, 3.9, 4.0, 4.7, 4.8, 5.2.**

THINK	WRITE
1. Identify the lowest score (X_{min}) of the data set.	Lowest score $= 2.1$
2. Identify the highest score (X_{max}) of the data set.	Highest score $= 5.2$
3. Write the rule for the range.	Range $= X_{\text{max}} - X_{\text{min}}$
4. Substitute the known values into the rule.	$= 5.2 - 2.1$
5. Evaluate and write the answer.	$= 3.1$

Interquartile range

- The interquartile range (IQR) is the spread of the middle 50% of all the scores in an ordered set. When calculating the interquartile range, the data are first organised into quartiles, each containing 25% of the data.
- The word 'quartile' comes from the word 'quarter'.

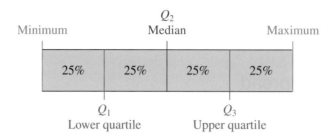

- The lower quartile (Q_1) is the median of the lower half of the data set.
- The upper quartile (Q_3) is the median of the upper half of the data set.

Calculating the IQR

Interquartile range (IQR) = upper quartile − lower quartile

$$= Q_{\text{upper}} - Q_{\text{lower}}$$

$$= Q_3 - Q_1$$

- The IQR is not affected by extremely large or extremely small data values (**outliers**), so in some circumstances the IQR is a better indicator of the spread of data than the range.

WORKED EXAMPLE 5 Calculating the IQR of a data set

Calculate the interquartile range (IQR) of the following set of data:

$$3, 2, 8, 6, 1, 5, 3, 7, 6.$$

THINK	WRITE
1. Arrange the scores in order.	$1, 2, 3, 3, 5, 6, 6, 7, 8$
2. Locate the median and use it to divide the data into two halves. *Note:* The median is the 5th score in this data set and should not be included in the lower or upper ends of the data.	$1, 2, 3, 3, \circledS, 6, 6, 7, 8$
3. Calculate Q_1, the median of the lower half of the data.	$Q_1 = \dfrac{2+3}{2}$ $= \dfrac{5}{2}$ $= 2.5$

4. Calculate Q_3, the median of the upper half of the data.

$$Q_3 = \frac{6+7}{2}$$
$$= \frac{13}{2}$$
$$= 6.5$$

5. Calculate the interquartile range.

$$\text{IQR} = Q_3 - Q_1$$
$$= 6.5 - 2.5$$

6. Write the answer.

$$= 4$$

| TI | THINK | DISPLAY/WRITE | CASIO | THINK | DISPLAY/WRITE |

TI | THINK

1. In a new problem, on a Lists & Spreadsheet page, label column A as 'xvalues'. Enter the values from the data set. Press ENTER after entering each value.

2. To find the summary statistics, open a Calculator page and press:
   • MENU
   • 6: Statistics
   • 1: Stat Calculations
   • 1: One-Variable Statistics
   Select 1 as the number of lists. Then on the One-Variable Statistics page select 'xvalues' as the X1 List and leave the Frequency as 1. Leave the remaining fields empty, TAB to OK and press ENTER. The summary statistics are shown.

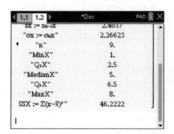

The IQR $= Q_3 - Q_1 = 6.5 - 2.5 = 4$

CASIO | THINK

1. On the Statistics screen, label list1 as 'x' and enter the values as shown in the table. Press EXE after entering each value.

2. To find the summary statistics, tap:
   • Calc
   • One-Variable
   Set values as:
   • XList: main\x
   • Freq: 1
   Tap OK.
   Calculate the IQR $= Q_3 - Q_1$

The IQR $= Q_3 - Q_1$
$= 6.5 - 2.5 = 4$

Determining the IQR from a graph

- When data are presented in a frequency distribution table, either ungrouped or grouped, the interquartile range is found by drawing an ogive.

WORKED EXAMPLE 6 Calculating the IQR from a graph

The following frequency distribution table gives the number of customers who order different volumes of concrete from a readymix concrete company during the course of a day. Calculate the interquartile range of the data.

Volume (m³)	Frequency
0.0 – < 0.5	15
0.5 – < 1.0	12
1.0 – < 1.5	10

Volume (m³)	Frequency
1.5 – < 2.0	8
2.0 – < 2.5	2
2.5 – < 3.0	4

THINK

1. To calculate the 25th and 75th percentiles from the ogive, first add a class centre column and a cumulative frequency column to the frequency distribution table and fill them in.

WRITE/DRAW

Volume	Class centre	f	cf
0.0 – < 0.5	0.25	15	15
0.5 – < 1.0	0.75	12	27
1.0 – < 1.5	1.25	10	37
1.5 – < 2.0	1.75	8	45
2.0 – < 2.5	2.25	2	47
2.5 – < 3.0	2.75	4	51

2. Draw the ogive. A percentage axis will be useful.

3. Identify the upper quartile (75th percentile) and lower quartile (25th percentile) from the ogive.

$Q_3 = 1.6 \, \text{m}^3$

$Q_1 = 0.4 \, \text{m}^3$

4. The interquartile range is the difference between the upper and lower quartiles.

$$IQR = Q_3 - Q_1$$
$$= 1.6 - 0.4$$
$$= 1.2 \, \text{m}^3$$

 Resources

eWorkbook   Topic 12 Workbook (worksheets, code puzzle and project) (ewbk-13302)

Interactivities Individual pathway interactivity: Measures of spread (int-4622)
Range (int-3822)
The interquartile range (int-4813)

12.3 Quick quiz on	12.3 Exercise

Individual pathways

■ PRACTISE	■ CONSOLIDATE	■ MASTER
1, 4, 7, 10, 13	2, 6, 8, 11, 14	3, 5, 9, 12, 15

Fluency

1. **WE4** Calculate the range for each of the following sets of data.

   a. 4, 3, 9, 12, 8, 17, 2, 16
   b. 49.5, 13.7, 12.3, 36.5, 89.4, 27.8, 53.4, 66.8
   c. $7\frac{1}{2}, 12\frac{3}{4}, 5\frac{1}{4}, 8\frac{2}{3}, 9\frac{1}{6}, 3\frac{3}{4}$

2. **WE5** Calculate the interquartile range (IQR) for the following sets of data.

   a. 3, 5, 8, 9, 12, 14
   b. 7, 10, 11, 14, 17, 23
   c. 66, 68, 68, 70, 71, 74, 79, 80
   d. 19, 25, 72, 44, 68, 24, 51, 59, 36

3. The following stem-and-leaf plot shows the mass of newborn babies (rounded to the nearest 100 g). Calculate:

   a. the range of the data
   b. the IQR of the data.

   Key: $1^* \mid 9 = 1.9\,\text{kg}$

Stem	Leaf
1*	9
2	2 4
2*	6 7 8 9
3	0 0 1 2 3 4
3*	5 5 6 7 8 8 8 9
4	0 1 3 4 4
4*	5 6 6 8 9
5	0 1 2 2

4. Use the ogive shown to calculate the interquartile range of the data.

5. **WE6** The following frequency distribution table gives the amount of time spent by 50 people shopping for Christmas presents.

Time (h)	$0 - <0.5$	$0.5 - <1$	$1 - <1.5$	$1.5 - <2$	$2 - <2.5$	$2.5 - <3$	$3 - <3.5$	$3.5 - <4$
Frequency	1	2	7	15	13	8	2	2

Estimate the IQR of the data.

6. **MC** Calculate the interquartile range of the following data:

$$17, 18, 18, 19, 20, 21, 21, 23, 25$$

A. 8   **B.** 18   **C.** 4   **D.** 20

Understanding

7. The following frequency distribution table shows the life expectancy in hours of 40 household batteries.

Life (h)	$50 - <55$	$55 - <60$	$60 - <65$	$65 - <70$	$70 - <75$	$75 - <80$
Frequency	4	10	12	8	5	1

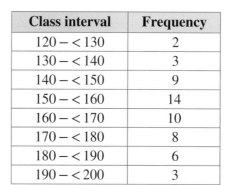

a. Draw an ogive curve that represents the data in the table above.
b. Use the ogive to answer the following questions.
   i. Calculate the median score.
   ii. Determine the upper and lower quartiles.
   iii. Calculate the interquartile range.
   iv. Identify the number of batteries that lasted less than 60 hours.
   v. Identify the number of batteries that lasted 70 hours or more.

8. Calculate the IQR for the following data.

Class interval	Frequency
$120 - <130$	2
$130 - <140$	3
$140 - <150$	9
$150 - <160$	14
$160 - <170$	10
$170 - <180$	8
$180 - <190$	6
$190 - <200$	3

9. For each of the following sets of data, state:
   i. the range
   ii. the IQR.
   a. 6, 9, 12, 13, 20, 22, 26, 29
   b. 7, 15, 2, 26, 47, 19, 9, 33, 38
   c. 120, 99, 101, 136, 119, 87, 123, 115, 107, 100

Reasoning

10. Explain what the measures of spread tell us about a set of data.

11. As newly appointed coach of Terrorolo's Meteors netball team, Kate decided to record each player's statistics for the previous season. The number of goals scored by the leading goal shooter was:

1,	3,	8,	18,	19,	23,	25,	25,	25,	26,	27,	28,
28,	28,	28,	29,	29,	30,	30,	33,	35,	36,	37,	40

a. Calculate the mean of the data.
b. Calculate the median of the data.
c. Calculate the range of the data.
d. Determine the interquartile range of the data.
e. There are three scores that are much lower than most. Explain the effect these scores have on the summary statistics.

12. The following back-to-back stem-and-leaf plot shows the ages of 30 pairs of men and women when entering their first marriage.

Key: $1\,|\,6 = 16$ years old

Leaf: Men	Stem	Leaf: Women
9 9 8	1	6 7 7 8 9
9 9 8 8 7 6 4 4 3 2 0	2	0 0 1 2 3 4 5 6 7 7 8 9
9 8 8 8 6 5 5 4 3 2	3	0 1 2 2 3 4 7 9
6 3 0 0	4	1 2 4 8
6 0	5	2

a. Determine the mean, median, range and interquartile range of each set.
b. Write a short paragraph comparing the two distributions.

Problem solving

13. Calculate the mean, median, mode, range and IQR of the following data collected when the temperature of the soil around 25 germinating seedlings was recorded:

28.9, 27.4, 23.6, 25.6, 21.1, 22.9, 29.6, 25.7, 27.4, 23.6, 22.4, 24.6, 21.8, 26.4, 24.9, 25.0, 23.5, 26.1, 23.6, 25.3, 29.5, 23.5, 22.0, 27.9, 23.6.

14. Four positive numbers a, b, c and d have a mean of 12, a median and mode of 9 and a range of 14. Determine the values of a, b, c and d.

15. A set of five positive integer scores have the following summary statistics:
- range $= 9$
- median $= 6$
- $Q_1 = 3$ and $Q_3 = 9$.

a. Explain whether the five scores could be $1, 3, 6, 9$ and 10.
b. A sixth score is added to the set. Determine whether there is a score that will maintain the summary statistics given above. Justify your answer.

LESSON
12.4 Box plots

LEARNING INTENTIONS

At the end of this lesson you should be able to:
- calculate the five-number summary for a set of data
- draw a box plot showing the five-number summary of a data set
- calculate outliers in a data set
- describe the skewness of distributions
- compare box plots to dot plots or histograms
- draw parallel box plots and compare sets of data.

12.4.1 Five-number summary

eles-4953

- A five-number summary is a list consisting of the lowest score (X_{min}), lower quartile (Q_1), median (Q_2), upper quartile (Q_3) and greatest score (X_{max}) of a set of data.

WORKED EXAMPLE 7 Calculations using the five-number summary

From the following five-number summary, calculate:
a. the interquartile range   b. the range.

X_{min}	Q_1	Median (Q_2)	Q_3	X_{max}
29	37	39	44	48

THINK

a. The interquartile range is the difference between the upper and lower quartiles.

b. The range is the difference between the greatest score and the lowest score.

WRITE

a. $IQR = Q_3 - Q_1$
$ = 44 - 37$
$ = 7$

b. $Range = X_{max} - X_{min}$
$ = 48 - 29$
$ = 19$

12.4.2 Box plots

eles-4954

- A **box plot** is a graph of the five-number summary.
- Box plots consist of a central divided box with attached whiskers.
- The box spans the interquartile range.
- The median is marked by a vertical line drawn inside the box.
- The whiskers indicate the range of scores:

The lowest score X_{min} The lower quartile Q_1 The median M The upper quartile Q_3 The largest score X_{max}

- Box plots are *always drawn to scale*.
- They are presented either with the five-number summary figures attached as labels (diagram at right) or with a scale presented alongside the box plot like the diagram below. They can also be drawn vertically.

Scale

Identification of extreme values or outliers

- If an extreme value or outlier occurs in a set of data, it can be denoted by a small cross on the box plot. The whisker is then shortened to the next largest, or smallest, value.
- The box plot below shows that the lowest score was 5. This was an extreme value as the rest of the scores were located within the range 15 to 42.
- Outliers are still included when calculating the range of the data.

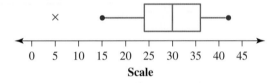

Scale

- Outliers sit more than $1.5 \times \text{IQR}$ away from Q_1 or Q_3.

Identifying outliers

$$\textbf{Lower limit} = Q_1 - 1.5 \times \textbf{IQR}$$

$$\textbf{Upper limit} = Q_3 + 1.5 \times \textbf{IQR}$$

Any scores that sit outside these limits are considered outliers.

Symmetry and skewness in distributions

- A **symmetrical** plot has data that are evenly spaced around a central point.

  Examples of a symmetrical stem-and-leaf plot and a symmetrical box plot are shown.

Stem	Leaf
26*	6
27	0 1 3
27*	5 6 8 9
28	0 1 1 1 2 4
28*	5 7 8 8
29	2 2 2
29*	5

- A **negatively skewed** plot has a higher frequency of data at the higher end. This is illustrated by the stem-and-leaf plot below where the leaves increase in length as the data increase in value.
  It is illustrated on the box plot when the median is much closer to the maximum value than the minimum value.

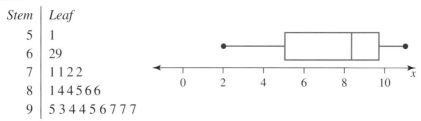

Stem	Leaf
5	1
6	2 9
7	1 1 2 2
8	1 4 4 5 6 6
9	5 3 4 4 5 6 7 7 7

- A **positively skewed** plot has a higher frequency of data at the lower end. This is illustrated on the stem-and-leaf plot below where the leaves increase in length as the data decrease in value.
  It is illustrated on the box plot when the median is much closer to the minimum value than the maximum value.

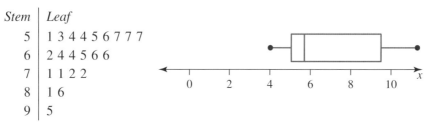

Stem	Leaf
5	1 3 4 4 5 6 7 7 7
6	2 4 4 5 6 6
7	1 1 2 2
8	1 6
9	5

WORKED EXAMPLE 8 Drawing a box plot

The following stem-and-leaf plot gives the speed of 25 cars caught by a roadside speed camera.

Key: $8 \mid 2 = 82\,\text{km/h}, 8^* \mid 6 = 86\,\text{km/h}$

Stem	Leaf
8	2 2 4 4 4 4
8*	5 5 6 6 7 9 9 9
9	0 1 1 2 4
9*	5 6 9
10	0 2
10*	
11	4

a. **Prepare a five-number summary of the data and draw a box plot to represent it.**
b. **Identify any outliers and redraw the box plot with outliers marked.**
c. **Describe the distribution of the data.**

THINK	WRITE
1. First identify the positions of the median and upper and lower quartiles. There are 25 data values. The median is the $\left(\dfrac{n+1}{2}\right)$ th score. The lower quartile is the median of the lower half of the data. The upper quartile is the median of the upper half of the data (each half contains 12 scores).	The median is the $\left(\dfrac{25+1}{2}\right)$ th score — that is, the 13th score. Q_1 is the $\left(\dfrac{12+1}{2}\right)$ th score in the lower half — that is, the 6.5th score. That is, halfway between the 6th and 7th scores. Q_3 is halfway between the 6th and 7th scores in the upper half of the data.

2. Mark the positions of the median and upper and lower quartiles on the stem-and-leaf plot.

Key: 8 | 2 = 82 km/h
      8* | 6 = 86 km/h

Stem	Leaf
8	2 2 4 4 4 4 *Q₁*
8*	5 5 6 6 7 9 ⑨ 9 Median
9	0 1 1 2 4
9*	5 6 9 *Q₃*
10	0 2
10*	
11	4

a. Write the five-number summary:
The lowest score is 82.
The lower quartile is between 84 and 85; that is, 84.5.
The median is 89.
The upper quartile is between 94 and 95; that is, 94.5.
The greatest score is 114.
Draw the box plot for this summary.

a. Five-number summary:

X_{min}	Q_1	Q_2	Q_3	X_{max}
82	84.5	89	94.5	114

Speed (km/h)

b. 1. Calculate the IQR.

b. $IQR = Q_3 - Q_1$
$= 94.5 - 84.5$
$= 10$

2. Calculate the lower and upper limits.

Lower limit $= 84.5 - 1.5 \times 10$
$= 69.5$
Upper limit $= 94.5 + 1.5 \times 10$
$= 109.5$

3. Identify the outliers.

114 is above the upper limit of 109.5, so it is an outlier.

4. Redraw the box plot, including the outlier marked as a cross. Draw the whisker to the next largest figure, 102.

Speed (km/h)

c. Describe the distribution.

c. The data is positively skewed with an outlier at 114.

TI	THINK	DISPLAY/WRITE	CASIO	THINK	DISPLAY/WRITE

a.

1. In a new problem, on a Lists & Spreadsheet page, label column A as 'cars' and enter the values from the stem-and-leaf plot. Press ENTER after each value.

2. To find a five-point summary of the data, on a Calculator page press:
- CATALOG
- 1
- F

Then use the down arrow to scroll down to FiveNumSummary and press ENTER.

3. Press VAR and select 'cars'. Complete the entry line as:
FiveNumSummary cars
Press ENTER.
Then press VAR, select 'stat.results' and press ENTER.

b.

To construct the box-and-whisker plot, open a Data & Statistics page. Press TAB to locate the label of the horizontal axis and select the variable 'cars'.
Then press:
- MENU
- 1: Plot Type
- 2: Box Plot

To change the colour, place the pointer over one of the data points. Then press CTRL MENU.
Then press:
- 6: Color
- 2: Fill Color.

Select whichever colour you like from the palette. Press ENTER.

The box-and-whisker plot is displayed. As you scroll over the box-and-whisker plot, the values of the five-number summary statistics are displayed. The data are skewed (positively).

a.

1. On the Statistics screen, label list1 as 'cars' and enter the values from the stem-and-leaf plot. Press EXE after entering each value.

2. To find the summary statistics, tap:
- Calc
- One-Variable

Set values as:
- XList: main\cars
- Freq: 1

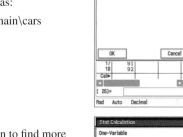

3. Tap OK.
Scroll down to find more statistics.
minX = 82
Q1 = 84.5
Med = 89
Q3 = 95.5
maxX = 114

b.

To construct the box-and-whisker plot, tap:
- SetGraph
- Setting...

Set:
- Type: MedBox
- XList: main\cars
- Freq: 1

Then tap the graphing icon.

The box-and-whisker plot is displayed.

12.4.3 Comparing different graphical representations

eles-4956

Box plots and dot plots

- A box plot can be directly related to a dot plot.
- **Dot plots** display each data value represented by a dot placed on a number line.
- The following data represent the amount of money (in $) that a group of 27 five-year-olds had with them on a day visiting the zoo with their parents.

0	0.85	0	1.8	1.65	8.45	3.75	0.55	4.1	2.4	2.15
1.2	1.35	0.9	3.45	1	0	0	1.45	1.25	1.7	2.65
1.85	4.75	3.9	1.15							

- The dot plot below and its comparative box plot show the distribution of the data.

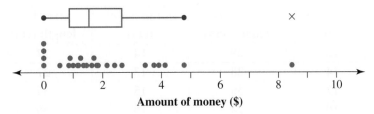

Amount of money ($)

- Both graphs indicate that the data is positively skewed and both graphs indicate the presence of the outlier. However, the box plot provides a more concise summary of the centre and spread of the distribution.

Box plots and histograms

- **Histograms** are graphs that graphically represent the distribution of numerical data. It is commonly used in statistics to visualize the frequency distribution of a set of continuous or discrete data.
- The following data are the number of minutes, rounded to the nearest minute, that forty Year 10 students take to travel to their school on a particular day.

15,	22,	14,	12,	21,	34,	19,	11,	13,	0,	16,
4,	23,	8,	12,	18,	24,	17,	14,	3,	10,	12,
9,	15,	20,	5,	19,	13,	17,	11,	16,	19,	24,
12,	7,	14,	17,	10,	14,	23				

- The data are displayed in the histogram and box plot shown.
- Both graphs indicate that the data is approximately symmetrical with an outlier. The histogram clearly shows the frequencies of each class interval. Neither graph displays the original values.
- The histogram does not give precise information about the centre, but the distribution of the data is visible.
- However, the box plot shows the presence of an outlier and provides a summary of the centre and spread of the distribution.

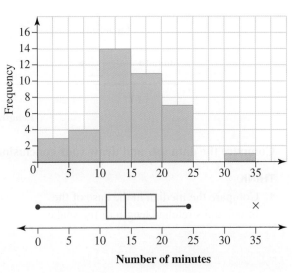

Number of minutes

Parallel box plots

- Parallel box plots are useful tools to compare two sets of data.
- Parallel box plots are two or more box plots displayed with the same scale.

WORKED EXAMPLE 9 Comparing two sets of data

Each member of a class was given a jelly snake to stretch. They each measured the initial length of their snake to the nearest centimetre and then slowly stretched the snake to make it as long as possible.

They then measured the maximum length of the snake by recording how far it had stretched at the time it broke. The results were recorded in the following table.

Initial length (cm)	Stretched length (cm)	Initial length (cm)	Stretched length (cm)
13	29	14	27
14	28	13	27
17	36	15	36
10	24	16	36
14	35	15	36
16	36	16	34
15	37	17	35
16	37	12	27
14	30	9	17
16	33	16	41
17	36	17	38
16	38	16	36
17	38	17	41
14	31	16	33
17	40	11	21

The above data was drawn on parallel box plots as shown below.

Length of snake (cm)

Compare the data sets and draw your conclusions.

THINK	WRITE
1. Compare the median in the case of the initial and stretched length of the snake.	The change in the length of the snake when stretched is evidenced by the increased median and spread shown on the box plots. The median snake length before being stretched was 15.5 cm, but the median snake length after being stretched was 35 cm.
2. Compare the spread for each box plot.	The range increased after stretching, as did the IQR.

Exercise 12.4 Box plots

learn on

12.4 Quick quiz on	12.4 Exercise

Individual pathways

■ PRACTISE	■ CONSOLIDATE	■ MASTER
1, 4, 6, 9, 12, 13, 16, 19	2, 5, 7, 10, 14, 17, 20	3, 8, 11, 15, 18, 21, 22

Fluency

1. **WE7** From the following five-number summary, calculate:

X_{min}	Q_1	Median	Q_3	X_{max}
6	11	13	16	32

   a. the interquartile range   b. the range.

2. From the following five-number summary, calculate:

X_{min}	Q_1	Median	Q_3	X_{max}
101	119	122	125	128

   a. the interquartile range   b. the range.

3. From the following five-number summary, calculate:

X_{min}	Q_1	Median	Q_3	X_{max}
39.2	46.5	49.0	52.3	57.8

   a. the interquartile range   b. the range.

4. The box plot shows the distribution of final points scored by a football team over a season's roster.

   a. Identify the team's greatest points score.
   b. Identify the team's least points score.
   c. Calculate the team's median points score.
   d. Calculate the range of points scored.
   e. Calculate the interquartile range of points scored.

5. The box plot shows the distribution of data formed by counting the number of gummy bears in each of a large sample of packs.

Number of gummy bears

   a. Identify the largest number of gummy bears in any pack.
   b. Identify the smallest number of gummy bears in any pack.
   c. Identify the median number of gummy bears in any pack.
   d. Calculate the range of numbers of gummy bears per pack.
   e. Calculate the interquartile range of gummy bears per pack.

Questions **6** to **8** refer to the following box plot.

Score

6. **MC** The median of the data is:

   A. 20   **B.** 23   **C.** 25   **D.** 31

7. **MC** The interquartile range of the data is:

   A. 23   **B.** 26   **C.** 5   **D.** 20 to 25

8. **MC** Select which of the following is not true of the data represented by the box plot.

   A. One-quarter of the scores are between 5 and 20.
   B. Half of the scores are between 20 and 25.
   C. Most of the data are contained between the scores of 5 and 20.
   D. One-third of the scores are between 5 and 20.

Understanding

9. The number of sales made each day by a salesperson is recorded over a 2-week period:

   $$25, 31, 28, 43, 37, 43, 22, 45, 48, 33$$

   a. Prepare a five-number summary of the data. (There is no need to draw a stem-and-leaf plot of the data. Just arrange them in order of size.)
   b. Draw a box plot of the data.

10. The data below show monthly rainfall in millimetres.

J	F	M	A	M	J	J	A	S	O	N	D
10	12	21	23	39	22	15	11	22	37	45	30

   a. Prepare a five-number summary of the data.
   b. Draw a box plot of the data.

11. **WE8** The stem-and-leaf plot shown details the age of 25 offenders who were caught during random breath testing.

   a. Prepare a five-number summary of the data.
   b. Draw a box plot of the data.
   c. Describe the distribution of the data.

Key: $1|8 = 18$ years

Stem	Leaf
1	8 8 9 9 9
2	0 0 0 1 1 3 4 6 9
3	0 1 2 7
4	2 5
5	3 6 8
6	6
7	4

12. The following stem-and-leaf plot details the price at which 30 blocks of land in a particular suburb were sold.

Key: 12|4 = $124 000

Stem	Leaf
12	4 7 9
13	0 0 2 5 5
14	0 0 2 3 5 5 7 9 9
15	0 0 2 3 7 7 8
16	0 2 2 5 8
17	5

a. Prepare a five-number summary of the data.
b. Draw a box plot of the data.

13. Prepare comparative box plots for the following dot plots (using the same axis) and describe what each plot reveals about the data.

a. Number of sick days taken by workers last year at factory A

b. Number of sick days taken by workers last year at factory B

14. An investigation into the transport needs of an outer suburb community recorded the number of passengers boarding a bus during each of its journeys, as follows.

12, 43,   76, 24, 46, 24, 21, 46, 54, 109, 87, 23, 78,
37, 22, 139, 65, 78, 89, 52, 23, 30,   54, 56, 32, 66, 49

Display the data by constructing a histogram using class intervals of 20 and a comparative box plot on the same axis.

15. **WE12** At a weight-loss clinic, the following weights (in kilograms) were recorded before and after treatment.

Before	75	80	75	140	77	89	97	123	128	95	152	92
After	69	66	72	118	74	83	89	117	105	81	134	85

Before	85	90	95	132	87	109	87	129	135	85	137	102
After	79	84	90	124	83	102	84	115	125	81	123	94

a. Prepare a five-number summary for weight before and after treatment.
b. Draw parallel box plots for weight before and after treatment.
c. Comment on the comparison of weights before and after treatment.

Reasoning

16. Explain the advantages and disadvantages of box plots as a visual form of representing data.

17. The following data detail the number of hamburgers sold by a fast food outlet every day over a 4-week period.

M	T	W	T	F	S	S
125	144	132	148	187	172	181
134	157	152	126	155	183	188
131	121	165	129	143	182	181
152	163	150	148	152	179	181

a. Prepare a stem-and-leaf plot of the data. (Use a class size of 10.)
b. Draw a box plot of the data.
c. Comment on what these graphs tell you about hamburger sales.

18. The following data show the ages of 30 mothers upon the birth of their first baby.

22, 21, 18, 33, 17, 23, 22, 24, 24, 20,
25, 29, 32, 18, 19, 22, 23, 24, 28, 20,
31, 22, 19, 17, 23, 48, 25, 18, 23, 20

a. Prepare a stem-and-leaf plot of the data. (Use a class size of 5.)
b. Draw a box plot of the data. Indicate any extreme values appropriately.
c. Describe the distribution in words. Comment on what the distribution says about the age that mothers have their first baby.

Problem solving

19. Sketch a histogram for the box plot shown.

20. Consider the box plot below, which shows the number of weekly sales of houses by two real estate agencies.

a. Determine the median number of weekly sales for each real estate agency.
b. State which agency had the greater range of sales. Justify your answer.
c. State which agency had the greater interquartile range of sales. Justify your answer.
d. State which agency performed better. Explain your answer.

21. Fifteen French restaurants were visited by three newspaper restaurant reviewers. The average price of a meal for a single person was investigated. The following box plot shows the results.

a. Identify the price of the cheapest meal.
b. Identify the price of the most expensive meal.
c. Identify the median cost of a meal.
d. Calculate the interquartile range for the price of a meal.
e. Determine the percentage of the prices that were below the median.

22. The following data give the box plots for three different age groups in a triathlon for under thirities.

a. Identify the slowest time for the 20–24-year-olds.
b. Estimate the difference in time between the fastest triathlete in:

  i. the under 20s and the 20–24 group
  ii. the under 20s and the 25–29 group
  iii. the under 20–24 group and the 25–29 group.
c. Comment on the overall performance of the three groups.

LESSON
12.5 The standard deviation (10A)

LEARNING INTENTIONS

At the end of this lesson you should be able to:
- calculate the standard deviation of a small data set by hand
- calculate the standard deviation using technology
- interpret the mean and standard deviation of data
- identify the effect of outliers on the standard deviation.

▶ 12.5.1 Standard deviation

eles-4958

- The **standard deviation** for a set of data is a measure of how far the data values are spread out (deviate) from the mean. The value of the standard deviation tells you the average deviation of the data from the mean.
- **Deviation** is the difference between each data value and the mean $(x - \bar{x})$. The standard deviation is calculated from the square of the deviations.

Standard deviation formula

- Standard deviation is denoted by the lowercase Greek letter sigma, σ, and can be calculated by using the following formula.

$$\sigma = \sqrt{\frac{\sum(x - \bar{x})^2}{n}}$$

where $\bar{x}$ is the mean of the data values and n is the number of data values.

- A low standard deviation indicates that the data values tend to be close to the mean.
- A high standard deviation indicates that the data values tend to be spread out over a large range, away from the mean.
- Standard deviation can be calculated using a scientific or graphics calculator, or it can be calculated from a frequency table by following the steps below.

  Step 1 Calculate the mean.
  Step 2 Calculate the deviations.
  Step 3 Square each deviation.
  Step 4 Sum the squares.
  Step 5 Divide the sum of the squares by the number of data values.
  Step 6 Take the square root of the result.

WORKED EXAMPLE 10 Calculating the standard deviation

The number of lollies in each of 8 packets is $11, 12, 13, 14, 16, 17, 18, 19$.
Calculate the mean and standard deviation correct to 2 decimal places. Interpret the result.

THINK

WRITE

1. Calculate the mean.

$$\bar{x} = \frac{11 + 12 + 13 + 14 + 16 + 17 + 18 + 19}{8}$$

$$= \frac{120}{8}$$

$$= 15$$

2. To calculate the deviations $(x - \bar{x})$, set up a frequency table as shown and complete.

No. of lollies (x)	$(x - \bar{x})$
11	$11 - 15 = -4$
12	-3
13	-2
14	-1
16	1
17	2
18	3
19	4
Total	

3. Add another column to the table to calculate the square of the deviations, $(x-\bar{x})^2$. Then sum the results: $\sum (x-\bar{x})^2$.

No. of lollies (x)	$(x-\bar{x})$	$(x-\bar{x})^2$
11	$11-15=-4$	16
12	-3	9
13	-2	4
14	-1	1
16	1	1
17	2	4
18	3	9
19	4	16
Total		$\sum (x-\bar{x})^2 = 60$

4. To calculate the standard deviation, divide the sum of the squares by the number of data values, then take the square root of the result.

$$\sigma = \sqrt{\frac{\sum (x-\bar{x})^2}{n}}$$

$$= \sqrt{\frac{60}{8}}$$

≈ 2.74 (correct to 2 decimal places)

5. Check the result using a calculator.

The calculator returns an answer of $\sigma_n = 2.73861$. The answer is confirmed.

6. Interpret the result.

The average (mean) number of lollies in each pack is 15 with a standard deviation of 2.74, which means that the number of lollies in each pack differs from the mean by an average of 2.74.

TI | THINK

1. In a new problem, on a Calculator page, complete the entry lines as shown. This stores the data values to the variable 'lollies'.
$lollies := \{11, 12, 13, 14, 15, 16, 17, 18, 19\}$
Although we can find many summary statistics, to calculate the mean only, open a Calculator page and press:
- MENU
- 6: Statistics
- 3: List Math
- 3: Mean
Press VAR and select 'lollies', then press ENTER.

DISPLAY/WRITE

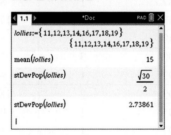

The mean number of lollies is 15 and the population standard deviation is $\sigma = 2.74$.

CASIO | THINK

On the Statistics screen, label list1 as 'x' and enter the values from the question. Press EXE after entering each value. To find the summary statistics, tap:
- Calc
- One-Variable
Set values as:
- XList: main\x
- Freq: 1
Tap OK.
The standard deviation is shown as σ_x and the mean is shown as $\bar{x}$.

DISPLAY/WRITE

The mean number of lollies is 15 with a standard deviation of 2.74.

2. To calculate the population standard deviation only, press:
   - MENU
   - 6: Statistics
   - 3: List Math
   - 9: Population standard deviation

Press VAR and select '*lollies*', then press ENTER. Press CTRL ENTER to get a decimal approximation.

- When calculating the standard deviation from a frequency table, the frequencies must be taken into account. Therefore, the following formula is used.

$$\sigma = \sqrt{\frac{\sum f(x - \bar{x})^2}{n}}$$

WORKED EXAMPLE 11 Calculating standard deviation from a frequency table

Lucy's scores in her last 12 games of golf were 87, 88, 88, 89, 90, 90, 90, 92, 93, 93, 95 and 97. Calculate the mean score and the standard deviation correct to 2 decimal places. Interpret your result.

THINK

WRITE

1. To calculate the mean, first set up a frequency table.

Golf score (x)	Frequency (f)	fx
87	1	87
88	2	176
89	1	89
90	3	270
92	1	92
93	2	186
95	1	95
97	1	97
Total	$\sum f = 12$	$\sum fx = 1092$

2. Calculate the mean.

$$\bar{x} = \frac{\sum fx}{\sum f}$$

$$= \frac{1092}{12}$$

$$= 91$$

3. To calculate the deviations $(x - \bar{x})$, add another column to the frequency table and complete.

Golf score (x)	Frequency (f)	fx	$(x - \bar{x})$
87	1	87	$87 - 91 = -4$
88	8	176	-3
89	1	89	-2
90	3	270	-1
92	1	92	1
93	2	186	2
95	1	95	4
97	1	97	6
Total	$\sum f = 12$	$\sum fx = 1092$	

4. Add another column to the table and multiply the square of the deviations, $(x - \bar{x})^2$, by the frequency $f(x - \bar{x})^2$. Then sum the results: $\sum f(x - \bar{x})^2$.

Golf score (x)	Frequency (f)	fx	$(x - \bar{x})$	$f(x - \bar{x})^2$
87	1	87	$87 - 91 = -4$	$1 \times (-4)^2 = 16$
88	8	176	-3	18
89	1	89	-2	4
90	3	270	-1	3
92	1	92	1	1
93	2	186	2	8
95	1	95	4	16
97	1	97	6	36
Total	$\sum f = 12$	$\sum fx = 1092$		102

5. Calculate the standard deviation using the formula.

$$\sigma = \sqrt{\frac{\sum f(x - \bar{x})^2}{n}}$$

$$= \sqrt{\frac{102}{12}}$$

$$\approx 2.92$$

(correct to 2 decimal places)

6. Check the result using a calculator.

The calculator returns an answer of $\sigma_n = 2.91548$.
The answer is confirmed.

7. Interpret the result.

The average (mean) score for Lucy is 91 with a standard deviation of 2.92, which means that her score differs from the mean by an average of 2.92.

TI	THINK	DISPLAY/WRITE	CASIO	THINK	DISPLAY/WRITE

TI | THINK

1. In a new problem, on a Lists & Spreadsheet page, label column A as 'score' and label column B as 'f'. Enter the values and the frequency corresponding to each score as shown in the table. Press ENTER after each value.

DISPLAY/WRITE

	A score	B f	C	D
1	87	1		
2	88	2		
3	89	1		
4	90	3		
5	92	1		

CASIO | THINK

1. On the Statistic screen, label list1 as 'score' and enter the values from the question. Press EXE after entering each value. To find the summary statistics, tap:
   • Calc
   • One-Variable
   Set values as:
   • XList: main\score
   • Freq: 1

DISPLAY/WRITE

2. To find all the summary statistics, open a Calculator page and press:
   • MENU
   • 6: Statistics
   • 1: Stat Calculations
   • 1: One Variable Statistics...
   Select 1 as the number of lists, then on the One-Variable Statistics page, select 'score' as the X1 List and 'f' as the Frequency List. Leave the next two fields empty, TAB to OK and press ENTER.

"Σx^2"	99474.
"SX := $S_{n-1}X$"	3.04512
"σX := $\sigma_n X$"	2.91548
"n"	12.
"MinX"	87.
"Q_1X"	88.5
"MedianX"	90.
"Q_3X"	93.
"MaxX"	97.
"SSX := $\Sigma(x-\bar{x})^2$"	102.

The mean is $\bar{x} = 91$ and the population standard deviation is $\sigma = 2.92$ correct to 2 decimal places.

2. Tap OK. The standard deviation is shown as σ_x and the mean is shown as $\bar{x}$.

The mean is $\bar{x} = 91$ and the population standard deviation is $\sigma = 2.92$ correct to 2 decimal places.

Why the deviations are squared

• When you take an entire data set, the sum of the deviations from the mean is zero, that is, $\sum(x - \bar{x}) = 0$.
• When the data value is less than the mean ($x < \bar{x}$), the deviation is negative.
• When the data value is greater than the mean ($x > \bar{x}$), the deviation is positive.
• The negative and positive deviations cancel each other out; therefore, calculating the sum and average of the deviations is not useful.
• By squaring all of the deviations, each deviation becomes positive, so the average of the deviations becomes meaningful.
   This explains why the standard deviation is calculated using the squares of the deviations, $(x - \bar{x})^2$, for all data values.

Standard deviations of populations and samples

• So far we have calculated the standard deviation for a population of data, that is, for complete sets of data. There is another formula for calculating standard deviation for samples of data, that is, data that have been randomly selected from a larger population.
• The sample standard deviation is more commonly used in day-to-day life, as it is usually impossible to collect data from an entire population.
   For example, if you wanted to know how much time Year 10 students across the country spend on social media, you would not be able to collect data from every student in the country. You would have to take a sample instead.
• The sample standard deviation is denoted by the letter s, and can be calculated using the following formula.

Sample standard deviation formula

$$s = \sqrt{\frac{\sum f(x - \bar{x})^2}{n - 1}}$$

- Calculators usually display both values for the standard deviation, so it is important to understand the difference between them.

12.5.2 Effects on standard deviation

eles-4959

- The standard deviation is affected by extreme values.

WORKED EXAMPLE 12 Interpreting the effects on standard deviation

On a particular day Lucy played golf brilliantly and scored 60.
The scores in her previous 12 games of golf were 87, 88, 88, 89, 90, 90, 90, 92, 93, 93, 95 and 97 (see Worked example 11).
Comment on the effect this latest score has on the standard deviation.

THINK	WRITE
1. Use a calculator to calculate the mean and the standard deviation.	$\bar{x} = 88.6154 \qquad \sigma = 8.7225$ $\phantom{\bar{x}} \approx 88.62 \qquad\quad\; \approx 8.7225$
2. Interpret the result and compare it to the results found in Worked example 11.	In the first 12 games Lucy's mean score was 91 with a standard deviation of 2.92. This implied that Lucy's scores on average were 2.92 either side of her average of 91. Lucy's latest performance resulted in a mean score of 88.62 with a standard deviation of 8.72. This indicates a slightly lower mean score, but the much higher standard deviation indicates that the data are now much more spread out.

12.5.3 Properties of standard deviation

eles-4961

- If a constant c is added to all data values in a set, the deviations $(x - \bar{x})$ will remain unchanged and consequently the standard deviation remains unchanged.
- If all data values in a set are multiplied by a constant k, the deviations $(x - \bar{x})$ will be multiplied by k, that is $k(x - \bar{x})$; consequently the standard deviation is increased by a factor of k.
- The standard deviation can be used to measure consistency.
- When the standard deviation is low we are able to say that the scores in the data set are more consistent with each other.

For the data 5, 9, 6, 11, 10, 7:
a. calculate the standard deviation
b. calculate the standard deviation if 4 is added to each data value. Comment on the effect.
c. calculate the standard deviation if all data values are multiplied by 2. Comment on the effect.

THINK	WRITE
a. 1. Calculate the mean.	a. $\bar{x} = \dfrac{5+9+6+11+10+7}{6}$
	$= 8$

2. Set up a frequency table and enter the squares of the deviations.

(x)	(x − x̄)	(x − x̄)²
5	5 − 8 = −3	9
6	−2	4
7	−1	1
9	1	1
10	2	4
11	3	9
Total		$\sum (x-\bar{x})^2 = 28$

3. To calculate the standard deviation, apply the formula for standard deviation.

$$\sigma = \sqrt{\dfrac{\sum (x-\bar{x})^2}{n}}$$

$$= \sqrt{\dfrac{28}{6}}$$

$$\approx 2.16 \text{ (correct to 2 decimal places)}$$

b. 1. Add 4 to each data value in the set.

b. 9, 13, 10, 15, 14, 11

2. Calculate the mean.

$$\bar{x} = \dfrac{9+13+10+15+14+11}{6}$$

$$= 12$$

3. Set up a frequency table and enter the squares of the deviations.

(x)	(x − x̄)	(x − x̄)²
9	9 − 12 = −3	9
10	−2	4
11	−1	1
13	1	1
14	2	4
15	3	9
Total		$\sum (x-\bar{x})^2 = 28$

4. To calculate the standard deviation, apply the formula for standard deviation.

$$\sigma = \sqrt{\dfrac{\sum (x-\bar{x})^2}{n}}$$

$$= \sqrt{\dfrac{28}{6}}$$

$$\approx 2.16 \text{ (correct to 2 decimal places)}$$

5. Comment on the effect of adding of 4 to each data value.

Adding 4 to each data value increased the mean but had no effect on the standard deviation, which remained at 2.16.

c. 1. Multiply each data value in the set by 2.

c. 10, 18, 12, 22, 20, 14

2. Calculate the mean.

$$\bar{x} = \frac{10 + 18 + 12 + 22 + 20 + 14}{6}$$

$$= 16$$

3. Set up a frequency table and enter the squares of the deviations.

(x)	$(x - \bar{x})$	$(x - \bar{x})^2$
10	$10 - 16 = -6$	36
12	-4	16
14	-2	4
18	2	4
20	4	16
22	6	36
Total		$\sum (x - \bar{x})^2 = 112$

4. To calculate the standard deviation, apply the formula for standard deviation.

$$\sigma = \sqrt{\frac{\sum (x - \bar{x})^2}{n}} = \sqrt{\frac{112}{6}}$$

$$\approx 4.32 \text{ (correct to 2 decimal places)}$$

5. Comment on the effect of multiplying each data value by 2.

Multiplying each data value by 2 doubled the mean and doubled the standard deviation, which changed from 2.16 to 4.32.

▶ 12.5.4 Average absolute deviations

eles-3219

- The average absolute deviation is a measure of spread that indicates the average absolute distance between data values and the centre of the distribution.
- The average absolute deviation may be calculated using the mean or the median.
- Begin by calculating the centre of the distribution and then calculate how far away each data point is from the centre using positive distances. Finally, calculate the centre of the deviations.

WORKED EXAMPLE 14 Calculating the average absolute deviation

For the data 5, 3, 17, 1, 4:
a. **calculate the median absolute deviation**
b. **calculate the mean absolute deviation.**

THINK

a. 1. Put the data in ascending order.

2. Locate the median.

WRITE

a. $1, 3, 4, 5, 17$

$1, 3, \mathbf{4}, 5, 17$

3. Calculate the absolute deviations by calculating how far away each data point is from the median, using positive distances.

x	$\lvert x - M \rvert$
1	3
3	1
4	0
5	1
17	13

4. Arrange the absolute deviations in ascending order.

0, 1, 1, 3, 13

5. Locate the median absolute deviation.

0, 1, **1**, 3, 13

6. Write the answer.

The median absolute deviation is 1.

b. 1. Calculate the mean of the data set.

$$\bar{x} = \frac{5 + 3 + 17 + 1 + 4}{5}$$

$$= \frac{30}{5}$$

$$= 6$$

2. Calculate the absolute deviations by calculating how far away each data point is from the mean, using positive distances.

x	$\lvert x - \bar{x} \rvert$
5	1
3	3
17	11
1	5
4	2

3. Calculate the mean absolute deviation.

$$MAD = \frac{1 + 3 + 11 + 5 + 2}{5}$$

$$= \frac{22}{5}$$

$$= 4.4$$

4. Write the answer.

The mean absolute deviation is 4.4.

 Resources

 eWorkbook Topic 12 Workbook (worksheets, code puzzle and project) (ewbk-13302)

Interactivity The standard deviation for a sample (int-4814)

Individual pathways

■ PRACTISE	■ CONSOLIDATE	■ MASTER
1, 2, 6, 10, 11, 12, 15	3, 4, 7, 9, 13, 16	5, 8, 14, 17

Fluency

1. **WE10** Calculate the standard deviation of each of the following data sets, correct to 2 decimal places.

   a. 3, 5, 8, 2, 7, 1, 6, 5
   b. 11, 8, 7, 12, 10, 11, 14
   c. 25, 15, 78, 35, 56, 41, 17, 24
   d. 5.2, 4.7, 5.1, 12.6, 4.8

2. **WE14a** Calculate the median absolute deviation for each of the data sets in question 1.

3. **WE14b** Calculate the mean absolute deviation for each of the following data sets, correct to 2 decimal places.

   a. 3, 5, 8, 2, 7, 1, 6, 5
   b. 25, 15, 78, 35, 56, 41, 17, 24

4. Calculate the standard deviation of each of the following data sets, correct to 2 decimal places.

a.

Score (x)	Frequency (f)
1	1
2	5
3	9
4	7
5	3

b.

Score (x)	Frequency (f)
16	15
17	24
18	26
19	28
20	27

c.

Score (x)	Frequency (f)
8	15
10	19
12	18
14	7
16	6
18	2

d.

Score (x)	Frequency (f)
65	15
66	15
67	16
68	17
69	16
70	15
71	15
72	12

5. Complete the following frequency distribution table and use it to calculate the standard deviation of the data set, correct to 2 decimal places.

Class	Class centre (x)	Frequency (f)
1–10		6
11–20		15
21–30		25
31–40		8
41–50		6

Understanding

6. **WE11** First-quarter profit increases for 8 leading companies are given below as percentages.

$$2.3, \quad 0.8, \quad 1.6, \quad 2.1, \quad 1.7, \quad 1.3, \quad 1.4, \quad 1.9$$

Calculate the mean score and the standard deviation for this set of data. Express your answers correct to 2 decimal places.

7. The heights in metres of a group of army recruits are given.

$$1.8, \quad 1.95, \quad 1.87, \quad 1.77, \quad 1.75, \quad 1.79, \quad 1.81, \quad 1.83, \quad 1.76, \quad 1.80, \quad 1.92, \quad 1.87, \quad 1.85, \quad 1.83$$

Calculate the mean score and the standard deviation for this set of data. Express your answers correct to 2 decimal places.

8. Times (to the nearest tenth of a second) for the heats in the open 100-m sprint at the school sports are given in the stem-and-leaf plot shown.
Calculate the standard deviation for this set of data and express your answer correct to 2 decimal places.

Key: $11|0 = 11.0$ s

Stem	Leaf
11	0
11	2 3
11	4 4 5
11	6 6
11	8 8 9
12	0 1
12	2 2 3
12	4 4
12	6
12	9

9. The number of outgoing phone calls from an office each day over a 4-week period is shown in the stem-and-leaf plot.

Key: $1|3 = 13$ calls

Stem	Leaf
0	8 9
1	3 4 7 9
2	0 1 3 7 7
3	3 4
4	1 5 6 7 8
5	3 8

Calculate the standard deviation for this set of data and express your answer correct to 2 decimal places.

10. **MC** A new legal aid service has been operational for only 5 weeks. The number of people who have made use of the service each day during this period is set out in the stem-and-leaf plot shown.

The standard deviation (to 2 decimal places) of these data is:

A. 6.00
B. 6.34
C. 6.47
D. 15.44

Key: 1|6 = 16 people

Stem	Leaf
0	2 4
0	7 7 9
1	0 1 4 4 4 4
1	5 6 6 7 8 8 9
2	1 2 2 3 3 3
2	7

11. **WE12** The speeds, in km/h, of the first 25 cars caught by a roadside speed camera on a particular day were:

82, 82, 84, 84, 84, 84, 85, 85, 85, 86, 86, 87, 89, 89, 89, 90, 91, 91, 92, 94, 95, 96, 99, 100, 102

The next car that passed the speed camera was travelling at 140 km/h.
Comment on the effect of the speed of this last car on the standard deviation for the data.

Reasoning

12. Explain what the standard deviation tells us about a set of data.

13. **WE13** For the data 1, 4, 5, 9, 11:
    a. calculate the standard deviation
    b. calculate the standard deviation if 7 is added to each data value. Comment on the effect.
    c. calculate the standard deviation if all data values are multiplied by 3. Comment on the effect.

14. Show using an example the effect, if any, on the standard deviation of adding a data value to a set of data that is equivalent to the mean.

Problem solving

15. If the mean for a set of data is 45 and the standard deviation is 6, determine how many standard deviations above the mean is a data value of 57.

16. Five numbers a, b, c, d and e have a mean of 12 and a standard deviation of 4.
    a. If each number is increased by 3, calculate the new mean and standard deviation.
    b. If each number is multiplied by 3, calculate the new mean and standard deviation.

17. Twenty-five students sat a test and the results for 24 of the students are given in the following stem-and-leaf plot.

Key: 1|2 = 12 marks

Stem	Leaf
0	8 9
1	1 2 3 7 8 9
2	2 3 5 6 8
3	0 1 2 4 6 8
4	0 2 5 6 8

a. If the average mark for the test was 27.84, determine the mark obtained by the 25th student.
b. Determine how many students scored higher than the median score.
c. Calculate the standard deviation of the marks, giving your answer correct to 2 decimal places.

LESSON
12.6 Comparing data sets

LEARNING INTENTIONS

At the end of this lesson you should be able to:
- choose an appropriate measure of centre and spread to analyse data
- interpret and make decisions based on measures of centre and spread.

▶ 12.6.1 Comparing data sets

- Decisions need to be made about which measure of centre and which measure of spread to use when analysing and comparing data.
- The mean is calculated using every data value in the set. The median is the middle score of an ordered set of data, so it is a more useful measure of centre when a set of data contains outliers.
- The range is determined by calculating the difference between the maximum and minimum data values, so it includes outliers. It is useful, however, when we are interested in extreme values such as high and low tides or maximum and minimum temperatures.
- The interquartile range is the difference between the upper and lower quartiles, so it does not include every data value in its calculation, but it will overcome the problem of outliers skewing data.
- The standard deviation is calculated using every data value in the set. It measures the spread around the mean.

WORKED EXAMPLE 15 Interpreting mean and standard deviation

For the two sets of data 6, 7, 8, 9, 10 and 12, 4, 10, 11, 3:
a. **calculate the mean**
b. **calculate the standard deviation**
c. **comment on the similarities and differences of each statistic.**

THINK	WRITE
a. 1. Calculate the mean of the first set of data.	a. $\bar{x}_1 = \dfrac{6+7+8+9+10}{5}$ $= 8$
2. Calculate the mean of the second set of data.	$\bar{x}_2 = \dfrac{12+4+10+11+3}{5}$ $= 8$
b. 1. Calculate the standard deviation of the first set of data.	b. $\sigma_1 = \sqrt{\dfrac{(6-8)^2+(7-8)^2+(8-8)^2+(9-8)^2+(10-8)^2}{5}}$ ≈ 1.41
2. Calculate the standard deviation of the second set of data.	$\sigma_2 = \sqrt{\dfrac{(12-8)^2+(4-8)^2+(10-8)^2+(11-8)^2+(3-8)^2}{5}}$ ≈ 3.74

---

858 Jacaranda Maths Quest 10 + 10A

c. Comment on the findings.　　c. For both sets of data the mean was the same, 8. However, the standard deviation for the second set (3.74) was much higher than the standard deviation of the first set (1.41), implying that the second set is more widely distributed than the first.

- When multiple data displays are used to display similar sets of data, comparisons and conclusions can then be drawn about the data.
- We can use **back-to-back stem-and-leaf plots** and **parallel box plots** to help compare statistics such as the median, range and interquartile range.

Parallel box plots

Back-to-back stem-and-leaf plots

Leaf	Stem	Leaf
5	0	8 9
6 6 3 2	1	4 5 6 8 9
8 4 1	2	0 4 5 7
9 8 5 3 0	3	1 6 9
7 1	4	1 3
	5	2 5

WORKED EXAMPLE 16 Comparing data sets

Below are the scores achieved by two students in eight Mathematics tests throughout the year.
John: 45, 62, 64, 55, 58, 51, 59, 62
Penny: 84, 37, 45, 80, 74, 44, 46, 50
a. Identify the student who performed better over the eight tests. Justify your answer.
b. Identify the student who was more consistent over the eight tests. Justify your answer.

THINK

a. Compare the mean for each student. The student with the higher mean performed better overall.

b. Compare the standard deviation for each student. The student with the lower standard deviation performed more consistently.

WRITE

a. John: $\bar{x} = 57$, $\sigma = 6$
Penny: $\bar{x} = 57.5$, $\sigma = 17.4$
Penny performed slightly better on average as her mean mark was higher than John's.

b. John was the more consistent student because his standard deviation was much lower than Penny's. This means that his test results were closer to his mean score than Penny's were to hers.

TI | THINK

a, b.
1. In a new problem, on a Lists & Spreadsheet page, label column A as 'john' and column B as 'penny'. Enter the data sets from the question. Press ENTER after each value.

DISPLAY/WRITE

	A john	B penny	C	D
=				
1	45	84		
2	62	37		
3	64	45		
4	55	80		
5	58	74		

CASIO | THINK

a, b.
1. On the Statistics screen, label list1 as 'John' and list2 as 'Penny'. Enter the data set as shown. Press EXE after each value.

DISPLAY/WRITE

2. To calculate only the mean and standard deviation of each data set, open a Calculator page and complete the entry lines as:
mean(*john*)
stDevPop(*john*)
mean(*penny*)
stDevPop(*penny*)
Press CTRL ENTER after each entry to get a decimal approximation.

mean(*john*)	57
stDevPop(*john*)	6
mean(*penny*)	57.5
stDevPop(*penny*)	17.4213

John: $\bar{x} = 57$, $\sigma = 6$
Penny: $\bar{x} = 57.5$, $\sigma = 17.4$ correct to 2 decimal places.

2. To calculate the mean and standard deviation of each data set, tap:
• Calc
• Two-Variable
Set values as:
• XList: main\John
• YList: main\Penny
• Freq: 1
Tap OK.
The *x*-values relate to John and the *y*-values to Penny. Scroll down to see all the statistics.

John: $\bar{x} = 57$, $\sigma = 6$
Penny: $\bar{x} = 57.5$, $\sigma = 17.4$ correct to 2 decimal places.

3. To draw the two box plots on the same Data & Statistics page, press TAB to locate the label of the horizontal axis and select the variable 'john'. Then press:
• MENU
• 1: Plot Type
• 2: Box Plot
Then press:
• MENU
• 2: Plot Properties
• 5: Add X-variable and select 'penny'.
To change the colour, place the pointer over one of the data points. Then press CTRL MENU. Then press:
• 6: Color
• 2: Fill Color
Select whichever colour you like from the palette for each of the box plots.

Penny performed slightly better overall as her mean mark was higher than John's; however, John was more consistent as his standard deviation was lower than Penny's.

To draw the two box-and-whisker plots on the same Statistics screen, tap:
• SetGraph
• Setting...
Set values as:
• Type: MedBox
• XList: main\John
• Freq: 1
Tap 2 in the row of numbers at the top of the screen.
Set values as:
• Type: MedBox
• XList: main\Penny
• Freq: 1
Tap Set.
Tap SetGraph and tick StatGraph1 and StatGraph2. Tap the graphing icon to display the graphs.

Penny performed slightly better overall as her mean mark was higher than John's; however, John was more consistent as his standard deviation was lower than Penny's.

 Resources

 eWorkbook   Topic 12 Workbook (worksheets, code puzzle and project) (ewbk-13302)

 Interactivities Individual pathway interactivity: Comparing data sets (int-4625)
          Back-to-back stem plots (int-6252)

Exercise 12.6 Comparing data sets

12.6 Quick quiz on	12.6 Exercise

Individual pathways

■ PRACTISE	■ CONSOLIDATE	■ MASTER
1, 5, 8, 10, 11, 17	2, 4, 6, 9, 12, 13, 18	3, 7, 14, 15, 16, 19, 20

Fluency

1. **WE15** For the two sets of data, 65, 67, 61, 63, 62, 60 and 56, 70, 65, 72, 60, 55:
   a. calculate the mean
   b. calculate the standard deviation
   c. comment on the similarities and differences.

2. A bank surveys the average morning and afternoon waiting times for customers. The figures were taken each Monday to Friday in the morning and afternoon for one month. The stem-and-leaf plot below shows the results.

   Key: 1|2 = 1.2 minutes

Leaf: Morning	Stem	Leaf: Afternoon
7	0	7 8 8
8 6 3 1 1	1	1 1 2 4 4 5 6 6 6 7
9 6 6 6 5 5 4 3 3 1	2	2 5 5 8
9 5 2	3	1 6
5	4	
	5	7

   a. Identify the median morning waiting time and the median afternoon waiting time.
   b. Calculate the range for morning waiting times and the range for afternoon waiting times.
   c. Use the information given in the display to comment about the average waiting time at the bank in the morning compared with the afternoon.

3. In a class of 30 students there are 15 boys and 15 girls. Their heights are measured in metres and are listed below.
   Boys: 1.65, 1.71, 1.59, 1.74, 1.66, 1.69 1.72, 1.66, 1.65, 1.64, 1.68, 1.74, 1.57, 1.59, 1.60
   Girls: 1.66, 1.69, 1.58, 1.55, 1.51, 1.56, 1.64, 1.69, 1.70, 1.57, 1.52, 1.58, 1.64, 1.68, 1.67

   Display this information in a back-to-back stem-and-leaf plot and comment on their height distribution.

4. The stem-and-leaf plot shown is used to display the number of vehicles sold by the Ford and Hyundai dealerships in a Sydney suburb each week for a three-month period.

   Key: 1|5 = 15 vehicles

Leaf: Ford	Stem	Leaf: Hyundai
7 4	0	3 9
9 5 2 2 1 0	1	1 1 1 6 6 8
8 5 4 4	2	2 2 7 9
0	3	5

   a. State the median of both distributions.
   b. Calculate the range of both distributions.
   c. Calculate the interquartile range of both distributions.
   d. Show both distributions on a box plot.

5. The box plot shown displays statistical data for two AFL teams over a season.

a. State the team that had the higher median score.
b. Determine the range of scores for each team.
c. For each team calculate the interquartile range.

Understanding

6. Tanya measures the heights (in m) of a group of Year 10 boys and girls and produces the following five-point summaries for each data set.

   Boys: 1.45, 1.56, 1.62, 1.70, 1.81
   Girls: 1.50, 1.55, 1.62, 1.66, 1.73

a. Draw a box plot for both sets of data and display them on the same scale.
b. Calculate the median of each distribution.
c. Calculate the range of each distribution.
d. Calculate the interquartile range for each distribution.
e. Comment on the spread of the heights among the boys and the girls.

7. The box plots show the average daily sales of cold drinks at the school canteen in summer and winter.

a. Calculate the range of sales in both summer and winter.
b. Calculate the interquartile range of the sales in both summer and winter.
c. Comment on the relationship between the two data sets, both in terms of measures of centre and measures of spread.

8. **MC** Andrea surveys the age of people at two movies being shown at a local cinema. The box plot shows the results.
   Select which of the following conclusions could be drawn based on the information shown in the box plot.

   A. Movie A attracts an older audience than Movie B
   B. Movie B attracts an older audience than Movie A.
   C. Movie A appeals to a wider age group than Movie B.
   D. Movie B appeals to a wider age group than Movie A.

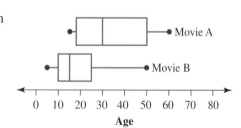

9. **MC** The figures below show the age of the first 10 men and women to finish a marathon.

Men: 28, 34, 25, 36, 25, 35, 22, 23, 40, 24

Women: 19, 27, 20, 26, 30, 18, 28, 25, 28, 22

Choose which of the following statements are correct.

Note: There may be more than one correct answer.

A. The mean age of the men is greater than the mean age of the women.

B. The range is greater among the men than among the women.

C. The interquartile range is greater among the men than among the women.

D. The standard deviation is greater among the men than among the women.

Reasoning

10. **WE16** Cory recorded his marks for each test that he did in English and Science throughout the year.

English: 55, 64, 59, 56, 62, 54, 65, 50

Science: 35, 75, 81 32, 37, 62, 77, 75

a. Identify the subject in which Cory received a better average. Justify your answer.

b. Identify the subject in which Cory performed more consistently. Justify your answer.

11. Draw an example of a graph that is:

a. symmetrical

b. positively skewed with one mode

c. negatively skewed with two modes.

12. The police set up two radar speed checks on a back street of Sydney and on a main road. In both places the speed limit is 60 km/h. The results of the first 10 cars that have their speed checked are given below.

Back street: 60, 62, 58, 55, 59, 56, 65, 70, 61, 64

Main road: 55, 58, 59, 50, 40, 90, 54, 62, 60, 60

a. Calculate the mean and standard deviation of the readings taken at each point.

b. Identify the road where drivers are generally driving faster. Justify your answer.

c. Identify the road where the spread of readings is greater. Justify your answer.

13. In boxes of Smarties it is advertised that there are 50 Smarties in each box. Two machines are used to distribute the Smarties into the boxes. The results from a sample taken from each machine are shown in the stem-and-leaf plot.

Key: $5|1 = 51$ $5*|6 = 56$

Leaf: Machine A	Stem	Leaf: Machine B
4	4	
9 9 8 7 7 6 6 5	4*	5 7 8 9 9 9 9 9 9 9
4 3 2 2 2 1 1 1 0 0 0 0 0 0	5	0 0 0 0 0 1 1 1 1 1 2 2 3
5 5	5*	9

a. Display the data from both machines on parallel box plots.

b. Calculate the mean and standard deviation of the number of Smarties distributed from both machines.

c. State which machine is the more dependable. Justify your answer.

14. Nathan and Timana are wingers in their local rugby league team. The number of tries they have scored in each season are listed below.

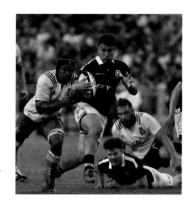

Nathan: 25, 23, 13, 36, 1, 8, 0, 9, 16, 20

Timana: 5, 10, 12, 14, 18, 11, 8, 14, 12, 19

a. Calculate the mean number of tries scored by each player.
b. Calculate the range of tries scored by each player. Justify your answer.
c. Calculate the interquartile range of tries scored by each player. Justify your answer.
d. State which player would you consider to be the more consistent. Justify your answer.

15. Year 10 students at Merrigong High School sit exams in Science and Maths. The results are shown in the table.

Mark	Number of students in Science	Number of students in Maths
51–60	7	6
61–70	10	7
71–80	8	12
81–90	8	9
91–100	2	6

a. Determine if either distribution is symmetrical.
b. If either distribution is not symmetrical, state whether it is positively or negatively skewed.
c. Discuss the possible reasons for any skewness.
d. State the modal class of each distribution.
e. Determine which subject has the greater standard deviation greater. Explain your answer.

16. A new drug for the relief of cold symptoms has been developed. To test the drug, 40 people were exposed to a cold virus. Twenty patients were then given a dose of the drug while another 20 patients were given a placebo.

(In medical tests a control group is often given a *placebo* drug. The subjects in this group believe that they have been given the real drug but in fact their dose contains no drug at all.)

All participants werc then asked to indicate the time when they first felt relief of symptoms. The number of hours from the time the dose was administered to the time when the patients first felt relief of symptoms are detailed below.

Group A (drug)

25,   29,   32,   45,   18,   21,   37,   42,   62,   13,
42,   38,   44,   42,   35,   47,   62,   17,   34,   32

Group B (placebo)

25,   17,   35,   42,   35,   28,   20,   32,   38,   35,
34,   32,   25,   18,   22,   28,   21,   24,   32,   36

a. Display the data on a back-to-back stem-and-leaf plot.
b. Display the data for both groups on a parallel box plot.
c. Make comparisons of the data. Use statistics in your answer.
d. Explain if the drug works. Justify your answer.
e. Determine other considerations that should be taken into account when trying to draw conclusions from an experiment of this type.

Problem solving

17. The heights of Year 10 and Year 12 students (to the nearest centimetre) are being investigated. The results of some sample data are shown in the table.

Year 10	160	154	157	170	167	164	172	158	177	180	175	168	159	155	163	163	169	173	172	170
Year 12	160	172	185	163	177	190	183	181	176	188	168	167	166	177	173	172	179	175	174	180

    a. Draw a back-to-back stem-and-leaf plot.

    b. Draw a parallel box plot.

    c. Comment on what the plots tell you about the heights of Year 10 and Year 12 students.

18. Kloe compares her English and Maths marks. The results of eight tests in each subject are shown below.

    English: 76, 64, 90, 67, 83, 60, 85, 37

    Maths: 80, 56, 92, 84, 65, 58, 55, 62

    a. Calculate Kloe's mean mark in each subject.

    b. Calculate the range of marks in each subject.

    c. Calculate the standard deviation of marks in each subject.

    d. Based on the above data, determine the subject that Kloe has performed more consistently in.

19. A sample of 50 students was surveyed on whether they owned an iPad or a mobile phone. The results showed that 38 per cent of the students owned both.

    Sixty per cent of the students owned a mobile phone and there were four students who had an iPad only.

    Evaluate the percentage of students that did not own a mobile phone or an iPad.

20. The life expectancy of non-Aboriginal and non–Torres Strait Islander people in Australian states and territories is shown on the box plot below.

Life expectancy of non-Aboriginal and non–Torres Strait Islander people in Australian states and territories

The life expectancies of Aboriginal and Torres Strait Islander people in each of the Australian states and territories are 56, 58.4, 51.3, 57.8, 53.9, 55.4 and 61.0.

    a. Draw parallel box plots on the same axes. Compare and comment on your results.

    b. Comment on the advantage and disadvantage of using a box plot.

LESSON
12.7 Populations and samples

LEARNING INTENTIONS

At the end of this lesson you should be able to:
- describe the difference between populations and samples
- recognise the difference between a census and a survey and identify a preferred method in different circumstances.

12.7.1 Populations

- The term **population** refers to a complete set of individuals, objects or events belonging to some category.
- When data are collected from a *whole* population, the process is known as a **census**.
  - It is often not possible, nor cost-effective, to conduct a census.
  - For this reason, **samples** have to be selected carefully from the population. A sample is a subset of a population.

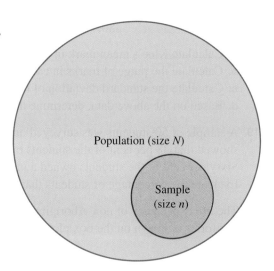

Population (size N)

Sample (size n)

WORKED EXAMPLE 17 Identifying problems with collecting data on populations

List some problems you might encounter in trying to collect data on the population of possums in a local area.

THINK	WRITE
Consider how the data might be collected and the problems in obtaining these data.	Firstly, it would be almost impossible to find all of the possums in a local area in order to count them. Secondly, possums are likely to stray into neighbouring areas, making it impossible to know if they belong to the area being observed. Thirdly, possums are more active at night-time, making it harder to detect their presence.

12.7.2 Samples

- Surveys are conducted using samples. Ideally the sample should reveal generalisations about the population.
- The sample selected to be surveyed should be chosen without **bias**, as this may result in a sample that is not representative of the whole population.

For example, the students conducting the investigation decide to choose a sample of 12 fellow students. Although it would be simplest to choose 12 of their friends as the sample, this would introduce bias since they would not be representative of the population as a whole.

- A random sample is generally accepted as being an ideal representation of the population from which it was drawn. However, it must be remembered that different random samples from the same population can produce different results.

  This means that we must be cautious about making predictions about a population, as results of surveys conducted using random samples may vary.

- A sample size must be sufficiently large. As a general rule, the sample size should be about $\sqrt{N}$, where N is the size of the population.

- If the sample size is too small, the conclusions that are drawn from the sample data may not reflect the population as a whole.

WORKED EXAMPLE 18 Determining statistics from samples

A die was rolled 50 times and the following results were obtained.

6, 5, 3, 1, 6, 2, 3, 6, 2, 5, 3, 4, 1, 3, 2, 6, 4, 5, 5, 4, 3, 1, 2, 1, 6,
4, 5, 2, 3, 6, 1, 5, 3, 3, 2, 4, 1, 4, 2, 3, 2, 6, 3, 4, 6, 2, 1, 2, 4, 2

a. **Determine the mean of the population (to 1 decimal place).**

b. **A suitable sample size for this population would be 7** $\left(\sqrt{50} \approx 7.1 \right)$.
   i. **Select a random sample of 7 scores, and determine the mean of these scores.**
   ii. **Select a second random sample of 7 scores, and determine the mean of these.**
   iii. **Select a third random sample of 20 scores, and determine the mean of these.**

c. **Comment on your answers to parts a and b.**

THINK	WRITE
a. Calculate the mean by first finding the sum of all the scores, then dividing by the number of scores (50).	a. Population mean $= \dfrac{\sum x}{n}$ $= \dfrac{169}{50}$ $= 3.4$
b. i. Use a calculator to randomly generate 7 scores from 1 to 50. Relate these numbers back to the scores, then calculate the mean.	b. i. The 7 scores randomly selected are numbers 17, 50, 40, 34, 48, 12, 19 in the set of 50 scores. These correspond to the scores: 4, 2, 3, 3, 2, 4, 5. The mean of these scores $= \dfrac{23}{7} \approx 3.3$.
ii. Repeat b i to obtain a second set of 7 randomly selected scores. This second set of random numbers produced the number 1 twice. Try again. Another attempt produced the number 14 twice. Try again. A third attempt produced 7 different numbers. This set of 7 random numbers will then be used to, again, calculate the mean of the scores.	ii. Ignore the second and third attempts to select 7 random numbers because of repeated numbers. The second set of 7 scores randomly selected is numbers 16, 49, 2, 42, 31, 11, 50 of the set of 50. These correspond to the scores: 6, 4, 5, 6, 1, 3, 2. The mean of these scores $= \dfrac{27}{7} \approx 3.9$.

iii. Repeat for a randomly selected 20 scores.

c. Comment on the results.

iii. The set of 20 randomly selected numbers produced a total of 68.

$$\text{Mean of 20 random scores} = \frac{68}{20} = 3.4$$

c. The population mean is 3.4.

The means of the two samples of 7 are 3.3 and 3.9. This shows that, even though the samples are randomly selected, their calculated means may be different.

The mean of the sample of 20 scores is 3.4.

This indicates that by using a bigger sample the result is more accurate than those obtained with the smaller samples.

TI	THINK	DISPLAY/WRITE

a.

1. In a new document, on a Lists & Spreadsheet page, label column A as 'die'. Enter the data from the question.

2. Although you can find many summary statistics, to find the mean only, open a Calculator page and press:
- MENU
- 6: Statistics
- 3: List Math
- 3: Mean

Press VAR and select 'die', then press CTRL ENTER to get a decimal approximation.

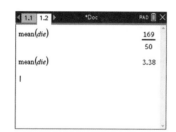

The mean of the 50 die rolls is 3.38.

CASIO	THINK	DISPLAY/WRITE

a.

1. On the Statistics screen, label list1 as 'die'. Enter the data from the question. Press EXE after each value.

2. To find the statistics summary, tap:
- Calc
- One-Variable
Set:
- XList: main\die
- Freq: 1
Tap OK.

The mean of the 50 die rolls is 3.38.

b.

i–ii. To generate a random sample of 7 scores, on the Calculator page, press:
- MENU
- 5: Probability
- 4: Random
- 2: Integer

Type the entry line as:
randInt (1,6,7)
Then press ENTER. This randomly generates 7 numbers between 1 and 6.
Press:
- MENU
- 6: Statistics
- 3: List Math
- 3: Mean

Complete the entry line as:
mean(ans)
Then press CTRL ENTER to get a decimal approximation of the mean.
Repeat this with another set of random numbers.

b.

i–ii.

The mean of the first sample of 7 rolls is 3.71, and the mean of the second sample of 7 rolls is 3.29, correct to 2 decimal places.

b.

i–ii. To generate a random sample of 7 dice rolls, on the Main screen, type the entry line as:
randList (7,1,6)
Then press EXE. The randList can be found in the Catalog on the Keyboard. This randomly generates 7 numbers between 1 and 6.

Tap:
- Action
- List
- Statistics
- Mean

Highlight the random list, including the brackets, and drag it to the 'mean('. Close the bracket and press EXE.
Repeat this with another set of random numbers.

b.

i–ii.

The mean of the first sample of 7 rolls is 4.71; the mean of the second sample of 7 rolls is also 4.71, correct to 2 decimal places.

iii. To repeat the procedure with 20 randomly generated values, change the first entry line to:
randList(1,6,20)
Follow the remaining steps to calculate the mean.

iii.

The mean of the third sample of 20 rolls is 3.15.

iii. To repeat the procedure with 20 randomly generated values, change the first entry line to:
randList(20,1,6)
Follow the remaining steps to calculate the mean.

iii.

The mean of the third sample of 20 rolls is 2.65.

c.

c. This indicates that the results obtained from a bigger sample are more accurate than those from smaller samples.

c.

c. This indicates that the results obtained from a bigger sample are more accurate than those from smaller samples.

▶ 12.7.3 To sample or to conduct a census?

eles-4948

- The particular circumstances determine whether data are collected from a population, or from a sample of the population.

  For example, suppose you collected data on the height of every Year 10 student in your class. If your class was the only Year 10 class in the school, your class would be the population. If, however, there were several Year 10 classes in your school, your class would be a sample of the Year 10 population.

- Worked example 18 showed that different random samples can produce different results. For this reason, it is important to acknowledge that there could be some uncertainty when using sample results to make predictions about the population.

WORKED EXAMPLE 19 Stating if the information was obtained by census or survey

For each of the following situations, state whether the information was obtained by census or survey. Justify why that particular method was used.

a. A roll call is conducted each morning at school to determine which students are absent.
b. TV ratings are collected from a selection of viewers to discover the popular TV shows.
c. Every hundredth light bulb off an assembly production line is tested to determine the life of that type of light bulb.
d. A teacher records the examination results of her class.

THINK	WRITE
a. Every student is recorded as being present or absent at the roll call.	a. This is a census. If the roll call only applied to a sample of the students, there would not be an accurate record of attendance at school. A census is essential in this case.
b. Only a selection of the TV audience contributed to these data.	b. This is a survey. To collect data from the whole viewer population would be time-consuming and expensive. For this reason, it is appropriate to select a sample to conduct the survey.
c. Only 1 bulb in every 100 is tested.	c. This is a survey. Light bulbs are tested to destruction (burn-out) to determine their life. If every bulb was tested in this way, there would be none left to sell! A survey on a sample is essential.
d. Every student's result is recorded.	d. This is a census. It is essential to record the result of every student.

---

on Resources

Exercise 12.7 Populations and samples

12.7 Quick quiz on   **12.7 Exercise**

Individual pathways

■ PRACTISE	■ CONSOLIDATE	■ MASTER
1, 4, 5, 8, 11	2, 6, 9, 12	3, 7, 10, 13

Fluency

1. **WE17** List some of the problems you might encounter in trying to collect data from the following populations.

   a. The life of a laptop computer battery
   b. The number of dogs in your neighbourhood
   c. The number of fish for sale at the fish markets
   d. The average number of pieces of popcorn in a bag of popcorn

2. **WE18** The data below show the results of the rolled die from Worked example 18.

   6, 5, 3, 1, 6, 2, 3, 6, 2, 5, 3, 4, 1, 3, 2, 6, 4, 5, 5, 4, 3, 1, 2, 1, 6,
   4, 5, 2, 3, 6, 1, 5, 3, 3, 2, 4, 1, 4, 2, 3, 2, 6, 3, 4, 6, 2, 1, 2, 4, 2

   The mean of the population is 3.4. Select your own samples for the following questions.

   a. Select a random sample of 7 scores, and determine the mean of these scores.
   b. Select a second random sample of 7 scores, and determine the mean of these.
   c. Select a third random sample of 20 scores, and determine the mean of these.
   d. Comment on your answers to parts a, b and c.

3. **WE19** In each of the following scenarios, state whether the information was obtained by census or survey. Justify why that particular method was used.

   a. Seating for all passengers is recorded for each aeroplane flight.
   b. Movie ratings are collected from a selection of viewers to discover the best movies for the week.
   c. Every hundredth soft drink bottle off an assembly production line is measured to determine the volume of its contents.
   d. A car driving instructor records the number of hours each learner driver has spent driving.

4. For each of the following, state whether a census or a survey has been used.

   a. Two hundred people in a shopping centre are asked to nominate the supermarket where they do most of their grocery shopping.
   b. To find the most popular new car on the road, new car buyers are asked what make and model they purchased.
   c. To find the most popular new car on the road, data are obtained from the transport department.
   d. Your Year 10 Maths class completed a series of questions on the amount of maths homework for Year 10 students.

Understanding

5. To conduct a statistical investigation, Gloria needs to obtain information from 630 students.

   a. Determine the appropriate sample size.
   b. Describe a method of generating a set of random numbers for this sample.

6. A local council wants the opinions of its residents regarding its endeavours to establish a new sporting facility for the community. It has specifically requested all residents over 10 years of age to respond to a set of on-line questions.

   a. State if this is a census or a survey.
   b. Determine what problems could be encountered when collecting data this way.

7. A poll was conducted at a school a few days before the election for Head Boy and Head Girl. After the election, it was discovered that the polls were completely misleading. Explain how this could have happened.

Reasoning

8. A sampling error is said to occur when results of a sample are different from those of the population from which the sample was drawn. Discuss some factors which could introduce sampling errors.

9. Since 1961, a census has been conducted in Australia every 5 years. Some people object to the census on the basis that their privacy is being invaded. Others say that the expense involved could be directed to a better cause. Others say that a sample could obtain statistics which are just as accurate.
   State your views on this. Justify your statements.

10. Australia has a very small population compared with other countries such as China and India. These are the world's most populous nations, so the problems we encounter in conducting a census in Australia would be insignificant compared with those encountered in those countries.
    Suggest what different problems authorities would come across when conducting a census in countries with large populations.

Problem solving

11. The game of Lotto involves picking the same 6 numbers in the range 1 to 45 as have been randomly selected by a machine containing 45 numbered balls. The balls are mixed thoroughly, then 8 balls are selected representing the 6 main numbers, plus 2 extra numbers, called supplementary numbers.
    Here is a list of the number of times each number had been drawn over a period of time, and also the number of weeks since each particular number has been drawn.

Number of weeks since each number drawn							
1	2	3	4	5	6	7	8
1	5	2	1	1	7	-	4
9	10	11	12	13	14	15	16
3	3	1	5	5	7	-	4
17	18	19	20	21	22	23	24
9	-	9	2	2	12	10	8
25	26	27	28	29	30	31	32
5	11	17	2	3	3	-	22
33	34	35	36	37	38	39	40
4	3	-	1	12	-	6	-
41	42	43	44	45			
6	1	7	-	31			

Number of times each number drawn since draw 413							
1	2	3	4	5	6	7	8
246	238	244	227	249	241	253	266
9	10	11	12	13	14	15	16
228	213	250	233	224	221	240	223
17	18	19	20	21	22	23	24
217	233	240	226	238	240	253	228
25	26	27	28	29	30	31	32
252	239	198	229	227	204	230	226
33	34	35	36	37	38	39	40
246	233	232	251	222	221	219	259
41	42	43	44	45			
245	242	237	221	224			

    If these numbers are randomly chosen, explain the differences shown in the tables.

12. A sample of 30 people was selected at random from those attending a local swimming pool. Their ages (in years) were recorded as follows.

19, 7, 58, 41, 17, 23,

62, 55, 40, 37, 32, 29,

21, 18, 16, 10, 40, 36,

33, 59, 65, 68, 15, 9,

20, 29, 38, 24, 10, 30

a. Calculate the mean and the median age of the people in this sample.
b. Group the data into class intervals (0–9 etc.) and complete the frequency distribution table.
c. Use the frequency distribution table to estimate the mean age.
d. Calculate the cumulative frequency and hence plot the ogive.
e. Estimate the median age from the ogive.
f. Compare the mean and median of the original data in part a with the estimates of the mean and the median obtained for the grouped data in parts c and e.
g. Were the estimates good enough? Explain your answer.

13. The typing speed (words per minute) was recorded for a group of Year 8 and Year 10 students. The results are displayed in this back-to-back stem plot.

Key: 2|6 = 26 wpm

Leaf: Year 8	Stem	Leaf: Year 10
9 9	0	
9 8 6 5 4 2 0	1	7 9
9 8 8 6 4 2 1 0 0	2	2 3 6 8 9
9 7 7 6 4 1 0	3	0 2 4 5 5 7 8 8
8 6 5 2 0	4	1 2 5 8 8 9 9
	5	0 3 5 7 8
	6	0 0 3

Write a report comparing the typing speeds of the two groups.

LESSON
12.8 Evaluating inquiry methods and statistical reports

LEARNING INTENTIONS

At the end of this lesson you should be able to:
- describe the difference between primary and secondary data collection
- identify misleading errors or graphical techniques in data
- evaluate the accuracy of a statistical report.

▶ 12.8.1 Data collection methods

eles-4963

- Data can be collected in different ways. The manner in which you collect data can affect the validity of the results you determine.
- **Primary data** is collected firsthand by observation, measurement, survey, experiment or simulation. and is owned by the data collector until it is published.
- **Secondary data** is obtained from external sources such as journals, newspapers, the internet, or any other previously collected data.

When collecting primary data through experiments, or using secondary data, considerations need to be given to:
- **Validity**. Results of valid investigations are supported by other investigations.
- **Reliability**. Data is reliable if the same data can be collected when the investigation is repeated. If the data cannot be repeated by other investigators, the data may not be valid or true.
- **Accuracy** and **precision**. When collecting primary data through experiments, the accuracy of the data is how close it is to a known value. Precise data is when multiple measurements of the same investigation are close to each other.
- **Bias**. When collecting primary data in surveys it is important to ensure your sample is representative of the population you are studying. When using secondary data you should consider who collected the data, and whether those researchers had any intentional or unintentional influence on the data.

WORKED EXAMPLE 20 Choosing appropriate collection methods

You have been given an assignment to investigate which year level uses the school library, after school, the most.

a. Explain whether it is more appropriate to use primary or secondary data in this case. Justify your choice.

b. Describe how the data could be collected. Discuss any problems which might be encountered.

THINK

a. No records have been kept on library use.

WRITE

a. Since records are not kept on the library use, secondary data is not an option.
Therefore, primary data is more appropriate to use in this case.

---

b. The data can be collected via a questionnaire or in person.

b. A questionnaire could be designed and distributed to a randomly-chosen sample. The problem here would be the non-return of the forms.
Observation could be used to personally interview students as they entered the library. This would take more time, but random interview times could be selected.

WORKED EXAMPLE 21 Choosing appropriate collection methods

State which method would be the most appropriate to collect the following data. Suggest an alternative method in each case.
a. The number of cars parked in the staff car park each day.
b. The mass of books students carry to school each day.
c. The length a spring stretches when weights are added to it.
d. The cost of mobile phone plans with various network providers.

THINK	WRITE
a. Observation	**a.** The best way would probably be observation by visiting the staff car park to count the number of cars there. An alternative method would be to conduct a census of all workers to ask if they parked in the staff car park. This method may be prone to errors as it relies on accurate reporting by many people.
b. Measurement	**b.** The mass of the books could be measured by weighing each student's pack on scales. A random sample would probably yield a reasonably accurate result.
c. Experiment	**c.** Conduct an experiment and measure the extension of the spring with various weights. There are probably no alternatives to this method as results will depend upon the type of spring used.
d. Internet search	**d.** An internet search would enable data to be collected. Alternatively, a visit to mobile phone outlets would yield similar results.

⏵ 12.8.2 Analysing statistical graphs

eles-4964

- Data can be graphed in a variety of ways — line graphs, bar graphs, histograms, stem plots, box plots, and so on.
- Because graphs give a quick visual impression, the temptation is to not look at them in great detail. Often graphs can be quite misleading.
- It is easy to manipulate a graph to give an impression which is supported by the creator of the graph. This is achieved by careful choice of scale on the horizontal and vertical axes.
  - Shortening the horizontal axis tends to highlight the increasing/decreasing nature of the trend of the graph. Lengthening the vertical axis tends to have the same effect.
  - Lengthening the horizontal and shortening the vertical axes tends to level out the trends.

The report shows the annual change in median house prices in the local government areas (LGA) of Queensland from 2019–20 to 2020–21.

a. Draw a bar graph which would give the impression that the percentage annual change was much the same throughout the whole state.

b. Construct a bar graph to give the impression that the percentage annual change in Brisbane was far greater than that in the other local government areas.

HOUSES Suburb/locality	Median house price		Annual change
	2020–21	2019–20	
Brisbane (LGA)	$700,000	$627,000	11.6%
Ipswich City (LGA)	$323,000	$310,000	4.2%
Redland City (LGA)	$467,500	$435,000	7.5%
Logan City (LGA)	$360,000	$340,000	5.9%
Moreton Bay (LGA)	$399,000	$372,000	7.3%
Gold Coast City (LGA)	$505,000	$465,000	8.6%
Toowoomba (LGA)	$360,000	$334,500	7.6%
Sunshine Coast (LGA)	$470,000	$445,000	5.6%
Fraser Coast (LGA)	$322,500	$312,500	3.2%
Bundaberg (LGA)	$282,000	$275,000	2.5%
Gladstone (LGA)	$286,000	$286,000	0.0%
Rockhampton (LGA)	$267,000	$254,000	5.1%
Mackay (LGA)	$398,000	$383,000	3.9%
Townsville City (LGA)	$375,000	$359,000	4.5%
Cairns (LGA)	$400,000	$389,000	2.8%

THINK

a. To flatten out trends, lengthen the horizontal axis and shorten the vertical axis.

WRITE/DRAW

a.

% house price changes in QLD 2019–20 to 2020-21

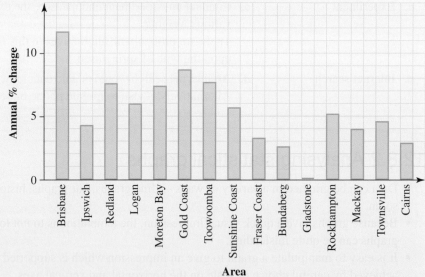

b. To accentuate trends, shorten the horizontal axis and lengthen the vertical axis.

b.

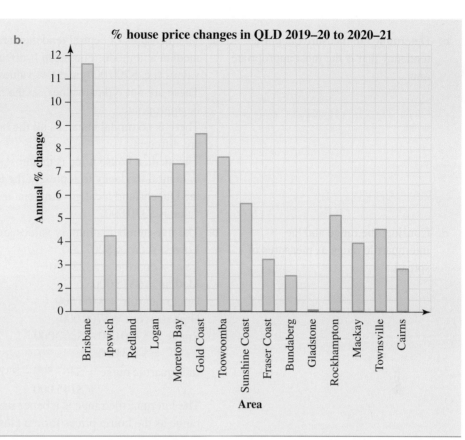

% house price changes in QLD 2019–20 to 2020–21

WORKED EXAMPLE 23 Comparing and choosing statistical measures

Consider the data displayed in the table of Worked example 22. Use the data collected for the median house prices in 2020–21.
a. Explain whether the data would be classed as primary or secondary data.
b. Explain why the data shows median house prices rather than the mean or modal house price.
c. Calculate a measure of central tendency for the data. Explain the reason for this choice.
d. Give a measure of spread of the data, giving a reason for the particular choice.
e. Display the data in a graphical form, explaining why this particular form was chosen.

THINK	WRITE
a. These are data that have been collected by someone else.	**a.** These are secondary data because they have been collected by someone else.
b. Median is the middle price, mean is the average price, and mode is the most frequently-occurring price.	**b.** The median price is the middle value. It is not affected by outliers as the mean is. The modal house price may only occur for two house sales with the same value. On the other hand, there may not be any mode. The median price is the most appropriate in this case.

c. Determine the measure of central tendency that is the most appropriate one.

c. The measures of central tendency are the mean, median and mode. The mean is affected by high values (i.e. $700\,000$) and low values (i.e. $282\,000$). These are not typical values, so the mean would not be appropriate.

There is no modal value, as all the house prices are different.

The median house price is the most suitable measure of central tendency to represent the house prices in the Queensland local government areas. The median value is $370\,000.

d. Consider the range and the interquartile range as measures of spread.

d. The five-number summary values are:

Lowest score $= \$267\,000$

Lowest quartile $= \$322\,500$

Median $= \$375\,000$

Upper quartile $= \$467\,500$

Highest score $= \$700\,000$

Range $= \$700\,000 - \$267\,000$

$\qquad = \$433\,000$

Interquartile range $= \$467\,500 - \$322\,500$

$\qquad\qquad = \$145\,000$

The interquartile range is a better measure for the range as the house prices form a cluster in this region.

e. Consider the graphing options.

e. Of all the graphing options, the box plot seems the most appropriate as it shows the spread of the prices as well as how they are grouped around the median price.

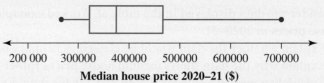

Median house price 2020–21 ($)

WORKED EXAMPLE 24 Analysing data and use of statistics to interpret results

The following data is the heights of the members of the Australian women's national basketball team (in metres):

$$1.73, 1.65, 1.8, 1.83, 1.96, 1.88, 1.63, 1.88, 1.83, 1.88, 1.8, 1.96$$

Provide calculations and explanations as evidence to verify or refute the following statements.

a. The mean height of the team is greater than their median height.

b. The range of the heights of the 12 players is almost 3 times their interquartile range.

c. Only 5 players are on the court at any one time. A team of 5 players can be chosen such that their mean, median and modal heights are all the same.

THINK

a. 1. Calculate the mean height of the 12 players.

WRITE

a. Mean $= \dfrac{\sum x}{n} = \dfrac{21.83}{12} = 1.82$ m

2. Order the heights to determine the median.

The heights of the players, in order, are: 1.63, 1.65, 1.73, 1.8, 1.8, 1.83, 1.83, 1.88, 1.88, 1.88, 1.96, 1.96. There are 12 scores, so the median is the average of the 6th and 7th scores.

$$\text{Median} = \frac{1.83 + 1.83}{2} = 1.83 \text{ m}$$

3. Comment on the statement.

The mean is 1.82 m, while the median is 1.83 m. This means that the mean is less than the median, so the statement is not true.

b. 1. Determine the range and the interquartile range of the 12 heights.

b. Range $= 1.96 - 1.63 = 0.33$ m
The lower quartile is the average of the 3rd and 4th scores.

$$\text{Lower quartile} = \frac{1.73 + 1.8}{2} = 1.765 \text{ m}$$

The upper quartile is average of the 3rd and 4th scores from the end.

$$\text{Upper quartile} = \frac{1.88 + 1.88}{2} = 1.88 \text{ m}$$

Interquartile range $= 1.88 - 1.765 = 0.115$ m

2. Compare the two values.

Range $= 0.33$ m

Interquartile range $= 0.115$ m

$$\frac{\text{Range}}{\text{Interquartile range}} = \frac{0.33}{0.115} = 2.9$$

3. Comment on the statement.

Range $= 2.9 \times$ interquartile range
This is almost 3 times, so the statement is true.

c. 1. Choose 5 players whose mean, median and modal heights are all equal. Trial and error is appropriate here. There may be more than one answer.

c. Three players have a height of 1.88 m. If a player shorter and one taller are chosen, both the same measurement from 1.88 m, this would make the mean, median and mode all the same. Choose players with heights:

1.8, 1.88, 1.88, 1.88, 1.96

$$\text{Mean} = \frac{9.4}{5} = 1.88 \text{ m}$$

$$\text{Median} = 3\text{rd score} = 1.88 \text{ m}$$

$$\text{Mode} = \text{most frequent score} = 1.88 \text{ m}$$

2. Comment on the statement.

The 5 players with heights, 1.8 m, 1.88 m, 1.88 m, 1.88 m and 1.96 m have a mean, median and modal height of 1.88 m. It is true that a team of 5 such players can be chosen.

| TI | THINK | DISPLAY/WRITE | CASIO | THINK | DISPLAY/WRITE |

a.

1. In a new problem, on a Lists & Spreadsheet page, label column A as 'heights'. Enter the data from the question.

a.

	A heights	B	C	D
=				
1	1.73			
2	1.65			
3	1.8			
4	1.83			
5	1.96			

a.

1. On the Statistics screen, label list1 as 'heights'. Enter the data in the table as shown. Press EXE after each value.

a.

2. Open a Calculator page and complete the entry lines as:
mean(*heights*)
median(*heights*)
Press ENTER after each entry.

mean(*heights*)   1.81917
median(*heights*)   1.83

The mean heights are less than the median heights, so the statement is false.

2. To find the statistics summary, tap:
• Calc
• One-Variable
Set values as:
• XList: main\heights
• Freq: 1
Tap OK.
The mean and median can be found in the list.

The mean heights are less than the median heights, so the statement is false.

b.

To find all the summary statistics, open the Calculator page and press:
• MENU
• 6: Statistics
• 1: Stat Calculations
• 1: One-Variable Statistics...
Select 1 as the number of lists. Then on the One-Variable Statistics page, select 'heights' as the X1 List and leave the frequency as 1. Leave the next two fields empty, TAB to OK and press ENTER.

b.

The range is max − min = $1.96 - 1.63 = 0.33$.
$Q_1 = 1.765$ and $Q_3 = 1.88$.
$IQR = Q_3 - Q_1$
$= 1.88 - 1.765 = 0.115$.
Now $2.9 \times IQR \approx$ range so the statement is true.

b.

More statistics can be found from the statistics summary.

b.

The range is max − min
$= 1.96 - 1.63 = 0.33$.
$Q_1 = 1.765$ and $Q_3 = 1.88$.
$IQR = Q_3 - Q_1$
$= 1.88 - 1.765 = 0.115$.
Now $2.9 \times IQR \approx$ range, so the statement is true.

⏵ 12.8.3 Statistical reports

eles-4955

• Reported data must not be simply taken at face value; all reports should be examined with a critical eye.

WORKED EXAMPLE 25 Analysing a statistical report

This is an excerpt from an article that appeared in a newspaper on Father's Day. It was reported to be national survey findings of a *Gallup Poll* of data from 1255 fathers of children aged 17 and under.

THE GREAT AUSSIE DADS SURVEY

Thinking about all aspects of your life, how happy would you say you are?

	%
I am very happy	26
I am fairly happy	49
Totally happy	**75**
Some days I'm happy and some days I'm not	21
I am fairly unhappy	3
I am very unhappy	1
Totally unhappy	4

How often, if ever, do you regret having children?

Every day	1
Most days	2
Some days	18
Never	79

Which one of these best describes the impact of having children on your relationship with your partner?

We're closer than ever	29
We don't spend as much time together as we should	40
We're more like friends now than lovers	21
We have drifted apart	6
None of the above	4

Which one of these best describes the allocation of cooking and cleaning duties in your household?

My partner does nothing/I do everything	1
I do most of it	11
We share the cooking and cleaning	42
My partner does most of it	41
I do nothing/my partner does everything	4
None of the above	1

Which of these aspects of your children's future do you have concerns about?

	%
Their safety	70
Being exposed to drugs	67
Their health	54
Bullying or cyber-bullying	50
Teenage violence	50
Their ability to afford a home	50
Alcohol consumption and binge drinking	47
Achieving academic success	47
Achieving academic success	47
Feeling pressured into sex	41
Being able to afford the lifestyle they expect to have	38
Climate change	23
Having them living with you in their mid 20s	14
None of the above	3

What is the best thing about being a dad?

The simple pleasures of family life	61
Enjoying the successes of your kids	24
The unpredictability it brings	9
The comfort of knowing that you will be looked after in later life	3
None of the above	3

Key findings

75% of Aussie dads are totally happy

79% have never regretted having children

67% are worried about their children being exposed to drugs

57% would like more intimacy with their partner

"Work–life balance is definitely an issue for dads in 2010."

David Briggs
Galaxy principal

Source: The Sunday Mail, 5 Sept. 2010, pp. 14–15.

a. **Comment on the sample chosen.**
b. **Discuss the percentages displayed.**
c. **Comment on the claim that 57% of dads would like more intimacy with their partner.**

THINK

a. How is the sample chosen? Is it truly representative of the population of Australian dads?

WRITE

a. The results of a national survey such as this should reveal the outlook of the whole nation's dads. There is no indication of how the sample was chosen, so without further knowledge we tend to accept that it is representative of the population. A sample of 1255 is probably large enough.

b. Look at the percentages in each of the categories.

b. For the first question regarding happiness, the percentages total more than 100%. It seems logical that, in a question such as this, the respondents would tick only one box, but obviously this has not been the case.

In the question regarding aspects of concern of 'your children's future', these percentages also total more than 100%. It seems appropriate here that dads would have more than one concerning area, so it is possible for the percentages to total more than 100%.

In each of the other three questions, the percentages total 100%, which is appropriate.

c. Look at the tables to try to find the source of this figure.

c. Examining the reported percentages in the question regarding 'relationship with your partner', there is no indication how a figure of 57% was determined.

Note: Frequently media reports make claims where the reader cannot confirm their truth.

WORKED EXAMPLE 26 Analysing a statistical report

This article appeared in a newspaper. Read the article, then answer the following questions.

SPONGES ARE TOXIC

Washing dishes can pose a serious health risk, with more than half of all kitchen sponges containing high levels of dangerous bacteria, research shows.

A new survey dishing the dirt on washing up shows more than 50 per cent of kitchen sponges have high levels of *E. coli*, which can cause severe cramps and diarrhoea, and *staphylococcus aureus*, which releases toxins that can lead to food poisoning or toxic shock syndrome.

Microbiologist Craig Andrew-Kabilafkas of Australian Food Microbiology said the Westinghouse study of more than 1000 households revealed germs can spread easily to freshly washed dishes.

The only way to safeguard homes from sickness was to wash utensils at very high temperatures in a dishwasher.

Source: The Sunday Mail, 5 Sept. 2010, p. 36.

a. Comment on the sample used in this survey.
b. Comment on the claims of the survey and identify any potential bias.
c. Is the heading of the article appropriate?

THINK

a. Look at sample size and selection of sample.

WRITE

a. The report claims that the sample size was more than 1000. There is no indication how the sample was selected.

The point to keep in mind is whether this sample is truly representative of the population consisting of all households. We have no way of knowing.

b. 1. Determine the results of the survey.

b. The survey claims that 50% of kitchen sponges have high levels of *E. coli*, which can cause severe medical problems.

2. Identify any potential bias.

The study was conducted by Westinghouse, so it is not surprising they recommend using a dishwasher. There is no detail of how the sample was selected.

c. Examine the heading in the light of the contents of the article.

c. The heading is quite shocking, designed to catch the attention of readers.

COLLABORATIVE TASK: Secondary data

Use the internet to source secondary data in an article that predicts the number of people likely to be infected with COVID-19. Reflect on the validity of the claims, consider the sample size used and discuss the ethical considerations of reporting this data to the wider public.

 Resources

📋 **eWorkbook**   Topic 12 Workbook (worksheets, code puzzle and project) (ewbk-13302)

🔧 **Interactivities** Individual pathway interactivity: Evaluating inquiry methods and statistical reports (int-4631)

Compare statistical reports (int-2790)

Exercise 12.8 Evaluating inquiry methods and statistical reports

learn on

12.8 Quick quiz on	**12.8 Exercise**

Individual pathways

■ PRACTISE	■ CONSOLIDATE	■ MASTER
1, 4, 6, 9, 12	2, 7, 10, 13	3, 5, 8, 11, 14

Fluency

1. You have been given an assignment to investigate which Year level has the greatest number of students who are driven to school each day by car.

a. Explain whether it is more appropriate to use primary or secondary data in this case. Justify your choice.

b. Describe how the data could be collected. Discuss any problems which might be encountered.

c. Explain whether an alternative method would be just as appropriate.

2. **WE22** You run a small company that is listed on the Australian Stock Exchange (ASX). During the past year you have given substantial rises in salary to3all your staff. However, profits have not been as spectacular as in the year before. This table gives the figures for the salary and profits for each quarter.

	1st quarter	2nd quarter	3rd quarter	4th quarter
Profits ($'000 000)	6	5.9	6	6.5
Salaries ($'000 000)	4	5	6	7

Draw two graphs, one showing profits, the other showing salaries, which will show you in the best possible light to your shareholders.

3. **WE23** The data below were collected from a real estate agent and show the sale prices of ten blocks of land in a new estate.
$150 000, $190 000, $175 000, $150 000, $650 000, $150 000, $165 000, $180 000, $160 000, $180 000

a. Calculate a measure of central tendency for the data. Explain the reason for this choice.
b. Give a measure of spread of the data, giving a reason for the particular choice.
c. Display the data in a graphical form, explaining why this particular form was chosen.
d. The real estate agent advertises the new estate land as:
   Own one of these amazing blocks of land for only $150 000 (average)!
   Comment on the agent's claims.

4. **WE24** The following data are the heights of the members of the Australian women's national basketball team (in metres):

$$1.73, 1.65, 1.8, 1.83, 1.96, 1.88, 1.63, 1.88, 1.83, 1.88, 1.8, 1.96$$

Provide calculations and explanations as evidence to verify or refute the following statements.

a. The mean height of the team is closer to the lower quartile than it is to the median.
b. Half the players have a height within the interquartile range.
c. Suggest which 5 players could be chosen to have the minimum range in heights.

5. The resting pulse of 20 female athletes was measured and is shown below.
   50   62   48   52   71   61   30   45   42   48   43   47   51   52   34   61   44   54   38   40

a. Represent the data in a distribution table using appropriate groupings.
b. Calculate the mean, median and mode of the data.
c. Comment on the similarities and differences between the three values.

6. The batting scores for two cricket players over six innings were recorded as follows.
   Player A: 31, 34, 42, 28, 30, 41
   Player B: 0, 0, 1, 0, 250, 0
   Player B was hailed as a hero for his score of 250.
   Comment on the performance of the two players.

Understanding

7. The table below shows the number of shoes of each size that were sold over a week at a shoe store.

Size	Number sold
4	5
5	7
6	19
7	24
8	16
9	8
10	7

a. Calculate the mean shoe size sold.
b. Determine the median shoe size sold.
c. Identify the modal shoe size sold.
d. Explain which measure of central tendency has the most meaning to the store proprietor.

8. **WE25&26** This report from Woolworths appeared in a newspaper.

IT'S A RECORD

- Woolworths posted 10.1% gain in annual profit to $2.02b
- 11th consecutive year of double-digit growth
- Flags 8% to 11% growth in the current financial year
- Sales rose 4.8% to $51.2b
- Wants to increase its share of the fresh food market
- Announced $700m off-market share buyback
- Final fully franked dividend 62% a share

Shares rebound

2.4% Yesterday

$
28.40
28.10
27.70
27.40
27.10
20.80
26.50
26.20
25.90
25.60

▲ May 26   ▲ Aug 26

Source: IRESS

Net profit

+$2.02b
+10.1%

+12.8%

+25.7%

+27.5%

+24.3%

$b
2
1.5
1
0.5
0

2006 2007 2008 2009 2010

Source: *The Courier Mail*, 27 Aug. 2010, pp. 40–1.

Comment on the report.

Reasoning

9. Explain the point of drawing a misleading graph in a report.

10. The graph shows the fluctuation in the Australian dollar in terms of the US dollar during the period 13 July to 13 September 2010.

    The higher the Australian dollar, the cheaper it is for Australian companies to import goods from overseas, and the cheaper they should be able to sell their goods to the Australian public.

    The manager of Company XYZ produced a graph to support his claim that, because there hasn't been much change in the Aussie dollar over that period, there hasn't been any change in the price he sells his imported goods to the Australian public.

    Draw a graph that would support his claim. Explain how you were able to achieve this effect.

AUSSIE

US 93.29¢

Source: IRESS

Source: *The Courier Mail*, 14 Sept. 2010, p. 25.

11. Two brands of light globes were tested by a consumer organisation. They obtained the following results.

    Brand A (hours lasted)
    385, 390, 425, 426, 570,
    640, 645, 730, 735, 760

    Brand B (hours lasted)
    500, 555, 560, 630, 720,
    735, 742, 770, 820, 860

    a. Complete a back-to-back stem plot for the data.
    b. State the brand that had the shortest lifetime. Justify your answer.
    c. State the brand that had the longest lifetime. Justify your answer.
    d. If you wanted to be certain that a globe you bought would last at least 500 hours, determine which brand would you buy. Show your working.

Problem solving

12. A small manufacturing plant employs 80 workers. The table below shows the structure of the plant.

Position	Salary ($)	Number of employees
Machine operator	18 000	50
Machine mechanic	20 000	15
Floor steward	24 000	10
Manager	62 000	4
Chief Executive Officer	80 000	1

a. Workers are arguing for a pay rise, but the management of the factory claims that workers are well paid because the mean salary of the factory is $22 100.
   Explain whether this is a sound argument.
b. Suppose that you were representing the factory workers and had to write a short submission in support of the pay rise.
   Explain the management's claim by providing some other statistics to support your case.

13. Look at the following bar charts and discuss why the one on the left is misleading and what characteristics the one on the right possesses that makes it acceptable.

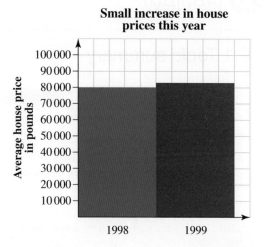

14. a. Determine what is wrong with this pie graph.

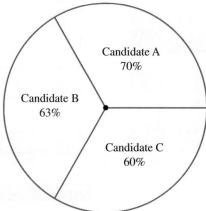

b. Explain why the following information is misleading.
Did scientists falsify research to support their own theories on global warming?
59% somewhat likely
35% very likely
26% not very likely

c. Discuss the implications of this falsification by statistics.

LESSON
12.9 Review

12.9.1 Topic summary

Populations and samples

- A population is the full set of people/things that you are collecting data on.
- A sample is a subset of a population.
- Samples must be randomly selected from a population in order for the results of the sample to accurately reflect the population.
- It is important to acknowledge that the results taken from a sample may not reflect the population as a whole. This is particularly true if the sample size is too small.
- The minimum sample size that should be used is approximately $\sqrt{N}$, where N is the size of the population.

Measures of spread

- Measures of spread describe how far the data values are spread from the centre or from each other.
- The range is the difference between the maximum and minimum data values.

Range = maxiumum value − minimum value

- The interquartile range (IQR) is the range of the middle 50% of the scores in an ordered set:

$$\text{IQR} = Q_3 - Q_1$$

where Q_1 and Q_3 are the first and third quartiles respectively.

Standard deviation

- The standard deviation is a more sophisticated measure of spread.
- The standard deviation measures how far, on average, each data value is away from the mean.
- The deviation of a data value is the difference between it and the mean $(x - \bar{x})$.
- The formula for the standard deviation is:

$$\sigma = \sqrt{\frac{\sum (x - \bar{x})^2}{n}}$$

- The standard deviation is always a positive number.

UNIVARIATE DATA

Measures of central tendency

- The three measures of central tendency are the mean, median and mode.
- The mean is the average of all values in a set of data. It is therefore affected by extreme values.

$$\bar{x} = \frac{\sum x}{n}$$

- The median is the middle value in an ordered set of data. It is located at the $\left(\frac{n+1}{2}\right)$th score.
- The mode is the most frequent value in a set of data.
- For the data set 2, 4, 5, 7, 7:

The mean is $\dfrac{2+4+5+7+7}{5} = 5$.

The median is the middle value, 5.
The mode is 7.

Box plots

- The five-number summary of a data set is a list containing:
  - the minimum value
  - the lower quartile, Q_1
  - the median
  - the upper quartile, Q_3
  - the maximum value.
- Boxplots are graphs of the five-number summary.

| The lowest score $X_{\min}$ (Lower extreme) | The lower quartile Q_1 | The median M | The upper quartile Q_3 | The greatest score $X_{\max}$ (Upper extreme) |

- Outliers are calculated and marked on the box plot. A score is considered an outlier if it falls outside the upper or lower boundary.
  Lower boundary = $Q_1 - 1.5 \times \text{IQR}$
  Upper boundary = $Q_3 + 1.5 \times \text{IQR}$

Comparing data sets

- Measures of centre and measures of spread are used to compare data sets.
- It is important to consider which measure of centre or spread is the most relevant when comparing data sets.
- For example, if there are outliers in a data set, the median will likely be a better measure of centre than the mean, and the IQR would be a better measure of spread than the range.
- Back-to-back stem-and-leaf plots can be used to compare two data sets.
- Parallel box plots can also be used to compare the spread of two or more data sets.

12.9.2 Success criteria

Tick the column to indicate that you have completed the lesson and how well you have understood it using the traffic light system.

(**Green:** I understand; **Yellow:** I can do it with help; **Red:** I do not understand.)

Lesson	Success criteria	🔘	⚪	⚫
12.2	I can calculate the mean, median and mode of data presented as ungrouped data, frequency distribution tables and grouped data.			
12.3	I can calculate the range and interquartile range of a data set.			
12.4	I can calculate the five-number summary for a set of data.			
	I can draw a box plot showing the five-number summary of a data set.			
	I can calculate outliers in a data set.			
	I can describe the skewness of distributions.			
	I can compare box plots to dot plots or histograms.			
	I can draw parallel box plots and compare sets of data.			
12.5	I can calculate the standard deviation of a small data set by hand.			
	I can calculate the standard deviation using technology.			
	I can interpret the mean and standard deviation of data.			
	I can identify the effect of outliers on the standard deviation.			
12.6	I can choose an appropriate measure of centre and spread to analyse data.			
	I can interpret and make decisions based on measures of centre and spread.			
12.7	I can describe the difference between populations and samples.			
	I can recognise the difference between a census and a survey and identify a preferred method in different circumstances.			
12.8	I can describe the difference between primary and secondary data collection.			
	I can identify misleading errors or graphical techniques in data.			
	I can evaluate the accuracy of a statistical report.			

12.9.3 Project

Cricket scores

Data are used to predict, analyse, compare and measure many aspects of the game of cricket. Attendance is tallied at every match. Players' scores are analysed to see if they should be kept on the team.

Comparisons of bowling and batting averages are used to select winners for awards. Runs made, wickets taken, no-balls bowled, the number of ducks scored in a game as well as the number of 4s and 6s are all counted and analysed after the game.

Data of all sorts are gathered and recorded, and measures of central tendency and spread are then calculated and interpreted.

Sets of data have been made available for you to analyse, and decisions based on the resultant measures can be made.

Batting averages

The following table shows the runs scored by four cricketers who are vying for selection to the state team.

Player	Runs in the last **25** matches	Mean	Median	Range	IQR
Will	13, 18, 23, 21, 9, 12, 31, 21, 20, 18, 14, 16, 28, 17, 10, 14, 9, 23, 12, 24, 0, 18, 14, 14, 20				
Rohit	2, 0, 112, 11, 0, 0, 8, 0, 10, 0, 56, 4, 8, 164, 6, 12, 2, 0, 5, 0, 0, 0, 8, 18, 0				
Marnus	12, 0, 45, 23, 0, 8, 21, 32, 6, 0, 8, 14, 1, 27, 23, 43, 7, 45, 2, 32, 0, 6, 11, 21, 32				
Ben	2, 0, 3, 12, 0, 2, 5, 8, 42, 0, 12, 8, 9, 17, 31, 28, 21, 42, 31, 24, 30, 22, 18, 20, 31				

1. Calculate the mean, median, range and IQR scored for each cricketer.
2. You need to recommend the selection of two of the four cricketers. For each player, write two points as to why you would or would not select them.
   Use statistics in your comments.

Bowling averages

The bowling average is the number of runs per wicket taken

$$\text{Bowling average} = \frac{\text{number of runs scored}}{\text{number of wicket taken}}$$

The smaller the average, the better the bowler has performed.

Josh and Ravi were competing for three bowling awards:
- Best in semifinal
- Best in final
- Best overall

The following table gives their scores.

	Semifinal		Final	
	Runs scored	Wickets taken	Runs scored	Wickets taken
Josh	12	5	28	6
Ravi	10	4	15	3

3. Calculate the bowling averages for the following and fill in the table below.
   • Semifinal
   • Final
   • Overall

	Semifinal average	Final average	Overall average
Josh			
Ravi			

4. Explain how Ravi can have the better overall average when Josh has the better average in both the semifinal and final.

 Resources

Exercise 12.9 Review questions

learn on

Fluency

1. List some problems you might encounter in trying to collect data from the following populations.
   a. The average number of mL in a can of soft drink
   b. The number of fish in a dam
   c. The number of workers who catch public transport to work each weekday morning

2. For each of the following investigations, state whether a census or a survey has been used.
   a. The average price of petrol in Canberra was estimated by averaging the price at 30 petrol stations in the area.
   b. The performance of a cricketer is measured by looking at his performance in every match he has played.
   c. Public opinion on an issue is sought by a telephone poll of 2000 homes.

3. Consider the box plot shown.
   a. Calculate the median.
   b. Calculate the range.
   c. Determine the interquartile range.

4. Calculate the mean, median and mode for each of the following sets of data:
   a. 7, 15, 8, 8, 20, 14, 8, 10, 12, 6, 19
   b. Key: 1|2 = 12

Stem	Leaf
1	2 6
2	1 7 8
3	0 3 3 4 6 8
4	0 1 1 5 9
5	1 3 6

   c.

Score (x)	Frequency (f)
70	2
71	6
72	9
73	7
74	4

5. For each of the following data sets, calculate the range.
   a. 4, 3, 6, 7, 2, 5, 8, 4, 3

   b.

x	13	14	15	16	17	18	19
f	3	6	7	12	6	7	8

   c. Key: 1|8 = 18

Stem	Leaf
1	7 8 8 9
2	1 2 4 4 5 7 7 7 8 9 9
3	0 0 0 1 3 4 7

6. For each of the following data sets, calculate the interquartile range.
   a. 18, 14, 15, 19, 20, 11 16, 19, 18, 19

   b. Key: 9|8 = 9.8

Stem	Leaf
8	7 8 8 9
9	0 2 4 4 5 7 7 7 8 9 9
10	0 1 1 1 3

7. The following back-to-back stem-and-leaf plot shows the typing speed in words per minute (wpm) of 30 Year 8 and Year 10 students.
   a. Calculate the mean, median, range, interquartile range and standard deviation of the Year 8 data set.
   b. Calculate the mean, median, range, interquartile range and standard deviation of the Year 10 data set.
   c. Compare the two distributions, using your answers to parts a and b.
   d. Using a calculator or otherwise, construct a pair of parallel box-and-whisker plots to represent the two sets of data.

   Key: 2|6 = 26 wpm

Leaf: Year 8	Stem	Leaf: Year 10
9 9	0	
9 8 6 5 4 2 0	1	7 9
9 8 8 6 4 2 1 0 0	2	2 3 6 8 9
9 7 7 6 4 1 0	3	0 2 4 5 5 7 8 8
8 6 5 2 0	4	1 2 5 8 8 9 9
	5	0 3 5 7 8
	6	0 0 3

8. The following data give the amount of cut meat (in kg) obtained from each of 20 lambs.

4.5	6.2	5.8	4.7	4.0	3.9	6.2	6.8	5.5	6.1
5.9	5.8	5.0	4.3	4.0	4.6	4.8	5.3	4.2	4.8

   a. Detail the data on a stem-and-leaf plot. (Use a class size of 0.5 kg.)
   b. Prepare a five-point summary of the data.
   c. Draw a box plot of the data.

9. Calculate the standard deviation of each of the following data sets correct to 1 decimal place.
   a. 58, 12, 98, 45, 60, 34, 42, 71, 90, 66

   b.

x	1	2	3	4	5
f	2	6	12	8	5

   c. Key: $1|4 = 14$

Stem	Leaf
0	1 3 4 4 5 7 8
1	0 0 0 1 2 2 4 5 7 8 9
2	0 2 2 3 5 7

10. **MC** The Millers obtained a number of quotes on the price of having their home painted. The quotes, to the nearest hundred dollars, were:

          4200,   5100,   4700,   4600,   4800,   5000,   4700,   4900

   The standard deviation for this set of data, to the nearest whole dollar, is:
   A. 260   B. 278
   C. 324   D. 325

11. **MC** The number of Year 12 students who, during semester 2, spent all their spare periods studying in the resource centre is shown on the stem-and-leaf plot.

Key: $2|5 = 25$ students

Stem	Leaf
0	8
1	
2	5 6 6 7
3	0 2 3 6 9
4	7 9
5	6
6	1

The standard deviation for this set of data, to the nearest whole number is:
   A. 12   B. 14
   C. 17   D. 35

12. Each week, varying amounts of a chemical are added to a filtering system. The amounts required (in mL) over the past 20 weeks are shown in the stem-and-leaf plot.

Key: 3|8 represents 0.38 ml

Stem	Leaf
2	1
2	2 2
2	4 4 4 5
2	6 6
2	8 8 9 9
3	0
3	2 2
3	4
3	6
3	8

Calculate to 2 decimal places the standard deviation of the amounts used.

13. Calculate the mean, median and mode of this data set: 2, 5, 6, 2, 5, 7, 8. Comment on the shape of the distribution.

14. The box plot shows the heights (in cm) of Year 12 students in a Maths class.

Height (cm)

a. State the median class height.
b. Calculate the range of heights.
c. Calculate the interquartile range of the heights.

Problem solving

15. **MC** A data set has a mean of 75 and a standard deviation of 5. Another score of 50 is added to the data set. Choose which of the following will occur.
A. The mean will increase and the standard deviation will increase.
B. The mean will increase and the standard deviation will decrease.
C. The mean will decrease and the standard deviation will increase.
D. The mean will decrease and the standard deviation will decrease.

16. **MC** A data set has a mean of 60 and a standard deviation of 10. A score of 100 is added to the data set. This score becomes the highest score in the data set. Choose which of the following will increase.
Note: There may be more than one correct answer.
A. Mean
B. Standard deviation
C. Range
D. Interquartile range

17. A sample of 30 people was selected at random from those attending a local swimming pool. Their ages (in years) were recorded as follows.

19, 7, 58, 41, 17, 23, 62, 55, 40, 37, 32, 29, 21, 18, 16,
10, 40, 36, 33, 59, 65, 68, 15, 9, 20, 29, 38, 24, 10, 30

a. Calculate the mean and the median age of the people in this sample.
b. Group the data into class intervals of 10 (0–9 etc.) and complete the frequency distribution table.
c. Use the frequency distribution table to estimate the mean age.
d. Calculate the cumulative frequency and hence plot the ogive.
e. Estimate the median age from the ogive.
f. Compare the mean and median of the original data in part a with the estimates of the mean and the median obtained for the grouped data in parts c and e.
g. Determine if the estimates were good enough. Explain your answer.

18. The table below shows the number of cars that are garaged at each house in a certain street each night.

Number of cars	Frequency
1	9
2	6
3	2
4	1
5	1

a. Show these data in a frequency histogram.
b. State if the data is positively or negatively skewed. Justify your answer.

19. Consider the data set represented by the frequency histogram shown.

a. Explain if the data is symmetrical.
b. State if the mean and median of the data can be seen. If so, determine their values.
c. Evaluate the mode of the data.

20. There are three m values in a data set for which $\bar{x} = m$ and $\sigma = \dfrac{m}{2}$.

a. Comment on the changes to the mean and standard deviation if each value of the data set is multiplied by m.
b. An additional value is added to the original data set, giving a new mean of $m + 2$. Evaluate the additional value.

21. The following data show the number of pets in each of the 12 houses in Coral Avenue, Rosebud.

$$2, 3, 3, 2, 2, 3, 2, 4, 3, 1, 1, 0$$

a. Calculate the mean and median number of pets.
b. The empty block of land at the end of the street was bought by a Cattery and now houses 20 cats. Recalculate the mean and median.
c. Explain why the answers are so different, and which measure of central tendency is best used for certain data.

22. The number of Year 10 students in all the 40 schools in the Northern District of the Education Department was recorded as follows.

56, 134, 93, 67, 123, 107, 167, 124, 108, 78, 89, 99, 103, 107, 110, 45, 112, 127, 106, 111, 127, 145, 87, 75, 90, 123, 100, 87, 116, 128, 131, 106, 123, 87, 105, 112, 145, 115, 126, 92

a. Using an interval of 10, produce a table showing the frequency for each interval.
b. Use the table to estimate the mean.
c. Calculate the mean of the ungrouped data.
d. Compare the results from parts **b** and **c** and explain any differences.

23. The following back-to-back stem-and-leaf plot shows the ages of a group of 30 males and 30 females as they enter hospital for the first time.

Key: $1 | 7 = 17$

Leaf: Male	Stem	Leaf: Female
9 8	0	5
9 9 8 8 8 6 3 2 1	1	7 7 8 9 9
8 7 7 6 4 3 2 0	2	0 0 1 2 4 5 5 6 7 9
8 6 3 1 0	3	0 1 3 3 5 8
7 5 2	4	2 3 6 8
5 3	5	1 3 4
	6	2
8	7	

a. Construct a pair of parallel box plots to represent the two sets of data, showing working out for the median and 1st and 3rd quartiles.
b. Calculate the mean, range and IQR for both sets of data.
c. Determine any outliers if they exist.
d. Write a short paragraph comparing the data.

24. The test scores, out of a total score of 50, for two classes A and B are shown in the back-to-back stem-and-leaf plot.

Key: $1 | 4 = 14$

Leaf: Class A	Stem	Leaf: Class B
5	0	1 2 4
9 7 5 3	1	1 4 5
9 7 7 5 4	2	0 0 5
8 8 6 5 5 1	3	1 5 5
3 2 0	4	1 5 7 7 8 9
0	5	0 0

a. Ms Vinculum teaches both classes and made the statement that 'Class A's performance on the test showed that the students' ability was more closely matched than the students' ability in Class B'. By calculating the measure of centre, first and third quartiles, and the measure of spread for the test scores for each class, explain if Ms Vinculum's statement was correct.
b. Would it be correct to say that Class A performed better on the test than Class B? Justify your answer by comparing the quartiles and median for each class.

25. The times, in seconds, of the duration of 20 TV advertisements shown in the 6–8 pm time slot are recorded below.

16, 60, 35, 23, 45, 15, 25, 55, 33, 20, 22, 30, 28, 38, 40, 18, 29, 19, 35, 75

a. From the data, determine:
  i. the mode
  ii. the median
  iii. the mean (write your answer correct to 2 decimal places)
  iv. the range
  v. the lower quartile
  vi. the upper quartile
  vii. the interquartile range.
b. Using your results from part **a**, construct a box plot for the time, in seconds, for the 20 TV advertisements in the 6–8 pm time slot.
c. From your box plot, determine:
  i. the percentage of advertisements that are more than 39 seconds in length
  ii. the percentage of advertisements that last between 21 and 39 seconds
  iii. the percentage of advertisements that are more than 21 seconds in length

The types of TV advertisements during the 6–8 pm time slot were categorised as Fast food, Supermarkets, Program information and Retail (clothing, sporting goods, furniture). A frequency table for the frequency of these advertisements is shown below.

Type	Frequency
Fast food	7
Supermarkets	5
Program information	3
Retail	5

d. State the type of data that has been collected in the table.
e. Determine the percentage of advertisements that are advertisements for fast food outlets.
f. Suggest a good option for a graphical representation of this type of data.

26. The speeds, in km/h, of 55 cars travelling along a major road are recorded in the following table.

Speed	Frequency
60–64	1
65–69	1
70–74	10
75–79	13
80–84	9
85–89	8
90–94	6
95–99	3
100–104	2
105–109	1
110–114	1
Total	55

a. By calculating the midpoint for each class interval, determine the mean speed, in km/h, of the cars travelling along the road.
Write your answer correct to 2 decimal places.

b. The speed limit along the road is 75 km/h. A speed camera is set to photograph the license plates of cars travelling 7% more than the speed limit.
A speeding fine is automatically sent to the owners of the cars photographed.
Based on the 55 cars recorded, determine the number of speeding fines that were issued.

c. Drivers of cars travelling 5 km/h up to 15 km/h over the speed limit are fined $135. Drivers of cars travelling more than 15 km/h and up to 25 km/h over the speed limit are fined $165 and drivers of cars recorded travelling more than 25 km/h and up to 35 km/h are fined $250.
Drivers travelling more than 35 km/h pay a $250 fine in addition to having their driver's license suspended.
Assume that this data is representative of the speeding habits of drivers along a major road and there are 30 000 cars travelling along this road on any given month.

  i. Determine the amount, in dollars, collected in fines throughout the month. Write your answer correct to the nearest cent.

  ii. Evaluate the number of drivers that would expect to have their licenses suspended throughout the month.

on   To test your understanding and knowledge of this topic, go to your learnON title at www.jacplus.com.au and complete the **post-test**.

Answers

Topic 12 Univariate data

12.1 Pre-test

1. 2
2. 29
3. 72
4. 29.8
5. subset
6. 22
7. 13
8. Positively skewed
9. 11
10. C
11. $\bar{x} = 184.42$ and $\sigma = 7.31$
12. D
13. B
14. The mean typing speed is 26.53 and IQR is 19 for Year 8. The mean typing speed is 40.53 and IQR is 20 for Year 10. This suggests, that the mean typing speed for Year 10 is greater than the Year 8 students. The interquartile range is not the same for both Year 8 and Year 10.
15. D

12.2 Measures of central tendency

1. a. 7   b. 8   c. 8
2. a. 6.875   b. 7   c. 4, 7
3. a. 39.125   b. 44.5   c. No mode
4. a. 4.857   b. 4.8   c. 4.8
5. a. 12   b. 12.625   c. 13.5
6. Science: mean $= 57.6$, median $= 57$, mode $= 42, 51$
   Maths: mean $= 69.12$, median $= 73$, mode $= 84$
7. a. 5.83   b. 6   c. 6
8. a. 14.425   b. 15   c. 15
9. a. Mean $= 2.5$, median $= 2.5$
   b. Mean $= 4.09$, median $= 3$
   c. Median
10. a. $70\dfrac{15}{16}$ b. $70 - {<}80$ c. $70 - {<}80$
11. 124.83
12. $65 - {<}70$
13. a. B   b. B
    c. C   d. D
14. a. Mean $= \$32.93$, median $= \$30$

b.

Class interval	Frequency	Cumulative frequency
0–9	5	5
10–19	5	10
20–29	5	15
30–39	3	18
40–49	5	23
50–59	3	26
60–69	3	29
70–79	1	30
Total	30	

Mean $= \$32.50$, median $= \$30$

c.

d. The mean is slightly underestimated; the median is exact. The estimate is good enough as it provides a guide only to the amount that may be spent by future customers.

15. a. 3
    b. 4, 5, 5, 5, 6 (one possible solution)
    c. One possible solution is to exchange 15 with 20.
16. a. Frequency column: 16, 6, 4, 2, 1, 1
    b. 6.8
    c. 0–4 hours
    d. 0–4 hours
17. a. Frequency column: 1, 13, 2, 0, 1, 8
    b.

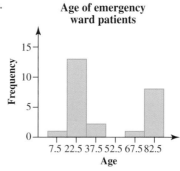

c. Asymmetrical or bimodal (as if the data come from two separate graphs)
d. 44.1
e. $15 - {<}30$
f. $15 - {<}30$

g.

h. 28

i. No

j. Sample responses can be found in the worked solutions in the online resources.

18. A

19. a. Player A median $= 34.33$, Player B median $= 41.83$

b. Player B

c. Player A median $= 32.5$, Player B median $= 0$

d. Player A

e. Player A is more consistent. One large score can distort the mean.

20. Sample responses can be found in the worked solutions in the online resources.

21. a. Frequency column: 3, 8, 5, 3, 1

b. 50.5

c. $40-<50$

d. $40-<50$

e.

f. Approximately 48 beats/min

22. Answers will vary. Sample responses include:

a. 3, 4, 5, 5, 8

b. 4, 4, 5, 7, 10

c. 2, 3, 6, 6, 12

23. 12

24. $\dfrac{2a+b}{3}$

25. 13, 31, 31, 47, 53, 59

12.3 Measures of spread

1. a. 15   b. 77.1   c. 9

2. a. 7   b. 7   c. 8.5   d. 39

3. a. 3.3 kg   b. 1.5 kg

4. 22 cm

5. 0.8

6. C

7. a.

b. i. 62.5

ii. $Q_1 = 58$, $Q_3 = 67$

iii. 9

iv. 14

v. 6

8. IQR $= 28$

9. a. i. Range $= 23$

ii. IQR $= 13.5$

b. i. Range $= 45$

ii. IQR $= 27.5$

c. i. Range $= 49$

ii. IQR $= 20$

10. Measures of spread tell us how far apart the values (scores) are from one another.

11. a. 25.5

b. 28

c. 39

d. 6

e. The three lower scores affect the mean but not the median or mode.

12. a.

Men:	mean $= 32.3$; median $= 32.5$; range $= 38$; IQR $= 14$
Women:	mean $= 29.13$; median $= 27.5$; range $= 36$; IQR $= 13$

b. Typically, women marry younger than men, although the spread of ages is similar.

13. Mean $= 25.036$, median $= 24.9$, mode $= 23.6$, range $= 8.5$, IQR $= 3.4$

14. $a = 22$, $b = 9$, $c = 9$ and $d = 8$

15. a. Yes

Range $= 10 - 1 = 9$

Median (middle score) $= 6$

b. 6

To maintain range, min $= 1$ and max $= 10$.
If median $= 6$, then 3rd and 4th scores must be 6.
Therefore, the 6th score must be 6. This will maintain the range, Q_1, Q_3 and median.

12.4 Box plots

1. a. 5   **b.** 26

2. a. 6   **b.** 27

3. a. 5.8   **b.** 18.6

4. a. 140   **b.** 55   **c.** 90   **d.** 85   **e.** 26

5. a. 58   **b.** 31   **c.** 43   **d.** 27   **e.** 15

6. B

7. C

8. C, D

9. a. 22, 28, 35, 43, 48

b.

Sales

10. a. $(10, 13.5, 22, 33.5, 45)$

b.

Rainfall (mm)

11. a. $(18, 20, 26, 43.5, 74)$

b.

Age

c. The distribution is positively skewed, with most of the offenders being young drivers.

12. a. $(124\,000, 135\,000, 148\,000, 157\,000, 175\,000)$

b.

($ × 1000$)

13. a.

Both graphs indicate that the data is slightly positively skewed. However, the box plot provides an excellent summary of the centre and spread of the distribution.

b.

Both graphs indicate that the data is slightly negatively skewed. However, the box plot provides an excellent summary of the centre and spread of the distribution.

14.

Number of passengers on bus journeys

Number of passengers

15. a.

	X_{min}	Q_1	Median	Q_3	X_{max}
Before	75	86	95	128.5	152
After	66	81	87	116	134

b.

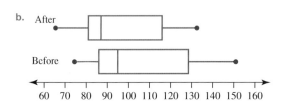

c. As a whole, the program was effective. The median weight dropped from 95 kg to 87 kg, a loss of 8 kg. A noticeable shift in the graph shows that after the program 50% of participants weighed between 66 and 87 kg, compared to 25% of participants weighing between 75 and 86 kg before they started.

Before the program, the range of weights was 77 kg (from 75 kg to 152 kg); after the program, the range had decreased to 68 kg. The IQR also diminished from 42.5 kg to 35 kg.

16. The advantages of box plots is that they are clear visual representations of 5-number summary, display outliers and can handle a large volume of data. The disadvantage is that individual scores are lost.

17. a. Key: 12|1 = 121

Stem	Leaf
12	1569
13	124
14	3488
15	022257
16	35
17	29
18	1112378

b.

120 140 160 180
Number sold

c. On most days the hamburger sales are less than 160. Over the weekend the sales figures spike beyond this.

18. a. Key: 1*|7 = 17 years

Stem	Leaf
1*	7 7 8 8 8 9 9
2	0 0 0 1 2 2 2 2 3 3 3 3 4 4 4
2*	5 5 8 9
3	1 2 3
3*	
4	
4*	8

b.

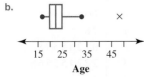

15   25   35   45
Age

c. The distribution is positively skewed, with first-time mothers being under the age of 30. There is one outlier (48) in this group.

19.

Size

20. a. HJ Looker: median = 5; Hane and Roarne: median = 6

b. HJ Looker

c. HJ Looker

d. Hane and Roarne had a higher median and a lower spread and so they appear to have performed better.

21. a. $50   **b.** $135   **c.** $100   **d.** $45   **e.** 50%

22. a. 9625 seconds

b. i. Under 20 − (20–24): 750 seconds difference

ii. Under 20 − (25–29): 500 seconds difference

iii. (20–24) − (25–29): 250 seconds difference

c. The under-20s performed best of the three groups, with the fastest time for each metric (minimum time, first

quartile, median, third quartile and maximum time). The next best performing group was the 25–29-year-olds. They had the same median as the 20–24-year-olds, but outperformed them in all of the other metrics. The 24–29-year-olds were the most consistent group, with a range of 1750 seconds compared to the range of 2000 seconds of the other groups.

12.5 The standard deviation (10A)

1. a. 2.29   **b.** 2.19   **c.** 20.17   **d.** 3.07

2. a. 2   **b.** 1   **c.** 12   **d.** 0.3

3. a. 1.97   **b.** 16.47

4. a. 1.03   **b.** 1.33   **c.** 2.67   **d.** 2.22

5. 10.82

6. Mean = 1.64%   Std dev. = 0.45%

7. Mean = 1.83   Std dev. = 0.06 m

8. 0.49 s

9. 15.10 calls

10. B

11. The mean of the first 25 cars is 89.24 km/h with a standard deviation of 5.60. The mean of the first 26 cars is 91.19 with a standard deviation of 11.20, indicating that the extreme speed of 140 km/h is an anomaly.

12. The standard deviation tells us how spread out the data is from the mean

13. a. $\sigma \approx 3.58$

b. The mean is increased by 7 but the standard deviation remains at $\sigma \approx 3.58$.

c. The mean is tripled and the standard deviation is tripled to $\sigma \approx 10.73$.

14. The standard deviation will decrease because the average distance to the mean has decreased.

15. 57 is two standard deviations above the mean.

16. a. New mean is the old mean increased by 3 (15) but no change to the standard deviation.

b. New mean is 3 times the old mean (36) and new standard deviation is 3 times the old standard deviation (12).

17. a. 43   **b.** 12   **c.** 12.19

12.6 Comparing data sets

1. a. The mean of the first set is 63. The mean of the second set is 63.

b. The standard deviation of the first set is. 2.38 The standard deviation of the second set is 6.53.

c. For both sets of data the mean is the same, 63. However, the standard deviation for the second set (6.53) is much higher than the standard deviation of the first set (2.38), implying that the second set is more widely distributed than the first. This is confirmed by the range, which is $67 − 60 = 7$ for the first set and $72 − 55 = 17$ for the second.

2. a. Morning: median = 2.45; afternoon: median = 1.6

b. Morning: range = 3.8; afternoon: range = 5

c. The waiting time is generally shorter in the afternoon. One outlier in the afternoon data causes the range to be larger. Otherwise the afternoon data are far less spread out.

3. Key: $16|1 = 1.61$ m

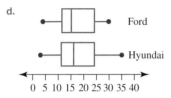

Leaf: Boys	Stem	Leaf: Girls
9 9 7	15	1 2 5 6 7 8 8
9 8 6 6 5 5 4 0	16	4 4 6 7 8 9 9
4 4 2 1	17	0

4. a. Ford: median $= 15$; Hyundai: median $= 16$
   b. Ford: range $= 26$; Hyundai: range $= 32$
   c. Ford: IQR $= 14$; Hyundai: IQR $= 13.5$

   d.

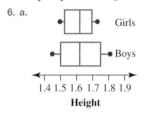

5. a. Brisbane Lions
   b. Brisbane Lions: range $= 65$;
      Sydney Swans: range $= 55$
   c. Brisbane Lions: IQR $= 40$;
      Sydney Swans: IQR $= 35$

6. a.

Girls
Boys
1.4 1.5 1.6 1.7 1.8 1.9
Height

   b. Boys: median $= 1.62$; girls: median $= 1.62$
   c. Boys: range $= 0.36$; girls: range $= 0.23$
   d. Boys: IQR $= 0.25$; girls: IQR $= 0.17$
   e. Although boys and girls have the same median height, the spread of heights is greater among boys as shown by the greater range and interquartile range.

7. a. Summer: range $= 23$; winter: range $= 31$
   b. Summer: IQR $= 14$; winter: IQR $= 11$
   c. There are generally more cold drinks sold in summer as shown by the higher median. The spread of data is similar as shown by the IQR although the range in winter is greater.

8. A

9. A, B, C, D

10. a. Cory achieved a better average mark in Science (59.25) than he did in English (58.125).
   b. Cory was more consistent in English ($\sigma = 4.9$) than he was in Science ($\sigma = 19.7$)

11. Sample responses can be found in the worked solutions in the online resources.

12. a. Back street: $\bar{x} = 61$, $\sigma = 4.3$;
      main road: $\bar{x} = 58.8$, $\sigma = 12.1$
   b. The drivers are generally driving faster on the back street.
   c. The spread of speeds is greater on the main road as indicated by the higher standard deviation.

13. a.

Machine A
Machine B
40 42 44 46 48 50 52 54 56 58 60
Number of Smarties in a box

   b. Machine A: mean $= 49.88$,
      standard deviation $= 2.87$;
      Machine B: mean $= 50.12$,
      standard deviation $= 2.44$
   c. Machine B is more reliable, as shown by the lower standard deviation and IQR. The range is greater on machine B only because of a single outlier.

14. a. Nathan: mean $= 15.1$; Timana: mean $= 12.3$
   b. Nathan: range $= 36$; Timana: range $= 14$
   c. Nathan: IQR $= 15$; Timana: IQR $= 4$
   d. Timana's lower range and IQR shows that he is the more consistent player.

15. a. Yes — Maths
   b. Science: positively skewed
   c. The Science test may have been more difficult.
   d. Science: 61–70, Maths: 71–80
   e. Maths has a greater standard deviation (12.6) compared to Science (11.9).

16. a. Key: $2|3 = 2.3$ hours

Leaf: Group A	Stem	Leaf: Group B
8 7 3	1	7 8
9 5 1	2	0 1 2 4 5 5 8 8
8 7 5 4 2 2	3	2 2 2 4 5 5 5 6 8
7 5 4 2 2 2	4	2
	5	
2 2	6	

   b. Five-point summary
      Group A: 13   27   36   43   62
      Group B: 17   23   30   35   42

   c. Sample responses can be found in the worked solutions in the online resources.
   d. Sample responses can be found in the worked solutions in the online resources.
   e. Sample responses can be found in the worked solutions in the online resources.

17. a.

Leaf: Year 10	Stem	Leaf: Year 11
9 8 7 5 4	15	
9 8 7 4 3 3 0	16	0 3 6 7 8
7 5 3 2 2 0 0	17	2 2 3 4 5 6 7 7 9
0	18	0 1 3 5 8
	19	0

b.

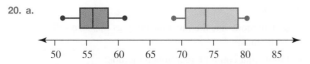

c. On average, the Year 12 students are about $6 - 10\,\text{cm}$ taller than the Year 10 students. The heights of the majority of Year 12 students are between $170\,\text{cm}$ and $180\,\text{cm}$, whereas the majority of the Year 10 students are between 160 and $172\,\text{cm}$ in height.

18. a. English: mean $= 70.25$; Maths: mean $= 69$

b. English: range $= 53$; Maths: range $= 37$

c. English: $\sigma = 16.1$; Maths: $\sigma = 13.4$

d. Kloe has performed more consistently in Maths as the range and standard deviation are both lower.

19. 32%

20. a.

The parallel box plots show a significant gap between the life expectancy of Aboriginal and Torres Strait Islander people and that of non-Aboriginal and non–Torres Strait Islander people. Even the maximum median age of Aboriginal and Torres Strait Islander people is much lower than the minimum of non-Aboriginal and non–Torres Strait Islander people.

b. The advantage of box plots is that it gives a clear graphical representation of the results and in this case shows a significant difference between the median life expectancy of Aboriginal and Torres Strait Islander people and non-Aboriginal and non–Torres Strait Islander people. The disadvantage is that we lose the data for individual states and territories.

12.7 Populations and samples

1. a. When was it first put into the machine? How old was the battery before being purchased? How frequently has the computer been used on battery?

b. Can't always see if a residence has a dog; a census is very time-consuming; perhaps could approach council for dog registrations.

c. This number is never constant with ongoing purchases, and continuously replenishing stock.

d. Would have to sample in this case as a census would involve opening every packet.

2. Sample responses can be found in the worked solutions in the online resources.

3. a. Census. The airline must have a record of every passenger on every flight.

b. Survey. It would be impossible to interview everyone.

c. Survey. A census would involve opening every bottle.

d. Census. The instructor must have an accurate record of each learner driver's progress.

4. a. Survey   b. Survey   c. Census   d. Survey

5. a. About 25

b. Drawing numbers from a hat, using a calculator.

6. a. The council is probably hoping it is a census, but it will probably be a survey because not all those over 10 will respond.

b. Residents may not all have internet access. Only those who are highly motivated are likely to respond.

7. The sample could have been biased. The questionnaire may have been unclear.

8. Sample size, randomness of sample

9. Sample responses can be found in the worked solutions in the online resources.

10. Sample response: A census of very large populations requires huge amounts of infrastructure and staff to collect the information for large numbers of people. These challenges could also be made harder because many people live in remote areas with poor transport access for census staff; forms may need to be created in multiple languages; and migrants who do not have residency permits may be unwilling to complete a census.

11. There is quite a variation in the frequency of particular numbers drawn. For example, the number 45 has not been drawn for 31 weeks, while most have been drawn within the last 10 weeks. In the long term, one should find the frequency of drawing each number is roughly the same. It may take a long time for this to happen, as only 8 numbers are drawn each week.

12. a. Mean $= 32.03$; median $= 29.5$

b.

Class interval	Frequency
0–9	2
10–19	7
20–29	6
30–39	6
40–49	3
50–59	3
60–69	3
Total	30

c. Mean $= 31.83$

d.

(cumulative frequency graph plotting Cumulative frequency against Age)

e. Median $= 30$

f. Estimates from parts c and e were fairly accurate.

g. Yes, they were fairly close to the mean and median of the raw data.

13. Year 8: mean = 26.83, median = 27, range = 39, IQR = 19
Year 10: mean = 40.7, median = 39.5, range = 46, IQR = 20
The typing speed of Year 10 students is about 13 to 14 wpm faster than that of Year students. The spread of data in Year 8 is slightly less than the spread in Year 10.

12.8 Evaluating inquiry methods and statistical reports

1. a. Primary. There is probably no secondary data available.

b. Sample responses can be found in the worked solutions in the online resources.

c. Sample responses can be found in the worked solutions in the online resources.

2. Company profits

Company profits

Mean salaries

Company salaries

3. a. Mean = $21 5000, median = $170 000, mode = $150 000. The median best represents these land prices. The mean is inflated by one large score, and the mode is the lowest price.

b. Range = $500 000, interquartile range = $30 000. The interquartile range is the better measure of spread.

c. This dot plot shows how 9 of the scores are grouped close together, while the score of $650 000 is an outlier.

d. The agent is quoting the modal price, which is the lowest price. This is not a true reflection of the average price of these blocks of land.

4. a. False. Mean = 1.82 m, lower quartile = 1.765 m, median = 1.83 m

b. True. This is the definition of interquartile range.

c. Players with heights 1.83 m, 1.83 m, 1.88 m, 1.88 m, 1.88 m.

5. a.

Resting pulse	Frequency	Middle value, x	$x \times$ frequency	Cumulative frequency
30–39	3	34.5	103.5	3
40–49	8	44.5	356.0	11
50–59	5	54.5	272.5	16
60–69	3	64.5	193.5	19
70–79	1	74.5	74.5	20

b. Mean = 48.65
Median = 48
Mode = 48

c. From the table it can be clearly seen that the highest concentration of resting pulse readings was in the 40–49 and 50–59 groups. All three measures of central tendency fell within these two groupings.
Because of the higher frequency in the 40–49 group, it was not surprising that the mode and median were contained there also. The mean was slightly higher, and this would have been influenced by the one reading in the 70–79 group.

6. Player B appears to be the better player if the mean result is used. However, Player A is the more consistent player.

7. a. 7.1   **b.** 7   **c.** 7

d. The mode has the most meaning as this size sells the most.

8. Points that could be mentioned include:
10.1% is only just 'double digit' growth.
2006–08 showed mid to low 20% growth. Growth has been declining since 2008.
The share price has rebounded, but not to its previous high. The share price scale is not consistent. Most increments are 30c, except for $27.70 to $28.10?(40c increment). Note also the figure of 20.80 — probably a typo instead of 26.80.

9. A misleading graph steers/convinces the reader towards a particular opinion. It can be biased and not accurate.

10. Shorten the y-axis and expand the x-axis.

Aussie dollar

11. a. Key: 3|85 = 385 hours

Leaf: Brand B	Stem	Leaf: Brand A
	3	85 90
	4	25 26
60 55 00	5	70
30	6	40 45
70 42 35 20	7	30 35 60
60 20	8	

Brand A: mean = 570.6, median = 605
Brand B: mean = 689.2, median = 727.5

b. Brand A had the shortest mean lifetime.

c. Brand B had the longest mean lifetime.

d. Brand B

12. a. The statement is true, but misleading as most of the employees earn $18000.

  b. The median and modal salary is $18000 and only 15 out of 80 (less than 20%) earn more than the mean.

13. Left bar chart suggests prices have tripled in one year due to fact vertical axis does not start at zero. Bar chart on right is truly indicative of situation.

14. a. Percentages do not add to 100%.

  b. Percentages do not add to 100%.

  c. Such representation allows multiple choices to have closer percentages than really exist.

Project

1.

Player	Runs in the last 25 matches	Mean	Median	Range	IQR
Will	13, 18, 23, 21, 9, 12, 31, 21, 20, 18, 14, 16, 28, 17, 10, 14, 9, 23, 12, 24, 0, 18, 14, 14, 20	16.76	17	31	8.5
Rohit	2, 0, 112, 11, 0, 0, 8, 0, 10, 0, 56, 4, 8, 164, 6, 12, 2, 0, 5, 0, 0, 0, 8, 18, 0	17.04	4	164	10.5
Marnus	12, 0, 45, 23, 0, 8, 21, 32, 6, 0, 8, 14, 1, 27, 23, 43, 7, 45, 2, 32, 0, 6, 11, 21, 32	16.76	12	45	25.5
Ben	2, 0, 3, 12, 0, 2, 5, 8, 42, 0, 12, 8, 9, 17, 31, 28, 21, 42, 31, 24, 30, 22, 18, 20, 31	16.72	17	42	25

2. a. Will: has a similar mean and median, which shows he was fairly consistent. The range and IQR values are low indicating that his scores remain at the lower end with not much deviation for the middle 50% .

  b. Rohit: has the best average but a very low median indicating his scores are not consistent. The range is extremely high and the IQR very low in comparison showing he can score very well at times but is not a consistent scorer.

  c. Marnus: has a similar mean to Will and Ben but a lower median, indicating his scores are sometimes high but generally are lower than the average. The range and IQR show a consistent batting average and spread with only a few higher scores and some lower ones.

  d. Ben: has a similar mean and median which shows he was a consistent player. The range and IQR show a consistent batting average and spread.

Players to be selected:

Would recommend **Will** if the team needs someone with very consistent batting scores every game but no outstanding runs.

Would recommend **Rohit** if the team needs someone who might score very high occasionally but in general fails to score many runs.

Would recommend **Marnus** if the team needs someone who is fairly consistent but can score quite well at times and the rest of the time does OK.

Would recommend **Ben** if the team needs someone who is fairly consistent but can score quite well at times and the rest of the time has a better median than Glenn.

3.

	Semifinal average	Final average	Overall average
Josh	2.4	4.67	3.64
Ravi	2.5	5	3.57

4. In the final, wickets were more costly than in the semifinal. That is, Josh conceded many runs in getting his six wickets. This affected the overall mean. In reality Josh was the most valuable player overall, but this method of combining the data of the two matches led to this unexpected result.

12.9 Review questions

1. a. You would need to open every can to determine this.

  b. Fish are continuously dying, being born, being caught.

  c. Approaching work places and public transport offices.

2. a. Survey

  b. Census

  c. Survey

3. a. 20　　　　b. 24　　　　c. 8

4. a. Mean = 11.55; median = 10; mode = 8

  b. Mean = 36; median = 36; mode = 33, 41

  c. Mean = 72.18; median = 72; mode = 72

5. a. 6　　　　b. 6　　　　c. 20

6. a. 4　　　　b. 0.85

7. a. Year 8: mean = 26.83, median = 27, range = 39, IQR = 19, sd = 11.45

  b. Year 10: mean = 40.7, median = 39.5, range = 46, IQR = 20, sd = 12.98

  c. The typing speed of Year 10 students is about 13 to 14 wpm faster than that of Year 8 students. The spread of data in Year 8 is slightly less than in Year 10.

  d.

8. a. Key: $3*|9 = 3.9\,\text{kg}$

Stem	Leaf
3*	9
4	0 0 0 2 3
4*	5 6 7 8
5	0 3
5*	5 8 8 9
6	1 2 2
6*	8

b. (3.9, 4.4, 4.9, 5.85, 6.8)

c.

3.5 4.5 5.5 6.5   kg

9. a. 24.4   **b.** 1.1   **c.** 7.3

10. A

11. B

12. 0.05 ml

13. Mean = 5, median = 5, mode = 2 and 5.
The distribution is positively skewed and bimodal.

14. a. Median height = 167 cm
b. Range = 25 cm
c. IQR = 5 cm

15. C

16. A, B and C

17. a. Mean = 32.03; median = 29.5

b.

Class interval	Frequency
0–9	2
10–19	7
20–29	6
30–39	6
40–49	3
50–59	3
60–69	3
Total	30

c. Mean = 31.83

d.

e. Median = 30

f. Estimates from parts c and e were fairly accurate.

g. Yes, they were fairly close to the mean and median of the raw data.

18. a.

b. Positively skewed — a greater number of scores is distributed at the lower end of the distribution.

19. a. Yes   **b.** Yes. Both are 3.
c. 3

20. a. $\bar{x} = m^2$ $\sigma = \dfrac{m^2}{2}$ **b.** $7m + 2$

21. a. Mean = 2.17, median = 2

b. Mean = 3.54, median = 2

c. The median relies on the middle value of the data and won't change much if an extra value is added. The mean however has increased because this large value will change the average of the numbers. The mean is used as a measure of central tendency if there are no outliers or if the data are symmetrical. The median is used as a measure of central tendency if there are outliers or the data are skewed.

22. a.

Interval	Frequency (f)	Midpoint × (f)
40–49	1	44.5 × 1 = 44.5
50–59	1	54.5 × 1 = 54.5
60–69	1	64.5 × 1 = 64.5
70–79	2	74.5 × 2 = 149
80–89	4	84.5 × 4 = 338
90–99	4	94.5 × 4 = 378
100–109	8	104.5 × 8 = 836
110–119	6	114.5 × 6 = 687
120–129	8	124.5 × 8 = 996
130–139	2	134.5 × 2 = 269
140–149	2	144.5 × 2 = 289
150–159	0	154.5 × 0 = 0
160–169	1	164.5 × 1 = 164.5
Total	40	4270

b. 106.75

c. 107.15

d. The differences in this case were minimal; however, the grouped data mean is not based on the actual data but on the frequency in each interval and the interval midpoint. It is unlikely to yield an identical value to the actual mean. The spread of the scores within the class interval has a great effect on the grouped data mean.

23. a.

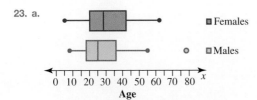

Age

b.

	Males	**Females**
Mean	28.2	31.1
Range	70	57
IQR	18	22

c. There is one outlier — a male aged 78.

d. Typically males seem to enter hospital for the first time at a younger age than females.

24. a. Class A: $Q_1 = 21.5$, median $= 30$, $Q_3 = 38$, IQR $= 16.5$
Class B: $Q_1 = 14.5$, median $= 33$, $Q_3 = 47$, IQR $= 32.5$
Based on the comparison between Class A's IQR (16.5) and Class B's IQR (32.5), Ms Vinculum was correct in her statement.

b. No, Class B has a higher median and upper quartile score than Class A, while Class A has a higher lower quartile. You can't confidently say that either class did better in the test than the other.

25. a. i. 35 s

ii. 29.5 s

iii. 33.05 s

iv. 60 s

v. 21 s

vi. 39 s

vii. 18 s

b.

c. i. 25%

ii. 50%

iii. 75%

d. Categorical

e. 35%

f. Pictogram, pie chart or bar chart.

26. a. 82.73 km/h

b. 30 cars

c. i. $2 607 272.73

ii. About 545

LESSON SEQUENCE

LESSON
13.1 Overview

Why learn this?

Bivariate data can be collected from all kinds of places. This includes data about the weather, data about athletic performance and data about the profitability of a business. By learning the tools you need to analyse bivariate data, you will be gaining skills that help you turn numbers (data) into information that can be used to make predictions (plots).

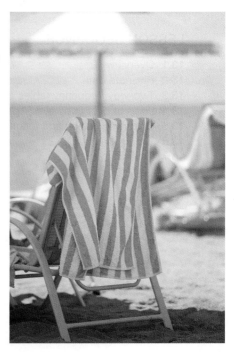

The use of bivariate data is not limited to the classroom; in fact, many professionals rely on bivariate data to help them make decisions. Examples of bivariate data in the real world include the following:

- When a new drug is created, scientists will run drug trials in which they collect bivariate data about how the drug works. When the drug is approved for use, the results of the scientific analysis help guide doctors, nurses, pharmacists and patients as to how much of the drug to use and how often.
- Manufacturers of products can use bivariate data about sales to help make decisions about when to make products and how many to make. For example, a beach towel manufacturer would know that they need to produce more towels for summer, and analysis of the data would help them decide how many to make.

By studying bivariate data, you can learn how to use data to make predictions. By studying and understanding how these predictions work, you will be able to understand the strengths and limitations of these types of predictions.

Hey students! Bring these pages to life online

▶ Watch videos 🧩 Engage with interactivities A⁺ Answer questions and check solutions

Find all this and MORE in jacPLUS ▶

Reading content and rich media, including interactivities and videos for every concept

Extra learning resources

Differentiated question sets

Questions with immediate feedback, and fully worked solutions to help students get unstuck

1. **MC** Choose the type of graph that shows whether there is a relationship between two variables, with each data value shown as a point on a Cartesian plane.
   A. Box plot
   B. Scatterplot
   C. Dot plot
   D. Ogive

2. **MC** Select the incorrect statement from the following.
   A. Bivariate data are data with two variables.
   B. Correlation describes the strength, direction and form of the relationship between two variables.
   C. The independent variable is placed on the y-axis and the dependent variable on the x-axis.
   D. The dependent variable is the one whose value depends on the other variable.

3. Data are compared from twenty students on the number of hours spent studying for an examination and the result of the examination.
   State if the number of hours spent studying is the independent or dependent variable.

4. Match the type of correlation with the data shown on the scatterplots.

Scatterplot	Type of correlation
a.	A. Strong negative linear correlation
b.	B. No correlation
c.	C. Weak positive linear correlation

5. **MC** The table shows the number of hours spent doing a problem-solving task for a subject and the corresponding total score for the task.

Number of hours spent on task	0	1.5	2	1	2	1.5	2.5	3	2	2.5
Task score (%)	20	50	60	45	80	70	75	97	85	20

Choose which data point is a possible outlier.

A. $(0, 20)$ **B.** $(1.5, 50)$ **C.** $(1.5, 70)$ **D.** $(2.5, 20)$

6. **MC** Each point on the scatterplot shown indicates the number of hours per week spent exercising by a person and their fitness level.

Choose the statement that best describes the scatterplot.
   A. The more time exercising, the worse the fitness level.
   B. The number of hours per week spent exercising is the independent variable.
   C. The correlation between the number of hours per week exercising and the fitness level is a weak positive non-linear correlation.
   D. Information has been collected from six people.

Exercising and fitness levels

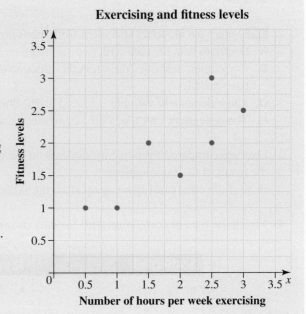

7. Select a term that describes a line of best fit being used to predict a value of a variable from within a given range from the following options: extrapolation, interpolation or regression.

8. Determine the gradient of the line of best fit shown in the scatterplot.

9. In time series data, explain whether time is the independent or dependent variable.

10. Select another term for an independent variable from the following options: a response variable or an explanatory variable.

11. **MC** Select the correct difference between a seasonal pattern and a cyclical pattern in a time series plot.
   A. A cyclical pattern shows upward trends, whereas a seasonal pattern shows only downward trends.
   B. A cyclical pattern displays fluctuations with no regular periods between peaks, whereas a seasonal pattern displays fluctuations that repeat at the same time each week, month, quarter or year.
   C. A cyclical pattern does not show any regular fluctuations, whereas a seasonal pattern does.
   D. A seasonal pattern displays fluctuations with no regular periods between peaks, whereas a cyclical pattern displays fluctuations that repeat at the same time each week, month, quarter or year.

12. **MC** The time series plot shown can be classified as:
   A. an upward trend
   B. a downward trend
   C. a seasonal pattern
   D. a cyclical pattern

13. Use the given scatterplot and the line of best fit to determine the value of x when $y = 2$.

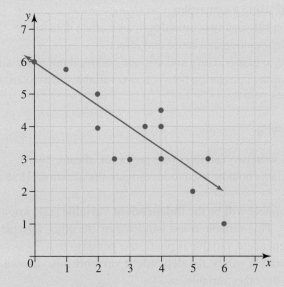

14. The table shows the number of new COVID-19 cases per month reported in Australia in 2020.

Month	March	April	May	June	July
New COVID-19 cases	11	304	14	9	86
Month	August	September	October	November	December
New COVID-19 cases	377	73	18	5	8

   a. Plot the time series.
   b. Interpret the trend in the data from March to December.

15. Use the given scatterplot and line of best fit to predict:
   a. the value of y when x = 4
   b. the value of x when y = 1.

LESSON
13.2 Bivariate data

13.2.1 Bivariate data

eles-4965

  • **Bivariate data** are data with two variables (the prefix 'bi' means 'two').
    For example, bivariate data could be used to investigate the question 'How are student marks affected by phone use?'
  • In bivariate data, one variable will be the **independent variable** (also known as the experimental variable or explanatory variable). This variable *is not impacted* by the other variable.
    In the example, the independent variable is phone use per day.
  • In bivariate data, one variable will be the **dependent variable** (also known as the response variable). This variable *is impacted* by the other variable.
    In the example, the dependent variable is average student marks.

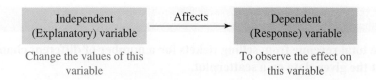

Change the values of this variable

To observe the effect on this variable

For example:

Change the level of education → To observe the effect on the salary

Change the daily temperature → To observe the effect on the sales of air conditioners

Change the amount of sunlight → To observe the effect on the height of a plant

Change the homework hours → To observe the effect on the test result

Scatterplots

- A **scatterplot** is a way of displaying bivariate data.
- In a scatterplot:
  - the independent variable is placed on the *x*-axis with a label and scale
  - the dependent variable is placed on the *y*-axis with a label and scale
  - the data points are shown on the plot.

Independent and dependent variables

- **The independent variable can explain changes in the dependent variable. It is placed on the horizontal axis of a scatterplot.**
- **The dependent variable changes in response to the independent variable. It is predicted using values of the independent variable. It is placed on the vertical axis of a scatterplot.**

Features of a scatterplot

WORKED EXAMPLE 1 Representing bivariate data on a scatterplot

The table shows the total revenue from selling tickets for a number of different chamber music concerts. Represent the given data on a scatterplot.

Number of tickets sold	400	200	450	350	250	300	500	400	350	250
Total revenue ($)	8000	3600	8500	7700	5800	6000	11 000	7500	6600	5600

THINK

1. Determine which is the dependent variable and which is the independent variable.

2. Draw a set of axes. Label the title of the graph. Label the horizontal axis 'Number of tickets sold' and the vertical axis 'Total revenue ($)'.

3. Use an appropriate scale on the horizontal and vertical axes.

4. Plot the points on the scatterplot.

WRITE/DRAW

The total revenue depends on the number of tickets being sold, so the total revenue is the dependent variable and the number of tickets is the independent variable.

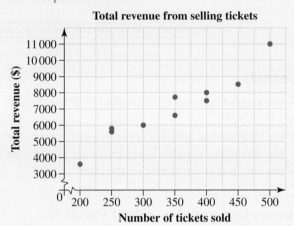

▶ 13.2.2 Correlation

eles-4966

- **Correlation** is a way of describing a connection between variables in a bivariate data set.

Describing correlation

Correlation between two variables will have:
- **a type (linear or non-linear)**

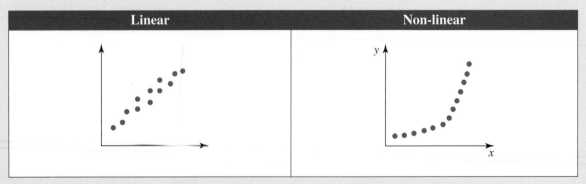

- a direction (positive or negative)

Positive	Negative
As the independent variable (*x*-axis) increases, the dependent variable (*y*-axis) also increases, forming an upward trend.	As the independent variable (*x*-axis) increases, the dependent variable (*y*-axis) decreases, forming a downward trend.

- a strength (strong, moderate or weak).

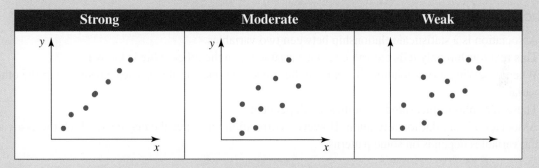

Data will have no correlation if the data are randomly spread out across the plot with no clear pattern, as shown in this example.

WORKED EXAMPLE 2 Describing the correlation

State the type of correlation between the variables *x* and *y* shown on the scatterplot.

THINK	WRITE
Carefully analyse the scatterplot and comment on its form, direction and strength.	The points on the scatterplot are close together and constantly increasing; therefore, the relationship is linear.
	The path is directed from the bottom left corner to the top right corner and the value of y increases as x increases. Therefore, the correlation is positive.
	The points are close together, so the correlation can be classified as strong.
	There is a linear, positive and strong relationship between x and y.

13.2.3 Association and causation

eles-6148

Association

- Association is a statistical relationship between two variables.
- This relationship only tells the value of one variable when the other value is known.
- When observing an association, it can never be stated that one variable causes a response in the other variable.
- These variables or measured quantities are dependent.
- Association is similar to correlation. However, correlation measures the degree to which the association of the variables depends on some patterns.

Causation

- Causation is defined as the relationship between two variables where one variable affects another.
- It is detected when there is an increase or decrease in the value of one variable as a result of the value of another (present) variable.
  For example: *If an employee works for extra hours a day, their income is likely to increase.* This means that one variable (extra hours) is causing an effect on another variable (income). This is a cause–effect or casual relationship.
- A correlation between variables, however, does not automatically mean that the change in one variable is the cause of the change in the values of the other variable.
- Causation requires that there is an association between two variables, but association does not necessarily imply causation.
  For example: *The summer season causes an increase in the sales of ice cream.* In this statement, the variables 'summer' and 'sales of ice cream' have a causal association.
  As the summer season causes the sales of ice cream to increase, we can say that one variable causes an effect in another variable and also the both of them are dependent.
  However, for example: *There is an increase in the sales of air-conditioning and ice cream.*
  In this statement, there is an association as 'summer season' is a common cause for the increase in both sales of air-conditioning and ice cream. However, there is no causation as the variables 'sales of air-conditioning' and 'sales of ice cream' have no cause–effect relationship between them — neither of them affects the other.

Experimentation

- One way to establish causation is to conduct experiments in which a control group is used.
  For example, scientists researching the effectiveness of a new drug designed to destroy a particular type of cancer cell would split the population (those participating in the study) into two groups.

- One group would be given the drug; the other group would be given a placebo. All other variables are the same across both groups.

  A study of the correlation between the administered drug and a decrease in the presence of the cancer cells could then establish causation.

  In this example, the administered drug and placebo are the independent variables, and the number of cancer cells present is the dependent variable.

WORKED EXAMPLE 3 Establishing causation

There is a strong positive association between rates of diabetes and the number of hip replacements. Does this mean that diabetes causes deterioration of the hip joint? Describe the common cause(s) that could link these two variables.

THINK	WRITE
1. Determine the independent and dependent variables.	Number of hip replacements (independent) Rates of diabetes (dependent)
2. Comment on the association.	There is a strong positive association between the two variables as given in the question.
3. There is no causation as the two variables 'number of hip replacements' and 'rates of diabetes' have no cause–effect relationship between them. Comment on the causation.	It cannot be concluded that diabetes causes deterioration of the hip joint. There may be other factors that need to be considered. Further investigation is needed before any conclusions can be drawn.
4. Write the common cause(s) that could link 'rates of diabetes' and 'number of hip replacements'.	Common causes could be age, over-weight, osteoarthritis, injuries, overuse, genetics and so on.

▶ 13.2.4 Drawing conclusions from correlation

eles-4967

- When drawing a conclusion from a scatterplot, state how the independent variable appears to affect the dependent variable and explain what that means.

  For the example of comparing time spent on a phone to average marks, a good conclusion for the graph shown in section 13.2.1 would be:

- Based on scatterplots, it is possible to draw conclusions about *correlation* but not *causation*.

  For example, for this graph:

  - it is correct to say that the amount of water drunk *appears to affect* the distance run
  - it is incorrect to say that drinking more water *causes* a person to run further. For example, someone might only be able to run 1 km, and even if they drink 3 litres of water, they will still only be able to run 1 km.

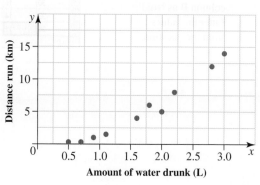

Mary sells business shirts in a department store. She always records the number of different styles of shirt sold during the day.

The table below shows her sales over one week.

Price ($)	14	18	20	21	24	25	28	30	32	35
Number of shirts sold	21	22	18	19	17	17	15	16	14	11

a. Construct a scatterplot of the data.

b. State the type of correlation between the two variables and, hence, draw a corresponding conclusion.

THINK

a. Draw the scatterplot showing 'Price ($)' (independent variable) on the horizontal axis and 'Number of shirts sold' (dependent variable) on the vertical axis.

WRITE/DRAW

a.

b. 1. Carefully analyse the scatterplot and comment on its form, direction and strength.

b. The points on the plot form a path that resembles a straight, narrow band directed from the top left corner to the bottom right corner. The points are close to forming a straight line. There is a linear, negative and strong correlation between the two variables.

2. Draw a conclusion corresponding to the analysis of the scatterplot.

The price of the shirt appears to affect the number sold; that is, the more expensive the shirt, the fewer sold.

TI | THINK

a, b.

1. In a new document, on a Lists & Spreadsheet page, label column A as 'price' and label column B as 'sold'. Enter the values from the question.

DISPLAY/WRITE

a, b.

CASIO | THINK

a, b.

1. On the Statistic screen, label list1 as 'Price' and list2 as 'Shirts', then enter the values from the question. Press EXE after entering each value.

DISPLAY/WRITE

a, b.

2. Open a Data & Statistics page. Press TAB to locate the label of the horizontal axis and select the variable 'price'. Press TAB again to locate the label of the vertical axis and select the variable 'sold'.

2. Tap:
   • SetGraph
   • Setting...
   Set values as shown in the screenshot. Then tap Set.

3. To change the colour of the scatterplot, place the pointer over one of the data points. Then press CTRL MENU. Press:
   • Colour
   • Fill Colour
   Select a colour from the palette for the scatterplot. Press ENTER.

The scatterplot is shown, using a suitable scale for both axes. The points are close to forming a straight line. There is a strong negative, linear correlation between the two variables. The trend indicates that the price of a shirt appears to affect the number sold; that is, the more expensive the shirt, the fewer are sold.

3. Tap the graphing icon and the scatterplot will appear.

The scatterplot is shown, using a suitable scale for both axes. The points are close to forming a straight line. There is a strong negative linear correlation between the two variables. The trend indicates that the price of a shirt appears to affect the number sold; that is, the more expensive the shirt, the fewer are sold.

DISCUSSION

How could you determine whether the change in one variable causes the change in another variable?

 Resources

 eWorkbook Topic 13 Workbook (worksheets, code puzzle and project) (ewbk-13304)

 Interactivity Individual pathway interactivity: Bivariate data (int-4626)

Exercise 13.2 Bivariate data

13.2 Quick quiz **on**   13.2 Exercise

Individual pathways

■ PRACTISE	■ CONSOLIDATE	■ MASTER
1, 4, 5, 7, 9, 13, 16	2, 6, 8, 11, 14, 17	3, 10, 12, 15, 18

Fluency

For questions **1** and **2**, decide which of the variables is independent and which is dependent.

1. **a.** Number of hours spent studying for a Mathematics test and the score on that test
   b. Daily amount of rainfall (in mm) and daily attendance at the Botanical Gardens
   c. Number of hours per week spent in a gym and number of visits per year to a doctor
   d. The amount of computer memory taken by an essay and the length of the essay (in words)

2. **a.** The cost of care in a childcare centre and attendance at the childcare centre
   b. The cost of a property (real estate) and the age of the property
   c. The entry requirements for a certain tertiary course and the number of applications for that course
   d. The heart rate of a runner and their running speed

3. **WE1** The table shows the cost of a wedding reception at 10 different venues. Represent the data on a scatterplot.

No. of guests	30	40	50	60	70	80	90	100	110	120
Total cost $(\times \$1000)$	1.5	1.8	2.4	2.3	2.9	4	4.3	4.5	4.6	4.6

4. **WE2** State the type of relationship between x and y for each of the following scatterplots.

a.

b.

c.

d.

e.
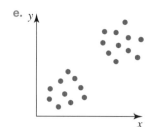

5. State the type of relationship between x and y for each of the following scatterplots.

a.

b.

c.

d.

e.

6. State the type of relationship between x and y for each of the following scatterplots.

a.

b.

c.

d.

e.

Understanding

7. **WE3** A strong positive association has been found between weight gain and the length of holidays.
   Does this mean that holidays cause weight gain?
   Describe a common cause that could link these variables.

8. **WE3** A strong positive association has been found between carbon dioxide concentration and the change in the Earth's average surface temperature.
   Does this mean that carbon dioxide in the atmosphere causes the change in the average Earth's average surface temperature?
   Describe a common cause that could link these variables.

9. **WE4** Eugene is selling leather bags at the local market. During the day he keeps records of his sales. The table below shows the number of bags sold over one weekend and their corresponding prices (to the nearest dollar).

Price of a bag ($)	30	35	40	45	50	55	60	65	70	75	80
Number of bags sold	10	12	8	6	4	3	4	2	2	1	1

a. Construct a scatterplot of the data.
b. State the type of correlation between the two variables and hence draw a corresponding conclusion.

10. The table below shows the number of bedrooms and the price of each of 30 houses.

Number of bedrooms	Price ($'000)	Number of bedrooms	Price ($'000)	Number of bedrooms	Price ($'000)
2	180	3	279	3	243
2	160	2	195	3	198
3	240	6	408	3	237
2	200	4	362	2	226
2	155	2	205	4	359
4	306	7	420	4	316
3	297	5	369	2	200
5	383	1	195	2	158
2	212	3	265	1	149
4	349	2	174	3	286

a. Construct a scatterplot of the data.

b. State the type of correlation between the number of bedrooms and the price of the house and hence draw a corresponding conclusion.

c. Suggest other factors that could contribute to the price of the house.

11. The table shows the number of questions solved by each student on a test and the corresponding total score on that test.

Number of questions	2	0	7	10	5	2	6	3	9	4	8	3	6
Total score (%)	22	39	69	100	56	18	60	36	87	45	84	32	63

a. Construct a scatterplot of the data.

b. Suggest the type of correlation shown in the scatterplot.

c. Give a possible explanation as to why the scatterplot is not perfectly linear.

12. A sample of 25 drivers who had obtained a full licence within the last month were asked to recall the approximate number of driving lessons they had taken (to the nearest 5) and the number of accidents they had while being on P plates. The results are summarised in the table that follows.

a. Represent these data on a scatterplot.

b. Specify the relationship suggested by the scatterplot.

c. Suggest some reasons why this scatterplot is not perfectly linear.

Number of lessons	5	20	15	25	10	35	5	15	10	20	40	25	10
Number of accidents	6	2	3	3	4	0	5	1	3	1	2	2	5
Number of lessons	5	20	40	25	30	15	35	5	30	15	20	10	
Number of accidents	5	3	0	4	1	4	1	4	0	2	3	4	

Reasoning

13. **MC** The scatterplot that best represents the relationship between the amount of water consumed daily by a certain household for a number of days in summer and the daily temperature is:

A.

B.

C.

D.

14. **MC** The scatterplot shows the number of sides and the sum of interior angles for a number of polygons.
Select the statement that is NOT true from the following statements.

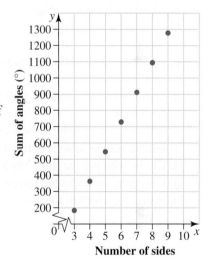

A. The correlation between the number of sides and the angle sum of the polygon is perfectly linear.

B. The increase in the number of sides causes the increase in the size of the angle sum.

C. The number of sides depends on the sum of the angles.

D. The correlation between the two variables is positive.

15. **MC** After studying a scatterplot, it was concluded that there was evidence that the greater the level of one variable, the smaller the level of the other variable. The scatterplot must have shown:

A. a strong, positive correlation

B. a strong, negative correlation

C. a moderate, positive correlation

D. a moderate, negative correlation

Problem solving

16. The table shown gives the number of kicks and handballs obtained by the top 8 players in an AFL game.

Player	A	B	C	D	E	F	G	H
Number of kicks	20	27	21	19	17	18	21	22
Number of handballs	11	3	11	6	5	1	9	7

a. Represent this information on a scatterplot by using the *x*-axis as the number of kicks and the *y*-axis as the number of handballs.

b. State whether the scatterplot supports the claim that the more kicks a player obtains, the more handballs they give.

17. Each point on the scatterplot shows the time (in weeks) spent by a person on a healthy diet and the corresponding mass lost (in kilograms).
Study the scatterplot and state whether each of the following statements is true or false.

a. The number of weeks that the person stays on a diet is the independent variable.

b. The y-coordinates of the points represent the time spent by a person on a diet.

c. There is evidence to suggest that the longer the person stays on a diet, the greater the loss in mass.

d. The time spent on a diet is the only factor that contributes to the loss in mass.

e. The correlation between the number of weeks on a diet and the number of kilograms lost is positive.

18. The scatterplot shown gives the marks obtained by students in two Mathematics tests. Mardi's score in the tests is represented by M. Determine which point represents each of the following students.

a. Mandy, who got the highest mark in both tests

b. William, who got the top mark in test 1 but not in test 2

c. Charlotte, who did better on test 1 than Mardi but not as well in test 2

d. Dario, who did not do as well as Charlotte in both tests

e. Edward, who got the same mark as Mardi in test 2 but did not do so well in test 1

f. Cindy, who got the same mark as Mardi for test 1 but did better than her for test 2

g. Georgina, who was the lowest in test 1

h. Harrison, who had the greatest discrepancy between his two marks

LESSON
13.3 Two-way frequency tables and Likert scales

13.3.1 Two-way frequency tables

eles-6149

- **Two-way tables** are used to examine the relationship between two related categorical variables.
- As the entries in these tables are frequency counts, they are also called **two-way frequency tables**.

- The table shown is a two-way table that relates two variables.

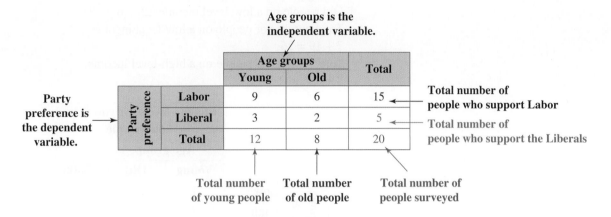

Age groups is the
independent variable.

Party
preference is
the dependent
variable.

Total number of
people who support Labor

Total number of
people who support the Liberals

Total number
of young people

Total number
of old people

Total number of
people surveyed

- The table displays the relationship between age groups and party preference. 'Age group' is the independent variable and 'party preference' is the dependent variable.

WORKED EXAMPLE 5 Displaying data in a two-way frequency table

A sample of 40 people was surveyed about their income level (low or high). The sample consisted of 15 young people (25–45 years old) and 25 old people (above 65 years old). Seven young people were on a low-level income and 6 old people were on a high-level income.
Display this data in a two-way table and determine the number of young people on a high-level income.

THINK	WRITE
1. State the two categorical variables.	Age group: young people and old people Income level low and high
2. Decide where the two variables are placed.	The independent variable is 'Age group' — columns. The dependent variable is 'Income level' — rows.
3. Construct the two-way table and label both the columns and the rows.	

Income levels for young and old people

		Age group		
		Young	Old	Total
Income level	Low			
	High			
	Total			

4. Fill in the known data:
   - 7 young people are on a low-level income.
   - 6 old people are on a high-level income.
   - There are 15 young people in total and 25 old people in total.
   - The total number of people in the sample is 40.

Income levels for young and old people

		Age group		
		Young	Old	Total
Income level	Low	7		
	High		6	
	Total	15	25	40

5. Determine the unknown data.

Young people on a high-level income: $15 - 7 = 8$
Old people on a low-level income: $25 - 6 = 19$
Total number of people on a low-level income:
$7 + 19 = 26$
Total number of people on a high-level income:
$8 + 6 = 14$

6. Fill in the table.

		Income levels for young and old people		
		Age group		
		Young	Old	Total
Income level	Low	7	19	26
	High	8	6	14
	Total	15	25	40

7. Write the answer as a sentence.

The total number of young people on a high-level income is 8.

13.3.2 Two-way relative frequency tables

eles-6150

- Two-way relative frequency tables display percentages or proportions (ratios) (called **relative frequencies**) rather than frequency counts.
- The relative frequencies may be shown as a proportion, a decimal (to the nearest hundredth) or a percent (to the nearest percent).
- Two-way relative frequency tables are good for seeing if there is an association between two variables.

Formula for relative frequency

The frequency given as a proportion is called relative frequency.

$$\text{relative frequency} = \frac{\text{score}}{\text{total number of scores}}$$

- Consider the party preference table shown. As the numbers of young people and old people are different, it is difficult to compare the data.

		Party preference among young and old people		
		Age group		
		Young	Old	Total
Party preference	Labor	9	6	15
	Liberal	3	2	5
	Total	12	8	20

- Nine young people and six old people prefer Labor; however, there are more young people in the sample than old people.

- To accurately compare the data, we need to convert these values in a more meaningful way as proportions or percentages.
- The relative frequency table for the party preference data is:

		Party preference among young and old people		
		Age group		Total
		Young	Old	
Party preference	Labor	$\dfrac{9}{20}$	$\dfrac{6}{20}$	$\dfrac{15}{20}$
	Liberal	$\dfrac{3}{20}$	$\dfrac{2}{20}$	$\dfrac{5}{20}$
	Total	$\dfrac{12}{20}$	$\dfrac{8}{20}$	$\dfrac{20}{20}$

- In simplified form using decimal numbers, the relative frequency table for party preference becomes:

		Party preference among young and old people		
		Age group		Total
		Young	Old	
Party preference	Labor	0.45	0.30	0.75
	Liberal	0.15	0.10	0.25
	Total	0.60	0.40	1.00

Note: In a relative frequency table, the total value is always 1.

- The frequency written as a percentage is called **percentage relative frequency**.

Formula for percentage relative frequency

We can use the following formula to calculate percentage relative frequency:

$$\text{percentage relative frequency} = \frac{\text{score}}{\text{total number of scores}} \times 100\%$$

or

$$\text{percentage relative frequency} = \text{relative frequency} \times 100\%$$

Consider the party preference data discussed previously with the relative frequencies shown.

Party preference among young and old people			
	Age group		
	Young	Old	Total
Labor	0.45	0.30	0.75
Liberal	0.15	0.10	0.25
Total	0.60	0.40	1.00

(Left-hand side labelled: **Party preference**)

Calculate the percentage relative frequency for all entries in the table out of the total number of people. Discuss your findings.

THINK

1. Write the formula for percentage relative frequency.

2. Substitute all scores into the percentage relative frequency formula.

WRITE

Percentage relative frequency = relative frequency × 100%

Party preference among young and old people			
	Age group		
	Young (%)	Old (%)	Total (%)
Labor	0.45×100	0.30×100	0.75×100
Liberal	0.15×100	0.10×100	0.25×100
Total	0.60×100	0.40×100	1.00×100

Party preference

3. Simplify.

Party preference among young and old people			
	Age group		
	Young (%)	Old (%)	Total (%)
Labor	45	30	75
Liberal	15	10	25
Total	60	40	100

(Left-hand side labelled: **Party preference**)

4. Discuss your findings.

Of the total number of people, 75% prefer Labor and 25% prefer the Liberals. 60% of people are young and 40% of people are old. 45% of people who prefer Labor are young and 30% of people who prefer the Liberals are old.

15% of people who prefer the Liberals are young and 10% of people who prefer the Liberals are old.

13.3.3 Two-way tables and Likert scales

eles-6151

- The five-point Likert scale is a type of rating scale used to measure attitudes, opinions and perceptions.
- It typically consists of five response options ranging from strongly agree to strongly disagree, with a neutral option in the middle.
- The response options are usually numbered 1 to 5 or labeled with descriptors such as 'strongly agree', 'agree', 'neutral', 'disagree', and 'strongly disagree'.
- Respondents are asked to choose the response that best represents their level of agreement or disagreement with a statement or question.

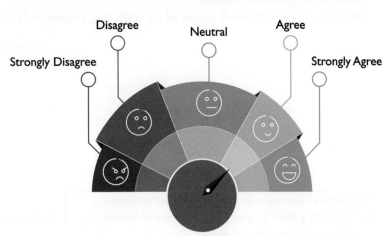

Likert Scale

For example, a five-point Likert scale question might be: 'How strongly do you agree or disagree with the following statement: "I enjoy spending time outdoors"?' The response options could be:

1. Strongly disagree
2. Disagree
3. Neither agree nor disagree
4. Agree
5. Strongly agree

- Researchers and survey designers often use the 5-point Likert scale because it allows for a relatively fine-grained measurement of attitudes and opinions while remaining simple and easy to use.
- Two-way tables can be used to investigate and compare survey responses to questions involving five-point Likert scales.
- This type of analysis can provide valuable insights into the attitudes and opinions of the survey respondents.

COLLABORATIVE TASK: Five-point Likert survey

The following collaborative task not only engages students in critical thinking and decision-making but also fosters teamwork and communication skills. It also allows for a more personalized and relevant Likert scale survey that the students can relate to.

1. Divide the students into small groups of 3 or 4.
2. Choose a topic for the Likert scale, such as 'How much do you enjoy playing video games?'
3. Have each group brainstorm 5–10 statements related to the topic that could be rated on a 5-point scale, for example 'I enjoy playing video games alone', 'I enjoy playing video games with friends', 'I get frustrated easily while playing video games'
4. Once each group has generated their statements, have them share with the rest of the class.
5. As a class, select the top 5–10 statements from all groups and create a final Likert scale survey.
6. Have each student individually rate each statement on the survey using a 5-point Likert scale.
7. Collect the surveys and analyse the results as a class, discussing the trends and patterns that emerge.

Students at Wiley High School were surveyed and asked to rate their agreement with the statement *'Maths is my favourite subject.'*

The following results were obtained by collecting responses from 100 students:

	Students	
	Junior	Senior
Strongly disagree	8	10
Disagree	16	22
Neutral	12	18
Agree	42	30
Strongly agree	22	20

a. Convert the data in the table to percentages.

b. Identify patterns and relationships between responses by the senior and junior students. Comment on your findings.

THINK

a. To calculate the percentage relative frequency for each cell in the table, we need to divide the frequency of each cell by the total number of respondents (100) and multiply by 100.

WRITE

a.

	Students	
	Junior	Senior
Strongly disagree	$\frac{8}{100} \times 100\% = 8\%$	$\frac{10}{100} \times 100\% = 10\%$
Disagree	$\frac{16}{100} \times 100\% = 16\%$	$\frac{22}{100} \times 100\% = 22\%$
Neutral	$\frac{12}{100} \times 100\% = 12\%$	$\frac{18}{100} \times 100\% = 18\%$
Agree	$\frac{42}{100} \times 100\% = 42\%$	$\frac{30}{100} \times 100\% = 30\%$
Strongly agree	$\frac{22}{100} \times 100\% = 22\%$	$\frac{20}{100} \times 100\% = 20\%$

b. From the relative percentage frequency table we can see that both junior students (42 respondents) and senior students (30 respondents) students agree with the statement.

b. More students (both junior and senior) either agree or strongly agree with the statement *'Maths is my favourite subject.'*

Specifically, 64% of juniors agreed or strongly agreed with the statement, while only 52% of seniors agreed or strongly agreed with the statement. More senior students disagreed with the statement than junior students.

  Resources

13.3 Quick quiz on	13.3 Exercise

Individual pathways

■ PRACTISE	■ CONSOLIDATE	■ MASTER
1, 4, 7, 10, 13	2, 5, 8, 11, 14	3, 6, 9, 12, 15

Fluency

1. For each of the following two-way frequency tables, complete the missing entries.

a.

		Grades		Total
		Year 12	Year 11	
Transport preference	Tram	25	i	47
	Train	ii	iii	iv
	Total	51	v	92

b.

		Voters' ages		
		18 to 45	over 45
Party preference	Labor	i	42%
	Liberal	53%	ii
	Total	iii	iv

2. Determine the missing values in the two-way table shown.

		Parking place and car theft		
		Parking type		Total
		In driveway	On the street	
Theft	**Car theft**		37	
	No car theft	16		
	Total		482	500

3. A survey of 263 students found that 58 students who owned a mobile phone also owned a tablet, while 11 students owned neither. The total number of students who owned a tablet was 174.
   Construct a complete two-way table to display this data and calculate all the unknown entries.

4. **WE5** The numbers of the Year 11 Maths students at Senior Secondary College who passed and failed their exams are displayed in the table shown.

		Year 11 students' Mathematics results		
		Result		Total
		Pass	Fail	
Subject	**Essential Mathematics**		23	
	General Mathematics		49	74
	Mathematical Methods	36	62	
	Specialist Mathematics	7		25
	Total	80		

a. Complete the two-way table.
b. Determine the total number of Year 11 students who study Mathematics at Senior Secondary College.
c. Calculate the number of students enrolled in Essential Mathematics.

5. George is a hairdresser who keeps a record of all his customers and their hairdressing requirements. He made a two-way table for last month and recorded the numbers of customers who wanted a haircut only, and those who wanted a haircut and colour.

a. Design a two-way table that George could use to record the numbers of young and old customers who wanted a haircut only, and those who wanted a haircut and colour.
b. George had 56 old customers in total and 2 young customers who wanted a haircut and colour. Place these values in the table from part a.
c. George had 63 customers during last month, with 30 old customers requiring a haircut and colour. Fill in the rest of the two-way table.

Understanding

6. A group of 650 people were asked their preferences on soft drink. The results are shown in the two-way table.

		Age group		Total
		Teenagers	Adults	
Drink types	Pepsi	221	155	376
	Coke	166	108	274
	Total	387	263	650

Convert the numbers in this table to percentages.

7. **WE6** Consider the data given in the two-way table shown.

		Cat and dog owners		
		Dog owners		Total
		Own	Do not own	
Cat owners	Own	12	78	90
	Do not own	151	64	215
	Total	163	142	305

a. Calculate the relative frequencies for all the entries in the two-way table given. Give your answers correct to 3 decimal places or as fractions.
b. Calculate the percentage relative frequencies for all the entries in the two-way table out of the total number of pet owners.

8. Using the two-way table from question 4, construct:
a. the relative frequency table, correct to 4 decimal places
b. the percentage relative frequency table.

9. **WE7** Students at Jacaranda Secondary School were surveyed and asked to rate their agreement with the statement '*How often do you feel stressed about schoolwork?*'
The following results were attained by collecting responses from 100 students.

	Students	
	Junior	**Senior**
Never	35	20
Rarely	30	25
Sometimes	20	30
Often	10	15
Very often	5	10

a. Convert the data in this table to percentages.
b. Identify patterns and relationships between responses by the senior and junior students. Comment on your findings.

Reasoning

10. The two-way-table shows the reactions of administrative staff and technical staff to an upgrade of the computer systems at a large corporation.

		Staff type		Total
		Administrative	**Technical**	
Attitude	For	53	98	151
	Against	37	31	68
	Total	90	129	219

a. **MC** From the table, we can conclude that:

A. 53% of administrative staff were for the upgrade.
B. 37% of administrative staff were for the upgrade.
C. 37% of administrative staff were against the upgrade.
D. 59% of administrative staff were for the upgrade.

b. **MC** From the table, we can conclude that:

A. 98% of technical staff were for the upgrade.
B. 65% of technical staff were for the upgrade.
C. 76% of technical staff were for the upgrade.
D. 31% of technical staff were against the upgrade.

11. Ninety-three people less than 40 years of age and 102 people aged 40 and over were asked where their priority financially was, given the three options 'mortgage', 'superannuation' or 'investing'.
Eighteen people in the 40 and over category and 42 people in the less than 40 category identified mortgage as their priority, whereas 21 people under 40 years of age and 33 people aged 40 and over said investment was most important.
The rest suggested superannuation was their priority.

a. Present these data in percentage form in a two-way table.
b. Compare the financial priorities of the under 40s to the people aged 40 and over.

12. Delegates at the respective Liberal Party and Australian Labor Party conferences were surveyed on whether or not they believed that marijuana should be legalised. Sixty-two Liberal delegates were surveyed and 40 of them were against legalisation. Seventy-one Labor delegates were surveyed and 43 were against legalisation.

a. Present the data in percentage form in a two-way frequency table.
b. Comment on any differences between the reactions of the Liberal and Labor delegates.

Problem solving

13. The manager at the local cinema recorded the movie genres and the ages of the audience at the Monday matinee movies. The total audience was 156 people. A quarter of the audience were children under the age of 10, 21 were teenagers and the rest were adults. There were 82 people who watched the drama movie, 16 teenagers who watched comedy, and 11 people under the age of 10 who watched the drama movie.

a. Construct a complete two-way table.
b. Determine the total number of adults who watched the comedy movie.
c. Calculate the total number of children who watched a Monday matinee movie.

14. Consider the data given in the two-way table shown.

| | | Swimming attendance | | |
| | | People | | |
		Adults	Children	Total
Day of the week	Monday	14	31	45
	Tuesday	9	17	26
	Wednesday	10	22	32
	Thursday	15	38	53
	Friday	18	45	63
	Saturday	27	59	86
	Sunday	32	63	95
	Total	125	275	400

a. Calculate the relative frequencies for all the entries for adults out of the total number of adults.
b. Calculate the percentage relative frequencies for all the entries for adults out of the total number of adults.

15. A ballroom studio runs classes for adults, teenagers and children in Latin, modern and contemporary dance. The enrolments in these classes are displayed in the two-way table shown.

| | | Dance class enrolments | | | |
| | | People | | | |
		Adults	Teenagers	Children	
Dance	Latin			23	75
	Modern		34	49	96
	Contemporary	36		62	
	Total	80			318

a. Complete the two-way table.
b. Construct a relative frequency table for this data set.
c. Construct a percentage relative frequency table for this data set.

LESSON
13.4 Lines of best fit by eye

▶ 13.4.1 Lines of best fit by eye

eles-4968

- A **line of best fit** is a line that follows the trend of the data in a scatterplot.

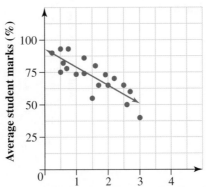

- A line of best fit is most appropriate for data with strong or moderate linear correlation.
- Drawing lines of best fit by eye is done by placing a line that:
  - represents the data trend
  - has an equal number of points above and below the line.

Determine the equation of a line of best fit by eye

To determine the equation of the line of best fit, follow these steps:
1. **Choose two points on the line.**
   Note: **It is best to use two data points on the line if possible.**
2. **Write the points in the coordinate form (x_1, y_1) and (x_2, y_2).**
3. **Calculate the gradient using $m = \dfrac{y_2 - y_1}{x_2 - x_1}$.**
4. **Write the equation in the form $y = mx + c$ using the m found in step 3.**
5. **Substitute one coordinate into the equation and rearrange to find c.**
6. **Write the final equation, replacing x and y if needed.**

The data in the table shows the cost of using the internet at a number of different internet cafes based on hours used per month.

Hours used per month	10	12	20	18	10	13	15	17	14	11
Total monthly cost ($)	15	18	30	32	18	20	22	23	22	18

a. Construct a scatterplot of the data.
b. Draw the line of best fit by eye.
c. Determine the equation of the line of best fit in terms of the variables n (number of hours) and C (monthly cost).

THINK

a. Draw the scatterplot placing the independent variable (hours used per month) on the horizontal axis and the dependent variable (total monthly cost) on the vertical axis. Label the axes.

b. 1. Carefully analyse the scatterplot.

2. Position the line of best fit so there is approximately an equal number of data points on either side of the line and so that all points are close to the line.
Note: With the line of best fit, there is no single definite solution.

c. 1. Select two points on the line that are not too close to each other.

2. Calculate the gradient of the line.

3. Write the rule for the equation of a straight line.

WRITE/DRAW

a.

b.

c. Let $(x_1, y_1) = (13, 20)$ and $(x_2, y_2) = (20, 30)$.

$$m = \frac{y_2 - y_1}{x_2 - x_1}$$

$$m = \frac{30 - 20}{20 - 13}$$

$$= \frac{10}{7}$$

$y = mx + c$

4. Substitute the known values into the equation.

$$y = \frac{10}{7}x + c$$

5. Substitute one pair of coordinates, say $(13, 20)$, into the equation to evaluate c.

$$20 = \frac{10}{7}(13) + c$$

$$c = 20 - \frac{130}{7}$$

$$= \frac{140 - 130}{7}$$

$$= \frac{10}{7}$$

6. Write the equation.
Note: The values of c and m are the same in this example. This is not always the case.

$$y = \frac{10}{7}x + \frac{10}{7}$$

7. Replace x with n (number of hours used) and y with C (the total monthly cost) as required.

$$C = \frac{10}{7}n + \frac{10}{7}$$

13.4.2 Predictions using lines of best fit

eles-4969

- Predictions can be made by using the line of best fit.
- To make a prediction, use one coordinate and determine the other coordinate by using either:
  - the line of best fit, or
  - the equation of the line of best fit.
- Predictions will be made using:
  - **interpolation** if the prediction sits within the given data
  - **extrapolation** if the prediction sits outside the given data.

Interpolation versus extrapolation

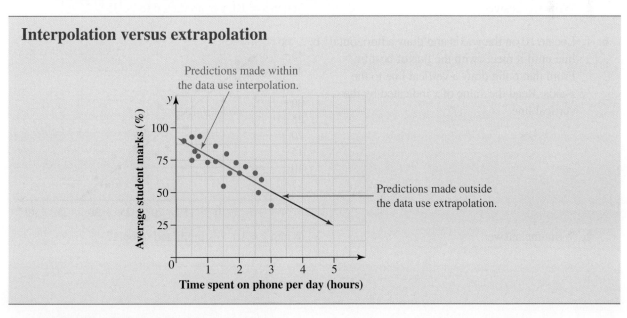

- Predictions will be reliable if they are made:
  - using interpolation
  - from data with a strong correlation
  - from a large number of data.

WORKED EXAMPLE 9 Making predictions using the line of best fit

Use the given scatterplot and line of best fit to predict:
a. the value of y when x = 10
b. the value of x when y = 10.

THINK

WRITE/DRAW

a. 1. Locate 10 on the *x*-axis and draw a vertical line until it meets with the line of best fit. From that point, draw a horizontal line to the *y*-axis. Read the value of *y* indicated by the horizontal line.

a.

2. Write the answer.

When x = 10, y is predicted to be 35.

b. 1. Locate 10 on the *y*-axis and draw a horizontal line until it meets with the line of best fit. From that point draw a vertical line to the *x*-axis. Read the value of *x* indicated by the vertical line.

b.

2. Write the answer.

When y = 10, x is predicted to be 27.

The table shows the number of boxes of tissues purchased by hayfever sufferers and the number of days affected by hayfever during the blooming season in spring.

Number of days affected by hayfever (d)	3	12	14	7	9	5	6	4	10	8
Total number of boxes of tissues purchased (T)	1	4	5	2	3	2	2	2	3	3

a. Construct a scatterplot of the data and draw a line of best fit.
b. Determine the equation of the line of best fit.
c. Interpret the meaning of the gradient.
d. Use the equation of the line of best fit to predict the number of boxes of tissues purchased by people suffering from hayfever over a period of:
   i. 11 days   ii. 15 days.

THINK

a. 1. Draw the scatterplot showing the independent variable (number of days affected by hayfever) on the horizontal axis and the dependent variable (total number of boxes of tissues purchased) on the vertical axis.

   2. Position the line of best fit on the scatterplot so that there is approximately an equal number of data points on either side of the line.

b. 1. Select two points on the line that are not too close to each other.

   2. Calculate the gradient of the line.

WRITE/DRAW

a.

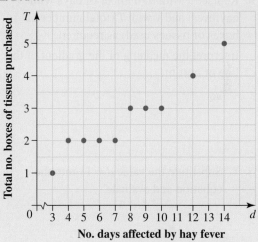

No. days affected by hay fever

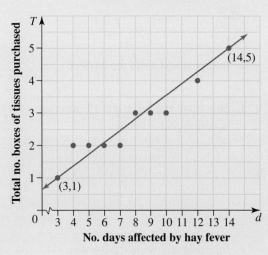

No. days affected by hay fever

b.   Let $(x_1, y_1) = (3, 1)$ and $(x_2, y_2) = (14, 5)$.

$$m = \frac{y_2 - y_1}{x_2 - x_1}$$

$$m = \frac{5 - 1}{14 - 3} = \frac{4}{11}$$

3. Write the rule for the equation of a straight line.

$$y = mx + c$$

4. Substitute the known values into the equation, say $(3, 1)$, into the equation to calculate c.

$$y = \frac{4}{11}x + c$$

$$1 = \frac{4}{11}(3) + c$$

$$c = 1 - \frac{12}{11}$$

$$= \frac{-1}{11}$$

5. Write the equation.

$$y = \frac{4}{11}x - \frac{1}{11}$$

6. Replace x with d (number of days with hayfever) and y with T (total number of boxes of tissues used) as required.

$$T = \frac{4}{11}d - \frac{1}{11}$$

c. Interpret the meaning of the gradient of the line of best fit.

c. The gradient indicates an increase in sales of tissues as the number of days affected by hayfever increases. A hayfever sufferer is using on average $\frac{4}{11}$ (or about 0.36) of a box of tissues per day.

d. i. 1. Substitute the value $d = 11$ into the equation and evaluate.

d. i. When $d = 11$,

$$T = \frac{4}{11} \times 11 - \frac{1}{11}$$

$$= 4 - \frac{1}{11}$$

$$= 3\frac{10}{11}$$

2. Interpret and write the answer.

In 11 days the hayfever sufferer will need 4 boxes of tissues.

ii. 1. Substitute the value $d = 15$ into the equation and evaluate.

ii. When $d = 15$,

$$T = \frac{4}{11} \times 15 - \frac{1}{11}$$

$$= \frac{60}{11} - \frac{1}{11}$$

$$= 5\frac{4}{11}$$

2. Interpret and write the answer.

In 15 days the hayfever sufferer will need about 6 boxes of tissues.

Reliability of predictions

- The more pieces of data there are in a set, the better the line of best fit you will be able to draw. More data points allow more reliable predictions.
- Interpolation closer to the centre of a data set will be more reliable than interpolation closer to the edge.
- A strong correlation between the points of data will give a more reliable line of good fit. This is shown when all of the points appear close to the line of best fit.
- The more points that appear far away from the line of best fit, the less reliable other predictions will be.
- When we use extrapolation, we are assuming that the association shown in the scatterplot continues beyond the data. However, for most sets of data, there will be a point where predictions made through extrapolation end up being nonsensical.
- Extrapolation is considered to be an unreliable prediction in most cases.

DISCUSSION

Why is extrapolation not considered to be reliable?

 Resources

 eWorkbook   Topic 13 Workbook (worksheets, code puzzle and project) (ewbk-13304)

 Interactivities Individual pathway interactivity: Lines of best fit (int-4627)
Lines of best fit (int-6180)
Interpolation and extrapolation (int-6181)

Exercise 13.4 Lines of best fit by eye   learn on

13.4 Quick quiz on	13.4 Exercise

Individual pathways

■ PRACTISE 1, 2, 5, 8, 11	■ CONSOLIDATE 3, 6, 9, 12	■ MASTER 4, 7, 10, 13

Fluency

1. **WE8** The data in the table shows the distances travelled by 10 cars and the amount of petrol used for their journeys (to the nearest litre).

Distance travelled, d (km)	52	36	83	12	44	67	74	23	56	95
Petrol used, P (L)	7	5	9	2	7	9	12	3	8	14

   a. Construct a scatterplot of the data and draw the line of best fit.
   b. Determine the equation of the line of best fit in terms of the variables d (distance travelled) and P (petrol used).

2. **WE9** Use the given scatterplot and line of best fit to predict:

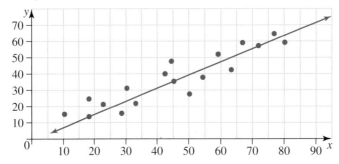

a. the value of y when $x = 45$
b. the value of x when $y = 15$.

3. Analyse the graph shown.

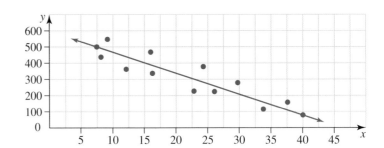

a. Use the line of best fit to estimate the value of y when the value of x is:

i. 7   ii. 22   iii. 36.

b. Use the line of best fit to estimate the value of x when the value of y is:

i. 120   ii. 260   iii. 480.

c. Determine the equation of the line of best fit, if it is known that it passes through the points $(5, 530)$ and $(40, 75)$.

d. Use the equation of the line to verify the values obtained from the graph in parts a and b.

4. A sample of ten Year 10 students who have part-time jobs was randomly selected. Each student was asked to state the average number of hours they work per week and their average weekly earnings (to the nearest dollar). The results are summarised in the table below.

Hours worked, h	4	8	15	18	10	5	12	16	14	6
Weekly earnings, E ($)	23	47	93	122	56	33	74	110	78	35

a. Construct a scatterplot of the data and draw the line of best fit.
b. Write the equation of the line of best fit in terms of the variables h (hours worked) and E (weekly earnings).
c. Interpret the meaning of the gradient.

Understanding

5. **WE10** The following table shows the average weekly expenditure on food for households of various sizes.

Number of people in a household	1	2	4	7	5	4	3	5
Cost of food ($ per week)	70	100	150	165	150	140	120	155
Number of people in a household	2	4	6	5	3	1	4	
Cost of food ($ per week)	90	160	160	160	125	75	135	

a. Construct a scatterplot of the data and draw in the line of best fit.
b. Determine the equation of the line of best fit. Write it in terms of the variables n (for the number of people in a household) and C (weekly cost of food).
c. Interpret the meaning of the gradient.
d. Use the equation of the line of best fit to predict the weekly food expenditure for a family of:

i. 8   ii. 9   iii. 10.

6. The number of hours spent studying, and the percentage marks obtained by a group of students on a test are shown in this table.

Hours spent studying	45	30	90	60	105	65	90	80	55	75
Marks obtained	40	35	75	65	90	50	90	80	45	65

a. State the values for marks obtained that can be used for interpolation.
b. State the values for hours spent studying that can be used for interpolation.

7. The following table shows the gestation time and the birth mass of 10 babies.

Gestation time (weeks)	31	32	33	34	35	36	37	38	39	40
Birth mass (kg)	1.080	1.470	1.820	2.060	2.230	2.540	2.750	3.110	3.080	3.370

a. i. Construct a scatterplot of the data.
   ii. Suggest the type of correlation shown by the scatterplot.
b. Draw in the line of best fit and determine its equation. Write it in terms of the variables t (gestation time) and M (birth mass).
c. Determine what the value of the gradient represents.
d. Although full term for gestation is considered to be 40 weeks, some pregnancies last longer. Use the equation obtained in part **b** to predict the birth mass of babies born after 41 and 42 weeks of gestation.
e. Many babies are born prematurely. Using the equation obtained in part **b**, predict the birth mass of a baby whose gestation time was 30 weeks.
f. Calculate a baby's gestation time (to the nearest week) if their birth mass was 2.390 kg.

Reasoning

8. **MC** Consider the scatterplot shown.

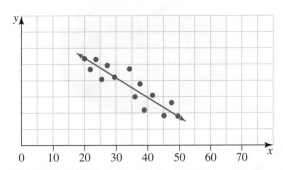

The line of best fit on the scatterplot is used to predict the values of y when $x = 15$, $x = 40$ and $x = 60$.

a. Interpolation would be used to predict the value of y when the value of x is:

    A. 15 and 40
    B. 15 and 60
    C. 15 only
    D. 40 only

b. The prediction of the y-value(s) can be considered reliable when:

    A. $x = 15$ and $x = 40$
    B. $x = 15$, $x = 40$ and $x = 60$
    C. $x = 40$
    D. $x = 40$ and $x = 60$

9. **MC** The scatterplot below is used to predict the value of y when $x = 300$. This prediction is:

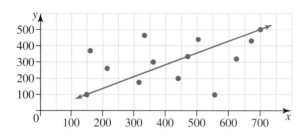

   A. reliable, because it is obtained using interpolation
   B. not reliable, because it is obtained using extrapolation
   C. not reliable, because only x-values can be predicted with confidence
   D. not reliable, because there is no correlation between x and y

10. As a part of her project, Rachel is growing a crystal. Every day she measures the crystal's mass using special laboratory scales and records it. The table below shows the results of her experiment.

Day number	1	2	3	4	5	8	9	10	11	12	15	16
Mass (g)	2.5	3.7	4.2	5.0	6.1	8.4	9.9	11.2	11.6	12.8	16.1	17.3

Measurements on days 6, 7, 13 and 14 are missing, since these were 2 consecutive weekends and hence Rachel did not have a chance to measure her crystal, which is kept in the school laboratory.

   a. Construct the scatterplot of the data and draw in the line of best fit.
   b. Determine the equation of the line of best fit. Write the equation using the variables d (day of the experiment) and M (mass of the crystal).
   c. Interpret the meaning of the gradient.
   d. For her report, Rachel would like to fill in the missing measurements (that is the mass of the crystal on days 6, 7, 13 and 14).
      Use the equation of the line of best fit to help Rachel determine these measurements.
      Explain whether this is an example of interpolation or extrapolation.
   e. Rachel needed to continue her experiment for 2 more days, but she fell ill and had to miss school.
      Help Rachel to predict the mass of the crystal on those two days (that is days 17 and 18) using the equation of the line of best fit.
      Explain whether these predictions are reliable.

Problem solving

11. Ari was given a baby rabbit for his birthday. To monitor the rabbit's growth, Ari decided to measure it once a week.

The table shows the length of the rabbit for various weeks.

Week number, n	1	2	3	4	6	8	10	13	14	17	20
Length (cm), l	20	21	23	24	25	30	32	35	36	37	39

a. Construct a scatterplot of the data.
b. Draw a line of best fit and determine its equation.
c. As can be seen from the table, Ari did not measure his rabbit on weeks 5, 7, 9, 11, 12, 15, 16, 18 and 19. Use the equation of the line of best fit to predict the length of the rabbit for those weeks.
d. Explain whether the predictions made in part c were an example of interpolation or extrapolation.
e. Predict the length of the rabbit in the next three weeks (that is weeks 21–23) using the line of best fit from part c.
f. Explain whether the predictions that have been made in part e are reliable.

12. Laurie is training for the long jump, hoping to make the Australian Olympic team. His best jump each year is shown in the table below.

Age (a)	Best jump (B) (metres)
8	4.31
9	4.85
10	5.29
11	5.74
12	6.05
13	6.21
14	—
15	6.88
16	7.24
17	7.35
18	7.57

a. Plot the points generated by the table on a scatterplot.
b. Join the points generated with straight line segments.
c. Draw a line of best fit and determine its equation.
d. The next Olympic Games will occur when Laurie is 20 years old.
   Use the equation of the line of best fit to estimate Laurie's best jump that year and whether it will pass the qualifying mark of 8.1 metres.
e. Explain whether a line of best fit is a good way to predict future improvement in this situation.
   State the possible problems there are with using a line of best fit.
f. The Olympic Games will also be held when Laurie is 24 years old and 28 years old. Using extrapolation, what length would you predict Laurie could jump at these two ages?
   Discuss whether this is realistic.

g. When Laurie was 14, he twisted a knee in training and did not compete for the whole season. In that year, a national junior championship was held. The winner of that championship jumped 6.5 metres.
   Use your line of best fit to predict whether Laurie would have won that championship.

13. Sam has a mean score of 88 per cent for his first nine tests of the semester. In order to receive an A+ his score must be 90 per cent or higher. There is one test remaining for the semester.
   Explain whether it is possible for him to receive an A+.

LESSON
13.5 Linear regression and technology (10A)

LEARNING INTENTIONS

At the end of this lesson you should be able to:
- display a scatterplot using technology
- determine the equation of a regression line (line of best fit) using technology
- display a scatterplot with its regression line using technology
- use a regression line to make predictions.

13.5.1 Scatterplots using technology

eles-4970

- Scatterplots can be displayed using graphics calculators and spreadsheets.
- To display a scatterplot using technology:
  - first input the data with the independent variable in the first column and the dependent variable in the second column
  - then use the technology to draw the plot.

WORKED EXAMPLE 11 Displaying a scatterplot using technology

The table shows the amount of time (hours) and the amount of distance walked (km) on a bushwalk. Display the data on a scatterplot using technology.

Time, hours (x)	1	2	3	4
Distance, km (y)	3.11	4.73	6.08	7.54

THINK

1. Determine which data will go in which column by identifying the independent and dependent variable.

2. Input the data into the spreadsheet or calculator.

3. Use the spreadsheet or calculator to create a scatterplot. Type your data into the spreadsheet and highlight all your data (including headings).
   For Google Sheets:
   Go to **Insert** and select **Chart**.
   For Excel:
   Go to **Insert**, select the scatterplot icon ⊾ and choose the **Scatter** option.

WRITE

Time is the independent variable — it will go in the first column. Distance is the dependent variable — it will go in the second column.

TI	THINK	DISPLAY/WRITE	CASIO	THINK	DISPLAY/WRITE

TI | THINK

1. In a new problem, on a Lists & Spreadsheet page, label column A as 'time' and B as 'distance'. Enter the data from the question.

DISPLAY/WRITE

2. Add a page by pressing CTRL then DOC and selecting:
   • 2: Add Graphs
   In the graphs page, press:
   • MENU
   • 3: Graph Entry/Edit
   • 6: Scatter Plot
   On the first line (next to x ←), press VAR and select 'time'. On the second line (next to y ←), press VAR and select 'distance'. Then press ENTER.
   To adjust the screen, press:
   • MENU
   • 4: Window / Zoom
   • 9: Zoom – Data

CASIO | THINK

1. From the menu, select Spreadsheet. Enter the data from the question in the columns A and B.

2. Highlight both columns A and B and tap:
   • Graph
   • Scatter

DISPLAY/WRITE

13.5.2 Regression lines using technology

eles-4971

- **Regression lines** are another name for lines of best fit.
- Using technology, it is possible to calculate and sketch a regression line with more accuracy compared to using a line of best fit by eye.
- Technology can be used to:
  - draw a regression line
  - determine the equation of the line.
- Note that regression lines are **only valid** if the independent and dependent variables have a connection.

WORKED EXAMPLE 12 Displaying a regression line using technology

The following data shows the amount of time (hours) and the amount of distance walked (km) on a bushwalk.
Determine the equation of the regression line and display the regression line using a spreadsheet.

Time, hours (x)	1	2	3	4
Distance, km (y)	3.11	4.73	6.08	7.54

THINK	WRITE
1. Use the scatterplot from Worked example 11. Start by sketching the regression line using the spreadsheet option called trendline.	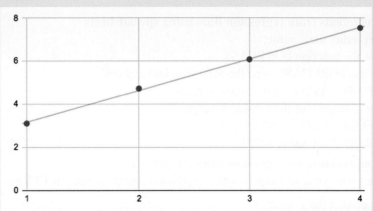
2. On the spreadsheet, display the equation by using the spreadsheet option to show equations on the graph.	The equation is $y = 1.46x + 1.71$.

TI | THINK	DISPLAY/WRITE	CASIO | THINK	DISPLAY/WRITE
1. Start with the scatterplot from Worked example 11. To determine the equation of the regression line, return to the spreadsheet page and click into the third column. Press: • MENU • 4: Statistics • 1: Stat Calculations • 3. Linear Regression (mx + b) In X List select 'time' and in Y List select 'distance', then click OK.		Start with the scatterplot from Worked example 11. To determine the equation of the regression line, tap: • Calc • Regression • Linear Reg	
2. To plot the regression line, return to graph page 1.2 and press: • MENU • 3: Graph Entry/Edit • 1: Function Press the up arrow to see f1 (x) = 1.464x + 1.705 Then press ENTER.			

13.5.3 Using regression lines to make predictions

eles-4972

- The regression line can be used to make predictions by using technology.
- Predictions will be reliable if they are made using interpolation, from data with a strong correlation and from a large number of data (see section 13.4.2).

Digital technology

1. **Predictions from regression lines using spreadsheets**
   To predict the y-value:
   Use the function FORECAST.
   To use FORECAST, type the following into any cell:
   = FORECAST(x-value, y-data, x-data)
   For example, for the x-value 3.5 type:
   = FORECAST(3.5, B2:B6, A2:A6)
   and find the y-value is 2.59.

2. **Predictions from regression lines using CAS**
   Start with your scatterplot with a regression line from section 13.5.2.

   To predict the y-value:
   Press **menu** then **3: Graph Entry/Edit**. Choose
   6: Scatterplot and press up to see **s1**. In **s1** untick the blue box. Press
   menu then **5: Trace** and select **1: Graph Trace**. Type the x-value and
   press **enter.** The y-coordinate will appear.

   For this example, when x is 3.5, y is 2.59.

WORKED EXAMPLE 13 Using a regression line to make predictions

For the following data:
a. **use technology to predict the distance after 2.2 hours**
b. **use technology to predict the distance after 6.4 hours**
c. **explain whether these predictions are reliable.**

Time, hours (x)	1	2	3	4
Distance, km (y)	3.11	4.73	6.08	7.54

THINK

a. 1. Use the spreadsheet or CAS pages set up in Worked example 12.

   2. We need to calculate the distance value, which is the y-value.
   For a spreadsheet, use the function FORECAST.
   OR
   For CAS, start on the graph page and use the Trace tool.

WRITE

a. See Worked example 12.

   In a spreadsheet, type:
   = FORECAST (2.2, B2:B5, A2:A5)
   Note the order B2:B5 first and A2:A5 second.
   Distance = 4.93 km
   OR
   Using CAS, select Trace, type the value 2.2 and press Enter.
   Note: The scatterplot must be turned off.
   Distance = 4.93 km

b. We need to calculate the distance value, which is the *y*-value.
For a spreadsheet, use the function FORECAST
OR
For CAS, start on the graph page and use the Trace tool.

b. In a spreadsheet, type:
= FORECAST(6.4, B2:B5, A2:A5)
Note the order B2:B5 first and A2:A5 second.
Distance = 11.07 km
OR
Using CAS, select Trace, type the value 6.4 and press enter.
Note: The scatterplot must be turned off.
Distance = 11.07 km

c. Predictions that use interpolation are reliable and predictions that use extrapolation are not reliable.

c. Answer **a** is reliable because it uses interpolation.
Answer **b** is not reliable because it uses extrapolation.

---

on **Resources**

eWorkbook Topic 13 Workbook (worksheets, code puzzle and project) (ewbk-13304)

Exercise 13.5 Linear regression and technology (10A)   **learn** **on**

13.5 Quick quiz **on**	13.5 Exercise

Individual pathways

■ PRACTISE	■ CONSOLIDATE	■ MASTER
1, 2, 3, 6, 11, 14	4, 7, 9, 12, 15	5, 8, 10, 13, 16

Fluency

1. **WE11** The table shows the amount of time athletes spent training in preparation for a marathon and their finishing position in the race. Display the data on a scatterplot using technology.

Time, hours (*x*)	25	30	35
Finishing position (*y*)	15	11	8

2. **WE12** The table shows the number of visitors to a store in a day and the profit of the store that day. Determine the equation of the regression line and display the regression line using either a spreadsheet or a CAS calculator.

Number of visitors (*x*)	80	85	94	101
Profit, dollars (*y*)	152	164	180	200

3. **WE13** Use the data from question **2** to answer the following questions.

   a. Use technology to predict the profit if there were 90 visitors.
   b. Use technology to predict the profit if there were 200 visitors. Round your answer to the nearest dollar.
   c. Explain if these predictions are reliable.

4. The table shows how far away students live from school in kilometres and the hours those students spend in a car per week.

Distance from school, kilometres (x)	2	2.5	3	5
Hours spent in a car each week (y)	2.2	2.8	2.9	3.4

a. Use technology to display the data on a scatterplot.
b. Use technology to determine the equation of the regression line and display the regression line.
c. Use technology to predict the hours spent in a car each week for a student who lives 6 km from school. Give your answer to 1 decimal place.
d. Use technology to predict the hours spent in a car each week for a student who lives 2.2 km from school. Give your answer to 1 decimal place.
e. Explain if these predictions are reliable.

5. The table shows how many thousands of bees are in a hive and the amount of honey produced in that hive per year:

Number of bees ($\times 1000$) (x)	11	16	24	31
Honey produced, kilograms per year (y)	15	18	22	24

a. Use technology to display the data on a scatterplot.
b. Use technology to determine the equation of the regression line and display the regression line.
c. Use technology to predict the honey produced per year if a hive had 35 000 bees. Round the prediction to the nearest whole number.
d. Use technology to predict the honey produced per year if a hive had 18 000 bees. Round the prediction to the nearest whole number.
e. Explain if these predictions are reliable.

Understanding

6. The following data shows the temperature on certain days of the year. The days have been numbered like this: 1 January is 1, 2 January is 2 and so on for 365 days. Assume it is a non-leap year.

Day of the year (x)	1	5	12	20
Maximum temperature, °C (y)	32	33	38	42

a. Use technology to determine the equation of the regression line.
b. Use technology to predict the temperature on 15 March (day 75 of the year). Give your answer to 1 decimal place.
c. Explain if your answer to part b makes sense. Within your answer use the word *extrapolation*.

7. Chantal is a big fan of the Dugongs baseball team. The following data shows the number of games Chantal watched per year and the games won by the Dugongs per year.

Number of games watched per year (x)	8	12	16	20
Number of games won by the Dugongs (y)	10	11	15	16

a. Use technology to determine the equation of the regression line.
b. Use technology to predict the number of games won if Chantal watches 15 games. Give your answer to the nearest whole number.
c. Explain if your answer to part b makes sense. Explain your answer in terms of whether a fan watching a sports game has a connection to the outcome.

8. The following data from four lawn mowing companies shows the yard size and the cost of their most recent lawn mowing jobs:

Fred's Mowing: $200\,m^2$ yard for $80
Dial-a-gardner: $150\,m^2$ yard for $75
Chopper Chops Limited: $50\,m^2$ yard for $60
Landscapes-r-Us: $400\,m^2$ yard for $120

a. Organise the data into a table and assign the x- and y-values.
b. Explain how you have assigned the x- and y-values.
c. Use technology to determine the equation of the regression line for this data.
d. Based on your equation for the regression line from part c, estimate the call-out fee for lawn mowing.

9. **MC** Sally Miles is a world-famous pop star. By analysing Sally's tour data, the equation for a regression line is found that relates numbers of fans at a concert (x) to Sally's earnings (y). The regression line equation is $y = 22x + 25144$. In the equation of the regression line, the number 22 represents:

A. the amount Sally earns per fan at the concert.
B. how much Sally earns per year.
C. how much Sally would earn if she had 10 fans at her concert.
D. the number of songs in Sally's playlist.

10. **MC** Regression lines are only valid for data where the independent variable has a connection to the dependent variable. Select which of the following would NOT have a valid regression line.

A. Number of growing days and height of a sunflower
B. Average top speed of cars and years since cars were invented
C. Number of ice creams purchased and the price of ice cream
D. Amount of cheese eaten per capita and the number of injuries in an AFL season

Reasoning

11. Two friends, Yousef and Gavin, were having an eating competition. In the competition they both ate one, then two, then three apples and recorded their time. The results are shown in the following tables.
Yousef

Number of apples eaten (x)	1	2	3
Time taken, seconds (y)	46	67	124

Gavin

Number of apples eaten (x)	1	2	3
Time taken, seconds (y)	38	75	112

a. Use technology to determine the equations of the regression lines for each set of data.
b. Identify the gradients for each set of data.
c. Compare the gradients. Explain what this information shows.
d. Using the data, explain who won the apple-eating contest.

12. The table shows the time it took for five people to complete one lap of a BMX track.

Age, years (x)	13	14	15	16	25
Time, minutes (y)	2.5	2.2	1.9	1.8	1.2

a. Determine the equation of the regression line for this data.

b. The outlier in this data is the point $(25, 1.2)$. Remove this piece of data and find the equation of the new regression line.

c. Explain the impact on the equation of the regression line after removing an outlier. In your answer refer to the gradient and y-intercept.

13. The table shows the time it took for packages of different weights to arrive in the post:

Weight, kilograms (x)	0.5	1.1	1.7	2.5	18
Time, days (y)	4	3	3	2	4

a. Determine the equation of the regression line for this data.

b. Use technology to draw the scatterplot for this data. Explain whether there is a correlation. Determine the type, direction and strength of the correlation.

c. There is an outlier in this data. Remove the outlier and determine the new regression line.

d. Use technology to draw the scatterplot for the data with the outlier removed. Explain whether there is a correlation; if so, describe the type, direction and strength.

e. Explain which regression line more accurately represents the data.

Problem solving

14. At the school athletics carnival Mr Wall was in charge of recording the student year levels and jump heights for the winning jumps.

Year level (x)	7	8	9	10	11	12
Winning jump height, cm (y)	110	122	126		130	139

Mr Wall knows the regression line for this data is $y = 4.91x + 79.3$. Calculate the missing jump height.

15. Use the data shown to answer the following questions.

x	1	2	3	4
y	3	5	9	11

a. Determine the equation of the regression line for this data.

b. Determine what happens to the equation of the regression line if you double all the y-values.

c. Determine the number that could be added to each y-value in the original data so that the regression line becomes $y = 2.8x + 5$. Show your working.

16. Answer the following questions with full working.

a. Determine four data points that give a regression line equation of $y = 8x + 3$.

b. Now determine four different data points that still give a regression line equation of $y = 8x + 3$.

LESSON
13.6 Time series

LEARNING INTENTIONS

At the end of this lesson you should be able to:
- describe time series data using trends and patterns
- draw a line of best fit by eye and use it to make predictions for a time series

---

▶ 13.6.1 Describing time series

- **Time series** are a type of bivariate data with time as the independent variable. In other words, time series show time on the x-axis.
- To describe time series data use *trends* and *patterns*.
- Time series trends can be:
  - increasing or decreasing
  - linear or non-linear.

Increasing linear time series	Decreasing non-linear time series
An upwards slope from left to right with data that is approximately a straight line	A downwards slope from left to right with data that is not a straight line

- Time series patterns can be:
  - seasonal
  - cyclical
  - random.

Seasonal	The pattern repeats over a period of time such as a day, week, month or year.	

Cyclical	Rises and falls happen over different periods of time.	
Random	No regular pattern and caused by unpredictable events	

WORKED EXAMPLE 14 Classifying a time series trend

Classify the trend suggested by the time series graph shown as being linear or non-linear, and upward, downward or no trend.

THINK

Carefully analyse the given graph and comment on whether the graph resembles a straight line or not, and whether the values of y increase or decrease over time.

WRITE

The time series graph does not resemble a straight line and overall the level of the variable, y, decreases over time. The time series graph suggests a non-linear downward trend.

WORKED EXAMPLE 15 Commenting on a time series trend

The data below show the average daily mass of a person (to the nearest 100 g) recorded over a 28-day period.

63.6, 63.8, 63.5, 63.7, 63.2, 63.0, 62.8, 63.3, 63.1, 62.7, 62.6, 62.5, 62.9, 63.0,
63.1, 62.9, 62.6, 62.8, 63.0, 62.6, 62.5, 62.1, 61.8, 62.2, 62.0, 61.7, 61.5, 61.2

a. Plot the data as a time series graph.

b. Comment on the trend.

THINK	WRITE/DRAW
a. 1. Draw the points on a scatterplot with time on the horizontal axis and mass on the vertical axis. **2.** Join the points with straight line segments to create a time series plot.	**a.**
b. Carefully analyse the given graph and comment on whether the graph resembles a straight line or not, and whether the values of *y* (in this case, mass) increase or decrease over time.	**b.** The graph resembles a straight line that slopes downwards from left to right (that is, mass decreases with increase in time). Although a person's mass fluctuates daily, the time series graph suggests a downward trend. That is, overall, the person's mass has decreased over the 28-day period.

13.6.2 Time series lines of best fit by eye

eles-4974

- It is possible to draw lines of best fit by eye for time series (see lesson 13.4).
- Lines of best fit can be used to make predictions.
  - For interpolations, the predictions are reliable.
  - For extrapolations, the predictions are not very reliable since there is an assumption that the trend will continue.

WORKED EXAMPLE 16 Making predictions using a line of best fit

The graph shows the average cost of renting a one-bedroom flat, as recorded over a 10-year period.

a. If appropriate, draw in a line of best fit and comment on the type of the trend.

b. Assuming that the current trend will continue, use the line of best fit to predict the cost of rent in 5 years' time.

THINK

a. 1. Analyse the graph and observe what occurs over a period of time. Draw a line of best fit.

WRITE/DRAW

a.

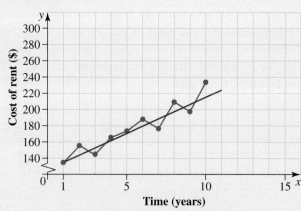

2. Comment on the type of trend observed.

The graph illustrates that the cost of rent increases steadily over the years. The time series graph indicates an upward linear trend.

b. 1. Extend the line of best fit drawn in part **a**. The last entry corresponds to the 10th year and we need to predict the cost of rent in 5 years' time, that is in the 15th year.

2. Locate the 15th year on the time axis and draw a vertical line until it meets with the line of best fit. From the trend line (line of best fit), draw a horizontal line to the cost axis.

b.

3. Read the cost from the vertical axis.

Cost of rent = $260

4. Write the answer.

Assuming that the cost of rent will continue to increase at the present rate, in 5 years we can expect the cost of rent to reach $260 per week.

DISCUSSION

Why are predictions in the future appropriate for time series even though they involve extrapolation?

 Resources

13.6 Quick quiz on

13.6 Exercise

Individual pathways

■ PRACTISE	■ CONSOLIDATE	■ MASTER
1, 4, 8, 11	2, 5, 7, 9, 12	3, 6, 10, 13

Fluency

1. **WE14** For questions **1** and **2**, classify the trend suggested by each time series graph as being linear or non-linear, and upward, downward or stationary in the mean (no trend).

   a.

   b.

   c.

   d.

2. a.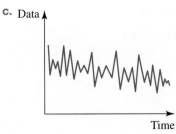

   b. Data

   c. Data

   d.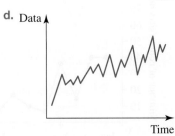

3. **WE15** The following data set shows the average daily temperatures recorded in June.

   17.6, 17.4, 18.0, 17.2, 17.5, 16.9, 16.3, 17.1, 16.9, 16.2, 16.0, 16.6, 16.1, 15.4, 15.1, 15.5, 16.0, 16.0, 15.4, 15.2, 15.0, 15.5, 15.1, 14.8, 15.3, 14.9, 14.6, 14.4, 15.0, 14.2

   a. Plot these temperatures as a time series graph.
   b. Comment on the trend.

Understanding

4. The table shows the quarterly sales (in thousands of dollars) recorded by the owner of a sheepskin product store over a period of 4 years.

Quarter	2006	2007	2008	2009
1	57	59	50	52
2	100	102	98	100
3	125	127	120	124
4	74	70	72	73

  a. Plot the time series.
  b. The time series plot displays seasonal fluctuations of period 4 (since there are four quarters). Explain in your own words what this means. Also write one or two possible reasons for the occurrence of these fluctuations.
  c. Determine if the time series plot indicates an upward trend, a downward trend or no trend.

5. The table shows the total monthly revenue (in thousands of dollars) obtained by the owners of a large reception hall. The revenue comes from rent and catering for various functions over a period of 3 years.

	Jan.	Feb.	Mar.	Apr.	May	June	July	Aug.	Sept.	Oct.	Nov.	Dec.
2007	60	65	40	45	40	50	45	50	55	50	55	70
2008	70	65	60	65	55	60	60	65	70	75	80	85
2009	80	70	65	70	60	65	70	75	80	85	90	100

  a. Construct a time series plot for this data.
  b. Describe the graph (peaks and troughs, long-term trend, and any other patterns).
  c. Suggest possible reasons for monthly fluctuations.
  d. Explain if the graph shows seasonal fluctuations over 12 months. Discuss any patterns that repeat from year to year.

6. The owner of a motel and caravan park in a small town keeps records of the total number of rooms and total number of camp sites occupied per month. The time series plots based on his records are shown below.

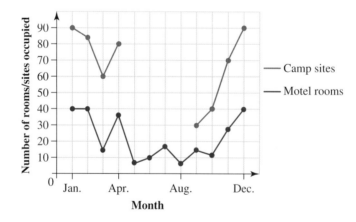

  a. Describe each graph, discussing the general trend, peaks and troughs, and so on. Explain particular features of the graphs and give possible reasons.
  b. Compare the two graphs and write a short paragraph commenting on any similarities and differences between them.

7. **WE16** The graph shows enrolments in the Health and Nutrition course at a local college over a 10-year period.

   a. If appropriate, draw in a line of best fit and comment on the type of the trend.

   b. Assuming that the trend will continue, use the line of best fit to predict the enrolment for the course in 5 years' time, that is in the 15th year.

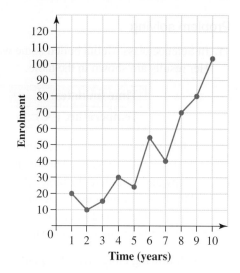

Reasoning

8. In June a new childcare centre was opened. The number of children attending full time (according to the enrolment at the beginning of each month) during the first year of operation is shown in the table.

June	July	Aug.	Sept.	Oct.	Nov.	Dec.	Jan.	Feb.	Mar.	Apr.	May
6	8	7	9	10	9	12	10	11	13	12	14

   a. Plot this time series (*Hint:* Let June = 1, July = 2 etc.)
   b. Justify if the childcare business is going well.
   c. Draw a line of best fit.
   d. Use your line of best fit to predict the enrolment in the centre during the second year of operation at the beginning of:
      i. August
      ii. January.

   e. State any assumptions that you have made.

9. The graph shows the monthly sales of a certain book since its publication. Explain in your own words why linear trend forecasting of the future sales of this book is not appropriate.

10. In the world of investing this phrase is commonly used when talking about investments:
    'Past performance is not an indicator of future returns.'

    a. Explain what this phrase means.
    b. Explain why this phrase is true using the term *extrapolation*.

Problem solving

11. In Science class Melita boiled some water and then recorded the temperature of the water over ten minutes. Her results are shown in the table.

Time (minutes)	0	1.5	3	4.5	6	7.5	10
Temperature (°C)	100	95	88	74	65	60	52

a. Melita wants to convert the time from minutes into seconds. She starts by converting 1.5 minutes to 150 seconds. Explain what she did wrong and find the correct number in seconds.
b. Copy and complete the table, changing the time in minutes to time in seconds.
c. Draw a scatterplot using seconds as the time scale.
d. Draw a line of best fit and use it to predict the time, in seconds, when the water will reach 20 °C.
e. Convert your answer from part e back to minutes.

12. The following table gives the quarterly sales figures for a second-hand car dealer over a three-year period.

Year	Q1	Q2	Q3	Q4
2012	75	65	92	99
2013	91	79	115	114
2014	93	85	136	118

a. Represent the data on a time series plot.
b. Briefly describe how the car sales have altered over the time period.
c. Discuss if it appears that the car dealer can sell more cars in a particular period each year.

13. Jasper owns an ice-cream truck.
    - In summer 2020–21 he sold 1536 ice-creams.
    - In autumn 2021 he sold one-quarter of that number.
    - In winter 2021 he sold one-eighth of that number
    - In spring 2021 he sold one-third of that number
    - From summer 2021–22 until spring 2022 his sales doubled from the season in the previous year.

a. Represent his sales from summer 2020–21 to spring 2022 on a scatterplot.
b. Describe the trend and the patterns in the data.

LESSON
13.7 Review

13.7.1 Topic summary

BIVARIATE DATA

Representing the data

- **Scatter plots** can be created by hand or using technology (CAS or Excel).
- The independent variable is placed on the x-axis and the dependent variable on the y-axis.

Properties of bivariate data

- **Bivariate data** can be displayed, analysed and used to make predictions.
- **Types of variables:**
  - Independent (experimental or explanatory variable): not impacted by the other variable.
  - Dependent (response variable): impacted by the other variable.

Correlations from scatter plots

- Correlation is a way of describing a connection between variables in a bivariate data set.
- Correlation between the two variables will have:
  - a type (linear or non-linear)
  - a direction (positive or negative)
  - a strength (strong, moderate or weak).
- There is no correlation if the data are spread out across the plot with no clear pattern.

Time series scatter plots

- A specific type of bivariate data where time is the independent variable.
- Describe time series by:
  - trends
  - patterns.
- Time series patterns can be:
  - seasonal
  - cyclical
  - random.

Line of best fit

- A line that follows the trend of the data in a scatter plot.
- It is most appropriate for data with strong or moderate linear correlation.
- Can be sketched as a line of best fit by eye or a regression line using technology.
  - The equation for the line can be found by using the gradient and equation of the straight line.
  - The line can be used to make predictions.
- Regression lines are only valid if the independent and dependent variables have a connection.

Association and cause-effect

- Association is a statistical relationship between two variables.
- Causation is defined as the relationship between two variables where one variable affects another. For example, *If an employee works for extra hours a day, their income is likely to increase.* This means, one variable (extra hours) is causing an effect on another variable (income). This is a cause-effect or casual relationship.
- Causation requires that there is an association between two variables, but association does not necessarily imply causation.

Two-way frequency tables and Likert scale

- Two-way tables are used to examine the relationship between two related categorical variables.
- Two-way relative frequency tables display percentages or proportions (ratios) (called relative frequencies) rather than frequency counts.
- They are good for seeing if there is an association between two variables.

Relative frequency $= \dfrac{\text{score}}{\text{total number of scores}}$

- The frequency written as a percentage is called percentage relative frequency.

Percentage relative frequency $= \dfrac{\text{score}}{\text{total number of scores}} \times 100\%$

$= \text{relative frequency} \times 100\%$

- The five-point Likert scale is a type of rating scale used to measure attitudes, opinions, and perceptions.

- Two-way tables can be to investigate and compare the survey responses to questions involving five-point Likert scale.

Interpolation and extrapolation

- Interpolation and extrapolation can be used to make predictions.
- **Interpolation:**
  - is more reliable from a large number of data
  - used if the prediction sits within the given data.
- **Extrapolation:**
  - assumes the trend will continue
  - used if the prediction sits outside the given data.

13.7.2 Success criteria

Tick the column to indicate that you have completed the lesson and how well you have understood it using the traffic light system.

(**Green:** I understand; **Yellow:** I can do it with help; **Red:** I do not understand)

Lesson	Success criteria	⬤	◯	⬤
13.2	I can recognise the independent and dependent variables in bivariate data.			
	I can represent bivariate data using a scatterplot.			
	I can describe the correlation between two variables in a bivariate data set.			
	I can differentiate between association and cause and effect.			
	I can use statistical evidence to make, justify and critique claims about association between variables.			
	I can draw conclusions about the correlation between two variables in a bivariate data set.			
13.3	I can construct two-way frequency tables and convert the numbers in a table to percentages			
	I can use percentages and proportions to identify patterns and associations in the data.			
	I can use two-way tables to investigate and compare survey responses to questions involving five-point Likert scales.			
13.4	I can draw a line of best fit by eye.			
	I can determine the equation of the line of best fit.			
	I can use the line of best fit to make interpolation or extrapolation predictions.			
13.5	I can display a scatterplot using technology.			
	I can determine the equation of a regression line using technology.			
	I can display a scatterplot with its regression line (line of best fit) using technology.			
	I can use a regression line to make predictions.			
13.6	I can describe time series data using trends and patterns.			
	I can draw a line of best fit by eye and use it to make predictions for a time series.			

13.7.3 Project

Collecting, recording and analysing data over time

A time series is a sequence of measurements taken at regular intervals (daily, weekly, monthly and so on) over a certain period of time. Time series are best represented using time-series plots, which are line graphs with the time plotted on the horizontal axis.

Examples of time series include daily temperature, monthly unemployment rates and daily share prices.

When data are recorded on a regular basis, the value of the variable may go up and down in what seems to be an erratic pattern. These are called fluctuations. However, over a long period of time, the time series usually suggests a certain trend. These trends can be classified as being linear or non-linear, and upward, downward or stationary (no trend).

Time series are often used for forecasting, that is making predictions about the future. The predictions made with the help of time series are always based on the assumption that the observed trend will continue in the future.

1. Choose a subject that is of interest to you and that can be observed and measured during one day or over the period of a week or more. (Suitable subjects are shown in the list below.)
2. Prepare a table for recording your results. Select appropriate regular time intervals. An example is shown.

Time	8 am	9 am	10 am	11 am	12 pm	1 pm	2 pm	3 pm	4 pm	5 pm
Pulse rate										

3. Take your measurements at the selected time intervals and record them in the table.
4. Use your data to plot the time series. You can use software such as Excel or draw the scatterplot by hand.
5. Describe the graph and comment on its trend.
6. If appropriate, draw a line of best fit and predict the next few data values.
7. Take the actual measurements during the hours you have made predictions for. Compare the predictions with the actual measurements. Were your predictions good? Give reasons.

Here are some suitable subjects for data observation and recording:
- minimum and maximum temperatures each day for 2 weeks (use the TV news or online data as resources)
- the value of a stock on the share market (e.g. Telstra, Wesfarmers and Rio Tinto)
- your pulse over 12 hours (ask your teacher how to do this or check on the internet)
- the value of sales each day at the school canteen
- the number of students absent each day
- the position of a song in the Top 40 over a number of weeks
- petrol prices each day for 2 weeks
- the relationship between vaccines and immunity from a virus
- world population statistics over time.

Resources

eWorkbook   Topic 13 Workbook (worksheets, code puzzle and project) (ewbk-13304)

Interactivities Crossword (int-9185)
                 Sudoku puzzle (int-3600)

Exercise 13.7 Review questions

learnon

Fluency

1. As preparation for a Mathematics test, 20 students were given a revision sheet containing 60 questions. The table shows the number of questions from the revision sheet successfully completed by each student and the mark, out of 100, gained by that student on the test.

Number of questions	9	12	37	60	55	40	10	25	50	48	60
Test result	18	21	52	95	100	67	15	50	97	85	89
Number of questions	50	48	35	29	19	44	49	20	16	58	52
Test result	97	85	62	54	30	70	82	37	28	99	80

   a. State which of the variables is dependent and which is independent.
   b. Construct a scatterplot of the data.
   c. State the type of correlation between the two variables suggested by the scatterplot and draw a corresponding conclusion.
   d. Suggest why the relationship is not perfectly linear.

2. Use the line of best fit shown on the graph to answer the following questions.

   a. Predict the value of y when the value of x is:
      i. 10   ii. 35.
   b. Predict the value of x when the value of y is:
      i. 15   ii. 30.
   c. Determine the equation of a line of best fit if it is known that it passes through the points $(5, 5)$ and $(20, 27)$.
   d. Use the equation of the line to algebraically verify the values obtained from the graph in parts **a** and **b**.

3. The graph shows the number of occupants of a large nursing home over the last 14 years.

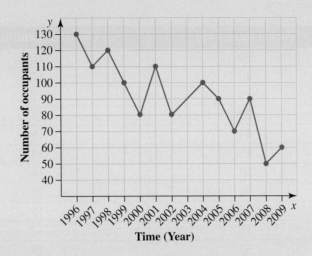

a. Comment on the type of trend displayed.
b. Explain why it is appropriate to draw in a line of best fit.
c. Draw a line of best fit and use it to predict the number of occupants in the nursing home in 3 years' time.
d. State the assumption that has been made when predicting figures for part **c.**

4. The table shows the advertised sale price ($'000) and the land size (m^2) for ten vacant blocks of land.

Land size (m^2)	Sale price ($'000)
632	36
1560	58
800	40
1190	44
770	41
1250	52
1090	43
1780	75
1740	72
920	43

a. Construct a scatterplot and determine the equation of the line of best fit.
b. State what the gradient represents.
c. Using the line of best fit, predict the approximate sale price, to the nearest thousand dollars, for a block of land with an area of $1600 \, m^2$.
d. Using the line of best fit, predict the approximate land size, to the nearest 10 square metres, that you could purchase with $50 000.

5. The table shows, for fifteen students, the amount of pocket money they receive and spend at the school canteen in an average week.

Pocket money ($)	Canteen spending ($)
30	16
40	17
15	12
25	14
40	16
15	14
30	16
30	17
25	15
15	13
50	19
20	14
35	17
20	15
10	13

a. Construct a scatterplot and determine the equation of the line of best fit.
b. State what the gradient represents.
c. Using your line of best fit, predict the amount of money spent at the canteen by a student receiving $45 pocket money a week.
d. Using your line of best fit, predict the amount of money spent at the canteen by a student who receives $100 pocket money each week.
   Explain if this seems reasonable.

6. The table shows, for 10 ballet students, the number of hours a week spent training and the number of pirouettes in a row they can complete.

Training (hours)	11	11	2	8	4	16	11	16	5	3
Number of pirouettes	15	13	3	12	7	17	13	16	8	5

a. Construct a scatterplot and determine the equation of the line of best fit.
b. State what the gradient represents.
c. Using your line of best fit, predict the number of pirouettes that could be completed if a student undertakes 14 hours of training.
d. Professional ballet dancers may undertake up to 30 hours of training a week. Using your line best fit, predict the number of pirouettes they should be able to do in a row.
   Comment on your findings.

7. Use the information in the data table to answer the following questions.

Age in years (x)	7	11	8	16	9	8	14	19	17	10	20	15
Hours of television watched in a week (y)	20	19	25	55	46	50	53	67	59	25	70	58

a. Use technology to determine the equation of the line of best fit for the following data.
b. Use technology to predict the value of the number of hours of television watched by a person aged 15.

8. MC A set of data was gathered from a large group of parents with children under 5 years of age. They were asked the number of hours they worked per week and the amount of money they spent on child care. The results were recorded, and a strong positive association was found between the number of hours worked and the amount of money spent on child care.
Determine which of the following statements is *not* true.
A. There is a positive correlation between the number of working hours and the amount of money spent on child care.
B. The correlation between the number of working hours and the amount of money spent on child care can be classified as strong.
C. The relationship between the number of working hours and the amount spent on child care is linear.
D. The increase in the number of hours worked causes the increase in the amount of money spent on child care.

9. Data showing reactions of junior staff and senior staff to a relocation of offices are given in the contingency table shown.

		Staff type		
		Junior	Senior	Total
Attitude	For	23	14	37
	Against	31	41	72
	Total	54	55	109

a. MC From this table, we can conclude that:
A. 23% of junior staff were for the relocation.
B. 42.6% of junior staff were for the relocation.
C. 31% of junior staff were against the relocation.
D. 62.1% of junior staff were for the relocation.
b. MC From this table, we can conclude that:
A. 14% of senior staff were for the relocation.
B. 37.8% of senior staff were for the relocation.
C. 12.8% of senior staff were for the relocation.
D. 74.5% of senior staff were against the relocation.

10. **MC** In a group of 165 people aged 16–30 years, 28 people aged 23–30 years have a learner driver licence while 15 people aged 16–22 years have a probationary driver licence. Out of the 117 people who have a licence, 59 are learners.

		Type of driver licence for people aged 16–30 years		
		Age groups		
		16–22 years	23–30 years	Total
Driver licence	L	31	28	59
	P	15	43	58
	None	27	21	48
	Total	73	92	165

If the total number of people aged 16–22 years surveyed was 73, the number of people aged 23–30 who have a probationary driver licence is:

A. 92   B. 58   C. 64   D. 43

11. One hundred teenagers were surveyed about their favourite type of music genre. The data were organised into a frequency table.

Music genre	Frequency
Hip hop/rap	28
Pop/R&B/soul	27
Rock	26
Country	3
Blues/jazz	4
Classical	2
Alternative	10

Calculate what percentage of teenagers preferred rock music.

Problem solving

12. Describe the trends present in the following time series data that show the mean monthly daily hours of sunshine in Melbourne from January to December.

Month	1	2	3	4	5	6	7	8	9	10	11	12
Daily hours of sunshine	8.7	8.0	7.5	6.4	4.8	4.0	4.5	5.5	6.3	7.3	7.5	8.3

13. The existence of the following situations is often considered an obstacle to making estimates from data.
    a. Outliers
    b. Extrapolation
    c. A small range of data
    d. A small number of data points
    Explain why each of these situations is considered an obstacle to making estimates of data and how each might be overcome.

14. Students from a Year 10 class were asked about their favourite subjects and the data were recorded in the frequency table shown.

Subject	Frequency
Maths	4
Art	4
Humanities	3
PE	6
Science	5
English	3
Total	**25**

a. Calculate the percentage of students who preferred Maths.

b. State the most popular subject.

c. Calculate the percentage of students who preferred this subject.

15. The table shows the heights of 10 students and the distances along the ground between their feet as they attempt to do the splits.

Height (cm)	Distance stretched (cm)
134.5	150
156	160
133.5	147
145	160
160	162
135	149
163	163
138	149
152	158
159	160

Using the data, estimate the distance that a person 1.8 m tall can achieve when attempting the splits.
Write a detailed analysis of your result.
Include:
- an explanation of the method(s) used
- any plots or formula generated
- comments on validity of the estimate
- any ways the validity of the estimate could be improved.

To test your understanding and knowledge of this topic, go to your learnON title at www.jacplus.com.au and complete the **post-test**.

Answers

Topic 13 Bivariate data

13.1 Pre-test

1. B
2. C
3. Independent variable
4. a. B   b. C   c. A
5. D
6. B
7. Interpolation
8. The gradient of the line is $\dfrac{16}{11}$.
9. Independent variable
10. Explanatory variable
11. B
12. C
13. $x = 6$
14. a. See the figure at the bottom of the page.*
    b. The number of COVID-19 cases started rising in March and peaked in April, then started to decline until June. There was an increase in cases in July and the cases reached peak again in August. Cases then started to decline again until December.
15. a. $y = 14$ b. $x = 12.5$

13.2 Bivariate data

1.

	Independent	Dependent
a.	Number of hours spent studying	Test results
b.	Rainfall	Attendance
c.	Hours in gym	Visits to the doctor
d.	Lengths of essay	Memory taken

2.

	Independent	Dependent
a.	Cost of care	Attendance
b.	Age of property	Cost of property
c.	Number of applicants	Entry requirements
d.	Running speed	Heart rate

3.

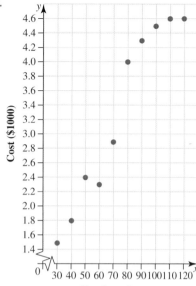

4. a. Perfectly linear, positive
   b. No correlation
   c. Non-linear, negative, moderate
   d. Strong, positive, linear
   e. No correlation
5. a. Non-linear, positive, strong
   b. Strong, linear, negative
   c. Non-linear, moderate, negative
   d. Weak, negative, linear
   e. Non-linear, moderate, positive
6. a. Positive, moderate, linear
   b. Non-linear, strong, negative
   c. Strong, negative, linear
   d. Weak, positive, linear
   e. Non-linear, moderate, positive
7. It cannot be concluded that holidays cause weight gain. The common cause could be any or a combination of the following: amount of exercise, alcohol consumption or diet.

*14. a.

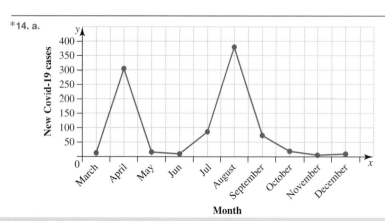

8. It cannot be concluded that carbon dioxide in the atmosphere causes the change in the average Earth's surface temperature.
The common cause could be any or a combination of the following: human activities, volcanic eruptions, the Sun's energy output and so on.

9. a.
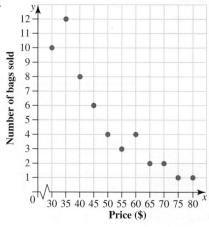

b. Negative, linear, moderate. The price of the bag appeared to affect the numbers sold; that is, the more expensive the bag, the fewer sold.

10. a.

b. Moderate positive linear correlation. There is evidence to show that the larger the number of bedrooms, the higher the price of the house.

c. Various answers; location, age, number of people interested in the house and so on.

11. a.

b. Strong, positive, linear correlation

c. Various answers; some students are of different ability levels and they may have attempted the questions but had incorrect answers.

12. a.
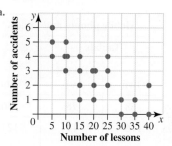

b. Weak, negative, linear relation

c. Various answers are possible: some drivers are better than others, live in lower traffic areas, traffic conditions etc.

13. B

14. C

15. D

16. a. See the figure at the bottom of the page.*

b. This scatterplot does not support the claim.

*16. a.
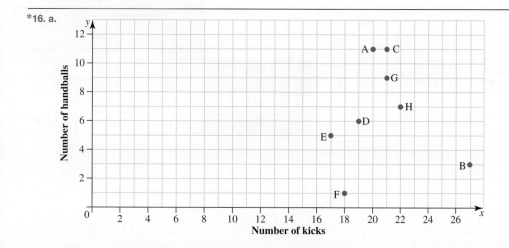

17. a. T   **b.** F   **c.** T
d. F   **e.** T

18. a. Mandy (iii)   **b.** William (iv)
c. Charlotte (viii)   **d.** Dario (vii)
e. Edward (vi)   **f.** Cindy (v)
g. Georgina (i)   **h.** Harrison (ii)

13.3 Two-way frequency tables and Likert scales

1. a. i. 22   ii. 26   iii. 19   iv. 45   v. 41

    b. i. 47%   ii. 58%   iii. 100%   iv. 100%

2.

Parking place and car theft			
	Parking type		
	In driveway	**On the street**	**Total**
Car theft	2	37	39
No car theft	16	445	461
Total	18	482	500

(Row label: **Theft**)

3.

Mobile phones and tablets			
	Tablets		
	Own	**Do not own**	**Total**
Own	58	78	136
Do not own	116	11	127
Total	174	89	263

(Row label: **Mobile phones**)

4. a.

Year 11 students' Mathematics results			
	Result		
	Pass	**Fail**	**Total**
Essential Mathematics	12	23	35
General Mathematics	25	49	74
Mathematical Methods	36	62	98
Specialist Mathematics	7	18	25
Total	80	152	232

(Row label: **Subject**)

    b. 232 students

    c. 35 students

5. a.

Hairdressing requirements by age group			
	Age group		
	Young	**Old**	**Total**
Haircut			
Haircut and colour			
Total			

(Row label: **Hairdressing requirement**)

b.

Hairdressing requirements by age group			
	Age group		
	Young	**Old**	**Total**
Haircut			
Haircut and colour	2		
Total		56	

(Row label: **Hairdressing requirement**)

c.

Hairdressing requirements by age group			
	Age group		
	Young	**Old**	**Total**
Haircut only	5	26	31
Haircut and colour	2	30	32
Total	7	56	63

(Row label: **Hairdressing requirement**)

6.

	Age group	
	Teenagers	**Adults**
Coke	57.1%	58.9%
Pepsi	42.9%	41.1%
Total	100%	100%

(Row label: **Drink**)

7. a. Relative frequencies

Cat and dog owners			
	Dog owners		
	Own	**Do not own**	**Total**
Own	0.039	0.256	0.295
Do not own	0.495	0.210	0.705
Total	0.534	0.466	1.000

(Row label: **Cat owners**)

b. Percentage relative frequencies

	Cat and dog owners (%)		
	Dog owners		
	Own (%)	Do not own (%)	Total
Cat owners Own (%)	3.9	25.6	29.5
Do not own (%)	49.5	21.0	70.5
Total	53.4	46.6	100.0

8. a. Relative frequency table

		Year 11 students' Mathematics results		
		Result		
		Pass	Fail	Total
Subject	Essential Mathematics	0.0517	0.0991	0.1509
	General Mathematics	0.1078	0.2112	0.3190
	Mathematical Methods	0.1552	0.2672	0.4224
	Specialist Mathematics	0.0302	0.0776	0.1078
	Total	0.3448	0.6552	1.0000

b. Percentage relative frequency table

		Year 11 students' Mathematics results		
		Result		
		Pass (%)	Fail (%)	Total (%)
Subject	Essential Mathematics (%)	5.17	9.91	15.09
	General Mathematics (%)	10.78	21.12	31.90
	Mathematical Methods (%)	15.52	26.72	42.24
	Specialist Mathematics (%)	3.02	7.76	10.78
	Total (%)	34.48	65.52	100.00

Note: Due to rounding, some of the percentages written do not add up to the actual values in some of the total columns. To have the correct values in the total columns, ensure that the actual values (no rounding) are summed up.

9. a.

	Students	
	Junior (%)	Senior (%)
Never	35	20
Rarely	30	25
Sometimes	20	30
Often	10	15
Very often	5	10

b. Please see the worked solution in the online resources.

10. a. D
b. C

11.

		Age group	
		< 40 (%)	40 + (%)
Choice	Mortgage	45	18
	Superannuation	32	50
	Investment	23	32
	Total	100	100

The under 40s have a focus on their mortgage, whereas the 40 and overs prioritise their superannuation.

12. a.

		Delegates	
		Liberal (%)	Labor (%)
Attitude	For	35.5	39.4
	Against	64.5	60.6
	Total	100.0	100.0

b. There is not a lot of difference in the reactions.

13. a.

		Monday matinee audience per movie genre		
		Movie genres		
		Drama	Comedy	Total
Audience type	Adults	66	30	96
	Teenagers	5	16	21
	Children	11	28	39
	Total	82	74	156

b. 30 adults
c. 39 children

14. a. Relative frequencies

Swimming attendance	
Day of the week	**Adults**
Monday	0.112
Tuesday	0.072
Wednesday	0.080
Thursday	0.120
Friday	0.144
Saturday	0.216
Sunday	0.256
Total	1.000

b. Percentage relative frequencies

Swimming attendance	
Day of the week	**Adults (%)**
Monday	11.2
Tuesday	7.2
Wednesday	8.0
Thursday	12.0
Friday	14.4
Saturday	21.6
Sunday	25.6
Total	100.0

Note: Due to rounding, some of the percentages written do not add up to the actual values in some of the total columns. To have the correct values in the total columns, ensure that the actual values (no rounding) are summed up.

15. a.

	Dance class enrolments			
	People			
	Adults	**Teenagers**	**Children**	**Total**
Latin	31	21	23	75
Modern	13	34	49	96
Contemporary	36	49	62	147
Total	80	104	134	318

Dance

b. Relative frequency table

	Dance class enrolments			
	People			
	Adults	**Teenagers**	**Children**	**Total**
Latin	0.10	0.07	0.07	0.24
Modern	0.04	0.11	0.15	0.30
Contemporary	0.11	0.15	0.19	0.46
Total	0.25	0.33	0.42	1.00

Dance

Note: Due to rounding, some of the percentages written do not add up to the actual values in some of the total columns. To have the correct values in the total columns, ensure that the actual values (no rounding) are summed up.

c. Percentage relative frequency table

	Dance class enrolments (%)			
	People			
	Adults (%)	**Teenagers (%)**	**Children (%)**	**Total**
Latin (%)	10	7	7	24
Modern (%)	4	11	15	30
Contemporary (%)	11	15	19	46
Total (%)	25	33	42	100

Dance

13.4 Lines of best fit by eye

Note: Answers may vary slightly depending on the line of best fit drawn.

1. a.

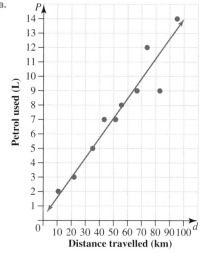

Distance travelled (km)

b. Using $(23, 3)$ and $(56, 8)$, the equation is $P = \dfrac{5}{33}d - \dfrac{16}{33}$.

2. a. 38   **b.** 18

3. a. i. 510   **ii.** 315   **iii.** 125

   b. i. 36.5   **ii.** 26   **iii.** 8

   c. $y = -13x + 595$

   d. y-values (a):

     i. 504

     ii. 309

     iii. 127

     x-values (b):

     i. 36.54

     ii. 25.77

     iii. 8.85

4. a.

b. Using $(8, 47)$ and $(12, 74)$, the equation is $E = 6.75h - 7$.

c. On average, students were paid $6.75 per hour.

5. a.

b. Using $(1, 75)$ and $(5, 150)$, the equation is
$C = 18.75n + 56.25$

c. On average, weekly cost of food increases by $18.75 for every extra person.

d. i. $206.25   ii. $225.00   iii. $243.75

6. a. 35 to 90

   b. 30 to 105

7. a. i.

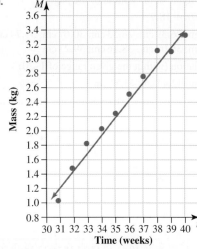

ii. Positive, strong, linear correlation

b. Using $(32, 1.470)$ and $(35, 2.230)$, $M = 0.25t - 6.5$.

c. With every week of gestation the mass of the baby increases by approximately. 250 g.

d. 3.75 kg; 4 kg

e. Approximately 1 kg

f. Between 35 and 36 weeks

8. a. D   **b.** C

9. D

10. a.

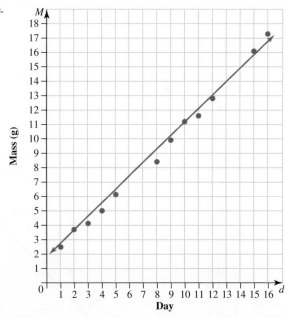

b. Using $(2, 3.7)$ and $(10, 11.2)$, $M = 0.88d + 1.94$.

c. Each day Rachel's crystal gains 0.88 g in mass. The line of best fit appears appropriate.

d. 7.22 g; 8.10 g; 13.38 g and 14.26 g; interpolation (within the given range of 1–16)

e. 16.9 g and 17.78 g; predictions are not reliable, since they were obtained using extrapolation.

11. a. See the figure at the bottom of the page.*

b. $L = 1.07n + 18.9$

c. 24.25 cm; 26.39 cm; 28.53 cm; 30.67 cm; 31.74 cm; 34.95 cm; 36.02 cm; 38.16 cm; 39.23 cm

d. Interpolation (within the given range of 1–20)

e. 41.37 cm; 42.44 cm; 43.51 cm

f. Not reliable, because extrapolation has been used.

12. a.

b.

c.

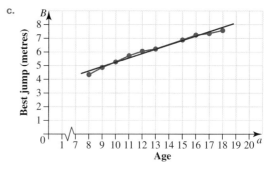

d. Yes. Using points (9, 4.85) and (16, 7.24), $B = 0.34a + 1.8$; estimated best jump $= 8.6$ m.

e. No, trends work well over the short term but in the long term are affected by other variables.

f. 24 years old: 9.97 m; 28 years old: 11.33 m. It is unrealistic to expect his jumping distance to increase indefinitely.

g. Equal first

13. No. He would have to get 108%, which would be impossible on a test.

13.5 Linear regression and technology (10A)

1. Sample responses can be found in the worked solutions in the online resources.

2. Sample responses can be found in the worked solutions in the online resources.

*11. a.

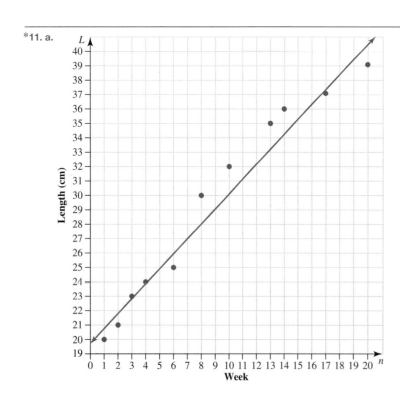

3. a. $174   b. $417
   c. a. Reliable, but b. not reliable.

4. a. Sample responses can be found in the worked solutions in the online resources.
   b. Sample responses can be found in the worked solutions in the online resources.
   c. 3.8. hours   d. 2.5 hours
   e. c. not reliable, d. reliable

5. a. Sample responses can be found in the worked solutions in the online resources.
   b. Sample responses can be found in the worked solutions in the online resources.
   c. 26 kg   d. 19 kg
   e. c. not reliable, d. reliable

6. a. $y = 0.553x + 31$ b. 72.5 °C
   c. Not possible. Extrapolation is not reliable.

7. a. $y = 0.55x + 5.3$ b. 14
   c. No, because there is no connection.

8. a. Sample responses can be found in the worked solutions in the online resources.
   b. Yard size: independent, price: dependent
   c. $y = 0.173x + 49.1$ d. $50

9. A

10. D

11. a. Yousef: $y = 39x + 1$, Gavin: $y = 37x + 1$
    b. Yousef 39, Gavin 37
    c. Time to eat each apple
    d. Gavin

12. a. $y = -0.094x + 3.35$ b. $y = -0.24x + 5.58$
    c. Sample responses can be found in the worked solutions in the online resources.

13. a. $y = 0.0508x + 2.96$
    b. No correlation
    c. $y = -0.913x + 4.32$
    d. Correlation is linear, negative and moderate/strong.
    e. Regression line from part c

14. 128 cm

15. a. $y = 2.8x$ b. The gradient doubles.
    c. 5

16. a. Sample responses can be found in the worked solutions in the online resources.
    b. Sample responses can be found in the worked solutions in the online resources.

13.6 Time series

1. a. Linear, downward
   b. Non-linear, upward
   c. Non-linear, stationary in the mean
   d. Linear, upward

2. a. Non-linear, downward
   b. Non-linear, downward
   c. Non-linear, downward
   d. Linear, upward

3. a. See the figure at the bottom of the page.*
   b. Linear downward trend

---

*3. a.

May temperature

4. a.

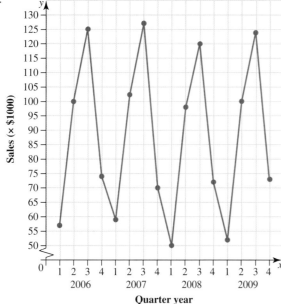

b. Sheepskin products are more popular in the third quarter (presumably winter) — discount sales, increase in sales, and so on.

c. No trend

5. a. See the figure at the bottom of the page.*

b. General upward trend with peaks around December and troughs around April

c. Peaks around Christmas where people have lots of parties, troughs around April where weather gets colder and people less inclined to go out

d. Yes; peaks in December, troughs in April

6. a. Peaks around Christmas holidays and a minor peak at Easter; no camping in colder months

b. Sample responses can be found in the worked solutions in the online resources.

7. a.

Upward linear.

b. In the 15th year the expected amount $= 122$.

8. a.

b. Yes, the graph shows an upward trend.

c. $y = \dfrac{4}{7}x + \dfrac{45}{7}$

d. i. 15   **ii.** 18

e. The assumption made was that business will continue on a linear upward trend.

9. The trend is non-linear; therefore, unable to forecast future sales.

---

*5. a.

10. a. Sample responses can be found in the worked solutions in the online resources.

   b. Extrapolation is not reliable.

11. a. 90 seconds

   b. Sample responses can be found in the worked solutions in the online resources.

   c. Sample responses can be found in the worked solutions in the online resources.

   d. Approximately 920 seconds

   e. Approximately 15.3 minutes

12. a. See the figure at the bottom of the page.*

   b. Secondhand car sales per quarter have shown a general upward trend but with some major fluctuations.

   c. More cars are sold in the third and fourth quarters compared to the first and second quarters.

13. a. Sample responses can be found in the worked solutions in the online resources.

   b. Trend: non-linear, increasing; pattern: seasonal

Project

1. Students could choose any subject given in the list that can be observed and measure for one day or over the period of a week or more.

2. Students need to create a data table for their recording. Students should use appropriate regular time intervals.

3. For a selected subject, student's need to take their measurements at the selected time intervals and record them in the table.

4. Students could use Excel or CAS to plot the time series.

5. Students should describe their graph and comment on its trend.

6. Students should draw a line of best fit and predict the next few data values.

7. Students should take the actual measurements during the hours they have made predictions for and then compare the predictions with the actual measurements. They should also comment on the accuracy of their predictions.

13.7 Review questions

1. a. Number of questions: independent; test result: dependent

   b.

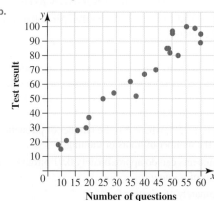

   c. Strong, positive, linear correlation; the larger the number of completed revision questions, the higher the mark on the test.

   d. Different abilities of the students

2. a. i. 12.5   ii. 49

   b. i. 12   ii. 22.5

   c. $y = \dfrac{22}{15}x - \dfrac{7}{3}$

   d. i. 12.33   ii. 49
       i. 11.82   ii. 22.05

3. a. Linear downwards

   b. The trend is linear.

   c. About 60–65 occupants

   d. Assumes that the current trend will continue.

*12. a.

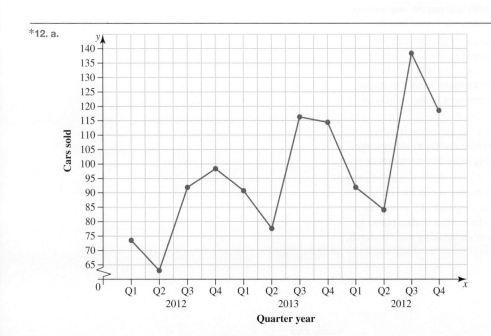

Quarter year

4. a. $P = 31.82a + 13070.4$, where P is the sale price and a is the land area.

b. The price of land is approximately \$31.82 per square metre.

c. \$64 000

d. $1160\,\text{m}^2$

5. a. $C = 0.15p + 11.09$, where C is the money spent at the canteen and p represents the pocket money received.

b. Students spend 15 cents at the canteen per dollar received for pocket money.

c. \$18

d. \$26. This involves extrapolation, which is considered unreliable. It does not seem reasonable that, if a student receives more money, they will eat more or have to purchase more than any other student.

6. a. $P = 0.91t + 2.95$, where P is the number of pirouettes and t is the number of hours of training.

b. Ballet students can do approximately 0.91 pirouettes for each hour of training.

c. Approximately 16 pirouettes

d. Approximately 30 pirouettes. This estimate is based on extrapolation, which is considered unreliable. To model this data linearly as the number of hours of training becomes large is unrealistic.

7. a. $y = 3.31x + 3.05$

b. Approximately 53 hours.

8. D

9. a. B

b. D

10. D

11. 26%

12. Overall the data appears to be following a seasonal trend, with peaks at either end of the year and a trough in the middle.

13. a. Outliers can unfairly skew data and as such dramatically alter the line of best fit. Identify and remove any outliers from the data before determining the line of best fit.

b. Extrapolation involves making estimates outside the data range and this is considered unreliable. When extrapolation is required, consider the data and the likelihood that the data would remain linear if extended. When giving results, make comment on the validity of the estimation.

c. A small range may not give a fair indication if a data set shows a strong linear correlation. Try to increase the range of the data set by taking more measurements or undertaking more research.

d. A small number of data points may not be able to establish with confidence the existence of a strong linear correlation. Try to increase the number of data points by taking more measurements or undertaking more research.

14. a. 16%   **b.** PE   **c.** 24%

15. About 170 cm; the data were first plotted as a scatterplot. $(145, 160)$ was identified as an outlier and removed from the data set. A line of best fit was then fitted to the remaining data and its equation determined as $d = 0.5h + 80$, where d is the distance stretched and h is the height. Substitution was used to obtain the estimate.

The estimation requires extrapolation and cannot be considered reliable. The presence of the outlier may indicate variation in flexibility rather than a strong linear correlation between the data. The estimate is based on a small set of data.

14 Circle geometry (10A)

LESSON
14.1 Overview

Why learn this?

For thousands of years, humans have been fascinated by circles. Since they first looked upwards towards the sun and moon, which, from a distance at least, looked circular, artists have created circular monuments to nature. The most famous circular invention, one that has been credited as the most important invention of all, is the wheel. The potter's wheel can be traced back to around 3500 BCE, approximately 300 years before wheels were used on chariots for transportation. Our whole transportation system revolves around wheels — bicycles, cars, trucks, trains and planes. Scholars as early as Socrates and Plato, Greek philosophers of the fourth century BCE, have been fascinated with the sheer beauty of the properties of circles. Many scholars made a life's work out of studying them, most famously Euclid, a Greek mathematician. It is in circle geometry that the concepts of congruence and similarity, studied earlier, have a powerful context.

Today, we see circles in many different areas. Some buildings are now constructed based on circular designs. Engineers, designers and architects understand the various properties of circles. Road systems often have circular interchanges, and amusement parks usually include ferris wheels. As with the simple rectangle, circles are now part of our everyday life. Knowing the various properties of circles helps with our understanding and appreciation of this simple shape.

Hey students! Bring these pages to life online

▶ Watch videos

🧩 Engage with interactivities

A+ Answer questions and check solutions

Find all this and MORE in jacPLUS

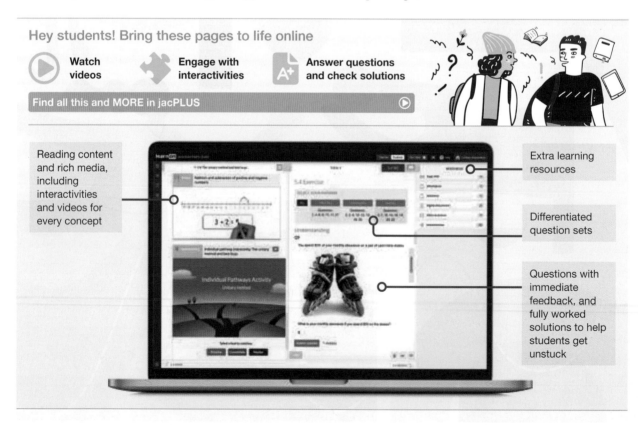

Reading content and rich media, including interactivities and videos for every concept

Extra learning resources

Differentiated question sets

Questions with immediate feedback, and fully worked solutions to help students get unstuck

1. State the name of the area of the circle between a chord and the circumference.

2. **MC** Select the name of a line that touches the circumference of a circle at one point only.

   A. Secant   **B.** Radius   **C.** Chord   **D.** Tangent

3. Calculate the angle of x within the circle shown.

4. **MC** Select the correct angles for y and z in the circle shown.

   A. $y = 40°$ and $z = 80°$ **B.** $y = 20°$ and $z = 40°$

   C. $y = 40°$ and $z = 80°$ **D.** $y = 40°$ and $z = 20°$

5. Determine the value of x in the circle shown.

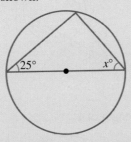

6. Determine the value of y in the shape shown.

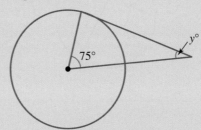

7. **MC** Select the correct values of x, y and z from the following list.

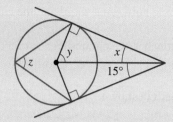

    A. $x = 15°$, $y = 75°$ and $z = 150°$
    B. $x = 15°$, $y = 30°$ and $z = 150°$
    C. $x = 15°$, $y = 15°$ and $z = 37.5°$
    D. $x = 15°$, $y = 75°$ and $z = 75°$

8. Calculate the length of p in the circle.

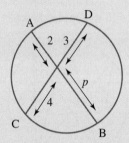

9. Calculate the length of m, correct to 1 decimal place.

10. **MC** Choose the correct value for the length of x.

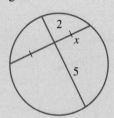

    A. $\sqrt{10}$
    B. $\sqrt{5}$
    C. 2
    D. 10

11. Determine the values of m and p.

12. **MC** Select the correct values for x and y.

A. $x = 110°$ and $y = 83°$
C. $x = 70°$ and $y = 97°$

B. $x = 83°$ and $y = 110°$
D. $x = 97°$ and $y = 70°$

13. **MC** Choose which of the following correctly states the relationships between x, y and z in the diagram.

A. $x = 2y$ and $z + y = 180°$

C. $x = 2y$ and $z + x = 180°$

B. $y = \dfrac{x}{2}$ and $z + y = 180°$

D. $y = 2x$ and $z + x = 180°$

14. **MC** Choose the correct value for x.

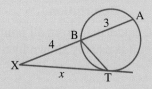

A. $2\sqrt{5}$ **B.** 28   **C.** 7   **D.** $2\sqrt{7}$

15. **MC** $AB = 9$ and P divides AB in the ratio $2 : 3$.
If $PO = 3$ where O is the centre of the circle, select the exact length of the diameter of the circle.

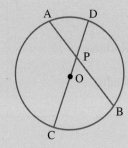

A. $\dfrac{6\sqrt{79}}{5}$

B. $\dfrac{4\sqrt{79}}{2}$

C. $\dfrac{3\sqrt{79}}{5}$

D. $\dfrac{6\sqrt{81}}{5}$

LESSON
14.2 Angles in a circle

LEARNING INTENTIONS

At the end of this lesson you should be able to:
- determine relationships between the angles at the centre and the circumference of a circle
- understand and use the angle in a semicircle being a right angle
- apply the relationships between tangents and radii of circles.

▶ 14.2.1 Circles

eles-4995

- A **circle** is a connected set of points that lie a fixed distance (the **radius**) from a fixed point (the **centre**).
- In circle geometry, there are many theorems that can be used to solve problems. It is important that we are also able to prove these theorems.

Steps to prove a theorem

Step 1. State the aim of the proof.

Step 2. Use given information and previously established theorems to establish the result.

Step 3. Give a reason for each step of the proof.

Step 4. State a clear conclusion.

Parts of a circle

Part (name)	Description	Diagram
Centre	The middle point, equidistant from all points on the circumference. It is usually shown by a dot and labelled O.	
Circumference	The outside length or the boundary forming the circle. It is the circle's perimeter.	
Radius Radii	A straight line from the centre to any point on the circumference Plural of radius	

Part (name)	Description	Diagram
Diameter	A straight line from one point on the circumference to another, passing through the centre	
Chord	A straight line from one point on the circumference to another	
Segment	The area of the circle between a chord and the circumference. The smaller segment is called the minor segment and the larger segment is the major segment.	
Sector	An area of a circle enclosed by 2 radii and the circumference	
Arc	A portion of the circumference	
Tangent	A straight line that touches the circumference at one point only	
Secant	A chord extended beyond the circumference on one side	

Angles in a circle

- In the diagram shown, chords AC and BC form the angle ACB. Arc AB has **subtended** angle ACB.

- **Theorem 1 Code**

  The angle subtended at the centre of a circle is twice the angle subtended at the circumference standing on the same arc.

  Proof:

  Let $\angle PRQ = x$ and $\angle QRO = y$

  $\qquad RO = PO = QO$ (radii of the same circle are equal)

  $\qquad \angle RPO = x$

  and $\angle RQO = y$

  $\qquad \angle POM = 2x$ (exterior angle of a triangle)

  and $\angle QOM = 2y$ (exterior angle of a triangle)

  $\qquad \angle POQ = 2x + 2y$

  $\qquad\qquad = 2(x + y)$

  which is twice the size of $\angle PRQ = x + y$.

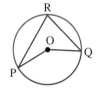

The angle subtended at the centre of a circle is twice the angle subtended at the circumference, standing on the same arc.

- **Theorem 2 Code**

  All angles that have their vertex on the circumference and are subtended by the same arc are equal.

  Proof:

  Join P and Q to O, the centre of the circle.

  Let $\angle PSQ = x$

  $\qquad \angle POQ = 2x$ (angle at the centre is twice the angle at the circumference)

  $\qquad \angle PRQ = x$ (angle at the circumference is half the angle of the centre)

  $\qquad \angle PSQ = \angle PRQ$.

  Angles at the circumference subtended by the same arc are equal.

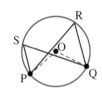

WORKED EXAMPLE 1 Determining angles in a circle

Determine the values of the pronumerals x and y in the diagram, giving reasons for your answers.

THINK	WRITE
1. Angles x and $46°$ are angles subtended by the same arc and both have their vertex on the circumference.	$x = 46°$
2. Angles y and $46°$ stand on the same arc. The $46°$ angle has its vertex on the circumference and y has its vertex at the centre. The angle at the centre is twice the angle at the circumference.	$y = 2 \times 46°$ $= 92°$

- **Theorem 3 Code**

Angles subtended by the diameter, that is angles in a semicircle, are right angles.

In the diagram, PQ is the diameter. Angles a, b and c are right angles. This theorem is in fact a special case of Theorem 1.

Proof:

$\angle POQ = 180°$ (straight line)

Let S refer to the angle at the circumference subtended by the diameter. In the diagram, S could be at the points where a, b and c are represented on the diagram.

$\angle PSQ = 90°$ (angle at the circumference is half the angle at the centre)

Angles subtended by a diameter are right angles.

▶ 14.2.2 Tangents to a circle

eles-4996

- In the diagram, the tangent touches the circumference of the circle at the point of contact.

- **Theorem 4 Code**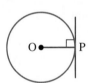

A tangent to a circle is perpendicular to the radius of the circle at the point of contact on the circumference.

In the diagram, the radius is drawn to a point, P, on the circumference. The tangent to the circle is also drawn at P. The radius and the tangent meet at right angles; that is, the angle at P equals 90°.

WORKED EXAMPLE 2 Determining angles when tangents are drawn to a circle

Determine the values of the pronumerals in the diagram, giving a reason for your answer.

THINK

1. Angle z is subtended by the diameter. Use an appropriate theorem to state the value of z.

2. Angle s is formed by a tangent and a radius is drawn to the point of contact. Apply the corresponding theorem to find the value of s.

WRITE

$z = 90°$

$s = 90°$

- **Theorem 5 Code**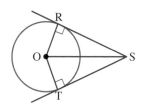

  The angle formed by two tangents meeting at an external point is bisected by a straight line joining the centre of the circle to that external point.

  Proof:

  Consider $\triangle$SOR and $\triangle$SOT.

  OR = OT (radii of the same circle are equal)

  OS is common.

  $\angle$ORS = $\angle$OTS = 90° (angle between a tangent and radii is 90°)

  $\therefore \triangle$SOR $\cong \triangle$SOT (RHS)

  So $\angle$ROS = $\angle$TOS and $\angle$OSR = $\angle$OST (corresponding angles in congruent triangles are equal).

  The angle formed by two tangents meeting at an external point is bisected by a straight line joining the centre of the circle to the external point.

WORKED EXAMPLE 3 Determining angles using properties of tangents

Given that BA and BC are tangents to the circle, determine the values of the pronumerals in the diagram. Give reasons for your answers.

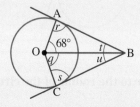

THINK	WRITE
1. Angles r and s are angles formed by the tangent and the radius drawn to the same point on the circle. State their size.	$s = r = 90°$
2. In the triangle ABO, two angles are already known, so angle t can be found using our knowledge of the sum of the angles in a triangle.	$\triangle$ABO: $t + 90° + 68° = 180°$ $t + 158° = 180°$ $t = 22°$
3. $\angle$ABC is formed by the two tangents, so the line BO, which joins the vertex B with the centre of the circle, bisects this angle. This means that angles t and u are equal.	$\angle$ABO = $\angle$CBO $\angle$ABO = $t = 22°$, $\angle$CBO = u $u = 22°$
4. All angles in a triangle have a sum of 180°. $\triangle$AOB and $\triangle$COB are congruent triangles using RHS, they are both right-angled triangles, the hypotenuse is common and OA = OB (both are radii).	In $\triangle$AOB and $\triangle$COB: $r + t + 68° = 180°$ $s + u + q = 180°$ $r = s = 90°$ (proved previously) $t = u = 22°$ (proved previously) $\therefore q = 68°$

Exercise 14.2 Angles in a circle

learn on

| 14.2 Quick quiz on | 14.2 Exercise |

Individual pathways

■ PRACTISE	■ CONSOLIDATE	■ MASTER
1, 4, 6, 8, 13, 17	2, 5, 9, 11, 14, 18	3, 7, 10, 12, 15, 16, 19

Fluency

WE1 For questions **1** to **3**, calculate the values of the pronumerals in each of the following diagrams, giving reasons for your answers.

1. a.   b.   c.

2. a.   b.   c.

3. a.   b.   c.

4. **WE2** Determine the values of the pronumerals in each of the following diagrams, giving reasons for your answers.

a.   b.   c.

5. Calculate the values of the pronumerals in each of the following diagrams, giving reasons for your answers.

a.

b.

c.

Understanding

6. **WE3** Given that AB and DB are tangents, determine the values of the pronumerals in each of the following, giving reasons for your answers.

a.

b.

c.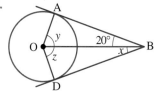

7. Given that AB and DB are tangents, determine the values of the pronumerals in each of the following, giving reasons for your answers.

a.

b.

c.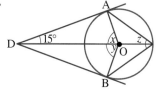

8. **MC** The value of x in the diagram is:

   A. 240°
   C. 90°

   B. 120°
   D. 60°

9. **MC** The value of x in the diagram is:

   A. 50°
   C. 100°

   B. 90°
   D. 80°

10. **MC** Choose which of the following statements is true for this diagram.

    A. ∠ACB = 2 × ∠ADB
    C. ∠ACB = ∠ADB

    B. ∠AEB = ∠ACB
    D. ∠AEB = ∠ADB

    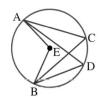

11. **MC** In the diagram shown, determine which angle is subtended by the same arc as ∠APB.
Note: There may be more than one correct answer.

    A. ∠APC   **B.** ∠BPC

    C. ∠ABP   **D.** ∠ADB

12. **MC** For the diagram shown, determine which of the statements is true.
Note: There may be more than one correct answer.

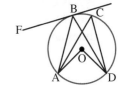

    A. $2\angle AOD = \angle ABD$ **B.** $\angle AOD = 2\angle ACD$

    C. $\angle ABF = \angle ABD$ **D.** $\angle ABD = \angle ACD$

Reasoning

13. Values are suggested for the pronumerals in the diagram shown. AB is a tangent to a circle and O is the centre.
In each case give reasons to justify the suggested values.

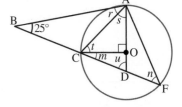

    a. $s = t = 45°$ **b.** $r = 45°$

    c. $u = 65°$ **d.** $m = 25°$

    e. $n = 45°$

14. Set out below is the proof of this result: The angle at the centre of a circle is twice the angle at the circumference standing on the same arc.
Copy and complete the following to show that $\angle POQ = 2 \times \angle PRQ$.
Construct a diameter through R. Let the opposite end of the diameter be S.
Let $\angle ORP = x$ and $\angle ORQ = y$.

OR = OP (_____)

$\angle OPR = x$ (_____)

$\angle SOP = 2x$ (exterior angle equals _____)

OR = OQ (_____)

$\angle OQR =$ _____ (_____)

$\angle SOQ =$ _____ (_____)

Now $\angle PRQ =$ _____ and $\angle POQ =$ _____.

Therefore, $\angle POQ = 2 \times \angle PRQ$.

15. Prove that the segments formed by drawing tangents from an external point to a circle are equal in length.

16. Use the diagram shown to prove that angles subtended by the same arc are equal.

Problem solving

17. Determine the value of x in each of the following diagrams.

    a.

    b.

18. Use your knowledge of types of triangles, angles in triangles and the fact that the radius of a circle meets the tangent to the circle at right angles *to prove* the following theorem:

    The angle formed between two tangents meeting at an external point is bisected by a line from the centre of the circle to the external point.

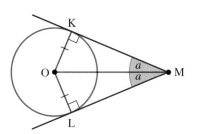

19. WX is the diameter of a circle with centre at O. Y is a point on the circle and WY is extended to Z so that OY = YZ. Prove that angle ZOX is three times angle YOZ.

LESSON
14.3 Intersecting chords, secants and tangents

LEARNING INTENTIONS

At the end of this lesson you should be able to:
- apply the intersecting chords, secants and tangents theorems
- recognise that a radius will bisect a chord at right angles
- understand the concept of the circumcentre of a triangle.

---

eles-4997

⊙ 14.3.1 Intersecting chords

- In the diagram shown, chords PQ and RS intersect at X.

- **Theorem 6 Code** ⊗

  If the two chords intersect inside a circle, then the point of intersection divides each chord into two segments so that the product of the lengths of the segments for both chords is the same.

$$PX \times QX = RX \times SX$$
$$\text{or } a \times b = c \times d$$

Proof:

Join PR and SQ.

Consider $\triangle$PRX and $\triangle$SQX.

$\angle$PXR = $\angle$SXQ (vertically opposite angles are equal)

$\angle$RSQ = $\angle$RPQ (angles at the circumference standing on the same arc are equal)

$\angle$PRS = $\angle$PQS (angles at the circumference standing on the same arc are equal)

$\triangle$PRX ~ $\triangle$SQX (**equiangular**)

$\dfrac{\text{PX}}{\text{SX}} = \dfrac{\text{RX}}{\text{QX}}$ (ratio of sides in similar triangles is equal)

or, PX $\times$ QX = RX $\times$ SX

WORKED EXAMPLE 4 Determining values using intersecting chords

Determine the value of the pronumeral *m*.

THINK	WRITE
1. Chords AB and CD intersect at X. Point X divides each chord into two parts so that the products of the lengths of these parts are equal. Write this as a mathematical statement.	AX $\times$ BX = CX $\times$ DX ⊗
2. Identify the lengths of the line segments.	AX = 4, BX = m, CX = 6, DX = 5
3. Substitute the given lengths into the formula and solve for *m*.	$4m = 6 \times 5$ $m = \dfrac{30}{4}$ $= 7.5$

⊙ 14.3.2 Intersecting secants

eles-4998

- In the diagram shown, chords CD and AB are extended to form secants CX and AX respectively. They intersect at X.

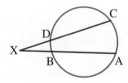

- **Theorem 7 Code** ⊸
  If two secants intersect outside the circle as shown, then the following relationship is always true:

$$\text{AX} \times \text{XB} = \text{XC} \times \text{DX}$$
$$\text{or } a \times b = c \times d.$$

Proof:

Join D and A to O, the centre of the circle.

Let ∠DCA = *x*

∠DOA = 2*x* (angle at the centre is twice the angle at the circumference standing on the same arc)

Reflex ∠DOA = 360° − 2*x* (angles in a revolution add to 360°)

∠DBA = 180° − *x* (angle at the centre is twice the angle at the circumference standing on the same arc)

∠DBX = *x* (angle sum of a straight line is 180°)

∠DCA = ∠DBX

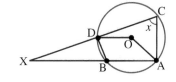

Consider ΔBXD and ΔCXA.

∠BXD is common.

∠DCA = ∠DBX (shown previously)

∠XAC = ∠XDB (angle sum of a triangle is 180°)

∠AXC ~ ΔDXB (equiangular)

$$\frac{AX}{DX} = \frac{XC}{XB}$$

or AX × XB = XC × DX

WORKED EXAMPLE 5 Determining pronumerals using intersecting secants

Determine the value of the pronumeral *y*.

THINK	WRITE
1. Secants XC and AX intersect outside the circle at X. Write the rule connecting the lengths of XC, DX, AX and XB.	XC × DX = AX × XB
2. State the length of the required line segments.	XC = *y* + 6 DX = 6 AX = 7 + 5 XB = 7 = 12
3. Substitute the length of the line segments and solve the equation for *y*.	(*y* + 6) × 6 = 12 × 7 6*y* + 36 = 84 6*y* = 48 *y* = 8

▶ ## 14.3.3 Intersecting tangents

eles-4999

- In the diagram, the tangents AC and BC intersect at C.

- **Theorem 8 Code**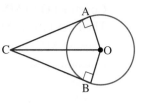
  If two tangents meet outside a circle, then the lengths from the external point to where they meet the circle are equal.
  Proof:
  Join A and B to O, the centre of the circle.
  Consider $\triangle OCA$ and $\triangle OCB$.
  OC is common.
  $\qquad$ OA = OB (radii of the same circle are equal)
  $\angle OAC = \angle OBC$ (radius is perpendicular to tangent through the point of contact)
  $\triangle OCA \cong \triangle OCB$ (RHS)
  $\qquad$ AC = BC (corresponding sides of congruent triangles are equal)

  If two tangents meet outside a circle, the lengths from the external point to the point of contact are equal.

WORKED EXAMPLE 6 Determining the pronumeral using lengths of tangents

Determine the value of the pronumeral m.

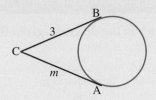

THINK	WRITE
1. BC and AC are tangents intersecting at C. State the rule that connects the lengths BC and AC.	AC = BC
2. State the lengths of BC and AC.	AC = m, BC = 3
3. Substitute the required lengths into the equation to find the value of m.	$m = 3$

14.3.4 Chords and radii

eles-5000

- In the diagram shown, the chord AB and the radius OC intersect at X at 90°; that is, $\angle OXB = 90°$. OC bisects the chord AB; that is, AX = XB.

- **Theorem 9 Code**
  If a radius and a chord intersect at right angles, then the radius bisects the chord.
  Proof:
  Join OA and OB.
  Consider $\triangle OAX$ and $\triangle OBX$.
  OA = OB (radii of the same circle are equal)
  $\angle OXB = \angle OXA$ (given)
  OX is common.
  $\triangle OAX \cong \triangle OBX$ (RHS)
  AX = BX (corresponding sides in congruent triangles are equal)
  If a radius and a chord intersect at right angles, then the radius bisects the chord.

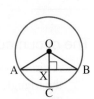

- The converse is also true:
  If a radius bisects a chord, the radius and the chord meet at right angles.

- **Theorem 10 Code**

  Chords equal in length are equidistant from the centre.

  This theorem states that if the chords MN and PR are of equal length, then OD = OC.

  Proof:

  Construct OA⊥MN and OB⊥PR.

  Then OA bisects MN and OB bisects PR (Theorem 9).

  Because MN = PR, MD = DN = PC = CR.

  Construct OM and OP, and consider ∆ODM and ∆OCP.

  MD = PC (shown above)

  OM = OP (radii of the same circle are equal)

  ∠ODM = ∠OCP = 90° (by construction)

  ∆ODM ≅ ∆OCP (RHS)

  So OD = OC (corresponding sides in congruent triangles are equal).

  Chords equal in length are equidistant from the centre.

WORKED EXAMPLE 7 Determining pronumerals using theorems on chords

Determine the values of the pronumerals in the diagram, given that AB = CD.

THINK	WRITE
1. Since the radius OG is perpendicular to the chord AB, the radius bisects the chord.	AE = EB
2. State the lengths of AE and EB.	AE = m, EB = 3
3. Substitute the lengths into the equation to find the value of m.	$m = 3$
4. AB and CD are chords of equal length and OE and OF are perpendicular to these chords. This implies that OE and OF are equal in length.	OE = OF
5. State the lengths of OE and OF.	OE = n, OF = 2.5
6. Substitute the lengths into the equation to find the value of n.	$n = 2.5$

The circumcentre of a triangle

- A circle passing through the three vertices of a triangle is called the **circumcircle** of the triangle.
- The centre of this circle is called the **circumcentre** of the triangle.

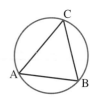

- This means OA = OB = OC, by congruent triangles.
  The circumcentre is located by the following procedure:
  Step 1: Draw any triangle ABC and label the vertices.

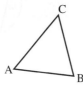

  Step 2: Construct the perpendicular bisectors of the three sides.
  Step 3: Let the bisectors intersect at O.
- A circle, centre O, can be drawn through the vertices A, B and C.
- The point O is the circumcentre of the triangle.

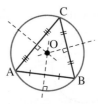

DISCUSSION

What techniques will you use to prove circle theorem?

Exercise 14.3 Intersecting chords, secants and tangents   **learn** on

14.3 Quick quiz on	14.3 Exercise

Individual pathways

■ PRACTISE	■ CONSOLIDATE	■ MASTER
1, 4, 9, 12	2, 5, 7, 10, 13	3, 6, 8, 11, 14

Fluency

1. **WE4** Determine the value of the pronumeral in each of the following diagrams.

   a.

   b.

   c.

2. **WE5** Determine the value of the pronumeral in each of the following diagrams.

a.

b.

3. Determine the value of the pronumeral in each of the following diagrams.

a.

b.

4. **WE6** Determine the value of the pronumerals in each of the following diagrams.

a.

b.

c.

5. **WE7** Determine the value of the pronumeral in each of the following diagrams.

a.

b.

6. Determine the value of the pronumeral in each of the following diagrams.

a.

b.

Understanding

7. **MC** Select which of the following diagrams allows the value of m to be determined by solving a linear equation.

   Note: There may be more than one correct answer.

A.

B.

C.

D.

8. Calculate the length ST in the diagram.

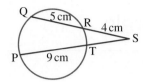

Reasoning

9. Prove the result: If a radius bisects a chord, then the radius meets the chord at right angles. Remember to provide reasons for your statements.

10. Prove the result: Chords that are an equal distance from the centre are equal in length. Provide reasons for your statements.

11. Prove that the line joining the centres of two intersecting circles bisects their common chord at right angles. Provide reasons for your statements.

Problem solving

12. Determine the value of the pronumeral in each of the following diagrams.

a.

b.

c.

13. AOB is the diameter of the circle. CD is a chord perpendicular to AB and meeting AB at M.

  a. Explain why M is the midpoint of CD.

  b. If $CM = c$, $AM = a$ and $MB = b$, prove that $c^2 = ab$.

  c. Explain why the radius of the circle is equal to $\dfrac{a+b}{2}$.

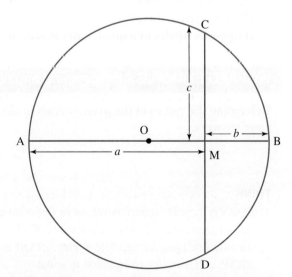

14. An astroid is the curve traced by a point on the circumference of a small circle as it rolls around the inside circumference of a circle that is four times larger than it. Draw the shape of an astroid.

LESSON
14.4 Cyclic quadrilaterals

14.4.1 Cyclic quadrilaterals

eles-5001

- A **cyclic quadrilateral** has all four vertices on the circumference of a circle; that is, the quadrilateral is inscribed in the circle.
- In the diagram shown, points A, B, C and D lie on the circumference; hence, ABCD is a cyclic quadrilateral.
- It can also be said that points A, B, C and D are **concyclic**; that is, the circle passes through all the points.

- **Theorem 11 Code**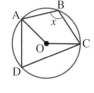
  The opposite angles of a cyclic quadrilateral are supplementary (add to 180°).
  Proof:
  Join A and C to O, the centre of the circle.
  Let $\angle ABC = x$.
  Reflex $\angle AOC = 2x$ (angle at centre is twice angle at circumference standing on same arc)
  Reflex $\angle AOC = 360° - 2x$ (angles revolution 360°)
  $\angle ADC = 180° - x$ (angle at centre is twice angle at circumference standing on same arc)
  $\angle ABC + \angle ADC = 180°$
  Similarly, $\angle DAB + \angle DCB = 180°$.
  Opposite angles in a cyclic quadrilateral are supplementary.

- The converse is also true:
  If opposite angles of a quadrilateral are supplementary, then the quadrilateral is cyclic.

WORKED EXAMPLE 8 Determining angles in a cyclic quadrilateral

Determine the values of the pronumerals in the diagram shown. Give reasons for your answers.

THINK	WRITE
1. PQRS is a cyclic quadrilateral, so its opposite angles are supplementary. Calculate the value of x by considering a pair of opposite angles, $\angle PQR$ and $\angle RSP$, and forming an equation to solve.	$\angle PQR + \angle RSP = 180°$ (The opposite angles of a cyclic quadrilateral are supplementary.) $\angle PQR = 75°$, $\angle RSP = x$ $x + 75° = 180°$ $x = 105°$

2. Calculate the value of y by considering the other pair of opposite angles ($\angle SPQ$ and $\angle QRS$).

$\angle SPQ + \angle QRS = 180°$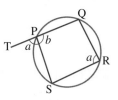
$\angle SPQ = 120°, \angle QRS = y$
$y + 120° = 180°$
$y = 60°$

- **Theorem 12 Code**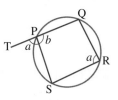
  The exterior angle of a cyclic quadrilateral is equal to the interior opposite angle.
  Proof:
  $\angle QPS + \angle QRS = 180°$ (opposite angles of a cyclic quadrilateral)
  $\angle QPS + \angle SPT = 180°$ (adjacent angles on a straight line)
  Therefore, $\angle SPT = \angle QRS$.
  The exterior angle of a cyclic quadrilateral is equal to the interior opposite angle.

WORKED EXAMPLE 9 Determining pronumerals in a cyclic quadrilateral

Determine the value of the pronumerals x and y in the diagram shown.

THINK

1. ABCD is a cyclic quadrilateral. The exterior angle, x, is equal to its interior opposite angle, $\angle DAB$.

2. The exterior angle, 100°, is equal to its interior opposite angle, $\angle ADC$.

WRITE

$x = \angle DAB, \angle DAB = 50°$
So $x = 50°$.

$\angle ADC = 100°, \angle ADC = y$
So $y = 100°$.

DISCUSSION

What is a cyclic quadrilateral?

 Resources

 eWorkbook   Topic 14 Workbook (worksheets, code puzzle and project) (ewbk-13306)

 Interactivities Circle theorem 11 (int-6228)
Circle theorem 12 (int-6229)

Exercise 14.4 Cyclic quadrilaterals

14.4 Quick quiz on	14.4 Exercise

Individual pathways

■ PRACTISE	■ CONSOLIDATE	■ MASTER
1, 4, 5, 10, 13, 16	2, 3, 6, 11, 14, 17	7, 8, 9, 12, 15, 18

Fluency

WE8 In questions **1** and **2**, calculate the values of the pronumerals in the diagrams shown.

1. a.

b.

c.

2. a.

b.

c.

WE9 In questions **3** to **6**, calculate the values of the pronumerals in the diagrams shown.

3. a.

b.

c.

4. a.

b.

c.

5. a.

b.

6. a.

b.

c.

d.

e.

f.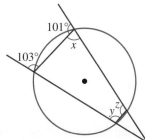

Understanding

7. **MC** Which of the following statements is true for this diagram?

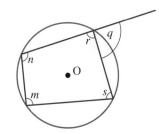

A. $r = q$ B. $r + n = 180°$

C. $m + n = 180°$ D. $r + m = 180°$

8. **MC** Which of the following statements is *not* true for this diagram?

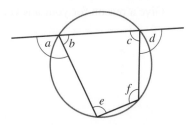

A. $b + f = 180°$ B. $a = f$

C. $e = d$ D. $c + f = 180°$

9. **MC** Choose which of the following correctly states the relationship between x, y and z in the diagram shown.

A. $x = y$ and $x = 2z$ B. $x = 2y$ and $y + z = 180°$

C. $z = 2x$ and $y = 2z$ D. $x + y = 180°$ and $z = 2x$

10. Follow the steps below to set out the proof that the opposite angles of a cyclic quadrilateral are equal.

    a. Calculate the size of ∠DOB.
    b. Calculate the size of the reflex angle DOB.
    c. Calculate the size of ∠BCD.
    d. Calculate ∠DAB + ∠BCD.

11. **MC** Choose which of the following statements is always true for the diagram shown.
    Note: There may be more than one correct answer.

    A. $r = t$ **B.** $r = p$

    C. $r = q$ **D.** $r = s$

12. **MC** Choose which of the following statements is correct for the diagram shown.
    Note: There may be more than one correct answer.

    A. $r + p = 180°$ **B.** $q + s = 180°$

    C. $t + p = 180°$ **D.** $t = r$

Reasoning

13. Prove that the exterior angle of a cyclic quadrilateral is equal to the interior opposite angle.

14. Determine the value of the pronumerals in the diagram shown.
    Give a reason for your answer.

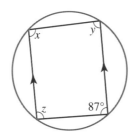

15. Determine the value of the pronumeral x in the diagram shown.
    Give a reason for your answer.

Problem solving

16. Determine the value of the pronumeral x in the diagram shown.

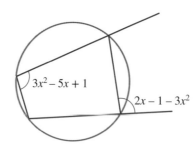

17. Determine the value of each pronumeral in the diagram shown.

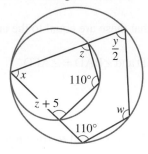

18. $\angle FAB = 70°$, $\angle BEF = a°$, $\angle BED = b°$ and $\angle BCD = c°$.

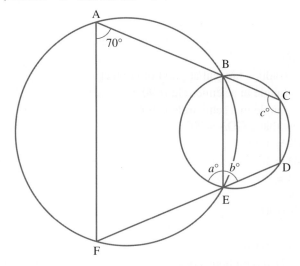

a. Calculate the values of a, b and c.
b. Prove that CD is parallel to AF.

LESSON
14.5 Tangents, secants and chords

LEARNING INTENTIONS

At the end of this lesson you should be able to:
- apply the alternate segment theorem to evaluate pronumerals
- determine the lengths of tangents and secants when they intersect.

▶ 14.5.1 The alternate segment theorem

eles-5002

- Consider the diagram shown. Line BC is a tangent to the circle at the point A.
- A line is drawn from A to a point D anywhere on the circumference.
  The angle $\angle BAD$ defines a segment (the shaded area).
  The unshaded part of the circle is called the **alternate segment** to $\angle BAD$.

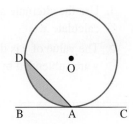

- Now consider angles subtended by the chord AD in the alternate segment, such as the angles marked in pink and blue.
- The alternate segment theorem states that these are equal to the angle that made the segment, namely:

$$\angle BAD = \angle AED \text{ and } \angle BAD = \angle AFD$$

- Theorem 13 Code

  The angle between a tangent and a chord is equal to the angle in the alternate segment.

  Proof:
  We are required to prove that $\angle BAD = \angle AFD$.
  Construct the diameter from A through O, meeting the circle at G.
  Join G to the points D and F.
  $\angle BAG = \angle CAG = 90°$ (radii $\perp$ tangent at point of contact)
  $\angle GFA = 90°$ (angle in a semicircle is 90°)
  $\angle GDA = 90°$ (angle in a semicircle is 90°)
  Consider $\triangle GDA$. We know that $\angle GDA = 90°$.
  $\angle GDA + \angle DAG + \angle AGD = 180°$
      $90° + \angle DAG + \angle AGD = 180°$
             $\angle DAG + \angle AGD = 90°$
  $\angle BAG$ is also a right angle.
  $\angle BAG = \angle BAD + \angle DAG = 90°$
  Equate the two results.
  $\angle DAG + \angle AGD = \angle BAD + \angle DAG$
  Cancel the equal angles ($\angle DAG$) on both sides.
  $\angle AGD = \angle BAD$
  Now consider the fact that both triangles DAG and DAF are subtended from the same chord (DA).
  $\angle AGD = \angle AFD$ (angles in the same segment standing on the same arc are equal)
  Equate the two equations.
  $\angle AFD = \angle BAD$

WORKED EXAMPLE 10 Determining pronumerals using the alternate segment theorem

Determine the values of the pronumerals x and y in the diagram shown, giving reasons.

THINK	WRITE
1. Use the alternate segment theorem to calculate x.	$x = 62°$ (angle between a tangent and a chord is equal to the angle in the alternate segment)
2. The value of y is the same as x because x and y are subtended by the same chord BT.	$y = 62°$ (angles in the same segment standing on the same arc are equal)

14.5.2 Tangents and secants

eles-5003

- **Theorem 14 Code**

  If a tangent and a secant intersect as shown, the following relationship is always true:

  $$XA \times XB = (XT)^2 \text{ or } a \times b = c^2.$$

  Proof:

  Join BT and AT.

  Consider $\triangle TXB$ and $\triangle AXT$.

  $\angle TXB$ is common.

  $\angle XTB = \angle XAT$ (angle between a tangent and a chord is equal to the angle
  in the alternate segment)

  $\angle XBT = \angle XTA$ (angle sum of a triangle is $180°$)

  $\triangle TXB \sim \triangle AXT$ (equiangular)

  So, $\dfrac{XB}{XT} = \dfrac{XT}{XA}$

  or, $XA \times XB = (XT)^2.$

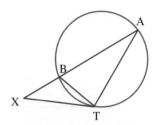

WORKED EXAMPLE 11 Determining pronumerals with intersecting tangents and secants

Determine the value of the pronumeral m in the diagram shown.

THINK	WRITE
1. Secant XA and tangent XT intersect at X. Write the rule connecting the lengths of XA, XB and XT.	$XA \times XB = (XT)^2$
2. State the values of XA, XB and XT.	$XA = m + 5,\ XB = 5,\ XT = 8$
3. Substitute the values of XA, XB and XT into the equation and solve for m.	$(m+5) \times 5 = 8^2$ $5m + 25 = 64$ $5m = 39$ $m = 7.8$

DISCUSSION

Describe the alternate segment of a circle.

 Resources

eWorkbook   Topic 14 Workbook (worksheets, code puzzle and project) (ewbk-13306)

Interactivities Circle theorem 13 (int-6230)
                    Circle theorem 14 (int-6231)

| 14.5 Quick quiz on | 14.5 Exercise |

Individual pathways

■ PRACTISE	■ CONSOLIDATE	■ MASTER
1, 2, 3, 9, 13, 14, 18, 22	4, 5, 6, 10, 15, 16, 17, 19, 23	7, 8, 11, 12, 20, 21, 24

Fluency

1. **WE10** Determine the values of the pronumerals in the following diagrams.

   a.

   b.

2. **WE11** Calculate the values of the pronumerals in the following diagrams.

   a.

   b.

   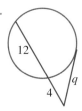

3. Line AB is a tangent to the circle as shown in the diagram. Calculate the values of the angles labelled x and y.

   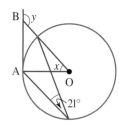

Questions **4** to **6** refer to the diagram shown. The line MN is a tangent to the circle, and EA is a straight line. The circles have the same radius.

4. **MC** If $\angle DAC = 20°$, then $\angle CFD$ and $\angle FDG$ are respectively:

   A. 70° and 50°   B. 70° and 40°
   C. 40° and 70°   D. 70° and 70°

5. **MC** A triangle similar to FDA is:

   A. FDG   B. FGB   C. EDA   D. GDE

6. State six different right angles.

7. Calculate the values of the angles x and y in the diagram shown.

   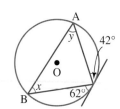

Understanding

8. Show that if the sum of the two given angles in question **7** is 90°, then the line AB must be a diameter.

9. Calculate the value of x in the diagram shown, given that the line underneath the circle is a tangent.

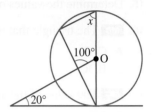

10. In the diagram shown, express x in terms of a and b. This is the same diagram as in question **9**.

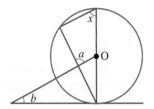

11. Two tangent lines to a circle meet at an angle y, as shown in the diagram. Determine the values of the angles x, y and z.

12. Solve question **11** in the general case (see the diagram) and show that $y = 2a$. This result is important for space navigation (imagine the circle to be the Earth) in that an object at y can be seen by people at x and z at the same time.

13. In the diagram shown, determine the values of the angles x, y and z.

14. **MC** Examine the diagram shown. The angles x and y (in degrees) are respectively:

   A. 51 and 99
   B. 51 and 129
   C. 39 and 122
   D. 51 and 122

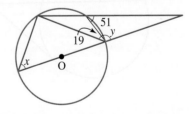

Questions **15** to **17** refer to the diagram shown. The line BA is a tangent to the circle at point B. Line AC is a chord that meets the tangent at A.

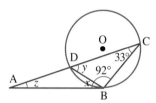

15. Determine the values of the angles x and y.

16. **MC** The triangle that is similar to triangle BAD is:

    A. CAB   **B.** BCD

    C. BDC   **D.** AOB

17. **MC** The value of the angle z is:

    A. 50°   **B.** 85°

    C. 95°   **D.** 100°

Reasoning

18. Determine the values of the angles x, y and z in the diagram shown. The line AB is a tangent to the circle at B.

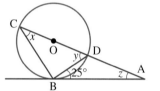

19. Calculate the values of the angles x, y and z in the diagram shown. The line AB is a tangent to the circle at B. The line CD is a diameter.

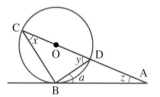

20. Solve question **19** in the general case; that is, express angles x, y and z in terms of a (see the diagram).

21. Prove that, when two circles touch, their centres and the point of contact are collinear.

Problem solving

22. Calculate the value of the pronumerals in the following diagrams.

    a.

    b.

    c.

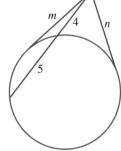

23. Determine the value of the pronumerals in the following diagrams.

a.

b.

c.

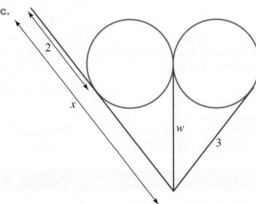

24. Calculate the values of a, b and c in each case.

a. $\angle BCE = 50°$ and $\angle ACE = c$

b.

LESSON
14.6 Review

14.6.1 Topic summary

Cyclic quadrilaterals

- **Opposite angles** of cyclic quadrilateral are supplementary.
  e.g. $a + b = 180°$
- The **exterior angle** of a cyclic quadrilateral equals the interior opposite angle.

CIRCLE GEOMETRY

Angles in a circle

- The angle at the centre is twice the angles at the circumference standing on the same arc.
  e.g. $b = 2a$

- Angles at the circumference standing on the same arc are equal.
  $a = b = c$

- The angle in a semicircle is 90°.
  $a = 90°, b = 180°$

Tangents to a circle

- Tangent is perpendicular to the radius at the point of contact.
- Lengths of tangents from an external point are equal.
  $a = b$

- A line from the centre bisects the external angle between two tangents.

Chords and secants

- When chords intersect inside a circle, the product of the lengths of the segments are equal.
  $a \times b = c \times d$

- When chords intersect outside a circle, the following is true:
  $a \times b = c \times d$

- The perpendicular bisector of a chord passes through the centre of the circle.

Tangents and secants

- The angle between a tangent and a chord equals the angle in the alternate segment.
- If a tangent and a segment intersect, then $a \times b = c^2$.

14.6.2 Success criteria

Tick the column to indicate that you have completed the lesson and how well you have understood it using the traffic light system.

(**Green:** I understand; **Yellow:** I can do it with help; **Red:** I do not understand)

Lesson	Success criteria	○	○	●
14.2	I can determine relationships between the angles at the centre and the circumference of a circle.			
	I can understand and use the angle in a semicircle being a right angle.			
	I can apply the relationships between tangents and radii of circles.			
14.3	I can apply the intersecting chords, secants and tangents theorems.			
	I can recognise that a radius will bisect a chord at right angles.			
	I can understand the concept of the circumcentre of a triangle.			
14.4	I can recognise a cyclic quadrilateral.			
	I can determine pronumerals knowing the opposite angles of a cyclic quadrilateral are supplementary.			
	I can determine pronumerals knowing the exterior angle of a cyclic quadrilateral equals the interior opposite angle.			
14.5	I can apply the alternate segment theorem to evaluate pronumerals.			
	I can determine the lengths of tangents and secants when they intersect.			

14.6.3 Project

Variation of distance

The Earth approximates the shape of a sphere. Lines of longitude travel between the North and South Poles, and lines of latitude travel east–west, parallel to the equator. Although the lines of longitude are all approximately the same length, this is not the case with lines of latitude. The line of latitude at the equator is the maximum length, and these lines decrease in length on approaching both the North and South Poles.

This investigation looks at how the distance between points on two given lines of longitude and the same line of latitude changes as we move from the equator to the pole.

Consider two lines of longitude, 0° and 100°E. Take two points, P_1 and P_2, lying on the equator on lines of longitude 0° and 100°E respectively.

The distance (in km) between two points on the same line of latitude is given by the formula:

Distance = angle sector between the two points × 111 × cos(degree of latitude)

1. The size of the angle sector between P_1 and P_2 is 100°, and these two points lie on 0° latitude. The distance between the points would be calculated as $100 \times 111 \times \cos(0°)$.
   Determine this distance.

2. Move the two points to the 10° line of latitude. Calculate the distance between P_1 and P_2 in this position.
   Round your answer to the nearest kilometre.

3. Complete the following table showing the distance (rounded to the nearest kilometre) between the points P_1 and P_2 as they move from the equator towards the pole.

Latitude	Distance between P_1 and P_2 (km)
0°	
10°	
20°	
30°	
40°	
50°	
60°	
70°	
80°	
90°	

4. Describe what happens to the distance between P_1 and P_2 as we move from the equator to the pole. Is there a constant change?
   Explain your answer.

5. You would perhaps assume that, at a latitude of 45°, the distance between P_1 and P_2 is half the distance between the points at the equator. This is not the case.
   At what latitude does this occur?

6. Using grid paper, sketch a graph displaying the change in distance between the points in moving from the equator to the pole.

7. Consider the points P_1 and P_2 on lines of longitude separated by 1°. On what line of latitude (to the nearest degree) would the points be 100 km apart?

8. Keeping the points P_1 and P_2 on the same line of latitude and varying their lines of longitude, investigate the rate that the distance between them changes from the equator to the pole.
   Explain whether it is more or less rapid in comparison to what you found earlier.

Exercise 14.6 Review questions

learnon

Fluency

For questions **1** to **3**, determine the value of the pronumeral in each of the diagrams.

1. a.

b.

c.

d.

e.

f.

2. a.

b.

c.

d.

e.

f.

3. a.

b.

c.

d.
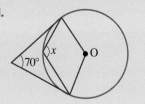

4. Determine the value of *m* in each of the following.

a.

b.

c.

d.

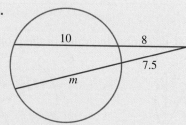

5. **MC** Choose the diagram(s) for which it is possible to get a reasonable value for the pronumeral.
Note: There may be more than one correct answer.

A.

B.

C.

D.

6. **MC** State which of the following statements is true for the diagram shown.
Note: There may be more than one correct answer.

A. $AO = BO$

B. $AC = BC$

C. $\angle OAC = \angle OBC$

D. $\angle AOC = 90°$

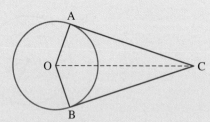

7. Determine the values of the pronumerals in the following diagrams.

a.

b.

c.

d.

8. **MC** State which of the following statements is *not* always true for the diagram shown.

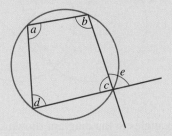

A. $\angle a + \angle c = 180°$

B. $\angle b + \angle d = 180°$

C. $\angle e + \angle c = 180°$

D. $\angle a + \angle e = 180°$

Problem solving

9. Determine the values of the pronumerals in the following diagrams.

a.

b.

c.

10. Two chords, AB and CD, intersect at E as shown. If AE = CE, prove that EB = ED.

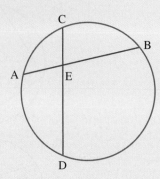

11. Two circles intersect at X and Y. Two lines, AXB AXB and CXD, intersect one circle at A and C, and the other at B and D, as shown. Prove that ∠AYC = ∠BYD.

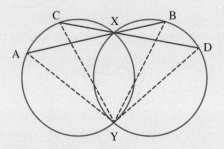

12. Name at least five pairs of equal angles in the diagram shown.

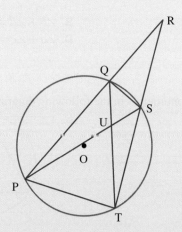

on To test your understanding and knowledge of this topic, go to your learnON title at www.jacplus.com.au and complete the **post-test**.

Answers

Topic 14 Circle geometry (10A)

14.1 Pre-test

1. Segment
2. D
3. $35°$
4. D
5. $65°$
6. $15°$
7. D
8. 6
9. 1.5
10. A
11. $m = 55°$, $p = 50°$
12. C
13. D
14. D
15. A

14.2 Angles in a circle

1. a. $x = 30°$ (theorem 2)
   b. $x = 25°$, $y = 25°$ (theorem 2 for both angles)
   c. $x = 32°$ (theorem 2)
2. a. $x = 40°$, $y = 40°$ (theorem 2 for both angles)
   b. $x = 60°$ (theorem 1)
   c. $x = 40°$ (theorem 1)
3. a. $x = 84°$ (theorem 1)
   b. $x = 50°$ (theorem 2); $y = 100°$ (theorem 1)
   c. $x = 56°$ (theorem 1)
4. a. $s = 90°$, $r = 90°$ (theorem 3 for both angles)
   b. $u = 90°$ (theorem 4); $t = 90°$ (theorem 3)
   c. $m = 90°$, $n = 90°$ (theorem 3 for both angles)
5. a. $x = 52°$ (theorem 3 and angle sum in a triangle $= 180°$)
   b. $x = 90°$ (theorem 4)
   c. $x = 90°$ (theorem 4); $y = 15°$ (angle sum in a triangle $= 180°$)
6. a. $x = z = 90°$ (theorem 4); $y = w = 20°$ (theorem 5 and angle sum in a triangle $= 180°$)
   b. $s = r = 90°$ (theorem 4); $t = 140°$ (angle sum in a quadrilateral $= 360°$)
   c. $x = 20°$ (theorem 5); $y = z = 70°$ (theorem 4 and angle sum in a triangle $= 180°$)
7. a. $s = y = 90°$ (theorem 4); $x = 70°$ (theorem 5); $r = z = 20°$ (angle sum in a triangle $= 180°$)
   b. $x = 70°$ (theorem 4 and angle sum in a triangle $= 180°$); $y = z = 20°$ (angle sum in a triangle $= 180°$)
   c. $x = y = 75°$ (theorem 4 and angle sum in a triangle $= 180°$); $z = 75°$ (theorem 1)
8. B
9. C
10. C
11. D
12. B, D
13. a. Base angles of a right-angled isosceles triangle
   b. $r + s = 90°$, $s = 45° \Rightarrow r = 45°$
   c. u is the third angle in $\triangle$ ABD, which is right-angled.
   d. m is the third angle in $\triangle$ OCD, which is right-angled.
   e. $\angle$AOC and $\angle$AFC stand on the same arc with $\angle$AOC at the centre and $\angle$AFC at the circumference.
14. OR $=$ OP (radii of the circle)
   $\angle$OPR $= x$ (equal angles lie opposite equal sides)
   $\angle$SOP $= 2x$ (exterior angle equals the sum of the two interior opposite angles)
   OR $=$ OQ (radii of the circle)
   $\angle$OQR $= y$ (equal angles lie opposite equal sides)
   $\angle$SOQ $= 2y$ (exterior angle equals the sum of the two interior opposite angles)
   Now $\angle$PRQ $= x + y$ and $\angle$POQ $= 2x + 2y = 2(x + y)$.
   Therefore, $\angle$POQ $= 2 \times \angle$PRQ.
15, 16. Sample responses can be found in the worked solutions in the online resources.
17. a. $16°$ b. $20°$
18, 19. Sample responses can be found in the worked solutions in the online resources.

14.3 Intersecting chords, secants and tangents

1. a. $m = 3$ b. $m = 3$ c. $m = 6$
2. a. $n = 1$ b. $m = 7.6$
3. a. $n = 13$ b. $m = 4$
4. a. $x = 5$ b. $m = 7$ c. $x = 2.5$, $y = 3.1$
5. a. $x = 2.8$ b. $x = 3.3$
6. a. $x = 5.6$ b. $m = 90°$
7. B, C, D
8. ST $= 3$ cm
9–11. Sample responses can be found in the worked solutions in the online resources.
12. a. $x = 3\sqrt{2}$ b. $x = 6$ c. $x = 3$, $y = 12$
13. a. The line from the centre perpendicular to the chord bisects the chord, giving M as the midpoint.
    b, c. Sample responses can be found in the worked solutions in the online resources.
14.

14.4 Cyclic quadrilaterals

1. a. $x = 115°$, $y = 88°$ b. $m = 85°$
   c. $n = 25°$
2. a. $x = 130°$ b. $x = y = 90°$ c. $x = 45°$, $y = 95°$
3. a. $x = 85°$, $y = 80°$
   b. $x = 110°$, $y = 115°$
   c. $x = 85°$

4. a. $x = 150°$ **b.** $x = 90°$, $y = 120°$
   c. $m = 120°$, $n = 130°$

5. a. $a = 89°$, $b = 45°$
   b. $a = 120°$, $b = 91°$, $c = 89°$

6. a. $x = 102°$, $y = 113°$
   b. $x = 95°$, $y = 85°$, $z = 95°$
   c. $x = 126°$, $y = 54°$
   d. $x = 60°$, $y = 120°$
   e. $x = 54°$, $y = 72°$
   f. $x = 79°$, $y = 101°$, $z = 103°$

7. D

8. D

9. D

10. a. $2x$ **b.** $360° - 2x$
  c. $180° - x$ **d.** $180°$

11. A

12. A, B, C, D

13. Sample responses can be found in the worked solutions in the online resources.

14. $x = 93°$, $y = 87°$, $z = 93°$

15. $x = -2°$ or $\dfrac{5°}{2}$

16. $x = \dfrac{2°}{3}$ or $\dfrac{1°}{2}$

17. $w = 110°$, $x = 70°$, $y = 140°$, $z = 87.5°$

18. a. $a = 110°$, $b = 70°$ and $c = 110°$
   b. Sample responses can be found in the worked solutions in the online resources.

14.5 Tangents, secants and chords

1. a. $x = 70°$ **b.** $x = 47°$, $y = 59°$

2. a. $p = 6$ **b.** $q = 8$

3. $x = 42°$, $y = 132°$

4. B

5. D

6. MAC, NAC, FDA, FBA, EDG, EBG.

7. $x = 42°$, $y = 62°$

8. Sample responses can be found in the worked solutions in the online resources.

9. $60°$

10. $x = 180° - a - b$

11. $x = 80°$, $y = 20°$, $z = 80°$

12. Sample responses can be found in the worked solutions in the online resources.

13. $x = 85°$, $y = 20°$, $z = 85°$

14. D

15. $x = 50°$, $y = 95°$

16. A

17. C

18. $x = 33°$, $y = 55°$, $z = 22°$

19. $x = 25°$, $y = 65°$, $z = 40°$

20. $x = a$, $y = 90° - a$, $z = 90° - 2a$

21. Sample responses can be found in the worked solutions in the online resources.

22. a. $x = 5$ **b.** $k = 12$ **c.** $m = 6$, $n = 6$

23. a. $x = 7$ **b.** $b = 4$, $a = 2$ **c.** $w = 3$, $x = 5$

24. a. $a = 50°$, $b = 50°$ and $c = 80°$
   b. $a = 50°$, $b = 70°$ and $c = 70°$

Project

1. $11\,100$ km

2. $10\,931$ km

3.

Latitude	Distance between P_1 and P_2 (km)
0°	11 100
10°	10 931
20°	10 431
30°	9613
40°	8503
50°	7135
60°	5550
70°	3796
80°	1927
90°	0

4. The distance between P_1 and P_2 decreases from 11 100 km at the equator to 0 km at the pole. The change is not constant. The distance between the points decreases more rapidly on moving towards the pole.

5. Latitude 60°

6.

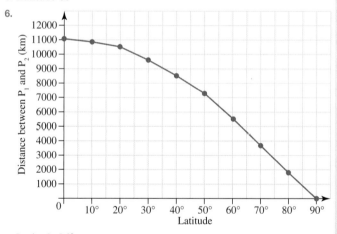

7. Latitude 26°

8. Sample responses can be found in the worked solutions in the online resources. Students need to investigate the rate at which the distance between the points changes from the equator to the pole and also compare with the earlier values.

14.6 Review questions

1. a. $x = 50°$ **b.** $x = 48°$, $y = 25°$
   c. $x = y = 28°$, $z = 56°$ **d.** $x = 90°$
   e. $y = 90°$ **f.** $y = 140°$

2. a. $x = 55°$ **b.** $x = 125°$
   c. $x = 70°$ **d.** $x = 100°$
   e. $m = 40°$ **f.** $x = 90°$, $y = 60°$, $z = 40°$

3. a. $x = 90°$
   b. $x = 20°$
   c. $x = 55°$
   d. $x = 125°$

4. a. $m = 3$
   b. $m = 12$
   c. $m = 9$
   d. $m = 11.7$

5. A, B, D

6. A, B, C

7. a. $x = 95°$, $y = 80°$
   b. $x = 99°$
   c. $x = 78°$, $y = 92°$
   d. $x = 97°$, $y = 92°$

8. D

9. a. $x = 42°$ b. $y = 62°$ c. $p = 65°$

10. $CE \times ED = AE \times EB$ ⊘

    $AE = CE$ (given)
    $\therefore ED = EB$

11. $\angle AYC = \angle AXC$ ⊗

    $\angle BXD = \angle BYD$ ⊗

    But $\angle AXC = \angle BXD$ ⋈

    $\Rightarrow \angle AYC = \angle BYD$

12. $\angle PQT$ and $\angle PST$, $\angle PTS$ and $\angle RQS$, $\angle TPQ$ and $\angle QSR$,
    $\angle QPS$ and $\angle QTS$, $\angle TPS$ and $\angle TQS$, $\angle PQS$ and $\angle PTS$,
    $\angle PUT$ and $\angle QUS$, $\angle PUQ$ and $\angle TUS$

15 Trigonometry II (10A)

LESSON
15.1 Overview

Why learn this?

Trigonometry is the branch of mathematics that describes the relationship between angles and side lengths in triangles. The ability to calculate distances using angles has long been critical. As early as the third century BCE, trigonometry was being used in the study of astronomy. Early explorers, using rudimentary calculations and the stars, were able to navigate their way around the world. They were even able to map coastlines along the way. Cartographers use trigonometry when they are making maps. It is essential to be able to calculate distances that can't be physically measured. Astronomers use trigonometry to calculate distances such as that from a particular planet to Earth. Our explorations have now turned towards the skies and outer space.

Scientists design and launch space shuttles and rockets to explore our universe. By applying trigonometry, they can approximate the distances to other planets. As well as in astronomy and space exploration, trigonometry is widely used in many other areas. Surveyors use trigonometry in setting out a land subdivision. Builders, architects and engineers use angles, lengths and forces in the design and construction of all types of buildings, both domestic and industrial. In music, a single note is a sine wave. Sound engineers manipulate sine waves to create sound effects. Trigonometry has many real-life applications.

Hey students! Bring these pages to life online

Watch videos

Engage with interactivities

Answer questions and check solutions

Find all this and MORE in jacPLUS

Reading content and rich media, including interactivities and videos for every concept

Extra learning resources

Differentiated question sets

Questions with immediate feedback, and fully worked solutions to help students get unstuck

1. **MC** From the following options select the exact value of sin(30°).

   A. $\dfrac{1}{\sqrt{2}}$

   B. $\dfrac{1}{2}$

   C. $\dfrac{\sqrt{3}}{2}$

   D. 1

2. Solve for x, correct to 2 decimal places.

3. Solve for y, correct to 2 decimal places.

4. **MC** Choose the values of the angles B and B' in the below triangle, to the nearest degree. (Assume $BC = B'C$.)

   A. 47° and 133°

   B. 46° and 134°

   C. 47° and 153°

   D. 25° and 155°

5. Calculate the perimeter of the following triangle, correct to 2 decimal places.

6. Solve for x, correct to 1 decimal place.

7. MC Calculate the area of the triangle shown.

    A. 13.05 cm^2 **B.** 18.64 cm^2

    C. 22.75 cm^2 **D.** 26.1 cm^2

8. State in which quadrant of the unit circle is the angle $203°$ located.

9. Determine the value of $\cos(60°)$ using part of the unit circle.

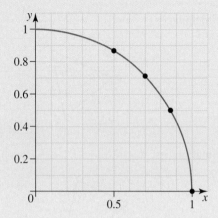

10. MC If $\cos(x°) = p$ for $0 \le x \le 90°$, then $\sin(180 - x°)$ in terms of p is:

    A. p **B.** $180 - p$ **C.** $1 - p$ **D.** $\sqrt{1 - p^2}$

11. MC Select the amplitude and period, respectively, of $y = -2\sin(2x)$ from the following options.

    A. $-2, 360°$ **B.** $-2, 180°$ **C.** $2, 2$ **D.** $2, 180°$

12. MC Select the correct equation for the graph shown.

    A. $y = 4\cos(2x)$

    B. $y = -4\cos(2x)$

    C. $y = -4\sin(2x)$

    D. $y = 4\sin(2x)$

13. MC Select the correct solutions for the equation $\sin(x) = \dfrac{1}{2}$ for x over the domain $0 \le x \le 360°$.

    A. $x = 30°$ and $x = 150°$ **B.** $x = 30°$ and $x = 210°$

    C. $x = 60°$ and $x = 120°$ **D.** $x = 60°$ and $x = 240°$

14. MC Select the correct solutions for the equation $\cos(2x) = -\dfrac{\sqrt{2}}{2}$ for x over the domain $0 \le x \le 360°$.

    A. $x = 22.5°$ and $x = 337.5°$ **B.** $x = 45°$ and $x = 315°$

    C. $x = 67.5°$ and $x = 112.5°$ **D.** $x = 157.5°$ and $x = 202.5°$

15. Using the graph shown, solve the equation $7\sin(x) = -7$ for $0 \le x \le 360°$.

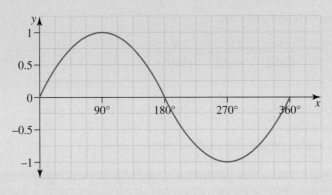

LESSON
15.2 The sine rule

LEARNING INTENTIONS

At the end of this lesson you should be able to:
- apply the sine rule to evaluate angles and sides of triangles
- recognise when the ambiguous case of the sine rule exists.

▶ 15.2.1 The sine rule

- In any triangle, label the angles are named by the vertices A, B and C and the corresponding opposite sides as a, b and c as shown in the diagram.
- Let BD be the perpendicular line from B to AC, of length h, giving two right-angled triangles, $\triangle ADB$ and $\triangle CDB$.

Using $\triangle ADB$:	Using $\triangle CDB$:

$$\sin(A) = \frac{h}{c} \qquad\qquad \sin(C) = \frac{h}{a}$$

$$h = c\sin(A) \qquad\qquad h = a\sin(C)$$

Equating the values of h:

$$c\sin(A) = a\sin(C)$$

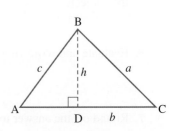

$$\frac{h}{c} = \sin(A) \text{ and } \frac{h}{a} = \sin(C)$$

giving:

$$\frac{c}{\sin(C)} = \frac{a}{\sin(A)}$$

- Similarly, if a perpendicular line is drawn from vertex A to BC, then:

$$\frac{c}{\sin(C)} = \frac{b}{\sin(B)}$$

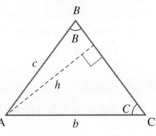

$$h = c\sin(B) \text{ and } h = b\sin(C)$$

Sine rule

- **The sine rule for any triangle ABC is:**

$$\frac{a}{\sin(A)} = \frac{b}{\sin(B)} = \frac{c}{\sin(C)}$$

- The sine rule can be used to solve non-right-angled triangles if we are given:
    1. two angles and one side
    2. two sides and an angle opposite one of these sides.

WORKED EXAMPLE 1 Determining unknown angles and sides of a given triangle

In the triangle ABC, $a = 4\,\text{m}$, $b = 7\,\text{m}$ and $B = 80°$. Calculate the values of A, C and c.

THINK	WRITE/DRAW
1. Draw a labelled diagram of the triangle ABC and fill in the given information.	
2. Check that one of the criteria for the sine rule has been satisfied.	The sine rule can be used since two side lengths and an angle opposite one of these side lengths have been given.
3. Write down the sine rule to calculate A.	To calculate angle A: $$\frac{a}{\sin(A)} = \frac{b}{\sin(B)}$$
4. Substitute the known values into the rule.	$$\frac{4}{\sin(A)} = \frac{7}{\sin(80°)}$$
5. Transpose the equation to make $\sin(A)$ the subject.	$4\sin(80°) = 7\sin(A)$ $$\sin(A) = \frac{4\sin(80°)}{7}$$
6. Evaluate and write your answer.	$$A = \sin^{-1}\left(\frac{4\sin(80°)}{7}\right)$$ $\approx 34.246\,004\,71°$
7. Round off the answer to degrees and minutes.	$\approx 34°15'$
8. Determine the value of angle C using the fact that the angle sum of any triangle is 180°.	$C \approx 180° - (80° + 34°15')$ $= 65°45'$
9. Write down the sine rule to calculate the value of c.	To calculate side length c: $$\frac{c}{\sin(C)} = \frac{b}{\sin(B)}$$
10. Substitute the known values into the rule.	$$\frac{c}{\sin(65°45')} = \frac{7}{\sin(80°)}$$

11. Transpose the equation to make c the subject.

$$c = \frac{7 \sin(65°45')}{\sin(80°)}$$

12. Evaluate. Round off the answer to 2 decimal places and include the appropriate unit.

$\approx 6.48\,\text{m}$

**TI	THINK**	**DISPLAY/WRITE**

TI | THINK

Open a new document and a Calculator page. Ensure your calculator is set to the degree and approximate mode, as shown for the next set of examples.
To do this, press:
- HOME
- 5: Settings
- 2: Document Settings
For Display Digits select 'Fix 2'.
TAB to Angle and select 'Degree'.
TAB to Calculation Mode and select 'Approximate'.
TAB to OK and press ENTER.

DISPLAY/WRITE

1. In a new problem, on the Calculator page, press TRIG to access and select the appropriate trigonometric ratio $(\sin^{-1})$. Complete the entry line as:

$$\sin^{-1}\left(\frac{4\sin(80)}{7}\right)$$

Then press ENTER.
To convert the decimal degree into degrees, minutes and seconds, press:
- CATALOG
- 1
- D
Scroll to and select ▶
DMS then press ENTER.
Complete the entry line as: $180 - (\text{ans} + 80)$
Then press ENTER.
Repeat the above process to convert to degrees, minutes and seconds.

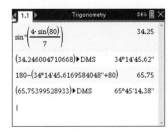

$A = 34°15'$ and $C = 65°45'$

CASIO | THINK

Ensure your calculator is set to the degree and decimal modes. To do this, at the bottom of the Main screen, tap on the mode options until you have Decimal and Deg.

DISPLAY/WRITE

1. On the Main screen, complete the entry line as:

$$\sin^{-1}\left(\frac{4\sin(80)}{7}\right)$$

$$180 - \left(80 + \sin^{-1}\left(\frac{4\sin(80)}{7}\right)\right)$$

Press EXE after each entry.
To convert the decimal degree into degrees, minutes and seconds, tap:
- Action
- Transformation
- DMS
- toDMS
Highlight and drag each of the decimal answers into the entry line and press EXE.

$A = 34°15'$ and $C = 65°45'$ rounded to the nearest minute.

2. To find the value of c, complete the entry line as:

$$\frac{7 \sin (\text{ans})}{\sin 80}$$

Then press ENTER.

$c = 6.48 \text{ m}$

2. To calculate the value of c, complete the entry lines as:

dms(65, 45)

$$\frac{7 \sin (65, 75)}{\sin (80)}$$

Press EXE after each entry line.

$c = 6.48 \text{ m}$

15.2.2 The ambiguous case

eles-5005

- If two side lengths and an angle opposite one of these side lengths are given, then two different triangles may be drawn.
- For example, if $a = 10$, $c = 6$ and $C = 30°$, two possible triangles could be created.
- In the first case angle A is an acute angle, and in the second case angle A is an obtuse angle.
- When using the sine rule to determine an angle, the inverse sine function is used.
- In lesson 15.5, you will see that the sine of an angle between $0°$ and $90°$ has the same value as the sine of its supplement.
  For example, $\sin 40° \approx 0.6427$ and $\sin 140° \approx 0.6427$.

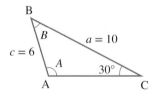

WORKED EXAMPLE 2 Solving triangles and checking for the ambiguous case

In the triangle ABC, $a = 10 \text{ m}$, $c = 6 \text{ m}$ and $C = 30°$. Determine two possible values of A, and hence two possible values of B and b.

Case 1

THINK

1. Draw a labelled diagram of the triangle ABC and fill in the given information.

2. Check that one of the criteria for the sine rule has been satisfied.

3. Write down the sine rule to determine A.

4. Substitute the known values into the rule.

WRITE/DRAW

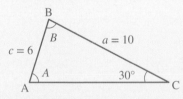

The sine rule can be used since two side lengths and an angle opposite one of these side lengths have been given.

To determine angle A: $\dfrac{a}{\sin(A)} = \dfrac{c}{\sin(C)}$

$$\frac{10}{\sin(A)} = \frac{6}{\sin(30°)}$$

$$10 \sin(30°) = 6 \sin(A)$$

5. Transpose the equation to make $\sin(A)$ the subject.

$$\sin(A) = \frac{10\sin(30°)}{6}$$

6. Evaluate angle A.

$$A = \sin^{-1}\left(\frac{10\sin(30°)}{6}\right)$$

$$\approx 56.442\,690\,24°$$

7. Round off the answer to degrees and minutes.

$$A = 56°27'$$

8. Determine the value of angle B, using the fact that the angle sum of any triangle is $180°$.

$$B \approx 180° - (30° + 56°27')$$

$$= 93°33'$$

9. Write down the sine rule to calculate b.

To calculate side length b:

$$\frac{b}{\sin(B)} = \frac{c}{\sin(C)}$$

10. Substitute the known values into the rule.

$$\frac{b}{\sin(93°33')} = \frac{6}{\sin(30°)}$$

11. Transpose the equation to make b the subject.

$$b = \frac{6\sin(93°33')}{\sin(30°)}$$

12. Evaluate. Round off the answer to 2 decimal places and include the appropriate unit.

$$\approx 11.98\,\text{m}$$

Note: The values we have just obtained are only one set of possible answers for the given dimensions of the triangle ABC.

We are told that $a = 10\,\text{m}$, $c = 6\,\text{m}$ and $C = 30°$. Since side a is larger than side c, it follows that angle A will be larger than angle C. Angle A must be larger than $30°$; therefore it may be an acute angle or an obtuse angle.

Case 2

THINK

1. Draw a labelled diagram of the triangle ABC and fill in the given information.

WRITE/DRAW

2. Write down the alternative value for angle A. Simply subtract the value obtained for A in case 1 from $180°$.

To determine the alternative angle A:
If $\sin A = 0.8333$, then A could also be:
$$A \approx 180° - 56°27'$$

$$= 123°33'$$

3. Determine the alternative value of angle B, using the fact that the angle sum of any triangle is $180°$.

$$B \approx 180° - (30° + 123°33')$$

$$= 26°27'$$

4. Write down the sine rule to determine the alternative b.

To calculate side length b:

$$\frac{b}{\sin(B)} = \frac{c}{\sin(C)}$$

5.	Substitute the known values into the rule.	$\dfrac{b}{\sin(26°27')} = \dfrac{6}{\sin(30°)}$
6.	Transpose the equation to make b the subject.	$b = \dfrac{6\sin(26°27')}{\sin(30°)}$
7.	Evaluate. Round off the answer to 2 decimal places and include the appropriate unit.	$\approx 5.34\,\text{m}$

TI | THINK	DISPLAY/WRITE	CASIO | THINK	DISPLAY/WRITE
1. In a new problem, on a Calculator page, complete the entry line as: solve $\left(\sin(a) = \dfrac{10\sin(30)}{6}, a\right) \mid$ $0 < a < 180$ Press the up arrow and highlight the answer shown. Press ENTER to bring this answer down into the new line. Convert these angles to degrees and minutes as shown in Worked Example 1 by completing the entry as shown. Press ENTER after each entry.	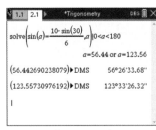 $A = 56°27'$ or $123°34'$	1. On the Main screen, complete the entry line as: solve $\left(\sin(a) = \dfrac{10\sin(30)}{6}, a\right) \mid$ $0 < a < 180$ Convert these angles to degrees and minutes as shown. Press EXE after each entry.	 $A = 56°27'$ or $123°33'$ rounded to the nearest minute.
2. Solve for the two values of B as shown in the screenshot. Instead of typing the angle manually, press the up arrow to highlight the previous answer you want, then press ENTER. Press ENTER after each entry. Convert these angles to degrees and minutes.	 $B = 93°33'$ or $26°27'$	2. Solve for the two values of B as shown at right. [This uses $B = 180 - (30 + A)$.] Press EXE after each entry, and convert these angles to degrees and minutes.	 $B = 93°33'$ or $26°27'$
3. To solve for the two values of b, complete the entry lines as: $\dfrac{6\sin(93°34')}{\sin(30)}$ $\dfrac{6\sin(26°26')}{\sin(30)}$ Press ENTER after each entry.	 $b = 11.98\,\text{m}$ or $5.34\,\text{m}$	3. To solve for the two values of b, complete the entry lines as: $\dfrac{6\sin(93.55730976)}{\sin(30)}$ $\dfrac{6\sin(26.4426902)}{\sin(30)}$ Press after each entry.	 $b = 11.98\,\text{m}$ or $5.34\,\text{m}$

- In Worked example 2 there were two possible solutions, as shown by the diagrams below.

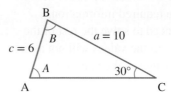

- The ambiguous case does not apply to every question.

Consider Worked example 1.
- Since $\angle A = 34°15'$, it could also have been $\angle A = 145°45'$, the supplementary angle.
- If $\angle A = 34°15'$ and $\angle B = 80°$, then
  $\angle C = 65°45'$ (angle sum of triangle).
- If $\angle A = 145°45'$ and $\angle B = 80°$, then
  $\angle C = 180° - (145°45' + 80°)$ which is not possible.
  $\angle C = -45°45'$
Hence, for Worked example 1, only one possible solution exists.
- The ambiguous case may exist if the angle found is opposite the larger given side.

WORKED EXAMPLE 3 Calculating heights given angles of elevation

To calculate the height of a building, Kevin measures the angle of elevation to the top as 52°. He then walks 20 m closer to the building and measures the angle of elevation as 60°. Calculate the height of the building.

THINK	WRITE/DRAW
1. Draw a labelled diagram of the situation and fill in the given information.	
2. Check that one of the criteria for the sine rule has been satisfied for triangle ABC.	The sine rule can be used for triangle ABC since two angles and one side length have been given.
3. Calculate the value of angle ACB, using the fact that the angle sum of any triangle is 180°.	$\angle ACB = 180° - (52° + 120°)$ $= 8°$
4. Write down the sine rule to calculate b (or AC).	To calculate side length b of triangle ABC: $$\frac{b}{\sin(B)} = \frac{c}{\sin(C)}$$
5. Substitute the known values into the rule.	$$\frac{b}{\sin(120°)} = \frac{20}{\sin(8°)}$$
6. Transpose the equation to make b the subject.	$$b = \frac{20\sin(120°)}{\sin(8°)}$$
7. Evaluate. Round off the answer to 2 decimal places and include the appropriate unit.	$\approx 124.45\ \text{m}$

8. Draw a diagram of the situation, that is triangle ADC, labelling the required information.
Note: There is no need to solve the rest of the triangle in this case as the values will not assist in calculating the height of the building.

9. Write down what is given for the triangle.

Have: angle and hypotenuse

10. Write down what is needed for the triangle.

Need: opposite side

11. Determine which of the trigonometric ratios is required (SOH − CAH − TOA).

$$\sin(\theta) = \frac{O}{H}$$

12. Substitute the given values into the appropriate ratio.

$$\sin(52°) = \frac{h}{124.45}$$

13. Transpose the equation and solve for h.

$$124.45 \sin(52°) = h$$
$$h = 124.45 \sin(52°)$$

14. Round off the answer to 2 decimal places.

$h \approx 98.07$

15. Write the answer.

The height of the building is 98.07 m.

DISCUSSION

Discuss a real-life situation in which the sine rule can be used.

 Resources

 eWorkbook   Topic 15 Workbook (worksheets, code puzzle and project) (ewbk-13308)

    Interactivities   The ambiguous case (int-4818)

                 The sine rule (int-6275)

Exercise 15.2 The sine rule

learn on

15.2 Quick quiz on	15.2 Exercise

Individual pathways

■ PRACTISE	■ CONSOLIDATE	■ MASTER
1, 2, 6, 7, 11, 13, 17, 18, 21	3, 4, 8, 9, 12, 14, 16, 19, 22	5, 10, 15, 20, 23, 24, 25

Where appropriate in this exercise, write your angles correct to the nearest minute and side lengths correct to 2 decimal places.

Fluency

1. **WE1** In the triangle ABC, $a = 10$, $b = 12$ and $B = 58°$. Calculate A, C and c.

2. In the triangle ABC, $c = 17.35$, $a = 26.82$ and $A = 101°47'$. Calculate C, B and b.

3. In the triangle ABC, $a = 5$, $A = 30°$ and $B = 80°$. Calculate C, b and c.

4. In the triangle ABC, $c = 27$, $C = 42°$ and $A = 105°$. Calculate B, a and b.

5. In the triangle ABC, $a = 7$, $c = 5$ and $A = 68°$. Determine the perimeter of the triangle.

6. Calculate all unknown sides and angles for the triangle ABC, given $A = 57°$, $B = 72°$ and $a = 48.2$.

7. Calculate all unknown sides and angles for the triangle ABC, given $a = 105$, $B = 105°$ and $C = 15°$.

8. Calculate all unknown sides and angles for the triangle ABC, given $a = 32$, $b = 51$ and $A = 28°$.

9. Calculate the perimeter of the triangle ABC if $a = 7.8$, $b = 6.2$ and $A = 50°$.

10. **MC** In a triangle ABC, $B = 40°$, $b = 2.6$ and $c = 3$. Identify the approximate value of C.
    Note: There may be more than one correct answer.
    A. $47°$ **B.** $48°$ **C.** $132°$ **D.** $133°$

Understanding

11. **WE2** In the triangle ABC, $a = 10$, $c = 8$ and $C = 50°$. Determine two possible values of A and hence two possible values of b.

12. In the triangle ABC, $a = 20$, $b = 12$ and $B = 35°$. Determine two possible values for the perimeter of the triangle.

13. Calculate all unknown sides and angles for the triangle ABC, given $A = 27°$, $B = 43°$ and $c = 6.4$.

14. Calculate all unknown sides and angles for the triangle ABC, given $A = 100°$, $b = 2.1$ and $C = 42°$.

15. Calculate all unknown sides and angles for the triangle ABC, given $A = 25°$, $b = 17$ and $a = 13$.

16. **WE3** To calculate the height of a building, Kevin measures the angle of elevation to the top as $48°$. He then walks 18 m closer to the building and measures the angle of elevation as $64°$. Calculate the height of the building.

Reasoning

17. Calculate the value of h, correct to 1 decimal place. Show the full working.

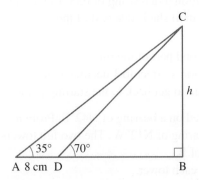

18. A boat sails on a bearing of N15°E for 10 km and then on a bearing of S85°E until it is due east of the starting point. Determine the distance from the starting point to the nearest kilometre. Show all your working.

19. A hill slopes at an angle of $30°$ to the horizontal. A tree that is 8 m tall and leaning downhill is growing at an angle of $10°$ m to the vertical and is part-way up the slope. Evaluate the vertical height of the top of the tree above the slope. Show all your working.

20. A cliff is 37 m high. The rock slopes outward at an angle of 50° to the horizontal and then cuts back at an angle of 25° to the vertical, meeting the ground directly below the top of the cliff. Carol wishes to abseil from the top of the cliff to the ground as shown in the diagram.

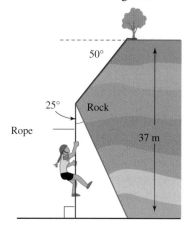

Her climbing rope is 45 m long and is secured to a tree at the top of the cliff. Determine if the rope will be long enough to allow her to reach the ground.

Problem solving

21. A river has parallel banks that run directly east–west. From the south bank, Kylie takes a bearing to a tree on the north side. The bearing is 047°T. She then walks 10 m due east and takes a second bearing to the tree. This is 305°T.
    Determine:
    a. her distance from the second measuring point to the tree
    b. the width of the river, to the nearest metre.

22. A ship sails on a bearing of S20°W for 14 km; then it changes direction, sails for 20 km and drops anchor. Its bearing from the starting point is now N65°W.
    a. Determine the distance of the ship from its starting point.
    b. Calculate the bearing on which the ship sails for the 20 km leg.

23. A cross-country runner runs at 8 km/h on a bearing of 150°T for 45 minutes; then she changes direction to a bearing of 053°T and runs for 80 minutes at a different speed until she is due east of the starting point.
    a. Calculate the distance of the second part of the run.
    b. Calculate her speed for this section, correct to 1 decimal place.
    c. Evaluate how far she needs to run to get back to the starting point.

24. From a fire tower, A, a fire is spotted on a bearing of N42°E. From a second tower, B, the fire is on a bearing of N12°W. The two fire towers are 23 km apart, and A is N63°W of B.
    Determine how far the fire is from each tower.

25. A yacht sets sail from a marina and sails on a bearing of 065°T for 3.5 km. It then turns and sails on a bearing of 127°T for another 5 km.
    a. Evaluate the distance of the yacht from the marina, correct to 1 decimal place.
    b. If the yacht was to sail directly back to the marina, on what bearing should it travel?
    Give your answer rounded to the nearest minute.

LESSON
15.3 The cosine rule

15.3.1 The cosine rule

eles-5006

- In triangle ABC, let BD be the perpendicular line from B to AC, of length h, giving two right-angled triangles, $\triangle$ADB and $\triangle$CDB.
- Let the length of AD $= x$, then DC $= (b - x)$.
- Using triangle ADB and Pythagoras' theorem, we obtain:

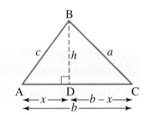

$$c^2 = h^2 + x^2 \qquad [1]$$

Using triangle CDB and Pythagoras' theorem, we obtain:

$$a^2 = h^2 + (b - x)^2 \qquad [2]$$

Expanding the brackets in equation [2]:

$$a^2 = h^2 + b^2 - 2bx + x^2$$

Rearranging equation [2] and using $c^2 = h^2 + x^2$ from equation [1]:

$$a^2 = h^2 + x^2 + b^2 - 2bx$$
$$= c^2 + b^2 - 2bx$$
$$= b^2 + c^2 - 2bx$$

From triangle ABD, $x = c \cos(A)$; therefore, $a^2 = b^2 + c^2 - 2bx$ becomes
$$a^2 = b^2 + c^2 - 2bc \cos(A).$$

Cosine rule

- **The cosine rule for any triangle ABC is:**

$$a^2 = b^2 + c^2 - 2bc \cos(A)$$
$$b^2 = a^2 + c^2 - 2ac \cos(B)$$
$$c^2 = a^2 + b^2 - 2ab \cos(C)$$

- The cosine rule can be used to solve non-right-angled triangles if we are given:
    1. three sides
    2. two sides and the included angle.

    Note: Once the third side has been calculated, the sine rule can be used to determine other angles if necessary.

Calculating angles by transposing the cosine rule

- **If three sides of a triangle are known, an angle can be found by transposing the cosine rule to make $\cos(A)$, $\cos(B)$ or $\cos(C)$ the subject.**

$$a^2 = b^2 + c^2 - 2bc\cos(A) \Rightarrow \cos(A) = \frac{b^2 + c^2 - a^2}{2bc}$$

$$b^2 = a^2 + c^2 - 2ac\cos(B) \Rightarrow \cos(B) = \frac{a^2 + c^2 - b^2}{2ac}$$

$$c^2 = a^2 + b^2 - 2ab\cos(C) \Rightarrow \cos(C) = \frac{a^2 + b^2 - c^2}{2ab}$$

WORKED EXAMPLE 4 Calculating sides using the cosine rule

Calculate the third side of triangle ABC given $a = 6$, $c = 10$ and $B = 76°$.

THINK	WRITE/DRAW
1. Draw a labelled diagram of the triangle ABC and fill in the given information.	
2. Check that one of the criteria for the cosine rule has been satisfied.	Yes, the cosine rule can be used since two side lengths and the included angle have been given.
3. Write down the appropriate cosine rule to calculate side b.	To calculate side b: $b^2 = a^2 + c^2 - 2ac\cos(B)$
4. Substitute the given values into the rule.	$= 6^2 + 10^2 - 2 \times 6 \times 10 \times \cos(76°)$
5. Evaluate.	≈ 106.9693725 $b \approx \sqrt{106.9693725}$
6. Round off the answer to 2 decimal places.	≈ 10.34

WORKED EXAMPLE 5 Calculating angles using the cosine rule

Calculate the smallest angle in the triangle with sides 4 cm, 7 cm and 9 cm.

THINK	WRITE/DRAW
1. Draw a labelled diagram of the triangle, call it ABC and fill in the given information. *Note:* The smallest angle will correspond to the smallest side.	 Let $a = 4, b = 7, c = 9$
2. Check that one of the criteria for the cosine rule has been satisfied.	The cosine rule can be used since three side lengths have been given.

3. Write down the appropriate cosine rule to calculate angle A.

$$\cos(A) = \frac{b^2 + c^2 - a^2}{2bc}$$

4. Substitute the given values into the rearranged rule.

$$= \frac{7^2 + 9^2 - 4^2}{2 \times 7 \times 9}$$

5. Evaluate.

$$= \frac{114}{126}$$

6. Transpose the equation to make A the subject by taking the inverse cos of both sides.

$$A = \cos^{-1}\left(\frac{114}{126}\right)$$

$$\approx 25.208\,765\,3°$$

7. Round off the answer to degrees and minutes.

$$\approx 25°13'$$

TI	THINK	DISPLAY/WRITE	CASIO	THINK	DISPLAY/WRITE

In a new problem, on a Calculator page, complete the entry lines as:
$$\frac{b^2 + c^2 - a^2}{2bc} \mid a = 4 \text{ and } b = 7$$
and $c = 9$
$\cos^{-1}(\text{ans})$
Convert the angle to DMS as shown previously.
Press ENTER after each entry.

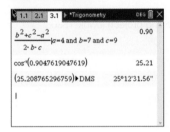

The smallest angle is $25°13'$ rounded up to the nearest minute.

On the Main screen, complete the entry lines as:
$$\frac{b^2 + c^2 - a^2}{2b \times c} \mid a = 4 \mid b = 7 \mid c = 9$$
$\cos^{-1}(0.9047619048)$
Convert the angle to DMS as shown previously.
Press EXE after each entry.

The smallest angle, rounded up to the nearest minute, is $25°13'$.

WORKED EXAMPLE 6 Applying the cosine rule to solve problems

Two rowers, Harriet and Kate, set out from the same point. Harriet rows N70°E for 2000 m and Kate rows S15°W for 1800 m.

Calculate the distance between the two rowers, correct to 2 decimal places.

THINK

1. Draw a labelled diagram of the triangle, call it ABC and fill in the given information.

2. Check that one of the criteria for the cosine rule has been satisfied.

3. Write down the appropriate cosine rule to calculate side c.

WRITE/DRAW

The cosine rule can be used since two side lengths and the included angle have been given.

To calculate side c:
$$c^2 = a^2 + b^2 - 2ab\cos(C)$$

4. Substitute the given values into the rule.	$= 2000^2 + 1800^2 - 2 \times 2000 \times 1800 \cos(125°)$
5. Evaluate.	$\approx 11\,369\,750.342$
	$c \approx \sqrt{11\,369\,750.342}$
6. Round off the answer to 2 decimal places.	≈ 3371.91
7. Write the answer.	The rowers are 3371.91 m apart.

DISCUSSION

In what situations would you use the sine rule rather than the cosine rule?

 Resources

 eWorkbook Topic 15 Workbook (worksheets, code puzzle and project) (ewbk-13308)

 Interactivity The cosine rule (int-6276)

Exercise 15.3 The cosine rule

learnon

15.3 Quick quiz on	15.3 Exercise

Individual pathways

■ PRACTISE	■ CONSOLIDATE	■ MASTER
1, 2, 3, 4, 7, 9, 14, 17	5, 6, 8, 10, 15, 18	11, 12, 13, 16, 19

Where appropriate in this exercise, write your angles correct to the nearest minute and side lengths correct to 2 decimal places.

Fluency

1. **WE4** Calculate the third side of triangle ABC given $a = 3.4$, $b = 7.8$ and $C = 80°$.

2. In triangle ABC, $b = 64.5$, $c = 38.1$ and $A = 58°34'$. Calculate the value of a.

3. In triangle ABC, $a = 17$, $c = 10$ and $B = 115°$. Calculate the value of b, and hence calculate the values of A and C.

4. **WE5** Calculate the size of the smallest angle in the triangle with sides 6 cm, 4 cm and 8 cm.
   (*Hint:* The smallest angle is opposite the smallest side.)

5. In triangle ABC, $a = 356$, $b = 207$ and $c = 296$. Calculate the size of the largest angle.

6. In triangle ABC, $a = 23.6$, $b = 17.3$ and $c = 26.4$. Calculate the size of all the angles.

7. **WE6** Two rowers set out from the same point. One rows N30°E for 1500 m and the other rows S40°E for 1200 m. Calculate the distance between the two rowers, correct to the nearest metre.

8. Maria cycles 12 km in a direction N68°W and then 7 km in a direction of N34°E.

   a. Calculate her distance from the starting point.
   b. Determine the bearing of the starting point from her finishing point.

Understanding

9. A garden bed is in the shape of a triangle with sides of length 3 m, 4.5 m and 5.2 m.

   a. Calculate the size of the smallest angle.
   b. Hence, calculate the area of the garden, correct to 2 decimal places.
      (*Hint:* Draw a diagram with the longest length as the base of the triangle.)

10. A hockey goal is 3 m wide. When Sophie is 7 m from one post and 5.2 m from the other, she shoots for goal. Determine within what angle, to the nearest degree, the shot must be made if it is to score a goal.

11. An advertising balloon is attached to two ropes 120 m and 100 m long. The ropes are anchored to level ground 35 m apart. Calculate the height of the balloon when both ropes are taut.

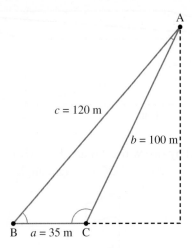

12. A plane flies in a direction of N70°E for 80 km and then on a bearing of S10°W for 150 km.

    a. Calculate the plane's distance from its starting point, correct to the nearest km.
    b. Calculate the plane's direction from its starting point.

13. Ship A is 16.2 km from port on a bearing of 053°T and ship B is 31.6 km from the same port on a bearing of 117°T.
    Calculate the distance between the two ships in km, correct to 1 decimal place.

Reasoning

14. A plane takes off at 10:00 am from an airfield and flies at 120 km/h on a bearing of N35°W. A second plane takes off at 10:05 am from the same airfield and flies on a bearing of S80°E at a speed of 90 km/h. Determine how far apart the planes are at 10:25 am in km, correct to 1 decimal place.

15. Three circles of radii 5 cm, 6 cm and 8 cm are positioned so that they just touch one another. Their centres form the vertices of a triangle.
    Determine the largest angle in the triangle. Show your working.

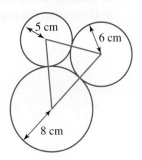

16. For the shape shown, determine:

    a. the length of the diagonal

    b. the magnitude (size) of angle B

    c. the length of x.

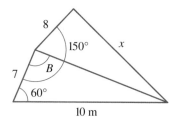

Problem solving

17. From the top of a vertical cliff 68 m high, an observer notices a yacht at sea. The angle of depression to the yacht is 47°. The yacht sails directly away from the cliff, and after 10 minutes the angle of depression is 15°. Determine the speed of the yacht in km/h, correct to 2 decimal places.

18. Determine the size of angles CAB, ABC and BCA.
Give your answers in degrees correct to 2 decimal places.

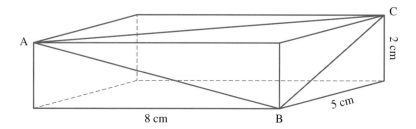

19. A vertical flag pole DB is supported by two wires AB and BC. AB is 5.2 metres long, BC is 4.7 metres long and B is 3.7 metres above ground level. Angle ADC is a right angle.

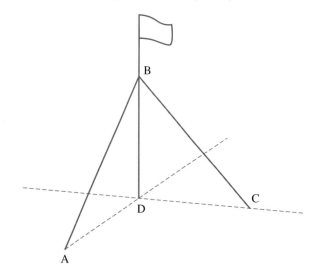

    a. Evaluate the distance from A to C in metres, correct to 4 decimal places.

    b. Determine the angle between AB and BC in degrees, correct to 2 decimal places.

LESSON
15.4 Area of triangles

15.4.1 Areas of triangles when two sides and the included angle are known

eles-5007

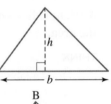

- The area of any triangle is given by the rule area $= \dfrac{1}{2}bh$, where b is the base and h is the perpendicular height of the triangle.
- In the triangle ABC, b is the base and h is the perpendicular height of the triangle.
- Using the trigonometric ratio for sine:

$$\sin(A) = \frac{h}{c}$$

Transposing the equation to make h the subject, we obtain:

$$h = c\sin(A)$$

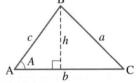

Area of a triangle

- **The area of triangle ABC can be found using the sine ratio:**

$$\textbf{Area} = \frac{1}{2}\textbf{\textit{bc}} \ \textbf{sin}(\textbf{\textit{A}})$$

- **Depending on how the triangle is labelled, the formula could read:**

$$\textbf{Area} = \frac{1}{2}\textbf{\textit{ab}} \ \textbf{sin}(\textbf{\textit{C}}) \quad \textbf{Area} = \frac{1}{2}\textbf{\textit{ac}} \ \textbf{sin}(\textbf{\textit{B}}) \quad \textbf{Area} = \frac{1}{2}\textbf{\textit{bc}} \ \textbf{sin}(\textbf{\textit{A}})$$

- The area formula may be used on any triangle provided that two sides of the triangle and the included angle (that is, the angle between the two given sides) are known.

WORKED EXAMPLE 7 Calculating the area of a triangle

Calculate the area of the triangle shown.

7 cm   120°   9 cm

THINK	WRITE/DRAW
1. Draw a labelled diagram of the triangle, label it ABC and fill in the given information.	 Let $a = 9$ cm, $c = 7$ cm, $B = 120°$.

2. Check that the criterion for the area rule has been satisfied.

The area rule can be used since two side lengths and the included angle have been given.

3. Write down the appropriate rule for the area.

$\text{Area} = \frac{1}{2}ac\sin(B)$

4. Substitute the known values into the rule.

$= \frac{1}{2} \times 9 \times 7 \times \sin 120°$

5. Evaluate. Round off the answer to 2 decimal places and include the appropriate unit.

$\approx 27.28 \text{ cm}^2$

WORKED EXAMPLE 8 Determining angles in a triangle and its area

A triangle has known dimensions of $a = 5$ cm, $b = 7$ cm and $B = 52°$. Determine A and C and hence the area.

THINK

1. Draw a labelled diagram of the triangle, label it ABC and fill in the given information.

2. Check whether the criterion for the area rule has been satisfied.

3. Write down the sine rule to calculate A.

4. Substitute the known values into the rule.

5. Transpose the equation to make $\sin(A)$ the subject.

6. Evaluate.

7. Round off the answer to degrees and minutes.

8. Determine the value of the included angle, C, using the fact that the angle sum of any triangle is 180°.

9. Write down the appropriate rule for the area.

WRITE/DRAW

Let $a = 5$, $b = 7$, $B = 52°$.

The area rule cannot be used since the included angle has not been given.

To calculate angle A:

$$\frac{a}{\sin(A)} = \frac{b}{\sin(B)}$$

$$\frac{5}{\sin(A)} = \frac{7}{\sin(52°)}$$

$5\sin(52°) = 7\sin(A)$

$$\sin(A) = \frac{5\sin(52°)}{7}$$

$$A = \sin^{-1}\left(\frac{5\sin(52°)}{7}\right)$$

$\approx 34.254\,151\,87°$

$\approx 34°15'$

$C \approx 180° - (52° + 34°15')$

$= 93°45'$

$\text{Area} = \frac{1}{2}ab\sin(C)$

10. Substitute the known values into the rule.

$$\approx \frac{1}{2} \times 5 \times 7 \times \sin(93°45')$$

11. Evaluate. Round off the answer to 2 decimal places and include the appropriate unit.

$$\approx 17.46 \, \text{cm}^2$$

TI	THINK	DISPLAY/WRITE

TI | THINK

In a new problem, open a Calculator page. To calculate the angle A, complete the entry line as:

$$\sin^{-1}\left(\frac{5\sin(52)}{7}\right)$$

Note that you can leave the angle in decimal degrees and work with this value. Determine the value of C as shown in the screenshot. Then calculate the area by completing the entry line as:

$$\frac{1}{2} \times 5 \times 7 \sin(93.75)$$

Press ENTER after each entry.

DISPLAY/WRITE

$A = 34.25°$
$C = 93.75°$
The area of the triangle is $17.46 \, \text{cm}^2$.

CASIO | THINK

To calculate the angle A, on the Main screen, complete the entry line as:

$$\sin^{-1}\left(\frac{5\sin(52)}{7}\right)$$

Note that we can leave the angle in decimal degrees and work with this value. Determine the value of C as shown. Then calculate the area by completing the entry line as:

$$\frac{1}{2} \times 5 \times 7 \sin(93.74584813)$$

Press EXE after each entry.

DISPLAY/WRITE

$A = 34.25°$
$C = 93.75°$
The area of the triangle is $17.46 \, \text{cm}^2$.

15.4.2 Areas of triangles when three sides are known

eles-3216

- If the lengths of all the sides of the triangle are known but none of the angles are known, apply the cosine rule to calculate one of the angles.
- Substitute the values of the angle and the two adjacent sides into the area of a triangle formula $\text{Area} = \frac{1}{2}bc\sin(A)$.

WORKED EXAMPLE 9 Calculating the area of a triangle given the lengths of the three sides

Calculate the area of the triangle with sides of 4 cm, 6 cm and 8 cm.

THINK

1. Draw a labelled diagram of the triangle, call it ABC and fill in the given information.

2. Choose an angle and apply the cosine rule to calculate angle A.

WRITE/DRAW

Let $a = 4$, $b = 6$, $c = 8$.

$$\cos(A) = \frac{b^2 + c^2 - a^2}{2bc}$$

$$= \frac{6^2 + 8^2 - 4^2}{2 \times 6 \times 8}$$

$$= \frac{7}{8}$$

$$A = \cos^{-1}\left(\frac{7}{8}\right)$$

$$\approx 28.955\,024\,37°$$

3. Write down the appropriate rule for the area.

$$\text{Area} = \frac{1}{2}bc \sin(A)$$

4. Substitute the known values into the rule.

$$\text{Area} = \frac{1}{2} \times 6 \times 8 \times \sin(28.955\,024\,37°)$$

5. Evaluate. Round off the answer to 2 decimal places and include the appropriate unit.

$$\approx 11.618\,95$$

The area is approximately 11.62 cm^2.

DISCUSSION

List three formulas for calculating the area of a triangle.

 Resources

 eWorkbook Topic 15 Workbook (worksheets, code puzzle and project) (ewbk-13308)

Interactivity Area of triangles (int-6483)

Exercise 15.4 Area of triangles

learn on

15.4 Quick quiz on	15.4 Exercise

Individual pathways

■ PRACTISE	■ CONSOLIDATE	■ MASTER
1, 4, 7, 10, 12, 15, 18, 21	2, 5, 8, 11, 13, 16, 19, 22	3, 6, 9, 14, 17, 20, 23, 24

Where appropriate in this exercise, write your angles correct to the nearest minute and other measurements correct to 2 decimal places.

Fluency

1. **WE7** Calculate the area of the triangle ABC with $a = 7$, $b = 4$ and $C = 68°$.

2. Calculate the area of the triangle ABC with $a = 7.3$, $c = 10.8$ and $B = 104°40'$.

3. Calculate the area of the triangle ABC with $b = 23.1$, $c = 18.6$ and $A = 82°17'$.

4. **WE8** A triangle has $a = 10$ cm, $c = 14$ cm and $C = 48°$. Determine A and B and hence the area.

5. A triangle has $a = 17$ m, $c = 22$ m and $C = 56°$. Determine A and B and hence the area.

6. A triangle has $b = 32$ mm, $c = 15$ mm and $B = 38°$. Determine A and C and hence the area.

7. **MC** In a triangle, $a = 15$ m, $b = 20$ m and $B = 50°$. The area of the triangle is:
   - **A.** 86.2 m^2
   - **B.** 114.9 m^2
   - **C.** 149.4 m^2
   - **D.** 172.4 m^2

8. **WE9** Calculate the area of the triangle with sides of 5 cm, 6 cm and 8 cm.

9. Calculate the area of the triangle with sides of 40 mm, 30 mm and 5.7 cm.

10. Calculate the area of the triangle with sides of 16 mm, 3 cm and 2.7 cm.

11. **MC** A triangle has sides of length 10 cm, 14 cm and 20 cm. The area of the triangle is:

   A. 41 cm^2 **B.** 65 cm^2 **C.** 106 cm^2 **D.** 137 cm^2

Understanding

12. A piece of metal is in the shape of a triangle with sides of length 114 mm, 72 mm and 87 mm. Calculate its area.

13. A triangle has the largest angle of 115°. The longest side is 62 cm and another side is 35 cm. Calculate the area of the triangle to the nearest whole number.

14. A triangle has two sides of 25 cm and 30 cm. The angle between the two sides is 30°. Determine:

   a. its area   **b.** the length of its third side

15. The surface of a fish pond has the shape shown in the diagram.

   Calculate how many goldfish the pond can support if each fish requires 0.3 m^2 surface area of water.

16. **MC** A parallelogram has sides of 14 cm and 18 cm and an angle between them of 72°.
   The area of the parallelogram is:

   A. 118.4 cm^2 **B.** 172.4 cm^2 **C.** 239.7 cm^2 **D.** 252 cm^2

17. **MC** An advertising hoarding is in the shape of an isosceles triangle with sides of length 15 m, 15 m and 18 m. It is to be painted with two coats of purple paint. If the paint covers 12 m^2 per litre, the amount of paint needed, to the nearest litre, would be:

   A. 9 L   **B.** 18 L   **C.** 24 L   **D.** 36 L

Reasoning

18. A parallelogram has diagonals of length 10 cm and 17 cm. An angle between them is 125°. Determine:

   a. the area of the parallelogram   **b.** the dimensions of the parallelogram.

19. A lawn is to be made in the shape of a triangle with sides of length 11 m, 15 m and 17.2 m. Determine how much grass seed, to the nearest kilogram, needs to be purchased if it is sown at the rate of 1 kg per 5 m^2.

20. A bushfire burns out an area of level grassland shown in the diagram.

   (*Note:* This is a sketch of the area and is not drawn to scale.)
   Evaluate the area, in hectares correct to 1 decimal place, of the land that is burned.

Problem solving

21. An earth embankment is 27 m long and has a vertical cross-section as shown in the diagram. Determine the volume of earth needed to build the embankment, correct to the nearest cubic metre.

22. Evaluate the area of this quadrilateral.

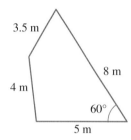

23. A surveyor measured the boundaries of a property as shown. The side AB could not be measured because it crossed through a marsh.
The owner of the property wanted to know the total area and the length of the side AB.
Give all lengths correct to 2 decimal places and angles to the nearest degree.

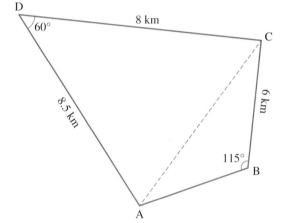

   a. Calculate the area of the triangle ACD.
   b. Calculate the distance AC.
   c. Calculate the angle CAB.
   d. Calculate the angle ACB.
   e. Calculate the length AB.
   f. Determine the area of the triangle ABC.
   g. Determine the area of the property.

24. A regular hexagon has sides of length 12 centimetres. It is divided into six smaller equilateral triangles. Evaluate the area of the hexagon, giving your answer correct to 2 decimal places.

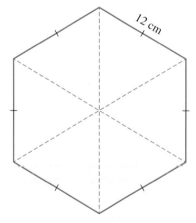

LESSON
15.5 The unit circle

LEARNING INTENTIONS

At the end of this lesson you should be able to:
- determine in which quadrant an angle lies
- use and interpret the relationship between a point on the unit circle and the angle made with the positive *x*-axis
- use the unit circle to determine approximate trigonometric ratios for angles greater than 90°
- use and interpret the relationships between supplementary and complementary angles.

15.5.1 The unit circle

eles-5009

- A **unit circle** is a circle with a radius of 1 unit.
- The unit circle is divided into 4 quadrants, numbered in an anticlockwise direction, as shown in the diagram.
- *Positive angles* are measured anticlockwise from 0°.
- *Negative angles* are measured clockwise from 0°.

WORKED EXAMPLE 10 Identifying where an angle lies on the unit circle

State the quadrant of the unit circle in which each of the following angles is found.
a. **145°**   b. **282°**

THINK

a. The given angle is between 90° and 180°. State the appropriate quadrant.

b. The given angle is between 270° and 360°. State the appropriate quadrant.

WRITE

a. 145° is in quadrant 2.

b. 282° is in quadrant 4.

- Consider the unit circle with point P(*x*, *y*) making the right-angled triangle OPN as shown in the diagram.
- Using the trigonometric ratios:

$$\frac{x}{1} = \cos(\theta), \quad \frac{y}{1} = \sin(\theta), \quad \frac{y}{x} = \tan(\theta)$$

where θ is measured anticlockwise from the positive *x*-axis.

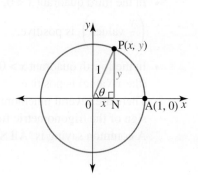

Calculating values of sine, cosine and tangent

To calculate the value of sine, cosine or tangent of any angle θ from the unit circle:

$$\cos(\theta) = x$$

$$\sin(\theta) = y$$

$$\tan(\theta) = \frac{y}{x} = \frac{\sin(\theta)}{\cos(\theta)}$$

▶ 15.5.2 The four quadrants of the unit circle

eles-5010

- Approximate values for sine, cosine and tangent of an angle can be found from the unit circle using the following steps, as shown in the diagram.
  Step 1: Draw a unit circle and label the x- and y-axes.
  Step 2: Mark the angles $0°$, $90°$, $180°$, $270°$ and $360°$.
  Step 3: Draw the given angle θ.
  Step 4: Mark $x = \cos(\theta)$, $y = \sin(\theta)$.
  Step 5: Approximate the values of x and y and equate to give the values of $\cos(\theta)$ and $\sin(\theta)$.
- Where the angle lies in the unit circle determines whether the trigonometric ratio is positive or negative.

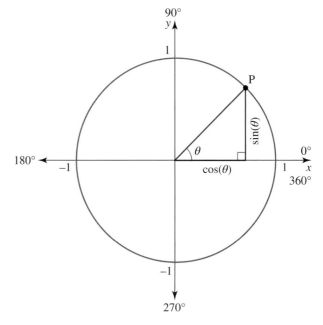

Sign of the trigonometric functions

- Consider the following.
  - In the first quadrant $x > 0$, $y > 0$; therefore, **All** trig ratios are positive.
  - In the second quadrant $x < 0$, $y > 0$; therefore, **Sine** (the y-value) is positive.
  - In the third quadrant $x < 0$, $y < 0$; therefore, **Tangent** $\left(\dfrac{y}{x} \text{ values}\right)$ is positive.

  - In the fourth quadrant $x > 0$, $y < 0$; therefore, **Cosine** (the x-value) is positive.
  - There are several mnemonics for remembering the sign of the trigonometric functions.
    A common saying is 'All Stations To Central'.

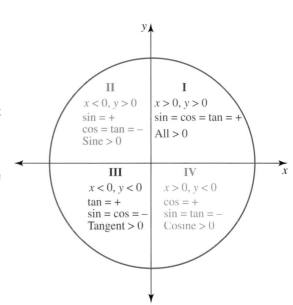

Determine the approximate value of each of the following using the unit circle.

a. $\sin(200°)$ b. $\cos(200°)$ c. $\tan(200°)$

THINK   **WRITE/DRAW**

Draw a unit circle and construct an angle of 200°.
Label the point corresponding to the angle of 200° on the
circle P. Highlight the lengths representing the x- and
y-coordinates of point P.

a. The sine of the angle is given by the y-coordinate   a. $\sin(200°) = -0.3$
of P. Determine the y-coordinate of P by measuring
the distance along the y-axis. State the value of
$\sin(200°)$. (*Note:* The sine value will be negative
as the y-coordinate is negative.)

b. The cosine of the angle is given by the x-coordinate   b. $\cos(200°) = -0.9$
of P. Determine the x-coordinate of P by measuring
the distance along the x-axis. State the value
of $\cos(200°)$. (*Note:* Cosine is also negative in
quadrant 3, as the x-coordinate is negative.)

c. $\tan(200°) = \dfrac{\sin(200°)}{\cos(200°)}$ c. $\dfrac{-0.3}{-0.9} = \dfrac{1}{3} = 0.3333$

- The approximate results obtained in Worked example 11 can be verified with the aid of a calculator:

$$\sin(200°) = -0.342\,020\,143, \cos(200°) = -0.939\,692\,62 \text{ and } \tan(200°) = 0.3640.$$

Rounding these values to 1 decimal place would give -0.3, -0.9 and 0.4 respectively, which match the
values obtained from the unit circle.

15.5.3 The relationship between supplementary and
complementary angles

eles-3217

Supplementary angles

- Consider the special relationship between the sine, cosine and tangent of supplementary angles, say $A°$ and
$(180 - A)°$.
In the diagram, the y-axis is an axis of symmetry.
  - The y-values of points C and E are the same.
    That is, $\sin(A°) = \sin(180 - A)°$.
  - The x-values of points C and E are opposites in value.
    That is, $\cos(A°) = -\cos(180 - A)°$

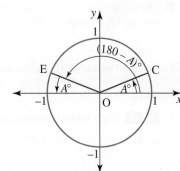

Thus: $\sin(180 - A)° = \sin(A°)$

$\cos(180 - A)° = -\cos(A°)$

$\tan(180 - A)° = \dfrac{\sin(180 - A)°}{\cos(180 - A)°} = \dfrac{\sin(A°)}{-\cos(A°)} = -\tan(A°).$

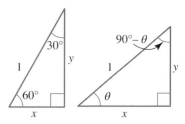

$$\textbf{Supplementary angles } A° \textbf{ and } (180° - A) \textbf{ where } 0° \le A \le 90°$$

$$\sin(180° - A) = \sin(A)$$

$$\cos(180° - A) = -\cos(A)$$

$$\tan(180° - A) = -\tan(A)$$

Complementary angles

- Complementary angles add to 90°. Therefore, 30° and 60° are complementary angles, and θ and $(90° - \theta)$ are also complementary angles.
- The sine of an angle is equal to the cosine of its complement. Therefore, $\sin(60°) = \cos(30°)$.
- We say that sine and cosine are complementary functions.

$$\sin A = \cos(90° - A)$$
$$\cos A = \sin(90° - A)$$

$$\textbf{Complementary angles where } 0° \le A \le 90°$$

$$\sin A = \cos(90° - A)$$

$$\cos A = \sin(90° - A)$$

on Resources

 eWorkbook Topic 15 Workbook (worksheets, code puzzle and project) (ewbk-13308)

Exercise 15.5 The unit circle

learn on

| 15.5 Quick quiz **on** | 15.5 Exercise |

Individual pathways

■ PRACTISE	■ CONSOLIDATE	■ MASTER
1, 2, 6, 8, 12, 17, 20	3, 4, 7, 9, 11, 13, 18, 21	5, 10, 14, 15, 16, 19, 22

Where appropriate in this exercise, give answers correct to 2 decimal places.

Fluency

1. **WE10** State which quadrant of the unit circle each of the following angles is in.

    a. 60°   b. 130°   c. 310°   d. 260°   e. 100°   f. 185°

2. **MC** If $\theta = 43°$, the triangle drawn to show this would be in:

   A. quadrant 1 **B.** quadrant 2 **C.** quadrant 3 **D.** quadrant 4

3. **MC** If $\theta = 295°$, the triangle drawn to show this would be in:

   A. quadrant 1 **B.** quadrant 2 **C.** quadrant 3 **D.** quadrant 4

4. **WE11** Determine the approximate value of each of the following using the unit circle.

   a. $\sin(20°)$ **b.** $\cos(20°)$ **c.** $\cos(100°)$ **d.** $\sin(100°)$

5. Determine the approximate value of each of the following using the unit circle.

   a. $\sin(320°)$ **b.** $\cos(320°)$ **c.** $\sin(215°)$ **d.** $\cos(215°)$

6. Use the unit circle to determine the approximate value of each of the following.

   a. $\sin(90°)$ **b.** $\cos(90°)$ **c.** $\sin(180°)$ **d.** $\cos(180°)$

7. Use the unit circle to determine the approximate value of each of the following.

   a. $\sin(270°)$ **b.** $\cos(270°)$ **c.** $\sin(360°)$ **d.** $\cos(360°)$

Understanding

8. On the unit circle, use a protractor to measure an angle of 30° from the positive x-axis. Mark the point P on the circle. Use this point to construct a triangle in quadrant 1 as shown.

   a. Calculate the value of $\cos(30°)$.
   (Remember that the length of the adjacent side of the triangle is $\cos(30°)$.)
   b. Calculate the value of $\sin(30°)$.
   (This is the length of the opposite side of the triangle.)

   Check your answers in **a** and **b** by calculating these values with a calculator.

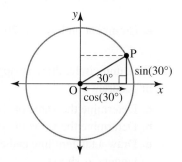

9. Using a graph of the unit circle, measure 150° with a protractor and mark the point P on the circle. Use this point to draw a triangle in quadrant 2 as shown.

   a. Determine the angle the radius OP makes with the negative x-axis.
   b. Remembering that $x = \cos(\theta)$, use your circle to determine the value of $\cos(150°)$.
   c. Comment on how $\cos(150°)$ compares to $\cos(30°)$.
   d. Remembering that $y = \sin(\theta)$, use your circle to determine the value of $\sin(150°)$.
   e. Comment on how $\sin(150°)$ compares with $\sin(30°)$.

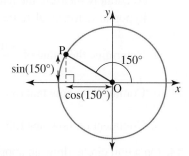

10. On a unit circle, measure 210° with a protractor and mark the point P on the circle.
    Use this point to draw a triangle in quadrant 3 as shown.

    a. Determine the angle the radius OP makes with the negative x-axis.
    b. Use your circle to determine the value of $\cos(210°)$.
    c. Comment on how $\cos(210°)$ compares to $\cos(30°)$.
    d. Use your circle to determine the value of $\sin(210°)$.
    e. Comment on how $\sin(210°)$ compares with $\sin(30°)$.

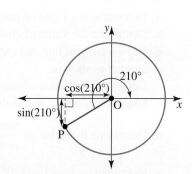

11. On a unit circle, measure 330° with a protractor and mark the point P on the circle. Use this point to draw a triangle in quadrant 4 as shown.

   a. Determine the angle the radius OP makes with the positive *x*-axis.
   b. Use your circle to determine the value of cos(330°).
   c. Comment on how cos(330°) compares to cos(30°).
   d. Use your circle to determine the value of sin(330°).
   e. Comment on how sin(330°) compares with sin(30°).

12. On a unit circle, draw an appropriate triangle for the angle of 20° in quadrant 1.

   a. Determine the value of sin(20°).
   b. Determine the value of cos(20°).
   c. Draw a tangent line and extend the hypotenuse of the triangle to meet the tangent as shown.
   Accurately measure the length of the tangent between the *x*-axis and the point where it meets the hypotenuse, and hence state the value of tan(20°).

   d. Determine the value of $\dfrac{\sin(20°)}{\cos(20°)}$.

   e. Comment on how tan(20°) compares with $\dfrac{\sin(20°)}{\cos(20°)}$.

13. On a unit circle, draw an appropriate triangle for the angle of 135° in quadrant 2.

   a. Determine the value of sin(135°) using sin(45°).
   b. Determine the value of cos(135°) using cos(45°).
   c. Draw a tangent line and extend the hypotenuse of the triangle to meet the tangent as shown.
   Accurately measure the length of the tangent to where it meets the hypotenuse to calculate the value of tan(135°).

   d. Determine the value of $\dfrac{\sin(135°)}{\cos(135°)}$.

   e. Comment on how tan(135°) compares with $\dfrac{\sin(135°)}{\cos(135°)}$.

   f. Comment on how tan(135°) compares with tan(45°).

14. On a unit circle, draw an appropriate triangle for the angle of 220° in quadrant 3.

   a. Determine the value of sin(220°).
   b. Determine the value of cos(220°).
   c. Draw a tangent line and extend the hypotenuse of the triangle to meet the tangent as shown.
   Determine the value of tan(220°) by accurately measuring the length of the tangent to where it meets the hypotenuse.

   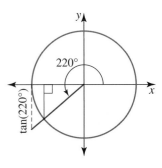

   d. Determine the value of $\dfrac{\sin(220°)}{\cos(220°)}$.

   e. Comment on how tan(220°) compares with $\dfrac{\sin(220°)}{\cos(220°)}$.

   f. Comment on how tan(220°) compares with tan(40°). (Use a calculator.)

15. On a unit circle, draw an appropriate triangle for the angle of 300° in quadrant 4.

  a. Determine the value of sin(300°).
  b. Determine the value of cos(300°).
  c. Draw a tangent line and extend the hypotenuse of the triangle to meet the tangent as shown. Determine the value of tan(300°) by accurately measuring the length of the tangent to where it meets the hypotenuse.

  d. Determine the value of the value of $\dfrac{\sin(300°)}{\cos(300°)}$.

  e. Comment on how tan(300°) compares with $\dfrac{\sin(300°)}{\cos(300°)}$.

  f. Comment on how tan(300°) compares with tan(60°). (Use a calculator.)

16. **MC** In a unit circle, the length of the radius is equal to:

  A. $\sin(\theta)$ **B.** $\cos(\theta)$ **C.** $\tan(\theta)$ **D.** 1

Reasoning

17. Show that $\sin^2(\alpha°) + \cos^2(\alpha°) = 1$.

18. Show that $1 - \sin^2(180 - \alpha)° = \cos^2(180 - \alpha)°$.

19. Show that $1 + \tan^2(\alpha°) = \sec^2(\alpha°)$, where $\sec(\alpha°) = \dfrac{1}{\cos(\alpha°)}$.

Problem solving

20. If $\sin(x°) = p$, $0 \le x \le 90°$, write each of the following in terms of p.

  a. $\cos(x°)$
  b. $\sin(180 - x)°$
  c. $\cos(180 - x)°$
  d. $\cos(90 - x)°$

21. Simplify $\sin(180 - x)° - \sin(x°)$.

22. Simplify $\cos(180 - x)° + \cos(x°)$, where $0 < x° < 90°$.

LESSON
15.6 Trigonometric functions

LEARNING INTENTIONS

At the end of this lesson you should be able to:
- sketch the graphs of the sine, cosine and tangent functions
- determine the amplitude of a given trigonometric function
- determine the period of a given trigonometric function.

▶ 15.6.1 Sine, cosine and tangent graphs

eles-5011

- The graphs of $y = \sin(x)$, $y = \cos(x)$ and $y = \tan(x)$ are shown below.

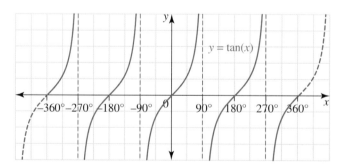

- Trigonometric graphs repeat themselves continuously in cycles; hence, they are called **periodic functions**.
- The **period** of the graph is the horizontal distance between repeating peaks or troughs. The period between the repeating peaks for $y = \sin(x)$ and $y = \cos(x)$ is 360°. The period of the graph of $y = \tan(x)$ is 180°, and **asymptotes** occur at $x = 90°$ and intervals of 180°.
- The **amplitude** of a periodic graph is half the distance between the maximum and minimum values of the function.
  Amplitude can also be described as the amount by which the graph goes above and below its mean value, which is the x-axis for the graphs of $y = \sin(x)$, $y = \cos(x)$ and $y = \tan(x)$.

- The following can be summarised from the graphs of the trigonometric functions.

Graph	Period	Amplitude
$y = \sin(x)$	360°	1
$y = \cos(x)$	360°	1
$y = \tan(x)$	180°	Undefined

- A scientific calculator can be used to check significant points such as 0°, 90°, 180°, 270° and 360° for the shape of a sin, cos or tan graph.

Transformations of trigonometric graphs

- **The sine, cosine and tangent graphs can be dilated, translated and reflected in the same way as other functions.**
- **These transformations are summarised in the table below.**

Graph	Period	Amplitude
$y = a\sin(nx)$	$\dfrac{360°}{n}$	a
$y = a\cos(nx)$	$\dfrac{360°}{n}$	a
$y = a\tan(nx)$	$\dfrac{180°}{n}$	Undefined

- **If $a < 0$, the graph is reflected in the x-axis. The amplitude is always the positive value of a.**

WORKED EXAMPLE 12 Sketching periodic functions

Sketch the graphs of a $y = 2\sin(x)$ and b $y = \cos(2x)$ for $0° \leq x \leq 360°$.

THINK

a. 1. The graph must be drawn from 0° to 360°.

2. Compared to the graph of $y = \sin(x)$ each value of $\sin(x)$ has been multiplied by 2; therefore, the amplitude of the graph must be 2.

3. Label the graph $y = 2\sin(x)$.

b. 1. The graph must be drawn from 0° to 360°.

2. Compared to the graph of $y = \cos(x)$, each value of x has been multiplied by 2; therefore, the period of the graph must become 180°.

3. Label the graph $y = \cos(2x)$.

WRITE/DRAW

a.

b.

TI	THINK	DISPLAY/WRITE	CASIO	THINK	DISPLAY/WRITE

TI | THINK

a.

1. In a new problem, on a Graphs page, ensure the Graphs & Geometry Settings are set to the degrees mode, as shown in the screenshot.
To do this, press:
- MENU
- 9: Settings

For Display Digits select 'Fix 2'.
TAB to Graphing Angle and select 'Degree'. TAB to OK and press ENTER.

2. To set an appropriate viewing window, press:
- MENU
- 4: Window/Zoom
- 1: Window Settings...

Select the values as shown in the screenshot.
TAB to OK and press ENTER.

DISPLAY/WRITE

a.

CASIO | THINK

On a Graph & Table screen, set an appropriate viewing window as shown.

a.
Complete the function entry line as:
$y1 = 2\sin(x) \mid 0 \le x \le 360$
Press EXE.
Tap the graphing icon and the graph is displayed as required only for $0° \le x \le 360°$.

DISPLAY/WRITE

a.

b.
Complete the function entry line as:
$y1 = \cos(2x) \mid 0 \le x \le 360$
Press EXE.
Tap the graphing icon and the graph is displayed as required only for $0° \le x \le 360°$.

b.

TI | THINK

3. Complete the function entry line as:
$f1(x) = 2\sin(x) \mid 0 \le x \le 360$
Press ENTER. The graph is displayed as required for $0° \le x \le 360°$.

DISPLAY/WRITE

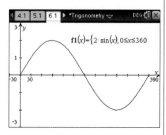

b.
Complete the function entry line as:
$f2(x) = \cos(2x) \mid 0 \le x \le 360$
Press ENTER, and the graph is displayed, as required only for $0° \le x \le 360°$.

b.

WORKED EXAMPLE 13 Stating the amplitude and period of periodic functions

For each of the following graphs, state:
 i. the amplitude
 ii. the period.

a. $y = 2\sin(3x)$

b. $y = \cos\left(\dfrac{x}{3}\right)$

c. $y = \tan(2x)$

THINK	WRITE
a. The value of a is 2.	a. i. Amplitude = 2
The period is $\dfrac{360°}{n}$.	ii. Period = $\dfrac{360}{3} = 120°$
b. The value of a is 1.	b. i. Amplitude = 1
The period is $\dfrac{360°}{n}$.	ii. Period = $\dfrac{360}{\frac{1}{3}} = 1080°$
c. The tangent curve has an undefined amplitude.	c. i. Amplitude = undefined
The period is $\dfrac{180°}{2}$.	ii. Period = $\dfrac{180°}{2} = 90°$

DISCUSSION

For the graph of $y = a\tan(nx)$, what would be the period and amplitude?

 Resources

Exercise 15.6 Trigonometric functions

15.6 Exercise

Individual pathways

■ PRACTISE	■ CONSOLIDATE	■ MASTER
1, 4, 8, 12, 16, 20, 25, 26, 28, 29, 32	3, 5, 7, 10, 13, 14, 17, 21, 23, 27, 30, 33	2, 6, 9, 11, 15, 18, 19, 22, 24, 31, 34

Fluency

1. Using your calculator (or the unit circle if you wish), complete the following table.

x	0°	30°	60°	90°	120°	150°	180°	210°	240°	270°	300°	330°	360°
$\sin(x)$													
x	390°	420°	450°	480°	510°	540°	570°	600°	630°	660°	690°	720°	
$\sin(x)$													

For questions 2 to 7, using graph paper, rule x- and y-axes and carefully mark a scale along each axis.

2. Use 1 cm = 30° on the x-axis to show x-values from 0° to 720°.
   Use 2 cm = 1 unit along the y-axis to show y-values from -1 to 1.
   Carefully plot the graph of $y = \sin(x)$ using the values from the table in question 1.

3. State how long it takes for the graph of $y = \sin(x)$ to complete one full cycle.

4. From your graph of $y = \sin(x)$, estimate to 1 decimal place the value of y for each of the following.
   a. $x = 42°$
   b. $x = 130°$
   c. $x = 160°$
   d. $x = 200°$

5. From your graph of $y = \sin(x)$, estimate to 1 decimal place the value of y for each of the following.
   a. $x = 180$
   b. $x = 70°$
   c. $x = 350°$
   d. $x = 290°$

6. From your graph of $y = \sin(x)$, estimate to the nearest degree a value of x for each of the following.
   a. $y = 0.9$
   b. $y = -0.9$
   c. $y = 0.7$

7. From your graph of $y = \sin(x)$, estimate to the nearest degree a value of x for each of the following.
   a. $y = -0.5$
   b. $y = -0.8$
   c. $y = 0.4$

8. Using your calculator (or the unit circle if you wish), complete the following table.

x	0°	30°	60°	90°	120°	150°	180°	210°	240°	270°	300°	330°	360°
$\cos(x)$													
x	390°	420°	450°	480°	510°	540°	570°	600°	630°	660°	690°	720°	
$\cos(x)$													

For questions 9 to 14, using graph paper, rule x- and y-axes and carefully mark a scale along each axis.

9. Use 1 cm = 30° on the x-axis to show x-values from 0° to 720°.
   Use 2 cm = 1 unit along the y-axis to show y-values from -1 to 1.
   Carefully plot the graph of $y = \cos(x)$ using the values from the table in question 8.

10. If you were to continue the graph of $y = \cos(x)$, state what shape you would expect it to take.

11. State whether the graph of $y = \cos(x)$ is the same as the graph of $y = \sin(x)$.
    Explain how it differs. State what features are the same.

12. Using the graph of $y = \cos(x)$, estimate to 1 decimal place the value of y for each of the following.

    a. $x = 48°$ b. $x = 155°$ c. $x = 180°$ d. $x = 340°$

13. Using the graph of $y = \cos(x)$, estimate to 1 decimal place the value of y for each of the following.

    a. $x = 240°$ b. $x = 140°$ c. $x = 40°$ d. $x = 200°$

14. Using the graph of $y = \cos(x)$, estimate to the nearest degree a value of x for each of the following.

    a. $y = -0.5$ b. $y = 0.8$ c. $y = 0.7$

15. Using the graph of $y = \cos(x)$, estimate to the nearest degree a value of x for each of the following.

    a. $y = -0.6$ b. $y = 0.9$ c. $y = -0.9$

16. Using your calculator (or the unit circle if you wish), complete the following table.

x	0°	30°	60°	90°	120°	150°	180°	210°	240°	270°	300°	330°	360°
$\tan(x)$													
x	390°	420°	450°	480°	510°	540°	570°	600°	630°	660°	690°	720°	
$\tan(x)$													

For questions 17 to 22, using graph paper, rule x- and y-axes and carefully mark a scale along each axis.

17. Use 1 cm = 30° on the x-axis to show x-values from 0° to 720°.
    Use 2 cm = 1 unit along the y-axis to show y-values from −2 to 2.
    Carefully plot the graph of $y = \tan(x)$ using the values from the table in question 16.

18. If you were to continue the graph of $y = \tan(x)$, state what shape would you expect it to take.

19. State whether the graph of $y = \tan(x)$ is the same as the graphs of $y = \sin(x)$ and $y = \cos(x)$.
    Explain how it differs. State what features are the same.

20. Using the graph of $y = \tan(x)$, estimate to 1 decimal place the value of y for each of the following.

    a. $x = 60°$ b. $x = 135°$ c. $x = 310°$ d. $x = 220°$

21. Using the graph of $y = \tan(x)$, determine the value of y for each of the following.

    a. $x = 500°$ b. $x = 590°$ c. $x = 710°$ d. $x = 585°$

22. Using the graph of $y = \tan(x)$, estimate to the nearest degree a value of x for each of the following.

    a. $y = 1$ b. $y = 1.5$ c. $y = -0.4$
    d. $y = -2$ e. $y = 0.2$ f. $y = -1$

WE12&13 For each of the graphs in questions 23 and 24:
  i. state the period   **ii.** state the amplitude   **iii.** sketch the graph.

23. a. $y = \cos(x)$, for $x \in [-180°, 180°]$
    b. $y = \sin(x)$, for $x \in [0°, 720°]$

24. a. $y = \sin(2x)$, for $x \in [0°, 360°]$
    b. $y = 2\cos(x)$, for $x \in [-360°, 0°]$

25. For each of the following, state:
  i. the period   **ii.** the amplitude.

    a. $y = 3\cos(2x)$ b. $y = 4\sin(3x)$ c. $y = 2\cos\left(\dfrac{x}{2}\right)$

    d. $y = \dfrac{1}{2}\sin\left(\dfrac{x}{4}\right)$ e. $y = -\sin(x)$ f. $y = -\cos(2x)$

Understanding

26. **MC** Use the graph shown to answer the following.

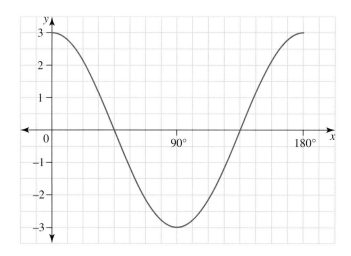

a. The amplitude of the graph is:

    A. $180°$ **B.** $90°$

    C. 3 **D.** -3

b. The period of the graph is:

    A. -3 **B.** 3

    C. $90°$ **D.** $360°$

c. The equation of the graph could be:

    A. $y = \cos(x)$ **B.** $y = \sin(x)$

    C. $y = 3\cos\left(\dfrac{x}{3}\right)$ **D.** $y = 3\cos(2x)$

27. Sketch each of the following graphs, stating the period and amplitude of each.

    a. $y = 2\cos\left(\dfrac{x}{3}\right)$, for $x \in [0°, 1080°]$

    b. $y = -3\sin(2x)$, for $x \in [0°, 360°]$

    c. $y = 3\sin\left(\dfrac{x}{2}\right)$, for $x \in [-180°, 180°]$

    d. $y = -\cos(3x)$, for $x \in [0°, 360°]$

    e. $y = 5\cos(2x)$, for $x \in [0°, 180°]$

    f. $y = -\sin(4x)$, for $x \in [0°, 180°]$

28. Use technology to sketch the graphs of each of the following for $0° \le x \le 360°$.

    a. $y = \cos(x) + 1$

    b. $y = \sin(2x) - 2$

    c. $y = \cos\left(\dfrac{\pi}{180}(x - 60)\right)$

    d. $y = 2\sin(4x) + 3$

Reasoning

29. **a.** Sketch the graph of $y = \cos(2x)$ for $x \in [0°, 360°]$.
    i. State the minimum value of y for this graph.
    ii. State the maximum value of y for this graph.

    b. Using the answers obtained in part **a**, write down the maximum and minimum values of $y = \cos(2x) + 2$.
    c. Determine what would be the maximum and minimum values of the graph of $y = 2\sin(x) + 3$. Explain how you obtained these values.

30. **a.** Complete the table below by filling in the exact values of $y = \tan(x)$

x	0°	30°	60°	90°	120°	150°	180°
$y = \tan(x)$							

    b. Sketch the graph of $y = \tan(x)$ for $[0°, 180°]$.
    c. Determine what happens at $x = 90°$.
    d. For the graph of $y = \tan(x)$, $x = 90°$ is called an asymptote. Write down when the next asymptote would occur.
    e. Determine the period and amplitude of $y = \tan(x)$.

31. Answer the following questions.
    a. Sketch the graph of $y = \tan(2x)$ for $[0°, 180°]$.
    b. Determine when the asymptotes occur.
    c. State the period and amplitude of $y = \tan(2x)$.

Problem solving

32. The height, h metres, of the tide above the mean sea level on the first day of the month is given by the rule

$$h = 3\sin(30t°)$$

where t is the time in hours since midnight.

    a. Sketch the graph of h versus t.
    b. Determine the height of the high tide.
    c. Calculate the height of the tide at 8 am.

33. The height, h metres, of the tide on the first day of January at Trig Cove is given by the rule

$$h = 6 + 4\sin(30t°)$$

where t is the time in hours since midnight.

    a. Sketch the graph of h versus t, for $0 \leq t \leq 24$.
    b. Determine the height of the high tide.
    c. Determine the height of the low tide.
    d. Calculate the height of the tide at 10 am, correct to the nearest centimetre.

34. The temperature, T, inside a house t hours after 3 am is given by the rule

$$T = 22 - 2\cos(15t°) \text{ for } 0 \leq t \leq 24$$

where T is the temperature in degrees Celsius.

    a. Determine the temperature inside the house at 9 am.
    b. Sketch the graph of T versus t.
    c. Determine the warmest and coolest temperatures that it gets inside the house over the 24-hour period.

LESSON
15.7 Solving trigonometric equations

LEARNING INTENTIONS

At the end of this lesson you should be able to:
- apply the exact values of sin, cos and tan for 30°, 45° and 60° angles
- solve trigonometric equations graphically for a given domain
- solve trigonometric equations algebraically, using exact values, for a given domain.

▶ 15.7.1 Exact values of trigonometric functions

eles-3215

- Most of the trigonometric values that we will deal with in this topic are approximations.
- However, angles of 30°, 45° and 60° have exact values of sine, cosine and tangent.
- Consider an equilateral triangle, ABC, of side length 2 cm.
  Let BD be the perpendicular bisector of AC, then:
  $\triangle ABD \cong \triangle CBD$ (using RHS)
  giving:
  $$AD = CD = 1 \text{ cm}$$
  $$\angle ABD = \angle CBD = 30°$$
  and:
  $$(AB)^2 = (AD)^2 + (BD)^2 \quad \text{(using Pythagoras' theorem)}$$
  $$2^2 = 1^2 + (BD)^2$$
  $$BD = \sqrt{3}$$

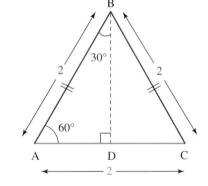

- Using $\triangle ABD$, the following exact values are obtained:

$$\sin(A) = \frac{\text{opp}}{\text{hyp}} \Rightarrow \sin(60°) = \frac{\sqrt{3}}{2} \qquad \sin(B) = \frac{\text{opp}}{\text{hyp}} \Rightarrow \sin(30°) = \frac{1}{2}$$

$$\cos(A) = \frac{\text{adj}}{\text{hyp}} \Rightarrow \cos(60°) = \frac{1}{2} \qquad \cos(B) = \frac{\text{adj}}{\text{hyp}} \Rightarrow \cos(30°) = \frac{\sqrt{3}}{2}$$

$$\tan(A) = \frac{\text{opp}}{\text{adj}} \Rightarrow \tan(60°) = \frac{\sqrt{3}}{1} \text{ or } \sqrt{3} \qquad \tan(B) = \frac{\text{opp}}{\text{adj}} \Rightarrow \tan(30°) = \frac{1}{\sqrt{3}} \text{ or } \frac{\sqrt{3}}{3}$$

- Consider a right-angled isosceles $\triangle EFG$ with equal sides of 1 unit.

$$(EG)^2 = (EF)^2 + (FG)\,2 \quad \text{(using Pythagoras' theorem)}$$
$$(EG)^2 = 1^2 + 1^2$$
$$EG = \sqrt{2}$$

- Using $\triangle EFG$, the following exact values are obtained:

$$\sin(E) = \frac{\text{opp}}{\text{hyp}} \Rightarrow \sin(45°) = \frac{1}{\sqrt{2}} \text{ or } \frac{\sqrt{2}}{2}$$

$$\cos(E) = \frac{\text{adj}}{\text{hyp}} \Rightarrow \cos(45°) = \frac{1}{\sqrt{2}} \text{ or } \frac{\sqrt{2}}{2}$$

$$\tan(E) = \frac{\text{opp}}{\text{adj}} \Rightarrow \tan(45°) = \frac{1}{1} \text{ or } 1$$

Summary of exact values

θ	30°	45°	60°
$\sin(\theta)$	$\dfrac{1}{2}$	$\dfrac{1}{\sqrt{2}} = \dfrac{\sqrt{2}}{2}$	$\dfrac{\sqrt{3}}{2}$
$\cos(\theta)$	$\dfrac{\sqrt{3}}{2}$	$\dfrac{1}{\sqrt{2}} = \dfrac{\sqrt{2}}{2}$	$\dfrac{1}{2}$
$\tan(\theta)$	$\dfrac{1}{\sqrt{3}} = \dfrac{\sqrt{3}}{3}$	1	$\sqrt{3}$

⏵ 15.7.2 Solving trigonometric equations

eles-5012

Solving trigonometric equations graphically

- Because of the periodic nature of circular functions, there are infinitely many solutions to unrestricted trigonometric equations.
- Equations are usually solved within a particular domain (x-values), to restrict the number of solutions.
- The sine graph shown displays the solutions between 0° and 360° for the equation $\sin(x) = 0.6$.

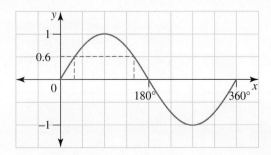

In the graph, it can clearly be seen that there are two solutions to this equation, which are approximately $x = 37°$ and $x = 143°$.

- It is difficult to obtain accurate answers from a graph. More accurate answers can be obtained using technology.

Solving trigonometric equations algebraically

• Exact values can be found for some trigonometric equations using the table in section 15.7.1.

WORKED EXAMPLE 14 Solving trigonometric equations using exact values

Solve the following equations.

a. $\sin(x) = \dfrac{\sqrt{3}}{2}, \; x \in [0°, 360°]$

b. $\cos(2x) = -\dfrac{1}{\sqrt{2}}, \; x \in [0°, 360°]$

THINK

WRITE

a. 1. The inverse operation of sine is $\sin^{-1}$.

a. $x = \sin^{-1}\left(\dfrac{\sqrt{3}}{2}\right)$

2. The first solution in the given domain from the table in subsection 15.7.1 is $x = 60°$. Since sine is positive in the first and second quadrants, another solution must be $x = 180° - 60° = 120°$.

There are two solutions in the given domain, $x = 60°$ and $x = 120°$.

b. 1. The inverse operation of cosine is $\cos^{-1}$.

b. $2x = \cos^{-1}\left(\dfrac{-1}{\sqrt{2}}\right)$

2. From the table of values, $\cos^{-1}\left(\dfrac{1}{\sqrt{2}}\right) = 45°$.

$2x = 135°, \, 225°$

Cosine is negative in the second and third quadrants, which gives the first two solutions to the equation as $180° - 45°$ and $180° + 45°$.

3. Solve for x by dividing by 2.

$x = 67.5°, \, 112.5°$

4. Since the domain in this case is $[0°, 360°]$ and the period has been halved, there must be four solutions altogether.
The other two solutions can be found by adding the period onto each solution.

$\text{Period} = \dfrac{360°}{2} = 180°$

$x = 67.5° + 180°, \, 112.5° + 180°$

$x = 67.5°, \, 112.5°, \, 247.5°, \, 292.5°$

TI	THINK	DISPLAY/WRITE	CASIO	THINK	DISPLAY/WRITE

a.

In a new problem, on a Calculator page, complete the entry line as:

$$\text{solve}\left(\sin(x) = \frac{\sqrt{3}}{2}, x\right)|$$

$0 \leq x \leq 360$

Then press ENTER.
Note that the calculator is set to the degrees mode.

a.

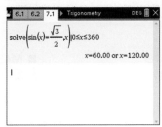

$$\sin(x) = \frac{\sqrt{3}}{2} \text{ for } x \in [0°, 360°]$$
$$\Rightarrow x = 60° \text{ or } 120°$$

a.

On the Main screen, complete the entry line as:

$$\text{solve}\left(\sin(x) = \frac{\sqrt{3}}{2}, x\right)|$$

$0 \leq x \leq 360$

Then press EXE.
Note that the calculator is set to the degrees mode.

a.

$$\sin(x) = \frac{\sqrt{3}}{2} \text{ for }$$
$$x \in [0°, 360°]$$
$$\Rightarrow x = 60° \text{ or } 120°$$

b.

On a Calculator page, complete the entry line as:

$$\text{solve}\left(\cos(2x) = -\frac{1}{\sqrt{2}}, x\right)|$$

$0 \leq x \leq 360$

Then press ENTER.
Note that the calculator is set to the degrees mode.

b.

$$\cos(2x) = -\frac{1}{\sqrt{2}} \text{ for }$$
$$x \in [0°, 360°]$$
$$\Rightarrow x = 67.5°, 112.5°, 247.5° \text{ or } 292.5°$$

b.

On the Main screen, complete the entry line as:

$$\text{solve}\left(\cos(2x) = -\frac{1}{\sqrt{2}}, x\right)|$$

$0 \leq x \leq 360$

Then press EXE.
Note that the calculator is set to the degrees mode.

b.

$$\cos(2x) = -\frac{1}{\sqrt{2}} \text{ for }$$
$$x \in [0°, 360°]$$
$$\Rightarrow x = 67.5°, 112.5°,$$
$$247.5° \text{ or } 292.5°$$

DISCUSSION

Explain why sine and cosine functions can be used to model situations that occur in nature such as tide heights and sound waves.

 Resources

Exercise 15.7 Solving trigonometric equations

15.7 Quick quiz on	15.7 Exercise

Individual pathways

■ PRACTISE	■ CONSOLIDATE	■ MASTER
1, 3, 6, 9, 13, 16	2, 4, 7, 10, 14, 17	5, 8, 11, 12, 15, 18

Fluency

For questions **1** and **2**, use the graph to determine approximate answers to the equations for the domain $0 \le x \le 360°$. Check your answers using a calculator.

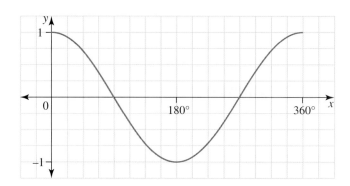

1. a. $\cos(x) = 0.9$ b. $\cos(x) = 0.3$

2. a. $\cos(x) = -0.2$ b. $\cos(x) = -0.6$

For questions **3** to **8**, solve the equations for the domain $0° \le x \le 360°$.

3. a. $\sin(x) = \dfrac{1}{2}$ b. $\sin(x) = \dfrac{\sqrt{3}}{2}$

4. a. $\cos(x) = -\dfrac{1}{2}$ b. $\cos(x) = -\dfrac{1}{\sqrt{2}}$

5. a. $\sin(x) = 1$ b. $\cos(x) = -1$

6. a. $\sin(x) = -\dfrac{1}{2}$ b. $\sin(x) = -\dfrac{1}{\sqrt{2}}$

7. a. $\cos(x) = \dfrac{\sqrt{3}}{2}$ b. $\cos(x) = -\dfrac{\sqrt{3}}{2}$

8. a. $\sin(x) = 1$ b. $\cos(x) = 0$

Understanding

WE14 For questions **9** to **12**, solve the equations for the given values of x.

9. **a.** $\sin(2x) = \dfrac{\sqrt{3}}{2}, x \in [0°, 360°]$

   b. $\cos(2x) = -\dfrac{\sqrt{3}}{2}, x \in [0°, 360°]$

10. **a.** $\tan(2x) = \dfrac{1}{\sqrt{3}}, x \in [0°, 360°]$

   b. $\sin(3x) = -\dfrac{1}{2}, x \in [0°, 180°]$

11. **a.** $\sin(4x) = -\dfrac{1}{2}, x \in [0°, 180°]$

   b. $\sin(3x) = -\dfrac{1}{\sqrt{2}}, x \in [-180°, 180°]$

12. **a.** $\tan(3x) = -1, x \in [0°, 90°]$

   b. $\cos(3x) = 0, x \in [0°, 360°]$

Reasoning

13. Solve the following equations for $x \in [0°, 360°]$.

   a. $2\sin(x) - 1 = 0$

   b. $2\cos(x) = \sqrt{3}$

14. Solve the following equations for $x \in [0°, 360°]$.

   a. $\sqrt{2}\cos(x) - 1 = 0$

   b. $\tan(x) + 1 = 0$

15. Sam measured the depth of water at the end of the Intergate jetty at various times on Thursday 13 August 2020. The table below provides her results.

Time	6 am	7	8	9	10	11	12 pm	1	2	3	4	5	6	7	8	9
Depth (m)	1.5	1.8	2.3	2.6	2.5	2.2	1.8	1.2	0.8	0.5	0.6	1.0	1.3	1.8	2.2	2.5

a. Plot the data.

b. Determine:
   i. the period
   ii. the amplitude.

c. Sam fishes from the jetty when the depth is a maximum. Specify these times for the next 3 days.

d. Sam's mother can moor her yacht when the depth is above 1.5 m. Determine during what periods she can moor the yacht on Sunday 16 January.

Problem solving

16. Solve:

   a. $\sqrt{3}\sin(x°) = \cos(x°)$ for $0° \le x \le 360°$

   b. $2\sin(x°) + \cos(x°) = 0$ for $0° \le x \le 360°$.

17. Solve $2\sin^2(x°) + 3\sin(x°) - 2 = 0$ for $0° \le x \le 360°$.

18. The grad (^{g}) is another measurement used when measuring the size of angles. A grad is equivalent to $\dfrac{1}{400}$ of a full circle. Write each of the following as grads (1 grad is written as 1^g).

   a. $90°$

   b. $180°$

   c. $270°$

   d. $360°$

LESSON
15.8 Review

15.8.1 Topic summary

Sine rule
- Connects two sides with the two opposite angles in a triangle.
- $\dfrac{a}{\sin A} = \dfrac{b}{\sin B} = \dfrac{c}{\sin C}$
- Used to solve triangles:
  - Two angles and one side.
  - Two sides and an angle opposite one of these sides.

Area of a triangle
- Given two sides and the included angle:
  $$\text{Area} = \frac{1}{2}\, ab\, \sin(C)$$
- If the lengths of all the sides of the triangle are known but none of the angles, apply the cosine rule to calculate one of the angles.
- Substitute the values of the angle and the two adjacent sides into the area of a triangle formula.

Cosine rule
- Connects three sides and one angle of a triangle.
  $$a^2 = b^2 + c^2 - 2bc\, \cos(A)$$
  Used to solve triangles given:
  - three sides, or
  - two sides and the included angle.

Ambiguous case
- When using the sine rule to calculate an angle, there may be two answers.
- The ambiguous case may occur when determining the angle opposite the larger side.
- Always check that the three angles add to 180°.
- e.g. In the triangle ABC, $a = 10$, $b = 6$ and $B = 30°$, using the sine rule, $A = 56°$ or $(180 - 56)°$.
  Angles in triangle would be:
  $A = 56°$, $B = 30°$ giving $C = 94°$
  $A = 124°$, $B = 30°$ giving $C = 26°$
  Two triangles are possible, so the ambiguous case exists.

TRIGONOMETRY II (10A)

Trigonometric equations
- Infinite number of solutions, so domain is normally restricted:
  e.g. $0° \leq x \leq 180°$
- Equations can be solved:
  - graphically — not very accurate
  - using technology
  - algebraically, using the exact values.
- e.g. $\sin \alpha = -\dfrac{1}{2}$, $0° \leq x \leq 360°$
  from exact values: $\sin 30° = \dfrac{1}{2}$
  sine is negative in 3rd and 4th quadrants and the angle is from the x-axis
  $\alpha = (180 + 30)°$ or $(360 - 30)°$
  $\alpha = 210°$ or $330°$

Exact values
- Exact trig ratios can be found using triangles.
- For 30° and 60° use the equilateral triangle.
- For 45°, use a right-angled isosceles triangle.
  e.g. $\sin 30° = \dfrac{1}{2}$
  $\tan 45° = 1$
  $\cos 60° = \dfrac{1}{2}$

Trigonometric graphs
- Trigonometric graphs repeat themselves continuously in cycles.
- Period: horizontal distance between repeating peaks or troughs.
- Amplitude: half the distance between the maximum and minimum values.

$y = \sin x$, Period = 360°, Amplitude = 1

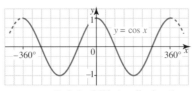

$y = \cos x$, Period = 360°, Amplitude = 1

$y = \tan x$, Period = 180°, Amplitude = undefined.

Unit circle
- Equation of the unit circle: $x^2 + y^2 = 1$
- Radius of length 1 unit.
- For any point on circumference:
  $x = \cos \theta$
  $y = \sin \theta$
- $\tan \theta = \dfrac{y}{x} = \dfrac{\sin \theta}{\cos \theta}$
- $\sin(180 - A)° = \sin A°$
  $\cos(180 - A)° = -\cos A°$
  $\tan(180 - A)° = -\tan A°$
- $\sin A = \cos(90° - A)$
  $\cos A = \sin(90° - A)$
- Quadrants are positive for:

II	I
$x < 0$, $y > 0$	$x > 0$, $y > 0$
$\sin = +$	$\sin = \cos = +$
$\cos = \tan = -$	$\tan = +$
Sin > 0	All > 0
III	**IV**
$x < 0$, $y < 0$	$x > 0$, $y < 0$
$\tan = +$	$\cos = +$
$\sin = \cos = -$	$\sin = \tan = -$
Tan > 0	Cos > 0

15.8.2 Success criteria

Tick the column to indicate that you have completed the lesson and how well you have understood it using the traffic light system.

(**Green:** I understand; **Yellow:** I can do it with help; **Red:** I do not understand)

Lesson	Success criteria	⬤	◯	⬤
15.2	I can apply the sine rule to evaluate angles and sides of triangles.			
	I can recognise when the ambiguous case of the sine rule exists.			
15.3	I can apply the cosine rule to calculate a side of a triangle.			
	I can apply the cosine rule to calculate the angles of a triangle.			
15.4	I can calculate the area of a triangle, given two sides and the included angle.			
	I can calculate the area of a triangle, given the three sides.			
15.5	I can determine in which quadrant an angle lies.			
	I can use and interpret the relationship between a point on the unit circle and the angle made with the positive x-axis.			
	I can use the unit circle to determine approximate trigonometric ratios for angles greater than 90°.			
	I can use and interpret the relationships between supplementary and complementary angles			
15.6	I can sketch the graphs of the sine, cosine and tangent graphs.			
	I can determine the amplitude of a given trigonometric function.			
	I can determine the period of a given trigonometric function.			
15.7	I can apply the exact values of sin, cos and tan for 30°, 45° and 60° angles.			
	I can solve trigonometric equations graphically for a given domain.			
	I can solve trigonometric equations algebraically, using exact values, for a given domain.			

15.8.3 Project

What's an arbelos?

As an introduction to this task, you are required to complete the following construction. The questions that follow require the application of measurement formulas and an understanding of semicircles related to this construction.

1. **Constructing an arbelos**
   - Rule a horizontal line AB 8 cm long.
   - Determine the midpoint of the line and construct a semicircle on top of the line with AB as the diameter.
   - Mark Y as a point on AB such that AY = 6 cm.
   - Determine the midpoint of AY and draw a small semicircle inside the larger semicircle with AY as the diameter.
   - Determine the midpoint of YB and construct a semicircle (also inside the larger semicircle) with a diameter YB.

The shape enclosed by the three semicircles is known as an arbelos. The word, in Greek, means 'shoemaker's knife' as it resembles the blade of a knife used by cobblers. The point Y is not fixed and can be located anywhere along the diameter of the larger semicircle, which can also vary in size.

2. **Perimeter of an arbelos**
   The perimeter of an arbelos is the sum of the arc length of the three semicircles. Perform the following calculations, leaving each answer in terms of π.
   a. Calculate the arc length of the semicircle with diameter AB.
   b. Calculate the arc length of the semicircle with diameter AY.
   c. Calculate the arc length of the semicircle on diameter YB.
   d. Compare the largest arc length with the two smaller arc lengths. What do you conclude?
3. We can generalise the arc length of an arbelos. The point Y can be located anywhere on the line AB, which can also vary in length.
   Let the diameter AB be d cm, AY be d_1 cm and YB be d_2 cm.
   Prove that your conclusion from question **2d** holds true for any value of d, where $d_1 + d_2 = d$.

4. Area of an arbelos

The area of an arbelos may be treated as the area of a composite shape.

a. Using your original measurements, calculate the area of the arbelos you drew in question 1. Leave your answer in terms of π.

The area of the arbelos can also be calculated using another method. We can draw the common tangent to the two smaller semicircles at their point of contact and extend this tangent to the larger semicircle. It is said that the area of the arbelos is the same as the area of the circle constructed on this common tangent as diameter.

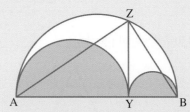

YZ is the common tangent.

Triangles AYZ, BYZ and AZB are all right-angled triangles. We can use Pythagoras' theorem, together with a set of simultaneous equations, to determine the length of the tangent YZ.

b. Complete the following.

In $\triangle$AYZ, $\qquad AZ^2 = AY^2 + YZ^2$

$\qquad\qquad\qquad\qquad = 6^2 + YZ^2$

In $\triangle$BYZ, $\qquad BZ^2 = BY^2 + YZ^2$

$\qquad\qquad\qquad\qquad = \dots\dots\dots\dots\dots + YZ^2$

Adding these two equations,

$\qquad\qquad AZ^2 + BZ^2 = \dots\dots\dots\dots\dots \ + \ \dots\dots\dots\dots\dots$

$\qquad\qquad AZ^2 + BZ^2 = AB^2$

But, in $\triangle$AZB $\qquad\qquad\quad = \dots\dots\dots\dots\dots$

$\dots\dots\dots\dots\dots \ + \ \dots\dots\dots\dots\dots$

So, $\qquad\qquad\qquad YZ = \dots\dots\dots\dots\dots$ (Leave your answer in surd form.)

c. Now calculate the area of the circle with diameter YZ.
Is your answer the same as that calculated in part a?

The area of an arbelos can be generalised.
Let the radii of the two smaller semicircles be r_1 and r_2.

5. Develop a formula for the area of the arbelos in terms of r_1 and r_2.

Demonstrate the use of your formula by checking your answer to question 4a.

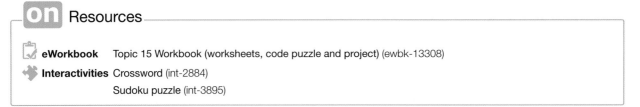
Exercise 15.8 Review questions

learn on

Fluency

1. Calculate the value of x, correct to 2 decimal places.

2. Calculate the value of θ, correct to the nearest minute.

3. Determine all unknown sides (correct to 2 decimal places) and angles (correct to the nearest minute) of triangle ABC, given $a = 25$ m, $A = 120°$ and $B = 50°$.

4. Calculate the value of x, correct to 2 decimal places.

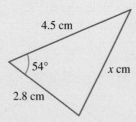

5. Calculate the value of θ, correct to the nearest degree.

6. A triangle has sides of length 12 m, 15 m and 20 m. Calculate the magnitude (size) of the largest angle, correct to the nearest minute.

7. A triangle has two sides of 18 cm and 25 cm. The angle between the two sides is 45°. Calculate, correct to 2 decimal places:
   a. its area
   b. the length of its third side.

8. If an angle of $\theta = 290°$ was represented on the unit circle, state which quadrant the triangle to show this would be drawn in.

9. On the unit circle, draw an appropriate triangle for the angle 110° in quadrant 2.
   a. Determine the value of sin(110°) and cos(110°), correct to 2 decimal places.
   b. Determine the value of tan(110°), correct to 2 decimal places.

10. **MC** The value of sin(53°) is equal to:
    A. cos(53°)   **B.** cos(37°)   **C.** sin(37°)   **D.** tan(53°)

11. Simplify $\dfrac{\sin(53°)}{\sin(37°)}$.

12. Draw a sketch of $y = \sin(x)$ from $0° \leq x \leq 360°$.

13. Draw a sketch of $y = \cos(x)$ from $0° \leq x \leq 360°$.

14. Draw a sketch of $y = \tan(x)$ from $0° \leq x \leq 360°$.

15. Label the triangle shown so that $\dfrac{x}{\sin(46°)} = \dfrac{y}{\sin(68°)}$.

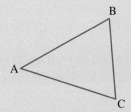

16. State the period and amplitude of each of the following graphs.
    a. $y = 2\sin(3x)$
    b. $y = -3\cos(2x)$
    c.

17. Sketch the following graphs.
    a. $y = 2\sin(x)$, $x \in [0°, 360°]$
    b. $y = \cos(2x)$, $x \in [-180°, 180°]$

18. Use technology to write down the solutions to the following equations for the domain $0° \le x \le 360°$ to 2 decimal places.
    a. $\sin(x) = -0.2$
    b. $\cos(2x) = 0.7$
    c. $3\cos(x) = 0.1$
    d. $2\tan(2x) = 0.5$

19. Solve each of the following equations.
    a. $\sin(x) = \dfrac{1}{2}$, $x \in [0°, 360°]$
    b. $\cos(x) = \dfrac{\sqrt{3}}{2}$, $x \in [0°, 360°]$

    c. $\cos(x) = \dfrac{1}{\sqrt{2}}$, $x \in [0°, 360°]$
    d. $\sin(x) = \dfrac{1}{\sqrt{2}}$, $x \in [0°, 360°]$

20. **MC** The equation that represents the graph shown could be:

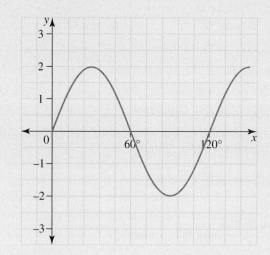

    A. $y = 3\sin(2x)$
    B. $y = 2\cos(3x)$
    C. $y = 3\cos(2x)$
    D. $y = 2\sin(3x)$

21. a. Use technology to help sketch the graph of $y = 2\sin(2x) - 3$.
    b. Write down the period and the amplitude of the graph in part **a**.

Problem solving

22. Sketch the graphs of each of the following, stating:
    i. the period
    ii. the amplitude.
    a. $y = 2\cos(2x)$, $x \in [0°, 360°]$
    b. $y = 3\sin(4x)$, $x \in [0°, 180°]$
    c. $y = -2\cos(3x)$, $x \in [-60°, 60°]$
    d. $y = 4\sin(2x)$, $x \in [-90°, 90°]$

23. Solve each of the following equations for the given values of x.
    a. $\cos(2x) = \dfrac{\sqrt{3}}{2}$, $x \in [0°, 360°]$
    b. $\sin(3x) = \dfrac{1}{2}$, $x \in [-90°, 90°]$

c. $\sin(2x) = \dfrac{1}{\sqrt{2}}, x \in [0°, 360°]$ **d.** $\cos(3x) = -\dfrac{1}{\sqrt{2}}, x \in [0°, 360°]$

e. $\sin(4x) = 0, x \in [0°, 180°]$ **f.** $\tan(4x) = -1, x \in [0°, 180°]$

24. Solve the following for $x \in [0°, 360°]$.

    a. $2\cos(x) - 1 = 0$ **b.** $2\sin(x) = -\sqrt{3}$ **c.** $-\sqrt{2}\cos(x) + 1 = 0$ **d.** $\sqrt{2}\sin(x) + 1 = 0$

25. Sketch the graph of $y = \tan(2x), x \in [0°, 180°]$. Write down the period, amplitude and the equations of any asymptotes.

26. A satellite dish is placed on top of an apartment building as shown in the diagram. Determine the height of the satellite dish in metres, correct to 2 decimal places.

27. Australian power points supply voltage, V, in volts, where $V = 240(\sin 18\,000t)$ and t is measured in seconds.

t	V
0.000	
0.005	
0.010	
0.015	
0.020	
0.025	
0.030	
0.035	
0.040	

    a. Copy and complete the table and sketch the graph, showing the fluctuations in voltage over time.
    b. State the times at which the maximum voltage output occurs.
    c. Determine how many seconds there are between times of maximum voltage output.
    d. Determine how many periods (or cycles) there are per second.

on To test your understanding and knowledge of this topic, go to your learnON title at www.jacplus.com.au and complete the **post-test**.

Answers

Topic 15 Trigonometry II (10A)

15.1 Pre-test

1. B
2. 9.06 m
3. 9.35 cm
4. A
5. 19.67
6. 7.2 m
7. B
8. 3rd quadrant
9. 0.5
10. D
11. D
12. B
13. A
14. C
15. 270°

15.2 The sine rule

1. $44°58', 77°2', 13.79$
2. $39°18', 38°55', 17.21$
3. $70°, 9.85, 9.40$
4. $33°, 38.98, 21.98$
5. 19.12
6. $C = 51°, b = 54.66, c = 44.66$
7. $A = 60°, b = 117.11, c = 31.38$
8. $B = 48°26', C = 103°34', c = 66.26$; or $B = 131°34'$, $C = 20°26', c = 23.8$
9. 24.17
10. B, C
11. $A = 73°15', b = 8.73$; or $A = 106°45', b = 4.12$
12. 51.90 or 44.86
13. $C = 110°, a = 3.09, b = 4.64$
14. $B = 38°, a = 3.36, c = 2.28$
15. $B = 33°33', C = 121°27', c = 26.24$; or $B = 146°27'$, $C = 8°33', c = 4.57$
16. 43.62 m
17. $h = 7.5$ cm
18. 113 km
19. 8.68 m
20. Yes, she needs 43 m altogether.
21. a. 6.97 m   b. 4 m
22. a. 13.11 km   b. N20°47′W
23. a. 8.63 km   b. 6.5 km/h   c. 9.90 km
24. 22.09 km from A and 27.46 km from B.
25. a. 7.3 km   b. 282°3′

15.3 The cosine rule

1. 7.95
2. 55.22
3. $23.08, 41°53', 23°7'$
4. 28°57′
5. 88°15′
6. $A = 61°15', B = 40°, C = 78°45'$
7. 2218 m
8. a. 12.57 km   b. S35°1′E
9. a. 35°6′   b. 6.73 m^2
10. 23°
11. 89.12 m
12. a. 130 km   b. S22°12′E
13. 28.5 km
14. 74.3 km
15. 70°49′
16. a. 8.89 m   b. 77°0′   c. $x = 10.07$ m
17. 1.14 km/h
18. $\angle CAB = 34.65°, \angle ABC = 84.83°$ and $\angle BCA = 60.52°$
19. a. 4.6637 m   b. 55.93°

15.4 Area of triangles

1. 12.98
2. 38.14
3. 212.88
4. $A = 32°4', B = 99°56',$ area $= 68.95$ cm^2
5. $A = 39°50', B = 84°10',$ area $= 186.03$ m^2
6. $A = 125°14', B = 16°46',$ area $= 196.03$ mm^2
7. C
8. 14.98 cm^2
9. 570.03 mm^2
10. 2.15 cm^2
11. B
12. 3131.41 mm^2
13. 610 cm^2
14. a. 187.5 cm^2 b. 15.03 cm^2
15. 17 goldfish
16. C
17. B
18. a. Area $= 69.63$ cm^2
    b. The dimensions are 12.08 cm and 6.96 cm.
19. 17 kg
20. 52.2 hectares
21. 175 m^3
22. 22.02 m^2
23. a. 29.44 km^2 b. 8.26 km   c. 41°
    d. 24°   e. 3.72 km   f. 10.11 km^2
    g. 39.55 km^2
24. 374.12 cm^2

15.5 The unit circle

1. a. 1st b. 2nd c. 4th d. 3rd e. 2nd f. 3rd

2. A

3. D

4. a. 0.34 b. 0.94 c. −0.17 d. 0.98

5. a. −0.64 b. 0.77 c. −0.57 d. −0.82

6. a. 1 b. 0 c. 0 d. −1

7. a. −1 b. 0 c. 0 d. 1

8. a. 0.87 b. 0.50

9. a. 30°
   b. −0.87
   c. $\cos(150°) = -\cos(30°)$
   d. 0.5
   e. $\sin(150°) = \sin(30°)$

10. a. 30°
    b. −0.87
    c. $\cos(210°) = -\cos(30°)$
    d. −0.50
    e. $\sin(210°) = -\sin(30°)$

11. a. 30°
    b. 0.87
    c. $\cos(330°) = -\cos(30°)$
    d. −0.50
    e. $\sin(330°) = -\sin(30°)$

12. a. 0.34 b. 0.94
    c. 0.36 d. 0.36
    e. They are equal.

13. a. 0.71 b. −0.71
    c. −1 d. −1
    e. They are equal. f. $\tan(135°) = -\tan(45°)$

14. a. −0.64 b. −0.77
    c. 0.84 d. 0.83
    e. They are approximately equal.
    f. $\tan(220°) = \tan(40°)$

15. a. −0.87 b. 0.5
    c. −1.73 d. −1.74
    e. They are approximately equal.
    f. $\tan(300°) = -\tan(60°)$

16. D

17, 18, 19. Sample responses can be found in the worked solutions in the online resources.

20. a. $\sqrt{1-p^2}$ b. p
    c. $-\sqrt{1-p^2}$ d. p

21. 0

22. 0

15.6 Trigonometric functions

1. See the table at the bottom of the page.*

2.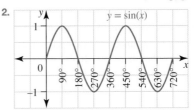

3. 360°

4. a. 0.7 b. 0.8 c. 0.3 d. −0.3

5. a. 0 b. 0.9 c. −0.2 d. −0.9

6. a. 64°, 116°, 424°, 476°
   b. 244°, 296°, 604°, 656°
   c. 44°, 136°, 404°, 496°

7. a. 210°, 330°, 570°, 690°
   b. 233°, 307°, 593°, 667°
   c. 24°, 156°, 384°, 516°

8. See the table at the bottom of the page.*

9.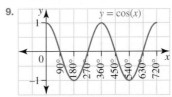

10. The graph would continue with the cycle.

11. It is a very similar graph with the same shape; however, the sine graph starts at $(0, 0)$, whereas the cosine graph starts at $(0, 1)$.

12. a. 0.7 b. −0.9 c. −1 d. 0.9

13. a. −0.5 b. −0.8 c. 0.8 d. −0.9

14. a. 120°, 240°, 480°, 600°
    b. 37°, 323°, 397°, 683°
    c. 46°, 314°, 406°, 674°

15. a. 127°, 233°, 487°, 593°
    b. 26°, 334°, 386°, 694°
    c. 154°, 206°, 514°, 566°

*1.

x	0°	30°	60°	90°	120°	150°	180°	210°	240°	270°	300°	330°	360°
$\sin(x)$	0	0.5	0.87	1	0.87	0.5	0	−0.5°	−0.87	−1	−0.87	−0.5	0
x	390°	420°	450°	480°	510°	540°	570°	600°	630°	660°	690°	720°	
$\sin(x)$	0.5	0.87	1	0.87	0.5	0	−0.5	−0.87	−1	−0.87	−0.5	0	

*8.

x	0°	30°	60°	90°	120°	150°	180°	210°	240°	270°	300°	330°	360°
$\cos(x)$	1	0.87	0.5	0	−0.5	−0.87	−1	−0.87	−0.5	0	0.5	0.87	1
x	390°	420°	450°	480°	510°	540°	570°	600°	630°	660°	690°	720°	
$\cos(x)$	0.87	0.5	0	−0.5	−0.87	−1	−0.87	−0.5	0	0.5	0.87	1	

16. See the table at the bottom of the page.*

17. $y = \tan(x)$

18. The graph would continue repeating every 180° as above.

19. Quite different. $y = \tan(x)$ has undefined values (asymptotes) and repeats every 180° rather than 360°. It also gives all y-values, rather than just values between -1 and 1.

20. a. 1.7　　b. -1　　c. -1.2　　d. 0.8

21. a. -0.8　　b. 1.2　　c. -0.2　　d. 1

22. a. 45°, 225°, 405°, 585°
    b. 56°, 236°, 416°, 596°
    c. 158°, 338°, 518°, 698°
    d. 117°, 297°, 477°, 657°
    e. 11°, 191°, 371°, 551°
    f. 135°, 315°, 495°, 675°

23. a. i. 360°
       ii. 1
       iii.

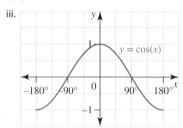

    b. i. 360°
       ii. 1
       iii.

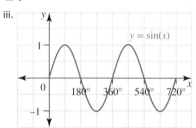

24. a. i. 180°
       ii. 1
       iii.

b. i. 360°
   ii. 2
   iii.

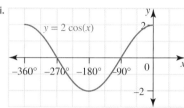

25. a. i. 180°　　　　　　ii. 3
    b. i. 120°　　　　　　ii. 4
    c. i. 720°　　　　　　ii. 2
    d. i. 1440°　　　　　　ii. $\dfrac{1}{2}$
    e. i. 360°　　　　　　ii. 1
    f. i. 180°　　　　　　ii. 1

26. a. C　　　　b. A　　　　c. D

27. a.

Period = 1080°
Amplitude = 2

b.

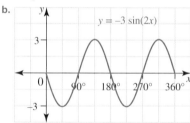

Period = 180°
Amplitude = 3

c.

Period = 720°
Amplitude = 3

*16.

x	0°	30°	60°	90°	120°	150°	180°	210°	240°	270°	300°	330°	360°
$\tan(x)$	0	0.58	1.73	undef.	-1.73	-0.58	0	0.58	1.73	undef.	-1.73	-0.58	0
x	390°	420°	450°	480°	510°	540°	570°	600°	630°	660°	690°	720°	
$\tan(x)$	0.58	1.73	undef.	-1.73	-0.58	0	0.58	1.73	undef.	-1.73	-0.58	0	

d.

Period = 120°
Amplitude = 1

e.

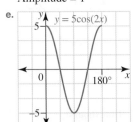

Period = 180°
Amplitude = 5

f.

Period = 90°
Amplitude = 1

28. a.

b.

c.

d.

29. a.

i. −1   ii. 1

b. Max value = 3, min value = 1

c. Max value of $\sin(x) = 1$; hence, max value of
$y = 2 \times 1 + 3 = 5$
Min value of $\sin(x) = -1$; hence, min value of
$y = 2 \times -1 + 3 = 1$

30. a.

x	0	30°	60°	90°	120°	150°	180°
y	0	$\dfrac{\sqrt{3}}{3}$	$\sqrt{3}$	undef.	$-\sqrt{3}$	$-\dfrac{\sqrt{3}}{3}$	0

b.

c. At $x = 90°$, y is undefined.

d. $x = 270°$

e. The period = 180° and the amplitude is undefined.

31. a.

b. $x = 45°$ and $x = 135°$

c. The period = 90° and the amplitude is undefined.

32. a.

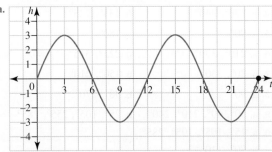

b. 3 metres

c. −2.6 metres

33. a.

b. 10 metres

c. 2 metres

d. 2.54 metres

34. a. 22 °C

b.

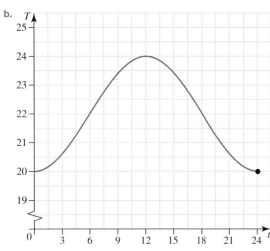

c. Coolest 20 °C, warmest 24 °C

15.7 Solving trigonometric equations

1. Calculator answers
   a. 25.84°, 334.16°   **b.** 72.54°, 287.46°

2. **a.** 101.54°, 258.46°   **b.** 126.87°, 233.13°

3. **a.** 30°, 150°   **b.** 60°, 120°

4. **a.** 120°, 240°   **b.** 135°, 225°

5. **a.** 90°   **b.** 180°

6. **a.** 210°, 330°   **b.** 225°, 315°

7. **a.** 30°, 330°   **b.** 150°, 210°

8. **a.** 90°   **b.** 90°, 270°

9. **a.** 30°, 60°, 210°, 240°   **b.** 75°, 105°, 255°, 285°

10. **a.** 15°, 105°, 195°, 285°
    b. 70°, 110°

11. **a.** 52.5°, 82.5°, 142.5°, 172.5°
    b. −165°, −135°, −45°, −15°, 75°, 105°

12. **a.** 45°
    b. 30°, 90°, 150°, 210°, 270°, 330°

13. **a.** 30°, 150°   **b.** 30°, 330°

14. **a.** 45°, 315°   **b.** 135°, 315°

15. a.

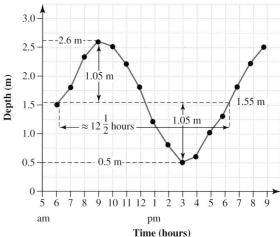

b. i. $12\dfrac{1}{2}$ hours   **ii.** 1.05 m

c. 10 : 00 am, 10 : 30 pm, 11 : 00 am, 11 : 30 pm, noon.

d. Until 1 : 45 am Sunday, 8 am to 2 : 15 pm and after 8 : 30 pm.

16. a. $x = 30°, 210°$ **b.** $x = 153.43°, 333.43°$

17. $x = 30°, 150°$

18. a. 100^g **b.** 200^g **c.** 300^g **d.** 400^g

Project

1. Follow the given instructions.

2. **a.** 4π cm
   b. 3π cm
   c. π cm
   d. The largest arc length equals the sum of the two smaller arc lengths.

3. Sample responses can be found in the worked solutions in the online resources.

4. **a.** 3π cm^2
   b. In $\triangle AYZ$:
   $$AZ^2 = AY^2 + YZ^2$$
   $$= 6^2 + YZ^2$$
   In $\triangle BYZ$:
   $$BZ^2 = BY^2 + YZ^2$$
   $$= 2^2 + YZ^2$$
   Adding these equations: $AZ^2 + BZ^2 = 6^2 + YZ^2 + 2^2 + YZ^2$
   But in $\triangle AZB$: $AZ^2 + BZ^2 = AB^2$
   $$6^2 + YZ^2 + 2^2 + YZ^2 = 8^2$$
   $$2\,YZ^2 = 64{-}36{-}4$$
   $$YZ^2 = 12$$
   $$YZ = \pm\sqrt{12}$$
   But YZ > 0 as it is a length: $YZ = 2\sqrt{3}$

   c. 3π cm^2
   Yes, same area

5. Area of the arbelos $= \pi r_1 r_2$
   Sample responses can be found in the worked solutions in the online resources.

15.8 Review questions

1. 14.15 cm

2. 20°31′

3. $b = 22.11\,\text{m}, c = 5.01\,\text{m}, C = 10°$

4. 3.64 cm

5. 34°

6. 94°56′

7. a. 159.10 cm²   b. 17.68 cm

8. 4th quadrant

9. a. 0.94, −0.34   b. −2.75

10. B

11. tan(53°)

12.

13.

14.

15.

16. a. Period = 120°, amplitude = 2
   b. Period = 180°, amplitude = 3
   c. Period = 180°, amplitude = 1

17. a.

   b.

18. a. $x = 191.54, 348.46$
   b. $x = 22.79, 157.21, 202.79, 337.21$
   c. $x = 88.09, 271.91$
   d. $x = 7.02, 97.02, 187.02, 277.02$

19. a. 30°, 150°   b. 30°, 330°
   c. 45°, 315°   d. 45°, 135°

20. D

21. a.

   b. Period = 180, amplitude = 2

22. a.

   i. Period = 180°
   ii. Amplitude = 2

   b.

   i. Period = 90°
   ii. Amplitude = 3

   c. $y = -2\cos(3x)$

   i. Period = 120°
   ii. Amplitude = 2

   d. $y = 4\sin(2x)$

   i. Period = 180°
   ii. Amplitude = 4

23. a. 15°, 165°, 195°, 345°
  b. −70°, 10°, 50°
  c. 22.5°, 67.5°, 202.5°, 247.5°
  d. 45°, 75°, 165°, 195°, 285°, 315°
  e. 0°, 45°, 90°, 135°, 180°
  f. 33.75°, 78.75°, 123.75°, 168.75°

24. a. 60°, 300° **b.** 240°, 300°
  c. 45°, 315° **d.** 225°, 315°

25.

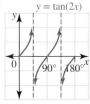

$y = \tan(2x)$

Period = 90°; amplitude is undefined.
Asymptotes are at $x = 45°$ and $x = 135°$.

26. 3.92 m

27. a.

t	V
0.000	0
0.005	240
0.010	0
0.015	−240
0.020	0
0.025	240
0.030	0
0.035	−240
0.040	0

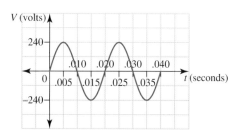

b. Maximum voltage occurs at $t = 0.005$ s, 0.025 s.

c. 0.02 s

d. 50 cycles per second

16 Polynomials (10A)

LESSON SEQUENCE

LESSON
16.1 Overview

Why learn this?

Just as your knowledge of numbers is learned in stages, so too are graphs. You have been building your knowledge of graphs and functions over time. First, you encountered linear functions. You saw how straight lines are everywhere in our daily lives. Then you learned about quadratic functions, or parabolas. Again, you saw, in everyday situations, how bridges and arches can be based on quadratic or parabolic functions. Circles and hyperbolas are other functions that you have studied. A polynomial is an algebraic expression with integer powers that are greater than or equal to zero, such as a parabola. Polynomial functions are represented by smooth and continuous curves. They can be used to model situations in many different fields, including business, science, architecture, design and engineering. An engineer and designer would use polynomials to model the curves on a rollercoaster. Economists use polynomials to model changes and fluctuations in the stock market. Scientists and researchers use polynomials when looking at changes in the behaviour of objects in different circumstances. Designers and architects incorporate polynomial functions in many areas of their designs in buildings and in landscaping. This topic introduces the building blocks of polynomials.

Reading content and rich media, including interactivities and videos for every concept

Extra learning resources

Differentiated question sets

Questions with immediate feedback, and fully worked solutions to help students get unstuck

1. State the degree of the following polynomials.

   a. $3x^2 - 5x + 1$ **b.** $2x^3 - 4x^2 + 3x$ **c.** $\dfrac{1}{2}x + 7$ **d.** 4

2. **MC** Choose which of the following is a polynomial.

   A. $\dfrac{x^3 + x}{2}$ **B.** $x^{-2} - \sqrt{7}$ **C.** $-x^2 + 3\sqrt{x}$ **D.** $6x^4 - \dfrac{1}{x}$

3. If $x^2 + 4x - 2 \equiv (x+1)^2 + a(x-1) + b$, determine the values of a and b.

4. **MC** Select the expanded and simplified expression of $(2x)(-3x)(x+1)$.

   A. $-6x^3 + 1$ **B.** $-6x^2 + 1$ **C.** $-6x^2 - 6x$ **D.** $-6x^3 - 6x^2$

5. **MC** Select the expanded and simplified expression for $(1-2x)(3-x)(4x+1)(x-5)$.

   A. $8x^4 - 66x^3 + 135x^2 - 22x - 15$ **B.** $8x^4 - 66x^3 + 135x^2 - 12x - 10$

   C. $9x^4 - 66x^2 + 135x^2 - 8x - 10$ **D.** $8x^3 - 66x^2 + 135x^2 - 22x - 15$

6. Consider the polynomial $-3x^3 + 2x^2 + 4x - 1$. State:

   a. the degree of the polynomial   **b.** the leading coefficient

   c. the value of the constant term   **d.** the coefficient of x^2.

7. **MC** Select the simplified expression for $(x^6 + 2x^3 + 3x + 1) - (x^5 - 2x^2 + 4x - 2)$.

   A. $x^6 - x^5 + 2x^3 + 2x^2 - x + 3$ **B.** $2x^3 + 2x^2 - 2x + 3$

   C. $x^6 - x^5 + 4x^3 - x + 3$ **D.** $x^6 - x^5 + 2x^3 + 2x^2 - 3x$

8. Calculate the quotient when $x^3 + 2x^2 + x - 1$ is divided by $x + 1$.

9. Determine the remainder when $x^3 - x^2 + 4x - 3$ is divided by $x - 2$.

10. If $P(x) = 3x^4 - 2x^3 + x^2 - 4x + 8$, calculate $P(-1)$.

11. **MC** If $P(x) = x^3 - x^2 + x + 1$, select the value of $P(x+1)$:

    A. $x^3 + 2x^2 + 5x + 4$ **B.** $2x + 3$

    C. $x^3 + 4x^2 + 5x + 4$ **D.** $x^3 + 2x^2 + 2x + 2$

12. When $x^3 - 2x^2 + bx + 3$ is divided by $x - 1$ the remainder is 4. Calculate the value of b.

13. **MC** Select the correct factor for $x^3 - x^2 - 5x - 3$:

    A. x **B.** $x + 1$ **C.** $x - 1$ **D.** $x + 3$

14. **MC** Select the factorised from of $x^3 + 4x^2 + x - 6$.

    A. $(x+1)(x-2)(x+3)$ **B.** $(x+1)(x-2)(x-3)$

    C. $(x-1)(x+2)(x+3)$ **D.** $(x-1)(x-2)(x-3)$

15. Evaluate the correct value of x, for $2x^3 + 15x^2 + 19x + 6 = 0$.

LESSON
16.2 Polynomials

▶ 16.2.1 Polynomials

eles-4975

A **polynomial** in x, sometimes denoted $P(x)$, is an expression containing only non-negative integer powers of x.

- The **degree** of a polynomial in x is the highest power of x in the expression.

  For example:

$$3x + 1 \qquad \text{is a polynomial of degree 1, or linear polynomial.}$$

$$x^2 + 4x - 7 \qquad \text{is a polynomial of degree 2, or quadratic polynomial.}$$

$$-5x^3 + \frac{x}{2} \qquad \text{is a polynomial of degree 3, or cubic polynomial.}$$

$$10 \qquad \text{is a polynomial of degree 0 (think of 10 as } 10x^0\text{).}$$

- Expressions containing a term similar to any of the following terms are **not** polynomials:

$$\frac{1}{x}, \qquad x^{-2}, \qquad \sqrt{x}, \qquad 2^x, \qquad \sin x, \qquad \text{and so on.}$$

  For example, the following are not polynomials.

$$3x^2 - 4x + \frac{2}{x} \qquad -5x^4 + x^3 - 2\sqrt{x} \qquad x^2 + \sin x + 1$$

- In the expression $6x^3 + 13x^2 - x + 1$

  x is the *variable*.
  6 is the *coefficient* of x^3.
  13 is the *coefficient* of x^2.
  -1 is the *coefficient* of x.
  $6x^3$, $13x^2$, $-x$ and $+1$ are all *terms*.
  The *constant* term is $+1$.
  The *degree* of the polynomial is 3.

- The **leading term** is $6x^3$ because it is the term that contains the highest power of x.
- The **leading coefficient** is 6.
- Any polynomial with a leading coefficient of 1 is called **monic**.

WORKED EXAMPLE 1 Identifying degrees and variables of polynomials

Answer the following questions.
i. State the degree of each of the following polynomials.
ii. State the variable for each of the following polynomials.
a. $x^3 + 3x^2 + x - 5$ b. $y^4 + 4y^3 - 8y^2 + 2y - 8$ c. $a^3 + 34a^6 - 12a - 72$

THINK	WRITE
a. i. Determine the highest power of x in the expression.	a. i. In the expression $x^3 + 3x^2 + x - 5$, the highest power of x is 3. Therefore, this polynomial is of degree 3.
ii. Determine the variable (unknown quantity) in the expression.	ii. x is the variable in this expression.
b. i. Determine the highest power of y in the expression.	b. i. In the expression $y^4 + 4y^3 - 8y^2 + 2y - 8$ the highest power of y is 4. Therefore, this polynomial is of degree 4.
ii. Determine the variable (unknown quantity) in the expression.	ii. y is the variable in this expression.
c. i. Determine the highest power of a in the expression.	c. i. In the expression $a^3 + 34a^6 - 12a - 72$, the highest power of a is 6. Therefore, this polynomial is of degree 6.
ii. Determine the variable (unknown quantity) in the expression.	ii. a is the variable in this expression.

WORKED EXAMPLE 2 Determining coefficients and terms of polynomials

Consider the polynomial $P(x) = 3x^4 - 5x^3 + 2x^2 + x - 12$
a. **State the degree and variable of the polynomial.**
b. **State the coefficient of x^3.**
c. **State the value of the constant term.**
d. **Determine the term that has a coefficient of 2.**
e. **Determine the leading term.**

THINK	WRITE
a. Determine the highest power of x and the variable (unknown quantity) in the expression.	a. The highest power of x is 4 and therefore, the degree is 4. x is the variable in this expression.
b. Determine the term with x^3 in the expression.	b. The coefficient in the term $-5x^3$ is -5.
c. Determine the term without variable in the expression.	c. The value of the constant term is -12.
d. Determine the term that has a coefficient of 2 in the expression.	d. The term that has a coefficient of 2 is $2x^2$.
e. Determine the term that contains the highest power of x in the expression.	e. The leading term is $3x^4$.

An example where polynomials are useful is in the construction of a greenhouse. The surface area, S, of a greenhouse of length L and height x can be approximated by the polynomial $S(x) = \pi x^2 + L\pi x - 4$.

on Resources

📋 **eWorkbook**   Topic 16 Workbook (worksheets, code puzzle and project) (ewbk-2040)

🧩 **Interactivity**   Degrees of polynomials (int-6203)

Exercise 16.2 Polynomials

learnon

| 16.2 Quick quiz   on | 16.2 Exercise |

Individual pathways

■ PRACTISE	■ CONSOLIDATE	■ MASTER
1, 4, 5, 8, 11, 14	2, 6, 9, 12, 15	3, 7, 10, 13, 16

Fluency

WE1 For questions **1** to **3**, answer the following questions.

i. State the degree of each of the following polynomials.

ii. State the variable for each of the following polynomials.

1. **a.** $x^3 - 9x^2 + 19x + 7$ **b.** $65 + 2x^7$ **c.** $3x^2 - 8 + 2x$

2. **a.** $x^6 - 3x^5 + 2x^4 + 6x + 1$ **b.** $y^8 + 7y^3 - 5$ **c.** $\dfrac{1}{2}u^5 - \dfrac{u^4}{3} + 2u - 6$

3. **a.** $18 - \dfrac{e^5}{6}$ **b.** $2g - 3$ **c.** $1.5f^6 - 800f$

4. Identify the polynomials in questions 1 to 3 that are:

    a. linear   b. quadratic   c. cubic   d. monic.

For questions 5 to 7, state whether each of the following is a polynomial (P) or not (N).

5. a. $7x + 6x^2 + \dfrac{5}{x}$ b. $33 - 4p$ c. $\dfrac{x^2}{9} + x$

6. a. $3x^4 - 2x^3 - 3\sqrt{x} - 4$ b. $k^{-2} + k - 3k^3 + 7$ c. $5r - r^9 + \dfrac{1}{3}$

7. a. $\dfrac{4c^6 - 3c^3 + 1}{2}$ b. $2^x - 8x + 1$ c. $\sin x + x^2$

Understanding

8. **WE2** Consider the polynomial $P(x) = -2x^3 + 4x^2 + 3x + 5$.

    a. State the degree of the polynomial.   b. State the variable.

    c. State the coefficient of x^2.   d. State the value of the constant term.

    e. State the term that has a coefficient of 3.   f. Determine the leading term.

9. Consider the polynomial $P(w) = 6w^7 + 7w^6 - 9$.

    a. State the degree of the polynomial.   b. State the variable.

    c. State the coefficient of w^6.   d. Determine the coefficient of w.

    e. State the value of the constant term.   f. State the term that has a coefficient of 6.

10. Consider the polynomial $f(x) = 4 - x^2 + x^4$.

    a. State the degree of the polynomial.   b. State the coefficient of x^4.

    c. Determine the leading term.   d. State the leading coefficient.

Reasoning

11. Write the following polynomials as simply as possible, arranging terms in descending powers of x.

    a. $7x + 2x^2 - 8x + 15 + 4x^3 - 9x + 3$

    b. $x^2 - 8x^3 + 3x^4 - 2x^2 + 7x + 5x^3 - 7$

    c. $x^3 - 5x^2 - 11x - 1 + 4x^3 - 2x + x^2 - 5$

12. A sports scientist determines the following equation for the velocity of a breaststroke swimmer during one complete stroke:

$$v(t) = 63.876t^6 - 247.65t^5 + 360.39t^4 - 219.41t^3 + 53.816t^2 + 0.4746t.$$

    a. Determine the degree of the polynomial.

    b. State the variable.

    c. State the number of terms that are there.

    d. Use a graphics calculator or graphing software to draw the graph of this polynomial.

    e. Match what happens during one complete stroke with points on the graph.

13. The distance travelled by a body after t seconds is given by $d(t) = t^3 + 2t^2 - 4t + 5$. Using a graphing calculator or suitable computer software, draw a graph of the above motion for $0 \le t \le 3$.
Use the graph to help you answer the following:

    a. State what information the constant term gives.

    b. Evaluate the position of the body after 1 second.

    c. Describe in words the motion in the first 2 second.

Problem solving

14. If $x^2 - 3x + 5 = x^2 + (a + b)x + (a - b)$, determine the values of a and b.

15. If $x^2 + 2x - 1 \equiv (x - 1)^2 + a(x + 1) + b$, evaluate a and b.

16. If $x^3 + 9x^2 + 12x + 7 \equiv x^3 + (ax + b)^2 + 3$, evaluate a and b.

LESSON
16.3 Adding, subtracting and multiplying polynomials

LEARNING INTENTION

At the end of this lesson you should be able to:
- add, subtract and multiply polynomial expressions.

▶ 16.3.1 Operations with polynomials

eles-4976

- To add or subtract polynomials, simply add or subtract any like terms in the expressions.

WORKED EXAMPLE 3 Simplifying polynomial expressions

Simplify each of the following.
a. $(5x^3 + 3x^2 - 2x - 1) + (x^4 + 5x^2 - 4)$
        b. $(5x^3 + 3x^2 - 2x - 1) - (x^4 + 5x^2 - 4)$

THINK	WRITE
a. 1. Write the expression.	**a.** $(5x^3 + 3x^2 - 2x - 1) + (x^4 + 5x^2 - 4)$
2. Remove any grouping symbols, watching any signs.	$= 5x^3 + 3x^2 - 2x - 1 + x^4 + 5x^2 - 4$
3. Re-order the terms with descending degrees of x.	$= x^4 + 5x^3 + 3x^2 + 5x^2 - 2x - 1 - 4$
4. Simplify by collecting like terms.	$= x^4 + 5x^3 + 8x^2 - 2x - 5$
b. 1. Write the expression.	**b.** $(5x^3 + 3x^2 - 2x - 1) - (x^4 + 5x^2 - 4)$
2. Remove any grouping symbols, watching any signs.	$= 5x^3 + 3x^2 - 2x - 1 - x^4 - 5x^2 + 4$
3. Re-order the terms with descending degrees of x.	$= -x^4 + 5x^3 + 3x^2 - 5x^2 - 2x - 1 + 4$
4. Simplify by collecting like terms.	$= -x^4 + 5x^3 - 2x^2 - 2x + 3$

TI | THINK	DISPLAY/WRITE	CASIO | THINK	DISPLAY/WRITE
a–b.	**a–b.**	**a–b.**	**a–b.**
In a new document, on a Calculator page, press: • MENU • 1: Actions • 1: Define Complete the entry lines as: Define $p1(x) = 5x^3 + 3x^2 - 2x - 1$ Define $p2(x) = x^4 + 5x^2 - 4$ $p1(x) + p2(x)$ $p1(x) - p2(x)$ Press ENTER after each entry.	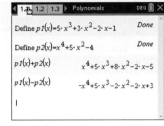 $p1(x) + p2(x)$ $= x^4 + 5x^3 + 8x^2 - 2x - 5$ $p1(x) - p2(x)$ $= -x^4 + 5x^3 - 2x^2 - 2x + 3$	On the Main screen, tap: • Action • Transformation • simplify Complete the entry lines as: simplify $((5x^3 + 3x^2 - 2x - 1)$ $+(x^4 + 5x^2 - 4))$ simplify $((5x^3 + 3x^2 - 2x - 1)$ $-(x^4 + 5x^2 - 4))$ Press EXE after each entry.	 $x^4 + 5x^3 + 8x^2 - 2x - 5$ $-x^4 + 5x^3 - 2x^2 - 2x + 3$

- To expand linear factors, for example $(x + 1)(x + 2)(x - 7)$, use FOIL from quadratic expansions.

WORKED EXAMPLE 4 Expanding polynomial expressions

Expand and simplify:

a. $x(x + 2)(x - 3)$

b. $(x - 1)(x + 5)(x + 2).$

THINK	WRITE
a. 1. Write the expression.	**a.** $x(x + 2)(x - 3)$
2. Expand the last two linear factors, using FOIL and simplify	$= x(x^2 - 3x + 2x - 6)$ $= x(x^2 - x - 6)$
3. Multiply the expression in brackets by x.	$= x^3 - x^2 - 6x$
b. 1. Write the expression.	**b.** $(x - 1)(x + 5)(x + 2)$
2. Expand the last two linear factors, using FOIL, and simplify	$= (x - 1)(x^2 + 2x + 5x + 10)$
3. Multiply the expression in the second bracket by x and then by -1.	$= (x - 1)(x^2 + 7x + 10)$ $= x^3 + 7x^2 + 10x - x^2 - 7x - 10$
4. Collect like terms.	$= x^3 + 6x^2 + 3x - 10$

TI	THINK	DISPLAY/WRITE	CASIO	THINK	DISPLAY/WRITE

TI | THINK

a–b.

On a Calculator page, press:
- MENU
- 3: Algebra
- 3: Expand

Complete the entry lines as:
expand
$(x \times (x+2) \times (x-3))$
expand
$(x \times (x-1) \times (x+5) \times (x+2))$
Press ENTER after each entry.
On the TI, the multiplication sign is displayed as a dot.

DISPLAY/WRITE

a–b.

$x(x+2)(x-3) = x^3 - x^2 - 6x$

$(x-1)(x+5)(x+2) = x^3 + 6x^2 + 3x - 10$

CASIO | THINK

a–b.

On the Main screen, tap:
- Action
- Transformation
- expand

Complete the entry lines as:
expand
$(x \times (x+2) \times (x-3))$
expand
$((x-1) \times (x+5) \times (x+2))$
Press EXE after each entry.

DISPLAY/WRITE

a–b.

$x(x+2)(x-3) = x^3 - x^2 - 6x$

$(x-1)(x+5)(x+2) = x^3 + 6x^2 + 3x - 10$

---

on Resources

 eWorkbook Topic 16 Workbook (worksheets, code puzzle and project) (ewbk-2040)

 Interactivity Adding and subtracting polynomials (int-6204)

Exercise 16.3 Adding, subtracting and multiplying polynomials **learn**on

16.3 Quick quiz on	16.3 Exercise

Individual pathways

■ PRACTISE	■ CONSOLIDATE	■ MASTER
1, 3, 5, 7, 9, 13, 16	2, 4, 6, 10, 14, 17	8, 11, 12, 15, 18

Fluency

1. **WE3a** Simplify each of the following.
   a. $(x^4 + x^3 - x^2 + 4) + (x^3 - 14)$
   b. $(x^6 + x^4 - 3x^3 + 6x^2) + (x^4 + 3x^2 + 5)$
   c. $(x^3 + x^2 + 2x - 4) + (4x^3 - 6x^2 + 5x - 9)$
   d. $(2x^4 - 3x^3 + 7x^2 + 9) + (6x^3 + 5x^2 - 4x + 5)$
   e. $(15x^4 - 3x^2 + 4x - 7) + (x^5 - 2x^4 + 3x^2 - 4x - 3)$

2. **WE3b** Simplify each of the following.
   a. $(x^4 + x^3 + 4x^2 + 5x + 5) - (x^3 + 2x^2 + 3x + 1)$
   b. $(x^6 + x^3 + 1) - (x^5 - x^2 - 1)$
   c. $(5x^7 + 6x^5 - 4x^3 + 8x^2 + 5x - 3) - (6x^5 + 8x^2 - 3)$
   d. $(10x^4 - 5x^2 + 16x + 11) - (2x^2 - 4x + 6)$
   e. $(6x^3 + 5x^2 - 7x + 12) - (4x^3 - x^2 + 3x - 3)$

3. **WE4a** Expand and simplify each of the following.

   a. $x(x+6)(x+1)$

   b. $x(x-9)(x+2)$

   c. $x(x-3)(x+11)$

   d. $2x(x+2)(x+3)$

   e. $-3x(x-4)(x+4)$

4. Expand and simplify each of the following.

   a. $5x(x+8)(x+2)$ b. $x^2(x+4)$ c. $-2x^2(7-x)$

   d. $(5x)(-6x)(x+9)$ e. $-7x(x+4)^2$

WE4b For questions **5** to **10**, expand and simplify each of the following.

5. a. $(x+7)(x+2)(x+3)$ b. $(x-2)(x+4)(x-5)$ c. $(x-1)(x-4)(x+8)$

   d. $(x-1)(x-2)(x-3)$ e. $(x+6)(x-1)(x+1)$

6. a. $(x-7)(x+7)(x+5)$ b. $(x+11)(x+5)(x-12)$ c. $(x+5)(x-1)^2$

   d. $(x+2)(x-7)^2$ e. $(x+1)(x-1)(x+1)$

7. a. $(x-2)(x+7)(x+8)$ b. $(x+5)(3x-1)(x+4)$ c. $(4x-1)(x+3)(x-3)(x+1)$

   d. $(5x+3)(2x-3)(x-4)$ e. $(1-6x)(x+7)(x+5)$

8. a. $3x(7x-4)(x-4)(x+2)$ b. $-9x(1-2x)(3x+8)$ c. $(6x+5)(2x-7)^2$

   d. $(3-4x)(2-x)(5x+9)(x-1)$ e. $2(7+2x)(x+3)(x+4)$

Understanding

9. a. $(x+2)^3$ b. $(x+5)^3$ c. $(x-1)^3$

10. a. $(x-3)^4$ b. $(2x-6)^3$ c. $(3x+4)^4$

11. Simplify the expression $2(ax+b)-5(c-bx)$.

12. Expand and simplify the expression $(x+a)(x-b)(x^2-3bx+2a)$.

Reasoning

13. If $(x-3)^4 = ax^4+bx^3+cx^2+dx+e$, determine the values of a, b, c, d and e. Show your working.

14. Simplify the expression $(2x-3)^3-(4-3x)^2$.

15. Determine the difference in volume between a cube of side $\dfrac{3(x-1)}{2}$ and a cuboid whose sides are x, $(x+1)$ and $(2x+1)$. Show your working.

Problem solving

16. Determine the values of the pronumerals a and b if: $\dfrac{5x+1}{(x-1)(x+2)} \equiv \dfrac{a}{(x-1)} + \dfrac{b}{(x+2)}$

17. Evaluate the constants a, b and c if: $\dfrac{5x-7}{(x-1)(x+1)(x-2)} \equiv \dfrac{a}{(x-1)} + \dfrac{b}{(x+1)} + \dfrac{c}{(x-2)}$.

18. Write $\dfrac{3x-5}{(x^2+1)(x-1)}$ in the form $\dfrac{ax+b}{(x^2+1)} + \dfrac{c}{(x-1)}$ and hence determine the values of a, b and c.

LESSON
16.4 Long division of polynomials

LEARNING INTENTIONS

At the end of this lesson you should be able to:
- divide polynomials by linear expressions using long division
- determine the quotient and remainder when dividing polynomials by linear expressions.

▶ 16.4.1 Long division of polynomials

eles-4977

- The reverse of expanding is factorising (expressing a polynomial as a product of its linear factors).
- To factorise polynomials we need to use a form of long division.
- The following steps show how to divide a polynomial by a linear factor using long division.

Consider $(x^3 + 2x^2 - 13x + 10) \div (x - 3)$.		
Step 1	Write the division out using long division notation.	$x - 3 \overline{\smash{\big)}\, x^3 + 2x^2 - 13x + 10}$
Step 2	Consider the leading terms only. Determine how many times x goes into x^3.	$x - 3 \overline{\smash{\big)}\, x^3 + 2x^2 - 13x + 10}$
Step 3	x into x^3 goes x^2 times. Write x^2 above the x^2 term of the polynomial.	$\begin{array}{r} x^2 \\ x - 3 \overline{\smash{\big)}\, x^3 + 2x^2 - 13x + 10} \end{array}$
Step 4	Multiply the term at the top (x^2) by the linear factor $(x - 3)$: $x^2 \times (x - 3) = x^3 - 3x^2$. Write the result beneath the first two terms of the polynomial.	$\begin{array}{r} x^2 \\ x - 3 \overline{\smash{\big)}\, x^3 + 2x^2 - 13x + 10} \\ x^3 - 3x^2 \end{array}$
Step 5	Subtract the first two terms of the polynomial by the terms written below them. $x^3 - x^3 = 0$ and $2x^2 - (-3x^2) = 5x^2$	$\begin{array}{r} x^2 \\ x - 3 \overline{\smash{\big)}\, x^3 + 2x^2 - 13x + 10} \\ \underline{x^3 - 3x^2} \\ 5x^2 \end{array}$
Step 6	Bring the next term of the polynomial down to sit next to $5x^2$.	$\begin{array}{r} x^2 \\ x - 3 \overline{\smash{\big)}\, x^3 + 2x^2 - 13x + 10} \\ \underline{x^3 - 3x^2} \\ 5x^2 - 13x \end{array}$

- The process now restarts, looking at the newly created $5x^2 - 13x$ expression.

Step 7	Consider the leading terms only. Determine how many times x goes into $5x^2$.	$\begin{array}{r} x^2 \\ x - 3 \overline{\smash{\big)}\, x^3 + 2x^2 - 13x + 10} \\ \underline{x^3 - 3x^2} \\ 5x^2 - 13x \end{array}$
Step 8	x into $5x^2$ goes $5x$ times. Write $5x$ above the x term of the polynomial.	$\begin{array}{r} x^2 + 5x \\ x - 3 \overline{\smash{\big)}\, x^3 + 2x^2 - 13x + 10} \\ \underline{x^3 - 3x^2} \\ 5x^2 - 13x \end{array}$

Step 9	Multiply the term at the top ($5x$) by the linear factor ($x - 3$): $5x \times (x - 3) = 5x^2 - 15x$. Write the result beneath the two terms written in step 6.	$$\begin{array}{r} x^2 + 5x \\ x - 3 \overline{)x^3 + 2x^2 - 13x + 10} \\ \underline{x^3 - 3x^2} \\ 5x^2 - 13x \\ 5x^2 - 15x \end{array}$$
Step 10	Subtract the two terms of the polynomial by the terms written below them. $5x^2 - 5x^2 = 0$ and $(-13x) - (-15x) = 2x$	$$\begin{array}{r} x^2 + 5x \\ x - 3 \overline{)x^3 + 2x^2 - 13x + 10} \\ \underline{x^3 - 3x^2} \\ 5x^2 - 13x \\ \underline{5x^2 - 15x} \\ 2x \end{array}$$
Step 11	Bring the next term of the polynomial down to sit next to $2x$.	$$\begin{array}{r} x^2 + 5x \\ x - 3 \overline{)x^3 + 2x^2 - 13x + 10} \\ \underline{x^3 - 3x^2} \\ 5x^2 - 13x \\ \underline{5x^2 - 15x} \\ 2x + 10 \end{array}$$

- Once again the process restarts, looking at the newly created $2x + 10$ expression.

Step 12	Consider the leading terms only. Determine how many times x goes into $2x$.	$$\begin{array}{r} x^2 + 5x \\ x - 3 \overline{)x^3 + 2x^2 - 13x + 10} \\ \underline{x^3 - 3x^2} \\ 5x^2 - 13x \\ \underline{5x^2 - 15x} \\ 2x + 10 \end{array}$$
Step 13	x into $2x$ goes 2 times. Write 2 above the constant term of the polynomial.	$$\begin{array}{r} x^2 + 5x + 2 \\ x - 3 \overline{)x^3 + 2x^2 - 13x + 10} \\ \underline{x^3 - 3x^2} \\ 5x^2 - 13x \\ \underline{5x^2 - 15x} \\ 2x + 10 \end{array}$$
Step 14	Multiply the term at the top (2) by the linear factor ($x - 3$): $2 \times (x - 3) = 2x - 6$. Write the result beneath the two terms written in step 11.	$$\begin{array}{r} x^2 + 5x + 2 \\ x - 3 \overline{)x^3 + 2x^2 - 13x + 10} \\ \underline{x^3 - 3x^2} \\ 5x^2 - 13x \\ \underline{5x^2 - 15x} \\ 2x + 10 \\ 2x - 6 \end{array}$$

(continued)

(continued)

Step 15	Subtract the two terms of the polynomial by the terms written below them. $2x - 2x = 0$ and $10 - (-6) = 16$	$$\begin{array}{r} x^2 + 5x + 2 \\ x - 3 \overline{)\, x^3 + 2x^2 - 13x + 10\,} \\ \underline{x^3 - 3x^2} \\ 5x^2 - 13x \\ \underline{5x^2 - 15x} \\ 2x + 10 \\ \underline{2x - 6} \\ 16 \end{array}$$
Step 16	The division is now complete! The top line is the quotient (Q), and the bottom number is the remainder (R).	$$\begin{array}{r} x^2 + 5x + 2 \quad \longleftarrow \text{Quotient} \\ x - 3 \overline{)\, x^3 + 2x^2 - 13x + 10\,} \\ \underline{x^3 - 3x^2} \\ 5x^2 - 13x \\ \underline{5x^2 - 15x} \\ 2x + 10 \\ \underline{2x - 6} \\ 16 \longleftarrow \text{Remainder} \end{array}$$
Write the answer: $(x^3 + 2x^2 - 13x + 10) \div (x - 3) = x^2 + 5x + 2$ remainder 16		

WORKED EXAMPLE 5 Performing long division of cubic polynomials

Perform the following long divisions and state the quotient and remainder.

a. $\left(x^3 + 3x^2 + x + 9\right) \div (x + 2)$ b. $\left(x^3 - 4x^2 - 7x - 5\right) \div (x - 1)$

c. $\left(2x^3 + 6x^2 - 3x + 2\right) \div (x - 6)$

THINK

a. 1. Write the question in long division format.
   2. Perform the long division process.

WRITE

a.
$$\begin{array}{r} x^2 + x - 1 \longleftarrow Q \\ x + 2 \overline{)\, x^3 + 3x^2 + x + 9\,} \\ \underline{x^3 + 2x^2} \\ x^2 + x \\ \underline{x^2 + 2x} \\ -x + 9 \\ \underline{-x - 2} \\ 11 \longleftarrow R \end{array}$$

   3. Write the quotient and remainder.

Quotient is $x^2 + x - 1$; remainder is 11.

b. 1. Write the question in long division format.
   2. Perform the long division process.

b.
$$\begin{array}{r} x^2 - 3x - 10 \longleftarrow Q \\ x - 1 \overline{)\, x^3 - 4x^2 - 7x - 5\,} \\ \underline{x^3 - x^2} \downarrow \\ -3x^2 - 7x \\ \underline{-3x^2 + 3x} \downarrow \\ -10x - 5 \\ \underline{-10x + 10} \\ -15 \longleftarrow R \end{array}$$

   3. Write the quotient and remainder.

Quotient is $x^2 - 3x - 10$; remainder is -15.

c. 1. Write the question in long division format.

2. Perform the long division process.

c.

$$
\begin{array}{r}
2x^2 + 18x + 105 \longleftarrow Q \\
x - 6 \overline{\smash{)}\,2x^3 + 6x^2 - 3x + 2} \\
\underline{2x^3 - 12x^2} \\
18x^2 - \downarrow 3x \\
\underline{18x^2 - 108x} \\
105x + 2 \\
\underline{105x - 630} \\
632 \longleftarrow R
\end{array}
$$

3. Write the quotient and remainder.

Quotient is $2x^2 + 18x + 105$; remainder is 632.

WORKED EXAMPLE 6 Determining the quotient and remainder of a degree 4 polynomial

Determine the quotient and the remainder when $x^4 - 3x^3 + 2x^2 - 8$ is divided by the linear expression $x + 2$.

THINK	WRITE
1. Set out the long division with each polynomial in descending powers of x. If one of the powers of x is missing, include it with 0 as the coefficient.	$x + 2 \overline{\smash{)}\,x^4 - 3x^3 + 2x^2 + 0x - 8}$
2. Divide x into x^4 and write the result above.	$\begin{array}{r} x^3 \\ x + 2 \overline{\smash{)}\,x^4 - 3x^3 + 2x^2 + 0x - 8} \end{array}$
3. Multiply the result x^3 by $x + 2$ and write the result underneath.	$\begin{array}{r} x^3 \\ x + 2 \overline{\smash{)}\,x^4 - 3x^3 + 2x^2 + 0x - 8} \\ x^4 + 2x^3 \end{array}$
4. Subtract and bring down the remaining terms to complete the expression.	$\begin{array}{r} x^3 \\ x + 2 \overline{\smash{)}\,x^4 - 3x^3 + 2x^2 + 0x - 8} \\ \underline{x^4 + 2x^3} \\ -5x^3 + 2x^2 + 0x - 8 \end{array}$
5. Divide x into $-5x^3$ and write the result above. **6.** Continue this process to complete the long division.	$\begin{array}{r} x^3 - 5x^2 + 12x - 24 \\ x + 2 \overline{\smash{)}\,x^4 - 3x^3 + 2x^2 + 0x - 8} \\ \underline{x^4 + 2x^3} \\ -5x^3 + 2x^2 + 0x - 8 \\ \underline{-5x^3 - 10x^2} \\ 12x^2 + 0x - 8 \\ \underline{12x^2 + 24x} \\ -24x - 8 \\ \underline{-24x - 48} \\ 40 \end{array}$
7. The polynomial $x^3 - 5x^2 + 12x - 24$, at the top, is the quotient.	The quotient is $x^3 - 5x^2 + 12x - 24$.
8. The result of the final subtraction, 40, is the remainder.	The remainder is 40.

Exercise 16.4 Long division of polynomials   learn

| **16.4 Quick quiz**   on | **16.4 Exercise** |

Individual pathways

■ PRACTISE	■ CONSOLIDATE	■ MASTER
1, 3, 5, 9, 14, 17	2, 4, 7, 10, 12, 15, 18	6, 8, 11, 13, 16, 19

Fluency

1. **WE5a** Perform the following long divisions and state the quotient and remainder.
   a. $(x^3 + 4x^2 + 4x + 9) \div (x + 2)$
   b. $(x^3 + 2x^2 + 4x + 1) \div (x + 1)$
   c. $(x^3 + 6x^2 + 3x + 1) \div (x + 3)$
   d. $(x^3 + 3x^2 + x + 3) \div (x + 4)$

2. Perform the following long divisions and state the quotient and remainder.
   a. $(x^3 + 6x^2 + 2x + 2) \div (x + 2)$
   b. $(x^3 + x^2 + x + 3) \div (x + 1)$
   c. $(x^3 + 8x^2 + 5x + 4) \div (x + 8)$
   d. $(x^3 + x^2 + 4x + 1) \div (x + 2)$

3. **WE5b** State the quotient and remainder for each of the following.
   a. $(x^3 + 2x^2 - 5x - 9) \div (x - 2)$
   b. $(x^3 + x^2 + x + 9) \div (x - 3)$
   c. $(x^3 + x^2 - 9x - 5) \div (x - 2)$
   d. $(x^3 - 4x^2 + 10x - 2) \div (x - 1)$

4. State the quotient and remainder for each of the following.
   a. $(x^3 - 5x^2 + 3x - 8) \div (x - 3)$
   b. $(x^3 - 7x^2 + 9x - 7) \div (x - 1)$
   c. $(x^3 + 9x^2 + 2x - 1) \div (x - 5)$
   d. $(x^3 + 4x^2 - 5x - 4) \div (x - 4)$

WE5c For questions 5 to 8, divide the first polynomial by the second and state the quotient and remainder.

5. a. $3x^3 - x^2 + 6x + 5, x + 2$
   b. $4x^3 - 4x^2 + 10x - 4, x + 1$
   c. $2x^3 - 7x^2 + 9x + 1, x - 2$

6. a. $2x^3 + 8x^2 - 9x - 1, x + 4$
   b. $4x^3 - 10x^2 - 9x + 8, x - 3$
   c. $3x^3 + 16x^2 + 4x - 7, x + 5$

7. a. $6x^3 - 7x^2 + 4x + 4, 2x - 1$
   b. $6x^3 + 23x^2 + 2x - 31, 3x + 4$
   c. $8x^3 + 6x^2 - 39x - 13, 2x + 5$

8. a. $2x^3 - 15x^2 + 34x - 13, 2x - 7$
   b. $3x^3 + 5x^2 - 16x - 23, 3x + 2$
   c. $9x^3 - 6x^2 - 5x + 9, 3x - 4$

Understanding

For questions 9 to 11, state the quotient and remainder for each of the following.

9. a. $\dfrac{-x^3 - 6x^2 - 7x - 16}{x + 1}$
   b. $\dfrac{-3x^3 + 7x^2 + 10x - 15}{x - 3}$
   c. $\dfrac{-2x^3 + 9x^2 + 17x + 15}{2x + 1}$
   d. $\dfrac{4x^3 - 20x^2 + 23x - 2}{-2x + 3}$

10. a. $(x^3 - 3x + 1) \div (x + 1)$
    b. $(x^3 + 2x^2 - 7) \div (x + 2)$
    c. $(x^3 - 5x^2 + 2x) \div (x - 4)$
    d. $(-x^3 - 7x + 8) \div (x - 1)$

11. **a.** $(5x^2 + 13x + 1) \div (x + 3)$
    b. $(2x^3 + 8x^2 - 4) \div (x + 5)$
    c. $(-2x^3 - x + 2) \div (x - 2)$
    d. $(-4x^3 + 6x^2 + 2x) \div (2x + 1)$

12. **WE6** Determine the quotient and the remainder when each polynomial is divided by the linear expression given.

    a. $x^4 + x^3 + 3x^2 - 7x$, $x - 1$
    b. $x^4 - 13x^2 + 36$, $x - 2$
    c. $x^5 - 3x^3 + 4x + 3$, $x + 3$

13. Determine the quotient and the remainder when each polynomial is divided by the linear expression given.

    a. $2x^6 - x^4 + x^3 + 6x^2 - 5x$, $x + 2$
    b. $6x^4 - x^3 + 2x^2 - 4x$, $x - 3$
    c. $3x^4 - 6x^3 + 12x$, $3x + 1$

Reasoning

14. Determine the quotient and remainder when $3x^4 - 6x^3 + 12x + a$ is divided by $3x + 6$. Show your working.

15. Determine the quotient and remainder when $ax^2 + bx + c$ is divided by $(x - d)$. Show your working.

16. A birthday cake in the shape of a cube had side length $(x + p)$ cm. The cake was divided between $(x - p)$ guests. The left-over cake was used for lunch the next day. There were q^3 guests for lunch the next day and each received c^3 cm^3 of cake, which was then all finished.
    Determine an expression for q in terms of p and c. Show your working.

Problem solving

17. When $x^3 - 2x^2 + 4x + a$ is divided by $x - 1$ the remainder is zero. Use long division to determine the value of a.

18. When $x^3 + 3x^2 + a$ is divided by $x + 1$, the remainder is 8. Use long division to determine the value of a.

19. When $2x^2 + ax + b$ is divided by $x - 1$ the remainder is zero but when $2x^2 + ax + b$ is divided by $x - 2$ the remainder is 9. Use long division to determine the value of the pronumerals a and b.

LESSON
16.5 Polynomial values

16.5.1 Polynomial values

eles-4978

• Consider the polynomial $P(x) = x^3 - 5x^2 + x + 1$.
• The value of the polynomial when $x = 3$ is denoted by $P(3)$ and is found by substituting $x = 3$ into the equation in place of x, as shown.

$$P(3) = (3)^3 - 5(3)^2 + (3) + 1$$
$$P(3) = 27 - 5(9) + 3 + 1$$
$$P(3) = 27 - 45 + 4$$
$$P(3) = -14$$

WORKED EXAMPLE 7 Evaluating polynomials for values of x

If $P(x) = 2x^3 + x^2 - 3x - 4$, determine the value of:

a. $P(1)$ b. $P(-2)$ c. $P(a)$ d. $P(2b)$ e. $P(x+1)$.

THINK

a. 1. Write the expression.

   2. Replace x with 1.

   3. Simplify and write the answer.

b. 1. Write the expression.

   2. Replace x with -2.

   3. Simplify and write the answer.

c. 1. Write the expression.

   2. Replace x with a.

   3. No further simplification is possible.

d. 1. Write the expression.

   2. Replace x with $2b$.

   3. Simplify and write the answer.

e. 1. Write the expression.

   2. Replace x with $(x+1)$.

   3. Expand the right-hand side and collect like terms.

   4. Write the answer.

WRITE

a. $P(x) = 2x^3 + x^2 - 3x - 4$

$P(1) = 2(1)^3 + (1)^2 - 3(1) - 4$

$\begin{aligned} &= 2 + 1 - 3 - 4 \\ &= -4 \end{aligned}$

b. $P(x) = 2x^3 + x^2 - 3x - 4$

$P(-2) = 2(-2)^3 + (-2)^2 - 3(-2) - 4$

$\begin{aligned} &= 2(-8) + (4) + 6 - 4 \\ &= -16 + 4 + 6 - 4 \\ &= -10 \end{aligned}$

c. $P(x) = 2x^3 + x^2 - 3x - 4$

$P(a) = 2a^3 + a^2 - 3a - 4$

d. $P(x) = 2x^3 + x^2 - 3x - 4$

$P(2b) = 2(2b)^3 + (2b)^2 - 3(2b) - 4$

$\begin{aligned} &= 2\left(8b^3\right) + 4b^2 - 6b + 4 \\ &= 16b^3 + 4b^2 - 6b + 4 \end{aligned}$

e. $P(x) = 2x^3 + x^2 - 3x - 4$

$P(x+1) = 2(x+1)^3 + (x+1)^2 - 3(x+1) - 4$

$\begin{aligned} &= 2(x+1)(x+1)(x+1) + (x+1)(x+1) - 3(x+1) - 4 \\ &= 2(x+1)\left(x^2 + 2x + 1\right) + x^2 + 2x + 1 - 3x - 3 - 4 \\ &= 2\left(x^3 + 2x^2 + x + x^2 + 2x + 1\right) + x^2 - x - 6 \\ &= 2\left(x^3 + 3x^2 + 3x + 1\right) + x^2 - x - 6 \\ &= 2x^3 + 6x^2 + 6x + 2 + x^2 - x - 6 \end{aligned}$

$= 2x^3 + 7x^2 + 5x - 4$

TI	THINK	DISPLAY/WRITE	CASIO	THINK	DISPLAY/WRITE

a–e.

On a Calculator page, press:
- MENU
- 1: Actions
- 1: Define

Complete the entry lines as:
Define
$p(x) = 2x^3 + x^2 - 3x - 4$
$p(1)$
$p(-2)$
$p(a)$
$p(2b)$
$p(x + 1)$
Press ENTER after each entry.

a–e.

$P(1) = -4$

$P(-2) = -10$

$P(a) = 2a^3 + a^2 - 3a - 4$

$P(2b) = 16b^3 + 4b^2 - 6b - 4$

$P(x + 1) = 2x^3 + 7x^2 + 5x - 4$

a–e.

On the Main screen, tap:
- Action
- Command
- Define

Complete the entry lines as:
Define
$p(x) = 2x^3 + x^2 - 3x - 4$
$p(1)$
$p(-2)$
$p(a)$
$p(2b)$
$p(x + 1)$
expand $(p(x + 1))$
Press EXE after each entry.

a–e.

$P(1) = -4$
$P(-2) = -10$

$P(a) = 2a^3 + a^2 - 3a - 4$

$P(2b) = 16b^3 + 4b^2 - 6b - 4$

$P(x + 1) = 2x^3 + 7x^2 + 5x - 4$

 Resources

 eWorkbook Topic 16 Workbook (worksheets, code puzzle and project) (ewbk-2040)

Exercise 16.5 Polynomial values

learn

16.5 Quick quiz **on**	16.5 Exercise

Individual pathways

■ PRACTISE	■ CONSOLIDATE	■ MASTER
1, 3, 7, 11, 12, 15	2, 5, 8, 9, 13, 16	4, 6, 10, 14, 17

Fluency

WE7 For questions 1 to 6, $P(x) = 2x^3 - 3x^2 + 2x + 10$. Calculate the following.

1. a. $P(0)$　　　　　　　　　　**b.** $P(1)$

2. a. $P(2)$　　　　　　　　　　**b.** $P(3)$

3. a. $P(-1)$　　　　　　　　　**b.** $P(-2)$

4. a. $P(-3)$　　　　　　　　　**b.** $P(a)$

5. a. $P(2b)$　　　　　　　　　**b.** $P(x + 2)$

6. a. $P(x - 3)$　　　　　　　　**b.** $P(-4y)$

Understanding

7. For the polynomial $P(x) = x^3 + x^2 + x + 1$, calculate the following showing your full working.
   a. $P(1)$
   b. $P(2)$
   c. $P(-1)$
   d. The remainder when $P(x)$ is divided by $(x - 1)$.
   e. The remainder when $P(x)$ is divided by $(x - 2)$.
   f. The remainder when $P(x)$ is divided by $(x + 1)$.

8. For the polynomial $P(x) = x^3 + 2x^2 + 5x + 2$, calculate the following showing your full working.
   a. $P(1)$
   b. $P(2)$
   c. $P(-2)$
   d. The remainder when $P(x)$ is divided by $(x - 1)$.
   e. The remainder when $P(x)$ is divided by $(x - 2)$.
   f. The remainder when $P(x)$ is divided by $(x + 2)$.

9. For the polynomial $P(x) = x^3 - x^2 + 4x - 1$, calculate the following showing your full working.
   a. $P(1)$
   b. $P(2)$
   c. $P(-2)$
   d. The remainder when $P(x)$ is divided by $(x - 1)$.
   e. The remainder when $P(x)$ is divided by $(x - 2)$.
   f. The remainder when $P(x)$ is divided by $(x + 2)$.

10. For the polynomial $P(x) = x^3 - 4x^2 - 7x + 3$, calculate the following showing your full working.
    a. $P(1)$
    b. $P(-1)$
    c. $P(-2)$
    d. The remainder when $P(x)$ is divided by $(x - 1)$.
    e. The remainder when $P(x)$ is divided by $(x + 1)$.
    f. The remainder when $P(x)$ is divided by $(x + 2)$.

Reasoning

11. Copy and complete:
    a. A quick way of determining the remainder when $P(x)$ is divided by $(x + 8)$ is to calculate _____.
    b. A quick way of determining the remainder when $P(x)$ is divided by $(x - 7)$ is to calculate _____.
    c. A quick way of determining the remainder when $P(x)$ is divided by $(x - a)$ is to calculate _____.

12. If $P(x) = 2(x - 3)^5 + 1$, determine:
    a. $P(2)$
    b. $P(-2)$
    c. $P(a)$

13. If $P(x) = -2x^3 - 3x^2 + x + 3$, evaluate:
    a. $P(a) + 1$
    b. $P(a + 1)$.

14. When $x^2 + bx + 2$ is divided by $(x - 1)$, the remainder is $b^2 - 4b + 7$. Determine the possible values of b.

Problem solving

15. If $P(x) = 2x^3 - 3x^2 + 4x + c$, determine the value of c if $P(2) = 20$.

16. If $P(x) = 3x^3 - 2x^2 - x + c$ and $P(2) = 8P(1)$, calculate the value of c.

17. If $P(x) = 5x^2 + bx + c$ and $P(-1) = 12$ while $P(2) = 21$, determine the values of b and c.

LESSON
16.6 The remainder and factor theorems

LEARNING INTENTION

At the end of this lesson you should be able to:
- identify factors of polynomials using the factor theorem.

▶ 16.6.1 The remainder theorem

eles-4979

- In the previous exercise, you may have noticed that:
  the remainder when P(x) is divided by (x − a) is equal to P(a).
- This fact is summarised in the **remainder theorem**.

The remainder theorem

When $P(x)$ is divided by $(x - a)$, the remainder $R = P(a)$.

- If $P(x) = x^3 + x^2 + x + 1$ is divided by $(x - 2)$, the quotient is $x^2 + 3x + 7$ and the remainder is $P(2)$, which equals 15. That is:

$$\left(x^3 + x^2 + x + 1\right) \div (x - 2) = x^2 + 3x + 7 + \frac{15}{x - 2}$$

$$\text{and } \left(x^3 + x^2 + x + 1\right) = \left(x^2 + 3x + 7\right)(x - 2) + 15$$

Dividing a polynomial by a linear factor

If $P(x)$ is divided by $(x - a)$, the quotient is $Q(x)$ and the remainder is $R = P(a)$, we can write:

$$P(x) \div (x - a) = Q(x) + \frac{R}{(x - a)}$$

$$\Rightarrow P(x) = (x - a)\,Q(x) + R$$

WORKED EXAMPLE 8 Calculating remainders using the remainder theorem

Without actually dividing, determine the remainder when $x^3 - 7x^2 - 2x + 4$ is divided by:
a. $x - 3$ **b. $x + 6$.**

THINK	WRITE
a. 1. Name the polynomial.	**a.** Let $P(x) = x^3 - 7x^2 - 2x + 4$.
2. The remainder when $P(x)$ is divided by $(x - 3)$ is equal to $P(3)$.	$R = P(3)$ $= 3^3 - 7(3)^2 - 2(3) + 4$ $= 27 - 7(9) - 6 + 4$ $= 27 - 63 - 6 + 4$
3. Write the remainder.	$= -38$

b. 1. The remainder when $P(x)$ is divided by $(x+6)$ is equal to $P(-6)$.

b. $R = P(-6)$

$$= (-6)^3 - 7(-6)^2 - 2(-6) + 4$$
$$= -216 - 7(36) + 12 + 4$$
$$= 216 - 252 + 12 + 4$$

2. Write the remainder.

$$= -452$$

▶ 16.6.2 The factor theorem

eles-4980

- The remainder when 12 is divided by 4 is zero, since 4 is a factor of 12.
- Similarly, if the remainder (R) when $P(x)$ is divided by $(x-a)$ is zero, then $(x-a)$ is a factor of $P(x)$.
- Since $R = P(a)$, determine the value of a that makes $P(a) = 0$, then $(x-a)$ is a factor.

This is summarised in the **factor theorem**.

> ### The factor theorem
>
> **If $P(a) = 0$, then $(x-a)$ is a factor of $P(x)$.**

- $P(x)$ could be factorised as follows:

$$P(x) = (x-a)Q(x), \text{ where } Q(x) \text{ is 'the other' factor of } P(x).$$

WORKED EXAMPLE 9 Applying the factor theorem to determine constants

$(x-2)$ **is a factor of** $x^3 + kx^2 + x - 2$**. Determine the value of** k**.**

THINK	WRITE
1. Name the polynomial.	Let $P(x) = x^3 + kx^2 + x - 2$.
2. The remainder when $P(x)$ is divided by $(x-2)$ is equal to $P(2) = 0$.	$0 = P(2)$ $= 2^3 + k(2)^2 + 2 - 2$ $0 = 8 + 4k$
3. Solve for k and write its value.	$4k = -8$ $k = -2$

 Resources

 eWorkbook Topic 16 Workbook (worksheets, code puzzle and project) (ewbk-2040)

Exercise 16.6 The remainder and factor theorems

16.6 Quick quiz **on**	16.6 Exercise

Individual pathways

■ PRACTISE	■ CONSOLIDATE	■ MASTER
1, 3, 5, 8, 9, 16, 18, 21	2, 4, 6, 10, 11, 14, 17, 19, 22	7, 12, 13, 15, 20, 23, 24

Fluency

WE6 For questions **1** and **2**, without actually dividing, determine the remainder when $x^3 + 3x^2 - 10x - 24$ is divided by:

1. a. $x - 1$ b. $x + 2$ c. $x - 3$ d. $x + 5$

2. a. $x - 0$ b. $x - k$ c. $x + n$ d. $x + 3c$

For questions **3** to **7**, determine the remainder when the first polynomial is divided by the second without performing long division.

3. a. $x^3 + 2x^2 + 3x + 4$, $(x - 3)$ b. $x^3 - 4x^2 + 2x - 1$, $(x + 1)$

4. a. $x^3 + 3x^2 - 3x + 1$, $(x + 2)$ b. $x^3 - x^2 - 4x - 5$, $(x - 1)$

5. a. $x^3 + x^2 + 8$, $(x - 5)$ b. $-3x^3 - 2x^2 + x + 6$, $(x + 1)$

6. a. $-x^3 + 8$, $(x + 3)$ b. $x^3 - 3x^2 - 2$, $(x - 2)$

7. a. $2x^3 + 3x^2 + 6x + 3$, $(x + 5)$ b. $x^3 + 2x^2$, $(x - 7)$

Understanding

8. **WE9** The remainder when $x^3 + kx + 1$ is divided by $(x + 2)$ is -19. Calculate the value of k.

9. The remainder when $x^3 + 2x^2 + mx + 5$ is divided by $(x - 2)$ is 27. Determine the value of m.

10. The remainder when $x^3 - 3x^2 + 2x + n$ is divided by $(x - 1)$ is 1. Calculate the value of n.

11. The remainder when $ax^3 + 4x^2 - 2x + 1$ is divided by $(x - 3)$ is -23. Determine the value of a.

12. The remainder when $x^3 - bx^2 - 2x + 1$ is divided by $(x + 1)$ is 0. Calculate the value of b.

13. The remainder when $-4x^2 + 2x + 7$ is divided by $(x - c)$ is -5. Determine a possible whole number value of c.

14. The remainder when $x^2 - 3x + 1$ is divided by $(x + d)$ is 11. Calculate the possible values of d.

15. The remainder when $x^3 + ax^2 + bx + 1$ is divided by $(x - 5)$ is -14. When the cubic polynomial is divided by $(x + 1)$, the remainder is -2. Determine the values of a and b.

16. **MC** Answer the following. *Note:* There may be more than one correct answer.

    a. When $x^3 + 2x^2 - 5x - 5$ is divided by $(x + 2)$, the remainder is:

    A. -5 **B.** -2 **C.** 2   **D.** 5

    b. Choose a factor of $2x^3 + 15x^2 + 22x - 15$ from the following.

    A. $(x - 1)$ **B.** $(x - 2)$ **C.** $(x + 3)$ **D.** $(x + 5)$

    c. When $x^3 - 13x^2 + 48x - 36$ is divided by $(x - 1)$, the remainder is:

    A. -3 **B.** -2 **C.** -1 **D.** 0

    d. Select a factor of $x^3 - 5x^2 - 22x + 56$ from the following.

    A. $(x - 2)$ **B.** $(x + 2)$ **C.** $(x - 7)$ **D.** $(x + 4)$

17. Determine one factor of each of the following cubic polynomials.

a. $x^3 - 3x^2 + 3x - 1$

b. $x^3 - 7x^2 + 16x - 12$

c. $x^3 + x^2 - 8x - 12$

d. $x^3 + 3x^2 - 34x - 120$

Reasoning

For questions 18 and 19, without actually dividing, show that the first polynomial is exactly divisible by the second (that is, the second polynomial is a factor of the first).

18. a. $x^3 + 5x^2 + 2x - 8$, $(x - 1)$

b. $x^3 - 7x^2 - x + 7$, $(x - 7)$

c. $x^3 - 7x^2 + 4x + 12$, $(x - 2)$

d. $x^3 + 2x^2 - 9x - 18$, $(x + 2)$

19. a. $x^3 + 3x^2 - 9x - 27$, $(x + 3)$

b. $-x^3 + x^2 + 9x - 9$, $(x - 1)$

c. $-2x^3 + 9x^2 - x - 12$, $(x - 4)$

d. $3x^3 + 22x^2 + 37x + 10$, $(x + 5)$

20. Prove that each of the following is a linear factor of $x^3 + 4x^2 - 11x - 30$ by substituting values into the cubic function: $(x + 2)$, $(x - 3)$, $(x + 5)$.

Problem solving

21. When $(x^3 + ax^2 - 4x + 1)$ and $(x^3 - ax^2 + 8x - 7)$ are each divided by $(x - 2)$, the remainders are equal. Determine the value of a.

22. When $x^4 + ax^3 - 4x^2 + b$ and $x^3 - ax^2 - 7x + b$ are each divided by $(x - 2)$, the remainders are 26 and 8 respectively. Calculate the values of a and b.

23. Both $(x - 1)$ and $(x - 2)$ are factors of $P(x) = x^4 + ax^3 - 7x^2 + bx - 30$. Determine the values of a and b and the remaining two linear factors.

24. The remainder when $2x - 1$ is divided into $6x^3 - x^2 + 3x + k$ is the same as when it is divided into $4x^3 - 8x^2 - 5x + 2$. Calculate the value of k.

LESSON
16.7 Factorising polynomials

LEARNING INTENTIONS

At the end of this lesson you should be able to:
- factorise polynomials using long division
- factorise polynomials using short division or inspection.

▶ 16.7.1 Using long division

eles-4981

- Once one factor of a polynomial has been found (using the factor theorem as in the previous section), long division may be used to find other factors.
- In the case of a cubic polynomial, one — possibly two — other factors may be found.

WORKED EXAMPLE 10 Factorising polynomials using long division

Use long division to factorise the following.

a. $x^3 - 5x^2 - 2x + 24$ **b.** $x^3 - 19x + 30$ **c.** $-2x^3 - 8x^2 + 6x + 4$

THINK

WRITE

a. 1. Name the polynomial.

a. $P(x) = x^3 - 5x^2 - 2x + 24$

2. Look for a value of x such that $P(x) = 0$. For cubics containing a single x^3, try a factor of the constant term (24 in this case).
Try $P(1)$.
$P(1) \neq 0$, so $(x - 1)$ is not a factor.
Try $P(2)$.

$$P(1) = 1^3 - 5 \times 1^2 - 2 \times 1 + 24$$
$$= 1 - 5 - 2 + 24$$
$$= 18$$
$$\neq 0$$
$$P(2) = 2^3 - 5 \times 2^2 - 2 \times 2 + 24$$
$$= 8 - 20 - 4 + 24$$
$$\neq 0$$

$P(2) \neq 0$, so $(x - 2)$ is not a factor.
Try $P(-2)$.

$$P(-2) = (-2)^3 - 5 \times (-2)^2 - 2 \times (-2) + 24$$
$$= -8 - 20 + 4 + 24$$
$$= -28 + 28$$
$$= 0$$
So, $(x + 2)$ is a factor.

$P(-2)$ does equal 0, so $(x + 2)$ is a factor.

3. Divide $(x + 2)$ into $P(x)$ using long division to determine a quadratic factor.

$$\begin{array}{r} x^2 - 7x + 12 \\ x+2\overline{)x^3 - 5x^2 - 2x + 24} \\ \underline{x^3 + 2x^2} \\ -7x^2 - 2x \\ \underline{-7x^2 - 14x} \\ 12x + 24 \\ \underline{12x + 24} \\ 0 \end{array}$$

4. Write $P(x)$ as a product of the two factors found so far.

$P(x) = (x + 2)(x^2 - 7x + 12)$

5. Factorise the quadratic factor if possible.

$P(x) = (x + 2)(x - 3)(x - 4)$

b. 1. Name the polynomial.
Note: There is no x^2 term, so include $0x^2$.

b. $P(x) = x^3 - 19x + 30$
$P(x) = x^3 + 0x^2 - 19x + 30$

2. Look at the last term in $P(x)$, which is 30. This suggests it is worth trying $P(5)$ or $P(-5)$. Try $P(-5)$. $P(-5) = 0$ so $(x + 5)$ is a factor.

$$P(-5) = (-5)^3 - 19 \times (-5) + 30$$
$$= -125 + 95 + 30$$
$$= 0$$
So, $(x + 5)$ is a factor.

3. Divide $(x + 5)$ into $P(x)$ using long division to find a quadratic factor.

$$\begin{array}{r} x^2 - 5x + 6 \\ x+5\overline{)x^3 + 0x^2 - 19x + 30} \\ \underline{x^3 + 5x^2} \\ -5x^2 - 19x \\ \underline{-5x^2 - 25x} \\ 6x + 30 \\ \underline{6x + 30} \\ 0 \end{array}$$

4. Write $P(x)$ as a product of the two factors found so far.

$P(x) = (x + 5)(x^2 - 5x + 6)$

▶

5. Factorise the quadratic factor if possible. $\qquad$ $P(x) = (x+5)(x-2)(x-3)$

c. 1. Write the given polynomial. $\qquad$ **c.** Let $P(x) = -2x^3 - 8x^2 + 6x + 4$

2. Take out a common factor of -2. $\qquad\qquad\qquad$ $= -2(x^3 + 4x^2 - 3x - 2)$
(We could take out $+2$ as the common factor, but taking out -2 results in a positive leading term in the part still to be factorised.)

3. Let $Q(x) = x^3 + 4x^2 - 3x - 2$ $\qquad\qquad$ Let $Q(x) = x^3 + 4x^2 - 3x - 2$
(We have already used P earlier.)

4. Evaluate $Q(1)$. $\qquad\qquad\qquad\qquad\qquad$ $Q(1) = 1 + 4 - 3 - 2$
$Q(1) = 0$, so $(x-1)$ is a factor. $\qquad\qquad\qquad = 0$
$\qquad\qquad\qquad\qquad\qquad\qquad\qquad\qquad$ So, $(x-1)$ is a factor.

5. Divide $(x-1)$ into $Q(x)$ using long division to determine a quadratic factor.

$$
\begin{array}{r}
x^2 + 5x + 2 \\
x - 1 \overline{)\, x^3 + 4x^2 - 3x - 2} \\
\underline{x^3 - x^2} \\
5x^2 - 3x \\
\underline{5x^2 - 5x} \\
2x - 2 \\
\underline{2x - 2} \\
0
\end{array}
$$

6. Write the original polynomial $P(x)$ as a product of the factors found so far. $\qquad$ $P(x) = -2(x-1)(x^2 + 5x + 2)$

In this case, it is not possible to further factorise $P(x)$.

- *Note:* In some of these examples, $P(x)$ could have been factorised without long division by finding all three values of x that make $P(x) = 0$ (and hence the three factors).

▶ 16.7.2 Using short division, or by inspection

eles-4982

- Short division, or factorising by inspection, is a quicker method than long division.
- Consider $P(x) = x^3 + 2x^2 - 13x + 10$
  Using the factor theorem:
  $P(1) = 0$ so $(x-1)$ is a factor.
  $\therefore P(x) = (x-1)Q(x)$, where $Q(x)$ is the quadratic quotient.
  so, $P(x) = (x-1)(ax^2 + bx + c)$

Expanding and equating:

Term in x^3: $x \times ax^2$	$(x-1)(ax^2 + bx + c)$	**Equating with $P(x)$:** $ax^3 = x^3$ $\therefore a = 1$
Terms in x^2: $-1 \times ax^2 + bx^2$	$(x-1)(ax^2 + bx + c)$	**Equating with $P(x)$:** $-1 \times 1x^2 + bx^2 = 2x^2$ $-1 + b = 2$ $\therefore b = 3$
Constant term: $-1 \times c$	$(x-1)(ax^2 + bx + c)$	**Equating with $P(x)$:** $-c = 10$ $\therefore c = -10$

$$\therefore P(x) = (x-1)\left(x^2 + 3x - 10\right)$$

Factorising the quadratic gives:

$$P(x) = (x-1)(x+5)(x-2)$$

- *Note*: In this example, the values of a and c can be seen simply by inspecting $P(x)$.
  Hence, $P(x) = (x-1)\left(x^2 + bx - 10\right)$, leaving only the value of b unknown.
- The following worked example is a repeat of a previous one, but explains the use of short, rather than long, division.

WORKED EXAMPLE 11 Factorising using short division

Use short division to factorise $x^3 - 5x^2 - 2x + 24$.

THINK	WRITE
1. Name the polynomial.	Let $P(x) = x^3 - 5x^2 - 2x + 24$
2. Look for a value of x such that $P(x) = 0$. Try $P(-2)$.	$P(-2) = (-2)^3 - 5 \times (-2)^2 - 2 \times (-2) + 24$ $= -8 - 20 + 4 + 24$ $= -28 + 28$ $= 0$
$P(-2)$ does equal 0, so $(x+2)$ is a factor.	So, $(x+2)$ is a factor.
3. Look again at the original and equate the factorised form to the expanded form. The values of a and c can be determined simply by inspection. Since the coefficient of the x^3 term is 1, $a = 1$. Since the constant term is 24, $c = 12$.	$x^3 - 5x^2 - 2x + 24 = (x+2)\left(ax^2 + bx + c\right)$ $= (x+2)\left(x^2 + bx + 12\right)$
4. Expand the brackets and equate the coefficients of the x^2 terms. We can then solve for b.	$x^3 - 5x^2 - 2x + 24 = x^3 + bx^2 + 12x + 2x^2 + 2bx + 24$ $-5x^2 = bx^2 + 2x^2$ $-5x^2 = (b+2)x^2$ $b + 2 = -5$ $b = -7$ $P(x) = (x+2)\left(x^2 - 7x + 12\right)$
5. Factorise the expression in the second pair of brackets if possible.	$P(x) = (x+2)(x-3)(x-4)$

TI | THINK	DISPLAY/WRITE	CASIO | THINK	DISPLAY/WRITE

TI | THINK

On a Calculator page, press:
- MENU
- 2: Number
- 3: Factor

Complete the entry lines as:
factor $(x^3 - 5x^2 - 2x + 24)$
Then press ENTER.

DISPLAY/WRITE

$$x^3 - 5x^2 - 2x + 24$$
$$= (x - 4)(x - 3)(x + 2)$$

CASIO | THINK

On the Main screen, tap:
- Action
- Transformation
- factor
- factor

Complete the entry line as:
factor
$(x^3 - 5x^2 - 2x + 24)$
Then press EXE.

DISPLAY/WRITE

$$x^3 - 5x^2 - 2x + 24$$
$$= (x + 2)(x - 3)(x - 4)$$

 Resources

📋 **eWorkbook** Topic 16 Workbook (worksheets, code puzzle and project) (ewbk-2040)

Exercise 16.7 Factorising polynomials

learnon

16.7 Quick quiz on	**16.7 Exercise**

Individual pathways

■ PRACTISE	■ CONSOLIDATE	■ MASTER
1, 2, 8, 11, 15, 18	3, 4, 5, 9, 12, 16, 19	6, 7, 10, 13, 14, 17, 20

Fluency

WE10 For question **1** to **7** apply long division to factorise each dividend.

1. a. $x + 1 \overline{)x^3 + 10x^2 + 27x + 18}$
   b. $x + 2 \overline{)x^3 + 8x^2 + 17x + 10}$

2. a. $x + 9 \overline{)x^3 + 12x^2 + 29x + 18}$
   b. $x + 1 \overline{)x^3 + 8x^2 + 19x + 12}$

3. a. $x + 3 \overline{)x^3 + 14x^2 + 61x + 84}$
   b. $x + 7 \overline{)x^3 + 12x^2 + 41x + 42}$

4. a. $x + 2 \overline{)x^3 + 4x^2 + 5x + 2}$
   b. $x + 3 \overline{)x^3 + 7x^2 + 16x + 12}$

5. a. $x + 5 \overline{)x^3 + 14x^2 + 65x + 100}$
   b. $x \overline{)x^3 + 13x^2 + 40x}$

6. a. $x \overline{\smash{)}x^3 + 7x^2 + 12x}$ **b.** $x + 5 \overline{\smash{)}x^3 + 10x^2 + 25x}$

7. a. $x + 1 \overline{\smash{)}x^3 + 6x^2 + 5x}$ **b.** $x + 6 \overline{\smash{)}x^3 + 6x^2}$

WE11 For questions **8** to **10**, factorise the following as fully as possible.

8. a. $x^3 + x^2 - x - 1$ **b.** $x^3 - 2x^2 - x + 2$
 c. $x^3 + 7x^2 + 11x + 5$ **d.** $x^3 + x^2 - 8x - 12$

9. a. $x^3 + 9x^2 + 24x + 16$ **b.** $x^3 - 5x^2 - 4x + 20$
 c. $x^3 + 2x^2 - x - 2$ **d.** $x^3 - 7x - 6$

10. a. $x^3 + 8x^2 + 17x + 10$ **b.** $x^3 + x^2 - 9x - 9$
 c. $x^3 - x^2 - 8x + 12$ **d.** $x^3 + 9x^2 - 12x - 160$

Understanding

For questions **11** to **14**, factorise as fully as possible.

11. a. $2x^3 + 5x^2 - x - 6$
 b. $3x^3 + 14x^2 + 7x - 4$
 c. $3x^3 + 2x^2 - 12x - 8$
 d. $4x^3 + 35x^2 + 84x + 45$

12. a. $x^3 + x^2 + x + 1$
 b. $4x^3 + 16x^2 + 21x + 9$
 c. $6x^3 - 23x^2 + 26x - 8$
 d. $7x^3 + 12x^2 - 60x + 16$

13. a. $3x^3 - x^2 - 10x$ **b.** $4x^3 + 2x^2 - 2x$
 c. $3x^3 - 6x^2 - 24x$ **d.** $-2x^3 - 12x^2 - 18x$

14. a. $-x^3 - 7x^2 - 12x$ **b.** $-x^3 - 3x^2 + x + 3$
 c. $-2x^3 + 10x^2 - 12x$ **d.** $-5x^3 + 24x^2 - 36x + 16$

Reasoning

15. Factorise $x^4 - 9x^2 - 4x + 12$.

16. Factorise $-x^5 + 6x^4 + 11x^3 - 84x^2 - 28x + 240$.

17. Two of the factors of $x^3 + px^2 + qx + r$ are $(x + a)$ and $(x + b)$. Determine the third factor.

Problem solving

18. Factorise $x^5 - 5x^4 + 5x^3 + 5x^2 - 6x$.

19. $(x - 1)$ and $(x - 2)$ are known to be factors of $x^5 + ax^4 - 2x^3 + bx^2 + x - 2$. Determine the values of a and b and hence fully factorise this fifth-degree polynomial.

20. The polynomial $x^4 - 6x^3 + 13x^2 - 12x - 32$ has three factors, one of which is $x^2 - 3x + 8$. Evaluate the other two factors.

LESSON
16.8 Solving polynomial equations

LEARNING INTENTION

At the end of this lesson you should be able to:
- solve polynomial equations by applying the Null Factor Law.

16.8.1 Solving polynomial equations

eles-4983

- To solve the polynomial equation of the form $P(x) = 0$:

  Step 1: factorise $P(x)$
  Step 2: apply the Null Factor Law
  Step 3: state the solutions.

- The Null Factor Law applies to polynomial equations just as it does for quadratics.
- If $P(x)$ is of degree n, then $P(x) = 0$ has up to n solutions.
  Solving each of these equations produces the solutions (roots).

$$x = a \qquad x = b \qquad x = c.$$

- If $P(x) = k(lx - a)(mx - b)(nx - c) = 0$, then the solutions can be found as follows. Let each factor $= 0$:

$$lx - a = 0 \qquad mx - b = 0 \qquad nx - c = 0$$

Solving each of these equations produces the solutions.
Note: The coefficient k used in this example does not produce a solution because $k \neq 0$.

$$x = \frac{a}{l} \qquad x = \frac{b}{m} \qquad x = \frac{c}{n}.$$

WORKED EXAMPLE 12 Solving polynomial equations

Solve:
a. $x^3 = 9x$ b. $-2x^3 + 4x^2 + 70x = 0$ c. $2x^3 - 11x^2 + 18x - 9 = 0.$

THINK	WRITE
a. 1. Write the equation.	a. $x^3 = 9x$
2. Rearrange so all terms are on the left.	$x^3 - 9x = 0$
3. Take out a common factor of x.	$x(x^2 - 9) = 0$
4. Factorise the quadratic expression using the difference of two squares.	$x(x + 3)(x - 3) = 0$
5. Use the Null Factor Law to solve.	$x = 0,\ x + 3 = 0$ or $x - 3 = 0$
6. Write the values of x.	$x = 0,\ x = -3$ or $x = 3$
b. 1. Write the equation.	b. $-2x^3 + 4x^2 + 70x = 0$
2. Take out a common factor of $-2x$.	$-2x(x^2 - 2x - 35) = 0$
3. Factorise the quadratic expression.	$-2x(x - 7)(x + 5) = 0$
4. Use the Null Factor Law to solve.	$-2x = 0,\ x - 7 = 0$ or $x + 5 = 0$
5. Write the values of x.	$x = 0,\ x = 7$ or $x = -5$

---

c. **1.** Name the polynomial.

2. Use the factor theorem to determine a factor (search for a value a such that $P(a) = 0$). Consider factors of the constant term (that is, factors of 9 such as 1 and 3). The simplest value to try is 1.

3. Use long or short division to determine another factor of $P(x)$.

4. Factorise the quadratic factor.

5. Use the Null Factor Law to solve.

6. Write the values of x.

c. Let $P(x) = 2x^3 - 11x^2 + 18x - 9$.

$P(1) = 2 - 11 + 18 - 9$
$\qquad = 0$

So $(x - 1)$ is a factor.

$$
\begin{array}{r}
2x^2 - 9x + 9 \\
x - 1 \overline{)\,2x^3 - 11x^2 + 18x - 9} \\
\underline{2x^3 - 2x^2} \\
-9x^2 + 18x \\
\underline{-9x^2 + 9x} \\
9x - 9 \\
\underline{9x - 9} \\
0
\end{array}
$$

$P(x) = (x - 1)(2x^2 + 9x - 9)$
$P(x) = (x - 1)(2x - 3)(x - 3)$
$(x - 1)(2x - 3)(x - 3) = 0$

$x - 1 = 0,\ 2x - 3 = 0 \text{ or } x - 3 = 0$

$x = 1,\ x = \dfrac{3}{2} \text{ or } x = 3$

 Resources

 eWorkbook Topic 16 Workbook (worksheets, code puzzle and project) (ewbk-2040)

Exercise 16.8 Solving polynomial equations

learnon

| **16.8 Quick quiz** on | **16.8 Exercise** |

Individual pathways

■ PRACTISE	■ CONSOLIDATE	■ MASTER
1, 5, 8, 10, 14, 17	2, 6, 9, 11, 12, 15, 18	3, 4, 7, 13, 16, 19

Fluency

 For questions **1** to **4**, solve the following.

1. **a.** $x^3 - 4x = 0$ **b.** $x^3 - 16x = 0$ **c.** $2x^3 - 50x = 0$

2. **a.** $-3x^3 + 81 = 0$ **b.** $x^3 + 5x^2 = 0$ **c.** $x^3 - 2x^2 = 0$

3. a. $-4x^3 + 8x = 0$ **b.** $12x^3 + 3x^2 = 0$ **c.** $4x^2 - 20x^3 = 0$

4. a. $x^3 - 5x^2 + 6x = 0$ **b.** $x^3 - 8x^2 + 16x = 0$ **c.** $x^3 + 6x^2 = 7x$

WE12c For questions 5 to 7, apply the factor theorem to solve the following.

5. a. $x^3 - x^2 - 16x + 16 = 0$ **b.** $x^3 - 6x^2 - x + 30 = 0$
   c. $x^3 - x^2 - 25x + 25 = 0$ **d.** $x^3 + 4x^2 - 4x - 16 = 0$

6. a. $x^3 - 4x^2 + x + 6 = 0$ **b.** $x^3 - 4x^2 - 7x + 10 = 0$
   c. $x^3 + 6x^2 + 11x + 6 = 0$ **d.** $x^3 - 6x^2 - 15x + 100 = 0$

7. a. $x^3 - 3x^2 - 6x + 8 = 0$ **b.** $x^3 + 2x^2 - 29x + 42 = 0$
   c. $2x^3 + 15x^2 + 19x + 6 = 0$ **d.** $-4x^3 + 16x^2 - 9x - 9 = 0$

8. MC *Note:* There may be more than one correct answer.
   Select a solution to $x^3 - 7x^2 + 2x + 40 = 0$ from the following.

   A. 5 **B.** -4 **C.** -2 **D.** 1

9. MC A solution of $x^3 - 9x^2 + 15x + 25 = 0$ is $x = 5$. Select the number of other (distinct) solutions there are.

   A. 0 **B.** 1 **C.** 2 **D.** 3

Understanding

10. Solve $P(x) = 0$ for each of the following.

   a. $P(x) = x^3 + 4x^2 - 3x - 18$ **b.** $P(x) = 3x^3 - 13x^2 - 32x + 12$
   c. $P(x) = -x^3 + 12x - 16$ **d.** $P(x) = 8x^3 - 4x^2 - 32x - 20$

11. Solve $P(x) = 0$ for each of the following.

   a. $P(x) = x^4 + 2x^3 - 13x^2 - 14x + 24$ **b.** $P(x) = -72 - 42x + 19x^2 + 7x^3 - 2x^4$
   c. $P(x) = x^4 + 2x^3 - 7x^2 - 8x + 12$ **d.** $P(x) = 4x^4 + 12x^3 - 24x^2 - 32x$

12. Solve each of the following equations.

   a. $x^3 - 3x^2 - 6x + 8 = 0$ **b.** $x^3 + x^2 - 9x - 9 = 0$ **c.** $3x^4 + 3x^3 - 18x = 0$

13. Solve each of the following equations.

   a. $2x^4 + 10x^3 - 4x^2 - 48x = 0$ **b.** $2x^4 + x^3 - 14x^2 - 4x + 24 = 0$ **c.** $x^4 - 2x^3 + 1 = 0$

Reasoning

14. Solve for a if $x = 2$ is a solution of $ax^3 - 6x^2 + 3x - 4 = 0$.

15. Solve for p if $x = \dfrac{p}{2}$ is a solution of $x^3 - 5x^2 + 2x + 8 = 0$.

16. Show that it is possible for a cuboid of side lengths x cm, $(x - 1)$ cm and $(x + 2)$ cm to have a volume that is 4 cm³ less than twice the volume of a cube of side length x cm. Comment on the shape of such a cuboid.

Problem solving

17. Solve the following equation for x.

$$x^3 + 8 = x(5x - 2)$$

18. Solve the following equation for x.

$$2(x^3 + 5) = 13x(x - 1)$$

19. Solve the following equation for z.

$$z(z - 1)^3 = -2(z^3 - 5z^2 + z + 3)$$

LESSON
16.9 Review

16.9.1 Topic summary

Polynomials

- Polynomials are expressions with only non-negative integer powers.
- The degree of a polynomial is the highest power of the variable that it contains.
- The leading term is the term with highest power of the variable.
- Monic polynomials have a leading coefficient of 1.
- Polynomials are often denoted $P(x)$.
- The value of a polynomial can be determined by substituting the x-value into the expression.
  e.g. $P(x) = x^4 - 3x^2 + 8$ is a monic polynomial of degree 4. The coefficient of the x^2 term is -3 and the constant term is 8.

The remainder theorem

- When $P(x)$ is divided by $(x - a)$, the remainder, R, is given by:
  $$R = P(a)$$
  e.g.
  The remainder when $P(x) = 2x^3 + 3x^2 - 4x - 5$ is divided by $(x + 2)$ is $P(-2)$.
  $$\begin{aligned} P(-2) &= 2(-2)^3 + 3(-2)^2 - 4(-2) - 5 \\ &= 2(-8) + 3(4) + 8 - 5 \\ &= -16 + 12 + 8 - 5 \\ &= -1 \end{aligned}$$

POLYNOMIALS (10A)

Operations on polynomials

- To add or subtract polynomials simply add or subtract like terms.
  e.g.
  $$\begin{aligned} &(2x^3 - 5x + 1) + (-6x^3 + 8x^2 + 3x - 11) \\ &= (2x^3 - 6x^3) + 8x^2 + (-5x + 3x) + (1 - 11) \\ &= -4x^3 + 8x^2 - 2x - 10 \end{aligned}$$
- To multiply polynomials use the same methods as with quadratic expressions. Use FOIL and then simplify.
- For polynomials of degree 3 and higher you may need to use FOIL multiple times.
  e.g.
  $$\begin{aligned} (x + 3)(x - 1)(2x + 4) &= (x + 3)(2x^2 + 4x - 2x - 4) \\ &= (x + 3)(2x^2 + 2x - 4) \\ &= 2x^3 + 2x^2 - 4x + 6x^2 + 6x - 12 \\ &= 2x^3 + 8x^2 + 2x - 12 \end{aligned}$$

Factorising polynomials

- The factor theorem states the following:
  If $P(a) = 0$, then $(x - a)$ is a factor of $P(x)$.
- This can be used to find a factor, and then other factors can be found using the methods used to factorise quadratics.
  e.g.
  $$P(x) = x^3 - 2x^2 - 5x + 6$$
  $$\begin{aligned} P(1) &= (1)^3 - 2(1)^2 - 5(1) + 6 \\ &= 1 - 2 - 5 + 6 \\ &= 0 \end{aligned}$$
  Therefore, $(x - 1)$ is a factor of $P(x)$.
  $$\begin{aligned} P(x) &= (x - 1)(x^2 - x - 6) \\ &= (x - 1)(x - 3)(x + 2) \end{aligned}$$

Long division

- Polynomials can be divided using long division.
- The example below shows the division of $P(x) = x^3 + 2x^2 - 13x + 10$ by $x - 3$.

$$
\begin{array}{r}
x^2 + 5x + 2 \quad \leftarrow \text{Quotient} \\
x - 3 \overline{) \, x^3 + 2x^2 - 13x + 10} \\
\underline{x^3 - 3x} \\
5x^2 - 13x \\
\underline{5x^2 - 15x} \\
2x + 10 \\
\underline{2x - 6} \\
16 \quad \leftarrow \text{Remainder}
\end{array}
$$

- The result is:
  $$P(x) = (x - 3)(x^2 + 5x + 2) + 16$$

Solving polynomial equations

- To solve a polynomial equation:
  1. express in the form $P(x) = 0$
  2. factorise $P(x)$
  3. solve using the Null Factor Law.
  e.g.
  $$2x^3 + 7x^2 = 9$$
  $$2x^3 + 7x^2 - 9 = 0$$
  Let $P(x) = 2x^3 + 7x^2 - 9$.
  $$\begin{aligned} P(1) &= 2(1)^3 + 7(1)^2 - 9 \\ &= 2 + 7 - 9 \\ &= 0 \end{aligned}$$
  Therefore, $(x - 1)$ is a factor of $P(x)$.
  $$\begin{aligned} P(x) &= (x - 1)(2x^2 + 9x + 9) \\ &= (x - 1)(2x - 3)(x + 3) \end{aligned}$$
  Using the Null Factor Law:
  $$x = 1, x = -\frac{3}{2} \text{ or } x = -3$$

16.9.2 Success criteria

Tick the column to indicate that you have completed the lesson and how well you have understood it using the traffic light system.

(**Green:** I understand; **Yellow:** I can do it with help; **Red:** I do not understand)

Lesson	Success criteria	○	○	●
16.2	I can identify polynomial expressions.			
	I can state the degree, leading term and leading coefficient of a polynomial.			
16.3	I can add, subtract and multiply polynomial expressions.			
16.4	I can divide polynomials by linear expressions using long division.			
	I can determine the quotient and remainder when dividing polynomials by linear expressions.			
16.5	I can determine the value of a polynomial for a given value.			
16.6	I can identify factors of polynomials using the factor theorem.			
16.7	I can factorise polynomials using long division.			
	I can factorise polynomials using short division or inspection.			
16.8	I can solve polynomial equations by applying the Null Factor Law.			

16.9.3 Project

Investigating polynomials

A polynomial is a function involving the sum of integer powers of a variable (for example, $y = -4x^3 + 3x^2 - 4$). The highest power of the variable determines the degree of the polynomial. In the case of the given example, the degree is 3.

A polynomial of the first degree is a linear function (for example, $y = 3x - 8$), and a second-degree function is a quadratic (for example, $y = 5x^2 - 6x + 7$). Let us investigate how the degree of a polynomial affects the shape of its graph.

In order to simplify the graphing of these functions, the polynomials will be expressed in factor form. A graphics calculator or some other digital technology will make the graphing process less tedious.

It will be necessary to adjust the window of the calculator from time to time in order to capture the relevant features of the graph.

1. Consider the following polynomials.
   a. $y_1 = (x+1)$
   b. $y_2 = (x+1)(x-2)$
   c. $y_3 = (x+1)(x-2)(x+3)$
   d. $y_4 = (x+1)(x-2)(x+3)(x-4)$
   e. $y_5 = (x+1)(x-2)(x+3)(x-4)(x+5)$
   f. $y_6 = (x+1)(x-2)(x+3)(x-4)(x+5)(x-6)$

   For each of the functions:
   i. give the degree of the polynomial
   ii. sketch the graph, marking in the x-intercepts
   iii. describe how the degree of the polynomial affects the shape of the graph.
   Complete question 1 on a separate sheet of paper.

2. Let us now look at the effect that the exponent of each factor has on the shape of the graph of the polynomial. Consider the following functions.
   a. $y_1 = (x+1)(x-2)(x+3)$
   b. $y_2 = (x+1)^2(x-2)(x+3)$
   c. $y_3 = (x+1)^2(x-2)^2(x+3)$
   d. $y_4 = (x+1)^2(x-2)(x+3)^3$
   e. $y_5 = (x+1)^3(x-2)(x+3)^4$
   f. $y_6 = (x+1)^5(x-2)^3(x+3)^2$

   i. On a separate sheet of paper, draw a sketch of each of the polynomials, marking in the x-intercepts.
   ii. Explain how the power of the factor affects the behaviour of the graph at the x-intercept.

3. Create and draw a sketch of polynomials with the following given characteristics. Complete your graphs on a separate sheet of paper.
   a. A first-degree polynomial that:
      i. crosses the x-axis
      ii. does not cross the x-axis.

   b. A second-degree polynomial that:
      i. crosses the x-axis twice
      ii. touches the x-axis at one and only one point.

   c. A third-degree polynomial that crosses the x-axis:
      i. three times
      ii. twice
      iii. once.

   d. A fourth-degree polynomial that crosses the x-axis:
      i. four times
      ii. three times
      iii. twice
      iv. once.

4. Considering the powers of factors of polynomials, write a general statement outlining the conditions under which the graph of a polynomial will pass through the x-axis or just touch the x-axis.

Resources

Fluency

1. **MC** Select which of the following is *not* a polynomial.

   A. $x^3 - \dfrac{x^2}{3} + 7x - 1$

   B. $a^4 + 4a^3 + 2a + 2$

   C. $\sqrt{x^2 + 3x + 2}$

   D. 5

2. Consider the polynomial $y = -\dfrac{1}{7}x^4 + x^5 + 3$.

   a. State the degree of y.
   b. State the coefficient of x^4.
   c. State the constant term.
   d. Determine the leading term.

3. **MC** The expansion of $(x + 5)(x + 1)(x - 6)$ is:

   A. $x^3 - 30$
   B. $x^3 + 12x^2 - 31x + 30$
   C. $x^3 - 31x - 30$
   D. $x^3 + 5x^2 - 36x - 30$

4. **MC** $x^3 + 5x^2 + 3x - 9$ is the expansion of:

   A. $(x + 3)^3$
   B. $x(x + 3)(x - 3)$
   C. $(x - 1)(x + 3)^2$
   D. $(x - 1)(x + 1)(x + 3)$

5. Expand each of the following.

   a. $(x - 2)^2(x + 10)$
   b. $(x + 6)(x - 1)(x + 5)$
   c. $(x - 7)^3$
   d. $(5 - 2x)(1 + x)(x + 2)$

6. **MC** Consider the following long division.

$$
\begin{array}{r}
x^2 + x + 2 \\
x - 4 \overline{\smash{)}\,x^3 + 5x^2 + 6x - 1} \\
\underline{x^3 + 4x^2} \\
x^2 + 6x \\
\underline{x^2 + 4x} \\
2x - 1 \\
\underline{2x + 8} \\
-9
\end{array}
$$

   a. The quotient is:

   A. -9 B. 9 C. $x + 4$ D. $x^2 + x + 2$

   b. The remainder is:

   A. -9 B. 2 C. 4 D. $2x - 1$

7. Determine the quotient and remainder when the first polynomial is divided by the second in each case.

   a. $x^3 + 2x^2 - 16x - 3, (x + 2)$ b. $x^3 + 3x^2 - 13x - 7, (x - 3)$ c. $-x^3 + x^2 + 4x - 7, (x + 1)$

8. **MC** If $P(x) = x^3 - 3x^2 + 7x + 1$, then $P(-2)$ equals:

   A. -34 B. -33 C. -9 D. 9

9. If $P(x) = -3x^3 + 2x^2 + x - 4$, calculate:

    a. $P(1)$ **b.** $P(-4)$ **c.** $P(2a)$

10. Without dividing, determine the remainder when $x^3 + 3x^2 - 16x + 5$ is divided by $(x - 1)$.

11. Show that $(x + 3)$ is a factor of $x^3 - 2x^2 - 29x - 42$.

12. Factorise $x^3 + 4x^2 - 100x - 400$.

13. Solve:

    a. $(2x + 1)(x - 3)^2 = 0$ **b.** $x^3 - 9x^2 + 26x - 24 = 0$ **c.** $x^4 - 4x^3 - x^2 + 16x - 12 = 0$

Problem solving

14. Let $P(x) = a_n x^n + a_{n-1} x^{n-1} + \ldots + a_1 x + a_0$ be a polynomial where the coefficients are integers. Also let $P(w) = 0$ where w is an integer. Show that w is a factor of a_0.

15. Evaluate the area of a square whose sides are $(2x - 3)$ cm. Expand and simplify your answer. If the area is $16 \, \text{cm}^2$, determine the value of x.

16. A window is in the shape of a semicircle above a rectangle. The height of the window is $(6x + 1)$ cm and its width is $(2x + 2)$ cm.

    a. Evaluate the total area of the window.

    b. Expand and simplify your answer.

    c. Determine the perimeter of the window.

17. Answer the following questions.

    a. Determine the volume of a cube of side $(x + 4)$ cm.

    b. Evaluate the surface area of the cube.

    c. Determine the value of x for which the volume and surface are numerically equal.

    d. Calculate the value of x if the numerical value of the volume is 5 less than the numerical value of the surface area.

18. Determine the quotient and remainder when $mx^2 + nx + q$ is divided by $(x - p)$.

19. When $P(x)$ is divided by $(x - n)$, the quotient is $x^2 - 2x + n$ and the remainder is $(n + 1)$. Evaluate the value of $P(x)$.

on To test your understanding and knowledge of this topic, go to your learnON title at www.jacplus.com.au and complete the **post-test**.

Answers

Topic 16 Polynomials (10A)

16.1 Pre-test

1. a. 2 b. 3 c. 1 d. 0
2. A
3. $a = 2, b = -1$
4. D
5. A
6. a. 3 b. -3 c. -1 d. 2
7. A
8. $x^2 + x$
9. 9
10. 18
11. D
12. $b = 2$
13. B
14. C
15. $x = -1, -\dfrac{1}{2}$ or -6

16.2 Polynomials

1. a. i. 3 ii. x
   b. i. 7 ii. x
   c. i. 2 ii. x
2. a. i. 6 ii. x
   b. i. 8 ii. y
   c. i. 5 ii. u
3. a. i. 5 ii. e
   b. i. 1 ii. g
   c. i. 6 ii. f
4. a. Polynomial 3b b. Polynomial 1c
   c. Polynomial 1a d. Polynomials 1a, 2a and 2b
5. a. N b. P c. P
6. a. N b. N c. P
7. a. P b. N c. N
8. a. 3 b. x c. 4
   d. 5 e. $3x$ f. $-2x^3$
9. a. 7 b. w c. 7
   d. 0 e. -9 f. $6w^7$
10. a. 4 b. 1 c. x^4 d. 1
11. a. $4x^3 + 2x^2 - 10x + 18$ b. $3x^4 - 3x^3 - x^2 + 7x - 7$
    c. $5x^3 - 4x^2 - 13x - 6$
12. a. 6
    b. t
    c. 6
    d. Sample responses can be found in the worked solutions in the online resources.
    e. Sample responses can be found in the worked solutions in the online resources.

13. a. 5 units to the right of the origin
    b. 4 units to the right of the origin
    c. The body moves towards the origin, then away.
14. $a = 1, b = -4$
15. $a = 4, b = -6$
16. $a = \pm 3, b = \pm 2$

16.3 Adding, subtracting and multiplying polynomials

1. a. $x^4 + 2x^3 - x^2 - 10$
   b. $x^6 + 2x^4 - 3x^3 + 9x^2 + 5$
   c. $5x^3 - 5x^2 + 7x - 13$
   d. $2x^4 + 3x^3 + 12x^2 - 4x + 14$
   e. $x^5 + 13x^4 - 10$
2. a. $x^4 + 2x^2 + 2x + 4$ b. $x^6 - x^5 + x^3 + x^2 + 2$
   c. $5x^7 - 4x^3 + 5x$ d. $10x^4 - 7x^2 + 20x + 5$
   e. $2x^3 + 6x^2 - 10x + 15$
3. a. $x^3 + 7x^2 + 6x$ b. $x^3 - 7x^2 - 18x$
   c. $x^3 + 8x^2 - 33x$ d. $2x^3 + 10x^2 + 12x$
   e. $48x - 3x^3$
4. a. $5x^3 + 50x^2 + 80x$ b. $x^3 + 4x^2$
   c. $2x^3 - 14x^2$ d. $-30x^3 - 270x^2$
   e. $-7x^3 - 56x^2 - 112x$
5. a. $x^3 + 12x^2 + 41x + 42$ b. $x^3 - 3x^2 - 18x + 40$
   c. $x^3 + 3x^2 - 36x + 32$ d. $x^3 - 6x^2 + 11x - 6$
   e. $x^3 + 6x^2 - x - 6$
6. a. $x^3 + 5x^2 - 49x - 245$ b. $x^3 + 4x^2 - 137x - 660$
   c. $x^3 + 3x^2 - 9x + 5$ d. $x^3 - 12x^2 + 21x + 98$
   e. $x^3 + x^2 - x - 1$
7. a. $x^3 + 13x^2 + 26x - 112$
   b. $3x^3 + 26x^2 + 51x - 20$
   c. $4x^4 + 3x^3 - 37x^2 - 27x + 9$
   d. $10x^3 - 49x^2 + 27x + 36$
   e. $-6x^3 - 71x^2 - 198x + 35$
8. a. $21x^4 - 54x^3 - 144x^2 + 96x$
   b. $54x^3 + 117x^2 - 72x$
   c. $24x^3 - 148x^2 + 154x + 245$
   d. $20x^4 - 39x^3 - 50x^2 + 123x - 54$
   e. $4x^3 + 42x^2 + 146x + 168$
9. a. $x^3 + 6x^2 + 12x + 8$ b. $x^3 + 15x^2 + 75x + 125$
   c. $x^3 - 3x^2 + 3x - 1$
10. a. $x^4 - 12x^3 + 54x^2 - 108x + 81$
    b. $8x^3 - 72x^2 + 216x - 216$
    c. $81x^4 + 432x^3 + 864x^2 + 768x + 256$
11. $(2a + 5b)x + (2b - 5c)$
12. $x^4 + (a - 4b)x^3 + (2a - 4ab + 3b^2)x^2 +$
    $(2a^2 - 2ab + 3ab^2)x - 2a^2b$
13. $a = 1, b = -12, c = 54, d = -108, e = 81$
14. $8x^3 - 45x^2 + 78x - 43$
15. $\dfrac{1}{8}(11x^3 - 105x^2 + 73x - 27)$
16. $a = 2, b = 3$

17. $a = 1$, $b = -2$ and $c = 1$

18. $a = 1$, $b = 4$ and $c = -1$

16.4 Long division of polynomials

1. a. $x^2 + 2x$, 9 b. $x^2 + x + 3$, -2
   c. $x^2 + 3x - 6$, 19 d. $x^2 - x + 5$, -17

2. a. $x^2 + 4x - 6$, 14 b. $x^2 + 1$, 2
   c. $x^2 + 5$, -36 d. $x^2 - x + 6$, -11

3. a. $x^2 + 4x + 3$, -3 b. $x^2 + 4x + 13$, 48
   c. $x^2 + 3x - 3$, -11 d. $x^2 - 3x + 7$, 5

4. a. $x^2 - 2x - 3$, -17 b. $x^2 - 6x + 3$, -4
   c. $x^2 + 14x + 72$, 359 d. $x^2 + 8x + 27$, 104

5. a. $3x^2 - 7x + 20$, -35 b. $4x^2 - 8x + 18$, -22
   c. $2x^2 - 3x + 3$, 7

6. a. $2x^2 - 9$, 35 b. $4x^2 + 2x - 3$, -1
   c. $3x^2 + x - 1$, -2

7. a. $3x^2 - 2x + 1$, 5 b. $2x^2 + 5x - 6$, -7
   c. $4x^2 - 7x - 2$, -3

8. a. $x^2 - 4x + 3$, 8 b. $x^2 + x - 6$, -11
   c. $3x^2 + 2x + 1$, 13

9. a. $-x^2 - 5x - 2$, -14 b. $-3x^2 - 2x + 4$, -3
   c. $-x^2 + 5x + 6$, 9 d. $-2x^2 + 7x - 1$, 1

10. a. $x^2 - x - 2$, 3 b. x^2, -7
    c. $x^2 - x - 2$, -8 d. $-x^2 - x - 8$, 0

11. a. $5x - 2$, 7 b. $2x^2 - 2x + 10$, -54
    c. $-2x^2 - 4x - 9$, -16 d. $-2x^2 + 4x - 1$, 1

12. a. $x^3 + 2x^2 + 5x - 2$, -2
    b. $x^3 + 2x^2 - 9x - 18$, 0
    c. $x^4 - 3x^3 + 6x^2 - 18x + 58$, -171

13. a. $2x^5 - 4x^4 + 7x^3 - 13x^2 + 32x - 69$, 138
    b. $6x^3 + 17x^2 + 53x + 155$, 465
    c. $x^3 - \dfrac{7}{3}x^2 + \dfrac{7}{9}x + 3\dfrac{20}{27}$, $-3\dfrac{20}{27}$

14. Quotient: $x^3 - 4x^2 + 8x - 12$
    Remainder: $(a + 72)$

15. Quotient $= ax + (b + ad)$
    Remainder $= Rc + d(b + ad)$

16. $q = \dfrac{2p}{c}$

17. $a = -3$

18. $a = 6$

19. $a = 3$, $b = -5$

16.5 Polynomial values

1. a. 10 b. 11

2. a. 18 b. 43

3. a. 3 b. -22

4. a. -77 b. $2a^3 - 3a^2 + 2a + 10$

5. a. $16b^3 - 12b^2 + 4b + 10$
   b. $2x^3 + 9x^2 + 14x + 18$

6. a. $2x^3 - 21x^2 + 74x - 77$
   b. $-128y^3 - 48y^2 - 8y + 10$

7. a. 4 b. 15 c. 0
   d. 4 e. 15 f. 0

8. a. 10 b. 28 c. -8
   d. 10 e. 28 f. -8

9. a. 3 b. 11 c. -21
   d. 3 e. 11 f. -21

10. a. -7 b. 5 c. -7
    d. -7 e. 5 f. -7

11. a. $P(-8)$ b. $P(7)$ c. $P(a)$

12. a. -1 b. -6249 c. $2(a - 3)^5 + 1$

13. a. $-2a^3 - 3a^2 + a + 4$ b. $-2a^3 - 9a^2 - 11a - 1$

14. $b = 1, 4$

15. $c = 8$

16. $c = 2$

17. $b = -2$, $c = 5$

16.6 The remainder and factor theorems

1. a. -30 b. 0 c. 0 d. -24

2. a. -24 b. $k^3 + 3k^2 - 10k - 24$
   c. $-n^3 + 3n^2 + 10n - 24$ d. $-27c^3 + 27c^2 + 30c - 24$

3. a. 58 b. -8

4. a. 11 b. -9

5. a. 158 b. 6

6. a. 35 b. -6

7. a. -202 b. 441

8. 6

9. 3

10. 1

11. -2

12. 2

13. 2

14. $-5, 2$

15. $a = -5, b = -3$

16. a. D   b. C, D   c. D   d. A, C, D

17. a. $(x - 1)$ b. $(x - 3)$ or $(x - 2)$
    c. $(x - 3)$ or $(x + 2)$ d. $(x - 6)$ or $(x + 4)$ or $(x + 5)$

18. a–d. Sample responses can be found in the worked solutions in the online resources.

19. a–d. Sample responses can be found in the worked solutions in the online resources.

20. Sample responses can be found in the worked solutions in the online resources.

21. $a = 2$

22. $a = 3, b = 2$

23. $a = -5, b = 41$, $(x + 3)$ and $(x - 5)$

24. $k = -4$

16.7 Factorising polynomials

1. a. $(x + 1)(x + 3)(x + 6)$ b. $(x + 1)(x + 2)(x + 5)$

2. a. $(x + 1)(x + 2)(x + 9)$ b. $(x + 1)(x + 3)(x + 4)$

3. a. $(x + 3)(x + 4)(x + 7)$ b. $(x + 2)(x + 3)(x + 7)$

4. a. $(x+1)^2(x+2)$ b. $(x+2)^2(x+3)$

5. a. $(x+4)(x+5)^2$ b. $x(x+5)(x+8)$

6. a. $x(x+3)(x+4)$ b. $x(x+5)^2$

7. a. $x(x+1)(x+5)$ b. $x^2(x+6)$

8. a. $(x-1)(x+1)^2$ b. $(x-2)(x-1)(x+1)$
   c. $(x+1)^2(x+5)$ d. $(x-3)(x+2)^2$

9. a. $(x+1)(x+4)^2$ b. $(x-5)(x-2)(x+2)$
   c. $(x-1)(x+1)(x+2)$ d. $(x-3)(x+1)(x+2)$

10. a. $(x+1)(x+2)(x+5)$ b. $(x-3)(x+1)(x+3)$
    c. $(x-2)^2(x+3)$ d. $(x-4)(x+5)(x+8)$

11. a. $(2x+3)(x-1)(x+2)$ b. $(3x-1)(x+1)(x+4)$
    c. $(3x+2)(x-2)(x+2)$ d. $(4x+3)(x+3)(x+5)$

12. a. $(x+1)(x^2+1)$ b. $(x+1)(2x+3)^2$
    c. $(x-2)(2x-1)(3x-4)$ d. $(7x-2)(x-2)(x+4)$

13. a. $x(x-2)(3x+5)$ b. $2x(x+1)(2x-1)$
    c. $3x(x-4)(x+2)$ d. $-2x(x+3)^2$

14. a. $-x(x+4)(x+3)$ b. $-(x-1)(x+1)(x+3)$
    c. $-2x(x-3)(x-2)$ d. $-(x-2)^2(5x-4)$

15. $(x-1)(x+2)(x+2)(x-3)$

16. $-(x-2)(x+2)(x+3)(x-4)(x-5)$

17. $(x-p+(a+b))$

18. $x(x-1)(x+1)(x-2)(x-3)$

19. $a=-2, b=4, (x-1)^2(x+1)^2(x-2)$

20. The other two factors are $(x-4)$ and $(x+1)$.

16.8 Solving polynomial equations

1. a. $-2, 0, 2$ b. $-4, 0, 4$ c. $-5, 0, 5$

2. a. 3 b. $-5, 0$ c. $0, 2$

3. a. $-\sqrt{2}, 0, \sqrt{2}$ b. $-\dfrac{1}{4}, 0$ c. $0, \dfrac{1}{5}$

4. a. $0, 2, 3$ b. $0, 4$ c. $-7, 0, 1$

5. a. $-4, 1, 4$ b. $-2, 3, 5$
   c. $-5, 1, 5$ d. $-4, -2, 2$

6. a. $-1, 2, 3$ b. $-2, 1, 5$
   c. $-3, -2, -1$ d. $-4, 5$

7. a. $-2, 1, 4$ b. $-7, 2, 3$
   c. $-6, -\dfrac{1}{2}, -1$ d. $-\dfrac{1}{2}, \dfrac{3}{2}, 3$

8. A, C

9. B

10. a. $-3, 2$ b. $-2, \dfrac{1}{3}, 6$

    c. $-4, 2$ d. $-1, \dfrac{5}{2}$

11. a. $-4, -2, 1, 3$ b. $-2, -\dfrac{3}{2}, 3, 4$

    c. $-3, -2, 1, 2$ d. $-4, -1, 0, 2$

12. a. $-2, 1, 4$ b. $-3, -1, 3$ c. $-3, 0, 2$

13. a. $-4, -3, 0, 2$ b. $-2, \dfrac{3}{2}, 2$ c. $-1, 1$

14. 2.75

15. $-2, 4, 8$

16. $x = 1.48$ (to 2 decimal places)

17. $x = -1, 4$ and 2

18. $x = \dfrac{-1}{2}, 2, 5$

19. $z = -1, 1, -2$ and 3

Project

1. a. i. 1
   ii.

iii. The graph is linear and crosses the x-axis once (at $x = -1$).

b. i. 2
   ii.

iii. The graph is quadratic and crosses the x-axis twice (at $x = -1$ and $x = 2$).

c. i. 3
   ii.

iii. The graph is a curve and crosses the x-axis 3 times (at $x = -1$, $x = 2$ and $x = -3$).

d. i. 4
   ii.

iii. The graph is a curve and crosses the x-axis 4 times (at $x = -1$, $x = 2$, $x = -3$ and $x = 4$).

e. i. 5
   ii.

iii. The graph is a curve and crosses the x-axis 5 times (at $x = -1$, $x = 2$, $x = -3$, $x = 4$ and $x = -5$).

f. i. 6

ii.

iii. The graph is a curve and crosses the x-axis 6 times (at $x = -1$, $x = 2$, $x = -3$, $x = 4$, $x = -5$ and $x = 6$).

2. a. i.

ii. Each factor is raised to the power 1. The polynomial is of degree 3 and the graph crosses the x-axis in 3 places(-3, -1 and 2).

b. i.

ii. The factor $(x + 1)$ is raised to the power 2 while the other two factors are raised to the power 1. The power 2 causes the curve not to cross the x-axis at $x = -1$ but to be curved back on itself.

c. i.

ii. The power 2 on the two factors $(x + 1)$ and $(x - 2)$ causes the curve to be directed back on itself and not to cross the x-axis at those two points ($x = -1$ and $x = 2$).

d. i.

ii. The power 3 on the factor $(x + 3)$ causes the curve to run along the axis at that point then to cross the axis (at $x = -3$).

e. i.

ii. The power 3 on the factor $(x + 1)$ causes the curve to run along the axis at $x = -1$, then cross the axis. The power 4 on the factor $(x + 3)$ causes the curve to

be directed back on itself without crossing the axis at $x = -3$.

f. i.

ii. The power 5 on the factor $(x + 1)$ causes the curve to run along the axis at $x = -1$, then cross the axis.

3. Answers will vary. Possible answers could be as follows.

a. i. $y = 3x + 2$

ii. $y = 4$

b. i. $y = (x + 1)(x + 2)$

ii. $y = (x + 1)^2$

c. i. $y = (x + 1)(x + 2)(x + 3)$

ii. Not possible

iii. $y = (x + 1)^2(x + 2)$

d. i. $y = (x + 1)(x + 2)(x + 3)(x + 4)$

ii. Not possible

iii. $y = (x + 1)^2(x + 2)(x + 3)$, $y = (x + 1)^3(x + 2)$

iv. Not possible

4. If the power of the factor of a polynomial is an odd integer, the curve will pass through the x-axis. If the power is 1, the curve passes straight through. If the power is 3, 5. . ., the curve will run along the x-axis before passing through it. On the other hand, an even power of a factor causes the curve to just touch the x-axis then move back on the same side of the x-axis.

16.9 Review questions

1. C

2. a. 5   **b.** $-\dfrac{1}{7}$ **c.** 3   **d.** x^5

3. C

4. C

5. a. $x^3 + 6x^2 - 36x + 40$

b. $x^3 + 10x^2 + 19x - 30$

c. $x^3 - 21x^2 + 147x - 343$

d. $-2x^3 - x^2 + 11x + 10$

6. a. D   **b.** A

7. a. $x^2 - 16, 29$ **b.** $x^2 + 6x + 5, 8$ **c.** $-x^2 + 2x + 2, -9$

8. B

9. a. -4

b. 216

c. $-24a^3 + 8a^2 + 2a - 4$

10. -7

11. Sample responses can be found in the worked solutions in the online resources.

12. $(x - 10)(x + 4)(x + 10)$

13. a. $-\dfrac{1}{2}, 3$ **b.** 2, 3, 4   **c.** $-2, 1, 2, 3$

14. For example, given $P(x) = x^3 - x^2 - 34x - 56$ and $P(7) = 0 \Rightarrow (x - 7)$ is a factor and 7 is a factor of 56.

15. $4x^2 - 12x + 9; x = -\dfrac{1}{2}, \dfrac{7}{2}$

16. a. Area $= (\dfrac{1}{2}\pi + 10)x^2 + (\pi + 10)x + \dfrac{\pi}{2}$

b. Area $= (\dfrac{1}{2}\pi + 10)x^2 + (\pi + 10)x + \dfrac{\pi}{2}$

c. Perimeter $= (12 + \pi)x + (2 + \pi)$

17. a. $(x + 4)^3$

b. $6(x + 4)^2$

c. $x = 2$

d. $-3, \dfrac{-3 + 3\sqrt{5}}{2}$

18. $mx + (n + mp); q + p(n + mp)$

19. $x^3 - (2 + n)x^2 + 3nx - (n^2 - n - 1)$

17 Functions and relations (10A)

LESSON SEQUENCE

LESSON
17.1 Overview

Why learn this?

Functions and relations are broad and interesting topics of study. They are topics with many real-world applications and are very important topics to understand as you head towards higher studies in mathematics. You will have already seen some functions and relations in your maths classes; linear equations, quadratics and polynomials are all examples of functions, and circles are examples of relations.

In your previous study of quadratics you learned about graphs with an x^2 term, but have you wondered what a graph would look like if it had an x^3 term or an x^4 term? You will be learning about these and other graphs in this topic. Have you ever heard the phrases 'exponential growth' or 'exponential decay'? In this topic you will also learn exactly what these phrases mean and how to graph and interpret various exponential situations.

An understanding of how to apply and use functions and relations is relevant to many professionals. Medical teams working to map the spread of diseases, engineers designing complicated structures such as the Sydney Opera House, graphic designers creating a new logo, video game designers developing a new map for their game — all require the use and understanding of functions and relations.

This topic builds on what you already know and extends it into new areas of mathematics. By the end of this topic you will know all about different types of functions and relations, and how to graph them, interpret them and transform them.

Hey students! Bring these pages to life online

▶ Watch videos

🧩 Engage with interactivities

A+ Answer questions and check solutions

Find all this and MORE in jacPLUS ▶

Reading content and rich media, including interactivities and videos for every concept

Extra learning resources

Differentiated question sets

Questions with immediate feedback, and fully worked solutions to help students get unstuck

1. **MC** Choose the type of relation that the graph represents.
   A. One-to-one relation
   B. One-to-many relation
   C. Many-to-one relation
   D. Many-to-many relation

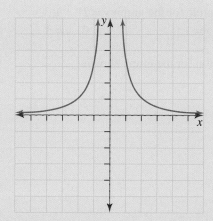

2. **MC** A function is a relation that is one-to-one or:
   A. many-to-one
   B. many-to-many
   C. one-to-many
   D. one-to-two

3. The graph below is a function.

State whether this is true or false.

4. **MC** Select the correct domain of the relation shown in the graph.
   A. $x \in R$
   B. $x \in [-1, 2]$
   C. $x \in [2, 8]$
   D. $x \in [0, 8]$

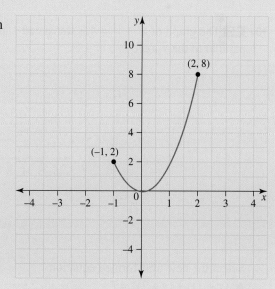

5. **MC** Select the correct range of the function $f(x) = \dfrac{1}{x-2} + 1$.

A. $y \in R$

B. $y \in R \backslash \{1\}$

C. $y \in R \backslash \{-1\}$

D. $y \in R \backslash \{2\}$

6. **MC** Select the correct equation of the inverse function $f(x) = (x+1)^2 - 2, x \le -1$.

A. $f^{-1}(x) = \sqrt{x+2} - 1, x \ge -1$

B. $f^{-1}(x) = \sqrt{x+2} - 1, x \le -2$

C. $f^{-1}(x) = -\sqrt{x+2} - 1, x \ge -1$

D. $f^{-1}(x) = -\sqrt{x+2} - 1, x \ge -2$

7. **MC** For a function to have an inverse function, it must be:

A. one-to-one

B. one-to-many

C. many-to-one

D. many-to-many

8. **MC** Select the correct asymptote for the graph $y = 2^{x-1} + 3$.

A. $y = -1$

B. $x = 1$

C. $x = 3$

D. $y = 3$

9. **MC** Select the correct equation for the graph shown.

A. $y = x(x-2)^2$

B. $y = x(x-2)$

C. $y = x(x+2)^2$

D. $y = x^2(x-2)$

10. **MC** The graph of $x^2 + y^2 = 4$ is translated 1 unit to the left parallel to the x-axis and 2 units upwards, parallel to the y-axis.

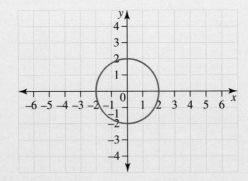

Select the new transformed equation.

A. $(x+1)^2 + (y+2)^2 = 7$

B. $(x+1)^2 + (y-2)^2 = 7$

C. $(x+1)^2 + (y+2)^2 = 4$

D. $(x+1)^2 + (y-2)^2 = 4$

11. **MC** Consider the function $f(x) = (x^2 - 9)(x - 2)(1 - x)$.
The graph of $f(x)$ is best represented by:

A.

B.

C.

D.

12. **MC** The quartic function has two x-intercepts at -1 and 4 and passes through the point $(0, -8)$.
Select the equation that best represents the function.

A. $f(x) = -\dfrac{1}{2}(x + 1)^2(x - 4)^2$

B. $f(x) = \dfrac{1}{2}(x + 1)^2(x - 4)^2$

C. $f(x) = -\dfrac{1}{2}(x - 1)^2(x + 4)^2$

D. $f(x) = -2(x + 1)^2(x - 4)^2$

13. **MC** If the graph shown is represented by the equation $y = P(x)$, select the correct graph for the equation $y = P(x - 1)$.

A.

B.

C.

D.

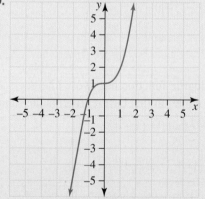

14. **MC** If the graph shown is represented by the equation $y = P(x)$, select the correct graph for the equation $y = -P(x) + 2$.

A.

B.

C.

D.

15. **MC** Consider the sketch of $y = P(x)$ and the graph of a transformation of $y = P(x)$.

Select the possible equation in terms of $P(x)$ for the transformation of $y = P(x)$.

A. $y = P(x) + 2$

B. $y = P(x - 2)$

C. $y = P(x + 2)$

D. $y = -P(x) + 2$

LESSON
17.2 Functions and relations

LEARNING INTENTIONS

At the end of this lesson you should be able to:
- identify the type of a relation
- determine the domain and range of a function or relation
- identify the points of intersection between two functions
- determine the inverse function of a one-to-one function.

17.2.1 Types of relations

eles-4984

- A **relation** is defined as a set of ordered pairs (x, y) which are related by a rule expressed as an algebraic equation. Examples of relations include $y = 3x$, $x^2 + y^2 = 4$ and $y = 2^x$.
- There are four types of relations, which are defined as follows.

Types of relations	Definition	Example
One-to-one relations	• A **one-to-one relation** exists if for any x-value there is only one corresponding y-value and vice versa.	
One-to-many relations	• A **one-to-many relation** exists if for any x-value there is more than one y-value, but for any y-value there is only one x-value.	
Many-to-one relations	• A **many-to-one relation** exists if there is more than one x-value for any y-value but for any x-value there is only one y-value.	
Many-to-many relations	• A **many-to-many relation** exists if there is more than one x-value for any y-value and vice versa.	

Determining the type of a relation

To determine the type of a relation:
- **Draw a horizontal line through the graph so that it cuts the graph the maximum number of times. Determine whether the number of cuts is one or many.**
- **Draw a vertical line through the graph so that it cuts the graph the maximum number of times. Determine whether the number of cuts is one or many.**

WORKED EXAMPLE 1 Identifying the types of relations

State the type of relation that each graph represents.

a.

b.

c.

THINK	WRITE
a. 1. Draw a horizontal line through the graph. The line cuts the graph **one** time.	a. One-to-_____ relation
2. Draw a vertical line through the graph. The line cuts the graph **many** times.	 One-to-many relation
b. 1. Draw a horizontal line through the graph. The line cuts the graph **one** time.	b. One-to-_____ relation
2. Draw a vertical line through the graph. The line cuts the graph **one** time.	 One-to-one relation
c. 1. Draw a horizontal line through the graph. The line cuts the graph **many** times.	c. Many-to-_____ relation

2. Draw a vertical line through the graph.
The line cuts the graph **one** time.

Many-to-many relation

17.2.2 Functions

- Relations that are one-to-one or many-to-one are called **functions**. That is, a function is a relation in which for any *x*-value there is at most one *y*-value.

Vertical line test

- To determine if a graph is a function, a vertical line is drawn anywhere on the graph. If it does not intersect with the curve more than once, then the graph is a function.
  For example, in each of the two graphs below, each vertical line intersects the graph only once.

1.

2.

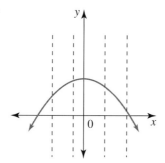

WORKED EXAMPLE 2 Identifying whether a relation is a function

State whether or not each of the following relations are functions.

a.

b.

THINK

a. It is possible for a vertical line to intersect with the curve more than once.

b. It is not possible for any vertical line to intersect with the curve more than once.

WRITE

a. Not a function

b. Function

Function notation

- Consider the relation $y = 2x$, which is a function.
  The y-values are determined from the x-values, so we say 'y is a function of x', which is abbreviated to $y = f(x)$.
  So, the rule $y = 2x$ can also be written as $f(x) = 2x$.
- For a given function $y = f(x)$, the value of y when $x = 1$ is written as $f(1)$, the value of y when $x = 5$ is written as $f(5)$, the value of y when $x = a$ as $f(a)$, etc.
- For the function $f(x) = 2x$:

  when $x = 1$, $y = f(1)$
  $$= 2 \times 1$$
  $$= 2.$$
  when $x = 2$, $y = f(2)$
  $$= 2 \times 2$$
  $$= 4, \text{ and so on.}$$

Domain and range

- The **domain** of a function is the set of all allowable values of x. It is sometimes referred to as the **maximal domain**.
- The **range** of a function is the set of y-values produced by the function.
- The following examples show how to determine the domain and range of some graphs.

Graph	Domain	Range
	The domain is all x values except 0. Domain: $x \in R \setminus \{0\}$	The range is all y values except 0. Range: $y \in R \setminus \{0\}$
	The domain is all x values. Domain: $x \in R$	The range is all y values that are greater than or equal to -3. Range: $y \geq -3$

If $f(x) = x^2 - 3$, calculate:

a. $f(1)$ b. $f(a)$ c. $3f(2a)$ d. $f(a) + f(b)$ e. $f(a+b)$.

THINK	WRITE
a. 1. Write the rule.	a. $f(x) = x^2 - 3$
2. Substitute $x = 1$ into the rule.	$f(1) = 1^2 - 3$
3. Simplify and write the answer.	$= 1 - 3$ $= -2$
b. 1. Write the rule.	b. $f(x) = x^2 - 3$
2. Substitute $x = a$ into the rule.	$f(a) = a^2 - 3$
c. 1. Write the rule.	c. $f(x) = x^2 - 3$
2. Substitute $x = 2a$ into the rule and simplify.	$f(2a) = (2a)^2 - 3$ $= 2^2 a^2 - 3$ $= 4a^2 - 3$
3. Multiply the answer by 3 and simplify.	$3f(2a) = 3(4a^2 - 3)$
4. Write the answer.	$= 12a^2 - 9$
d. 1. Write the rule.	d. $f(x) = x^2 - 3$
2. Evaluate $f(a)$.	$f(a) = a^2 - 3$
3. Evaluate $f(b)$.	$f(b) = b^2 - 3$
4. Evaluate $f(a) + f(b)$.	$f(a) + f(b) = a^2 - 3 + b^2 - 3$
5. Write the answer.	$= a^2 + b^2 - 6$
e. 1. Write the rule	e. $f(x) = x^2 - 3$
2. Evaluate $f(a+b)$.	$f(a+b) = (a+b)^2 - 3$ $= (a+b)(a+b) - 3$
3. Write the answer.	$= a^2 + 2ab + b^2 - 3$

TI	THINK	DISPLAY/WRITE	CASIO	THINK	DISPLAY/WRITE

a-e.

In a new document, on a Calculator page, press:
- MENU
- 1: Actions
- 1: Define

Complete the entry lines as:
Define $f(x) = x^2 - 3$
$f(1)$
$f(a)$
$3f(2a)$
$f(a) + f(b)$
$f(a + b)$

Press ENTER after each entry.

a-e.

$f(1) = -2$

$f(a) = a^2 - 3$

$3f(2a) = 3\left(4a^2 - 3\right)$

$f(a) + f(b) = a^2 + b^2 - 6$

$f(a + b) = a^2 + 2ab + b^2 - 3$

a-e.

On the Main screen, tap:
- Action
- Command
- Define

Complete the entry lines as:
Define $f(x) = x^2 - 3$
$f(1)$
$f(a)$
$3f(2a)$
$f(a) + f(b)$
$f(a + b)$
expand $\left((a + b)^2 - 3\right)$

Press EXE after each entry.

a-e.

$f(1) = -2$

$f(a) = a^2 - 3$

$3f(2a) = \left(12a^2 - 9\right)$

$f(a) + f(b) = a^2 + b^2 - 6$

$f(a + b) = a^2 + 2ab + b^2 - 3$

17.2.3 Identifying features of functions

eles-4986

Behaviour of functions as they approach extreme values

- We can identify features of certain functions by observing what happens to the function value (y value) when x approaches a very small value such as 0 ($x \to 0$) or a very large value such as ∞ ($x \to \infty$).

WORKED EXAMPLE 4 Identifying end behaviour of a function

Describe what happens to these functions as the value of x increases, that is, as $x \to \infty$.

a. $y = x^2$

b. $f(x) = 2^{-x}$

c. $f(x) = \dfrac{1}{x} + 1$

THINK	WRITE
a. 1. Write the function.	**a.** $y = x^2$
2. Substitute large x values into the function, such as $x = 10\,000$ and $x = 1\,000\,000$.	$f(10\,000) = 100\,000\,000$ $f(1\,000\,000) = 1 \times 10^{12}$
3. Write a conclusion.	As $x \to \infty$, $f(x)$ also increases; that is, $f(x) \to \infty$.
b. 1. Write the function.	**b.** $f(x) = 2^{-x}$
2. Substitute large x values into the function, such as $x = 10\,000$ and $x = 1\,000\,000$.	$f(10\,000) \approx 0$ $f(1\,000\,000) \approx 0$
3. Write a conclusion.	As $x \to \infty$, $f(x) \to 0$.
c. 1. Write the function.	**c.** $f(x) = \dfrac{1}{x} + 1$
2. Substitute large x values into the function, such as $x = 10\,000$ and $x = 1\,000\,000$.	$f(10\,000) = 1.0001$ $f(1\,000\,000) = 1.000\,001$
3. Write a conclusion.	As $x \to \infty$, $f(x) \to 1$.

Points of intersection

- A **point of intersection** between two functions is a point at which the two graphs cross paths.
- To determine points of intersection, equate the two graphs and solve to calculate the coordinates of the points of intersection.

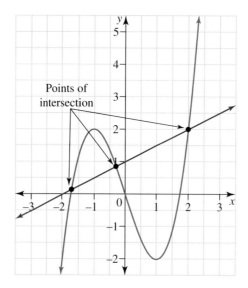

WORKED EXAMPLE 5 Determining points of intersection

Determine any points of intersection between $f(x) = 2x + 1$ and $g(x) = \dfrac{1}{x}$.

THINK	WRITE
1. Write the two equations.	$f(x) = 2x + 1$ $g(x) = \dfrac{1}{x}$
2. Points of intersection are common values between the two curves. To solve the equations simultaneously, equate both functions.	For points of intersection: $2x + 1 = \dfrac{1}{x}$
3. Rearrange the resulting equation and solve for x.	$2x^2 + x = 1$ $2x^2 + x - 1 = 0$ $(2x - 1)(x + 1) = 0$ $x = \dfrac{1}{2} \text{ or } -1$
4. Substitute the x values into either function to calculate the y values.	$f\left(\dfrac{1}{2}\right) = 2 \times \dfrac{1}{2} + 1 = 2$ $f(-1) = 2 \times -1 + 1 = -1$
5. Write the coordinates of the two points of intersection.	The points of intersection are $\left(\dfrac{1}{2}, 2\right)$ and $(-1, -1)$.

TI | THINK	DISPLAY/WRITE	CASIO | THINK	DISPLAY/WRITE

1. In a new problem, on a Calculator page, press:
- MENU
- 1: Actions
- 1: Define

Complete the entry lines as:
Define $f1(x) = 2x + 1$
Define $f2(x) = \dfrac{1}{x}$

Press ENTER after each entry. Then press:
- MENU
- 3: Algebra
- 1: Solve

Complete the entry lines as:
solve $(f1(x) = f2(x), x)$
$f1(-1)$
$f2\left(\dfrac{1}{2}\right)$

Press ENTER after each entry.

The points of intersection are $(-1, -1)$ and $\left(\dfrac{1}{2}, 2\right)$.

1. On the Main screen, complete the entry line as:

solve $\left(2x + 1 = \dfrac{1}{x},\ x\right)$

$2x + 1 \mid x = -1$

$2x + 1 \mid x = \dfrac{1}{2}$

Press EXE after each entry.

The points of intersection are $(-1, -1)$ and $\left(\dfrac{1}{2}, 2\right)$.

2. Alternatively, open a Graphs page in the current document. Since the functions have already been entered, just select the functions and press ENTER.
The graphs will be displayed.

2. Alternatively, tap the graphing icon.
Highlight each side of the equation separately and drag it down to the axis below.

3. To locate the points of intersection between the two functions, press:
- MENU
- 6: Analyze Graph
- 4: Intersection

Move the cursor to the left of the intersection point and press ENTER. Then move the cursor to the right of the intersection point and press ENTER. The intersection point is displayed. Repeat for the other intersection point.

The points of intersection are $(-1, -1)$ and $(0.5, 2)$.

3. To locate the points of intersection, tap:
- Analysis
- G-Solve
- Intersection

To find the other intersection point, tap the right arrow.

The points of intersection are $(-1, -1)$ and $(0.5, 2)$.

17.2.4 Inverse functions

eles-4987

- An **inverse function** is when a function is reflected across the line $y = x$.
- The inverse of $f(x)$ is written as $f^{-1}(x)$.
- To determine the equation of the inverse function, swap x and y and rearrange to make y the subject. For example, for $f(x) = 2x$:
  - rewrite the equation as $y = 2x$
  - swap x and y to get $x = 2y$
  - rearrange to make y the subject to get $y = \dfrac{1}{2}x$
  - rewrite the inverse in function notation $f^{-1}(x) = \dfrac{1}{2}x$.

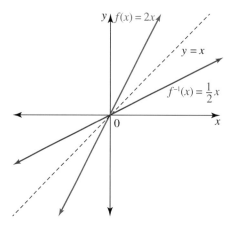

- Not all functions have inverses that are functions.
  For example, the inverse of the function $y = x^2$ is $x = y^2$ or $y = \pm\sqrt{x}$ which is a one-to-many relation.

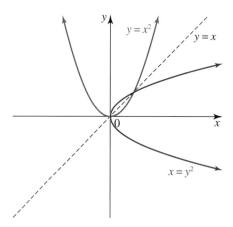

- For a function to have an inverse function it must be a one-to-one function.
- To determine whether a function has an inverse which is a function we can use the horizontal line test.
- Only when a function is one-to-one, will it have an inverse function. This can be tested by drawing a horizontal line to see if it cuts the graph once (has an inverse function) or more than once (does not have an inverse function).
- If a graph fails the horizontal line test its domain can be restricted so that it passes the test. The function with the restricted domain will then have an inverse function.

Graph	Result
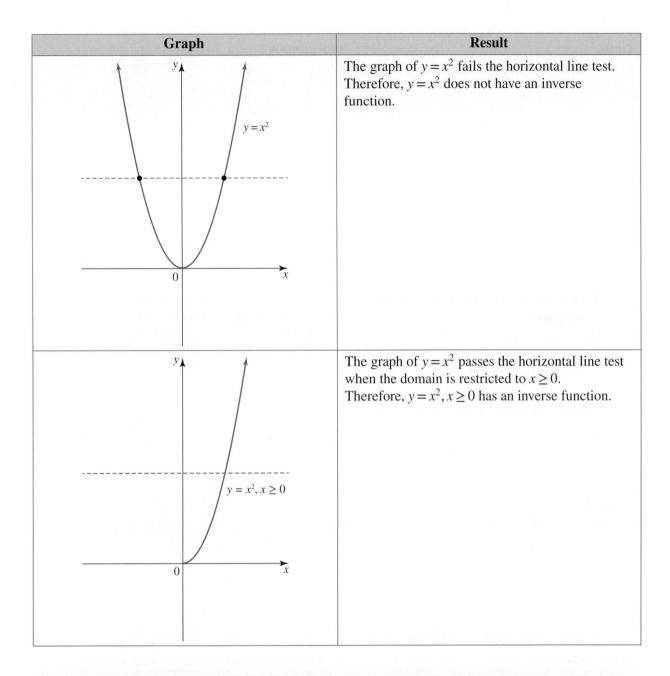	The graph of $y = x^2$ fails the horizontal line test. Therefore, $y = x^2$ does not have an inverse function.
	The graph of $y = x^2$ passes the horizontal line test when the domain is restricted to $x \geq 0$. Therefore, $y = x^2, x \geq 0$ has an inverse function.

WORKED EXAMPLE 6 Determining inverse functions of functions

Answer the following questions.
a. i. **Show that the function** $f(x) = x(x - 5)$ **will have not have an inverse function.**
   ii. **Suggest a restriction that would result in an inverse function.**
b. i. **Show that the function** $f(x) = x^2 + 4$, $x \geq 0$ **will have an inverse function.**
   ii. **Determine the equation of the inverse function.**

THINK	WRITE/DRAW

a. i. 1. Sketch the graph of $f(x) = x(x-5)$.

a. i.

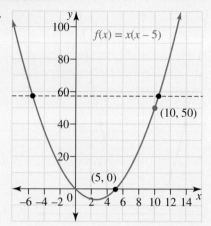

2. Draw a dotted horizontal line(s) through the graph.

The graph does not satisfy the horizontal line test, so the function $f(x) = x(x-5)$ will not have an inverse function.

ii. Apply a restriction to the function so that it will have an inverse.

ii. An inverse function will exist if
$f(x) = x(x-5), x \leq 2.5$ or
$f(x) = x(x-5), x \geq 2.5$.

b. i. 1. Sketch the graph of $f(x) = x^2 + 4, x \geq 0$.

b. i.

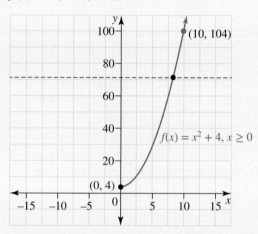

2. Draw a dotted horizontal line through the graph.

The graph satisfies the horizontal line test, so the function $f(x) = x^2 + 4, x \geq 0$ has an inverse function.

ii. 1. Determine the equation of the inverse function by swapping x and y.

ii. Let $y = x^2 + 4, x \geq 0$.
Swap x and y.
$x = y^2 + 4$
Make y the subject.
$$x = y^2 + 4$$
$$x - 4 = y^2$$
$$\sqrt{x-4} = y$$
$$y = \sqrt{x-4}$$

2. Write the answer in correct form, noting the domain.

The inverse of $f(x) = x^2 + 4$ is
$f^{-1}(x) = \sqrt{x-4}, x \geq 4$.

Exercise 17.2 Functions and relations   **learn**on

Individual pathways

■ PRACTISE	■ CONSOLIDATE	■ MASTER
1, 4, 7, 10, 12, 15, 17, 22	2, 5, 8, 11, 13, 18, 19, 23	3, 6, 9, 14, 16, 20, 21, 24

Fluency

WE1 For questions **1** to **3**, state the type of relation that each graph represents.

1. a.   b.   c.   d.

2. a.   b.   c.   d.

3. a.   b.   c.   d.

WE2 For questions **4** to **6**, answer the following questions.

4. a. Use the vertical line test to determine which of the relations in question **1** are functions.
    b. Determine which of these functions have inverses that are also functions.

5. a. Use the vertical line test to determine which of the relations in question **2** are functions.
    b. Determine which of these functions have inverses that are also functions.

6. **a.** Use the vertical line test to determine which of the relations in question **3** are functions.
   b. Determine which of these functions have inverses that are also functions.

7. **WE3** If $f(x) = 3x + 1$, calculate:
   a. $f(0)$ **b.** $f(2)$ **c.** $f(-2)$ **d.** $f(5)$

8. If $g(x) = \sqrt{x+4}$, calculate:
   a. $g(0)$ **b.** $g(-3)$ **c.** $g(5)$ **d.** $g(-4)$

9. If $g(x) = 4 - \dfrac{1}{x}$, calculate:

   a. $g(1)$ **b.** $g\left(\dfrac{1}{2}\right)$ **c.** $g\left(-\dfrac{1}{2}\right)$ **d.** $g\left(-\dfrac{1}{5}\right)$

10. If $f(x) = (x+3)^2$, calculate:
   a. $f(0)$ **b.** $f(-2)$ **c.** $f(1)$ **d.** $f(a)$

11. If $h(x) = \dfrac{24}{x}$, calculate:
   a. $h(2)$ **b.** $h(4)$ **c.** $h(-6)$ **d.** $h(12)$

Understanding

12. State which of the following relations are functions.
   a. **b.** $x^2 + y^2 = 9$ **c.** $y = 8x - 3$ **d.**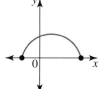

13. State which of the following relations are functions.
   a. $y = 2x + 1$ **b.** $y = x^2 + 2$ **c.** $y = 2^x$
   d. $x^2 + y^2 = 25$ **e.** $x^2 + 4x + y^2 + 6y = 14$ **f.** $y = -4x$

14. Given that $f(x) = \dfrac{10}{x} - x$, determine:

   a. $f(2)$ **b.** $f(-5)$ **c.** $f(2x)$
   d. $f(x^2)$ **e.** $f(x+3)$ **f.** $f(x-1)$

15. Calculate the value (or values) of x for which each function has the value given.
   a. $f(x) = 3x - 4, f(x) = 5$ **b.** $g(x) = x^2 - 2, g(x) = 7$ **c.** $f(x) = \dfrac{1}{x}, f(x) = 3$

16. Calculate the value (or values) of x for which each function has the value given.
   a. $h(x) = x^2 - 5x + 6, h(x) = 0$ **b.** $g(x) = x^2 + 3x, g(x) = 4$ **c.** $f(x) = \sqrt{8 - x}, f(x) = 3$

Reasoning

17. **WE4** Describe what happens to:

   a. $f(x) = x^2 + 3$ as $x \to \infty$ **b.** $f(x) = 2^x$ as $x \to -\infty$ **c.** $f(x) = \dfrac{1}{x}$ as $x \to \infty$

   d. $f(x) = x^3$ as $x \to -\infty$ **e.** $f(x) = -5^x$ as $x \to -\infty$

18. **WE5** Determine any points of intersection between the following curves.

a. $f(x) = 2x - 4$ and $g(x) = x^2 - 4$

b. $f(x) = -3x + 1$ and $g(x) = -\dfrac{2}{x}$

c. $f(x) = x^2 - 4$ and $g(x) = 4 - x^2$

d. $f(x) = \dfrac{3}{4}x - 6\dfrac{1}{4}$ and $x^2 + y^2 = 25$

19. Determine the equation of the inverse function of each of the following, placing restrictions on the original values of x as required.

a. $f(x) = 2x - 1$

b. $f(x) = x^2 - 3$

c. $f(x) = (x - 2)^2 + 4$

20. **WE6** Answer the following questions with full working.

a. Show that the function $f(x) = x(x - 2)$ will not have an inverse function.

b. Suggest a restriction that would result in an inverse function.

21. Answer the following questions with full working.

a. Show that the function $f(x) = -x^2 + 4, x \le 0$ will have an inverse function.

b. Determine the equation of the inverse function.

Problem solving

22. Determine the value(s) of for which:

a. $f(x) = x^2 + 7$ and $f(x) = 16$

b. $g(x) = \dfrac{1}{x - 2}$ and $g(x) = 3$

c. $h(x) = \sqrt{8 + x}$ and $h(x) = 6$.

23. Compare the graphs of the inverse functions $y = a^x$ and $y = \log_a x$, choosing various values for a. Explain why these graphs are inverses.

24. Consider the function defined by the rule $f : R \to R, \ f(x) = (x - 1)^2 + 2$.

a. State the range of the function.

b. Determine the type of mapping for the function.

c. Sketch the graph of the function stating where it cuts the y-axis and its turning point.

d. Select a domain where x is positive such that f is a one-to-one function.

e. Determine the inverse function. Give the domain and range of the inverse function.

f. Sketch the graph of the inverse function on the same set of axes used for part c.

g. Determine where f and the function $g(x) = x + 3$ intersect each other.

LESSON
17.3 Exponential functions

LEARNING INTENTIONS

At the end of this lesson you should be able to:
- sketch graphs of exponential functions
- determine the equation of an exponential function.

▶ 17.3.1 Exponential functions

eles-4988

- **Exponential functions** can be used to model many real situations involving natural growth and decay.

Exponential growth	Exponential decay
Exponential growth is when a quantity grows by a constant percentage in each fixed period of time. Examples of exponential growth include growth of investment at a certain rate of compound interest and growth in the number of cells in a bacterial colony.	**Exponential decay** is when a quantity decreases by a constant percentage in each fixed period of time. Examples of exponential decay include yearly loss of value of an item (called depreciation) and radioactive decay.
	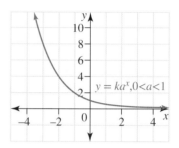

- Both exponential growth and decay can be modelled by exponential functions of the type $y = ka^x$ ($y = k \times a^x$). The difference is in the value of the base a. When $a > 1$, there is exponential growth and when $0 < a < 1$ there is exponential decay.

  When x is used to represent time, the value of k corresponds to the initial quantity that is growing or decaying.

WORKED EXAMPLE 7 Modelling bacterial growth using exponential functions

The number of bacteria, N, in a Petri dish after x hours is given by the equation $N = 50 \times 2^x$.
a. Identify the initial number of bacteria in the Petri dish.
b. Determine the number of bacteria in the Petri dish after 3 hours.
c. Draw the graph of the function of N against x.
d. Use the graph to estimate the length of time it will take for the initial number of bacteria to treble.

THINK

a. 1. Write the equation.

   2. Substitute $x = 0$ into the given formula and evaluate. (Notice that this is the value of x for equations of the form $y = k \times a^x$.)

   3. Write the answer in a sentence.

b. 1. Substitute $x = 3$ into the formula and evaluate.

   2. Write the answer in a sentence.

WRITE/DRAW

a. $N = 50 \times 2^x$

When $x = 0$, $N = 50 \times 2^0$
$= 50 \times 1$
$= 50$

The initial number of bacteria in the Petri dish is 50.

b. When $x = 3$, $N = 50 \times 2^3$
$= 50 \times 8$
$= 400$

After 3 hours there are 400 bacteria in the Petri dish.

c. 1. Calculate the value of N when $x = 1$ and $x = 2$.

c. At $x = 1$, $N = 50 \times 2^1$
$$= 50 \times 2$$
$$= 100$$

At $x = 2$, $N = 50 \times 2^2$
$$= 50 \times 4$$
$$= 200$$

2. Draw a set of axes, labelling the horizontal axis as x and the vertical axis as N.

3. Plot the points generated by the answers to parts **a**, **b** and **c 1**.

4. Join the points plotted with a smooth curve.

5. Label the graph.

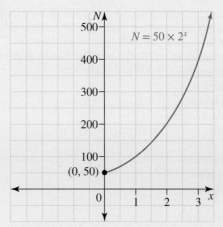

d. 1. Determine the number of bacteria required.

d. Number of bacteria $= 3 \times 50$
$$= 150$$

2. Draw a horizontal line from $N = 150$ to the curve and from this point draw a vertical line to the x-axis. This will help us to the estimate the time taken for the number of bacteria to treble.

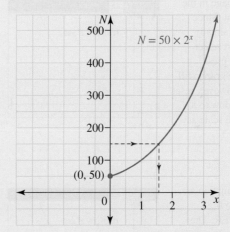

3. Write the answer in a sentence.

The time taken will be approximately 1.6 hours.

TI | THINK	**DISPLAY/WRITE**	**CASIO | THINK**	**DISPLAY/WRITE**

a-b.

In a new problem, on a Calculator page, press:
• MENU
• 1: Actions
• 1: Define

Complete the function entry line as:
Define
$f1(x) = 50 \times 2^x | x \geq 0$
Then press ENTER.
Note that the $x \geq 0$ needs to be included as the graph is only sketched for $x \geq 0$.
To determine the initial number of bacteria, complete the entry line as:
$f1(0)$
To determine the number of bacteria after 3 hours, complete the entry line as:
$f1(3)$
Press ENTER after each entry.

a-b.

Initially there are 50 bacteria present, and after 3 hours there are 400 bacteria present.

a-b.

On the Main screen, complete the entry line as:
Define
$f1(x) = 50 \times 2^x | x \geq 0$
Then press EXE.
Note that we need to include the $x \geq 0$ as we want to sketch only the graph for $x \geq 0$.
To determine the initial number of bacteria, type:
$f1(0)$
To determine the number of bacteria after 3 hours, type:
$f1(3)$

a-b.

Initially there are 50 bacteria present, and after 3 hours there are 400 bacteria present.

c.

Open a Graphs page in the current document. Since the function has already been entered, just select the function and press ENTER, and the graph will be displayed. However, reset the viewing window to a more appropriate scale as shown.

c.

c.

On the Graph & Table screen, complete the function entry line as:
$y1 = 50 \times 2^x | x \geq 0$
Tick the $y1$ box and tap the graphing icon.
Reset the viewing window to a more appropriate scale as shown.

c.

d.

Now enter the function as:
$f2(x) = 150 | x \geq 0$
Then press ENTER. The graph will be displayed. To find the point of intersection between the two graphs, press:
• MENU
• 6: Analyze Graph
• 4: Intersection
Move the cursor to the left of the intersection point and press ENTER. Then move the cursor to the right of the intersection point and press ENTER. The intersection point is displayed.

d.

The point of intersection is at $(1.58, 150)$. The initial number of bacteria will treble after 1.58 hours.

d.

Enter the equation by typing the entry line:
$y2 = 150 | x \geq 0$
Tick the $y2$ box and tap the graphing icon.
To find the point of intersection, tap:
• Analysis
• G-Solve
• Intersection

d.

The point of intersection is at $(1.58, 150)$. The initial number of bacteria will treble after 1.58 hours.

WORKED EXAMPLE 8 Determining the equation of an exponential function

A new computer costs \$3000. It is estimated that each year it will be losing 12% of the previous year's value.

a. Calculate the value, \$V, of the computer after the first year.
b. Calculate the value of the computer after the second year.
c. Determine the equation that relates the value of the computer to the number of years, n, it has been used.
d. Use your equation to calculate the value of the computer in 10 years' time.

THINK	WRITE
a. 1. State the original value of the computer.	**a.** $V_0 = 3000$
2. Since 12% of the value is being lost each year, the value of the computer will be 88% or $(100 - 12)\%$ of the previous year's value. Therefore, the value after the first year (V_1) is 88% of the original cost.	$V_1 = 88\%$ of 3000 $= 0.88 \times 3000$ $= 2640$
3. Write the answer in a sentence.	The value of the computer after 1 year is \$2640.
b. 1. The value of the computer after the second year, V_2, is 88% of the value after the first year.	**b.** $V_2 = 88\%$ of 2640 $= 0.88 \times 2640$ $= 2323.2$
2. Write the answer in a sentence.	The value of the computer after the second year is \$2323.20.
c. 1. The original value is V_0.	**c.** $V_0 = 3000$
2. The value after the first year, V_1, is obtained by multiplying the original value by 0.88.	$V_1 = 3000 \times 0.88$
3. The value after the second year, V_2, is obtained by multiplying V_1 by 0.88, or by multiplying the original value, V_0, by $(0.88)^2$.	$V_2 = (3000 \times 0.88) \times 0.88$ $= 3000 \times (0.88)^2$
4. The value after the third year, V_3, is obtained by multiplying V_2 by 0.88, or V_0 by $(0.88)^3$.	$V_3 = (3000 \times 0.88)^2 \times 0.88$ $= 3000 \times (0.88)^3$
5. By observing the pattern we can generalise as follows: the value after the nth year, V_n, can be obtained by multiplying the original value, V_0, by 0.88 n times; that is, by $(0.88)^n$.	$V_n = 3000 \times (0.88)^n$
d. 1. Substitute $n = 10$ into the equation obtained in part **c** to calculate the value of the computer after 10 years.	**d.** When $n = 10$, $V_{10} = 3000 \times (0.88)^{10}$ $= 835.50$
2. Write the answer in a sentence.	The value of the computer after 10 years is \$835.50.

17.3.2 Determining the equation of an exponential function using data

- If the initial value, k, is known, it is possible to substitute in a data point to determine the value of a and thereby find the equation of the function.
- Sometimes the relationship between the two variables closely resembles an exponential pattern, but cannot be described exactly by an exponential function. In such cases, part of the data are used to model the relationship with exponential growth or the decay function.

WORKED EXAMPLE 9 Determining the equation of an exponential function using data

The population of a certain city is shown in the table below.

Year	2000	2005	2010	2015	2020
Population (×1000)	128	170	232	316	412

Assume that the relationship between the population, P, and the year, x, can be modelled by the function $P = ka^x$, where x is the number of years after 2000.
Multiply the value of P by 1000 to determine the actual population.
a. State the value of k, which is the population, in thousands, at the start of the period.
b. Use a middle point in the data set to calculate the value of a, correct to 2 decimal places.
Hence, write the formula, connecting the population, P, with the number of years, x, since 2000.
c. For the years given, calculate the size of the population using the formula obtained in part b.
Compare it with the actual size of the population in those years.
d. Predict the population of the city in the years 2030 and 2035.

THINK

a. From the given table, state the value of k that corresponds to the population of the city in the year 2000.

b. 1. Write the given formula for the population of the city.

2. Replace the value of k with the value found in a.

3. Using a middle point of the data, replace x with the number of years since 1985 and P with the corresponding value

4. Solve the equation for a.

5. Round the answer to 2 decimal places.

6. Rewrite the formula with this value of a.

c. 1. Draw a table of values and enter the given years, the number of years since 2000, x, and the population for each year, P. Round values of P to the nearest whole number.

2. Comment on the closeness of the fit.

WRITE

a. $k = 128$

b. $P = ka^x$

$P = 128 \times a^x$

Middle point is (2010, 232).
When $x = 10, P = 232$, so $232 = 128 \times a^{10}$.

$a^{10} = \dfrac{232}{128}$

$a^{10} = 1.8125$

$a = \sqrt[10]{1.8125}$

$a = 1.0613...$

$a \approx 1.06$

So, $P = 128 \times (1.06)^x$.

c.

Year	2000	2005	2010	2015	2020
x	0	5	10	15	20
P	128	171	229	307	411

The values for the population obtained using the formula closely resemble the actual data.

1160 Jacaranda Maths Quest 10 + 10A

d. 1. Determine the value of x, the number of years after 2000.

d. For the year, $x = 30$.

2. Substitute this value of x into the formula and evaluate.

$P = 128 \times (1.06)^{30}$
$= 735.166\,87...$

3. Round to the nearest whole number.

$P \approx 735$

4. Write the answer in a sentence.

The predicted population for 2030 is 735 000.

5. Repeat for the year 2035.

For the year 2035, $x = 35$.
$P = 128 \times (1.06)^{35}$
$= 983.819...$
$P \approx 984$

6. Write the answer in a sentence.

The predicted population for 2035 is 984 000.

 Resources

 eWorkbook Topic 17 Workbook (worksheets, code puzzle and project) (ewbk-2041)

 Interactivity Exponential growth and decay (int-6211)

Exercise 17.3 Exponential functions

learnon

17.3 Quick quiz on	17.3 Exercise

Individual pathways

■ PRACTISE	■ CONSOLIDATE	■ MASTER
1, 4, 7, 10, 13	2, 5, 8, 11, 14	3, 6, 9, 12, 15

Fluency

1. **WE7** The number of micro-organisms, N, in a culture dish after x hours is given by the equation $N = 2000 \times 3^x$.

   a. Identify the initial number of micro-organisms in the dish.
   b. Determine the number of micro-organisms in a dish after 5 hours.
   c. Draw the graph of N against x.
   d. Use the graph to estimate the number of hours needed for the initial number of micro-organisms to quadruple.

2. The value of an investment (in dollars) after n years is given by $A = 5000 \times (1.075)^n$.

   a. Identify the size of the initial investment.
   b. Determine the value of the investment (to the nearest dollar) after 6 years.
   c. Draw the graph of A against n.
   d. Use the graph to estimate the number of years needed for the initial investment to double.

3. **MC** **a.** The function $P = 300 \times (0.89)^n$ represents an:

 A. exponential growth with the initial amount of 300
 B. exponential growth with the initial amount of 0.89
 C. exponential decay with the initial amount of 300
 D. exponential decay with the initial amount of 0.89

b. The relationship between two variables, A and t, is described by the function $A = 45 \times (1.095)^t$, where t is the time, in months, and A is the amount, in dollars. This function indicates:

 A. a monthly growth of $45
 B. a monthly growth of 9.5 cents
 C. a monthly growth of 1.095%
 D. a monthly growth of 9.5%

Understanding

4. **WE8** A new washing machine costs $950. It is estimated that each year it will be losing 7% of the previous year's value.

 a. Calculate the value of the machine after the first year.
 b. Calculate the value of the machine after the second year.
 c. Determine the equation that relates the value of the machine, $V, to the number of years, n, that it has been used.
 d. Use your equation to find the value of the machine in 12 years' time.

5. A certain radioactive element decays in such a way that every 50 years the amount present decreases by 15%. In 1900, 120 mg of the element was present.

 a. Calculate the amount present in 1950.
 b. Calculate the amount present in the year 2000.
 c. Determine the rule that connects the amount of the element present, A, with the number of 50-year intervals, t, since 1900.
 d. Calculate the amount present in the year 2010. Round your answer to 3 decimal places.
 e. Graph the function of A against t.
 f. Use the graph to estimate the half-life of this element (that is, the number of years needed for half the initial amount to decay).

6. When a shirt made of a certain fabric is washed, it loses 2% of its colour.

 a. Determine the percentage of colour that remains after:

 i. two washes
 ii. five washes.

 b. Write a function for the percentage of colour, C, remaining after w washings.
 c. Draw the graph of C against w.
 d. Use the graph to estimate the number of washes after which there is only 85% of the original colour left.

7. **WE9** The population of a certain country is shown in the table.

Year	Population (in millions)
2000	118
2005	130
2010	144
2015	160
2020	178

Assume that the relationship between the population, P and the year, n can be modelled by the formula $P = ka^n$, where n is the number of years since 2000.

a. State the value of k.

b. Use the middle point of the data set to calculate the value of a rounded to 2 decimal places. Hence, write the formula that connects the two variables, P and n.

c. For the years given in the table, determine the size of the population, using your formula. Compare the numbers obtained with the actual size of the population.

d. Predict the population of the country in the year 2045.

8. The temperature in a room (in degrees Celsius), recorded at 10-minute intervals after the air conditioner was turned on, is shown in the table below.

Time (min)	0	10	20	30	40
Temperature (°C)	32	26	21	18	17

Assume that the relationship between the temperature, T, and the time, t, can be modelled by the formula $T = ca^t$, where t is the time, in minutes, since the air conditioner was turned on.

a. State the value of c.

b. Use the middle point in the data set to determine the value of a to 2 decimal places.

c. Write the rule connecting T and t.

d. Using the rule, calculate the temperature in the room, 10, 20, 30 and 40 minutes after the air conditioner was turned on and compare your numbers with the recorded temperature. Comment on your findings. (Give answers correct to 1 decimal place.)

9. The population of a species of dogs (D) increases exponentially and is described by the equation $D = 60\left(1 - 0.6^t\right) + 3$, where t represents the time in years.

a. Calculate the initial number of dogs.

b. Calculate the number of dogs after 1 year.

c. Determine the time taken for the population to reach 50 dogs.

Reasoning

10. Fiona is investing \$20 000 in a fixed term deposit earning 6% p.a. interest. When Fiona has \$30 000 she intends to put a deposit on a house.

a. Determine an exponential function that will model the growth of Fiona's investment.

b. Graph this function.

c. Determine the length of time (correct to the nearest year) that it will take for Fiona's investment to grow to \$30 000.

d. Suppose Fiona had been able to invest at 8% p.a. Explain how much quicker Fiona's investment would have grown to the \$30 000 she needs.

e. Alvin has \$15 000 to invest. Determine the interest rate at which Alvin must invest his money, if his investment is to grow to \$30 000 in less than 8 years.

11. A Petri dish containing a bacteria colony was exposed to an antiseptic. The number of bacteria within the colony, B, over time, t, in hours is shown in the graph.

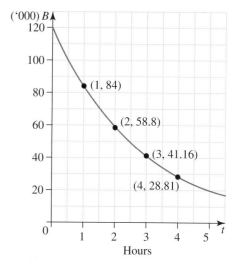

a. Using the graph, predict the number of bacteria in the Petri dish after 5 hours.

b. Using the points from the graph, show that if B (in thousands) can be modelled by the function $B = ab^t$, then $a = 120$ and $b = 0.7$.

c. After 8 hours, another type of antiseptic was added to the Petri dish. Within three hours, the number of bacteria in the Petri dish had decreased to 50.

If the number of bacteria decreased at a constant rate, show that the total of number of bacteria that had decreased within two hours was approximately 6700.

12. One hundred people were watching a fireworks display at a local park. As the fireworks were set off, more people started to arrive to see the show. The number of people, P, at time, t, after the start of the fireworks display, can be modelled by the function, $P = ab^t$.

a. If after 5 minutes there were approximately 249 people, show that the number of people arriving at the park to watch the fireworks increased by 20% each minute.

b. The fireworks display lasted for 40 minutes. After 40 minutes, people started to leave the park. The number of people leaving the park could be modelled by an exponential function.
15 minutes after the fireworks ceased there were only 700 people in the park.
Derive an exponential function that can determine the number of people, N, remaining in the park after the fireworks had finished at any time, m, in minutes.

Problem solving

13. A hot plate used as a camping stove is cooling down. The formula that describes this cooling pattern is $T = 500 \times 0.5^t$ where T is the temperature in degrees Celsius and t is the time in hours.

a. Identify the initial temperature of the stove.
b. Determine the temperature of the stove after 2 hours.
c. Decide when the stove will be cool enough to touch and give reasons.

14. The temperature in a greenhouse is monitored when the door is left open. The following measurements are taken.

Time (min)	0	5	10	15	20
Temperature (°C)	45	35	27	21	16

a. State the initial temperature of the greenhouse.
b. Determine an exponential equation to fit the collected data.
c. Evaluate the temperature after 30 minutes.
d. Explain whether the temperature will ever reach 0°C.

15. Carbon-14 decomposes in such a way that the amount present can be calculated using the equation $Q = Q_0(1 - 0.038)^t$, where Q is measured in milligrams and t in centuries.

    a. If there is 40 mg present initially, evaluate how much is present in:

      i. 10 years' time   **ii.** 2000 years' time.

    b. Determine how many years will it take for the amount to be less than 10 mg.

LESSON
17.4 Cubic functions

LEARNING INTENTIONS

At the end of this lesson you should be able to:
- plot the graph of a cubic function using a table of values
- sketch the graph of a cubic function by calculating its intercepts
- determine the equation of a cubic function by inspection.

17.4.1 Cubic functions

eles-4989

- **Cubic functions** are **polynomials** where the highest power of the variable is three or the product of pronumeral makes up three.
- Some examples of cubic functions are $y = x^3$, $y = (x + 1)(x - 2)(x + 3)$ and $y = 2x^2(4x - 1)$.
- The following worked examples show how the graphs of cubic functions can be created by plotting points.

WORKED EXAMPLE 10 Plotting a cubic function using a table of values

Plot the graph of $y = x^3 - 1$ by completing a table of values.

THINK

1. Prepare a table of values, taking x-values from -3 to 3. Fill in the table by substituting each x-value into the given equation to determine the corresponding y-value.

2. Draw a set of axes and plot the points from the table. Join them with a smooth curve.

WRITE/DRAW

x	-3	-2	-1	0	1	2	3
y	-28	-9	-2	-1	0	7	26

Plot the curve of $y = x(x-2)(x+2)$ by completing a table of values.

THINK

1. Prepare a table of values, taking x-values from -3 to 3. Fill in the table by substituting each x-value into the given equation.

2. Draw a set of axes and plot the points from the table. Join them with a smooth curve.

WRITE/DRAW

x	-3	-2	-1	0	1	2	3
y	-15	0	3	0	-3	0	15

17.4.2 Sketching cubic functions

eles-4990

- Graphs of cubic functions have either *two turning points* or *one point of inflection*.
  - These two types of graphs are shown. Note that a point of inflection is a point where the gradient of the graph changes from decreasing to increasing or vice versa.
  - For the purposes of this topic, we will only consider the points of inflection where the graph momentarily flattens out.

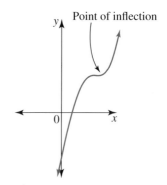

- Cubic functions can be positive or negative.
  - A positive cubic function will have a **positive** x^3 term and will have a general **upward** slope.
  - A negative cubic function will have a **negative** x^3 term and will have a general **downward** slope.

Example of a positive cubic function
$y = (x+2)(x-1)(x+3)$

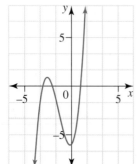

Example of a negative cubic function
$y = (x+2)(1-x)(x+3)$

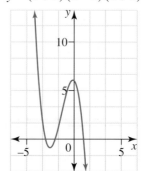

- To sketch the graph of a cubic function:
    1. determine the y-intercept by setting $x = 0$ and solving for y
    2. determine the x-intercepts by setting $y = 0$ and solving for x (you will need to use the Null Factor Law)
    3. draw the intercepts on the graph, then use those points to sketch the cubic graph.
- There are two special cases when sketching a cubic graph in factorised form:
    - Cubic functions of the form $y = (x - a)(x - b)^2$ will have a turning point on the x-axis at the point $(0, b)$.
    - Cubic functions of the form $y = (x - a)^3$ will have a point of inflection on the x-axis at the point $(0, a)$.

WORKED EXAMPLE 12 Sketching a cubic graph by determining intercepts

Sketch the following, showing all intercepts.

a. $y = (x - 2)(x - 3)(x + 5)$ b. $y = (x - 6)^2 (4 - x)$ c. $y = (x - 2)^3$

THINK

a. 1. Write the equation.

   2. The y-intercept occurs where $x = 0$. Substitute $x = 0$ into the equation.

   3. Solve $y = 0$ to calculate the x-intercepts.

   4. Combine the above steps to sketch.

b. 1. Write the equation.

   2. Substitute $x = 0$ to calculate the y-intercept.

   3. Solve $y = 0$ to calculate the x-intercepts.

   4. Combine all information and sketch the graph.
Note: The curve just touches the x-axis at $x = 6$. This occurs with a double factor such as $(x - 6)^2$.

WRITE/DRAW

a. $y = (x - 2)(x - 3)(x + 5)$

y-intercept: if $x = 0$,
$y = (-2)(-3)(5)$
$\quad = 30$
Point: $(0, 30)$

x-intercepts: if $y = 0$,
$x - 2 = 0, x - 3 = 0$ or $x + 5 = 0$
$x = 2, x = 3$ or $x = -5$
Points: $(2, 0), (3, 0), (-5, 0)$

b. $y = (x - 6)^2 (4 - x)$

y-intercept: if $x = 0$,
$y = (-6)^2 (4)$
$\quad = 144$
Point: $(0, 144)$

x-intercept: if $y = 0$,
$x - 6 = 0$ or $4 - x = 0$
$x = 6$ or $x = 4$
Point: $(6, 0), (4, 0)$

c. **1.** Write the equation.

c. $y = (x - 2)^3$

2. Substitute $x = 0$ to calculate the y-intercept.

y-intercept: if $x = 0$,
$y = (-2)^3$
$= -8$

3. Solve $y = 0$ to calculate the x-intercepts.

x-intercept: if $x = 0$,

$x - 2 = 0$
$x = 2$

4. Combine all information and sketch the graph.
Note: The point of inflection is at $x = 2$. This occurs with a triple factor such as $(x - 2)^3$.

| TI | THINK | DISPLAY/WRITE | CASIO | THINK | DISPLAY/WRITE |

a.

1. In a new problem, on a Calculator page, complete the entry lines as:
Define
$f1(x) = (x - 2)(x - 3)(x + 5)$
$f1(0)$
solve $(f1(x) = 0, x)$
Press ENTER after each entry.
Remember to include the implied multiplication sign between the brackets.

a.

2. Open a Graphs page in the current document. Since the function has already been entered, just select the function and press ENTER and the graph will be displayed. Reset the viewing window to a more appropriate scale as shown. This graph does cross the x-axis at three distinct points. Find all the axial intercepts as described earlier.

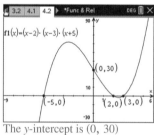

The y-intercept is $(0, 30)$ and the x-intercepts are $(-5, 0), (2, 0)$ and $(3, 0)$.

a.

1. On the Graph & Table screen, complete the function entry line as:
$y1 = (x - 2)(x - 3)(x + 5)$
Touch the graphing icon and resize the graph. Touch the Y = 0 icon for the x-intercepts. Use the right arrow to move between the x-intercepts.

a.

2. For the y-intercept, tap:
• Analysis
• G-Solve
• y-intercept

The y-intercept is $(0, 30)$ and the x-intercepts are $(-5, 0), (2, 0)$ and $(3, 0)$.

b.

1. On a Calculator page, complete the entry lines as:

Define

$f1(x) = (x-6)^2 (4-x)$

$f1(0)$

solve $(f1(x) = 0, x)$

Then press ENTER. Remember to include the implied multiplication sign between the brackets.

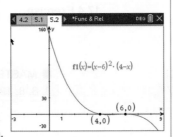

b.

1. On the Graph & Table screen, complete the function entry line as:

$y1 = (x-6)^2 (4-x)$

Touch the graphing icon and resize the graph. Touch the Y = 0 icon for the x-intercepts. Use the right arrow to move between the x-intercepts.

b.

2. Open a Graphs page in the current document. Since the function has already been entered, just select the function and press ENTER and the graph will be displayed. Reset the viewing window to a more appropriate scale as shown. This graph does cross the x-axis at two distinct points; however, this is not clear from the graph shown. Find all the axial intercepts as described earlier.

The y-intercept is $(0, 144)$ and the x-intercepts are $(4, 0)$ and $(6, 0)$.

2. For the y-intercept, tap:
- Analysis
- G-Solve
- y-intercept

The y-intercept is $(0, 144)$ and the x-intercepts are $(4, 0)$ and $(6, 0)$.

c.

1. In a new problem, on a Calculator page, complete the entry lines as:

Define $f1(x) = (x-2)^3$

$f1(0)$

solve $(f1(x) = 0, x)$

Press ENTER after each entry.

c.

1. On the Graph & Table screen, complete the function entry as:

$y1 = (x-2)^3$

Touch the graphing icon and resize the graph. Touch the Y = 0 icon for the x-intercepts. Use the right arrow to move between the x-intercepts.

c.

2. Open a Graphs page in the current document. Since the function has already been entered, just select the function and press ENTER and the graph will be displayed. Reset the viewing window to a more appropriate scale as shown. Find all the axial intercepts as described earlier.

The y-intercept is $(0, -8)$ and the x-intercept is $(2, 0)$. For this example there is only one x-intercept as it is a triple factor; this point is called a point of inflection.

2. For the y-intercept, tap:
- Analysis
- G-Solve
- y-intercept

The y-intercept is $(0, -8)$ and the x-intercept is $(2, 0)$. For this example there is only one x-intercept as it is a triple factor; this point is called a point of inflection.

Exercise 17.4 Cubic functions

learn on

17.4 Quick quiz on	17.4 Exercise

Individual pathways

■ PRACTISE	■ CONSOLIDATE	■ MASTER
1, 4, 9, 10, 13, 16	2, 5, 7, 11, 14, 17	3, 6, 8, 12, 15, 18

Fluency

WE10, 11, 12 For questions **1** to **3**, sketch the following, showing all intercepts.

1. **a.** $y = (x-1)(x-2)(x-3)$
   c. $y = (x+6)(x+1)(x-7)$
   b. $y = (x-3)(x-5)(x+2)$
   d. $y = (x+4)(x+9)(x+3)$

2. **a.** $y = (x+8)(x-11)(x+1)$
   c. $y = (2x-5)(x+4)(x-3)$
   b. $y = (2x-6)(x-2)(x+1)$
   d. $y = (3x+7)(x-5)(x+6)$

3. **a.** $y = (4x-3)(2x+1)(x-4)$
   c. $y = (x-3)^2(x-6)$
   b. $y = (2x+1)(2x-1)(x+2)$
   d. $y = (x+2)(x+5)^2$

For questions **4** to **6**, sketch the following (a mixture of positive and negative cubics).

4. **a.** $y = (2-x)(x+5)(x+3)$
   c. $y = (x+8)(x-8)(2x+3)$
   b. $y = (1-x)(x+7)(x-2)$
   d. $y = (x-2)(2-x)(x+6)$

5. **a.** $y = x(x+1)(x-2)$
   c. $y = 3(x+1)(x+10)(x+5)$
   b. $y = -2(x+3)(x-1)(x+2)$
   d. $y = -3x(x-4)^2$

6. **a.** $y = 4x^2(x+8)$
   c. $y = (6x-1)^2(x+7)$
   b. $y = (5-3x)(x-1)(2x+9)$
   d. $y = -2x^2(7x+3)$

7. **MC** Select a reasonable sketch of $y = (x+2)(x-3)(2x+1)$ from the following.

A.

B.

C.

D.

8. **MC** The graph shown could be that of:

A. $y = x^2(x+2)$
B. $y = (x+2)^3$
C. $y = (x-2)(x+2)^2$
D. $y = (x-2)^2(x+2)$

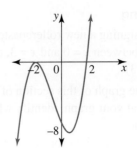

9. **MC** The graph shown has the equation:

A. $y = (x+1)(x+2)(x+3)$
B. $y = (x+1)(x-2)(x+3)$
C. $y = (x-1)(x+2)(x+3)$
D. $y = (x-1)(x+2)(x-3)$

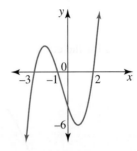

10. **MC** If a, b and c are positive numbers, the equation of the graph shown could be:

A. $y = (x-a)(x-b)(x-c)$
B. $y = (x+a)(x-b)(x+c)$
C. $y = (x+a)(x+b)(x-c)$
D. $y = (x-a)(x+b)(x-c)$

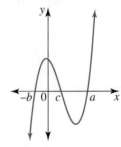

Understanding

11. Sketch the graph of each of the following.

a. $y = x(x-1)^2$
b. $y = -(x+1)^2(x-1)$

12. Sketch the graph of each of the following.

a. $y = (2-x)(x^2-9)$
b. $y = -x(1-x^2)$

Reasoning

13. For the graph shown, explain whether:

a. the gradient is positive, negative or zero to the left of the point of inflection.
b. the gradient is positive, negative or zero to the right of the point of inflection.
c. the gradient is positive, negative or zero at the point of inflection.
d. this is a positive or negative cubic graph.

14. The function $f(x) = x^3 + ax^2 + bx + 4$ has x-intercepts at $(1, 0)$ and $(-4, 0)$. Determine the values of a and b. Show full working.

15. The graphs of the functions $f(x) = x^3 + (a+b)x^2 + 3x - 4$ and $g(x) = (x-3)^3 + 1$ touch. Express a in terms of b.

Problem solving

16. Susan is designing a new rollercoaster ride using maths. For the section between $x = 0$ and $x = 3$, the equation of the ride is $y = x(x - 3)^2$.

    a. Sketch the graph of this section of the ride.
    b. Looking at your graph, identify where the ride touches the ground.
    c. The maximum height for this section is reached when $x = 1$. Use algebra to calculate the maximum height.

17. Determine the rule for the cubic function shown.

    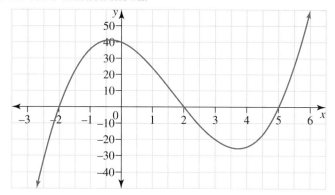

18. A girl uses 140 cm of wire to make a frame of a cuboid with a square base as shown.

    The base length of the cuboid is x cm and the height is h cm.

    a. Explain why the volume cm^3 is given by $V = 35x^2 - 2x^3$.
    b. Determine possible values that x can assume.
    c. Evaluate the volume of the cuboid when the base area is 81 cm^2.
    d. Sketch the graph of V versus x.
    e. Use technology to determine the coordinates of the maximum turning point. Explain what these coordinates mean.

LESSON
17.5 Quartic functions

▶ 17.5.1 Quartic functions

eles-4991

- **Quartic function** are polynomials where the highest power of the variable is 4 or the product of pronumeral makes up four.
- Some examples of quartic functions are $y = x^4$ and $y = (x + 1)(x - 2)(x + 3)(x - 4)$.
- There are three types of quartic functions: those with *one turning point*, those with *three turning points* and those with *one turning point and one point of inflection*.
- The table below includes the standard types of positive quartic equations.

Type of quartic function	Standard positive quartic graphs and equations with $a > 0$
One turning point	
Three turning points	
One turning point and one point of inflection	

For the "One turning point" row: $y = ax^4$ and $y = x^2(ax^2 + c),\ c \geq 0$

For the "Three turning points" row: $y = ax^2(x - b)(x - c)$, $y = a(x - b)^2(x - c)^2$, and $y = a(x - b)(x - c)(x - d)(x - e)$

For the "One turning point and one point of inflection" row: $y = a(x - b)(x - c)^3$

- There are also negative equivalents to all of the above graphs when $a < 0$. The negative graphs have the same shape, but are reflected across the x-axis. For example, the graph of $y = ax^4$ for $a < 0$ is shown.

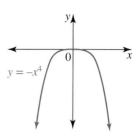

$y = -x^4$

17.5.2 Sketching quartic functions

- To sketch a quartic function:
    1. factorise the equation so that the equation is in the form matching one of the standard quartics. To factorise, use the **factor theorem** and long division. (If the equation is already factorised, skip this step.)
    2. calculate all x- and y-intercepts
    3. draw the intercepts on the graph, then use those points to sketch the cubic graph.

WORKED EXAMPLE 13 Factorising then sketching the graph of a quartic function

Sketch the graph of $y = x^4 - 2x^3 - 7x^2 + 8x + 12$, showing all intercepts.

THINK	WRITE/DRAW
1. Calculate the y-intercept.	When $x = 0$, $y = 12$. The y-intercept is 12.
2. Let $P(x) = y$.	Let $P(x) = x^4 - 2x^3 - 7x^2 + 8x + 12$.
3. Determine two linear factors of the quartic expressions, if possible, using the factor theorem.	$P(1) = (1)^4 - 2(-1)^3 - 7(1)^2 + 8(1) + 12$ $\quad = 12$ $\quad \neq 0$ $P(-1) = (-1)^4 - 2(-1)^3 - 7(-1)^2 + 8(-1) + 12$ $\quad = 0$ $(x + 1)$ is a factor. $P(2) = (2)^4 - 2(2)^3 - 7(2)^2 + 8(2) + 12$ $\quad = 0$ $(x - 2)$ is a factor.
4. Calculate the product of the two linear factors	$(x + 1)(x - 2) = x^2 - x - 2$
5. Use long division to divide the quartic by the quadratic factor $x^2 - x - 2$.	$$\begin{array}{r} x^2 - x - 6 \\ x^2 - x - 2 \overline{)x^4 - 2x^3 - 7x^2 + 8x + 12} \\ \underline{x^4 - x^3 - 2x^2} \\ -x^3 - 5x^2 + 8x \\ \underline{-x^3 + x^2 + 2x} \\ -6x^2 + 6x + 12 \\ \underline{-6x^2 + 6x + 12} \\ 0 \end{array}$$
6. Express the quartic in factorised form.	$y = (x + 1)(x - 2)(x^2 - x - 6)$ $\quad = (x + 1)(x - 2)(x - 3)(x + 2)$

7. To calculate the *x*-intercepts, solve $y = 0$.

If $0 = (x+1)(x-2)(x-3)(x+2)$
$x = -1, 2, 3, -2.$

8. State the *x*-intercepts.

The *x*-intercepts are $-2, -1, 2, 3$.

9. Sketch the graph of the quartic.

TI | THINK

1. In a new problem, on a Calculator page, complete the entry lines as:
Define $f1(x) = x^4 - 2x^3 - 7x^2 + 8x + 12$
$f1(0)$
solve $(f1(x) = 0, x)$
Press ENTER after each entry.

DISPLAY/WRITE

2. Open a Graphs page in the current document. Since the function has already been entered, just select the function and press ENTER and the graph will be displayed. Reset the viewing window to a more appropriate scale as shown. This graph does cross the *x*-axis at four distinct points. Find all the axial intercepts as described earlier.

The *y*-intercept is $(0, 12)$ and the *x*-intercepts are $(-2, 0), (-1, 0),$ $(2, 0)$ and $(3, 0)$.

CASIO | THINK

1. On the Graph & Table screen, complete the function entry line as:
$y1 = x^4 - 2x^3 - 7x^2 + 8x + 12$
Touch the graphing icon and resize the graph. Touch the $Y = 0$ icon for the *x*-intercepts. Use the right arrow to move between the *x*-intercepts.

DISPLAY/WRITE

2. For the *y*-intercept, tap:
• Analysis
• G-Solve
• *y*-intercept

The *y*-intercept is $(0, 12)$ and the *x*-intercepts are $(-2, 0), (-1, 0), (2, 0)$ and $(3, 0)$.

---

 on Resources

 eWorkbook Topic 17 Workbook (worksheets, code puzzle and project) (ewbk-2041)

Interactivity Quartic functions (int-6213)

| 17.5 Quick quiz on | 17.5 Exercise |

Individual pathways

■ PRACTISE	■ CONSOLIDATE	■ MASTER
1, 4, 7, 10, 11, 16	2, 5, 9, 12, 13, 17	3, 6, 8, 14, 15, 18

Fluency

WE13 For questions **1** to **3**, sketch the graph of each of the following showing all intercepts. You may like to verify the shape of the graph using a graphics calculator or another form of digital technology.

1. a. $y = (x-2)(x+3)(x-4)(x+1)$
   b. $y = (x^2-1)(x+2)(x-5)$

2. a. $y = 2x^4 + 6x^3 - 16x^2 - 24x + 32$
   b. $y = x^4 + 4x^3 - 11x^2 - 30x$

3. a. $y = x^4 - 4x^2 + 4$
   b. $y = 30x - 37x^2 + 15x^3 - 2x^4$

For questions **4** to **6**, sketch each of the following.

4. a. $y = x^2(x-1)^2$
   b. $y = -(x+1)^2(x-4)^2$

5. a. $y = -x(x-3)^3$
   b. $y = (2-x)(x-1)(x+1)(x-4)$

6. $y = (x-a)(b-x)(x+c)(x+d)$, where $a, b, c, d > 0$

Understanding

7. **MC** A quartic touches the x-axis at $x = -3$ and $x = 2$. It crosses the y-axis at $y = -9$. A possible equation is:

   A. $y = \dfrac{1}{4}(x+3)^2(x-2)^2$

   B. $y = -\dfrac{1}{6}(x+3)^3(x-2)$

   C. $y = -\dfrac{3}{8}(x+3)(x-2)^3$

   D. $y = -\dfrac{1}{4}(x+3)^2(x-2)^2$

8. **MC** Consider the function $f(x) = x^4 - 8x^2 - 16$. When factorised, $f(x)$ is equal to:

   A. $(x+2)(x-2)(x-1)(x+4)$
   B. $(x+3)(x-2)(x-1)(x+1)$
   C. $(x-2)^3(x+2)$
   D. $(x-2)^2(x+2)^2$

9. **MC** Consider the function $f(x) = x^4 - 8x^2 - 16$.
   The graph of $f(x)$ is best represented by:

   A.

   B.

   C.

   D.

   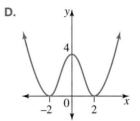

Reasoning

For questions **10** to **12**, sketch the graph of each of the following functions. Verify your answers using a graphics calculator.

10. **a.** $y = x(x-1)^3$ **b.** $y = (2-x)(x^2-4)(x+3)$

11. **a.** $y = (x+2)^3(x-3)$ **b.** $y = 4x^2 - x^4$

12. **a.** $y = -(x-2)^2(x+1)^2$ **b.** $y = x^4 - 6x^2 - 27$

13. The functions $y = (a-2b)x^4 - 3x - 2$ and $y = x^4 - x^3 + (a+5b)x^2 - 5x + 7$ both have an x-intercept of 1. Determine the value of a and b. Show your working.

14. Sketch the graph of each of the following functions. Verify your answers using a graphics calculator.
    a. $y = x^4 - x^2$ **b.** $y = 9x^4 - 30x^3 + 13x^2 + 20x + 4$

15. Patterns emerge when we graph polynomials with repeated factors, that is, polynomials of the form $P(x) = (x-a)^n, n > 1$. Discuss what happens if:
    a. n is even   **b.** n is odd.

Problem solving

16. The function $f(x) = x^4 + ax^3 - 4x^2 + bx + 6$ has x-intercepts $(2,0)$ and $(-3,0)$. Determine the values of a and b.

17. A carnival ride has a piece of the track modelled by the rule

    $$h = -\frac{1}{300}x(x-12)^2(x-20) + 15, 0 \le x \le 20$$

    where x metres is the horizontal displacement from the origin and h metres is the vertical displacement of the track above the horizontal ground.

    a. Determine how high above the ground level the track is at the origin.
    b. Use technology to sketch the function. Give the coordinates of any stationary points (that is, turning points or points of inflection).
    c. Evaluate how high above ground level the track is when $x = 3$.

18. Determine the rule for the quartic function shown.

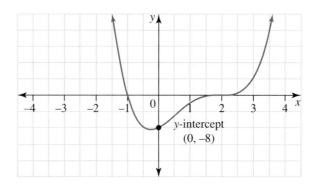

y-intercept
(0, –8)

LESSON
17.6 Transformations

17.6.1 General transformations

eles-4993

- When the graph of a function has been moved, stretched and/or flipped, this is called a **transformation**.
- There are three types of transformations:
  - **translations** are movements of graphs left, right, up or down
  - **dilations** are stretches of graphs to make them thinner or wider
  - **reflections** are when a graph is flipped in the x- or y-axis.
- The following table summarises transformations of the general polynomial $P(x)$ with examples given for transformations of the basic quadratic function $y = x^2$ shown here:

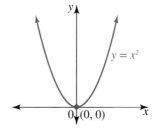

Transformation	Equation and explanation	Example(s)
Vertical translation	$y = P(x) + c$ This is a vertical translation of c units. If c is positive the translation is up, if c is negative the translation is down.	—— original function —— translated function $y = x^2 + 2$ $y = x^2$ (0, 2) $y = x^2$ $y = x^2 - 3$ (0, –3)

Horizontal translation	$y = P(x - b)$ This is a horizontal translation of b units. If b is positive the translation is right, if b is negative the translation is left.	*Graph showing $y = x^2$ and $y = (x-2)^2$ with points (0, 4) and (2, 0)*	*Graph showing $y = (x+1)^2$ and $y = x^2$ with points (0, 1) and (−1, 0)*
Dilation	$y = aP(x)$ This is a dilation by a factor of a in the x direction. If $a > 1$ the graph becomes thinner, if $0 < a < 1$ the graph becomes wider.	*Graph showing $y = 2x^2$ and $y = x^2$ with point (0, 0)*	*Graph showing $y = \frac{1}{4}x^2$ and $y = x^2$ with point (0, 0)*
Reflection	$y = -P(x)$ This is a reflection in the x-axis. $y = P(-x)$ This is a reflection in the y-axis.	*Graph showing $y = x^2$ and $y = -x^2$ with point (0, 0)*	

- Note that the graph $y = x^2$ does not change when reflected across the y-axis because it is symmetrical about the y-axis.
- With knowledge of the transformations discussed in this section, it is possible to generate many other graphs without knowing the equation of the original function.

WORKED EXAMPLE 14 Sketching transformations

Use the sketch of $y = P(x)$ shown to sketch:

a. $y = P(x) + 1$ b. $y = P(x) - 1$ c. $y = -P(x)$.

THINK	**WRITE/DRAW**
a. 1. Sketch the original $y = P(x)$.	**a.**
2. Look at the equation $y = P(x) + 1$. This is a translation of one unit in the vertical direction — one unit up.	
	$y = P(x)$
3. Sketch the graph of $y = P(x) + 1$ using a similar scale to the original.	
	$y = P(x) + 1$
b. 1. Sketch the original $y = P(x)$.	**b.**
2. Look at the equation $y = P(x) - 1$. This is a translation of negative one unit in the vertical direction; that is, one unit down.	
	$y = P(x)$
3. Sketch the graph of $y = P(x) - 1$ using a similar scale to the original.	
	$y = P(x) - 1$

c. **1.** Sketch the original $y = P(x)$.

2. Look at the equation $y = -P(x)$. This is a reflection across the x-axis.

c.

$y = P(x)$

3. Sketch the graph of $y = -P(x)$ using a similar scale to the original.

$y = -P(x)$

 ## 17.6.2 Transformations of hyperbolas, exponential functions and circles

eles-4994

- For transformations of hyperbolas, start with the standard hyperbola $y = \dfrac{1}{x}$ then transform using

$y = \dfrac{a}{x - b} + c$ where:

  - translations are c units vertically and b units horizontally
  - dilations are by a factor of a
  - reflections across the x-axis happen when there is a negative at the front, for example $y = -\dfrac{a}{x - b} + c$;

    and reflections across the y-axis happen when there is x is negative, for example $y = \dfrac{a}{-x - b} + c$.

Standard hyperbola	Transformed hyperbola
$y = \dfrac{1}{x}$	$y = \dfrac{2}{(x - 1)} + 1$

- For transformations of exponentials, start with the standard exponential $y = a^x$ where $a \neq 1$ then transform using $y = ka^{(x-b)} + c$ where:
  - translations are c units vertically and b units horizontally.
  - dilations are by a factor of k.
  - reflections across the x-axis happen when there is a negative out the front, $y = -\left(ka^{(x-b)} + c\right)$ and reflections across the y-axis happen when there is a negative on the x, $y = ka^{(-x-b)} + c$.

Standard exponential graph	Transformed exponential graph

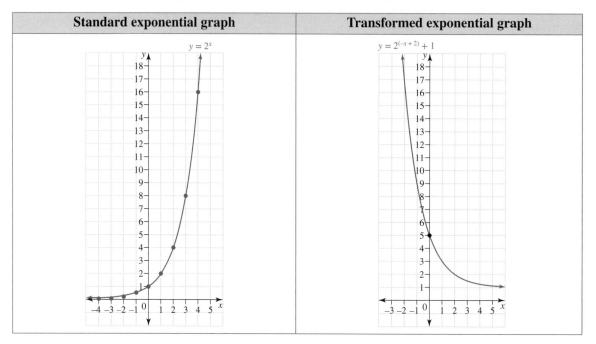

- For transformations of circles, start with the standard circle $x^2 + y^2 = r^2$ then transform using $(x - h)^2 + (y - k)^2 = r^2$ where:
  - translations are h units horizontally and k units vertically.

Standard circle	Transformed circle

WORKED EXAMPLE 15 Describing a transformation

Describe the transformations for the following graphs.

a. A transformation of the graph $x^2 + y^2 = 4$ to the graph $(x - 3)^2 + (y + 7)^2 = 4$

b. A transformation of the graph $y = \dfrac{1}{x}$ to the graph $y = \dfrac{2}{-x - 4} + 5$

THINK

a. 1. Using the standard translated formula $(x - h)^2 + (y - k)^2 = r^2$ identify the value of h for the horizontal translation and the value of k for the vertical translation.

WRITE

a. $h = 3, k = -7$.

The graph is translated 3 units in the horizontal direction and -7 units in the vertical direction.

b. 1. Using the standard translated formula
$y = \dfrac{a}{x-b} + c$ identify the value of c for the vertical translation and the value of b for the horizontal translation.

b. $b = 4, c = 5.$
The graph is translated 5 units vertically and 4 units horizontally.

2. Using the standard translated formula
$y = \dfrac{a}{x-b} + c$ identify the value of a for the dilation factor.

$a = 2.$
The graph is dilated by a factor of 2.

3. Look for $y = -P(x)$ (reflection in the x-axis) and/or $y = P(-x)$ (reflection in the y-axis). This equation has a negative on the x-value so it is $y = P(-x).$

The graph is reflected in the y-axis.

Resources

📋 **eWorkbook** Topic 17 Workbook (worksheets, code puzzle and project) (ewbk-2041)

🔧 **Interactivities** Horizontal translations of parabolas (int-6054)
Vertical translations of parabolas (int-6055)
Dilation of parabolas (int-6096)
Exponential functions (int-5959)
Reflection of parabolas (int-6151)
Hyperbolas (int-6155)
Translations of circles (int-6214)
Transformations of exponentials (int-6216)
Transformations of cubics (int-6217)
The rectangular hyperbola (int-2573)

Exercise 17.6 Transformations

learnon

| **17.6 Quick quiz** on | **17.6 Exercise** |

Individual pathways

■ PRACTISE	■ CONSOLIDATE	■ MASTER
1, 4, 7, 10, 13	2, 5, 8, 11, 14	3, 6, 9, 12, 15

Fluency

1. **WE14** Use the sketch of $y = P(x)$ shown to sketch:

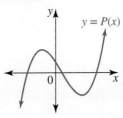

a. $y = P(x) + 1$

b. $y = -P(x)$

2. Use the sketch of $y = P(x)$ shown to sketch:

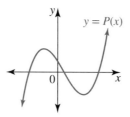

a. $y = P(x) - 2$
b. $y = 2P(x)$.

3. Consider the sketch of $y = P(x)$ shown. Sketch:

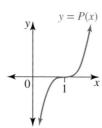

a. $y = P(x) + 1$
b. $y = -P(x)$
c. $y = P(x + 2)$.

4. **WE15** Describe the transformations for the following graphs.

   a. A transformation of the graph $x^2 + y^2 = 9$ to the graph $(x - 2)^2 + (y - 1)^2 = 9$

   b. A transformation of the graph $y = \dfrac{1}{x}$ to the graph $y = \dfrac{3}{x - 2} + 7$

5. Describe the transformations for the following graphs.

   a. A transformation of the graph $y = \dfrac{1}{x}$ to the graph $y = \dfrac{5}{x - 1} + 2$

   b. A transformation of the graph $y = 3^x$ to the graph $y = 3^{(-x+2)} + 4$

6. Describe the transformations for the following graphs.

   a. A transformation of the graph $y = \dfrac{1}{x}$ to the graph $y = \dfrac{4}{-x + 3} + 10$

   b. A transformation of the graph $y = 5^x$ to the graph $y = -\left(5^{(-x+7)} - 6\right)$

Understanding

7. Draw any polynomial $y = P(x)$. Discuss the similarities and differences between the graphs of $y = P(x)$ and $y = -P(x)$.

8. Draw any polynomial $y = P(x)$. Discuss the similarities and differences between the graphs of $y = P(x)$ and $y = 2P(x)$.

9. Draw any polynomial $y = P(x)$. Discuss the similarities and differences between the graphs of $y = P(x)$ and $y = P(x) - 2$.

Reasoning

10. Consider the sketch of $y = P(x)$ shown below.

Give a possible equation for each of the following in terms of $P(x)$.

a.

b.

c.

11. $y = x(x-2)(x-3)$ and $y = -2x(x-2)(x-3)$ are graphed on the same set of axes. Describe the relationship between the two graphs using the language of transformations.

12. If $y = -hr^{-(x+p)} - r$, explain what translations take place from the original graph, $y = r^x$.

Problem solving

13. a. Sketch the graph of $y = 3 \times 2^x$.
    b. If the graph of $y = 3 \times 2^x$ is transformed into the graph $y = 3 \times 2^x + 4$, describe the transformation.
    c. Sketch the graph of $y = 3 \times 2^x + 4$.
    d. Determine the coordinates of the y-intercept of $y = 3 \times 2^x + 4$.

14. The graph of $y = \dfrac{1}{x}$ is reflected in the y-axis, dilated by a factor of 2 parallel to the x-axis, translated 2 units to the left and up 1 unit. Determine the equation of the resultant curve. Give the equations of any asymptotes.

15. The graph of an exponential function is shown.

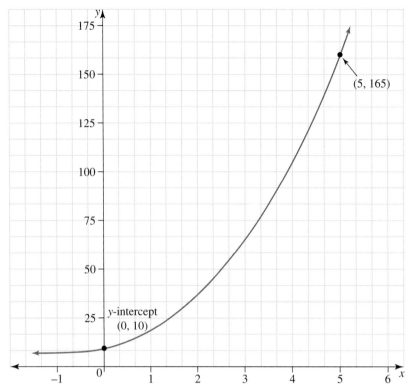

Its general rule is given by $y = a(2^x) + b$.

a. Determine the values of a and b.

b. Describe any transformations that had to be applied to the graph of $y = 2^x$ to achieve this graph.

LESSON
17.7 Review

17.7.1 Topic summary

Relations

- There are four types of relations: one-to-one, one-to-many, many-to-one and many-to-many.
- Relations that are one-to-one or many-to-one are called functions.

Domain and range of functions

- The domain of a function is the set of allowable values of x.
- The range of a function is the set of y-values it produces.
- Consider the function $y = x^2 - 1$. The domain of the function is $x \in R$. The range of the function is $y \geq -1$.

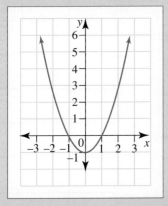

Inverse functions

- Functions have an inverse function if they are one-to-one.
- The inverse of a function $f(x)$ is denoted $f^{-1}(x)$.
- To determine the equation of the inverse function:
  1. Let $y = f(x)$.
  2. Switch x and y in the equation.
  3. Rearrange for y.

e.g.
$$f(x) = 2x^3 - 4$$
$$y = 2x^3 - 4$$
$$x = 2y^3 - 4$$
$$x + 4 = 2y^3$$
$$\frac{x+4}{2} = y^3$$
$$y = \sqrt[3]{\frac{x+4}{2}}$$
$$f^{-1}(x) = \sqrt[3]{\frac{x+4}{2}}$$

- Sometimes a function needs to have its domain restricted for the inverse to exist. For example, $f(x) = x^2$ is many-to-one, so does not have an inverse; but $f(x) = x^2$, $x \geq 0$ is one-to-one, and therefore does have an inverse function.

FUNCTIONS AND RELATIONS

Function notation

- Functions are denoted $f(x)$.
- The value of a function at a point can be determined by substituting the x-value into the equation.

e.g.
$$f(x) = 2x^3 - 3x + 5$$
The value of the function when $x = 2$ is:
$$f(2) = 2(2)^3 - 3(2) + 5$$
$$= 2(8) - 6 + 5$$
$$= 15$$

Transformations

- There are 3 types of transformations that can be applied to functions and relations:
  - Translations
  - Dilations
  - Reflections
- The equation of a hyperbola $\left(f(x) = \dfrac{1}{x}\right)$ when transformed is
$$f(x) = \frac{a}{x - b} + c.$$
- The equation of an exponential $(f(x) = a^x)$ when transformed is
$$f(x) = ka^{(x-b)} + c.$$
- The equation of a circle $(x^2 + y^2 = r^2)$ when transformed is
$$(x - h)^2 + (y - k)^2 = r^2.$$

Types of functions

- Cubic functions are functions where the highest power of x is 3. Cubic functions have 2 turning points, or 1 point of inflection.
- Quartic functions are functions where the highest power of x is 4. Quartic functions have 1 turning point, 3 turning points or 1 turning point and 1 point of inflection.
- Exponential functions are of the form $f(x) = k \times a^x$. The y-intercept is $y = k$.
- Examples of a cubic (top), a quartic (middle) and an exponential (bottom) function are shown below.

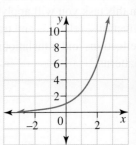

17.7.2 Success criteria

Tick the column to indicate that you have completed the lesson and how well you have understood it using the traffic light system.

(**Green:** I understand; **Yellow:** I can do it with help; **Red:** I do not understand)

Lesson	Success criteria			
17.2	I can identify the type of a relation.			
	I can determine the domain and range of a function or relation.			
	I can identify the points of intersection between two functions.			
	I can determine the inverse function of a one-to-one function.			
17.3	I can sketch graphs of exponential functions.			
	I can determine the equation of an exponential function.			
17.4	I can plot the graph of a cubic function using a table of values.			
	I can sketch the graph of a cubic function by calculating its intercepts			
	I can determine the equation of a cubic function by inspection.			
17.5	I can factorise the equation of a quartic function.			
	I can sketch the graph of a quartic function from a factorised equation.			
17.6	I can sketch the graph of a function that has undergone some transformations.			
	I can describe a transformation using the words translation, dilation and reflection.			

17.7.3 Project

Shaping up!

Many beautiful patterns are created by starting with a single function or relation and transforming and repeating it over and over.

In this task you will apply what you have learned about functions, relations and transformations (dilations, reflections and translations) to explore mathematical patterns.

Exploring patterns using transformations

1. **a.** On the same set of axes, draw the graphs of:
    i. $y = x^2 - 4x + 1$
    ii. $y = x^2 - 3x + 1$
    iii. $y = x^2 - 2x + 1$
    iv. $y = x^2 + 2x + 1$
    v. $y = x^2 + 3x + 1$
    vi. $y = x^2 + 4x + 1$

   b. Describe the pattern formed by your graphs. Use mathematical terms such as intercepts, turning point, shape and transformations.

What you have drawn is referred to as a family of curves — curves in which the shape of the curve changes if the values of a, b and c in the general equation $y = ax^2 + bx + c$ change.

   c. Explore the family of parabolas formed by changing the values of a and c. Comment on your findings.
   d. Explore exponential functions belonging to the family of curves with equation $y = ka^x$, families of cubic functions with equations $y = ax^3$ or $y = ax^3 + bx^2 + cx + d$, and families of quartic functions with equations $y = ax^4$ or $y = ax^4 + bx^3 + cx^2 + dx + e$. Comment on your findings.
   e. Choose one of the designs shown below and recreate it (or a simplified version of it). Record the mathematical equations used to complete the design.

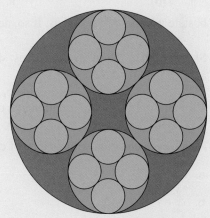

Coming up with your design

2. Use what you know about transformations to functions and relations to create your own design from a basic graph. You could begin with a circle, add some line segments and then repeat the pattern with some change. Record all the equations and restrictions you use.
It may be helpful to apply your knowledge of inverse functions too.
A digital technology will be very useful for this task.
Create a poster of your design to share with the class.

Fluency

1. State which of the following are functions.

   a.

   b.

2. Identify which of the following are functions. For each identified as functions, state the equation of the inverse function, if it exists.

   a. $y = 2x - 7$ b. $x^2 + y^2 = \sqrt{30}$ c. $y = 2^x$ d. $y = \dfrac{1}{x+1}$

3. If $f(x) = \sqrt{4 - x^2}$:
   a. calculate:

      i. $f(0)$ ii. $f(1)$ iii. $f(2)$

   b. state whether $f^{-1}(x)$ exists. If so, determine its equation.

4. Sketch each of the following curves, showing all intercepts.

   a. $y = (x - 1)(x + 2)(x - 3)$ b. $y = (2x + 1)(x + 5)^2$

5. Give an example of the equation of a cubic that would just touch the x-axis and cross it at another point.

6. Match each equation with its type of curve.

   a. $y = x^2 + 2$ A. circle

   b. $x^2 + y^2 = 9$ B. cubic

   c. $f(x) = \dfrac{2}{x + 2}$ C. exponential

   d. $g(x) = 6^{-x}$ D. parabola

   e. $h(x) = (x + 1)(x - 3)(x + 5)$ E. hyperbola

7. **MC** The equation for the graph shown could be:
   A. $y = (x - 5)(x + 1)(x + 3)$ B. $y = (x - 3)(x - 1)(x + 5)$
   C. $y = (x - 3)(x + 1)(x + 5)$ D. $y = (5 - x)(1 + x)(3 + x)$

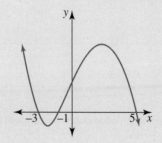

8. **MC** Select which of the following shows the graph of $y = -2(x+5)^3 - 12$.

A.
(−5, −12)

B.
(−5, 12)

C.
(5, −12)

D.
(5, 12)

9. Sketch the following functions:

   a. $y = x(x-2)(x+11)$ b. $y = x^3 + 6x^2 - 15x + 8$ c. $y = -2x^3 + x^2$

10. **MC** The rule for the graph shown could be:

    A. $f(x) = x(x+2)^3$

    C. $f(x) = x^2(x-2)^2$

    B. $f(x) = -x(x-2)^2$

    D. $f(x) = x(x-2)^3$

11. **MC** The graph of $y = (x+3)^2(x-1)(x-3)$ is best represented by:

A.

B.

C.

D.

Problem solving

12. Sketch the graph of $y = x^4 - 7x^3 + 12x^2 + 4x - 16$, showing all intercepts.

13. Consider the sketch of $y = P(x)$ shown. Sketch $y = -P(x)$.

14. Draw any polynomial $y = P(x)$. Discuss the similarities and differences between the graphs of $y = P(x)$ and $y = P(x) + 3$.

15. Describe what happens to $f(x) = -2^x$ as $x \to \infty$ and $x \to -\infty$.

16. Determine any points of intersection between $f(x) = x^2 - 4$ and $g(x) = x^3 + x^2 - 12$.

17. The concentration of alcohol (mg/L) in a bottle of champagne is modelled by $C = C_0 \times 0.33^{kt}$ where t represents the time in days after the bottle is opened.

If the initial concentration is 80 mg/L and the concentration after 1 day is 70 mg/L, evaluate the concentration remaining after:

   a. 3 days
   b. 1 week
   c. 18 hours.

18. The number of hyenas, H, in the zoo is given by $H = 20(10^{0.1t})$, where t is the number of years since counting started. At the same time, the number of dingoes, D, is given by $D = 25(10^{0.05t})$.

   a. Calculate the number of:
      i. hyenas
      ii. dingoes when counting began.
   b. Calculate the numbers of each after:
      i. 1 year
      ii. 18 months.
   c. Which of the animals is the first to reach a population of 40 and by how long?
   d. After how many months are the populations equal and what is this population?

19. a. Consider the equation $f(x) = a(x - h)^2 + k$. By restricting the x-values, find the equation of the inverse function.
   b. Show that the function $f(x) = \dfrac{a}{x} + b$ and its inverse function intersect on the line $y = x$.

20. A shend is a type of tropical pumpkin grown by the people of Outer Thrashia. The diameter (Dm) of a shend increases over a number of months (m) according to the rule $D = 0.25 \times (10)^{0.01m}$.
   a. Determine the diameter of the shend after 4 months.
   b. If the shend is not harvested it will explode when it reaches a critical diameter of 0.5 metres. Show that it takes approximately 30 months for an unharvested shend to explode.

21. The surface area of a lake is evaporating at a rate of 5% per year due to climate change. To model this situation, the surface area of the lake (S km^2) over time is given by $S = 20\,000 \times 0.95^x$, where x is the time in years.
   a. Explain whether this is an exponential relationship.
   b. State the initial surface area of the lake.
   c. Determine the surface area of the lake after 10 years.
   d. Plot a graph for this relationship.
   e. Determine the surface area of the lake in 100 years.
   f. Explain whether this is a realistic model.

 To test your understanding and knowledge of this topic, go to your learnON title at www.jacplus.com.au and complete the **post-test**.

Answers

Topic 17 Functions and relations (10A)

17.1 Pre-test

1. B
2. A
3. True
4. B
5. B
6. D
7. A
8. D
9. D
10. D
11. B
12. A
13. B
14. B
15. D

17.2 Functions and relations

1. a. One-to-many   b. Many-to-one
   c. Many-to-one   d. One-to-one

2. a. One-to-one   b. Many-to-one
   c. Many-to-many   d. Many-to-one

3. a. One-to-one   b. Many-to-one
   c. One-to-one   d. Many-to-one

4. a. b, c, d   b. d

5. a. a, b, d   b. a

6. a. a, b, c, d   b. a, c

7. a. 1   b. 7   c. −5   d. 16

8. a. 2   b. 1   c. 3   d. 0

9. a. 3   b. 2   c. 6   d. 9

10. a. 0   b. 1
    c. 16   d. $a^2 + 6a + 9$

11. a. 12   b. 6   c. −4   d. 2

12. a, c, d

13. a, b, c, f

14. a. 3   b. 3   c. $\dfrac{5}{x} - 2x$

    d. $\dfrac{10}{x^2} - x^2$ e. $\dfrac{10}{x+3} - x - 3$ f. $\dfrac{10}{x-1} - x + 1$

15. a. 3   b. −3 or 3   c. $\dfrac{1}{3}$

16. a. 2 or 3   b. −4 or 1   c. −1

17. a. $f(x) \to \infty$ b. $f(x) \to 0$ c. $f(x) \to 0$
    d. $f(x) \to -\infty$ e. $f(x) \to 0$

18. a. $(0, -4), (2, 0)$ b. $(1, -2), \left(-\dfrac{2}{3}, 3\right)$

    c. $(2, 0), (-2, 0)$ d. $(3, -4)$

19. a. $f^{-1}(x) = \dfrac{x+1}{2}$

    b. $f^{-1}(x) = \sqrt{x+3}$ for original $x > 0$.

    c. $f^{-1}(x) = \sqrt{x-4} + 2$ for original $x > 2$

20. a. The horizontal line test fails.

    b. An inverse function will exist for $f(x) = x(x-2)$, $x \le 1$
       or $f(x) = x(x-2)$, $x \ge 1$.

21. a. The horizontal line test is upheld.

    b. $f^{-1}(x) = -\sqrt{4-x}$, $x \le 4$.

22. a. $x = \pm 3$ b. $x = 2\dfrac{1}{3}$ c. $x = 28$

23. These graphs are inverse because they are the mirror images
    of each other through the line $y = x$.

24. a. Ran $= [2, \infty)$

    b. Many-to-one

    c. and f.

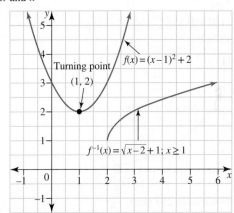

    d. Dom $= [1, \infty)$

    e. $f^{-1}(x) = \sqrt{x-2} + 1$, Dom $= [2, \infty)$, Ran $= [1, \infty)$

    g. $(0, 3)$ and $(3, 6)$

17.3 Exponential functions

1. a. 2000

    b. 486 000

    c.

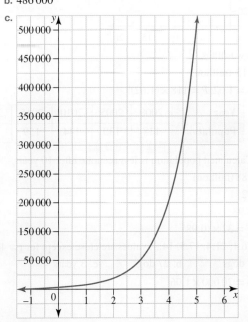

    d. 1.26 hours

2. a. $5000

b. $7717

c.
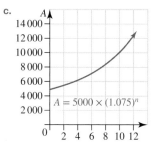
$A = 5000 \times (1.075)^n$

d. 10 years

3. a. C   **b.** D

4. a. $883.50   **b.** $821.66

  c. $V = 950 \times (0.93)^n$ **d.** $397.67

5. a. 102 mg

b. 86.7 mg

c. $A = 120 \times (0.85)^t$

d. 83.927 mg

e.
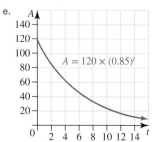
$A = 120 \times (0.85)^t$

f. Approximately 213 years

6. a. i. 96.04%   **ii.** 90.39%

b. $C = 100(0.98)^w$

c.
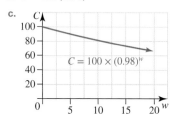
$C = 100 \times (0.98)^w$

d. 8 washings

7. a. 118 (million)

b. $a = 1.02; P = 118 \times (1.02)^n$

c.

Year	2000	2005	2010	2015	2020
Population	118	130	144	159	175

Calculated population is less accurate after 10 years.

d. 288 (million)

8. a. 32

b. 0.98

c. $T = 32 \times (0.98)^t$

d. 26.1, 21.4, 17.5, 14.3; values are close except for $t = 40$.

9. a. 3 dogs   **b.** 27 dogs   **c.** 3 years

10. a. $A = 20\,000 \times 1.06^x$

b.
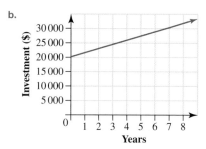

c. 7 years

d. 6 years —1 year quicker

e. 9.05% p.a.

11. a. Approximately 20 000

b, c. Sample responses can be found in the worked solutions in the online resources.

12. a. $a = 100, b = 1.20$, increase $= 20\%/\min$

b. $N = 14\,6977 \times 0.70^m$

13. a. 500 °C

b. 125 °C.

c. Between 5 and 6 hours once it has cooled to below 15 °C.

14. a. 45 °C

b. $T = 45 \times 0.95^t$

c. 10 °C

d. No. The line $T = 0$ is an asymptote.

15. a. 1. 39.85 mg   **2.** 18.43 mg

b. More than 35.78 centuries.

17.4 Cubic functions

1. a.

b.

c.

d.

2. a.

b.

c.

d.

6. a.

b.

c.

d.

7. C

8. C

9. B

10. D

11. a.

b.

12. a.

3. a.

b.

c.

d.

4. a.

b.

c.

d.

5. a.

b.

c.

d.

b.

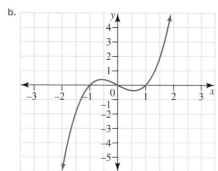

13. a. Positive   **b.** Positive
    c. Zero   **d.** Positive

14. $a = 2, b = -7$

15. $a = \dfrac{-(27 + 11\,b)}{11}$

16. a.

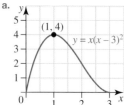

  b. $x = 0$ and $x = 3$

  c. 4 units

17. $y = 2(x + 2)(x - 2)(x - 5)$

18. a. Sample responses can be found in the worked solutions
     in the online resources.

  b. $0 < x < 17.5$

  c. 1377 cm^3

  d.

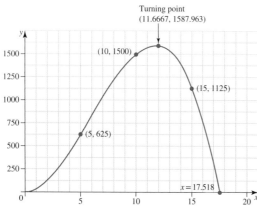

  e. $(11.6667, 1587.963)$; this is the value of x which creates
     the maximum volume.

17.5 Quartic functions

1. a.

  b.

2. a.

  b.

3. a.

  b.

4. a.

  b.

5. a.

b.

6.

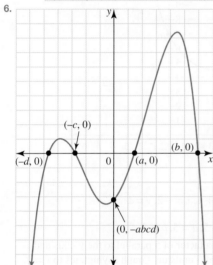

7. D

8. D

9. B

10. a.

b.

11. a.

b.

12. a.

b.

13. $a = 3, b = -1$

14. a.

b.

15. a. If n is even, the graph touches the x-axis.

b. If n is odd, the graph cuts the x-axis.

16. $a = 4, b = -19$

17. a. 15 m

b.

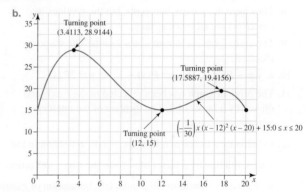

c. 28.77 m

18. $y = (x + 1)(x - 2)^3$

17.6 Transformations

1.

2.

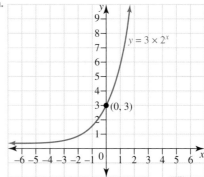

3.

$y = P(x+2)$ $y = P(x) + 1$ $y = P(x)$ $y = -P(x)$

4. a. The transformation is a translation 2 units horizontally and a translation 1 unit vertically.

   b. The transformation is a dilation by a factor of 3 followed by a translation 2 units horizontally and 7 units vertically.

5. a. The transformation is a dilation by a factor of 5 followed by a translation 1 unit horizontally and 2 units vertically.

   b. The transformation is a reflection in the y-axis followed by a translation -2 units horizontally and 4 units vertically.

6. a. The transformation is a dilation by a factor of 4 followed by a reflection in the y-axis and then a translation -3 units horizontally and 10 units vertically.

   b. The transformation is a reflection in the y-axis followed by a translation -7 units horizontally and -6 units vertically and then a reflection in the x-axis.

7. They have the same x-intercepts, but $y = -P(x)$ is a reflection of $y = P(x)$ in the x-axis.

8. They have the same x-intercepts, but the y-values in $y = 2P(x)$ are all twice as large.

9. The entire graph is moved down 2 units. The shape is identical.

10. a. $y = -P(x)$ **b.** $y = P(x) - 3$ **c.** $y = 2P(x)$

11. The original graph has been reflected in the x-axis and dilated by a factor of 2. The location of the intercepts remains unchanged.

12. Dilation by a factor of h from the x-axis, reflection in the x-axis, dilation by a factor of $\dfrac{1}{q}$ from the y-axis, reflection in the y-axis, translation of p units left, translation of r units down.

13. a.

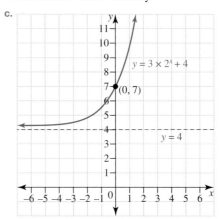

$y = 3 \times 2^x$ $(0, 3)$

   b. Translation 4 units vertically.

   c.

$y = 3 \times 2^x + 4$ $(0, 7)$ $y = 4$

   d. $(0, 7)$

14. $y = -\dfrac{2}{(x+2)} + 1, x = -2, y = 1$

15. a. $y = 5(2^x) + 5, a = 5, b = 5$

   b. Dilation by a factor of 5 parallel to the y-axis and translation of 5 units up. Graph asymptotes to $y = 5$.

Project

1. a. See figure at the bottom of the page.*

   b. All of the graphs have a y-intercept of $y = 1$. The graphs which have a positive coefficient of x have two positive x-intercepts. The graphs which have a negative coefficient

---

*1. a.

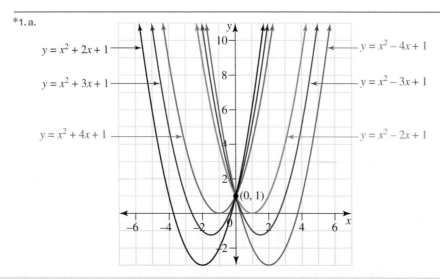

$y = x^2 + 2x + 1$ $y = x^2 - 4x + 1$

$y = x^2 + 3x + 1$ $y = x^2 - 3x + 1$

$y = x^2 + 4x + 1$ $y = x^2 - 2x + 1$

$(0, 1)$

of x have two negative x-intercepts. As the coefficient of x decreases the x-intercepts move further left.

c. The value of a affects the width of the graph. The greater the value, the narrower the graph. When a is negative, the graph is reflected in the x-axis. The value of c affects the position of the graph vertically. As the value of c increases, the graph moves upwards. The value of c does not affect the shape of the graph.

d. Students may find some similarities in the effect of the coefficients on the different families of graphs. For example, the constant term always affects the vertical position of the graph, not its shape and the leading coefficient always affects the width of the graph.

e. Student responses will vary. Students should aim to recreate the pattern as best as possible, and describe the process in mathematical terms. Sample responses can be found in the worked solutions in the online resources.

2. Student responses will vary. Students should try to use their imagination and create a pattern that they find visually beautiful.

17.7 Review questions

1. a

2. a, c, d, $\dfrac{x+7}{2}, \log_2 x, \dfrac{1}{x} - 1$

3. a. i. 2   ii. $\sqrt{3}$ iii. 0

    b. No.

4. a.

$y = (x-1)(x+2)(x-3)$

    b.

$y = (2x+1)(x+5)^2$

5. Sample responses can be found in the worked solutions in the online resources. One possible answer is $y = (x-1)(x-2)^2$.

6. a. D   b. A   c. E
    d. C   e. B

7. D

8. A

9. a.

b.

c.

10. D

11. A

12.

13.

14. The entire graph is moved up 3 units. The shape is identical.

15. As $x \to \infty, f(x) \to -\infty$
    As $x \to -\infty, f(x) \to 0$

16. $(2, 0)$

17. a. 53.59 mg/L   b. 31.42 mg/L   c. 72.38 mg/L

18. a. i. 20   ii. 25

    b. i. $H = 25; D = 28$ ii. $H = 28; D = 30$

    c. The hyenas are the first species to reach a population of 40 by 1 year.

    d. After about 23 months, 21 animals

19. a. $f^{-1}(x) = \sqrt{\dfrac{x-k}{a}} + h, x \geq \dfrac{k}{a}$

    b. Sample responses can be found in the worked solutions in the online resources.

20. a. 0.27 m

    b. Sample responses can be found in the worked solutions in the online resources.

21. a. Yes, because the relationship involves a variable as an exponent.

    b. 20 000 km^2

    c. 11 975 km^2

d.

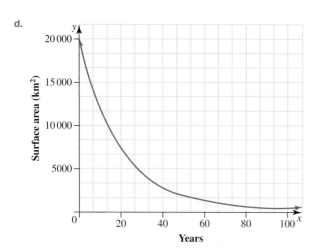

e. $118 \, \text{km}^2$

f. No, this is not a realistic model as is it does not take into account changes to climate, rain, runoff from mountains, glaciers and so on.

18 Algorithms and pseudocode

LESSON
18.1 Overview

Why learn this?

Learning about algorithms and pseudocode can offer several benefits. Here are some reasons why it is important to study these concepts:

- **Problem-solving skills:** Algorithms are step-by-step procedures or formulas for solving problems, which teaches you how to approach and break down complex problems into smaller, more manageable tasks.
- **Logical thinking:** Algorithmic thinking involves breaking down problems into smaller steps and understanding the logical flow of a solution.
- **Computational thinking:** Algorithms are fundamental to computer science and involve understanding how computers process information and solve problems, which is valuable in our increasingly digital world.
- **Coding skills:** Learning about algorithms often goes hand-in-hand with learning how to code.
- **Critical thinking:** Algorithmic problem solving requires you to analyse problems, evaluate potential solutions and choose the most effective approach.
- **Understanding algorithms:** Pseudocode provides a way to describe algorithms in a more readable and understandable form. It helps you focus on the logic and structure of an algorithm without getting bogged down by specific programming syntax.
- **Preparation for future technologies:** As technology continues to advance, understanding algorithms becomes increasingly important. Whether you're interested in artificial intelligence, machine learning or other cutting-edge technologies, a strong foundation in algorithms will be beneficial.

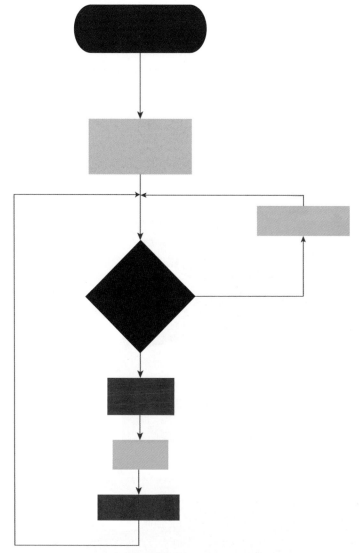

In summary, learning about algorithms and pseudocode is not just about computer science; it's about developing essential skills that promote clear thinking, problem solving and effective communication of ideas.

Reading content and rich media, including interactivities and videos for every concept

Extra learning resources

Differentiated question sets

Questions with immediate feedback, and fully worked solutions to help students get unstuck

Exercise 18.1 Pre-test

1. Algorithms are a set of rules or steps to solve a specific problem. True or false?

2. Pseudocode is a formal programming language. True or false?

3. Algorithms can only be used in computers. True or false?

4. Sorting is an example of an algorithm. True or false?

5. Algorithms are only used by computer programmers. True or false?

6. Algorithms are always written in programming languages. True or false?

7. Pseudocode is similar to a programming language but less formal and more readable. True or false?

8. An algorithm can have multiple correct solutions. True or false?

9. Searching for a specific item in a list is an example of a search algorithm. True or false?

10. Algorithms are only used in mathematical problems. True or false?

11. Computers can only follow algorithms; they cannot create them. True or false?

12. Algorithms are used in everyday activities, such as following a recipe. True or false?

13. Algorithms must be written in a specific order, and changing the order won't affect the result. True or false?

14. Pseudocode is a formal programming language used to write executable code. True or false?

15. Pseudocode is primarily used for communication between computers and programming languages. True or false?

LESSON
18.2 Algorithms

LEARNING INTENTION

At the end of this lesson you should be able to:
- understand how algorithms can be used to solve problems.

18.2.1 Algorithms

An algorithm is a set of **rules** that must be followed when solving a particular problem.

That is, given a number or set of numbers (the input), apply a set of rules to achieve a result (the output).

In mathematics an algorithm may be a set of processes or calculations to solve a particular problem, such as the method we use to solve a long division problem, but algorithms occur everywhere in day-to-day life. For example, the recipe for baking a cake can be thought of as an algorithm.

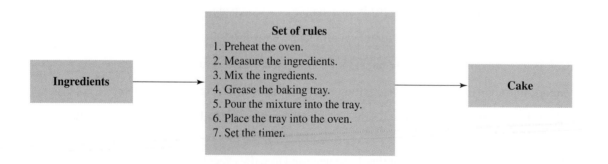

Flowcharts

Flowcharts can be used to step through an algorithm and are helpful in making sure that no steps are missed. Flowcharts typically show Begin and End elements.

Taking the example of baking the cake, this algorithm may be written as a flowchart.

The next example shows a very simple flowchart that demonstrates the importance of capturing all steps in an algorithm to get the desired result.

WORKED EXAMPLE 1 Writing an algorithm and drawing a flowchart

a. Write an algorithm for the division of two numbers.
b. Draw a flowchart representing the algorithm from part **a.**

THINK

a. 1. Draw a diagram to represent the algorithm.

WRITE/DRAW

a.

2. Write the algorithm as a series of steps or rules.

Step 1. Input two numbers, x and y.
Step 2. Read the numbers x and y.
Step 3. Calculate $z = \dfrac{x}{y}$.
Step 4. Write the output, z.

b. Draw a flowchart showing the steps.

b.

Taking another example, let's write an algorithm for checking whether a number is odd or even.

The steps involved require you to know what problem is being solved.

We need to think about this problem in a way that gives us a set of rules; that is, we need to think algorithmically. We know that when even numbers are divided by 2, there will be no remainder, but when odd numbers are divided by 2, there will be a remainder of 1.

For example:

$$\frac{3}{2} = 1 \text{ with remainder } 1$$

$$\frac{4}{2} = 2 \text{ with remainder } 0$$

$$\frac{5}{2} = 2 \text{ with remainder } 1$$

$$\frac{6}{2} = 3 \text{ with remainder } 0$$

$$\frac{7}{2} = 3 \text{ with remainder } 1$$

$$\frac{8}{2} = 4 \text{ with remainder } 0$$

We can represent this algorithm as follows.

A possible flowchart to represent this information is shown below.

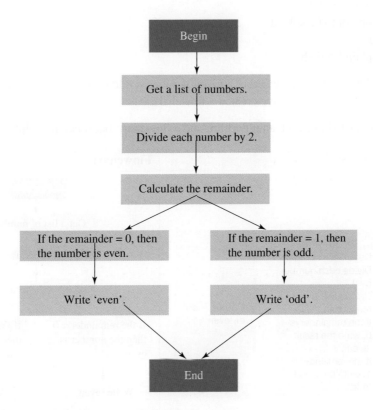

This is a slightly more complex example of an algorithm than the previous examples and illustrates that as processes become more complex, we need better ways to work with them.

Complex problems are better solved using computers, but we need to instruct the computer on what to do by writing computer programs. To make sure the computer can understand the program, it must be written in a specific computer language. There are many computer languages around, such as Python, Java, C++, Lua and older languages such as Pascal, Fortran, Cobol and Basic. In this topic we will not delve into any particular language. Instead, we will use pseudocode, which avoids the specific rules and notations used in each of these computer languages.

18.2.2 Pseudocode

In the next lesson, you will learn more deeply about pseudocode, but let's introduce it here as it helps to express algorithms and assists with algorithmic thinking.

Pseudocode is a plain English description of the step-by-step sequence used in solving an algorithm. Pseudocode often uses some of the conventions of programming language but is intended for human reading rather than machine reading, and is independent of the particular syntax, that is the rules and requirements of a specific programming language.

Computer scientists who solve problems usually start with pseudocode before writing the code for a specific programming language. It helps them to create the outline or structure for the code, very similar to writing the outline or plan for an essay before writing the full essay.

Each person will write pseudocode in their own style.

To write pseudocode, you need to:
- understand the problem to be solved
- have an algorithm
- write the code in plain English.

Taking our earlier problem of checking whether a given number is odd or even, we can write a pseudocode. So far, we have written the algorithm and have created a flowchart.

Note that it's not necessary to have both an algorithm and a flowchart, but it can be helpful.

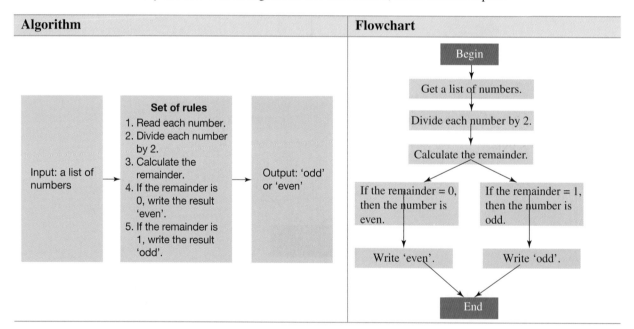

Algorithm	Flowchart

Algorithm:

Input: a list of numbers →

Set of rules
1. Read each number.
2. Divide each number by 2.
3. Calculate the remainder.
4. If the remainder is 0, write the result 'even'.
5. If the remainder is 1, write the result 'odd'.

→ Output: 'odd' or 'even'

Flowchart:

Begin → Get a list of numbers. → Divide each number by 2. → Calculate the remainder. → If the remainder = 0, then the number is even. / If the remainder = 1, then the number is odd. → Write 'even'. / Write 'odd'. → End

---

WORKED EXAMPLE 2 Writing pseudocode

Write a pseudocode to check whether a given number is odd or even.

THINK	WRITE
1. Start each pseudocode with 'Begin'.	Begin
2. The first requirement would be to input a list of numbers so that we can test whether each number is odd or even.	Get a list of numbers.

3. Next, we need to read each number in the list separately. This can also be described as iterating or looping through each number.	Iterate through each number in the list.
4. Examine each input number one at a time. Assign a variable to represent each number; let's use x.	For each number, x, in the list,
5. Divide each number by 2 and record the remainder, y, which will be either 0 or 1.	$\dfrac{x}{2}$ has a remainder of 0 or 1. Let the remainder equal y.
6. Check if $y = 0$.	Check if $y = 0$, then the number x is even.
7. Check if $y = 1$.	Check if $y = 1$, then the number x is odd.
8. Once all numbers in the list are iterated through, then the code can end.	End

The pseudocode above is an example of turning an algorithm into plain English that can later be converted into a specific programming language. The focus of pseudocode is on solving the problem and being able to articulate the solution to the problem. All programmers need to be able to think systematically; pseudocode frees them up from having to worry about the syntax or machine code of a particular programming language, which can come later.

Equally important is the fact that anybody should be able to read pseudocode and be able to understand it. The reader doesn't need to understand the technical syntax of programming.

Exercise 18.2 Algorithms

learn on

18.2 Quick quiz on	18.2 Exercise

Individual pathways

■ PRACTISE	■ CONSOLIDATE	■ MASTER
1, 4, 5, 9	2, 6, 8, 10	3, 7, 11

Fluency

1. Why is it important to break problems down into smaller steps when designing an algorithm?

2. How does pseudocode help in the transition from problem solving to actual coding?

3. In what real-life scenarios can algorithms be applied?

4. Why is it essential to express algorithms clearly and unambiguously?

Understanding

5. **WE1** Write an algorithm to find the maximum of two numbers.

6. **WE2** Create a pseudocode for checking if a given number is even or odd. *Note:* Try to write a version different to the Worked example.

7. Develop an algorithm to calculate the average of three numbers.

Reasoning

8. Write a pseudocode to check if a given year is a leap year. Provide an explanation.

9. Design an algorithm to swap the values of two variables. Provide an explanation.

Problem solving

10. Develop an algorithm to check if a given word is a palindrome. Provide an explanation.
    Note: A palindrome is a word, phrase, number or other sequence of characters that reads the same forward and backward. For example, the word DEED is a palindrome.

11. Design an algorithm to convert degrees Celsius to degrees Fahrenheit.

LESSON
18.3 Pseudocode

LEARNING INTENTION

At the end of this lesson you should be able to:
* write pseudocode to perform simple calculations and tasks.

18.3.1 Pseudocode and the concept of programming

In the previous lesson, you were introduced to pseudocode as part of the work on algorithms. In this lesson, you will learn more deeply about pseudocode.

Pseudocode helps us to understand the **concepts** of programming, some of which are:
* iterations
* loops
* conditions (e.g. if this, then that).

At its simplest, the structure of pseudocode is as follows.

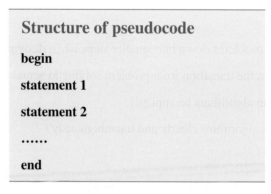

Write a pseudocode to calculate the perimeter and area of a rectangle.

THINK	WRITE
1. Start the code with begin. Determine which input variables and which output variables are needed. In this situation we need input variables *length* and *width* and output variables *perimeter* and *area*. These are all numeric variables, so write them out on the second line, separated by commas as shown.	begin numeric length, width, perimeter, area
2. We need the user to input the values of the input variables. Use 'display' to give some context to the user about which variable they are inputting.	display 'Enter the length.' input length display 'Enter the width.' input width
3. Perform the required calculations. Calculate the perimeter and area of the rectangle from the input variables. Note that we use * to indicate multiplication.	perimeter = 2 * length + 2 * width area = length * width
4. We need to output the results of our calculated output variables.	display 'The perimeter of the rectangle is', perimeter display 'The area of the rectangle is', area
5. End the code with end.	end

The complete pseudocode is shown below. Note the indentation to make the blocks of code easier to read.

```
begin
        numeric length, width, perimeter, area
        display 'Enter the length.'
        input length
        display 'Enter the width.'
        input width
                perimeter = 2 * length + 2 * width
                area = length * width
        display 'The perimeter of the rectangle is', perimeter
        display 'The area of the rectangle is', area
end
```

18.3.2 Selection or conditional constructs

For Worked example 3, the statements are executed line by line in order. Often in programming or pseudocode we need to decide which set of statements or blocks of statements are selected or repeated. One way of achieving this will depend on a statement that is a proposition that evaluates to a Boolean expression of 'true' or 'false'. This is done using the if ... (Boolean expression) ... then ... endif construct.

Selection constructs

if (Boolean expression) then

> {some lines of code that will be executed if the Boolean expression in the above line is true}

endif

Note: If the Boolean expression is false, the lines of code are not executed and the pseudocode moves on to the following lines.

WORKED EXAMPLE 4 Determining the greatest of two numbers

Write a pseudocode to determine which of two numbers is greater.

THINK	WRITE
1. Start the code with begin. Only two input variables are need in this problem, and there are no other output variables. We will call them num1 and num2.	begin numeric num1, num2
2. Display and prompt for the two input variables.	display 'Enter the first number.' input num1 display 'Enter the second number.' input num2

3. Use the if statement to decide if the first number is the largest of the two numbers.	if (num1 > num2) then
	    display 'The first number', num1, 'is the largest.'
	endif

4. Use the if statement to decide if the second number is the largest of the two numbers.	if (num2 > num1) then
	    display 'The second number', num2, 'is the largest.'
	endif

5. If neither of the above two conditions are met, test to check if the two numbers are equal. Note that the two equal signs are used to test if the values are equal, not to assign the value of num1 to the value of num2.	if (num1 == num2) then
	    display 'The two numbers are equal.'
	endif

6. End the code with end.	end

The complete pseudocode is shown below.

> begin
>
>   numeric num1, num2
>
>   display 'Enter the first number.'
>
>   input num1
>
>   display 'Enter the second number.'
>
>   input num2

```
                if (num1 > num2) then

                        display 'The first number', num1, 'is the largest.'

                endif

                if (num2 > num2) then

                        display 'The second number', num2, 'is the largest.'

                endif

                if (num1 == num2) then

                        display 'The two numbers are equal.'

                endif

        end
```

18.3.3 Extended if then elseif

Sometimes there will be more than one case to consider, or the cases are mutually exclusive, that is if one occurs then the other can't. In these situations we use the if... then... elseif... else... endif constructs. Also, we can combine Boolean expressions using 'and', 'or' to get true or false values.

if then elseif

if (Boolean expression) then

    {execute these lines of pseudocode if the above Boolean expression is true}

elseif (Boolean expression)

    {execute these lines of pseudocode if the above Boolean expression is true}

elseif (Boolean expression)

    {execute these lines of pseudocode if the above Boolean expression is true}

else

    {execute these lines of pseudocode if none of the above Boolean expressions are true}

endif

Note that pseudocode is not necessarily unique; there are many ways to write code and to achieve the same outputs. For example, the pseudocode for Worked example 4 could also be written as:

```
        begin

                numeric num1, num2

                display 'Enter the first number.'

                input num1

                display 'Enter the second number.'

                input num2

                if (num1 == num2) then                               ▶
```

```
                    display 'The numbers are equal.'

            elseif (num1 > num2) then

                    display 'The first number', num1, 'is the largest.'

            else

                    display 'The second number', num2, 'is the largest.'

            endif

    end
```

WORKED EXAMPLE 5 Determining the maximum of three numbers

Write a pseudocode to determine the maximum of three numbers.

THINK	WRITE
1. Start the code with begin. Three input variables are needed in this problem, and one output variable to store the information about which variable is the maximum. We will call the input variables num1, num2 and num3, and the output variable maxx. (Sometimes in programming languages the word max is a reserved word and cannot be used as a variable name.)	begin numeric num1, num2, num3, maxx
2. Display and prompt for the three input variables.	display 'Enter the first number.' input num1 display 'Enter the second number.' input num2 display 'Enter the third number.' input num3

3. Use the if statement to decide the maximum of the first two numbers and assign the variable maxx to one of these.	if (num1 < num2) then
	    maxx = num2
	else
	    maxx = num1
	endif

4. Use the if statement to decide if the third number is greater than the maximum of the other two numbers.	if (num3 > maxx) then
	    maxx = num3
	endif

5. Output the largest of the three numbers.	display 'The maximum of the three numbers is', maxx
6. End the code with end.	end

The complete pseudocode is shown below.

```
begin
        numeric num1, num2, num3, maxx
        display 'Enter the first number.'
        input num1
        display 'Enter the second number.'
        input num2
        display 'Enter the third number.'
        input num3
        if (num1 < num2) then
            maxx = num2
        else
            maxx = num1
        endif
        if (num3 > maxx) then
            maxx = num3
        endif
        display 'The maximum of the three numbers is', maxx
end
```

WORKED EXAMPLE 6 Displaying different messages based on the time

Write a pseudocode to take the time from a 24-hour clock and output 'Good morning' between 12 am and 12 pm, 'Good afternoon' between 12 pm and 6 pm, 'Have a good night' between 6 pm and 10 pm, or 'Time to go to bed' between 10 pm and 12 am.

THINK

1. Start the code with begin.
   Only one input is needed in this problem, the current time.

WRITE

begin

    numeric time

2. Display and prompt for the time.

display 'Using a 24-hour clock, enter the time as a decimal. For example, 13:27 = 13.27.'

input time

3. Use the if statement to decide whether the time is between 12 am and 12 pm.

if ((time ≥ 0) and (time < 12)) then
    display 'Good morning'

4. Use the elseif statement to decide whether the time is between 12 pm and 6 pm.

elseif ((time ≥ 12) and (time < 18)) then
    display 'Good afternoon'

5. Use the elseif statement to decide whether the time is between 6 pm and 10 pm.

elseif ((time ≥ 18) and (time < 22)) then
    display 'Have a good night'

6. Use the elseif statement to decide whether the time is between 10 pm and 12 am.

elseif ((time ≥ 22) and (time < 24)) then
    display 'Time to go to bed'

7. If none of the above conditions are met, the time must be invalid. End the if...then...elseif block with endif.

else
    display 'Invalid time'
endif

8. End the code with end.

end

The complete pseudocode is shown below.

```
begin

    numeric time

    display 'Using a 24-hour clock, enter the time as a decimal. For example, 13:27 = 13.27.'

    input time

if ((time ≥ 0) and (time < 12)) then

    display 'Good morning'

elseif ((time ≥ 12) and (time < 18)) then

    display 'Good afternoon'

elseif ((time ≥ 18) and (time < 22)) then

    display 'Have a good night'

elseif ((time ≥ 22) and (time < 24)) then

    display 'Time to go to bed'

else

    display 'Invalid time'

endif

end
```

18.3.4 Programming using CAS calculators

CAS calculators are capable of creating and running programs such as the ones we have been modelling throughout this topic. The language and methodology differs depending on the CAS calculator used; however, they all use similar steps to those we have been using in the pseudocodes throughout this topic. The following Worked example is a repeat of the previous one, using the TI-Nspire and CASIO ClassPad CAS calculators.

Write a program on your CAS calculator to determine the maximum of three numbers.

TI	THINK	DISPLAY/WRITE	CASIO	THINK	DISPLAY/WRITE

1. Open a new TI-Nspire document and choose 9: Add Program Editor and 1: New... to open a new program.

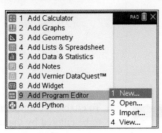

1. Open the Program application found on the second page of the Menu.

2. Type in a name for the program, call it 'maxof3' and tab and press OK.

2. Select Edit in the toolbar and then choose New File.

3. Some of the code is already present. Press MENU option 3: Define Variables and 1: Local.

3. Type in a name for the program. This example will be called 'maxof3'. Tap OK.

4. Type in the variable names we will use. Notice the keywords are coloured.

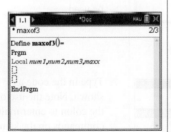

4. Define the variables to use by first selecting Misc > Variable > Local and then typing in the variable names.

5. To obtain the input, press MENU option 6: I/O and 3: Request.

6. Complete and repeat for the I/O (input output) for the other variables as shown.

7. To obtain other constructs to change the flow of the program, press MENU option 4: Control and 3: If…Then…Else…Endif.

5. To enable the numbers to be entered by the user, select the Input command in the I/O sub-menu and complete the coding as shown.

6. To obtain other command constructs to change the flow of the program, use the options available in the Ctrl menu. In this example the items of If…Then…IfEnd will be used.

7. Type in the code as shown. Note the use of the colon to enter multiple commands on the same line.

8. Repeat for If...Then...EndIf and type in the rest of the code as shown. Notice again the colour coding and the automatic indentation.

8. For the output display, select the menu I/O. Commonly used options include ClrText (clears any previous output from the screen) and Print (prints text and results of the program).

9. For the output, press MENU option 6: I/O and 1: Disp.

9. Complete the program as shown. This example will print a statement, the three numbers that were used and then the maximum of these numbers.

10. Complete the program as shown.

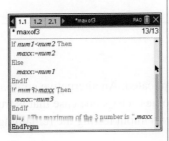

10. The program will test the syntax when Edit > Save File > Save is selected. If no errors are detected, the program will save. If not, an error screen will appear.

11. To check the syntax, press MENU option 2: Check Syntax & Store and 1: Check Syntax & Store (or the shortcut key Ctrl + B).
Provided there are no errors, that is the code is syntax free, you will get a message stating 'maxof3' stored successfully.

11. To run the program, select the 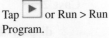 icon to open the Program Loader screen. Select the program to load by tapping the Name: down arrow button and selecting 'maxof3'.
Tap ▶ or Run > Run Program.

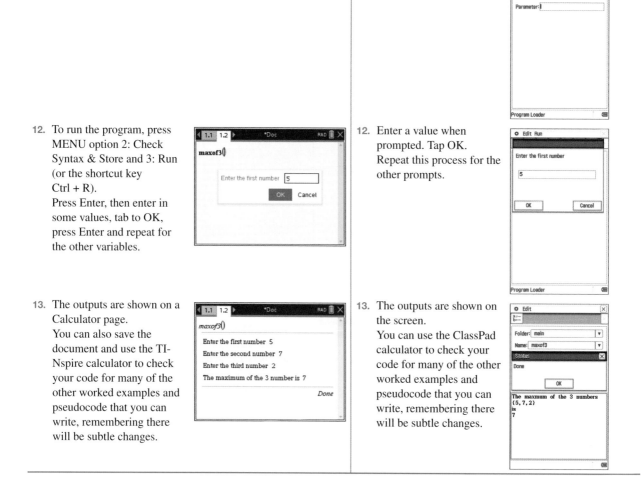

12. To run the program, press MENU option 2: Check Syntax & Store and 3: Run (or the shortcut key Ctrl + R).
Press Enter, then enter in some values, tab to OK, press Enter and repeat for the other variables.

12. Enter a value when prompted. Tap OK. Repeat this process for the other prompts.

13. The outputs are shown on a Calculator page.
You can also save the document and use the TI-Nspire calculator to check your code for many of the other worked examples and pseudocode that you can write, remembering there will be subtle changes.

13. The outputs are shown on the screen.
You can use the ClassPad calculator to check your code for many of the other worked examples and pseudocode that you can write, remembering there will be subtle changes.

18.3.5 Case constructs

Often 'if then else' constructs can become quite complicated. An alternative approach is case constructs. A case construct tests specific cases for a variable. Once it comes across one case that is true, it executes the following lines of code and then breaks out of the case construct without testing the remaining cases.

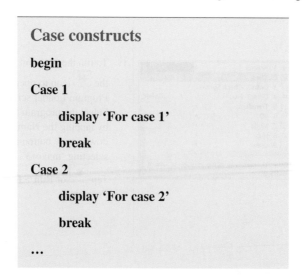

Case constructs

begin

Case 1

    display 'For case 1'

    break

Case 2

    display 'For case 2'

    break

...

```
default
        display 'Out of range'
break
endcase
end
```

Note: The case command is not available in the programming function on the TI-Nspire CAS calculator.

WORKED EXAMPLE 8 Writing a pseudocode using a case command

Write pseudocode to list the day of the week, with Monday being 1 and Sunday being 7, and the day number entered as input.

THINK	WRITE
1. Start the code with begin. Only one input variable is needed in the problem, no output variables. We will call it daycode.	begin numeric daycode
2. Display and prompt for the daycode.	display 'Enter the day code.' input daycode
3. As 1 represents Monday, use the case statement to display the corresponding weekday name and break out of the case.	begin case 1 display 'Monday' break
4. As 2 represents Tuesday, use the case statement to display the corresponding weekday name.	case 2 display 'Tuesday' break
5. Repeat for the other days of the week.	case 3 display 'Wednesday' break case 4 display 'Thursday' break case 5 display 'Friday' break

case 6

    display 'Saturday'

    break

case 7

    display 'Sunday'

    break

6. Generate an error message if the daycode is invalid. The message 'Value not valid' will be displayed if the number entered is not an integer between 1 and 7 inclusive.

default

    display 'Value not valid'

    break

7. End the cases.

endcase

8. End the code with end.

end

18.3.6 Repetition constructs

When a block of statements needs to be repeated a fixed known number of times, we use the for loop. These repetition constructs are also known as iterations. A for loop has the form:

for (index variable name = initial value, final value, increment value)

The index variable name that is commonly used for this variable is i, j or counter; the initial value is usually 1; and the incremental value again is usually 1, and if omitted will be assumed to be 1. Note that again we will use indentation and finish the block with endfor to indicate that all of these statements within this loop will be executed and repeated.

Repetition

for (index variable name = initial value, final value, increment value)

    statement 1

    statement 2

    ...

endfor

WORKED EXAMPLE 9 Creating a list of numbers and their squares (1)

Write a pseudocode using a for loop to print out the first 10 natural numbers and their squares.

THINK	WRITE
1. Start the code with begin. No input variables are needed in the problem, only output variables, which are the counter and the value squared.	begin numeric counter, squaredvalue
2. We need a heading indicating what will be displayed.	display 'Number Square'
3. Write the for loop.	for (counter = 1, 10)

4. We need to calculate and display the output, and end the for loop.	squaredvalue = counter * counter display counter, squaredvalue endfor
5. End the code with end.	end

The complete pseudocode is shown below.

```
begin
        display 'Number Square'
        numeric counter, valuesquared
        for (counter = 1, 10)
                square = counter * counter
                display counter, square
        endfor
end
```

Note that the output from this code will be:

Number	Square
1	1
2	4
3	9
4	16
5	25
6	36
7	49
8	64
9	81
10	100

In the Worked example above the statements are executed a fixed number of times that is known in advance. We could have an input to determine the number of times the loop would be executed. Sometimes we may need to make decisions when a variable value has changed.

In mathematics a statement such as $x = x + 1$ is meaningless and has no solution, but in coding it is often used. What it is really saying is increment or increase the value of x by 1. Using this we can use the while (Boolean expression is true)... endwhile construct as a loop that is pre-tested at the beginning of the block. When the conditional Boolean expression is true, repeat the block of statements up to the endwhile, but when the condition is false, the block of statements is not executed. Within the block of code we need to change a value so that the Boolean expression that is tested every time must eventually become false, otherwise the loop and the code will never terminate. Note that if the condition of the while loop is initially false, then the while loop will not be executed at all.

Often code can be written in many ways. In the next example we repeat the last example using the while... endwhile block.

WORKED EXAMPLE 10 Creating a list of numbers and their squares (2)

Write a pseudocode using a while loop to print out the first 10 natural numbers and their squares.

THINK	WRITE
1. Start the code with begin. No input variables are needed in the problem, only output variables, which are the counter and the value squared.	begin numeric counter, squaredvalue
2. We need to initialise the value of counter before the while loop starts.	counter = 1
3. We need a heading indicating what will be displayed.	display 'Number Square'
4. Write the while loop.	while (counter <=10)
5. We need to calculate and display the output, and increment the value of counter.	squaredvalue = counter * counter display counter, squarevalue counter = counter + 1
6. End the while loop.	endwhile
7. End the code with end.	end

The complete pseudocode is shown below and the output is the same as from Worked example 9.

The statements inside the while loop are executed and tested. As soon as the variable counter becomes greater than 10, the condition is false and the loop is terminated.

```
begin
        numeric counter, squaredvalue
        counter = 1
        display 'Number Square'
        while (counter <=10)
                squaredvalue = counter * counter
                display counter, squarevalue
                counter = counter + 1
        endwhile
end
```

18.3.7 Using functions

Often when writing pseudocode we may need to use other built-in mathematical functions and mathematical notations. Some of these are described in this section.

For example, we will assume the value of pi (π) is built in and has the value 3.14159.

When raising number to powers with indices, we write for example x^n. In pseudocode we can write this as $x{**}n$, and for square roots, $\sqrt{x}$, we can write sqrt(x). Often we may need the Boolean expression 'not equal to', for which we will use !=.

Operations on integers are useful as well, for example mod$(14, 3) = 2$, so that mod gives the remainder of the division. Other functions to obtain random numbers include randInt$(1, n)$, which gives a random integer between 1 and n inclusive. We will also assume that functions such as sin(x) and cos(x) are available and can be used when required.

Although the above notes are only an introduction to writing pseudocode, the core of writing pseudocode (and programming in general) is the ability to represent algorithms using constructs such as 'sequence', 'if … then … else', 'case', 'for loop' and 'while'. These constructs are also called keywords and are used to describe the flow of the algorithm. Note that variable names cannot contain any of these special words.

Exercise 18.3 Pseudocode

learn on

18.3 Quick quiz on	18.3 Exercise

Individual pathways

■ PRACTISE	■ CONSOLIDATE	■ MASTER
1, 6, 11, 12, 15, 19, 22	2, 3, 7, 8, 13, 14, 16, 20, 23	4, 5, 9, 10, 17, 18, 21, 24, 25

Fluency

For questions 1–5, write a pseudocode to determine each of the following.

1. **WE3** The area and circumference of a circle

2. The area and perimeter of a square

3. The average of two numbers

4. The volume of a cone

5. The area of a parallelogram

6. **WE4** Write a pseudocode to output a 'pass' if a student scores 50% or more on a test, or a 'fail' if a student scores less than 50% on the test. Both the student's mark and the total marks for the test will need to be given as inputs.

7. **WE5** Write a pseudocode to determine the minimum of three numbers.

8. Write a pseudocode to determine whether a number is even or odd.

9. **WE6** Write a pseudocode to give the grades of students' results; that is, output 'A' for marks 80% or greater, 'B' for marks 70–79, 'C' for marks 60–69, 'D' for marks 50–59 and 'E' for marks less than 50. Assume the value entered is a percentage rounded to the nearest whole number.

10. Write a pseudocode to input your age in years and if you are less than 12 years old, output 'Only a child'; if your age is between 13 and 19, output 'Just a teenager'; if your age is 20 to 29, output 'In your twenties'; if your age is 30 to 39, output 'In your thirties'; if your age is 40 to 49, output 'In your forties'; if your age is 50 to 59, output 'In your fifties'; if your age is 60 to 99, output 'You can retire'; otherwise, output 'Wow over 100'.

Understanding

For questions 11–13, write a pseudocode for each of the following.

11. **WE7** To list the months of the year according to the number value (assume that January corresponds to 1 and December corresponds to 12)

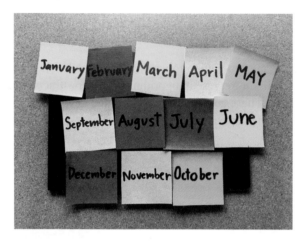

12. To decide if a letter of the alphabet is a vowel

13. To add, subtract, multiply or divide two numbers entered when prompted for the operation, if 1 is addition, 2 is subtraction, 3 is multiplication and 4 is division

14. **WE8** Write a pseudocode using for loops to print out the first 10 natural numbers and their square roots.

15. Write a pseudocode using for loops to sum all the numbers from 1 up to 100 inclusive.

16. Write a pseudocode using for loops to sum all the odd numbers from 1 up to n, when n is entered as input.

17. The factorial for a positive number or integer n, denoted or represented as $n!$, is the product of all the positive numbers preceding, for example $5! = 5 \times 4 \times 3 \times 2 \times 1 = 120$.
    Write a pseudocode to determine the value of the factorial of an input when entered.

18. The Fibonacci sequence of numbers is given by 1, 1, 2, 3, 5, 8, 13, ... where each number is the sum of the two previous numbers, the first term being 1, the second term being 1 and so on.
    Write a pseudocode to output all the Fibonacci numbers up to the nth value where the value of n is to be inputted.

Reasoning

19. **WE9** Write a pseudocode using while loops to print out the first 10 natural numbers and their square roots.

20. Write a pseudocode using while loops to sum all the numbers from 1 up to 100 inclusive.

21. Write a pseudocode using while loops to sum all the odd numbers from 1 up to n, when n is entered as input.

Problem solving

22. Write a pseudocode using while loops to determine the value of a factorial.

23. Write a pseudocode using while loops to determine the mean of a list of numbers that are inputted. The inputs continue until the number -1 is entered; this is a trigger to stop inputting numbers.

24. 'Guess the number game': Write a pseudocode to generate a random number between 1 and 100 inclusive, and let a user try to guess the number. The outputs are 'Too high', 'Too low' or 'Correct number', depending on the guess.
    Guesses continue until the correct number is guessed; the code then states the total number of guesses needed to guess the number.

25. Write a pseudocode to sort three numbers in ascending order.

LESSON
18.4 Review

18.4.1 Topic summary

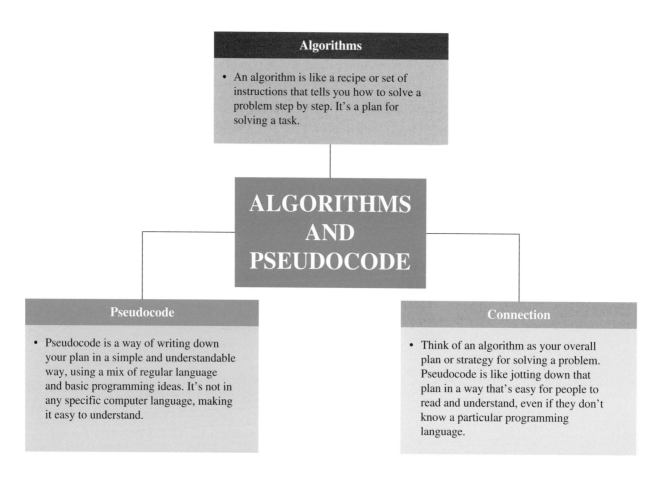

Algorithms

- An algorithm is like a recipe or set of instructions that tells you how to solve a problem step by step. It's a plan for solving a task.

ALGORITHMS AND PSEUDOCODE

Pseudocode

- Pseudocode is a way of writing down your plan in a simple and understandable way, using a mix of regular language and basic programming ideas. It's not in any specific computer language, making it easy to understand.

Connection

- Think of an algorithm as your overall plan or strategy for solving a problem. Pseudocode is like jotting down that plan in a way that's easy for people to read and understand, even if they don't know a particular programming language.

18.4.2 Success criteria

Tick the column to indicate that you have completed the lesson and how well you have understood it using the traffic light system.

(**Green:** I understand; **Yellow:** I can do it with help; **Red:** I do not understand)

Lesson	Success criteria	○	○	●
18.2	I understand how algorithms can be used to solve problems.			
18.3	I can write pseudocode to perform simple calculations and tasks.			

Fluency

1. Write an algorithm for cooking a steak.

Problem solving

2. Write a pseudocode for cooking a steak.

3. Create a pseudocode for a simple guessing game where the user has to guess a randomly chosen number between 1 and 100.

4. Write an algorithm expressed in pseudocode for the addition of two proper fractions.

5. Write an algorithm expressed in pseudocode for simplifying a ratio.

6. Write a pseudocode for factorising a quadratic expression in the form $ax^2 + bx + c$.

on To test your understanding and knowledge of this topic, go to your learnON title at www.jacplus.com.au and complete the **post-test**.

Answer

Topic 18 Algorithms and pseudocode

18.1 Pre-test

1. True
2. False
3. False
4. True
5. False
6. False
7. True
8. True
9. True
10. False
11. False
12. True
13. False
14. False
15. False

18.2 Algorithms

1–11. Please see the Worked solutions.

18.3 Pseudocode

1–25. Please see the Worked solutions.

18.4 Review questions

1–6. Please see the Worked solutions.

Semester review 2

The learnON platform is a powerful tool that enables students to complete revision independently and allows teachers to set mixed and spaced practice with ease.

Student self-study

Review the **Course Content** to determine which topics and subtopics you studied throughout the year. Notice the green bubbles showing which elements were covered.

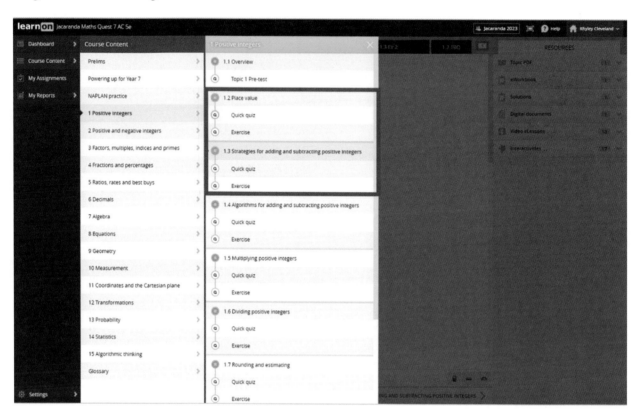

Review your results in **My Reports** and highlight the areas where you may need additional practice.

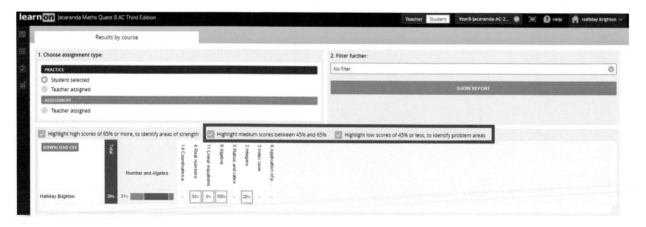

Use these and other tools to help identify areas of strengths and weakness and target those areas for improvement.

Teachers

It is possible to set questions that span multiple topics. These assignments can be given to individual students, to groups or to the whole class in a few easy steps.

Go to **Menu** and select **Assignments** and then **Create Assignment**. You can select questions from one or many topics simply by ticking the boxes as shown below.

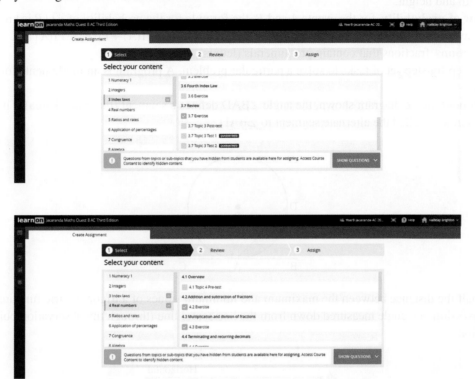

Once your selections are made, you can assign to your whole class or subsets of your class, with individualised start and finish times. You can also share with other teachers.

More instructions and helpful hints are available at www.jacplus.com.au.

GLOSSARY

3-dimensional a shape that occupies space (a solid); that is, one that has dimensions in three directions — length, width and height.

Addition Law of probability for the events A and B, the formula for the probability of $A \cup B$ is known as the Addition Law of probability and is given by the formula: $\Pr(A \cup B) = \Pr(A) + \Pr(B) - \Pr(A \cap B)$.

algebraic fractions fractions that contain pronumerals (letters).

algorithm a step-by-step set of tasks to solve a particular problem. A program is an implementation of an algorithm.

alternate segment in the diagram shown, the angle $\angle BAD$ defines a segment (the shaded area). The unshaded part of the circle is called the alternate segment to $\angle BAD$.

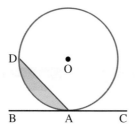

amplitude half the distance between the maximum and minimum values of a trigonometric function.

angle of depression the angle measured down from the horizontal line (through the observation point) to the line of vision.

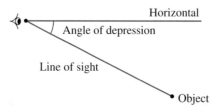

angle of elevation the angle measured up from the horizontal line (through the observation point) to the line of vision.

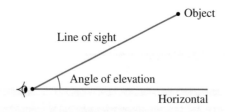

area the amount of flat surface enclosed by the shape. It is measured in square units, such as square metres, m^2, or square kilometres, km^2.

Associative Law a method of combining two numbers or algebraic expressions is associative if the result of the combination of these objects does not depend on the way in which the objects are grouped. Addition and multiplication obey the Associative Law, but subtraction and division are not associative.

asymptote a line that a graph approaches but never meets.

axis of symmetry a line through a shape so that each side is a mirror image.

back-to-back stem-and-leaf plot a method for comparing two data distributions by attaching two sets of 'leaves' to the same 'stem' in a stem-and-leaf plot; for example, comparing the pulse rate before and after exercise. A key should always be included.

Key: 8 | 6 = 86

Leaf Before exercise	Stem	Leaf After exercise
9 8 8 8	6	
8 6 6 4 1 1 0	7	
8 8 6 2	8	6 7 8 8
6 0	9	0 2 2 4 5 8 9 9
4	10	0 4 4
0	11	8
	12	4 4
	13	
	14	6

base the digit at the bottom of a number written in exponent form. For example, in 6^4, the base is 6. This tells us that 6 is multiplied by itself four times.

bias designing a questionnaire or choosing a method of sampling that would not be representative of the population as a whole.

bivariate data sets of data where each piece is represented by two variables.

Boolean a JavaScript data type with two possible values: true or false. JavaScript Booleans are used to make logical decisions.

boundary line indicates whether the points on a line satisfy the inequality.

box plot a graphical representation of the 5-number summary; that is, the lowest score, lower quartile, median, upper quartile and highest score, for a particular set of data. Two or more box plots can be drawn on the same scale to visually compare the five-number summaries of the data sets. These are called parallel box plots.

bridge an edge that disconnects the graph when removed

canvas a defined area on a web page where graphics can be drawn with JavaScript.

capacity the maximum amount of fluid that can be contained in an object. It is usually applied to the measurement of liquids and is measured in units such as millilitres (mL), litres (L) and kilolitres (kL).

Cartesian plane the area formed by a horizontal line with a scale (*x*-axis) joined to a vertical line with a scale (*y*-axis). The point of intersection of the lines is called the origin.

census collection of data from a population (e.g. all Year 10 students) rather than a sample.

centre middle point of a circle, equidistant (equal in distance) from all points on its circumference.

centre of enlargement the point from which the enlargement of an image is measured.

character in programming, a string of length 1. A JavaScript character is used to represent a letter, digit or symbol.

circle the general equation of a circle with centre (0, 0) and radius *r* is $x^2 + y^2 = r^2$.

circuit a trail beginning and ending at the same vertex.

circumcentre the centre of a circle drawn so that it touches all three vertices of a triangle.

circumcircle a circle drawn so that it touches all three vertices of a triangle.

clinometer a tool used to measure angles of inclination.

Closure Law when an operation is performed on an element (or elements) of a set, the result produced must also be an element of that set.

coincident lines that lie on top of each other.

collinear points points that all lie on the same straight line.

combination a group of objects where order is not important.

Commutative Law a method of combining two numbers or algebraic expressions is commutative if the result of the combination does not depend on the order in which the objects are given. For example, the addition of 2 and 3 is commutative, since $2 + 3 = 3 + 2$. However, subtraction is not commutative, since $2 - 3 \neq 3 - 2$.

compass bearing directions measured in degrees from the north–south line in either a clockwise or anticlockwise direction. To write the compass bearing we need to state whether the angle is measured from the north or south, the size of the angle and whether the angle is measured in the direction of east or west; for example, N27°W, S32°E.

complement the complement of a set, A, written A', is the set of elements that are in ξ but not in A.

complementary angles two angles that add to 90°; for example, 24° and 66° are complementary angles.

complementary events events that have no common elements and together make up the sample space. If A and A' are complementary events, then $\Pr(A) + \Pr(A') = 1$.

completing the square the process of writing a general quadratic expression in turning point form.

composite figure a figure made up of more than one basic shape.

compound interest interest that is paid on the sum of the principal plus preceding interest over time; $A = P(1 + r)^n$.

compounded value the value of an investment with accrued interest included.

compounding period the period of time over which compound interest is calculated.

concave polygon a polygon with at least one reflex interior angle.

concyclic points that lie on the circumference of a circle.

conditional probability where the probability of an event is conditional (depends) on another event occurring first. For two events A and B, the conditional probability of event B, given that event A occurs, is denoted by $P(B|A)$ and can be calculated using the formula: $P(B|A) = \dfrac{P(A \cap B)}{P(A)}, P(A) \neq 0$.

congruent triangles there are five standard congruence tests for triangles: SSS (side, side, side), SAS (side, included angle, side), ASA (two angles and one side), AAS (two angles and a non-included side) and RHS (right angle, hypotenuse, side).

conjugate surds surds that, when multiplied together, result in a rational number. For example, $\left(\sqrt{a} + \sqrt{b}\right)$ and $\left(\sqrt{a} - \sqrt{b}\right)$ are conjugate surds, because $\left(\sqrt{a} + \sqrt{b}\right) \times \left(\sqrt{a} - \sqrt{b}\right) = a - b$.

connected graph a graph in which it is possible to reach every vertex by moving along the edges.

console a special region in a web browser for monitoring the running of JavaScript programs.

constant of proportionality used to prove that a proportionality relationship (direct or inverse) exists between two or more variables (or quantities).

convex polygon a polygon with no interior reflex angles.

coordinates a pair of values (typically x and y) that represent a point on the screen.

correlation a measure of the relationship between two variables. Correlation can be classified as linear, non-linear, positive, negative, weak, moderate or strong.

cosine ratio the ratio of the adjacent side to the hypotenuse in a right-angled triangle; $\cos \theta = \dfrac{\text{adjacent}}{\text{hypotenuse}}$.

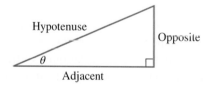

cosine rule in any triangle ABC, $a^2 = b^2 + c^2 - 2bc \cos(A)$.

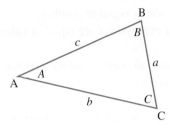

cubic function a polynomial in which the highest power of the variable is 3. The basic form of a cubic function is $y = ax^3$. These functions can have 1, 2 or 3 roots.

cumulative frequency curve a line graph that is formed when the cumulative frequencies of a set of data are plotted against the end points of their respective class intervals and then joined up by straight-line segments. It is also called an ogive.

cumulative frequency the total of all frequencies up to and including the frequency for a particular score in a frequency distribution.

cycle a path beginning and ending at the same vertex.

cyclic quadrilateral a quadrilateral that has all four vertices on the circumference of a circle. That is, the quadrilateral is inscribed in the circle.

data various forms of information.

degree (geometry) a unit used to measure the size of an angle; (functions and relations) the degree of a polynomial in x is the highest power of x in the expression; (network theory) the total number of edges directly connected to a vertex, except that a loop at a vertex contributes twice to the degree of that vertex.

degrees a unit used to measure the size of an angle.

denominator the lower number of a fraction that represents the number of equal fractional parts a whole has been divided into.

dependent events successive events in which one event affects the occurrence of the next.

dependent variable this variable is graphed on the y-axis.

deviation the difference between a data value and the mean.

digraphs *see* **directed graphs**

dilation occurs when a graph or figure is made thinner or wider.

dimension lines thin continuous lines showing the dimensions of lines on scale drawings.

directed graphs graphs that only allow movement in a specified direction; also called digraphs.

disconnected graph a graph in which it is not possible to reach every vertex by moving along the edges.

discriminant referring to the quadratic equation $ax^2 + bx + c = 0$, the discriminant is given by $\Delta = b^2 - 4ac$. It is the expression under the square-root sign in the quadratic formula and can be used to determine the number and type of solutions of a quadratic equation.

domain the set of all allowable values of x for a function or relation.

dot plot this graphical representation uses one dot to represent a single observation. Dots are placed in columns or rows, so that each column or row corresponds to a single category or observation.

edge a connection in a network, represented by a line.

elimination method a method used to solve simultaneous equations. This method combines the two equations into a third equation involving only one of the variables.

enlargement a scaled-up (or down) version of a figure in which the transformed figure is in proportion to the original figure; that is, the two figures are similar.

equate the process of writing one expression as equal to another.

equation a statement that asserts that two expressions are equal in value. An equation must have an equal sign. For example, $x + 4 = 12$.

equiangular when two or more shapes have all corresponding angles equal.

equilateral triangle a triangle with all sides equal in length, and all angles equal to 60°.

Euler trail a trail in which every edge is used once.

Eulerian circuit (or Euler circuit) an Euler trail that starts and ends at the same vertex.

event space a list of all the possible outcomes obtained from a probability experiment. It is written as ξ or S, and the list is enclosed in a pair of curled brackets { }. It is also called the sample space.

experimental probability the probability of an event based on the outcomes of experiments, simulations or surveys.

explanatory variable this variable *is not impacted* by the other variable. This is the x-axis (or horizontal) variable. Also known as the independent variable.

exponent the number that indicates how many times the base is being multiplied by itself when an expression is written in exponent form, also known as an *index* or *power*.

exponent form the form a^x expressing a number in terms of a base, a, and an index x.

exponential decay a quantity that decreases by a constant percentage in each fixed period of time. This growth can be modelled by exponential functions of the type $y = ka^x$, where $a > 1$.

exponential function relationship of the form $y = ka^x$, where $a \neq 1$, are called exponential functions with base a.

exponential growth a quantity that grows by a constant percentage in each fixed period of time. This growth can be modelled by exponential functions of the type $y = ka^x$, where $0 < a < 1$.

extrapolation the process of predicting a value of a variable outside the range of the data.

factor theorem If $P(x)$ is a polynomial, and $P(a) = 0$ for some number a, then $P(x)$ is divisible by $(x - a)$.

factorial $n!$ (n factorial) is the product of each of the integers from n down to 1.

FOIL a diagrammatic method of expanding a pair of brackets. The letters in FOIL represent the order of the expansion: First, Outer, Inner and Last.

frequency the number of times a particular score appears.

function a process that takes a set of x-values and produces a related set of y-values. For each distinct x-value, three is only one related y-value. They are usually defined by a formula for $f(x)$ in terms of x; for example $f(x) = x^2$.

future value the future value of a loan or investment.

half plane a region that represents all the points that satisfy an inequality.

Hamiltonian cycle a Hamiltonian path that starts and ends at the same vertex.

Hamiltonian path a path that passes through every vertex exactly once.

Heron's formula this formula is used to calculate the area of a triangle when all three sides are known. The formula is $A = \sqrt{s(s-a)(s-b)(s-c)}$, where a, b and c are the lengths of the sides and s is the semi-perimeter or $s = \dfrac{a+b+c}{2}$.

histogram graph that displays continuous numerical variables and does not retain all original data.

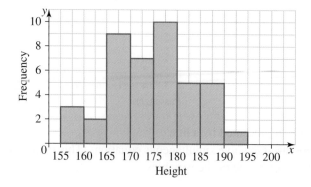

hyperbola the graph of $y = \dfrac{1}{x}$ is a rectangular hyperbola with asymptotes on the x- and y-axes.

hypotenuse the longest side of a right-angled triangle. It is the side opposite the right angle.

Identity Law when 0 is added to an expression or the expression is multiplied by 1, the value of the expression does not change. For example, $x + 0 = x$ and $x \times 1 = x$.

image the enlarged (or reduced) figure produced.

independent events successive events that have no effect on each other.

independent variable this variable *is not impacted* by the other variable. This is the x-axis (or horizontal) variable. Also known as the explanatory variable.

index (algorithms) an integer that points to a particular item in an array.

inequality when one algebraic expression or one number is greater than or less than another.

inequation similar to equations, but contain an inequality sign instead of an equal sign. For example, $x = 3$ is an equation, but $x = 3$ is an inequation.

integers these include the positive and negative whole numbers, as well as zero; that is, $\dots, -3, -2, -1, 0, 1, 2, \dots$

interpolation the process of predicting a value of a variable from within the range of the data.

interquartile range (IQR) the difference between the upper (or third) quartile, Q_{upper} (or Q_3), and the lower (or first) quartile, Q_{lower} (or Q_1); that is, $IQR = Q_{upper} - Q_{lower} = Q_3 - Q_1$. It is the range of approximately the middle half of the data.

inverse function the reflection of a function across the line $y = x$.

Inverse Law when the additive inverse of a number or pronumeral is added to itself, the sum is equal to 0. When the multiplicative inverse of a number or pronumeral is multiplied by itself, the product is equal to 1. So, $x + (-x) = 0$ and $x \times \dfrac{1}{x} = 1$.

inverse proportion describes a particular relationship between two variables (or quantities); that is, as one variable increases, the other decreases. The rule used to relate the two variables is $y = \dfrac{k}{x}$.

irrational number numbers that cannot be written as fractions. Examples of irrational numbers include surds, π and non-terminating, non-recurring decimals.

isomorphic graphs graphs that represent exactly the same information. Isomorphic graphs have the same number of vertices and edges, with corresponding vertices having identical degrees and connections.

isosceles triangle a triangle with exactly two sides equal in length.

leading coefficient the coefficient of the leading term in a polynomial.

leading term the term in a polynomial that contains the highest power of x.

line of best fit a straight line that best fits the data points of a scatterplot that appear to follow a linear trend. It is positioned on the scatterplot so that there is approximately an equal number of data points on either side of the line, and so that all the points are as close to the line as possible.

line segment a line segment or interval is a part of a line with end points.

linear graph consist of an infinite number of points that can be joined to form a straight line.

linked list a list of objects. Each object stores data and points to the next object in the list. The last object points to a terminator to indicate the end of the list.

literal equation an equation that includes two or more pronumerals or variables.

logarithm the power to which a given positive number b, called the base, must be raised in order to produce the number x. The logarithm of x, to the base b, is denoted by $\log_b (x)$. Algebraically: $\log_b (x) = y \leftrightarrow b^y = x$; for example, $\log_{10} (100) = 2$ because $10^2 = 100$.

logarithmic equation an equation that requires the application of the laws of indices and logarithms to solve.

loop in JavaScript, a process that executes the same code many times with different data each time; in network theory, an edge that connects a vertex to itself.

many-to-many relation a relation in which one range value may yield more than one domain value and vice versa.

many-to-one relation a function or mapping that takes the same value for at least two different elements of its domain.

matrix a rectangular array of numbers arranged in rows and columns.

maximal domain the limit of the x-values that a function can have.

mean one measure of the centre of a set of data. It is given by mean $= \dfrac{\text{sum of all scores}}{\text{number of scores}}$ or $\bar{x} = \dfrac{\sum x}{n}$. When data are presented in a frequency distribution table, $\bar{x} = \dfrac{\sum(f \times x)}{n}$.

measures of central tendency mean, median and mode.

measures of spread range, interquartile range, standard deviation.

median one measure of the centre of a set of data. It is the middle score for an odd number of scores arranged in numerical order. If there is an even number of scores, the median is the mean of the two middle scores when they are ordered. Its location is determined by the rule $\dfrac{n+1}{2}$. For example, the median value of the set 1 3 3 4 5 6 8 9 9 is 5, while the median value for the set 1 3 3 4 5 6 8 9 9 10 is the mean of 5 and 6 (5.5).

midpoint the midpoint of a line segment is the point that divides the segment into two equal parts. The coordinates of the midpoint M between the two points $P(x_1, y_1)$ and $Q(x_2, y_2)$ is given by the formula $\left(\dfrac{x_1 + x_2}{2}, \dfrac{y_1 + y_2}{2} \right)$.

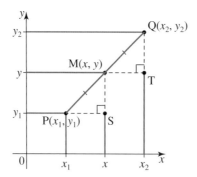

minimum spanning tree a spanning tree where the total distance along the tree branches is as small as possible; that is, it has the lowest total weighting.

mode one measure of the centre of a set of data. It is the score that occurs most often. There may be no mode, one mode or more than one mode (two or more scores occur equally frequently).

monic any polynomial with a leading coefficient of 1.

Multiplication Law of probability if events A and B are independent, then: $\Pr(A \text{ and } B) = \Pr(A) \times \Pr(B)$ or $\Pr(A \cap B) = \Pr(A) \times \Pr(B)$.

multiplication principle If there are n ways of performing operation A and m ways of performing operation B, then there are $n \times m$ ways of performing A *and* B in the order AB.

mutually exclusive events that cannot occur together. On a Venn diagram, two mutually exclusive events will appear as disjoint sets.

natural numbers the set of positive integers, or counting numbers; that is, the set 1, 2, 3, …

negatively skewed showing larger amounts of data as the values of the data increase.

nested loop a loop within a loop. The outer loop contains an inner loop. The first iteration of the outer loop triggers a full cycle of the inner loop until the inner loop completes. This triggers the second iteration of the outer loop, which triggers a full cycle of the inner loop again. This process continues until the outer loop finishes a full cycle.

node an object in a network, represented by a point; also known as a vertex.

number a JavaScript data type that represents a numerical value.

object a general JavaScript data type that can have many properties. Each property is a name–value pair so that the property has a name to reference a value.

ogive a graph formed by joining the top right-hand corners of the columns of a cumulative frequency histogram.

one-dimensional array a simple array of values in which the values can be of any type except for another array.

one-to-many relation a relation in which there may be more than one range value for one domain value but only one domain value for each range value.

one-to-one relation a relation in which every element of the first set corresponds to one and only one element of the second set.

outlier a piece of data that is considerably different from the rest of the values in a set of data; for example, 24 is the outlier in the set of ages {12, 12, 13, 13, 13, 13, 13, 14, 14, 24}. Outliers sit $1.5 \times$ IQR or greater away from Q_1 or Q_3.

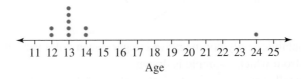

parabola the graph of a quadratic function has the shape of a parabola. For example, the typical shape is that of the graph of $y = x^2$.

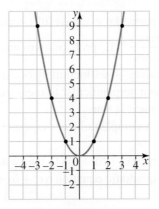

parallel parallel lines in a plane never meet, no matter how far they are extended. Parallel lines have the same gradient.

parallelogram a quadrilateral with both pairs of opposite sides parallel.

path a walk with no repeated vertices and no repeated edges.

percentile the value below which a given percentage of all scores lie. For example, the 20 th percentile is the value below which 20% of the scores in the set of data lie.

period the distance between repeating peaks or troughs of periodic functions.

periodic function a function that has a graph that repeats continuously in cycles, for example, graphs of $y = \sin(x)$ and $y = \cos(x)$.

permutation an arrangement of objects in a particular order.

perpendicular perpendicular lines are at right angles to each other. The product of the gradients of two perpendicular lines is -1.

pi (π) the Greek letter π represents the ratio of the circumference of any circle to its diameter. The number π is irrational, with an approximate value of $\frac{22}{7}$, and a decimal value of $\pi = 3.141\ 59 \ldots$.

planar network a network or graph in which the edges do not cross each other.

point of intersection a point where two or more lines intersect. The solution to a pair of simultaneous equations can be found by graphing the two equations and identifying the coordinates of the point of intersection.

pointer in JavaScript, a variable that points to a JavaScript object or array. Multiple pointers can point to the same object or array.

polygon a plane figure bounded by line segments.

polynomial an expression containing only non-negative integer powers of a variable.

population the whole group from which a sample is drawn.

positively skewed showing smaller amounts of data as the values of the data decrease.

Prim's algorithm a set of logical steps that can be used to identify the minimum spanning tree for a weighted connected graph.

primary data data collected by the user.

probability the likelihood or chance of a particular event (result) occurring.
$$\text{Pr(event)} = \frac{\text{number of favourable outcomes}}{\text{number of possible outcomes}}.$$ The probability of an event occurring ranges from 0 (impossible — will not occur) to 1 (certainty — will definitely occur).

projection lines thin continuous lines drawn perpendicular to measurements on scale drawings.

proof an argument that shows why a statement is true.

property references a value on an object. A complex object may have many properties. Each property has a unique name on the object.

quadratic equation an equation in the form $ax^2 + bx + c = 0$, where a, b and c are numbers.

quadratic formula gives the roots of the quadratic equation $ax^2 + bx + c = 0$. It is expressed as
$$x = \frac{-b \pm \sqrt{b^2 - 4ac}}{2a}.$$

quadratic function a polynomial in which the highest power of the variable is 2.

quantile percentiles expressed as decimals. For example, the 95th percentile is the same as the 0.95 quantile.

quartic function a polynomial in which the highest power of the variable is 4. The basic form of a quadratic function is $y = ax^4$. If the value of a is positive, the curve is upright, whereas a negative value of a results in an inverted graph. A maximum of 4 roots can result.

quartile values that divide an ordered set into four (approximately) equal parts. There are three quartiles — the first (or lower) quartile Q_1, the second quartile (or median) Q_2 and the third (or upper) quartile Q_3.

radius the straight line from a circle's centre to any point on its circumference.

range (functions and relations) the set of y-values produced by the function; (statistics) the difference between the highest and lowest scores in a set of data; that is, range = highest score − lowest score.

rational number numbers that can be written as fractions, where the denominator is not zero.

real numbers rational and irrational numbers combine to form the set of real numbers.

reciprocal when a number is multiplied by its reciprocal, the result is 1.

recurring decimal a decimal which has one or more digits repeated continuously; for example, 0.999 …. They can be expressed exactly by placing a dot or horizontal line over the repeating digits; for example,
$8.343\,434 = 8.\overset{..}{3}\overset{}{4}$ or $8.\overline{34}$.

reflection when a graph or figure is flipped in the x- or y-axis.

regression line a line of best fit that is created using technology.

regular polygon a polygon with sides of the same length and interior angles of the same size.

relation a set of ordered pairs.

relative frequency represents the frequency of a particular score divided by the total sum of the frequencies. It is given by the rule: relative frequency of a score $= \dfrac{\text{frequency of the score}}{\text{total sum of frequencies}}$.

remainder theorem if a polynomial $P(x)$ is divided by $x - a$, where a is any real number, the remainder is $P(a)$.

required region the region that contains the points that satisfy an inequality.

response variable another way to reference the dependent variable; this variable is graphed on the y-axis.

rhombus a parallelogram with all sides equal.

sample part of a population chosen so as to give information about the population as a whole.

sample space a list of all the possible outcomes obtained from a probability experiment. It is written as ξ or S, and the list is enclosed in a pair of curled brackets {}. It is also called the event space.

scale drawing a drawing on paper of a real-life object.

scale factor the ratio of the corresponding sides in similar figures, where the enlarged (or reduced) figure is referred to as the image and the original figure is called the object. scale factor $= \dfrac{\text{image length}}{\text{object length}}$

scatterplot a graphical representation of bivariate data that displays the degree of correlation between two variables. Each piece of data on a scatterplot is shown by a point. The x-coordinate of this point is the value of the independent variable and the y-coordinate is the corresponding value of the dependent variable.

secondary data data collected by others.

similar describes figures that have the same shape but different size.

similar triangles triangles that have similar shape but different size. There are four standard tests to determine whether two triangles are similar: AAA (angle, angle, angle), SAS (side, angle, side), SSS (side, side, side) and RHS (right angle, hypotenuse side).

simple graph a graph in which pairs of vertices are connected by at most one edge.

simulations probability experiments used to model real-life situations.

simultaneous equations the equations of two (or more) linear graphs that have the same solution.

simultaneous occurring at the same time.

sine ratio the ratio of the opposite side to the hypotenuse in a right-angled triangle; $\sin \theta = \dfrac{\text{opposite}}{\text{hypotenuse}}$.

sine rule in any triangle ABC, $\dfrac{a}{\sin A} = \dfrac{b}{\sin B} = \dfrac{c}{\sin C}$.

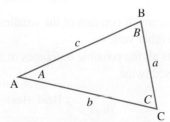

spanning trees trees that include all the vertices of the original graph.

standard deviation a measure of the variability of spread of a data set. It gives an indication of the degree to which the individual data values are spread around the mean.

string a JavaScript data type that represents text.

subject of an equation the variable that is expressed in terms of the other variables. In the equation $y = 3x + 4$, the variable y is the subject.

substitution method a method used to solve simultaneous equations. It is useful when one (or both) of the equations has one of the variables as the subject.

subtended In the diagram shown, chords AC and BC form the angle ACB. Arc AB has subtended angle ACB.

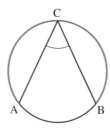

supplementary angles angles that add to 180°.

surd roots of numbers that do not have an exact answer, so they are irrational numbers. Surds themselves are exact numbers; for example, $\sqrt{6}$ or $\sqrt[3]{5}$.

symmetrical the identical size, shape and arrangement of parts of an object on opposite sides of a line or plane.

system of equations a set of two or more equations with the same variables.

tangent ratio the ratio of the opposite side to the adjacent side in a right-angled triangle; $\tan \theta = \dfrac{\text{opposite}}{\text{adjacent}}$.

terminating decimal a decimal which has a fixed number of places; for example, 0.6 and 2.54.

theorem rules or laws.

theoretical probability given by the rule $\text{Pr(event)} = \dfrac{\text{number of favourable outcomes}}{\text{number of possible outcomes}}$ or $\text{Pr}(E) = \dfrac{n(E)}{n(S)}$, where $n(E) =$ number of times or ways an event, E, can occur and $n(S) =$ number of elements in the sample space or number of ways all outcomes can occur, given all the outcomes are equally likely.

three-step chance experiment a probability experiment that involves three trials.

time series a sequence of measurements taken at regular intervals (daily, weekly, monthly and so on) over a certain period of time. They are used for analysing general trends and making predictions for the future.

total surface area (TSA) the area of the outside surface of a 3-dimensional figure.

trail a walk with no repeated edges.

transformation changes that occur to a graph or figure in order to obtain another graph or figure. Examples of transformations are translations, reflections or dilations.

translation movement of a graph left, right, up or down.

transversal a line that meets two or more other lines in a plane.

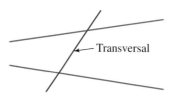

tree a simple connected graph with no circuits. It consists of the smallest number of edges necessary so that each vertex is connected to at least one other vertex.

tree diagram branching diagram that lists all the possible outcomes of a probability experiment. This diagram shows the outcomes when a coin is tossed twice.

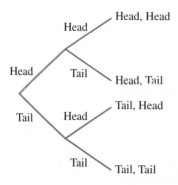

trial the number of times a probability experiment is conducted.

trigonometric ratios three different ratios of one side of a triangle to another. The three ratios are the sine, cosine and tangent.

true bearing a direction which is written as the number of degrees (3 digits) from north in a clockwise direction, followed by the word *true* or T; for example, due east would be 090° true or 090°T.

turning point a point at which a graph changes direction (either up or down).

two-dimensional array an array of one-dimensional arrays.

two-step chance experiment a probability experiment that involves two trials.

two-way table tables that list all the possible outcomes of a probability experiment in a logical manner.

Hair colour	Hair type		Total
Red	1	1	2
Brown	8	4	12
Blonde	1	3	4
Black	7	2	9
Total	17	10	27

undirected graphs graphs in which it is possible to move along edges in both directions.

unit circle a circle with its centre at the origin and having a radius of 1 unit.

univariate data data relating to a single variable.

variable a named container or memory location that holds a value.

vertex (pl. **vertices**) in geometry, the point at which two lines meet, forming an angle; in graphs, the point at which the graph of a quadratic function (parabola) changes direction (either up or down); in network theory, an object in a network represented by a point, also known as a node.

vertically opposite angles when two lines intersect, four angles are formed at the point of intersection, and two pairs of vertically opposite angles result. Vertically opposite angles are equal.

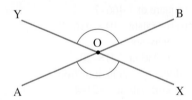

volume the amount of space a 3-dimensional object occupies. The units used are cubic units, such as cubic centimetres (cm³) and cubic metres (m³).

walk any route taken through a network, including routes that repeat edges and vertices.

weighted graphs graphs in which edges have values.

x-intercept the point where a graph intersects the *x*-axis.

y-intercept the point where a graph intersects the *y*-axis. In the equation of a straight line, $y = mx + c$, the constant term, c, represents the *y*-intercept of that line.

INDEX